# *Your* Guide to Success in Math

**Complete Step 0 as soon as you begin your math cou~~rse.~~**

## STEP 0:  Plan Your Semester

☐ Register for the online part of the course (if there is one) as soon as possible.

☐ Fill in your Course and Contact information on this pull-out card.

☐ Write important dates from your syllabus on the Semester Organizer section
on this pull-out card.

**Follow Steps 1–3 during your course. Your instructor will tell you which resources
to use—and when—in the textbook or eText, *MyMathGuide* workbook, videos, and
MyLab Math. Use these resources for extra help and practice.**

## STEP 1:  PREPARE: Studying the Concepts

☐ Read the textbook or eText, listen to your instructor's lecture, and/or watch the section
videos. You can work in *MyMathGuide* as you do this, saving all your work to review
throughout the course.

☐ Work the Skill to Review exercises and/or watch the videos in MyLab Math in
each section.

☐ Stop and do the Margin Exercises, including the Guided Solution Exercises, as directed.

## STEP 2:  PARTICIPATE: Making Connections
##            through Active Exploration

☐ Explore the concepts using the Animations in MyLab Math.

☐ Work the Visualizing for Success or Translating for Success exercises in the text and/or
in MyLab Math.

☐ Answer the Check Your Understanding exercises in the Section Exercises in the
textbook and/or in MyLab Math.

## STEP 3:  PRACTICE: Reinforcing Understanding

☐ Complete your assigned homework from the textbook and/or in MyLab Math.

    ☐ When doing homework from the textbook, use the answer section to check your work.

    ☐ When doing homework in MyLab Math, use the Learning Aids, such as Help Me
Solve This and View an Example, as needed, working toward being able to complete
exercises without the aids.

☐ Work the exercises in the Mid-Chapter Review.

☐ Read the Study Guide and work the Review Exercises in the Chapter Summary
and Review.

☐ Take the Chapter Test as a practice exam. To watch an instructor solve each problem,
go to the Chapter Test Prep Videos in MyLab Math.

Use the ***Studying for Success*** tips in the text and the ***MyLab Math Study Skills*** modules
(with videos) to help you develop effective time-management, note-taking, test-prep, and
other skills.

# Student Organizer

## Course Information

Course Number: _____ Name: _____

Location: _____ Days/Time: _____

## Contact Information

| Contact | Name | Email | Phone | Office Hours | Location |
|---|---|---|---|---|---|
| Instructor | | | | | |
| Tutor | | | | | |
| Math Lab | | | | | |
| Classmate | | | | | |
| Classmate | | | | | |

## Semester Organizer

| Week | Homework | Quizzes and Tests | Other |
|---|---|---|---|
| | | | |
| | | | |
| | | | |
| | | | |
| | | | |
| | | | |
| | | | |
| | | | |
| | | | |
| | | | |
| | | | |
| | | | |
| | | | |
| | | | |
| | | | |
| | | | |
| | | | |
| | | | |

# Prealgebra

EDITION 8

## Marvin L. Bittinger
*Indiana University Purdue University Indianapolis*

## David J. Ellenbogen
*Community College of Vermont*

## Barbara L. Johnson
*Indiana University Purdue University Indianapolis*

| | |
|---|---|
| *Director, Courseware Portfolio Management:* | Michael Hirsch |
| *Courseware Portfolio Management Assistant:* | Shannon Slocum |
| *Managing Producer:* | Karen Wernholm |
| *Content Producer:* | Ron Hampton |
| *Producer:* | Erin Carreiro |
| *Manager, Courseware QA:* | Mary Durnwald |
| *Manager, Content Development:* | Eric Gregg |
| *Field Marketing Managers:* | Jennifer Crum; Lauren Schur |
| *Marketing Manager:* | Kyle DiGiannantonio |
| *Marketing Assistant:* | Brooke Imbornone |
| *Senior Author Support/ Technology Specialist:* | Joe Vetere |
| *Manager, Rights and Permissions:* | Gina Cheselka |
| *Manufacturing Buyer:* | Carol Melville, LSC Communications |
| *Associate Director of Design:* | Blair Brown |
| *Text Design:* | Geri Davis/The Davis Group, Inc. |
| *Editorial and Production Service:* | Jane Hoover/Lifland et al., Bookmakers |
| *Composition:* | Pearson CSC |
| *Illustration:* | Network Graphics; William Melvin |
| *Cover Design:* | Pearson CSC |
| *Cover Image:* | Peter Dazeley/Getty Images |

**Library of Congress Cataloging-in-Publication Data is on file with the publisher.**

ISBN 13: 978-0-13-518256-7
ISBN 10: 0-13-518256-5

2 2019

# Contents

## 1 WHOLE NUMBERS      1

## 2 INTRODUCTION TO INTEGERS AND ALGEBRAIC EXPRESSIONS      85

## 3 FRACTION NOTATION: MULTIPLICATION AND DIVISION      151

## 4 FRACTION NOTATION: ADDITION, SUBTRACTION, AND MIXED NUMERALS      221

# Index of Activities

# Index of Animations

| Section | Title |
|---------|-------|
| 1.9b | Exponential Notation |
| 2.2b | Order on the Number Line |
| 3.4b | Multiplying Fractions |
| 4.2b | Adding Fractions |
| 5.1d | Rounding Decimal Notation |
| 6.5a | Similar Triangles |
| 7.5a | Percent Increase and Percent Decrease |
| 8.6c | Histograms |
| 9.4a | Volume of Rectangular Solids |
| 10.1b | Negative Exponents |

# Photo Credits

CHAPTER 1: p. 1 (bowl with hands) Suzanne Tucker/Shutterstock; ("Stop Hunger" and background) Jacek Dudzinski/123RF. p. 3 Tina Manley/Alamy Stock Photo. p. 8 (left) Anton Balazh/Shutterstock; (right) NASA. p. 55 Courtesy of Barbara Johnson. p. 66 (left) meunierd/Shutterstock; (right) Courtesy of Geri Davis. p. 82 Sebastian Duda/Fotolia. CHAPTER 2: p. 85 (Dead Sea resort and background) Anna Johnson; (road sign) Rob Swanson/Shutterstock. p. 90 (left) chuyu/Shutterstock; (right) Carlos Villoch/MagicSea/Alamy. p. 99 Mellowbox/Fotolia. p. 103 Serge Black/Fotolia. p. 146 Robert Nyholm/123RF. CHAPTER 3: p. 151 (t-shirts and background) khatawut Chaemchamras /123RF; ("Made in China") Dawid Zagorsk /123RF. p. 172 cynoclub/Fotolia. p. 177 Petr84/Shutterstock. p. 178 Larry Roberg/Fotolia. p. 196 (left) apidach jansawang/123RF; (right) Gloriole/Shutterstock. CHAPTER 4: p. 221 (honeybee and background) Bohdan Hetman/123RF; (honeycomb graphic) Alexander Parenkin/123RF. p. 237 (left) 3000ad/Shutterstock; (right) Jim West/Alamy Stock Photo. p. 264 Keystone Pictures USA/Alamy Stock Photo. p. 265 Ruud Morijn/Fotolia. p. 276 Maridav/Fotolia. p. 280 dpa picture alliance/Alamy Stock Photo. p. 281 WENN Ltd/Alamy Stock Photo. p. 290 Tom Sears. p. 295 MPH Photos/Shutterstock. p. 299 alisonhancock/Fotolia. CHAPTER 5: p. 301 (food display and background) puhhha/Shutterstock; (menu) Victor Metelskiy/Shutterstock. p. 302 ZUMA Press, Inc./Alamy Stock Photo. p. 304 (left) Foto Arena LTDA/Alamy Stock Photo; (right) David Chedgy/Alamy Stock Photo. p. 309 (left) CHEN WS/ Shutterstock; (right) digitalskillet/Shutterstock. p. 323 (top) Trong Nguyen/Shutterstock; (bottom) iQoncept/Shutterstock. p. 324 feverpitched/123RF. p. 325 Martin Valigursky/Fotolia. p. 327 Natthapong Khromkrathok/123RF. p. 348 Rtimages/Shutterstock. p. 363 rido/123RF. p. 364 (top) Marmaduke St. John/Alamy Stock Photo; (bottom) Simon Mayer/Shutterstock. p. 368 Kotangens/ Fotolia. p. 372 (left) Pictorial Press Ltd/Alamy Stock Photo; (right) Boston Globe/Getty Images. p. 375 (left) wusuowei/Fotolia; (right) karrapavan/Fotolia. p. 376 candan/Fotolia. p. 379 Marcel de Grijs/123RF. CHAPTER 6: p. 387 (stranded sea mammal and background) Lano Lan/Shutterstock; ("The Orcas Rescue") Dima Gorohow/Shutterstock. p. 392 (left) artem evdokimov/ Shutterstock; (right) pasiphae/123RF. p. 395 Budimir Jevtic/Shutterstock. p. 397 kidia/123RF. p. 399 (upper left) Randy Duchaine/ Alamy Stock Photo; (upper right) Katarzyna Bialasiewicz/123RF; (lower left) Science Picture Co/Alamy Stock Photo; (lower right) jose garcia/Fotolia. p. 408 (left) FOTOimage Montreal/Shutterstock; (right) Geri Lynn Smith/Shutterstock. p. 412 M. Timothy O'Keefe/Alamy Stock Photo. p. 415 Ariel Bravy/Shutterstock. p. 417 (left) Pictorial Press Ltd/Alamy Stock Photo; (right) Kim D. French/Fotolia. p. 418 (left) Duplass/Shutterstock; (right) sondoggie/Shutterstock. p. 420 (left) Stephen Meese/Fotolia; (right) mattjeppson/Fotolia. p. 421 (left) PACIFIC PRESS/Alamy; (right) dpa picture alliance/Alamy. p. 424 (middle) A Periam Photography/Shutterstock; (lower left) Courtesy of Geri Davis. p. 427 (both) Courtesy of Geri Davis. p. 432 CountryStock/Alamy. p. 433 (top) Marek Uliasz/Alamy Stock Photo; (bottom) Rawpixel.com/Shutterstock. CHAPTER 7: p. 439 (transplant scene and background) Michelle Del Guercio/Science Source; (donor card) PA Images/Alamy Stock Photo. p. 441 Jenoche/Shutterstock. p. 444 scyther5/123RF. p. 448 (top right) pio3/Shutterstock; (bottom left) Tom Oliveira/Shutterstock; (bottom right) Matyas Rehak/ Shutterstock. p. 449 (left) om12/123RF; (right) Andriy Popov/123RF. p. 454 mikekiev/123RF. p. 455 (top) Brent Hofacker/123RF; (bottom) Todd Arena/123RF. p. 459 mark adams/123RF. p. 469 jewhyte/123RF. p. 471 (left) fsstock/123RF; (right) OSDG/ Shutterstock. p. 474 Bart Sadowski/Shutterstock. p. 475 stockyimages/Shutterstock; p. 476 DUANGJAN JITMART/Shutterstock; p. 484 welcomia/Shutterstock; p. 485 Barry Blackburn/123RF; p. 487 (left) Patti McConville/Alamy Stock Photo; (right) Africa Studio/Shutterstock. p. 488 (left) auremar/Fotolia; (right) Christina Richards/Shutterstock. p. 493 ocusfocus/123RF. p. 496 (left) Dmitry Vereshchagin/Fotolia; (right) Andres Rodriguez/Fotolia. p. 508 Anne Rippy/Alamy Stock Photo. CHAPTER 8: p. 509 (traffic and background) chuyu/123RF; ("Expect Delays") Mr Doomits/Alamy Stock Photo. p. 515 Roman Tiraspolsky/ Shutterstock. p. 550 dolgachov/123RF. p. 552 Cal Sport Media/Alamy Stock Photo. p. 566 Bruce Leighty/Alamy Stock Photo. p. 567 Leonard Zhukovsky/Shutterstock. p. 568 Ian Allenden/123RF. p. 575 vasin leenanuruksa/123RF. p. 576 sashafolly/ Shutterstock. p. 597 Elffle.95/Shutterstock. CHAPTER 9: p. 599 (tunnel and background) PA Images/Alamy Stock Photo; (St. Gotthard sign) BRIAN HARRIS/Alamy Stock Photo. p. 602 mandritoiu/Shutterstock. p. 604 golddc/Shutterstock. p. 607 Klaus Rademaker/Shutterstock. p. 608 PA Images/Alamy Stock Photo. p. 617 (left) Hemis/Alamy Stock Photo; (right) Jerry Ballard/ Alamy Stock Photo. p. 628 (left) Deatonphotos/Fotolia; (right) Wisconsin DNR. p. 640 (left) Patti McConville/Alamy Stock Photo; (right) Lev1977/Fotolia. p. 641 (left) imageBROKER/Alamy Stock Photo; (right) Natural History Library/Alamy Stock Photo. p. 672 (left) Victoria Field/Shutterstock; (right) Arina P. Habich/123RF. p. 674 Aurora Photos/Alamy Stock Photo. p. 676 Neal Hamberg/Associated Press. p. 679 belchonock/123RF. p. 690 Monkey Business Images/Shutterstock. p. 693 Michael Zysman/123rf. CHAPTER 10: p. 695 (octopus and background) Michal Adamczyk/123RF; ("Have no fear ...") Vozzy/Shutterstock. p. 707 (top) Andrea Izzotti/123RF; (bottom) cooldesign/123RF. p. 712 (left) Terry Leung/Pearson Education Asia Ltd; (right) auremar/123RF. p. 713 (left) Melissa Burovac/Shutterstock; (right) Michal Adamczyk/123RF. p. 714 (left) Vadim Fedotov/123RF; (right) royaltystockphoto/123RF. p. 720 sirtravelalot/Shutterstock. p. 723 (left) Brian Buckland; (right) Buzz Pictures/Alamy Stock Photo. p. 735 dolgachov/123RF. p. 738 Mariusz Prusaczyk/Fotolia. p. 741 NASA.

# Preface

**Math doesn't change, but students' needs—and the way students learn—do.**

With this in mind, *Prealgebra*, 8th edition, continues the Bittinger tradition of objective-based, guided learning, while integrating many updates with the proven pedagogy. These updates are motivated by feedback that we received from students and instructors, as well as our own experience in the classroom. In this edition, our focus is on guided learning and retention: helping each student (and instructor) get the most out of all the available program resources—wherever and whenever they engage with the math.

We believe that student success in math hinges on four key areas: **Foundation, Engagement, Application,** and **Retention**. In the 8th edition, we have added key new program features (highlighted below, for quick reference) in each area to make it easier for each student to personalize his or her learning experience. In addition, you will recognize many proven features and presentations from the previous edition of the program.

## FOUNDATION
### Studying the Concepts

Students can learn the math concepts by reading the textbook or the eText, participating in class, watching the videos, working in the *MyMathGuide* workbook—or using whatever combination of these course resources works best for them.

> In order to understand new math concepts, students must recall and use skills and concepts previously studied.
>
> ☐ *New!*  **Skill Review,** in nearly every section of the text and the eText, reviews a previously presented skill at the objective level where it is key to learning the new material. This feature offers students two practice exercises with answers. In MyLab Math, new **Skill Review Videos,** created by the Bittinger author team, offer a concise, step-by-step solution for each Skill Review exercise.

**Margin Exercises with Guided Solutions,** with fill-in blanks at key steps in the problem-solving process, appear in nearly every text section and can be assigned in MyLab Math.

*Prealgebra* **Video Program,** our comprehensive program of objective-based, interactive videos, can be used hand-in-hand with our *MyMathGuide* workbook. **Interactive Your Turn exercises** in the videos prompt students to solve problems and receive instant feedback. These videos can be accessed at the section, objective, and example levels.

*MyMathGuide* offers students a guided, hands-on learning experience. This objective-based workbook (available in print and in MyLab Math) includes vocabulary, skill, and concept review—as well as problem-solving practice with space for students to fill in the answers and stepped-out solutions to problems, to show (and keep) their work, and to write notes. Students can use *MyMathGuide* while watching the videos, listening to the instructor's lecture, or reading the text or the eText, in order to reinforce and self-assess their learning.

**Studying for Success** sections are checklists of study skills designed to ensure that students develop the skills they need to succeed in math, school, and life. They are available at the beginning of selected sections.

☐ *New!* **Expanded Statistics Content** Chapters 6–8 have been revised and reordered to allow for expansion of the material on statistics, as well as coverage of percents prior to the statistics material. Chapter 8, Data, Graphs, and Statistics, has been revised and expanded. Beginning with tables and graphs and continuing with discussions of one-variable statistics, frequency distributions, and probability, this chapter provides students with an introduction to foundational concepts of statistics. New to this edition is coverage of measures of spread, quartiles, frequency distributions and tables, stem-and-leaf plots, and construction of histograms and tree diagrams. Students completing this chapter will be better equipped to understand and analyze the data and graphs they encounter, as well as to enter an introductory statistics course.

## ENGAGEMENT
### Making Connections Through Active Exploration

Since understanding the big picture is key to student success, we offer many active learning opportunities for the practice, review, and reinforcement of important concepts and skills.

☐ *New!* **Chapter Opener Applications** with infographics use current data and applications to present the math in context. Each application is related to exercises in the text to help students model, visualize, learn, and retain the math.

☐ *New!* **Student Activities,** included with each chapter, have been developed as multistep, data-based activities for students to apply the math in the context of an authentic application. Student Activities are available in *MyMathGuide* and in MyLab Math.

☐ *New!* **Interactive Animations** can be manipulated by students in MyLab Math through guided and open-ended exploration to further solidify their understanding of important concepts.

**Translating for Success** offers extra practice with the important first step of the process for solving applied problems. This activity is available in the text and in MyLab Math.

**Calculator Corner** is an optional feature throughout the text that helps students use a calculator to perform calculations and to visualize concepts.

**Learning Catalytics** uses students' mobile devices for an engagement, assessment, and classroom intelligence system that gives instructors real-time feedback on student learning.

## APPLICATION
### Reinforcing Understanding

As students explore the math, they have frequent opportunities to apply new concepts, practice, self-assess, and reinforce their understanding.

**Margin Exercises,** labeled "Do Exercise . . . ," give students frequent opportunities to apply concepts just discussed by solving problems that parallel text examples.

**Exercise Sets** in each section offer abundant opportunity for practice and review in the text and in MyLab Math. The Section Exercises are grouped by objective for ease of use, and each set includes the following special exercise types:

☐ *New!* **Check Your Understanding** with **Reading Check** and **Concept Check** exercises, at the beginning of each exercise set, gives students the opportunity to assess their grasp of the skills and concepts before moving on to the objective-based section exercises. In MyLab Math, many of these exercises use drag-and-drop functionality.

☐ **Skill Maintenance Exercises** offer a thorough review of the math in the preceding sections of the text.

☐ **Synthesis Exercises** help students develop critical-thinking skills by requiring them to use what they know in combination with content from the current and previous sections.

## RETENTION
### Carrying Success Forward

Because continual practice and review is so important to retention, we have integrated both throughout the program in the text and in MyLab Math.

☐ *New!* **Skill Builder Adaptive Practice,** built into MyLab Math, offers each student a personalized learning experience. When a student struggles with the assigned homework, Skill Builder exercises offer just-in-time additional adaptive practice. The adaptive engine tracks student performance and delivers to each individual questions that are appropriate for his or her level of understanding. When the system has determined that the student has a high probability of successfully completing the assigned exercise, it suggests that the student return to the assigned homework.

**Mid-Chapter Review** offers an opportunity for active review midway through each chapter. This review offers four types of practice problems: **Concept Reinforcement, Guided Solutions, Mixed Review,** and **Understanding Through Discussion and Writing.**

**Summary and Review** is a comprehensive learning and review section at the end of each chapter. Each of the five sections—**Vocabulary Reinforcement** (fill-in-the-blank), **Concept Reinforcement** (true/false), **Study Guide** (examples with stepped-out solutions paired with similar practice problems), **Review Exercises,** and **Understanding Through Discussion and Writing**—includes references to the section in which the material was covered to facilitate review.

**Chapter Test** offers students the opportunity for comprehensive review and reinforcement prior to taking their instructor's exam. **Chapter Test Prep Videos** in MyLab Math show step-by-step solutions to the questions on the chapter test.

**Cumulative Review** follows each chapter beginning with Chapter 2. These reviews revisit skills and concepts from all preceding chapters to help students retain previously presented material.

# Resources for Success

## MyLab Math Online Course for Bittinger, Ellenbogen, and Johnson, *Prealgebra*, 8th Edition

(access code required or visit www.mylabmath.com)

MyLab™ Math is available to accompany Pearson's market-leading text offerings. To give students a consistent tone, voice, and teaching method, the pedagogical approach of the text is tightly integrated throughout the accompanying MyLab Math course, making learning the material as seamless as possible.

### UPDATED! Learning Path

Structured, yet flexible, the updated learning path highlights author-created, faculty-vetted content—giving students what they need exactly when they need it. The learning path directs students to resources such as two new types of video: **Just-in-Time Review** (concise presentations of key topics from previous courses) and **Skill Review** (author-created exercises with step-by-step solutions that reinforce previously presented skills), both available in the Multimedia Library and assignable in MyLab Math.

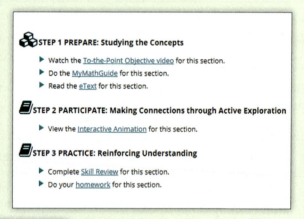

**STEP 1 PREPARE: Studying the Concepts**

▶ Watch the To-the-Point Objective video for this section.
▶ Do the MyMathGuide for this section.
▶ Read the eText for this section.

**STEP 2 PARTICIPATE: Making Connections through Active Exploration**

▶ View the Interactive Animation for this section.

**STEP 3 PRACTICE: Reinforcing Understanding**

▶ Complete Skill Review for this section.
▶ Do your homework for this section.

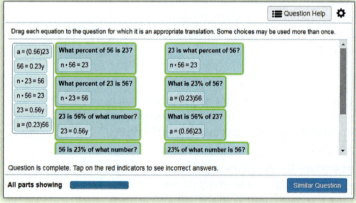

### NEW! Drag-and-Drop Exercises

Drag-and-drop exercises are now available in MyLab Math. This new assignment type allows students to drag answers and values within a problem, providing a new and engaging way to test students' concept knowledge.

### NEW and UPDATED! Animations

New animations encourage students to learn key concepts through guided and open-ended exploration. Animations are available through the learning path and multimedia library, and they can be assigned within MyLab Math.

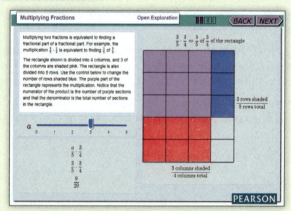

**pearson.com/mylab/math**

# Resources for Success

## Instructor Resources

Additional resources can be downloaded from **www.pearsonhighered.com** or hardcopy resources can be ordered from your sales representative.

### Annotated Instructor's Edition
ISBN: 0135183308
- Answers to all text exercises.
- Helpful teaching tips, including suggestions for incorporating Student Activities in the course.

### Instructor's Resource Manual with Tests and Mini-lectures
(download only)
ISBN: 0135230063
- Resources designed to help both new and experienced instructors with course preparation and class management.
- Chapter teaching tips and support for media supplements.
- Multiple versions of multiple-choice and free-response chapter tests, as well as final exams.

### Instructor's Solutions Manual
(download only)
By Judy Penna
ISBN: 0135183316
The *Instructor's Solutions Manual* includes brief solutions for the even-numbered exercises in the exercise sets and fully worked-out annotated solutions for all the exercises in the Mid-Chapter Reviews, the Summary and Reviews, the Chapter Tests, and the Cumulative Reviews.

### PowerPoint® Lecture Slides
(download only)
- Editable slides present key concepts and definitions from the text.
- Available to both instructors and students.
- Fully accessible.

### TestGen®
TestGen enables instructors to build, edit, print, and administer tests using a computerized test bank of questions developed to cover all the objectives of the text. (www.pearsoned.com/testgen)

## Student Resources

### Prealgebra Lecture Videos
- Concise, interactive, and objective-based videos.
- View a whole section, choose an objective, or go straight to an example.
- Available in MyLab Math.

### Chapter Test Prep Videos
- Step-by-step solutions for every problem in the chapter tests.
- Available in MyLab Math.

### Skill Review Videos
Students can review previously presented skills at the objective level with two practice exercises before moving forward in the content. Videos include a step-by-step solution for each exercise.
- Available in MyLab Math.

### MyMathGuide: Notes, Practice, and Video Path
ISBN: 0135184487
- Guided, hands-on learning in a workbook format with space for students to show their work and record their notes and questions.
- Highlights key concepts, skills, and definitions; offers quick reviews of key vocabulary terms with practice problems, examples with guided solutions, similar Your Turn exercises, and practice exercises with readiness checks.
- Includes student activities utilizing real data.
- Available in MyLab Math and as a printed manual.

### Student's Solutions Manual
ISBN: 0135183391
By Judy Penna
- Includes completely worked-out annotated solutions for odd-numbered exercises in the text, as well as all the exercises in the Mid-Chapter Reviews, the Summary and Reviews, the Chapter Tests, and the Cumulative Reviews.
- Available in MyLab Math and as a printed manual.

**pearson.com/mylab/math**

# Acknowledgments

Our deepest appreciation to all the instructors and students who helped to shape this revision of our program by reviewing our texts and courses, providing feedback, and sharing their experiences with us at conferences and on campus. In particular, we would like to thank the following for reviewing the titles in our worktext program for this revision:

Amanda L. Blaker, *Gallatin College*
Jessica Bosworth, *Nassau Community College*
Judy G. Burn, *Trident Technical College*
Dr. Abushieba A. Ibrahim, *Broward College*
Laura P. Kyser, *Savannah Technical College*
Dr. David Mandelbaum, *Nova Southeastern University*

An outstanding team of professionals was involved in the production of this text. We want to thank Judy Penna for creating the new Skill Review videos and for writing the *Student's Solutions Manual* and the *Instructor's Solutions Manual.* We also thank Laurie Hurley for preparing *MyMathGuide*, and Tom Atwater for supporting and overseeing new videos. Accuracy checkers Judy Penna and Laurie Hurley contributed immeasurably to the quality of the text.

Jane Hoover, of Lifland et al., Bookmakers, provided editorial and production services of the highest quality, and Geri Davis, of The Davis Group, performed superb work as designer, art editor, and photo researcher. Their countless hours of work and consistent dedication have led to products of which we are immensely proud.

In addition, a number of people at Pearson, including the Developmental Math Team, have contributed in special ways to the development and production of our program. Special thanks are due to Cathy Cantin, Courseware Portfolio Manager, for her visionary leadership and development support. In addition, Ron Hampton, Content Producer, contributed invaluable coordination for all aspects of the project. We also thank Erin Carreiro, Producer, and Kyle DiGiannantonio, Marketing Manager, for their exceptional support.

Our goal in writing this textbook was to make mathematics accessible to every student. We want you to be successful in this course and in the mathematics courses you take in the future. Realizing that your time is both valuable and limited, and that you learn in a uniquely individual way, we employ a variety of pedagogical and visual approaches to help you learn in the best and most efficient way possible. We wish you a positive and successful learning experience.

*Marv Bittinger*
*David Ellenbogen*
*Barbara Johnson*

# Index of Applications

## Transportation

Stop Hunger

# Whole Numbers

Many people around the world lack access to clean water or to sufficient food. Approximately 663 million people, or 9% of the world's population, drink water that is not clean.

Hunger is a reality for even more people: 10% of the world's population lacks sufficient nourishment. As the graph indicates, some regions of the world are more affected by undernourishment than others.

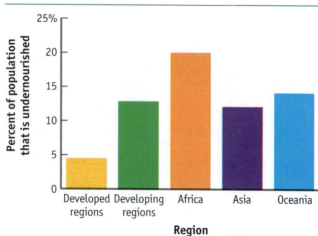

**World Undernourishment**

DATA: worldhunger.org

*DATA: charitywater.org; actionagainsthunger.org; stopthehunger.com*

In Example 8 and Margin Exercise 9 of Section 1.1, we will examine the numbers of families helped by one charity that seeks to alleviate hunger.

# 1.1

## OBJECTIVES

**a** Give the meaning of digits in standard notation.

**b** Convert from standard notation to expanded notation.

**c** Convert between standard notation and word names.

# Standard Notation

We study mathematics in order to be able to solve problems. In this section, we study how numbers are named. We begin with the concept of place value.

## a  PLACE VALUE

Attendance at various types of Broadway performances in New York City for the 2016–2017 season is given in the following table.

| TYPE OF PERFORMANCE | ATTENDANCE |
|---|---|
| Musicals | 11,362,732 |
| Plays | 1,798,723 |
| Specials | 109,797 |

DATA: The Broadway League

A **digit** is a number 0, 1, 2, 3, 4, 5, 6, 7, 8, or 9 that names a place-value location. For large numbers, digits are separated by commas into groups of three, called **periods**. Each period has a name: *ones, thousands, millions, billions, trillions,* and so on. To understand the number of people attending Broadway musicals in the table above, we can use a **place-value chart**, as shown below.

| PLACE-VALUE CHART | | | | | | | | | | | | | | |
|---|---|---|---|---|---|---|---|---|---|---|---|---|---|---|
| Periods → | Trillions | | | Billions | | | Millions | | | Thousands | | | Ones | | |
| | | | | | | | | 1 | 1 | 3 | 6 | 2 | 7 | 3 | 2 |
| | Hundreds | Tens | Ones | Hundreds | Tens | Ones | Hundreds | Tens | Ones | Hundreds | Tens | Ones | Hundreds | Tens | Ones |

11 millions   362 thousands   732 ones

**EXAMPLES**  In each of the following numbers, what does the digit 8 mean?

1. 278,342          8 thousands
2. 872,342          8 hundred thousands
3. 28,343,399,223   8 billions
4. 98,413,099       8 millions
5. 6328             8 ones

**Do Exercises 1–6 (in the margin at right).** ▶

**EXAMPLE 6**  *Websites.*  In July 2017, the total number of active websites on the world wide web was 1,225,423,079. What digit names the number of ten millions?

**Data:** internetlivestats.com

Ten millions
1,225,423,079

The digit 2 is in the ten millions place, so 2 names the number of ten millions.

**Do Exercise 7.** ▶

**b  CONVERTING FROM STANDARD NOTATION TO EXPANDED NOTATION**

Heifer International is a charitable organization whose mission is to work with communities to end hunger and poverty and care for the earth by providing farm animals to impoverished families around the world. Consider the data in the following table.

| GEOGRAPHICAL AREAS OF NEED | NUMBER OF FAMILIES ASSISTED DIRECTLY AND INDIRECTLY BY HEIFER INTERNATIONAL IN 2016 |
|---|---|
| Africa | 959,734 |
| Americas | 640,604 |
| Asia, South Pacific | 1,699,836 |
| Central and Eastern Europe | 254,427 |

DATA: *Heifer International 2016 Annual Report*

What does the digit 2 mean in each number?

1. 526,555          2. 265,789

3. 42,789,654       4. 24,789,654

5. 8924             6. 5,643,201

7. *Government Payroll.*  In 2015, the total payroll for all full-time federal employees in the United States was $19,369,134,421. What digit names the number of ten billions?

   **Data:** U.S. Census Bureau

**Answers**

1. 2 ten thousands   2. 2 hundred thousands
3. 2 millions   4. 2 ten millions   5. 2 tens
6. 2 hundreds   7. 1

Write expanded notation.

**8.** 2718 mi, the length of the Congo River in Africa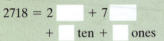

$$2718 = 2\ \boxed{\phantom{xx}} + 7\ \boxed{\phantom{xx}}$$
$$+\ \boxed{\phantom{xx}}\ \text{ten} + \boxed{\phantom{xx}}\ \text{ones}$$

**9.** 254,427, the number of families in Central and Eastern Europe assisted by Heifer International in 2016

**10.** 1670 ft, the height of the Taipei 101 Tower in Taiwan

**11.** 104,094 square miles, the area of Colorado

The number of families assisted in Africa was 959,734. This number is expressed in **standard notation**. We write **expanded notation** for 959,734 as follows:

959,734 = 9 hundred thousands + 5 ten thousands
+ 9 thousands + 7 hundreds
+ 3 tens + 4 ones.

**EXAMPLE 7**   Write expanded notation for 1776 ft, the height of One World Trade Center in New York City.

1776 = 1 thousand + 7 hundreds + 7 tens + 6 ones

**EXAMPLE 8**   Write expanded notation for 640,604, the number of families in the Americas assisted by Heifer International in 2016.

640,604 = 6 hundred thousands + 4 ten thousands
+ 0 thousands + 6 hundreds + 0 tens + 4 ones

or

6 hundred thousands + 4 ten thousands + 6 hundreds + 4 ones

◀ **Do Exercises 8–11.**

### c   CONVERTING BETWEEN STANDARD NOTATION AND WORD NAMES

We often use **word names** for numbers. When we pronounce a number, we are speaking its word name. Russia won 56 medals in the 2016 Summer Olympics in Rio de Janeiro, Brazil. A word name for 56 is "fifty-six." Word names for some two-digit numbers like 36, 51, and 72 use hyphens. Others, like that for 17, use only one word, "seventeen."

**2016 Summer Olympics Medal Count**

| COUNTRY | GOLD | SILVER | BRONZE | TOTAL |
|---------|------|--------|--------|-------|
| United States of America | 46 | 37 | 38 | 121 |
| Great Britain | 27 | 23 | 17 | 67 |
| People's Republic of China | 26 | 18 | 26 | 70 |
| Russia | 19 | 18 | 19 | 56 |
| Germany | 17 | 10 | 15 | 42 |

DATA: espn.com

*Answers*

**8.** 2 thousands + 7 hundreds + 1 ten + 8 ones
**9.** 2 hundred thousands + 5 ten thousands
+ 4 thousands + 4 hundreds + 2 tens + 7 ones
**10.** 1 thousand + 6 hundreds + 7 tens
+ 0 ones, or 1 thousand + 6 hundreds + 7 tens
**11.** 1 hundred thousand + 0 ten thousands
+ 4 thousands + 0 hundreds + 9 tens
+ 4 ones, or 1 hundred thousand
+ 4 thousands + 9 tens + 4 ones

*Guided Solution:*
**8.** thousands, hundreds, 1, 8

**EXAMPLES**  Write a word name.

**9.** 46, the number of gold medals won by the United States

Forty-six

**10.** 15, the number of bronze medals won by Germany

Fifteen

**11.** 121, the total number of medals won by the United States

One hundred twenty-one

**Do Exercises 12–14.** ▶

For word names for larger numbers, we begin at the left with the largest period. The number named in the period is followed by the name of the period; then a comma is written and the next number and period are named. Note that the name of the ones period is not included in the word name for a whole number.

**EXAMPLE 12**  Write a word name for 46,605,314,732.

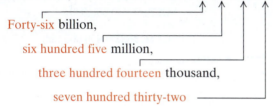

Forty-six billion,

six hundred five million,

three hundred fourteen thousand,

seven hundred thirty-two

The word "and" *should not* appear in word names for whole numbers. Although we commonly hear such expressions as "two hundred *and* one," the use of "and" is not, strictly speaking, correct in word names for whole numbers. For decimal notation, it is appropriate to use "and" for the decimal point. For example, 317.4 is read as "three hundred seventeen *and* four tenths."

**Do Exercises 15–18.** ▶

**EXAMPLE 13**  Write standard notation.

Five hundred six million,

three hundred forty-five thousand,

two hundred twelve

Standard notation is 506,345,212.

**Do Exercise 19.** ▶

---

Write a word name. (Refer to the chart on the previous page.)

**12.** 67, the total number of medals won by Great Britain

**13.** 18, the number of silver medals won by the People's Republic of China

**14.** 38, the number of bronze medals won by the United States

Write a word name.

**15.** 204

**16.** 10,336, the number of state parks in the United States

**Data:** stateparks.org

**GS** **17.** 1,879,204

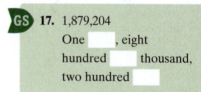

One ____ , eight

hundred ____ thousand,

two hundred ____

**18.** 7,401,989,718, the world population in 2017

**Data:** U.S. Census Bureau

**19.** Write standard notation.

Two hundred thirteen million, one hundred five thousand, three hundred twenty-nine

**Answers**

**12.** Sixty-seven  **13.** Eighteen
**14.** Thirty-eight  **15.** Two hundred four
**16.** Ten thousand, three hundred thirty-six
**17.** One million, eight hundred seventy-nine thousand, two hundred four
**18.** Seven billion, four hundred one million, nine hundred eighty-nine thousand, seven hundred eighteen
**19.** 213,105,329

**Guided Solution:**
**17.** Million, seventy-nine, four

✓ **Check Your Understanding**

**Reading Check** Complete each statement with the correct word from the following list.

digit          expanded          period          standard

**RC1.** In 983, the _____ 9 represents 9 hundreds.

**RC2.** In 615,702, the number 615 is in the thousands _____.

**RC3.** The phrase "3 hundreds + 2 tens + 9 ones" is _____ notation for 329.

**RC4.** The number 721 is written in _____ notation.

**Concept Check** Write a word name.

**CC1.** 5,000,000         **CC2.** 42,000,000         **CC3.** 3,000,000,000

**CC4.** 18,000,000,000      **CC5.** 7,000,000,000,000     **CC6.** 40,000,000,000,000

**a**   What does the digit 5 mean in each number?

**1.** 235,888         **2.** 253,777         **3.** 1,488,526         **4.** 500,736

*Broadway Shows.*   In the 2016–2017 season, Broadway shows grossed $1,449,321,564. What digit names the number of:

**5.** thousands?        **6.** millions?        **7.** ten millions?        **8.** hundred thousands?

**b**   Write expanded notation.

*Radio and Television Stations.*   The figure below shows the number of AM radio, FM radio, and full-power television stations in the United States. In Exercises 9–12, write expanded notation for the given number of stations.

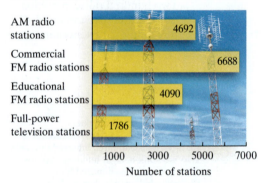

**Radio and Television Stations**

DATA: radiosurvivor.com; Federal Communications Commission

**9.** 4692 AM radio stations

**10.** 6688 commercial FM radio stations

**11.** 4090 educational FM radio stations

**12.** 1786 full-power television stations

**13.** 93,986

**14.** 38,453

**15.** 401,690

**16.** 135,080

*Population.* The table below shows the populations of four countries in 2016. In Exercises 17–20, write expanded notation for the given population.

**Four Most Populous Countries in the World**

| COUNTRY | POPULATION, 2016 |
|---------|------------------|
| China | 1,373,541,278 |
| India | 1,266,883,598 |
| United States | 323,995,528 |
| Indonesia | 258,316,051 |

DATA: *The CIA World Factbook*

**17.** 1,373,541,278 for China

**18.** 1,266,883,598 for India

**19.** 258,316,051 for Indonesia

**20.** 323,995,528 for the United States

**C** Write a word name.

**21.** 85

**22.** 48

**23.** 88,000

**24.** 45,987

**25.** 123,765

**26.** 111,013

**27.** 7,754,211,577

**28.** 43,550,651,808

**29.** *Airports.* In 2017, the world's busiest airport, Hartsfield-Jackson Atlanta International Airport, scheduled 394,249 departures. Write a word name for 394,249.

**Data:** U.S. Bureau of Transportation Statistics

**30.** *NASCAR Racing.* The average attendance at a NASCAR race is 99,853. Write a word name for 99,853.

**Data:** statisticbrain.com

**31.** *Sports Salaries.* The average annual salary over the life of his contract for Major League Baseball player Clayton Kershaw is $30,714,286. Write a word name for 30,714,286.

**Data:** *USA Today*

**32.** *Do Not Call Registry.* The number of active registrations in the National Do Not Call Registry in 2016 was 226,001,288. Write a word name for 226,001,288.

**Data:** Federal Trade Commission

Write each number in standard notation.

**33.** Six hundred thirty-two thousand, eight hundred ninety-six

**34.** Three hundred fifty-four thousand, seven hundred two

**35.** Fifty thousand, three hundred twenty-four

**36.** Seventeen thousand, one hundred twelve

**37.** Two million, two hundred thirty-three thousand, eight hundred twelve

**38.** Nineteen million, six hundred ten thousand, four hundred thirty-nine

**39.** Eight billion

**40.** Seven hundred million

**41.** Forty million

**42.** Twenty-six billion

**43.** Thirty million, one hundred three

**44.** Two hundred thousand, seventeen

Write standard notation for the number in each sentence.

**45.** *Pacific Ocean.* The area of the Pacific Ocean is sixty-four million, one hundred eighty-six thousand square miles.

**46.** The average distance from the sun to Neptune is two billion, seven hundred ninety-three million miles.

# Synthesis

*To the student and the instructor*: The Synthesis exercises found at the end of every exercise set challenge students to combine concepts or skills studied in the section or in preceding parts of the text. Exercises marked with a 🖩 symbol are meant to be solved using a calculator.

**47.** How many whole numbers between 100 and 400 contain the digit 2 in their standard notation?

**48.** 🖩 What is the largest number that you can name on your calculator using standard notation? How many digits does that number have? How many periods?

# Addition

## a ADDITION OF WHOLE NUMBERS

To answer questions such as "How many?", "How much?", and "How tall?", we often use whole numbers. The set, or collection, of **whole numbers** is

$$0, 1, 2, 3, 4, 5, 6, 7, 8, 9, 10, 11, 12, \ldots .$$

The set goes on indefinitely. There is no largest whole number, and the smallest whole number is 0. Each whole number can be named using various notations. The set $1, 2, 3, 4, 5, \ldots$, without 0, is called the set of **natural numbers**.

Addition of whole numbers corresponds to combining things together.

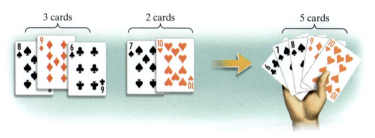

3 cards     2 cards     5 cards

We say that the **sum** of 3 and 2 is 5. The numbers added are called **addends**. The addition that corresponds to the figure above is

$$3 \quad + \quad 2 \quad = \quad 5.$$

    Addend    Addend    Sum

To add whole numbers, we add the ones digits first, then the tens, then the hundreds, then the thousands, and so on.

**EXAMPLE 1** Add: $878 + 995$.

Place values are lined up in columns.

$$
\begin{array}{r}
  \overset{1}{8}\ 7\ 8 \\
+ \ 9\ 9\ 5 \\
\hline
  3
\end{array}
$$

Add ones. We get 13 ones, or 1 ten + 3 ones. Write 3 in the ones column and 1 above the tens. This is called *carrying*, or *regrouping*.

$$
\begin{array}{r}
  \overset{1}{8}\ \overset{1}{7}\ 8 \\
+ \ 9\ 9\ 5 \\
\hline
  7\ 3
\end{array}
$$

Add tens. We get 17 tens, so we have 10 tens + 7 tens. This is also 1 hundred + 7 tens. Write 7 in the tens column and 1 above the hundreds.

$$
\begin{array}{r}
  \overset{1}{8}\ \overset{1}{7}\ 8 \\
+ \ 9\ 9\ 5 \\
\hline
1\ 8\ 7\ 3
\end{array}
$$

Add hundreds. We get 18 hundreds.

We show you these steps for explanation. You need write only this.

$$
\begin{array}{r}
  \overset{1}{8}\ \overset{1}{7}\ 8 \\
+ \ 9\ 9\ 5 \\
\hline
1\ 8\ 7\ 3
\end{array}
$$

← Addends

← Sum

## TO THE INSTRUCTOR AND THE STUDENT

This section presents a review of addition of whole numbers. Students who are successful should go on to Section 1.3. Those who have trouble should study developmental unit A near the back of this text and then repeat Section 1.2.

How do we perform an addition of three numbers, like $2 + 3 + 6$? We could do it by adding 3 and 6, and then 2. We can show this with parentheses:

$$2 + (3 + 6) = 2 + 9 = 11. \qquad \text{Parentheses tell what to do first.}$$

We could also add 2 and 3, and then 6:

$$(2 + 3) + 6 = 5 + 6 = 11.$$

Either way the result is 11. It does not matter how we group the numbers. This illustrates the **associative law of addition**, $a + (b + c) = (a + b) + c$. We can also add whole numbers in any order. That is, $2 + 3 = 3 + 2$. This illustrates the **commutative law of addition**, $a + b = b + a$. Together, the commutative and associative laws tell us that to add more than two numbers, we can use any order and grouping we wish. Adding 0 to a number does not change the number: $a + 0 = 0 + a = a$. That is, $6 + 0 = 0 + 6 = 6$, or $198 + 0 = 0 + 198 = 198$. We say that 0 is the **additive identity**.

**EXAMPLE 2** Add: $391 + 1276 + 789 + 5498$.

$$
\begin{array}{r}
\overset{\phantom{0}\overset{2}{\phantom{0}}}{3\ 9\ 1} \\
1\ 2\ 7\ 6 \\
7\ 8\ 9 \\
+\ 5\ 4\ 9\ 8 \\
\hline
4
\end{array}
$$

Add ones. We get 24, so we have 2 tens + 4 ones. Write 4 in the ones column and 2 above the tens.

$$
\begin{array}{r}
\overset{3\ 2}{3\ 9\ 1} \\
1\ 2\ 7\ 6 \\
7\ 8\ 9 \\
+\ 5\ 4\ 9\ 8 \\
\hline
5\ 4
\end{array}
$$

Add tens. We get 35 tens, so we have 30 tens + 5 tens. This is also 3 hundreds + 5 tens. Write 5 in the tens column and 3 above the hundreds.

$$
\begin{array}{r}
\overset{1\ 3\ 2}{3\ 9\ 1} \\
1\ 2\ 7\ 6 \\
7\ 8\ 9 \\
+\ 5\ 4\ 9\ 8 \\
\hline
9\ 5\ 4
\end{array}
$$

Add hundreds. We get 19 hundreds, so we have 10 hundreds + 9 hundreds. This is also 1 thousand + 9 hundreds. Write 9 in the hundreds column and 1 above the thousands.

$$
\begin{array}{r}
\overset{1\ 3\ 2}{3\ 9\ 1} \\
1\ 2\ 7\ 6 \\
7\ 8\ 9 \\
+\ 5\ 4\ 9\ 8 \\
\hline
7\ 9\ 5\ 4
\end{array}
$$

Add thousands. We get 7 thousands.

◀ **Do Exercises 1–4.**

Add.

**1.** $6203 + 3542$

**2.**
$$
\begin{array}{r}
7\ 9\ 6\ 8 \\
+\ 5\ 4\ 9\ 7 \\
\hline
\end{array}
$$

$$
\begin{array}{r}
\overset{1\ 1\ \phantom{0}}{7\ 9\ 6\ 8} \\
+\ \ 5\ 4\ 9\ 7 \\
\hline
\square\ \square\ ,\ \square\ \ 5
\end{array}
$$

**3.**
$$
\begin{array}{r}
9\ 8\ 0\ 4 \\
+\ 6\ 3\ 7\ 8 \\
\hline
\end{array}
$$

**4.**
$$
\begin{array}{r}
1\ 9\ 3\ 2 \\
6\ 7\ 2\ 3 \\
9\ 8\ 7\ 8 \\
+\ 8\ 9\ 4\ 1 \\
\hline
\end{array}
$$

*Answers*

**1.** 9745  **2.** 13,465  **3.** 16,182
**4.** 27,474

*Guided Solution:*

**2.**
$$
\begin{array}{r}
\overset{1\ 1\ 1}{7\ 9\ 6\ 8} \\
+\ 5\ 4\ 9\ 7 \\
\hline
13{,}4\ 6\ 5
\end{array}
$$

---

 **CALCULATOR CORNER**

*Adding Whole Numbers*  This is the first of a series of *optional* discussions on using a calculator. A calculator is *not* a requirement for this text. Check with your instructor about whether you are allowed to use a calculator in the course.

There are many kinds of calculators and different instructions for their usage. Be sure to consult your user's manual.

To add whole numbers on a calculator, we use the ⊞ and ⊜ keys. After we press ⊜, the sum appears on the display.

**EXERCISES:**  Use a calculator to find each sum.

**1.** $73 + 48$  **2.** $925 + 677$  **3.** $826 + 415 + 691$  **4.** $253 + 490 + 121$

## b FINDING PERIMETER

Addition can be used when finding perimeter.

> ### PERIMETER
>
> The distance around an object is its **perimeter**.

**EXAMPLE 3**  Find the perimeter of the figure.

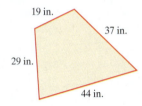

We add the lengths of the sides:

$$\text{Perimeter} = 29 \text{ in.} + 19 \text{ in.} + 37 \text{ in.} + 44 \text{ in.}$$
$$= 129 \text{ in.}$$

The perimeter of the figure is 129 in. (inches).

**Do Exercises 5 and 6.** ▶

**EXAMPLE 4**  Lucas Oil Stadium in Indianapolis has a unique retractable roof. When the roof is opened (retracted) in good weather to create an open-air stadium, the opening approximates a rectangle 588 ft long and 300 ft wide. Find the perimeter of the opening.

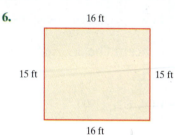

Opposite sides of a rectangle have equal lengths, so this rectangle has two sides of length 588 ft and two sides of length 300 ft.

$$\text{Perimeter} = 588 \text{ ft} + 300 \text{ ft} + 588 \text{ ft} + 300 \text{ ft}$$
$$= 1776 \text{ ft}$$

The perimeter of the opening is 1776 ft.

**Do Exercise 7.** ▶

Find the perimeter of each figure.

**GS 5.**

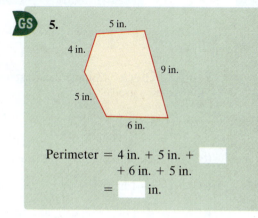

$$\text{Perimeter} = 4 \text{ in.} + 5 \text{ in.} + \boxed{\phantom{xx}}$$
$$+ 6 \text{ in.} + 5 \text{ in.}$$
$$= \boxed{\phantom{xx}} \text{ in.}$$

**6.**

(square: 16 ft, 15 ft, 15 ft, 16 ft)

**7. Index Cards.**  Two standard sizes for index cards are 3 in. by 5 in. and 5 in. by 8 in. Find the perimeter of each type of card.

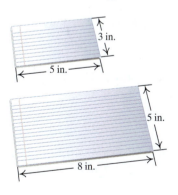

**Answers**

**5.** 29 in.   **6.** 62 ft   **7.** 16 in.; 26 in.
**Guided Solution:**
**5.** 9 in., 29

## ✔ Check Your Understanding

**Reading Check** Complete each statement with the appropriate word or number from the following list. Not every choice will be used.

| | | | |
|---|---|---|---|
| 0 | addends | law | product |
| 1 | factors | perimeter | sum |

**RC1.** In the addition 5 + 2 = 7, the numbers 5 and 2 are _____.

**RC2.** In the addition 5 + 2 = 7, the number 7 is the _____.

**RC3.** The sum of _____ and any number *a* is *a*.

**RC4.** The distance around an object is its _____.

**Concept Check** Add.

**CC1.** 20 + 30

**CC2.** 20 + 90

**CC3.** 300 + 500

**CC4.** 800 + 900

**CC5.** 5000 + 1000

**CC6.** 1000 + 9000

**a** Add.

**1.**
```
   3 6 4
 +   2 3
```

**2.**
```
   1 5 2 1
 +   3 4 8
```

**3.**
```
   8 6
 + 7 8
```

**4.**
```
   7 3
 + 6 9
```

**5.**
```
   1 7 1 6
 + 3 4 8 2
```

**6.**
```
   7 5 0 3
 + 2 6 8 3
```

**7.**
```
   9 9
 +   1
```

**8.**
```
   9 9 9
 +   1 1
```

**9.** 8113 + 390

**10.** 271 + 3338

**11.** 356 + 4910

**12.** 280 + 34,902

**13.** 3870 + 92 + 7 + 497

**14.** 10,120 + 12,989 + 5738

**15.**
```
   4 8 2 5
 + 1 7 8 3
```

**16.**
```
   3 6 5 4
 + 2 7 0 0
```

**17.**
```
   2 3,4 4 3
 + 1 0,9 8 9
```

**18.**
```
   4 5,8 7 9
 + 2 1,7 8 6
```

**19.**
```
   7 7,5 4 3
 + 2 3,7 6 7
```

**20.**
```
   9 9,9 9 9
 +     1 1 2
```

**21.**
```
   4 5
   2 5
   3 6
   4 4
 + 8 0
```

**22.**
```
   3 8
   2 7
   3 2
   1 4
 + 7 6
```

**23.**
```
  1 2,0 7 0
     2,9 5 4
+    3,4 0 0
```

**24.**
```
   4 2,4 8 7
   8 3,1 4 1
+  3 6,7 1 2
```

**25.**
```
  4 8 3 5
    7 2 9
  9 2 0 4
  8 9 8 6
+ 7 9 3 1
```

**26.**
```
  9 8 9
  5 6 6
  8 3 4
  9 2 0
+ 7 0 3
```

**b** Find the perimeter of each figure.

**27.**

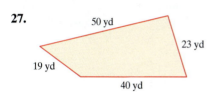

**28.**

**29.**

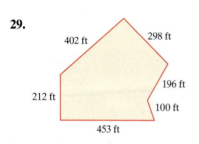

**30.**

**31.** Find the perimeter of a standard hockey rink.

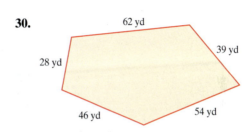

**32.** In Major League Baseball, how far does a batter travel when circling the bases after hitting a home run?

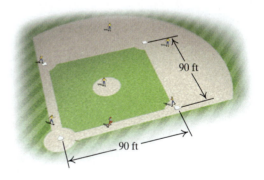

## Skill Maintenance

The exercises that follow begin an important feature called *Skill Maintenance exercises.* These exercises provide an ongoing review of topics previously covered in the text. You will see them in virtually every exercise set. It has been found that this kind of continuing review can significantly improve your performance on a final examination.

**33.** What does the digit 8 mean in 486,205?  [1.1a]

**34.** Write a word name for 9,346,399,468.  [1.1c]

## Synthesis

**35.** A fast way to add all the numbers from 1 to 10 inclusive is to pair 1 with 9, 2 with 8, and so on. Use a similar approach to add all numbers from 1 to 100 inclusive.

# 1.3

## OBJECTIVE

**a** Subtract whole numbers.

# Subtraction

## **a** SUBTRACTION OF WHOLE NUMBERS

**SKILL REVIEW**

*Give the meaning of digits in standard notation.* [1.1a]

Consider the number 328,974.

1. What digit names the number of hundreds?

2. What digit names the number of ones?

**Answers: 1.** 9 **2.** 4

MyLab Math
**VIDEO**

Subtraction is finding the difference of two numbers. Suppose you purchase 6 tickets for a concert and give 2 to a friend.

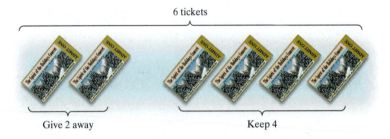

6 tickets

Give 2 away          Keep 4

The subtraction that represents this situation is

$$6 \quad - \quad 2 \quad = \quad 4.$$

Minuend   Subtrahend   Difference

The **minuend** is the number from which another number is being subtracted. The **subtrahend** is the number being subtracted. The **difference** is the result of subtracting the subtrahend from the minuend.

In the subtraction above, note that the difference, 4, is the number we add to 2 to get 6. This illustrates the relationship between addition and subtraction and leads us to the following definition of subtraction.

---

### SUBTRACTION

The difference $a - b$ is that unique whole number $c$ for which $a = c + b$.

---

We see that $6 - 2 = 4$ because $4 + 2 = 6$.

To subtract whole numbers, we subtract the ones digits first, then the tens digits, then the hundreds, then the thousands, and so on.

**EXAMPLE 1** Subtract: 9768 − 4320.

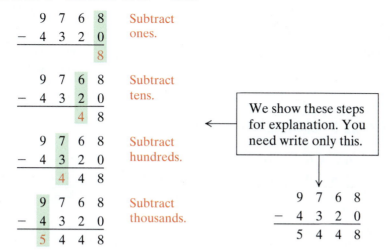

We can use addition to *check* subtraction.

*Subtraction:*

```
  9 7 6 8
− 4 3 2 0
  5 4 4 8
```

*Check by Addition:*

```
  5 4 4 8
+ 4 3 2 0
  9 7 6 8
```

**Do Exercise 1.** ▶

Sometimes we need to rename or regroup in order to perform a subtraction. This is also called *borrowing*.

**EXAMPLE 2** Subtract: 348 − 165.

First, we subtract the ones.

```
  3 4 8     Subtract ones.
− 1 6 5
      3
```

We cannot subtract the tens because there is no whole number that when added to 6 gives 4. To complete the subtraction, we must *borrow* 1 hundred from 3 hundreds and regroup it with the 4 tens. Then we can do the subtraction 14 tens − 6 tens = 8 tens.

```
  2  14
  3  4  8     Borrow one hundred. That is, 1 hundred = 10 tens, and
− 1  6  5     10 tens + 4 tens = 14 tens. Write 2 above the hundreds
        3     column and 14 above the tens.
```

```
  2  14
  3  4  8     Subtract tens; subtract hundreds.
− 1  6  5
  1  8  3
```

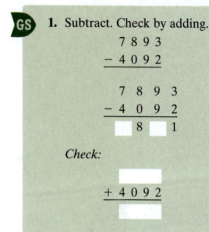

GS **1.** Subtract. Check by adding.

```
  7 8 9 3
− 4 0 9 2
```

```
  7 8 9 3
− 4 0 9 2
  □ 8 □ 1
```

*Check:*

```
  □ □ □ □
+ 4 0 9 2
  □ □ □ □
```

**CALCULATOR CORNER**

***Subtracting Whole Numbers***
To subtract whole numbers on a calculator, we use the ⊟ and ⊟ keys.

**EXERCISES:** Use a calculator to perform each subtraction. Check by adding.

**1.** 57 − 29

**2.** 81 − 34

**3.** 145 − 78

**4.** 612 − 493

**5.** 
```
  4 9 7 6
− 2 8 4 8
```

**6.** 
```
  1 2, 4 0 6
−  9, 8 1 3
```

*Answer*
**1.** 3801
*Guided Solution:*
**1.** 3, 0; 3801, 7893

This is what you should write.

$$
\begin{array}{r}
\overset{2\ \ 14}{3\ 4\ 8} \\
-\ 1\ 6\ 5 \\
\hline
1\ 8\ 3
\end{array}
\quad \textit{Check:} \quad
\begin{array}{r}
\overset{1}{1\ 8\ 3} \\
+\ 1\ 6\ 5 \\
\hline
3\ 4\ 8
\end{array}
$$

The answer checks because this is the top number in the subtraction.

**EXAMPLE 3**  Subtract: 6246 − 1879.

$$
\begin{array}{r}
6\ 2\ \overset{3\ \ 16}{4\ 6} \\
-\ 1\ 8\ 7\ 9 \\
\hline
7
\end{array}
$$

We cannot subtract 9 ones from 6 ones, but we can subtract 9 ones from 16 ones. We borrow 1 ten to get 16 ones.

$$
\begin{array}{r}
6\ 2\ \overset{\overset{13}{3}\ 16}{4\ 6} \\
-\ 1\ 8\ 7\ 9 \\
\hline
6\ 7
\end{array}
$$

We cannot subtract 7 tens from 3 tens, but we can subtract 7 tens from 13 tens. We borrow 1 hundred to get 13 tens.

$$
\begin{array}{r}
\overset{5\ \ 11\ 13}{6\ 2\ 4\ 6} \\
-\ 1\ 8\ 7\ 9 \\
\hline
4\ 3\ 6\ 7
\end{array}
$$

We cannot subtract 8 hundreds from 1 hundred, but we can subtract 8 hundreds from 11 hundreds. We borrow 1 thousand to get 11 hundreds. Finally, we subtract the thousands.

This is what you should write.

$$
\begin{array}{r}
\overset{5\ \ 11\ 13}{6\ 2\ 4\ 6} \\
-\ 1\ 8\ 7\ 9 \\
\hline
4\ 3\ 6\ 7
\end{array}
\quad \textit{Check:} \quad
\begin{array}{r}
\overset{1\ 1\ 1}{4\ 3\ 6\ 7} \\
+\ 1\ 8\ 7\ 9 \\
\hline
6\ 2\ 4\ 6
\end{array}
$$

The answer checks because this is the top number in the subtraction.

◀ **Do Exercises 2 and 3.**

**EXAMPLE 4**  Subtract: 902 − 477.

$$
\begin{array}{r}
\overset{8\ \ 9\ \ 12}{9\ 0\ 2} \\
-\ 4\ 7\ 7 \\
\hline
4\ 2\ 5
\end{array}
$$

We cannot subtract 7 ones from 2 ones. We have 9 hundreds, or 90 tens. We borrow 1 ten to get 12 ones. We then have 89 tens.

◀ **Do Exercises 4 and 5.**

**EXAMPLE 5**  Subtract: 8003 − 3667.

$$
\begin{array}{r}
\overset{7\ \ 9\ \ 9\ \ 13}{8\ 0\ 0\ 3} \\
-\ 3\ 6\ 6\ 7 \\
\hline
4\ 3\ 3\ 6
\end{array}
$$

We have 8 thousands, or 800 tens. We borrow 1 ten to get 13 ones. We then have 799 tens.

**EXAMPLES**

**6.** Subtract: 6000 − 3762.

$$
\begin{array}{r}
\overset{5\ \ 9\ \ 9\ \ 10}{6\ 0\ 0\ 0} \\
-\ 3\ 7\ 6\ 2 \\
\hline
2\ 2\ 3\ 8
\end{array}
$$

**7.** Subtract: 6024 − 2968.

$$
\begin{array}{r}
\overset{5\ \ 9\ \ 11\ \ 14}{6\ 0\ 2\ 4} \\
-\ 2\ 9\ 6\ 8 \\
\hline
3\ 0\ 5\ 6
\end{array}
$$

◀ **Do Exercises 6–9.**

---

Subtract. Check by adding.

**2.**
$$
\begin{array}{r}
8\ 6\ 8\ 6 \\
-\ 2\ 3\ 5\ 8 \\
\hline
\end{array}
$$

**3.**
$$
\begin{array}{r}
7\ 1\ 4\ 5 \\
-\ 2\ 3\ 9\ 8 \\
\hline
\end{array}
$$

Subtract.

**4.**
$$
\begin{array}{r}
7\ 0 \\
-\ 1\ 4 \\
\hline
\end{array}
$$

**5.**
$$
\begin{array}{r}
5\ 0\ 3 \\
-\ 2\ 9\ 8 \\
\hline
\end{array}
$$

$$
\begin{array}{r}
5\ \overset{\square\square\ \ \overset{13}{\phantom{0}}}{0\ 3} \\
-\ 2\ 9\ 8 \\
\hline
\square\ \square\ 0
\end{array}
$$

Subtract.

**6.**
$$
\begin{array}{r}
7\ 0\ 0\ 7 \\
-\ 6\ 3\ 4\ 9 \\
\hline
\end{array}
$$

**7.**
$$
\begin{array}{r}
6\ 0\ 0\ 0 \\
-\ 3\ 1\ 4\ 9 \\
\hline
\end{array}
$$

**8.**
$$
\begin{array}{r}
9\ 0\ 3\ 5 \\
-\ 7\ 4\ 8\ 9 \\
\hline
\end{array}
$$

**9.**
$$
\begin{array}{r}
2\ 0\ 0\ 1 \\
-\ \ \ 1\ 2\ 4 \\
\hline
\end{array}
$$

**Answers**

**2.** 6328  **3.** 4747  **4.** 56  **5.** 205  **6.** 658
**7.** 2851  **8.** 1546  **9.** 1877

**Guided Solution:**

**5.**
$$
\begin{array}{r}
\overset{4\ \ 9\ \ 13}{5\ 0\ 3} \\
-\ 2\ 9\ 8 \\
\hline
2\ 0\ 5
\end{array}
$$

✓ **Check Your Understanding**

**Reading Check**   Match each word or phrase from the following list with the indicated part of the subtraction sentence.

difference          minuend          subtraction symbol          subtrahend

**RC1.** A _____

**RC2.** B _____

**RC3.** C _____

**RC4.** D _____

(A)   (B)   (C)   (D)
↓      ↓      ↓      ↓
97   −   51   =   26

**Concept Check**   Subtract.

**CC1.** $10 - 8$

**CC2.** $100 - 7$

**CC3.** $100 - 93$

**CC4.** $1000 - 400$

**CC5.** $1000 - 5$

**CC6.** $1000 - 999$

**a**   Subtract. Check by adding.

**1.**
```
  6 5
− 2 1
```

**2.**
```
  8 7
− 3 4
```

**3.**
```
  8 6 6
− 3 3 3
```

**4.**
```
  5 2 6
− 3 2 3
```

**5.** $86 - 47$

**6.** $73 - 28$

**7.** $51 - 37$

**8.** $64 - 19$

**9.**
```
  5 6 3
− 1 9 4
```

**10.**
```
  7 9 5
− 3 9 8
```

**11.**
```
  3 9 1
− 3 6 5
```

**12.**
```
  3 1 6
− 2 4 7
```

**13.** $981 - 747$

**14.** $887 - 698$

**15.** $683 - 266$

**16.** $342 - 217$

**17.**
```
  7 7 6 9
− 2 3 8 7
```

**18.**
```
  6 4 3 1
− 2 8 9 6
```

**19.**
```
  4 5 1 2
− 1 7 3 4
```

**20.**
```
  8 3 6 4
− 5 3 7 5
```

**21.** 5318 − 2249

**22.** 9241 − 5643

**23.** 3947 − 2858

**24.** 7583 − 3641

**25.**  12,647
− 4,899

**26.**  16,222
− 5,888

**27.**  51,342
− 47,198

**28.**  32,194
− 29,236

**29.**  80
− 24

**30.**  90
− 78

**31.**  690
− 236

**32.**  803
− 418

**33.**  6808
− 3059

**34.**  9405
− 258

**35.**  2300
− 109

**36.**  7500
− 3604

**37.** 90,237 − 47,209

**38.** 84,703 − 298

**39.** 101,734 − 5760

**40.** 15,017 − 7809

**41.**  6007
− 1589

**42.**  8003
− 599

**43.**  39,000
− 37,695

**44.**  17,000
− 11,598

**45.** 10,008 − 19

**46.** 40,006 − 147

**47.** 50,001 − 1984

**48.** 30,004 − 6749

## Skill Maintenance

Add.  [1.2a]

**49.** 567 + 778

**50.** 901 + 23

**51.** 12,885 + 9807

**52.** 9909 + 1011

**53.** Write a word name for 6,375,602.  [1.1c]

**54.** Write expanded notation for 9103.  [1.1b]

## Synthesis

**55.** Fill in the missing digits to make the subtraction true:
9,☐48,621 − 2,097,☐81 = 7,251,140.

**56.** ▦ Subtract: 3,928,124 − 1,098,947.

# Multiplication

## a MULTIPLICATION OF WHOLE NUMBERS

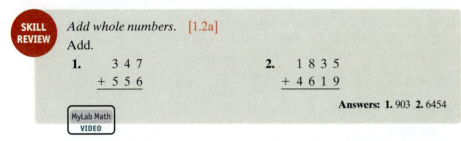

**SKILL REVIEW**

*Add whole numbers.* [1.2a]
Add.

| | |
|---|---|
| **1.**     3 4 7 <br>  + 5 5 6 | **2.**   1 8 3 5 <br>   + 4 6 1 9 |

**Answers: 1.** 903 **2.** 6454

MyLab Math
**VIDEO**

When you write a multiplication corresponding to a real-world situation, you should think of either a rectangular array or repeated addition. In some cases, it may help to think both ways.

### Repeated Addition

The multiplication $3 \times 5$ corresponds to this repeated addition.

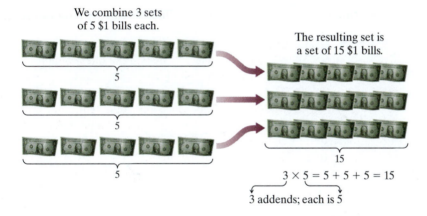

We combine 3 sets of 5 $1 bills each.

The resulting set is a set of 15 $1 bills.

$$3 \times 5 = 5 + 5 + 5 = 15$$

3 addends; each is 5

### Rectangular Arrays

Multiplications can also be thought of as rectangular arrays. Each of the following corresponds to the multiplication $3 \times 5$.

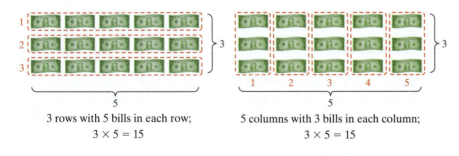

3 rows with 5 bills in each row;
$3 \times 5 = 15$

5 columns with 3 bills in each column;
$3 \times 5 = 15$

The numbers that we multiply are called **factors**. The result of the multiplication is called a **product**.

$$3 \quad \times \quad 5 \quad = \quad 15$$

Factor   Factor   Product

We have used an "×" to denote multiplication. A dot " · " is also commonly used. (Use of the dot is attributed to the German mathematician Gottfried Wilhelm von Leibniz over three centuries ago.) Parentheses are also used to denote multiplication. For example,

$$3 \times 5 = 3 \cdot 5 = (3)(5) = 3(5) = 15.$$

**EXAMPLE 1** Multiply: $5 \times 734$.

We have

```
        7 3 4
    ×       5
    ─────────
        2 0  ←── Multiply the 4 ones by 5: 5 × 4 = 20.
      1 5 0  ←── Multiply the 3 tens by 5: 5 × 30 = 150.
    3 5 0 0  ←── Multiply the 7 hundreds by 5: 5 × 700 = 3500.
    ─────────
    3 6 7 0  ←── Add.
```

Instead of writing each product on a separate line, we can use a shorter form.

```
      2
    7 3 4
  ×     5
  ───────
        0
```
Multiply the 4 ones by 5: $5 \cdot (4 \text{ ones}) = 20$ ones = 2 tens + 0 ones. Write 0 in the ones column and 2 above the tens.

```
    1 2
    7 3 4
  ×     5
  ───────
      7 0
```
Multiply the 3 tens by 5 and add 2 tens: $5 \cdot (3 \text{ tens}) = 15$ tens, 15 tens + 2 tens = 17 tens = 1 hundred + 7 tens. Write 7 in the tens column and 1 above the hundreds.

```
    1 2
    7 3 4
  ×     5
  ───────
  3 6 7 0
```
Multiply the 7 hundreds by 5 and add 1 hundred: $5 \cdot (7 \text{ hundreds}) = 35$ hundreds, 35 hundreds + 1 hundred = 36 hundreds.

```
    1 2
    7 3 4
  ×     5
  ───────
  3 6 7 0
```
You should write only this.

◀ **Do Exercises 1–4.**

Multiplication of whole numbers is based on a property called the **distributive law**. It says that to multiply a number by a sum, $a \cdot (b + c)$, we can multiply each addend by $a$ and then add like this: $(a \cdot b) + (a \cdot c)$. Thus, $a \cdot (b + c) = (a \cdot b) + (a \cdot c)$. For example, consider the following.

$$4 \cdot (2 + 3) = 4 \cdot 5 = 20 \qquad \text{Adding first; then multiplying}$$

$$4 \cdot (2 + 3) = (4 \cdot 2) + (4 \cdot 3) = 8 + 12 = 20 \qquad \text{Multiplying first; then adding}$$

The results are the same, so $4 \cdot (2 + 3) = (4 \cdot 2) + (4 \cdot 3)$.

---

Multiply.

**1.**
```
    5 8
  ×   2
```

**2.**
```
    3 7
  ×   4
```

**3.**
```
    8 2 3
  ×     6
```

**4.** GS
```
    1 3 4 8
  ×       5
```

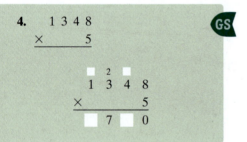

```
      2
  □     □
  1 3 4 8
  ×     5
  ─────────
  □ 7 □ 0
```

**Answers**

**1.** 116  **2.** 148  **3.** 4938  **4.** 6740

**Guided Solution:**

**4.**
```
    1 2 4
    1 3 4 8
  ×       5
  ─────────
    6 7 4 0
```

Let's find the product $51 \times 32$. Since $32 = 2 + 30$, we can think of this product as

$$51 \times 32 = 51 \times (2 + 30) = (51 \times 2) + (51 \times 30).$$

That is, we multiply 51 by 2, then we multiply 51 by 30, and finally we add. We can write our work in columns.

```
      5 1
   ×  3 2
   ─────────
    1 0 2      Multiplying by 2
  1 5 3 0      Multiplying by 30. (We write a 0 and then multiply 51 by 3.)
```

We add to obtain the product.

```
      5 1
   ×  3 2
   ─────────
    1 0 2
  1 5 3 0
  ─────────
  1 6 3 2      Adding to obtain the product
```

**EXAMPLE 2**   Multiply: $457 \times 683$.

```
          5 2
        6 8 3
   ×    4 5 7
   ───────────
      4 7 8 1      Multiplying 683 by 7
```

```
        4 1
        5 2
        6 8 3
   ×    4 5 7
   ───────────
      4 7 8 1
    3 4 1 5 0      Multiplying 683 by 50
```

```
        3 1
        4 1
        5 2
        6 8 3
   ×    4 5 7
   ───────────
      4 7 8 1
    3 4 1 5 0
  2 7 3 2 0 0      Multiplying 683 by 400. (We write 00 and then multiply 683 by 4.)
  ───────────
  3 1 2 , 1 3 1    Adding
```

Do Exercises 5–8. ▶

**EXAMPLE 3**   Multiply: $306 \times 274$.

Note that $306 = 3$ hundreds $+ 6$ ones.

```
      2 7 4
   ×  3 0 6
   ───────────
    1 6 4 4        Multiplying by 6
  8 2 2 0 0        Multiplying by 3 hundreds. (We write 00 and then multiply 274 by 3.)
  ───────────
  8 3 , 8 4 4      Adding
```

Do Exercises 9–12. ▶

**CALCULATOR CORNER**

*Multiplying Whole Numbers*
To multiply whole numbers on a calculator, we use the ⊠ and ⊟ keys.

**EXERCISES:**   Use a calculator to find each product.

1. $56 \times 8$
2. $845 \times 26$
3. $5 \cdot 1276$
4. $126(314)$
5.     $\begin{array}{r} 3\,7\,6\,0 \\ \times \quad\;\; 4\,8 \end{array}$
6.     $\begin{array}{r} 5\,2\,1\,8 \\ \times \quad 4\,5\,3 \end{array}$

Multiply.

5.  $\begin{array}{r} 4\,5 \\ \times\, 2\,3 \end{array}$       6. $48 \times 63$

7.  $\begin{array}{r} 7\,4\,6 \\ \times \;\; 6\,2 \end{array}$       8. $245 \times 837$

Multiply.

9.  $\begin{array}{r} 4\,7\,2 \\ \times\, 3\,0\,6 \end{array}$       10. $408 \times 704$

11.  $\begin{array}{r} 2\,3\,4\,4 \\ \times\, 6\,0\,0\,5 \end{array}$       12. $\begin{array}{r} 1\,0\,0\,6 \\ \times \;\; 7\,0\,3 \end{array}$

*Answers*

5. 1035   6. 3024   7. 46,252   8. 205,065
9. 144,432   10. 287,232   11. 14,075,720
12. 707,218

**Multiply.**

**13.**
```
    4 7 2
  × 8 3 0
```

**14.**
```
    2 3 4 4
  × 7 4 0 0
```

**15.** $100 \times 562$

**16.** $1000 \times 562$

**17. a)** Find $23 \cdot 47$.

  **b)** Find $47 \cdot 23$.

  **c)** Compare your answers to parts (a) and (b).

**Multiply.**

**18.** $5 \cdot 2 \cdot 4$

**19.** $4 \cdot 2 \cdot 6$

**EXAMPLE 4** Multiply: $360 \times 274$.

Note that $360 = 3$ hundreds $+ 6$ tens.

```
        2 7 4        ┌─ Multiplying by 6 tens. (We write 0 and
      ×   3 6 0      │   then multiply 274 by 6.)
      1 6 4 4 0  ◄───┘
      8 2 2 0 0  ◄── Multiplying by 3 hundreds. (We write 00
      ─────────      and then multiply 274 by 3.)
      9 8,6 4 0      Adding
```

◄ **Do Exercises 13–16.**

When we multiply two numbers, we can change the order of the numbers without changing their product. For example, $3 \cdot 6 = 18$ and $6 \cdot 3 = 18$. This illustrates the **commutative law of multiplication:** $a \cdot b = b \cdot a$.

◄ **Do Exercise 17.**

To multiply three or more numbers, we group them so that we multiply two at a time. Consider $2 \cdot 3 \cdot 4$. We can group these numbers as $2 \cdot (3 \cdot 4)$ or as $(2 \cdot 3) \cdot 4$. The parentheses tell what to do first:

$$2 \cdot (3 \cdot 4) = 2 \cdot (12) = 24. \qquad \text{We multiply 3 and 4 and then multiply that product by 2.}$$

We can also multiply 2 and 3 and then multiply that product by 4:

$$(2 \cdot 3) \cdot 4 = (6) \cdot 4 = 24.$$

Either way we get 24. It does not matter how we group the numbers. This illustrates the **associative law of multiplication:** $a \cdot (b \cdot c) = (a \cdot b) \cdot c$.

◄ **Do Exercises 18 and 19.**

Two more properties of multiplication involve multiplying by 0 and by 1. The product of 0 and any whole number is 0: $0 \cdot a = a \cdot 0 = 0$. For example, $0 \cdot 3 = 3 \cdot 0 = 0$. Multiplying a number by 1 does not change the number: $1 \cdot a = a \cdot 1 = a$. For example, $1 \cdot 3 = 3 \cdot 1 = 3$. We say that 1 is the **multiplicative identity**.

## b FINDING AREA

We can think of the area of a rectangular region as the number of square units needed to fill it. Here is a rectangle 4 cm (centimeters) long and 3 cm wide. It takes 12 square centimeters (sq cm) to fill it.

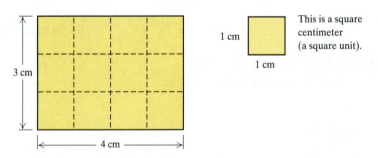

In this case, we have a rectangular array of 3 rows, each of which contains 4 squares. The number of square units is given by $3 \cdot 4$, or 12. That is, $A = l \cdot w = 3 \text{ cm} \cdot 4 \text{ cm} = 12 \text{ sq cm}$.

*Answers*

**13.** 391,760  **14.** 17,345,600
**15.** 56,200  **16.** 562,000
**17. (a)** 1081; **(b)** 1081; **(c)** same
**18.** 40  **19.** 48

**EXAMPLE 5** *Professional Pool Table.* The playing area of a pool table used in professional tournaments is 50 in. by 100 in. (There are 6-in.-wide rails on the outside that are not included in the playing area.) Determine the playing area.

100 in.

50 in.

If we think of filling the rectangle with square inches, we have a rectangular array. The length $l = 100$ in. and the width $w = 50$ in. Thus the area $A$ is given by the formula

$$A = l \cdot w = 100 \text{ in.} \cdot 50 \text{ in.} = 5000 \text{ sq in.}$$

**Do Exercise 20.** ▶

**GS 20. Painting a Room.** Ben and Elizabeth plan to paint one wall of a bedroom in a dark accent color. The wall is a rectangle 12 ft long and 8 ft high. Determine the area of the wall.

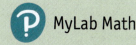

$$A = l \cdot w$$
$$= 12 \text{ ft} \cdot \boxed{\phantom{xx}}$$
$$= \boxed{\phantom{xx}} \text{ sq ft}$$

*Answer*
20. 96 sq ft
*Guided Solution:*
20. 8 ft, 96

---

| 1.4 | **Exercise Set** | FOR EXTRA HELP | Ⓟ MyLab Math |

### ✓ Check Your Understanding

**Reading Check** Complete each statement with the appropriate word or number from the following list. Not every choice will be used.

| 0 | addends | product |
|---|---------|---------|
| 1 | factors | sum |

**RC1.** In the multiplication $4 \times 3 = 12$, 4 and 3 are _____.

**RC2.** In the multiplication $4 \times 3 = 12$, 12 is the _____.

**RC3.** The product of _____ and any number $a$ is 0.

**RC4.** The product of _____ and any number $a$ is $a$.

**Concept Check** Multiply.

**CC1.** $20 \times 40$

**CC2.** $60 \times 80$

**CC3.** $300 \times 20$

**CC4.** $400 \times 200$

**CC5.** $70 \times 900$

**CC6.** $500 \times 800$

Multiply.

**1.**  6 5
    × 8

**2.**  8 7
    × 4

**3.**  9 4
    × 6

**4.**  7 6
    × 9

**5.** 3 · 509

**6.** 7 · 806

**7.** 7(9229)

**8.** 4(7867)

**9.** 90(53)

**10.** 60(78)

**11.** (47)(85)

**12.** (34)(87)

**13.**  8 7
    × 1 0

**14.**  2 3 4 0
    × 1 0 0 0

**15.**  9 6
    × 2 0

**16.**  8 0 0
    × 7 0 0

**17.**  6 4 3
    × 7 2

**18.**  7 7 7
    × 7 7

**19.**  4 4 4
    × 3 3

**20.**  5 4 9
    × 8 8

**21.**  5 6 4
    × 4 5 8

**22.**  4 3 2
    × 3 7 5

**23.**  8 5 3
    × 9 3 6

**24.**  3 4 6
    × 6 5 9

**25.**  6 4 2 8
    × 3 2 2 4

**26.**  8 9 2 8
    × 3 1 7 2

**27.**  3 4 8 2
    × 1 0 4

**28.**  6 4 0 8
    × 6 0 6 4

**29.**  8 7 6
    × 3 4 5

**30.**  3 5 5
    × 2 9 9

**31.**  7 8 8 9
    × 6 2 2 4

**32.**  6 5 2 1
    × 3 4 4 9

**33.**
$$
\begin{array}{r}
5\ 6\ 0\ 8 \\
\times\ 4\ 5\ 0\ 0 \\
\hline
\end{array}
$$

**34.**
$$
\begin{array}{r}
4\ 5\ 0\ 6 \\
\times\ 7\ 8\ 0\ 0 \\
\hline
\end{array}
$$

**35.**
$$
\begin{array}{r}
5\ 0\ 0\ 6 \\
\times\ 4\ 0\ 0\ 8 \\
\hline
\end{array}
$$

**36.**
$$
\begin{array}{r}
6\ 0\ 0\ 9 \\
\times\ 2\ 0\ 0\ 3 \\
\hline
\end{array}
$$

**b**   Find the area of each region.

**37.**

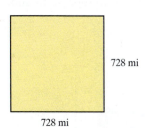

728 mi

728 mi

**38.**

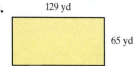

129 yd

65 yd

**39.** Find the area of the region formed by the base lines on a Major League Baseball diamond.

90 ft

90 ft

**40.** Find the area of a standard-sized hockey rink.

85 ft

200 ft

## Skill Maintenance

Add. [1.2a]

**41.**
$$
\begin{array}{r}
4\ 9\ 0\ 8 \\
5\ 6\ 6\ 7 \\
+\ 2\ 1\ 1\ 0 \\
\hline
\end{array}
$$

**42.**
$$
\begin{array}{r}
9\ 8\ 7\ 6 \\
8\ 7\ 6 \\
7\ 6 \\
+\ \ \ \ \ 6 \\
\hline
\end{array}
$$

Subtract. [1.3a]

**43.**
$$
\begin{array}{r}
9\ 8\ 7\ 6 \\
-\ \ \ 9\ 8\ 7 \\
\hline
\end{array}
$$

**44.**
$$
\begin{array}{r}
3\ 4\ 0\,,7\ 9\ 8 \\
-\ \ \ 8\ 6\,,6\ 7\ 9 \\
\hline
\end{array}
$$

**45.** What does the digit 4 mean in 9,482,157? [1.1a]

**46.** What digit in 38,026 names the number of hundreds? [1.1a]

**47.** Write expanded notation for 12,847. [1.1b]

**48.** Write a word name for 7,432,000. [1.1c]

## Synthesis

**49.** ▦ An 18-story office building is box-shaped. Each floor measures 172 ft by 84 ft with a 20-ft by 35-ft rectangular area lost to an elevator and a stairwell. How much area is available as office space?

# Division

a  **DIVISION OF WHOLE NUMBERS**

## Repeated Subtraction

Division of whole numbers applies to two kinds of situations. The first is repeated subtraction. Suppose we have 20 doughnuts and we want to find out how many sets of 5 there are. One way to do this is to repeatedly subtract sets of 5 as follows.

20 doughnuts

How many sets of 5 doughnuts each?

Since there are 4 sets of 5 doughnuts each, we have

20  ÷  5  =  4.

Dividend   Divisor   Quotient

The division $20 \div 5$ is read "20 divided by 5." The **dividend** is 20, the **divisor** is 5, and the **quotient** is 4. We divide the *dividend* by the *divisor* to get the *quotient*. We can also express the division $20 \div 5 = 4$ as

$$\frac{20}{5} = 4 \quad \text{or} \quad 5\overline{)20}.$$

## Rectangular Arrays

We can also think of division in terms of rectangular arrays. Consider again the 20 doughnuts and division by 5. We can arrange the doughnuts in a rectangular array with 5 rows and ask, "How many are in each row?"

We can also consider a rectangular array with 5 doughnuts in each column and ask, "How many columns are there?" The answer is still 4.

In each case, we are asking, "What do we multiply 5 by in order to get 20?"

Missing factor                Quotient

$$5 \cdot \square = 20 \qquad 20 \div 5 = \square$$

This leads us to the following definition of division.

## DIVISION

The quotient $a \div b$, where $b$ is not 0, is that unique number $c$ for which $a = b \cdot c$.

This definition shows the relation between division and multiplication. We see, for instance, that

$$20 \div 5 = 4 \text{ because } 20 = 5 \cdot 4.$$

This relation allows us to use multiplication to check division.

**EXAMPLE 1** Divide. Check by multiplying.

**a)** $16 \div 8$      **b)** $\dfrac{36}{4}$      **c)** $7\overline{)56}$

We do so as follows.

**a)** $16 \div 8 = 2$    *Check:* $8 \cdot 2 = 16.$

**b)** $\dfrac{36}{4} = 9$    *Check:* $4 \cdot 9 = 36.$

**c)** $\begin{array}{r} 8 \\ 7\overline{)56} \end{array}$    *Check:* $7 \cdot 8 = 56.$

**Do Exercises 1–3.** ▶

Divide. Check by multiplying.

**1.** $9\overline{)45}$

**2.** $27 \div 3$

**3.** $\dfrac{48}{6}$

Let's consider some basic properties of division.

## DIVIDING BY 1

Any number divided by 1 is that same number: $a \div 1 = \dfrac{a}{1} = a.$

For example, $6 \div 1 = 6$ and $\dfrac{15}{1} = 15.$

## DIVIDING A NUMBER BY ITSELF

Any nonzero number divided by itself is 1: $a \div a = \dfrac{a}{a} = 1, \quad a \neq 0.$

For example, $7 \div 7 = 1$ and $\dfrac{22}{22} = 1.$

## DIVIDENDS OF 0

Zero divided by any nonzero number is 0: $0 \div a = \dfrac{0}{a} = 0, \quad a \neq 0.$

For example, $0 \div 14 = 0$ and $\dfrac{0}{3} = 0.$

**Do Exercises 4–7.** ▶

Divide.

**4.** $\dfrac{9}{9}$      **5.** $\dfrac{8}{1}$

**6.** $\dfrac{0}{12}$      **7.** $\dfrac{28}{28}$

*Answers*
**1.** 5 **2.** 9 **3.** 8 **4.** 1 **5.** 8 **6.** 0 **7.** 1

Why can't we divide by 0? Suppose the number 4 could be divided by 0. Then if $\square$ were the answer, we would have

$$4 \div 0 = \square,$$

and since 0 times any number is 0, we would have

$$4 = \square \cdot 0 = 0. \qquad \text{False!}$$

Thus, the only possible number that could be divided by 0 would be 0 itself. But such a division would give us any number we wish. For instance,

$$\left. \begin{array}{lll} 0 \div 0 = 8 & \text{because} & 0 = 8 \cdot 0; \\ 0 \div 0 = 3 & \text{because} & 0 = 3 \cdot 0; \\ 0 \div 0 = 7 & \text{because} & 0 = 7 \cdot 0. \end{array} \right\} \qquad \text{All true!}$$

We avoid the preceding difficulties by agreeing to exclude division by 0.

◀ **Do Exercises 8–9.**

**8.** Divide if possible: $0 \div 2$. If not possible, write "not defined." **GS**

$0 \div 2$ means ☐ divided by ☐ .

Since zero divided by any non-zero number is 0, $0 \div 2 = $ ☐ .

**9.** Divide if possible: $7 \div 0$. If not possible, write "not defined."

$7 \div 0$ means ☐ divided by ☐ .

Since division by 0 is not defined, $7 \div 0$ is ☐ .

## Division with a Remainder

Suppose everyone in a group of 22 people wants to ride a roller coaster. If each car on the ride holds 6 people, the group will fill 3 cars and there will be 4 people left over.

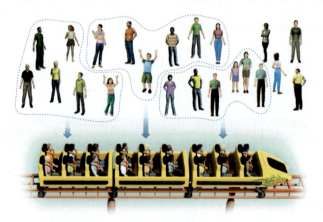

We can think of this situation as the following division. The people left over are the **remainder**.

$$
\begin{array}{r}
3 \;\leftarrow \text{Quotient} \\
6 \,\overline{)\, 2\,2} \\
\underline{1\,8} \\
4 \;\; < \;\; \text{Remainder}
\end{array}
$$

We express the result as

$$22 \div 6 = 3 \text{ R } 4.$$

Dividend    Divisor    Quotient    Remainder

Note that

Quotient · Divisor + Remainder = Dividend.

Thus we have

$3 \cdot 6 = 18$      Quotient · Divisor

and   $18 + 4 = 22.$     Adding the remainder. The result is the dividend.

We now show a procedure for dividing whole numbers.

**EXAMPLE 2** Divide and check: $4 \overline{)3\ 4\ 5\ 7}$.

First, we try to divide the first digit of the dividend, 3, by the divisor, 4. Since $3 \div 4$ is not a whole number, we consider the first *two* digits of the dividend.

$$
\begin{array}{r}
8\phantom{\ \ \ \ } \\
4\overline{)3\ 4\ 5\ 7} \\
3\ 2\phantom{\ \ \ \ } \\
\hline
2\phantom{\ \ \ \ }
\end{array}
$$

Since $4 \cdot 8 = 32$ and 32 is smaller than 34, we write an 8 in the quotient above the 4. We also write 32 below 34 and subtract.

What if we had chosen a number other than 8 for the first digit of the quotient? Suppose we had used 7 instead of 8 and subtracted $4 \cdot 7$, or 28, from 34. The result would have been $34 - 28$, or 6. Because 6 is larger than the divisor, 4, we know that there is at least one more factor of 4 in 34, and thus 7 is too small. If we had used 9 instead of 8, then we would have tried to subtract $4 \cdot 9$, or 36, from 34. That difference is not a whole number, so we know 9 is too large. When we subtract, the difference must be smaller than the divisor.

Let's continue dividing.

$$
\begin{array}{r}
8\ 6\phantom{\ \ } \\
4\overline{)3\ 4\ 5\ 7} \\
3\ 2\phantom{\ \ \ \ } \\
\hline
2\ 5\phantom{\ \ } \\
2\ 4\phantom{\ \ } \\
\hline
1\phantom{\ \ }
\end{array}
$$

Now we bring down the 5 in the dividend and consider $25 \div 4$. Since $4 \cdot 6 = 24$ and 24 is smaller than 25, we write 6 in the quotient above the 5. We also write 24 below 25 and subtract. The difference, 1, is smaller than the divisor, so we know that 6 is the correct choice.

$$
\begin{array}{r}
8\ 6\ 4 \\
4\overline{)3\ 4\ 5\ 7} \\
3\ 2\phantom{\ \ \ \ } \\
\hline
2\ 5\phantom{\ \ } \\
2\ 4\phantom{\ \ } \\
\hline
1\ 7 \\
1\ 6 \\
\hline
1
\end{array}
$$

We bring down the 7 and consider $17 \div 4$. Since $4 \cdot 4 = 16$ and 16 is smaller than 17, we write 4 in the quotient above the 7. We also write 16 below 17 and subtract.

$1 \leftarrow$ The remainder is 1.

Check: $864 \cdot 4 = 3456$ and $3456 + 1 = 3457$.

The answer is 864 R 1.

**Do Exercises 10–12.** ▶

Divide and check.

**10.** $3 \overline{)2\ 3\ 9}$

**11.** $5 \overline{)5\ 8\ 6\ 4}$

**12.** $6 \overline{)3\ 8\ 5\ 5}$

**EXAMPLE 3** Divide: $8904 \div 42$.

Because 42 is close to 40, we think of the divisor as 40 when we make our choices of digits in the quotient.

$$
\begin{array}{r}
2\phantom{000} \\
42\overline{)8904} \\
84\phantom{0}\downarrow \\
\hline
5\,0
\end{array}
$$
← *Think:* $89 \div 40$. We try 2. Multiply $42 \cdot 2$ and subtract. Then bring down the 0.

$$
\begin{array}{r}
2\;1\phantom{00} \\
42\overline{)8904} \\
84\phantom{00} \\
\hline
5\,0\phantom{0} \\
4\,2\downarrow \\
\hline
8\,4
\end{array}
$$
← *Think:* $50 \div 40$. We try 1. Multiply $42 \cdot 1$ and subtract. Then bring down the 4.

$$
\begin{array}{r}
2\;1\;2 \\
42\overline{)8904} \\
84\phantom{00} \\
\hline
5\,0\phantom{0} \\
4\,2\phantom{0} \\
\hline
8\,4 \\
8\,4 \\
\hline
0
\end{array}
$$
← *Think:* $84 \div 40$. We try 2. Multiply $2 \cdot 42$ and subtract.

The remainder is 0, so the answer is 212.

◀ **Do Exercises 13 and 14.**

Divide.

**13.** $45\overline{)6030}$

**14.** $52\overline{)3288}$

..................................... **Caution!** .....................................

Be careful to keep the digits lined up correctly when you divide.

...........................................................................................................................

---

**CALCULATOR CORNER**

*Dividing Whole Numbers* To divide whole numbers on a calculator, we use the ÷ and = keys.

When we enter $453 \div 15$, the display reads ⬚30.2⬚. Note that the result is not a whole number. This tells us that there is a remainder. The number 30.2 is expressed in decimal notation. The symbol "." is called a decimal point. The number to the right of the decimal point is not the remainder, although it is possible to use that number to find the remainder. We will not do so here.

**EXERCISES:** Use a calculator to perform each division.

**1.** $19\overline{)532}$

**2.** $7\overline{)861}$

**3.** $9367 \div 29$

**4.** $12{,}276 \div 341$

---

*Answers*

**13.** 134   **14.** 63 R 12

## Zeros in Quotients

**EXAMPLE 4**   Divide: 6341 ÷ 7.

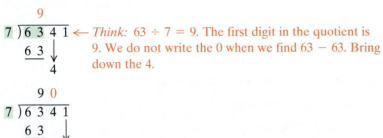

$$
\begin{array}{r}
9 \\
7\,\overline{)6\,3\,4\,1} \\
6\,3 \\
\hline
4
\end{array}
$$
← *Think:* 63 ÷ 7 = 9. The first digit in the quotient is 9. We do not write the 0 when we find 63 − 63. Bring down the 4.

$$
\begin{array}{r}
9\ 0 \\
7\,\overline{)6\,3\,4\,1} \\
6\,3 \\
\hline
4\,1
\end{array}
$$
← *Think:* 4 ÷ 7. If we subtract a group of 7's, such as 7, 14, 21, etc., from 4, we do not get a whole number, so the next digit in the quotient is 0. Bring down the 1.

$$
\begin{array}{r}
9\ 0\ 5 \\
7\,\overline{)6\,3\,4\,1} \\
6\,3 \\
\hline
4\,1 \\
3\,5 \\
\hline
6
\end{array}
$$
← *Think:* 41 ÷ 7. We try 5. Multiply 7 · 5 and subtract.

← The remainder is 6.

The answer is 905 R 6.

**Do Exercises 15 and 16.** ▶

**EXAMPLE 5**   Divide: 8169 ÷ 34.

Because 34 is close to 30, we think of the divisor as 30 when we make our choices of digits in the quotient.

$$
\begin{array}{r}
2 \\
34\,\overline{)8\,1\,6\,9} \\
6\,8 \\
\hline
1\,3\,6
\end{array}
$$
← *Think:* 81 ÷ 30. We try 2. Multiply 34 · 2 and subtract. Then bring down the 6.

$$
\begin{array}{r}
2\ 4 \\
34\,\overline{)8\,1\,6\,9} \\
6\,8 \\
\hline
1\,3\,6 \\
1\,3\,6 \\
\hline
9
\end{array}
$$
← *Think:* 136 ÷ 30. We try 4. Multiply 34 · 4 and subtract. The difference is 0, so we do not write it. Bring down the 9.

$$
\begin{array}{r}
2\ 4\ 0 \\
34\,\overline{)8\,1\,6\,9} \\
6\,8 \\
\hline
1\,3\,6 \\
1\,3\,6 \\
\hline
9 \\
0 \\
\hline
9
\end{array}
$$
*Think:* 9 ÷ 34. If we subtract a group of 34's, such as 34 or 68, from 9, we do not get a whole number, so the last digit in the quotient is 0.
← The remainder is 9.

The answer is 240 R 9.

**Do Exercises 17 and 18.** ▶

### TO THE INSTRUCTOR AND THE STUDENT

This section presents a review of division of whole numbers. Students who are successful should go on to Section 1.6. Those who have trouble should study developmental unit D near the back of this text and then repeat Section 1.5.

Divide.

**15.**  $6\,\overline{)4\,8\,4\,6}$

**16.**  $7\,\overline{)7\,6\,1\,6}$

Divide.

**17.**  $27\,\overline{)9\,7\,2\,4}$

**18.**  $56\,\overline{)4\,4,\!8\,4\,7}$

***Answers***
**15.** 807 R 4   **16.** 1088   **17.** 360 R 4
**18.** 800 R 47

✓ **Check Your Understanding**

**Reading Check** Match each word from the following list with the indicated part of the division.

dividend          divisor          quotient          remainder

**RC1.** A _____

**RC2.** B _____

**RC3.** C _____

**RC4.** D _____

$$\begin{array}{r} 2\ 9 \leftarrow Ⓐ \\ Ⓓ \rightarrow 8\overline{)2\ 3\ 5} \leftarrow Ⓑ \\ \underline{1\ 6} \\ 7\ 5 \\ \underline{7\ 2} \\ 3 \leftarrow Ⓒ \end{array}$$

**Concept Check** Divide.

**CC1.** $3\overline{)7}$

**CC2.** $4\overline{)15}$

**CC3.** $5\overline{)64}$

**CC4.** $2\overline{)97}$

**a**  Divide, if possible. If not possible, write "not defined."

**1.** $72 \div 6$

**2.** $54 \div 9$

**3.** $\dfrac{23}{23}$

**4.** $\dfrac{37}{37}$

**5.** $22 \div 1$

**6.** $\dfrac{56}{1}$

**7.** $\dfrac{0}{7}$

**8.** $\dfrac{0}{32}$

**9.** $\dfrac{16}{0}$

**10.** $74 \div 0$

**11.** $\dfrac{48}{8}$

**12.** $\dfrac{20}{4}$

Divide.

**13.** $277 \div 5$

**14.** $699 \div 3$

**15.** $864 \div 8$

**16.** $869 \div 8$

**17.** $4\overline{)1\ 2\ 2\ 8}$

**18.** $3\overline{)2\ 1\ 2\ 4}$

**19.** $6\overline{)4\ 5\ 2\ 1}$

**20.** $9\overline{)9\ 1\ 1\ 0}$

**21.** 297 ÷ 4

**22.** 389 ÷ 2

**23.** 738 ÷ 8

**24.** 881 ÷ 6

**25.** 5 ) 8 5 1 5

**26.** 3 ) 6 0 2 7

**27.** 9 ) 8 8 8 8

**28.** 8 ) 4 1 3 9

**29.** 127,000 ÷ 10

**30.** 127,000 ÷ 100

**31.** 127,000 ÷ 1000

**32.** 4260 ÷ 10

**33.** 7 0 ) 3 6 9 2

**34.** 2 0 ) 5 7 9 8

**35.** 3 0 ) 8 7 5

**36.** 4 0 ) 9 8 7

**37.** 852 ÷ 21

**38.** 942 ÷ 23

**39.** 8 5 ) 7 6 7 2

**40.** 5 4 ) 2 7 2 9

**41.** 1 1 1 ) 3 2 1 9

**42.** 1 0 2 ) 5 6 1 2

**43.** 8 ) 8 4 3

**44.** 7 ) 7 4 9

**45.** 5 ) 8 0 4 7

**46.** 9 ) 7 2 7 3

**47.** 5 ) 5 0 3 6

**48.** 7 ) 7 0 7 4

**49.** 1058 ÷ 46

**50.** 7242 ÷ 24

**51.** 3425 ÷ 32

**52.** 4 8 ) 4 8 9 9

**53.** $24\overline{)8880}$  **54.** $36\overline{)7563}$  **55.** $28\overline{)17,067}$  **56.** $36\overline{)28,929}$

**57.** $80\overline{)24,320}$  **58.** $90\overline{)88,560}$  **59.** $285\overline{)999,999}$

**60.** $306\overline{)888,888}$  **61.** $456\overline{)3,679,920}$  **62.** $803\overline{)5,622,606}$

## Skill Maintenance

Subtract.  [1.3a]

**63.**  $\begin{array}{r} 4\,9\,0\,8 \\ -\ 3\,6\,6\,7 \\ \hline \end{array}$

**64.**  $\begin{array}{r} 8\,8,7\,7\,7 \\ -\ 2\,2,3\,3\,3 \\ \hline \end{array}$

Multiply.  [1.4a]

**65.**  $\begin{array}{r} 1\,9\,8 \\ \times\ 1\,0\,0 \\ \hline \end{array}$

**66.**  $\begin{array}{r} 2\,6\,8 \\ \times\ \ \ 3\,5 \\ \hline \end{array}$

Use the following figure for Exercises 67 and 68.

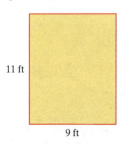

11 ft

9 ft

**67.** Find the perimeter of the figure.  [1.2b]

**68.** Find the area of the figure.  [1.4b]

## Synthesis

**69.** Complete the following table.

| $a$ | $b$ | $a \cdot b$ | $a + b$ |
|-----|-----|-------------|---------|
|     | 68  | 3672        |         |
| 84  |     |             | 117     |
|     |     | 32          | 12      |

**70.** Find a pair of factors whose product is 36 and:

  **a)** whose sum is 13.
  **b)** whose difference is 0.
  **c)** whose sum is 20.
  **d)** whose difference is 9.

**71.** A group of 1231 college students is going to use buses to take a field trip. Each bus can hold 42 students. How many buses are needed?

**72.** ▦ Fill in the missing digits to make the equation true:
$$34,584,132 \div 76\square = 4\square,386.$$

# Mid-Chapter Review

## Concept Reinforcement

Determine whether each statement is true or false.

_____ **1.** If $a - b = c$, then $b = a + c$.   [1.3a]

_____ **2.** We can think of the multiplication $4 \times 3$ as a rectangular array containing 4 rows with 3 items in each row.   [1.4a]

_____ **3.** We can think of the multiplication $4 \times 3$ as a rectangular array containing 3 columns with 4 items in each column.   [1.4a]

_____ **4.** The product of two whole numbers is always greater than either of the factors.   [1.4a]

_____ **5.** Zero divided by any nonzero number is 0.   [1.5a]

_____ **6.** Any number divided by 1 is the number 1.   [1.5a]

## Guided Solutions

**GS**  Fill in each blank to create a correct statement or solution.

**7.** Write a word name for 95,406,237.   [1.1c]

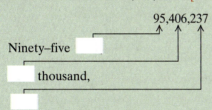

**8.** Subtract: $604 - 497$.   [1.3a]

$$\begin{array}{r} 6\ \ 0\ \ 4 \\ -\ 4\ \ 9\ \ 7 \\ \hline \end{array}$$

## Mixed Review

In each of the following numbers, what does the digit 6 mean?   [1.1a]

**9.** 2698

**10.** 61,204

**11.** 146,237

**12.** 586

Consider the number 306,458,129. What digit names the number of:   [1.1a]

**13.** tens?

**14.** millions?

**15.** ten thousands?

**16.** hundreds?

Write expanded notation.   [1.1b]

**17.** 5602

**18.** 69,345

Write a word name.   [1.1c]

**19.** 136

**20.** 64,325

Write standard notation.   [1.1c]

**21.** Three hundred eight thousand, seven hundred sixteen

**22.** Four million, five hundred sixty-seven thousand, two hundred sixteen

Add. [1.2a]

23. 
$$\begin{array}{r} 3\ 1\ 6 \\ +\ 4\ 8\ 2 \\ \hline \end{array}$$

24. 
$$\begin{array}{r} 5\ 9\ 3 \\ +\ 4\ 3\ 7 \\ \hline \end{array}$$

25. 
$$\begin{array}{r} 2\ 6\ 3\ 8 \\ +\ 5\ 2\ 8\ 4 \\ \hline \end{array}$$

26. 
$$\begin{array}{r} 4\ 6\ 1\ 7 \\ 2\ 4\ 3\ 6 \\ +\ \ \ \ 4\ 8\ 1 \\ \hline \end{array}$$

Subtract. [1.3a]

27. 
$$\begin{array}{r} 7\ 8\ 6 \\ -\ 3\ 2\ 1 \\ \hline \end{array}$$

28. 
$$\begin{array}{r} 6\ 2\ 4 \\ -\ 2\ 8\ 5 \\ \hline \end{array}$$

29. 
$$\begin{array}{r} 3\ 6\ 0\ 2 \\ -\ 1\ 7\ 4\ 8 \\ \hline \end{array}$$

30. 
$$\begin{array}{r} 5\ 0\ 0\ 4 \\ -\ \ \ 6\ 7\ 6 \\ \hline \end{array}$$

Multiply. [1.4a]

31. 
$$\begin{array}{r} 3\ 6 \\ \times\ \ \ 6 \\ \hline \end{array}$$

32. 
$$\begin{array}{r} 5\ 6\ 7 \\ \times\ \ \ 2\ 8 \\ \hline \end{array}$$

33. 
$$\begin{array}{r} 4\ 0\ 7 \\ \times\ 3\ 2\ 5 \\ \hline \end{array}$$

34. 
$$\begin{array}{r} 9\ 4\ 3\ 5 \\ \times\ \ \ 6\ 0\ 2 \\ \hline \end{array}$$

Divide. [1.5a]

35. $4\overline{)1\ 0\ 1\ 2}$

36. $38\overline{)4\ 2\ 6\ 1}$

37. $60\overline{)1\ 3\ 9\ 9}$

38. $56\overline{)8\ 0\ 9\ 5}$

39. Find the perimeter of the figure (m stands for "meters"). [1.2b]

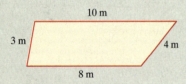

40. Find the area of the region. [1.4b]

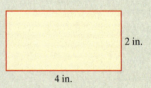

## Understanding Through Discussion and Writing

*To the student and the instructor:* The Discussion and Writing exercises are meant to be answered with one or more sentences. They can be discussed and answered collaboratively by the entire class or by small groups.

41. Explain in your own words what the associative law of addition means. [1.2a]

42. Is subtraction commutative? That is, is there a commutative law of subtraction? Why or why not? [1.3a]

43. Describe a situation that corresponds to each multiplication: $4 \cdot \$150$;  $\$4 \cdot 150$. [1.4a]

44. Suppose a student asserts that "$0 \div 0 = 0$ because nothing divided by nothing is nothing." Devise an explanation to persuade the student that the assertion is false. [1.5a]

# Rounding and Estimating; Order

## 1.6

### a ROUNDING

**SKILL REVIEW**

*Give the meaning of digits in standard notation.*  [1.1a]
In the number 145,627, what digit names the number of:

1. Tens?                    2. Thousands?

Answers: **1.** 2  **2.** 5

MyLab Math
VIDEO

### OBJECTIVES

**a** Round to the nearest ten, hundred, or thousand.

**b** Estimate sums, differences, products, and quotients by rounding.

**c** Use < or > for ☐ to write a true sentence in a situation like 6 ☐ 10.

We round numbers in various situations when we do not need an exact answer. For example, we might round to see if we are being charged the correct amount in a store. We might also round to check if an answer to a problem is reasonable or to check a calculation done by hand or on a calculator.

To understand how to round, we first look at some examples using the number line. The number line displays numbers at equally spaced intervals.

**EXAMPLE 1**   Round 47 to the nearest ten.

47 is between 40 and 50. Since 47 is closer to 50, we round up to 50.

**EXAMPLE 2**   Round 42 to the nearest ten.

42 is between 40 and 50. Since 42 is closer to 40, we round down to 40.

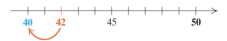

**EXAMPLE 3**   Round 45 to the nearest ten.

45 is halfway between 40 and 50. We could round 45 down to 40 or up to 50. We agree to round up to 50.

Do Exercises 1–7. ▶

When a number is halfway between rounding numbers, round up.

Round to the nearest ten.

1. 37

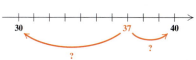

2. 52                    3. 35

4. 73                    5. 75

6. 88                    7. 64

**Answers**
**1.** 40  **2.** 50  **3.** 40  **4.** 70  **5.** 80
**6.** 90  **7.** 60

Based on these examples, we can state a rule for rounding whole numbers.

---

**ROUNDING WHOLE NUMBERS**

To round to a certain place:

**a)** Locate the digit in that place.

**b)** Consider the next digit to the right.

**c)** If the digit to the right is 5 or higher, round up. If the digit to the right is 4 or lower, round down.

**d)** Change all digits to the right of the rounding location to zeros.

---

**EXAMPLE 4**  Round 6485 to the nearest ten.

**a)** Locate the digit in the tens place, 8.

6 4 8 5

**b)** Consider the next digit to the right, 5.

6 4 8 5

**c)** Since that digit, 5, is 5 or higher, round 8 tens up to 9 tens.

**d)** Change all digits to the right of the tens digit to zeros.

6 4 9 0 ⟵ This is the answer.

◀ Do Exercises 8–11.

**EXAMPLE 5**  Round 6485 to the nearest hundred.

**a)** Locate the digit in the hundreds place, 4.

6 4 8 5

**b)** Consider the next digit to the right, 8.

6 4 8 5

**c)** Since that digit, 8, is 5 or higher, round 4 hundreds up to 5 hundreds.

**d)** Change all digits to the right of hundreds to zeros.

6 5 0 0 ⟵ This is the answer.

◀ Do Exercises 12–15.

**EXAMPLE 6**  Round 6485 to the nearest thousand.

**a)** Locate the digit in the thousands place, 6.

6 4 8 5

**b)** Consider the next digit to the right, 4.

6 4 8 5

**c)** Since that digit, 4, is 4 or lower, round down, meaning that 6 thousands stays as 6 thousands.

**d)** Change all digits to the right of thousands to zeros.

6 0 0 0 ⟵ This is the answer.

◀ Do Exercises 16–19.

Round to the nearest ten.

**8.** 137

**9.** 473

**10.** 235

**11.** 285

Round to the nearest hundred.

**12.** 641

**13.** 759

**14.** 1871

**15.** 9325

············· **Caution!** ·············

It is incorrect in Example 6 to round from the ones digit over, as follows:

6485 → 6490 → 6500 → 7000.

Note that 6485 is closer to 6000 than it is to 7000.

Round to the nearest thousand.

**16.** 7896

**17.** 8459

**18.** 19,343

**19.** 68,500

*Answers*

**8.** 140  **9.** 470  **10.** 240  **11.** 290
**12.** 600  **13.** 800  **14.** 1900  **15.** 9300
**16.** 8000  **17.** 8000  **18.** 19,000  **19.** 69,000

Sometimes rounding involves changing more than one digit in a number.

**EXAMPLE 7**   Round 78,595 to the nearest ten.

**a)** Locate the digit in the tens place, 9.

7 8 , 5 9 5
          ↑

**b)** Consider the next digit to the right, 5.

7 8 , 5 9 5

**c)** Since that digit, 5, is 5 or higher, round 9 tens to 10 tens. We think of 10 tens as 1 hundred + 0 tens and increase the hundreds digit by 1, to get 6 hundreds + 0 tens. We then write 6 in the hundreds place and 0 in the tens place.

**d)** Change the digit to the right of the tens digit to zero.

7 8 , 6 0 0 ←— This is the answer.

Note that if we round this number to the nearest hundred, we get the same answer.

**Do Exercises 20 and 21.** ▶

## b   ESTIMATING

Estimating can be done in many ways. In general, an estimate rounded to the nearest ten is more accurate than one rounded to the nearest hundred, an estimate rounded to the nearest hundred is more accurate than one rounded to the nearest thousand, and so on.

**EXAMPLE 8**   Estimate this sum by first rounding to the nearest ten:

78 + 49 + 31 + 85.

We round each number to the nearest ten. Then we add.

```
 7 8        8 0
 4 9        5 0
 3 1        3 0
+8 5      + 9 0
          2 5 0  ←— Estimated answer
```

**Do Exercises 22 and 23.** ▶

**EXAMPLE 9**   Estimate the difference by first rounding to the nearest thousand: 9324 − 2849.

We have

```
 9 3 2 4      9 0 0 0
−2 8 4 9    − 3 0 0 0
            6 0 0 0  ←— Estimated answer
```

**Do Exercises 24 and 25.** ▶

**20.** Round 48,968 to the nearest ten, hundred, and thousand.

**21.** Round 269,582 to the nearest ten, hundred, and thousand.

**22.** Estimate the sum by first rounding to the nearest ten. Show your work.

```
   7 4
   2 3
   3 5
 + 6 6
```

**23.** Estimate the sum by first rounding to the nearest hundred. Show your work.

```
   6 5 0
   6 8 5
   2 3 8
 + 1 6 8
```

**24.** Estimate the difference by first rounding to the nearest hundred. Show your work.

```
   9 2 8 5
 − 6 7 3 9
```

**25.** Estimate the difference by first rounding to the nearest thousand. Show your work.

```
   2 3,2 7 8
 − 1 1,6 9 8
```

*Answers*
**20.** 48,970; 49,000; 49,000
**21.** 269,580; 269,600; 270,000
**22.** 70 + 20 + 40 + 70 = 200
**23.** 700 + 700 + 200 + 200 = 1800
**24.** 9300 − 6700 = 2600
**25.** 23,000 − 12,000 = 11,000

In the sentence $7 - 5 = 2$, the equals sign indicates that $7 - 5$ is the *same* as 2. When we round to make an estimate, the outcome is rarely the same as the exact result. Thus we cannot use an equals sign when we round. Instead, we use the symbol $\approx$. This symbol means "**is approximately equal to.**" In Example 9, for instance, we could have written

$$9324 - 2849 \approx 6000.$$

**EXAMPLE 10**  Estimate the following product by first rounding to the nearest ten. Then estimate the product by first rounding to the nearest hundred: $683 \times 457$.

*Nearest ten:*

```
      6 8 0      683 ≈ 680
   ×  4 6 0      457 ≈ 460
   ─────────
    4 0 8 0 0
  2 7 2 0 0 0
  ───────────
  3 1 2,8 0 0
```

*Nearest hundred:*

```
      7 0 0      683 ≈ 700
   ×  5 0 0      457 ≈ 500
   ─────────
  3 5 0,0 0 0
```

*Exact:*

```
      6 8 3
   ×  4 5 7
   ─────────
      4 7 8 1
    3 4 1 5 0
  2 7 3 2 0 0
  ───────────
  3 1 2,1 3 1
```

We see that rounding to the nearest ten gives a better estimate than rounding to the nearest hundred.

◀ **Do Exercise 26.**

**EXAMPLE 11**  Estimate the following quotient by first rounding to the nearest ten. Then estimate the quotient by first rounding to the nearest hundred: $12{,}238 \div 175$.

*Nearest ten:*  *Nearest hundred:*

```
             6 8                    6 1
  1 8 0 ) 1 2,2 4 0      2 0 0 ) 1 2,2 0 0
         1 0 8 0                 1 2 0 0
         ───────                 ───────
           1 4 4 0                  2 0 0
           1 4 4 0                  2 0 0
           ───────                  ─────
                 0                      0
```

The exact answer is 69 R 163. Again we see that rounding to the nearest ten gives a better estimate than rounding to the nearest hundred.

◀ **Do Exercise 27.**

---

**26.** Estimate the product by rounding first to the nearest ten. Then estimate the product by rounding first to the nearest hundred. Show your work.

```
    8 3 7
  × 2 4 5
```

*Nearest ten:*

```
          8 4 0
      ×
      ─────────
      4 2 0 0 0
          0 0 0
    ,0 0 0
```

*Nearest hundred:*

```
          8 0 0
      ×
      ─────────
      0,0 0 0
```

**27.** Estimate the quotient by rounding first to the nearest hundred. Show your work.

$$64{,}534 \div 349$$

**Answers**

**26.** $840 \times 250 = 210{,}000$;
    $800 \times 200 = 160{,}000$

**27.** $64{,}500 \div 300 = 215$

*Guided Solution:*
**26.** Nearest ten: 250, 1, 6, 8, 2, 1, 0;
Nearest hundred: 200, 1, 6

The next two examples show how estimating can be used in making financial decisions.

**EXAMPLE 12** *Tuition.* Ellen plans to take 12 credit hours of classes next semester. If she takes the courses on campus, the cost per credit hour is $248. Estimate, by rounding to the nearest ten, the total cost of tuition.

We have

$$
\begin{array}{r}
2\ 5\ 0 \\
\times \quad 1\ 0 \\
\hline
2\ 5\ 0\ 0.
\end{array}
$$

The tuition will cost about $2500.

**Do Exercise 28.** ▶

**EXAMPLE 13** *Purchasing a New Vehicle.* Jenn is considering buying a Ford F-150 XL truck. She has a budget of $35,000. The base price of the truck is $27,110. A number of options are available at additional cost. Some of these are listed in the following table.

Estimate, by rounding to the nearest hundred, the cost of the F-150 XL with the SuperCab, 5.0L V8 engine, 4-wheel drive, and off-road package. Then determine whether this will fit within Jenn's budget.

| FORD F-150 XL | PRICE |
|---|---|
| Base price | $27,110 |
| SuperCab | $4,085 |
| 8-foot bed | $300 |
| 2.7L EcoBoost engine | $995 |
| 5.0L V8 engine | $1,795 |
| 4-wheel drive | $4,645 |
| Equipment upgrade package | $2,255 |
| Off-road package | $770 |
| Trailer tow package | $1,145 |
| Spray bedliner | $495 |

DATA: Ford

First, we list the base price of the truck and then the cost of each of the options. We then round each number to the nearest hundred and add.

$$
\begin{array}{rr}
2\,7,1\,1\,0 & 2\,7,1\,0\,0 \\
4,0\,8\,5 & 4,1\,0\,0 \\
1,7\,9\,5 & 1,8\,0\,0 \\
4,6\,4\,5 & 4,6\,0\,0 \\
+\quad 7\,7\,0 & +\quad 8\,0\,0 \\
\hline
& 3\,8,4\,0\,0 \leftarrow \text{Estimated cost}
\end{array}
$$

The estimated cost is $38,400. This exceeds Jenn's budget of $35,000, so she will have to forgo at least one option.

**Do Exercises 29 and 30.** ▶

**28.** *Tuition.* If Ellen takes courses online, the cost per credit hour is $198. Estimate, by rounding to the nearest ten, the total cost of 12 credit hours of classes.

Refer to the table in Example 13 to do Margin Exercises 29 and 30.

**29.** By eliminating one option, show how Jenn can buy an F-150 XL and stay within her budget.

**30.** Logan is also considering buying an F-150 XL. Estimate, by rounding to the nearest hundred, the cost of this truck with all the listed options except the 2.7L EcoBoost engine.

*Answers*

**28.** $2000   **29.** She can eliminate either the 4-wheel drive or the SuperCab. (If she eliminates the 4-wheel drive, the off-road package is no longer an option.)   **30.** $42,600

## C ORDER

A sentence like $8 + 5 = 13$ is called an **equation**. It is a *true* equation. The equation $4 + 8 = 11$ is a *false* equation.

A sentence like $7 < 11$ is called an **inequality**. The sentence $7 < 11$ is a *true* inequality. The sentence $23 > 69$ is a *false* inequality.

Some common **inequality symbols** follow.

---

### INEQUALITY SYMBOLS

$<$  means "is less than"

$>$  means "is greater than"

$\neq$  means "is not equal to"

---

We know that 2 is not the same as 5. We express this by the sentence $2 \neq 5$. We also know that 2 is less than 5. We symbolize this by the expression $2 < 5$. We can see this order on the number line: 2 is to the left of 5. The number 0 is the smallest whole number.

---

### ORDER OF WHOLE NUMBERS

For any whole numbers $a$ and $b$:

1. $a < b$ (read "$a$ is less than $b$") is true when $a$ is to the left of $b$ on the number line.

2. $a > b$ (read "$a$ is greater than $b$") is true when $a$ is to the right of $b$ on the number line.

---

**EXAMPLE 14**  Use $<$ or $>$ for ☐ to write a true sentence: $7\ \square\ 11$.

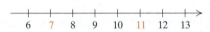

Since 7 is to the left of 11 on the number line, $7 < 11$.

**EXAMPLE 15**  Use $<$ or $>$ for ☐ to write a true sentence: $92\ \square\ 87$.

Since 92 is to the right of 87 on the number line, $92 > 87$.

◀ **Do Exercises 31–36.**

---

Use $<$ or $>$ for ☐ to write a true sentence. Draw the number line if necessary.

**31.** $8\ \square\ 12$

Since 8 is to the ☐ of 12 on the number line, $8\ \square\ 12$.

**32.** $12\ \square\ 8$

**33.** $76\ \square\ 64$

**34.** $64\ \square\ 76$

**35.** $217\ \square\ 345$

**36.** $345\ \square\ 217$

---

*Answers*

**31.** $<$  **32.** $>$  **33.** $>$  **34.** $<$  **35.** $<$

**36.** $>$

*Guided Solution:*

**31.** left, $<$

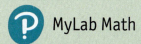

## ✔ Check Your Understanding

**Reading Check**　Determine whether each statement is true or false.

_____ **RC1.** When rounding to the nearest hundred, if the digit in the tens place is 5 or higher, we round up.

_____ **RC2.** When rounding 3500 to the nearest thousand, we should round down.

_____ **RC3.** An estimate made by rounding to the nearest thousand is more accurate than an estimate made by rounding to the nearest ten.

_____ **RC4.** Since 78 rounded to the nearest ten is 80, we can write 78 ≈ 80.

**Concept Check**　Use the graph of each of the following numbers to round the number to the nearest hundred.

**CC1.** 147                          **CC2.** 961                          **CC3.** 650

---

**a**　Round to the nearest ten.

**1.** 48                **2.** 532              **3.** 463              **4.** 8945

**5.** 731              **6.** 54               **7.** 895              **8.** 798

Round to the nearest hundred.

**9.** 146              **10.** 874             **11.** 957             **12.** 650

**13.** 9079            **14.** 4645            **15.** 32,839          **16.** 198,402

Round to the nearest thousand.

**17.** 5876            **18.** 4500            **19.** 7500            **20.** 2001

**21.** 45,340          **22.** 735,562         **23.** 373,405         **24.** 6,713,255

**b**　Estimate each sum or difference by first rounding to the nearest ten. Show your work.

**25.** 　 7 8
　　 + 9 2

**26.** 　 6 2
　　 　 9 7
　　 　 4 6
　　 + 8 1

**27.** 　 8 0 7 4
　　 − 2 3 4 7

**28.** 　 6 7 3
　　 　 − 2 8

Estimate each sum by first rounding to the nearest ten. State if the given sum seems to be incorrect when compared to the estimate.

**29.**
```
    4 5
    7 7
    2 5
+   5 6
─────────
  3 4 3
```

**30.**
```
    4 1
    2 1
    5 5
+   6 0
─────────
  1 7 7
```

**31.**
```
  6 2 2
    7 8
    8 1
+ 1 1 1
─────────
  9 3 2
```

**32.**
```
  8 3 6
  3 7 4
  7 9 4
+ 9 3 8
─────────
3 9 4 7
```

Estimate each sum or difference by first rounding to the nearest hundred. Show your work.

**33.**
```
  7 3 4 8
+ 9 2 4 7
```

**34.**
```
    5 6 8
    4 7 2
    9 3 8
+   4 0 2
```

**35.**
```
  6 8 5 2
− 1 7 4 8
```

**36.**
```
  9 4 3 8
− 2 7 8 7
```

Estimate each sum by first rounding to the nearest hundred. State if the given sum seems to be incorrect when compared to the estimate.

**37.**
```
    2 1 6
      8 4
    7 4 5
+   5 9 5
─────────
  1 6 4 0
```

**38.**
```
    4 8 1
    7 0 2
    6 2 3
+ 1 0 4 3
─────────
  1 8 4 9
```

**39.**
```
    7 5 0
    4 2 8
      6 3
+   2 0 5
─────────
  1 4 4 6
```

**40.**
```
    3 2 6
    2 7 5
    7 5 8
+   9 4 3
─────────
  2 3 0 2
```

Estimate each sum or difference by first rounding to the nearest thousand. Show your work.

**41.**
```
  9 6 4 3
  4 8 2 1
  8 9 4 3
+ 7 0 0 4
```

**42.**
```
  7 6 4 8
  9 3 4 8
  7 8 4 2
+ 2 2 2 2
```

**43.**
```
  9 2,1 4 9
− 2 2,5 5 5
```

**44.**
```
  8 4,8 9 0
− 1 1,1 1 0
```

Estimate each product by first rounding to the nearest ten. Show your work.

**45.**
```
    4 5
×   6 7
```

**46.**
```
    5 1
×   7 8
```

**47.**
```
    3 4
×   2 9
```

**48.**
```
    6 3
×   5 4
```

Estimate each product by first rounding to the nearest hundred. Show your work.

**49.**
```
    8 7 6
×   3 4 5
```

**50.**
```
    3 5 5
×   2 9 9
```

**51.**
```
    4 3 2
×   1 9 9
```

**52.**
```
    7 8 9
×   4 3 4
```

Estimate each quotient by first rounding to the nearest ten. Show your work.

**53.** 347 ÷ 73       **54.** 454 ÷ 87       **55.** 8452 ÷ 46       **56.** 1263 ÷ 29

Estimate each quotient by first rounding to the nearest hundred. Show your work.

**57.** 1165 ÷ 236       **58.** 3641 ÷ 571       **59.** 8358 ÷ 295       **60.** 32,854 ÷ 748

*Planning a Vacation.* Planning a trip to a theme park resort involves making lodging, ticket, and dining decisions. The following table lists the prices of several options for a three-day trip to a theme park resort. Use the data for Exercises 61–66.

| HOTEL OPTIONS | PRICE |
| --- | --- |
| Family Value | $498 |
| Wilderness Lodge | $921 |
| Beachfront | $1122 |
| Spa and Resort | $1722 |

| PARK TICKETS | PRICE |
| --- | --- |
| Theme Park | $289 |
| Water Park | $62 |
| Fast-Lane Upgrade | $159 |

| MEAL PLANS | PRICE |
| --- | --- |
| Fast Food | $145 |
| Full Service | $208 |
| Extended Dining | $320 |

**61.** Estimate the total price of a three-day trip with a beachfront hotel room, a theme park ticket, a fast-lane upgrade, and an extended dining meal plan. Round each price to the nearest hundred dollars.

**62.** Estimate the total price of a three-day trip with a family value hotel room, a theme park ticket, and a fast food meal plan. Round each price to the nearest hundred dollars.

**63.** Marcus has a budget of $1500 for a three-day theme park vacation. He would like to stay in a wilderness lodge, visit both the theme park and the water park, and have an extended dining meal plan. Estimate the total price of this vacation by rounding each price to the nearest hundred dollars. Can he afford his choices?

**64.** Alyssa has a budget of $1500 for a three-day theme park vacation. She is planning to book a family value hotel room. She would like to visit both the theme park and the water park, purchase a fast-lane upgrade, and have an extended dining meal plan. Estimate the total price of this vacation by rounding each price to the nearest hundred dollars. Does her budget cover her choices?

**65.** If you were going on a three-day theme park vacation and had a budget of $1500, what options would you choose? Decide on options that fit your budget and estimate the total price by rounding each price to the nearest hundred dollars.

**66.** If you were going on a three-day theme park vacation and had a budget of $2000, what options would you choose? Decide on options that fit your budget and estimate the total price by rounding each price to the nearest hundred dollars.

**67. Mortgage Payments.** To pay for their new home, Tim and Meribeth will make 360 payments of $751.55. In addition, they must add an escrow amount of $112.67 to each payment for insurance and taxes.

a) Estimate the total amount they will pay by rounding the number of payments, the amount of each payment, and the escrow amount to the nearest ten.

b) Estimate the total amount they will pay by rounding the number of payments, the amount of each payment, and the escrow amount to the nearest hundred.

**68. Conference Expenses.** The cost to attend a three-day teachers' conference is $245, and a hotel room costs $169 a night. One year, 489 teachers attended the conference, and 315 rooms were rented for two nights each.

a) Estimate the total amount spent by the teachers by rounding the cost of attending the conference, the nightly cost of a hotel room, the number of teachers, and the number of rooms to the nearest ten.

b) Estimate the total amount spent by the teachers by rounding the cost of attending the conference, the nightly cost of a hotel room, the number of teachers, and the number of rooms to the nearest hundred.

**69. Banquet Attendance.** Tickets to the annual awards banquet for the Riviera Swim Club cost $28 each. Ticket sales for the banquet totaled $2716. Estimate the number of people who attended the banquet by rounding the cost of a ticket to the nearest ten and the total sales to the nearest hundred.

**70. School Fundraiser.** For a school fundraiser, Charlotte sells trash bags at a price of $11 per roll. If her sales totaled $2211, estimate the number of rolls she sold by rounding the price per roll to the nearest ten and the total sales to the nearest hundred.

**c**    Use < or > for ☐ to write a true sentence. Draw the number line if necessary.

**71.** 0 ☐ 17        **72.** 32 ☐ 0        **73.** 34 ☐ 12        **74.** 28 ☐ 18

**75.** 1000 ☐ 1001        **76.** 77 ☐ 117        **77.** 133 ☐ 132        **78.** 999 ☐ 997

**79.** 460 ☐ 17        **80.** 345 ☐ 456        **81.** 37 ☐ 11        **82.** 12 ☐ 32

*Postsecondary Education.* The number of degrees awarded by postsecondary institutions in the United States in 2014–2015 is shown in the table below. Use this table to do Exercises 83 and 84.

| TYPE OF DEGREE | NUMBER OF DEGREES |
|---|---|
| Associate's | 1,014,023 |
| Bachelor's | 1,894,934 |
| Master's | 758,708 |
| Doctorate | 178,547 |

DATA: National Center for Education Statistics

**83.** Write an inequality to compare the numbers of associate's degrees and bachelor's degrees.

**84.** Write an inequality to compare the numbers of associate's degrees and master's degrees.

**85.** *Animal Species.* The number of species for categories of vertebrate animals is shown in the following figure. Write an inequality to compare the number of species of birds and the number of species of reptiles.

**Vertebrate Animal Species**

| | Number of species |
|---|---|
| Mammals | 5,513 |
| Birds | 10,425 |
| Reptiles | 10,038 |
| Amphibians | 7,302 |
| Fishes | 32,900 |

10,000  20,000  30,000  40,000
Number of species

DATA: currentresults.com

**86.** *City Population.* The populations of several cities in the United States are shown in the following table. Write an inequality to compare the populations of Indianapolis, Indiana, and Jacksonville, Florida.

| CITY | POPULATION |
|---|---|
| Austin, Texas | 885,400 |
| Columbus, Ohio | 822,553 |
| Indianapolis, Indiana | 843,393 |
| Jacksonville, Florida | 842,583 |
| San Francisco, California | 837,442 |

DATA: ballotpedia.com

## Skill Maintenance

Add.  [1.2a]

**87.**   6 7,7 8 9
     + 1 8,9 6 5

**88.**   9 0 0 2
     + 4 5 8 7

Subtract.  [1.3a]

**89.**   6 7,7 8 9
     − 1 8,9 6 5

**90.**   9 0 0 2
     − 4 5 8 7

Multiply.  [1.4a]

**91.**   4 6
     × 3 7

**92.**   3 0 6
     ×   5 8

Divide.  [1.5a]

**93.** 328 ÷ 6

**94.** 4784 ÷ 23

## Synthesis

**95.–98.** ▦ Use a calculator to find the sums and the differences in each of Exercises 41–44. Then compare your answers with those found using estimation. Even when using a calculator it is possible to make an error if you press the wrong buttons, so it is a good idea to check by estimating.

Find a number that makes each sentence true.

**1.** $8 = 1 + \square$

**2.** $\square + 2 = 7$

**3.** Determine whether 7 is a solution of $\square + 5 = 9$.

**4.** Determine whether 4 is a solution of $\square + 5 = 9$.

# Solving Equations

## a SOLUTIONS BY TRIAL

Let's find a number that we can put in the blank to make this sentence true:

$$9 = 3 + \square.$$

We are asking "9 is 3 plus what number?" The answer is 6.

$$9 = 3 + 6$$

◀ **Do Exercises 1 and 2.**

A sentence with = is called an **equation**. A **solution** of an equation is a number that makes the sentence true. Thus, 6 is a solution of

$$9 = 3 + \square \quad \text{because} \quad 9 = 3 + 6 \text{ is true.}$$

However, 7 is not a solution of

$$9 = 3 + \square \quad \text{because} \quad 9 = 3 + 7 \text{ is false.}$$

◀ **Do Exercises 3 and 4.**

We can use a letter in an equation instead of a blank:

$$9 = 3 + n.$$

We call $n$ a **variable** because it can represent any number. If a replacement for a variable makes an equation true, the replacement is a solution of the equation.

---

### SOLUTIONS OF AN EQUATION

A **solution of an equation** is a replacement for the variable that makes the equation true. When we find all the solutions, we say that we have **solved** the equation.

---

**EXAMPLE 1** Solve $y + 12 = 27$ by trial.

We replace $y$ with several numbers.

If we replace $y$ with 13, we get a false equation: $13 + 12 = 27$.
If we replace $y$ with 14, we get a false equation: $14 + 12 = 27$.
If we replace $y$ with 15, we get a true equation: $15 + 12 = 27$.

No other replacement makes the equation true, so the solution is 15.

Solve by trial.

**5.** $n + 3 = 8$

**6.** $x - 2 = 8$

**7.** $45 \div 9 = y$

**8.** $10 + t = 32$

**EXAMPLES** Solve.

**2.** $7 + n = 22$
(7 plus what number is 22?)
The solution is 15.

**3.** $63 = 3 \cdot x$
(63 is 3 times what number?)
The solution is 21.

◀ **Do Exercises 5–8.**

---

**Answers**

**1.** 7  **2.** 5  **3.** No  **4.** Yes  **5.** 5
**6.** 10  **7.** 5  **8.** 22

## b SOLVING EQUATIONS

*Divide whole numbers.* [1.5a]
Divide.

**1.** $1008 \div 36$        **2.** $675 \div 15$

**Answers: 1.** 28  **2.** 45

MyLab Math
VIDEO

We now begin to develop more efficient ways to solve certain equations. When an equation has a variable alone on one side and a calculation on the other side, we can find the solution by carrying out the calculation.

**EXAMPLE 4**  Solve: $x = 245 \times 34$.

To solve the equation, we carry out the calculation.

$$\begin{array}{r} 2\ 4\ 5 \\ \times\ \ 3\ 4 \\ \hline 9\ 8\ 0 \\ 7\ 3\ 5\ 0 \\ \hline 8\ 3\ 3\ 0 \end{array}$$

$x = 245 \times 34$
$x = 8330$

The solution is 8330.

**Do Exercises 9–12.** ▶

Look at the equation

$$x + 12 = 27.$$

We can get $x$ alone by subtracting 12 *on both sides*. Thus,

$x + 12 - 12 = 27 - 12$   Subtracting 12 on both sides
$x + 0 = 15$    Carrying out the subtraction
$x = 15.$    We can "see" that the solution is 15: $15 + 12 = 27$.

---

### SOLVING $x + a = b$

To solve $x + a = b$, subtract $a$ on both sides.

---

If we can get an equation in a form with the variable alone on one side, we can "see" the solution.

**EXAMPLE 5**  Solve: $t + 28 = 54$.

We have

$t + 28 = 54$
$t + 28 - 28 = 54 - 28$   Subtracting 28 on both sides
$t + 0 = 26$
$t = 26.$

Solve.

**9.** $346 \times 65 = y$

**10.** $x = 2347 + 6675$

**11.** $4560 \div 8 = t$

**12.** $x = 6007 - 2346$

*Answers*
**9.** 22,490  **10.** 9022  **11.** 570  **12.** 3661

Solve. Be sure to check.

**13.** $x + 9 = 17$  **GS**

$x + 9 - \boxed{\phantom{0}} = 17 - 9$

$x = \boxed{\phantom{0}}$

Check:  $x + 9 = 17$

$\boxed{\phantom{0}} + 9 \ \overset{?}{?} \ 17$

$\boxed{\phantom{0}}$

Since $17 = 17$ is $\boxed{\phantom{0}}$, the answer checks.
The solution is $\boxed{\phantom{0}}$.

**14.** $77 = m + 32$

**15.** Solve: $155 = t + 78$. Be sure to check.

Solve. Be sure to check.

**16.** $4566 + x = 7877$

**17.** $8172 = h + 2058$

---

To check the answer, we substitute 26 for $t$ in the original equation.

Check:  $t + 28 = 54$

$26 + 28 \ \overset{?}{?} \ 54$

$54$   TRUE   Since $54 = 54$ is true, 26 checks.

The solution is 26.

◀ **Do Exercises 13 and 14.**

**EXAMPLE 6**   Solve: $182 = 65 + n$.

We have

$182 = 65 + n$

$182 - 65 = 65 + n - 65$    Subtracting 65 on both sides

$117 = 0 + n$    65 plus $n$ minus 65 is $0 + n$.

$117 = n$.

Check:  $182 = 65 + n$

$182 \ \overset{?}{?} \ 65 + 117$

$182$    TRUE

The solution is 117.

◀ **Do Exercise 15.**

**EXAMPLE 7**   Solve: $7381 + x = 8067$.

We have

$7381 + x = 8067$

$7381 + x - 7381 = 8067 - 7381$    Subtracting 7381 on both sides

$x = 686$.

Check:  $7381 + x = 8067$

$7381 + 686 \ \overset{?}{?} \ 8067$

$8067$    TRUE

The solution is 686.

◀ **Do Exercises 16 and 17.**

We now learn to solve equations like $8 \cdot n = 96$. We can get $n$ alone by dividing by 8 *on both sides*:

$8 \cdot n = 96$

$\dfrac{8 \cdot n}{8} = \dfrac{96}{8}$    Dividing by 8 on both sides

$n = 12$.    8 times $n$ divided by 8 is $n$.

To check the answer, we substitute 12 for $n$ in the original equation.

Check:  $8 \cdot n = 96$

$8 \cdot 12 \ \overset{?}{?} \ 96$

$96$    TRUE

Since $96 = 96$ is a true equation, 12 is the solution of the equation.

*Answers*

**13.** 8  **14.** 45  **15.** 77
**16.** 3311  **17.** 6114

*Guided Solution:*
**13.** 9, 8, 8, 17; true; 8

To solve $a \cdot x = b$, divide by $a$ on both sides.

**EXAMPLE 8**   Solve: $10 \cdot x = 240$.

We have

$$10 \cdot x = 240$$

$$\frac{10 \cdot x}{10} = \frac{240}{10} \qquad \text{Dividing by 10 on both sides}$$

$$x = 24.$$

Check:   $\dfrac{10 \cdot x = 240}{10 \cdot 24 \; ? \; 240}$
          $\phantom{10 \cdot 24} 240 \; | \qquad$ TRUE

The solution is 24.

**EXAMPLE 9**   Solve: $5202 = 9 \cdot t$.

We have

$$5202 = 9 \cdot t$$

$$\frac{5202}{9} = \frac{9 \cdot t}{9} \qquad \text{Dividing by 9 on both sides}$$

$$578 = t.$$

Check:   $\dfrac{5202 = 9 \cdot t}{5202 \; ? \; 9 \cdot 578}$
          $\phantom{5202 \; ? \; } | \; 5202 \qquad$ TRUE

The solution is 578.

Do Exercises 18–20. ▶

**EXAMPLE 10**   Solve: $14 \cdot y = 1092$.

We have

$$14 \cdot y = 1092$$

$$\frac{14 \cdot y}{14} = \frac{1092}{14} \qquad \text{Dividing by 14 on both sides}$$

$$y = 78.$$

The check is left to the student. The solution is 78.

**EXAMPLE 11**   Solve: $n \cdot 56 = 4648$.

We have

$$n \cdot 56 = 4648$$

$$\frac{n \cdot 56}{56} = \frac{4648}{56} \qquad \text{Dividing by 56 on both sides}$$

$$n = 83.$$

The check is left to the student. The solution is 83.

Do Exercises 21 and 22. ▶

Solve. Be sure to check.

**18.** $8 \cdot x = 64$

**GS** **19.** $144 = 9 \cdot n$

$$\frac{144}{9} = \frac{9 \cdot n}{\boxed{\phantom{x}}}$$

$$\boxed{\phantom{x}} = n$$

Check:   $\dfrac{144 = 9 \cdot n}{144 \; ? \; 9 \cdot \boxed{\phantom{x}}}$
          $\phantom{144 \; ? \; 9 \cdot} | \; \boxed{\phantom{x}}$

Since $144 = 144$ is $\boxed{\phantom{xx}}$, the answer checks. The solution is $\boxed{\phantom{x}}$.

**20.** $5152 = 8 \cdot t$

Solve. Be sure to check.

**21.** $18 \cdot y = 1728$

**22.** $n \cdot 48 = 4512$

*Answers*
**18.** 8   **19.** 16   **20.** 644   **21.** 96   **22.** 94
*Guided Solution:*
**19.** 9, 16, 16, 144; true; 16

## ✓ Check Your Understanding

**Reading Check** Match each word with its definition from the list on the right.

RC1. Equation ____

RC2. Solution ____

RC3. Solved ____

RC4. Variable ____

a) A replacement for the variable that makes an equation true

b) A letter that can represent any number

c) A sentence containing $=$

d) Found all the solutions for

**Concept Check** For each equation, determine whether the given number is a solution of the equation.

CC1. $x + 27 = 65$; 92

CC2. $672 = 6 \cdot n$; 112

CC3. $6516 \div 543 = n$; 12

CC4. $5462 = 3189 + t$; 2327

### a Solve by trial.

**1.** $x + 0 = 14$

**2.** $x - 7 = 18$

**3.** $y \cdot 17 = 0$

**4.** $56 \div m = 7$

### b Solve. Be sure to check.

**5.** $x = 12{,}345 + 78{,}555$

**6.** $t = 5678 + 9034$

**7.** $908 - 458 = p$

**8.** $9007 - 5667 = m$

**9.** $16 \cdot 22 = y$

**10.** $34 \cdot 15 = z$

**11.** $t = 125 \div 5$

**12.** $w = 256 \div 16$

**13.** $13 + x = 42$

**14.** $15 + t = 22$

**15.** $12 = 12 + m$

**16.** $16 = t + 16$

**17.** $10 + x = 89$

**18.** $20 + x = 57$

**19.** $61 = 16 + y$

**20.** $53 = 17 + w$

**21.** $3 \cdot x = 24$

**22.** $6 \cdot x = 42$

**23.** $112 = n \cdot 8$

**24.** $162 = 9 \cdot m$

**25.** $3 \cdot m = 96$

**26.** $4 \cdot y = 96$

**27.** $715 = 5 \cdot z$

**28.** $741 = 3 \cdot t$

**29.** $8322 + 9281 = x$

**30.** $9281 - 8322 = y$

**31.** $47 + n = 84$

**32.** $56 + p = 92$

**33.** $45 \cdot 23 = x$

**34.** $23 \cdot 78 = y$

**35.** $x + 78 = 144$

**36.** $z + 67 = 133$

**37.** $6 \cdot p = 1944$

**38.** $4 \cdot w = 3404$

**39.** $567 + x = 902$

**40.** $438 + x = 807$

**41.** $234 \cdot 78 = y$

**42.** $10{,}534 \div 458 = q$

**43.** $18 \cdot x = 1872$

**44.** $19 \cdot x = 6080$

**45.** $40 \cdot x = 1800$

**46.** $20 \cdot x = 1500$

**47.** $2344 + y = 6400$

**48.** $9281 = 8322 + t$

**49.** $m = 7006 - 4159$

**50.** $n = 3004 - 1745$

**51.** $165 = 11 \cdot n$

**52.** $660 = 12 \cdot n$

**53.** $58 \cdot m = 11{,}890$

**54.** $233 \cdot x = 22{,}135$

**55.** $491 - 34 = y$

**56.** $512 - 63 = z$

## Skill Maintenance

Divide.  [1.5a]

**57.** $1283 \div 9$

**58.** $1278 \div 9$

**59.** $1\,7 \overline{)\,5\,6\,7\,8}$

**60.** $1\,7 \overline{)\,5\,6\,8\,9}$

Use > or < for ☐ to write a true sentence.  [1.6c]

**61.** $123 \ \square \ 789$

**62.** $342 \ \square \ 339$

**63.** $688 \ \square \ 0$

**64.** $0 \ \square \ 11$

**65.** Round 6,375,602 to the nearest thousand.  [1.6a]

**66.** Round 6,375,602 to the nearest ten.  [1.6a]

## Synthesis

Solve.

**67.** 🖩 $23{,}465 \cdot x = 8{,}142{,}355$

**68.** 🖩 $48{,}916 \cdot x = 14{,}332{,}388$

## OBJECTIVE

a Solve applied problems involving addition, subtraction, multiplication, or division of whole numbers.

# Applications and Problem Solving

## a A PROBLEM-SOLVING STRATEGY

**SKILL REVIEW**

*Estimate sums, differences, products, and quotients by rounding.*   [1.6b]

Estimate each sum or difference by first rounding to the nearest thousand. Show your work.

**1.**   3 6 7,9 8 2
      +   4 3,4 9 5

**2.**   9 2 8 7
      −  3 5 0 2

**Answers: 1.** 368,000 + 43,000 = 411,000
**2.** 9000 − 4000 = 5000

One of the most important ways in which we use mathematics is as a tool in solving problems. To solve a problem, we use the following five-step strategy.

---

**FIVE STEPS FOR PROBLEM SOLVING**

1. **Familiarize** yourself with the problem situation.
2. **Translate** the problem to an equation using a variable.
3. **Solve** the equation.
4. **Check** to see whether your possible solution actually fits the problem situation and is thus really a solution of the problem.
5. **State** the answer clearly using a complete sentence and appropriate units.

---

The first of these five steps, becoming familiar with the problem, is probably the most important.

---

**THE FAMILIARIZE STEP**

- If the problem is presented in words, read and reread it carefully until you understand what you are being asked to find.
- Make a drawing, if it makes sense to do so.
- Write a list of the known facts and a list of what you wish to find out.
- Assign a letter, or *variable*, to the unknown.
- Organize the information in a chart or a table.
- Find further information. Look up a formula, consult a reference book or an expert in the field, or do research on the Internet.
- Guess or estimate the answer and check your guess or estimate.

---

**EXAMPLE 1** *Video Games.* The following table shows the number of games released for eight popular video game platforms sold worldwide, as of July 2017. Three of these platforms are made by PlayStation®. Find the total number of games released for the PlayStation platforms.

| PLATFORM | NUMBER OF GAMES |
|---|---|
| Nintendo DS | 4009 |
| Xbox 360 | 3676 |
| PlayStation 2 | 3549 |
| PlayStation 3 | 3315 |
| Nintendo Wii | 2808 |
| PlayStation | 2680 |
| Nintendo Game Boy | 1608 |

DATA: vgchartz.com

1. **Familiarize.** First, we assign a letter, or variable, to the number we wish to find. We let $p$ = the total number of video games released for the PlayStation platforms. Since we are combining numbers, we will add.

2. **Translate.** We translate to an equation:

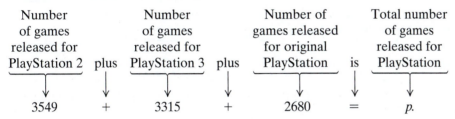

| Number of games released for PlayStation 2 | plus | Number of games released for PlayStation 3 | plus | Number of games released for original PlayStation | is | Total number of games released for PlayStation |
|---|---|---|---|---|---|---|
| 3549 | + | 3315 | + | 2680 | = | $p$. |

3. **Solve.** We solve the equation by carrying out the addition.

$$3549 + 3315 + 2680 = p$$
$$9544 = p$$

$$\begin{array}{r} 3\,5\,4\,9 \\ 3\,3\,1\,5 \\ +\,2\,6\,8\,0 \\ \hline 9\,5\,4\,4 \end{array}$$

4. **Check.** We check our result by rereading the original problem and seeing if 9544 answers the question. Since we are looking for a total, we could repeat the addition calculation. We could also check whether the answer is reasonable. In this case, since the total is greater than any of the three separate game numbers, the result seems reasonable. Another way to check is to estimate the expected result and compare the estimate with the calculated result. If we round each number of PlayStation games to the nearest thousand and add, we have

$$4000 + 3000 + 3000 = 10,000.$$

Since $9544 \approx 10,000$, our result again seems reasonable.

5. **State.** The total number of games released for the three PlayStation platforms listed is 9544.

**Do Exercises 1–3.** ▶

Refer to the table in Example 1 to do Margin Exercises 1–3.

1. Find the total number of games released for the Nintendo platforms listed.

2. Find the total number of games released for the four game platforms listed in the table with the most games released.

3. Find the total number of games released for all the game platforms listed in the table.

**EXAMPLE 2** *Travel Distance.* Abigail is driving from Indianapolis to Salt Lake City to attend a family reunion. The distance from Indianapolis to Salt Lake City is 1634 mi. In the first two days, she travels 1154 mi to Denver. How much farther must she travel?

1. **Familiarize.** We first make a drawing or at least visualize the situation. We let $d$ = the remaining distance to Salt Lake City.

2. **Translate.** We want to determine how many more miles Abigail must travel. We translate to an equation:

| Distance already traveled | plus | Distance to go | is | Total distance of trip |
|:---:|:---:|:---:|:---:|:---:|
| ↓ | ↓ | ↓ | ↓ | ↓ |
| 1154 | + | $d$ | = | 1634. |

3. **Solve.** To solve the equation, we subtract 1154 on both sides.

$$1154 + d = 1634$$
$$1154 + d - 1154 = 1634 - 1154$$
$$d = 480$$

$$\begin{array}{r} {}^{5\ 13} \\ 1\,6\,3\,4 \\ -\,1\,1\,5\,4 \\ \hline 4\,8\,0 \end{array}$$

4. **Check.** We check our answer of 480 mi in the original problem. This number should be less than the total distance, 1634 mi, and it is. We can add the distance traveled, 1154, and the distance left to go, 480: $1154 + 480 = 1634$. We can also estimate:

$$1634 - 1154 \approx 1600 - 1200$$
$$= 400 \approx 480.$$

The answer, 480 mi, checks.

5. **State.** Abigail must travel 480 mi farther to Salt Lake City.

◀ **Do Exercise 4.**

---

4. **Reading Assignment.** William 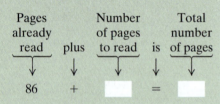 has been assigned 234 pages of reading for his history class. He has read 86 pages. How many more pages does he have to read?

1. **Familiarize.** Let $p$ = the number of pages William still has to read.

2. **Translate.**

| Pages already read | plus | Number of pages to read | is | Total number of pages |
|:---:|:---:|:---:|:---:|:---:|
| ↓ | ↓ | ↓ | ↓ | ↓ |
| 86 | + | ☐ | = | ☐ |

3. **Solve.**

$$86 + p = 234$$
$$86 + p - \boxed{\phantom{00}} = 234 - 86$$
$$p = \boxed{\phantom{00}}$$

4. **Check.** If William reads 148 more pages, he will have read a total of $86 + 148$ pages, or ☐ pages.

5. **State.** William has ☐ more pages to read.

---

*Answer*
**4.** 148 pages
*Guided Solution:*
**4.** $p$, 234; 86, 148; 234; 148

**56**   CHAPTER 1   Whole Numbers

**EXAMPLE 3** *Total Cost of Chairs.*   What is the total cost of 6 Adirondack chairs if each one costs $169?

**1. Familiarize.**   We make a drawing and let $C$ = the cost of 6 chairs.

$169       $169       $169       $169       $169       $169

**2. Translate.**   We translate to an equation:

$$\underbrace{\text{Number of chairs}}_{6} \; \underbrace{\text{times}}_{\times} \; \underbrace{\text{Cost of each chair}}_{169} \; \underbrace{\text{is}}_{=} \; \underbrace{\text{Total cost}}_{C.}$$

**3. Solve.**   This sentence tells us what to do. We multiply.

$$6 \times 169 = C$$
$$1014 = C$$

$$\begin{array}{r} 1\ 6\ 9 \\ \times \qquad 6 \\ \hline 1\ 0\ 1\ 4 \end{array}$$

**4. Check.**   We have an answer, 1014, that is greater than the cost of any one chair, which is reasonable. We can also check by estimating:

$$6 \times 169 \approx 6 \times 170 = 1020 \approx 1014.$$

The answer checks.

**5. State.**   The total cost of 6 chairs is $1014.

**Do Exercise 5.** ▶

**EXAMPLE 4** *Area of an Oriental Rug.*   The dimensions of the rectangular oriental rug in the Fosters' front hallway are 42 in. by 66 in. What is the area of the rug?

**1. Familiarize.**   We let $A$ = the area of the rug and use the formula for the area of a rectangle, $A$ = length $\cdot$ width = $l \cdot w$. Since we usually consider length to be larger than width, we will let $l$ = 66 in. and $w$ = 42 in.

42 in.

66 in.

**2. Translate.**   We substitute in the formula:

$$A = l \cdot w = 66 \cdot 42.$$

**5.** *Total Cost of Gas Grills.*
What is the total cost of 14 gas grills, each with 520 sq in. of total cooking surface, if each one costs $398?

*Answer*
**5.** $5572

**3. Solve.** We carry out the multiplication.

$$A = 66 \cdot 42$$
$$A = 2772$$

$$\begin{array}{r} 6\ 6 \\ \times\ 4\ 2 \\ \hline 1\ 3\ 2 \\ 2\ 6\ 4\ 0 \\ \hline 2\ 7\ 7\ 2 \end{array}$$

**4. Check.** We can repeat the calculation. We can also round and estimate:

$$66 \times 42 \approx 70 \times 40 = 2800 \approx 2772.$$

The answer checks.

**5. State.** The area of the rug is 2772 sq in.

◀ **Do Exercise 6.**

**6. Bed Sheets.** The dimensions of a flat sheet for a queen-size bed are 90 in. by 102 in. What is the area of the sheet?

**EXAMPLE 5** *Packages of Gum.* A candy company produces 3304 sticks of gum. How many 12-stick packages can be filled? How many sticks will be left over?

**1. Familiarize.** We make a drawing to visualize the situation and let $n =$ the number of 12-stick packages that can be filled. The problem can be considered as repeated subtraction, taking successive sets of 12 sticks and putting them into $n$ packages.

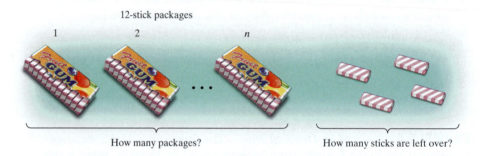

12-stick packages

1    2    n

How many packages?          How many sticks are left over?

**2. Translate.** We translate to an equation:

| Number of sticks | divided by | Number in each package | is | Number of packages |
|---|---|---|---|---|
| ↓ | ↓ | ↓ | ↓ | ↓ |
| 3304 | ÷ | 12 | = | $n$. |

**3. Solve.** We solve the equation by carrying out the division.

$$3304 \div 12 = n$$
$$275 \text{ R } 4 = n$$

$$\begin{array}{r} 2\ 7\ 5 \\ 1\ 2\ \overline{)\ 3\ 3\ 0\ 4} \\ 2\ 4\phantom{\ 0\ 0} \\ \hline 9\ 0\phantom{\ 0} \\ 8\ 4\phantom{\ 0} \\ \hline 6\ 4 \\ 6\ 0 \\ \hline 4 \end{array}$$

**Answer**

6. 9180 sq in.

**4. Check.** We can check by multiplying the number of packages by 12 and adding the remainder, 4:

$$12 \cdot 275 = 3300,$$
$$3300 + 4 = 3304.$$

**5. State.** Thus, 275 twelve-stick packages of gum can be filled. There will be 4 sticks left over.

**Do Exercise 7.** ▶

**EXAMPLE 6** *Automobile Mileage.* A 2017 Honda Accord Sport gets 26 miles per gallon (mpg) in city driving. How many gallons will it use in 4758 mi of city driving?
**Data:** Honda

**1. Familiarize.** We make a drawing and let $g$ = the number of gallons of gasoline used in 4758 mi of city driving.

**2. Translate.** Repeated addition or multiplication applies here.

| Number of miles per gallon | times | Number of gallons used | is | Number of miles driven |
|:---:|:---:|:---:|:---:|:---:|
| 26 | · | $g$ | = | 4758 |

**3. Solve.** To solve the equation, we divide by 26 on both sides.

$$26 \cdot g = 4758$$
$$\frac{26 \cdot g}{26} = \frac{4758}{26}$$
$$g = 183$$

```
        1 8 3
   2 6 ) 4 7 5 8
        2 6
        2 1 5
        2 0 8
            7 8
            7 8
               0
```

**4. Check.** To check, we multiply 183 by 26.

```
      1 8 3
   ×    2 6
    1 0 9 8
    3 6 6 0
    4 7 5 8
```

The answer checks.

**5. State.** The 2017 Honda Accord Sport will use 183 gal of gasoline.

**Do Exercise 8.** ▶

**7.** *Packages of Gum.* The candy company in Example 5 also produces 6-stick packages. How many 6-stick packages can be filled with 2269 sticks of gum? How many sticks will be left over?

**8.** *Automobile Mileage.* A 2017 Honda Accord Sport gets 34 miles per gallon (mpg) in highway driving. How many gallons will it use in 2686 mi of highway driving?
**Data:** Honda

*Answers*
**7.** 378 packages with 1 stick left over  **8.** 79 gal

## Multistep Problems

Sometimes we must use more than one operation to solve a problem, as in the following example.

**EXAMPLE 7** *Weight Loss.* To lose one pound, you must burn about 3500 calories in excess of what you already burn doing your regular daily activities. The following chart shows how long a person must engage in several types of exercise in order to burn 100 calories. For how long would a person have to run at a brisk pace in order to lose one pound?

To burn 100 calories, you must:

- Run for 8 minutes at a brisk pace, or
- Swim for 2 minutes at a brisk pace, or
- Bicycle for 15 minutes at 9 mph, or
- Do aerobic exercises for 15 minutes, or
- Golf, walking, for 20 minutes, or
- Play tennis, singles, for 11 minutes

1. **Familiarize.** This is a multistep problem. We will first find how many hundreds are in 3500. This will tell us how many times a person must run for 8 min in order to lose one pound. Then we will find the total number of minutes required for the weight loss.

   We let $x$ = the number of hundreds in 3500 and $t$ = the time it takes to lose one pound.

2. **Translate.** We translate to two equations.

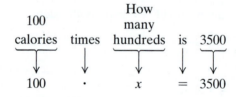

$$100 \quad \cdot \quad x \quad = \quad 3500$$

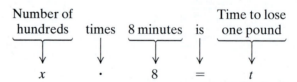

$$x \quad \cdot \quad 8 \quad = \quad t$$

**3. Solve.** We divide by 100 on both sides of the first equation to find $x$.

$$100 \cdot x = 3500$$

$$\frac{100 \cdot x}{100} = \frac{3500}{100}$$

$$x = 35$$

$$\begin{array}{r} 3\ 5 \\ 1\ 0\ 0\ )\overline{3\ 5\ 0\ 0} \\ \underline{3\ 0\ 0} \\ 5\ 0\ 0 \\ \underline{5\ 0\ 0} \\ 0 \end{array}$$

Then we use the fact that $x = 35$ to find $t$.

$$x \cdot 8 = t$$

$$35 \cdot 8 = t$$

$$280 = t$$

$$\begin{array}{r} 3\ 5 \\ \times\ \ \ 8 \\ \hline 2\ 8\ 0 \end{array}$$

**4. Check.** Suppose you run for 280 min. For every 8 min of running, you burn 100 calories. Since $280 \div 8 = 35$, there are 35 groups of 8 min in 280 min, so you will burn $35 \times 100 = 3500$ calories.

**5. State.** You must run for 280 min, or 4 hr 40 min, at a brisk pace in order to lose one pound.

**Do Exercise 9.** ▶

The key words, phrases, and concepts in the following table are useful when translating problems to equations.

**Key Words, Phrases, and Concepts**

| ADDITION (+) | SUBTRACTION (−) | MULTIPLICATION (×) | DIVISION (÷) |
|---|---|---|---|
| add | subtract | multiply | divide |
| added to | subtracted from | multiplied by | divided by |
| sum | difference | product | quotient |
| total | minus | times | repeated |
| plus | less than | of |    subtraction |
| more than | decreased by | repeated | missing factor |
| increased by | take away |    addition | finding equal |
| | how much more | rectangular arrays |    quantities |

The following tips are also helpful in problem solving.

---

**PROBLEM-SOLVING TIPS**

1. Look for patterns when solving problems.
2. When translating in mathematics, consider the dimensions of the variables and constants in the equation. The variables that represent length should all be in the same unit, those that represent money should all be in dollars or all in cents, and so on.
3. Make sure that units appear in the answer whenever appropriate and that you completely answer the original problem.

---

 **9. Weight Loss.** Use the information in Example 7 to determine how long an individual must swim at a brisk pace in order to lose one pound.

**1. Familiarize.** Let $x =$ the number of hundreds in 3500. Let $t =$ the time it takes to lose one pound.

**2. Translate.**

$$100 \cdot x = \boxed{\phantom{000}}$$

$$x \cdot \boxed{\phantom{00}} = t$$

**3. Solve.** From Example 7, we know that $x = \boxed{\phantom{00}}$.

$$x \cdot 2 = t$$

$$\boxed{\phantom{00}} \cdot 2 = t$$

$$\boxed{\phantom{00}} = t$$

**4. Check.** Since $70 \div 2 = 35$, there are $\boxed{\phantom{00}}$ groups of 2 min in 70 min. Thus, you will burn $35 \times 100 = \boxed{\phantom{000}}$ calories.

**5. State.** You must swim for $\boxed{\phantom{00}}$ min, or 1 hr $\boxed{\phantom{00}}$ min, in order to lose one pound.

*Answer*

**9.** 70 min, or 1 hr 10 min

*Guided Solution:*

**9.** 3500, 2; 35, 35, 70; 35, 3500; 70, 10

# Translating for Success

1. **Brick-Mason Expense.** A commercial contractor is building 30 two-unit condominiums in a retirement community. The brick-mason expense for each building is $10,860. What is the total cost of bricking the buildings?

2. **Heights.** Dean's sons are on the high school basketball team. Their heights are 73 in., 69 in., and 76 in. How much taller is the tallest son than the shortest son?

3. **Account Balance.** James has $423 in his checking account. Then he deposits $73 and uses his debit card for purchases of $76 and $69. How much is left in the account?

4. **Purchasing a Computer.** A computer is on sale for $423. Jenny has only $69. How much more does she need to buy the computer?

5. **Purchasing Coffee Makers.** Sara purchases 8 coffee makers for the newly remodeled bed-and-breakfast that she manages. If she pays $52 for each coffee maker, what is the total cost of her purchase?

*The goal of these matching questions is to practice step (2), Translate, of the five-step problem-solving process. Translate each word problem to an equation and select a correct translation from equations A–0.*

A. $8 \cdot 52 = n$

B. $69 \cdot n = 76$

C. $73 - 76 - 69 = n$

D. $423 + 73 - 76 - 69 = n$

E. $30 \cdot 10{,}860 = n$

F. $15 \cdot n = 195$

G. $69 + n = 423$

H. $n = 10{,}860 - 300$

I. $n = 423 \div 69$

J. $30 \cdot n = 10{,}860$

K. $15 \cdot 195 = n$

L. $n = 52 - 8$

M. $69 + n = 76$

N. $15 \div 195 = n$

O. $52 + n = 60$

*Answers on page A-2*

6. **Hourly Rate.** Miller Auto Repair charges $52 per hour for labor. Jackson Auto Care charges $60 per hour. How much more does Jackson charge than Miller?

7. **College Band.** A college band with 195 members marches in a 15-row formation in the homecoming halftime performance. How many members are in each row?

8. **Shoe Purchase.** A college football team purchases 15 pairs of shoes at $195 a pair. What is the total cost of this purchase?

9. **Loan Payment.** Kendra's uncle loans her $10,860, interest free, to buy a car. The loan is to be paid off in 30 payments. How much is each payment?

10. **College Enrollment.** At the beginning of the fall term, the total enrollment in Lakeview Community College was 10,860. By the end of the first two weeks, 300 students had withdrawn. How many students were then enrolled?

FOR EXTRA HELP   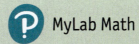 MyLab Math

## ✓ Check Your Understanding

**Reading Check**   List the steps of the problem-solving strategy in order, using the choices given below. The last step is already listed.

    Check          Familiarize          Solve          Translate

**RC1. 1.** _____.

**RC2. 2.** _____.

**RC3. 3.** _____.

**RC4. 4.** _____.

   **5.** State.

**Concept Check**   Choose from the list on the right the best translation of each problem.

**CC1.** Jessica placed 500 pieces of candy into 125 giveaway bags for children. If she put the same number of pieces of candy in each bag, how many pieces did she put in each bag?

**CC2.** Joseph drove 500 miles on Wednesday and 125 miles on Thursday. How much farther did he drive on Wednesday than on Thursday?

**CC3.** Hayley mows a rectangular field that is 500 ft long and 125 feet wide. What is the area of the field?

**CC4.** Alexis spent \$500 for groceries in May and \$125 for groceries in June. How much did she spend in total for groceries for the two months?

**a)**  $500 + 125 = x$
**b)**  $500 = x + 125$
**c)**  $x = 500 \cdot 125$
**d)**  $x \cdot 125 = 500$

---

**a**   Solve.

*Waste.*   The following figure shows waste generation data for the four countries with the highest per capita generation of municipal solid waste. Use the information for Exercises 1–4.

**Municipal Solid Waste**

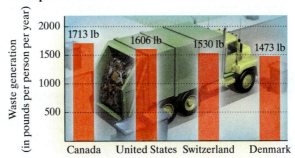

DATA: conferenceboard.ca; U.S. Environmental Protection Agency; European Environment Agency

**1.** How much more waste is generated annually per capita in Canada than in the United States?

**2.** How much more waste is generated annually per capita in the United States than in Denmark?

**3.** Switzerland generates 981 lb more waste annually per capita than does Romania. How much waste per capita is generated annually in Romania?

**4.** Canada generates 531 lb more waste annually per capita than does Iceland. How much waste per capita is generated annually in Iceland?

**5.** A carpenter drills 216 holes in a rectangular array to construct a pegboard. There are 12 holes in each row. How many rows are there?

**6.** Lou arranges 504 entries on a spreadsheet in a rectangular array that has 36 rows. How many entries are in each row?

**7.** *Olympics.* There were 306 events in the 2016 Summer Olympics in Rio de Janeiro, Brazil. This was 263 more events than there were in the first modern Olympic games in Athens, Greece, in 1896. How many events were there in 1896?

**Data:** *USA Today*; thedailybeast.com

**8.** *Olympics.* Athletes from 206 countries participated in the 2016 Summer Olympics in Rio de Janeiro, Brazil. The number of countries represented in the 2016 Olympics was 192 more than the number of countries represented in the first modern Olympic games in Athens, Greece, in 1896. How many countries were represented in 1896?

**Data:** history.com; thedailybeast.com

**9.** *Boundaries between Countries.* The boundary between mainland United States and Canada including the Great Lakes is 3987 mi long. The length of the boundary between the United States and Mexico is 1933 mi. How much longer is the Canadian border?

**Data:** U.S. Geological Survey

**10.** *Longest Rivers.* The longest river in the world is the Nile in Africa at about 4135 mi. The longest river in the United States is the Missouri–Mississippi at about 3860 mi. How much longer is the Nile?

**11.** *Pixels.* A high-definition television (HDTV) screen consists of small rectangular dots called *pixels*. How many pixels are there on a screen that has 1080 rows with 1920 pixels in each row?

**12.** *Crossword Puzzle.* The *USA Today* crossword puzzle is a rectangle containing 15 rows with 15 squares in each row. How many squares does the puzzle have altogether?

13. **Caffeine Content.** An 8-oz serving of Red Bull energy drink contains 76 milligrams of caffeine. An 8-oz serving of brewed coffee contains 19 more milligrams of caffeine than the energy drink. How many milligrams of caffeine does the 8-oz serving of coffee contain?

   **Data:** The Mayo Clinic

14. **Caffeine Content.** Hershey's 6-oz milk chocolate almond bar contains 25 milligrams of caffeine. A 20-oz bottle of Coca-Cola has 32 more milligrams of caffeine than the Hershey bar. How many milligrams of caffeine does the 20-oz bottle of Coca-Cola have?

   **Data:** *National Geographic*, "Caffeine," by T. R. Reid, January 2005

15. There are 24 hr in a day and 7 days in a week. How many hours are there in a week?

16. There are 60 min in an hour and 24 hr in a day. How many minutes are there in a day?

**Housing Costs.** The graph below shows the average monthly rent for a one-bedroom apartment in several cities in June 2017. Use this graph to do Exercises 17–22.

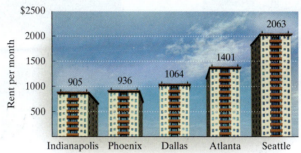

**Average Rent for a One-Bedroom Apartment**

DATA: rentjungle.com

17. How much higher is the average monthly rent in Seattle than in Dallas?

18. How much lower is the average monthly rent in Indianapolis than in Atlanta?

19. Phil, Scott, and Julio plan to rent a one-bedroom apartment in Atlanta immediately after graduation, sharing the rent equally. What average monthly rent can each of them expect to pay?

20. Maria and her sister Theresa plan to share a one-bedroom apartment in Phoenix, dividing the monthly rent equally between them. About how much can each of them expect to pay?

21. On average, how much rent would a tenant pay for a one-bedroom apartment in Phoenix during a 12-month period?

22. On average, how much rent would a tenant pay for a one-bedroom apartment in Seattle during a 6-month period?

23. **Colonial Population.** Before the establishment of the U.S. Census in 1790, it was estimated that the colonial population in 1780 was 2,780,400. This was an increase of 2,628,900 from the population in 1680. What was the colonial population in 1680?

   **Data:** *Time Almanac*

24. **Interstate Speed Limits.** The speed limit for passenger cars on interstate highways in rural areas in Montana is 75 mph. This is 10 mph faster than the speed limit for trucks on the same roads. What is the speed limit for trucks?

**25.** *Yard-Sale Profit.* Ruth made $312 at her yard sale and divided the money equally among her four grandchildren. How much did each child receive?

**26.** *Paper Measures.* A quire of paper consists of 25 sheets, and a ream of paper consists of 500 sheets. How many quires are in a ream?

**27.** *Parking Rates.* The most expensive parking in the world is found in New York City, where the average rate is $606 per month. This is $216 per month more than in Sydney, Australia. What is the average monthly parking rate in Sydney?

**Data:** autoguide.com

**28.** *Trade Balance.* In 2016, international visitors spent $153,700,000,000 traveling in the United States, while Americans spent $110,500,000,000 traveling abroad. How much more was spent by visitors to the United States than by Americans traveling abroad?

**Data:** U.S. Travel Association

**29.** *Refrigerator Purchase.* Gourmet Deli has a chain of 24 restaurants. It buys a commercial refrigerator for each store at a cost of $1019 each. Determine the total cost of the purchase.

**30.** *Microwave Purchase.* Each room in the new dorm at Bridgeway College has a small kitchen. To furnish the kitchens, the college buys 96 microwave ovens at $88 each. Determine the total cost of the purchase.

**31.** *Seinfeld.* A local television station plans to air the 177 episodes of the long-running comedy series *Seinfeld*. If the station airs 5 episodes per week, how many full weeks will pass before it must begin re-airing previously shown episodes? How many unaired episodes will be shown the following week before the previously aired episodes are rerun?

**32.** *Everybody Loves Raymond.* The popular television comedy series *Everybody Loves Raymond* had 208 scripted episodes and 2 additional episodes consisting of clips from previous shows. A local television station plans to air the 208 scripted episodes, showing 5 episodes per week. How many full weeks will pass before it must begin re-airing episodes? How many unaired episodes will be shown the following week before the previously aired episodes are rerun?

**33.** *Crossword Puzzle.* The *Los Angeles Times* crossword puzzle is a rectangle containing 441 squares arranged in 21 rows. How many columns does the puzzle have?

**34.** *Mailing Labels.* A box of mailing labels contains 750 labels on 25 sheets. How many labels are on each sheet?

**35. *Automobile Mileage.*** The 2017 Ford Focus FWD with a manual transmission gets 30 miles per gallon (mpg) in city driving. How many gallons will it use in 7080 mi of city driving?

**Data:** fueleconomy.gov

**36. *Automobile Mileage.*** The 2017 Chevrolet Tahoe C1500 gets 23 miles per gallon (mpg) in highway driving. How many gallons will it use in 3795 mi of highway driving?

**Data:** fueleconomy.gov

**37. *High School Court.*** The standard basketball court used by high school players has dimensions of 50 ft by 84 ft.
  **a)** What is its area?
  **b)** What is its perimeter?

**38. *College Court.*** The standard basketball court used by college players has dimensions of 50 ft by 94 ft.
  **a)** What is its area?
  **b)** What is its perimeter?
  **c)** How much greater is the area of a college court than that of a high school court? (See Exercise 37.)

**39.** Copies of this book are usually shipped from the warehouse in cartons containing 24 books each. How many cartons are needed to ship 1344 books?

**40.** The H. J. Heinz Company ships 16-oz bottles of ketchup in cartons containing 12 bottles each. How many cartons are needed to ship 528 bottles of ketchup?

**41. *Map Drawing.*** A map has a scale of 215 mi to the inch. How far apart *in reality* are two cities that are 3 in. apart on the map? How far apart *on the map* are two cities that, in reality, are 1075 mi apart?

**42. *Map Drawing.*** A map has a scale of 288 mi to the inch. How far apart *on the map* are two cities that, in reality, are 2016 mi apart? How far apart *in reality* are two cities that are 8 in. apart on the map?

**43. Loan Payments.** Dana borrows $5928 for a used car. The loan is to be paid off in 24 equal monthly payments. How much is each payment (excluding interest)?

**44. Home Improvement Loan.** The Van Reken family borrows $7824 to build a detached garage next to their home. The loan is to be paid off in equal monthly payments of $163 (excluding interest). How many months will it take to pay off the loan?

Refer to the information in Example 7 to do Exercises 45 and 46.

**45.** For how long must you do aerobic exercises in order to lose one pound?

**46.** For how long must you golf, walking, in order to lose one pound?

**New Jobs.** Many of the fastest-growing occupations in the United States require education beyond a high school diploma. The following table lists some of these and gives the projected numbers of new jobs expected to be created between 2014 and 2024. Use the information in the table for Exercises 47 and 48.

**New Jobs Created, 2014–2024**

| JOB | NUMBER |
| --- | --- |
| Registered nurse | 439,300 |
| Postsecondary teacher | 177,000 |
| Accountant | 142,400 |
| Physical therapist | 71,800 |
| Web developer | 39,500 |
| Marketing manager | 19,700 |

DATA: U.S. Bureau of Labor Statistics

**47.** The U.S. Bureau of Labor Statistics predicts that between 2014 and 2024, there will be 411,800 more new jobs created for registered nurses and physical therapists than there will be for physicians. How many new jobs will be created for physicians between 2014 and 2024?

**48.** The U.S. Bureau of Labor Statistics predicts that between 2014 and 2024, there will be 143,100 more new jobs created for marketing managers and accountants than there will be for sales managers. How many new jobs will be created for sales managers between 2014 and 2024?

**49. Seating Configuration.** The seats in the Boeing 737-700 airplanes in United Airlines' North American fleet are configured with 3 rows of 4 seats across in first class, 17 rows of 6 seats across in economy class, and one exit row of 4 seats across. Determine the total seating capacity of one of these planes.

**Data:** United Airlines

Economy class: 17 rows of 6 seats; 1 row of 4 seats    First class: 3 rows of 4 seats

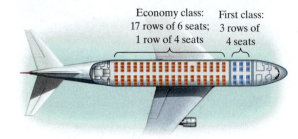

**50. Seating Configuration.** The seats in the Airbus 320 airplanes in United Airlines' North American fleet are configured with 3 rows of 4 seats across in first class and 23 rows of 6 seats across in economy class. Determine the total seating capacity of one of these planes.

**Data:** United Airlines

Economy class: 23 rows of 6 seats    First class: 3 rows of 4 seats

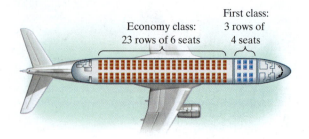

**51.** Elena buys 5 video games at $64 each and pays for them with $10 bills. How many $10 bills does it take?

**52.** Pedro buys 5 video games at $64 each and pays for them with $20 bills. How many $20 bills does it take?

**53.** The balance in Meg's bank account is $568. She uses her debit card for purchases of $46, $87, and $129. Then she deposits $94 in the account after returning a textbook. How much is left in her account?

**54.** The balance in Dylan's bank account is $749. He uses his debit card for purchases of $34 and $65. Then he makes a deposit of $123 from his paycheck. What is the new balance?

**55.** *Bones in the Hands and Feet.* There are 27 bones in each human hand and 26 bones in each human foot. How many bones are there in all in the hands and feet?

**56.** An office for adjunct instructors at a community college has 6 bookshelves, each of which is 3 ft wide. The office is moved to a new location that has dimensions of 16 ft by 21 ft. Is it possible for the bookshelves to be put side by side on the 16-ft wall?

## Skill Maintenance

**57.** Add: [1.2a]

```
  6 2 5 4
  1 5 3 7
+   4 8 2
```

**58.** Subtract: [1.3a]

```
  9 6 0 2
− 1 8 4 3
```

**59.** Multiply: [1.4a]

```
  3 4 0 5
×   2 3 7
```

**60.** Divide: [1.5a]

```
3 2 ) 4 7 0 8
```

**61.** Find the perimeter of the figure. [1.2b]

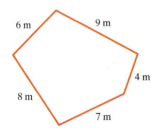

**62.** Find the area of the region. [1.4b]

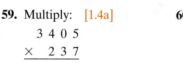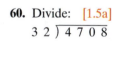

211 ft

46 ft

**63.** Estimate $238 \times 596$ by rounding to the nearest hundred. [1.6b]

**64.** Solve: $x + 15 = 81$. [1.7b]

## Synthesis

**65.** 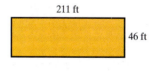 *Speed of Light.* Light travels about 186,000 miles per second (mi/sec) in a vacuum such as in outer space. In ice it travels about 142,000 mi/sec, and in glass it travels about 109,000 mi/sec. In 18 sec, how many more miles will light travel in a vacuum than in ice? than in glass?

**66.** Carney Community College has 1200 students. Each instructor teaches 4 classes, and each student takes 5 classes. There are 30 students and 1 instructor in each classroom. How many instructors are there at Carney Community College?

# Exponential Notation and Order of Operations

**OBJECTIVES**

**a** Write exponential notation for products such as $4 \cdot 4 \cdot 4$.

**b** Evaluate exponential notation.

**c** Simplify expressions using the rules for order of operations.

**d** Remove parentheses within parentheses.

## a WRITING EXPONENTIAL NOTATION

Consider the product $3 \cdot 3 \cdot 3 \cdot 3$. Such products occur often enough that mathematicians have found it convenient to create a shorter notation, called **exponential notation**, for them. For example,

$3 \cdot 3 \cdot 3 \cdot 3$ is shortened to $3^4$. ← exponent

4 factors     base

We read exponential notation as follows.

| NOTATION | WORD DESCRIPTION |
|---|---|
| $3^4$ | "three to the fourth power," or "the fourth power of three" |
| $5^3$ | "five cubed," or "the cube of five," or "five to the third power," or "the third power of five" |
| $7^2$ | "seven squared," or "the square of seven," or "seven to the second power," or "the second power of seven" |

The wording "seven squared" for $7^2$ is derived from the fact that a square with side $s$ has area $A$ given by $A = s^2$.

$s$

$A = s^2$     $s$

An expression like $3 \cdot 5^2$ is read "three times five squared," or "three times the square of five."

**EXAMPLE 1** Write exponential notation for $10 \cdot 10 \cdot 10 \cdot 10 \cdot 10$.

Exponential notation is $10^5$.     5 is the *exponent*.
10 is the *base*.

**EXAMPLE 2** Write exponential notation for $2 \cdot 2 \cdot 2$.

Exponential notation is $2^3$.

◀ **Do Exercises 1–4.**

Write exponential notation.

**1.** $5 \cdot 5 \cdot 5 \cdot 5$

**2.** $5 \cdot 5 \cdot 5 \cdot 5 \cdot 5 \cdot 5$

**3.** $10 \cdot 10$

**4.** $10 \cdot 10 \cdot 10 \cdot 10$

*Answers*

**1.** $5^4$  **2.** $5^6$  **3.** $10^2$  **4.** $10^4$

## b  EVALUATING EXPONENTIAL NOTATION

MyLab Math
ANIMATION

SKILL REVIEW

*Multiply whole numbers.*  [1.4a]
Multiply.

1. $5 \times 5 \times 5$     2. $2 \times 2 \times 2 \times 2 \times 2$

Answers: 1. 125  2. 32

MyLab Math
VIDEO

We evaluate exponential notation by rewriting it as a product and then computing the product.

**EXAMPLE 3**  Evaluate: $10^3$.

$$10^3 = 10 \cdot 10 \cdot 10 = 1000$$

·········  **Caution!**  ·········

$10^3$ does not mean $10 \cdot 3$.

**EXAMPLE 4**  Evaluate: $5^4$.

$$5^4 = 5 \cdot 5 \cdot 5 \cdot 5 = 625$$

Do Exercises 5–8. ▶

GS  **5.** Evaluate: $10^4$.

$$10^4 = \boxed{\phantom{0}} \cdot \boxed{\phantom{0}} \cdot \boxed{\phantom{0}} \cdot \boxed{\phantom{0}} = \boxed{\phantom{0}}$$

Evaluate.

**6.** $10^2$     **7.** $8^3$

**8.** $2^5$

## c  SIMPLIFYING EXPRESSIONS

Suppose we have a calculation like the following:

$$3 + 4 \cdot 8.$$

How do we find the answer? Do we add 3 to 4 and then multiply by 8, or do we multiply 4 by 8 and then add 3? In the first case, the answer is 56. In the second, the answer is 35. We agree to compute as in the second case:

$$3 + 4 \cdot 8 = 3 + 32 = 35.$$

The following rules are an agreement regarding the order in which we perform operations. These are the rules that computers and most scientific calculators use to do computations.

---

**RULES FOR ORDER OF OPERATIONS**

1. Do all calculations within parentheses ( ), brackets [ ], or braces { } before operations outside.
2. Evaluate all exponential expressions.
3. Do all multiplications and divisions in order from left to right.
4. Do all additions and subtractions in order from left to right.

---

**EXAMPLE 5**  Simplify: $16 \div 8 \cdot 2$.

There are no parentheses or exponents, so we begin with the third step.

$$\left.\begin{array}{l} 16 \div 8 \cdot 2 = 2 \cdot 2 \\ \qquad\qquad = 4 \end{array}\right\}$$

Doing all multiplications and divisions in order from left to right

**CALCULATOR CORNER**

***Exponential Notation***  Many calculators have a $\boxed{y^x}$ or $\boxed{\wedge}$ key for raising a base to a power. To find $16^3$, for example, we press $\boxed{1}\,\boxed{6}\,\boxed{y^x}\,\boxed{3}\,\boxed{=}$ or $\boxed{1}\,\boxed{6}\,\boxed{\wedge}\,\boxed{3}\,\boxed{=}$. The result is 4096.

**EXERCISES:**  Use a calculator to find each of the following.

1. $3^5$
2. $5^6$
3. $12^4$
4. $2^{11}$

*Answers*
**5.** 10,000  **6.** 100  **7.** 512  **8.** 32
*Guided Solution:*
**5.** 10, 10, 10, 10, 10,000

Simplify.

**9.** $93 - 14 \cdot 3$

**10.** $104 \div 4 + 4$

**11.** $25 \cdot 26 - (56 + 10)$

**12.** $75 \div 5 + (83 - 14)$

Simplify and compare.

**13.** $64 \div (32 \div 2)$ and $(64 \div 32) \div 2$

**14.** $(28 + 13) + 11$ and $28 + (13 + 11)$

---

**15. Simplify:**
$9 \times 4 - (20 + 4) \div 8 - (6 - 2).$
$9 \times 4 - (20 + 4) \div 8 - (6 - 2)$
$= 9 \times 4 - \boxed{\phantom{0}} \div 8 - \boxed{\phantom{0}}$
$= \boxed{\phantom{0}} - 24 \div 8 - 4$
$= 36 - \boxed{\phantom{0}} - 4$
$= \boxed{\phantom{0}} - 4$
$= \boxed{\phantom{0}}$

GS

---

Simplify.

**16.** $5 \cdot 5 \cdot 5 + 26 \cdot 71$
$\quad -(16 + 25 \cdot 3)$

**17.** $30 \div 5 \cdot 2 + 10 \cdot 20 + 8 \cdot 8$
$\quad - 23$

**18.** $95 - 2 \cdot 2 \cdot 2 \cdot 5 \div (24 - 4)$

Simplify.

**19.** $5^3 + 26 \cdot 71 - (16 + 25 \cdot 3)$

**20.** $(1 + 3)^3 + 10 \cdot 20 + 8^2 - 23$

**21.** $81 - 3^2 \cdot 2 \div (12 - 9)$

*Answers*

**9.** 51 **10.** 30 **11.** 584 **12.** 84 **13.** 4; 1
**14.** 52; 52 **15.** 29 **16.** 1880 **17.** 253
**18.** 93 **19.** 1880 **20.** 305 **21.** 75

*Guided Solution:*
**15.** 24, 4, 36, 3, 33, 29

---

**EXAMPLE 6** Simplify: $7 \cdot 14 - (12 + 18)$.

$$7 \cdot 14 - (12 + 18) = 7 \cdot 14 - 30 \quad \text{Carrying out operations inside parentheses}$$
$$= 98 - 30 \quad \text{Doing all multiplications and divisions}$$
$$= 68 \quad \text{Doing all additions and subtractions}$$

◀ **Do Exercises 9–12.**

**EXAMPLE 7** Simplify and compare: $23 - (10 - 9)$ and $(23 - 10) - 9$.
We have
$$23 - (10 - 9) = 23 - 1 = 22;$$
$$(23 - 10) - 9 = 13 - 9 = 4.$$

We can see that $23 - (10 - 9)$ and $(23 - 10) - 9$ represent different numbers. Thus subtraction is not associative.

◀ **Do Exercises 13 and 14.**

**EXAMPLE 8** Simplify: $7 \cdot 2 - (12 + 0) \div 3 - (5 - 2)$.

$7 \cdot 2 - (12 + 0) \div 3 - (5 - 2)$
$= 7 \cdot 2 - 12 \div 3 - 3 \quad \text{Carrying out operations inside parentheses}$
$= 14 - 4 - 3 \quad \text{Doing all multiplications and divisions in order from left to right}$
$\left. \begin{array}{l} = 10 - 3 \\ = 7 \end{array} \right\} \quad \text{Doing all additions and subtractions in order from left to right}$

**EXAMPLE 9** Simplify: $15 \div 3 \cdot 2 \div (10 - 8)$.

$15 \div 3 \cdot 2 \div (10 - 8)$
$= 15 \div 3 \cdot 2 \div 2 \quad \text{Carrying out operations inside parentheses}$
$\left. \begin{array}{l} = 5 \cdot 2 \div 2 \\ = 10 \div 2 \\ = 5 \end{array} \right\} \quad \text{Doing all multiplications and divisions in order from left to right}$

◀ **Do Exercises 15–18.**

**EXAMPLE 10** Simplify: $4^2 \div (10 - 9 + 1)^3 \cdot 3 - 5$.

$4^2 \div (10 - 9 + 1)^3 \cdot 3 - 5$
$= 4^2 \div (1 + 1)^3 \cdot 3 - 5 \quad \text{Subtracting inside parentheses}$
$= 4^2 \div 2^3 \cdot 3 - 5 \quad \text{Adding inside parentheses}$
$= 16 \div 8 \cdot 3 - 5 \quad \text{Evaluating exponential expressions}$
$\left. \begin{array}{l} = 2 \cdot 3 - 5 \\ = 6 - 5 \end{array} \right\} \quad \text{Doing all multiplications and divisions in order from left to right}$
$= 1 \quad \text{Subtracting}$

◀ **Do Exercises 19–21.**

**EXAMPLE 11**   Simplify: $2^9 \div 2^6 \cdot 2^3$.

$2^9 \div 2^6 \cdot 2^3 = 512 \div 64 \cdot 8$   Since there are no parentheses, we evaluate the exponential expressions.

$\left. \begin{array}{l} = 8 \cdot 8 \\ = 64 \end{array} \right\}$   Doing all multiplications and divisions in order from left to right

**Do Exercise 22.** ▶

**22.** Simplify: $2^3 \cdot 2^8 \div 2^9$.

## Averages

In order to find the average of a set of numbers, we use addition and then division. For example, the average of 2, 3, 6, and 9 is found as follows.

The number of addends is 4.

$$\text{Average} = \frac{2 + 3 + 6 + 9}{4} = \frac{20}{4} = 5$$

Divide by 4.

The fraction bar acts as a grouping symbol, so

$$\frac{2 + 3 + 6 + 9}{4} \text{ is equivalent to } (2 + 3 + 6 + 9) \div 4.$$

Thus we are using order of operations when we compute an average.

---

**AVERAGE**

The **average** of a set of numbers is the sum of the numbers divided by the number of addends.

---

**EXAMPLE 12**   *National Parks.*   Since 2000, four national parks have been established in the United States. The sizes of these parks are shown in the figure below. Determine the average size of the four parks.

**U.S. National Parks**

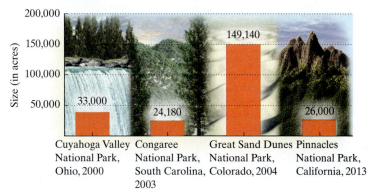

Cuyahoga Valley National Park, Ohio, 2000 — 33,000

Congaree National Park, South Carolina, 2003 — 24,180

Great Sand Dunes National Park, Colorado, 2004 — 149,140

Pinnacles National Park, California, 2013 — 26,000

Size (in acres)

DATA: National Geographic

*Answer:*

**22.** 4

**23. *Average Number of Career Hits.*** The numbers of career hits of five Hall of Fame baseball players are given in the graph below. Find the average number of career hits of all five.

**Career Hits**

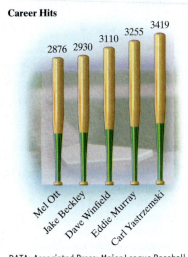

2876  2930  3110  3255  3419

Mel Ott  Jake Beckley  Dave Winfield  Eddie Murray  Carl Yastrzemski

DATA: Associated Press; Major League Baseball

Simplify.

**24.** $9 \times 5 + \{6 \div [14 - (5 + 3)]\}$

**25.** $[18 - (2 + 7) \div 3]$ **GS**
$\qquad - (31 - 10 \times 2)$
$= [18 - \boxed{\phantom{0}} \div 3] - (31 - 10 \times 2)$
$= [18 - \boxed{\phantom{0}}] - (31 - \boxed{\phantom{0}})$
$= \boxed{\phantom{0}} - \boxed{\phantom{0}}$
$= \boxed{\phantom{0}}$

The average is given by

$$\frac{33{,}000 + 24{,}180 + 149{,}140 + 26{,}000}{4} = \frac{232{,}320}{4} = 58{,}080.$$

The average size of the four national parks is 58,080 acres.

◀ **Do Exercise 23.**

**d  REMOVING PARENTHESES WITHIN PARENTHESES**

When parentheses occur within parentheses, we can make them different shapes, such as [ ] (called "brackets") and { } (called "braces"). All of these have the same meaning. When parentheses occur within parentheses, computations in the innermost ones are to be done first.

**EXAMPLE 13**  Simplify: $[25 - (4 + 3) \cdot 3] \div (11 - 7)$.

$[25 - (4 + 3) \cdot 3] \div (11 - 7)$

$= [25 - 7 \cdot 3] \div (11 - 7)$   Doing the calculations in the innermost parentheses first

$= [25 - 21] \div (11 - 7)$   Doing the multiplication in the brackets

$= 4 \div 4$   Subtracting

$= 1$   Dividing

**EXAMPLE 14**  Simplify: $16 \div 2 + \{40 - [13 - (4 + 2)]\}$.

$16 \div 2 + \{40 - [13 - (4 + 2)]\}$

$= 16 \div 2 + \{40 - [13 - 6]\}$   Doing the calculations in the innermost parentheses first

$= 16 \div 2 + \{40 - 7\}$   Again, doing the calculations in the innermost brackets

$= 16 \div 2 + 33$   Subtracting inside the braces

$= 8 + 33$   Dividing

$= 41$   Adding

◀ **Do Exercises 24 and 25.**

*Answers*
**23.** 3118 hits  **24.** 46  **25.** 4
*Guided Solution:*
**25.** 9, 3, 20, 15, 11, 4

### ✓ Check Your Understanding

**Reading Check**  Complete each statement by choosing the correct word or number from below the blank.

**RC1.** In the expression $5^3$, the number 3 is the _____.
base/exponent

**RC2.** The expression $9^2$ can be read "nine _____."
cubed/squared

**RC3.** To calculate $10 - 4 \cdot 2$, we perform the _____ first.
multiplication/subtraction

**RC4.** To find the average of 7, 8, and 9, we add the numbers and divide the sum by _____.
2/3

**Concept Check**  Choose from the list on the right the operation that should be performed first to evaluate each given expression. Choices may be used more than once or not at all.

**CC1.** $10 + 2 \cdot 5$

**CC2.** $100 \div 10 \cdot 2$

**CC3.** $18 - 3 + 5$

**CC4.** $20 - (4 \cdot 3)$

**CC5.** $40 - 5 \cdot 3^2$

Addition
Subtraction
Multiplication
Division
Exponentiation

---

**a**  Write exponential notation.

**1.** $3 \cdot 3 \cdot 3 \cdot 3$

**2.** $2 \cdot 2 \cdot 2 \cdot 2 \cdot 2$

**3.** $5 \cdot 5$

**4.** $13 \cdot 13 \cdot 13$

**5.** $7 \cdot 7 \cdot 7 \cdot 7 \cdot 7$

**6.** $9 \cdot 9$

**7.** $10 \cdot 10 \cdot 10$

**8.** $1 \cdot 1 \cdot 1 \cdot 1$

**b**  Evaluate.

**9.** $7^2$

**10.** $5^3$

**11.** $9^3$

**12.** $8^2$

**13.** $12^4$

**14.** $10^5$

**15.** $3^5$

**16.** $2^6$

**c**  Simplify.

**17.** $12 + (6 + 4)$

**18.** $(12 + 6) + 18$

**19.** $52 - (40 - 8)$

**20.** $(52 - 40) - 8$

**21.** $1000 \div (100 \div 10)$

**22.** $(1000 \div 100) \div 10$

**23.** $(256 \div 64) \div 4$

**24.** $256 \div (64 \div 4)$

**25.** $(2 + 5)^2$

**26.** $2^2 + 5^2$

**27.** $(11 - 8)^2 - (18 - 16)^2$

**28.** $(32 - 27)^3 + (19 + 1)^3$

**29.** $16 \cdot 24 + 50$

**30.** $23 + 18 \cdot 20$

**31.** $83 - 7 \cdot 6$

**32.** $10 \cdot 7 - 4$

**33.** $10 \cdot 10 - 3 \cdot 4$

**34.** $90 - 5 \cdot 5 \cdot 2$

**35.** $4^3 \div 8 - 4$

**36.** $8^2 - 8 \cdot 2$

**37.** $17 \cdot 20 - (17 + 20)$

**38.** $1000 \div 25 - (15 + 5)$

**39.** $6 \cdot 10 - 4 \cdot 10$

**40.** $3 \cdot 8 + 5 \cdot 8$

**41.** $300 \div 5 + 10$

**42.** $144 \div 4 - 2$

**43.** $3 \cdot (2 + 8)^2 - 5 \cdot (4 - 3)^2$

**44.** $7 \cdot (10 - 3)^2 - 2 \cdot (3 + 1)^2$

**45.** $4^2 + 8^2 \div 2^2$

**46.** $6^2 - 3^4 \div 3^3$

**47.** $10^3 - 10 \cdot 6 - (4 + 5 \cdot 6)$

**48.** $7^2 + 20 \cdot 4 - (28 + 9 \cdot 2)$

**49.** $6 \cdot 11 - (7 + 3) \div 5 - (6 - 4)$

**50.** $8 \times 9 - (12 - 8) \div 4 - (10 - 7)$

**51.** $120 - 3^3 \cdot 4 \div (5 \cdot 6 - 6 \cdot 4)$

**52.** $80 - 2^4 \cdot 15 \div (7 \cdot 5 - 45 \div 3)$

**53.** $2^3 \cdot 2^8 \div 2^6$

**54.** $2^7 \div 2^5 \cdot 2^4 \div 2^2$

**55.** Find the average of $64, $97, and $121.

**56.** Find the average of four test grades of 86, 92, 80, and 78.

**57.** Find the average of 320, 128, 276, and 880.

**58.** Find the average of $1025, $775, $2062, $942, and $3721.

**59.** $8 \times 13 + \{42 \div [18 - (6 + 5)]\}$

**60.** $72 \div 6 - \{2 \times [9 - (4 \times 2)]\}$

**61.** $[14 - (3 + 5) \div 2] - [18 \div (8 - 2)]$

**62.** $[92 \times (6 - 4) \div 8] + [7 \times (8 - 3)]$

**63.** $(82 - 14) \times [(10 + 45 \div 5) - (6 \cdot 6 - 5 \cdot 5)]$

**64.** $(18 \div 2) \cdot \{[(9 \cdot 9 - 1) \div 2] - [5 \cdot 20 - (7 \cdot 9 - 2)]\}$

**65.** $4 \times \{(200 - 50 \div 5) - [(35 \div 7) \cdot (35 \div 7) - 4 \times 3]\}$

**66.** $15(23 - 4 \cdot 2)^3 \div (3 \cdot 25)$

**67.** $\{[18 - 2 \cdot 6] - [40 \div (17 - 9)]\} + \{48 - 13 \times 3 + [(50 - 7 \cdot 5) + 2]\}$

**68.** $(19 - 2^4)^5 - (141 \div 47)^2$

## Skill Maintenance

Solve. [1.7b]

**69.** $x + 341 = 793$

**70.** $4197 + x = 5032$

**71.** $7 \cdot x = 91$

**72.** $1554 = 42 \cdot y$

**73.** $6000 = 1102 + t$

**74.** $10,000 = 100 \cdot t$

Solve. [1.8a]

**75.** *Colorado.* The state of Colorado is roughly the shape of a rectangle that is 273 mi by 382 mi. What is its area?

**76.** On a long four-day trip, a family bought the following amounts of gasoline for their motor home: 23 gal, 24 gal, 26 gal, and 25 gal. How much gasoline did they buy in all?

## Synthesis

Each of the answers in Exercises 77–79 is incorrect. First find the correct answer. Then place as many parentheses as needed in the expression in order to make the incorrect answer correct.

**77.** $1 + 5 \cdot 4 + 3 = 36$

**78.** $12 \div 4 + 2 \cdot 3 - 2 = 2$

**79.** $12 \div 4 + 2 \cdot 3 - 2 = 4$

**80.** Use one occurrence each of 1, 2, 3, 4, 5, 6, 7, 8, and 9, in order, and any of the symbols $+, -, \cdot, \div$, and ( ) to represent 100.

## Vocabulary Reinforcement

In each of Exercises 1–8, fill in the blank with the correct term from the given list. Some of the choices may not be used and some may be used more than once.

1. The distance around an object is its _____. [1.2b]

2. The _____ is the number from which another number is being subtracted. [1.3a]

3. For large numbers, _____ are separated by commas into groups of three, called _____. [1.1a]

4. In the sentence $28 \div 7 = 4$, the _____ is 28. [1.5a]

5. In the sentence $10 \times 1000 = 10,000$, 10 and 1000 are called _____ and 10,000 is called the _____. [1.4a]

6. The number 0 is called the _____ identity. [1.2a]

7. The sentence $3 \times (6 \times 2) = (3 \times 6) \times 2$ illustrates the _____ law of multiplication. [1.4a]

8. We can use the following statement to check division: quotient · _____ + _____ = _____. [1.5a]

associative

commutative

addends

factors

area

perimeter

minuend

subtrahend

product

digits

periods

additive

multiplicative

dividend

quotient

remainder

divisor

## Concept Reinforcement

Determine whether each statement is true or false.

_____ 1. $a > b$ is true when $a$ is to the right of $b$ on the number line. [1.6c]

_____ 2. Any nonzero number divided by itself is 1. [1.5a]

_____ 3. For any whole number $a$, $a \div 0 = 0$. [1.5a]

_____ 4. Every equation is true. [1.7a]

_____ 5. The rules for order of operations tell us to multiply and divide before adding and subtracting. [1.9c]

_____ 6. The average of three numbers is the middle number. [1.9c]

# Study Guide

**Objective 1.1a**  Give the meaning of digits in standard notation.

**Example**  What does the digit 7 mean in 2,379,465?

2, 3 7 9, 4 6 5

7 means 7 ten thousands.

**Practice Exercise**

1. What does the digit 2 mean in 432,079?

---

**Objective 1.2a**  Add whole numbers.

**Example**  Add: 7368 + 3547.

$$
\begin{array}{r}
\overset{1}{7}\,\overset{1}{3}\,6\,8 \\
+\ 3\,5\,4\,7 \\
\hline
1\,0,9\,1\,5
\end{array}
$$

**Practice Exercise**

2. Add: 36,047 + 29,255.

---

**Objective 1.3a**  Subtract whole numbers.

**Example**  Subtract: 8045 − 2897.

$$
\begin{array}{r}
{}^{7}\,{}^{9}\,{}^{\overset{13}{3}}\,{}^{15} \\
8\,0\,4\,5 \\
-\ 2\,8\,9\,7 \\
\hline
5\,1\,4\,8
\end{array}
$$

**Practice Exercise**

3. Subtract: 4805 − 1568.

---

**Objective 1.4a**  Multiply whole numbers.

**Example**  Multiply: 57 × 315.

$$
\begin{array}{r}
\overset{1}{}\,\overset{3}{}\,\ \ \ \\
\overset{}{}\,\overset{2}{}\,\ \ \ \\
3\,1\,5 \\
\times\ \ \ 5\,7 \\
\hline
2\,2\,0\,5 \leftarrow 315 \times 7 \\
1\,5\,7\,5\,0 \leftarrow 315 \times 50 \\
\hline
1\,7,9\,5\,5
\end{array}
$$

**Practice Exercise**

4. Multiply: 329 × 684.

---

**Objective 1.5a**  Divide whole numbers.

**Example**  Divide: 6463 ÷ 26.

$$
\begin{array}{r}
2\,4\,8 \\
26\overline{)6\,4\,6\,3} \\
\underline{5\,2\ \ \ \ } \\
1\,2\,6\ \ \\
\underline{1\,0\,4\ \ } \\
2\,2\,3 \\
\underline{2\,0\,8} \\
1\,5
\end{array}
$$

The answer is 248 R 15.

**Practice Exercise**

5. Divide: 8519 ÷ 27.

**Objective 1.6a** Round to the nearest ten, hundred, or thousand.

**Example** Round to the nearest thousand.

6 4 7 1

The digit 6 is in the thousands place. We consider the next digit to the right. Since the digit, 4, is 4 or lower, we round down, meaning that 6 thousands stays as 6 thousands. Change all digits to the right of the thousands digit to zeros. The answer is 6000.

**Practice Exercise**

**6.** Round 36,468 to the nearest thousand.

---

**Objective 1.6c** Use < or > for ☐ to write a true sentence in a situation like 6 ☐ 10.

**Example** Use < or > for ☐ to write a true sentence:

34 ☐ 29.

Since 34 is to the right of 29 on the number line,

34 > 29.

**Practice Exercise**

**7.** Use < or > for ☐ to write a true sentence:

78 ☐ 81.

---

**Objective 1.7b** Solve equations like $t + 28 = 54$, $28 \cdot x = 168$, and $98 \cdot 2 = y$.

**Example** Solve: $y + 12 = 27$.
$$y + 12 = 27$$
$$y + 12 - 12 = 27 - 12$$
$$y + 0 = 15$$
$$y = 15$$
The solution is 15.

**Practice Exercise**

**8.** Solve: $24 \cdot x = 864$.

---

**Objective 1.9b** Evaluate exponential notation.

**Example** Evaluate: $5^4$.
$$5^4 = 5 \cdot 5 \cdot 5 \cdot 5 = 625$$

**Practice Exercise**

**9.** Evaluate: $6^3$.

---

# Review Exercises

The review exercises that follow are for practice. Answers are given at the back of the book. If you miss an exercise, restudy the objective indicated in red next to the exercise or on the direction line that precedes it.

**1.** What does the digit 8 mean in 4,678,952? [1.1a]

**2.** In 13,768,940, what digit tells the number of millions? [1.1a]

Write expanded notation. [1.1b]

**3.** 2793

**4.** 56,078

**5.** 4,007,101

Write a word name. [1.1c]

**6.** 67,819

**7.** 2,781,427

Write standard notation. [1.1c]

**8.** Four hundred seventy-six thousand, five hundred eighty-eight

**9.** *Candy.* About one billion, five hundred million marshmallow peeps are produced in the United States each year.

Data: *USA Today*

Add.  [1.2a]

**10.** 7304 + 6968

**11.** 27,609 + 38,415

**12.** 2703 + 4125 + 6004 + 8956

**13.**
$$\begin{array}{r} 9\,1,4\,2\,6 \\ +\quad 7,4\,9\,5 \\ \hline \end{array}$$

Subtract.  [1.3a]

**14.** 8045 − 2897

**15.** 9001 − 7312

**16.** 6003 − 3729

**17.**
$$\begin{array}{r} 3\,7,4\,0\,5 \\ -\,1\,9,6\,4\,8 \\ \hline \end{array}$$

Multiply.  [1.4a]

**18.** 17,000 · 300

**19.** 7846 · 800

**20.** 726 · 698

**21.** 587 · 47

**22.**
$$\begin{array}{r} 8\,3\,0\,5 \\ \times\quad 6\,4\,2 \\ \hline \end{array}$$

Divide.  [1.5a]

**23.** 63 ÷ 5

**24.** 80 ÷ 16

**25.** 7 ) 6 3 9 4

**26.** 3073 ÷ 8

**27.** 6 0 ) 2 8 6

**28.** 4266 ÷ 79

**29.** 3 8 ) 1 7,1 7 6

**30.** 1 4 ) 7 0,1 1 2

**31.** 52,668 ÷ 12

Round 345,759 to the nearest:  [1.6a]

**32.** Hundred.

**33.** Ten.

**34.** Thousand.

**35.** Hundred thousand.

Use < or > for ☐ to write a true sentence.  [1.6c]

**36.** 67 ☐ 56

**37.** 1 ☐ 23

Estimate each sum, difference, or product by first rounding to the nearest hundred. Show your work.  [1.6b]

**38.** 41,348 + 19,749

**39.** 38,652 − 24,549

**40.** 396 · 748

Solve.  [1.7b]

**41.** $46 \cdot n = 368$

**42.** $47 + x = 92$

**43.** $1 \cdot y = 58$

**44.** $24 = x + 24$

**45.** Write exponential notation: 4 · 4 · 4.  [1.9a]

Evaluate.  [1.9b]

**46.** $10^4$

**47.** $6^2$

Simplify.  [1.9c, d]

**48.** $8 \cdot 6 + 17$

**49.** $10 \cdot 24 - (18 + 2) \div 4 - (9 - 7)$

**50.** $(80 \div 16) \times [(20 - 56 \div 8) + (8 \cdot 8 - 5 \cdot 5)]$

**51.** Find the average of 157, 170, and 168.  [1.9c]

Solve.

**52.** *Computer Purchase.*   Natasha has $196 and wants to buy a computer for $698. How much more does she need?  [1.8a]

**53.** Toni has $406 in her checking account. She is paid $78 for a part-time job and deposits that in her checking account. How much is then in her account?  [1.8a]

**54.** *Lincoln-Head Pennies.*   In 1909, the first Lincoln-head pennies were minted. Seventy-three years later, these pennies were first minted with a decreased copper content. In what year was the copper content reduced?  [1.8a]

**55.** A beverage company packed 228 cans of soda into 12-can cartons. How many cartons were filled?  [1.8a]

**56.** An apartment builder bought 13 gas stoves at $425 each and 13 refrigerators at $620 each. What was the total cost?   [1.8a]

**57.** An apple farmer keeps bees in her orchard to help pollinate the apple blossoms. The bees from an average beehive can pollinate 30 surrounding trees during one growing season. The farmer has 420 trees. How many beehives does she need to pollinate all of them?   [1.8a]

**Data:** Jordan Orchards, Westminster, PA

**58.** *Olympic Trampoline.*   Shown below is an Olympic trampoline. Determine the area and the perimeter of the trampoline.   [1.2b], [1.4b]

**Data:** International Trampoline Industry Association, Inc.

**59.** A chemist has 2753 mL of alcohol. How many 20-mL beakers can be filled? How much will be left over?   [1.8a]

**60.** A family budgeted $7825 a year for food and clothing and $2860 for entertainment. The yearly income of the family was $38,283. How much of this income remained after these two allotments?   [1.8a]

**61.** Simplify: $7 + (4 + 3)^2$.   [1.9c]
  **A.** 32                **B.** 56
  **C.** 151             **D.** 196

**62.** Simplify: $7 + 4^2 + 3^2$.   [1.9c]
  **A.** 32                **B.** 56
  **C.** 130             **D.** 196

**63.** $[46 - (4 - 2) \cdot 5] \div 2 + 4$   [1.9d]
  **A.** 6                 **B.** 20
  **C.** 114             **D.** 22

## Synthesis

**64.** 🖩 Determine the missing digit $d$.   [1.4a]

$$\begin{array}{r} 9\ d \\ \times\ \ d\ 2 \\ \hline 8\ 0\ 3\ 6 \end{array}$$

**65.** 🖩 Determine the missing digits $a$ and $b$.   [1.5a]

$$2\ b\ 1\ \overline{)\ 2\ 3\ 6{,}4\ 2\ 1}\quad^{9\ a\ 1}$$

**66.** A mining company estimates that a crew must tunnel 2000 ft into a mountain to reach a deposit of copper ore. Each day, the crew tunnels about 500 ft. Each night, about 200 ft of loose rocks roll back into the tunnel. How many days will it take the mining company to reach the copper deposit?   [1.8a]

## Understanding Through Discussion and Writing

**1.** Is subtraction associative? Why or why not?   [1.3a]

**2.** Explain how estimating and rounding can be useful when shopping for groceries.   [1.6b]

**3.** Write a problem for a classmate to solve. Design the problem so that the solution is "The driver still has 329 mi to travel."   [1.8a]

**4.** Consider the expressions $9 - (4 \cdot 2)$ and $(3 \cdot 4)^2$. Are the parentheses necessary in each case? Explain.   [1.9c]

CHAPTER

**1**    **Test**

For
Extra
Help

For step-by-step test solutions, access the Chapter Test Prep Videos
in MyLab Math.

**1.** In the number 546,789, which digit tells the number of hundred thousands?

**2.** Write expanded notation: 8843.

**3.** Write a word name: 38,403,277.

Add.

**4.**
```
  6 8 1 1
+ 3 1 7 8
```

**5.**
```
  4 5,8 8 9
+ 1 7,9 0 2
```

**6.**
```
  1 2 3 9
    8 4 3
    3 0 1
+   7 8 2
```

**7.**
```
  6 2 0 3
+ 4 3 1 2
```

Subtract.

**8.**
```
  7 9 8 3
− 4 3 5 3
```

**9.**
```
  2 9 7 4
− 1 9 3 5
```

**10.**
```
  8 9 0 7
− 2 0 5 9
```

**11.**
```
  2 3,0 6 7
− 1 7,8 9 2
```

Multiply.

**12.**
```
  4 5 6 8
×       9
```

**13.**
```
  8 8 7 6
×     6 0 0
```

**14.**
```
    6 5
×   3 7
```

**15.**
```
    6 7 8
×   7 8 8
```

Divide.

**16.** $15 \div 4$

**17.** $420 \div 6$

**18.** $89 \overline{)8633}$

**19.** $44 \overline{)35,428}$

Round 34,528 to the nearest:

**20.** Thousand.

**21.** Ten.

**22.** Hundred.

Estimate each sum, difference, or product by first rounding to the nearest hundred. Show your work.

**23.**
```
  2 3,6 4 9
+ 5 4,7 4 6
```

**24.**
```
  5 4,7 5 1
− 2 3,6 4 9
```

**25.**
```
    8 2 4
×   4 8 9
```

Use < or > for ☐ to write a true sentence.

**26.** 34 ☐ 17

**27.** 117 ☐ 157

Solve.

**28.** $28 + x = 74$

**29.** $169 \div 13 = n$

**30.** $38 \cdot y = 532$

**31.** $381 = 0 + a$

Solve.

**32.** *Calorie Content.* An 8-oz serving of whole milk contains 146 calories. This is 63 calories more than the number of calories in an 8-oz serving of skim milk. How many calories are in an 8-oz serving of skim milk?

**Data:** *American Journal of Clinical Nutrition*

**33.** A box contains 5000 staples. How many staplers can be filled from the box if each stapler holds 250 staples?

**34.** *Largest States.* The following table lists the five largest states in terms of their land area. Find the total land area of these states.

| STATE | AREA (in square miles) |
|---|---|
| Alaska | 571,951 |
| Texas | 261,797 |
| California | 155,959 |
| Montana | 145,552 |
| New Mexico | 121,356 |

DATA: U.S. Department of Commerce; U.S. Census Bureau

**35.** *Pool Tables.* The Bradford™ pool table made by Brunswick Billiards comes in three sizes, with playing area dimensions of 50 in. by 100 in., 44 in. by 88 in., and 38 in. by 76 in.

**Data:** Brunswick Billiards

a) Determine the perimeter and the playing area of each table.
b) By how much does the area of the largest table exceed the area of the smallest table?

**36.** *Hostess Ding Dongs®.* Hostess packages its Ding Dong snack cakes in 12-packs. How many 12-packs can it fill with 22,231 cakes? How many will be left over?

**37.** *Office Supplies.* Morgan manages the office of a small graphics firm. He buys 3 black inkjet cartridges at $15 each and 2 photo inkjet cartridges at $25 each. How much does the purchase cost?

**38.** Write exponential notation: $12 \cdot 12 \cdot 12 \cdot 12$.

Evaluate.

**39.** $7^3$

**40.** $10^5$

Simplify.

**41.** $35 - 1 \cdot 28 \div 4 + 3$

**42.** $10^2 - 2^2 \div 2$

**43.** $(25 - 15) \div 5$

**44.** $2^4 + 24 \div 12$

**45.** $8 \times \{(20 - 11) \cdot [(12 + 48) \div 6 - (9 - 2)]\}$

**46.** Find the average of 97, 99, 87, and 89.

**A.** 93      **B.** 124      **C.** 186      **D.** 372

## Synthesis

**47.** An open cardboard container is 8 in. wide, 12 in. long, and 6 in. high. How many square inches of cardboard are used?

**48.** Use trials to find the single-digit number $a$ for which
$$359 - 46 + a \div 3 \times 25 - 7^2 = 339.$$

**49.** Cara spends $229 a month to repay her student loan. If she has already paid $9160 on the 10-year loan, how many payments remain?

# Introduction to Integers and Algebraic Expressions

Elevation of land is measured in terms of distance above sea level. The highest elevations on each continent vary from the top of Mt. Everest in Asia, at an elevation of 29,035 ft, to the top of Mt. Kosciuszko in Australia, at an elevation of 7,310 ft. Some land is actually below sea level and thus has a negative elevation. The largest city in the world that is partially below sea level is New Orleans, Louisiana, where some neighborhoods have elevations below −7 ft. The accompanying graph shows the lowest elevation on each continent.

*DATA: new.uno.edu; infoplease.com*

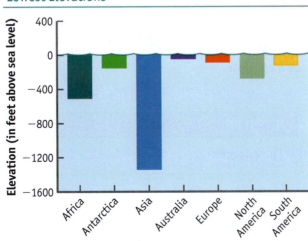

**Lowest Elevations**

DATA: infoplease.com; worldatlas.com

In Example 1 of Section 2.1 and in Exercises 73 and 80 of Exercise Set 2.3, we will write elevations below sea level and find the difference between elevations.

# 2.1

# Integers and the Number Line

## OBJECTIVES

**a** State the integer that corresponds to a real-world situation.

**b** Form a true sentence using < or >.

**c** Find the absolute value of any integer.

**d** Find the opposite of any integer.

In this section, we extend the set of whole numbers to form the set of *integers*. You have probably already used negative numbers. For example, the outside temperature could drop to *negative five* degrees, or a credit card statement could indicate activity of *negative forty-eight* dollars.

To describe integers, we start with the whole numbers, 0, 1, 2, 3, and so on. For each number 1, 2, 3, and so on, we obtain a new number that is the same number of units to the left of zero on the number line.

For the number 1, there is the *opposite* number $-1$ (negative 1).

For the number 2, there is the *opposite* number $-2$ (negative 2).

For the number 3, there is the *opposite* number $-3$ (negative 3), and so on.

The **integers** consist of the whole numbers and these new numbers. We picture them on the number line as follows.

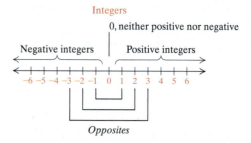

We call the integers to the left of zero **negative integers**. Those to the right of zero are called **positive integers**. Zero is neither positive nor negative. We call $-1$ and 1 **opposites** of each other. Similarly, $-2$ and 2 are opposites, $-3$ and 3 are opposites, $-100$ and 100 are opposites, and 0 is its own opposite.

---

**INTEGERS**

The **integers**: $\ldots, -5, -4, -3, -2, -1, 0, 1, 2, 3, 4, 5, \ldots$

---

## a  INTEGERS AND THE REAL WORLD

Integers correspond to many real-world problems and situations. The following examples will help you get ready to translate problem situations to mathematical language.

**EXAMPLE 1**   Tell which integer corresponds to this situation: Baku, the capital of Azerbaijan, lies on the Caspian Sea. Its elevation is 28 m below sea level.

**Data:** elevationmap.net

The integer −28 corresponds to the situation. The elevation is −28 m.

**EXAMPLE 2**   *Stock Price Changes.*   Tell which integers correspond to the stock price changes. Hal owns a stock whose price decreased from $27 per share to $11 per share over a recent time period. He owns another stock whose price increased from $20 per share to $22 per share over the same time period.

The integer −16 corresponds to the decrease in the value of the first stock. The integer 2 represents the increase in the value of the second stock.

**Do Exercises 1–5.** ▶

## b   ORDER ON THE NUMBER LINE

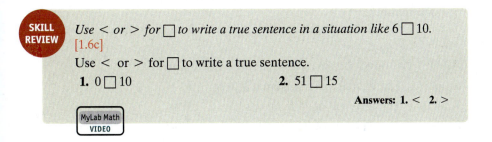

**SKILL REVIEW**

*Use < or > for ☐ to write a true sentence in a situation like 6 ☐ 10.*
[1.6c]

Use < or > for ☐ to write a true sentence.

**1.** 0 ☐ 10                **2.** 51 ☐ 15

Answers: **1.** <  **2.** >

**MyLab Math VIDEO**

Numbers are written in order on the number line, increasing as we move from left to right. For any two numbers on the line, the one to the left is *less than* the one to the right.

Since the symbol < means "is less than," the sentence −5 < 9 is read "−5 is less than 9." The symbol > means "is greater than," so the sentence −4 > −8 is read "−4 is greater than −8."

Tell which integers correspond to each situation.

1. *High and Low Temperatures.* The highest recorded temperature in Illinois was 117°F on July 14, 1954, in East St. Louis. The lowest recorded temperature in Illinois was 36°F below zero on January 5, 1999, in Congerville.

   **Data:** National Climate Data Center, NESDIS, NOAA, U.S. Dept. of Commerce

2. *Stock Decrease.*   The price of a stock decreased from $41 per share to $38 per share over a recent period.

3. At 10 sec before liftoff, ignition occurs. At 148 sec after liftoff, the first stage is detached from the rocket.

4. The halfback gained 8 yd on first down. The quarterback was sacked for a 5-yd loss on second down.

5. A submarine dived 120 ft, rose 50 ft, and then dived 80 ft.

**MyLab Math ANIMATION**

*Answers*

**1.** 117; −36   **2.** The integer −3 corresponds to the decrease in the value of the stock.   **3.** −10; 148   **4.** 8; −5   **5.** −120; 50; −80

Use either < or > for □ to write a true sentence.

**6.** −3 □ 7

**7.** −8 □ −5

**8.** 7 □ −10

**9.** −4 □ −20

◀ Do Exercises 6–9.

**EXAMPLES**   Use either < or > for □ to form a true sentence.

**3.** −9 □ 2      Since −9 is to the left of 2, we have −9 < 2.

**4.** 4 □ −6      Since 4 is to the right of −6, we have 4 > −6.

**5.** −8 □ −1     Since −8 is to the left of −1, we have −8 < −1.

## c   ABSOLUTE VALUE

From the number line, we see that some integers, like 4 and −4, are the same distance from zero. We call the distance of a number from zero the **absolute value** of the number. Since distance is always a nonnegative number, absolute value is always nonnegative.

> ### ABSOLUTE VALUE
>
> The **absolute value** of a number is its distance from zero on the number line. We use the symbol $|x|$ to represent the absolute value of a number $x$.

The distance of −4 from 0 is 4.      The distance of 4 from 0 is 4.
The absolute value of −4 is 4.        The absolute value of 4 is 4.

$|-4| = 4$                            $|4| = 4$

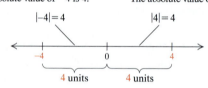

4 units        4 units

> To find the absolute value of a number:
>
> **a)** If a number is negative, its absolute value is its opposite.
>
> **b)** If a number is positive or zero, its absolute value is the same as the number.

Find the absolute value.

**10.** $|18|$
The distance of 18 from 0 is ___, so $|18| = $ ___.

**GS**

**EXAMPLES**   Find the absolute value of each number.

**6.** $|-3|$   The distance of −3 from 0 is 3, so $|-3| = 3$.

**7.** $|25|$   The distance of 25 from 0 is 25, so $|25| = 25$.

**8.** $|0|$   The distance of 0 from 0 is 0, so $|0| = 0$.

◀ Do Exercises 10–13.

**11.** $|-9|$        **12.** $|-29|$

**13.** $|52|$

## d   OPPOSITES

Given a number on one side of 0 on the number line, we can get a number on the other side by *reflecting* the number across zero. For example, the *reflection* of 2 is −2. We can read −2 as "negative 2" or "the opposite of 2."

The opposite of a number is also called its *additive inverse*.

*Answers*

**6.** <   **7.** <   **8.** >   **9.** >   **10.** 18
**11.** 9   **12.** 29   **13.** 52

*Guided Solution:*

**10.** 18, 18

## NOTATION FOR OPPOSITES

The **opposite** of a number $x$ is written $-x$ (read "the opposite of $x$").

**EXAMPLE 9**  If $x$ is $-3$, find $-x$.

To find the opposite of $x$ when $x$ is $-3$, we reflect $-3$ to the other side of 0.

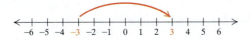

When $x = -3, -x = -(-3)$. We have $-(-3) = 3$. The opposite of $-3$ is 3.

When we replace a variable with a number and find the value of an expression, we say that we are **evaluating** the expression.

**EXAMPLE 10**  Evaluate $-x$ when $x$ is 4.

We have $-(4) = -4$. The opposite of 4 is $-4$.

**Do Exercises 14–16.** ▶

A negative number is sometimes said to have a negative *sign*. A positive number is said to have a positive sign, even though it is rarely written. Replacing a number with its opposite, or additive inverse, is sometimes called *changing the sign*.

**EXAMPLES**  Change the sign (find the opposite, or additive inverse) of each number.

**11.** $-6$    $-(-6) = 6$       **12.** $-10$    $-(-10) = 10$

**13.** $0$    $-(0) = 0$      **14.** $14$    $-(14) = -14$

**Do Exercises 17–20.** ▶

**EXAMPLE 15**  If $x$ is 2, find $-(-x)$.

If $x = 2$, then $-(-x) = -(-2) = 2$.    The opposite of the opposite of 2 is 2.

**EXAMPLE 16**  Evaluate $-(-x)$ for $x = -4$.

If $x = -4$, then $-(-x) = -(-(-4)) = -(4) = -4$.    The opposite of the opposite of $-4$ is $-4$.

**When we change a number's sign twice, we return to the original number. That is, $-(-x) = x$.**

**Do Exercises 21 and 22.** ▶

**EXAMPLE 17**  Evaluate $-|-x|$ for $x = 2$.

If $x = 2$, then $-|-x| = -|-2| = -2$.    The absolute value of $-2$ is 2; $-(2) = -2$.

Note that $-(-2) = 2$, whereas $-|-2| = -2$.

**Do Exercises 23 and 24.** ▶

---

In each case, draw a number line, if necessary.

**14.** Find $-x$ when $x$ is 1.

**15.** Find $-x$ when $x$ is 0.

**16.** Evaluate $-x$ when $x$ is $-2$.

Change the sign. (Find the opposite, or additive inverse.)

**17.** $-4$            **18.** $-13$

**19.** $39$            **20.** $0$

**21.** If $x$ is 7, find $-(-x)$.

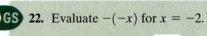
**GS 22.** Evaluate $-(-x)$ for $x = -2$.

$$-(-x) = -(-(\quad))$$
$$= -(\quad) = \boxed{\phantom{xx}}$$

**23.** Find $-|-7|$.

**24.** Find $-|-39|$.

*Answers*
**14.** $-1$   **15.** 0   **16.** 2   **17.** 4   **18.** 13
**19.** $-39$   **20.** 0   **21.** 7   **22.** $-2$   **23.** $-7$
**24.** $-39$
*Guided Solution:*
**22.** $-2, 2, -2$

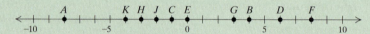
## ✓ Check Your Understanding

**Reading Check**   Determine whether each statement is true or false.

_____ **RC1.** If $x$ is a negative integer, then $x$ is less than any positive integer.

_____ **RC2.** The absolute value of an integer is never negative.

_____ **RC3.** The opposite of an integer is never positive.

**Concept Check**   Use the number line below, on which the letters name numbers, for Exercises CC1–CC8.

Determine whether each statement is true or false.

_____ **CC1.** $K < B$    _____ **CC2.** $H < B$    _____ **CC3.** $E < C$    _____ **CC4.** $J > D$

_____ **CC5.** $|K| = 4$    _____ **CC6.** $|H| = |B|$    _____ **CC7.** $A = -F$    _____ **CC8.** $-G = H$

---

**a**   State the integers that correspond to each situation.

**1.** On Wednesday, the temperature was 24° above zero. On Thursday, it was 2° below zero.

**2.** A student deposited her tax refund of $750 in a savings account. Two weeks later, she withdrew $125 to pay technology fees.

**3.** *Temperature Extremes.* The highest temperature ever created in a lab is 7,200,000,000,000°F. The lowest temperature ever created is approximately 460°F below zero.

**Data:** *Live Science; The Guinness Book of World Records*

**4.** *Extreme Climate.* Verkhoyansk, a river port in northeast Siberia, has the most extreme climate on the planet. Its average monthly winter temperature is 58.5°F below zero, and its average monthly summer temperature is 56.5°F.

**Data:** *The Guinness Book of World Records*

**5.** *Architecture.* The Shanghai Tower in Shanghai, China, has a total height of 2073 ft. The foundation depth is 282 ft below ground level.

**Data:** travelchinaguide.com

**6.** *Sunken Ships.* There are numerous sunken ships to explore near Bermuda. One of the most frequently visited sites is the Hermes, a decommissioned freighter that was sunk in 1985 to create an artificial reef. This ship is 80 ft below the surface.

**Data:** skin-diver.com

**b** Use either $<$ or $>$ for ☐ to form a true sentence.

**7.** $-8 \,☐\, 0$      **8.** $7 \,☐\, 0$      **9.** $9 \,☐\, 0$      **10.** $-7 \,☐\, 0$      **11.** $8 \,☐\, -8$

**12.** $6 \,☐\, -6$      **13.** $-6 \,☐\, -4$      **14.** $-1 \,☐\, -7$      **15.** $-8 \,☐\, -5$      **16.** $-5 \,☐\, -3$

**17.** $-13 \,☐\, -9$      **18.** $-5 \,☐\, -11$      **19.** $-3 \,☐\, -4$      **20.** $-6 \,☐\, -5$

**c** Find the absolute value.

**21.** $|57|$      **22.** $|11|$      **23.** $|0|$      **24.** $|-4|$      **25.** $|-24|$

**26.** $|-36|$      **27.** $|53|$      **28.** $|54|$      **29.** $|-8|$      **30.** $|-79|$

**d** Find $-x$ when $x$ is each of the following.

**31.** $-7$      **32.** $-6$      **33.** $7$      **34.** $6$      **35.** $0$      **36.** $-1$

Change the sign. (Find the opposite, or additive inverse.)

**37.** $-21$      **38.** $-67$      **39.** $53$      **40.** $0$      **41.** $-1$      **42.** $16$

Evaluate $-(-x)$ when $x$ is each of the following.

**43.** $7$      **44.** $-8$      **45.** $-9$      **46.** $3$      **47.** $-17$      **48.** $-19$

**49.** $23$      **50.** $0$      **51.** $-1$      **52.** $73$      **53.** $85$      **54.** $-37$

Evaluate $-|-x|$ when $x$ is each of the following.

**55.** $345$      **56.** $729$      **57.** $0$      **58.** $1$      **59.** $-8$      **60.** $-3$

## Skill Maintenance

**61.** Add: $327 + 498$.   [1.2a]

**62.** Evaluate: $5^3$.   [1.9b]

**63.** Multiply: $209 \cdot 34$.   [1.4a]

**64.** Solve: $300 \cdot x = 1200$.   [1.7b]

**65.** Simplify: $7(9 - 3)$.   [1.9c]

**66.** Simplify: $9^2 - 3[2 + (10 - 8)]$.   [1.9d]

## Synthesis

Use either $<$, $>$, or $=$ for ☐ to write a true sentence.

**67.** $|-5| \,☐\, |-2|$      **68.** $|4| \,☐\, |-7|$      **69.** $|-8| \,☐\, |8|$

Solve. Consider only integer replacements.

**70.** $|x| = 7$      **71.** $|x| < 2$

**72.** Simplify $-(-x)$, $-(-(-x))$, and $-(-(-(-x)))$.

**73.** List these integers in order from least to greatest.
$2^{10}$, $-5$, $|-6|$, $4$, $|3|$, $-100$, $0$, $2^7$, $7^2$, $10^2$

# Addition of Integers

**a  ADDITION**

To explain addition of integers, we can use the number line. Once our understanding is developed, we will streamline our approach.

### Addition on the Number Line

To find $a + b$, we start at $a$ and then move according to $b$.

**a)** If $b$ is positive, we move from $a$ to the right.

**b)** If $b$ is negative, we move from $a$ to the left.

**c)** If $b$ is 0, we stay at $a$.

Add using the number line.

**1.** $1 + (-4)$

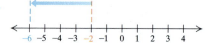

**2.** $-3 + (-2)$

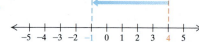

**3.** $-3 + 7$

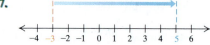

**4.** $-5 + 5$

For each illustration, write a corresponding addition sentence.

**5.**

**6.**

**7.**

**EXAMPLE 1**  Add: $2 + (-5)$.

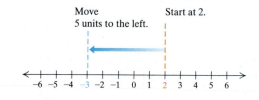

Start at 2.

Move 5 units to the left.

$2 + (-5) = -3$

**EXAMPLE 2**  Add: $-1 + (-3)$.

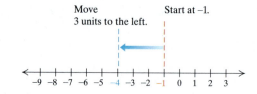

Start at –1.

Move 3 units to the left.

$-1 + (-3) = -4$

**EXAMPLE 3**  Add: $-4 + 9$.

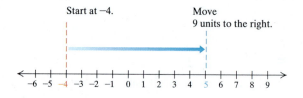

Start at –4.

Move 9 units to the right.

$-4 + 9 = 5$

◀ **Do Exercises 1–7.**

You may have noticed a pattern in Example 2 and Margin Exercises 2 and 6. When two negative integers are added, the result is negative.

---

**ADDING NEGATIVE INTEGERS**

To add two negative integers, add their absolute values and change the sign (making the answer negative).

---

*Answers*

**1.** $-3$  **2.** $-5$  **3.** 4  **4.** 0
**5.** $4 + (-5) = -1$  **6.** $-2 + (-4) = -6$
**7.** $-3 + 8 = 5$

**EXAMPLES** Add.

**4.** $-5 + (-7) = -12$    *Think*: Add the absolute values: $5 + 7 = 12$. Make the answer negative, $-12$.

**5.** $-8 + (-2) = -10$

**Do Exercises 8–11.** ▶

When the number 0 is added to any number, that number remains unchanged. For this reason, the number 0 is referred to as the **additive identity**.

**EXAMPLES** Add.

**6.** $-4 + 0 = -4$      **7.** $0 + (-9) = -9$      **8.** $17 + 0 = 17$

**Do Exercises 12–14.** ▶

When we add a positive integer and a negative integer, as in Examples 1 and 3, the sign of the number with the greater absolute value is the sign of the answer.

---

### ADDING POSITIVE AND NEGATIVE INTEGERS

To add a positive integer and a negative integer, find the difference of their absolute values.

**a)** If the negative integer has the greater absolute value, the answer is negative.

**b)** If the positive integer has the greater absolute value, the answer is positive.

**c)** If the integers have the same absolute value, the answer is 0.

---

**EXAMPLES** Add.

**9.** $3 + (-5) = -2$    *Think*: The absolute values are 3 and 5. The difference is 2. Since the negative number has the larger absolute value, the answer is *negative*, $-2$.

**10.** $11 + (-8) = 3$    *Think*: The absolute values are 11 and 8. The difference is 3. The positive number has the larger absolute value, so the answer is *positive*, 3.

**11.** $-7 + 4 = -3$
**12.** $-6 + 10 = 4$
**13.** $9 + (-9) = 0$

**Do Exercises 15–19.** ▶

Sometimes $-a$ is referred to as the **additive inverse** of $a$ because adding any number to its additive inverse always results in the additive identity, 0.

---

### OPPOSITES, OR ADDITIVE INVERSES

For any integer $a$, the **opposite**, or **additive inverse**, of $a$, denoted $-a$, is such that

$$a + (-a) = (-a) + a = 0.$$

---

Add. Do not use the number line except as a check.

**8.** $-5 + (-6)$

**9.** $-9 + (-3)$

**10.** $-20 + (-14)$

**11.** $-11 + (-11)$

Add.
**12.** $0 + (-17)$

**13.** $49 + 0$

**14.** $-56 + 0$

Add, using the number line only as a check.

**15.** $-4 + 6$

**16.** $-7 + 3$

**17.** $5 + (-7)$

**18.** $10 + (-7)$

**GS** **19.** $-12 + 12$

     $-12$ and 12 have the same  value. The answer is ☐.

Add, using the number line only as a check.

**20.** $5 + (-5)$

**21.** $-6 + 6$

**22.** $89 + (-89)$

---

**CALCULATOR CORNER**

*Negative Numbers* On many calculators, we can enter negative numbers using the $\boxed{+/-}$ key. To enter $-8$, for example, we press $\boxed{8}\ \boxed{+/-}$. To find the sum $-14 + (-9)$, we press $\boxed{1}\ \boxed{4}$ $\boxed{+/-}\ \boxed{+}\ \boxed{9}\ \boxed{+/-}\ \boxed{=}$. The result is $-23$. Note that it is not necessary to use parentheses when entering this expression. On some calculators, the $\boxed{+/-}$ key is labeled $\boxed{(-)}$ and is pressed *before* the number.

**EXERCISES:** Add.

**1.** $-4 + 17$

**2.** $3 + (-11)$

---

Add.

**23.** $-5 + (-10)$

**24.** $18 + (-11)$

**25.** $-13 + 13$

**26.** $-20 + 7$

---

Add.

**27.** $-15 + (-5) + 25 + (-9) + 10 + (-14)$

GS

Add the positive numbers:
$$25 + \boxed{\phantom{00}} = \boxed{\phantom{00}}.$$

Add the negative numbers:
$$-15 + (-5) + (-9) + (\boxed{\phantom{00}}) = \boxed{\phantom{00}}.$$

Finally, add the results:
$$35 + (\boxed{\phantom{00}}) = \boxed{\phantom{00}}.$$

---

*Answers*

**20.** 0  **21.** 0  **22.** 0  **23.** $-15$  **24.** 7
**25.** 0  **26.** $-13$  **27.** $-8$

*Guided Solution:*
**27.** 10, 35, $-14$, $-43$, $-43$, $-8$

---

**EXAMPLES** Add.

**14.** $-8 + 8 = 0$          **15.** $14 + (-14) = 0$

◀ **Do Exercises 20–22.**

In summary, to add integers, look first at the signs of the numbers you are adding. This tells you whether you should add or subtract to find the sum.

---

**RULES FOR ADDITION OF INTEGERS**

1. *Positive numbers*: Add the same way as you add arithmetic numbers. The answer is positive.

2. *Negative numbers*: Add absolute values. The answer is negative.

3. *A positive and a negative number*: Subtract absolute values.
   a) If the positive number has the greater absolute value, the answer is positive.
   b) If the negative number has the greater absolute value, the answer is negative.
   c) If the numbers have the same absolute value, they are additive inverses and the answer is 0.

4. *One number is zero*: The sum is the other number.

---

Rule 4 is known as the **identity property of 0**. It says that for any real number $a$, $a + 0 = 0 + a = a$.

**EXAMPLES** Add.

**16.** $-12 + (-7) = -19$    Two negative numbers; add absolute values. The answer is negative.

**17.** $-20 + 36 = 16$    A positive and a negative number; subtract absolute values. The answer is positive.

◀ **Do Exercises 23–26.**

Suppose we wish to add several numbers, some positive and some negative, as in $15 + (-2) + 7 + 14 + (-5) + (-12)$. Because of the commutative and associative laws for addition, we can group the positive numbers together and the negative numbers together and add them separately. Then we add the two results.

**EXAMPLE 18**  Add: $15 + (-2) + 7 + 14 + (-5) + (-12)$.

First add the positive numbers: $15 + 7 + 14 = 36$.

Then add the negative numbers: $-2 + (-5) + (-12) = -19$.

Finally, add the results: $36 + (-19) = 17$.

We can also add in any other order we wish—say, from left to right:

$$
\begin{aligned}
15 + (-2) + 7 + 14 + (-5) + (-12) &= 13 + 7 + 14 + (-5) + (-12) \\
&= 20 + 14 + (-5) + (-12) \\
&= 34 + (-5) + (-12) \\
&= 29 + (-12) \\
&= 17.
\end{aligned}
$$

◀ **Do Exercise 27.**

---

## ✓ Check Your Understanding

**Reading Check** Choose the correct word or words from the list on the right to complete each sentence. Words may be used more than once or not at all.

**RC1.** To add $-3 + (-6)$, _____ 3 and 6 and make the answer _____.

**RC2.** To add $-11 + 5$, _____ 5 from 11 and make the answer _____.

**RC3.** The sum of two numbers that are _____ is 0.

**RC4.** The addition $-7 + 0 = -7$ illustrates the _____ property of 0.

add
subtract
opposites
identity
positive
negative

**Concept Check** Fill in each blank with either "left" or "right" so that the statement describes the steps for adding the given numbers on the number line.

**CC1.** To add $7 + 2$, start at 7 and then move 2 units _____. The sum is 9.

**CC2.** To add $-3 + (-5)$, start at $-3$ and then move 5 units _____. The sum is $-8$.

**CC3.** To add $4 + (-6)$, start at 4 and then move 6 units _____. The sum is $-2$.

**CC4.** To add $-8 + 3$, start at $-8$ and then move 3 units _____. The sum is $-5$.

---

**a**   Add using the number line.

**1.** $-7 + 2$     **2.** $1 + (-5)$     **3.** $-9 + 5$     **4.** $8 + (-3)$     **5.** $-3 + 9$

**6.** $9 + (-9)$     **7.** $-7 + 7$     **8.** $-8 + (-5)$     **9.** $-3 + (-1)$     **10.** $-2 + (-9)$

Add. Use the number line only as a check.

**11.** $-3 + (-9)$     **12.** $-3 + (-7)$     **13.** $-6 + (-5)$     **14.** $-10 + (-14)$

**15.** $-15 + 0$     **16.** $0 + (-11)$     **17.** $0 + 42$     **18.** $27 + 0$

**19.** $9 + (-4)$     **20.** $-7 + 8$     **21.** $-10 + 6$     **22.** $6 + (-13)$

**23.** $5 + (-5)$     **24.** $10 + (-10)$     **25.** $-2 + 2$     **26.** $-3 + 3$

**27.** $-4 + (-5)$     **28.** $10 + (-12)$     **29.** $13 + (-6)$     **30.** $-3 + 14$

**31.** $-25 + 25$     **32.** $40 + (-40)$     **33.** $63 + (-18)$     **34.** $85 + (-65)$

**35.** $-11 + 8$

**36.** $0 + (-34)$

**37.** $-19 + 19$

**38.** $-10 + 3$

**39.** $-16 + 6$

**40.** $-15 + 5$

**41.** $-17 + (-7)$

**42.** $-15 + (-5)$

**43.** $11 + (-16)$

**44.** $-8 + 14$

**45.** $-15 + (-6)$

**46.** $-8 + 8$

**47.** $-15 + (-15)$

**48.** $-25 + (-25)$

**49.** $-11 + 17$

**50.** $19 + (-19)$

**51.** $-15 + (-7) + 1$

**52.** $23 + (-5) + 4$

**53.** $30 + (-10) + 5$

**54.** $40 + (-8) + 5$

**55.** $-23 + (-9) + 15$

**56.** $-25 + 25 + (-9)$

**57.** $40 + (-40) + 6$

**58.** $63 + (-18) + 12$

**59.** $12 + (-65) + (-12)$

**60.** $-35 + (-63) + 35$

**61.** $-24 + (-37) + (-19) + (-45) + (-35)$

**62.** $75 + (-14) + (-17) + (-5)$

**63.** $28 + (-44) + 17 + 31 + (-94)$

**64.** $27 + (-54) + (-32) + 65 + 46$

**65.** $-19 + 73 + (-23) + 19 + (-73)$

**66.** $35 + (-51) + 29 + 51 + (-35)$

## Skill Maintenance

**67.** Add: $587 + 6094$.   [1.2a]

**68.** Subtract: $3046 - 2973$.   [1.3a]

**69.** Write in expanded notation: 39,417.   [1.1b]

**70.** Multiply: $42 \cdot 56$.   [1.4a]

**71.** Divide: $288 \div 9$.   [1.5a]

**72.** Round to the nearest ten: 3496.   [1.6a]

## Synthesis

Add.

**73.** $-|27| + (-|-13|)$

**74.** $|-32| + (-|15|)$

**75.** ▦ $-3496 + (-2987)$

**76.** ▦ $497 + (-3028)$

**77.** For what numbers $x$ is $-x$ positive?

**78.** For what numbers $x$ is $-x$ negative?

Tell whether each sum is positive, negative, or zero.

**79.** If $n$ is positive and $m$ is negative, then $-n + m$ is
_____.

**80.** If $n = m$ and $n$ is negative, then $-n + (-m)$ is
_____.

**81.** If $n$ is negative and $m$ is less than $n$, then $n + m$ is
_____.

**82.** If $n$ is positive and $m$ is greater than $n$, then $n + m$ is
_____.

# Subtraction of Integers

## a SUBTRACTION

OBJECTIVES

a Subtract integers, and simplify combinations of additions and subtractions.

b Solve applied problems involving addition and subtraction of integers.

We now consider subtraction of integers. To find the difference $a - b$, we look for a number to add to $b$ that gives us $a$.

---

**THE DIFFERENCE**

The difference $a - b$ is the number that when added to $b$ gives $a$.

---

For example, $45 - 17 = 28$ because $17 + 28 = 45$. Let's consider an example in which the answer is a negative number.

**EXAMPLE 1**  Subtract: $5 - 8$.

*Think*: $5 - 8$ is the number that when added to 8 gives 5. What number can we add to 8 to get 5? The number must be negative. The number is $-3$:

$$5 - 8 = -3.$$

That is, $5 - 8 = -3$ because $8 + (-3) = 5$.

**Do Exercises 1–3.** ▶

The definition of $a - b$ above does not always provide the most efficient way to subtract. To understand a faster way to subtract, consider finding $5 - 8$ using the number line. We start at 5. Then we move 8 units to the *left* to do the subtracting. Note that this is the same as adding the opposite of 8, or $-8$, to 5.

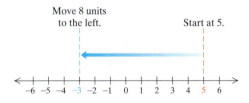

Move 8 units to the left.  Start at 5.

$$5 - 8 = -3$$

Look for a pattern in the following table.

| SUBTRACTIONS | ADDING AN OPPOSITE |
|---|---|
| $5 - 8 = -3$ | $5 + (-8) = -3$ |
| $-6 - 4 = -10$ | $-6 + (-4) = -10$ |
| $-7 - (-10) = 3$ | $-7 + 10 = 3$ |
| $-7 - (-2) = -5$ | $-7 + 2 = -5$ |

**Do Exercises 4–7.** ▶

Perhaps you have noticed that we can subtract by adding the opposite of the number being subtracted. This can always be done.

Subtract.

1. $-6 - 4$
   *Think*: What number can be added to 4 to get $-6$:
   $\square + 4 = -6$?

2. $-7 - (-10)$
   *Think*: What number can be added to $-10$ to get $-7$:
   $\square + (-10) = -7$?

3. $-7 - (-2)$
   *Think*: What number can be added to $-2$ to get $-7$:
   $\square + (-2) = -7$?

Complete the addition and compare with the subtraction.

4. $4 - 6 = -2$;
   $4 + (-6) = $ _____

5. $-3 - 8 = -11$;
   $-3 + (-8) = $ _____

6. $-5 - (-9) = 4$;
   $-5 + 9 = $ _____

7. $-5 - (-3) = -2$;
   $-5 + 3 = $ _____

*Answers*
1. $-10$  2. 3  3. $-5$  4. $-2$  5. $-11$
6. 4  7. $-2$

## SUBTRACTING BY ADDING THE OPPOSITE

To subtract, add the opposite, or additive inverse, of the number being subtracted:

$$a - b = a + (-b).$$

This is the method generally used for quick subtraction of integers.

**EXAMPLES** Write each subtraction as a corresponding addition. Then write the equation in words.

**2.** $-12 - 30$

$-12 - 30 = -12 + (-30)$      Adding the opposite of 30

Negative twelve minus thirty is negative twelve plus negative thirty.

**3.** $-20 - (-17)$

$-20 - (-17) = -20 + 17$      Adding the opposite of $-17$

Negative twenty minus negative seventeen is negative twenty plus seventeen.

◄ **Do Exercises 8–10.**

**EXAMPLES** Subtract.

**4.** $2 - 6 = 2 + (-6)$      The opposite of 6 is $-6$. We change the subtraction to addition and add the opposite. Instead of subtracting 6, we add $-6$.

$\phantom{2 - 6} = -4$

**5.** $4 - (-9) = 4 + 9$      The opposite of $-9$ is 9. We change the subtraction to addition and add the opposite. Instead of subtracting $-9$, we add 9.

$\phantom{4 - (-9)} = 13$

**6.** $-4 - 8 = -4 + (-8)$      We change the subtraction to addition and add the opposite. Instead of subtracting 8, we add $-8$.

$\phantom{-4 - 8} = -12$

**7.** $10 - 7 = 10 + (-7)$      We change the subtraction to addition and add the opposite. Instead of subtracting 7, we add $-7$.

$\phantom{10 - 7} = 3$

**8.** $-4 - (-9) = -4 + 9$      Instead of subtracting $-9$, we add 9.

$\phantom{-4 - (-9)} = 5$      To check, note that $5 + (-9) = -4$.

**9.** $-7 - (-3) = -7 + 3$      Instead of subtracting $-3$, we add 3.

$\phantom{-7 - (-3)} = -4$      *Check*: $-4 + (-3) = -7$.

◄ **Do Exercises 11–16.**

---

Write each subtraction as a corresponding addition. Then write the equation in words.

**8.** $3 - 10$

**9.** $-12 - (-9)$

**10.** $-12 - 10$

Subtract.

**11.** $2 - 8$     **GS**

$= 2 + \boxed{\phantom{xx}} = \boxed{\phantom{xx}}$

**12.** $-6 - 10$

**13.** $13 - 8$

**14.** $-7 - (-9)$

**15.** $-8 - (-2)$

**16.** $5 - (-8)$

*Answers*

**8.** $3 - 10 = 3 + (-10)$; three minus ten is three plus negative ten.
**9.** $-12 - (-9) = -12 + 9$; negative twelve minus negative nine is negative twelve plus nine.
**10.** $-12 - 10 = -12 + (-10)$; negative twelve minus ten is negative twelve plus negative ten.  **11.** $-6$
**12.** $-16$  **13.** 5  **14.** 2  **15.** $-6$  **16.** 13

*Guided Solution:*
**11.** $(-8)$, $-6$

When several additions and subtractions occur together, we can make them all additions. The commutative law for addition can then be used.

**EXAMPLE 10**  Simplify: $-3 - (-5) - 9 + 4 - (-6)$.

$$-3 - (-5) - 9 + 4 - (-6) = -3 + 5 + (-9) + 4 + 6 \qquad \text{Adding opposites}$$

$$= -3 + (-9) + 5 + 4 + 6 \qquad \text{Using a commutative law}$$

$$= -12 + 15$$

$$= 3$$

**Do Exercises 17 and 18.** ▶

Simplify.

**GS** **17.** $-6 - (-2) - (-4) - 12 + 3$

$= -6 + 2 + \boxed{\phantom{x}} + (-12) + 3$

$= -6 + (-12) + 2 + \boxed{\phantom{x}} + 3$

$= -18 + \boxed{\phantom{x}}$

$= \boxed{\phantom{x}}$

**18.** $9 - (-6) + 7 - 9 - 8 - (-20)$

## b  APPLICATIONS AND PROBLEM SOLVING

We use addition and subtraction of real numbers to solve a variety of applied problems.

**EXAMPLE 11**  *Surface Temperatures on Mars.*  Surface temperatures on Mars vary from $-128°C$ during the polar night to $27°C$ at the equator at midday when Mars is at the point in its orbit closest to the sun. Find the difference between the highest value and the lowest value in this temperature range.

**Data:** Mars Institute

We let $D =$ the difference in the temperatures. Then the problem translates to the following subtraction:

Difference in temperature  is  Highest temperature  minus  Lowest temperature

$$D \qquad = \qquad 27 \qquad - \qquad (-128).$$

We then solve the equation: $D = 27 - (-128) = 27 + 128 = 155$.

The difference in the temperatures is $155°C$.

**Do Exercise 19.** ▶

**19.** *Temperature Extremes.*  The highest temperature ever recorded in the United States was $134°F$ in Greenland Ranch, California, on July 10, 1913. The lowest temperature ever recorded was $-80°F$ in Prospect Creek, Alaska, on January 23, 1971. How much higher was the temperature in Greenland Ranch than the temperature in Prospect Creek?

**Data:** National Oceanographic and Atmospheric Administration

**Answers**
**17.** $-9$  **18.** $25$  **19.** $214°F$
***Guided Solution:***
**17.** $4, 4, 9, -9$

## ✓ Check Your Understanding

**Reading Check** Choose from the list on the right the word that makes each statement true. Words may be used more than once or not at all.

**RC1.** The number 3 is the _____ of −3.

**RC2.** To subtract, we add the _____ of the number being subtracted.

**RC3.** The word _____ usually translates to subtraction.

difference
opposite
reciprocal
sum

**Concept Check** Match each expression with an expression from the column on the right that names the same number.

**CC1.** 18 − 6

**CC2.** −18 − (−6)

**CC3.** −18 − 6

**CC4.** 18 − (−6)

**a)** 18 + 6
**b)** −18 + 6
**c)** 18 + (−6)
**d)** −18 + (−6)

**a** Subtract.

**1.** 3 − 7

**2.** 5 − 10

**3.** 0 − 7

**4.** 0 − 8

**5.** −8 − (−2)

**6.** −6 − (−8)

**7.** −10 − (−10)

**8.** −8 − (−8)

**9.** 12 − 16

**10.** 14 − 19

**11.** 20 − 27

**12.** 26 − 7

**13.** −9 − (−3)

**14.** −6 − (−9)

**15.** −11 − (−11)

**16.** −14 − (−14)

**17.** 7 − 7

**18.** 9 − 9

**19.** 7 − (−7)

**20.** 4 − (−4)

**21.** 8 − (−3)

**22.** −7 − 4

**23.** −6 − 8

**24.** 6 − (−10)

**25.** −4 − (−9)

**26.** −14 − 2

**27.** 2 − 9

**28.** 2 − 8

**29.** −6 − (−5)

**30.** −4 − (−3)

**31.** 8 − (−10)

**32.** 5 − (−6)

**33.** 0 − 10

**34.** 0 − 23

**35.** −5 − (−2)

**36.** −3 − (−1)

**37.** −7 − 14

**38.** −9 − 16

**39.** 0 − (−5)

**40.** 0 − (−1)

**41.** −8 − 0

**42.** −9 − 0

**43.** 7 − (−5)

**44.** 7 − (−4)

**45.** 6 − 25

**46.** 18 − 63

**47.** −42 − 26

**48.** −18 − 63

**49.** −72 − 9

**50.** −49 − 3

**51.** 24 − (−92)

**52.** 48 − (−73)

**53.** −50 − (−50)

**54.** −70 − (−70)

**55.** −30 − (−85)

**56.** −25 − (−15)

Simplify.

**57.** 7 − (−5) + 4 − (−3)

**58.** −5 − (−8) + 3 − (−7)

**59.** −31 + (−28) − (−14) − 17

**60.** −43 − (−19) − (−21) + 25

**61.** −34 − 28 + (−33) − 44

**62.** 39 + (−88) − 29 − (−83)

**63.** −93 − (−84) − 41 − (−56)

**64.** 84 + (−99) + 44 − (−18) − 43

**65.** −5 − (−30) + 30 + 40 − (−12)

**66.** 14 − (−50) + 20 − (−32)

**67.** 132 − (−21) + 45 − (−21)

**68.** 81 − (−20) − 14 − (−50) + 53

 Solve.

**69.** *Offshore Oil.* In 1998, the world's deepest deepwater well was drilled at an elevation of −7718 ft. In 2016, the Raya-1 exploratory well (which proved to be dry) was drilled at an elevation 3438 ft deeper. What was the elevation of the Raya-1 well?

**Data:** www.deepwater.com; gcaptain.com

**70.** *Oceanography.* The deepest point in the Pacific Ocean is the Marianas Trench, with a depth of 11,033 m. The deepest point in the Atlantic Ocean is the Puerto Rico Trench, with a depth of 8648 m. What is the difference in the elevation of the two trenches?

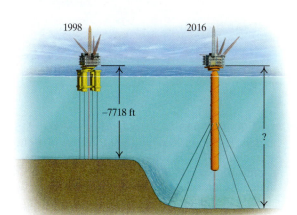

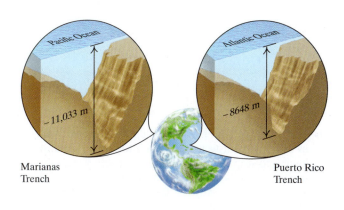

**71.** Through exercise, Rod went from 8 lb above his "ideal" body weight to 9 lb below it. How many pounds did Rod lose?

**72.** Laura has a charge of $477 on her credit card, but she then returns a sweater that cost $129. How much does she now owe on her credit card?

**73. Difference in Elevation.** The highest elevation in Japan is 3776 m above sea level at Fujiyama. The lowest elevation in Japan is 4 m below sea level at Hachirogata. Find the difference in the elevations.

**Data:** *The CIA World Factbook*

**74. Copy-Center Account.** Rachel's copy-center bill for July was $327. She made a payment of $200 and then made $48 worth of copies in August. How much did she then owe on her account?

**75. Temperature Changes.** One day the temperature in Lawrence, Kansas, is 32° at 6:00 A.M. It rises 15° by noon, but falls 50° by midnight after a cold front moves in. What is the final temperature?

**76. Stock Price Changes.** On a recent day, the price of a stock opened at $61. It rose $5, dropped $7, and rose $4. Find the price of the stock at the end of the day.

**77. Points per Game Differential.** A basketball team's point differential is the difference between that team's score in a game and its opponent's score. The points per game differential is the sum of a team's point differentials divided by the number of games it has played. At one point in a recent season, the Indiana Pacers had a points per game differential of +9, and the Milwaukee Bucks had a points per game differential of −9. How much higher was the Pacers' points per game differential?

**78. Golf.** As a result of coaching, Cedric's average golf score improved from 3 over par to 2 under. By how many strokes did his score change?

**79. "Flipping" Houses.** Buying run-down houses, fixing them up, and reselling them is referred to as "flipping" houses. Charlie and Sophia flipped four houses in a recent year. The profits and losses are shown in the following bar graph. Find the sum of the profits and losses.

**Flipping Houses**

**80. Elevations in Asia.** The elevation of the highest point in Asia, Mt. Everest, on the border between Nepal and Tibet, is 29,035 ft. The lowest elevation, at the Dead Sea, between Israel and Jordan, is −1348 ft. What is the difference in the elevations of the two locations?

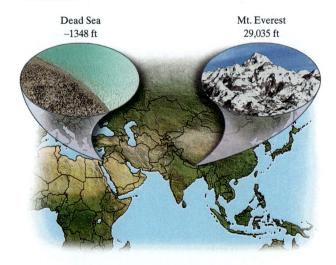

Dead Sea
−1348 ft

Mt. Everest
29,035 ft

**81. Toll Roads.** The E-Z Pass program allows drivers in the Northeast to travel certain toll roads without having to stop to pay. Instead, a transponder attached to the vehicle is scanned as the vehicle rolls through a toll booth. Recently the Ramones began a trip to New York City with a balance of $12 in their E-Z Pass account. They accumulated $15 in tolls on their trip, and because they overspent their balance, the Ramones had to pay $80 in administrative fees. By how much were the Ramones in debt as a result of their travel on toll roads?

**82. Toll Roads.** The Murrays began a trip with $13 in their E-Z Pass account. (See Exercise 81.) They accumulated $20 in tolls and had to pay $80 in administrative fees. By how much were the Murrays in debt as a result of their travel on toll roads?

## Skill Maintenance

Evaluate.

**83.** $4^3$  [1.9b]

**84.** $68 \cdot 72$  [1.4a]

**85.** $1^7$  [1.9b]

**86.** $143 \cdot 29$  [1.4a]

**87.** How many 12-oz cans of soda can be filled with 96 oz of soda?  [1.8a]

**88.** A case of soda contains 24 bottles. If each bottle contains 12 oz, how many ounces of soda are in the case?  [1.8a]

Simplify.  [1.9c]

**89.** $5 + 4^2 + 2 \cdot 7$

**90.** $45 \div (2^2 + 11)$

**91.** $(9 + 7)(9 - 7)$

**92.** $(13 - 2)(13 + 2)$

## Synthesis

Subtract.

**93.** ▦ $123{,}907 - 433{,}789$

**94.** ▦ $23{,}011 - (-60{,}432)$

For Exercises 95–100, tell whether each statement is true or false for all integers $a$ and $b$. If false, give an example to show why.

**95.** $a - 0 = 0 - a$

**96.** $0 - a = a$

**97.** If $a \neq b$, then $a - b \neq 0$.

**98.** If $a = -b$, then $a + b = 0$.

**99.** If $a + b = 0$, then $a$ and $b$ are opposites.

**100.** If $a - b = 0$, then $a = -b$.

**101.** If $a - 54$ is $-37$, find the value of $a$.

**102.** If $x - 48$ is $-15$, find the value of $x$.

**103.** Maureen kept track of the weekly changes in the stock market over a period of 5 weeks. By how many points (pts) had the market risen or fallen over this time?

| WEEK 1 | WEEK 2 | WEEK 3 | WEEK 4 | WEEK 5 |
|--------|--------|--------|--------|--------|
| Down 13 pts | Down 16 pts | Up 36 pts | Down 11 pts | Up 19 pts |

**104.** *Blackjack Counting System.*   Players of the casino game of blackjack make use of many card-counting systems in which a negative count means that a player has an advantage. One such system, called *High–Low*, was developed by Harvey Dubner in 1963. Each card already showing counts as $-1, 0,$ or $1$ as follows:

    $2, 3, 4, 5, 6$           count as $+1$;

    $7, 8, 9$               count as $0$;

    $10, J, Q, K, A$     count as $-1$.

**Data:** Patterson, Jerry L. *Casino Gambling*. New York: Perigee, 1982

**a)** Find the total count on this sequence of cards:

    K, A, 2, 4, 5, 10, J, 8, Q, K, 5.

**b)** Does the player have an advantage?

# 2.4

## OBJECTIVES

**a** Multiply integers.

**b** Find products of three or more integers, and simplify powers of integers.

# Multiplication of Integers

## a MULTIPLICATION

Multiplication of integers is like multiplication of whole numbers. The difference is that we must determine whether the answer is positive or negative.

### Multiplication of a Positive Integer and a Negative Integer

To see how to multiply a positive integer and a negative integer, consider the following pattern.

This number decreases by 1 each time.

$$4 \cdot 5 = 20$$
$$3 \cdot 5 = 15$$
$$2 \cdot 5 = 10$$
$$1 \cdot 5 = 5$$
$$0 \cdot 5 = 0$$
$$-1 \cdot 5 = -5$$
$$-2 \cdot 5 = -10$$
$$-3 \cdot 5 = -15$$

This number decreases by 5 each time.

**1.** Complete, as in the example in the text.

$$4 \cdot 10 = 40$$
$$3 \cdot 10 = 30$$
$$2 \cdot 10 =$$
$$1 \cdot 10 =$$
$$0 \cdot 10 =$$
$$-1 \cdot 10 =$$
$$-2 \cdot 10 =$$
$$-3 \cdot 10 =$$

◀ **Do Exercise 1.**

According to this pattern, it looks as though the product of a negative integer and a positive integer is negative. To confirm this, use repeated addition:

$$-1 \cdot 5 = 5 \cdot (-1) = -1 + (-1) + (-1) + (-1) + (-1) = -5$$
$$-2 \cdot 5 = 5 \cdot (-2) = -2 + (-2) + (-2) + (-2) + (-2) = -10$$
$$-3 \cdot 5 = 5 \cdot (-3) = -3 + (-3) + (-3) + (-3) + (-3) = -15$$

---

**MULTIPLYING A POSITIVE INTEGER AND A NEGATIVE INTEGER**

To multiply a positive integer and a negative integer, multiply their absolute values. The answer is negative.

---

Multiply.

**2.** $-3 \cdot 6$     **3.** $20 \cdot (-5)$

**4.** $12(-1)$

**EXAMPLES** Multiply.

**1.** $8(-5) = -40$     **2.** $50(-1) = -50$     **3.** $-7 \cdot 6 = -42$

◀ **Do Exercises 2–4.**

**5.** Complete, as in the example in the text.

$$3 \cdot (-10) = -30$$
$$2 \cdot (-10) = -20$$
$$1 \cdot (-10) =$$
$$0 \cdot (-10) =$$
$$-1 \cdot (-10) =$$
$$-2 \cdot (-10) =$$
$$-3 \cdot (-10) =$$

### Multiplication of Two Negative Integers

How do we multiply two negative integers? Again we look for a pattern.

This number decreases by 1 each time.

$$4 \cdot (-5) = -20$$
$$3 \cdot (-5) = -15$$
$$2 \cdot (-5) = -10$$
$$1 \cdot (-5) = -5$$
$$0 \cdot (-5) = 0$$
$$-1 \cdot (-5) = 5$$
$$-2 \cdot (-5) = 10$$
$$-3 \cdot (-5) = 15$$

This number increases by 5 each time.

*Answers*

**1.** 20; 10; 0; −10; −20; −30   **2.** −18
**3.** −100   **4.** −12   **5.** −10; 0; 10; 20; 30

◀ **Do Exercise 5.**

According to the pattern, the product of two negative integers is positive. This leads to the second part of the rule for multiplying integers.

---

**MULTIPLYING TWO NEGATIVE INTEGERS**
..............................................................................

To multiply two negative integers, multiply their absolute values. The answer is positive.

---

**EXAMPLES** Multiply.

**4.** $(-2)(-4) = 8$

**5.** $(-10)(-7) = 70$

> **Do Exercises 6–8.** ▶

The following is another way to state the rules for multiplication.

---

**MULTIPLYING TWO NONZERO INTEGERS**
..............................................................................

To multiply two nonzero integers:

**a)** Multiply the absolute values.

**b)** If the signs are the same, the answer is positive.

**c)** If the signs are different, the answer is negative.

---

**EXAMPLES** Multiply.

**6.** $(-3)(-5) = 15$

**7.** $8(-10) = -80$

> **Do Exercises 9–12.** ▶

## Multiplication by 0

No matter how many times 0 is added to itself, the answer is 0.

---

The product of 0 and any integer is 0:

$$a \cdot 0 = 0.$$

---

**EXAMPLES** Multiply.

**8.** $-19 \cdot 0 = 0$

**9.** $0(-7) = 0$

> **Do Exercises 13 and 14.** ▶

## b MULTIPLICATION OF MORE THAN TWO INTEGERS

**SKILL REVIEW**

*Evaluate exponential notation.* [1.9b]

Evaluate.

**1.** $5^3$

**2.** $10^4$

Answers: **1.** 125 **2.** 10,000

MyLab Math
VIDEO

---

Multiply.

**6.** $(-3)(-4)$

**7.** $-9(-5)$

**8.** $(-1)(-6)$

Multiply.

**GS** **9.** $3(-6)$

Multiply absolute values:

$$3 \cdot 6 = \boxed{\phantom{00}}.$$

The signs are different, so the answer is $\boxed{\phantom{00}}$.

$$3(-6) = \boxed{\phantom{00}}$$

**10.** $(-6)(-5)$

**11.** $(-1)(-50)$

**12.** $(-8) \cdot 11$

Multiply.

**13.** $0(-5)$

**14.** $-23 \cdot 0$

When multiplying more than two integers, we can choose order and grouping as we please, using the commutative and associative laws.

**EXAMPLES** Multiply.

**10. a)** $-8 \cdot 2(-3) = -16(-3)$     Multiplying the first two numbers
$= 48$     Multiplying the results

**b)** $-8 \cdot 2(-3) = 24 \cdot 2$     Multiplying the negative numbers
$= 48$     The result is the same as above.

**11.** $7(-1)(-4)(-2) = (-7)8$     Multiplying the first two numbers and
$= -56$     the last two numbers

**12.** $-5 \cdot (-2) \cdot (-3) \cdot (-6) = 10 \cdot 18$     Each pair of negative numbers
$= 180$     gives a positive product.

**13.** $(-3)(-5)(-2)(-3)(-1) = 15 \cdot 6 \cdot (-1)$
$= 90(-1) = -90$

We can see the following pattern in the results of Examples 12 and 13.

> The product of an even number of negative integers is positive.
> The product of an odd number of negative integers is negative.

Multiply.

**15.** $5 \cdot (-3) \cdot 2$

**16.** $-2 \cdot (-5) \cdot (-4) \cdot (-3)$

**17.** $(-4)(-5)(-2)(-3)(-1)$     **GS**
$= \boxed{\phantom{00}} \cdot 6 \cdot (-1)$
$= \boxed{\phantom{00}} \cdot (-1)$
$= \boxed{\phantom{00}}$

**18.** $(-1)(-1)(-2)(-3)(-1)(-1)$

◀ **Do Exercises 15–18.**

## Powers of Integers

A positive number raised to any exponent is positive. When a negative number is raised to an integer exponent, the sign of the result depends upon whether the exponent is even or odd.

**EXAMPLES** Simplify.

**14.** $(-7)^2 = (-7)(-7) = 49$     The result is positive.

**15.** $(-4)^3 = (-4)(-4)(-4)$
$= 16(-4)$
$= -64$     The result is negative.

**16.** $(-3)^4 = (-3)(-3)(-3)(-3)$
$= 9 \cdot 9$
$= 81$     The result is positive.

**17.** $(-2)^5 = (-2)(-2)(-2)(-2)(-2)$
$= 4 \cdot 4 \cdot (-2)$
$= 16(-2)$
$= -32$     The result is negative.

*Answers*
**15.** $-30$ **16.** $120$ **17.** $-120$ **18.** $6$
*Guided Solution:*
**17.** $20, 120, -120$

Perhaps you noted the following.

> When a negative number is raised to an even exponent, the result is positive. When a negative number is raised to an odd exponent, the result is negative.

**Do Exercises 19–22.** ▶

When an integer is multiplied by $-1$, the result is the opposite of that integer.

> For any integer $a$,
> $$-1 \cdot a = -a.$$

**EXAMPLE 18** Simplify: $-7^2$.

In the expression $-7^2$, the base is 7, not $-7$. We can regard $-7^2$ as $-1 \cdot 7^2$:

$-7^2 = -1 \cdot 7^2$   The rules for order of operations tell us to square first: $7^2 = 7 \cdot 7$.

$= -1 \cdot 49$

$= -49.$

Compare Examples 14 and 18 and note that $(-7)^2 \neq -7^2$.

**Do Exercises 23 and 24.** ▶

Simplify.

**19.** $(-2)^3$     **20.** $(-9)^2$

**21.** $(-1)^9$     **22.** $2^5$

**23.** Simplify: $-5^2$.

**24.** Simplify: $(-5)^2$.

**Answers**
**19.** $-8$  **20.** $81$  **21.** $-1$  **22.** $32$
**23.** $-25$  **24.** $25$

---

**CALCULATOR CORNER**

**Exponential Notation** When using a calculator to calculate expressions like $(-39)^4$ or $-39^4$, it is important to use the correct sequence of keystrokes.

**Calculators with** $\boxed{+/-}$ **key:** To calculate $(-39)^4$, we press $\boxed{3}\,\boxed{9}\,\boxed{+/-}\,\boxed{x^y}\,\boxed{4}\,\boxed{=}$.

To calculate $-39^4$, we must first raise 39 to the power 4. Then the sign of the result must be changed. This can be done with the keystrokes $\boxed{3}\,\boxed{9}\,\boxed{x^y}\,\boxed{4}\,\boxed{=}\,\boxed{+/-}$, or by multiplying $39^4$ by $-1$ using the keystrokes $\boxed{3}\,\boxed{9}\,\boxed{x^y}\,\boxed{4}\,\boxed{=}\,\boxed{\times}\,\boxed{1}\,\boxed{+/-}\,\boxed{=}$.

**Calculators with** $\boxed{(-)}$ **key:** On some calculators, the $\boxed{(-)}$ key is pressed before a number to indicate that the number is negative. This is similar to the way the expression is written on paper. With these calculators, $(-39)^4$ is found by pressing $\boxed{(}\,\boxed{(-)}$ $\boxed{3}\,\boxed{9}\,\boxed{)}\,\boxed{\wedge}\,\boxed{4}\,\boxed{ENTER}$, and $-39^4$ is found by pressing $\boxed{(-)}\,\boxed{3}\,\boxed{9}\,\boxed{\wedge}\,\boxed{4}\,\boxed{ENTER}$.

You can either experiment or consult a user's manual if you are unsure of the proper keystrokes for your calculator.

**EXERCISES:** Use a calculator to determine each of the following.

**1.** $(-23)^6$                 **2.** $(-17)^5$

**3.** $(-104)^3$              **4.** $(-4)^{10}$

**5.** $-9^6$                     **6.** $-7^6$

**7.** $-6^5$                     **8.** $-3^9$

✓ **Check Your Understanding**

**Reading Check** Fill in the blank with either "positive" or "negative."

**RC1.** To multiply a positive number and a negative number, multiply their absolute values. The answer is _____.

**RC2.** To multiply two negative numbers, multiply their absolute values. The answer is _____.

**RC3.** The product of an even number of negative numbers is _____.

**RC4.** The product of an odd number of negative numbers is _____.

**Concept Check** Evaluate.

**CC1.** $(-1)(-1)$    **CC2.** $(-1)(-1)(-1)$    **CC3.** $(-1)(-1)(-1)(-1)$    **CC4.** $(-1)(-1)(-1)(-1)(-1)$

**a** Multiply.

**1.** $-2 \cdot 8$

**2.** $-7 \cdot 3$

**3.** $10 \cdot (-6)$

**4.** $12 \cdot (-2)$

**5.** $8 \cdot (-6)$

**6.** $8 \cdot (-3)$

**7.** $-10 \cdot 3$

**8.** $-9 \cdot 8$

**9.** $-3(-5)$

**10.** $-8 \cdot (-2)$

**11.** $-9 \cdot (-2)$

**12.** $(-8)(-9)$

**13.** $(-6)(-7)$

**14.** $-8 \cdot (-3)$

**15.** $-10(-2)$

**16.** $-9(-8)$

**17.** $12(-10)$

**18.** $15(-8)$

**19.** $-23 \cdot 0$

**20.** $-38 \cdot 0$

**21.** $(-72)(-1)$

**22.** $41(-3)$

**23.** $(-20)17$

**24.** $(-1)(-43)$

**25.** $-8(-50)$

**26.** $(-25) \cdot 8$

**27.** $0(-14)$

**28.** $0(-38)$

Multiply.

**29.** $3 \cdot (-8) \cdot (-1)$

**30.** $(-7) \cdot (-4) \cdot (-1)$

**31.** $7(-4)(-3)5$

**32.** $9(-2)(-6)7$

**33.** $-2(-5)(-7)$

**34.** $(-2)(-5)(-3)(-5)$

**35.** $(-5)(-2)(-3)(-1)$

**36.** $-6(-5)(-9)$

**37.** $(-15)(-29) \cdot 0 \cdot 8$

**38.** $19(-7)(-8) \cdot 0 \cdot 6$

**39.** $(-7)(-1)(7)(-6)$

**40.** $(-5)6(-4)5$

Simplify.

**41.** $(-6)^2$

**42.** $(-8)^2$

**43.** $(-5)^3$

**44.** $(-2)^4$

**45.** $(-10)^4$

**46.** $(-1)^5$

**47.** $-2^4$

**48.** $(-2)^6$

**49.** $(-3)^5$

**50.** $-10^4$

**51.** $(-1)^{12}$

**52.** $(-1)^{13}$

**53.** $-11^2$

**54.** $-2^6$

**55.** $-4^3$

**56.** $-2^5$

## Skill Maintenance

**57.** Round 532,451 to the nearest hundred.  [1.6a]

**58.** Write standard notation for sixty million.  [1.1c]

**59.** Divide: $2880 \div 36$.  [1.5a]

**60.** Multiply: $75 \times 34$.  [1.4a]

**61.** Simplify: $10 - 2^3 + 6 \div 2$.  [1.9c]

**62.** Simplify: $2 \cdot 5^2 - 3 \cdot 2^3 \div (3 + 3^2)$.  [1.9c]

**63.** A rectangular rug measures 5 ft by 8 ft. What is the area of the rug?  [1.4b], [1.8a]

**64.** An elevator can hold 16 people and 50 people are waiting to go up. How many trips will be required to transport all of them?  [1.8a]

## Synthesis

Simplify.

**65.** $(-3)^5(-1)^{379}$

**66.** $(-2)^3 \cdot [(-1)^{29}]^{46}$

**67.** $-9^4 + (-9)^4$

**68.** $-5^2(-1)^{29}$

**69.** $|(-2)^5 + 3^2| - (3 - 7)^2$

**70.** $|-12(-3)^2 - 5^3 - 6^2 - (-5)^2|$

**71.** ▦ $-47^2$

**72.** ▦ $-53^2$

**73.** ▦ $(-19)^4$

**74.** ▦ $(-23)^4$

**75.** ▦ $(73 - 86)^3$

**76.** ▦ $(-49 + 34)^3$

**77.** ▦ $-935(238 - 243)^3$

**78.** ▦ $(-17)^4(129 - 133)^5$

**79.** Jo had a balance of $68 in her account and wrote seven checks for $13 each. What was her balance after writing the checks?

**80.** After diving 95 m below the surface, a diver rises at a rate of 7 meters per minute for 9 min. What is the diver's new elevation?

**81.** What must be true of $m$ and $n$ if $[(-5)^m]^n$ is to be **(a)** negative? **(b)** positive?

**82.** What must be true of $m$ and $n$ if $-mn$ is to be **(a)** positive? **(b)** zero? **(c)** negative?

## 2.5

### OBJECTIVES

**a** Divide integers.

**b** Use the rules for order of operations with integers.

# Division of Integers and Order of Operations

We now consider division of integers. Because of the way in which division is defined, its sign rules are similar to those for multiplication.

## a  DIVISION OF INTEGERS

### DIVISION

The quotient $\dfrac{a}{b}$ ( or $a \div b$, or $a/b$ ) is the number, if there is one, that when multiplied by $b$ gives $a$.

Let's use the definition to divide integers.

**EXAMPLES**  Divide, if possible. Check each answer.

**1.** $14 \div (-7) = -2$     *Think*: What number multiplied by $-7$ gives 14? The number is $-2$. *Check*: $(-2)(-7) = 14$.

**2.** $\dfrac{-32}{-4} = 8$     *Think*: What number multiplied by $-4$ gives $-32$? The number is 8. *Check*: $8(-4) = -32$.

**3.** $-21 \div 7 = -3$     *Think*: What number multiplied by 7 gives $-21$? The number is $-3$. *Check*: $(-3) \cdot 7 = -21$.

**4.** $\dfrac{0}{-5} = 0$     *Think*: What number multiplied by $-5$ gives 0? The number is 0. *Check*: $0(-5) = 0$.

**5.** $\dfrac{-5}{0}$ is **not defined**.     *Think*: What number multiplied by 0 gives $-5$? There is no such number because the product of 0 and *any* number is 0.

The rules for determining the sign of a quotient are the same as those for determining the sign of a product. We state them together.

### MULTIPLYING OR DIVIDING TWO INTEGERS

To multiply or divide two integers:

**a)** Multiply or divide the absolute values.

**b)** If the signs are the same, the answer is positive.

**c)** If the signs are different, the answer is negative.

◀ Do Exercises 1–6.

### Dividing by 0

Recall that, in general, $a \div b$ and $b \div a$ are different numbers. In Example 4, we divided *into* 0. In Example 5, we attempted to divide *by* 0. Since any number times 0 gives 0, not $-5$, we say that $-5 \div 0$ is **not defined** or is **undefined**. Also, since *any* number times 0 gives 0, $0 \div 0$ is also not defined.

Divide.

**1.** $6 \div (-2)$
*Think:* What number multiplied by $-2$ gives 6?

**2.** $\dfrac{-15}{-3}$
*Think:* What number multiplied by $-3$ gives $-15$?

**3.** $-24 \div 8$
*Think:* What number multiplied by 8 gives $-24$?

**4.** $\dfrac{0}{-4}$     **5.** $\dfrac{30}{-5}$

**6.** $\dfrac{-45}{9}$

**Answers**
**1.** $-3$  **2.** 5  **3.** $-3$  **4.** 0
**5.** $-6$  **6.** $-5$

## EXCLUDING DIVISION BY 0

Division by 0 is not defined:

$a \div 0$, or $\dfrac{a}{0}$, is undefined for all real numbers $a$.

## Dividing 0 by Other Numbers

Note that $0 \div 8 = 0$ because $0 = 0 \cdot 8$.

## DIVIDENDS OF 0

Zero divided by any nonzero real number is 0:

$\dfrac{0}{a} = 0, \quad a \neq 0.$

**EXAMPLES**   Divide.

**6.** $0 \div (-6) = 0$      **7.** $\dfrac{0}{12} = 0$      **8.** $\dfrac{-3}{0}$ is undefined.

**Do Exercises 7–9.** ▶

Divide, if possible.

**7.** $34 \div 0$

**8.** $0 \div (-4)$

**9.** $-52 \div 0$

## b   RULES FOR ORDER OF OPERATIONS

*Simplify expressions using the rules for order of operations.*   [1.9d]
Simplify.

**1.** $5^2 - 2(10 - 3)$      **2.** $2[21 - (11 - 3)]$

**Answers: 1.** 11  **2.** 26

When more than one operation appears in a calculation or problem, we apply the rules for order of operations.

## RULES FOR ORDER OF OPERATIONS

1. Do all calculations within grouping symbols, including parentheses, brackets, braces, and absolute-value symbols, and within numerators or denominators.
2. Evaluate all exponential expressions.
3. Do all multiplications and divisions in order from left to right.
4. Do all additions and subtractions in order from left to right.

*Answers*
**7.** Undefined  **8.** 0  **9.** Undefined

Simplify.

**10.** $5 - (-7)(-3)^2$

**11.** $(-2) \cdot |3 - 2^2| + 5$
$= (-2) \cdot |3 - \square| + 5$
$= (-2) \cdot |\square| + 5$
$= (-2) \cdot \square + 5$
$= \square + 5$
$= \square$ **GS**

**12.** $52 \cdot 5 + 5^3 - (4^2 - 48 ÷ 4)$

**13.** $\dfrac{(-5)(-9)}{1 - 2 \cdot 2}$

**EXAMPLE 9** Simplify: $17 - 10 ÷ 2 \cdot 4$.

There are no grouping symbols or exponents, so we begin with the third rule.

$$\left.\begin{aligned} 17 - 10 ÷ 2 \cdot 4 &= 17 - 5 \cdot 4 \\ &= 17 - 20 \end{aligned}\right\}$$ Carrying out all multiplications and divisions in order from left to right
$$= -3$$

**EXAMPLES** Simplify.

**10.** $|(-2)^3 ÷ 4| - 5(-2)$

We first simplify within the absolute-value symbols.

$$\begin{aligned} |(-2)^3 ÷ 4| - 5(-2) &= |-8 ÷ 4| - 5(-2) \quad && (-2)(-2)(-2) = -8 \\ &= |-2| - 5(-2) \quad && \text{Dividing} \\ & && |-2| = 2 \\ &= 2 - 5(-2) \quad && \text{Multiplying} \\ &= 2 - (-10) \quad && \text{Subtracting;} \\ &= 12 \quad && 2 - (-10) = 2 + 10 \end{aligned}$$

**11.** $2^4 + 51 \cdot 4 - (37 + 23 \cdot 2)$
$2^4 + 51 \cdot 4 - (37 + 23 \cdot 2)$

$= 2^4 + 51 \cdot 4 - (37 + 46)$ — Carrying out all operations inside parentheses first, following the rules for order of operations within the parentheses

$= 2^4 + 51 \cdot 4 - 83$ — Adding inside parentheses

$= 16 + 51 \cdot 4 - 83$ — Evaluating exponential expressions

$= 16 + 204 - 83$ — Doing all multiplications

$= 220 - 83$ — Doing all additions and subtractions in order from left to right

$= 137$

Always regard a fraction bar as a grouping symbol. It separates any calculations in the numerator from those in the denominator.

**EXAMPLE 12** Simplify: $\dfrac{5 - (-3)^2}{8 - 10}$.

$$\left.\begin{aligned} \frac{5 - (-3)^2}{8 - 10} &= \frac{5 - 9}{-2} \\ &= \frac{-4}{-2} \end{aligned}\right\}$$ Calculating within the numerator and within the denominator: $(-3)^2 = (-3)(-3) = 9$, $8 - 10 = -2$, and $5 - 9 = -4$

$$= 2$$ Dividing

◀ **Do Exercises 10–13.**

✓ **Check Your Understanding**

**Reading Check** Determine whether each statement is true or false.

_____ **RC1.** The product of $-10$ and $-2$ has the same sign as the quotient of $-10$ and $-2$.

_____ **RC2.** Absolute-value symbols are grouping symbols.

_____ **RC3.** It is impossible to divide zero by any number.

**Concept Check** Tell whether each division results in 0 or is undefined.

**CC1.** $0 \div 17$   **CC2.** $(-3) \div 0$   **CC3.** $\dfrac{132}{0}$   **CC4.** $\dfrac{0}{-74}$

**a**    Divide, if possible, and check each answer by multiplying. If an answer is undefined, state so.

**1.** $36 \div (-6)$

**2.** $\dfrac{42}{-7}$

**3.** $\dfrac{26}{-2}$

**4.** $24 \div (-12)$

**5.** $\dfrac{-16}{8}$

**6.** $-22 \div (-2)$

**7.** $\dfrac{-48}{-12}$

**8.** $-72 \div (-9)$

**9.** $\dfrac{-72}{8}$

**10.** $\dfrac{-50}{25}$

**11.** $-100 \div (-50)$

**12.** $\dfrac{-400}{8}$

**13.** $-108 \div 9$

**14.** $\dfrac{-128}{8}$

**15.** $\dfrac{200}{-25}$

**16.** $-651 \div (-31)$

**17.** $\dfrac{-56}{0}$

**18.** $\dfrac{0}{-5}$

**19.** $\dfrac{88}{-11}$

**20.** $\dfrac{-145}{-5}$

**21.** $-\dfrac{276}{12}$

**22.** $-\dfrac{217}{7}$

**23.** $\dfrac{0}{-2}$

**24.** $\dfrac{-13}{0}$

**25.** $\dfrac{19}{-1}$

**26.** $\dfrac{-17}{1}$

**27.** $-41 \div 1$

**28.** $23 \div (-1)$

**b** Simplify, if possible. If an answer is undefined, state so.

**29.** $8 - 2 \cdot 3 - 9$

**30.** $8 - (2 \cdot 3 - 9)$

**31.** $(8 - 2 \cdot 3) - 9$

**32.** $(8 - 2)(3 - 9)$

**33.** $16 \cdot (-24) + 50$

**34.** $10 \cdot 20 - 15 \cdot 24$

**35.** $40 - 3^2 - 2^3$

**36.** $2^4 + 2^2 - 10$

**37.** $4 \cdot (6 + 8)/(4 + 3)$

**38.** $4^3 + 10 \cdot 20 + 8^2 - 23$

**39.** $4 \cdot 5 - 2 \cdot 6 + 4$

**40.** $5^3 + 4 \cdot 9 - (8 + 9 \cdot 3)$

**41.** $1 - (-2)^2 \cdot 3 \div 6$

**42.** $-6 + (-3)^2 + 6 \div (-2)$

**43.** $18 - (-3)^3 - 3^2 \cdot 5$

**44.** $9 - (-2)^3 - 50 \div 2$

**45.** $\dfrac{9^2 - 1}{1 - 3^2}$

**46.** $\dfrac{100 - 6^2}{(-5)^2 - 3^2}$

**47.** $8(-7) + 6(-5)$

**48.** $10(-5) \div 1(-1)$

**49.** $20 \div 5(-3) + 3$

**50.** $14 \div 2(-6) + 7$

**51.** $18 - 0(3^2 - 5^2 \cdot 7 - 4)$

**52.** $9 \cdot 0 \div 5 \cdot 4$

**53.** $-4^2 + 6$

**54.** $-5^2 + 7$

**55.** $-8^2 - 3$

**56.** $-9^2 - 11$

**57.** $4 \cdot 5^2 \div 10$

**58.** $(2 - 5)^2 \div (-9)$

**59.** $(3 - 8)^2 \div (-1)$

**60.** $3 - 3^2$

**61.** $12 - 20^3$

**62.** $20 + 4^3 \div (-8)$

**63.** $2 \times 10^3 - 5000$

**64.** $-7(3^4) + 18$

**65.** $6[9 - (3 - 4)]$

**66.** $8[(6 - 13) - 11]$

**67.** $-1000 \div (-100) \div 10$

**68.** $256 + (-32) \div (-4)$

**69.** $-7 - 3[-80 \div (2 - 10)]$

**70.** $-1 - 5[3 - (7 - 4^2)]$

**71.** $-2[3 - (7 - 9)^3]$

**72.** $-10[(2 - 8)^2 - 6]$

**73.** $8 - |7 - 9| \cdot 3$

**74.** $|8 - 7 - 9| \cdot 2 + 1$

**75.** $9 - |7 - 3^2|$

**76.** $9 - |5 - 7|^3$

**77.** $\dfrac{6^3 - 7 \cdot 3^4 - 2^5 \cdot 9}{(1 - 2^3)^3 + 7^3}$

**78.** $\dfrac{6 \div 2 \cdot 4^2 - 7^2 + 1}{(7 - 4)^3 - 2 \cdot 5 - 4}$

**79.** $\dfrac{2 \cdot 3^2 \div (3^2 - (2 + 1))}{5^2 - 6^2 - 2^2(-3)}$

**80.** $\dfrac{5 \cdot 6^2 \div (2^2 \cdot 5) - 7^2}{3^2 - 4^2 - (-2)^3 - 2}$

**81.** $\dfrac{(-5)^3 + 17}{10(2 - 6) - 2(5 + 2)}$

**82.** $\dfrac{(3 - 5)^2 - (7 - 13)}{(2 - 5)3 + 2 \cdot 4}$

**83.** $\dfrac{2 \cdot 4^3 - 4 \cdot 32}{19^3 - 17^4}$

**84.** $\dfrac{-16 \cdot 28 \div 2^2}{5 \cdot 25 - 5^3}$

## Skill Maintenance

**85.** Fabrikant Fine Diamonds ran a 4-in. by 7-in. advertisement in the *New York Times*. Find the area of the ad. [1.4b], [1.8a]

**86.** A hotel has 4 floors with 62 rooms on each floor. How many rooms are there in the hotel? [1.8a]

**87.** Cindi's Ford Focus gets 32 mpg (miles per gallon). How many gallons will it take to travel 384 mi? [1.8a]

**88.** Craig's Chevy Blazer gets 14 mpg. How many gallons will it take to travel 378 mi? [1.8a]

**89.** A 7-oz bag of tortilla chips contains 1050 calories. How many calories are in a 1-oz serving? [1.8a]

**90.** A 7-oz bag of tortilla chips contains 8 g (grams) of fat per ounce. How many grams of fat are in a carton containing 12 bags of chips? [1.8a]

**91.** There are 18 sticks in a large pack of Trident gum. If 4 people share the pack equally, how many whole pieces will each person receive? How many extra pieces will remain? [1.8a]

**92.** A bag of Ricola cough drops contains 24 drops. If 5 people share the bag equally, how many drops will each person receive? How many extra drops will remain? [1.8a]

## Synthesis

Simplify, if possible.

**93.** $\dfrac{9 - 3^2}{2 \cdot 4^2 - 5^2 \cdot 9 + 8^2 \cdot 7}$

**94.** $\dfrac{7^3 \cdot 9 - 6^2 \cdot 8 + 4^3 \cdot 6}{5^2 - 25}$

**95.** $\dfrac{(25 - 4^2)^3}{17^2 - 16^2} \cdot ((-6)^2 - 6^2)$

**96.** $\dfrac{(7 - 8)^{37}}{7^2 - 8^2} \cdot (98 - 7^2 \cdot 2)$

**97.** ▦ $\dfrac{19 - 17^2}{13^2 - 34}$

**98.** ▦ $\dfrac{195 + (-15)^3}{195 - 7 \cdot 5^2}$

**99.** ▦ $28^2 - 36^2/4^2 + 17^2$

**100.** ▦ $9^3 - 36^3/12^2 + 9^2$

**101.** ▦ Write down the keystrokes needed to calculate $\dfrac{15^2 - 5^3}{3^2 + 4^2}$.

**102.** ▦ Write down the keystrokes needed to calculate $\dfrac{16^2 - 24 \cdot 23}{3 \cdot 4 + 5^2}$.

**103.** Evaluate the expression for which the keystrokes are as follows: $\boxed{4}\ \boxed{-}\ \boxed{1}\ \boxed{0}\ \boxed{\div}\ \boxed{2}\ \boxed{+}\ \boxed{6}\ \boxed{=}$.

**104.** Evaluate the expression for which the keystrokes are $\boxed{4}\ \boxed{-}\ \boxed{1}\ \boxed{6}\ \boxed{\div}\ \boxed{(}\ \boxed{2}\ \boxed{+}\ \boxed{6}\ \boxed{)}\ \boxed{=}$.

Determine the sign of each expression if $m$ is negative and $n$ is positive.

**105.** $\dfrac{-n}{m}$

**106.** $\dfrac{-n}{-m}$

**107.** $-\left(\dfrac{-n}{m}\right)$

**108.** $-\left(\dfrac{n}{-m}\right)$

**109.** $-\left(\dfrac{-n}{-m}\right)$

# Mid-Chapter Review

## Concept Reinforcement

Determine whether each statement is true or false.

_____ **1.** Every integer is either positive or negative. [2.1a]

_____ **2.** If $a > b$, then $a$ lies to the left of $b$ on the number line. [2.1b]

_____ **3.** The absolute value of $a$ number is always nonnegative. [2.1c]

## Guided Solutions

 Fill in each blank with the number that creates a correct statement or solution.

**4.** Evaluate $-x$ and $-(-x)$ when $x = -4$. [2.1d]

$-x = -(\boxed{\phantom{00}}) = \boxed{\phantom{00}}$ ;

$-(-x) = -(-(\boxed{\phantom{00}})) = -(\boxed{\phantom{00}}) = \boxed{\phantom{00}}$

Subtract. [2.3a]

**5.** $5 - 13 = 5 + (\boxed{\phantom{00}}) = \boxed{\phantom{00}}$

**6.** $-6 - (-7) = -6 + \boxed{\phantom{00}} = \boxed{\phantom{00}}$

## Mixed Review

**7.** State the integers that correspond to this situation.

Jerilyn deposited \$450 in her checking account. Later that week she used her debit card for a \$79 purchase. [2.1a]

**8.** Change the sign of 9. [2.1d]

Use either $<$ or $>$ for $\boxed{\phantom{0}}$ to write a true sentence. [2.1b]

**9.** $-6 \,\boxed{\phantom{0}}\, 6$

**10.** $-5 \,\boxed{\phantom{0}}\, -3$

**11.** $-10 \,\boxed{\phantom{0}}\, 0$

**12.** $-20 \,\boxed{\phantom{0}}\, -30$

Find the absolute value. [2.1c]

**13.** $|38|$

**14.** $|-18|$

**15.** $|0|$

**16.** $|-12|$

Find the opposite, or additive inverse, of the number. [2.1d]

**17.** $-56$

**18.** $3$

**19.** $0$

**20.** $-49$

**21.** Find $-x$ when $x$ is $-19$. [2.1d]

**22.** Evaluate $-(-x)$ when $x$ is 23. [2.1d]

Compute and simplify.   [2.2a], [2.3a], [2.4a, b], [2.5a, b]

**23.** $7 + (-9)$

**24.** $-6 + (-10)$

**25.** $36 + (-36)$

**26.** $-8 + (-9)$

**27.** $-9 + 10$

**28.** $19 + (-17)$

**29.** $2 - 28$

**30.** $-8 - (-4)$

**31.** $-3 - 10$

**32.** $5 - (-11)$

**33.** $0 - (-6)$

**34.** $12 - 24$

**35.** $-12 \cdot 3$

**36.** $6(-9)$

**37.** $(-13)(-2)$

**38.** $(-2)(-41)$

**39.** $(-9)^2$

**40.** $-9^2$

**41.** $-75 \div (-3)$

**42.** $-20 \div 4$

**43.** $17 - (-25) + 15 - (-18)$

**44.** $-9 + (-3) + 16 - (-10)$

**45.** $(-7)(-2)(-1)(-3)$

**46.** $3 - 6 \cdot 5 - 11$

**47.** $-5^2 + 6[1 - (3 - 4)]$

**48.** $\dfrac{6^2 - 3(5 - 9)}{7^2 - (-5)^2}$

Solve.   [2.3b]

**49.** *Temperature Change.*   In a chemistry lab, Ben works with a substance whose initial temperature is 25°C. During an experiment, the temperature falls to −8°C. Find the difference between the two temperatures.

**50.** *Stock Price Change.*   The price of a stock opened at $56. During the day, it dropped $6, then rose $2, and dropped $8. Find the value of the stock at the end of the day.

## Understanding Through Discussion and Writing

**51.** A student states "−45 is bigger than −21." What mistake do you think the student is making?   [2.1b]

**52.** Is subtraction of integers associative? Why or why not?   [2.3a]

**53.** Explain in your own words why the sum of two negative numbers is always negative.   [2.2a]

**54.** If a negative number is subtracted from a positive number, will the result always be positive? Why or why not?   [2.3a]

**STUDYING FOR SUCCESS**   *Your Textbook as a Resource*

☐   Study any drawings. Observe details in any sketches or graphs that accompany the explanations.

☐   Note the careful use of color to indicate substitutions and highlight steps in a multi-step solution.

## 2.6

### OBJECTIVES

**a**  Evaluate an algebraic expression by substitution.

**b**  Use the distributive law to find equivalent expressions.

# Introduction to Algebra and Expressions

In this section, we will write *equivalent expressions* by making use of the *distributive law*.

## a   EVALUATING ALGEBRAIC EXPRESSIONS

In arithmetic, we work with expressions such as

$$37 + 86, \qquad 7 \cdot 8, \qquad 19 - 7, \quad \text{and} \quad \frac{3}{8}.$$

In algebra, we can use both numbers and letters and work with *algebraic expressions* such as

$$x + 86, \qquad 7 \cdot t, \qquad 19 - y, \quad \text{and} \quad \frac{a}{b}.$$

When a letter can stand for various numbers, we call the letter a **variable**. A number or a letter that stands for just one number is called a **constant**. Let $b =$ your birth year. Then $b$ is a constant. Let $a =$ your age. Then $a$ is a variable since the value of $a$ changes from year to year.

An **algebraic expression** consists of variables, constants, numerals, and operation signs. When we replace a variable with a number, we say that we are **substituting** for the variable. Carrying out the **operations** of addition, subtraction, and so on, is called **evaluating the expression**.

**EXAMPLE 1**   Evaluate $x + y$ for $x = 37$ and $y = 29$.

We substitute 37 for $x$ and 29 for $y$ and carry out the addition:

$$x + y = 37 + 29 = 66.$$

The number 66 is called the **value** of the expression.   ■

Algebraic expressions involving multiplication can be written in several ways. For example, "8 times $a$" can be written as $8 \times a$, $8 \cdot a$, $8(a)$, or simply $8a$. Two letters written together without an operation symbol, such as $ab$, also indicate multiplication.

**1.** Evaluate $a + b$ for $a = 38$ and $b = 26$.

**2.** Evaluate $x - y$ for $x = 57$ and $y = 29$.

**3.** Evaluate $5t$ for $t = -14$.

**EXAMPLE 2**   Evaluate $3y$ for $y = -14$.

$$3y = 3(-14) = -42 \qquad \text{Parentheses are required here.}$$

◀ **Do Exercises 1–3.**

Algebraic expressions involving division can also be written several ways.

For example, "8 divided by $t$" can be written as $8 \div t$, $8/t$, or $\dfrac{8}{t}$.

*Answers*

**1.** 64  **2.** 28  **3.** −70

**EXAMPLE 3** Evaluate $\dfrac{a}{b}$ and $\dfrac{-a}{-b}$ for $a = 35$ and $b = 7$.

We substitute 35 for $a$ and 7 for $b$:

$$\frac{a}{b} = \frac{35}{7} = 5; \qquad \frac{-a}{-b} = \frac{-35}{-7} = 5.$$

**EXAMPLE 4** Evaluate $-\dfrac{a}{b}, \dfrac{-a}{b}$, and $\dfrac{a}{-b}$ for $a = 15$ and $b = 3$.

We substitute 15 for $a$ and 3 for $b$:

$$-\frac{a}{b} = -\frac{15}{3} = -5; \qquad \frac{-a}{b} = \frac{-15}{3} = -5; \qquad \frac{a}{-b} = \frac{15}{-3} = -5.$$

Examples 3 and 4 illustrate the following.

---

$\dfrac{-a}{-b}$ and $\dfrac{a}{b}$ represent the same number.

$-\dfrac{a}{b}, \dfrac{-a}{b}$, and $\dfrac{a}{-b}$ all represent the same number.

---

**Do Exercises 4–7.** ▶

**EXAMPLE 5** Evaluate $\dfrac{9C}{5} + 32$ for $C = 20$.

This expression can be used to find the Fahrenheit temperature that corresponds to 20 degrees Celsius.

$$\frac{9C}{5} + 32 = \frac{9 \cdot 20}{5} + 32 = \frac{180}{5} + 32 = 36 + 32 = 68.$$

**Do Exercise 8.** ▶

**EXAMPLE 6** Evaluate $5x^2$ for $x = 3$ and $x = -3$.

The rules for order of operations specify that the replacement for $x$ be squared. That result is then multiplied by 5.

$$\left.\begin{array}{l} 5x^2 = 5(3)^2 = 5(9) = 45; \\ 5x^2 = 5(-3)^2 = 5(9) = 45 \end{array}\right\}$$

You can always use parentheses when substituting. They are usually necessary when substituting a negative number.

**Do Exercises 9 and 10.** ▶

**EXAMPLE 7** Evaluate $(-x)^2$ and $-x^2$ for $x = 7$.

$$(-x)^2 = (-7)^2 = (-7)(-7) = 49 \qquad \text{Substitute 7 for } x. \text{ Then evaluate the power.}$$

To evaluate $-x^2$, we again substitute 7 for $x$. We must recall that taking the opposite of a number is the same as multiplying that number by $-1$.

$$\begin{aligned} -x^2 &= -1 \cdot x^2 & &-a = -1 \cdot a \\ -7^2 &= -1 \cdot 7^2 & &\text{Substituting 7 for } x \\ &= -1 \cdot 49 = -49 & &\text{Using the rules for order of operations} \end{aligned}$$

**Do Exercises 11–13.** ▶

---

For each expression, find two equivalent expressions with the negative sign in different places.

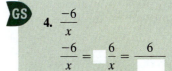

**GS** **4.** $\dfrac{-6}{x}$

$$\frac{-6}{x} = \frac{\boxed{\phantom{6}}\,6}{x} = \frac{6}{\boxed{\phantom{x}}}$$

**5.** $-\dfrac{m}{n}$

**6.** $\dfrac{r}{-4}$

**7.** Evaluate $\dfrac{a}{-b}, \dfrac{-a}{b}$, and $-\dfrac{a}{b}$ for $a = 28$ and $b = 4$.

**8.** Find the Fahrenheit temperature that corresponds to 10 degrees Celsius. (See Example 5.)

**9.** Evaluate $3x^2$ for $x = 4$ and $x = -4$.

**10.** Evaluate $a^4$ for $a = 3$ and $a = -3$.

**11.** Evaluate $(-x)^2$ and $-x^2$ for $x = 3$.

**12.** Evaluate $(-x)^2$ and $-x^2$ for $x = 2$.

**13.** Evaluate $x^5$ for $x = 2$ and $x = -2$.

***Answers***
**4.** $-\dfrac{6}{x}; \dfrac{6}{-x}$  **5.** $\dfrac{-m}{n}; \dfrac{m}{-n}$  **6.** $\dfrac{-r}{4}; -\dfrac{r}{4}$
**7.** $-7; -7; -7$  **8.** 50  **9.** 48; 48
**10.** 81; 81  **11.** 9; $-9$  **12.** 4; $-4$
**13.** 32; $-32$

***Guided Solution:***
**4.** $-, -x$

Examples 6 and 7 illustrate the following.

$$(-x)^2 = (x)^2$$
$$(-x)^2 \neq -x^2$$

## b EQUIVALENT EXPRESSIONS AND THE DISTRIBUTIVE LAW

Some pairs of algebraic expressions are *equivalent*.

**EXAMPLE 8**   Evaluate $x + x$ and $2x$ for $x = 3$ and $x = -5$.

We substitute 3 for $x$ in $x + x$ and again in $2x$:

$$x + x = 3 + 3 = 6; \qquad 2x = 2 \cdot 3 = 6.$$

Next we repeat the procedure, substituting $-5$ for $x$:

$$x + x = -5 + (-5) = -10; \qquad 2x = 2(-5) = -10.$$

The results can be shown in a table. It appears that $x + x$ and $2x$ represent the same number.

|        | $x + x$ | $2x$ |
|--------|---------|------|
| $x = 3$  | 6       | 6    |
| $x = -5$ | $-10$   | $-10$ |

◀ **Do Exercises 14 and 15.**

Example 8 suggests that $x + x$ and $2x$ represent the same number for any replacement of $x$. This is in fact true, so we can say that $x + x$ and $2x$ are **equivalent expressions**.

### EQUIVALENT EXPRESSIONS

Two expressions that have the same value for all allowable replacements are called **equivalent**.

In Examples 3 and 7 we saw that the expressions $\dfrac{-a}{-b}$ and $\dfrac{a}{b}$ are equivalent but that the expressions $(-x)^2$ and $-x^2$ are *not* equivalent.

An important concept, known as the **distributive law**, is useful for finding equivalent algebraic expressions. The distributive law involves two operations: multiplication and either addition or subtraction.

To understand how the distributive law works, consider the following:

$$
\begin{array}{r}
4\,5 \\
\times\ \ 7 \\
\hline
3\,5 \\
2\,8\,0 \\
\hline
3\,1\,5
\end{array}
$$

← This is $7 \cdot 5$.
← This is $7 \cdot 40$.
← This is the sum $7 \cdot 40 + 7 \cdot 5$.

To carry out the multiplication, we actually added two products. That is,

$$7 \cdot 45 = 7(40 + 5) = 7 \cdot 40 + 7 \cdot 5.$$

The distributive law says that if we want to multiply a sum of several numbers by a number, either we can add within the grouping symbols and then multiply or we can multiply each of the terms separately and then add.

---

Complete each table by evaluating each expression for the given values.

**14.**

|        | $3x + 2x$ | $5x$ |
|--------|-----------|------|
| $x = 4$  |           |      |
| $x = -2$ |           |      |
| $x = 0$  |           |      |

**15.**

|        | $4x - x$ | $3x$ |
|--------|----------|------|
| $x = 2$  |          |      |
| $x = -2$ |          |      |
| $x = 0$  |          |      |

**CALCULATOR CORNER**

*Evaluating Powers*   To evaluate an expression like $-x^3$ for $x = -14$ with a calculator, we must keep in mind the rules for order of operations. On some calculators, this expression is evaluated by pressing $\boxed{1}\,\boxed{4}\,\boxed{+/-}\,\boxed{x^y}$ $\boxed{3}\,\boxed{=}\,\boxed{+/-}$. Other calculators use the keystrokes $\boxed{(-)}\,\boxed{(}\,\boxed{(-)}$ $\boxed{1}\,\boxed{4}\,\boxed{)}\,\boxed{\wedge}\,\boxed{3}\,\boxed{\text{ENTER}}$. The result should be 2744. Consult your owner's manual or an instructor, or simply experiment if your calculator behaves differently.

**EXERCISES:**   Evaluate.

1. $-a^5$ for $a = -3$
2. $-x^5$ for $x = -4$
3. $-x^5$ for $x = 2$
4. $-x^5$ for $x = 5$

---

**Answers**

**14.** 20, 20; $-10, -10$; 0, 0
**15.** 6, 6; $-6, -6$; 0, 0

## THE DISTRIBUTIVE LAW

For any numbers $a$, $b$, and $c$,

$$a(b + c) = ab + ac.$$

**EXAMPLE 9**  Evaluate $a(b + c)$ and $ab + ac$ for $a = 3$, $b = 4$, and $c = 2$.

We have

$$a(b + c) = 3(4 + 2) = 3 \cdot 6 = 18$$

and    $ab + ac = 3 \cdot 4 + 3 \cdot 2 = 12 + 6 = 18.$

The parentheses in the statement of the distributive law tell us to multiply both $b$ and $c$ by $a$. Without the parentheses, we would have $ab + c$.

**EXAMPLE 10**  Use the distributive law to write an expression equivalent to $2(l + w)$.

$2(l + w) = 2 \cdot l + 2 \cdot w$    Note that the $+$ sign between $l$ and $w$ now appears between $2 \cdot l$ and $2 \cdot w$.

$\qquad = 2l + 2w.$    Try to go directly to this step.

**Do Exercises 16 and 17.** ▶

Since subtraction can be regarded as addition of the opposite, it follows that the distributive law is true for subtraction as well as addition.

**EXAMPLE 11**  Use the distributive law to write an expression equivalent to each of the following:

**a)** $9(x - 5)$;  **b)** $(a - 7)b$;  **c)** $-4(x - 2y + 3z)$;  **d)** $-(2x - 3y)$.

**a)** $9(x - 5) = 9x - 9(5)$

$\qquad\qquad = 9x - 45$    Try to go directly to this step.

**b)** $(a - 7)b = b(a - 7)$    Using a commutative law

$\qquad\qquad = b \cdot a - b \cdot 7$    Using the distributive law

$\qquad\qquad = ab - 7b$    Using a commutative law to write $ba$ alphabetically and $b \cdot 7$ with the constant first

**c)** $-4(x - 2y + 3z) = -4 \cdot x - (-4)(2y) + (-4)(3z)$    Using the distributive law

$\qquad\qquad = -4x - (-4 \cdot 2)y + (-4 \cdot 3)z$    Using an associative law (twice)

$\qquad\qquad = -4x - (-8y) + (-12z)$

$\qquad\qquad = -4x + 8y - 12z$

**d)** $-(2x - 3y) = -1(2x - 3y)$    Finding the opposite of a number is the same as multiplying by $-1$.

$\qquad\qquad = -1 \cdot 2x - (-1)(3y)$    Using the distributive law

$\qquad\qquad = -2x - (-3y)$    Using an associative law (twice)

$\qquad\qquad = -2x + 3y$

**Do Exercises 18–22.** ▶

---

Use the distributive law to write an equivalent expression.

**16.** $5(a + b)$

**GS 17.** $6(x + y + z)$
$= 6 \cdot \square + 6 \cdot \square + 6 \cdot \square$
$= 6x + \square + 6z$

Use the distributive law to write an equivalent expression.

**18.** $4(x - y)$

**19.** $3(a - b + c)$

**20.** $(m - 4)6$

**21.** $-8(2a - b + 3c)$

**22.** $-(c - 8d)$

***Answers***

**16.** $5a + 5b$  **17.** $6x + 6y + 6z$
**18.** $4x - 4y$  **19.** $3a - 3b + 3c$
**20.** $6m - 24$  **21.** $-16a + 8b - 24c$
**22.** $-c + 8d$

***Guided Solution:***
**17.** $x, y, z, 6y$

## ✔ Check Your Understanding

**Reading Check**   Classify each algebraic expression as involving either multiplication or division.

**RC1.** $3/q$ _____

**RC2.** $3q$ _____

**RC3.** $3 \cdot q$ _____

**RC4.** $\dfrac{3}{q}$ _____

**Concept Check**   Substitute the given values for the variables in each expression. Do not evaluate.

**CC1.** $3 - 2x; x = 5$

**CC2.** $3(a + b); a = 4, b = -7$

**CC3.** $\dfrac{5m}{n}; m = -4, n = 2$

**CC4.** $-x^2; x = -3$

---

**a**   Evaluate.

**1.** $10n$, for $n = 2$
(The cost, in dollars, of buying 2 appetizers)

**2.** $99n$, for $n = 2$
(The cost, in cents, of downloading 2 songs)

**3.** $\dfrac{x}{y}$, for $x = 6$ and $y = -3$

**4.** $\dfrac{m}{n}$, for $m = 18$ and $n = 2$

**5.** $\dfrac{2d}{c}$, for $c = 6$ and $d = 3$

**6.** $\dfrac{5y}{z}$, for $y = 15$ and $z = -25$

**7.** $\dfrac{72}{r}$, for $r = 4$
(The approximate doubling time, in years, for an investment earning 4% interest per year)

**8.** $\dfrac{72}{i}$, for $i = 2$
(The approximate doubling time, in years, for an investment earning 2% interest per year)

**9.** $3 - 5 \cdot x$, for $x = 2$

**10.** $9 - 2 \cdot x$, for $x = 5$

**11.** $2l + 2w$, for $l = 3$ and $w = 4$
(The perimeter, in feet, of a 3-ft by 4-ft rectangle)

**12.** $3(a + b)$, for $a = 2$ and $b = 4$

**13.** $2(l + w)$, for $l = 3$ and $w = 4$
(The perimeter, in feet, of a 3-ft by 4-ft rectangle)

**14.** $3a + 3b$, for $a = 2$ and $b = 4$

**15.** $7a - 7b$, for $a = -1$ and $b = 2$

**16.** $4x - 4y$, for $x = -5$ and $y = 1$

**17.** $7(a - b)$, for $a = -1$ and $b = 2$

**18.** $4(x - y)$, for $x = -5$ and $y = 1$

**19.** $16t^2$, for $t = 5$
(The distance, in feet, that an object falls in 5 sec)

**20.** $\dfrac{49t^2}{10}$, for $t = 10$
(The distance, in meters, that an object falls in 10 sec)

**21.** $a + (b - a)^2$, for $a = 6$ and $b = 10$

**22.** $(x + y)^2$, for $x = 2$ and $y = 10$

**23.** $9a + 9b$, for $a = 13$ and $b = -13$

**24.** $8x + 8y$, for $x = 17$ and $y = -17$

**25.** $\dfrac{n^2 - n}{2}$, for $n = 9$
(For determining the number of handshakes possible among 9 people)

**26.** $\dfrac{5(F - 32)}{9}$, for $F = 50$
(For converting 50 degrees Fahrenheit to degrees Celsius)

**27.** $1 - x^2$, for $x = -2$

**28.** $4 - y^2$, for $y = -1$

**29.** $m^2 - n^2$, for $m = 6$ and $n = 5$

**30.** $a^2 + b^2$, for $a = 3$ and $b = 4$

**31.** $a^3 - a^2$, for $a = -10$

**32.** $x^2 - x^3$, for $x = -10$

For each expression, write two equivalent expressions with the negative sign in different places.

**33.** $-\dfrac{5}{t}$

**34.** $\dfrac{7}{-x}$

**35.** $\dfrac{-n}{b}$

**36.** $-\dfrac{3}{r}$

**37.** $\dfrac{9}{-p}$

**38.** $\dfrac{-u}{5}$

**39.** $\dfrac{-14}{w}$

**40.** $\dfrac{-23}{m}$

Evaluate $\dfrac{-a}{b}, \dfrac{a}{-b}$, and $-\dfrac{a}{b}$ for the given values.

**41.** $a = 45, b = 9$

**42.** $a = 40, b = 2$

**43.** $a = 81, b = 3$

**44.** $a = 56, b = 7$

Evaluate.

**45.** $(-3x)^2$ and $-3x^2$, for $x = 2$

**46.** $(-2x)^2$ and $-2x^2$, for $x = 3$

**47.** $5x^2$, for $x = 3$ and $x = -3$

**48.** $2x^2$, for $x = 5$ and $x = -5$

**49.** $x^3$, for $x = 6$ and $x = -6$

**50.** $x^6$, for $x = 2$ and $x = -2$

**51.** $x^8$, for $x = 1$ and $x = -1$

**52.** $x^5$, for $x = 3$ and $x = -3$

**53.** $a^5$, for $a = 2$ and $a = -2$

**54.** $a^7$, for $a = 1$ and $a = -1$

**b**   Use the distributive law to write an equivalent expression.

**55.** $5(a + b)$

**56.** $7(x + y)$

**57.** $4(x + 1)$

**58.** $6(a + 1)$

**59.** $2(b + 5)$

**60.** $3(x - 6)$

**61.** $7(1 - t)$

**62.** $4(1 - y)$

**63.** $6(5x - 2)$

**64.** $9(6m - 7)$

**65.** $8(x + 7 + 6y)$

**66.** $4(5x + 8 + 3p)$

**67.** $-7(y - 2)$

**68.** $-9(y - 7)$

**69.** $(x + 2)3$

**70.** $(x + 4)2$

**71.** $-4(x - 3y - 2z)$

**72.** $8(2x - 5y - 8z)$

**73.** $8(a - 3b + c)$

**74.** $-6(a + 2b - c)$

**75.** $4(x - 3y - 7z)$

**76.** $5(9x - y + 8z)$

**77.** $(4a - 5b + c - 2d)5$

**78.** $(9a - 4b + 3c - d)7$

**79.** $-1(3m + 2n)$

**80.** $-1(6a + 7b)$

**81.** $-1(2a - 3b + 4)$

**82.** $-1(7x - 8y - 9)$

**83.** $-(x - y - z)$

**84.** $-(a - b - c)$

## Skill Maintenance

**85.** Write a word name for 23,043,921.   [1.1c]

**86.** Multiply: $17 \cdot 53$.   [1.4a]

**87.** Estimate by rounding to the nearest ten. Show your work.   [1.6b]

$$\begin{array}{r} 5\ 2\ 8\ 3 \\ -\ 2\ 4\ 7\ 5 \\ \hline \end{array}$$

**88.** Divide: $2982 \div 3$.   [1.5a]

**89.** On March 9, it snowed 12 in., but on March 10, the sun melted 7 in. How much snow remained?   [1.8a]

**90.** For Tania's graduation party, her husband ordered three buckets of chicken wings at \$12 apiece and 3 trays of nachos at \$9 a tray. How much did he pay for the wings and nachos?   [1.8a]

## Synthesis

**91.** A car's catalytic converter works most efficiently after it is heated to about 370°C. To what Fahrenheit temperature does this correspond? (*Hint:* See Example 5.)

**92.** Evaluate $\dfrac{9C}{5} + 32$ for $C = 10$ and for $C = 20$.

(See Example 5 and Margin Exercise 8.) When the Celsius temperature is doubled, is the corresponding Fahrenheit temperature also doubled?

Evaluate.

**93.** ▦ $a - b^3 + 17a$, for $a = 19$ and $b = -16$

**94.** ▦ $x^2 - 23y + y^3$, for $x = 18$ and $y = -21$

**95.** ▦ $r^3 + r^2t - rt^2$, for $r = -9$ and $t = 7$

**96.** ▦ $a^3b - a^2b^2 + ab^3$, for $a = -8$ and $b = -6$

**97.** $a^{1996} - a^{1997}$, for $a = -1$

**98.** $x^{1493} - x^{1492}$, for $x = -1$

**99.** $(m^3 - mn)^m$, for $m = 4$ and $n = 6$

**100.** $5a^{3a-4}$, for $a = 2$

Replace each blank with $\boxed{+}$, $\boxed{-}$, $\boxed{\times}$, or $\boxed{\div}$ to make each statement true.

**101.** ▦ $-32 \ \square \ (88 \ \square \ 29) = \ -1888$

**102.** ▦ $59 \ \square \ 17 \ \square \ 59 \ \square \ 8 = 1475$

Classify each statement as true or false. If false, write an example showing why.

**103.** For any choice of $x$, $x^2 = (-x)^2$.

**104.** For any choice of $x$, $x^3 = -x^3$.

**105.** For any choice of $x$, $x^6 + x^4 = (-x)^6 + (-x)^4$.

**106.** For any choice of $x$, $(-3x)^2 = 9x^2$.

## 2.7

### OBJECTIVES

a  Combine like terms.

b  Determine the perimeter of a polygon.

# Like Terms and Perimeter

One common way in which equivalent expressions are formed is by *combining like terms*.

## a  COMBINING LIKE TERMS

A **term** is a number, a variable, a product of numbers and/or variables, or a quotient of numbers and/or variables. Terms are separated by addition signs. If there are subtraction signs, we can find an equivalent expression that uses addition signs.

**EXAMPLE 1**   What are the terms of $3xy - 4y + \dfrac{2}{z}$?

$$3xy - 4y + \dfrac{2}{z} = 3xy + (-4y) + \dfrac{2}{z} \quad \text{Separating parts with } + \text{ signs}$$

The terms are $3xy$, $-4y$, and $\dfrac{2}{z}$.

What are the terms of each expression?

**1.** $5x - 4y + 3$

**2.** $-4y - 2x + \dfrac{x}{y}$

◀ Do Exercises 1 and 2.

Terms in which the variable factors are exactly the same, such as $9x$ and $-4x$, are called **like**, or **similar**, **terms**. For example, $3y^2$ and $7y^2$ are like terms, but $5x$ and $6x^2$ are not. Constants, like 7 and 3, are also like terms.

**EXAMPLES**   Identify the like terms.

**2.** $7x + 5x^2 + 2x + 8 + 5x^3 + 1$

$7x$ and $2x$ are like terms;   8 and 1 are like terms.

**3.** $5ab + a^3 - a^2b - 2ab + 7a^3$

$5ab$ and $-2ab$ are like terms;   $a^3$ and $7a^3$ are like terms.

Identify the like terms.

**3.** $9a^3 + 4ab + a^3 + 3ab + 7$

**4.** $3xy - 5x^2 + y^2 - 4xy + y$

◀ Do Exercises 3 and 4.

When an algebraic expression contains like terms, an equivalent expression can be formed by **combining**, or **collecting**, **like terms**. To combine like terms, we use the distributive law.

**EXAMPLE 4**   Combine like terms to form an equivalent expression.

**a)** $4x + 3x$

**b)** $6mn - 7mn$

**c)** $7y - 2 - 6y + 5$

**d)** $2a^5 + 9ab + 3 + a^5 - 7 - 4ab$

**a)** $4x + 3x = (4 + 3)x$    Using the distributive law (in "reverse")

$\qquad\qquad = 7x$          We usually go directly to this step.

**b)** $6mn - 7mn = (6 - 7)mn$    Try to do this mentally.

$\qquad\qquad\qquad = -1mn$, or simply $-mn$

**c)** $7y - 2 - 6y + 5 = 7y + (-2) + (-6y) + 5$   Rewriting as addition

$\qquad\qquad\qquad\qquad = 7y + (-6y) + (-2) + 5$   Using a commutative law

$\qquad\qquad\qquad\qquad = 1y + 3$, or simply $y + 3$

**Answers**

**1.** $5x$; $-4y$; 3   **2.** $-4y$; $-2x$; $\dfrac{x}{y}$

**3.** $9a^3$ and $a^3$; $4ab$ and $3ab$   **4.** $3xy$ and $-4xy$

**d)** $2a^5 + 9ab + 3 + a^5 - 7 - 4ab$

$= 2a^5 + 9ab + 3 + a^5 + (-7) + (-4ab)$

$= 2a^5 + a^5 + 9ab + (-4ab) + 3 + (-7)$     Rearranging terms

$= 3a^5 + 5ab + (-4)$     Think of $a^5$ as $1a^5$; $2a^5 + a^5 = 3a^5$

$= 3a^5 + 5ab - 4$

Do Exercises 5–7. ▶

## b   PERIMETER

---

### PERIMETER OF A POLYGON

A **polygon** is a closed geometric figure with three or more sides. The **perimeter** of a polygon is the distance around it, or the sum of the lengths of its sides.

---

**EXAMPLE 5**   Find the perimeter of this polygon.

> A polygon with five sides is called a *pentagon*.

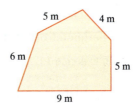

We add the lengths of all sides. Since all the units are the same, we are effectively combining like terms.

$\text{Perimeter} = 6\,\text{m} + 5\,\text{m} + 4\,\text{m} + 5\,\text{m} + 9\,\text{m}$

$= (6 + 5 + 4 + 5 + 9)\,\text{m}$     Using the distributive law

$= 29\,\text{m}$     Try to go directly to this step.

∙∙∙∙∙∙∙∙∙∙∙∙∙∙∙∙∙∙∙∙∙∙∙∙∙∙∙∙∙ **Caution!** ∙∙∙∙∙∙∙∙∙∙∙∙∙∙∙∙∙∙∙∙∙∙∙∙∙∙∙∙∙

When units of measurement are given in the statement of a problem, as in Example 5, the solution should also contain units of measurement.

∙∙∙∙∙∙∙∙∙∙∙∙∙∙∙∙∙∙∙∙∙∙∙∙∙∙∙∙∙∙∙∙∙∙∙∙∙∙∙∙∙∙∙∙∙∙∙∙∙∙∙∙∙∙∙∙∙∙∙∙∙∙∙∙∙∙∙∙∙∙∙∙∙∙∙∙∙∙∙∙∙∙∙∙∙∙∙∙∙∙∙

Do Exercises 8 and 9. ▶

A **rectangle** is a polygon with four sides and four 90° angles. Opposite sides of a rectangle have the same measure. The symbol ⌐ or ¬ indicates a 90° angle. A 90° angle is often referred to as a **right angle**.

**EXAMPLE 6**   Find the perimeter of a rectangle that is 3 cm by 4 cm.

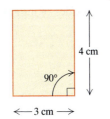

$\text{Perimeter} = 3\,\text{cm} + 3\,\text{cm} + 4\,\text{cm} + 4\,\text{cm}$

$= (3 + 3 + 4 + 4)\,\text{cm}$

$= 14\,\text{cm}$

Do Exercise 10. ▶

---

Combine like terms to form an equivalent expression.

**5.** $2a + 7a$

**6.** $5x^2 + 9 - 4x^2 + 3$

**GS** **7.** $4m - 2n^2 + 5 + n^2 + m - 9$

The like terms are

$4m$ and ☐,

$-2n^2$ and ☐,

$5$ and ☐.

$4m + (-2n^2) + 5 + n^2 + m + (-9)$

$= 4m + m + (-2n^2) + ☐$

$\qquad\qquad\qquad + 5 + (-9)$

$= 5m + (\;\;☐\;\;) + (-4)$

$= 5m - ☐ - 4$

Find the perimeter of each polygon.

**8.**

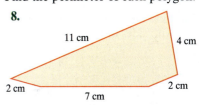

**9.**

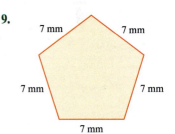

**10.** Find the perimeter of a rectangle that is 4 cm by 2 cm.

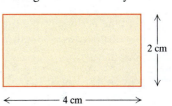

***Answers***
**5.** $9a$   **6.** $x^2 + 12$   **7.** $5m - n^2 - 4$
**8.** 26 cm   **9.** 35 mm   **10.** 12 cm
***Guided Solution:***
**7.** $m, n^2, -9, n^2, -n^2, n^2$

The perimeter of the rectangle in Example 6 is $2 \cdot 3 \text{ cm} + 2 \cdot 4 \text{ cm}$, or equivalently $2(3 \text{ cm} + 4 \text{ cm})$. This can be generalized, as follows.

---

### PERIMETER OF A RECTANGLE

The **perimeter $P$ of a rectangle** of length $l$ and width $w$ is given by

$$P = 2l + 2w, \quad \text{or} \quad P = 2 \cdot (l + w).$$

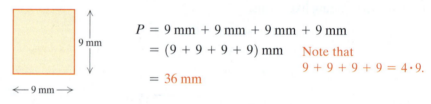

---

**EXAMPLE 7** A common door size is 7 ft by 3 ft. Find the perimeter of such a door.

$$\begin{aligned} P &= 2l + 2w & &\text{We could also use } P = 2(l + w). \\ &= 2 \cdot 7 \text{ ft} + 2 \cdot 3 \text{ ft} \\ &= (2 \cdot 7) \text{ ft} + (2 \cdot 3) \text{ ft} & &\text{Try to do this mentally.} \\ &= 14 \text{ ft} + 6 \text{ ft} \\ &= 20 \text{ ft} & &\text{Combining like terms} \end{aligned}$$

The perimeter of the door is 20 ft.

◀ **Do Exercise 11.**

**11.** Find the perimeter of a 4-ft by 8-ft sheet of plywood.

A **square** is a rectangle in which all sides have the same length.

**EXAMPLE 8** Find the perimeter of a square with sides of length 9 mm.

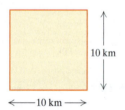

$$\begin{aligned} P &= 9 \text{ mm} + 9 \text{ mm} + 9 \text{ mm} + 9 \text{ mm} \\ &= (9 + 9 + 9 + 9) \text{ mm} & &\text{Note that} \\ & & &9 + 9 + 9 + 9 = 4 \cdot 9. \\ &= 36 \text{ mm} \end{aligned}$$

**12.** Find the perimeter of a square with sides of length 10 km.

◀ **Do Exercise 12.**

---

### PERIMETER OF A SQUARE

The **perimeter $P$ of a square** is four times $s$, the length of a side:

$$\begin{aligned} P &= s + s + s + s \\ &= 4s. \end{aligned}$$

---

**13.** Find the perimeter of a square book with sides of length 9 in.

$$\begin{aligned} P &= 4s \\ &= 4 \cdot \boxed{\phantom{0}} \text{ in.} \\ &= \boxed{\phantom{00}} \text{ in.} \end{aligned}$$

**EXAMPLE 9** Find the perimeter of a square room with sides of length 12 ft.

$$\begin{aligned} P &= 4s \\ &= 4 \cdot 12 \text{ ft} \\ &= 48 \text{ ft} \end{aligned}$$

The perimeter of the room is 48 ft.

◀ **Do Exercise 13.**

*Answers*
**11.** 24 ft   **12.** 40 km   **13.** 36 in.
*Guided Solution:*
**13.** 9, 36

# Translating for Success

1. **Wood Costs.** It costs $8 for the wood for each birdhouse that Annette builds. If she used $120 worth of wood, how many birdhouses did she build?

2. **Elevation.** Genine started hiking a 10-mi trail at an elevation that was 150 ft below sea level. At the end of the trail, she was 75 ft above sea level. How many feet higher was she at the end of the trail than at the beginning?

3. **Community Service.** In order to fulfill the requirements for a sociology class, Glen must log 120 hr of community service. So far, he has spent 75 hr volunteering at a youth center. How many more hours must he serve?

4. **Disaster Relief.** Each package that is prepared for a disaster relief effort contains 15 meal bars. How many packages can be filled from a donation of 750 bars?

5. **Perimeter.** A rectangular building lot is 75 ft wide and 150 ft long. What is the perimeter of the lot?

*The goal of these matching questions is to practice step (2), Translate, of the five-step problem-solving process. Translate each word problem to an equation and select a correct translation from equations A–O.*

**A.** $75 + x = 120$

**B.** $15 \div 750 = x$

**C.** $-10 - 15 = x$

**D.** $8 \cdot 120 = x$

**E.** $150 - 75 = x$

**F.** $15 \cdot 750 = x$

**G.** $75 - (-150) = x$

**H.** $2 \cdot 150 + 2 \cdot 75 = x$

**I.** $-10 - (-15) = x$

**J.** $8 \cdot x = 120$

**K.** $750 \div 15 = x$

**L.** $15 - (-10) = x$

**M.** $75 + 120 = x$

**N.** $75 - 150 = x$

**O.** $75 = 120 + x$

*Answers on page A-4*

6. **Account Balance.** Lorenzo had $75 in his checking account. He then wrote a check for $150. What was the balance in his account?

7. **Laptop Computers.** Great Graphics purchased a laptop computer for each of its 15 employees. If each laptop cost $750, how much did the computers cost?

8. **Basketball.** A basketball team scored 75 points in one game. In the next game, the team scored a record 120 points. How many points did the team score in the two games?

9. **Pizza Sales.** A youth club sold 120 pizzas for a fundraiser. If each pizza sold for $8, how much money was taken in?

10. **Temperature.** The temperature in Fairbanks was $-10°$ at 6:00 p.m. and fell another 15° by midnight. What was the temperature at midnight?

## ✓ Check Your Understanding

**Reading Check** Complete each statement with the correct word from the list on the right. A word may be used more than once or not at all.

**RC1.** A polygon is a(n) _____ figure with three or more sides.

**RC2.** The distance around a polygon is its _____.

**RC3.** The formula $P = 2l + 2w$ gives the _____ of a rectangle.

**RC4.** The perimeter of a(n) _____ is given by the formula $P = 4s$.

closed
open
perimeter
polygon
rectangle
square

**Concept Check** List each pair of like terms.

**CC1.** $5a + 5b + a$

**CC2.** $4 + x + 5x + 7$

**CC3.** $x - 6x - 3y + 2y$

**CC4.** $7a - 2ab + 3b + 4ab$

**CC5.** $2t + t^2 + 7t^3 - t^2$

**CC6.** $6xy^2 + 9x^2y - x^2y - 8xy^2$

**a** List the terms of each expression.

**1.** $2a + 5b - 7c$

**2.** $4x - 6y + 7z$

**3.** $mn - 6n + 8$

**4.** $7rs - s - 5$

**5.** $3x^2y - 4y^2 - 2z^3$

**6.** $4a^3b + ab^2 - 9b^3$

Combine like terms to form an equivalent expression.

**7.** $5x + 9x$

**8.** $9a + 7a$

**9.** $10a - 15a$

**10.** $-17x + x$

**11.** $2x + 6y + x$

**12.** $3t - y + 7t$

**13.** $27a + 70 - 40a - 8$

**14.** $42x - 6 - x + 2$

**15.** $9 + 5t + 7y - t - y - 13$

**16.** $8 - 4a + 9b + 5a - 3b - 15$

**17.** $a + 3b + 5a - 2 + b$

**18.** $x + 7y + 5 - 2y + 3x$

**19.** $-8 + 11a - 5b - 10a - 7b + 7$

**20.** $8x - 5x + 6 + 3y - y - 4$

**21.** $8x^2 + 3y - x^2$

**22.** $8y^3 - 3z + 4y^3$

**23.** $11x^4 + 2y^3 - 4x^4 - y^3$

**24.** $13a^5 + 9b^4 - 2a^5 - 4b^4$

**25.** $9a^2 - 4a + a - 3a^2$

**26.** $3a^2 + 7a^3 - a^2 + 5 + a^3$

**27.** $x^3 - 5x^2 + 2x^3 - 3x^2 + 4$

**28.** $9xy + 4y^2 - 2xy + 2y^2 - 1$

**29.** $9x^3y + 4xy^3 - 5xy^3 + 3xy$

**30.** $8a^2b - 3ab^2 - 7a^2b + 2ab$

**31.** $3a^6 - b^4 + 2a^6b^4 - 7a^6 - 2b^4$

**32.** $3x^4 - 2y^4 + 8x^4y^4 - 7x^4 + y^4$

**b**    Find the perimeter of each polygon.

**33.**

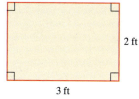

2 ft

3 ft

**34.**

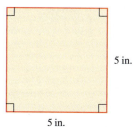

5 in.

5 in.

**35.**

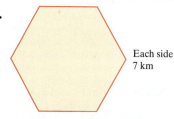

Each side
7 km

**36.**

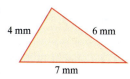

4 mm    6 mm

7 mm

**37.**

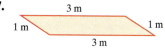

3 m

1 m        1 m

3 m

**38.**

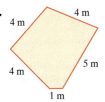

4 m    4 m

5 m

4 m

1 m

*Tennis Court.*   A tennis court contains many rectangles. Use the diagram of a regulation tennis court to calculate the perimeters in Exercises 39–42.

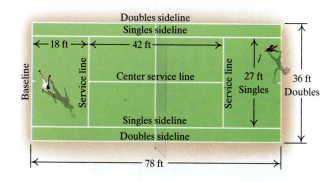

**39.** The perimeter of a singles court

**40.** The perimeter of a doubles court

**41.** The perimeter of the rectangle formed by the service lines and the singles sidelines

**42.** The perimeter of the rectangle formed by a service line, a baseline, and the singles sidelines

**43.** Find the perimeter of a rectangular 8-ft by 10-ft bedroom.

**44.** Find the perimeter of a rectangular 3-ft by 4-ft doghouse.

**45.** Find the perimeter of a checkerboard that is 14 in. on each side.

**46.** Find the perimeter of a square skylight that is 2 m on each side.

**47.** Find the perimeter of a square frame that is 65 cm on each side.

**48.** Find the perimeter of a square garden that is 12 yd on each side.

**49.** Find the perimeter of a 12-ft by 20-ft rectangular deck.

**50.** Find the perimeter of a 40-ft by 35-ft rectangular backyard.

## Skill Maintenance

**51.** A box of Shaw's Corn Flakes contains 510 grams (g) of corn flakes. A serving of corn flakes weighs 30 g. How many servings are in one box?   [1.8a]

**52.** Estimate the difference by rounding to the nearest ten.   [1.6b]

$$\begin{array}{r} 7\ 0\ 4 \\ -\ 4\ 8\ 6 \\ \hline \end{array}$$

Solve.   [1.7b]

**53.** $25 = t + 9$

**54.** $19 = x + 6$

**55.** $45 = 3x$

**56.** $2t = 50$

**57.** $25 + n = 400$

**58.** $25 \cdot n = 400$

## Synthesis

Simplify. (Multiply and then combine like terms.)

**59.** $5(x + 3) + 2(x - 7)$

**60.** $3(a - 7) + 7(a + 4)$

**61.** $2(3 - 4a) + 5(a - 7)$

**62.** $7(2 - 5x) + 3(x - 8)$

**63.** $-5(2 + 3x + 4y) + 7(2x - y)$

**64.** $3(4 - 2x) + 5(9x - 3y + 1)$

**65.** In order to save energy, Andrea plans to run a bead of caulk sealant around 3 exterior doors and 13 windows. Each window measures 3 ft by 4 ft, each door measures 3 ft by 7 ft, and there is no need to caulk the bottom of each door. If each cartridge of caulk seals 56 ft and costs $9, how much will it cost Andrea to seal the windows and doors?

**66.** Jacie is adding borders to small rectangular bulletin boards that are 5 ft by 7 ft and to larger bulletin boards that are 7 ft by 9 ft. If the border material costs $5 per yard, how much will Jacie spend to do 4 small bulletin boards and 3 large bulletin boards?

**67.** 🔲 A square wooden rack is used to store the 15 numbered pool balls as well as the cue ball. If a pool ball has a diameter of 57 mm, find the inside perimeter of the storage rack.

**68.** A rectangular box is used to store six Christmas ornaments. Find the perimeter of such a box if each ornament has a diameter of 72 mm.

# Solving Equations

A *solution* of an equation is a replacement for the variable that makes the equation true. To *solve* an equation means to find all of its solutions.

## a EQUIVALENT EQUATIONS

The solution of the equation $x = 10$ is $10$. The solution of the equation $x + 5 = 15$ is also $10$, since $10 + 5 = 15$ is true. Because their solutions are identical, $x = 10$ and $x + 5 = 15$ are said to be **equivalent equations**.

a Distinguish between equivalent expressions and equivalent equations.

b Use the addition principle to solve equations.

c Use the division principle to solve equations.

d Decide which principle should be used to solve an equation.

e Solve equations that require use of both the addition principle and the division principle.

> ### EQUIVALENT EQUATIONS
>
> Equations with the same solutions are called **equivalent equations**.

It is important to be able to distinguish between equivalent *expressions* and equivalent *equations*.

- $6a$ and $4a + 2a$ are equivalent *expressions* because, for any replacement of $a$, both expressions represent the same number.
- $3x = 15$ and $4x = 20$ are equivalent *equations* because any solution of one equation is also a solution of the other equation.

**EXAMPLE 1** Classify each pair as either equivalent equations or equivalent expressions:

**a)** $5x + 1$; $2x - 4 + 3x + 5$    **b)** $x = -7$; $x + 2 = -5$.

**a)** First note that these are expressions, not equations. To see if they are equivalent, we combine like terms in the second expression:

$$2x - 4 + 3x + 5 = (2 + 3)x + (-4 + 5) \quad \text{Regrouping and using the distributive law}$$

$$= 5x + 1.$$

We see that $2x - 4 + 3x + 5$ and $5x + 1$ are *equivalent expressions*.

**b)** Both $x = -7$ and $x + 2 = -5$ are equations. The solution of $x = -7$ is $-7$. We substitute to see if $-7$ is also the solution of $x + 2 = -5$:

$$x + 2 = -5$$
$$-7 + 2 = -5 \quad \text{TRUE}$$

Since $x = -7$ and $x + 2 = -5$ have the same solution, they are *equivalent equations*.

**Do Exercises 1 and 2.** ▶

Classify each pair as equivalent expressions or equivalent equations.

**1.** $a - 5 = -3$; $a = 2$

**2.** $a - 9 + 6a$; $7a - 9$

*Answers*

**1.** Equivalent equations
**2.** Equivalent expressions

## b THE ADDITION PRINCIPLE

We now begin to consider principles that enable us to start with one equation and create an equivalent equation similar to $x = 15$, for which the solution is more easily seen. One such principle, the *addition principle*, tells us that we can add the same number to both sides of an equation without changing the solutions of the equation.

---

**THE ADDITION PRINCIPLE**

For any numbers $a$, $b$, and $c$,

$$a = b \quad \text{is equivalent to} \quad a + c = b + c.$$

---

**EXAMPLE 2**   Solve: $x - 7 = -2$.

We have

$$
\begin{aligned}
x - 7 &= -2 \\
x - 7 + 7 &= -2 + 7 \quad &&\text{Using the addition principle:} \\
&&&\text{adding 7 to both sides} \\
x + 0 &= 5 \quad &&\text{Adding 7 "undoes" the subtraction of 7.} \\
x &= 5. \quad &&\text{This equation has the same solution as} \\
&&&x - 7 = -2.
\end{aligned}
$$

The solution appears to be 5. To check, we use the original equation.

Check:    $\dfrac{x - 7 = -2}{5 - 7\ ?\ -2}$

                     $-2$     |     TRUE

The solution is 5.

◀ **Do Exercises 3 and 4.**

We can subtract by adding the opposite of the number being subtracted. Because of this, the addition principle allows us to subtract the same number from both sides of an equation.

**EXAMPLE 3**   Solve: $23 = t + 7$.

We have

$$
\begin{aligned}
23 &= t + 7 \\
23 - 7 &= t + 7 - 7 \quad &&\text{Using the addition principle to add } -7 \\
&&&\text{or to subtract 7 on both sides} \\
16 &= t + 0 \quad &&\text{Subtracting 7 "undoes" the addition of 7.} \\
16 &= t. \quad &&\text{The solution of } 23 = t + 7 \text{ is also 16.}
\end{aligned}
$$

The solution is 16. The check is left to the student.

◀ **Do Exercises 5 and 6.**

Solve.

**3.** $x - 5 = 19$

**4.** $x - 9 = -12$

Solve.

**5.** $42 = x + 17$

**6.** $a + 8 = -6$

**Answers**

**3.** 24   **4.** $-3$   **5.** 25   **6.** $-14$

To visualize the addition principle, think of a jeweler's balance. When both sides of the balance hold equal amounts of weight, the balance is level. If weight is added or removed, equally, on both sides, the balance remains level.

## c THE DIVISION PRINCIPLE

We can solve $8n = 96$ by dividing both sides by 8:

$8 \cdot n = 96$

$\dfrac{8 \cdot n}{8} = \dfrac{96}{8}$    Dividing both sides by 8

$n = 12.$    8 times $n$, divided by 8, is $n$. $96 \div 8$ is 12.

Both $8n = 96$ and $n = 12$ have the solution 12. We can divide both sides of an equation by any nonzero number in order to find an equivalent equation.

---

**THE DIVISION PRINCIPLE**

For any numbers $a$, $b$, and $c$ ($c \neq 0$),

$a = b$   is equivalent to   $\dfrac{a}{c} = \dfrac{b}{c}$.

---

After we have discussed multiplication of fractions, we can use an equivalent form of this principle: the multiplication principle.

**EXAMPLE 4**   Solve: $9x = 63$.

We have

$9x = 63$

$\dfrac{9x}{9} = \dfrac{63}{9}$    Using the division principle to divide both sides by 9

$x = 7.$

Check:    $\dfrac{9x = 63}{9 \cdot 7 \;?\; 63}$    Checking in the original equation

$\phantom{9 \cdot 7 \;?\;} 63 \mid$    TRUE

The solution is 7.

**Do Exercises 7 and 8.** ▶

**SKILL REVIEW** *Solve equations like* $28 \cdot x = 168$. [1.7b]

Solve.
1. $17 \cdot y = 357$
2. $480 = 32 \cdot x$

**Answers: 1.** 21 **2.** 15

MyLab Math
VIDEO

Solve.

**7.** $7x = 42$

**8.** $-24 = 3t$

*Answers*

**7.** 6 **8.** $-8$

**EXAMPLE 5** Solve: $48 = -8n$.

To undo multiplication by $-8$, we use the division principle:

$$48 = -8n$$

$$\frac{48}{-8} = \frac{-8n}{-8} \qquad \text{Dividing both sides by } -8$$

$$-6 = n.$$

Check: 
$$\frac{48 = -8n}{48 \; ? \; -8(-6)}$$
$$48 \qquad \text{TRUE}$$

The solution is $-6$.

◀ **Do Exercises 9 and 10.**

Our goal in equation solving is to have the variable by itself on one side of the equation. In an equation like $-x = 7$, the variable is not by itself—there is a negative sign in front of $x$. We can solve an equation like $-x = 7$ by dividing or by multiplying both sides of the equation by $-1$.

**EXAMPLE 6** Solve: $-x = 7$.

One way to solve this equation is to note that $-x = -1 \cdot x$. Then we can divide both sides by $-1$:

$$-x = 7$$
$$-1 \cdot x = 7$$
$$\frac{-1 \cdot x}{-1} = \frac{7}{-1} \qquad \text{Using the division principle}$$
$$x = -7.$$

Check: 
$$\frac{-x = 7}{-(-7) \; ? \; 7} \qquad \text{Be sure to check in the original equation.}$$
$$7 \; | \; \text{TRUE}$$

The solution is $-7$.

Another way to solve the equation in Example 6 is to remember that when an expression is multiplied or divided by $-1$, its sign is changed. Here we multiply on both sides by $-1$ to change the sign of $-x$:

$$-x = 7$$
$$(-1)(-x) = (-1) \cdot 7 \qquad \text{Multiplying both sides by } -1$$
$$x = -7. \qquad \text{Note that } (-1)(-x) \text{ is the same as } (-1)(-1)x.$$

◀ **Do Exercises 11 and 12.**

**d  SELECTING THE CORRECT APPROACH**

You can determine which principle should be used to solve a particular equation by thinking of undoing an operation.

It is important to distinguish between addition of a negative number, as in $39 = -3 + t$, and multiplication by a negative number, as in $39 = -3t$. We undo addition by subtracting on both sides, and we undo multiplication by dividing on both sides.

---

Solve

**9.** $63 = -7n$

**10.** $-6x = 72$

Solve.

**11.** $-x = 23$

**12.** $-t = -3$

GS

$$\square \cdot t = -3$$
$$\frac{-1 \cdot t}{\square} = \frac{-3}{\square}$$
$$t = \square$$

*Answers*

**9.** $-9$  **10.** $-12$  **11.** $-23$  **12.** $3$

*Guided Solution:*

**12.** $-1, -1, -1, 3$

**EXAMPLE 7**  Solve: $39 = -3 + t$.

Note that $-3$ is added to $t$. To undo addition of $-3$, we subtract $-3$ or simply add 3 on both sides:

$$3 + 39 = 3 + (-3) + t \qquad \text{Using the addition principle}$$
$$42 = 0 + t$$
$$42 = t.$$

Check:
$$\begin{array}{c|c} 39 = -3 + t \\ \hline 39 \ ? \ -3 + 42 \\ \mid \quad 39 \qquad \text{TRUE} \end{array}$$

The solution is 42.

**EXAMPLE 8**  Solve: $39 = -3t$.

Here $t$ is multiplied by $-3$. To undo multiplication by $-3$, we divide by $-3$ on both sides:

$$39 = -3t$$
$$\frac{39}{-3} = \frac{-3t}{-3} \qquad \text{Using the division principle}$$
$$-13 = t.$$

Check:
$$\begin{array}{c|c} 39 = -3t \\ \hline 39 \ ? \ -3(-13) \\ \mid \quad 39 \qquad \text{TRUE} \end{array}$$

The solution is $-13$.

**Do Exercises 13–15.** ▶

### e   USING THE PRINCIPLES TOGETHER

Suppose we want to determine whether 7 is the solution of $5x - 8 = 27$. To check, we replace $x$ with 7 and simplify.

Check:
$$\begin{array}{c|c} 5x - 8 = 27 \\ \hline 5 \cdot 7 - 8 \ ? \ 27 \\ 35 - 8 \mid \\ 27 \mid \quad \text{TRUE} \end{array}$$

This shows that 7 is the solution.

**Do Exercises 16 and 17.** ▶

In the check above, note that the rules for order of operations require that we multiply before we subtract (or add). Thus, to evaluate $5x - 8$,

we *select* a value:     $x$

then *multiply* by 5:     $5x$

and then *subtract* 8:     $5x - 8$.

In Example 9, which follows, these steps are reversed to solve for $x$:

we *add* 8:     $5x - 8 + 8$

then *divide* by 5:     $\dfrac{5x}{5}$

and then *isolate* $x$:     $x$.

Solve.

**13.** $-2x = -52$

**14.** $-2 + x = -52$

**15.** $x \cdot 7 = -28$

**16.** Determine whether $-9$ is the solution of $7x + 8 = -55$.

**17.** Determine whether $-6$ is the solution of $4x + 3 = -25$.

In general, the *last* step performed when calculating is the *first* step to be reversed when finding a solution.

When $x$ is by itself on one side of an equation, we say that it is *isolated*. Our goal in the remaining examples in this section will be to isolate the term containing the variable (using the addition principle) and then to isolate the variable itself (using the division principle).

**EXAMPLE 9**  Solve: $5x - 8 = 27$.

We first note that the term containing $x$ is $5x$. To isolate $5x$, we add 8 on both sides:

$$5x - 8 = 27$$
$$5x - 8 + 8 = 27 + 8 \qquad \text{Using the addition principle}$$
$$5x + 0 = 35 \qquad \text{Try to do this step mentally.}$$
$$5x = 35. \qquad \text{We have isolated } 5x.$$

Next, we isolate $x$ by dividing by 5 on both sides:

$$5x = 35$$
$$\frac{5x}{5} = \frac{35}{5} \qquad \text{Using the division principle}$$
$$1x = 7 \qquad \text{Try to do this step mentally.}$$
$$x = 7. \qquad \text{We have isolated } x.$$

The check was performed on the previous page. The solution is 7.

◀ **Do Exercise 18.**

**18.** Solve: $2x - 9 = 43$.

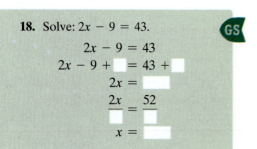

$$2x - 9 = 43$$
$$2x - 9 + \boxed{\phantom{0}} = 43 + \boxed{\phantom{0}}$$
$$2x = \boxed{\phantom{0}}$$
$$\frac{2x}{\boxed{\phantom{0}}} = \frac{52}{\boxed{\phantom{0}}}$$
$$x = \boxed{\phantom{0}}$$

**EXAMPLE 10**  Solve: $38 = -9t + 2$.

We first isolate $-9t$ by subtracting 2 on both sides:

$$38 = -9t + 2$$
$$38 - 2 = -9t + 2 - 2 \qquad \text{Subtracting 2 (or adding } -2 \text{) on both sides}$$
$$36 = -9t + 0 \qquad \text{Try to do this step mentally.}$$
$$36 = -9t.$$

Now that we have isolated $-9t$ on one side of the equation, we can divide by $-9$ to isolate $t$:

$$36 = -9t$$
$$\frac{36}{-9} = \frac{-9t}{-9} \qquad \text{Dividing both sides by } -9$$
$$-4 = t. \qquad \text{Simplifying}$$

Check:
$$38 = -9t + 2$$
$$38 \; ? \; -9 \cdot (-4) + 2$$
$$36 + 2$$
$$38 \qquad \text{TRUE}$$

The solution is $-4$.

◀ **Do Exercise 19.**

**19.** Solve: $-3n + 2 = 47$.

---

*Answers*

**18.** 26  **19.** $-15$

*Guided Solution:*

**18.** 9, 9, 52, 2, 2, 26

✓ **Check Your Understanding**

**Reading Check**  Choose from the list on the right the word that makes each statement true. Words may be used more than once.

**RC1.** Equivalent expressions represent the same number for any replacement of the _____(s).

solution

variable

**RC2.** Equivalent equations have the same _____(s).

**RC3.** We can add the same number to both sides of an equation without changing the _____(s) of the equation.

**RC4.** In the equation $7a - 3 = a + 5$, the _____ is $a$.

**Concept Check**  Match each equation with the correct first step for solving it from the list on the right.

**CC1.** $7x = 28$  _____

**a)** Add 7 to both sides.

**CC2.** $x - 7 = 13$  _____

**b)** Add $-7$ to both sides.

**CC3.** $-7x = -14$  _____

**c)** Divide both sides by 7.

**CC4.** $x + 7 = 1$  _____

**d)** Divide both sides by $-7$.

Ⓐ  Classify each pair as either equivalent expressions or equivalent equations.

**1.** $x = -1$; $x + 5 = 4$

**2.** $4x + 1$; $6 + 4x - 5$

**3.** $7a - 3$; $13 + 7a - 16$

**4.** $t = -2$; $5t = -10$

**5.** $4r + 3$; $9 - r + 5r - 6$

**6.** $2r - 7$; $r - 10 + r + 3$

**7.** $x - 9 = 8$; $x = 20 - 3$

**8.** $t + 4 = 19$; $t = 9 + 6$

**9.** $3(t + 2)$; $5 + 3t + 1$

**10.** $2x = -14$; $x - 2 = -9$

**11.** $x + 4 = -8$; $2x = -24$

**12.** $4(x - 7)$; $3x - 28 + x$

Ⓑ  Solve.

**13.** $x - 6 = -9$

**14.** $x - 5 = -7$

**15.** $x - 4 = -12$

**16.** $x - 7 = 5$

**17.** $a + 7 = 25$

**18.** $x + 9 = -3$

**19.** $-8 = n + 7$

**20.** $38 = a + 12$

**21.** $24 = t - 8$

**22.** $-9 = x + 3$

**23.** $-12 = x + 5$

**24.** $17 = n - 6$

**25.** $-5 + a = 12$

**26.** $3 = 17 + x$

**27.** $-8 = -8 + t$

**28.** $-7 + t = -7$

**c** Solve.

**29.** $6x = 60$

**30.** $-8t = 40$

**31.** $-3t = 42$

**32.** $3x = 24$

**33.** $-7n = -35$

**34.** $64 = -2t$

**35.** $0 = 6x$

**36.** $-5n = -65$

**37.** $55 = -5t$

**38.** $-x = 83$

**39.** $-x = 56$

**40.** $-2x = 0$

**41.** $n(-4) = -48$

**42.** $-x = -475$

**43.** $-x = -390$

**44.** $n(-7) = 42$

**d** Solve.

**45.** $t - 6 = -2$

**46.** $3t = -45$

**47.** $6x = -54$

**48.** $x + 9 = -15$

**49.** $15 = -x$

**50.** $-13 = x - 4$

**51.** $-21 = x + 5$

**52.** $-42 = -x$

**53.** $35 = -7t$

**54.** $7 + t = -18$

**55.** $-17x = 68$

**56.** $-34 = x - 10$

**57.** $12 + t = -160$

**58.** $-48 = t(-12)$

**59.** $-27 = x - 23$

**60.** $-135 = -9t$

Solve.

**61.** $5x - 1 = 34$

**62.** $7x - 3 = 25$

**63.** $4t + 2 = 14$

**64.** $3t + 5 = 26$

**65.** $6a + 1 = -17$

**66.** $8a + 3 = -37$

**67.** $2x - 9 = -23$

**68.** $3x - 5 = -35$

**69.** $-2x + 1 = 17$

**70.** $-4t + 3 = -17$

**71.** $-8t - 3 = -67$

**72.** $-7x - 4 = -46$

**73.** $-x + 9 = -15$

**74.** $-x - 6 = 8$

**75.** $7 = 2x - 5$

**76.** $9 = 4x - 7$

**77.** $13 = 3 + 2x$

**78.** $33 = 5 - 4x$

**79.** $13 = 5 - x$

**80.** $12 = 7 - x$

## Skill Maintenance

Simplify. [1.9c]

**81.** $5 + 3 \cdot 2^3$

**82.** $(9 - 7)^4 - 3^2$

**83.** $12 \div 3 \cdot 2$

**84.** $27 \div 3(2 + 1)$

**85.** $15 - 3 \cdot 2 + 7$

**86.** $30 - 4^2 \div 8 \cdot 2$

**87.** $3(8 - 6) - (9 - 4)$

**88.** $2(10 - 6)^2 - (120 \div 6 - 3 \cdot 4)$

## Synthesis

Solve.

**89.** $2x - 7x = -40$

**90.** $9 + x - 5 = 23$

**91.** $2x - 7 + x = 5 - 12$

**92.** $3 - 6x + 5 = 2(4)$

**93.** $n + n = -2 - 3 \cdot 6 \div 2 + 1$

**94.** $10 \div 5 \cdot 2 + 3 = 5t - 4t$

**95.** $17 - 3^2 = 4 + t - 5^2$

**96.** $(-9)^2 = 2^3 t + (3 \cdot 6 + 1)t$

**97.** $(-7)^2 - 5 = t + 4^3$

**98.** ▦ $(-42)^3 = 14^2 t$

**99.** ▦ $x - (19)^3 = -18^3$

**100.** ▦ $23^2 = x + 22^2$

**101.** ▦ $35^3 = -125t$

**102.** ▦ $248 = 24 - 32x$

**103.** ▦ $529 - 143x = -1902$

## Vocabulary Reinforcement

Complete each statement with the appropriate word or phrase from the list on the right. Some of the choices may not be used.

1. The _____ are ..., $-3, -2, -1, 0, 1, 2, 3, \ldots$ .   [2.1a]

2. The _____ value of a number is its distance from zero on the number line.   [2.1c]

3. The number 3 is the _____ of $-3$.   [2.1d]

4. Division by 0 is _____.   [2.5a]

5. When we replace a variable with a number, we say that we are _____ for the variable.   [2.6a]

6. A letter that stands for just one number is called a _____.   [2.6a]

7. The _____ states that for any numbers $a$, $b$, and $c$, $a(b + c) = ab + ac$.   [2.6b]

8. The _____ for solving equations states that for any real numbers $a$, $b$, and $c$, $a = b$ is equivalent to $a + c = b + c$.   [2.8b]

9. Equations with the same solutions are called _____ equations.   [2.8a]

constant

variable

opposite

equivalent

distributive law

integers

substituting

absolute

not defined

zero

addition principle

division principle

distributive law

## Concept Reinforcement

Determine whether each statement is true or false.

_____ 1. The opposite of the opposite of a number is the original number.   [2.1d]

_____ 2. The product of an even number of negative numbers is positive.   [2.4b]

_____ 3. The expression $2(x + 3)$ is equivalent to the expression $2 \cdot x + 3$.   [2.6b]

_____ 4. $3y$ and $3y^2$ are like terms.   [2.7a]

## Study Guide

**Objective 2.1c**   Find the absolute value of any integer.

**Example**   Find the absolute value.

a) $|-97|$   b) $|35|$

a) The number is negative, so we write the opposite.

$|-97| = 97$

b) The number is positive, so the absolute value is the same as the number.

$|35| = 35$

**Practice Exercise**

1. Find the absolute value.
   a) $|-17|$   b) $|300|$

**Objective 2.2a**  Add integers without using the number line.

**Example**  Add without using the number line: $-15 + 9$.

To add a negative number and a positive number, subtract the absolute values: $15 - 9 = 6$. The negative number has the larger absolute value, so the answer is negative.

$$-15 + 9 = -6$$

**Example**  Add without using the number line: $-8 + (-9)$.

To add two negative numbers, add the absolute values: $8 + 9 = 17$. The answer is negative.

$$-8 + (-9) = -17$$

**Practice Exercise**

**2.** Add without using the number line: $37 + (-16)$.

---

**Objective 2.3a**  Subtract integers, and simplify combinations of additions and subtractions.

**Example**  Subtract: $8 - 12$.

We add the opposite of the number being subtracted.

$$8 - 12 = 8 + (-12)$$
$$= -4$$

**Practice Exercise**

**3.** Subtract: $6 - (-8)$.

---

**Objective 2.4a**  Multiply integers.

**Example**  Multiply: $-6(-4)$.

The signs are the same, so the answer is positive.

$$-6(-4) = 24$$

**Example**  Multiply: $-4(3)$.

The signs are different, so the answer is negative.

$$-4(3) = -12$$

**Practice Exercise**

**4.** Multiply: $6(-15)$.

---

**Objective 2.5a**  Divide integers.

**Example**  Divide: $-36 \div (-4)$.

The signs are the same, so the answer is positive.

$$-36 \div (-4) = 9$$

**Example**  Divide: $-30 \div 10$.

The signs are different, so the answer is negative.

$$-30 \div 10 = -3$$

**Practice Exercise**

**5.** Divide: $99 \div (-9)$.

---

**Objective 2.5b**  Use the rules for order of operations with integers.

**Example**  Simplify: $3^2 - 24 \div 2 - (4 + 2 \cdot 8)$.

$$3^2 - 24 \div 2 - (4 + 2 \cdot 8)$$
$$= 3^2 - 24 \div 2 - (4 + 16)$$
$$= 3^2 - 24 \div 2 - 20$$
$$= 9 - 24 \div 2 - 20$$
$$= 9 - 12 - 20$$
$$= -3 - 20$$
$$= -23$$

**Practice Exercise**

**6.** Simplify: $4 - 8^2 \div (10 - 6)$.

**Objective 2.6b**  Use the distributive law to find equivalent expressions.

**Example**  Use the distributive law to write an expression equivalent to $-2(3x - 4)$.

$$-2(3x - 4) = -2(3x) - (-2)(4)$$
$$= -6x - (-8)$$
$$= -6x + 8$$

**Practice Exercise**

7. Use the distributive law to write an expression equivalent to $5(6x - 8y - z)$.

---

**Objective 2.7a**  Combine like terms.

**Example**  Combine like terms to form an expression equivalent to $6 - 30x + 12x - 7$.

$$6 - 30x + 12x - 7 = 6 + (-30x) + 12x + (-7)$$
$$= -30x + 12x + 6 + (-7)$$
$$= -18x - 1$$

**Practice Exercise**

8. Combine like terms to form an expression equivalent to $8a - b + 9a - 6b$.

---

**Objective 2.8e**  Solve equations that require use of both the addition principle and the division principle.

**Example**  Solve: $-3x + 1 = 16$.

$$-3x + 1 = 16$$
$$-3x + 1 - 1 = 16 - 1$$
$$-3x = 15$$
$$\frac{-3x}{-3} = \frac{15}{-3}$$
$$x = -5$$

The solution is $-5$.

**Practice Exercise**

9. Solve: $-19 = 5x + 11$.

---

# Review Exercises

1. Tell which integers correspond to this situation:

   David has a debt of $45 and Joe has $72 in his savings account.   [2.1a]

Use either $<$ or $>$ for ☐ to form a true statement.   [2.1b]

2. $0$ ☐ $-5$        3. $-7$ ☐ $6$        4. $-4$ ☐ $-19$

Find the absolute value.   [2.1c]

5. $|-39|$          6. $|23|$              7. $|0|$

8. Find $-x$ when $x = -72$.   [2.1d]

9. Find $-(-x)$ when $x = 59$.   [2.1d]

Compute and simplify.

10. $-14 + 5$   [2.2a]

11. $-5 + (-6)$   [2.2a]

12. $14 + (-8)$   [2.2a]

13. $0 + (-24)$   [2.2a]

14. $17 - 29$   [2.3a]

15. $9 - (-14)$   [2.3a]

16. $-8 - (-7)$   [2.3a]

17. $-3 - (-3)$   [2.3a]

18. $-3 + 7 + (-8)$   [2.3a]

19. $8 - (-9) - 7 + 2$   [2.3a]

**20.** $-23 \cdot (-4)$   [2.4a]

**21.** $7(-12)$   [2.4a]

**22.** $2(-4)(-5)(-1)$   [2.4b]

**23.** $15 \div (-5)$   [2.5a]

**24.** $\dfrac{-55}{11}$   [2.5a]

**25.** $\dfrac{0}{7}$   [2.5a]

Simplify.   [2.5b]

**26.** $625 \div (-25) \div 5$

**27.** $-16 \div 4 - 30 \div (-5)$

**28.** $9[(7 - 14) - 13]$

**29.** $(-3)|4 - 3^2| - 5$

**30.** $[-12(-3) - 2^3] - (-9)(-10)$

**31.** Evaluate $3a + b$ for $a = 4$ and $b = -5$.   [2.6a]

**32.** Evaluate $\dfrac{-x}{y}, \dfrac{x}{-y}$, and $-\dfrac{x}{y}$ for $x = 30$ and $y = 5$.   [2.6a]

Use the distributive law to write an equivalent expression.   [2.6b]

**33.** $4(5x + 9)$          **34.** $3(2a - 4b + 5)$

**35.** $-10(2x + y)$

Combine like terms.   [2.7a]

**36.** $5a + 12a$          **37.** $-7x + 13x$

**38.** $9m + 14 - 12m - 8$

**39.** Find the perimeter of a rectangular frame that is 8 in. by 10 in.   [2.7b]

**40.** Find the perimeter of a square pane of glass that is 25 cm on each side.   [2.7b]

Solve.   [2.8b, c, d, e]

**41.** $x - 9 = -17$          **42.** $-4t = 36$

**43.** $13 = -x$          **44.** $56 = 6x - 10$

**45.** $-x + 3 = -12$          **46.** $18 = 4 - 2x$

Solve.

**47.** On the first, second, and third downs, a football team had these gains and losses: 5-yd gain, 12-yd loss, and 15-yd gain, respectively. Find the total gain (or loss). [2.3b]

**48.** Kaleb's total assets are $170. He borrows $300. What are his total assets now? [2.3b]

**49.** Evaluate $-|-x|$ when $x = -10$. [2.1c, d]
**A.** $-10$          **B.** $-20$
**C.** $100$          **D.** $10$

**50.** Simplify: $-3 \cdot 4 - 12 \div 4$. [2.5b]
**A.** $-16$         **B.** $-15$
**C.** $0$            **D.** $6$

## Synthesis

**51.** The sum of two numbers is 800. The difference is 6. Find the numbers. [2.3b]

**52.** The following are examples of consecutive integers: 4, 5, 6, 7, 8; $-13, -12, -11, -10$. [2.2a], [2.4b]
  **a)** Express the number 8 as the sum of 16 consecutive integers.
  **b)** Find the product of the 16 consecutive integers in part (a).

Simplify. [2.5b]

**53.** ▦ $87 \div 3 \cdot 29^3 - (-6)^6 + 1957$

**54.** ▦ $1969 + (-8)^5 - 17 \cdot 15^3$

**55.** ▦ $\dfrac{113 - 17^3}{15 + 8^3 - 507}$

**56.** For what values of $x$ will $8 + x^3$ be negative? [2.6a]

**57.** For what values of $x$ is $|x| > x$? [2.1b, c]

## Understanding Through Discussion and Writing

**1.** What rule have we developed that would tell you the sign of $(-7)^8$ and $(-7)^{11}$ without doing the computations? Explain. [2.4b]

**2.** Does $-x$ always represent a negative number? Why or why not? [2.1d]

**3.** Jake enters $18/2 \cdot 3$ on his calculator and expects the result to be 3. What mistake is he making? [2.5b]

**4.** Does $-x^2$ always represent a negative number? Why or why not? [2.6a]

**CHAPTER**

**2** **Test**

For
Extra
Help

For step-by-step test solutions, access the Chapter Test Prep Videos
in MyLab Math.

1. Tell which integers correspond to this situation:

   The Tee Shop sold 542 fewer muscle shirts than expected in January and 307 more than expected in February.

2. Use either $<$ or $>$ for $\square$ to form a true statement.

   $$-14 \ \square \ -21$$

3. Find the absolute value: $|-739|$.

4. Find $-(-x)$ when $x = -19$.

Compute and simplify.

5. $6 + (-17)$

6. $-9 + (-12)$

7. $-8 + 17$

8. $0 - 12$

9. $7 - 22$

10. $-5 - 19$

11. $-8 - (-27)$

12. $31 - (-3) - 5 + 9$

13. $(-4)^3$

14. $27(-10)$

15. $-9 \cdot 0$

16. $-72 \div (-9)$

17. $\dfrac{-56}{7}$

18. $8 \div 2 \cdot 2 - 3^2$

19. $29 - (3 - 5)^2$

**20.** *Antarctica Highs and Lows.* The continent of Antarctica, which lies in the southern hemisphere, experiences winter in July. The average high temperature is $-67°F$ and the average low temperature is $-81°F$. How much higher is the average high than the average low?

**Data:** National Climatic Data Center

**21.** Evaluate $\dfrac{a - b}{6}$ for $a = -8$ and $b = 10$.

**22.** Use the distributive law to write an expression equivalent to $7(2x + 3y - 1)$.

**23.** Combine like terms.
$9x - 14 - 5x - 3$

**24.** Find the perimeter of a square garden that is 5 ft on each side.

Solve.

**25.** $-7x = -35$

**26.** $a + 9 = -3$

**27.** $95 = -x$

**28.** $3t - 7 = 5$

**29.** Use the distributive law to write an expression equivalent to $-2(n - 6m)$.

   **A.** $-2n - 6m$        **B.** $-2n - 12m$
   **C.** $-2n + 12m$       **D.** $2n + 12m$

## Synthesis

**30.** Monty plans to attach trim around the doorway and along the base of the walls in a 12-ft by 14-ft room. If the doorway is 3 ft by 7 ft, how many feet of trim are needed? (Only three sides of a doorway get trim.)

Simplify.

**31.** $9 - 5[x + 2(3 - 4x)] + 14$

**32.** $15x + 3(2x - 7) - 9(4 + 5x)$

**33.**  ▦  $49 \cdot 14^3 \div 7^4 + 1926^2 \div 6^2$

**34.**  ▦  $3487 - 16 \div 4 \cdot 4 \div 2^8 \cdot 14^4$

1. Write standard notation for the number written in words in the following sentence: In 2017, the total number of AP exams taken in U.S. History was five hundred five thousand, three hundred two.

   **Data:** collegeboard.org

2. Write a word name for 5,380,001,437.

Add.

3.   $\begin{array}{r} 1\,5{,}8\,9\,2 \\ +\ 2{,}9\,3\,5 \end{array}$

4.   $\begin{array}{r} 7\,9\,8\,9 \\ 7\,8\,9 \\ +\quad 7\,9 \end{array}$

5. $-16 + 72$

6. $-30 + (-72)$

Subtract.

7.   $\begin{array}{r} 8\,2\,7\,6 \\ -\quad 4\,3\,0 \end{array}$

8.   $\begin{array}{r} 3\,0\,0\,6 \\ -\quad 5\,7\,8 \end{array}$

9. $7 - (-7)$

10. $-18 - 25$

Multiply.

11.   $\begin{array}{r} 6\,2\,1 \\ \times\quad 2\,7 \end{array}$

12.   $\begin{array}{r} 2\,5\,0\,5 \\ \times\ 3\,3\,0\,0 \end{array}$

13. $43 \cdot (-8)$

14. $-12(-6)$

Divide.

15. $63\,)\overline{6\,5\,5\,2}$

16. $62\,)\overline{3\,8\,4\,4}$

17. $0 \div (-67)$

18. $60 \div (-12)$

19. Round 427,931 to the nearest thousand.

20. Round 5309 to the nearest hundred.

Estimate each sum or product by rounding to the nearest hundred. Show your work.

21.   $\begin{array}{r} 7\,4\,9{,}5\,5\,9 \\ +\ 3\,0\,1{,}3\,6\,2 \end{array}$

22.   $\begin{array}{r} 7\,4\,9 \\ \times\ 5\,3\,1 \end{array}$

23. Use $<$ or $>$ for ☐ to form a true sentence:
    $-26$ ☐ $19$.

24. Find the absolute value: $|-279|$.

Simplify.

**25.** $35 - 25 \div 5 + 2 \times 3$

**26.** $\{17 - [8 - (5 - 2 \times 2)]\} \div (3 + 12 \div 6)$

**27.** $10 \div 1(-5) - 6^2$

**28.** $5^3$

**29.** Evaluate $\dfrac{x - y}{5}$ for $x = 11$ and $y = -4$.

**30.** Evaluate $7x^2$ for $x = -2$.

Use the distributive law to write an equivalent expression.

**31.** $-2(x + 5)$

**32.** $6(3x - 2y + 4)$

Simplify.

**33.** $12 + (-14)$      **34.** $(-2)^3$

**35.** $23 - 38$      **36.** $-64 \div (-2)$

**37.** $-12 - (-25)$      **38.** $(-2)(-3)(-5)$

**39.** $3 - (-8) + 2 - (-3)$      **40.** $16 \div 2(-8) + 7$

Solve.

**41.** $x + 8 = 35$      **42.** $-12t = 36$

**43.** $6 - x = -9$      **44.** $-39 = 4x - 7$

Solve.

**45.** In the movie *Little Big Man,* Dustin Hoffman plays a character who ages from 17 to 121. This represents the greatest age range depicted by one actor in one film. How many years does Hoffman's character age?

**Data:** Guinness Book of World Records

**46.** There are four national parks in Colorado: Black Canyon of the Gunnison, Great Sand Dunes, Mesa Verde, and Rocky Mountain. These contain 30,045 acres, 149,137 acres, 52,074 acres, and 265,795 acres, respectively. What is the total number of acres in the four parks?

**Data:** nps.gov; nationalgeographic.com

**47.** Surface temperatures on the moon vary from $-184°C$ to $101°C$. Find the difference between the highest value and the lowest value in this temperature range.

**Data:** northern-stars.com/solar_system_info.htm

**48.** Eastside Appliance sells a refrigerator for $900 and $30 tax with no delivery charge. Westside Appliance sells the same model for $860 and $28 tax plus a $25 delivery charge. Which is the better buy?

**49.** Write an equivalent expression by combining like terms: $7x - 9 + 3x - 5$.

## Synthesis

**50.** A soft-drink distributor has 166 loose cans of cola. The distributor wishes to form as many 24-can cases as possible and then, with any remaining cans, as many six-packs as possible. How many cases will be filled? How many six-packs? How many loose cans will remain?

**51.** Simplify: $a - \{3a - [4a - (2a - 4a)]\}$.

**52.** ▦ Simplify: $37 \cdot 64 \div 4^2 \cdot 2 - (7^3 - (-4)^5)$.

**53.** Find two solutions of $5|x| - 2 = 13$.

MADE IN CHINA

# Fraction Notation: Multiplication and Division

Each year over 2 billion t-shirts are sold worldwide. T-shirt exports by all countries totaled $43,100,000,000 in 2016. The table below lists the sales, in dollars, for the top six countries. The most common type of fabric used to make t-shirts is cotton, for example, combed, organic, pima, or slub cotton. Two other fabrics often used to make t-shirts are linen and polyester.

*DATA: www.worldstopexports.com, "T-shirt Exports by Country" by Daniel Workman, June 29, 2017; www.huffingtonpost.com; theadairgroup.com*

## T-Shirt Exports

| Country | Export Sales (in U.S. dollars) |
|---|---|
| China | $8,700,000,000 |
| Bangladesh | 5,600,000,000 |
| Turkey | 2,900,000,000 |
| India | 2,800,000,000 |
| Germany | 1,900,000,000 |
| Vietnam | 1,500,000,000 |

DATA: worldstopexports.com

In Exercise 55 in Exercise Set 3.6, we will calculate the dollar value of worldwide exports of cotton t-shirts.

# 3.1

## OBJECTIVES

**a** Find some multiples of a number, and determine whether a number is divisible by another number.

**b** Test to see if a number is divisible by 2, 3, 5, 6, 9, or 10.

# Multiples and Divisibility

In this section, we discuss *multiples* and *divisibility* in order to be able to simplify fractions like $\frac{117}{225}$.

## a   MULTIPLES AND DIVISIBILITY

**SKILL REVIEW**

*Divide whole numbers.*   [1.5a]
Divide.

1. $329 \div 8$

2. $2\,3\,)\overline{1\,0\,8\,1}$

Answers: **1.** 41 R 1 **2.** 47

MyLab Math
VIDEO

A **multiple** of a number is a product of that number and an integer. For example, some multiples of 2 are

2   (because $2 = 1 \cdot 2$);

4   (because $4 = 2 \cdot 2$);

6   (because $6 = 3 \cdot 2$);

8   (because $8 = 4 \cdot 2$).

We can also find multiples of 2 by counting by twos: 2, 4, 6, 8, and so on.

**EXAMPLE 1**   Show that each of the numbers 3, 6, 9, and 15 is a multiple of 3.

$3 = 1 \cdot 3;$     $6 = 2 \cdot 3;$     $9 = 3 \cdot 3;$     $15 = 5 \cdot 3.$

◀ Do Exercises 1 and 2.

**EXAMPLE 2**   Multiply by 1, 2, 3, and so on, to find eight multiples of six.

$1 \cdot 6 = 6$     $5 \cdot 6 = 30$

$2 \cdot 6 = 12$     $6 \cdot 6 = 36$

$3 \cdot 6 = 18$     $7 \cdot 6 = 42$

$4 \cdot 6 = 24$     $8 \cdot 6 = 48$

◀ Do Exercise 3.

---

**DIVISIBILITY**

A number $b$ is said to be **divisible** by another number $a$ if $b$ is a multiple of $a$.

---

1. Show that each of the numbers 5, 45, and 100 is a multiple of 5.

2. Show that each of the numbers 10, 60, and 110 is a multiple of 10.

3. Multiply by 1, 2, 3, and so on, to find ten multiples of 5.

**Answers**

**1.** $5 = 1 \cdot 5; 45 = 9 \cdot 5; 100 = 20 \cdot 5$
**2.** $10 = 1 \cdot 10; 60 = 6 \cdot 10; 110 = 11 \cdot 10$
**3.** 5, 10, 15, 20, 25, 30, 35, 40, 45, 50

Thus,

6 is divisible by 2 because 6 is a multiple of 2 ($6 = 3 \cdot 2$), and

100 is divisible by 25 because 100 is a multiple of 25 ($100 = 4 \cdot 25$).

Saying that $b$ is divisible by $a$ means that if we divide $b$ by $a$, the remainder is 0. When this happens, we sometimes say that $a$ divides $b$ "evenly."

**EXAMPLE 3** Determine **(a)** whether 45 is divisible by 9 and **(b)** whether 45 is divisible by 4.

**a)** We divide 45 by 9.

$$\begin{array}{r} 5 \\ 9)\overline{45} \\ \underline{45} \\ 0 \end{array} \leftarrow \text{Remainder is 0.}$$

Because the remainder is 0, 45 is divisible by 9.

**b)** We divide 45 by 4.

$$\begin{array}{r} 11 \\ 4)\overline{45} \\ \underline{4} \\ 5 \\ \underline{4} \\ 1 \end{array} \leftarrow \text{Not 0}$$

Since the remainder is not 0, 45 is not divisible by 4.

**Do Exercises 4–6.** ▶

## b  TESTS FOR DIVISIBILITY

We now look at quick ways of checking for divisibility by 2, 3, 5, 6, 9, and 10 without actually performing long division. Tests do exist for divisibility by 4, 7, and 8, but they can be as difficult to perform as the actual long division.

To test for divisibility by 2, 5, and 10, we examine the ones digit.

### Divisibility by 2

All even numbers are divisible by 2.

---

**BY 2**

A number is **divisible by 2** (is *even*) if it has a ones digit of 0, 2, 4, 6, or 8 (that is, it has an even ones digit).

---

To see why this test works, start counting by twos: 2, 4, 6, 8, 10, 12, 14, 16, 18, 20, 22,. . . . Note that the ones digit will always be 0, 2, 4, 6, or 8, no matter how high we count.

**EXAMPLES** Determine whether each number is divisible by 2.

**4.** 35**5** *is not* divisible by 2 because **5** is not even.

**5.** 478**6** *is* divisible by 2 because **6** is even.

**6.** 899**0** *is* divisible by 2 because **0** is even.

**7.** 426**1** *is not* divisible by 2 because **1** is not even.

**Do Exercises 7–10.** ▶

 **4.** Determine whether 16 is divisible by 2.

$$\begin{array}{r} 8 \\ \boxed{\phantom{0}})\overline{16} \\ \underline{16} \\ \boxed{\phantom{0}} \end{array}$$

Since the remainder is $\boxed{\phantom{0}}$, 16 $\boxed{\phantom{0}}$ divisible by 2.
  $\underset{\text{is/is not}}{}$

**5.** Determine whether 125 is divisible by 5.

**6.** Determine whether 125 is divisible by 6.

---

**CALCULATOR CORNER**

***Divisibility*** We can use a calculator to determine whether one number is divisible by another number. For example, to determine whether 387 is divisible by 9, we press $\boxed{3}\boxed{8}\boxed{7}\boxed{\div}\boxed{9}\boxed{=}$. The display is $\boxed{\phantom{00}43}$.

Since 43 contains no digits to the right of the decimal point, we know that 387 is divisible by 9. On the other hand, since $387 \div 10 = 38.7$, we know that 387 is *not* divisible by 10.

**EXERCISES:** For each pair of numbers, determine whether the first number is divisible by the second number.

**1.** 731; 17

**2.** 1502; 79

**3.** 1053; 36

**4.** 4183; 47

---

Determine whether each number is divisible by 2.

**7.** 84      **8.** 59

**9.** 998      **10.** 2225

*Answers*
**4.** Yes   **5.** Yes   **6.** No   **7.** Yes
**8.** No   **9.** Yes   **10.** No
*Guided Solution:*
**4.** 2, 0; 0, is

## Divisibility by 5

To determine the test for divisibility by 5, we start counting by fives: 5, 10, 15, 20, 25, 30, 35, . . . . Note that the ones digit will always be 5 or 0, no matter how high we count.

| BY 5 |
| --- |
| A number is **divisible by 5** if its ones digit is 0 or 5. |

**EXAMPLES** Determine whether each number is divisible by 5.

**8.** 22**0**   *is* divisible by 5 because the ones digit is 0.

**9.** 47**5**   *is* divisible by 5 because the ones digit is 5.

**10.** 651**4**   *is not* divisible by 5 because the ones digit is neither 0 nor 5.

◀ **Do Exercises 11–14.**

Determine whether each number is divisible by 5.

**11.** 5780      **12.** 3427

**13.** 34,678      **14.** 7775

## Divisibility by 10

| BY 10 |
| --- |
| A number is **divisible by 10** if its ones digit is 0. |

We know that this test works because the product of 10 and *any* number has a ones digit of 0.

**EXAMPLES** Determine whether each number is divisible by 10.

**11.** 344**0**   *is* divisible by 10 because the ones digit is 0.

**12.** 344**7**   *is not* divisible by 10 because the ones digit is not 0.

◀ **Do Exercises 15–18.**

Determine whether each number is divisible by 10.

**15.** 305      **16.** 847

**17.** 300      **18.** 8760

Determine whether each number is divisible by 3.

**19.** 111      **20.** 1111

**21.** 309

## Divisibility by 3

To test for divisibility by 3, we examine the sum of a number's digits.

| BY 3 |
| --- |
| A number is **divisible by 3** if the sum of its digits is divisible by 3. |

An explanation of why this test works is outlined in Exercise 69 at the end of this section.

**EXAMPLES** Determine whether each number is divisible by 3.

**13.** 18      $1 + 8 = 9$

**14.** 93      $9 + 3 = 12$   Each *is* divisible by 3 because the sum of its digits *is* divisible by 3.

**15.** 201      $2 + 0 + 1 = 3$

**16.** 256      $2 + 5 + 6 = 13$   The sum of the digits, 13, *is not* divisible by 3, so 256 *is not* divisible by 3.

◀ **Do Exercises 19–22.**

**22.** 17,216      **GS**

Add the digits:

$1 + 7 + \boxed{\phantom{0}} + 1 + 6 = \boxed{\phantom{00}}$ .

Since 17 ___ divisible by 3,
                is/is not

the number 17,216 ___
                       is/is not

divisible by 3.

**Answers**

**11.** Yes  **12.** No  **13.** No  **14.** Yes
**15.** No  **16.** No  **17.** Yes  **18.** Yes
**19.** Yes  **20.** No  **21.** Yes  **22.** No
*Guided Solution:*
**22.** 2, 17; is not, is not

# Divisibility by 9

The test for divisibility by 9 is similar to the test for divisibility by 3.

> ### BY 9
>
> A number is **divisible by 9** if the sum of its digits is divisible by 9.

**EXAMPLES** Determine whether each number is divisible by 9.

**17.** 6984

Because $6 + 9 + 8 + 4 = 27$ and 27 is divisible by 9, 6984 *is* divisible by 9.

**18.** 322

Because $3 + 2 + 2 = 7$ and 7 is not divisible by 9, 322 *is not* divisible by 9.

**Do Exercises 23–26.** ▶

Determine whether each number is divisible by 9.

**23.** 16  **24.** 117

**25.** 309  **26.** 29,223

# Divisibility by 6

A number divisible by 6 is a multiple of 6. But $6 = 2 \cdot 3$, so the number is also a multiple of 2 and 3. Since 2 and 3 have no factors in common, a number is divisible by 6 if it is divisible by 2 *and* by 3.

> ### BY 6
>
> A number is **divisible by 6** if its ones digit is 0, 2, 4, 6, or 8 (is even) and the sum of its digits is divisible by 3.

**EXAMPLES** Determine whether each number is divisible by 6.

**19.** 720

Because 720 is even, it is divisible by 2. Also, $7 + 2 + 0 = 9$ and 9 is divisible by 3, so 720 is divisible by 3. Thus, 720 *is* divisible by 6.

$$72\underset{\uparrow}{0} \qquad 7 + 2 + 0 = \underset{\uparrow}{9}$$

Even  Divisible by 3

**20.** 73

73 *is not* divisible by 6 because it is not even.

**21.** 256

Although 256 is even, it *is not* divisible by 6 because the sum of its digits, $2 + 5 + 6$, or 13, is not divisible by 3.

**Do Exercises 27–30.** ▶

Determine whether each number is divisible by 6.

**27.** 420  **28.** 106

**29.** 321  **30.** 444

*Answers*

**23.** No  **24.** Yes  **25.** No  **26.** Yes
**27.** Yes  **28.** No  **29.** No  **30.** Yes

## 3.1 Exercise Set

### ✓ Check Your Understanding

**Reading and Concept Check**  Match the beginning of each divisibility test with the appropriate ending from the list on the right.

**RC1.** A number is divisible by 2 if _____.

**RC2.** A number is divisible by 3 if _____.

**RC3.** A number is divisible by 5 if _____.

**RC4.** A number is divisible by 6 if _____.

**RC5.** A number is divisible by 9 if _____.

**RC6.** A number is divisible by 10 if _____.

**a)** the sum of its digits is divisible by 3

**b)** the sum of its digits is divisible by 9

**c)** it has an even ones digit

**d)** its ones digit is 0 or 5

**e)** its ones digit is 0

**f)** it has an even ones digit and the sum of its digits is divisible by 3

**a**  Multiply by 1, 2, 3, and so on, to find ten multiples of each number.

**1.** 7     **2.** 4     **3.** 20     **4.** 50     **5.** 3     **6.** 8

**7.** 12     **8.** 15     **9.** 10     **10.** 11     **11.** 25     **12.** 100

**13.** Determine whether 83 is divisible by 3.

**14.** Determine whether 29 is divisible by 2.

**15.** Determine whether 525 is divisible by 7.

**16.** Determine whether 346 is divisible by 8.

**17.** Determine whether 8127 is divisible by 9.

**18.** Determine whether 4144 is divisible by 4.

**b**  For Exercises 19–30, answer "Yes" or "No" and give a reason based on the tests for divisibility.

**19.** Determine whether 84 is divisible by 3.

**20.** Determine whether 467 is divisible by 9.

**21.** Determine whether 5553 is divisible by 5.

**22.** Determine whether 2004 is divisible by 6.

**23.** Determine whether 671,500 is divisible by 10.

**24.** Determine whether 6120 is divisible by 5.

**25.** Determine whether 1773 is divisible by 9.

**26.** Determine whether 3286 is divisible by 3.

**27.** Determine whether 21,687 is divisible by 2.

**28.** Determine whether 64,091 is divisible by 10.

**29.** Determine whether 32,109 is divisible by 6.

**30.** Determine whether 9840 is divisible by 2.

For Exercises 31–38, test each number for divisibility by 2, 3, 5, 6, 9, and 10.

**31.** 6825

**32.** 12,600

**33.** 119,117

**34.** 2916

**35.** 127,575

**36.** 25,088

**37.** 9360

**38.** 143,507

To answer Exercises 39–44, consider the following numbers. Use the tests for divisibility.

| | | | |
|---|---|---|---|
| 46 | 300 | 85 | 256 |
| 224 | 36 | 711 | 8064 |
| 19 | 45,270 | 13,251 | 1867 |
| 555 | 4444 | 254,765 | 21,568 |

**39.** Which of the above are divisible by 3?

**40.** Which of the above are divisible by 2?

**41.** Which of the above are divisible by 10?

**42.** Which of the above are divisible by 5?

**43.** Which of the above are divisible by 6?

**44.** Which of the above are divisible by 9?

To answer Exercises 45–50, consider the following numbers.

| | | | |
|---|---|---|---|
| 56 | 200 | 75 | 35 |
| 324 | 42 | 812 | 402 |
| 784 | 501 | 2345 | 111,111 |
| 55,555 | 3009 | 2001 | 1005 |

**45.** Which of the above are divisible by 2?

**46.** Which of the above are divisible by 3?

**47.** Which of the above are divisible by 5?

**48.** Which of the above are divisible by 10?

**49.** Which of the above are divisible by 9?

**50.** Which of the above are divisible by 6?

# Skill Maintenance

Solve.

**51.** $16 \cdot t = 848$   [1.7b], [2.8c]

**52.** $m + 9 = 14$   [1.7b], [2.8b]

**53.** $23 + x = 15$   [1.7b], [2.8b]

**54.** $24 \cdot m = -576$   [1.7b], [2.8c]

Solve.   [1.8a]

**55.** Marty's automobile has a 5-speed transmission and gets 33 mpg in city driving. How many gallons of gas will it use in 1485 mi of city driving?

**56.** There are 60 min in 1 hr. How many minutes are there in 72 hr?

Evaluate.   [1.9b], [2.4b]

**57.** $5^3$

**58.** $(-2)^4$

Write in exponential notation.   [1.9a]

**59.** $9 \cdot 9 \cdot 9$

**60.** $3 \cdot 3 \cdot 3 \cdot 3 \cdot 3 \cdot 3$

# Synthesis

**61.** ▦ Find the largest five-digit number that is divisible by 47.

**62.** ▦ Find the largest six-digit number that is divisible by 53.

Find the smallest number that is simultaneously a multiple of the given numbers.

**63.** 2, 3, and 5

**64.** 3, 5, and 7

**65.** 6, 10, and 14

**66.** ▦ 17, 43, and 85

**67.** 30, 70, and 120

**68.** 25, 100, and 175

**69.** To help see why the tests for divisibility by 3 and 9 work, note that any four-digit number *abcd* can be rewritten as $1000 \cdot a + 100 \cdot b + 10 \cdot c + d$, or $999a + 99b + 9c + a + b + c + d$.

   **a)** Explain why $999a + 99b + 9c$ is divisible by both 9 and 3 for all choices of *a*, *b*, *c*, and *d*.

   **b)** Explain why the four-digit number *abcd* is divisible by 9 if $a + b + c + d$ is divisible by 9 and is divisible by 3 if $a + b + c + d$ is divisible by 3.

**70.** A passenger in a taxicab asks for the cab number. The driver says abruptly, "Sure—it's the smallest multiple of 11 that, when divided by 2, 3, 4, 5, or 6, has a remainder of 1." What is the number?

Following are the tests for divisibility by 4, 8, 7, and 11. Use these for Exercises 71 and 72.

| NUMBER | TEST | TEST 23,904,328 FOR DIVISIBILITY |
|---|---|---|
| 4 | A number is divisible by 4 if the number named by its last two digits is divisible by 4. | The number named by the last two digits is 28. Since 28 is divisible by 4, 23,904,328 is divisible by 4. |
| 8 | A number is divisible by 8 if the number named by its last three digits is divisible by 8. | The number named by the last three digits is 328. Since 328 is divisible by 8, 23,904,328 is divisible by 8. |
| 7 | Divide the number into groups of three digits, starting at the right. Start with the group at the right, subtract the next group to the left, add the next group to the left, and so on. If the resulting number is divisible by 7, the original number is divisible by 7. | The number is already divided into groups of three by commas. Calculate $328 - 904 + 23 = -553$. Since $-553$ is divisible by 7, 23,904,328 is divisible by 7. |
| 11 | Find the sum of the odd-numbered digits (the 1st digit plus the 3rd plus the 5th and so on). Find the sum of the even-numbered digits. Subtract these sums. If the difference is divisible by 11, then the original number is divisible by 11. | Add the odd-numbered digits: $2 + 9 + 4 + 2 = 17$. Add the even-numbered digits: $3 + 0 + 3 + 8 = 14$. Subtract: $17 - 14 = 3$. Since 3 is *not* divisible by 11, 23,904,328 is not divisible by 11. |

**71.** Test 332,986,412 for divisibility by 4, 8, 7, and 11.

**72.** Test 6,637,105,860 for divisibility by 4, 8, 7, and 11.

# Factorizations

## 3.2

### OBJECTIVES

a Find the factors of a number.

b Given a number from 1 to 100, tell whether it is prime, composite, or neither.

c Find the prime factorization of a composite number.

When we express a number as a product, we say that we have *factored* the original number. The numbers in the product are called *factors*. Thus "factor" can be used as either a noun or a verb. The ability to factor is an important skill needed for a solid understanding of fractions.

## a FACTORS AND FACTORIZATIONS

Here we consider only the natural numbers: 1, 2, 3, and so on. From the equation $3 \cdot 4 = 12$, we can say that 3 and 4 are **factors** of 12. Since $12 = 12 \cdot 1$, we know that 12 and 1 are also factors of 12.

---

### FACTORS AND FACTORIZATIONS

- In the product $a \cdot b$, $a$ and $b$ are called **factors**.
- A number $c$ is a **factor** of $a$ if $a$ is divisible by $c$.
- A **factorization** of $a$ expresses $a$ as a product of two or more numbers.

---

Note that saying that 3 and 4 are factors of 12 is the same as saying that 12 is a multiple of 3 and that 12 is a multiple of 4. Each of the following gives a factorization of 12.

$12 = 4 \cdot 3$ $\leftarrow$ This factorization shows that 4 and 3 are factors of 12.

$12 = 12 \cdot 1$ $\leftarrow$ This factorization shows that 12 and 1 are factors of 12.

$12 = 6 \cdot 2$ $\leftarrow$ This factorization shows that 6 and 2 are factors of 12.

$12 = 2 \cdot 3 \cdot 2$ $\leftarrow$ This factorization shows that 2 and 3 are factors of 12.

Thus, 1, 2, 3, 4, 6, and 12 are all factors of 12. Note that since $n = n \cdot 1$, every number has a factorization, and every number has itself and 1 as factors.

**EXAMPLE 1** List all the factors of 18.

Beginning at 1, we check all positive integers to see if they are factors of 18. If they are, we write the factorization. We stop when we have already included the next integer in a factorization.

| | |
|---|---|
| 1 is a factor of every number. | $1 \cdot 18$ |
| 2 is a factor of 18. | $2 \cdot 9$ |
| 3 is a factor of 18. | $3 \cdot 6$ |
| 4 is *not* a factor of 18. | |
| 5 is *not* a factor of 18. | |

The next integer is 6, but we have already listed 6 as a factor in the product $3 \cdot 6$. We need check no additional numbers, because any integer greater than 6 must be paired with a factor less than 6.

We now write the factors of 18 beginning with 1, going down the list of factorizations writing each first factor, then up the list of factorizations writing each second factor:

1, 2, 3, 6, 9, 18.

**Do Exercises 1–4.** ▶

---

List all the factors of each number.

**1.** 10     **2.** 62     **3.** 24

GS **4.** 45

1 is a factor of 45.    $1 \cdot 45$

2 ___ a factor of 45.

3 ___ a factor of 45. 3 · ___

4 ___ a factor of 45.

5 ___ a factor of 45. 5 · ___

6 is not a factor of 45.

7 is not a factor of 45.

8 is not a factor of 45.

Factors of 45: 1, 3, 5, ___, ___, 45.

---

***Answers***

**1.** 1, 2, 5, 10     **2.** 1, 2, 31, 62
**3.** 1, 2, 3, 4, 6, 8, 12, 24    **4.** 1, 3, 5, 9, 15, 45
***Guided Solution:***
**4.** is not; is, 15; is not; is, 9; 9, 15

## b PRIME AND COMPOSITE NUMBERS

---

**PRIME AND COMPOSITE NUMBERS**

A natural number that has exactly two *different* factors, only itself and 1, is called a **prime number**.

- The number 1 is *not* prime.
- A natural number, other than 1, that is not prime is **composite**.

---

**EXAMPLE 2**  Determine which of the numbers 1, 2, 7, 8, 9, 11, 18, 27, 39, 43, 56, 59, and 77 are prime, which are composite, and which are neither.

The number 1 is not prime. It does not have *two* different factors.

The number 2 is prime. It has only the factors 2 and 1.

The numbers 7, 11, 43, and 59 are prime. Each has only two factors, itself and 1.

The number 8 is not prime. It has the factors 1, 2, 4, and 8 and is composite.

The numbers 9, 18, 27, 39, 56, and 77 are composite. Each has more than two factors.

Thus, we have

Prime:　　　2, 7, 11, 43, 59;

Composite:　8, 9, 18, 27, 39, 56, 77;

Neither:　　1.

We can make several observations about prime numbers.

- Because 0 is not a natural number, it is neither prime nor composite.
- The number 1 is not prime because it does not have two different factors.
- The number 2 is the smallest prime and the only even prime, since 2 is a factor of all even numbers.
- To determine whether an odd number is prime, check divisibility by prime numbers beginning with 3 and 5. If you reach a point where the quotient is less than the divisor and none of the primes up to that point are factors, the number you are checking is prime.

**5.** Classify each number as prime, composite, or neither.

1, 2, 6, 12, 13, 23, 41, 65, 73, 99

◀ **Do Exercise 5.**

The following is a table of the prime numbers from 2 to 157. Although you need not memorize the entire list, remembering at least the first nine or ten is important.

---

**A TABLE OF PRIMES FROM 2 TO 157**

2, 3, 5, 7, 11, 13, 17, 19, 23, 29, 31, 37, 41, 43, 47, 53, 59, 61, 67, 71, 73, 79, 83, 89, 97, 101, 103, 107, 109, 113, 127, 131, 137, 139, 149, 151, 157

---

There are infinitely many prime numbers. It takes extensive computer operations to determine whether very large numbers are prime. In 2018, the largest known prime was $2^{77232917} - 1$. This number has over 23 million digits!

**Data:** mersenne.org

*Answer*

**5.** 2, 13, 23, 41, 73 are prime; 6, 12, 65, 99 are composite; 1 is neither

## C   PRIME FACTORIZATIONS

A factorization like $2 \cdot 2 \cdot 5 \cdot 11$ could also be expressed in other forms, such as $5 \cdot 2 \cdot 2 \cdot 11$ or $2 \cdot 5 \cdot 11 \cdot 2$ or $2^2 \cdot 5 \cdot 11$ or $11 \cdot 2^2 \cdot 5$. The prime factors are the same in each case. For this reason, we agree that any of these may be considered "the" prime factorization of 220.

When we factor a composite number into a product of primes, we find the **prime factorization** of the number. We can do this by making a series of successive divisions or by using a *factor tree.*

To use division, we consider the prime numbers 2, 3, 5, 7, 11, 13, and so on, and determine whether a given number is divisible by the primes.

**EXAMPLE 3**   Find the prime factorization of 39.

We check for divisibility by the first prime, 2. Since 39 is not even, 2 is not a factor of 39. Since the sum of the digits in 39 is 12 and 12 is divisible by 3, we know that 39 is divisible by 3. We then perform the division.

$$\begin{array}{r} 13 \\ 3\overline{)39} \end{array} \quad R = 0 \qquad \text{A remainder of 0 confirms that 3 is a factor of 39.}$$

Because 13 is prime, we can now write the prime factorization:

$$39 = 3 \cdot 13.$$

**EXAMPLE 4**   Find the prime factorization of 220.

We consider the first prime, 2. Since 220 is even, it must have 2 as a factor.

$$\begin{array}{r} 110 \\ 2\overline{)220} \end{array} \quad R = 0 \qquad 220 = 2 \cdot 110$$

Because 110 is also even, we divide again by 2.

$$\begin{array}{r} 55 \\ 2\overline{)110} \end{array} \quad R = 0 \qquad 220 = 2 \cdot 2 \cdot 55$$

Since 55 is odd, it is not divisible by 2. We move to the next prime, 3. The sum of the digits is 10 and 10 is not divisible by 3, so we move to the next prime, 5. Since the ones digit of 55 is 5, we divide by 5.

$$\begin{array}{r} 11 \\ 5\overline{)55} \end{array} \quad R = 0 \qquad 220 = 2 \cdot 2 \cdot 5 \cdot 11$$

Because 11 is prime, we are finished. The prime factorization is

$$220 = 2 \cdot 2 \cdot 5 \cdot 11.$$

We abbreviate our procedure as follows.

$$\begin{array}{r} 11 \\ 5\overline{)55} \\ 2\overline{)110} \\ 2\overline{)220} \end{array}$$

$$220 = 2 \cdot 2 \cdot 5 \cdot 11$$

> Every composite number has just one (unique) prime factorization.

This result is sometimes called the Fundamental Theorem of Arithmetic.

**EXAMPLE 5**  Find the prime factorization of 187.

We check for divisibility by 2, 3, and 5, and find that 187 is not divisible by any of these numbers. The next prime number, 7, does not divide 187 evenly. However, when we divide by 11, the remainder is 0, so 11 is a factor of 187.

$$\begin{array}{r} 17 \\ 11\overline{)187} \end{array}$$    We can write $187 = 11 \cdot 17$.

Because 17 is prime, we can factor no further. The complete factorization is

$187 = 11 \cdot 17.$    All factors are prime.

**EXAMPLE 6**  Find the prime factorization of 72.

$$\begin{array}{r} 3 \\ 3\overline{)\,9} \\ 2\overline{)18} \\ 2\overline{)36} \\ 2\overline{)72} \end{array}$$    ← 3 is prime, so we stop dividing.

← Begin here and work upward.

$72 = 2 \cdot 2 \cdot 2 \cdot 3 \cdot 3$

Another way to find the prime factorization of 72 is to use a **factor tree** as follows. Begin by determining any factorization you can, and then continue factoring until all of the factors are prime numbers. Each of the following trees gives the same prime factorization.

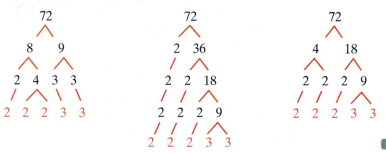

**EXAMPLE 7**  Find the prime factorization of 130.

We can use a string of divisions or a factor tree.

$$\begin{array}{r} 13 \\ 5\overline{)65} \\ 2\overline{)130} \end{array}$$

$130 = 2 \cdot 5 \cdot 13$

◀ **Do Exercises 6–13.**

Finding a number's prime factorization can be quite challenging, especially when the prime factors themselves are large. This difficulty is used worldwide as a way of securing transactions over the Internet.

Find the prime factorization of each number.

**6.** 6        **7.** 12

**8.** 45

**9.** 98

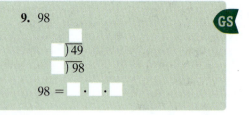

**10.** 126        **11.** 144

**12.** 91        **13.** 1925

*Answers*
**6.** $2 \cdot 3$  **7.** $2 \cdot 2 \cdot 3$  **8.** $3 \cdot 3 \cdot 5$
**9.** $2 \cdot 7 \cdot 7$  **10.** $2 \cdot 3 \cdot 3 \cdot 7$
**11.** $2 \cdot 2 \cdot 2 \cdot 2 \cdot 3 \cdot 3$  **12.** $7 \cdot 13$
**13.** $5 \cdot 5 \cdot 7 \cdot 11$
*Guided Solution:*
**9.** $\begin{array}{r} 7 \\ 7\overline{)49} \\ 2\overline{)98} \end{array}$; 2, 7, 7

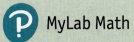
✓ **Check Your Understanding**

**Reading Check** Fill in the blank with a word from the list on the right that completes each statement. Each word is used only once.

**RC1.** The _____ factorization of 42 is $2 \cdot 3 \cdot 7$.

**RC2.** 6 is a _____ of 42.

**RC3.** 42 is _____ by 14.

**RC4.** 42 is a _____ of 3.

**RC5.** 42 is a _____ number.

**RC6.** One _____ of 42 is $2 \cdot 21$.

composite
divisible
factor
factorization
multiple
prime

**Concept Check** Determine whether each statement is true or false.

_____ **CC1.** One factorization of 20 is $4 \cdot 5$.

_____ **CC2.** One factor of 15 is 30.

_____ **CC3.** The smallest prime number is 1.

_____ **CC4.** A prime number has exactly two different factors.

_____ **CC5.** The prime factorization of 30 is $3 \cdot 10$.

**a** Determine whether the second number is a factor of the first.

**1.** 52; 14
**2.** 52; 13
**3.** 625; 25
**4.** 680; 16

List all the factors of each number.

**5.** 18
**6.** 16
**7.** 54
**8.** 48

**9.** 4
**10.** 9
**11.** 1
**12.** 13

**13.** 98
**14.** 100
**15.** 255
**16.** 120

**b** Classify each number as prime, composite, or neither.

**17.** 19
**18.** 24
**19.** 22
**20.** 31

**21.** 48
**22.** 43
**23.** 53
**24.** 54

**25.** 1
**26.** 2
**27.** 81
**28.** 37

**29.** 47
**30.** 51
**31.** 29
**32.** 49

**c** Find the prime factorization of each number.

**33.** 27  **34.** 16  **35.** 14  **36.** 15  **37.** 80

**38.** 32  **39.** 25  **40.** 75  **41.** 62  **42.** 169

**43.** 100  **44.** 110  **45.** 143  **46.** 40  **47.** 121

**48.** 170  **49.** 273  **50.** 675  **51.** 175  **52.** 196

**53.** 209  **54.** 217  **55.** 1200  **56.** 1800  **57.** 693

**58.** 2370  **59.** 2884  **60.** 484  **61.** 1122  **62.** 6435

## Skill Maintenance

Multiply.

**63.** $-2 \cdot 13$  [2.4a]  **64.** $(-8)(-32)$  [2.4a]

Add.

**65.** $-17 + 25$  [2.2a]  **66.** $-9 + (-14)$  [2.2a]

Divide

**67.** $53 \div 53$  [1.5a]  **68.** $-98 \div 1$  [2.5a]  **69.** $0 \div 22$  [1.5a]  **70.** $0 \div (-42)$  [2.5a]

## Synthesis

Find the prime factorization of each number.

**71.** ▦ 136,097  **72.** ▦ 102,971  **73.** ▦ 473,073,361

**74.** Describe an arrangement of 54 objects that corresponds to the factorization $54 = 6 \times 9$.

**75.** Describe an arrangement of 24 objects that corresponds to the factorization $24 = 2 \cdot 3 \cdot 4$.

**76.** Two numbers are **relatively prime** if there is no prime number that is a factor of both numbers. For example, 10 and 21 are relatively prime but 15 and 18 are not. List five pairs of composite numbers that are relatively prime.

**77.** *Factors and Sums.* In the table below, the top number in each column can be factored in such a way that the sum of the factors is the bottom number in the column. For example, in the first column, 56 has been factored as $7 \cdot 8$, and $7 + 8 = 15$, the bottom number. (Such thinking will be important in understanding the meaning of a factor and in algebra.) Fill in the missing numbers in the table.

| PRODUCT | 56 | 63 | 36 | 72 | 140 | 96 | 48 | 168 | 110 | 90 | 432 | 63 |
|---------|----|----|----|----|-----|----|----|-----|-----|----|-----|----|
| FACTOR  | 7  |    |    |    |     |    |    |     |     |    |     |    |
| FACTOR  | 8  |    |    |    |     |    |    |     |     |    |     |    |
| SUM     | 15 | 16 | 20 | 38 | 24  | 20 | 14 | 29  | 21  | 19 | 42  | 24 |

# Fractions and Fraction Notation

### OBJECTIVES

a Identify the numerator and the denominator of a fraction, and write fraction notation for part of an object or part of a set of objects and as a ratio.

b Simplify fraction notation like $n/n$ to 1, $0/n$ to 0, and $n/1$ to $n$.

The study of arithmetic begins with the set of whole numbers. But we also need to be able to use fractional parts of numbers such as halves, thirds, fourths, and so on. Here is an example.

Households in auto-dependent locations spend about $\frac{1}{4}$ of their income on transportation costs, while location-efficient households (those with easy access to public transportation) can hold transportation costs to $\frac{1}{10}$ of their income.

**Auto-Dependent Households**          **Location-Efficient Households**

$\frac{1}{4}$ Transportation    $\frac{3}{4}$ Remaining      $\frac{1}{10}$ Transportation    $\frac{9}{10}$ Remaining

DATA: U.S. Department of Transportation, Federal Highway Administration

## a FRACTIONS AND THE REAL WORLD

Expressions like those below are written in **fraction notation**. The top number is called the **numerator** and the bottom number is called the **denominator**.

$$\frac{1}{2}, \frac{7}{8}, \frac{-8}{5}, \frac{x}{y}, -\frac{4}{25}, \frac{2a}{7b}$$

**EXAMPLE 1**  Identify the numerator and the denominator.

$\dfrac{7}{8}$ ← Numerator
   ← Denominator

**Do Exercises 1–4.** ▶

Let's look at various situations that involve fractions.

### Fractions as a Partition of an Object Divided into Equal Parts

Consider a candy bar divided into 5 equal sections. If you eat 2 sections, you have eaten $\frac{2}{5}$ of the candy bar. The denominator 5 tells us the unit, $\frac{1}{5}$. The numerator 2 tells us the number of equal parts we are considering, 2.

For each fraction, identify the numerator and the denominator.

1. About $\frac{4}{5}$ of the parts on a Toyota Camry were produced in the United States.

   **Data:** "Made in America: Which Car Creates the Most Jobs?" by David Muir and Sharyn Alfonsi, on abcnews.go.com

2. It is projected that $\frac{19}{100}$ of the U.S. population in 2050 will be foreign-born.

   **Data:** Pew Research Center

3. $\dfrac{5a}{7m}$          4. $\dfrac{-22}{3}$

*Answers*

1. Numerator: 4; denominator: 5
2. Numerator: 19; denominator: 100
3. Numerator: 5a; denominator: 7m
4. Numerator: −22; denominator: 3

**EXAMPLE 2** What part is shaded?

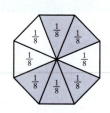

There are 8 equal parts. This tells us the unit, $\frac{1}{8}$. The *denominator* is 8. We have 5 of the units shaded. This tells us the *numerator*, 5. Thus,

$$\frac{5}{8} \begin{array}{l} \leftarrow \text{5 units are shaded.} \\ \leftarrow \text{The unit is } \frac{1}{8}. \end{array}$$

is shaded.

**EXAMPLE 3** What part of the measuring cup is filled?

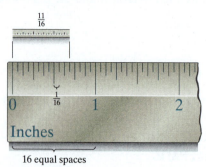

The measuring cup is divided into 3 parts of the same size, and 2 of them are filled. This is $2 \cdot \frac{1}{3}$, or $\frac{2}{3}$. Thus, $\frac{2}{3}$ (read *two-thirds*) of the cup is filled.

◀ **Do Exercises 5–8.**

The markings on a ruler use fractions.

**EXAMPLE 4** What part of an inch is indicated?

Each inch on the ruler shown above is divided into 16 equal parts. The marked section extends to the 11th mark. Thus, $\frac{11}{16}$ (read *eleven–sixteenths*) of an inch is indicated.

◀ **Do Exercise 9.**

---

What part is shaded?

**5.**

**6.**

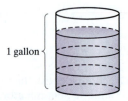

**7.**

**8.**

**9.** What part of an inch is indicated?

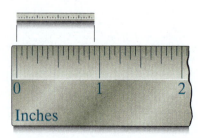

*Answers*

**5.** $\frac{5}{6}$ **6.** $\frac{1}{3}$ **7.** $\frac{3}{4}$ **8.** $\frac{8}{15}$ **9.** $\frac{15}{16}$

Fractions greater than or equal to 1, such as $\frac{24}{24}$, $\frac{10}{3}$, and $\frac{5}{4}$, correspond to situations like the following.

**EXAMPLE 5** What part is shaded?

a)

The rectangle is divided into 24 equal parts. Thus, the unit is $\frac{1}{24}$. The denominator is 24. All 24 equal parts are shaded. This tells us that the numerator is 24. Thus, $\frac{24}{24}$ is shaded.

b)

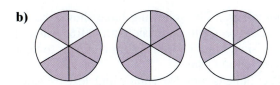

Each circle is divided into 6 parts. Thus, the unit is $\frac{1}{6}$. The denominator is 6. We see that 11 of the equal units are shaded. This tells us that the numerator is 11. Thus, $\frac{11}{6}$ is shaded.

**EXAMPLE 6** *Ice-Cream Roll-up Cake.* What part of an ice-cream roll-up cake is dark brown?

3 ice cream roll-up cakes

Each cake is divided into 6 equal slices. The unit is $\frac{1}{6}$. The *denominator* is 6. We see that 13 of the slices are shaded dark brown. This tells us that the *numerator* is 13. Thus, $\frac{13}{6}$ is shaded.

**Do Exercises 10–12.** ▶

Fractions larger than or equal to 1, such as $\frac{13}{6}$ or $\frac{9}{9}$, are sometimes referred to as "improper" fractions. We will not use this terminology because notation such as $\frac{27}{8}$, $\frac{11}{3}$, and $\frac{4}{4}$ is quite "proper" and very common in algebra.

## Fractions as Ratio

A **ratio** is a quotient of two quantities. We can express a ratio with fraction notation. (We will consider ratios in more detail in Chapter 6.)

What part is shaded?

10.

11.

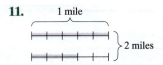

1 mile

2 miles

**GS** 12.

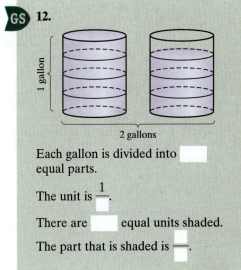

1 gallon

2 gallons

Each gallon is divided into ☐ equal parts.

The unit is $\frac{1}{\Box}$.

There are ☐ equal units shaded.

The part that is shaded is $\frac{\Box}{\Box}$.

*Answers*

10. $\frac{15}{15}$  11. $\frac{8}{5}$  12. $\frac{7}{4}$

*Guided Solution:*

12. 4, 4, 7, $\frac{7}{4}$

Morocco
Egypt
Mali
Ethiopia
Yaounde
Equator
Angola
Zambia
Botswana

**13.** What part of the set of countries in Example 7 is west of Yaounde?

| CENTRAL | W | L | Pct. | Home | Road |
|---|---|---|---|---|---|
| Cleveland Indians | 94 | 67 | .584 | 53–28 | 41–39 |
| Detroit Tigers | 86 | 75 | .534 | 45–35 | 41–40 |
| Kansas City Royals | 81 | 81 | .500 | 47–34 | 34–47 |
| Chicago White Sox | 78 | 84 | .481 | 45–36 | 33–48 |
| Minnesota Twins | 59 | 103 | .364 | 30–51 | 29–52 |

DATA: Major League Baseball

**14.** *Baseball Standings.* Refer to the table in Example 8. The Minnesota Twins finished fifth in the American League Central division in 2016. Find the Twins' ratio of wins to losses, of wins to total games, and of losses to total games.

**Data:** Major League Baseball

**EXAMPLE 7** *Countries of Africa.* What part of this group of countries is north of the equator? south of the equator?

| | |
|---|---|
| Angola | Mali |
| Botswana | Morocco |
| Egypt | Zambia |
| Ethiopia | |

There are 7 countries in the set, and 4 of them—Egypt, Ethiopia, Mali, and Morocco—are north of the equator. Thus, 4 of 7, or $\frac{4}{7}$, are north of the equator. The 3 remaining countries are south of the equator. Thus, $\frac{3}{7}$ are south of the equator.

◀ **Do Exercise 13.**

**EXAMPLE 8** *Baseball Standings.* The following table shows the final standings in the American League Central division for 2016, when the Cleveland Indians won the division. Find the Indians' ratio of wins to losses, wins to total games, and losses to total games.

The Indians won 94 games and lost 67 games. They played a total of 94 + 67, or 161, games. Thus, we have the following.

The ratio of wins to losses is $\frac{94}{67}$.

The ratio of wins to total games is $\frac{94}{161}$.

The ratio of losses to total games is $\frac{67}{161}$.

◀ **Do Exercise 14.**

**b** **SOME FRACTION NOTATION FOR WHOLE NUMBERS**

### Fraction Notation for 1

The number 1 corresponds to situations like those shown here. If we divide an object into *n* parts and take *n* of them, we get all of the object (1 whole object).

Since a negative number divided by a negative number is a positive number, we state the following for all nonzero integers.

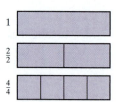

**THE NUMBER 1 IN FRACTION NOTATION**

$\frac{n}{n} = 1$, for any integer *n* that is not 0.

*Answers*

**13.** $\frac{2}{7}$ **14.** $\frac{59}{103}, \frac{59}{162}, \frac{103}{162}$

**EXAMPLES** Simplify. Assume any variables are nonzero.

**9.** $\dfrac{5}{5} = 1$   **10.** $\dfrac{-9}{-9} = 1$   **11.** $\dfrac{23x}{23x} = 1$

**Do Exercises 15–20.** ▶

## Fraction Notation for 0

Consider the fraction $\frac{0}{4}$. This corresponds to dividing an object into 4 parts and taking none of them. We get 0.

> **THE NUMBER 0 IN FRACTION NOTATION**
>
> $\dfrac{0}{n} = 0$, for any integer $n$ that is not 0.

**EXAMPLES** Simplify.

**12.** $\dfrac{0}{1} = 0$   **13.** $\dfrac{0}{9} = 0$

**14.** $\dfrac{0}{-23} = 0$   **15.** $\dfrac{0}{5a} = 0$, assuming that $a \neq 0$

Fraction notation with a denominator of 0, such as $n/0$, does not represent a number because we cannot speak of an object being divided into *zero* parts. (If it is not divided at all, then we say it is undivided and remains in one part.)

> **A DENOMINATOR OF 0**
>
> $\dfrac{n}{0}$ is not defined. We say that $\dfrac{n}{0}$ is *undefined*.

**Do Exercises 21–26.** ▶

## Other Integers

Consider the fraction $\frac{4}{1}$. This corresponds to taking 4 objects and dividing each into 1 part. (In other words, we do not divide them.) We have 4 objects.

> **ANY INTEGER IN FRACTION NOTATION**
>
> Any integer divided by 1 is the original integer. That is,
>
> $\dfrac{n}{1} = n$, for any integer $n$.

**EXAMPLES** Simplify.

**16.** $\dfrac{2}{1} = 2$   **17.** $\dfrac{-9}{1} = -9$   **18.** $\dfrac{3x}{1} = 3x$

**Do Exercises 27–30.** ▶

Simplify.

**15.** $\dfrac{1}{1}$   **16.** $\dfrac{4}{4}$

**17.** $\dfrac{a}{a}$   **18.** $\dfrac{-100}{-100}$

**19.** $\dfrac{2347}{2347}$   **20.** $\dfrac{54n}{54n}$

Simplify, if possible. Assume that $x \neq 0$.

**21.** $\dfrac{0}{1}$   **22.** $\dfrac{0}{-6}$

**23.** $\dfrac{0}{4x}$

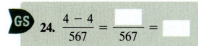

**GS** **24.** $\dfrac{4-4}{567} = \dfrac{\boxed{\phantom{0}}}{567} = \boxed{\phantom{0}}$

**25.** $\dfrac{15}{0}$   **26.** $\dfrac{-4}{3-3}$

Simplify.

**27.** $\dfrac{8}{1}$   **28.** $\dfrac{-10}{1}$

**29.** $\dfrac{-346}{1}$   **30.** $\dfrac{24-1}{23-22}$

*Answers*

**15.** 1   **16.** 1   **17.** 1   **18.** 1   **19.** 1
**20.** 1   **21.** 0   **22.** 0   **23.** 0   **24.** 0
**25.** Undefined   **26.** Undefined   **27.** 8
**28.** −10   **29.** −346   **30.** 23

*Guided Solution:*
**24.** 0, 0

## ✓ Check Your Understanding

**Reading Check** Match each expression with the appropriate description or value from the list on the right.

**RC1.** The 3 in $\frac{3}{4}$ _____

**RC2.** The 4 in $\frac{3}{4}$ _____

**RC3.** The fraction $\frac{3}{4}$ _____

**RC4.** $\frac{0}{1}$ _____

**RC5.** $\frac{n}{0}$ _____

**RC6.** $\frac{n}{1}$ _____

**a)** $n$

**b)** 0

**c)** a ratio

**d)** a denominator

**e)** a numerator

**f)** not defined

**Concept Check** Illustrate each fraction by shading parts of the figure.

**CC1.** $\frac{11}{16}$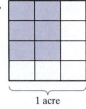

**CC2.** $\frac{4}{9}$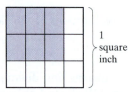

---

**a** Identify the numerator and the denominator.

**1.** $\frac{3}{4}$

**2.** $\frac{-9}{10}$

**3.** $\frac{7}{-9}$

**4.** $\frac{15}{8}$

**5.** $\frac{2x}{3y}$

**6.** $\frac{9a}{2b}$

What part of each object or set of objects is shaded?

**7.**

1 acre

**8.**

1 square inch

**9.**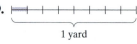

1 yard

**10.**

1 mile

**11.**

**12.**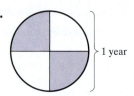

1 year

**13.**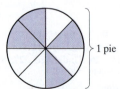

1 pie

**14.**

**15.**

**16.**

**17.** 1 gold bar / 2 gold bars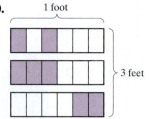

**18.** 1 spool

**19.**  1 quart / 2 quarts

**20.** 1 foot / 3 feet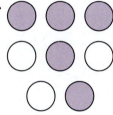

**21.** 1 window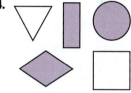

**22.** 1 quart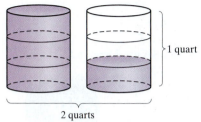

**23.**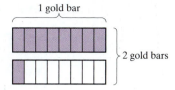

**24.**

**25.**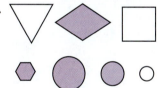

**26.**

What part of an inch is indicated?

**27.**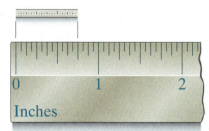

**28.**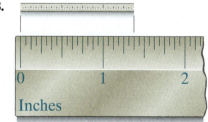

**29.**

**30.**

For each of Exercises 31–34, give fraction notation for the amount of gas **(a)** in the tank and **(b)** used from a full tank.

**31.**    **32.**    **33.**    **34.**

**35.** For the following set of animals, what is the ratio of:
   **a)** puppies to the total number of animals?
   **b)** puppies to kittens?
   **c)** kittens to the total number of animals?
   **d)** kittens to puppies?

**36.** For the following set of sports equipment, what is the ratio of:
   **a)** basketballs to footballs?
   **b)** footballs to basketballs?
   **c)** basketballs to the total number of balls?
   **d)** total number of balls to basketballs?

**37.** Bryce delivers car parts to auto service centers. On Thursday he had 15 deliveries scheduled. By noon he had delivered only 4 orders. What is the ratio of:
   **a)** orders delivered to total number of orders?
   **b)** orders delivered to orders not delivered?
   **c)** orders not delivered to total number of orders?

**38.** *Gas Mileage.* A Volkswagen Passat TDI® SE will travel 473 mi on 11 gal of gasoline in highway driving. What is the ratio of:
   **a)** miles driven to gasoline used?
   **b)** gasoline used to miles driven?

   **Data:** vw.com

For Exercises 39 and 40, use the following bar graph, which shows the number of police officers per 10,000 residents in each of twelve cities.

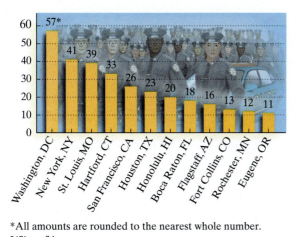

*All amounts are rounded to the nearest whole number.
DATA: ucr.fbi.gov

**39.** What is the ratio of police officers to 10,000 residents in the given city?
  **a)** Washington, DC        **b)** St. Louis, Missouri
  **c)** Hartford, Connecticut  **d)** Honolulu, Hawaii
  **e)** Flagstaff, Arizona     **f)** Rochester, Minnesota

**40.** What is the ratio of police officers to 10,000 residents in the given city?
  **a)** New York, New York     **b)** San Francisco, California
  **c)** Houston, Texas         **d)** Boca Raton, Florida
  **e)** Fort Collins,          **f)** Eugene, Oregon
     Colorado

For Exercises 41 and 42, use the following set of states, as illustrated on the map.

| Alabama | Nebraska | West Virginia |
| Arkansas | South Dakota | Wisconsin |
| Illinois | | |

Mississippi River

**41.** What part of this group of states is east of the Mississippi River?

**42.** What part of this group of states is north of Nashville, Tennessee?

**b**  Simplify. Assume that all variables are nonzero.

**43.** $\dfrac{0}{8}$

**44.** $\dfrac{8}{8}$

**45.** $\dfrac{8-1}{9-8}$

**46.** $\dfrac{16}{1}$

**47.** $\dfrac{20}{20}$

**48.** $\dfrac{-20}{1}$

**49.** $\dfrac{-45}{-45}$

**50.** $\dfrac{11-1}{10-9}$

**51.** $\dfrac{0}{-238}$

**52.** $\dfrac{19x}{19x}$

**53.** $\dfrac{19x}{1}$

**54.** $\dfrac{0}{-16}$

**55.** $\dfrac{13t}{13t}$

**56.** $\dfrac{-27m}{1}$

**57.** $\dfrac{-87}{1}$

**58.** $\dfrac{-98}{-98}$

**59.** $\dfrac{0}{2a}$

**60.** $\dfrac{0}{18}$

**61.** $\dfrac{52}{0}$

**62.** $\dfrac{8-8}{1247}$

**63.** $\dfrac{7n}{1}$

**64.** $\dfrac{1317}{0}$

**65.** $\dfrac{5}{6-6}$

**66.** $\dfrac{13}{10-10}$

## Skill Maintenance

Multiply.

**67.** $-7(30)$  [2.4a]

**68.** $23 \cdot (-14)$  [2.4a]

**69.** $(-71)(-12)0$  [2.4b]

**70.** $32(-29)0$  [2.4b]

Solve.  [1.8a]

**71.** *Libraries.*  There are 9082 public library systems in the United States and 3793 academic libraries. How many more public libraries are there than academic libraries?

**Data:** American Library Association

**72.** *Gas Mileage.*  The 2018 Subaru Impreza gets 38 mpg in highway driving. How many gallons will it use in 2698 mi of highway driving?

**Data:** fueleconomy.gov

## Synthesis

What part of each object is shaded?

**73.**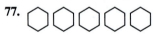

**74.**

**75.**

**76.**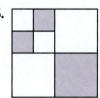

Shade or mark each figure to show $\frac{3}{5}$. Answers may vary.

**77.**

**78.**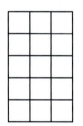

**79.**

**80.**

**81.** The year 2017 began on a Sunday. What fractional part of the days in 2017 were Mondays?

**82.** The year 2018 began on a Monday. What fractional part of the days in 2018 were Mondays?

**83.** The surface of Earth is 3 parts water and 1 part land. What fractional part of Earth is water? land?

**84.** A couple had 3 sons, each of whom had 3 daughters. If each daughter gave birth to 3 sons, what fractional part of the couple's descendants is female?

# Multiplication and Applications

## a MULTIPLICATION BY AN INTEGER

**SKILL REVIEW**

*Multiply whole numbers.* [1.4a]
Multiply.

**1.** $24 \cdot 17$     **2.** $5(13)$

**Answers: 1.** 408   **2.** 65

MyLab Math
VIDEO

We can find $3 \cdot \frac{1}{4}$ by thinking of repeated addition. We add three $\frac{1}{4}$'s.

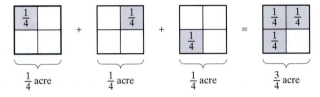

We see that $3 \cdot \frac{1}{4} = \frac{1}{4} + \frac{1}{4} + \frac{1}{4} = \frac{3}{4}$.

**Do Exercises 1 and 2.** ▶

**1.** Find $2 \cdot \frac{1}{3}$.

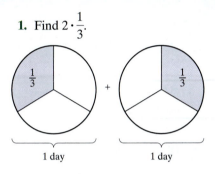

1 day          1 day

**2.** Find $5 \cdot \frac{1}{8}$.

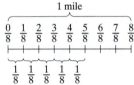

To multiply a fraction by an integer,

**a)** multiply the top number (the numerator) by the integer and

**b)** keep the same denominator.

$$6 \cdot \frac{4}{5} = \frac{6 \cdot 4}{5} = \frac{24}{5}$$

**EXAMPLES** Multiply.

**1.** $5 \times \frac{3}{8} = \frac{5 \times 3}{8} = \frac{15}{8}$     We generally replace the $\times$ symbol with $\cdot$.

Skip this step when you feel comfortable doing so.

**2.** $\frac{2}{5} \cdot 13 = \frac{2 \cdot 13}{5} = \frac{26}{5}$

**3.** $-10 \cdot \frac{1}{3} = \frac{-10}{3}$, or $-\frac{10}{3}$     Recall that $\frac{-a}{b} = -\frac{a}{b}$.

**4.** $a \cdot \frac{4}{7} = \frac{4a}{7}$     Recall that $a \cdot 4 = 4 \cdot a$.

**Do Exercises 3–6.** ▶

Multiply.

**3.** $7 \times \frac{2}{3}$     **4.** $(-11) \cdot \frac{3}{10}$

**5.** $34 \cdot \frac{2}{5}$     **6.** $x \cdot \frac{4}{9}$

***Answers***

**1.** $\frac{2}{3}$  **2.** $\frac{5}{8}$  **3.** $\frac{14}{3}$  **4.** $-\frac{33}{10}$, or $\frac{-33}{10}$

**5.** $\frac{68}{5}$  **6.** $\frac{4x}{9}$

**7.** Draw diagrams like the ones at right to illustrate $\frac{1}{4}$ and $\frac{1}{2} \cdot \frac{1}{4}$.

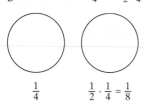

$\frac{1}{4}$

$\frac{1}{2} \cdot \frac{1}{4} = \frac{1}{8}$

## b  MULTIPLICATION USING FRACTION NOTATION

To illustrate the meaning of an expression like $\frac{1}{2} \cdot \frac{1}{3}$, we first represent $\frac{1}{3}$ and then shade half of that region. Note that $\frac{1}{2} \cdot \frac{1}{3}$ is the same as $\frac{1}{2}$ of $\frac{1}{3}$.

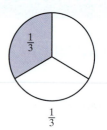
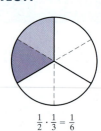

$\frac{1}{3}$

$\frac{1}{2} \cdot \frac{1}{3} = \frac{1}{6}$

◀ **Do Exercise 7.**

To visualize $\frac{2}{5} \cdot \frac{3}{4}$, we first represent $\frac{3}{4}$. This is shown by the shading on the left below. Next, we divide the shaded area into 5 equal parts (using horizontal lines) and take 2 of them. That is shown by the darker shading on the right below. Note that $\frac{2}{5} \cdot \frac{3}{4}$ is the same as $\frac{2}{5}$ of $\frac{3}{4}$.

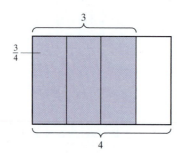
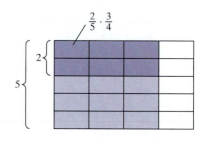

**8.** Draw diagrams like the ones at right to illustrate $\frac{1}{3}$ and $\frac{4}{5} \cdot \frac{1}{3}$.

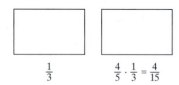

$\frac{1}{3}$

$\frac{4}{5} \cdot \frac{1}{3} = \frac{4}{15}$

The entire object has now been divided into 20 parts, and we have shaded 6 of them twice. Thus,

$$\frac{2}{5} \cdot \frac{3}{4} = \frac{6}{20}.$$

← This is the product of the numerators.
← This is the product of the denominators.

◀ **Do Exercise 8.**

Notice that the product of two fractions is the product of the numerators over the product of the denominators.

> To multiply a fraction by a fraction,
>
> **a)** multiply the numerators and
>
> $$\frac{9}{7} \cdot \frac{3}{4} = \frac{9 \cdot 3}{7 \cdot 4} = \frac{27}{28}$$
>
> **b)** multiply the denominators.

**EXAMPLES**  Multiply.

**5.** $\frac{5}{6} \cdot \frac{7}{4} = \frac{5 \cdot 7}{6 \cdot 4} = \frac{35}{24}$

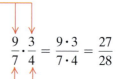

Skip this step when you feel comfortable doing so.

**6.** $\frac{3}{5} \cdot \frac{7}{8} = \frac{3 \cdot 7}{5 \cdot 8} = \frac{21}{40}$

*Answers*

**7.**
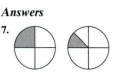

**8.**
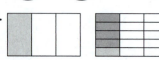

**7.** $\left(-\dfrac{4}{x}\right)\left(-\dfrac{y}{7}\right) = \dfrac{4y}{7x}$   A negative times a negative is positive.

**8.** $(-6)\left(-\dfrac{4}{5}\right) = \dfrac{-6}{1} \cdot \dfrac{-4}{5} = \dfrac{24}{5}$   We can always write $n$ as $\dfrac{n}{1}$.

Do Exercises 9–12. ▶

## c  APPLICATIONS AND PROBLEM SOLVING

Many problems that can be solved by multiplying fractions can be thought of in terms of rectangular arrays.

**EXAMPLE 9**   A real estate developer owns a plot of land and plans to use $\frac{4}{5}$ of the plot for a small strip mall and parking lot. Of this, $\frac{2}{3}$ will be needed for the parking lot. What part of the plot will be used for parking?

1. **Familiarize.**   We first make a drawing to help familiarize ourselves with the problem. The land may not be rectangular, but we can think of it as a rectangle. The strip mall, including the parking lot, uses $\frac{4}{5}$ of the plot. We shade $\frac{4}{5}$ as shown on the left below. The parking lot alone uses $\frac{2}{3}$ of the part we just shaded. We shade that as shown on the right below.

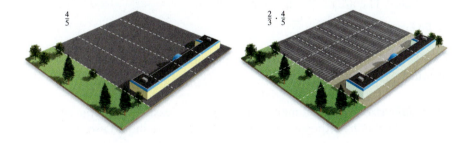

2. **Translate.**   We let $n =$ the part of the plot that is used for parking. We are taking "two-thirds of four-fifths." The word "of" corresponds to multiplication. Thus, the following multiplication sentence corresponds to the situation:

$$\frac{2}{3} \cdot \frac{4}{5} = n.$$

3. **Solve.**   The number sentence tells us what to do. We multiply:

$$\frac{2}{3} \cdot \frac{4}{5} = \frac{2 \cdot 4}{3 \cdot 5} = \frac{8}{15}.$$

Thus, $\dfrac{8}{15} = n.$

4. **Check.**   We can do a partial check by noting that the answer is a fraction less than 1, which we expect since the developer is using only part of the original plot of land. Thus, $\frac{8}{15}$ is a reasonable answer. We can also check this in the figure above, where we see that 8 of 15 parts represent the parking lot.

5. **State.**   The parking lot takes up $\frac{8}{15}$ of the plot of land.

Do Exercise 13. ▶

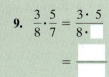
**GS** Multiply.

**9.** $\dfrac{3}{8} \cdot \dfrac{5}{7} = \dfrac{3 \cdot 5}{8 \cdot \boxed{\phantom{0}}}$

$= \dfrac{\boxed{\phantom{0}}}{\boxed{\phantom{0}}}$

**10.** $\dfrac{4}{3} \cdot \dfrac{8}{5}$

**11.** $\left(-\dfrac{3}{10}\right)\left(-\dfrac{1}{10}\right)$

**12.** $(-7)\dfrac{a}{b}$

**13.** A developer plans to set aside $\frac{3}{4}$ of the land in a housing development as open (undeveloped) space. Of this, $\frac{1}{2}$ will be green (natural) space. What part of the land will be green space?

*Answers*

**9.** $\dfrac{15}{56}$   **10.** $\dfrac{32}{15}$   **11.** $\dfrac{3}{100}$

**12.** $-\dfrac{7a}{b}$, or $\dfrac{-7a}{b}$   **13.** $\dfrac{3}{8}$

*Guided Solution:*

**9.** $7, \dfrac{15}{56}$

**14. *Area of a Ceramic Tile.*** The length of a rectangular ceramic tile inlaid on a countertop is $\frac{4}{9}$ ft. The width is $\frac{2}{9}$ ft. What is the area of the tile?

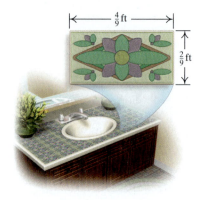

**15.** Of the students at Overton Junior College, $\frac{1}{8}$ participate in sports and $\frac{3}{5}$ of these play football. What fractional part of the students play football?

*Answers*

**14.** $\frac{8}{81}$ ft² **15.** $\frac{3}{40}$

The area of a rectangular region is found by multiplying length by width. That is true whether length and width are whole numbers or not. Remember, the area of a rectangular region is given by the formula

$$A = l \cdot w \quad (Area = length \cdot width).$$

**EXAMPLE 10** *Area of a Cranberry Bog.* The length of a rectangular cranberry bog is $\frac{9}{16}$ mi. The width is $\frac{3}{8}$ mi. What is the area of the bog?

1. **Familiarize.** Recall that area is length times width. We let $A =$ the area of the cranberry bog.

2. **Translate.** Next, we translate:

$$\underbrace{\text{Area}}_{A} \quad \underbrace{\text{is}}_{=} \quad \underbrace{\text{Length}}_{\frac{9}{16}} \quad \underbrace{\text{times}}_{\times} \quad \underbrace{\text{Width}}_{\frac{3}{8}}.$$

3. **Solve.** The sentence tells us what to do. We multiply:

$$A = \frac{9}{16} \cdot \frac{3}{8} = \frac{9 \cdot 3}{16 \cdot 8} = \frac{27}{128}.$$

4. **Check.** We check by repeating the calculation. This is left to the student.

5. **State.** The area is $\frac{27}{128}$ square mile (mi²).

◀ **Do Exercise 14.**

**EXAMPLE 11** A recipe for oatmeal chocolate chip cookies calls for $\frac{3}{4}$ cup of rolled oats. Monica is making $\frac{1}{2}$ of the recipe. How much oats should she use?

1. **Familiarize.** We first make a drawing or at least visualize the situation. We let $n =$ the amount of oats that Monica should use.

$\frac{3}{4}$ cup in the recipe    $\frac{1}{2} \cdot \frac{3}{4}$ cup in $\frac{1}{2}$ the recipe

2. **Translate.** We are finding $\frac{1}{2}$ of $\frac{3}{4}$, so the multiplication sentence $\frac{1}{2} \cdot \frac{3}{4} = n$ corresponds to the situation.

3. **Solve.** We carry out the multiplication:

$$\frac{1}{2} \cdot \frac{3}{4} = \frac{1 \cdot 3}{2 \cdot 4} = \frac{3}{8}.$$

Thus, $\frac{3}{8} = n$.

4. **Check.** We check by repeating the calculation. This is left to the student.

5. **State.** Monica should use $\frac{3}{8}$ cup of oats.

◀ **Do Exercise 15.**

## ✓ Check Your Understanding

**Reading Check**  Determine whether each statement is true or false.

_____ **RC1.** Multiplying $\frac{1}{2} \cdot \frac{3}{5}$ is the same as finding $\frac{1}{2}$ of $\frac{3}{5}$.

_____ **RC2.** When we multiply two fractions, the new numerator is the product of the numerators in the two fractions.

_____ **RC3.** The whole number 6 can be written $\frac{6}{1}$.

_____ **RC4.** The product of two fractions can be smaller than either of the two fractions.

**Concept Check**  Draw a diagram to show the multiplication.

**CC1.** $\frac{2}{3} \cdot \frac{2}{5}$

**CC2.** $\frac{1}{4} \cdot \frac{1}{6}$

---

**a**  Multiply.

**1.** $3 \cdot \frac{1}{8}$

**2.** $2 \cdot \frac{1}{5}$

**3.** $(-5) \times \frac{1}{6}$

**4.** $(-4) \times \frac{1}{7}$

**5.** $\frac{2}{3} \cdot 7$

**6.** $\frac{2}{5} \cdot 7$

**7.** $(-1)\frac{7}{9}$

**8.** $(-1)\frac{4}{11}$

**9.** $\frac{5}{6} \cdot x$

**10.** $\left(-\frac{7}{8}\right)y$

**11.** $\frac{2}{5}(-3)$

**12.** $\frac{3}{5}(-4)$

**13.** $a \cdot \frac{2}{7}$

**14.** $b \cdot \frac{3}{8}$

**15.** $-17 \times \frac{m}{6}$

**16.** $\frac{n}{7} \cdot 30$

**17.** $-3 \cdot \frac{-2}{5}$

**18.** $-4 \cdot \frac{-5}{7}$

**19.** $-\frac{2}{7}(-x)$

**20.** $-\frac{3}{4}(-a)$

---

**b**  Multiply.

**21.** $\frac{2}{5} \cdot \frac{2}{3}$

**22.** $\frac{3}{4} \cdot \frac{3}{5}$

**23.** $\left(-\frac{1}{4}\right) \times \frac{1}{10}$

**24.** $\left(-\frac{1}{3}\right) \times \frac{1}{10}$

**25.** $\frac{2}{3} \times \frac{1}{5}$

**26.** $\frac{3}{5} \times \frac{1}{5}$

**27.** $\frac{2}{y} \cdot \frac{x}{9}$

**28.** $\left(-\frac{3}{4}\right)\left(-\frac{3}{5}\right)$

**29.** $\left(-\frac{3}{4}\right)\left(-\frac{3}{4}\right)$

**30.** $\frac{3}{b} \cdot \frac{a}{7}$

31. $\dfrac{2}{3} \cdot \dfrac{7}{13}$

32. $\dfrac{3}{11} \cdot \dfrac{4}{5}$

33. $\dfrac{1}{10}\left(\dfrac{-3}{5}\right)$

34. $\dfrac{3}{10}\left(\dfrac{-7}{5}\right)$

35. $\dfrac{7}{8} \cdot \dfrac{a}{8}$

36. $\dfrac{4}{5} \cdot \dfrac{7}{x}$

37. $\dfrac{1}{y} \cdot 100$

38. $\dfrac{b}{10} \cdot 13$

39. $\dfrac{-21}{4} \cdot \dfrac{7}{5}$

40. $\dfrac{-8}{3} \cdot \dfrac{20}{9}$

 Solve.

41. **Hair Bows.** It takes $\frac{5}{3}$ yd of ribbon to make a hair bow. How much ribbon is needed to make 8 bows?

42. **Gasoline Can Capacity.** A gasoline can holds $\frac{5}{2}$ gal. How much will the can hold when it is $\frac{1}{2}$ full?

43. **Women's Basketball: High School to Pro.** One of 26 girls who play high school basketball also plays college basketball. One of 474 women who play college basketball also plays professional basketball. What fractional part of female high school basketball players play professional basketball?

Data: NCAA, March 10, 2017

44. **Men's Soccer: High School to Pro.** One of 18 boys who play high school soccer also plays college soccer. One of 330 men who play college soccer also plays professional soccer. What fractional part of male high school soccer players play professional soccer?

Data: NCAA, March 10, 2017

45. **Serving of Cheesecake.** At the Cheesecake Factory, a piece of cheesecake is $\frac{1}{12}$ of a cheesecake. How much of the cheesecake is $\frac{1}{2}$ piece?

Data: The Cheesecake Factory

46. **Tossed Salad.** The recipe for a tossed salad calls for $\frac{3}{4}$ cup of sliced almonds. How much is needed to make $\frac{1}{2}$ of the recipe?

47. **Floor Tiling.** The floor of a room is being covered with tile. An area $\frac{3}{5}$ of the length and $\frac{3}{4}$ of the width is covered. What fraction of the floor has been tiled?

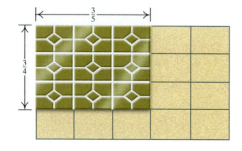

48. **Table Top Size.** A rectangular table top measures $\frac{4}{5}$ m long by $\frac{3}{5}$ m wide. What is its area?

## Skill Maintenance

Simplify.  [1.9c]

49. $8 \cdot 12 - (63 \div 9 + 13 \cdot 3)$

50. $(10 - 3)^4 + 10^3 \cdot 4 - 10 \div 5$

## Synthesis

Multiply. Write each answer using fraction notation.

51. ▦ $\dfrac{341}{517} \cdot \dfrac{209}{349}$

52. ▦ $\left(\dfrac{57}{61}\right)^3$

53. $\left(\dfrac{2}{5}\right)^3\left(-\dfrac{7}{9}\right)$

54. $\left(-\dfrac{1}{2}\right)^5\left(\dfrac{3}{5}\right)$

55. A chain saw holds $\frac{1}{5}$ gal of fuel. Chain saw fuel is $\frac{1}{16}$ two-cycle oil and $\frac{15}{16}$ unleaded gasoline. How much two-cycle oil is in a freshly filled chain saw?

56. Evaluate $-\frac{2}{3}xy$ for $x = \frac{2}{5}$ and $y = -\frac{1}{7}$.

# Simplifying

## a  MULTIPLYING BY 1

Recall the following:

$$1 = \frac{1}{1} = \frac{2}{2} = \frac{3}{3} = \frac{4}{4} = \frac{10}{10} = \frac{45}{45} = \frac{100}{100} = \frac{n}{n}.$$

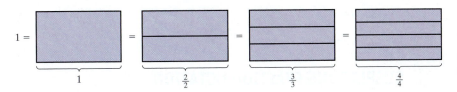

### OBJECTIVES

a  Multiply by 1 to find an equivalent expression with a specified denominator.

b  Simplify fraction notation.

c  Use the test for equality to determine whether two fractions are equivalent.

The multiplicative identity states that for any number $a$, $1 \cdot a = a \cdot 1 = a$. Since any nonzero number divided by itself is 1, we can state the multiplicative identity using fraction notation.

---

**MULTIPLICATIVE IDENTITY FOR FRACTIONS**

When we multiply a number by 1, we get the same number:

$$a = a \cdot 1 = a \cdot \frac{n}{n} = a.$$

---

For example, $\frac{3}{5} = \frac{3}{5} \cdot 1 = \frac{3}{5} \cdot \frac{4}{4} = \frac{12}{20}$. Since $\frac{3}{5} = \frac{12}{20}$, we say that $\frac{3}{5}$ and $\frac{12}{20}$ are **equivalent fractions.**

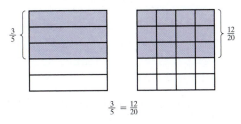

$$\frac{3}{5} = \frac{12}{20}$$

**Do Exercises 1–4.** ▶

Multiply.

1. $\dfrac{1}{2} \cdot \dfrac{8}{8}$    2. $\dfrac{3}{5} \cdot \dfrac{a}{a}$

3. $-\dfrac{13}{25} \cdot \dfrac{4}{4}$    4. $\dfrac{8}{3}\left(\dfrac{-2}{-2}\right)$

Suppose we want to find another name for $\frac{2}{3}$, one that has a denominator of 15. We can multiply by 1 to find equivalent fractions. Since $15 = 3 \cdot 5$, we choose $\frac{5}{5}$ for 1 in order to get a denominator of 15:

$$\frac{2}{3} = \frac{2}{3} \cdot 1 = \frac{2}{3} \cdot \frac{5}{5} = \frac{2 \cdot 5}{3 \cdot 5} = \frac{10}{15}.$$

**EXAMPLE 1**  Find a fraction equivalent to $\frac{1}{4}$ with a denominator of 24.

Since $4 \cdot 6 = 24$, we multiply by $\frac{6}{6}$:

$$\frac{1}{4} = \frac{1}{4} \cdot \frac{6}{6} = \frac{1 \cdot 6}{4 \cdot 6} = \frac{6}{24}.$$

The numbers $\frac{1}{4}$ and $\frac{6}{24}$ represent the same number. They are equivalent.

*Answers*

1. $\dfrac{8}{16}$   2. $\dfrac{3a}{5a}$   3. $-\dfrac{52}{100}$   4. $\dfrac{-16}{-6}$

Find an equivalent expression, using the denominator indicated. Use multiplication by 1.

**5.** $\dfrac{4}{3} = \dfrac{?}{15}$

$$\dfrac{4}{3} = \dfrac{4}{3} \cdot \dfrac{\square}{\square}$$

$$= \dfrac{4 \cdot 5}{3 \cdot 5}$$

$$= \dfrac{\square}{15}$$

**6.** $\dfrac{3}{4} = \dfrac{?}{-24}$

**7.** $\dfrac{9}{10} = \dfrac{?}{10x}$

**8.** $\dfrac{15}{1} = \dfrac{?}{2}$

**9.** $\dfrac{-8}{7} = \dfrac{?}{49}$

**10.** $\dfrac{1}{-2} = \dfrac{?}{-6}$

**Answers**

**5.** $\dfrac{20}{15}$   **6.** $\dfrac{-18}{-24}$   **7.** $\dfrac{9x}{10x}$
**8.** $\dfrac{30}{2}$   **9.** $\dfrac{-56}{49}$   **10.** $\dfrac{3}{-6}$

**Guided Solution:**

**5.** $\dfrac{5}{5}$, 20

**EXAMPLE 2** Find a number equivalent to $\frac{2}{5}$ with a denominator of $-35$.

Since $5(-7) = -35$, we multiply by 1, using $\frac{-7}{-7}$:

$$\dfrac{2}{5} = \dfrac{2}{5}\left(\dfrac{-7}{-7}\right) = \dfrac{2(-7)}{5(-7)} = \dfrac{-14}{-35}.$$

**EXAMPLE 3** Find an expression equivalent to $\frac{9}{8}$ with a denominator of $8a$.

Since $8 \cdot a = 8a$, we multiply by 1, using $\dfrac{a}{a}$:

$$\dfrac{9}{8} \cdot \dfrac{a}{a} = \dfrac{9a}{8a}.$$

◀ **Do Exercises 5–10.**

## b   SIMPLIFYING FRACTION NOTATION

All of the following are names for three-fourths:

$$\dfrac{3}{4}, \quad \dfrac{-3}{-4}, \quad \dfrac{6}{8}, \quad \dfrac{30}{40}, \quad \dfrac{-15}{-20}.$$

We say that $\frac{3}{4}$ is **simplest** because it has the smallest positive denominator. Note that 3 and 4 have no factor in common other than 1.

To simplify fraction notation, we reverse the process of multiplying by 1:

$$\dfrac{12}{18} = \dfrac{2 \cdot 6}{3 \cdot 6} \quad \begin{array}{l} \leftarrow \text{Factoring the numerator} \\ \leftarrow \text{Factoring the denominator} \end{array}$$

$$= \dfrac{2}{3} \cdot \dfrac{6}{6} \quad \text{Factoring the fraction}$$

$$= \dfrac{2}{3} \cdot 1 \quad\quad \dfrac{6}{6} = 1$$

$$= \dfrac{2}{3}. \quad\quad \text{Removing the factor 1: } \dfrac{2}{3} \cdot 1 = \dfrac{2}{3}$$

---

**SIMPLIFYING FRACTION NOTATION**

1. Factor the numerator and factor the denominator.
2. Identify any common factors in the numerator and denominator.
3. Use the common factors to remove a factor equal to 1.

---

**EXAMPLES** Simplify.

**4.** $\dfrac{-8}{20} = \dfrac{-2 \cdot 4}{5 \cdot 4} = \dfrac{-2}{5} \cdot \dfrac{4}{4} = \dfrac{-2}{5}$    Removing a factor equal to 1: $\dfrac{4}{4} = 1$

**5.** $\dfrac{2}{6} = \dfrac{1 \cdot 2}{3 \cdot 2} = \dfrac{1}{3} \cdot \dfrac{2}{2} = \dfrac{1}{3}$    Writing 1 allows for pairing of factors in the numerator and the denominator.

**6.** $\dfrac{30}{6} = \dfrac{5 \cdot 6}{1 \cdot 6} = \dfrac{5}{1} \cdot \dfrac{6}{6} = \dfrac{5}{1} = 5$ ← We could also simplify $\frac{30}{6}$ by doing the division $30 \div 6$. That is, $\frac{30}{6} = 30 \div 6 = 5$.

**7.** $-\dfrac{15}{10} = -\dfrac{3 \cdot 5}{2 \cdot 5} = -\dfrac{3}{2} \cdot \dfrac{5}{5} = -\dfrac{3}{2}$    Removing a factor equal to 1: $\dfrac{5}{5} = 1$

**8.** $\dfrac{4x}{15x} = \dfrac{4 \cdot x}{15 \cdot x} = \dfrac{4}{15} \cdot \dfrac{x}{x} = \dfrac{4}{15}$   Removing a factor equal to 1: $\dfrac{x}{x} = 1$

Do Exercises 11–16. ▶

The use of prime factorizations can be helpful for simplifying.

**EXAMPLE 9**   Simplify: $\dfrac{90}{84}$.

$$\dfrac{90}{84} = \dfrac{2 \cdot 3 \cdot 3 \cdot 5}{2 \cdot 2 \cdot 3 \cdot 7}$$   Factoring the numerator and the denominator into primes

$$= \dfrac{2 \cdot 3 \cdot 3 \cdot 5}{2 \cdot 3 \cdot 2 \cdot 7}$$   Changing the order so that like primes are above and below each other

$$= \dfrac{2}{2} \cdot \dfrac{3}{3} \cdot \dfrac{3 \cdot 5}{2 \cdot 7}$$   Factoring the fraction

$$= \dfrac{3 \cdot 5}{2 \cdot 7}$$   Removing factors of 1

$$= \dfrac{15}{14}$$

**EXAMPLE 10**   Simplify: $\dfrac{105}{135}$.

Since both 105 and 135 end in 5, we know that 5 is a factor of both the numerator and the denominator:

$$\dfrac{105}{135} = \dfrac{21 \cdot 5}{27 \cdot 5} = \dfrac{21}{27} \cdot \dfrac{5}{5} = \dfrac{21}{27}.$$   To find the 21, we divided 105 by 5. To find the 27, we divided 135 by 5.

A fraction is not "simplified" if common factors of the numerator and the denominator remain. Because 21 and 27 are both divisible by 3, we must simplify further:

$$\dfrac{105}{135} = \dfrac{21}{27} = \dfrac{7 \cdot 3}{9 \cdot 3} = \dfrac{7}{9} \cdot \dfrac{3}{3} = \dfrac{7}{9}.$$   To find the 7, we divided 21 by 3. To find the 9, we divided 27 by 3.

**EXAMPLE 11**   Simplify: $\dfrac{322}{434}$.

Since 322 and 434 are both even, we know that 2 is a common factor:

$$\dfrac{322}{434} = \dfrac{2 \cdot 161}{2 \cdot 217} = \dfrac{2}{2} \cdot \dfrac{161}{217} = \dfrac{161}{217}$$   Removing a factor equal to 1: $\dfrac{2}{2} = 1$

$$= \dfrac{7 \cdot 23}{7 \cdot 31} = \dfrac{7}{7} \cdot \dfrac{23}{31} = \dfrac{23}{31}.$$   7 is also a common factor; removing a factor equal to 1

We found the common factor 7 by focusing first on 161. After determining that 7 is a factor of 161, we checked to see if 7 is also a factor of 217.

Do Exercises 17–23. ▶

Simplify.

**11.** $\dfrac{8}{14}$

**12.** $\dfrac{-10}{12}$

**13.** $\dfrac{40}{8}$

**14.** $\dfrac{4a}{3a}$

**15.** $-\dfrac{50}{30}$

**16.** $\dfrac{x}{3x}$

Simplify.

**17.** $\dfrac{-35}{40}$

**18.** $\dfrac{801}{702}$

**19.** $\dfrac{24}{21}$

**20.** $\dfrac{429}{561}$

**21.** $\dfrac{280}{960}$

**22.** $\dfrac{1332}{2880}$

**23.** Simplify each fraction in this circle graph.

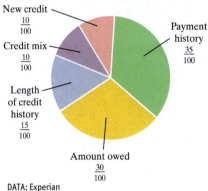

**Factors Used to Calculate FICO Scores**

New credit $\dfrac{10}{100}$

Credit mix $\dfrac{10}{100}$

Length of credit history $\dfrac{15}{100}$

Amount owed $\dfrac{30}{100}$

Payment history $\dfrac{35}{100}$

DATA: Experian

*Answers*

**11.** $\dfrac{4}{7}$   **12.** $\dfrac{-5}{6}$   **13.** 5   **14.** $\dfrac{4}{3}$   **15.** $-\dfrac{5}{3}$

**16.** $\dfrac{1}{3}$   **17.** $\dfrac{-7}{8}$   **18.** $\dfrac{89}{78}$   **19.** $\dfrac{8}{7}$   **20.** $\dfrac{13}{17}$

**21.** $\dfrac{7}{24}$   **22.** $\dfrac{37}{80}$

**23.** $\dfrac{35}{100} = \dfrac{7}{20}; \dfrac{30}{100} = \dfrac{3}{10}; \dfrac{15}{100} = \dfrac{3}{20};$

$\dfrac{10}{100} = \dfrac{1}{10}; \dfrac{10}{100} = \dfrac{1}{10}$

## Canceling

Canceling is a shortcut that you may have used for removing a factor that equals 1 when working with fraction notation. With concern, we mention it as a possibility for speeding up your work. Canceling may be done only when removing common factors in numerators and denominators. Each pair of common factors allows us to remove a factor equal to 1.

Our concern is that canceling be done with care and understanding. Generally, slashes are used to indicate factors equal to 1 that have been removed. For instance, Example 9 might have been done faster as follows:

$$\frac{90}{84} = \frac{2 \cdot 3 \cdot 3 \cdot 5}{2 \cdot 2 \cdot 3 \cdot 7} \qquad \text{Factoring the numerator and the denominator}$$

$$= \frac{2 \cdot 3 \cdot 3 \cdot 5}{2 \cdot 2 \cdot 3 \cdot 7} \qquad \begin{array}{l}\text{When a factor equal to 1 is noted,} \\ \text{it is canceled as shown: } \dfrac{2}{2} \cdot \dfrac{3}{3} = 1.\end{array}$$

$$= \frac{3 \cdot 5}{2 \cdot 7} = \frac{15}{14}.$$

············· **Caution!** ·············

The problem with canceling is that it is often applied incorrectly in situations like the following:

$$\frac{2 + 3}{2} = 3; \qquad \frac{4 + 1}{4 + 2} = \frac{1}{2}; \qquad \frac{15}{54} = \frac{1}{4}.$$

$$\quad\text{Wrong!} \qquad\qquad \text{Wrong!} \qquad\qquad \text{Wrong!}$$

The correct answers are

$$\frac{2 + 3}{2} = \frac{5}{2}; \qquad \frac{4 + 1}{4 + 2} = \frac{5}{6}; \qquad \frac{15}{54} = \frac{3 \cdot 5}{3 \cdot 18} = \frac{3}{3} \cdot \frac{5}{18} = \frac{5}{18}.$$

In each incorrect case, the numbers canceled did not form a factor equal to 1. Factors are parts of products. For example, in $2 \cdot 3$, the numbers 2 and 3 are factors, but in $2 + 3$, 2 and 3 are terms, not factors.

- **If you cannot factor, do not cancel! If in doubt, do not cancel!**
- **Only factors can be canceled, and factors are never separated by + or − signs.**

·····································

**c** **A TEST FOR EQUALITY**

When denominators are the same, we say that fractions have a **common denominator**. When fractions have a common denominator, we can compare them by comparing numerators. Suppose we want to compare $\frac{3}{6}$ and $\frac{2}{4}$. First, we find a common denominator. To do this, we multiply each fraction by 1, using the denominator of the other fraction to form the symbol for 1. We multiply $\frac{3}{6}$ by $\frac{4}{4}$ and $\frac{2}{4}$ by $\frac{6}{6}$:

$$\frac{3}{6} = \frac{3}{6} \cdot \frac{4}{4} = \frac{3 \cdot 4}{6 \cdot 4} = \frac{12}{24}; \qquad \text{Multiplying by } \frac{4}{4}$$

$$\frac{2}{4} = \frac{2}{4} \cdot \frac{6}{6} = \frac{2 \cdot 6}{4 \cdot 6} = \frac{12}{24}. \qquad \text{Multiplying by } \frac{6}{6}$$

Once we have a common denominator, 24, we compare the numerators. And since these numerators are both 12, the fractions are equal:

$$\frac{3}{6} = \frac{2}{4}.$$

The "key" to the above work is that $3 \cdot 4$ and $2 \cdot 6$ are equal. Had these products differed, we would have shown that $\frac{3}{6}$ and $\frac{2}{4}$ were *not* equal.

---

### A TEST FOR EQUALITY

Two fractions are equal if their cross products are equal.

We multiply these two numbers: $3 \cdot 4 = 12$.

We multiply these two numbers: $6 \cdot 2 = 12$.

$$\frac{3}{6} \,\square\, \frac{2}{4}$$

We call $3 \cdot 4$ and $6 \cdot 2$ **cross products**. Since the cross products are the same—that is, $3 \cdot 4 = 6 \cdot 2$—we know that $\dfrac{3}{6} = \dfrac{2}{4}$.

---

In the sentence $a \neq b$, the symbol $\neq$ means "is not equal to."

**EXAMPLE 12**   Use $=$ or $\neq$ for $\square$ to write a true sentence:

$$\frac{6}{7} \,\square\, \frac{7}{8}.$$

We multiply these two numbers: $6 \cdot 8 = 48$.

We multiply these two numbers: $7 \cdot 7 = 49$.

$$\frac{6}{7} \,\square\, \frac{7}{8}$$

Because $48 \neq 49$, $\frac{6}{7}$ and $\frac{7}{8}$ do not name the same number. Thus, $\frac{6}{7} \neq \frac{7}{8}$.

**EXAMPLE 13**   Use $=$ or $\neq$ for $\square$ to write a true sentence:

$$\frac{6}{10} \,\square\, \frac{3}{5}.$$

We multiply these two numbers: $6 \cdot 5 = 30$.

We multiply these two numbers: $10 \cdot 3 = 30$.

$$\frac{6}{10} \,\square\, \frac{3}{5}$$

Because the cross products are the same, we have $\frac{6}{10} = \frac{3}{5}$.

Remembering that $\dfrac{-a}{b}$, $\dfrac{a}{-b}$, and $-\dfrac{a}{b}$ all represent the same number can be helpful when checking for equality.

**EXAMPLE 14**   Use $=$ or $\neq$ for $\square$ to write a true sentence:

$$\frac{-6}{8} \,\square\, -\frac{9}{12}.$$

We rewrite $-\frac{9}{12}$ as $\frac{-9}{12}$ and then check cross products:

$$-6 \cdot 12 = -72 \qquad\qquad\qquad 8(-9) = -72$$

$$\frac{-6}{8} \,\square\, \frac{-9}{12}$$

Because the cross products are the same, we have $\frac{-6}{8} = -\frac{9}{12}$.

**Do Exercises 24–26.** ▶

---

Use $=$ or $\neq$ for $\square$ to write a true sentence.

**24.** $\dfrac{2}{6} \,\square\, \dfrac{3}{9}$

**GS** **25.** $\dfrac{2}{3} \,\square\, \dfrac{14}{20}$

$2 \cdot \blacksquare = 40 \qquad\qquad 3 \cdot \blacksquare = 42$

$$\frac{2}{3} \,\square\, \frac{14}{20}$$

Since $40 \neq 42$, $\dfrac{2}{3} \,\square\, \dfrac{14}{20}$.

**26.** $-\dfrac{10}{15} \,\square\, \dfrac{8}{-12}$

*Answers*
**24.** $=$   **25.** $\neq$   **26.** $=$
*Guided Solution:*
**25.** 20, 14, $\neq$

✓ **Check Your Understanding**

**Reading Check**   Complete each statement with the appropriate the word from the following list.

    common               cross               equivalent               simplify

**RC1.** _____ fractions name the same number.

**RC2.** To _____ a fraction, we find a fraction that names the same number and that has a numerator and a denominator with no common factor.

**RC3.** The fractions $\frac{2}{7}$ and $\frac{4}{7}$ have a _____ denominator.

**RC4.** Two fractions are equal if their _____ products are equal.

**Concept Check**   Determine if the fractions are equivalent.

**CC1.** $\frac{12}{22}, \frac{6}{11}$            **CC2.** $\frac{5}{5}, \frac{13}{13}$            **CC3.** $\frac{7}{2}, \frac{2}{7}$

**CC4.** $\frac{4}{9}, \frac{16}{27}$            **CC5.** $\frac{1}{5}, \frac{7}{35}$            **CC6.** $\frac{7}{6}, \frac{28}{24}$

**a**   Find an equivalent expression for each number, using the denominator indicated. Use multiplication by 1.

**1.** $\frac{1}{2} = \frac{?}{10}$      **2.** $\frac{1}{6} = \frac{?}{12}$      **3.** $\frac{3}{4} = \frac{?}{-48}$      **4.** $\frac{2}{9} = \frac{?}{-18}$

**5.** $\frac{7}{10} = \frac{?}{50}$      **6.** $\frac{3}{8} = \frac{?}{48}$      **7.** $\frac{11}{5} = \frac{?}{5t}$      **8.** $\frac{5}{3} = \frac{?}{3a}$

**9.** $\frac{5}{1} = \frac{?}{4}$      **10.** $\frac{7}{1} = \frac{?}{5}$      **11.** $-\frac{17}{18} = -\frac{?}{54}$      **12.** $-\frac{11}{16} = -\frac{?}{256}$

**13.** $\frac{3}{-8} = \frac{?}{-40}$      **14.** $\frac{7}{-8} = \frac{?}{-32}$      **15.** $\frac{-7}{22} = \frac{?}{132}$      **16.** $\frac{-10}{21} = \frac{?}{126}$

**17.** $\frac{1}{8} = \frac{?}{8x}$      **18.** $\frac{1}{3} = \frac{?}{3a}$      **19.** $\frac{-10}{7} = \frac{?}{7a}$      **20.** $\frac{-4}{3} = \frac{?}{3n}$

**21.** $\frac{4}{9} = \frac{?}{9ab}$      **22.** $\frac{8}{11} = \frac{?}{11xy}$      **23.** $\frac{4}{9} = \frac{?}{27b}$      **24.** $\frac{8}{11} = \frac{?}{55y}$

**b**   Simplify.

**25.** $\frac{2}{4}$      **26.** $\frac{3}{6}$      **27.** $-\frac{6}{9}$      **28.** $\frac{-9}{12}$      **29.** $\frac{10}{25}$      **30.** $\frac{8}{10}$

**31.** $\frac{24}{8}$      **32.** $\frac{36}{9}$      **33.** $\frac{27}{36}$      **34.** $\frac{30}{40}$      **35.** $-\frac{24}{14}$      **36.** $-\frac{16}{10}$

**37.** $\dfrac{3n}{4n}$    **38.** $\dfrac{7x}{8x}$    **39.** $\dfrac{-17}{51}$    **40.** $-\dfrac{13}{26}$    **41.** $\dfrac{-100}{20}$    **42.** $\dfrac{-150}{25}$

**43.** $\dfrac{420}{480}$    **44.** $\dfrac{180}{240}$    **45.** $\dfrac{-540}{810}$    **46.** $\dfrac{-1000}{1080}$    **47.** $\dfrac{12x}{30x}$    **48.** $\dfrac{54n}{90n}$

**49.** $\dfrac{153}{136}$    **50.** $\dfrac{117}{91}$    **51.** $\dfrac{132}{143}$    **52.** $\dfrac{91}{259}$    **53.** $\dfrac{221}{247}$    **54.** $\dfrac{299}{403}$

**55.** $\dfrac{3ab}{8ab}$    **56.** $\dfrac{6xy}{7xy}$    **57.** $\dfrac{9xy}{6x}$    **58.** $\dfrac{10ab}{15a}$    **59.** $\dfrac{-18a}{20ab}$    **60.** $\dfrac{-19x}{38xy}$

**c**   Use = or ≠ for ☐ to write a true sentence.

**61.** $\dfrac{3}{4} \,\square\, \dfrac{9}{12}$    **62.** $\dfrac{4}{8} \,\square\, \dfrac{3}{6}$    **63.** $\dfrac{1}{5} \,\square\, \dfrac{2}{9}$    **64.** $\dfrac{1}{4} \,\square\, \dfrac{2}{9}$

**65.** $\dfrac{3}{8} \,\square\, \dfrac{6}{16}$    **66.** $\dfrac{2}{6} \,\square\, \dfrac{6}{18}$    **67.** $\dfrac{2}{5} \,\square\, \dfrac{2}{7}$    **68.** $\dfrac{3}{10} \,\square\, \dfrac{3}{11}$

**69.** $\dfrac{-3}{10} \,\square\, \dfrac{-4}{12}$    **70.** $\dfrac{-2}{9} \,\square\, \dfrac{-8}{36}$    **71.** $-\dfrac{12}{9} \,\square\, \dfrac{-8}{6}$    **72.** $\dfrac{-8}{7} \,\square\, -\dfrac{16}{14}$

**73.** $\dfrac{5}{-2} \,\square\, -\dfrac{17}{7}$    **74.** $-\dfrac{10}{3} \,\square\, \dfrac{24}{-7}$    **75.** $\dfrac{305}{145} \,\square\, \dfrac{122}{58}$    **76.** $\dfrac{425}{165} \,\square\, \dfrac{130}{66}$

## Skill Maintenance

Solve.   [1.7b]

**77.** $5280 = 1760 + t$    **78.** $10{,}947 = 123 \cdot y$    **79.** $8797 = y + 2299$    **80.** $x \cdot 74 = 6290$

## Synthesis

Simplify. Use a list of prime numbers (see p. 160).

**81.** $\dfrac{391}{667}$    **82.** $\dfrac{209ab}{247ac}$    **83.** $-\dfrac{1073x}{555y}$    **84.** ▦ $\dfrac{3473}{3197}$

**85.** Sociologists have found that 4 of 10 people are shy. Write fraction notation for **(a)** the part of the population that is shy and **(b)** the part that is not shy. Simplify.

**86.** Sociologists estimate that 3 of 20 people are left-handed. In a crowd of 460 people, how many would you expect to be left-handed?

**87.** *Baseball Batting Averages.* For the 2017 season, José Altuve, of the Houston Astros, won the American League batting title with 204 hits in 590 times at bat. Charlie Blackmon, of the Colorado Rockies, won the National League title with 213 hits in 664 times at bat. Did they have the same fraction for hits per times at bat (batting average)? Why or why not?

**Data:** Major League Baseball

**88.** ▦ On a test with 82 questions, Taylor got 63 correct. On another test with 100 questions, she got 77 correct. Did she get the same portion of each test correct? Why or why not?

# Mid-Chapter Review

## Concept Reinforcement

Determine whether each statement is true or false.

_____ **1.** A number $a$ is divisible by another number $b$ if $b$ is factor of $a$.  [3.2a]

_____ **2.** If a number is not divisible by 6, then it is not divisible by 3.  [3.1b]

_____ **3.** The fraction $\frac{9}{4}$ is equal to the fraction $\frac{13}{6}$.  [3.5c]

_____ **4.** The number 1 is not prime.  [3.2b]

## Guided Solutions

 Fill in each blank with the number that creates a correct statement or solution.

**5.** $\dfrac{25}{\boxed{\phantom{0}}} = 1$  [3.3b]    **6.** $\dfrac{\boxed{\phantom{0}}}{9} = 0$  [3.3b]    **7.** $\dfrac{8}{\boxed{\phantom{0}}} = 8$  [3.3b]    **8.** $\dfrac{6}{13} = \dfrac{\boxed{\phantom{0}}}{39}$  [3.5a]

**9.** Simplify: $\dfrac{70}{225}$.  [3.5b]

$$\frac{70}{225} = \frac{2 \cdot \boxed{\phantom{0}} \cdot 7}{\boxed{\phantom{0}} \cdot 3 \cdot 5 \cdot \boxed{\phantom{0}}}$$  Factoring the numerator
Factoring the denominator

$$= \frac{5}{5} \cdot \frac{\boxed{\phantom{0}} \cdot 7}{3 \cdot \boxed{\phantom{0}} \cdot 5}$$  Factoring the fraction

$$= \boxed{\phantom{0}} \cdot \frac{\boxed{\phantom{0}}}{45} \qquad \frac{5}{5} = 1$$

$$= \frac{\boxed{\phantom{0}}}{\boxed{\phantom{0}}}$$  Removing the factor 1

## Mixed Review

To answer Exercises 10–14, consider the following numbers.  [3.1a, b]

| | | | |
|---|---|---|---|
| 84 | 132 | 594 | 350 |
| 300 | 500 | 120 | 14,850 |
| 17,576 | 180 | 1125 | 504 |
| 224 | 351 | 495 | 1632 |

**10.** Which of the above are divisible by 2 but not by 10?

**11.** Which of the above are divisible by 4 but not by 8?

**12.** Which of the above are divisible by 4 but not by 6?

**13.** Which of the above are divisible by 3 but not by 9?

**14.** Which of the above are divisible by 4, 5, and 6?

Determine whether each number is prime, composite, or neither.  [3.2b]

**15.** 61    **16.** 2    **17.** 91    **18.** 1

Find all the factors of each composite number. Then find the prime factorization of the number. [3.2a], [3.2c]

**19.** 160

**20.** 222

**21.** 98

**22.** 315

What part of each object or set of objects is shaded? [3.3a]

**23.**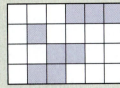

**24.**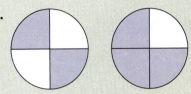

Multiply. [3.4a], [3.4b]

**25.** $7 \cdot \dfrac{1}{9}$

**26.** $\dfrac{4}{15} \cdot \dfrac{2}{3}$

**27.** $\dfrac{5}{11}(-8)$

Simplify. [3.3b], [3.5b]

**28.** $\dfrac{24}{60}$

**29.** $\dfrac{220n}{60n}$

**30.** $\dfrac{17x}{17x}$

**31.** $\dfrac{0}{-23}$

**32.** $\dfrac{54}{186}$

**33.** $\dfrac{-36}{20}$

**34.** $\dfrac{75}{630}$

**35.** $\dfrac{315}{435}$

**36.** $\dfrac{14}{0}$

Use = or ≠ for ☐ to write a true sentence. [3.5c]

**37.** $\dfrac{3}{-7}$ ☐ $\dfrac{-39}{91}$

**38.** $\dfrac{19}{3}$ ☐ $\dfrac{95}{18}$

**39.** *Job Applications.* Of every 200 online applications started, only 25 are reviewed by a hiring manager. What is the ratio of applications reviewed to applications started? [3.3a], [3.5b]

**Data:** Talent Function Group LLC, in "Your Résumé vs. Oblivion," wsj.com, 1/24/12

**40.** *Area of an Ice-Skating Rink.* The length of a rectangular ice-skating rink in the atrium of a shopping mall is $\frac{7}{100}$ mi. The width is $\frac{3}{100}$ mi. What is the area of the rink? [3.4c]

## Understanding Through Discussion and Writing

**41.** Explain a method for finding a composite number that contains exactly two factors other than itself and 1. [3.2b]

**42.** Which of the years from 2000 to 2020, if any, also happen to be prime numbers? Explain at least two ways in which you might go about solving this problem. [3.2b]

**43.** Explain in your own words when it *is* possible to "cancel" and when it *is not* possible to "cancel." [3.5b]

**44.** Can fraction notation be simplified if the numerator and the denominator are two different prime numbers? Why or why not? [3.5b]

# 3.6 Multiplying, Simplifying, and More with Area

## OBJECTIVES

**a** Multiply and simplify using fraction notation.

**b** Solve applied problems involving multiplication of fractions.

### a   SIMPLIFYING WHEN MULTIPLYING

It is often possible to simplify after we multiply. To make such simplifying easier, it is usually best not to carry out the products in the numerator and the denominator immediately, but to factor and simplify first. Consider the product

$$\frac{3}{8} \cdot \frac{4}{9}.$$

We proceed as follows:

$$\frac{3}{8} \cdot \frac{4}{9} = \frac{3 \cdot 4}{8 \cdot 9}$$   We write the products in the numerator and the denominator, but we do not carry out the multiplication.

$$= \frac{3 \cdot 2 \cdot 2}{2 \cdot 2 \cdot 2 \cdot 3 \cdot 3}$$   Factoring the numerator and the denominator

$$= \frac{3 \cdot 2 \cdot 2 \cdot 1}{2 \cdot 2 \cdot 2 \cdot 3 \cdot 3}$$   Using the identity property of 1 to insert the number 1 as a factor

$$= \frac{3 \cdot 2 \cdot 2}{3 \cdot 2 \cdot 2} \cdot \frac{1}{2 \cdot 3}$$   Factoring the fraction

$$= 1 \cdot \frac{1}{2 \cdot 3}$$

$$= \frac{1}{2 \cdot 3}$$   Removing a factor of 1

$$= \frac{1}{6}.$$

To multiply and simplify:

**a)** Write the products in the numerator and the denominator, but do not carry out the multiplication.

**b)** Factor the numerator and the denominator.

**c)** Factor the fraction to remove a factor equal to 1, if possible.

**d)** Carry out the remaining products.

**EXAMPLES**   Multiply and simplify.

**1.** $\dfrac{2}{3} \cdot \dfrac{9}{4} = \dfrac{2 \cdot 9}{3 \cdot 4} = \dfrac{2 \cdot 3 \cdot 3}{3 \cdot 2 \cdot 2} = \dfrac{2 \cdot 3}{2 \cdot 3} \cdot \dfrac{3}{2} = 1 \cdot \dfrac{3}{2} = \dfrac{3}{2}$

**2.** $\dfrac{6}{7} \cdot \dfrac{-5}{3} = \dfrac{3 \cdot 2 \cdot (-5)}{7 \cdot 3}$     Note that 3 is a common factor of 6 and 3.

$\qquad = \dfrac{3}{3} \cdot \dfrac{2(-5)}{7} = \dfrac{-10}{7}$, or $-\dfrac{10}{7}$     Removing a factor equal to 1: $\dfrac{3}{3} = 1$

**3.** $\dfrac{10}{21} \cdot \dfrac{14a}{15} = \dfrac{5 \cdot 2 \cdot 7 \cdot 2 \cdot a}{7 \cdot 3 \cdot 5 \cdot 3}$     Note that 5 is a common factor of 10 and 15.
Note that 7 is a common factor of 21 and 14a.

$\qquad = \dfrac{5 \cdot 7}{5 \cdot 7} \cdot \dfrac{2 \cdot 2 \cdot a}{3 \cdot 3}$

$\qquad = \dfrac{4a}{9}$     $\Big\}$ Removing a factor equal to 1: $\dfrac{5 \cdot 7}{5 \cdot 7} = 1$

**4.** $32 \cdot \dfrac{7}{8} = \dfrac{32}{1} \cdot \dfrac{7}{8} = \dfrac{8 \cdot 4 \cdot 7}{8 \cdot 1}$     Note that 8 is a common factor of 32 and 8.

$\qquad = \dfrac{8}{8} \cdot \dfrac{4 \cdot 7}{1} = 28$     Removing a factor equal to 1: $\dfrac{8}{8} = 1$

························· **Caution!** ·························

Canceling can be used as follows for these examples.

**1.** $\dfrac{2}{3} \cdot \dfrac{9}{4} = \dfrac{2 \cdot \cancel{3} \cdot 3}{\cancel{3} \cdot 2 \cdot 2} = \dfrac{3}{2}$     Removing a factor equal to 1: $\dfrac{2 \cdot 3}{2 \cdot 3} = 1$

**2.** $\dfrac{6}{7} \cdot \dfrac{-5}{3} = \dfrac{\cancel{3} \cdot 2(-5)}{7 \cdot \cancel{3}} = \dfrac{-10}{7}$     Removing a factor equal to 1: $\dfrac{3}{3} = 1$

**3.** $\dfrac{10}{21} \cdot \dfrac{14a}{15} = \dfrac{\cancel{5} \cdot \cancel{7} \cdot 2 \cdot a}{\cancel{7} \cdot 3 \cdot \cancel{5} \cdot 3} = \dfrac{4a}{9}$     Removing a factor equal to 1: $\dfrac{5 \cdot 7}{5 \cdot 7} = 1$

**4.** $32 \cdot \dfrac{7}{8} = \dfrac{8 \cdot 4 \cdot 7}{8 \cdot 1} = 28$     Removing a factor equal to 1: $\dfrac{8}{8} = 1$

**Remember, only factors can be canceled!**

·····································································

**Do Exercises 1–4.** ▶

Multiply and simplify.

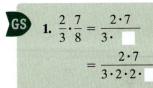

**GS 1.** $\dfrac{2}{3} \cdot \dfrac{7}{8} = \dfrac{2 \cdot 7}{3 \cdot \square}$

$\qquad = \dfrac{2 \cdot 7}{3 \cdot 2 \cdot 2 \cdot \square}$

$\qquad = \dfrac{2}{\square} \cdot \dfrac{7}{3 \cdot 2 \cdot 2}$

$\qquad = \dfrac{\square}{\square} \cdot \dfrac{7}{3 \cdot 2 \cdot 2}$

$\qquad = \dfrac{7}{\square}$

**2.** $\dfrac{4}{5} \cdot \dfrac{-5}{12}$     **3.** $16 \cdot \dfrac{3}{8}$

**4.** $\dfrac{5}{2x} \cdot 4$

## b  SOLVING PROBLEMS

**EXAMPLE 5** *Landscaping.*  Celina's Landscaping uses $\frac{2}{3}$ lb of peat moss when planting a rosebush. How much will be needed to plant 21 rosebushes?

**1. Familiarize.**  We let $n =$ the number of pounds of peat moss needed. Each rosebush requires $\frac{2}{3}$ lb of peat moss, so repeated addition, or multiplication, applies.

**2. Translate.**  The problem translates to the following equation:

$$n = 21 \cdot \frac{2}{3}.$$

**3. Solve.**  To solve the equation, we carry out the multiplication:

$$n = 21 \cdot \frac{2}{3} = \frac{21}{1} \cdot \frac{2}{3} = \frac{21 \cdot 2}{1 \cdot 3}$$     Multiplying

$$\qquad = \frac{3 \cdot 7 \cdot 2}{1 \cdot 3} = \frac{3}{3} \cdot \frac{7 \cdot 2}{1} = 14.$$     Removing the factor $\frac{3}{3}$ and simplifying

*Answers*

**1.** $\dfrac{7}{12}$  **2.** $-\dfrac{1}{3}$  **3.** 6  **4.** $\dfrac{10}{x}$

*Guided Solution:*
**1.** 8, 2, 2, 1, 12

**5. Candy.** Chocolate Delight sells $\frac{4}{5}$-lb boxes of truffles. How many pounds of truffles will be needed to fill 85 boxes?

85 boxes

$\frac{4}{5}$ pound of truffles in each box

**4. Check.** We check by repeating the calculation. (This is left to the student.) We can also ask if the answer seems reasonable. We are putting less than a pound of peat moss on each bush, so the answer should be less than 21. Since 14 is less than 21, we have a partial check.

A second partial check can be performed using the units:

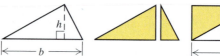

$$21 \text{ bushes} \cdot \frac{2}{3} \text{ pounds per bush}$$

$$= 21 \cdot \frac{2}{3} \cdot \text{bushes} \cdot \frac{\text{pounds}}{\text{bush}}$$

$$= 14 \text{ pounds}.$$

Since the resulting unit is pounds, we have another partial check.

**5. State.** Celina's Landscaping will need 14 lb of peat moss to plant 21 rosebushes.

◀ **Do Exercise 5.**

## Area

We multiply to find the area of a triangle. Consider a triangle with a base of length $b$ and a height of $h$. A rectangle can be formed by splitting and inverting a copy of this triangle.

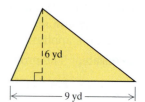

The rectangle's area, $b \cdot h$, is exactly twice the area of the triangle. We have the following result.

---

### AREA OF A TRIANGLE

The **area $A$ of a triangle** is half the length of the base $b$ times the height $h$:

$$A = \frac{1}{2} \cdot b \cdot h.$$

---

**EXAMPLE 6** Find the area of this triangle.

$$A = \frac{1}{2} \cdot b \cdot h$$

$$= \frac{1}{2} \cdot 9 \text{ yd} \cdot 6 \text{ yd}$$

$$= \frac{9 \cdot 6}{2} \text{ yd}^2$$

$$= \frac{54}{2} \text{ yd}^2$$

$$= 27 \text{ yd}^2, \text{ or } 27 \text{ square yards}$$

··············· **Caution!** ···············

Use square units for units of area.
·····································································

**EXAMPLE 7** Find the area of this triangle.

$$A = \frac{1}{2} \cdot b \cdot h$$

$$= \frac{1}{2} \cdot \frac{10}{3} \, cm \cdot 4 \, cm$$

$$= \frac{1 \cdot 10 \cdot 4}{2 \cdot 3} \, cm^2$$

$$= \frac{1 \cdot 2 \cdot 5 \cdot 4}{2 \cdot 3} \, cm^2 \qquad \text{Removing a factor equal to 1: } \frac{2}{2} = 1$$

$$= \frac{20}{3} \, cm^2$$

**Do Exercises 6 and 7.** ▶

**EXAMPLE 8** Find the area of this kite.

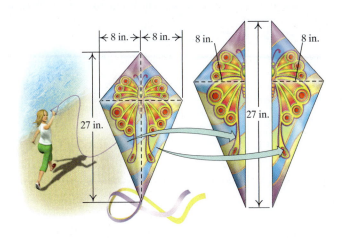

1. **Familiarize.** We look for figures with areas we can calculate using area formulas that we already know. We let $K = $ the kite's area.

2. **Translate.** The kite consists of two triangles, each with a base of 27 in. and a height of 8 in. We can apply the formula $A = \frac{1}{2} \cdot b \cdot h$ for the area of a triangle and then multiply by 2.

*Rephrase:*  Kite's area  is  twice  Area of long triangle

*Translate:*  $K$  $=$  $2$  $\cdot$  $\frac{1}{2}(27 \text{ in.}) \cdot (8 \text{ in.})$

3. **Solve.** We have

$$K = 2 \cdot \frac{1}{2} \cdot (27 \text{ in.}) \cdot (8 \text{ in.})$$
$$= 1 \cdot 27 \text{ in.} \cdot 8 \text{ in.} = 216 \text{ in}^2. \qquad 2 \cdot \frac{1}{2} = \frac{2}{2} = 1$$

4. **Check.** We can check by repeating the calculations. The unit, $in^2$, is appropriate for area.

5. **State.** The area of the kite is 216 $in^2$.

**Do Exercise 8.** ▶

---

Find the area.

**6.**

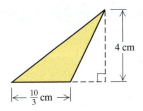

12 m
16 m

**GS 7.**

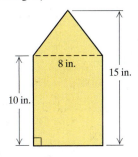

$\frac{12}{5}$ cm     11 cm

$$A = \frac{1}{2} \cdot b \cdot h$$
$$= \frac{1}{2} \cdot 11 \text{ cm} \cdot \square \text{ cm}$$
$$= \frac{1 \cdot 11 \cdot \square}{2 \cdot 5} \text{ cm}^2$$
$$= \frac{1 \cdot 11 \cdot 2 \cdot 2 \cdot \square}{2 \cdot 5} \text{ cm}^2$$
$$= \frac{\square}{5} \text{ cm}^2$$

**8.** Find the area. (*Hint:* The figure is made up of a rectangle and a triangle.)

8 in.
15 in.
10 in.

**Answers**

**6.** 96 $m^2$   **7.** $\frac{66}{5}$ $cm^2$

**8.** Rectangle: $(10 \text{ in.}) \cdot (8 \text{ in.}) = 80 \text{ in}^2$;

triangle: $\frac{1}{2}(8 \text{ in.}) \cdot (5 \text{ in.}) = 20 \text{ in}^2$;

$80 \text{ in}^2 + 20 \text{ in}^2 = 100 \text{ in}^2$

*Guided Solution:*

**7.** $\frac{12}{5}$, 12, 3, 66

✓ **Check Your Understanding**

**Reading Check** Complete each step in the process for multiplying and simplifying using fraction notation.

**RC1.** Write the _____ in the numerator and the denominator, but do not carry out the multiplication.

**RC2.** _____ the numerator and the denominator.

**RC3.** Factor the fraction to remove a factor equal to _____, if possible.

**RC4.** _____ the remaining products.

**Concept Check** Multiply and simplify by canceling and removing factors of 1. Indicate factors of 1 with slashes.

**CC1.** $\dfrac{15}{80} \cdot \dfrac{2}{3} = \dfrac{3 \cdot 5 \cdot 2 \cdot 1}{2 \cdot 2 \cdot 2 \cdot 2 \cdot 5 \cdot 3} =$

**CC2.** $\dfrac{70}{77} \cdot \dfrac{44}{210} = \dfrac{2 \cdot 5 \cdot 7 \cdot 2 \cdot 2 \cdot 11}{7 \cdot 11 \cdot 2 \cdot 3 \cdot 5 \cdot 7} =$

**a**   Multiply. Don't forget to simplify, if possible.

**1.** $\dfrac{2}{3} \cdot \dfrac{1}{2}$      **2.** $\dfrac{4}{5} \cdot \dfrac{1}{4}$      **3.** $\dfrac{7}{8} \cdot \dfrac{-1}{7}$      **4.** $\dfrac{5}{6} \cdot \dfrac{-1}{5}$

**5.** $\dfrac{2}{3} \cdot \dfrac{6}{7}$      **6.** $\dfrac{2}{5} \cdot \dfrac{3}{10}$      **7.** $\dfrac{2}{9} \cdot \dfrac{3}{10}$      **8.** $\dfrac{3}{5} \cdot \dfrac{10}{9}$

**9.** $\dfrac{9}{-5} \cdot \dfrac{12}{8}$      **10.** $\dfrac{16}{-15} \cdot \dfrac{5}{4}$      **11.** $\dfrac{5x}{9} \cdot \dfrac{4}{5}$      **12.** $\dfrac{25}{4a} \cdot \dfrac{4}{3}$

**13.** $9 \cdot \dfrac{1}{9}$      **14.** $4 \cdot \dfrac{1}{4}$      **15.** $\dfrac{7}{10} \cdot \dfrac{10}{7}$      **16.** $\dfrac{8}{9} \cdot \dfrac{9}{8}$

**17.** $\dfrac{1}{4} \cdot 12$      **18.** $\dfrac{1}{6} \cdot 12$      **19.** $21 \cdot \dfrac{1}{3}$      **20.** $18 \cdot \dfrac{1}{2}$

**21.** $-16\left(-\dfrac{3}{4}\right)$      **22.** $-24\left(-\dfrac{5}{6}\right)$      **23.** $\dfrac{3}{8} \cdot 8a$      **24.** $\dfrac{2}{9} \cdot 9x$

**25.** $\left(-\dfrac{3}{8}\right)\left(-\dfrac{8}{3}\right)$

**26.** $\left(-\dfrac{7}{9}\right)\left(-\dfrac{9}{7}\right)$

**27.** $\dfrac{a}{b} \cdot \dfrac{b}{a}$

**28.** $\dfrac{n}{m} \cdot \dfrac{m}{n}$

**29.** $\dfrac{4}{10} \cdot \dfrac{5}{10}$

**30.** $\dfrac{11}{24} \cdot \dfrac{3}{5}$

**31.** $\dfrac{8}{10} \cdot \dfrac{45}{100}$

**32.** $\dfrac{3}{10} \cdot \dfrac{8}{10}$

**33.** $\dfrac{1}{6} \cdot 360n$

**34.** $\dfrac{1}{3} \cdot 12y$

**35.** $20\left(\dfrac{1}{-6}\right)$

**36.** $35\left(\dfrac{1}{-10}\right)$

**37.** $-8x \cdot \dfrac{1}{-8x}$

**38.** $-5a \cdot \dfrac{1}{-5a}$

**39.** $\dfrac{2x}{9} \cdot \dfrac{27}{2x}$

**40.** $\dfrac{10a}{3} \cdot \dfrac{3}{5a}$

**41.** $\dfrac{7}{10} \cdot \dfrac{34}{150}$

**42.** $\dfrac{15}{22} \cdot \dfrac{4}{7}$

**43.** $\dfrac{36}{85} \cdot \dfrac{25}{-99}$

**44.** $\dfrac{-70}{45} \cdot \dfrac{50}{49}$

**45.** $\dfrac{-98}{99} \cdot \dfrac{27a}{175a}$

**46.** $\dfrac{70}{-49} \cdot \dfrac{63}{300x}$

**47.** $\dfrac{110}{33} \cdot \dfrac{-24}{25x}$

**48.** $\dfrac{-19}{130} \cdot \dfrac{65}{38x}$

**49.** $\left(-\dfrac{11}{24}\right)\dfrac{3}{5}$

**50.** $\left(-\dfrac{15}{22}\right)\dfrac{4}{7}$

**51.** $\dfrac{10a}{21} \cdot \dfrac{3}{8b}$

**52.** $\dfrac{17}{21y} \cdot \dfrac{3x}{5}$

**b**  Solve.

The *pitch* of a screw is the distance between its threads. With each complete rotation, the screw goes in or out a distance equal to its pitch. Use this information to answer Exercises 53 and 54.

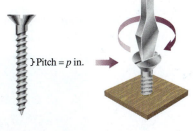

}Pitch = p in.

Each rotation moves the screw in or out p in.

**53.** The pitch of a screw is $\frac{1}{16}$ in. How far will it go into a piece of oak when it is turned 10 complete rotations clockwise?

**54.** The pitch of a screw is $\frac{3}{32}$ in. How far will it come out of a piece of plywood when it is turned 10 complete rotations counterclockwise?

**55. Cotton T-Shirt Exports.** Worldwide, t-shirt exports had a total value of $43,100,000,000 in 2016. Of this amount, approximately $\frac{2}{3}$ came from cotton t-shirts. Worldwide, what was the value of cotton t-shirts exported in 2016?

**Data:** www.worldstopexports.com

**56. Corn Exports.** In 2016, the United States exported 51,135,000 metric tons of corn. In that year, South Korea purchased $\frac{2}{25}$ of U.S. corn exports. How many metric tons of U.S. corn did South Korea purchase?

**Data:** *Outlook for U.S. Agriculture Trade*/AES-99/May 25, 2017, USDA's Economic Research Service and Foreign Agricultural Service; National Corn Growers Association

**57. Mailing-List Changes.** The United States Postal Service estimates that $\frac{4}{25}$ of the addresses on a mailing list will change in one year. A business has a mailing list of 3000 people. After one year, how many addresses on that list will be incorrect?

**Data:** usps.com

**58. Substitute Teaching.** After Vivian completes 60 hr of teacher training in college, she can earn $120 for working a full day as a substitute teacher. How much will she receive for working $\frac{3}{5}$ of a day?

**59. College Enrollment.** Approximately $\frac{2}{3}$ of all high school graduates in the United States in 2016 enrolled in college. Of these, $\frac{3}{4}$ enrolled in public colleges and universities. What fraction of the 2016 graduating class enrolled in public colleges and universities?

**Data:** thecollegesolution.com; U.S. Bureau of Labor Statistics

**60. Culinary Arts.** A recipe for piecrust calls for $\frac{2}{3}$ cup of flour. A chef is making $\frac{1}{2}$ of the recipe. How much flour should the chef use?

**61. Assessed Value.** A house worth $154,000 is assessed for $\frac{3}{4}$ of its value. What is the assessed value of the house?

**62. Student Loans.** Roxanne's tuition was $4600. A loan was obtained for $\frac{3}{4}$ of the tuition. How much was the loan?

**63. Map Scaling.** On a map, 1 in. represents 240 mi. How much does $\frac{2}{3}$ in. represent?

**64. Map Scaling.** On a map, 1 in. represents 120 mi. How much does $\frac{3}{4}$ in. represent?

**65.** *Household Budgets.* A family has an annual income of $42,000. Of this, $\frac{1}{5}$ is spent for food, $\frac{1}{4}$ for housing, $\frac{1}{10}$ for clothing, $\frac{1}{14}$ for savings, $\frac{1}{5}$ for taxes, and the rest for other expenses. How much is spent for each?

**66.** *Household Budgets.* A family has an annual income of $28,140. Of this, $\frac{1}{5}$ is spent for food, $\frac{1}{4}$ for housing, $\frac{1}{10}$ for clothing, $\frac{1}{14}$ for savings, $\frac{1}{5}$ for taxes, and the rest for other expenses. How much is spent for each?

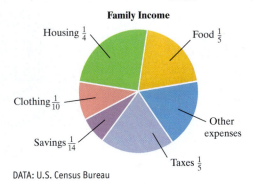

**Family Income**

Housing $\frac{1}{4}$    Food $\frac{1}{5}$

Clothing $\frac{1}{10}$    Other expenses

Savings $\frac{1}{14}$    Taxes $\frac{1}{5}$

DATA: U.S. Census Bureau

Find the area.

**67.**

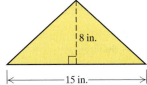

8 in.

15 in.

**68.**

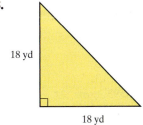

18 yd

18 yd

**69.**

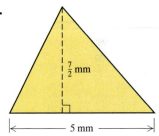

$\frac{7}{2}$ mm

5 mm

**70.**

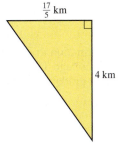

$\frac{17}{5}$ km

4 km

**71.**

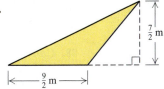

$\frac{7}{2}$ m

$\frac{9}{2}$ m

**72.**

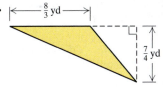

$\frac{8}{3}$ yd

$\frac{7}{4}$ yd

**73.**

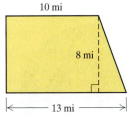

10 mi

8 mi

13 mi

**74.**

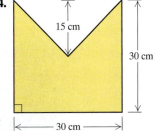

15 cm

30 cm

30 cm

**75.** *Quilting.*  Michelle is designing a quilt using hexagons. Each hexagon is made using six kite-shaped pieces of cloth, as shown below. Find the area of one of the kite-shaped pieces.

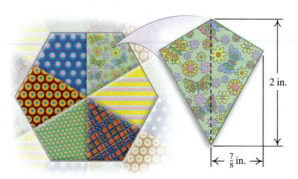

2 in.

$\frac{7}{8}$ in.

**76.** *Construction.*  Find the total area of the sides and ends of the town office building shown. Do not subtract for any windows, doors, or steps.

25 ft

11 ft

75 ft

50 ft

Dumont Town Offices

## Skill Maintenance

Solve.  [1.7b]

**77.** $48 \cdot t = 1680$

**78.** $74 \cdot x = 6290$

**79.** $3125 = 25 \cdot t$

**80.** $2880 = 24 \cdot y$

**81.** $t + 28 = 5017$

**82.** $456 + x = 9002$

**83.** $8797 = y + 98$

**84.** $10,000 = 3593 + m$

## Synthesis

Simplify. Use the list of prime numbers on p. 160.

**85.** ▦ $\frac{201}{535} \cdot \frac{4601}{6499}$

**86.** ▦ $\frac{667}{899} \cdot \frac{558}{621}$

**87.** *College Profile.*  Of students entering a college, $\frac{7}{8}$ have completed high school and $\frac{2}{3}$ are older than 20. If $\frac{1}{7}$ of all students are left-handed, what fraction of students entering the college are left-handed high school graduates over the age of 20?

**88.** *College Profile.*  Refer to the information in Exercise 87. If 480 students are entering the college, how many of them are left-handed high school graduates 20 years old or younger?

**89.** ▦ *Manufacturing.*  A candy box is triangular at each end, as shown below. Find the surface area of the box.

30 mm

30 mm

30 mm

26 mm

140 mm

**90.** ▦ *Painting.*  Shoreline Painting needs to determine the surface area of an octagonal steeple. Find the total area, if the dimensions are as shown below.

15 ft

6 ft

4 ft

# Reciprocals and Division

## a  RECIPROCALS

Each of the following products is 1:

$$8 \cdot \frac{1}{8} = \frac{8}{8} = 1; \qquad \frac{-2}{3} \cdot \frac{3}{-2} = \frac{-6}{-6} = 1.$$

### RECIPROCALS

If the product of two numbers is 1, we say that they are **reciprocals** of each other.* To find the reciprocal of a fraction, interchange the numerator and the denominator.

Number: $\frac{3}{4}$ ⟶ Reciprocal: $\frac{4}{3}$

**EXAMPLES**  Find the reciprocal.

1. The reciprocal of $\frac{4}{y}$ is $\frac{y}{4}$.  $\quad \frac{4}{y} \cdot \frac{y}{4} = \frac{4 \cdot y}{y \cdot 4} = \frac{4y}{4y} = 1$

2. The reciprocal of 8 is $\frac{1}{8}$.  $\quad$ Think of 8 as $\frac{8}{1}$: $\frac{8}{1} \cdot \frac{1}{8} = \frac{8}{8} = 1$.

3. The reciprocal of $\frac{1}{3}$ is 3.  $\quad \frac{1}{3} \cdot 3 = \frac{3}{3} = 1$

4. The reciprocal of $-\frac{5}{9}$ is $-\frac{9}{5}$.  Negative numbers have negative reciprocals: $\left(-\frac{5}{9}\right)\left(-\frac{9}{5}\right) = \frac{45}{45} = 1$.

**Do Exercises 1–5.** ▶

Find the reciprocal.

1. $\frac{2}{5}$    2. $\frac{-6}{x}$

3. 9    4. $\frac{1}{5}$

5. $\frac{-3}{10}$

Does 0 have a reciprocal? If it did, it would have to be a number $x$ such that $0 \cdot x = 1$. But 0 times any number is 0. Thus, we have the following.

### 0 HAS NO RECIPROCAL

The number 0, or $\frac{0}{n}$, has no reciprocal. $\left(\text{Recall that } \frac{n}{0} \text{ is not defined.}\right)$

## b  DIVISION

**SKILL REVIEW**

*Divide integers.*  [2.5a]

Divide.

1. $\frac{26}{-2}$    2. $\frac{-100}{-25}$

**Answers: 1.** $-13$ **2.** 4

MyLab Math
VIDEO

*A reciprocal is also called a *multiplicative inverse*.

*Answers*

1. $\frac{5}{2}$  2. $\frac{x}{-6}$  3. $\frac{1}{9}$  4. 5  5. $-\frac{10}{3}$

Consider the division $\frac{3}{4} \div \frac{1}{8}$. We are asking how many $\frac{1}{8}$'s are in $\frac{3}{4}$. From the figure below, we see that there are six $\frac{1}{8}$'s in $\frac{3}{4}$. Thus,

$$\frac{3}{4} \div \frac{1}{8} = 6.$$

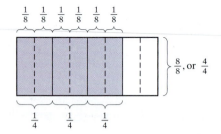

We can check this by multiplying:

$$6 \cdot \frac{1}{8} = \frac{6}{1} \cdot \frac{1}{8} = \frac{6}{8} = \frac{2 \cdot 3}{2 \cdot 4} = \frac{2}{2} \cdot \frac{3}{4} = \frac{3}{4}.$$

Here is a faster way to do this division:

$$\frac{3}{4} \div \frac{1}{8} = \frac{3}{4} \cdot \frac{8}{1} = \frac{3 \cdot 8}{4 \cdot 1} = \frac{24}{4} = 6. \qquad \text{Multiplying by the reciprocal of the divisor}$$

### DIVISION OF FRACTIONS

To divide by a fraction, multiply by its reciprocal:

$$\frac{a}{b} \div \frac{c}{d} = \frac{a}{b} \cdot \frac{d}{c}. \qquad \text{Multiply by the reciprocal of the divisor.}$$

**EXAMPLES** Divide and simplify.

5. $\dfrac{5}{6} \div \dfrac{2}{3} = \dfrac{5}{6} \cdot \dfrac{3}{2}$     Multiplying by the reciprocal of the divisor

$\qquad = \dfrac{5 \cdot 3}{3 \cdot 2 \cdot 2}$     Factoring and identifying a common factor

$\qquad = \dfrac{3}{3} \cdot \dfrac{5}{2 \cdot 2}$     Removing a factor equal to 1: $\dfrac{3}{3} = 1$

$\qquad = \dfrac{5}{4}$

6. $\dfrac{-3}{5} \div \dfrac{1}{2} = \dfrac{-3}{5} \cdot 2$     The reciprocal of $\dfrac{1}{2}$ is 2.

$\qquad = \dfrac{-3 \cdot 2}{5} = \dfrac{-6}{5}, \text{ or } -\dfrac{6}{5}$

7. $\dfrac{2a}{5} \div 7 = \dfrac{2a}{5} \cdot \dfrac{1}{7}$     The reciprocal of 7 is $\dfrac{1}{7}$.

$\qquad = \dfrac{2a \cdot 1}{5 \cdot 7} = \dfrac{2a}{35}$

**8.**
$$\cfrac{\frac{7}{10}}{-\frac{14}{15}} = \frac{7}{10} \div \left(-\frac{14}{15}\right)$$ 　The fraction bar indicates division.

$$= \frac{7}{10} \cdot \left(-\frac{15}{14}\right)$$ 　Multiplying by the reciprocal of the divisor

$$= \frac{7 \cdot 5(-3)}{2 \cdot 5 \cdot 7 \cdot 2}$$ 　Factoring and identifying common factors

$$= \frac{7 \cdot 5}{7 \cdot 5} \cdot \frac{-3}{4}$$ 　Removing a factor equal to 1: $\frac{7 \cdot 5}{7 \cdot 5} = 1$

$$= \frac{-3}{4}, \text{ or } -\frac{3}{4}$$

$\blacksquare$

.......................... **Caution!** ..........................

Canceling can be used as follows for Examples 5 and 8.

**5.** $\dfrac{5}{6} \div \dfrac{2}{3} = \dfrac{5}{6} \cdot \dfrac{3}{2} = \dfrac{5 \cdot 3}{6 \cdot 2} = \dfrac{5 \cdot \cancel{3}}{\cancel{3} \cdot 2 \cdot 2} = \dfrac{5}{2 \cdot 2} = \dfrac{5}{4}$

Removing a factor equal to 1: $\frac{3}{3} = 1$

**8.** $\dfrac{7}{10} \div \left(-\dfrac{14}{15}\right) = \dfrac{7}{10} \cdot \left(-\dfrac{15}{14}\right) = \dfrac{7 \cdot 5(-3)}{2 \cdot \cancel{5} \cdot \cancel{7} \cdot 2} = \dfrac{-3}{4}, \text{ or } -\dfrac{3}{4}$

Removing a factor equal to 1: $\frac{7 \cdot 5}{7 \cdot 5} = 1$

**Remember, if you can't factor, you can't cancel!**

..........................

**Do Exercises 6–10.** ▶

What is the explanation for multiplying by a reciprocal when dividing? Let's consider $\frac{2}{3} \div \frac{7}{5}$. We multiply by 1. The name for 1 that we will use is $(5/7)/(5/7)$; it comes from the reciprocal of $\frac{7}{5}$.

$$\frac{2}{3} \div \frac{7}{5} = \cfrac{\frac{2}{3}}{\frac{7}{5}} = \cfrac{\frac{2}{3}}{\frac{7}{5}} \cdot 1 = \cfrac{\frac{2}{3} \cdot \frac{5}{7}}{\frac{7}{5} \cdot \frac{5}{7}} = \cfrac{\frac{2}{3} \cdot \frac{5}{7}}{\frac{7}{5} \cdot \frac{5}{7}} = \frac{\frac{2}{3} \cdot \frac{5}{7}}{1} = \frac{2}{3} \cdot \frac{5}{7} = \frac{10}{21}$$

Thus,

$$\frac{2}{3} \div \frac{7}{5} = \frac{2}{3} \cdot \frac{5}{7} = \frac{10}{21}.$$

Expressions of the form $\cfrac{\frac{a}{b}}{\frac{c}{d}}$ are examples of *complex fractions*:

$$\cfrac{\frac{a}{b}}{\frac{c}{d}} = \frac{a}{b} \div \frac{c}{d}.$$

**Do Exercise 11.** ▶

---

Divide and simplify.

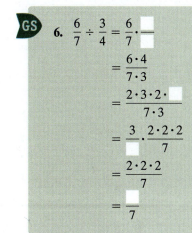

**GS** **6.** $\dfrac{6}{7} \div \dfrac{3}{4} = \dfrac{6}{7} \cdot \dfrac{\square}{\square}$

$$= \frac{6 \cdot 4}{7 \cdot 3}$$

$$= \frac{2 \cdot 3 \cdot 2 \cdot \square}{7 \cdot 3}$$

$$= \frac{3}{\square} \cdot \frac{2 \cdot 2 \cdot 2}{7}$$

$$= \frac{2 \cdot 2 \cdot 2}{7}$$

$$= \frac{\square}{7}$$

**7.** $\left(-\dfrac{2}{3}\right) \div \dfrac{1}{4}$ 　　**8.** $\dfrac{4}{5} \div 8$

**9.** $60 \div \dfrac{3a}{5}$ 　　**10.** $\cfrac{\frac{-6}{7}}{\frac{3}{5}}$

**11.** To remember *why* fractions are divided as they are, multiply by 1 to perform the following division, using the reciprocal of $\frac{4}{5}$ to write 1.

$$\cfrac{\frac{6}{7}}{\frac{4}{5}}$$

*Answers*

**6.** $\dfrac{8}{7}$ 　**7.** $-\dfrac{8}{3}$ 　**8.** $\dfrac{1}{10}$ 　**9.** $\dfrac{100}{a}$

**10.** $\dfrac{-10}{7}$, or $-\dfrac{10}{7}$ 　**11.** $\dfrac{15}{14}$

*Guided Solution:*

**6.** $\dfrac{4}{3}$, 2, 3, 8

## ✓ Check Your Understanding

**Reading Check** Determine whether each statement is true or false.

_____ **RC1.** The numbers $\frac{1}{7}$ and 7 are reciprocals.

_____ **RC2.** The number 1 has no reciprocal.

_____ **RC3.** To divide by a fraction, we multiply by its reciprocal.

_____ **RC4.** We can divide by $\frac{1}{2}$ by multiplying by 2.

**Concept Check** Choose an equivalent expression from the columns on the right. Not all choices will be used.

**CC1.** $\frac{3}{5} \div \frac{1}{5}$

**CC2.** $\frac{1}{5} \div \frac{3}{5}$

**CC3.** $5 \div \frac{1}{5}$

**CC4.** $\frac{1}{5} \div 5$

a) $5 \cdot \frac{1}{5}$

b) $\frac{1}{5} \cdot \frac{1}{5}$

c) $5 \cdot 5$

d) $\frac{1}{5} \cdot \frac{3}{5}$

e) $5 \cdot \frac{5}{3}$

f) $\frac{1}{5} \cdot \frac{5}{3}$

g) $\frac{3}{5} \cdot 5$

---

**a** Find the reciprocal.

**1.** $\frac{7}{3}$

**2.** $\frac{6}{5}$

**3.** 9

**4.** 3

**5.** $\frac{1}{7}$

**6.** $\frac{1}{4}$

**7.** $-\frac{8}{9}$

**8.** $-\frac{12}{5}$

**9.** $\frac{a}{c}$

**10.** $\frac{x}{y}$

**11.** $\frac{-3n}{m}$

**12.** $\frac{8t}{-7r}$

**13.** $\frac{8}{-15}$

**14.** $\frac{-6}{25}$

**15.** $7m$

**16.** $5n$

**17.** $\frac{1}{4a}$

**18.** $\frac{1}{9t}$

**19.** $-\frac{1}{3z}$

**20.** $-\frac{1}{2x}$

---

**b** Divide. Don't forget to simplify when possible. Assume that all variables are nonzero.

**21.** $\frac{3}{7} \div \frac{3}{4}$

**22.** $\frac{2}{3} \div \frac{3}{4}$

**23.** $\frac{3}{5} \div \frac{9}{4}$

**24.** $\frac{6}{7} \div \frac{3}{5}$

**25.** $\frac{4}{3} \div \frac{1}{3}$

**26.** $\frac{10}{9} \div \frac{1}{2}$

**27.** $\left(-\frac{1}{3}\right) \div \frac{1}{6}$

**28.** $\left(-\frac{1}{4}\right) \div \frac{1}{5}$

**29.** $\left(-\dfrac{10}{21}\right) \div \left(-\dfrac{2}{15}\right)$

**30.** $-\dfrac{15}{28} \div \left(-\dfrac{9}{20}\right)$

**31.** $\dfrac{3}{8} \div 3$

**32.** $\dfrac{5}{6} \div 5$

**33.** $\dfrac{12}{7} \div 16$

**34.** $\dfrac{18}{5} \div 27$

**35.** $(-12) \div \dfrac{3}{2}$

**36.** $(-24) \div \dfrac{3}{8}$

**37.** $\dfrac{x}{8} \div \dfrac{1}{4}$

**38.** $\dfrac{3}{4} \div \dfrac{2}{y}$

**39.** $\dfrac{2}{3} \div (6x)$

**40.** $\dfrac{12}{5} \div (4x)$

**41.** $28 \div \dfrac{4}{5a}$

**42.** $40 \div \dfrac{2}{3m}$

**43.** $\left(-\dfrac{5}{8}\right) \div \left(-\dfrac{5}{8}\right)$

**44.** $\left(-\dfrac{2}{5}\right) \div \left(-\dfrac{2}{5}\right)$

**45.** $\dfrac{-8}{15} \div \dfrac{4}{5}$

**46.** $\dfrac{6}{-13} \div \dfrac{3}{26}$

**47.** $\dfrac{77}{64} \div \dfrac{49}{18}$

**48.** $\dfrac{81}{42} \div \dfrac{33}{56}$

**49.** $120a \div \dfrac{45}{14}$

**50.** $360n \div \dfrac{27n}{8}$

**51.** $\dfrac{\frac{2}{5}}{\frac{3}{7}}$

**52.** $\dfrac{\frac{5}{6}}{\frac{2}{7}}$

**53.** $\dfrac{-\frac{7}{20}}{-\frac{8}{5}}$

**54.** $\dfrac{-\frac{8}{21}}{-\frac{6}{5}}$

**55.** $\dfrac{-\frac{15}{8}}{\frac{9}{10}}$

**56.** $\dfrac{-\frac{27}{10}}{\frac{21}{20}}$

## Skill Maintenance

**57.** Evaluate $-(-x)$ when $x = -9$.  [2.1d]

**58.** Evaluate $-x^2$ when $x = -9$.  [2.6a]

**59.** Evaluate $3x^2$ when $x = 5$.  [2.6a]

**60.** Evaluate $(-x)^2$ when $x = -9$.  [2.6a]

**61.** Evaluate $\dfrac{-a}{b}$ for $a = 20$ and $b = -10$.  [2.6a]

**62.** Evaluate $-\dfrac{a}{b}$ for $a = 20$ and $b = -10$.  [2.6a]

## Synthesis

Simplify.

**63.** $\left(\dfrac{4}{15} \div \dfrac{2}{25}\right)^2$

**64.** $\left(\dfrac{9}{10} \div \dfrac{12}{25}\right)^2$

**65.** $\left(\dfrac{9}{10} \div \dfrac{2}{5} \div \dfrac{3}{8}\right)^2$

**66.** $\dfrac{\left(-\frac{3}{7}\right)^2 \div \frac{12}{5}}{\left(\frac{-2}{9}\right)\left(\frac{9}{2}\right)}$

**67.** $\left(\dfrac{14}{15} \div \dfrac{49}{65} \cdot \dfrac{77}{260}\right)^2$

**68.** $\left(\dfrac{10}{9}\right)^2 \div \dfrac{35}{27} \cdot \dfrac{49}{44}$

Simplify. Use the list of prime numbers on p. 160.

**69.** ▦ $\dfrac{711}{1957} \div \dfrac{10,033}{13,081}$

**70.** ▦ $\dfrac{8633}{7387} \div \dfrac{485}{581}$

**71.** ▦ $\dfrac{451}{289} \div \dfrac{123}{340}$

**72.** ▦ $\dfrac{530}{490} \div \dfrac{1060}{980}$

# 3.8

## OBJECTIVES

a Use the multiplication principle to solve equations.

b Solve problems by using the multiplication principle.

**SKILL REVIEW**

*Use the division principle to solve equations.* [2.8c]

Solve.

1. $-8t = 32$  2. $-x = -9$

Answers: 1. $-4$  2. $9$

MyLab Math
VIDEO

Solve.

1. $\frac{2}{3}x = 8$  **GS**

$\frac{\square}{\square} \cdot \frac{2}{3}x = \frac{\square}{\square} \cdot 8$

$\square x = \frac{3 \cdot 8}{2}$

$x = \frac{3 \cdot 2 \cdot \square}{2}$

$x = \square$

2. $\frac{2}{7}a = -6$

**Answers**

1. 12  2. $-21$

**Guided Solution:**

1. $\frac{3}{2}, \frac{3}{2}, 1, 4, 12$

# Solving Equations: The Multiplication Principle

With fraction notation, we can solve equations like $a \cdot x = b$ by using multiplication.

## a THE MULTIPLICATION PRINCIPLE

To divide by a fraction, we multiply by the reciprocal of that fraction. This suggests that we restate the division principle in its more common form—the multiplication principle.

> **THE MULTIPLICATION PRINCIPLE**
>
> For any numbers $a$, $b$, and $c$, with $c \neq 0$,
>
> $a = b$ is equivalent to $a \cdot c = b \cdot c$.

**EXAMPLE 1** Solve: $\frac{3}{4}x = 15$.

We multiply both sides of the equation by the reciprocal of $\frac{3}{4}$.

$$\frac{3}{4}x = 15$$

$$\frac{4}{3} \cdot \frac{3}{4}x = \frac{4}{3} \cdot 15 \quad \text{Using the multiplication principle; note that } \frac{4}{3} \text{ is the reciprocal of } \frac{3}{4}.$$

$$\left(\frac{4}{3} \cdot \frac{3}{4}\right)x = \frac{4 \cdot 15}{3} \quad \text{Using an associative law; try to do this mentally.}$$

$$1x = 20 \quad \text{Multiplying; note that } \frac{4 \cdot 15}{3} = \frac{4 \cdot 3 \cdot 5}{3}.$$

$$x = 20 \quad \text{Remember that } 1x \text{ is } x.$$

To confirm that 20 is the solution, we perform a check.

Check: $\quad \dfrac{3}{4}x = 15$

$\dfrac{3}{4} \cdot 20 \; ? \; 15$

$\dfrac{3 \cdot 4 \cdot 5}{4}$ $\quad$ Removing a factor equal to 1: $\dfrac{4}{4} = 1$

$15 \mid 15$ $\quad$ TRUE

The solution is 20.

◀ **Do Exercises 1 and 2.**

In an expression like $\frac{3}{4}x$, the constant factor—in this case, $\frac{3}{4}$—is called the **coefficient**. In Example 1, we multiplied on both sides by $\frac{4}{3}$, the reciprocal of the coefficient of $x$. After we carried out the multiplication, the coefficient of $x$ was 1. Note that using the multiplication principle to multiply by $\frac{4}{3}$ on both sides is the same as using the division principle to divide by $\frac{3}{4}$ on both sides.

**EXAMPLE 2** Solve: $5a = -\dfrac{7}{3}$.

We have

$$5a = -\frac{7}{3}$$

$$\frac{1}{5} \cdot 5a = \frac{1}{5} \cdot \left(-\frac{7}{3}\right) \qquad \color{red}\text{Multiplying both sides by } \tfrac{1}{5}, \text{ the reciprocal of 5}$$

$$a = -\frac{1 \cdot 7}{5 \cdot 3} = -\frac{7}{15}.$$

We leave the check to the student. The solution is $-\frac{7}{15}$.

**EXAMPLE 3** Solve: $\dfrac{10}{3} = -\dfrac{4}{9}x$.

We have

$$\frac{10}{3} = -\frac{4}{9}x$$

$$-\frac{9}{4} \cdot \frac{10}{3} = -\frac{9}{4} \cdot \left(-\frac{4}{9}\right)x \qquad \color{red}\text{Multiplying both sides by } -\tfrac{9}{4}, \text{ the reciprocal of } -\tfrac{4}{9}$$

$$-\frac{3 \cdot 3 \cdot 2 \cdot 5}{2 \cdot 2 \cdot 3} = x$$

$$-\frac{15}{2} = x. \qquad \color{red}\text{Removing a factor equal to 1: } \tfrac{3 \cdot 2}{2 \cdot 3} = 1$$

We leave the check to the student. The solution is $-\frac{15}{2}$.

**Do Exercises 3 and 4.** ▶

Solve.

**3.** $-\dfrac{9}{8} = 4x$

**GS** **4.** $-\dfrac{6}{7}a = \dfrac{9}{14}$

$$-\frac{\square}{\square} \cdot \left(-\frac{6}{7}a\right) = -\frac{\square}{\square} \cdot \frac{9}{14}$$

$$\square = -\frac{7 \cdot 3 \cdot 3}{2 \cdot 3 \cdot 2 \cdot 7}$$

$$a = -\frac{3}{\square}$$

### b  APPLICATIONS AND PROBLEM SOLVING

**EXAMPLE 4**  *Doses of an Antibiotic.*  How many doses, each containing $\frac{15}{4}$ milliliters (mL), can be obtained from a bottle of a children's antibiotic that contains 60 mL?

**1. Familiarize.** We make a drawing, as shown at the right, and let $n =$ the number of doses in the bottle.

**2. Translate.** The problem can be translated to the equation

$$\frac{15}{4} \cdot n = 60.$$

**3. Solve.** To solve the equation, we use the multiplication principle.

$$\frac{4}{15} \cdot \frac{15}{4} \cdot n = \frac{4}{15} \cdot 60 \qquad \color{red}\text{Multiplying both sides by } \tfrac{4}{15}$$

$$1n = \frac{4 \cdot 60}{15}$$

$$n = \frac{2 \cdot 2 \cdot 2 \cdot 2 \cdot 3 \cdot 5}{1 \cdot 3 \cdot 5} = \frac{3 \cdot 5}{3 \cdot 5} \cdot \frac{2 \cdot 2 \cdot 2 \cdot 2}{1} = 16$$

$\frac{15}{4}$ milliliter in each dose

$n$ doses in the bottle

*Answers*

**3.** $-\dfrac{9}{32}$  **4.** $-\dfrac{3}{4}$

*Guided Solution:*

**4.** $\dfrac{7}{6}, \dfrac{7}{6}, a, 4$

**5.** Each loop in a spring uses $\frac{21}{8}$ in. of wire. How many loops can be made from 210 in. of wire?

**6.** *Servings of Cereal.* A box contains 12 cups of cereal. The nutrition label on the box states that the serving size is $\frac{3}{4}$ cup. How many servings of cereal are in the box?

**7.** *Sales Trip.* John Penna sells soybean seeds to seed companies. After he had driven 210 mi, $\frac{5}{6}$ of his sales trip was completed. How long was the total trip?

$\frac{5}{6}$ of the trip
210 mi

**4. Check.** We check by multiplying the number of doses by the size of the dose: $16 \cdot \frac{15}{4} = 60$. Note too that

$$\text{doses} \cdot \frac{\text{mL}}{\text{dose}} = \text{mL},$$

so the units also check.

**5. State.** There are 16 doses of the antibiotic in the 60-mL bottle.

◀ **Do Exercises 5 and 6.**

**EXAMPLE 5** *Bicycle Paths.* The city of Indianapolis has adopted the *Indianapolis Bicycle Master Plan* as a strategy for creating an environment where bicycling is a safe, practical, and enjoyable transportation choice. After the city finished constructing 60 mi of bike paths and on-road bike lanes, the master plan was $\frac{3}{10}$ complete. What is the total number of miles of bicycling surface that the city of Indianapolis plans to construct?

**Data:** *Indianapolis Bicycle Master Plan*, June 2012

**1. Familiarize.** We ask: "60 mi is $\frac{3}{10}$ of what length?" We make a drawing or at least visualize the problem. We let $b =$ the total number of miles of bicycling surface in the master plan.

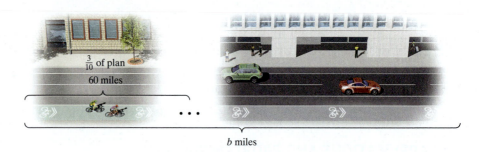

$\frac{3}{10}$ of plan

60 miles

$b$ miles

**2. Translate.** We translate to an equation:

| Fraction completed | of | Total miles planned | is | Amount completed |
|:---:|:---:|:---:|:---:|:---:|
| ↓ | ↓ | ↓ | ↓ | ↓ |
| $\frac{3}{10}$ | $\cdot$ | $b$ | $=$ | 60. |

**3. Solve.** We use the multiplication principle to solve the equation.

$$\frac{10}{3} \cdot \frac{3}{10} \cdot b = \frac{10}{3} \cdot 60 \qquad \text{Multiplying both sides by } \frac{10}{3}$$

$$1b = \frac{10 \cdot 60}{3}$$

$$b = \frac{10 \cdot 3 \cdot 20}{1 \cdot 3} = \frac{200}{1} = 200$$

**4. Check.** We determine whether $\frac{3}{10}$ of 200 is 60: $\frac{3}{10} \cdot 200 = 60$. The answer, 200, checks.

**5. State.** The *Indianapolis Bicycle Master Plan* calls for 200 mi of bicycling surface.

◀ **Do Exercise 7.**

# Translating for Success

The goal of these matching questions is to practice step (2), Translate, of the five-step problem-solving process. Translate each problem to an equation and select a correct translation from equations A–O.

1. **Boxes of Candy.** Jane's Fudge Shop is preparing gift boxes of fudge. How many pounds of fudge will be needed to fill 80 boxes if each box contains $\frac{5}{16}$ lb?

2. **Gallons of Gasoline.** On the third day of a business trip, a sales representative used $\frac{4}{5}$ of a tank of gasoline. If the tank holds 20 gal of gasoline, how many gallons were used on the third day?

3. **Purchasing a Shirt.** Tom received $36 for his birthday. If he spends $\frac{3}{4}$ of the gift on a new shirt, what is the cost of the shirt?

4. **Checkbook Balance.** The balance in Sam's checking account is $1456. He writes a check for $28 and makes a deposit of $52. What is the new balance?

5. **Boxes of Candy.** Jane's Fudge Shop prepared 80 lb of fudge for gift boxes. If each box contains $\frac{5}{16}$ lb, how many boxes can be filled?

A. $x = \frac{3}{4} \cdot 36$

B. $28 \cdot x = 52$

C. $x = 80 \cdot \frac{5}{16}$

D. $x = 1456 \div 28$

E. $x = 20 - \frac{4}{5}$

F. $20 = \frac{4}{5} \cdot x$

G. $x = 12 \cdot 28$

H. $x = \frac{4}{5} \cdot 20$

I. $\frac{3}{4} \cdot x = 36$

J. $x = 1456 - 52 - 28$

K. $x \div 28 = 1456$

L. $x = 52 - 28$

M. $x = 52 \cdot 28$

N. $x = 1456 - 28 + 52$

O. $\frac{5}{16} \cdot x = 80$

6. **Gasoline Tank.** A gasoline tank contains 20 gal when it is $\frac{4}{5}$ full. How many gallons can it hold when full?

7. **Knitting a Scarf.** It takes Rachel 36 hr to knit a scarf. She can knit only $\frac{3}{4}$ hr per day because she is taking 16 hr of college classes. How many days will it take her to knit the scarf?

8. **Bicycle Trip.** On a recent 52-mi bicycle trip, David stopped to make a cell-phone call after completing 28 mi. How many more miles did he bicycle after the call?

9. **Crème de Menthe Thins.** Andes Candies L.P. makes Crème de Menthe Thins. How many 28-piece packages can be filled with 1456 pieces?

10. **Cereal Donations.** The Williams family donates 28 boxes of cereal weekly to the local Family in Crisis Center. How many boxes does this family donate in one year?

*Answers on page A-6*

## ✓ Check Your Understanding

**Reading Check** Determine whether each statement is true or false.

_____ **RC1.** In the expression $\frac{7}{8}x$, the factor $\frac{7}{8}$ is the coefficient.

_____ **RC2.** To solve $6x = \frac{2}{3}$, we can either divide on both sides by 6 or multiply on both sides by $\frac{1}{6}$.

_____ **RC3.** The solution of $\frac{3}{4} = \frac{2}{3}x$ is $\frac{1}{2}$.

_____ **RC4.** If $-\frac{5}{9} = -\frac{7}{2}x$, then $x = \left(-\frac{2}{7}\right)\left(-\frac{5}{9}\right)$.

**Concept Check** From the list on the right, choose the fraction to use in solving each equation by multiplying on both sides.

**CC1.** $\frac{3}{4}x = 10$ _____

**CC2.** $-\frac{4}{9}x = \frac{3}{4}$ _____

**CC3.** $-7 = \frac{9}{4}x$ _____

**CC4.** $\frac{4}{9} = \frac{4}{3}x$ _____

a) $\frac{3}{4}$

b) $\frac{4}{3}$

c) $-\frac{9}{4}$

d) $\frac{4}{9}$

---

**a** Use the multiplication principle to solve each equation. Don't forget to check!

**1.** $\frac{4}{5}x = 12$

**2.** $\frac{4}{3}x = 20$

**3.** $\frac{7}{3}a = 21$

**4.** $\frac{4}{5}a = 24$

**5.** $\frac{2}{9}y = -10$

**6.** $\frac{3}{8}y = -21$

**7.** $6t = \frac{12}{17}$

**8.** $3t = \frac{15}{14}$

**9.** $\frac{1}{4}x = \frac{3}{5}$

**10.** $\frac{1}{6}x = \frac{2}{7}$

**11.** $\frac{3}{2}t = -\frac{8}{7}$

**12.** $\frac{4}{3}t = -\frac{5}{2}$

**13.** $\frac{4}{5} = -10a$

**14.** $\frac{6}{5} = -12a$

**15.** $x \cdot \frac{9}{5} = \frac{3}{10}$

**16.** $x \cdot \frac{10}{3} = \frac{8}{15}$

**17.** $-\frac{1}{10}x = 8$

**18.** $-\frac{1}{11}x = -5$

**19.** $a \cdot \frac{9}{7} = -\frac{3}{14}$

**20.** $a\left(-\frac{9}{4}\right) = -\frac{3}{10}$

**21.** $-x = \dfrac{7}{13}$

**22.** $-x = \dfrac{7}{11}$

**23.** $-x = -\dfrac{27}{31}$

**24.** $-x = -\dfrac{35}{39}$

**25.** $7t = 6$

**26.** $-6t = 1$

**27.** $-24 = -10a$

**28.** $-18 = -20a$

**29.** $-\dfrac{14}{9} = \dfrac{10}{3}t$

**30.** $-\dfrac{15}{7} = \dfrac{3}{2}t$

**31.** $n \cdot \dfrac{4}{15} = \dfrac{12}{25}$

**32.** $n \cdot \dfrac{5}{16} = \dfrac{15}{14}$

**33.** $-\dfrac{7}{20}x = -\dfrac{21}{10}$

**34.** $-\dfrac{7}{15}x = -\dfrac{21}{10}$

**35.** $-\dfrac{25}{17} = -\dfrac{35}{34}a$

**36.** $-\dfrac{49}{45} = -\dfrac{28}{27}a$

 Solve.

**37.** *Extension Cords.* An electrical supplier sells rolls of SJO 14-3 cable to a company that makes extension cords. It takes $\frac{7}{3}$ ft of cable to make each cord. How many extension cords can be made with a roll of cable containing 2240 ft of cable?

**38.** Benny uses $\frac{2}{5}$ gram (g) of toothpaste each time he brushes his teeth. If Benny buys a 30-g tube, how many times will he be able to brush his teeth?

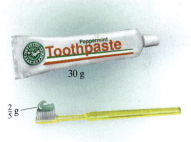

**39.** *Sewing.* A pair of basketball shorts requires $\frac{3}{4}$ yd of nylon. How many pairs of shorts can be made from 24 yd of nylon?

**40.** *Sewing.* A child's baseball shirt requires $\frac{5}{6}$ yd of fabric. How many shirts can be made from 25 yd of fabric?

**41.** How many $\frac{2}{3}$-cup sugar bowls can be filled from 16 cups of sugar?

**42.** For a party, Kyrsten makes an 8-ft submarine sandwich. If one serving is $\frac{2}{3}$ ft, how many servings does Kyrsten's sub contain?

**43.** A bucket had 12 L of water in it when it was $\frac{3}{4}$ full. How much could it hold when full?

**44.** A tank had 20 L of gasoline in it when it was $\frac{4}{5}$ full. How much could it hold when full?

**45. Packaging.** The South Shore Co-op prepackages cheddar cheese in $\frac{3}{4}$-lb packages. How many packages can be made from a 15-lb wheel of cheese?

**46. Meal Planning.** Ian purchased 6 lb of cold cuts for a luncheon. If Ian is to allow $\frac{3}{8}$ lb per person, how many people can attend the luncheon?

**47. Art Supplies.** The Ferristown School District purchased $\frac{3}{4}$ T (ton) of clay. The clay is to be shared equally among the district's 6 art departments. How much will each art department receive?

**48. Gardening.** The Bingham community garden is to be split into 16 equally sized plots. If the garden occupies $\frac{3}{4}$ acre of land, how large will each plot be?

**49. Honey.** A worker bee will produce $\frac{1}{12}$ tsp (teaspoon) of honey in her lifetime. How many worker bees does it take to produce $\frac{3}{4}$ tsp of honey?

**Data:** www.pbs.org/wgbh/nova/bees/buzz.html

**50. Gardening.** Large quantities of soil, gravel, or mulch are normally sold by the *yard* (yd). Although technically the unit for volume is the *cubic yard* ($yd^3$), in this context only the word *yard* is used. Green Season Gardening uses about $\frac{2}{3}$ yd of bark mulch per customer every spring. How many customers can they accommodate with one 30-yd load of bark mulch?

**51. Writing.** On Monday, Ayesha wrote 450 words of an essay that was due on Friday. She then calculated that she had written $\frac{3}{16}$ of the minimum number of words required. What was the minimum number of words required for the essay?

**52. Landscaping.** As part of a semester project for an urban planning course, Fridrik planted 240 perennials in the median strip of an avenue. If Fridrik planted $\frac{5}{32}$ of the total number of perennials planted, how many perennials were planted in all?

**53. Running.** Chad runs 12 mi three days a week. In bad weather, he runs on an indoor track that is $\frac{3}{8}$ mi long. How many laps must he complete in order to run 12 mi?

**54. Herbal Tea.** At Perfect Tea, Donna fills each bag of Sereni-Tea with $\frac{3}{5}$ g (gram) of chamomile. If she begins with 51 g of chamomile, how many tea bags can she fill?

**55.** Yoshi Teramoto sells tools to hardware stores. After driving 180 kilometers (km), he has completed $\frac{5}{8}$ of a sales trip. How long is the total trip? How many kilometers are left to drive?

**56.** A piece of coaxial cable $\frac{4}{5}$ meter (m) long is to be cut into 8 pieces of the same length. What is the length of each piece?

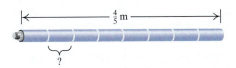

*Pitch of a Screw.* The pitch of a screw is the distance between its threads. With each complete rotation, the screw goes in or out a distance equal to its pitch. Use this information to do Exercises 57 and 58.

**57.** After a screw has been turned 8 complete rotations, it is extended $\frac{1}{2}$ in. into a piece of wallboard. What is the pitch of the screw?

**58.** The pitch of a screw is $\frac{3}{32}$ in. How many complete rotations are necessary to drive the screw $\frac{3}{4}$ in. into a piece of pine wood?

## Skill Maintenance

Simplify.

**59.** $-23 + 49$ [2.2a]

**60.** $-69 + 27$ [2.2a]

**61.** $-38 - 29$ [2.3a]

**62.** $-47 - 18$ [2.3a]

**63.** $36 \div (-3)^2 \times (7 - 2)$ [2.5b]

**64.** $(-37 - 12 + 1) \div (-2)^3$ [2.5b]

Form an equivalent expression by combining like terms. [2.7a]

**65.** $13x + 4x$

**66.** $9a - 5a$

**67.** $2a + 3 + 5a$

**68.** $3x - 7 + x$

## Synthesis

Solve.

**69.** $2x - 7x = -\dfrac{10}{9}$

**70.** $\left(-\dfrac{4}{7}\right)^2 = \left(\dfrac{2^3 - 9}{3}\right)^3 x$

Solve using the five-step problem-solving approach.

**71.** A package of coffee beans weighed $\frac{21}{32}$ lb when it was $\frac{3}{4}$ full. How much could the package hold when completely filled?

**72.** After swimming $\frac{2}{7}$ mi, Katie had swum $\frac{3}{4}$ of the race. How long a race was Katie competing in?

**73.** A block of Swiss cheese is 12 in. long. How many slices will it yield if half of the brick is cut by a slicer set for $\frac{3}{32}$-in. slices and half is cut by a slicer set for $\frac{5}{32}$-in. slices?

**74.** If $\frac{1}{3}$ of a number is $\frac{1}{4}$, what is $\frac{1}{2}$ of the number?

**75.** See Exercise 49. There are 3 teaspoons in a tablespoon and 4 tablespoons in $\frac{1}{4}$ cup. How many worker bees does it take to produce $\frac{1}{2}$ cup of honey?

**76.** ▦ See Exercise 50. If each customer of Green Season Gardening uses $\frac{3}{4}$ yd of bark mulch and Green Season charges each customer $45 for the mulch, how much will Green Season receive from a 25-yd load of mulch?

## Vocabulary Reinforcement

Fill in each blank with the correct term from the list at the right. Some of the choices may not be used.

1. For any number $a$, $a \cdot 1 = a$. The number 1 is the _____ identity.   [3.5a]

2. In the product $10 \cdot \frac{3}{4}$, 10 and $\frac{3}{4}$ are called _____.   [3.2a]

3. A natural number that has exactly two different factors, only itself and 1, is called a(n) _____ number.   [3.2b]

4. In the fraction $\frac{4}{17}$, we call 17 the _____.   [3.3a]

5. Since $\frac{2}{5}$ and $\frac{6}{15}$ are two names for the same number, we say that $\frac{2}{5}$ and $\frac{6}{15}$ are _____ fractions.   [3.5a]

6. The product of 6 and $\frac{1}{6}$ is 1. We say that 6 and $\frac{1}{6}$ are _____.   [3.7a]

7. Since $20 = 4 \cdot 5$, we say that $4 \cdot 5$ is a _____ of 20.   [3.2a]

8. Since $20 = 4 \cdot 5$, we say that 20 is a _____ of 5.   [3.1a]

| |
|---|
| equivalent |
| additive |
| multiplicative |
| reciprocals |
| factors |
| prime |
| composite |
| numerator |
| denominator |
| factorization |
| variables |
| multiple |

## Concept Reinforcement

Determine whether each statement is true or false.

_____ 1. For any natural number $n$, $\dfrac{n}{n} > \dfrac{0}{n}$.   [3.3b]

_____ 2. A number is divisible by 10 if its ones digit is 0 or 5.   [3.1b]

_____ 3. If a number is divisible by 9, then it is also divisible by 3.   [3.1b]

_____ 4. The fraction $\dfrac{-6}{8}$ is equivalent to the fraction $\dfrac{15}{-20}$.   [3.5c]

## Study Guide

**Objective 3.2a**   Find the factors of a number.

**Example**   Find the factors of 84.

We find as many "two-factor" factorizations as we can.

$1 \cdot 84$      $4 \cdot 21$

$2 \cdot 42$      $6 \cdot 14$

$3 \cdot 28$      $7 \cdot 12$ $\longleftarrow$ Since 8, 9, 10, and 11 are not factors, we are finished.

The factors are 1, 2, 3, 4, 6, 7, 12, 14, 21, 28, 42, and 84.

**Practice Exercise**

1. Find the factors of 104.

**Objective 3.2c**  Find the prime factorization of a composite number.

**Example**  Find the prime factorization of 84.

$$\begin{array}{r} 7 \\ 3\overline{)21} \\ 2\overline{)42} \\ 2\overline{)84} \end{array}$$

Thus, $84 = 2 \cdot 2 \cdot 3 \cdot 7$.

**Practice Exercise**

2. Find the prime factorization of 104.

---

**Objective 3.3b**  Simplify fraction notation like $n/n$ to 1, $0/n$ to 0, and $n/1$ to $n$.

**Example**  Simplify $\dfrac{6}{6}, \dfrac{0}{6},$ and $\dfrac{6}{1}$.

$$\dfrac{6}{6} = 1, \qquad \dfrac{0}{6} = 0, \qquad \dfrac{6}{1} = 6$$

**Practice Exercise**

3. Simplify $\dfrac{0}{18}, \dfrac{18}{18},$ and $\dfrac{18}{1}$.

---

**Objective 3.5b**  Simplify fraction notation.

**Example**  Simplify: $\dfrac{315}{1650}$.

Using the test for divisibility by 5, we see that both the numerator and the denominator are divisible by 5:

$$\dfrac{315}{1650} = \dfrac{5 \cdot 63}{5 \cdot 330} = \dfrac{5}{5} \cdot \dfrac{63}{330} = 1 \cdot \dfrac{63}{330}$$

$$= \dfrac{63}{330} = \dfrac{3 \cdot 21}{3 \cdot 110} = \dfrac{3}{3} \cdot \dfrac{21}{110} = 1 \cdot \dfrac{21}{110} = \dfrac{21}{110}.$$

**Practice Exercise**

4. Simplify: $\dfrac{100}{280}$.

---

**Objective 3.5c**  Test to determine whether two fractions are equivalent.

**Example**  Use $=$ or $\neq$ for $\square$ to write a true sentence:

$$\dfrac{10}{54} \,\square\, \dfrac{15}{81}.$$

We find the cross products: We multiply 10 and 81: $10 \cdot 81 = 810$. Then we multiply 54 and 15: $54 \cdot 15 = 810$. Because the cross products are the same, we have

$$\dfrac{10}{54} = \dfrac{15}{81}.$$

If the cross products had been different, the fractions would not be equal.

**Practice Exercise**

5. Use $=$ or $\neq$ for $\square$ to write a true sentence:

$$\dfrac{8}{48} \,\square\, \dfrac{6}{44}.$$

---

**Objective 3.6a**  Multiply and simplify using fraction notation.

**Example**  Multiply and simplify: $\dfrac{7}{16} \cdot \dfrac{40}{49}$.

$$\dfrac{7}{16} \cdot \dfrac{40}{49} = \dfrac{7 \cdot 40}{16 \cdot 49} = \dfrac{7 \cdot 2 \cdot 2 \cdot 2 \cdot 5}{2 \cdot 2 \cdot 2 \cdot 2 \cdot 7 \cdot 7}$$

$$= \dfrac{2 \cdot 2 \cdot 2 \cdot 7}{2 \cdot 2 \cdot 2 \cdot 7} \cdot \dfrac{5}{2 \cdot 7} = 1 \cdot \dfrac{5}{14} = \dfrac{5}{14}$$

**Practice Exercise**

6. Multiply and simplify: $\dfrac{80}{3} \cdot \dfrac{21}{72}$.

**Objective 3.7b** Divide and simplify using fraction notation.

**Example** Divide and simplify: $\dfrac{9}{20} \div \dfrac{18}{25}$.

$$\frac{9}{20} \div \frac{18}{25} = \frac{9}{20} \cdot \frac{25}{18} = \frac{9 \cdot 25}{20 \cdot 18} = \frac{3 \cdot 3 \cdot 5 \cdot 5}{2 \cdot 2 \cdot 5 \cdot 2 \cdot 3 \cdot 3}$$

$$= \frac{3 \cdot 3 \cdot 5}{3 \cdot 3 \cdot 5} \cdot \frac{5}{2 \cdot 2 \cdot 2} = 1 \cdot \frac{5}{8} = \frac{5}{8}$$

**Practice Exercise**

7. Divide and simplify: $\dfrac{9}{4} \div \dfrac{45}{14}$.

---

**Objective 3.8b** Solve problems by using the multiplication principle.

**Example** A rental car had 18 gal of gasoline when its gas tank was $\frac{6}{7}$ full. How much could the tank hold when full?

The equation that corresponds to the situation is

$$\frac{6}{7} \cdot g = 18.$$

We multiply both sides by $\frac{7}{6}$:

$$\frac{7}{6} \cdot \frac{6}{7} \cdot g = \frac{7}{6} \cdot 18$$

$$g = \frac{7}{6} \cdot \frac{18}{1} = \frac{7 \cdot 3 \cdot 6}{6 \cdot 1} = 21.$$

The rental car can hold 21 gal of gasoline.

**Practice Exercise**

8. A flower vase has $\frac{7}{4}$ cups of water in it when it is $\frac{3}{4}$ full. How much can it hold when full?

---

## Review Exercises

1. Multiply by 1, 2, 3, and so on, to find ten multiples of 8. [3.1a]

Use the tests for divisibility to answer Exercises 2–6. [3.1b]

2. Determine whether 3920 is divisible by 6.

3. Determine whether 68,537 is divisible by 3.

4. Determine whether 673 is divisible by 5.

5. Determine whether 4936 is divisible by 2.

6. Determine whether 5238 is divisible by 9.

Find all the factors of each number. [3.2a]

7. 60  8. 176

Classify each number as prime, composite, or neither. [3.2b]

9. 37  10. 1  11. 91

Find the prime factorization of each number. [3.2c]

12. 70  13. 72

14. 45  15. 150

16. 648  17. 1200

18. Identify the numerator and the denominator of $\dfrac{9}{7}$. [3.3a]

What part is shaded? [3.3a]

19. 

20.

**21.** For a committee in the United States Senate that consists of 3 Democrats and 5 Republicans, what is the ratio of: [3.3a]

  **a)** Democrats to Republicans?
  **b)** Republicans to Democrats?
  **c)** Democrats to the total number of members of the committee?

Simplify, if possible. Assume that all variables are nonzero.

**22.** $\dfrac{0}{6}$ [3.3b]   **23.** $\dfrac{74}{74}$ [3.3b]   **24.** $\dfrac{48}{1}$ [3.3b]

**25.** $\dfrac{7x}{7x}$ [3.3b]   **26.** $-\dfrac{10}{15}$ [3.5b]   **27.** $\dfrac{7}{28}$ [3.5b]

**28.** $\dfrac{-42}{42}$ [3.5b]   **29.** $\dfrac{9m}{12m}$ [3.5b]   **30.** $\dfrac{-12}{-30}$ [3.5b]

**31.** $\dfrac{-27}{0}$ [3.3b]   **32.** $\dfrac{140}{490}$ [3.5b]   **33.** $\dfrac{288}{2025}$ [3.5b]

Find an equivalent expression for each number, using the denominator indicated. Use multiplication by 1. [3.5a]

**34.** $\dfrac{5}{7} = \dfrac{?}{21}$   **35.** $\dfrac{-6}{11} = \dfrac{?}{55}$

**36.** Simplify, if possible, the fractions on this circle graph. [3.5b]

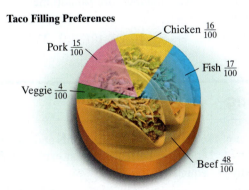

**Taco Filling Preferences**

Chicken $\frac{16}{100}$
Pork $\frac{15}{100}$
Fish $\frac{17}{100}$
Veggie $\frac{4}{100}$
Beef $\frac{48}{100}$

DATA: *Food Network Magazine,* May 2017

Use = or ≠ for ☐ to write a true sentence. [3.5c]

**37.** $\dfrac{3}{5} \,\square\, \dfrac{4}{6}$   **38.** $\dfrac{4}{-7} \,\square\, \dfrac{-8}{14}$

**39.** $\dfrac{4}{5} \,\square\, \dfrac{5}{6}$   **40.** $\dfrac{4}{3} \,\square\, \dfrac{28}{21}$

Find the reciprocal of each number. [3.7a]

**41.** $\dfrac{2}{13}$   **42.** $-7$

**43.** $\dfrac{1}{8}$   **44.** $\dfrac{3x}{5y}$

Perform the indicated operation and, if possible, simplify.

**45.** $\dfrac{2}{9} \cdot \dfrac{7}{5}$ [3.4b]   **46.** $\dfrac{3}{x} \cdot \dfrac{y}{7}$ [3.4b]

**47.** $\dfrac{3}{4} \cdot \dfrac{8}{9}$ [3.6a]   **48.** $-10 \cdot \dfrac{7}{5}$ [3.6a]

**49.** $\dfrac{11}{3} \cdot \dfrac{30}{77}$ [3.6a]   **50.** $\dfrac{4a}{7} \cdot \dfrac{7}{4a}$ [3.6a]

**51.** $\dfrac{6}{5} \cdot 20x$ [3.6a]   **52.** $\dfrac{3}{14} \div \dfrac{6}{7}$ [3.7b]

**53.** $20 \div \dfrac{3}{4}$ [3.7b]   **54.** $-\dfrac{5}{36} \div \left(-\dfrac{25}{12}\right)$ [3.7b]

**55.** $21 \div \dfrac{7}{2a}$ [3.7b]   **56.** $-\dfrac{23}{25} \div \dfrac{23}{25}$ [3.7b]

**57.** $\dfrac{\frac{21}{30}}{\frac{14}{15}}$ [3.7b]   **58.** $\dfrac{-\frac{2}{3}}{-\frac{3}{2}}$ [3.7b]

Solve. [3.8a]

**59.** $\dfrac{2}{3}x = 160$   **60.** $\dfrac{3}{8} = -\dfrac{5}{4}t$

**61.** $-\dfrac{1}{7}n = -4$   **62.** $y \cdot \dfrac{1}{2} = \dfrac{1}{3}$

Find the area. [3.6b]

**63.**

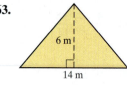

6 m
14 m

**64.**

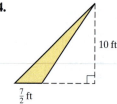

10 ft
$\frac{7}{2}$ ft

Solve. [3.8b]
**65.** A road crew repaves $\frac{1}{12}$ mi of road each day. How long will it take the crew to repave a $\frac{3}{4}$-mi stretch of road?

**66.** *Level of Education and Median Income.* The median yearly income of someone with an associate's degree is approximately $\frac{3}{5}$ of the median income of someone with a master's degree. If the median income for those with master's degrees is $101,323, what is the median income of those with associate's degrees?

**Data:** Statista

Associate's degree        Master's degree

**67.** After driving 600 km, the Buxton family has completed $\frac{3}{5}$ of their vacation. How long is the total trip?

**68.** Molly is making a pepper steak recipe that calls for $\frac{2}{3}$ cup of green bell peppers. How much would be needed to make $\frac{1}{2}$ recipe?

**69.** The Winchester swim team has 4 swimmers in a $\frac{2}{3}$-mi relay race. If each swims the same distance, how far will each person swim?

**70.** A book bag requires $\frac{4}{5}$ yd of fabric. How many bags can be made from 48 yd?

**71.** Solve: $\frac{2}{13} \cdot x = \frac{1}{2}$. [3.8a]

**A.** $\frac{1}{13}$   **B.** 13   **C.** $\frac{4}{13}$   **D.** $\frac{13}{4}$

**72.** Multiply and simplify: $\frac{15}{26} \cdot \frac{13}{90}$. [3.6a]

**A.** $\frac{195}{234}$   **B.** $\frac{1}{12}$   **C.** $\frac{3}{36}$   **D.** $\frac{13}{156}$

## Synthesis

**73.** Simplify: $\frac{15x}{14z} \cdot \frac{17yz}{35xy} \div \left(-\frac{3}{7}\right)^2$. [3.6a], [3.7b]

**74.** What digit(s) could be inserted in the ones place in

574 ☐

to make it divisible by 6? [3.1b]

**75.** 🖩 In the division below, find $a$ and $b$. [3.7b]

$$\frac{19}{24} \div \frac{a}{b} = \frac{187,853}{268,224}$$

**76.** A prime number that remains a prime number when its digits are reversed is called a **palindrome prime**. For example, 17 is a palindrome prime because both 17 and 71 are primes. Which of the following numbers are palindrome primes? [3.2b]

13, 19, 61, 11, 53, 41, 29, 101, 103, 37

# Understanding Through Discussion and Writing

**1.** A student incorrectly insists that $\frac{2}{5} \div \frac{3}{4}$ is $\frac{15}{8}$. What mistake is he probably making? [3.7b]

**2.** Use the number 9432 to explain why the test for divisibility by 9 works. [3.1b]

**3.** A student claims that "taking $\frac{1}{2}$ of a number is the same as dividing by $\frac{1}{2}$." Explain the error in this reasoning. [3.7b]

**4.** On p. 176 we explained, using words and pictures, why $\frac{2}{5} \cdot \frac{3}{4}$ equals $\frac{6}{20}$. Present a similar explanation of why $\frac{2}{3} \cdot \frac{4}{7}$ equals $\frac{8}{21}$. [3.4b]

**5.** Without performing the division, explain why $5 \div \frac{1}{7}$ is a greater number than $5 \div \frac{2}{3}$. [3.7b]

**6.** If a fraction's numerator and denominator have no factors (other than 1) in common, can the fraction be simplified? Why or why not? [3.5b]

CHAPTER

**3**   **Test**

For
Extra
Help

For step-by-step test solutions, access the Chapter Test Prep Videos
in MyLab Math.

**1.** Determine whether 5682 is divisible by 3. Do not use long division.

**2.** Determine whether 7018 is divisible by 5. Do not use long division.

**3.** Find all the factors of 90.

**4.** Determine whether 93 is prime, composite, or neither.

Find the prime factorization of each number.

**5.** 36

**6.** 60

**7.** Identify the numerator and the denominator of $\frac{4}{9}$.

**8.** What part is shaded?

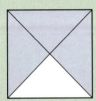

**9.** What part of the set is shaded?

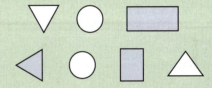

**10.** *Cholesterol.*   Morrison's cholesterol test showed that his total cholesterol level was 180 mg/dL, his HDL level was 47 mg/dL, and his LDL level was 93 mg/dL.
  **a)** What was the ratio of total cholesterol to HDL cholesterol?
  **b)** What was the ratio of HDL cholesterol to LDL cholesterol?

Simplify. if possible. Assume that all variables are nonzero.

**11.** $\frac{32}{1}$

**12.** $\frac{-12}{-12}$

**13.** $\frac{0}{16}$

**14.** $\frac{-8}{24}$

**15.** $\frac{42}{7}$

**16.** $\frac{9x}{45x}$

**17.** $\frac{-62}{0}$

**18.** $\frac{72}{108}$

Use = or ≠ for ☐ to write a true sentence.

**19.** $\frac{3}{4}$ ☐ $\frac{6}{8}$

**20.** $\frac{5}{4}$ ☐ $\frac{9}{7}$

**21.** Find an equivalent expression for $\frac{3}{8}$ with a denominator of 40.

Find the reciprocal.

**22.** $\frac{a}{42}$

**23.** −9

Perform the indicated operation. Simplify, if possible.

**24.** $\dfrac{2}{3} \cdot \dfrac{15}{4}$

**25.** $\dfrac{2}{11} \div \dfrac{3}{4}$

**26.** $3 \cdot \dfrac{x}{8}$

**27.** $\dfrac{\dfrac{4}{7}}{-\dfrac{8}{3}}$

**28.** $12 \div \dfrac{2}{3}$

**29.** $\dfrac{22c}{15} \cdot \dfrac{5}{33c}$

Solve.

**30.** A $\frac{3}{4}$-lb slab of cheese is shared equally by 5 people. How much does each person receive?

**31.** Monroe weighs $\frac{5}{7}$ of his dad's weight. If his dad weighs 175 lb, how much does Monroe weigh?

**32.** $\dfrac{7}{8} \cdot x = 56$

**33.** $\dfrac{7}{10} = \dfrac{-2}{5} \cdot t$

**34.** Find the area.

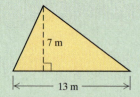

7 m

13 m

**35.** In which figure does the shaded part represent $\frac{7}{6}$ of the figure?

**A.**

**B.**

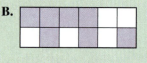

**C.**

**D.**

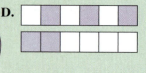

## Synthesis

**36.** Grandma Shelby left $\frac{2}{3}$ of her $\frac{7}{8}$-acre apple farm to Karl. Karl gave $\frac{1}{4}$ of his share to his oldest daughter, Shannon. How much land did Shannon receive?

**37.** Simplify: $\left(-\dfrac{3}{8}\right)^2 \div \dfrac{6}{7} \cdot \dfrac{2}{9} \div (-5)$.

**1.** Write a word name: 2,056,783.

Add.

**2.**
$$\begin{array}{r} 2\ 7\ 4\ 3 \\ +\ 8\ 2\ 3\ 9 \\ \hline \end{array}$$

**3.** $-29 + (-14)$

**4.** $-45 + 12$

Subtract.

**5.**
$$\begin{array}{r} 6\ 3\ 2\ 4 \\ -\ 4\ 1\ 9\ 5 \\ \hline \end{array}$$

**6.** $27 - 50$

**7.** $-12 - (-4)$

Multiply and, if possible, simplify.

**8.**
$$\begin{array}{r} 7\ 3\ 5 \\ \times\ \ 2\ 3 \\ \hline \end{array}$$

**9.** $-52 \cdot 6$

**10.** $\dfrac{6}{7} \cdot (-35x)$

**11.** $\dfrac{2}{9} \cdot \dfrac{21}{10}$

Divide and, if possible, simplify.

**12.** $1\,3\overline{)3\,0\,5\,8}$

**13.** $-85 \div 5$

**14.** $-16 \div \dfrac{4}{7}$

**15.** $\dfrac{3}{7} \div \dfrac{9}{14}$

**16.** Round 4509 to the nearest ten.

**17.** Estimate the product by rounding to the nearest hundred. Show your work.

$$\begin{array}{r} 9\ 2\ 1 \\ \times\ 4\ 5\ 3 \\ \hline \end{array}$$

**18.** Find the absolute value: $|-479|$.

**19.** Simplify: $10^2 \div 5(-2) - 8(2 - 8)$.

**20.** Determine whether 98 is prime, composite, or neither.

**21.** Evaluate $a - b^2$ for $a = -5$ and $b = 4$.

Solve.

**22.** $a + 24 = 49$

**23.** $7x = 49$

**24.** $\dfrac{2}{9} \cdot a = -10$

**25.** $48 = -4t$

**26.** $2x + 13 = 3$

**27.** $-x = -10$

Combine like terms.

**28.** $8 - 4x - 13 + 9x$

**29.** $-12x + 7y + 15x$

Simplify, if possible.

**30.** $\dfrac{97}{97}$

**31.** $\dfrac{0}{81}$

**32.** $\dfrac{63x}{1}$

**33.** $\dfrac{-10}{54}$

Find the reciprocal.

**34.** $\dfrac{2}{5}$

**35.** $57$

**36.** Find an equivalent expression for $\frac{3}{10}$ with a denominator of 70. Use multiplying by 1.

**37.** A 48-oz coffee pot is emptied into 6 mugs. How much will each mug hold if the coffee is poured out evenly?

**38.** A truck that gets 17 miles per gallon is traded in toward a van that gets 25 miles per gallon. How many more miles per gallon does the van get?

**39.** There are 7000 students at La Poloma College, and $\frac{5}{8}$ of them live in dorms. How many live in dorms?

**40.** A thermos of iced tea contained 3 qt of tea when it was $\frac{3}{5}$ full. How much tea could it hold when full?

**41.** Tony has jogged $\frac{2}{3}$ of a course that is $\frac{9}{10}$ of a mile long. How far has Tony gone?

## Synthesis

**42.** Evaluate $\dfrac{ab}{c}$ for $a = -\dfrac{2}{5}$, $b = \dfrac{10}{13}$, and $c = \dfrac{26}{27}$.

**43.** Evaluate $-|xy|^2$ for $x = -\dfrac{3}{5}$ and $y = \dfrac{1}{2}$.

**44.** Wayne and Patty each earn $85 a day, while Janet earns $90 a day. They decide to pool their earnings from three days and spend $\frac{2}{5}$ of that on entertainment and save the rest. How much will Wayne, Patty, and Janet end up saving?

# Fraction Notation: Addition, Subtraction, and Mixed Numerals

Living in colonies of up to 60,000 inhabitants, honeybees gather nectar from flowers during the summer and store it as honey, to serve as food during the winter. In healthy colonies, enough honey is produced for beekeepers to harvest an average of 30 lb of honey and leave sufficient honey for the bees for the winter. We depend on honeybees for much of our food. About $\frac{1}{3}$ of the food we eat depends on pollinators for its production, and honeybees are responsible for $\frac{4}{5}$ of that pollination. Agriculturists in the United States are concerned about the high percentage of colony losses in recent years, as shown in the accompanying graph.

*DATA: thoughtco.com; nbcnews.com*

### Honeybee Colony Loss

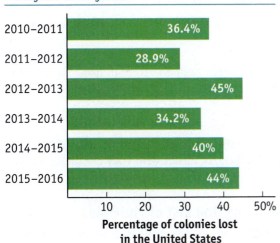

| Year | Percentage |
|------|------------|
| 2010–2011 | 36.4% |
| 2011–2012 | 28.9% |
| 2012–2013 | 45% |
| 2013–2014 | 34.2% |
| 2014–2015 | 40% |
| 2015–2016 | 44% |

**Percentage of colonies lost in the United States**

DATA: USDA

In Example 11 and Margin Exercise 14 of Section 4.3, we will compare the average lengths of types of honeybees.

# 4.1

## OBJECTIVE

a  Find the least common multiple, or LCM, of two or more numbers.

# Least Common Multiples

In order to add or subtract fractions, the fractions must share a common denominator. If necessary, we find equivalent fractions using the **least common denominator (LCD)**, or **least common multiple (LCM)** of the denominators.

## a  FINDING LEAST COMMON MULTIPLES

**SKILL REVIEW**

*Find some multiples of a number.*  [3.1a]

Multiply by 1, 2, 3, and so on, to find six multiples of each number.

**1.** 8                **2.** 25

Answers: **1.** 8, 16, 24, 32, 40, 48
**2.** 25, 50, 75, 100, 125, 150

MyLab Math
VIDEO

### LEAST COMMON MULTIPLE, LCM

The **least common multiple**, or LCM, of two natural numbers is the smallest number that is a multiple of both numbers.

**EXAMPLE 1**  Find the LCM of 20 and 30.

First, we list some multiples of 20 by multiplying 20 by 1, 2, 3, and so on:

20, 40, 60, 80, 100, 120, 140, 160, 180, 200, 220, 240, . . . .

Then we list some multiples of 30 by multiplying 30 by 1, 2, 3, and so on:

30, 60, 90, 120, 150, 180, 210, 240, . . . .

Now we determine the smallest number *common* to both lists. The LCM of 20 and 30 is 60.

◀ **Do Exercises 1 and 2.**

Next, we develop two more efficient methods for finding LCMs. You may choose to learn only one method. (Consult with your instructor.) If you are going to study algebra, you should definitely learn method 2.

## Method 1: Finding LCMs Using One List of Multiples

The first method for finding LCMs works especially well when the numbers are relatively small.

**1.** Find the LCM of 9 and 15 by examining lists of multiples.

**2.** Find the LCM of 8 and 14 by examining lists of multiples.

*Answers*

**1.** 45  **2.** 56

> *Method 1.* To find the LCM of a set of numbers using a list of multiples:
>
> **a)** Determine whether the largest number is a multiple of the others. If it is, it is the LCM. That is, if the largest number has the others as factors, the LCM is that number.
>
> **b)** If not, check multiples of the largest number until you get one that is a multiple of each of the others.

**EXAMPLE 2**   Find the LCM of 12 and 15.

**a)**   15 is the larger number, but it is not a multiple of 12.

**b)**   Check multiples of 15:

$$2 \cdot 15 = 30, \qquad \text{Not a multiple of 12}$$
$$3 \cdot 15 = 45, \qquad \text{Not a multiple of 12}$$
$$4 \cdot 15 = 60. \qquad \text{A multiple of 12: } 5 \cdot 12 = 60$$

The LCM = 60.

**EXAMPLE 3**   Find the LCM of 4 and 14.

**a)**   14 is the larger number, but it is not a multiple of 4.

**b)**   Check multiples of 14:

$$2 \cdot 14 = 28. \qquad \text{A multiple of 4}$$

The LCM = 28.

**EXAMPLE 4**   Find the LCM of 8 and 32.

32 is the larger number and 32 is a multiple of 8 ($4 \cdot 8 = 32$), so it is the LCM.

The LCM = 32.

**Do Exercises 3–6.** ▶

Find each LCM using one list of multiples.

**3.** 6, 9    **4.** 10, 12

**5.** 9, 36    **6.** 3, 20

**EXAMPLE 5**   Find the LCM of 10, 20, and 50.

**a)**   50 is a multiple of 10 but not a multiple of 20.

**b)**   Check multiples of 50:

$$2 \cdot 50 = 100. \qquad \text{A multiple of 10 and of 20: } 10 \cdot 10 = 100 \text{ and } 5 \cdot 20 = 100$$

The LCM = 100.

**Do Exercises 7 and 8.** ▶

Find each LCM using one list of multiples.

**7.** 30, 40, 50    **8.** 10, 20, 40

## Method 2: Finding LCMs Using Prime Factorizations

A second method for finding LCMs uses prime factorizations. Consider again 20 and 30. Their prime factorizations are $20 = 2 \cdot 2 \cdot 5$ and $30 = 2 \cdot 3 \cdot 5$. Any multiple of 20 will have to have *two* 2's as factors and *one* 5 as a factor. Any multiple of 30 will need to have *one* 2, *one* 3, and *one* 5 as factors.

The smallest number satisfying these conditions is

Two 2's, one 5; $2 \cdot 2 \cdot 3 \cdot 5$ is a multiple of 20.

$$2 \cdot 2 \cdot 3 \cdot 5.$$

One 2, one 3, one 5; $2 \cdot 2 \cdot 3 \cdot 5$ is a multiple of 30.

Thus, the LCM of 20 and 30 is $2 \cdot 2 \cdot 3 \cdot 5$, or 60. It has all the factors of 20 and all the factors of 30, but the factors are not repeated when they are common to both numbers.

Note that each prime factor is used the greatest number of times that it occurs in either of the individual factorizations.

---

*Method 2.* To find the LCM of two numbers using prime factorizations:

1. Write the prime factorization of each number.
2. Select one of the factorizations and see whether it contains the other.
   a) If it does, it is the LCM.
   b) If it does not, multiply that factorization by those prime factors of the other number that it lacks. The final product is the LCM.
3. As a check, make sure that the LCM includes each factor the greatest number of times that it occurs in either factorization.

---

**EXAMPLE 6**   Find the LCM of 18 and 21.

1. We begin by writing the prime factorization of each number:

   $18 = 2 \cdot 3 \cdot 3$  and  $21 = 3 \cdot 7$.

2. We select the factorization of 18: $2 \cdot 3 \cdot 3$.

   a) We note that $2 \cdot 3 \cdot 3$ does not contain the other factorization, $3 \cdot 7$.

   b) To find the LCM of 18 and 21, we multiply $2 \cdot 3 \cdot 3$ by the factor of 21 that it lacks, 7:

   18 is a factor.

   $$\text{LCM} = 2 \cdot 3 \cdot 3 \cdot 7.$$

   21 is a factor.

3. The greatest number of times that 2 occurs as a factor of 18 or 21 is **one** time; the greatest number of times that 3 occurs as a factor of 18 or 21 is **two** times; and the greatest number of times that 7 occurs as a factor of 18 or 21 is **one** time. To check, note that the LCM has exactly **one** 2, **two** 3's, and **one** 7. The LCM is $2 \cdot 3 \cdot 3 \cdot 7$, or 126.

**EXAMPLE 7**   Find the LCM of 7 and 21.

1. Because 7 is prime, we think of $7 = 7$ as a "factorization":

   $7 = 7$  and  $21 = 3 \cdot 7$.

2. One factorization, $3 \cdot 7$, contains the other. Thus, the LCM is $3 \cdot 7$, or 21.

**EXAMPLE 8**  Find the LCM of 24 and 36.

1. We write the prime factorization of each number:

   $$24 = 2 \cdot 2 \cdot 2 \cdot 3 \quad \text{and} \quad 36 = 2 \cdot 2 \cdot 3 \cdot 3.$$

2. We select the factorization of 24.

   **a)** The factorization of 24 does not contain the factorization of 36.

   **b)** To find the LCM of 24 and 36, we multiply the factorization of 24 by any prime factors of 36 that it lacks. We need another factor of 3.

   $$\text{LCM} = 2 \cdot 2 \cdot 2 \cdot 3 \cdot 3.$$

   24 is a factor.
   36 is a factor.

3. Note that the LCM includes 2 and 3 the greatest number of times that each appears as a factor of either 24 or 36. The LCM is $2 \cdot 2 \cdot 2 \cdot 3 \cdot 3$, or 72.

**Do Exercises 9–12.** ▶

Exponential notation is often helpful when writing least common multiples. Let's reconsider Example 8 using exponents. The largest exponents indicate the greatest number of times that 2 and 3 occur as factors.

$$24 = 2 \cdot 2 \cdot 2 \cdot 3 = 2^3 \cdot 3^1 \quad \color{red}{2^3 \text{ is the greatest power of 2}}$$
$$36 = 2 \cdot 2 \cdot 3 \cdot 3 = 2^2 \cdot 3^2 \quad \color{red}{3^2 \text{ is the greatest power of 3}}$$
$$\text{LCM} = 2 \cdot 2 \cdot 2 \cdot 3 \cdot 3 = 2^3 \cdot 3^2, \text{ or } 72.$$

Note that the greatest power of each factor is used to construct the LCM.

Lining up the different prime numbers in the factorizations can help us construct the LCM. This method also works well when finding the LCM of more than two numbers.

**EXAMPLE 9**  Find the LCM of 27, 90, and 84.

We find the prime factorization of each number and write the factorizations in exponential notation.

$$27 = 3 \cdot 3 \cdot 3 = 3^3$$
$$90 = 2 \cdot 3 \cdot 3 \cdot 5 = 2 \cdot 3^2 \cdot 5$$
$$84 = 2 \cdot 2 \cdot 3 \cdot 7 = 2^2 \cdot 3 \cdot 7$$

No one factorization contains the others. The prime numbers 2, 3, 5, and 7 appear as factors.

We write the factorizations, lining up all the powers of 2, the powers of 3, and so on.

$$27 = \phantom{2^2 \cdot} 3^3$$
$$90 = 2 \phantom{^2} \cdot 3^2 \cdot 5$$
$$84 = 2^2 \cdot 3 \phantom{^2} \cdot \phantom{5} 7$$

The LCM is formed by choosing the greatest power of each factor:

$$2^2 \cdot 3^3 \cdot 5 \cdot 7 = 3780.$$

The LCM of 27, 90, and 84 is 3780.

**Do Exercises 13 and 14.** ▶

---

Use prime factorizations to find the LCM.

**9.** 8, 10        **10.** 5, 30

**11.** 12, 48

 **12.** 18, 40

  **1.** $18 = 2 \cdot 3 \cdot \square$

     $40 = 2 \cdot 2 \cdot 2 \cdot \square$

  **2.** Select the factorization of 40:

    $2 \cdot 2 \cdot 2 \cdot 5.$

    This is not a multiple of 18. We need two factors of $\square$.

  **3.** $\text{LCM} = 2 \cdot 2 \cdot 2 \cdot 5 \cdot \square \cdot \square$

     $= \square$

Find the LCM.

**13.** 8, 18, 30        **14.** 10, 20, 25

*Answers*

**9.** 40  **10.** 30  **11.** 48  **12.** 360  **13.** 360
**14.** 100

*Guided Solution:*

**12.** 3, 5; 3; 3, 3, 360

**EXAMPLE 10**  Find the LCM of 8 and 25.

We write the prime factorization of each number in exponential notation.

$$8 = 2 \cdot 2 \cdot 2 = 2^3$$
$$25 = 5 \cdot 5 = 5^2$$

The prime numbers 2 and 5 appear as factors. We write the factorizations as

$$8 = 2^3$$
$$25 = \phantom{2^3} 5^2.$$

Note that the two numbers, 8 and 25, have no common prime factor. When this is the case, the LCM is just the product of the two numbers. Thus, the LCM is $2^3 \cdot 5^2 = 8 \cdot 25 = 200$.

◀ **Do Exercises 15 and 16.**

The same method works perfectly with variables.

**EXAMPLE 11**  Find the LCM of $7a^2b$ and $ab^3$.

We have the following factorizations:

$$7a^2b = 7 \cdot a \cdot a \cdot b \quad \text{and} \quad ab^3 = a \cdot b \cdot b \cdot b.$$

No one factorization contains the other.

Consider the factorization of $7a^2b$, which is $7 \cdot a \cdot a \cdot b$. Since $ab^3$ contains two more factors of $b$, we multiply the factorization of $7a^2b$ by $b \cdot b$.

$$7 \cdot a \cdot a \cdot b \cdot b \cdot b$$

$7a^2b$ is a factor.
$ab^3$ is a factor.

As a second approach, we find the greatest power of each factor using exponential notation.

$$7a^2b = 7 \cdot a^2 \cdot b$$
$$ab^3 = \phantom{7 \cdot} a \cdot b^3$$

The LCM is $7 \cdot a \cdot a \cdot b \cdot b \cdot b$, or $7a^2b^3$.

◀ **Do Exercises 17 and 18.**

**EXAMPLE 12**  Find the LCM of $12x^2y^3z$ and $18x^4z^3$.

We write the factorizations using exponential notation:

$$12x^2y^3z = 2^2 \cdot 3 \cdot x^2 \cdot y^3 \cdot z \quad \text{and} \quad 18x^4z^3 = 2 \cdot 3^2 \cdot x^4 \cdot z^3.$$

No one factorization contains the other.

We form the LCM using the greatest power of each factor.

$$12x^2y^3z = 2^2 \cdot 3 \phantom{^2} \cdot x^2 \cdot y^3 \cdot z$$
$$18x^4z^3 = 2 \phantom{^2} \cdot 3^2 \cdot x^4 \cdot \phantom{y^3 \cdot} z^3$$

The LCM is $2^2 \cdot 3^2 \cdot x^4 \cdot y^3 \cdot z^3$, or $36x^4y^3z^3$.

◀ **Do Exercise 19.**

Find the LCM.

**15.** $4, 9$

**16.** $5, 6, 7$

Find the LCM.

**17.** $xy, yz$

**18.** $5a^2, a^3b$

$$5a^2 = 5 \cdot \boxed{\phantom{xx}}$$
$$a^3b = \phantom{5 \cdot} a^3 \cdot \boxed{\phantom{xx}}$$
$$\text{LCM} = 5 \cdot a^3 \cdot \boxed{\phantom{xx}}, \text{or} \boxed{\phantom{xxx}}$$

GS

**19.** Find the LCM of $8a^3b^2$ and $10a^2c^4$.

*Answers*

**15.** 36  **16.** 210  **17.** $xyz$  **18.** $5a^3b$
**19.** $40a^3b^2c^4$

*Guided Solution:*
**18.** $a^2, b; b, 5a^3b$

## ✓ Check Your Understanding

**Reading Check** Determine whether each statement is true or false.

_____ **RC1.** Any two numbers have more than one common multiple.

_____ **RC2.** If one number is a multiple of a second number, the larger number is the LCM of the two numbers.

_____ **RC3.** If two numbers have no common prime factor, then the LCM of the numbers is their product.

_____ **RC4.** LCMs cannot be found using prime factorizations.

**Concept Check** Find the LCM of each set of numbers using the given prime factorizations.

**CC1.** $6 = 2 \cdot 3$,
$15 = 3 \cdot 5$

**CC2.** $20 = 2 \cdot 2 \cdot 5$,
$24 = 2 \cdot 2 \cdot 2 \cdot 3$

**CC3.** $6 = 2 \cdot 3$,
$55 = 5 \cdot 11$

**CC4.** $36 = 2 \cdot 2 \cdot 3 \cdot 3$,
$600 = 2 \cdot 2 \cdot 2 \cdot 3 \cdot 5 \cdot 5$

**CC5.** $12 = 2 \cdot 2 \cdot 3$,
$20 = 2 \cdot 2 \cdot 5$,
$75 = 3 \cdot 5 \cdot 5$

**CC6.** $15 = 3 \cdot 5$
$40 = 2 \cdot 2 \cdot 2 \cdot 5$,
$28 = 2 \cdot 2 \cdot 7$,
$125 = 5 \cdot 5 \cdot 5$

**a** Find the LCM of each set of numbers or expressions.

**1.** 2, 4

**2.** 3, 15

**3.** 10, 25

**4.** 10, 15

**5.** 20, 40

**6.** 8, 12

**7.** 18, 27

**8.** 9, 11

**9.** 30, 50

**10.** 8, 36

**11.** 30, 40

**12.** 21, 27

**13.** 18, 24

**14.** 12, 18

**15.** 60, 70

**16.** 35, 45

**17.** 16, 36

**18.** 24, 32

**19.** 18, 20

**20.** 36, 48

**21.** 2, 3, 7

**22.** 2, 5, 9

**23.** 3, 6, 15

**24.** 6, 12, 18

**25.** 24, 36, 12

**26.** 8, 16, 22

**27.** 5, 12, 15

**28.** 12, 18, 40

**29.** 9, 12, 6

**30.** 8, 16, 12

**31.** 180, 100, 450    **32.** 18, 30, 50, 48    **33.** 8, 48    **34.** 16, 32    **35.** 10, 21

**36.** 14, 15    **37.** 75, 100    **38.** 81, 90    **39.** 12, 15, 60    **40.** 24, 36, 72

**41.** $ab, bc$    **42.** $7x, xy$    **43.** $3x, 9x^2$    **44.** $10x^4, 5x^3$    **45.** $4x^3, x^2y$

**46.** $6ab^2, 9a^3b$    **47.** $6r^3st^4, 8rs^2t$    **48.** $3m^2n^4p^5, 9mn^2p^4$    **49.** $a^3b, b^2c, ac^2$    **50.** $x^2z^3, x^3y, y^2z$

*Applications of LCMs: Planet Orbits.* Jupiter, Saturn, and Uranus all revolve around the sun. Jupiter takes 12 yr, Saturn 30 yr, and Uranus 84 yr to make a complete revolution. On a certain night, you look at Jupiter, Saturn, and Uranus and wonder how many years it will take before they have the same position again. (*Hint*: The number of years is the LCM of 12, 30, and 84.)

**Data:** *The Handy Science Answer Book*

**51.** How often will Jupiter and Saturn appear in the same direction in the night sky as seen from the earth?

**52.** How often will Jupiter and Uranus appear in the same direction in the night sky as seen from the earth?

**53.** How often will Saturn and Uranus appear in the same direction in the night sky as seen from the earth?

**54.** How often will Jupiter, Saturn, and Uranus appear in the same direction in the night sky as seen from the earth?

## Skill Maintenance

Perform the indicated operation and, if possible, simplify.

**55.** $-38 + 52$  [2.2a]    **56.** $-18 \div \left(\dfrac{2}{3}\right)$  [3.7b]    **57.** $23 \cdot 345$  [1.4a]

**58.** $\dfrac{4}{5} \cdot \dfrac{10}{12}$  [3.6a]    **59.** $\dfrac{4}{5} \div \left(-\dfrac{7}{10}\right)$  [3.7b]    **60.** $382 - 549$  [2.3a]

## Synthesis

🖩 Use a calculator and the multiples method to find the LCM of each pair of numbers.

**61.** 288; 324    **62.** 2700; 7800    **63.** 7719; 18,011    **64.** 17,385; 24,339

**65.** The tables at a flea market are either 6 ft long or 8 ft long. Each row consists entirely of 6-ft tables or entirely of 8-ft tables, and all rows are the same length. What is the shortest possible length of the rows at the flea market?

*African Artistry.* In southern Africa, the design of every woven handbag, or *gipatsi* (plural *sipatsi*), is created by repeating two or more geometric patterns. Each pattern encircles the bag, sharing the strands of fabric with any pattern above or below. The length, or period, of each pattern is the number of strands required to construct the pattern. For a gipatsi to be considered beautiful, each individual pattern must fit a whole number of times around the bag.

**Data:** Gerdes, Paulus. *Women, Art and Geometry in Southern Africa.* Asmara, Eritrea: Africa World Press, Inc., p. 5.

**66.** A weaver is using two patterns to create a gipatsi. Pattern A is 10 strands long, and pattern B is 3 strands long. What is the smallest number of strands that can be used to complete the gipatsi?

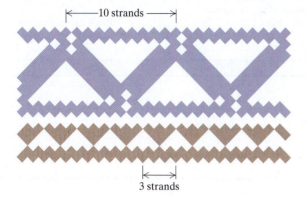

**67.** A weaver is using a four-strand pattern, a six-strand pattern, and an eight-strand pattern. What is the smallest number of strands that can be used to complete the gipatsi?

**68.** *Prescriptions.* Prescriptions for a 30-day supply of simvastatin and a 14-day supply of pain medication are filled at a pharmacy. Assuming the prescriptions are refilled regularly, how long will it be until they are both refilled on the same day?

**69.** Consider $a^3b^2$ and $a^2b^5$. Determine whether each of the following is the LCM of $a^3b^2$ and $a^2b^5$. Tell why or why not.
**a)** $a^3b^3$
**b)** $a^2b^5$
**c)** $a^3b^5$

**70.** Use Example 9 to help find the LCM of 27, 90, 84, 210, 108, and 50.

**71.** Use Examples 6 and 8 to help find the LCM of 18, 21, 24, 36, 63, 56, and 20.

**72.** Find three different pairs of numbers for which 56 is the LCM. Do not use 56 itself in any of the pairs.

**73.** Find three different pairs of numbers for which 54 is the LCM. Do not use 54 itself in any of the pairs.

## 4.2 Addition, Order, and Applications

### OBJECTIVES

**a** Add using fraction notation when denominators are the same.

**b** Add using fraction notation when denominators are different.

**c** Use < or > to form a true statement with fraction notation.

**d** Solve problems involving addition with fraction notation.

### a LIKE DENOMINATORS

Addition using fraction notation corresponds to combining or putting like things together, just as when we combined like terms. For example,

We combine two sets, each of which consists of equally sized parts of one object.

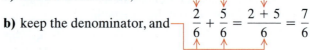

This is the resulting set.

$$\frac{2}{8} \qquad + \qquad \frac{3}{8} \qquad = \qquad \frac{5}{8}$$

2 eighths + 3 eighths = 5 eighths,

or $\quad 2 \cdot \frac{1}{8} + 3 \cdot \frac{1}{8} = 5 \cdot \frac{1}{8},\quad$ or $\quad \frac{2}{8} + \frac{3}{8} = \frac{5}{8}.$

◀ Do Exercise 1.

**1.** Find $\frac{1}{5} + \frac{3}{5}$.

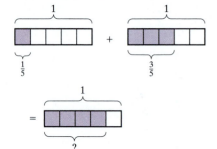

> To add when denominators are the same,
>
> **a)** add the numerators,
>
> **b)** keep the denominator, and $\quad \dfrac{2}{6} + \dfrac{5}{6} = \dfrac{2+5}{6} = \dfrac{7}{6}$
>
> **c)** simplify, if possible.

**EXAMPLES**  Add and, if possible, simplify.

**1.** $\dfrac{2}{4} + \dfrac{1}{4} = \dfrac{2+1}{4} = \dfrac{3}{4}$   No simplifying is possible.

**2.** $\dfrac{3}{12} + \dfrac{5}{12} = \dfrac{3+5}{12} = \dfrac{8}{12}$   Adding numerators; the denominator remains unchanged.

$= \dfrac{4}{4} \cdot \dfrac{2}{3} = \dfrac{2}{3}$   Simplifying by removing a factor equal to 1: $\frac{4}{4} = 1$

**3.** $\dfrac{-11}{6} + \dfrac{3}{6} = \dfrac{-11+3}{6} = \dfrac{-8}{6}$

$= \dfrac{2}{2} \cdot \dfrac{-4}{3} = \dfrac{-4}{3},$ or $-\dfrac{4}{3}$   Removing a factor equal to 1: $\frac{2}{2} = 1$

Add and, if possible, simplify.

**2.** $\dfrac{5}{13} + \dfrac{9}{13}$

**3.** $\dfrac{1}{3} + \dfrac{2}{3}$

**4.** $\dfrac{-5}{12} + \left(\dfrac{-1}{12}\right)$

**5.** $\dfrac{3}{x} + \dfrac{-7}{x}$

**4.** $-\dfrac{2}{a} + \left(-\dfrac{3}{a}\right) = \dfrac{-2}{a} + \dfrac{-3}{a}$   Recall that $-\dfrac{m}{n} = \dfrac{-m}{n}.$

$= \dfrac{-2 + (-3)}{a} = \dfrac{-5}{a},$ or $-\dfrac{5}{a}$

◀ Do Exercises 2–5.

**Answers**

**1.** $\dfrac{4}{5}$   **2.** $\dfrac{14}{13}$   **3.** 1   **4.** $-\dfrac{1}{2}$   **5.** $-\dfrac{4}{x}$

We may need to add fractions when combining like terms.

**EXAMPLE 5**  Simplify by combining like terms: $\dfrac{2}{7}x + \dfrac{3}{7}x$.

$$\dfrac{2}{7}x + \dfrac{3}{7}x = \left(\dfrac{2}{7} + \dfrac{3}{7}\right)x \qquad \text{Try to do this step mentally.}$$

$$= \dfrac{5}{7}x$$

**Do Exercises 6 and 7.** ▶

Simplify by combining like terms.

**6.** $\dfrac{3}{10}a + \dfrac{1}{10}a$   **7.** $-\dfrac{3}{4}x + \dfrac{1}{4}x$

## b  DIFFERENT DENOMINATORS

We cannot add $\frac{1}{2} + \frac{1}{3}$ by simply adding numerators. However, by rewriting $\frac{1}{2}$ as $\frac{1}{2} \cdot \frac{3}{3} = \frac{3}{6}$ and $\frac{1}{3}$ as $\frac{1}{3} \cdot \frac{2}{2} = \frac{2}{6}$, we can determine the sum.

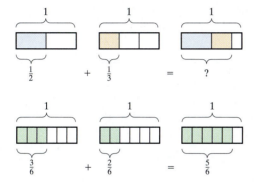

When denominators are different, we can find a common denominator by multiplying by 1. Consider the addition $\frac{3}{4} + \frac{1}{6}$ using two different common denominators.

**A.** We use 24 as a common denominator:

$$\dfrac{3}{4} + \dfrac{1}{6} = \dfrac{3}{4} \cdot \dfrac{6}{6} + \dfrac{1}{6} \cdot \dfrac{4}{4}$$

$$= \dfrac{18}{24} + \dfrac{4}{24} = \dfrac{22}{24} = \dfrac{11}{12}.$$

**B.** We use 12 as a common denominator:

$$\dfrac{3}{4} + \dfrac{1}{6} = \dfrac{3}{4} \cdot \dfrac{3}{3} + \dfrac{1}{6} \cdot \dfrac{2}{2}$$

$$= \dfrac{9}{12} + \dfrac{2}{12} = \dfrac{11}{12}.$$

We had to simplify at the end of (A), but not in (B). In (B), we used the *least* common multiple of the denominators, 12, as the common denominator. That number is called the **least common denominator**, or **LCD**. We may still need to simplify when using the LCD, but it is usually easier than when we use a larger denominator.

> To add when denominators are different:
>
> **a)** Find the least common multiple of the denominators. That number is the least common denominator, LCD.
> **b)** Multiply by 1, writing 1 in the form of $n/n$ to express each fraction in an equivalent form that contains the LCD.
> **c)** Add the numerators, keeping the same denominator.
> **d)** Simplify, if possible.

*Answers*

**6.** $\dfrac{2}{5}a$   **7.** $-\dfrac{1}{2}x$

**EXAMPLE 6** Add: $\dfrac{1}{8} + \dfrac{3}{4}$.

**a)** Since 4 is a factor of 8, the LCM of 4 and 8 is 8. Thus, the LCD is 8.

**b)** We need to find a fraction equivalent to $\frac{3}{4}$ with a denominator of 8:

$$\frac{1}{8} + \frac{3}{4} = \frac{1}{8} + \frac{3}{4} \cdot \frac{2}{2}. \qquad \textit{Think: } 4 \times \square = 8. \text{ The answer is 2, so we multiply by 1, using } \tfrac{2}{2}.$$

**c)** We add: $\dfrac{1}{8} + \dfrac{6}{8} = \dfrac{7}{8}$.     Adding numerators

**d)** No simplifying is possible. The sum is $\frac{7}{8}$.

In Examples 7–10, we follow the same steps without labeling them.

**EXAMPLE 7** Add: $\dfrac{5}{6} + \dfrac{1}{9}$.

The LCD is 18.     $6 = 2 \cdot 3$ and $9 = 3 \cdot 3$, so the LCM of 6 and 9 is $2 \cdot 3 \cdot 3$, or 18.

$$\frac{5}{6} + \frac{1}{9} = \frac{5}{6} \cdot \frac{3}{3} + \frac{1}{9} \cdot \frac{2}{2}$$

*Think*: $9 \times \square = 18$. The answer is 2, so we multiply by 1, using $\frac{2}{2}$.

*Think*: $6 \times \square = 18$. The answer is 3, so we multiply by 1, using $\frac{3}{3}$.

$$= \frac{15}{18} + \frac{2}{18} = \frac{17}{18}$$

◀ **Do Exercises 8 and 9.**

**EXAMPLE 8** Add: $\dfrac{3}{-5} + \dfrac{11}{10}$.

$$\frac{3}{-5} + \frac{11}{10} = \frac{-3}{5} + \frac{11}{10}$$

Recall that $\dfrac{m}{-n} = \dfrac{-m}{n}$. We generally avoid negative signs in the denominator. The LCD is 10.

$$= \frac{-3}{5} \cdot \frac{2}{2} + \frac{11}{10}$$

$$= \frac{-6}{10} + \frac{11}{10}$$

$$= \frac{5}{10} = \frac{1}{2}$$

We may still have to simplify, but simplifying is almost always easier if the LCD has been used.

◀ **Do Exercise 10.**

**EXAMPLE 9** Add: $\dfrac{5}{8} + 2$.

$$\frac{5}{8} + 2 = \frac{5}{8} + \frac{2}{1} \qquad \text{Rewriting 2 in fraction notation}$$

$$= \frac{5}{8} + \frac{2}{1} \cdot \frac{8}{8} \qquad \text{The LCD is 8.}$$

$$= \frac{5}{8} + \frac{16}{8} = \frac{21}{8}$$

◀ **Do Exercise 11.**

---

Add using the least common denominator.

**8.** $\dfrac{2}{3} + \dfrac{1}{6}$

**9.** $\dfrac{3}{8} + \dfrac{5}{6}$

The LCD is [ ].

$$\frac{3}{8} + \frac{5}{6} = \frac{3}{8} \cdot 1 + \frac{5}{6} \cdot 1$$

$$= \frac{3}{8} \cdot \frac{3}{3} + \frac{5}{6} \cdot \frac{\phantom{0}}{\phantom{0}}$$

$$= \frac{\phantom{00}}{24} + \frac{\phantom{00}}{24}$$

$$= \frac{\phantom{00}}{24}$$

**10.** Add: $\dfrac{1}{-6} + \dfrac{7}{18}$.

**11.** Add: $7 + \dfrac{3}{5}$.

**Answers**

**8.** $\dfrac{5}{6}$  **9.** $\dfrac{29}{24}$  **10.** $\dfrac{2}{9}$  **11.** $\dfrac{38}{5}$

**Guided Solution:**

**9.** 24; $\dfrac{4}{4}$, 9, 20, 29

**EXAMPLE 10** Add: $\frac{9}{70} + \frac{11}{21} + \frac{-4}{15}$.

We need to determine the LCM of 70, 21, and 15:

$$70 = 2 \cdot 5 \cdot 7, \qquad 21 = 3 \cdot 7, \qquad 15 = 3 \cdot 5.$$

The LCM is $2 \cdot 3 \cdot 5 \cdot 7$, or 210.

$$\frac{9}{70} + \frac{11}{21} + \frac{-4}{15} = \frac{9}{70} \cdot \frac{3}{3} + \frac{11}{21} \cdot \frac{2 \cdot 5}{2 \cdot 5} + \frac{-4}{15} \cdot \frac{7 \cdot 2}{7 \cdot 2}$$

In each case, we multiply by 1 to obtain the LCD. To form 1, look at the prime factorization of the LCD and use the factor(s) missing from each denominator.

$$= \frac{9 \cdot 3}{70 \cdot 3} + \frac{11 \cdot 10}{21 \cdot 10} + \frac{-4 \cdot 14}{15 \cdot 14}$$

$$= \frac{27}{210} + \frac{110}{210} + \frac{-56}{210}$$

$$= \frac{137 + (-56)}{210} = \frac{81}{210}$$

$$= \frac{3 \cdot 3 \cdot 3 \cdot 3}{2 \cdot 3 \cdot 5 \cdot 7} = \frac{3}{3} \cdot \frac{3 \cdot 3 \cdot 3}{2 \cdot 5 \cdot 7} = \frac{27}{70}$$

**Do Exercises 12 and 13.** ▶

Add.

**12.** $\frac{4}{10} + \frac{1}{100} + \frac{3}{1000}$

**13.** $\frac{7}{10} + \frac{-2}{21} + \frac{1}{7}$

## c ORDER

When two fractions share a common denominator, the larger number can be found by comparing numerators. For example, 4 is greater than 3, so $\frac{4}{5}$ is greater than $\frac{3}{5}$.

$$\frac{4}{5} > \frac{3}{5}$$

Similarly, because $-6$ is less than $-2$, we have

$$\frac{-6}{7} < \frac{-2}{7}, \quad \text{or} \quad -\frac{6}{7} < -\frac{2}{7}.$$

**Do Exercises 14–16.** ▶

Use $<$ or $>$ for ☐ to form a true sentence.

**14.** $\frac{3}{8} \ \square \ \frac{5}{8}$

**15.** $\frac{7}{10} \ \square \ \frac{6}{10}$     **16.** $\frac{-2}{9} \ \square \ \frac{-5}{9}$

**EXAMPLE 11** Use $<$ or $>$ for ☐ to form a true sentence:

$$\frac{5}{8} \ \square \ \frac{2}{3}.$$

You can confirm that the LCD is 24. We multiply by 1 to find two fractions equivalent to $\frac{5}{8}$ and $\frac{2}{3}$ with denominators the same:

$$\frac{5}{8} \cdot \frac{3}{3} = \frac{15}{24}, \qquad \frac{2}{3} \cdot \frac{8}{8} = \frac{16}{24}. \qquad \frac{5}{8} \ \square \ \frac{2}{3} \text{ is equivalent to } \frac{15}{24} \ \square \ \frac{16}{24}.$$

Since $15 < 16$, it follows that $\frac{15}{24} < \frac{16}{24}$. Thus,

$$\frac{5}{8} < \frac{2}{3}.$$

**EXAMPLE 12** Use < or > for ☐ to form a true sentence:

$$-\frac{89}{100} \; \square \; -\frac{9}{10}.$$

We rewrite $-\frac{9}{10}$ with a denominator of 100: $-\frac{9}{10} \cdot \frac{10}{10} = -\frac{90}{100}$. We then have

$$\frac{-89}{100} \; \square \; \frac{-90}{100}. \qquad \text{Recall that } -\frac{m}{n} = \frac{-m}{n}.$$

Since $-89 > -90$, it follows that $-\frac{89}{100} > -\frac{90}{100}$, so

$$-\frac{89}{100} > -\frac{9}{10}.$$

◀ **Do Exercises 17–19.**

Use < or > for ☐ to form a true sentence.

**17.** $\frac{2}{3} \square \frac{3}{4}$      **18.** $\frac{-3}{4} \square \frac{-8}{12}$

**19.** $\frac{7}{8} \square \frac{5}{6}$

---

## d   APPLICATIONS AND PROBLEM SOLVING

**EXAMPLE 13** *Construction.* A contractor uses two layers of subflooring under a ceramic tile floor. First, she installs a $\frac{3}{4}$-in. layer of oriented strand board (OSB). Then a $\frac{1}{2}$-in. sheet of cement board is mortared to the OSB. The mortar is $\frac{1}{8}$-in. thick. What is the total thickness of the two installed subfloors?

**1. Familiarize.** We first make a drawing. We let $T =$ the total thickness of the subfloors.

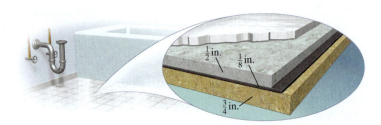

**2. Translate.** The problem can be translated to an equation as follows.

| OSB | plus | mortar | plus | cement board | is | total thickness |
|---|---|---|---|---|---|---|
| $\frac{3}{4}$ | $+$ | $\frac{1}{8}$ | $+$ | $\frac{1}{2}$ | $=$ | $T$ |

**3. Solve.** To solve the equation, we carry out the addition.

$$\frac{3}{4} + \frac{1}{8} + \frac{1}{2} = T \qquad \text{The LCM of the denominators is 8.}$$

$$\frac{3}{4} \cdot \frac{2}{2} + \frac{1}{8} + \frac{1}{2} \cdot \frac{4}{4} = T \qquad \text{Multiplying by 1 to obtain the LCD}$$

$$\frac{6}{8} + \frac{1}{8} + \frac{4}{8} = T$$

$$\frac{11}{8} = T$$

**4. Check.** We check by repeating the calculation.

**5. State.** The total thickness of the installed subfloors is $\frac{11}{8}$ in.

◀ **Do Exercise 20.**

---

**20.** *Catering.* **GS** A caterer prepares a mixed berry salad with $\frac{7}{8}$ qt of strawberries, $\frac{3}{4}$ qt of raspberries, and $\frac{5}{16}$ qt of blueberries. How many quarts of berries are in the salad?

**1. Familiarize.** Let $T =$ the total amount of berries in the salad.

**2. Translate.** To find the total amount, we add.

$$\frac{7}{\square} + \frac{3}{\square} + \frac{5}{\square} = T$$

**3. Solve.** The LCD is ☐.

$$\frac{7}{8} \cdot \frac{2}{2} + \frac{3}{4} \cdot \frac{\square}{\square} + \frac{5}{16} = T$$

$$\frac{\square}{16} + \frac{\square}{16} + \frac{5}{16} = T$$

$$\frac{\square}{16} = T$$

**4. Check.** The answer is reasonable because it is larger than any of the individual amounts.

**5. State.** There are ☐ qt of berries in the salad.

---

*Answers*

**17.** <   **18.** <   **19.** >   **20.** $\frac{31}{16}$ qt

*Guided Solution:*

**20.** 8, 4, 16; 16; $\frac{4}{4}$, 14, 12, 31; $\frac{31}{16}$

## ✓ Check Your Understanding

**Reading Check** Determine whether each statement is true or false.

_____ **RC1.** Before we can add two fractions, they must have the same denominator.

_____ **RC2.** To add fractions, we add numerators and add denominators.

_____ **RC3.** If we use the LCD to add fractions, we never need to simplify the result.

_____ **RC4.** Adding fractions with different denominators involves multiplying at least one fraction by 1.

**Concept Check** Rewrite the fractions in each addition as equivalent fractions with the given LCD. Do not perform the addition.

**CC1.** $\dfrac{7}{20} + \dfrac{3}{4}$; LCD = 20

**CC2.** $\dfrac{3}{8} + \dfrac{5}{6}$; LCD = 24

**CC3.** $\dfrac{7}{10} + \dfrac{11}{15} + \dfrac{1}{4}$; LCD = 60

**a** , **b**    Add and, if possible, simplify.

**1.** $\dfrac{4}{9} + \dfrac{1}{9}$

**2.** $\dfrac{3}{11} + \dfrac{5}{11}$

**3.** $\dfrac{4}{7} + \dfrac{3}{7}$

**4.** $\dfrac{7}{8} + \dfrac{1}{8}$

**5.** $\dfrac{7}{10} + \dfrac{3}{-10}$

**6.** $\dfrac{1}{-6} + \dfrac{5}{6}$

**7.** $\dfrac{9}{a} + \dfrac{4}{a}$

**8.** $\dfrac{2}{t} + \dfrac{3}{t}$

**9.** $\dfrac{-1}{4} + \dfrac{-1}{4}$

**10.** $\dfrac{7}{12} + \dfrac{-5}{12}$

**11.** $\dfrac{2}{9}x + \dfrac{5}{9}x$

**12.** $\dfrac{3}{11}a + \dfrac{2}{11}a$

**13.** $\dfrac{3}{32}t + \dfrac{13}{32}t$

**14.** $\dfrac{3}{25}x + \dfrac{12}{25}x$

**15.** $-\dfrac{2}{x} + \left(-\dfrac{7}{x}\right)$

**16.** $-\dfrac{7}{a} + \dfrac{5}{a}$

**17.** $\dfrac{1}{8} + \dfrac{1}{6}$

**18.** $\dfrac{1}{9} + \dfrac{1}{6}$

**19.** $\dfrac{-4}{5} + \dfrac{7}{10}$

**20.** $\dfrac{-3}{4} + \dfrac{-1}{12}$

**21.** $\dfrac{7}{12} + \dfrac{3}{8}$

**22.** $\dfrac{7}{8} + \dfrac{1}{16}$

**23.** $\dfrac{3}{20} + 4$

**24.** $\dfrac{2}{15} + 3$

**25.** $\dfrac{5}{-8} + \dfrac{5}{6}$

**26.** $\dfrac{5}{-6} + \dfrac{7}{9}$

**27.** $\dfrac{3}{10}x + \dfrac{7}{100}x$

**28.** $\dfrac{9}{20}a + \dfrac{3}{40}a$

**29.** $\dfrac{5}{12} + \dfrac{8}{15}$

**30.** $\dfrac{3}{16} + \dfrac{1}{12}$

**31.** $\dfrac{7}{8} + \dfrac{0}{1}$

**32.** $\dfrac{0}{6} + \dfrac{5}{3}$

**33.** $\dfrac{-7}{10} + \dfrac{-29}{100}$

**34.** $\dfrac{-3}{10} + \dfrac{-27}{100}$

**35.** $-\dfrac{1}{10}x + \dfrac{1}{15}x$

**36.** $-\dfrac{1}{6}x + \dfrac{1}{4}x$

**37.** $-5t + \dfrac{2}{7}t$

**38.** $-4x + \dfrac{3}{5}x$

**39.** $-\dfrac{5}{12} + \dfrac{7}{-24}$

**40.** $-\dfrac{1}{18} + \dfrac{5}{-12}$

**41.** $\dfrac{3}{16} + \dfrac{5}{16} + \dfrac{4}{16}$

**42.** $\dfrac{3}{8} + \dfrac{1}{8} + \dfrac{2}{8}$

**43.** $\dfrac{4}{10} + \dfrac{3}{100} + \dfrac{7}{1000}$

**44.** $\dfrac{7}{10} + \dfrac{2}{100} + \dfrac{9}{1000}$

**45.** $\dfrac{3}{10} + \dfrac{5}{12} + \dfrac{8}{15}$

**46.** $\dfrac{1}{2} + \dfrac{3}{8} + \dfrac{1}{4}$

**47.** $\dfrac{5}{6} + \dfrac{25}{52} + \dfrac{7}{4}$

**48.** $\dfrac{15}{24} + \dfrac{7}{36} + \dfrac{91}{48}$

**49.** $\dfrac{2}{9} + \dfrac{7}{10} + \dfrac{-4}{15}$

**50.** $\dfrac{5}{12} + \dfrac{-3}{8} + \dfrac{1}{10}$

**51.** $-\dfrac{3}{4} + \dfrac{1}{5} + \dfrac{-7}{10}$

**52.** $\dfrac{1}{3} + \dfrac{-7}{9} + \dfrac{-1}{2}$

**c** Use < or > for ☐ to form a true sentence.

**53.** $\dfrac{3}{8} \, \square \, \dfrac{2}{8}$

**54.** $\dfrac{7}{9} \, \square \, \dfrac{5}{9}$

**55.** $\dfrac{2}{3} \, \square \, \dfrac{5}{6}$

**56.** $\dfrac{11}{18} \, \square \, \dfrac{5}{9}$

**57.** $\dfrac{-2}{7} \, \square \, \dfrac{-5}{7}$

**58.** $\dfrac{-4}{5} \, \square \, \dfrac{-3}{5}$

**59.** $\dfrac{9}{15} \, \square \, \dfrac{7}{10}$

**60.** $\dfrac{5}{14} \, \square \, \dfrac{8}{21}$

**61.** $\dfrac{3}{4} \, \square \, -\dfrac{1}{5}$

**62.** $\dfrac{3}{8} \, \square \, -\dfrac{13}{16}$

**63.** $\dfrac{-7}{20} \, \square \, \dfrac{-6}{15}$

**64.** $\dfrac{-7}{12} \, \square \, \dfrac{-9}{16}$

Arrange each group of fractions from smallest to largest.

**65.** $\dfrac{3}{10}, \dfrac{5}{12}, \dfrac{4}{15}$

**66.** $\dfrac{5}{6}, \dfrac{19}{21}, \dfrac{11}{14}$

**d** Solve.

67. *Segway® Tour.* On her Segway® tour of Chicago, Alexis rode $\frac{5}{6}$ mi to the lakefront, then $\frac{3}{4}$ mi along the beach, and then $\frac{3}{2}$ mi through the park. How far did she ride the Segway?

68. *Volunteering.* For a community project, an earth science class volunteered one hour per day for three days to join the state highway beautification project. The students collected trash along a $\frac{4}{5}$-mi stretch of highway the first day, a $\frac{5}{8}$-mi stretch the second day, and a $\frac{1}{2}$-mi stretch the third day. How many miles along the highway did they clean?

69. *Caffeine.* To cut back on caffeine intake, Michelle and Gerry mix caffeinated and decaffeinated coffee beans before grinding for a customized mix. They mix $\frac{3}{16}$ lb of decaffeinated beans with $\frac{5}{8}$ lb of caffeinated beans. What is the total amount of coffee beans in the mixture?

70. *Purchasing Tea.* Alyse bought $\frac{1}{3}$ lb of orange pekoe tea and $\frac{1}{2}$ lb of English cinnamon tea. How many pounds of tea did she buy?

71. *Culinary Arts.* The campus culinary arts department is preparing brownies for the international student reception. Students in the catering program iced the $\frac{11}{16}$-in. $\left(\frac{11''}{16}\right)$ brownies with a $\frac{5}{32}$-in. $\left(\frac{5''}{32}\right)$ layer of butterscotch icing. What is the thickness of the iced brownies?

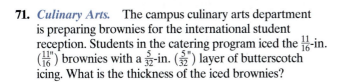

72. *Carpentry.* A carpenter glues two kinds of plywood together. He glues a $\frac{1}{4}$-in. $\left(\frac{1''}{4}\right)$ piece of cherry plywood to a $\frac{3}{8}$-in. $\left(\frac{3''}{8}\right)$ piece of less expensive plywood. What is the total thickness of these pieces?

73. *Baking.* A baker used $\frac{1}{2}$ lb of flour for rolls, $\frac{1}{4}$ lb for donuts, and $\frac{1}{3}$ lb for cookies. How much flour was used?

74. *Baking.* A recipe for muffins calls for $\frac{1}{2}$ qt (quart) of buttermilk, $\frac{1}{3}$ qt of skim milk, and $\frac{1}{16}$ qt of oil. How many quarts of liquid ingredients does the recipe call for?

75. *Meteorology.* On April 15, it rained $\frac{1}{2}$ in. in the morning and $\frac{3}{8}$ in. in the afternoon. How much did it rain altogether?

76. *Medication.* Janine took $\frac{1}{5}$ g of ibuprofen before lunch and $\frac{1}{2}$ g after lunch. How much did she take altogether?

77. *Hiking.* A park naturalist hiked $\frac{3}{5}$ mi to a lookout, another $\frac{3}{10}$ mi to an osprey's nest, and finally $\frac{3}{4}$ mi to a campsite. How far did the naturalist hike?

78. *Triathlon.* A triathlete runs $\frac{7}{8}$ mi, canoes $\frac{1}{3}$ mi, and swims $\frac{1}{6}$ mi. How many miles does the triathlete cover?

**79. *Culinary Arts.*** A recipe for strawberry punch calls for $\frac{1}{5}$ qt of ginger ale and $\frac{3}{5}$ qt of strawberry soda. How much liquid is needed? If the recipe is doubled, how much liquid is needed? If the recipe is halved, how much liquid is needed?

**80. *Construction.*** A cubic meter of concrete mix contains 420 kg (kilograms) of cement, 150 kg of stone, and 120 kg of sand. What is the total weight of a cubic meter of the mix? What fractional part is cement? stone? sand? Add these fractional amounts. What is the result?

## Skill Maintenance

Subtract. [2.3a]

**81.** $-7 - 6$

**82.** $-5 - (-9)$

**83.** $9 - 17$

**84.** $-8 - 23$

Evaluate. [2.6a]

**85.** $\dfrac{x - y}{3}$, for $x = 7$ and $y = -3$

**86.** $3(x + y)$ and $3x + 3y$, for $x = 5$ and $y = 9$

Solve.

**87.** $48 \cdot t = 1680$ [1.7b]

**88.** $10{,}000 = m + 3593$ [1.7b]

**89.** $3x - 8 = 25$ [2.8e]

**90.** $5x + 9 = 24$ [2.8e]

**91.** $\dfrac{2}{3}x = \dfrac{6}{7}$ [3.8a]

**92.** $-\dfrac{5}{8}t = \dfrac{1}{4}$ [3.8a]

## Synthesis

Add and, if possible, simplify.

**93.** $\dfrac{3}{10}t + \dfrac{2}{7} + \dfrac{2}{15}t + \dfrac{3}{5}$

**94.** $\dfrac{2}{9} + \dfrac{4}{21}x + \dfrac{4}{15} + \dfrac{3}{14}x$

**95.** $5t^2 + \dfrac{6}{a}t + 2t^2 + \dfrac{3}{a}t$

Use <, >, or = for ☐ to form a true sentence.

**96.** 🖩 $\dfrac{10}{97} + \dfrac{67}{137} \ \square\ \dfrac{8123}{13{,}289}$

**97.** 🖩 $\dfrac{12}{169} + \dfrac{53}{103} \ \square\ \dfrac{10{,}192}{17{,}407}$

**98.** 🖩 $\dfrac{37}{157} + \dfrac{20}{107} \ \square\ \dfrac{6942}{16{,}799}$

**99.** A guitarist's band is booked for Friday and Saturday nights at a local club. The guitarist's group is a trio on Friday and expands to a quintet on Saturday. Thus, the guitarist is paid one-third of one-half the weekend's pay for Friday and one-fifth of one-half the weekend's pay for Saturday. What fractional part of the total pay did the guitarist receive for the weekend's work? If the band was paid $1200, how much did the guitarist receive?

**100.** 🖩 Consider only the numbers 2, 3, 4, and 5. Assume each is placed in a blank in the following.

$$\dfrac{\square}{\square} + \dfrac{\square}{\square} = \ ?$$

What placement of the numbers in the blanks yields the largest sum?

**101.** 🖩 In the sum below, $a$ and $b$ are digits (so $1b$ is a two-digit number and $35a$ is a three-digit number). Find $a$ and $b$. (*Hint:* $a < 4$ and $b > 6$.)

$$\dfrac{a}{17} + \dfrac{1b}{23} = \dfrac{35a}{391}$$

**102.** 🖩 Use a standard calculator. Arrange the following in order from smallest to largest.

$$\dfrac{3}{4},\ \dfrac{17}{21},\ \dfrac{13}{15},\ \dfrac{7}{9},\ \dfrac{15}{17},\ \dfrac{13}{12},\ \dfrac{19}{22}$$

# Subtraction, Equations, and Applications

### a SUBTRACTION

**OBJECTIVES**

**a** Subtract using fraction notation.

**b** Solve equations of the type $x + a = b$ and $a + x = b$, where $a$ and $b$ may be fractions.

**c** Solve applied problems involving subtraction with fraction notation.

#### Like Denominators

Let's consider the difference $\frac{4}{8} - \frac{3}{8}$.

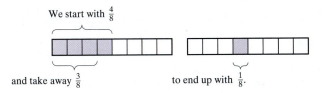

We start with $\frac{4}{8}$

and take away $\frac{3}{8}$     to end up with $\frac{1}{8}$.

We start with 4 eighths and take away 3 eighths:

4 eighths − 3 eighths = 1 eighth,

or $\quad 4 \cdot \frac{1}{8} - 3 \cdot \frac{1}{8} = \frac{1}{8}$, or $\quad \frac{4}{8} - \frac{3}{8} = \frac{1}{8}$.

---

To subtract when denominators are the same,

**a)** subtract the numerators,

**b)** keep the denominator, and

**c)** simplify, if possible.

$$\frac{7}{10} - \frac{4}{10} = \frac{7-4}{10} = \frac{3}{10}$$

---

**EXAMPLES** Subtract and, if possible, simplify.

**1.** $\dfrac{8}{13} - \dfrac{3}{13} = \dfrac{8-3}{13} = \dfrac{5}{13}$

**2.** $\dfrac{3}{35} - \dfrac{13}{35} = \dfrac{3-13}{35} = \dfrac{-10}{35} = \dfrac{5}{5} \cdot \dfrac{-2}{7} = \dfrac{-2}{7}$, or $-\dfrac{2}{7}$     Removing a factor equal to 1: $\frac{5}{5} = 1$

**3.** $\dfrac{13}{2a} - \dfrac{5}{2a} = \dfrac{13-5}{2a} = \dfrac{8}{2a} = \dfrac{2}{2} \cdot \dfrac{4}{a} = \dfrac{4}{a}$     Removing a factor equal to 1: $\frac{2}{2} = 1$

**4.** $-\dfrac{7}{t} - \dfrac{2}{t} = \dfrac{-7-2}{t} = \dfrac{-9}{t}$, or $-\dfrac{9}{t}$

**Do Exercises 1–4.** ▶

Subtract and, if possible, simplify.

**1.** $\dfrac{7}{8} - \dfrac{3}{8}$     **2.** $\dfrac{5}{9a} - \dfrac{1}{9a}$

**3.** $\dfrac{7}{10} - \dfrac{13}{10}$     **4.** $-\dfrac{2}{x} - \dfrac{4}{x}$

#### Different Denominators

---

To subtract when denominators are different:

**a)** Find the least common multiple of the denominators. That number is the least common denominator, LCD.

**b)** Multiply by 1, writing 1 in the form $n/n$, to express each fraction in an equivalent form that contains the LCD.

**c)** Subtract the numerators, keeping the same denominator.

**d)** Simplify, if possible.

---

**Answers**

**1.** $\dfrac{1}{2}$  **2.** $\dfrac{4}{9a}$  **3.** $-\dfrac{3}{5}$  **4.** $-\dfrac{6}{x}$

**EXAMPLE 5** Subtract: $\dfrac{2}{5} - \dfrac{3}{8}$.

a) The LCM of 5 and 8 is 40, so the LCD is 40.

b) We need to find numbers equivalent to $\frac{2}{5}$ and $\frac{3}{8}$ with denominators of 40:

$$\frac{2}{5} - \frac{3}{8} = \frac{2}{5} \cdot \frac{8}{8} - \frac{3}{8} \cdot \frac{5}{5}. \quad \longleftarrow$$

*Think*: $8 \times \square = 40$. The answer is 5, so we multiply by 1, using $\frac{5}{5}$.

*Think*: $5 \times \square = 40$. The answer is 8, so we multiply by 1, using $\frac{8}{8}$.

c) We subtract: $\dfrac{16}{40} - \dfrac{15}{40} = \dfrac{16 - 15}{40} = \dfrac{1}{40}$.

d) Since $\frac{1}{40}$ cannot be simplified, we are finished. The answer is $\frac{1}{40}$.

◀ **Do Exercise 5.**

**5.** Subtract: $\dfrac{3}{4} - \dfrac{2}{3}$.

**EXAMPLE 6** Subtract: $\dfrac{7}{12} - \dfrac{5}{6}$.

Since 12 is a multiple of 6, the LCM of 6 and 12 is 12. The LCD is 12.

$$\frac{7}{12} - \frac{5}{6} = \frac{7}{12} - \frac{5}{6} \cdot \frac{2}{2} \qquad \textit{Think: } 6 \times \square = 12. \text{ The answer is 2,}$$
so we multiply by 1, using $\frac{2}{2}$.

$$= \frac{7}{12} - \frac{10}{12}$$

$$= \frac{7 - 10}{12} = \frac{-3}{12} \qquad 7 - 10 = 7 + (-10) = -3$$

$$= \frac{3}{3} \cdot \frac{-1}{4} = \frac{-1}{4}, \text{ or } -\frac{1}{4} \qquad \begin{array}{l}\text{Simplifying by removing a} \\ \text{factor equal to 1: } \frac{3}{3} = 1\end{array}$$

**EXAMPLE 7** Subtract: $\dfrac{17}{24} - \dfrac{4}{15}$.

We need to find the LCM of 24 and 15:

$$\left.\begin{array}{l}24 = 2 \cdot 2 \cdot 2 \cdot 3, \\ 15 = 3 \cdot 5.\end{array}\right\} \quad \text{The LCM is } 2 \cdot 2 \cdot 2 \cdot 3 \cdot 5, \text{ or } 120.$$

Multiplying by 1 to obtain the LCD. To form 1, use the factors of the LCM that each denominator lacks. Note that $2 \cdot 2 \cdot 2 = 8$.

$$\frac{17}{24} - \frac{4}{15} = \frac{17}{24} \cdot \frac{5}{5} - \frac{4}{15} \cdot \frac{8}{8}$$

$$= \frac{85}{120} - \frac{32}{120} = \frac{85 - 32}{120} = \frac{53}{120}$$

◀ **Do Exercises 6–9.**

Subtract.

**6.** $\dfrac{5}{6} - \dfrac{1}{9}$    **(GS)**

The LCD is ▯.

$$\frac{5}{6} - \frac{1}{9} = \frac{5}{6} \cdot \frac{3}{3} - \frac{1}{9} \cdot \frac{\square}{\square}$$

$$= \frac{\square}{18} - \frac{\square}{18}$$

$$= \frac{\square}{18}$$

**7.** $\dfrac{2}{5} - \dfrac{7}{10}$    **8.** $\dfrac{2}{3} - \dfrac{5}{6}$

**9.** $\dfrac{11}{28} - \dfrac{5}{16}$

**10.** Simplify: $\dfrac{9}{10}x - \dfrac{3}{5}x$.

**EXAMPLE 8** Simplify by combining like terms: $\dfrac{7}{8}x - \dfrac{3}{4}x$.

$$\frac{7}{8}x - \frac{3}{4}x = \left(\frac{7}{8} - \frac{3}{4}\right)x \qquad \text{Try to do this step mentally.}$$

$$= \left(\frac{7}{8} - \frac{6}{8}\right)x = \frac{1}{8}x \qquad \text{Multiplying } \frac{3}{4} \text{ by } \frac{2}{2} \text{ and subtracting}$$

◀ **Do Exercise 10.**

**Answers**

**5.** $\dfrac{1}{12}$   **6.** $\dfrac{13}{18}$   **7.** $-\dfrac{3}{10}$   **8.** $-\dfrac{1}{6}$

**9.** $\dfrac{9}{112}$   **10.** $\dfrac{3}{10}x$

**Guided Solution:**

**6.** 18; $\dfrac{2}{2}$, 15, 2, 13

## b SOLVING EQUATIONS

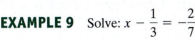

We can use the addition principle to solve equations containing fractions.

**EXAMPLE 9** Solve: $x - \dfrac{1}{3} = -\dfrac{2}{7}$.

$$x - \frac{1}{3} = -\frac{2}{7}$$

$$x - \frac{1}{3} + \frac{1}{3} = -\frac{2}{7} + \frac{1}{3}$$    Using the addition principle: adding $\frac{1}{3}$ to both sides

$$x + 0 = -\frac{2}{7} + \frac{1}{3}$$    Adding $\frac{1}{3}$ "undid" the subtraction of $\frac{1}{3}$ on the left-hand side of the equation.

$$x = -\frac{2}{7} \cdot \frac{3}{3} + \frac{1}{3} \cdot \frac{7}{7}$$    Multiplying by 1 to obtain the LCD, 21

$$x = -\frac{6}{21} + \frac{7}{21} = \frac{1}{21}$$    The solution appears to be $\frac{1}{21}$.

Check:
$$\frac{x - \dfrac{1}{3} = -\dfrac{2}{7}}{\dfrac{1}{21} - \dfrac{1}{3} \;\overset{?}{\vert}\; -\dfrac{2}{7}}$$

$$\frac{1}{21} - \frac{1}{3} \cdot \frac{7}{7}$$

$$\frac{1}{21} - \frac{7}{21}$$

$$\frac{-6}{21}$$

$$-\frac{2}{7} \cdot \frac{3}{3}$$

$$-\frac{2}{7} \quad \text{TRUE}$$

Our answer checks. The solution is $\frac{1}{21}$.

Recall that we can also create equivalent equations if we subtract the same number on both sides of an equation.

**EXAMPLE 10** Solve: $x + \dfrac{1}{4} = \dfrac{3}{5}$.

$$x + \frac{1}{4} - \frac{1}{4} = \frac{3}{5} - \frac{1}{4}$$    Using the addition principle: adding $-\frac{1}{4}$ to, or subtracting $\frac{1}{4}$ from, both sides

$$x + 0 = \frac{3}{5} \cdot \frac{4}{4} - \frac{1}{4} \cdot \frac{5}{5}$$    The LCD is 20. We multiply by 1 to get the LCD.

$$x = \frac{12}{20} - \frac{5}{20} = \frac{7}{20}$$

The solution is $\frac{7}{20}$. We leave the check to the student.

**Do Exercises 11–13.** ▶

Solve.

**11.** $x - \dfrac{2}{5} = \dfrac{1}{5}$

**12.** $x + \dfrac{2}{3} = \dfrac{5}{6}$

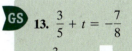

**GS** **13.** $\dfrac{3}{5} + t = -\dfrac{7}{8}$

$$\frac{3}{5} + t - \frac{\square}{\square} = -\frac{7}{8} - \frac{\square}{\square}$$

$$t + 0 = -\frac{7}{8} \cdot \frac{5}{\square} - \frac{3}{5} \cdot \frac{8}{\square}$$

$$t = -\frac{\square}{40} - \frac{\square}{40}$$

$$t = \frac{-35 - \square}{40}$$

$$t = \frac{-35 + (\square)}{40}$$

$$t = \frac{\square}{40} = -\frac{\square}{\square}$$

*Answers*

**11.** $\dfrac{3}{5}$   **12.** $\dfrac{1}{6}$   **13.** $-\dfrac{59}{40}$

*Guided Solution:*

**13.** $\dfrac{3}{5}, \dfrac{3}{5}, 5, 8, 35, 24, 24, -24, -59, \dfrac{59}{40}$

**APPLICATIONS AND PROBLEM SOLVING**

**EXAMPLE 11** *Honeybees.* A colony of honeybees is made up of one queen, hundreds of drones, and thousands of worker bees. Each type of honeybee looks different and is a different size. The average length of each type is illustrated in the graph below. How much longer, on average, is a queen bee than a drone?

1. **Familiarize.** From the graph, we see that the average length of a queen bee is $\frac{3}{4}$ in. and the average length of a drone is $\frac{2}{3}$ in.

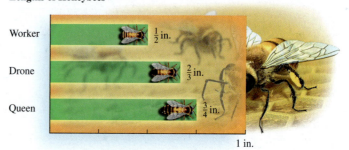

**Lengths of Honeybees**

Worker — $\frac{1}{2}$ in.

Drone — $\frac{2}{3}$ in.

Queen — $\frac{3}{4}$ in.

1 in.

Average length (in inches)

DATA: Washington State University

We let $b =$ the length by which a queen bee is longer than a drone.

2. **Translate.** We translate to an equation.

$$
\underbrace{\text{Length of drone}}_{\frac{2}{3}} \quad \underbrace{\text{plus}}_{+} \quad \underbrace{\text{Additional length}}_{b} \quad \underbrace{\text{is}}_{=} \quad \underbrace{\text{Length of queen}}_{\frac{3}{4}}
$$

3. **Solve.** To solve the equation, we subtract $\frac{2}{3}$ on both sides:

$$\frac{2}{3} + b = \frac{3}{4}$$

$$\frac{2}{3} + b - \frac{2}{3} = \frac{3}{4} - \frac{2}{3} \qquad \text{Subtracting } \frac{2}{3} \text{ on both sides}$$

$$b + 0 = \frac{3}{4} \cdot \frac{3}{3} - \frac{2}{3} \cdot \frac{4}{4} \qquad \text{The LCD is 12. We multiply by 1 to obtain the LCD.}$$

$$b = \frac{9}{12} - \frac{8}{12}$$

$$b = \frac{1}{12}.$$

4. **Check.** To check, we add $\frac{1}{12}$ to the length of a drone:

$$\frac{2}{3} + \frac{1}{12} = \frac{2}{3} \cdot \frac{4}{4} + \frac{1}{12} = \frac{8}{12} + \frac{1}{12} = \frac{9}{12} = \frac{3 \cdot 3}{4 \cdot 3} = \frac{3}{4} \cdot \frac{3}{3} = \frac{3}{4} \cdot 1 = \frac{3}{4}.$$

We get the length of a queen bee, so the answer checks.

5. **State.** On average, a queen bee is $\frac{1}{12}$ in. longer than a drone.

◀ **Do Exercise 14.**

**14. Honeybees.** Use the graph in Example 11 to find how much shorter, on average, a worker bee is than a drone.

*Answer*

**14.** $\frac{1}{6}$ in.

# Translating for Success

1. **Packaging.** One-Stop Postal Center orders bubble wrap in 64-yd rolls. On average, $\frac{3}{4}$ yd is used per small package. How many small packages can be prepared with 2 rolls of bubble wrap?

2. **Distance from College.** The post office is $\frac{7}{9}$ mi from the community college. The medical clinic is $\frac{2}{5}$ as far from the college as the post office is. How far is the clinic from the college?

3. **Swimming.** Andrew swims $\frac{7}{9}$ mi every day. One day he swims $\frac{2}{5}$ mi by 11:00 A.M. How much farther must Andrew swim to reach his daily goal?

4. **Tuition.** The average tuition at Waterside University is $12,000. If a loan is obtained for $\frac{1}{3}$ of the tuition, how much is the loan?

5. **Thermos Bottle Capacity.** A thermos bottle holds $\frac{11}{12}$ gal. How much is in the bottle when it is $\frac{4}{7}$ full?

*The goal of these matching questions is to practice step (2), Translate, of the five-step problem-solving process. Translate each word problem to an equation and select a correct translation from equations A–O.*

**A.** $\frac{3}{4} \cdot 64 = x$

**B.** $\frac{1}{3} \cdot 12,000 = x$

**C.** $\frac{1}{3} + \frac{2}{5} = x$

**D.** $\frac{2}{5} + x = \frac{7}{9}$

**E.** $\frac{2}{5} \cdot \frac{7}{9} = x$

**F.** $\frac{3}{4} \cdot x = 64$

**G.** $\frac{4}{7} = x + \frac{11}{12}$

**H.** $\frac{2}{5} = x + \frac{7}{9}$

**I.** $\frac{4}{7} \cdot \frac{11}{12} = x$

**J.** $\frac{3}{4} \cdot x = 128$

**K.** $\frac{1}{3} \cdot x = 12,000$

**L.** $\frac{1}{3} + \frac{2}{5} + x = 1$

**M.** $\frac{2}{5} = \frac{7}{9}x$

**N.** $\frac{4}{7} + x = \frac{11}{12}$

**O.** $\frac{1}{3} + x = \frac{2}{5}$

*Answers on page A-8*

6. **Cutting Rope.** A piece of rope $\frac{11}{12}$ yd long is cut into two pieces. One piece is $\frac{4}{7}$ yd long. How long is the other piece?

7. **Planting Corn.** Each year, Prairie State Farm plants 64 acres of corn. With good weather, $\frac{3}{4}$ of the planting can be completed by April 20. How many acres can be planted by April 20 with good weather?

8. **Painting Trim.** A painter used $\frac{1}{3}$ gal of white paint for the trim in the library and $\frac{2}{5}$ gal for the trim in the family room. How much paint was used for the trim in the two rooms?

9. **Lottery Winnings.** Sally won $12,000 in a state lottery and decided to give the net amount after taxes to three charities. One received $\frac{1}{3}$ of the net amount, and a second received $\frac{2}{5}$. What fractional part of the net amount did the third charity receive?

10. **Reading Assignment.** When Lowell had read 64 pages of his political science assignment, he had completed $\frac{3}{4}$ of his required reading. How many total pages were assigned?

✓ **Check Your Understanding**

**Reading Check** Complete each statement with the appropriate word or words from the following list. A word may be used more than once or not at all.

denominator          numerator

denominators          numerators

**RC1.** To subtract fractions with like denominators, we subtract the _____ and keep the _____.

**RC2.** Before we can subtract fractions, the _____ must be the same.

**RC3.** To subtract fractions when denominators are different, we find the LCM of the _____.

**RC4.** To subtract fractions when denominators are different, we multiply one or both fractions by 1 to make the _____ the same.

**Concept Check** Find the LCM of the denominators of each pair of fractions.

**CC1.** $\dfrac{3}{8}, \dfrac{1}{16}$          **CC2.** $\dfrac{7}{10}, \dfrac{5}{8}$          **CC3.** $\dfrac{16}{11}, \dfrac{7}{12}$          **CC4.** $\dfrac{17}{90}, \dfrac{109}{120}$

**a** Subtract and, if possible, simplify.

**1.** $\dfrac{5}{6} - \dfrac{1}{6}$

**2.** $\dfrac{7}{5} - \dfrac{2}{5}$

**3.** $\dfrac{9}{16} - \dfrac{13}{16}$

**4.** $\dfrac{5}{12} - \dfrac{7}{12}$

**5.** $\dfrac{8}{a} - \dfrac{6}{a}$

**6.** $\dfrac{4}{t} - \dfrac{9}{t}$

**7.** $-\dfrac{3}{8} - \dfrac{1}{8}$

**8.** $-\dfrac{3}{10} - \dfrac{1}{10}$

**9.** $\dfrac{3}{5a} - \dfrac{7}{5a}$

**10.** $\dfrac{2}{7t} - \dfrac{10}{7t}$

**11.** $\dfrac{10}{3t} - \dfrac{4}{3t}$

**12.** $\dfrac{9}{2a} - \dfrac{5}{2a}$

**13.** $\dfrac{7}{8} - \dfrac{1}{16}$

**14.** $\dfrac{4}{3} - \dfrac{5}{6}$

**15.** $\dfrac{7}{15} - \dfrac{4}{5}$

**16.** $\dfrac{3}{28} - \dfrac{3}{4}$

**17.** $\dfrac{3}{4} - \dfrac{1}{20}$

**18.** $\dfrac{3}{4} - \dfrac{4}{16}$

**19.** $\dfrac{2}{15} - \dfrac{5}{12}$

**20.** $\dfrac{11}{16} - \dfrac{9}{10}$

**21.** $\dfrac{7}{10} - \dfrac{23}{100}$

**22.** $\dfrac{9}{10} - \dfrac{3}{100}$

**23.** $\dfrac{7}{15} - \dfrac{3}{25}$

**24.** $\dfrac{18}{25} - \dfrac{4}{35}$

**25.** $\dfrac{-41}{100} - \dfrac{3}{10}$

**26.** $\dfrac{-13}{100} - \dfrac{7}{20}$

**27.** $\dfrac{2}{3} - \dfrac{1}{8}$

**28.** $\dfrac{3}{4} - \dfrac{1}{2}$

**29.** $-\dfrac{3}{10} - \dfrac{7}{25}$

**30.** $-\dfrac{5}{18} - \dfrac{2}{27}$

**31.** $\dfrac{3}{8} - \dfrac{5}{12}$

**32.** $\dfrac{2}{9} - \dfrac{7}{12}$

**33.** $\dfrac{-5}{18} - \dfrac{7}{24}$

**34.** $\dfrac{-7}{25} - \dfrac{2}{15}$

**35.** $\dfrac{13}{90} - \dfrac{17}{120}$

**36.** $\dfrac{8}{25} - \dfrac{29}{150}$

**37.** $\dfrac{2}{3}x - \dfrac{4}{9}x$

**38.** $\dfrac{7}{4}x - \dfrac{5}{12}x$

**39.** $\dfrac{2}{5}a - \dfrac{3}{4}a$

**40.** $\dfrac{4}{7}a - \dfrac{1}{3}a$

**b** Solve.

**41.** $x - \dfrac{4}{9} = \dfrac{3}{9}$

**42.** $x - \dfrac{3}{11} = \dfrac{7}{11}$

**43.** $a + \dfrac{2}{11} = \dfrac{6}{11}$

**44.** $a + \dfrac{4}{15} = \dfrac{13}{15}$

**45.** $y + \dfrac{1}{30} = \dfrac{1}{10}$

**46.** $y + \dfrac{1}{3} = \dfrac{5}{6}$

**47.** $a - \dfrac{3}{8} = \dfrac{3}{4}$

**48.** $x - \dfrac{3}{10} = \dfrac{2}{5}$

**49.** $\dfrac{2}{3} + x = \dfrac{4}{5}$

**50.** $\dfrac{4}{5} + x = \dfrac{6}{7}$

**51.** $\dfrac{3}{8} + a = \dfrac{1}{12}$

**52.** $\dfrac{5}{6} + a = \dfrac{2}{9}$

**53.** $n - \dfrac{3}{10} = -\dfrac{1}{6}$

**54.** $n - \dfrac{3}{4} = -\dfrac{5}{12}$

**55.** $x + \dfrac{3}{4} = -\dfrac{1}{2}$

**56.** $x + \dfrac{5}{6} = -\dfrac{11}{12}$

**c** Solve.

**57.** For a research paper, Kaitlyn spent $\frac{3}{4}$ hr searching on google.com and $\frac{1}{3}$ hr on yahoo.com. How much longer did she spend on google.com than on yahoo.com?

**58.** As part of a fitness program, Deb swims $\frac{1}{2}$ mi every day. One day she had already swum $\frac{1}{5}$ mi. How much farther did Deb need to swim?

**59.** The tread depth of an IRL Indy Car Series tire is $\frac{3}{32}$ in. Tires for a normal car have a tread depth of $\frac{5}{16}$ in. when new and are considered bald at $\frac{1}{16}$ in. How much deeper is the tread depth of an Indy Car tire than that of a bald tire for a normal car?

**Data:** Indy500.com; *Consumer Reports*

**60.** Ash uses $\frac{1}{3}$ lb of fresh mozzarella cheese and $\frac{1}{4}$ lb of grated Parmesan cheese on a homemade margherita pizza. How much more mozzarella cheese does he use than Parmesan cheese?

$\frac{3}{32}$ in.

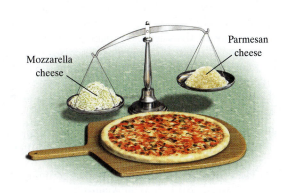

Parmesan cheese

Mozzarella cheese

**61.** From a $\frac{4}{5}$-lb wheel of cheese, a $\frac{1}{4}$-lb piece was served. How much cheese remained on the wheel?

**62.** A baker has a dispenser containing $\frac{15}{16}$ cup of icing and puts $\frac{1}{12}$ cup on a cinnamon roll. How much icing remains in the dispenser?

**63.** Jorge's $\frac{3}{4}$-hr drive to a job was part city and part country driving. If $\frac{2}{5}$ hr was city driving, how much time was spent on country driving?

**64.** Keri exercises $\frac{5}{6}$ of an hour every day. She jogs for $\frac{7}{12}$ of an hour and spends the remaining time warming up and cooling down. How much time does she spend warming up and cooling down?

**65.** *Woodworking.* Natalie is replacing a $\frac{3}{4}$-in.-thick shelf in her bookcase. If her replacement board is $\frac{15}{16}$ in. thick, how much must it be planed down before the repair can be completed?

**66.** *Furniture Cleaner.* A $\frac{2}{3}$-cup mixture of lemon juice and olive oil is a homemade cleaner for wood furniture. If the mixture contains $\frac{1}{4}$ cup of lemon juice, how much olive oil is in the cleaner?

$\frac{1}{4}$ cup $\quad + \quad$ ? $\quad = \quad$ $\frac{2}{3}$ cup

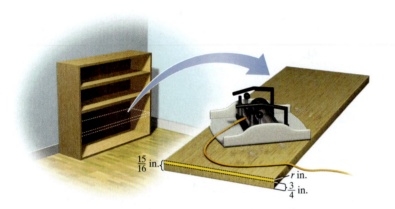

**67.** Blake used $\frac{1}{3}$ cup of maple syrup in preparing the batter for a batch of maple oatbran muffins. Sheila pointed out that the recipe actually calls for $\frac{5}{8}$ cup of syrup. How much more syrup should Blake add to the batter?

**68.** Amber added $\frac{1}{3}$ qt of two-cycle oil to a fuel mixture for her lawn mower. She then noticed that the owner's manual indicates $\frac{1}{2}$ qt should have been added. How much more two-cycle oil should Amber add to the mixture?

*Phases of the Moon.* The moon rotates in such a way that the same side always faces the earth. Throughout a lunar cycle, the portion of the moon that appears illuminated increases from nearly none (new moon) to nearly all (full moon), then decreases back to nearly none. These *phases* of the moon can be described by fractions between 0 and 1, indicating the portion of the moon illuminated. The partial calendar from August 2013 shows the fraction of the moon illuminated at midnight, Eastern Standard Time, for each day.

**69.** How much more of the moon appeared illuminated on August 18, 2013, than on August 15, 2013?

**70.** How much less of the moon appeared illuminated on August 31, 2013, than on August 23, 2013?

| 11 | 12 | 13 | 14 | | 15 | 16 | 17 |
|---|---|---|---|---|---|---|---|
| $\frac{1}{5}$ | $\frac{1}{4}$ | $\frac{9}{25}$ | First quarter | $\frac{1}{2}$ | $\frac{3}{5}$ | $\frac{7}{10}$ | $\frac{4}{5}$ |
| 18 | 19 | 20 | 21 | | 22 | 23 | 24 |
| $\frac{17}{20}$ | $\frac{19}{20}$ | $\frac{99}{100}$ | Full moon | 1 | $\frac{49}{50}$ | $\frac{19}{20}$ | $\frac{17}{20}$ |
| 25 | 26 | 27 | 28 | | 29 | 30 | 31 |
| $\frac{4}{5}$ | $\frac{7}{10}$ | $\frac{3}{5}$ | Last quarter | $\frac{1}{2}$ | $\frac{2}{5}$ | $\frac{1}{3}$ | $\frac{6}{25}$ |

DATA: Astronomical Applications Department, U.S. Naval Observatory, Washington, DC 20392-5420

## Skill Maintenance

Divide, if possible. If not possible, write "Not defined." [1.5a], [3.3b]

**71.** $\dfrac{38}{38}$

**72.** $\dfrac{38}{0}$

**73.** $\dfrac{124}{0}$

**74.** $\dfrac{124}{31}$

Divide and simplify. [3.7b]

**75.** $\dfrac{3}{7} \div \dfrac{9}{4}$

**76.** $\dfrac{9}{10} \div \dfrac{3}{5}$

**77.** $7 \div \dfrac{1}{3}$

**78.** $\dfrac{1}{4} \div 8$

## Synthesis

Simplify.

**79.** $\dfrac{7}{8} - \dfrac{3}{4} - \dfrac{1}{16}$

**80.** $\dfrac{9}{10} - \dfrac{1}{2} - \dfrac{2}{15}$

**81.** $\dfrac{2}{5} - \dfrac{1}{6}(-3)^2$

**82.** $\dfrac{7}{8} - \dfrac{1}{10}\left(-\dfrac{5}{6}\right)^2$

**83.** $-4 \cdot \dfrac{3}{7} - \dfrac{1}{7} \cdot \dfrac{4}{5}$

**84.** $\left(\dfrac{5}{6}\right)^2 - \left(\dfrac{3}{4}\right)^2$

**85.** $\left(-\dfrac{2}{5}\right)^3 - \left(-\dfrac{3}{10}\right)^3$

**86.** $\dfrac{3}{17} - \dfrac{2}{19} - \left(\dfrac{3}{17} - \dfrac{2}{19}\right)$

**87.** As part of a rehabilitation program, an athlete must swim and then walk a total of $\frac{9}{10}$ km each day. If one lap in the swimming pool is $\frac{3}{80}$ km, how far must the athlete walk after swimming 10 laps?

**88.** A small community garden was divided among four local residents. Based on the time they could spend on their garden sections and their individual crop plans, the residents each received a different-size plot to tend. One received $\frac{1}{4}$ of the garden, the second $\frac{1}{16}$, and the third $\frac{3}{8}$ of the garden. How much did the fourth gardener receive?

**89.** The circle graph below shows how long shoppers stay when visiting a mall. What portion of shoppers stay for 0–2 hr?

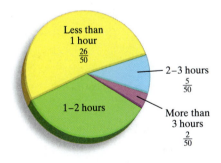

Less than 1 hour $\frac{26}{50}$

2–3 hours $\frac{5}{50}$

1–2 hours

More than 3 hours $\frac{2}{50}$

**90.** Four business partners plan to upgrade their computer system at a cost of \$12,600. Celia will pay $\frac{1}{2}$ of the cost, Reba will pay $\frac{1}{4}$ of the cost, Jon will pay $\frac{1}{6}$ of the cost, and Karl will pay the rest.

a) How much will Karl pay?
b) What fractional part will Karl pay?

**91.** ▦ Solve: $1x + \dfrac{16}{323} = \dfrac{10}{187}$.

**92.** ▦ Determine what whole number $a$ must be in order for the following to be true:

$$\dfrac{10 + a}{23} = \dfrac{330}{391} - \dfrac{a}{17}.$$

**93.** *Microsoft Interview.* The following is a question taken from an employment interview with Microsoft. "Given a gold bar that can be cut exactly twice and a contractor who must be paid one-seventh of a gold bar every day for seven days, how should the bar be cut?"

**Data:** *Fortune Magazine,* January 22, 2001

# Solving Equations: Using the Principles Together

## OBJECTIVES

**a** Solve equations that involve fractions and require use of both the addition principle and the multiplication principle.

**b** Solve equations by using the multiplication principle to clear fractions.

---

**SKILL REVIEW** *Use the multiplication principle to solve equations.* [3.8a]

Solve.

**1.** $\dfrac{2}{3}x = \dfrac{5}{6}$  **2.** $18 = -\dfrac{2}{3}t$

**Answers: 1.** $\dfrac{5}{4}$  **2.** $-27$

MyLab Math
VIDEO

---

We have used the multiplication principle to solve equations like

$$\frac{2}{3}x = \frac{5}{6} \quad \text{and} \quad 7 = \frac{5}{4}t,$$

and we have used the addition principle to solve equations like

$$\frac{4}{5} + x = \frac{1}{2} \quad \text{and} \quad \frac{7}{3} = t - \frac{2}{9}.$$

We are now ready to solve equations in which both principles are required.

### a   USING THE PRINCIPLES TOGETHER

Recall that we use the addition and multiplication principles to write equivalent equations. In the following steps, all five equations are equivalent:

$$5x - 2 = 43 \qquad \text{We first isolate } 5x.$$

$$5x - 2 + 2 = 43 + 2 \qquad \text{Using the addition principle}$$

$$5x = 45 \qquad \text{We now isolate } x.$$

$$\frac{1}{5} \cdot 5x = \frac{1}{5} \cdot 45 \qquad \text{Using the multiplication principle}$$

$$x = 9. \qquad \text{The solution of } x = 9 \text{ is the solution of } 5x - 2 = 43.$$

As a check, note that $5 \cdot 9 - 2 = 45 - 2 = 43$, as desired. The solution is 9.

**EXAMPLE 1**   Solve: $\dfrac{3}{4}x - \dfrac{1}{8} = \dfrac{1}{2}$.

We first isolate $\frac{3}{4}x$ by adding $\frac{1}{8}$ to both sides:

$$\frac{3}{4}x - \frac{1}{8} = \frac{1}{2}$$

$$\frac{3}{4}x - \frac{1}{8} + \frac{1}{8} = \frac{1}{2} + \frac{1}{8} \qquad \text{Using the addition principle}$$

$$\frac{3}{4}x + 0 = \frac{4}{8} + \frac{1}{8} \qquad \text{Writing with a common denominator}$$

$$\frac{3}{4}x = \frac{5}{8}.$$

Next, we isolate $x$ by multiplying both sides by $\frac{4}{3}$:

$$\frac{3}{4}x = \frac{5}{8} \qquad \text{Note that the reciprocal of } \tfrac{3}{4} \text{ is } \tfrac{4}{3}.$$

$$\frac{4}{3} \cdot \frac{3}{4}x = \frac{4}{3} \cdot \frac{5}{8} \qquad \text{Using the multiplication principle}$$

$$1x = \frac{20}{24}, \text{ or } \frac{5}{6}. \qquad \text{Simplifying; the solution appears to be } \tfrac{5}{6}.$$

Check:

$$\frac{3}{4}x - \frac{1}{8} = \frac{1}{2}$$

$$\frac{3}{4} \cdot \frac{5}{6} - \frac{1}{8} \;\overset{?}{\vert}\; \frac{1}{2}$$

$$\frac{3 \cdot 5}{4 \cdot 2 \cdot 3} - \frac{1}{8}$$      Removing a factor equal to 1: $\frac{3}{3} = 1$

$$\frac{5}{8} - \frac{1}{8}$$

$$\frac{1}{2}$$      **TRUE**

The solution is $\frac{5}{6}$.

**Do Exercises 1 and 2.** ▶

**EXAMPLE 2**   Solve: $5 + \dfrac{9}{2}t = -\dfrac{7}{2}$.

We first isolate $\frac{9}{2}t$ by subtracting 5 from both sides:

$$5 + \frac{9}{2}t = -\frac{7}{2}$$

$$5 + \frac{9}{2}t - 5 = -\frac{7}{2} - 5$$      Subtracting 5 from both sides

$$\frac{9}{2}t = -\frac{7}{2} - \frac{10}{2}$$      Writing 5 as $\frac{10}{2}$ to use the LCD

$$\frac{9}{2}t = -\frac{17}{2}$$      Note that the reciprocal of $\frac{9}{2}$ is $\frac{2}{9}$.

$$\frac{2}{9} \cdot \frac{9}{2}t = \frac{2}{9} \cdot \left(-\frac{17}{2}\right)$$      Multiplying both sides by $\frac{2}{9}$

$$1t = -\frac{2 \cdot 17}{9 \cdot 2}$$      Removing a factor equal to 1: $\frac{2}{2} = 1$

$$t = -\frac{17}{9}.$$

Check:

$$5 + \frac{9}{2}t = -\frac{7}{2}$$

$$5 + \frac{9}{2}\left(-\frac{17}{9}\right) \;\overset{?}{\vert}\; -\frac{7}{2}$$      Removing a factor equal to 1: $\frac{9}{9} = 1$

$$5 + \left(-\frac{17}{2}\right)$$

$$\frac{10}{2} + \left(\frac{-17}{2}\right)$$

$$\frac{10 - 17}{2}$$

$$\frac{-7}{2}$$      **TRUE**

The solution is $-\frac{17}{9}$.

**Do Exercises 3 and 4.** ▶

Sometimes the variable appears on the right side of the equation. The strategy for solving the equation remains the same.

Solve.

**1.** $\dfrac{3}{5}t - \dfrac{8}{15} = \dfrac{2}{15}$

**GS**  **2.** $\dfrac{1}{2}x - \dfrac{1}{5} = \dfrac{7}{10}$

$$\frac{1}{2}x - \frac{1}{5} + \frac{\square}{\square} = \frac{7}{10} + \frac{1}{5}$$

$$\frac{1}{2}x = \frac{7}{10} + \frac{\square}{10}$$

$$\frac{1}{2}x = \frac{\square}{10}$$

$$\square \cdot \frac{1}{2}x = 2 \cdot \frac{9}{10}$$

$$1x = \frac{2 \cdot 3 \cdot 3}{2 \cdot \square}$$

$$x = \frac{\square}{\square}$$

Solve.

**3.** $3 + \dfrac{14}{5}t = -\dfrac{21}{5}$

**4.** $2x + 4 = \dfrac{1}{2}$

**EXAMPLE 3**  Solve: $20 = 6 - \frac{2}{3}x$.

Our plan is to first use the addition principle to isolate $-\frac{2}{3}x$ and then use the multiplication principle to isolate $x$.

$$20 = 6 - \frac{2}{3}x$$

$$20 - 6 = 6 - \frac{2}{3}x - 6 \qquad \text{Subtracting 6 (or adding } -6\text{) on both sides}$$

$$14 = -\frac{2}{3}x$$

$$\left(-\frac{3}{2}\right)14 = \left(-\frac{3}{2}\right)\left(-\frac{2}{3}x\right) \qquad \text{Multiplying both sides by } -\frac{3}{2}$$

$$-\frac{3 \cdot 14}{2} = 1x$$

$$-\frac{3 \cdot 7 \cdot 2}{2} = 1x \qquad \text{Removing a factor equal to 1: } \frac{2}{2} = 1.$$

$$-21 = x$$

Check:  $20 = 6 - \frac{2}{3}x$

$$\begin{array}{c|c} 20 & 6 - \frac{2}{3}(-21) \\ & 6 + \frac{42}{3} \\ & 6 + 14 \\ & 20 \qquad \text{TRUE} \end{array}$$

The solution is $-21$.

◄ **Do Exercise 5.**

**5.** Solve: $9 - \frac{3}{4}x = 21$.

## b  CLEARING FRACTIONS

We now look at an alternative approach for solving Examples 1–3. Key to this approach is using the multiplication principle in the *first* step to produce an equivalent equation that is "cleared of fractions."

To "clear fractions," we identify the LCM of the denominators and use the multiplication principle. Because the LCM is a common multiple of the denominators, when both sides of the equation are multiplied by the LCM, the resulting terms can all be simplified. An equivalent equation can then be written without using fractions. We demonstrate this approach by solving Examples 1 and 2 by clearing fractions.

·········································  **Caution!**  ·········································

We can "clear fractions" in equations, not in expressions. Do not multiply to clear fractions when simplifying an expression.

·······························································································

Either of the methods discussed in this section can be used to solve equations that contain fractions, but it is important for students planning to continue in algebra to thoroughly understand *both* methods.

**EXAMPLE 4**  Solve Example 1 by clearing fractions:

$$\frac{3}{4}x - \frac{1}{8} = \frac{1}{2}.$$

The LCM of the denominators is 8, so we begin by multiplying both sides of the equation by 8:

$$\frac{3}{4}x - \frac{1}{8} = \frac{1}{2}$$

$$8\left(\frac{3}{4}x - \frac{1}{8}\right) = 8 \cdot \frac{1}{2}$$  Multiplying both sides by 8. We use parentheses when we are multiplying more than one term.

·············· **Caution!** ··············

$$\frac{8 \cdot 3}{4}x - 8 \cdot \frac{1}{8} = \frac{8}{2}$$  Use the distributive law carefully! Here we multiply every term inside the parentheses by 8.

$$\frac{4 \cdot 2 \cdot 3}{4}x - 1 = 4$$  Factoring and simplifying

$$6x - 1 = 4$$  The equation is now cleared of fractions. This is the advantage of using this method.

$$6x - 1 + 1 = 4 + 1$$  Adding 1 to both sides

$$6x = 5$$

$$\frac{6x}{6} = \frac{5}{6}$$  Dividing both sides by 6 (or multiplying both sides by $\frac{1}{6}$)

$$x = \frac{5}{6}.$$  Simplifying

Since $\frac{5}{6}$ was the solution in Example 1, we have a check. The solution is $\frac{5}{6}$.

**EXAMPLE 5**  Solve Example 2 by clearing fractions:

$$5 + \frac{9}{2}t = -\frac{7}{2}.$$

The LCM of the denominators is 2, so we begin by multiplying both sides of the equation by 2:

$$2\left(5 + \frac{9}{2}t\right) = 2\left(-\frac{7}{2}\right)$$  Using the multiplication principle

$$2 \cdot 5 + \frac{2 \cdot 9}{2}t = -\frac{2 \cdot 7}{2}$$  Using the distributive law; multiplying every term by 2

$$10 + 9t = -7$$  Simplifying and removing a factor equal to 1: $\frac{2}{2} = 1$. The equation is now cleared of fractions.

$$10 + 9t - 10 = -7 - 10$$  Subtracting 10 from both sides

$$9t = -17$$  Simplifying

$$\frac{9t}{9} = \frac{-17}{9}$$  Dividing both sides by 9 (or multiplying both sides by $\frac{1}{9}$)

$$t = -\frac{17}{9}.$$  Simplifying

Since the solution in Example 2 is also $-\frac{17}{9}$, we have a check. The solution is $-\frac{17}{9}$.

**Do Exercises 6 and 7.** ▶

  **6.** Solve Example 3 by clearing fractions:

$$20 = 6 - \frac{2}{3}x.$$

$$\boxed{\phantom{x}}(20) = 3\left(6 - \frac{2}{3}x\right)$$

$$\boxed{\phantom{x}} = 3 \cdot \boxed{\phantom{x}} - \frac{3 \cdot 2}{3}x$$

$$60 = \boxed{\phantom{x}} - \boxed{\phantom{x}}$$

$$60 - \boxed{\phantom{x}} = 18 - 2x - 18$$

$$\boxed{\phantom{x}} = -2x$$

$$\frac{42}{\boxed{\phantom{x}}} = \frac{-2x}{-2}$$

$$\boxed{\phantom{x}} = x$$

**7.** Solve Margin Exercise 1 by clearing fractions:

$$\frac{3}{5}t - \frac{8}{15} = \frac{2}{15}.$$

*Answers*

**6.** $-21$  **7.** $\dfrac{10}{9}$

*Guided Solution:*

**6.** 3, 60, 6, 18, 2x, 18, 42, −2, −21

## ✓ Check Your Understanding

**Reading Check** Determine whether each statement is true or false.

_____ **RC1.** We cannot use both the addition principle and the multiplication principle in the solution of an equation.

_____ **RC2.** To solve for a variable, it must appear on the left side of an equation.

_____ **RC3.** The first step in the solution of an equation always involves the addition principle.

_____ **RC4.** We clear fractions by multiplying both sides of an equation by the LCM of the denominators.

**Concept Check** Determine whether the given number is a solution of the equation.

**CC1.** $\frac{1}{3}t + \frac{1}{2} = \frac{4}{9}; -\frac{1}{6}$

**CC2.** $\frac{1}{4} - \frac{2}{5}x = \frac{17}{20}; -\frac{3}{2}$

**CC3.** $7 - \frac{2}{3}y = 3; -6$

---

**a** Solve using the addition principle and/or the multiplication principle. Don't forget to check!

**1.** $6x - 3 = 15$

**2.** $7x - 6 = 22$

**3.** $5x + 7 = 10$

**4.** $19 = 2x + 4$

**5.** $8 = 3x + 11$

**6.** $2a + 9 = -7$

**7.** $\frac{2}{3}y - 8 = 1$

**8.** $\frac{3}{10}y - 7 = 3$

**9.** $\frac{3}{2}t - \frac{1}{4} = \frac{1}{2}$

**10.** $\frac{1}{4}t + \frac{1}{8} = \frac{1}{2}$

**11.** $\frac{1}{5}x + \frac{3}{10} = \frac{3}{5}$

**12.** $\frac{4}{3}x - \frac{2}{15} = \frac{2}{15}$

**13.** $5 - \frac{3}{4}x = 6$

**14.** $3 - \frac{2}{5}x = 6$

**15.** $-1 + \frac{2}{5}t = -\frac{4}{5}$

**16.** $-2 + \frac{1}{6}t = -\frac{7}{4}$

**17.** $12 = 8 + \frac{7}{2}t$

**18.** $7 = 5 + \frac{3}{2}t$

**19.** $-11 = \frac{2}{3}x - 7$

**20.** $-10 = \frac{2}{5}x - 4$

**21.** $7 = a + \frac{14}{5}$

**22.** $9 = a + \frac{47}{10}$

**23.** $\frac{2}{5}t - 1 = \frac{7}{5}$

**24.** $-\frac{53}{4} = \frac{3}{2}a + 2$

**25.** $\frac{39}{8} = \frac{11}{4} - \frac{1}{2}x$

**26.** $\frac{7}{2} = \frac{13}{2} - \frac{1}{7}y$

**27.** $-\frac{13}{3}s + \frac{11}{2} = \frac{35}{4}$

**28.** $-\frac{11}{5}t + \frac{36}{5} = \frac{7}{2}$

**b** Solve by using the multiplication principle to clear fractions.

**29.** $\frac{1}{2}x - \frac{1}{4} = \frac{1}{2}$

**30.** $\frac{1}{3}x - \frac{1}{6} = \frac{2}{3}$

**31.** $7 = \frac{4}{9}t + 5$

**32.** $5 = \frac{4}{7}t + 3$

**33.** $-3 = \frac{3}{4}t - \frac{1}{2}$

**34.** $-2 = \frac{4}{3}t - \frac{5}{6}$

**35.** $\frac{4}{3} - \frac{5}{6}x = \frac{3}{2}$

**36.** $\frac{3}{2} - \frac{5}{3}x = \frac{5}{6}$

**37.** $-\frac{3}{4} = -\frac{5}{6} - \frac{1}{2}x$

**38.** $-\frac{1}{4} = -\frac{2}{3} - \frac{1}{6}x$

**39.** $\frac{4}{3} - \frac{1}{5}t = \frac{3}{4}$

**40.** $\frac{2}{5} - \frac{3}{4}t = \frac{4}{3}$

## Skill Maintenance

Divide.  [2.5a]

**41.** $39 \div (-3)$

**42.** $56 \div (-7)$

**43.** $(-72) \div (-4)$

**44.** $(-81) \div (-3)$

Solve.  [2.3b]

**45.** Jeremy withdraws $200 from his bank account, makes a $90 deposit, and then withdraws another $40. How much has Jeremy's account balance changed?

**46.** Animal Instinct, a pet supply store, makes a profit of $850 on Friday and a profit of $375 on Saturday, but suffers a loss of $45 on Sunday. Find the total profit or loss for the three days.

Divide and simplify.  [3.7b]

**47.** $\frac{10}{7} \div (2m)$

**48.** $45n \div \frac{9}{4}$

## Synthesis

Solve.

**49.** ▦ $\frac{553}{2451}a - \frac{13}{57} = \frac{29}{43}$

**50.** ▦ $\frac{1081}{3599}x - \frac{17}{61} = \frac{19}{59}$

**51.** ▦ $\frac{11}{17} = \frac{13}{41} - \frac{23}{29}t$

**52.** $-\frac{a}{5} + \frac{31}{4} = \frac{16}{3}$

**53.** $\frac{47}{5} - \frac{a}{4} = \frac{44}{7}$

**54.** $\frac{49}{8} + \frac{2x}{9} = 4$

**55.** The perimeter of the figure shown is 15 cm. Solve for $x$.

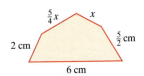

**56.** The perimeter of the figure is 15 cm. Solve for $n$.

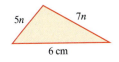

# Mixed Numerals

### OBJECTIVES

 a  Convert between mixed numerals and fraction notation.

b  Divide, writing the quotient as a mixed numeral.

## a  MIXED NUMERALS

> **SKILL REVIEW**  *Divide whole numbers.*  [1.5a]
> Divide.
> 1.  $598 \div 11$
> 2.  $15\overline{)326}$
>
> Answers: **1.** 54 R 4  **2.** 21 R 11
>
> MyLab Math
> VIDEO

The following figure illustrates the use of a **mixed numeral**. The bolt shown is $2\frac{3}{8}$ in. long. The length is given as a whole-number part, 2, and a fraction part less than 1, $\frac{3}{8}$. We can represent the measurement as $\frac{19}{8}$, but $2\frac{3}{8}$ makes the length easier to visualize and is thus more descriptive.

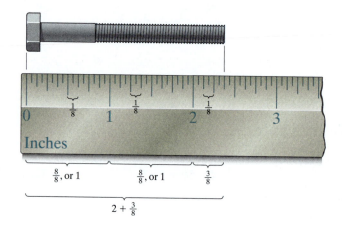

A mixed numeral $2\frac{3}{8}$ represents a sum:

$$2\frac{3}{8} \quad \text{means} \quad 2 + \frac{3}{8}.$$

This is a whole number.   This is a fraction less than 1.

**EXAMPLES**  Convert to a mixed numeral.

**1.**  $7 + \frac{2}{5} = 7\frac{2}{5}$   **2.**  $4 + \frac{3}{10} = 4\frac{3}{10}$

◀ **Do Exercises 1–3.**

The notation $2\frac{3}{4}$ has a plus sign left out. To aid in understanding, we sometimes write the missing plus sign: $2 + \frac{3}{4}$. We can think of a negative mixed numeral as having a missing minus sign, since $-5\frac{2}{3} = -(5 + \frac{2}{3}) = -5 - \frac{2}{3}$.

---

Convert to a mixed numeral.

**1.**  $1 + \frac{2}{3} = \square\frac{\square}{\square}$

**2.**  $2 + \frac{3}{4} = \square\frac{\square}{\square}$

**3.**  $12 + \frac{2}{7}$

---

*Answers*

**1.** $1\frac{2}{3}$  **2.** $2\frac{3}{4}$  **3.** $12\frac{2}{7}$

Mixed numerals can be displayed on the number line, as shown here.

**EXAMPLES**  Convert to fraction notation.

**3.** $2\dfrac{3}{4} = 2 + \dfrac{3}{4}$    Inserting the missing plus sign

$= \dfrac{2}{1} + \dfrac{3}{4}$    $2 = \dfrac{2}{1}$

$= \dfrac{2}{1} \cdot \dfrac{4}{4} + \dfrac{3}{4}$    Finding a common denominator

$= \dfrac{8}{4} + \dfrac{3}{4} = \dfrac{11}{4}$

**4.** $4\dfrac{3}{10} = 4 + \dfrac{3}{10} = \dfrac{4}{1} + \dfrac{3}{10} = \dfrac{4}{1} \cdot \dfrac{10}{10} + \dfrac{3}{10} = \dfrac{40}{10} + \dfrac{3}{10} = \dfrac{43}{10}$

**Do Exercises 4 and 5.** ▶

We can streamline the process of converting a mixed numeral to fraction notation.

> To convert from a mixed numeral like $4\frac{3}{10}$ to fraction notation:
>
> (a) Multiply the whole number by the denominator: $4 \cdot 10 = 40$.
>
> (b) Add the result to the numerator: $40 + 3 = 43$.
>
> (c) Keep the denominator.
>
>

**EXAMPLES**  Convert to fraction notation.

**5.** $6\dfrac{2}{3} = \dfrac{20}{3}$    $6 \cdot 3 = 18,\ 18 + 2 = 20$

**6.** $8\dfrac{2}{9} = \dfrac{74}{9}$    $8 \cdot 9 = 72,\ 72 + 2 = 74$

**Do Exercises 6–9.** ▶

To find the opposite of the number in Example 5, we can write either $-6\frac{2}{3}$ or $-\frac{20}{3}$. Thus, to convert a negative mixed numeral to fraction notation, we remove the negative sign for purposes of computation and then include it in the answer.

**EXAMPLES**  Convert to fraction notation.

**7.** $-5\dfrac{1}{3} = -\left(5 + \dfrac{1}{3}\right) = -\dfrac{16}{3}$    $5 \cdot 3 = 15;\ 15 + 1 = 16;$ include the negative sign

**8.** $-7\dfrac{5}{6} = -\left(7 + \dfrac{5}{6}\right) = -\dfrac{47}{6}$    $7 \cdot 6 = 42;\ 42 + 5 = 47$

**Do Exercises 10 and 11.** ▶

Convert to fraction notation.

**4.** $4\dfrac{2}{5}$    **5.** $6\dfrac{1}{10}$

Convert to fraction notation. Use the faster method.

**GS** **6.** $4\dfrac{5}{6}$

$4 \cdot 6 = \boxed{\phantom{0}}$

$24 + \boxed{\phantom{0}} = 29$

$4\dfrac{5}{6} = \dfrac{\boxed{\phantom{0}}}{6}$

**7.** $9\dfrac{1}{4}$    **8.** $20\dfrac{2}{3}$

**9.** $1\dfrac{9}{13}$

Convert to fraction notation.

**10.** $-6\dfrac{2}{5}$    **11.** $-7\dfrac{2}{9}$

***Answers***

**4.** $\dfrac{22}{5}$  **5.** $\dfrac{61}{10}$  **6.** $\dfrac{29}{6}$  **7.** $\dfrac{37}{4}$

**8.** $\dfrac{62}{3}$  **9.** $\dfrac{22}{13}$  **10.** $-\dfrac{32}{5}$  **11.** $-\dfrac{65}{9}$

***Guided Solution:***
**6.** 24, 5, 29

Convert to a mixed numeral.

12. $\dfrac{7}{3}$    13. $\dfrac{11}{10}$

14. $\dfrac{110}{6}$    15. $\dfrac{231}{18}$

## Writing Mixed Numerals

We can find a mixed numeral for $\frac{5}{3}$ as follows:

$$\frac{5}{3} = \frac{3}{3} + \frac{2}{3} = 1 + \frac{2}{3} = 1\frac{2}{3}.$$

In terms of objects, we can think of $\frac{5}{3}$ as $\frac{3}{3}$, or 1, plus $\frac{2}{3}$, as shown below.

$$\frac{5}{3} = \underbrace{\qquad}_{\frac{3}{3},\ \text{or}\ 1} + \underbrace{\qquad}_{\frac{2}{3}}$$

Fraction symbols like $\frac{5}{3}$ also indicate division; $\frac{5}{3}$ means $5 \div 3$. Let's divide the numerator by the denominator.

$$\begin{array}{r} 1 \\ 3\overline{)5} \\ \underline{3} \\ 2 \leftarrow 2 \div 3 = \frac{2}{3} \end{array}$$

Thus, $\frac{5}{3} = 1\frac{2}{3}$.

> To convert from fraction notation to a mixed numeral, divide.
>
> $$\frac{13}{5} \qquad \begin{array}{r} 2 \\ 5\overline{)13} \\ \underline{10} \\ 3 \end{array} \qquad 2\frac{3}{5}$$
>
> 2 — The quotient, 5)13 — The divisor, 3 — The remainder

**EXAMPLES** Convert to a mixed numeral.

9. $\dfrac{69}{10}$    $\begin{array}{r} 6 \\ 10\overline{)69} \\ \underline{60} \\ 9 \end{array}$    $\dfrac{69}{10} = 6\dfrac{9}{10}$

10. $\dfrac{122}{8}$    $\begin{array}{r} 15 \\ 8\overline{)122} \\ \underline{8} \\ 42 \\ \underline{40} \\ 2 \end{array}$    $\dfrac{122}{8} = 15\dfrac{2}{8} = 15\dfrac{1}{4}$

> Simplify the fraction part of a mixed numeral, if possible.

◀ **Do Exercises 12–15.**

A fraction larger than 1, such as $\frac{27}{8}$, is sometimes referred to as an "improper" fraction. However, the use of notation such as $\frac{27}{8}$, $\frac{11}{9}$, and $\frac{89}{10}$ is quite "proper" and very common in algebra.

*Answers*

**12.** $2\frac{1}{3}$   **13.** $1\frac{1}{10}$   **14.** $18\frac{1}{3}$   **15.** $12\frac{5}{6}$

The same procedure also works with negative numbers. Of course, the result will be a negative mixed numeral.

**EXAMPLE 11** Convert $\dfrac{-9}{4}$ to a mixed numeral.

Since $\begin{array}{r} 2 \\ 4\overline{)9} \\ 8 \\ \hline 1 \end{array}$, we have $\dfrac{9}{4} = 2\dfrac{1}{4}$. Thus, $\dfrac{-9}{4} = -2\dfrac{1}{4}$.

**Do Exercises 16 and 17.** ▶

## b  FINDING QUOTIENTS AND AVERAGES

It is quite common when performing long division to express the quotient as a mixed numeral. As in Examples 9–11, the remainder becomes the numerator of the fraction part of the mixed numeral.

**EXAMPLE 12** Divide. Write a mixed numeral for the quotient.

$7\overline{)6341}$

We first divide as usual.

$$\begin{array}{r} 905 \\ 7\overline{)6341} \\ 6300 \\ \hline 41 \\ 35 \\ \hline 6 \end{array}$$

$\dfrac{6341}{7} = 905\dfrac{6}{7}$

The answer is 905 R 6, or $905\dfrac{6}{7}$. Using fraction notation, we write $\dfrac{6341}{7} = 905\dfrac{6}{7}$.

**Do Exercises 18 and 19.** ▶

**EXAMPLE 13** *Dietary Fiber.* Each of the following five fruits is a good source of dietary fiber. This list gives the amount of fiber, in grams, contained in each serving of fruit. How much fiber is contained, on average, in these fruits?

| | |
|---|---|
| Raspberries (1 cup) | 8 g |
| Pear (1 medium) | 6 g |
| Blueberries (1 cup) | 4 g |
| Apple (1 medium) | 3 g |
| Banana (1 medium) | 3 g |

**Data:** NationalFiberCouncil.org

To find the *average* of a set of values, we add the values and divide the sum by the number of values being added.

$$\text{Average grams of fiber} = \frac{8\text{ g} + 6\text{ g} + 4\text{ g} + 3\text{ g} + 3\text{ g}}{5} = \frac{24\text{ g}}{5} = 4\frac{4}{5}\text{ g}$$

On average, these fruits contain $4\dfrac{4}{5}$ g of fiber per serving.

**Do Exercise 20.** ▶

---

Convert to a mixed numeral.

**GS 16.** $\dfrac{-17}{5}$

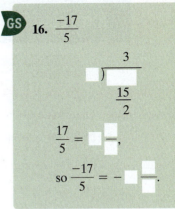

**17.** $-\dfrac{134}{12}$

Divide. Write a mixed numeral for the answer.

**18.** $6\overline{)4846}$

**19.** $45\overline{)6053}$

**20.** *Fitness.* Shelby recently added jogging to her exercise routine. The following list gives the total amount of time she jogged for each of the first four weeks of her new routine. Find the average number of minutes she jogged per week.

Week 1: 24 min
Week 2: 30 min
Week 3: 27 min
Week 4: 32 min

*Answers*

**16.** $-3\dfrac{2}{5}$  **17.** $-11\dfrac{1}{6}$  **18.** $807\dfrac{2}{3}$
**19.** $134\dfrac{23}{45}$  **20.** $28\dfrac{1}{4}$ min

*Guided Solution:*
**16.** $5, 17, 3\dfrac{2}{5}, 3\dfrac{2}{5}$

✓ **Check Your Understanding**

**Reading Check** Determine whether each statement is true or false.

_____ **RC1.** A mixed numeral consists of a whole-number part and a fraction less than 1.

_____ **RC2.** The mixed numeral $5\frac{1}{4}$ represents $5 + \frac{1}{4}$.

_____ **RC3.** It is never appropriate to use fraction notation such as $\frac{33}{25}$.

_____ **RC4.** When a quotient is written as a mixed numeral, the divisor is the denominator of the fraction part (assuming that the fraction has not been simplified).

**Concept Check** Use the results of the accompanying division to convert each fraction to a mixed numeral.

**CC1.** $\frac{37}{12} = $ _____

$$\begin{array}{r} 3 \\ 1\,2\,\overline{)3\,7} \\ 3\,6 \\ \hline 1 \end{array}$$

**CC2.** $\frac{99}{8} = $ _____

$$\begin{array}{r} 1\,2 \\ 8\,\overline{)9\,9} \\ 8 \\ \hline 1\,9 \\ 1\,6 \\ \hline 3 \end{array}$$

**CC3.** $\frac{619}{23} = $ _____

$$\begin{array}{r} 2\,6 \\ 2\,3\,\overline{)6\,1\,9} \\ 4\,6 \\ \hline 1\,5\,9 \\ 1\,3\,8 \\ \hline 2\,1 \end{array}$$

**a** Convert to fraction notation.

**1.** $7\frac{2}{3}$

**2.** $6\frac{2}{5}$

**3.** $6\frac{1}{4}$

**4.** $8\frac{1}{2}$

**5.** $-20\frac{1}{8}$

**6.** $-10\frac{1}{3}$

**7.** $5\frac{1}{10}$

**8.** $8\frac{1}{10}$

**9.** $20\frac{3}{5}$

**10.** $30\frac{4}{5}$

**11.** $-33\frac{1}{3}$

**12.** $-66\frac{2}{3}$

**13.** $1\frac{5}{8}$

**14.** $1\frac{3}{5}$

**15.** $-12\frac{3}{4}$

**16.** $-15\frac{2}{3}$

**17.** $5\frac{7}{10}$

**18.** $7\frac{3}{100}$

**19.** $-5\frac{7}{100}$

**20.** $-6\frac{4}{15}$

Convert to a mixed numeral.

**21.** $\frac{16}{3}$

**22.** $\frac{19}{8}$

**23.** $\frac{45}{6}$

**24.** $\frac{30}{9}$

**25.** $\frac{57}{10}$

**26.** $\frac{-89}{10}$

**27.** $\frac{65}{9}$

**28.** $\frac{65}{8}$

**29.** $\frac{-33}{6}$

**30.** $\frac{-50}{8}$

**31.** $\frac{46}{4}$

**32.** $\frac{39}{9}$

**33.** $\dfrac{-12}{8}$    **34.** $-\dfrac{57}{6}$    **35.** $\dfrac{307}{5}$    **36.** $\dfrac{227}{4}$    **37.** $-\dfrac{413}{50}$    **38.** $\dfrac{467}{100}$

**b**   Divide. Write a mixed numeral for the answer.

**39.** $8\overline{)869}$    **40.** $3\overline{)2126}$    **41.** $7\overline{)6345}$    **42.** $9\overline{)9110}$    **43.** $21\overline{)852}$

**44.** $85\overline{)7670}$    **45.** $-302 \div 15$    **46.** $-475 \div 13$    **47.** $471 \div (-21)$    **48.** $545 \div (-25)$

*Education.*   The non-profit organization School on Wheels provides one-on-one tutoring and educational advocacy for school-aged children experiencing homelessness in Indianapolis. The table below shows the number of children who received services each year since 2004–2005, as well as other related information.

**49.** What is the average number of backpacks with school supplies given to children for the school years 2007–2008 through 2012–2013?

**50.** What is the average number of homeless children served for the school years 2007–2008 through 2012–2013?

**51.** What is the average number of volunteer tutor hours for the school years 2004–2005 through 2009–2010?

**52.** What is the average number of volunteer tutor hours for the school years 2010–2011 through 2015–2016?

| SCHOOL YEAR | NUMBER OF HOMELESS CHILDREN SERVED | NUMBER OF VOLUNTEER TUTOR HOURS | NUMBER OF BACKPACKS WITH SCHOOL SUPPLIES GIVEN TO CHILDREN |
| --- | --- | --- | --- |
| 2015–2016 | 428 | 7972 | N/A |
| 2014–2015 | 363 | 7774 | 219 |
| 2013–2014 | 331 | 5885 | 198 |
| 2012–2013 | 359 | 5511 | 278 |
| 2011–2012 | 407 | 6018 | 310 |
| 2010–2011 | 415 | 6708 | 320 |
| 2009–2010 | 373 | 5965 | 242 |
| 2008–2009 | 365 | 4796 | 303 |
| 2007–2008 | 322 | 4027 | 201 |
| 2006–2007 | 363 | 4083 | 101 |
| 2005–2006 | 388 | 3385 | 105 |
| 2004–2005 | 375 | 3131 | 87 |

**Data:** indyschoolonwheels.org

## Skill Maintenance

Use the distributive law to write an equivalent expression.   [2.6b]

**53.** $6(3x + y - 4)$

**54.** $-3(a - 2b + 4c)$

## Synthesis

Write a mixed numeral for each number or sum listed.

**55.** ▦ $\dfrac{128{,}236}{541}$    **56.** ▦ $\dfrac{103{,}676}{349}$    **57.** $\dfrac{56}{7} + \dfrac{2}{3}$    **58.** $\dfrac{72}{12} + \dfrac{5}{6}$    **59.** $\dfrac{12}{5} + \dfrac{19}{15}$

**60.** There are $\frac{366}{7}$ weeks in a leap year.

**61.** There are $\frac{365}{7}$ weeks in a year that is not a leap year.

**62.** *Athletics.*   At a track and field meet, the hammer that is thrown has a wire length ranging from 3 ft $10\frac{1}{4}$ in. to 3 ft $11\frac{3}{4}$ in., a $4\frac{1}{8}$-in. grip, and a 16-lb ball with a diameter of $4\frac{3}{8}$ in. to $5\frac{1}{8}$ in. Give specifications for the wire length and diameter of an "average" hammer.

# Mid-Chapter Review

## Concept Reinforcement

Determine whether each statement is true or false.

_____ **1.** If $\dfrac{a}{b} > \dfrac{c}{b}, b \neq 0$, then $a > c$.  [4.2c]

_____ **2.** The mixed numeral $1\frac{1}{2}$ and the fraction $\frac{6}{4}$ represent the same number.  [4.5a]

_____ **3.** The least common multiple of two natural numbers is the smallest number that is a factor of both.  [4.1a]

_____ **4.** To add fractions when denominators are the same, we keep the numerator and add the denominators.  [4.2a]

## Guided Solutions

**GS** Fill in each blank with the number that creates a correct solution.

**5.** Subtract: $\dfrac{11}{42} - \dfrac{3}{35}$.  [4.3a]

$$\frac{11}{42} - \frac{3}{35} = \frac{11}{2 \cdot \boxed{\phantom{0}} \cdot 7} - \frac{3}{\boxed{\phantom{0}} \cdot 7} \qquad \text{Factoring the denominators}$$

$$= \frac{11}{2 \cdot 3 \cdot 7} \cdot \left(\frac{\boxed{\phantom{0}}}{\boxed{\phantom{0}}}\right) - \frac{3}{5 \cdot 7} \cdot \left(\frac{2 \cdot 3}{2 \cdot 3}\right) \qquad \text{Multiplying by 1 to get the LCD}$$

$$= \frac{11 \cdot \boxed{\phantom{0}}}{2 \cdot 3 \cdot 7 \cdot \boxed{\phantom{0}}} - \frac{3 \cdot 2 \cdot 3}{5 \cdot 7 \cdot 2 \cdot 3} \qquad \text{Multiplying}$$

$$= \frac{\boxed{\phantom{0}}}{2 \cdot 3 \cdot 5 \cdot 7} - \frac{\boxed{\phantom{0}}}{2 \cdot 3 \cdot 5 \cdot 7} \qquad \text{Simplifying}$$

$$= \frac{\boxed{\phantom{0}} - \boxed{\phantom{0}}}{2 \cdot 3 \cdot 5 \cdot 7} = \frac{\boxed{\phantom{0}}}{\boxed{\phantom{0}}} \qquad \text{Subtracting and simplifying}$$

**6.** Solve: $x + \dfrac{1}{8} = \dfrac{2}{3}$.  [4.3b]

$$x + \frac{1}{8} = \frac{2}{3}$$

$$x + \frac{1}{8} - \boxed{\phantom{0}} = \frac{2}{3} - \boxed{\phantom{0}} \qquad \text{Subtracting on both sides}$$

$$x + 0 = \frac{2}{3} \cdot \frac{\boxed{\phantom{0}}}{\boxed{\phantom{0}}} - \frac{1}{8} \cdot \frac{\boxed{\phantom{0}}}{\boxed{\phantom{0}}} \qquad \text{Multiplying by 1 to get the LCD}$$

$$x = \frac{\boxed{\phantom{0}}}{24} - \frac{\boxed{\phantom{0}}}{24} \qquad \text{Simplifying and multiplying}$$

$$x = \frac{\boxed{\phantom{0}}}{\boxed{\phantom{0}}} \qquad \text{Subtracting}$$

The solution is $\dfrac{\boxed{\phantom{0}}}{\boxed{\phantom{0}}}$.

## Mixed Review

**7.** Match each set of numbers in the first column with its least common multiple in the second column by drawing connecting lines.  [4.1a]

45 and 50

50 and 80          120

30 and 24          720

18, 24, and 80     400

30, 45, and 50     450

Calculate and simplify.  [4.2b], [4.3a]

**8.** $\dfrac{1}{5} + \dfrac{7}{45}$

**9.** $\dfrac{5}{6} + \dfrac{2}{3} + \dfrac{7}{12}$

**10.** $\dfrac{2}{9} - \dfrac{1}{6}$

**11.** $\dfrac{1}{15} - \dfrac{5}{18}$

**12.** $\dfrac{19}{48} - \dfrac{11}{30}$

**13.** $-\dfrac{3}{8}x + \dfrac{1}{12}x$

**14.** $\dfrac{-3}{40} + \dfrac{-5}{24}$

**15.** $\dfrac{8}{65} - \dfrac{2}{35}$

Solve.

**16.** Miguel jogs for $\frac{4}{5}$ mi, rests, and then jogs for another $\frac{2}{3}$ mi. How far does he jog in all?  [4.2d]

**17.** One weekend, Kirby spent $\frac{39}{5}$ hr playing two games—Brain Challenge and Scrabble. She spent $\frac{11}{4}$ hr playing Scrabble. How many hours did she spend playing Brain Challenge?  [4.3c]

**18.** Arrange in order from smallest to largest:

$\dfrac{4}{9}, \dfrac{3}{10}, \dfrac{2}{7},$ and $\dfrac{1}{5}.$  [4.2c]

**19.** Solve: $\dfrac{2}{5} + x = \dfrac{9}{16}.$  [4.3b]

**20.** Solve: $\dfrac{3}{4}x + 1 = \dfrac{1}{3}.$  [4.4a], [4.4b]

**21.** Divide: $15\overline{)263}$. Write a mixed numeral for the answer.  [4.5b]

**22.** Fraction notation for $9\frac{3}{8}$ is which of the following?  [4.5a]

**A.** $\dfrac{27}{8}$    **B.** $\dfrac{93}{8}$    **C.** $\dfrac{75}{8}$    **D.** $\dfrac{80}{3}$

**23.** Mixed numeral notation for $-\frac{39}{4}$ is which of the following?  [4.5a]

**A.** $-35\dfrac{1}{4}$    **B.** $-\dfrac{4}{39}$    **C.** $-9\dfrac{3}{4}$    **D.** $-36\dfrac{3}{4}$

## Understanding Through Discussion and Writing

**24.** Is the LCM of two numbers always larger than either number? Why or why not?  [4.1a]

**25.** Explain the role of multiplication when adding using fraction notation with different denominators.  [4.2b]

**26.** A student made the following error:

$$\dfrac{8}{5} - \dfrac{8}{2} = \dfrac{8}{3}.$$

Find at least two ways to convince him of the mistake.  [4.3a]

**27.** Are the numbers $2\frac{1}{3}$ and $2 \cdot \frac{1}{3}$ equal? Why or why not?  [4.5a]

# 4.6

## Addition and Subtraction Using Mixed Numerals; Applications

### OBJECTIVES

**a** Add using mixed numerals.

**b** Subtract using mixed numerals.

**c** Solve applied problems involving addition and subtraction with mixed numerals.

**d** Add and subtract using negative mixed numerals.

## a  ADDITION USING MIXED NUMERALS

**SKILL REVIEW**

*Simplify fraction notation.*  [3.5b]

Simplify.

1. $\dfrac{18}{32}$    2. $\dfrac{78}{117}$

Answers: 1. $\dfrac{9}{16}$  2. $\dfrac{2}{3}$

MyLab Math
VIDEO

To add mixed numerals, we first add the fractions. Then we add the whole numbers.

**EXAMPLE 1**  Add: $1\frac{5}{8} + 3\frac{1}{8}$. Write a mixed numeral for the answer.

$$
\begin{array}{r}
1\dfrac{5}{8} \\
+\,3\dfrac{1}{8} \\
\hline
\dfrac{6}{8}
\end{array}
=
\qquad
\begin{array}{r}
1\dfrac{5}{8} \\
+\,3\dfrac{1}{8} \\
\hline
4\dfrac{6}{8} = 4\dfrac{3}{4}
\end{array}
$$

Simplifying: $\dfrac{6}{8} = \dfrac{3}{4}$

↑ Add the fractions.   ↑ Add the whole numbers.

Sometimes we must write the fraction parts with a common denominator before we can add.

**EXAMPLE 2**  Add: $5\frac{2}{3} + 3\frac{5}{6}$. Write a mixed numeral for the answer.

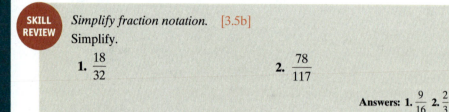

$$
\begin{array}{r}
5\dfrac{2}{3}\cdot\dfrac{2}{2} \;=\; 5\dfrac{4}{6} \\
+\,3\dfrac{5}{6} \;=\; +\,3\dfrac{5}{6} \\
\hline
8\dfrac{9}{6} = 8 + \dfrac{9}{6}
\end{array}
$$

The LCD is 6.

$$= 8 + 1\dfrac{1}{2} = 9\dfrac{1}{2} \qquad \dfrac{9}{6} = 1\dfrac{3}{6} = 1\dfrac{1}{2}$$

Add.

1.  $\begin{array}{r} 2\dfrac{3}{10} \\ +\,5\dfrac{1}{10} \\ \hline \end{array}$    2.  $\begin{array}{r} 8\dfrac{2}{5} \\ +\,3\dfrac{7}{10} \\ \hline \end{array}$

**Answers**

1. $7\dfrac{2}{5}$  2. $12\dfrac{1}{10}$

◀ Do Exercises 1 and 2.

The fraction part of a mixed numeral should always be less than 1.

**EXAMPLE 3** Add: $10\frac{5}{6} + 7\frac{3}{8}$.

The LCD is 24.

$$
\begin{array}{rcl}
10\dfrac{5}{6}\cdot\dfrac{4}{4} &=& 10\dfrac{20}{24} \\[2mm]
+\ 7\dfrac{3}{8}\cdot\dfrac{3}{3} &=& +\ 7\dfrac{9}{24} \\[2mm]
\hline
&& 17\dfrac{29}{24} = 17 + \dfrac{29}{24} \\[4mm]
&& \qquad\ \ = 17 + 1\dfrac{5}{24} \qquad \text{Writing } \tfrac{29}{24} \text{ as a mixed} \\
&& \qquad\qquad\qquad\qquad\quad \text{numeral, } 1\tfrac{5}{24} \\[3mm]
&& \qquad\ \ = 18\dfrac{5}{24}
\end{array}
$$

**Do Exercise 3.** ▶

**3.** Add.

$$
\begin{array}{r}
9\dfrac{3}{4} \\[2mm]
+\ 3\dfrac{5}{6} \\[1mm]
\hline
\end{array}
$$

---

**b** ## SUBTRACTION USING MIXED NUMERALS

**EXAMPLE 4** Subtract: $7\frac{3}{4} - 2\frac{1}{4}$.

$$
\begin{array}{rcl}
7\dfrac{3}{4} &=& 7\dfrac{3}{4} \\[2mm]
-\ 2\dfrac{1}{4} &=& -\ 2\dfrac{1}{4} \\[1mm]
\hline
\dfrac{2}{4} && 5\dfrac{2}{4} = 5\dfrac{1}{2}
\end{array}
$$

↑ Subtract the fractions.

↑ Subtract the whole numbers.

↑ Simplifying: $\dfrac{2}{4} = \dfrac{1}{2}$

**Subtract.**

**4.**
$$
\begin{array}{r}
10\dfrac{7}{8} \\[2mm]
-\ 9\dfrac{3}{8} \\[1mm]
\hline
\end{array}
$$

**GS** **5.**
$$
\begin{array}{rcl}
8\dfrac{2}{3} &=& 8\dfrac{\square}{6} \\[2mm]
-\ 5\dfrac{1}{2} &=& -\ 5\dfrac{\square}{6} \\[1mm]
\hline
&& 3\dfrac{\square}{6}
\end{array}
$$

**EXAMPLE 5** Subtract: $9\frac{4}{5} - 3\frac{1}{2}$.

The LCD is 10.

$$
\begin{array}{rcl}
9\dfrac{4}{5}\cdot\dfrac{2}{2} &=& 9\dfrac{8}{10} \\[2mm]
-\ 3\dfrac{1}{2}\cdot\dfrac{5}{5} &=& -\ 3\dfrac{5}{10} \\[1mm]
\hline
&& 6\dfrac{3}{10}
\end{array}
$$

**Do Exercises 4 and 5.** ▶

**GS** **6.** Subtract: $5 - 1\frac{1}{3}$.

$$
\begin{array}{rcl}
5 &=& 4\dfrac{\square}{\square} \\[2mm]
-\ 1\dfrac{1}{3} &=& -\ 1\dfrac{1}{3} \\[1mm]
\hline
&& 3\dfrac{\square}{3}
\end{array}
$$

**EXAMPLE 6** Subtract: $12 - 9\frac{3}{8}$.

$$
\begin{array}{rcl}
12 &=& 11\dfrac{8}{8} \qquad 12 = 11 + 1 = 11 + \tfrac{8}{8} = 11\tfrac{8}{8} \\[2mm]
-\ 9\dfrac{3}{8} &=& -\ 9\dfrac{3}{8} \\[1mm]
\hline
&& 2\dfrac{5}{8}
\end{array}
$$

**Do Exercise 6.** ▶

*Answers*

**3.** $13\dfrac{7}{12}$ **4.** $1\dfrac{1}{2}$ **5.** $3\dfrac{1}{6}$ **6.** $3\dfrac{2}{3}$

*Guided Solutions:*

**5.** 4, 3, 1 **6.** $\dfrac{3}{3}$, 2

**EXAMPLE 7**  Subtract: $7\frac{1}{6} - 2\frac{1}{4}$.

$$
\begin{array}{ll}
7\dfrac{1}{6}\cdot\dfrac{2}{2} = & 7\dfrac{2}{12}\\[2mm]
-\,2\dfrac{1}{4}\cdot\dfrac{3}{3} = & -\,2\dfrac{3}{12}
\end{array}
\Biggr\}
$$

← The LCD is 12.

$$
\begin{array}{rr}
7\dfrac{2}{12} = & 6\dfrac{14}{12}\\[2mm]
-\,2\dfrac{3}{12} = & -\,2\dfrac{3}{12}\\[2mm]
\hline
 & 4\dfrac{11}{12}
\end{array}
$$

We cannot subtract $\frac{3}{12}$ from $\frac{2}{12}$.
We borrow 1, or $\frac{12}{12}$, from 7:
$7\frac{2}{12} = 6 + 1 + \frac{2}{12} = 6 + \frac{12}{12} + \frac{2}{12} = 6\frac{14}{12}$.

◀ **Do Exercise 7.**

**7.** Subtract.

$$
\begin{array}{r}
8\dfrac{1}{9}\\[2mm]
-4\dfrac{5}{6}\\[1mm]
\hline
\end{array}
$$

**EXAMPLE 8**  Combine like terms: $9\frac{3}{4}x + 4\frac{1}{2}x$.

$$
9\frac{3}{4}x + 4\frac{1}{2}x = \left(9\frac{3}{4} + 4\frac{1}{2}\right)x
$$

Using the distributive law; this is often done mentally.

$$
= \left(9\frac{3}{4} + 4\frac{2}{4}\right)x
$$

The LCD is 4.

$$
= 13\frac{5}{4}x = 14\frac{1}{4}x
$$

Combine like terms.

**8.** $7\frac{1}{12}t + 1\frac{5}{12}t$

**9.** $7\frac{11}{12}x - 5\frac{2}{3}x$

**10.** $5\frac{11}{15}x + 8\frac{3}{10}x$

◀ **Do Exercises 8–10.**

### c  APPLICATIONS AND PROBLEM SOLVING

**EXAMPLE 9**  *Men's Long-Jump World Records.*  On October 18, 1968, Bob Beamon set a world record of $29\frac{3}{16}$ ft for the long jump, a record that was not broken for nearly 23 years. This record-setting jump was significantly longer than the previous one of $27\frac{19}{48}$ ft, accomplished on May 29, 1965, by Ralph Boston. How much longer was Beamon's jump than Boston's?

**Data:** thoughtco.com

1. **Familiarize.**  The phrase "how much longer" indicates subtraction. We let $w$ = the difference in the world records.

2. **Translate.**  We translate as follows:

| Beamon's jump | − | Boston's jump | = | Difference in length |
|:---:|:---:|:---:|:---:|:---:|
| ↓ | ↓ | ↓ | ↓ | ↓ |
| $29\frac{3}{16}$ | − | $27\frac{19}{48}$ | = | $w$. |

3. **Solve.**  To solve the equation, we carry out the subtraction. The LCD is 48.

$$
\begin{array}{rrrr}
29\dfrac{3}{16} = & 29\dfrac{3}{16}\cdot\dfrac{3}{3} = & 29\dfrac{9}{48} = & 28\dfrac{57}{48}\\[2mm]
-\,27\dfrac{19}{48} = & -\,27\dfrac{19}{48} = & -\,27\dfrac{19}{48} = & -\,27\dfrac{19}{48}\\[2mm]
\hline
 & & & 1\dfrac{38}{48} = 1\dfrac{19}{24}
\end{array}
$$

Thus, $w = 1\frac{19}{24}$.

*Answers*

**7.** $3\frac{5}{18}$  **8.** $8\frac{1}{2}t$  **9.** $2\frac{1}{4}x$  **10.** $14\frac{1}{30}x$

**4. Check.** To check, we add the difference, $1\frac{19}{24}$, to Boston's jump.

Since $27\frac{19}{48} + 1\frac{19}{24} = 29\frac{3}{16}$, the answer checks.

**5. State.** Beamon's jump was $1\frac{19}{24}$ ft longer than Boston's.

**Do Exercise 11.** ▶

**EXAMPLE 10** *Hair Donation.* Wigs for Kids is a non-profit organization that provides hairpieces for children who have lost their hair as a result of medical conditions. Karissa and Cayla allowed their hair to grow in order to donate it. The length cut from Karissa's hair was $15\frac{1}{4}$ in., and the length cut from Cayla's hair was $14\frac{1}{2}$ in. After the hair was cut, it was discovered that the ends of each lock had highlighting that needed to be trimmed. Because of the highlighting, $1\frac{1}{2}$ in. was cut from Karissa's lock of hair and $2\frac{3}{4}$ in. was cut from Cayla's. In all, what was the total usable length of hair that Karissa and Cayla donated?

**1. Familiarize.** We let $l$ = the total usable length of hair that Karissa and Cayla donated.

**2. Translate.** The length $l$ is the sum of the lengths that were cut, minus the sum of the lengths that were trimmed from the locks. Thus we have

$$l = \left(15\frac{1}{4} + 14\frac{1}{2}\right) - \left(1\frac{1}{2} + 2\frac{3}{4}\right).$$

**3. Solve.** This is a three-step problem.

**a)** We first add the two lengths $15\frac{1}{4}$ and $14\frac{1}{2}$.

$$\begin{array}{r} 15\frac{1}{4} = \phantom{+}15\frac{1}{4} \\ +\,14\frac{1}{2} = +\,14\frac{2}{4} \\ \hline 29\frac{3}{4} \end{array}$$

**b)** Next, we add the two lengths $1\frac{1}{2}$ and $2\frac{3}{4}$.

$$\begin{array}{r} 1\frac{1}{2} = \phantom{+}1\frac{2}{4} \\ +\,2\frac{3}{4} = +\,2\frac{3}{4} \\ \hline 3\frac{5}{4} = 4\frac{1}{4} \end{array}$$

**c)** Finally, we subtract $4\frac{1}{4}$ from $29\frac{3}{4}$.

$$\begin{array}{r} 29\frac{3}{4} \\ -\;4\frac{1}{4} \\ \hline 25\frac{2}{4} = 25\frac{1}{2} \end{array}$$

Thus, $l = 25\frac{1}{2}$.

**4. Check.** We can check by doing the problem a different way. We can subtract the trimmed length from each lock, then add the adjusted lengths together.

| Karissa's lock | Cayla's lock | Sum of lengths |
|---|---|---|
| $15\frac{1}{4} = \phantom{+}14\frac{5}{4}$ | $14\frac{1}{2} = \phantom{+}13\frac{6}{4}$ | $13\frac{3}{4}$ |
| $-\;1\frac{1}{2} = -\;1\frac{2}{4}$ | $-\;2\frac{3}{4} = -\;2\frac{3}{4}$ | $+\,11\frac{3}{4}$ |
| $13\frac{3}{4}$ | $11\frac{3}{4}$ | $24\frac{6}{4} = 25\frac{1}{2}$ |

We obtained the same answer, so our answer checks.

**5. State.** The sum of the usable lengths of hair donated was $25\frac{1}{2}$ in.

**Do Exercise 12.** ▶

**11.** *Travel Distance.* On a two-day business trip, Paul drove $213\frac{7}{10}$ mi the first day and $107\frac{5}{8}$ mi the second day. How much farther did Paul drive on the first day than on the second day?

$15\frac{1}{4}$ in.    $14\frac{1}{2}$ in.

$1\frac{1}{2}$ in.    $2\frac{3}{4}$ in.

**12.** *Liquid Fertilizer.* There is $283\frac{5}{8}$ gal of liquid fertilizer in a fertilizer application tank. After applying $178\frac{2}{3}$ gal to a soybean field, the farmer requests that Braden's Farm Supply deliver an additional 250 gal to the tank. How many gallons of fertilizer are in the tank after the delivery?

***Answers***

**11.** $106\frac{3}{40}$ mi  **12.** $354\frac{23}{24}$ gal

## d  NEGATIVE MIXED NUMERALS

Consider the numbers $5\frac{3}{4}$ and $-5\frac{3}{4}$ on the number line.

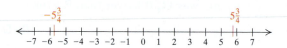

Note that just as $5\frac{3}{4}$ means $5 + \frac{3}{4}$, we can regard $-5\frac{3}{4}$ as $-5 - \frac{3}{4}$.

Consider the subtraction $4 - 4\frac{1}{2}$. We know that if we have \$4 and make a \$$4\frac{1}{2}$ purchase, we will owe half a dollar. Thus,

$$4 - 4\frac{1}{2} = -\frac{1}{2}.$$

We can subtract by rewriting the subtraction as addition:

$$4 - 4\frac{1}{2} = 4 + \left(-4\frac{1}{2}\right).$$

Because $-4\frac{1}{2}$ has the greater absolute value, the answer will be negative. The difference in absolute value is $4\frac{1}{2} - 4 = \frac{1}{2}$, so

$$4 - 4\frac{1}{2} = -\frac{1}{2}.$$

Another way to see this is to convert to fraction notation and subtract:

$$4 - 4\frac{1}{2} = \frac{8}{2} - \frac{9}{2} = \frac{8}{2} + \left(-\frac{9}{2}\right) = -\frac{1}{2}.$$

**EXAMPLE 11**  Subtract: $3\frac{2}{7} - 4\frac{2}{5}$.

Since $4\frac{2}{5}$ is greater than $3\frac{2}{7}$, the answer will be negative. We can also see this by rewriting the subtraction as $3\frac{2}{7} + (-4\frac{2}{5})$. The difference in absolute values is

$$
\begin{array}{rcccl}
4\frac{2}{5} & = & 4\,\dfrac{2}{5}\cdot\dfrac{7}{7} & = & 4\dfrac{14}{35} \\[2mm]
-3\frac{2}{7} & = & -3\,\dfrac{2}{7}\cdot\dfrac{5}{5} & = & -3\dfrac{10}{35} \\[2mm]
\hline
& & & & 1\dfrac{4}{35}.
\end{array}
$$

> Because $-4\frac{2}{5}$ has the larger absolute value, we make the answer negative.

Thus, $3\frac{2}{7} - 4\frac{2}{5} = -1\frac{4}{35}$. ◄

◀ **Do Exercises 13–15.**

Subtract.

**13.** $7 - 7\frac{3}{4}$

**14.** $5\frac{1}{2} - 9\frac{3}{4}$

**15.** $4\frac{2}{3} - 7\frac{1}{6}$

**EXAMPLE 12**  Subtract: $-6\frac{4}{5} - \left(-9\frac{3}{10}\right)$.

We write the subtraction as addition:

$$-6\frac{4}{5} - \left(-9\frac{3}{10}\right) = -6\frac{4}{5} + 9\frac{3}{10}.$$  Instead of subtracting, we add the opposite.

Since $9\frac{3}{10}$ has the greater absolute value, the answer will be positive. The difference in absolute values is

$$
\begin{array}{rcl}
9\frac{3}{10} = & 9\frac{3}{10} & = \quad 9\frac{3}{10} = \quad 8\frac{13}{10} \\
-\,6\frac{4}{5} = & -6\,\dfrac{4\cdot2}{5\cdot2} & = \quad -6\frac{8}{10} = -6\frac{8}{10} \\
\hline
& & \qquad\qquad 2\frac{5}{10} = 2\frac{1}{2}.
\end{array}
$$

Thus, $-6\frac{4}{5} - (-9\frac{3}{10}) = 2\frac{1}{2}$.

We can check by converting to fraction notation and redoing the calculation:

$$
\begin{aligned}
-6\frac{4}{5} - \left(-9\frac{3}{10}\right) &= -\frac{34}{5} - \left(-\frac{93}{10}\right) \\
&= -\frac{34}{5} + \frac{93}{10} \qquad &&\text{Adding the opposite} \\
&= -\frac{68}{10} + \frac{93}{10} \qquad &&\text{Writing with a common denominator} \\
&= \frac{25}{10} = \frac{5}{2} = 2\frac{1}{2}. \qquad &&\text{$-68 + 93 = 25$}
\end{aligned}
$$

**Do Exercises 16 and 17.** ▶

To add two negative numbers we add absolute values and make the answer negative.

**EXAMPLE 13**  Add: $-4\frac{1}{6} + \left(-5\frac{2}{9}\right)$.

We add the absolute values and make the answer negative.

$$
\begin{array}{rcl}
4\frac{1}{6} = & 4\,\dfrac{1\cdot3}{6\cdot3} = & 4\frac{3}{18} \qquad \text{Adding absolute values} \\
+\,5\frac{2}{9} = & +5\,\dfrac{2\cdot2}{9\cdot2} = & +5\frac{4}{18} \\
\hline
& & 9\frac{7}{18}
\end{array}
$$

Thus, $-4\frac{1}{6} + (-5\frac{2}{9}) = -9\frac{7}{18}$.

**Do Exercise 18.** ▶

---

Subtract.

**16.** $-7\frac{1}{3} - \left(-5\frac{1}{2}\right)$

**17.** $-4\frac{1}{10} - \left(-7\frac{2}{5}\right)$

**18.** Add: $-7\frac{1}{10} + \left(-6\frac{2}{15}\right)$.

***Answers***

**16.** $-1\frac{5}{6}$  **17.** $3\frac{3}{10}$  **18.** $-13\frac{7}{30}$

FOR EXTRA HELP ⓟ MyLab Math

## ✓ Check Your Understanding

**Reading Check** Match each addition or subtraction with the correct first step from the following list.

**a)** Add the fractions.

**b)** Write the fraction parts with a common denominator.

**c)** Rename 5 as $4\frac{9}{9}$.

**d)** Borrow 1 from 5 and add it to $\frac{1}{9}$.

_____ **RC1.** $5\frac{1}{9}$
$-3\frac{4}{9}$

_____ **RC2.** $5\frac{4}{9}$
$+3\frac{1}{9}$

_____ **RC3.** $5\frac{4}{9}$
$+3\frac{1}{18}$

_____ **RC4.** $5$
$-3\frac{1}{9}$

### Concept Check

**CC1.** Determine which of the following are equivalent to 20.

**a)** $19\frac{4}{4}$      **b)** $19\frac{8}{8}$      **c)** $19\frac{15}{15}$      **d)** $19\frac{1}{20}$

**CC2.** Determine which of the following are equivalent to $7\frac{1}{8}$.

**a)** $7\frac{2}{4}$      **b)** $6\frac{9}{8}$      **c)** $7\frac{2}{16}$      **d)** $6\frac{18}{16}$

**a , b** Perform the indicated operation. Write a mixed numeral for each answer.

**1.** $6$
$+5\frac{2}{5}$

**2.** $3$
$+6\frac{5}{7}$

**3.** $2\frac{7}{8}$
$+6\frac{5}{8}$

**4.** $2\frac{5}{6}$
$+5\frac{5}{6}$

**5.** $4\frac{1}{4}$
$+1\frac{1}{12}$

**6.** $4\frac{1}{12}$
$+5\frac{1}{6}$

**7.** $7\frac{3}{4}$
$+5\frac{5}{6}$

**8.** $4\frac{3}{8}$
$+6\frac{5}{12}$

**9.** $3\frac{2}{5}$
$+8\frac{7}{10}$

**10.** $5\frac{1}{2}$
$+3\frac{7}{10}$

**11.** $6\frac{3}{8}$
$+10\frac{5}{6}$

**12.** $5\frac{5}{8}$
$+1\frac{5}{6}$

**13.** $18\frac{4}{5}$
$+2\frac{7}{10}$

**14.** $15\frac{5}{8}$
$+11\frac{3}{4}$

**15.** $14\frac{5}{8}$
$+13\frac{1}{4}$

**16.** $16\frac{1}{4}$
$+15\frac{7}{8}$

**17.** $8\frac{9}{10}$
$-1\frac{7}{10}$

**18.** $6\frac{7}{8}$
$-5\frac{3}{8}$

**19.** $9\frac{3}{5}$
$-3\frac{1}{2}$

**20.** $8\frac{2}{3}$
$-7\frac{1}{2}$

**21.** $4\dfrac{1}{5}$

　$-2\dfrac{3}{5}$

**22.** $5\dfrac{1}{8}$

　$-2\dfrac{3}{8}$

**23.** $19$

　$-5\dfrac{3}{4}$

**24.** $17$

　$-3\dfrac{7}{8}$

**25.** $34$

　$-18\dfrac{5}{8}$

**26.** $23$

　$-19\dfrac{3}{4}$

**27.** $21\dfrac{1}{6}$

　$-13\dfrac{3}{4}$

**28.** $42\dfrac{1}{10}$

　$-23\dfrac{7}{12}$

**29.** $25\dfrac{1}{9}$

　$-13\dfrac{5}{6}$

**30.** $23\dfrac{5}{16}$

　$-14\dfrac{7}{12}$

Combine like terms.

**31.** $1\dfrac{1}{8}t + 7\dfrac{5}{8}t$

**32.** $6\dfrac{1}{4}x + 8\dfrac{3}{4}x$

**33.** $9\dfrac{1}{2}x - 7\dfrac{1}{2}x$

**34.** $7\dfrac{3}{4}x - 2\dfrac{1}{4}x$

**35.** $5\dfrac{9}{10}y + 2\dfrac{2}{5}y$

**36.** $9\dfrac{3}{4}t + 2\dfrac{3}{8}t$

**37.** $37\dfrac{5}{9}t - 25\dfrac{2}{3}t$

**38.** $23\dfrac{1}{6}t - 19\dfrac{2}{3}t$

**39.** $2\dfrac{5}{6}x + 7\dfrac{3}{8}x$

**40.** $7\dfrac{3}{20}t + 1\dfrac{2}{15}t$

**41.** $11a - 8\dfrac{2}{3}a$

**42.** $6a - 3\dfrac{7}{10}a$

 Solve.

**43.** *Widening a Driveway.* Sherry and Woody are widening their existing $17\frac{1}{4}$-ft driveway by adding $5\frac{9}{10}$ ft on one side. What is the new width of the driveway?

**44.** *Plumbing.* A plumber uses two pipes, each of length $51\frac{5}{16}$ in., and one pipe of length $34\frac{3}{4}$ in. when installing a shower. How much pipe is used in all?

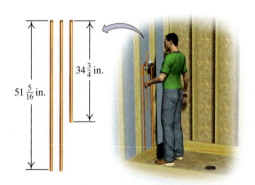

**45.** *Height.* Casey's beagle is $14\frac{1}{4}$ in. from shoulder to floor, and her basset hound is $13\frac{5}{16}$ in. from shoulder to floor. How much shorter is her basset hound?

**46.** *Winterizing a Swimming Pool.* To winterize their swimming pool, the Jablonskis are draining the water into a nearby field. The distance to the field is $103\frac{1}{2}$ ft. Because their only hose measures $62\frac{3}{4}$ ft, they need to buy an additional hose. How long must the new hose be?

**47.** *Fashion Design.* Jordan is designing two shoes. The Tower pump has a $4\frac{1}{10}$-in. heel, and the Classic pump has a $2\frac{4}{5}$-in. heel. How much taller is the heel of the Tower pump than the heel of the Classic pump?

**48.** *Guitar Design.* Josh's nylon-string guitar has a neck width of 2 in. His steel-string guitar has a neck width of $1\frac{13}{16}$ in. How much wider is the neck of his nylon-string guitar than the neck of his steel-string guitar?

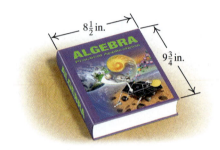

$1\frac{13}{16}$ in.

Steel string guitar

2 in.

Nylon string guitar

$4\frac{1}{10}$ in.

$2\frac{4}{5}$ in.

Tower pump          Classic pump

Find the perimeter of (distance around) each figure.

**49.**

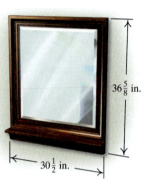

$36\frac{5}{8}$ in.

$30\frac{1}{2}$ in.

**50.**

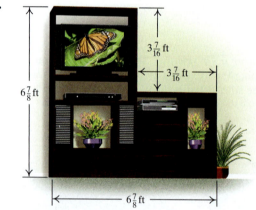

$8\frac{1}{2}$ in.

$9\frac{3}{4}$ in.

**51.**

$5\frac{3}{4}$ yd   $5\frac{3}{4}$ yd

$5\frac{3}{4}$ yd          $5\frac{3}{4}$ yd

$5\frac{3}{4}$ yd

**52.**

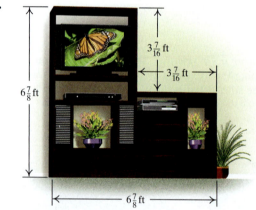

$3\frac{7}{16}$ ft

$3\frac{7}{16}$ ft

$6\frac{7}{8}$ ft

$6\frac{7}{8}$ ft

**53.**

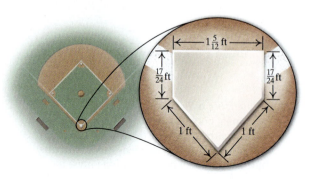

$1\frac{5}{12}$ ft

$\frac{17}{24}$ ft          $\frac{17}{24}$ ft

1 ft          1 ft

**54.**

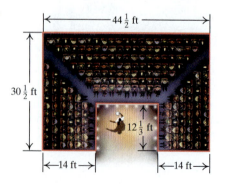

$44\frac{1}{2}$ ft

$30\frac{1}{2}$ ft

$12\frac{1}{3}$ ft

14 ft          14 ft

**55.** *Stone Bench.* Baytown Village Stone Creations is making a custom stone bench as shown below. The recommended height for the bench is 18 in. The depth of the stone bench is $3\frac{3}{8}$ in. Each of the two supporting legs is made up of three stacked stones. Two of the stones measure $3\frac{1}{2}$ in. and $5\frac{1}{4}$ in. How much must the third stone measure?

$3\frac{3}{8}$ in.
$3\frac{1}{2}$ in.
$5\frac{1}{4}$ in.
?
18 in.

**56.** *Window Dimensions.* The Sanchez family is replacing a window in their home. The original window measures $4\frac{5}{6}$ ft $\times$ $8\frac{1}{4}$ ft. The new window is $2\frac{1}{3}$ ft wider. What are the dimensions of the new window?

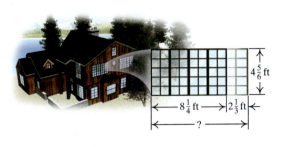

$4\frac{5}{6}$ ft
$8\frac{1}{4}$ ft
$2\frac{1}{3}$ ft
?

**57.** *Carpentry.* When cutting wood with a saw, a carpenter must take into account the thickness of the saw blade. Suppose that from a piece of wood 36 in. long, a carpenter cuts a $15\frac{3}{4}$-in. length with a saw blade that is $\frac{1}{8}$ in. thick. How long is the piece that remains?

**58.** *Cutco Cutlery.* The Essentials 5-piece set sold by Cutco contains three knives with different blade lengths: $7\frac{5}{8}''$ Petite Chef, $6\frac{3}{4}''$ Petite Carver, and $2\frac{3}{4}''$ Paring Knife. How much longer is the blade of the Petite Chef than that of the Petite Carver? than that of the Paring Knife?

**Data:** Cutco Cutlery Corporation

**59.** *Interior Design.* Eric worked $10\frac{1}{2}$ hr over a three-day period on an interior design project. If he worked $2\frac{1}{2}$ hr on the first day and $4\frac{1}{5}$ hr on the second, how many hours did Eric work on the third day?

**60.** *Painting.* Geri had $3\frac{1}{2}$ gal of paint. It took $2\frac{3}{4}$ gal to paint the family room. She estimated that it would take $2\frac{1}{4}$ gal to paint the living room. How much more paint did Geri need?

**61.** *Fly Fishing.* Bryan is putting together a fly fishing line and uses $58\frac{5}{8}$ ft of slow-sinking fly line and $8\frac{3}{4}$ ft of leader line. He uses $\frac{3}{8}$ ft of the slow-sinking fly line to connect the two lines. The knot used to connect the fly to the leader line uses $\frac{1}{6}$ ft of the leader line. How long is the finished fly fishing line?

**62.** Find the smallest length of a bolt that will pass through a piece of tubing with an outside diameter of $\frac{1}{2}$ in., a washer $\frac{1}{16}$ in. thick, a piece of tubing with a $\frac{3}{4}$-in. outside diameter, another washer, and a nut $\frac{3}{16}$ in. thick.

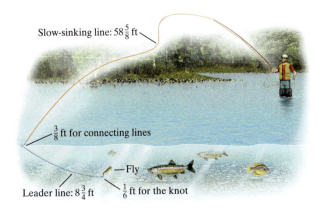

Slow-sinking line: $58\frac{5}{8}$ ft
$\frac{3}{8}$ ft for connecting lines
Fly
Leader line: $8\frac{3}{4}$ ft
$\frac{1}{6}$ ft for the knot

Find the length *d* in each figure.

**63.**

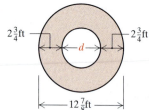

$2\frac{3}{4}$ft ——— ———$2\frac{3}{4}$ft

$d$

———$12\frac{7}{8}$ft———

**64.**

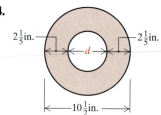

$2\frac{1}{5}$in.——— ———$2\frac{1}{5}$in.

$d$

———$10\frac{1}{2}$in.———

**d** Perform the indicated operation. Write a mixed numeral for each answer.

**65.** $9 - 9\frac{2}{5}$

**66.** $8 - 8\frac{3}{7}$

**67.** $3\frac{1}{2} - 6\frac{3}{4}$

**68.** $5\frac{1}{2} - 7\frac{3}{4}$

**69.** $3\frac{4}{5} + \left(-7\frac{2}{3}\right)$

**70.** $2\frac{3}{7} + \left(-5\frac{1}{2}\right)$

**71.** $-3\frac{1}{5} - 4\frac{2}{5}$

**72.** $-5\frac{3}{8} - 4\frac{1}{8}$

**73.** $-4\frac{1}{12} + 6\frac{2}{3}$

**74.** $-2\frac{3}{4} + 5\frac{3}{8}$

**75.** $-6\frac{1}{9} - \left(-4\frac{2}{9}\right)$

**76.** $-2\frac{3}{5} - \left(-1\frac{1}{5}\right)$

## Skill Maintenance

**77.** Write expanded notation for 38,125.  [1.1b]

**78.** Write a word name for 2,005,689.  [1.1c]

**79.** Write exponential notation for $9 \cdot 9 \cdot 9 \cdot 9$.  [1.9a]

**80.** Evaluate: $3^4$.  [1.9b]

Determine whether the first number is divisible by the second.  [3.1a], [3.1b]

**81.** 9993 by 3

**82.** 9993 by 9

**83.** 2345 by 9

**84.** 2345 by 5

**85.** 2335 by 10

**86.** 7764 by 6

**87.** 18,888 by 2

**88.** 18,888 by 6

## Synthesis

Calculate each of the following. Write the result as a mixed numeral.

**89.** ▦ $3289\frac{1047}{1189} + 5278\frac{32}{41}$

**90.** ▦ $4230\frac{19}{73} - 5848\frac{17}{29}$

Solve.

**91.** $35\frac{2}{3} + n = 46\frac{1}{4}$

**92.** $42\frac{7}{9} = x - 13\frac{2}{5}$

**93.** $-15\frac{7}{8} = 12\frac{1}{2} + t$

**94.** A post for a pier is 29 ft long. Half of the post extends above the water's surface and $8\frac{3}{4}$ ft of the post is buried in mud. How deep is the water at that location?

**95.** An algebra text is $1\frac{1}{8}$ in. thick, $9\frac{3}{4}$ in. long, and $8\frac{1}{2}$ in. wide. If the front, back, and spine of the book were unfolded, they would form a rectangle. What would the perimeter of that rectangle be?

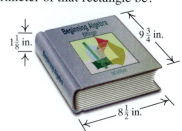

$1\frac{1}{8}$ in.

$9\frac{3}{4}$ in.

$8\frac{1}{2}$ in.

# Multiplication and Division Using Mixed Numerals; Applications

## OBJECTIVES

a   Multiply using mixed numerals.

b   Divide using mixed numerals.

c   Evaluate expressions using mixed numerals.

d   Solve applied problems involving multiplication and division with mixed numerals.

Carrying out addition and subtraction with mixed numerals is usually easier if the numbers are left as mixed numerals. With multiplication and division, however, it is easier to convert the numbers to fraction notation first.

## a   MULTIPLICATION USING MIXED NUMERALS

> **MULTIPLICATION USING MIXED NUMERALS**
>
> To multiply using mixed numerals, first convert to fraction notation and multiply. Then convert the answer to a mixed numeral, if appropriate.

**EXAMPLE 1**   Multiply: $6 \cdot 2\frac{1}{2}$.

$$6 \cdot 2\frac{1}{2} = \frac{6}{1} \cdot \frac{5}{2} = \frac{6 \cdot 5}{1 \cdot 2} = \frac{2 \cdot 3 \cdot 5}{2 \cdot 1} = 15$$

Removing a factor equal to 1: $\frac{2}{2} = 1$

Convert the numbers to fraction notation first.

**EXAMPLE 2**   Multiply: $3\frac{1}{2} \cdot \frac{3}{4}$.

$$3\frac{1}{2} \cdot \frac{3}{4} = \frac{7}{2} \cdot \frac{3}{4} = \frac{21}{8} = 2\frac{5}{8}$$

Recall that common denominators are *not* required when multiplying fractions.

**Do Exercises 1 and 2.** ▶

Multiply.

1. $8 \cdot 3\frac{1}{2}$

2. $5\frac{1}{2} \cdot \frac{3}{7}$

**EXAMPLE 3**   Multiply: $-10 \cdot 5\frac{2}{3}$.

$$-10 \cdot 5\frac{2}{3} = -\frac{10}{1} \cdot \frac{17}{3} = -\frac{170}{3} = -56\frac{2}{3}$$

Multiply.

**GS**   3.  $-2 \cdot 6\frac{2}{5} = -\frac{2}{1} \cdot \frac{\square}{5}$

$$= -\frac{\square}{5} = -\square\frac{\square}{\square}$$

**EXAMPLE 4**   Multiply: $2\frac{1}{4} \cdot 5\frac{2}{3}$.

$$2\frac{1}{4} \cdot 5\frac{2}{3} = \frac{9}{4} \cdot \frac{17}{3} = \frac{9 \cdot 17}{4 \cdot 3} = \frac{3 \cdot 3 \cdot 17}{2 \cdot 2 \cdot 3} = \frac{51}{4} = 12\frac{3}{4}$$

4. $3\frac{1}{3} \cdot 2\frac{1}{2}$

.......... **Caution!** ..........

Note that $2\frac{1}{4} \cdot 5\frac{2}{3} \neq 10\frac{2}{12}$. A common error is to multiply the whole numbers and then the fractions. The correct answer, $12\frac{3}{4}$, is found only after converting to fraction notation.

**Answers**

1. 28   2. $2\frac{5}{14}$   3. $-12\frac{4}{5}$   4. $8\frac{1}{3}$

**Guided Solution:**

3. 32, 64, $12\frac{4}{5}$

**Do Exercises 3 and 4.** ▶

## b DIVISION USING MIXED NUMERALS

The division $1\frac{1}{2} \div \frac{1}{6}$ is shown here. This division means "How many $\frac{1}{6}$'s are in $1\frac{1}{2}$?" We see that the answer is 9.

$\dfrac{1}{6}$ goes into $1\dfrac{1}{2}$ nine times.

When we divide using mixed numerals, we convert to fraction notation first. Recall that to divide by a fraction, we multiply by its reciprocal.

$$1\frac{1}{2} \div \frac{1}{6} = \frac{3}{2} \div \frac{1}{6} = \frac{3}{2} \cdot \frac{6}{1}$$

$$= \frac{3 \cdot 6}{2 \cdot 1} = \frac{3 \cdot 3 \cdot 2}{2 \cdot 1} = \frac{3 \cdot 3}{1} \cdot \frac{2}{2} = \frac{3 \cdot 3}{1} \cdot 1 = 9$$

---

### DIVISION USING MIXED NUMERALS

To divide using mixed numerals, first write fraction notation and divide. Then convert the answer to a mixed numeral, if appropriate.

---

**EXAMPLE 5** Divide: $32 \div 3\frac{1}{5}$.

$$32 \div 3\frac{1}{5} = \frac{32}{1} \div \frac{16}{5} \qquad \text{Writing fraction notation}$$

$$= \frac{32}{1} \cdot \frac{5}{16} = \frac{32 \cdot 5}{1 \cdot 16} = \frac{2 \cdot 16 \cdot 5}{1 \cdot 16} = \frac{16}{16} \cdot \frac{2 \cdot 5}{1} = 1 \cdot \frac{2 \cdot 5}{1} = 10$$

Remember to multiply by the reciprocal.

............................................................ **Caution!** ............................................................

The reciprocal of $3\frac{1}{5}$ is neither $5\frac{1}{3}$ nor $3\frac{5}{1}$!

..............................................................................................................................

**5.** Divide: $63 \div 5\dfrac{1}{4}$.

◀ Do Exercise 5.

**EXAMPLE 6** Divide: $2\frac{1}{3} \div 1\frac{3}{4}$.

$$2\frac{1}{3} \div 1\frac{3}{4} = \frac{7}{3} \div \frac{7}{4} = \frac{7}{3} \cdot \frac{4}{7} = \frac{7 \cdot 4}{7 \cdot 3} = \frac{4}{3} = 1\frac{1}{3} \qquad \text{Removing a factor equal to 1: } \frac{7}{7} = 1$$

*Answer*

**5.** 12

**EXAMPLE 7** Divide: $-1\frac{3}{5} \div (-3\frac{1}{3})$.

$$-1\frac{3}{5} \div \left(-3\frac{1}{3}\right) = -\frac{8}{5} \div \left(-\frac{10}{3}\right) = \frac{8}{5} \cdot \frac{3}{10}$$ The product or quotient of two negatives is positive.

$$= \frac{2 \cdot 4 \cdot 3}{5 \cdot 2 \cdot 5} = \frac{12}{25}$$ Removing a factor equal to 1: $\frac{2}{2} = 1$

**Do Exercises 6 and 7.** ▶

## c  EVALUATING EXPRESSIONS

Mixed numerals can appear in algebraic expressions.

**EXAMPLE 8** A train traveling $r$ miles per hour for $t$ hours travels a total of $rt$ miles. (*Remember*: Distance = Rate · Time.)

**a)** Find the distance traveled by a train moving at 60 mph in $2\frac{3}{4}$ hr.

**b)** Find the distance traveled if the speed of the train is $26\frac{1}{2}$ mph and the time is $2\frac{2}{3}$ hr.

**a)** We evaluate $rt$ for $r = 60$ and $t = 2\frac{3}{4}$:

$$rt = 60 \cdot 2\frac{3}{4}$$

$$= \frac{60}{1} \cdot \frac{11}{4}$$

$$= \frac{15 \cdot 4 \cdot 11}{1 \cdot 4} = 165.$$ Removing a factor equal to 1: $\frac{4}{4} = 1$

In $2\frac{3}{4}$ hr, a train moving at 60 mph travels 165 mi.

**b)** We evaluate $rt$ for $r = 26\frac{1}{2}$ and $t = 2\frac{2}{3}$:

$$rt = 26\frac{1}{2} \cdot 2\frac{2}{3}$$

$$= \frac{53}{2} \cdot \frac{8}{3} = \frac{53 \cdot 2 \cdot 4}{2 \cdot 3}$$ Removing a factor equal to 1: $\frac{2}{2} = 1$

$$= \frac{212}{3} = 70\frac{2}{3}.$$

In $2\frac{2}{3}$ hr, a train moving at $26\frac{1}{2}$ mph travels $70\frac{2}{3}$ mi.

**EXAMPLE 9** Evaluate $x + yz$ for $x = 7\frac{1}{3}$, $y = \frac{1}{3}$, and $z = 5$.

We substitute and follow the rules for order of operations:

$$x + yz = 7\frac{1}{3} + \frac{1}{3} \cdot 5$$

$$= 7\frac{1}{3} + \frac{1}{3} \cdot \frac{5}{1}$$ Multiply first; then add.

$$= 7\frac{1}{3} + \frac{5}{3}$$

$$\left.\begin{array}{l} = 7\frac{1}{3} + 1\frac{2}{3} \\ \\ = 8\frac{3}{3} = 9. \end{array}\right\}$$ Adding mixed numerals

**Do Exercises 8–10.** ▶

---

Divide.

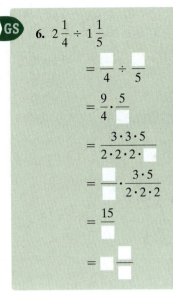

**GS** **6.** $2\frac{1}{4} \div 1\frac{1}{5}$

$$= \frac{\square}{4} \div \frac{\square}{5}$$

$$= \frac{9}{4} \cdot \frac{5}{\square}$$

$$= \frac{3 \cdot 3 \cdot 5}{2 \cdot 2 \cdot 2 \cdot \square}$$

$$= \frac{\square}{\square} \cdot \frac{3 \cdot 5}{2 \cdot 2 \cdot 2}$$

$$= \frac{15}{\square}$$

$$= \square\frac{\square}{\square}$$

**7.** $1\frac{3}{4} \div \left(-2\frac{1}{2}\right)$

Evaluate.

**8.** $rt$, for $r = 78$ and $t = 2\frac{1}{4}$

**9.** $7xy$, for $x = 9\frac{2}{5}$ and $y = 2\frac{3}{7}$

**10.** $x - y \div z$, for $x = 5\frac{7}{8}$, $y = \frac{1}{4}$, and $z = 2$

*Answers*

**6.** $1\frac{7}{8}$   **7.** $-\frac{7}{10}$   **8.** $175\frac{1}{2}$   **9.** $159\frac{4}{5}$   **10.** $5\frac{3}{4}$

*Guided Solution:*

**6.** $9, 6, 6, 3, \frac{3}{3}, 8, 1\frac{7}{8}$

## d APPLICATIONS AND PROBLEM SOLVING

**EXAMPLE 10** *Training Regimens.* Fitness trainers suggest training regimens for athletes who are preparing to run marathons and mini-marathons. One suggested twelve-week regimen combines days of short, easy running with other days of cross-training, rest, and long-distance running. During week nine, this regimen calls for a long-distance run of 10 mi, which is $2\frac{1}{2}$ times the length of the long-distance run recommended for week one. What is the length of the long-distance run recommended for week one?

**Data:** shape.com

1. **Familiarize.** We ask the question "10 is $2\frac{1}{2}$ times what number?" We let $r$ = the length of the long-distance run recommended for week one.

2. **Translate.** The problem can be translated to an equation.

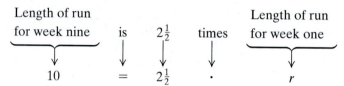

$$10 \quad = \quad 2\frac{1}{2} \quad \cdot \quad r$$

3. **Solve.** To solve the equation, we multiply on both sides.

$$10 = 2\frac{1}{2} \cdot r$$

$$10 = \frac{5}{2} \cdot r \qquad \text{Converting } 2\frac{1}{2} \text{ to fraction notation}$$

$$\frac{2}{5} \cdot 10 = \frac{2}{5} \cdot \frac{5}{2}r \qquad \text{Using the multiplication principle}$$

$$\frac{2 \cdot 10}{5} = 1 \cdot r \qquad \text{Multiplying}$$

$$4 = r \qquad \text{Simplifying: } \frac{2 \cdot 10}{5} = \frac{20}{5} = 4$$

4. **Check.** If the length of the long-distance run recommended for week one is 4 mi, we find the length of the run recommended for week nine by multiplying 4 by $2\frac{1}{2}$.

$$2\frac{1}{2} \cdot 4 = \frac{5}{2} \cdot 4 = \frac{20}{2} = 10$$

The answer checks.

5. **State.** The regimen recommends a long-distance run of 4 mi for week one.

◀ **Do Exercises 11 and 12.**

Solve.

11. Kyle's pickup truck travels on an interstate highway at 65 mph for $3\frac{1}{2}$ hr. How far does it travel?

12. Holly's minivan traveled 302 mi on $15\frac{1}{10}$ gal of gas. How many miles per gallon did it get?

*Answers*

**11.** $227\frac{1}{2}$ mi  **12.** 20 mpg

**EXAMPLE 11** *Flooring.* Ann and Tony plan to lay a hardwood floor only on the part of their living room not covered by a rug. If the room is $22\frac{1}{2}$ ft by $15\frac{1}{2}$ ft and the rug is 9 ft by 12 ft, how much hardwood flooring do they need? How much hardwood flooring would it take to cover the entire floor of the room?

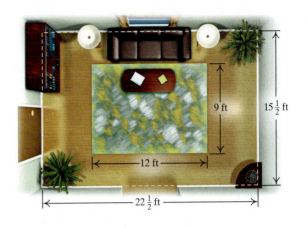

1. **Familiarize.** We draw a diagram and let $B$ = the area of the room, $R$ = the area of the rug, and $H$ = the area to be covered by hardwood flooring.

2. **Translate.** This is a multistep problem. We first find the area of the room, $B$, and the area of the rug, $R$. Then $H = B - R$. We find each area using the formula for the area of a rectangle: $A = l \times w$.

3. **Solve.** We carry out the calculations.

$$B = \text{length} \times \text{width}$$
$$= 22\frac{1}{2} \cdot 15\frac{1}{2}$$
$$= \frac{45}{2} \cdot \frac{31}{2}$$
$$= \frac{1395}{4} = 348\frac{3}{4} \text{ sq ft}$$

$$R = \text{length} \times \text{width}$$
$$= 12 \cdot 9$$
$$= 108 \text{ sq ft}$$

Then $H = B - R$

$$= 348\frac{3}{4} \text{ sq ft} - 108 \text{ sq ft}$$

$$= 240\frac{3}{4} \text{ sq ft}$$

4. **Check.** We can perform a check by repeating the calculations.

5. **State.** Ann and Tony will need $240\frac{3}{4}$ sq ft of hardwood flooring. It would take $348\frac{3}{4}$ sq ft of hardwood flooring to cover the entire floor of the room.

**Do Exercise 13.** ▶

**13.** *Koi Pond.* Colleen designed a koi fish pond for her backyard. Using the dimensions shown in the diagram below, determine the area of Colleen's backyard remaining after the pond was completed.

***Answer***
**13.** 945 sq yd

---

### CALCULATOR CORNER

***Operations on Fractions and Mixed Numerals***   Fraction calculators can add, subtract, multiply, and divide fractions and mixed numerals. The $\boxed{a^b/c}$ key is used to enter fractions and mixed numerals. The fraction $\frac{3}{4}$ is entered by pressing $\boxed{3}$ $\boxed{a^b/c}$ $\boxed{4}$, and it appears on the display as $\boxed{\qquad 3 \lrcorner 4}$. The mixed numeral $1\frac{5}{16}$ is entered by pressing $\boxed{1}$ $\boxed{a^b/c}$ $\boxed{5}$ $\boxed{a^b/c}$ $\boxed{1}$ $\boxed{6}$, and it is displayed as $\boxed{1 \lrcorner 5 \lrcorner 16}$. To express the result for $1\frac{5}{16}$ as a fraction, we press $\boxed{\text{SHIFT}}$ $\boxed{d/c}$. We get $\boxed{21 \lrcorner 16}$, or $\frac{21}{16}$. Some calculators display fractions and mixed numerals in the way in which we write them.

**EXERCISES:**  Perform each calculation. Give the answer in fraction notation.

**1.** $\dfrac{1}{3} + \dfrac{1}{4}$

**2.** $\dfrac{7}{5} - \dfrac{3}{10}$

**3.** $\dfrac{15}{4} \cdot \dfrac{7}{12}$

**4.** $-\dfrac{4}{5} \div \dfrac{8}{3}$

Perform each calculation. Give the answer as a mixed numeral.

**5.** $4\dfrac{1}{3} + 5\dfrac{4}{5}$

**6.** $9\dfrac{2}{7} - 8\dfrac{1}{4}$

**7.** $-2\dfrac{1}{3} \cdot \left(-4\dfrac{3}{5}\right)$

**8.** $10\dfrac{7}{10} \div 3\dfrac{5}{6}$

# Translating for Success

1. **Raffle Tickets.** At the Happy Hollow Camp Fall Festival, Rico and Becca, together, spent $270 on raffle tickets that sell for $\frac{9}{20}$ each. How many tickets did they buy?

2. **Irrigation Pipe.** Jed uses two pipes, one of which measures $5\frac{1}{3}$ ft, to repair the irrigation system in the Aguilars' lawn. The total length of the two pipes is $8\frac{7}{12}$ ft. How long is the other pipe?

3. **Vacation Days.** Together, Oscar and Claire have 36 vacation days a year. Oscar has 22 vacation days per year. How many does Claire have?

4. **Enrollment in Japanese Classes.** Last year at Lakeside Community College, 225 students enrolled in basic mathematics. This number is $4\frac{1}{2}$ times as many as the number who enrolled in Japanese. How many enrolled in Japanese?

5. **Bicycling.** Cole rode his bicycle $5\frac{1}{3}$ mi on Saturday and $8\frac{7}{12}$ mi on Sunday. How far did he ride on the weekend?

The goal of these matching questions is to practice step (2), Translate, of the five-step problem-solving process. Translate each word problem to an equation and select a correct translation from equations A–O.

**A.** $13\frac{11}{12} = x + 5\frac{1}{3}$

**B.** $\frac{3}{4} \cdot x = 1\frac{2}{3}$

**C.** $270 - \frac{20}{9} = x$

**D.** $225 = 4\frac{1}{2} \cdot x$

**E.** $98 \div 2\frac{1}{3} = x$

**F.** $22 + x = 36$

**G.** $x = 4\frac{1}{2} \cdot 225$

**H.** $x = 5\frac{1}{3} + 8\frac{7}{12}$

**I.** $22 \cdot x = 36$

**J.** $x = \frac{3}{4} \cdot 1\frac{2}{3}$

**K.** $5\frac{1}{3} + x = 8\frac{7}{12}$

**L.** $\frac{9}{20} \cdot 270 = x$

**M.** $1\frac{2}{3} + \frac{3}{4} = x$

**N.** $98 - 2\frac{1}{3} = x$

**O.** $\frac{9}{20} \cdot x = 270$

*Answers on page A-9*

6. **Deli Order.** For a promotional open house for contractors last year, the Bayside Builders Association ordered 225 turkey sandwiches. Because of increased registrations this year, $4\frac{1}{2}$ times as many sandwiches are needed. How many sandwiches should be ordered?

7. **Dog Ownership.** In Sam's community, $\frac{9}{20}$ of the households own at least one dog. There are 270 households. How many own dogs?

8. **Magic Tricks.** Samantha has 98 ft of rope and needs to cut it into $2\frac{1}{3}$-ft pieces to be used in a magic trick. How many pieces can be cut from the rope?

9. **Painting.** Laura needs $1\frac{2}{3}$ gal of paint to paint the ceiling of the exercise room and $\frac{3}{4}$ gal of the same paint for the bathroom. How much paint does Laura need?

10. **Chocolate Fudge Bars.** A recipe for chocolate fudge bars that serves 16 includes $1\frac{2}{3}$ cups of sugar. How much sugar is needed for $\frac{3}{4}$ of this recipe?

✓ **Check Your Understanding**

**Reading Check** Determine whether each statement is true or false.

_____ **RC1.** To multiply using mixed numerals, we first convert to fraction notation.

_____ **RC2.** To divide using mixed numerals, we first convert to fraction notation.

_____ **RC3.** The product of mixed numerals is generally written as a mixed numeral, unless it is an integer or between $-1$ and $1$.

_____ **RC4.** To divide fractions, we multiply by the reciprocal of the divisor.

**Concept Check** Rewrite each division as a product, using fraction notation. Do not carry out the calculations.

**CC1.** $2\frac{1}{3} \div 4\frac{3}{5}$ **CC2.** $7\frac{3}{10} \div 1\frac{7}{10}$ **CC3.** $8\frac{2}{9} \div \frac{1}{16}$ **CC4.** $6\frac{5}{7} \div 30$

**a** Multiply. Write a mixed numeral for each answer.

**1.** $8 \cdot 2\frac{5}{6}$

**2.** $10 \cdot 3\frac{3}{4}$

**3.** $6\frac{2}{3} \cdot \frac{1}{4}$

**4.** $-\frac{1}{3} \cdot 5\frac{2}{5}$

**5.** $20\left(-2\frac{5}{6}\right)$

**6.** $6\frac{3}{8} \cdot 4\frac{1}{3}$

**7.** $3\frac{1}{2} \cdot 4\frac{2}{3}$

**8.** $4\frac{1}{5} \cdot 5\frac{1}{4}$

**9.** $-2\frac{3}{10} \cdot 4\frac{2}{5}$

**10.** $4\frac{7}{10} \cdot 5\frac{3}{10}$

**11.** $\left(-6\frac{3}{10}\right)\left(-5\frac{7}{10}\right)$

**12.** $-20\frac{1}{2} \cdot \left(-10\frac{1}{5}\right)$

**b** Divide. Write a mixed numeral for each answer whenever possible.

**13.** $20 \div 3\frac{1}{5}$

**14.** $18 \div 2\frac{1}{4}$

**15.** $8\frac{2}{5} \div 7$

**16.** $3\frac{3}{8} \div 3$

**17.** $6\frac{1}{4} \div 3\frac{3}{4}$

**18.** $5\frac{4}{5} \div 2\frac{1}{2}$

**19.** $-1\frac{7}{8} \div 1\frac{2}{3}$

**20.** $-4\frac{3}{8} \div 2\frac{5}{6}$

**21.** $5\frac{1}{10} \div 4\frac{3}{10}$

**22.** $4\frac{1}{10} \div 2\frac{1}{10}$

**23.** $-20\frac{1}{4} \div (-90)$

**24.** $-12\frac{1}{2} \div (-50)$

**c** Evaluate.

**25.** $lw$, for $l = 2\frac{3}{5}$ and $w = 9$

**26.** $mv$, for $m = 7$ and $v = 3\frac{2}{5}$

**27.** $rs$, for $r = 5$ and $s = 3\frac{1}{7}$

**28.** $rt$, for $r = 5\frac{2}{3}$ and $t = 4\frac{1}{5}$

**29.** $mt$, for $m = 6\frac{2}{9}$ and $t = -4\frac{3}{8}$

**30.** $M \div NP$, for $M = 2\frac{1}{4}$, $N = -5$, and $P = 2\frac{1}{3}$

**31.** $R \cdot S \div T$, for $R = 4\frac{2}{3}$, $S = 1\frac{3}{7}$, and $T = -5$

**32.** $a - bc$, for $a = 18$, $b = 2\frac{1}{5}$, and $c = 3\frac{3}{4}$

**33.** $r + ps$, for $r = 5\frac{1}{2}$, $p = 3$, and $s = 2\frac{1}{4}$

**34.** $s + rt$, for $s = 3\frac{1}{2}$, $r = 5\frac{1}{2}$, and $t = 7\frac{1}{2}$

**35.** $m + n \div p$, for $m = 7\frac{2}{5}$, $n = 4\frac{1}{2}$, and $p = 6$

**36.** $x - y \div z$, for $x = 9$, $y = 2\frac{1}{2}$, and $z = 3\frac{3}{4}$

**d** Solve.

**37.** *Longest Tunnels.* The Gotthard Base Tunnel in Switzerland is the longest rail tunnel in the world. It is about $2\frac{1}{3}$ times as long as the Laerdal Tunnel in Norway, which is the longest road tunnel in the world. If the Laerdal Tunnel is 15 mi long, how long is the Gotthard Base Tunnel?

**38.** *Spreading Grass Seed.* Emily seeds lawns for Sam's Superior Lawn Care. When she walks at a rapid pace, the wheel on the broadcast spreader completes $150\frac{2}{3}$ revolutions per minute. How many revolutions does the wheel complete in 15 min?

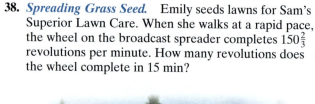

**39.** *Coffee Consumption.* On average, coffee drinkers in the United States drink $3\frac{1}{5}$ cups of coffee a day. At Northwest High School, there are 45 teachers who drink coffee. If their average coffee consumption is the same as the national consumption, how many cups of coffee do they drink each day?

**Data:** e-importz.com

**40.** *Population.* The population of Michigan is $1\frac{1}{2}$ times the population of Indiana. The population of Indiana is approximately 6,600,000. What is the population of Michigan?

**Data:** U.S. Census Bureau

**41. *Apple Net Income.*** Apple, Inc., reported net income of about $8,000,000,000 in fiscal year 2009. In fiscal year 2016, Apple's net income was about $5\frac{3}{4}$ times that amount. What was Apple's net income for fiscal year 2016?

**Data:** statista.com

**42. *Median Income.*** Median household income in the United States was about $12,000 in 1975. By 2016, median household income was $4\frac{2}{3}$ times that amount. What was the median household income in 2016?

**Data:** U.S. Census Bureau

**43. *Average Speed in Indianapolis 500.*** Tony Kanaan won the Indianapolis 500 in 2013 with a record average speed of about $187\frac{1}{2}$ mph. This record is $2\frac{1}{2}$ times the average speed of the first winner, Ray Harroun, in 1911. What was the average speed in the first Indianapolis 500?

**Data:** Indianapolis Motor Speedway

**44. *Population.*** The population of Cleveland is about $1\frac{1}{3}$ times the population of Cincinnati. In 2016, the population of Cleveland was approximately 385,800. What was the population of Cincinnati in 2016?

**Data:** U.S. Census Bureau

**45. *Sidewalk.*** A sidewalk alongside a garden at the conservatory is to be $14\frac{2}{5}$ yd long. Rectangular stone tiles that are each $1\frac{1}{8}$ yd long are used to form the sidewalk. How many tiles are used?

**46. *Aeronautics.*** Most space shuttles orbit the earth once every $1\frac{1}{2}$ hr. How many orbits are made every 24 hr?

**47. *Doubling a Recipe.*** The chef of a five-star hotel is doubling a recipe for chocolate cake. The original recipe requires $2\frac{3}{4}$ cups of flour and $1\frac{1}{3}$ cups of sugar. How much flour and sugar will she need?

**48. *Half of a Recipe.*** A caterer is following a salad dressing recipe that calls for $1\frac{7}{8}$ cups of mayonnaise and $1\frac{1}{6}$ cups of sugar. How much mayonnaise and sugar will he need if he prepares $\frac{1}{2}$ of the amount of salad dressing?

**49. *Mileage.*** A car traveled 213 mi on $14\frac{2}{10}$ gal of gas. How many miles per gallon did it get?

**50. *Mileage.*** A car traveled 385 mi on $15\frac{4}{10}$ gal of gas. How many miles per gallon did it get?

**51. *Mural.*** A student artist painted a mural on the wall under a bridge. The dimensions of the mural are $6\frac{2}{3}$ ft by $9\frac{3}{8}$ ft. What is the area of the mural?

**52. *Weight of Water.*** The weight of water is $62\frac{1}{2}$ lb per cubic foot. What is the weight of $2\frac{1}{4}$ cubic feet of water?

**53. Weight of Water.** The weight of water is $62\frac{1}{2}$ lb per cubic foot. How many cubic feet would be occupied by 25,000 lb of water?

**54. Weight of Water.** The weight of water is $8\frac{1}{3}$ lb per gallon. Harry rolls his lawn with an 800-lb capacity roller. Express the water capacity of the roller in gallons.

**55. Servings of Salmon.** A serving of filleted fish is generally considered to be about $\frac{1}{3}$ lb. How many servings can be prepared from $5\frac{1}{2}$ lb of salmon fillet?

**56. Servings of Tuna.** A serving of fish steak (cross section) is generally $\frac{1}{2}$ lb. How many servings can be prepared from a cleaned $18\frac{3}{4}$-lb tuna?

**57. Landscaping.** The previous owners of Ashley's new home had a large L-shaped vegetable garden consisting of a rectangle that was $15\frac{1}{2}$ ft by 20 ft adjacent to one that was $10\frac{1}{2}$ ft by $12\frac{1}{2}$ ft. Ashley wants to cover the garden with sod. What is the total area of the sod she must purchase?

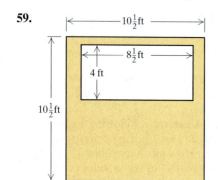

**58. Home Furnishings.** An L-shaped sunroom consists of a rectangle that is $9\frac{1}{2}$ ft by 12 ft adjacent to one that is $9\frac{1}{2}$ ft by 8 ft. What is the total area of a carpet that covers the floor?

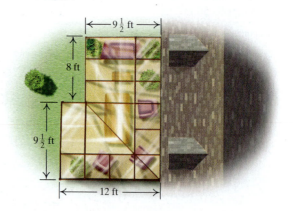

Find the area of each shaded region.

**59.**

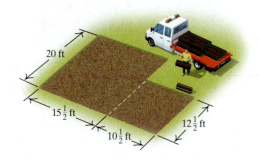

**60.**

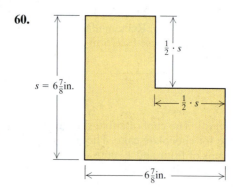

**61.**

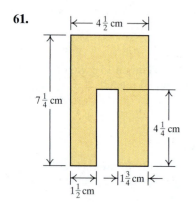

**62.**

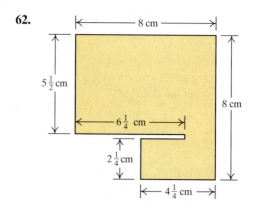

**63. *Temperature.*** Fahrenheit temperature can be obtained from Celsius (Centigrade) temperature by multiplying by $1\frac{4}{5}$ and adding $32°$. What Fahrenheit temperature corresponds to a Celsius temperature of $20°$?

**64. *Temperature.*** Fahrenheit temperature can be obtained from Celsius (Centigrade) temperature by multiplying by $1\frac{4}{5}$ and adding $32°$. What Fahrenheit temperature corresponds to the Celsius temperature of boiling water, $100°$?

**65. *Building a Ziggurat.*** The dimensions of all of the square bricks that King Nebuchadnezzar used over 2500 years ago to build ziggurats were $13\frac{1}{4}$ in. $\times$ $13\frac{1}{4}$ in. $\times$ $3\frac{1}{4}$ in. What are the perimeter and the area of the $13\frac{1}{4}$ in. $\times$ $13\frac{1}{4}$ in. side? of the $13\frac{1}{4}$ in. $\times$ $3\frac{1}{4}$ in. side?

**Data:** eartharchitecture.org

**66. *Word Processing.*** For David's design report, he needs to create a table containing two columns, each $1\frac{1}{2}$ in. wide, and five columns, each $\frac{3}{4}$ in. wide. Will this table fit on a piece of standard paper that is $8\frac{1}{2}$ in. wide? If so, how wide will each side margin be if the margins on each side are to be of equal width?

## Skill Maintenance

**67.** Find the perimeter of the figure. [1.2b]

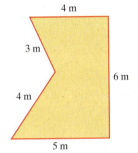

**68.** Find the area of the region. [1.4b]

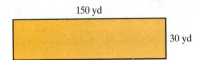

## Synthesis

Simplify. Write each answer as a mixed numeral whenever possible.

**69.** $-8 \div \frac{1}{2} + \frac{3}{4} + \left(-5 - \frac{5}{8}\right)^2$

**70.** $\left(\frac{5}{9} - \frac{1}{4}\right)(-12) + \left(-4 - \frac{3}{4}\right)^2$

**71.** $\frac{1}{3} \div \left(\frac{1}{2} - \frac{1}{5}\right) \times \frac{1}{4} + \frac{1}{6}$

**72.** $\frac{7}{8} - 1\frac{1}{8} \times \frac{2}{3} + \frac{9}{10} \div \frac{3}{5}$

**73.** Find $r$ if

$$\frac{1}{r} = \frac{1}{40} + \frac{1}{60} + \frac{1}{80}.$$

**74. *Heights.*** Find the average height of the following NBA players:

| | |
|---|---|
| LeBron James | 6 ft 8 in. |
| Dwayne Wade | 6 ft 4 in. |
| Dwight Howard | 6 ft 11 in. |
| Dirk Nowitzki | 7 ft 0 in. |
| Al Jefferson | 6 ft 10 in. |

**75. *Water Consumption.*** According to the U.S. Department of Energy, washing one load of clothes uses $2\frac{2}{3}$ times the amount of hot water required for the average shower. If the average shower uses 12 gal of hot water, how much hot water will two showers and two loads of wash require?

## Order of Operations and Complex Fractions

### OBJECTIVES

**a** Simplify expressions containing fraction notation using the rules for order of operations.

**b** Simplify complex fractions.

## a ORDER OF OPERATIONS; FRACTION NOTATION AND MIXED NUMERALS

Like expressions containing integers, expressions containing fraction notation follow the rules for order of operations.

**RULES FOR ORDER OF OPERATIONS**

1. Do all calculations within grouping symbols before operations outside.
2. Evaluate all exponential expressions.
3. Do all multiplications and divisions in order from left to right.
4. Do all additions and subtractions in order from left to right.

Simplify.

**1.** $\dfrac{2}{5} \cdot \dfrac{5}{8} + \dfrac{1}{4}$

**2.** $\dfrac{1}{3} \cdot \dfrac{3}{4} \div \dfrac{5}{8} - \dfrac{1}{10}$

**GS**

$= \dfrac{\square}{12} \div \dfrac{5}{8} - \dfrac{1}{10} = \dfrac{3}{12} \cdot \dfrac{\square}{5} - \dfrac{1}{10}$

$= \dfrac{3 \cdot 2 \cdot 2 \cdot \square}{3 \cdot 2 \cdot 2 \cdot 5} - \dfrac{1}{10} = \dfrac{\square}{5} - \dfrac{1}{10}$

$= \dfrac{\square}{10} - \dfrac{1}{10} = \dfrac{\square}{10}$

**EXAMPLE 1** Simplify: $\dfrac{1}{6} + \dfrac{2}{3} \div \dfrac{1}{2} \cdot \dfrac{5}{8}$.

$\dfrac{1}{6} + \dfrac{2}{3} \div \dfrac{1}{2} \cdot \dfrac{5}{8} = \dfrac{1}{6} + \dfrac{2}{3} \cdot \dfrac{2}{1} \cdot \dfrac{5}{8}$  
Doing the division first by multiplying by the reciprocal of $\frac{1}{2}$

$= \dfrac{1}{6} + \dfrac{2 \cdot 2 \cdot 5}{3 \cdot 1 \cdot 8}$  Doing the multiplications

$= \dfrac{1}{6} + \dfrac{2 \cdot 2 \cdot 5}{3 \cdot 1 \cdot 2 \cdot 2 \cdot 2}$  Factoring

$= \dfrac{1}{6} + \dfrac{5}{6}$  Removing a factor equal to 1: $\dfrac{2 \cdot 2}{2 \cdot 2} = 1$; simplifying

$= \dfrac{6}{6}$, or 1  Adding

◀ Do Exercises 1 and 2.

**EXAMPLE 2** Simplify: $\dfrac{2}{3} \cdot 24 - 11\dfrac{1}{2}$.

$\dfrac{2}{3} \cdot 24 - 11\dfrac{1}{2} = \dfrac{2 \cdot 24}{3 \cdot 1} - 11\dfrac{1}{2}$  Doing the multiplication first

$= \dfrac{2 \cdot 3 \cdot 8}{3 \cdot 1} - 11\dfrac{1}{2}$  Factoring

$= 2 \cdot 8 - 11\dfrac{1}{2}$  Removing a factor equal to 1: $\dfrac{3}{3} = 1$

$= 16 - 11\dfrac{1}{2}$  Multiplying

$= 4\dfrac{1}{2}$, or $\dfrac{9}{2}$  Subtracting

**3.** Simplify: $\dfrac{3}{4} \cdot 16 + 8\dfrac{2}{3}$.

◀ Do Exercise 3.

**Answers**

**1.** $\dfrac{1}{2}$  **2.** $\dfrac{3}{10}$  **3.** $20\dfrac{2}{3}$, or $\dfrac{62}{3}$

**Guided Solution:**
**2.** 3, 8, 2, 2, 4, 3

**EXAMPLE 3** Simplify.

**a)** $-\dfrac{1}{2} - \dfrac{2}{3}\left(-\dfrac{1}{2}\right)^2$

**b)** $1 - \dfrac{2}{3}\left(\dfrac{1}{3} - \dfrac{1}{2}\right)$

**a)** The parentheses here are not grouping symbols, so we begin by evaluating the exponential expression.

$$-\frac{1}{2} - \frac{2}{3}\left(-\frac{1}{2}\right)^2 = -\frac{1}{2} - \frac{2}{3}\cdot\frac{1}{4} \qquad \left(-\frac{1}{2}\right)^2 = \left(-\frac{1}{2}\right)\cdot\left(-\frac{1}{2}\right) = \frac{1}{4}$$

$$= -\frac{1}{2} - \frac{2\cdot 1}{3\cdot 2\cdot 2} \qquad \text{Multiplying and factoring}$$

$$= -\frac{1}{2} - \frac{1}{6} \qquad \text{Simplifying}$$

$$= -\frac{3}{6} - \frac{1}{6} \qquad \text{Writing with the LCD, 6, in order to subtract}$$

$$= -\frac{4}{6} = -\frac{2}{3} \qquad \text{Subtracting and simplifying}$$

**b)** We first subtract within the parentheses.

$$1 - \frac{2}{3}\left(\frac{1}{3} - \frac{1}{2}\right) = 1 - \frac{2}{3}\left(\frac{2}{6} - \frac{3}{6}\right) \qquad \text{Writing with the LCD, 6, in order to subtract}$$

$$= 1 - \frac{2}{3}\left(-\frac{1}{6}\right) \qquad \begin{array}{l}\text{Subtracting within} \\ \text{the parentheses:} \\ \dfrac{2}{6} - \dfrac{3}{6} = \dfrac{2}{6} + \left(-\dfrac{3}{6}\right) = -\dfrac{1}{6}\end{array}$$

$$= 1 - \left(-\frac{2\cdot 1}{3\cdot 2\cdot 3}\right) \qquad \begin{array}{l}\text{There are no exponential} \\ \text{expressions, so we multiply and} \\ \text{factor.}\end{array}$$

$$= 1 - \left(-\frac{1}{9}\right) \qquad \text{Simplifying}$$

$$= \frac{9}{9} + \frac{1}{9} = \frac{10}{9} \qquad \begin{array}{l}\text{Writing with the LCD, 9, and} \\ \text{adding the opposite of } -\dfrac{1}{9}\end{array}$$

**Do Exercises 4–6.** ▶

Simplify.

**4.** $\left(\dfrac{3}{4}\right)^2 - \dfrac{1}{2} \div \left(-\dfrac{4}{5}\right)$

**5.** $1 - \left(\dfrac{2}{3} - \dfrac{3}{4}\right)^2$

**6.** $\left(\dfrac{2}{3} + \dfrac{3}{4}\right) \div 2\dfrac{1}{3} - \left(\dfrac{1}{2}\right)^3$

## b  COMPLEX FRACTIONS

A **complex fraction** is a fraction in which the numerator and/or the denominator contain one or more fractions. The following are some examples of complex fractions.

$\dfrac{\frac{2}{3}}{7}$  ← The numerator contains a fraction.

$\dfrac{-\frac{1}{5}}{-\frac{9}{10}}$  ← The numerator contains a fraction.
← The denominator contains a fraction.

Since a fraction bar represents division, complex fractions can be rewritten using the division symbol ÷.

**EXAMPLE 4** Simplify.

a) $\dfrac{\frac{2}{3}}{7}$

b) $\dfrac{-\frac{1}{5}}{-\frac{9}{10}}$

a)
$$\dfrac{\frac{2}{3}}{7} = \frac{2}{3} \div 7 \qquad \text{Rewriting using a division symbol}$$

$$= \frac{2}{3} \cdot \left(\frac{1}{7}\right) \qquad \text{Multiplying by the reciprocal of the divisor}$$

$$= \frac{2 \cdot 1}{3 \cdot 7} \qquad \text{Multiplying numerators and multiplying denominators}$$

$$= \frac{2}{21} \qquad \text{This expression cannot be simplified.}$$

b)
$$\dfrac{-\frac{1}{5}}{-\frac{9}{10}} = -\frac{1}{5} \div \left(-\frac{9}{10}\right) \qquad \text{Rewriting using a division symbol}$$

$$= -\frac{1}{5} \cdot \left(-\frac{10}{9}\right) \qquad \text{Multiplying by the reciprocal of the divisor; the product will be positive.}$$

$$= \frac{1 \cdot 2 \cdot \cancel{5}}{\cancel{5} \cdot 3 \cdot 3} \qquad \text{Multiplying numerators and multiplying denominators; factoring}$$

$$= \frac{2}{9} \qquad \text{Removing a factor equal to 1 and simplifying}$$

◀ **Do Exercises 7 and 8.**

Complex fractions may contain variables.

**EXAMPLE 5** Simplify: $\dfrac{\frac{x}{24}}{\frac{15}{28}}$.

$$\dfrac{\frac{x}{24}}{\frac{15}{28}} = \frac{x}{24} \div \frac{15}{28} \qquad \text{Rewriting using a division symbol}$$

$$= \frac{x}{24} \cdot \frac{28}{15} \qquad \text{Multiplying by the reciprocal of the divisor}$$

$$= \frac{x \cdot 2 \cdot 2 \cdot 7}{2 \cdot 2 \cdot 2 \cdot 3 \cdot 3 \cdot 5} \qquad \text{Multiplying numerators and multiplying denominators; factoring}$$

$$= \frac{x \cdot 7}{2 \cdot 3 \cdot 3 \cdot 5} \qquad \text{Removing a factor equal to 1: } \frac{2 \cdot 2}{2 \cdot 2} = 1$$

$$= \frac{7x}{90} \qquad \text{Multiplying}$$

◀ **Do Exercises 9 and 10.**

Simplify.

**7.** $\dfrac{\frac{10}{5}}{\frac{5}{8}} = 10 \div \dfrac{\boxed{\phantom{0}}}{\boxed{\phantom{0}}}$

$$= 10 \cdot \dfrac{\boxed{\phantom{0}}}{\phantom{0}}$$

$$= \dfrac{10 \cdot \boxed{\phantom{0}}}{5}$$

$$= \dfrac{\boxed{\phantom{0}} \cdot \cancel{5} \cdot 8}{\cancel{5} \cdot 1}$$

$$= \boxed{\phantom{0}}$$

**8.** $\dfrac{\frac{7}{5}}{-\frac{10}{7}}$

Simplify.

**9.** $\dfrac{\frac{x}{20}}{\frac{3}{10}}$

**10.** $\dfrac{-\frac{7}{12}}{-\frac{x}{18}}$

**Answers**

**7.** 16  **8.** $-\dfrac{49}{50}$  **9.** $\dfrac{x}{6}$  **10.** $\dfrac{21}{2x}$

**Guided Solution:**

**7.** $\dfrac{5}{8}, \dfrac{8}{5}, 8, 2, 16$

When the numerator or denominator of a complex fraction consists of more than one term, first simplify the numerator and/or denominator separately.

**EXAMPLE 6** Simplify: $\dfrac{\frac{1}{2} - \frac{2}{3}}{1\frac{7}{8}}$.

$\dfrac{\frac{1}{2} - \frac{2}{3}}{1\frac{7}{8}} = \dfrac{\frac{3}{6} - \frac{4}{6}}{\frac{15}{8}}$    Writing the fractions in the numerator with a common denominator

Writing the mixed numeral in the denominator as a fraction

$= \dfrac{-\frac{1}{6}}{\frac{15}{8}}$    Subtracting in the numerator of the complex fraction

$= -\dfrac{1}{6} \div \dfrac{15}{8}$    Rewriting using a division symbol

$= -\dfrac{1}{6} \cdot \dfrac{8}{15}$    Multiplying by the reciprocal of the divisor

$= -\dfrac{1 \cdot 2 \cdot 2 \cdot 2}{2 \cdot 3 \cdot 3 \cdot 5}$    Multiplying numerators and multiplying denominators; factoring

$= -\dfrac{4}{45}$    Removing a factor equal to 1: $\dfrac{2}{2} = 1$

**Do Exercises 11 and 12.** ▶

Simplify.

**11.** $\dfrac{\frac{7}{12} + \frac{5}{6}}{\frac{4}{9}}$

**12.** $\dfrac{-\frac{3}{5}}{\frac{2}{3} - \frac{7}{10}}$

**EXAMPLE 7** *Harvesting Walnut Trees.* A woodland owner decided to harvest five walnut trees in order to improve the growing conditions of the remaining trees. The logs she sold measured $7\frac{5}{8}$ ft, $8\frac{1}{4}$ ft, $8\frac{3}{4}$ ft, $9\frac{1}{8}$ ft, and $10\frac{1}{2}$ ft. What is the average length of the logs?

Recall that to compute an average, we add the numbers and then divide the sum by the number of addends. We have

$\dfrac{7\frac{5}{8} + 8\frac{1}{4} + 8\frac{3}{4} + 9\frac{1}{8} + 10\frac{1}{2}}{5} = \dfrac{7\frac{5}{8} + 8\frac{2}{8} + 8\frac{6}{8} + 9\frac{1}{8} + 10\frac{4}{8}}{5}$

$= \dfrac{42\frac{18}{8}}{5} = \dfrac{42\frac{9}{4}}{5} = \dfrac{\frac{177}{4}}{5}$    Adding, simplifying, and converting to fraction notation

$= \dfrac{177}{4} \div 5$    Rewriting using a division symbol

$= \dfrac{177}{4} \cdot \dfrac{1}{5}$    Multiplying by the reciprocal of 5

$= \dfrac{177}{20} = 8\frac{17}{20}.$    Converting to a mixed numeral

**13.** Rachel has triplets. Their birth weights are $3\frac{1}{2}$ lb, $2\frac{3}{4}$ lb, and $3\frac{1}{8}$ lb. What is the average birth weight of her babies?

**14.** Find the average of $\dfrac{1}{2}$, $\dfrac{1}{3}$, and $\dfrac{5}{6}$.

**15.** Find the average of $\dfrac{3}{4}$ and $\dfrac{4}{5}$.

The average length of the logs is $8\frac{17}{20}$ ft.

**Do Exercises 13–15.** ▶

*Answers*

**11.** $\dfrac{51}{16}$, or $3\dfrac{3}{16}$   **12.** 18   **13.** $3\dfrac{1}{8}$ lb
**14.** $\dfrac{5}{9}$   **15.** $\dfrac{31}{40}$

✓ **Check Your Understanding**

**Reading Check** Match the beginning of each statement with the correct ending from the list at the right so that the rules for order of operations are listed in the correct order.

**RC1.** Do all _____.

**RC2.** Evaluate all _____.

**RC3.** Do all _____.

**RC4.** Do all _____.

**a)** multiplications and divisions in order from left to right.

**b)** additions and subtractions in order from left to right.

**c)** calculations within grouping symbols.

**d)** exponential expressions.

**Concept Check** Rewrite each complex fraction using a division symbol, ÷, and fraction notation.

**CC1.** $\dfrac{\frac{2}{3}}{\frac{1}{6}}$

**CC2.** $\dfrac{10}{\frac{3}{8}}$

**CC3.** $\dfrac{\frac{1}{4}}{4}$

**CC4.** $\dfrac{3\frac{1}{5}}{4\frac{1}{2}}$

**a** Simplify.

**1.** $\dfrac{1}{8} + \dfrac{1}{4} \cdot \dfrac{2}{3}$

**2.** $\dfrac{2}{5} - \dfrac{4}{5} \div \dfrac{2}{3}$

**3.** $-\dfrac{1}{6} - 3\left(-\dfrac{5}{9}\right)$

**4.** $1 + \dfrac{2}{3}\left(-\dfrac{6}{25}\right)$

**5.** $\dfrac{9}{10} - \left(\dfrac{2}{5} - \dfrac{3}{8}\right)$

**6.** $-\dfrac{5}{9} + \left(\dfrac{1}{3} - \dfrac{5}{6}\right)$

**7.** $\dfrac{5}{8} \div \dfrac{1}{4} - \dfrac{2}{3} \cdot \dfrac{4}{5}$

**8.** $\dfrac{4}{7} \cdot \dfrac{7}{15} + \dfrac{2}{3} \div 8$

**9.** $\dfrac{7}{8} \div \dfrac{1}{2} \cdot \dfrac{1}{4}$

**10.** $\dfrac{7}{10} \cdot \dfrac{4}{5} \div \dfrac{2}{3}$

**11.** $\dfrac{3}{4} - \dfrac{2}{3} \cdot \left(\dfrac{1}{2} + \dfrac{2}{5}\right)$

**12.** $\dfrac{3}{4} \div \dfrac{1}{2} \cdot \left(\dfrac{8}{9} - \dfrac{2}{3}\right)$

**13.** $\dfrac{4}{5} \div \left(\dfrac{2}{9} \cdot \dfrac{1}{2}\right) \cdot \left(-\dfrac{5}{6}\right)$

**14.** $-\dfrac{4}{9} \cdot \left(\dfrac{3}{8} \div \dfrac{1}{2}\right) \div \left(-\dfrac{2}{3}\right)$

**15.** $\left(\dfrac{2}{3}\right)^2 - \dfrac{1}{3} \cdot 1\dfrac{1}{4}$

**16.** $-1\dfrac{3}{5} - \dfrac{9}{10} + \left(-\dfrac{1}{2}\right)^2$

**17.** $-\dfrac{12}{25}\left(\dfrac{3}{4} - \dfrac{1}{2}\right)^2$

**18.** $\dfrac{2}{3}\left(\dfrac{1}{3} - \dfrac{1}{5}\right)^2$

**19.** $-\dfrac{3}{4} \div \left(\dfrac{2}{3} - \dfrac{1}{6}\right) + \dfrac{1}{2}$

**20.** $-6 \div \left(\dfrac{1}{2} - \dfrac{2}{3}\right) + \dfrac{1}{3}$

**21.** $\left(-\dfrac{3}{2}\right)^2 - 2\left(\dfrac{1}{4} - \dfrac{3}{2}\right)$

**22.** $\left(-\dfrac{1}{2}\right)^3 - \dfrac{3}{4}\left(\dfrac{1}{3} - \dfrac{1}{6}\right)$

**23.** $\dfrac{1}{2} - \left(\dfrac{1}{2}\right)^2 + \left(\dfrac{1}{2}\right)^3$

**24.** $1 + \dfrac{1}{4} + \left(\dfrac{1}{4}\right)^2 - \left(\dfrac{1}{4}\right)^3$

**25.** $\left(\dfrac{3}{5} - \dfrac{1}{2}\right) \div \left(\dfrac{3}{4} - \dfrac{3}{10}\right)$

**26.** $\left(\dfrac{2}{3} + \dfrac{3}{4}\right) \div \left(\dfrac{5}{6} - \dfrac{1}{3}\right)$

**b** Simplify.

**27.** $\dfrac{\frac{3}{8}}{\frac{11}{8}}$

**28.** $\dfrac{-\frac{1}{8}}{\frac{3}{4}}$

**29.** $\dfrac{\frac{-4}{6}}{7}$

**30.** $\dfrac{-\frac{3}{8}}{-12}$

**31.** $\dfrac{\frac{1}{40}}{-\frac{1}{50}}$

**32.** $\dfrac{\frac{7}{9}}{\frac{3}{9}}$

**33.** $\dfrac{-\frac{1}{10}}{-10}$

**34.** $\dfrac{\frac{28}{7}}{-\frac{7}{4}}$

**35.** $\dfrac{\frac{5}{18}}{-1\frac{2}{3}}$

**36.** $\dfrac{-2\frac{1}{5}}{\frac{7}{10}}$

**37.** $\dfrac{\frac{x}{28}}{\frac{5}{8}}$

**38.** $\dfrac{\frac{x}{15}}{\frac{3}{4}}$

**39.** $\dfrac{\frac{n}{14}}{-\frac{2}{3}}$

**40.** $\dfrac{-\frac{t}{10}}{\frac{4}{45}}$

**41.** $\dfrac{-\frac{3}{35}}{-\frac{x}{10}}$

**42.** $\dfrac{\frac{7}{40}}{-\frac{6}{x}}$

**43.** $\dfrac{-\frac{5}{8}}{\left(\frac{3}{2}\right)^2}$

**44.** $\dfrac{\left(-\frac{2}{3}\right)^2}{\left(\frac{9}{10}\right)^2}$

**45.** $\dfrac{\frac{1}{6} - \frac{5}{9}}{\frac{2}{3}}$

**46.** $\dfrac{\frac{7}{12}}{\frac{1}{4} - \frac{5}{8}}$

**47.** $\dfrac{\frac{1}{4} - \frac{3}{8}}{\frac{1}{2} - \frac{7}{8}}$

**48.** $\dfrac{\frac{1}{2} - \frac{3}{5}}{\frac{2}{5} - \frac{1}{2}}$

**49.** Find the average of $\dfrac{2}{3}$ and $\dfrac{7}{8}$.

**50.** Find the average of $\dfrac{1}{4}$ and $\dfrac{1}{5}$.

**51.** Find the average of $\dfrac{1}{6}, \dfrac{1}{8}$, and $\dfrac{3}{4}$.

**52.** Find the average of $\dfrac{4}{5}, \dfrac{1}{2}$, and $\dfrac{1}{10}$.

**53.** Find the average of $3\dfrac{1}{2}$ and $9\dfrac{3}{8}$.

**54.** Find the average of $10\dfrac{2}{3}$ and $24\dfrac{5}{6}$.

**55.** *Hiking the Appalachian Trail.* Ellen camped and hiked for three consecutive days along a section of the Appalachian Trail. The distances she hiked on the three days were $15\frac{5}{32}$ mi, $20\frac{3}{16}$ mi, and $12\frac{7}{8}$ mi. Find the average of these distances.

**56.** *Vertical Leaps.* Eight-year-old Zachary registered vertical leaps of $12\frac{3}{4}$ in., $13\frac{3}{4}$ in., $13\frac{1}{2}$ in., and 14 in. Find his average vertical leap.

**57. Black Bear Cubs.** Black bears typically have two cubs. In January 2007 in northern New Hampshire, a black bear sow gave birth to a litter of 5 cubs. This is so rare that Tom Sears, a wildlife photographer, spent 28 hr per week for six weeks watching for the perfect opportunity to photograph this family of six. At the time of this photo, an observer estimated that the cubs weighed $7\frac{1}{2}$ lb, 8 lb, $9\frac{1}{2}$ lb, $10\frac{5}{8}$ lb, and $11\frac{3}{4}$ lb. What was the average weight of the cubs?

**Data:** Andrew Timmins, New Hampshire Fish and Game Department, *Northcountry News*, Warren, NH; Tom Sears, photographer

**58. Acceleration.** The results of acceleration tests for five cars are given in the graph below. The test measures the time in seconds required to go from 0 mph to 60 mph. What was the average time?

**Acceleration: 0 mph to 60 mph**

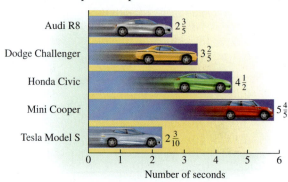

DATA: 0–60specs.com

## Skill Maintenance

Simplify.

**59.** $12 + 30 \div 3 - 2$ [1.9c]

**60.** $5 \cdot 2^2 \div 10$ [1.9c]

**61.** $10^2 - [3 \cdot 2^4 \div (10 - 2) + 5 \cdot 2]$ [1.9d]

**62.** $(10 + 3 \cdot 4 \div 6)^2 - 11 \cdot 2^2$ [1.9c]

**63.** List all the factors of 42. [3.2a]

**64.** Determine whether 114 is divisible by 7. [3.1a]

**65.** Classify the given numbers as prime, composite, or neither. [3.2b]

  1, 5, 7, 9, 14, 23, 43

**66.** Find the prime factorization of 150. [3.2c]

## Synthesis

Simplify.

**67.** $\left(1\frac{1}{2} - 1\frac{1}{3}\right)^2 \cdot 144 - \frac{9}{10} \div 4\frac{1}{5}$

**68.** 🖩 $\left(3\frac{1}{2} - 2\frac{1}{3}\right)^3 - 30 \cdot 2\frac{1}{2} \div (-2)^5$

**69.** $\dfrac{\frac{2}{3}x - \frac{1}{2}x}{\left(1\frac{1}{4} + \frac{1}{2}\right)^2}$

**70.** $\dfrac{\frac{x}{24}}{\frac{x}{48}}$

Estimate each of the following as $0, \frac{1}{2}$, or 1.

**71.** $\frac{2}{99}$

**72.** $\frac{19}{20}$

**73.** $\frac{13}{27}$

**74.** $\frac{101}{100}$

**75.** $\frac{215}{429}$

**76.** $\frac{1}{1000}$

## Vocabulary Reinforcement

Complete each statement with the correct term from the list to the right. Some of the choices may not be used and some may be used more than once.

1. The _____ of two numbers is the smallest number that is a multiple of both numbers. [4.1a]

2. A _____ represents a sum of a whole number and a fraction less than 1. [4.5a]

3. To multiply using mixed numerals, we first convert to _____ notation. [4.7a]

4. A _____ contains a fraction in its numerator and/or denominator. [4.8b]

5. To add fractions, the _____ of the fractions must be the same. [4.2a]

6. The least common denominator of two fractions is the _____ of the denominators of the fractions. [4.2b]

7. When finding the LCM of a set of numbers using prime factorizations, we use each prime number the _____ number of times that it appears in any one factorization. [4.1a]

8. To compare two fractions with a common denominator, we compare their _____ . [4.2c]

greatest

least

numerators

denominators

fraction

decimal

mixed numeral

complex fraction

least common multiple

greatest common factor

## Concept Reinforcement

Determine whether each statement is true or false.

_____ 1. The mixed numeral $5\frac{2}{3}$ can be represented by the sum $5 \cdot \frac{3}{3} + \frac{2}{3}$. [4.5a]

_____ 2. The least common multiple of two natural numbers is always larger than or equal to the larger number. [4.1a]

_____ 3. To clear fractions in an equation, multiply both sides by the LCM of all denominators in the equation. [4.4b]

_____ 4. The sum of any two mixed numerals is a mixed numeral. [4.6a]

## Study Guide

**Objective 4.1a** Find the least common multiple, or LCM, of two or more numbers.

**Example** Find the LCM of 105 and 90.

$$105 = 3 \cdot 5 \cdot 7,$$
$$90 = 2 \cdot 3 \cdot 3 \cdot 5;$$
$$LCM = 2 \cdot 3 \cdot 3 \cdot 5 \cdot 7 = 630$$

**Practice Exercise**

1. Find the LCM of 52 and 78.

**Objective 4.2b**  Add using fraction notation when denominators are different.

**Example**  Add: $\dfrac{5}{24} + \dfrac{7}{45}$.

$$\dfrac{5}{24} + \dfrac{7}{45} = \dfrac{5}{2\cdot2\cdot2\cdot3}\cdot\dfrac{3\cdot5}{3\cdot5} + \dfrac{7}{3\cdot3\cdot5}\cdot\dfrac{2\cdot2\cdot2}{2\cdot2\cdot2}$$

$$= \dfrac{75}{360} + \dfrac{56}{360} = \dfrac{131}{360}$$

**Practice Exercise**

2. Add: $\dfrac{19}{60} + \dfrac{11}{36}$.

---

**Objective 4.2c**  Use $<$ or $>$ to form a true statement with fraction notation.

**Example**  Use $<$ or $>$ for $\square$ to write a true sentence:

$$\dfrac{5}{12}\,\square\,\dfrac{9}{16}.$$

The LCD is 48. Thus, we have

$$\dfrac{5}{12}\cdot\dfrac{4}{4}\,\square\,\dfrac{9}{16}\cdot\dfrac{3}{3}, \quad\text{or}\quad \dfrac{20}{48}\,\square\,\dfrac{27}{48}.$$

Since $20 < 27$, $\dfrac{20}{48} < \dfrac{27}{48}$ and thus $\dfrac{5}{12} < \dfrac{9}{16}$.

**Practice Exercise**

3. Use $<$ or $>$ for $\square$ to write a true sentence:

$$\dfrac{3}{13}\,\square\,\dfrac{5}{12}.$$

---

**Objective 4.3a**  Subtract using fraction notation.

**Example**  Subtract: $\dfrac{7}{12} - \dfrac{11}{60}$.

$$\dfrac{7}{12} - \dfrac{11}{60} = \dfrac{7}{12}\cdot\dfrac{5}{5} - \dfrac{11}{60} = \dfrac{35}{60} - \dfrac{11}{60}$$

$$= \dfrac{35 - 11}{60} = \dfrac{24}{60} = \dfrac{2\cdot12}{5\cdot12} = \dfrac{2}{5}$$

**Practice Exercise**

4. Subtract: $\dfrac{29}{35} - \dfrac{5}{7}$.

---

**Objective 4.4a**  Solve equations that involve fractions and require use of both the addition principle and the multiplication principle.

**Example**  Solve: $\dfrac{1}{2}x + \dfrac{1}{6} = \dfrac{5}{8}$.

$$\dfrac{1}{2}x + \dfrac{1}{6} = \dfrac{5}{8}$$

$$\dfrac{1}{2}x + \dfrac{1}{6} - \dfrac{1}{6} = \dfrac{5}{8} - \dfrac{1}{6}$$

$$\dfrac{1}{2}x = \dfrac{5}{8}\cdot\dfrac{3}{3} - \dfrac{1}{6}\cdot\dfrac{4}{4}$$

$$\dfrac{1}{2}x = \dfrac{15}{24} - \dfrac{4}{24}$$

$$\dfrac{1}{2}x = \dfrac{11}{24}$$

$$\dfrac{2}{1}\left(\dfrac{1}{2}x\right) = \dfrac{2}{1}\left(\dfrac{11}{24}\right)$$

$$x = \dfrac{2\cdot11}{1\cdot2\cdot12} = \dfrac{11}{12}$$

The solution is $\dfrac{11}{12}$.

**Practice Exercise**

5. Solve: $\dfrac{2}{9} + \dfrac{2}{3}x = \dfrac{1}{6}$.

**Objective 4.5a** Convert between mixed numerals and fraction notation.

**Example** Convert $2\frac{5}{13}$ to fraction notation: $2\frac{5}{13} = \frac{31}{13}$.

**Example** Convert $\frac{40}{9}$ to a mixed numeral: $\frac{40}{9} = 4\frac{4}{9}$.

**Practice Exercises**

6. Convert $8\frac{2}{3}$ to fraction notation.

7. Convert $\frac{47}{6}$ to a mixed numeral.

---

**Objective 4.6b** Subtract using mixed numerals.

**Example** Subtract: $3\frac{3}{8} - 1\frac{4}{5}$.

$$3\frac{3}{8} = \quad 3\frac{15}{40} = \quad 2\frac{55}{40}$$

$$\underline{-1\frac{4}{5} = -1\frac{32}{40} = -1\frac{32}{40}}$$

$$1\frac{23}{40}$$

**Practice Exercise**

8. Subtract: $10\frac{5}{7} - 2\frac{3}{4}$.

---

**Objective 4.7a** Multiply using mixed numerals.

**Example** Multiply: $7\frac{1}{4} \cdot 5\frac{3}{10}$. Write a mixed numeral for the answer.

$$7\frac{1}{4} \cdot 5\frac{3}{10} = \frac{29}{4} \cdot \frac{53}{10}$$

$$= \frac{1537}{40} = 38\frac{17}{40}$$

**Practice Exercise**

9. Multiply: $4\frac{1}{5} \cdot 3\frac{7}{15}$.

---

**Objective 4.7d** Solve applied problems involving multiplication and division with mixed numerals.

**Example** The population of Chicago is $1\frac{4}{5}$ times that of Philadelphia. The population of Chicago is approximately 2,700,000. What is the population of Philadelphia?

Translate:

$$\underbrace{\text{Population of Chicago}} \quad \text{is} \quad 1\frac{4}{5} \quad \text{times} \quad \underbrace{\text{Population of Philadelphia}}$$

$$2{,}700{,}000 \quad = \quad 1\frac{4}{5} \quad \cdot \quad x.$$

Solve: $\quad 2{,}700{,}000 = \frac{9}{5} \cdot x$

$$\frac{5}{9} \cdot 2{,}700{,}000 = \frac{5}{9} \cdot \frac{9}{5} \cdot x$$

$$2{,}700{,}000 \cdot \frac{5}{9} = x$$

$$1{,}500{,}000 = x$$

The population of Philadelphia is about 1,500,000.

**Practice Exercise**

10. The population of Louisiana is $2\frac{1}{2}$ times the population of West Virginia. The population of West Virginia is approximately 1,800,000. What is the population of Louisiana?

**Objective 4.8a** Simplify expressions containing fraction notation using the rules for order of operations.

**Example** Simplify: $\left(\dfrac{4}{5}\right)^2 - \dfrac{1}{5} \cdot 2\dfrac{1}{8}$.

$$\left(\dfrac{4}{5}\right)^2 - \dfrac{1}{5} \cdot 2\dfrac{1}{8} = \dfrac{16}{25} - \dfrac{1}{5} \cdot \dfrac{17}{8}$$

$$= \dfrac{16}{25} - \dfrac{17}{40} = \dfrac{16}{25} \cdot \dfrac{8}{8} - \dfrac{17}{40} \cdot \dfrac{5}{5}$$

$$= \dfrac{128}{200} - \dfrac{85}{200} = \dfrac{43}{200}$$

**Practice Exercise**

**11.** Simplify: $\dfrac{3}{2} \cdot 1\dfrac{1}{3} \div \left(\dfrac{2}{3}\right)^2$.

## Review Exercises

Find the LCM.  [4.1a]

**1.** 12 and 18   **2.** 18 and 45

**3.** 3, 6, and 30   **4.** 26, 36, and 54

Perform the indicated operation and, if possible, simplify.  [4.2a, b], [4.3a]

**5.** $\dfrac{2}{9} + \dfrac{5}{9}$   **6.** $\dfrac{7}{x} + \dfrac{2}{x}$

**7.** $-\dfrac{6}{5} + \dfrac{11}{15}$   **8.** $\dfrac{5}{16} + \dfrac{3}{24}$

**9.** $\dfrac{7}{9} - \dfrac{5}{9}$   **10.** $\dfrac{1}{4} - \dfrac{3}{8}$

**11.** $\dfrac{10}{27} - \dfrac{2}{9}$   **12.** $\dfrac{5}{6} - \dfrac{7}{9}$

Use $<$ or $>$ for $\square$ to form a true sentence.  [4.2c]

**13.** $\dfrac{4}{7} \,\square\, \dfrac{5}{9}$   **14.** $-\dfrac{8}{9} \,\square\, -\dfrac{11}{13}$

Solve.  [4.3b], [4.4a]

**15.** $x + \dfrac{2}{5} = \dfrac{7}{8}$   **16.** $\dfrac{1}{2}a - 3 = \dfrac{5}{2}$

**17.** $5 + \dfrac{16}{3}x = \dfrac{5}{9}$   **18.** $\dfrac{22}{5} = \dfrac{16}{5} + \dfrac{5}{2}x$

Solve by using the multiplication principle to clear fractions.  [4.4b]

**19.** $\dfrac{5}{3}x + \dfrac{5}{6} = \dfrac{3}{2}$

Convert to fraction notation.  [4.5a]

**20.** $7\dfrac{1}{2}$   **21.** $8\dfrac{3}{8}$

**22.** $4\dfrac{1}{3}$   **23.** $-1\dfrac{5}{7}$

Convert to a mixed numeral.  [4.5a]

**24.** $\dfrac{7}{3}$   **25.** $\dfrac{-27}{4}$

**26.** $\dfrac{63}{5}$   **27.** $\dfrac{7}{2}$

**28.** Divide. Write a mixed numeral for the answer.
$7896 \div (-9)$  [4.5b]

**29.** Gina's golf scores were 80, 82, and 85. What was her average score?  [4.5b]

Perform the indicated operation. Write a mixed numeral for each answer.  [4.6a, b, d]

**30.** $\begin{array}{r} 7\dfrac{3}{5} \\ + 2\dfrac{4}{5} \\ \hline \end{array}$   **31.** $\begin{array}{r} 6\dfrac{1}{3} \\ + 5\dfrac{2}{5} \\ \hline \end{array}$

**32.** $-3\dfrac{5}{6} + \left(-5\dfrac{1}{6}\right)$   **33.** $-2\dfrac{3}{4} + 4\dfrac{1}{2}$

**34.** $\begin{array}{r} 14 \\ - 6\dfrac{2}{9} \\ \hline \end{array}$   **35.** $\begin{array}{r} 9\dfrac{3}{5} \\ - 4\dfrac{13}{15} \\ \hline \end{array}$

**36.** $4\dfrac{5}{8} - 9\dfrac{3}{4}$   **37.** $-7\dfrac{1}{2} - 6\dfrac{3}{4}$

Combine like terms. [4.2b], [4.6b]

**38.** $\dfrac{4}{9}x + \dfrac{1}{3}x$

**39.** $8\dfrac{3}{10}a - 5\dfrac{1}{8}a$

Perform the indicated operation. Write a mixed numeral or integer for each answer, unless the answer is less than 1. [4.7a, b]

**40.** $6 \cdot 2\dfrac{2}{3}$

**41.** $-5\dfrac{1}{4} \cdot \dfrac{2}{3}$

**42.** $2\dfrac{1}{5} \cdot 1\dfrac{1}{10}$

**43.** $2\dfrac{2}{5} \cdot 2\dfrac{1}{2}$

**44.** $-54 \div 2\dfrac{1}{4}$

**45.** $2\dfrac{2}{5} \div \left(-1\dfrac{7}{10}\right)$

**46.** $3\dfrac{1}{4} \div 26$

**47.** $4\dfrac{1}{5} \div 4\dfrac{2}{3}$

Evaluate. [4.7c]

**48.** $5x - y$, for $x = 3\dfrac{1}{5}$ and $y = 2\dfrac{2}{7}$

**49.** $2a \div b$, for $a = 5\dfrac{2}{11}$ and $b = 3\dfrac{4}{5}$

Solve. [4.6c], [4.7d]

**50.** *Sewing.* Kim wants to make slacks and a jacket. She needs $1\dfrac{5}{8}$ yd of 60-in. fabric for the slacks and $2\dfrac{5}{8}$ yd for the jacket. How many yards in all does Kim need to make the outfit?

**51.** *Party Planning.* The San Diaz drama club had $\dfrac{3}{8}$ of a vegetarian pizza, $1\dfrac{1}{2}$ cheese pizzas, and $1\dfrac{1}{4}$ pepperoni pizzas remaining after a cast party. How many pizzas remained altogether?

**52.** *Turkey Servings.* There are $1\dfrac{1}{3}$ servings per pound in a whole turkey. How many pounds are needed for 32 servings?

**53.** *Fundraisers.* Green River's Humane Society recently hosted its annual dessert social. Each of the 83 pies donated was cut into 6 pieces. At the end of the evening, 382 pieces of pie had been sold. How many pies were sold? How many were left over? Express your answers in mixed numerals.

**54.** *Running.* Janelle has mapped a $1\dfrac{1}{2}$-mi running route in her neighborhood. One Saturday, she ran this route $2\dfrac{1}{2}$ times. How many miles did she run?

**55.** What is the sum of the areas in the figure below?

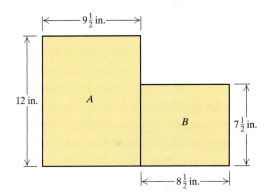

**56.** In the figure above, how much larger is the area of rectangle $A$ than the area of rectangle $B$?

**57.** *Painting a Border.* Katie hired an artist to paint a decorative border around the top of her son's bedroom. The artist charges $20 per foot. The room measures $11\dfrac{3}{4}$ ft $\times$ $9\dfrac{1}{2}$ ft. What is Katie's cost for the project?

$9\dfrac{1}{2}$ ft

$11\dfrac{3}{4}$ ft

**58.** *NCAA Football.* In college football, the distance between goalposts was reduced from $23\frac{1}{3}$ ft to $18\frac{1}{2}$ ft. By how much was it reduced?

**Data:** NCAA

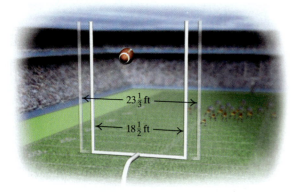

$23\frac{1}{3}$ ft

$18\frac{1}{2}$ ft

Simplify each expression using the rules for order of operations. [4.8a]

**59.** $\dfrac{1}{2} + \dfrac{1}{8} \div \dfrac{1}{4}$

**60.** $\dfrac{4}{5} - \dfrac{1}{2} \cdot \left( \dfrac{1}{4} - \dfrac{3}{5} \right)$

**61.** $20\dfrac{3}{4} - 1\dfrac{1}{2} \times 12 + \left( \dfrac{1}{2} \right)^2$

**62.** Find the average of $\dfrac{1}{2}, \dfrac{1}{4}, \dfrac{1}{3}$, and $\dfrac{1}{5}$. [4.8b]

Simplify. [4.8b]

**63.** $\dfrac{\dfrac{1}{3}}{\dfrac{5}{8} - 1}$

**64.** $\dfrac{-\dfrac{x}{7}}{\dfrac{3}{14}}$

**65.** Simplify: $\dfrac{1}{4} + \dfrac{2}{5} \div 5^2$. [4.8a]

  **A.** $\dfrac{133}{500}$  **B.** $\dfrac{3}{500}$

  **C.** $\dfrac{117}{500}$  **D.** $\dfrac{5}{2}$

**66.** Solve: $x + \dfrac{2}{3} = 5$. [4.3b]

  **A.** $\dfrac{15}{2}$  **B.** $5\dfrac{2}{3}$

  **C.** $\dfrac{10}{3}$  **D.** $4\dfrac{1}{3}$

## Synthesis

**67.** Find $r$ if
$$\dfrac{1}{r} = \dfrac{1}{100} + \dfrac{1}{150} + \dfrac{1}{200}.$$ [4.2b], [4.4b]

**68.** Place the numbers 3, 4, 5, and 6 in the boxes in order to make a true equation: [4.5a]
$$\dfrac{\square}{\square} + \dfrac{\square}{\square} = 3\dfrac{1}{4}.$$

**69.** Find the largest integer for which each fraction is greater than 1. [4.2c]

  a) $\dfrac{7}{\square}$  b) $\dfrac{11}{\square}$

  c) $\dfrac{\square}{-27}$  d) $\dfrac{\square}{-\frac{1}{2}}$

# Understanding Through Discussion and Writing

**1.** Is the sum of two mixed numerals always a mixed numeral? Why or why not? [4.6a]

**2.** Write a real-world problem for a classmate to solve. Design the problem so that its solution is found by performing the multiplication $4\frac{1}{2} \cdot 33\frac{1}{3}$. [4.7d]

**3.** A student insists that $3\frac{2}{5} \cdot 1\frac{3}{7} = 3\frac{6}{35}$. What mistake is he making and how should he have proceeded? [4.7a]

**4.** Discuss the role of least common multiples in adding and subtracting with fraction notation. [4.2b], [4.3a]

**5.** Find a real-world situation that fits this equation:
$$2 \cdot 15\frac{3}{4} + 2 \cdot 28\frac{5}{8} = 88\frac{3}{4}.$$ [4.6c], [4.7d]

**6.** A student insists that $5 \cdot 3\frac{2}{7} = (5 \cdot 3) \cdot (5 \cdot \frac{2}{7})$. What mistake is she making and how should she have proceeded? [4.7a]

CHAPTER
**4** **Test**

For
Extra
Help

For step-by-step test solutions, access the Chapter Test Prep Videos
in MyLab Math.

**1.** Find the LCM of 12 and 16.

Perform the indicated operation and, if possible, simplify.

**2.** $\dfrac{1}{2} + \dfrac{5}{2}$

**3.** $-\dfrac{7}{8} + \dfrac{2}{3}$

**4.** $\dfrac{5}{t} - \dfrac{3}{t}$

**5.** $\dfrac{5}{6} - \dfrac{3}{4}$

**6.** $\dfrac{5}{8} - \dfrac{17}{24}$

Solve.

**7.** $x + \dfrac{2}{3} = \dfrac{11}{12}$

**8.** $-5x - 3 = 9$

**9.** $\dfrac{3}{4} = \dfrac{1}{2} + \dfrac{5}{3}x$

**10.** Use $<$ or $>$ for $\square$ to form a true sentence.

$\dfrac{6}{7} \,\square\, \dfrac{21}{25}$

Convert to fraction notation.

**11.** $3\dfrac{1}{2}$

**12.** $-9\dfrac{3}{8}$

**13.** Convert to a mixed numeral:

$-\dfrac{74}{9}$.

**14.** Divide. Write a mixed numeral for the answer.

$11\overline{)1789}$

Perform the indicated operation. Write a mixed numeral for each answer.

**15.** $\quad 6\dfrac{2}{5}$
$\quad + 7\dfrac{4}{5}$
$\quad \overline{\phantom{xxxx}}$

**16.** $\quad 3\dfrac{1}{4}$
$\quad + 9\dfrac{1}{6}$
$\quad \overline{\phantom{xxxx}}$

**17.** $\quad 10\dfrac{1}{6}$
$\quad - 5\dfrac{7}{8}$
$\quad \overline{\phantom{xxxx}}$

**18.** $14 + \left(-5\dfrac{3}{7}\right)$

**19.** $3\dfrac{4}{5} - 9\dfrac{1}{2}$

Combine like terms.

**20.** $\dfrac{3}{8}x - \dfrac{1}{2}x$

**21.** $5\dfrac{2}{11}a - 3\dfrac{1}{5}a$

Perform the indicated operation.

**22.** $9 \cdot 4\frac{1}{3}$

**23.** $6\frac{3}{4} \cdot \left(-2\frac{2}{3}\right)$

**24.** $33 \div 5\frac{1}{2}$

**25.** $2\frac{1}{3} \div 1\frac{1}{6}$

Evaluate.

**26.** $\frac{2}{3}ab$, for $a = 7$ and $b = 4\frac{1}{5}$

**27.** $4 + mn$, for $m = 7\frac{2}{5}$ and $n = 3\frac{1}{4}$

Solve.

**28.** *Flying Speed.* At top speed, a red-tailed hawk can fly $2\frac{1}{5}$ times as fast as a California condor. If a red-tailed hawk can fly at 121 mph, how fast can a California condor fly?

**Data:** speedofanimals.com

**29.** *Book Order.* An order of books for a math course weighs 220 lb. Each book weighs $2\frac{3}{4}$ lb. How many books are in the order?

**30.** *Carpentry.* The following diagram shows a middle drawer support guide for a cabinet drawer. Find each of the following.

   **a)** The short length $a$ across the top

   **b)** The length $b$ across the bottom

**31.** *Carpentry.* In carpentry, some pieces of plywood that are called "$\frac{3}{4}$-inch" plywood are actually $\frac{11}{16}$ in. thick. How much thinner is such a piece than its name indicates?

**32.** *Women's Dunks.* The first three women in the history of college basketball able to dunk a basketball are listed below. Their names, heights, and universities are

   Michelle Snow, $6\frac{5}{12}$ ft, Tennessee;

   Charlotte Smith, $5\frac{11}{12}$ ft, North Carolina;

   Georgeann Wells, $6\frac{7}{12}$ ft, West Virginia.

Find the average height of these women.

**Data:** *USA Today*, 11/30/00, p. 3C

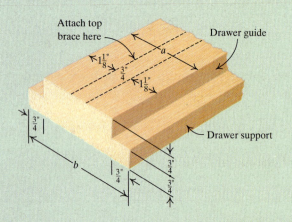

Simplify.

**33.** $\frac{2}{3} + 1\frac{1}{3} \cdot 2\frac{1}{8}$

**34.** $-1\frac{1}{2} - \frac{1}{2}\left(\frac{1}{2} \div \frac{1}{4}\right) + \left(\frac{1}{2}\right)^2$

**35.** $\dfrac{\frac{1}{3} - \frac{7}{9}}{\frac{1}{2} + \frac{1}{6}}$

**36.** Find the LCM of 12, 36, and 60.
   **A.** 6          **B.** 12
   **C.** 60          **D.** 180

## Synthesis

**37.** The students in a math class can be organized into study groups of 8 each so that no students are left out. The same class of students can also be organized into groups of 6 so that no students are left out.

   **a)** Find some class sizes for which this will work.

   **b)** Find the smallest such class size.

**38.** Rebecca walks 17 laps at her health club. Trent walks 17 laps at his health club. If the track at Rebecca's health club is $\frac{1}{7}$ mi long and the track at Trent's is $\frac{1}{8}$ mi long, who walks farther? How much farther?

Solve.

**1.** There are $20\frac{1}{3}$ gal of water in a rainbarrel; $5\frac{3}{4}$ gal are poured out and $8\frac{2}{3}$ gal are returned after a heavy rainfall. How many gallons of water are then in the barrel?

**2.** How many people can receive equal $16 shares from a total of $496?

**3.** A recipe calls for $\frac{4}{5}$ tsp of salt. How much salt should be used for $\frac{1}{2}$ recipe? for 5 recipes?

**4.** How many pieces, each $2\frac{3}{8}$ ft long, can be cut from a piece of wire 38 ft long?

**5.** An emergency food pantry fund contains $423. From this fund, $148 and $167 are withdrawn for expenses. How much is left in the fund?

**6.** In a walkathon, Jermaine walked $\frac{9}{10}$ mi and Oleta walked $\frac{3}{4}$ mi. What was the total distance they walked?

What part is shaded?

**7.**

**8.**

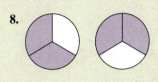

Calculate and simplify.

**9.**  $\begin{array}{r} 3\,7\,0\,4 \\ +\,5\,2\,7\,8 \end{array}$

**10.**  $\begin{array}{r} 7\,6\,0\,5 \\ -\,3\,0\,8\,7 \end{array}$

**11.**  $\begin{array}{r} 2\,7\,8 \\ \times\,1\,8 \end{array}$

**12.** $29(-5)$

**13.** $\frac{3}{8} + \frac{1}{24}$

**14.**  $\begin{array}{r} 2\frac{3}{4} \\ +\,5\frac{1}{2} \end{array}$

**15.** $\frac{4}{t} - \frac{9}{t}$

**16.**  $\begin{array}{r} 2\frac{1}{3} \\ -\,1\frac{1}{6} \end{array}$

**17.** $\frac{9}{10} \cdot \frac{5}{3}$

**18.** $18\left(-\frac{5}{6}\right)$

**19.** $-9 - (-25)$

**20.** $2\frac{1}{5} \div \frac{3}{10}$

Divide. If there is a remainder, write the answer in the form 34 R 7.

**21.** $6\overline{)4290}$

**22.** $45\overline{)2531}$

**23.** Write a mixed numeral for the answer in Exercise 22.

**24.** In the number 2753, what digit names tens?

**25.** *Room Carpeting.* The Chandlers are carpeting an L-shaped family room consisting of one rectangle that is $8\frac{1}{2}$ ft by 11 ft and another rectangle that is $6\frac{1}{2}$ ft by $7\frac{1}{2}$ ft.

a) Find the area of the carpet.

b) Find the perimeter of the carpet.

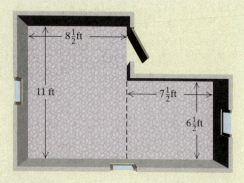

**26.** Round 38,478 to the nearest hundred.

**27.** Find the LCM of 18 and 24.

**28.** Simplify:

$$\left(\frac{1}{2} + \frac{2}{5}\right)^2 \div 3 + 6 \times \left(2 + \frac{1}{4}\right).$$

Use $<$, $>$, or $=$ for $\square$ to write a true sentence.

**29.** $\frac{4}{5} \square \frac{4}{6}$

**30.** $\frac{3}{13} \square \frac{9}{39}$

**31.** $-\frac{5}{12} \square -\frac{3}{7}$

**32.** Evaluate $\dfrac{t + p}{3}$ for $t = -4$ and $p = 16$.

Simplify.

**33.** $\dfrac{36}{45}$

**34.** $\dfrac{0}{27}$

**35.** $\dfrac{-320}{10}$

**36.** Convert to fraction notation: $4\frac{5}{8}$.

**37.** Convert to a mixed numeral: $-\dfrac{17}{3}$.

Solve.

**38.** $x + 24 = 117$

**39.** $x + \dfrac{7}{9} = \dfrac{4}{3}$

**40.** $\dfrac{7}{9} \cdot t = -\dfrac{4}{3}$

**41.** $\dfrac{5}{7} = \dfrac{1}{3} + 4a$

**42.** *Matching.* Match each item in the first column with the appropriate item in the second column by drawing connecting lines. There can be more than one correct correspondence for an item.

| | |
|---|---|
| Factors of 68 | 12, 54, 72, 300 |
| Factorization of 68 | 2, 3, 17, 19, 23, 31, 47, 101 |
| Prime factorization of 68 | $2 \cdot 2 \cdot 17$ |
| Numbers divisible by 6 | $2 \cdot 34$ |
| Numbers divisible by 8 | 8, 16, 24, 32, 40, 48, 64, 864 |
| Numbers divisible by 5 | 1, 2, 4, 17, 34, 68 |
| Prime numbers | 70, 95, 215 |

## Synthesis

**43.** Find the smallest prime number that is larger than 2000.

**44.** Solve: $7x - \dfrac{2}{3}(x - 6) = 6\frac{5}{7}$.

# Decimal Notation

The average American eats 1996.3 lb of food per year. The graph shown here gives a visual comparison of the numbers of pounds of certain foods consumed per year per person. The average American also consumes 53 gallons of soda, 29 lb of French fries, 23 lb of pizza, and 24 lb of ice cream per year.

*DATA: niftyhomestead.com; visualeconomics.com*

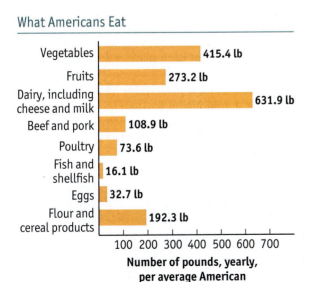

## What Americans Eat

| Food | Number of pounds, yearly, per average American |
|------|-----------------------------------------------|
| Vegetables | 415.4 lb |
| Fruits | 273.2 lb |
| Dairy, including cheese and milk | 631.9 lb |
| Beef and pork | 108.9 lb |
| Poultry | 73.6 lb |
| Fish and shellfish | 16.1 lb |
| Eggs | 32.7 lb |
| Flour and cereal products | 192.3 lb |

**Number of pounds, yearly, per average American**

DATA: visualeconomics.com

In Exercise 57 of Exercise Set 5.8, we will calculate the number of pounds of fruits and vegetables the average American eats per day.

# 5.1

# Decimal Notation, Order, and Rounding

## OBJECTIVES

**a** Given decimal notation, write a word name.

**b** Convert between decimal notation and fraction notation.

**c** Given a pair of numbers in decimal notation, tell which is larger.

**d** Round decimal notation to the nearest thousandth, hundredth, tenth, one, ten, hundred, or thousand.

The set of **rational numbers** consists of the **integers** $\ldots, -3, -2, -1, 0, 1, 2, 3, \ldots$, and fractions like $\frac{1}{2}, \frac{2}{3}, \frac{-7}{8}, \frac{17}{-10}$, and so on. In this chapter, we will use *decimal notation* to represent rational numbers. Using decimal notation, we can write 0.875 for $\frac{7}{8}$, for example, or 48.97 for $48\frac{97}{100}$. A number written in decimal notation is often simply referred to as a *decimal*.

The word *decimal* comes from the Latin word *decima*, meaning a *tenth part*. Since our usual counting system is based on tens, decimal notation is a natural extension of an already familiar system.

## a  DECIMAL NOTATION AND WORD NAMES

One model of the iRobot Roomba robot vacuum with Wi-Fi connectivity sells for \$697.99. The dot in \$697.99 is called a **decimal point**. Since \$0.99, or 99¢, is $\frac{99}{100}$ of a dollar, it follows that

$$\$697.99 = \$697 + \$0.99.$$

Also, since \$0.99, or 99¢, has the same value as 9 dimes + 9 pennies and 1 dime is $\frac{1}{10}$ of a dollar and 1 penny is $\frac{1}{100}$ of a dollar, we can write

$$697.99 = 6 \cdot 100 + 9 \cdot 10 + 7 \cdot 1 + 9 \cdot \frac{1}{10} + 9 \cdot \frac{1}{100}.$$

This is an extension of the expanded notation for whole numbers that we used in Chapter 1. The place values are 100, 10, 1, $\frac{1}{10}$, $\frac{1}{100}$, and so on. We can see this on a **place-value chart**. The value of each place is $\frac{1}{10}$ as large as that of the one to its left.

Let's consider decimal notation using a place-value chart to represent 3.46583 min, the winning time for a gold medal by the U.S. men's 4 × 100 meters medley relay team. Members of the swim team, pictured at left, were Nathan Adrian, Michael Phelps, Ryan Murphy, and Cody Miller.

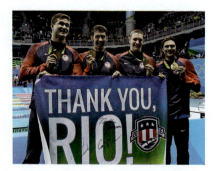

| PLACE-VALUE CHART | | | | | | | |
|---|---|---|---|---|---|---|---|
| Hundreds | Tens | Ones | Ten*ths* | Hundred*ths* | Thousand*ths* | Ten-Thousand*ths* | Hundred-Thousand*ths* |
| 100 | 10 | 1 | $\frac{1}{10}$ | $\frac{1}{100}$ | $\frac{1}{1000}$ | $\frac{1}{10,000}$ | $\frac{1}{100,000}$ |
| 3 . | 4 | 6 | 5 | 8 | 3 | | |

The decimal notation 3.46583 means

$$3 + \frac{4}{10} + \frac{6}{100} + \frac{5}{1000} + \frac{8}{10,000} + \frac{3}{100,000}, \quad \text{or} \quad 3\frac{46,583}{100,000}.$$

We read both 3.46583 and $3\frac{46,583}{100,000}$ as

"Three *and* forty-six thousand five hundred eighty-three hundred-thousandths."

We read the decimal point as "and." Note that the place values to the right of the decimal point always end in *th*. We can also read 3.46583 as "three *point* four six five eight three."

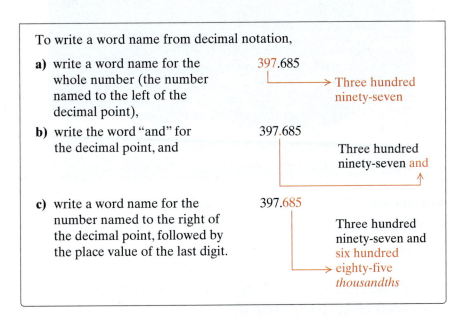

EXAMPLE 1 *Median Age.* The median age of residents in Florida is 41.8 years. The median age of residents in Alaska is 33.3 years. The median age in the United States is 37.8 years. Write word names for 41.8, 33.3, and 37.8.

**Data:** Statista; U.S. Census Bureau

Forty-one and eight tenths

Thirty-three and three tenths

Thirty-seven and eight tenths

1. **Life Expectancy.** The life expectancy at birth in South Africa in 2016 was 61.6 years for males and 64.6 years for females. Write word names for 61.6 and 64.6.

   **Data:** World Health Organization

2. **10,000-Meter Record.** Almaz Ayana of Ethiopia holds the women's world record for the 10,000-meter run: 29.2908 min. Write a word name for 29.2908.

   **Data:** *The World Almanac 2017*

Write a word name for each number.

3. 245.89

4. 34.0064

5. 31,079.756

**EXAMPLE 2** Write a word name for 19.3806.

Nineteen and three thousand eight hundred six ten-thousandths

**EXAMPLE 3** *Highest Auction Price.* A 1956 Aston Martin DBR1 sold for $22.55 million at the RM Sotheby's Monterey 2017 auction. This amount was the highest paid for a car sold at auction from January through August 2017. Write a word name for 22.55.

**Data:** classic-car-auctions.info, "2017 (January to August): Ten Most-Expensive Cars Sold at Public Auction" by Henk Bekker

Twenty-two and fifty-five hundredths

**EXAMPLE 4** Write a word name for 1788.045.

One thousand, seven hundred eighty-eight and forty-five thousandths

◀ Do Exercises 1–5.

## b  CONVERTING BETWEEN DECIMAL NOTATION AND FRACTION NOTATION

Given decimal notation, we can convert to fraction notation as follows:

$$9.875 = 9 + \frac{8}{10} + \frac{7}{100} + \frac{5}{1000}$$

$$= 9 \cdot \frac{1000}{1000} + \frac{8}{10} \cdot \frac{100}{100} + \frac{7}{100} \cdot \frac{10}{10} + \frac{5}{1000}$$

$$= \frac{9000}{1000} + \frac{800}{1000} + \frac{70}{1000} + \frac{5}{1000} = \frac{9875}{1000}.$$

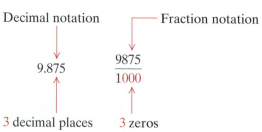

*Answers*

1. Sixty-one and six tenths; sixty-four and six tenths
2. Twenty-nine and two thousand nine hundred eight ten-thousandths
3. Two hundred forty-five and eighty-nine hundredths
4. Thirty-four and sixty-four ten-thousandths
5. Thirty-one thousand, seventy-nine and seven hundred fifty-six thousandths

To convert from decimal notation to fraction notation,

**a)** count the number of decimal places,

$$4.98$$
2 places

**b)** move the decimal point that many places to the right, and

$$4.98.$$ Move 2 places.

**c)** write the answer over a denominator of 1 followed by that number of zeros.

$$\frac{498}{100}$$ 2 zeros

For a number like 0.876, we generally write a 0 before the decimal point to call attention to the presence of the decimal point.

**EXAMPLE 5**  Write fraction notation for 0.876. Do not simplify.

$$0.876 \qquad 0.876. \qquad 0.876 = \frac{876}{1000}$$

3 places        3 zeros

Decimals greater than 1 or less than $-1$ can be written either as fractions or as mixed numerals.

**EXAMPLE 6**  Write 56.23 as a fraction and as a mixed numeral.

$$56.23 \qquad 56.23. \qquad 56.23 = \frac{5623}{100}, \qquad \text{and} \qquad 56.23 = 56\frac{23}{100}$$

2 places        2 zeros

As a check, note that both 56.23 and $56\frac{23}{100}$ are read as "fifty-six and twenty-three hundredths."

**EXAMPLE 7**  Write $-2.6073$ as a fraction and as a mixed numeral.

$$-2.6073. = -\frac{26{,}073}{10{,}000} \qquad \text{and} \qquad -2.6073 = -2\frac{6073}{10{,}000}$$

4 places        4 zeros

**Do Exercises 6–9.** ▶

To write $\frac{5328}{10}$ as a decimal, we can first divide to find an equivalent mixed numeral.

$$\frac{5328}{10} = 532\frac{8}{10}$$

Next note that

$$532\frac{8}{10} = 532 + \frac{8}{10}$$

$$= 532.8.$$

$$\begin{array}{r} 5\ 3\ 2 \\ 10\overline{)5\ 3\ 2\ 8} \\ \underline{5\ 0} \\ 3\ 2 \\ \underline{3\ 0} \\ 2\ 8 \\ \underline{2\ 0} \\ 8 \end{array}$$

Write fraction notation. Do not simplify.

**6.** 0.5491

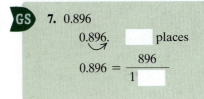

**GS** **7.** 0.896

$$0.896.$$ ⬚ places

$$0.896 = \frac{896}{1\ ⬚}$$

Write as a fraction and as a mixed numeral.

**8.** 75.069        **9.** $-312.9$

***Answers***

**6.** $\dfrac{5491}{10{,}000}$  **7.** $\dfrac{896}{1000}$  **8.** $\dfrac{75{,}069}{1000}; 75\dfrac{69}{1000}$

**9.** $-\dfrac{3129}{10}; -312\dfrac{9}{10}$

***Guided Solution:***

**7.** 3; 000

Thus, if fraction notation has a denominator that is a power of ten, such as 10, 100, 1000, and so on, we convert to decimal notation by reversing the procedure that we just used in Examples 5–7.

---

To convert from fraction notation to decimal notation when the denominator is 10, 100, 1000, and so on,

a) count the number of zeros in the denominator and

$$\frac{8679}{\underbrace{1000}_{\text{3 zeros}}}$$

b) move the decimal point that number of places to the left. Leave off the denominator.

8.679.   Move 3 places.

$$\frac{8679}{1000} = 8.679$$

---

**EXAMPLE 8**   Write decimal notation for $\frac{47}{10}$.

$$\frac{47}{\underset{\text{1 zero}}{10}} \qquad 4.7. \atop \text{1 place} \qquad \frac{47}{10} = 4.7$$

The decimal point is moved to the left.

**EXAMPLE 9**   Write decimal notation for $\frac{123{,}067}{10{,}000}$.

$$\frac{123{,}067}{\underset{\text{4 zeros}}{10{,}000}} \qquad 12.3067. \atop \text{4 places} \qquad \frac{123{,}067}{10{,}000} = 12.3067$$

To move the decimal point to the left, we may need to write extra zeros.

**EXAMPLE 10**   Write decimal notation for $-\frac{9}{100}$.

$$-\frac{9}{\underset{\text{2 zeros}}{100}} \qquad -0.09. \atop \text{2 places} \qquad -\frac{9}{100} = -0.09$$

◀ **Do Exercises 10–13.**

To convert fractions with denominators other than 10, 100, and so on to decimal notation, we will usually perform long division.

If a mixed numeral has a fraction part with a denominator that is a power of ten, such as 10, 100, or 1000, and so on, we first write the mixed numeral as a sum of a whole number and a fraction. Then we convert to decimal notation.

**EXAMPLE 11**   Write decimal notation for $23\frac{59}{100}$.

$$23\frac{59}{100} = 23 + \frac{59}{100} = 23 \text{ and } \frac{59}{100} = 23.59$$

◀ **Do Exercises 14 and 15.**

---

Write decimal notation for each number.

**10.** $\frac{743}{100}$ **GS**

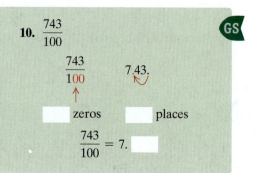
$$\frac{743}{100} \qquad 7.43.$$
$$\boxed{\phantom{0}}\text{ zeros} \qquad \boxed{\phantom{0}}\text{ places}$$
$$\frac{743}{100} = 7.\boxed{\phantom{0}}$$

**11.** $-\frac{73}{1000}$   **12.** $\frac{67{,}089}{10{,}000}$

**13.** $-\frac{9}{10}$

Write decimal notation for each number.

**14.** $-7\frac{3}{100}$   **15.** $23\frac{47}{1000}$

**Answers**

**10.** 7.43   **11.** −0.073   **12.** 6.7089
**13.** −0.9   **14.** −7.03   **15.** 23.047

**Guided Solution:**

**10.** 2, 2; 43

## c ORDER

To understand how to compare numbers in decimal notation, consider 0.85 and 0.9. First note that $0.9 = 0.90$ because $\frac{9}{10} = \frac{90}{100}$. Since $0.85 = \frac{85}{100}$, it follows that $\frac{85}{100} < \frac{90}{100}$ and $0.85 < 0.9$. This leads us to a quick way to compare two numbers in decimal notation.

> To compare two positive numbers in decimal notation, start at the left and compare corresponding digits, moving from left to right. When two digits differ, the number with the larger digit is the larger of the two numbers. Extra zeros can be written to the right of the last decimal place.

**EXAMPLE 12**   Which is larger: 2.109 or 2.1?

2.109          2.109          2.109          2.109
↕ The same     ↑ The same     ↑ The same     ↕ Different;
2.1            2.1            2.10           2.100   9 is larger
                                                     than 0.

Thus, 2.109 is larger than 2.1. In symbols, $2.109 > 2.1$.

**EXAMPLE 13**   Which is larger: 0.09 or 0.108?

0.09                0.09
↑ The same          ↕ Different; 1 is larger than 0.
0.108               0.108

Thus, 0.108 is larger than 0.09. In symbols, $0.108 > 0.09$.

**Do Exercises 16–21.** ▶

Which number is larger?

**16.** 2.04,   2.039

**17.** 0.06,   0.008

**18.** 0.5,   0.58

**19.** 1,   0.9999

**20.** 0.8989,   0.09898

**21.** 21.006,   21.05

We can use the number line to visualize order. We illustrate Examples 12 and 13 below. Larger numbers are always to the right.

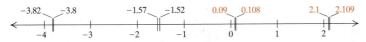

Note from the number line that $-2 < -1$. Similarly, $-1.57 < -1.52$.

> To compare two negative numbers in decimal notation, start at the left and compare corresponding digits, moving from left to right. When two digits differ, the number with the smaller digit is the larger of the two numbers.

**EXAMPLE 14**   Which is larger: −3.8 or −3.82?

−3.8                      −3.80
↕↕ The same               ↕ Different; 0 is smaller than 2.
−3.82                     −3.82

Thus, −3.8 is larger than −3.82. In symbols, $-3.8 > -3.82$. (See the number line above.)

**Do Exercises 22 and 23.** ▶

Which number is larger?

**22.** −34.01,   −34.008

**23.** −9.12,   −8.98

*Answers*

**16.** 2.04   **17.** 0.06   **18.** 0.58   **19.** 1
**20.** 0.8989   **21.** 21.05   **22.** −34.008
**23.** −8.98

MyLab Math

**ANIMATION**

Round to the nearest tenth.

**24.** 2.76  **25.** 13.85

**26.** −234.448  **27.** 7.009

Round to the nearest hundredth.

**28.** 0.6362  **29.** −7.8348

**30.** 34.67514  **31.** −0.02521

Round to the nearest thousandth:

**32.** 0.94347  **33.** −8.00382

**34.** −43.111943  **35.** 37.400526

Round 7459.3548 to the nearest:

**36.** Thousandth.

**37.** Hundredth.

**38.** Tenth.  **39.** One.

**40.** Ten. (*Caution:* "Tens" are not "tenths.")

**41.** Hundred.  **42.** Thousand.

*Answers*
**24.** 2.8 **25.** 13.9 **26.** −234.4 **27.** 7.0
**28.** 0.64 **29.** −7.83 **30.** 34.68
**31.** −0.03 **32.** 0.943 **33.** −8.004
**34.** −43.112 **35.** 37.401 **36.** 7459.355
**37.** 7459.35 **38.** 7459.4 **39.** 7459
**40.** 7460 **41.** 7500 **42.** 7000

---

## d | ROUNDING

We round decimals in much the same way that we round whole numbers. To see how, we use the number line.

**EXAMPLE 15**  Round 0.37 to the nearest tenth.

Here is part of the number line, magnified.

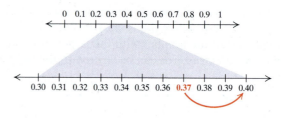

We see that 0.37 is closer to 0.40 than to 0.30. Thus, 0.37 rounded to the nearest tenth is 0.4.

---

To round to the nearest tenth, hundredth, thousandth, and so forth:

**a)** Locate the digit in that place.

**b)** Consider the next digit to the right.

**c)** If the digit to the right is 5 or greater, add 1 to the original digit. If the digit to the right is 4 or less, the original digit does not change. In either case, drop all numbers to the right of the original digit.

---

**EXAMPLE 16**  Round 72.3846 to the nearest hundredth.

**a)** Locate the digit in the hundredths place, 8.

$$72.3846$$

**b)** Consider the next digit to the right, 4.

**c)** Since that digit, 4, is less than 5, the original digit does not change. The rounded number is 72.38.

---

································ **Caution!** ································

72.39 is not a correct answer to Example 16. It is *incorrect* to round sequentially from right to left as follows: 72.3846, 72.385, 72.39.

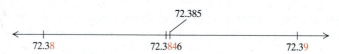

72.3846 is closer to 72.38 than to 72.39.

··························································································

**EXAMPLE 17**  Round −0.064 to the nearest tenth.

**a)** Locate the digit in the tenths place, 0.

$$-0.064$$

**b)** Consider the next digit to the right, 6.

**c)** Since that digit, 6, is greater than 5, round from −0.064 to −0.1.

The answer is −0.1. Since −0.1 < −0.064, we actually rounded *down*.

◀ **Do Exercises 24–42.**

## ✓ Check Your Understanding

**Reading Check**    Name the digit that represents each place value in the number 436.81205.

**RC1.** Hundred-thousandths ____

**RC2.** Thousandths ____

**RC3.** Tens ____

**RC4.** Ten-thousandths ____

**RC5.** Tenths ____

**RC6.** Hundreds ____

**RC7.** Hundredths ____

**RC8.** Ones ____

**Concept Check**    Arrange the numbers in order from smallest to largest.

**CC1.** 0.99, 0.099, 1, 0.9999, 0.89999, 1.00009, 0.909, 0.9889

**CC2.** 2.1, 2.109, 2.108, 2.018, 2.0119, 2.0302, 2.000001

**a**

**1.** *Currency Conversion.*    One U.S. dollar was worth about 17.7178 Mexican pesos recently. Write a word name for 17.7178.

**Data:** OANDA Rates™

**2.** *Currency Conversion.*    One U.S. dollar was worth 1.257 Canadian dollars recently. Write a word name for 1.257.

**Data:** OANDA Rates™

**3.** *Birth Rate.*    There were 103.6 triplet and higher-order multiple births per 100,000 live births in a recent year in the United States. Write a word name for 103.6.

**Data:** multiplesofamerica.org

**4.** *Rice Consumption.*    Annual per capita rice consumption in Indonesia is 324.5 lb. In the United States, annual per capita rice consumption is only 73.48 lb. Write word names for 324.5 and 73.48.

**Data:** uark.edu

**5.** *Pole Position at Indy 500.*    Scott Dixon won the pole position for the 2017 Indianapolis 500 with a speed of 232.164 mph. This was the fastest qualifying speed for four laps recorded in 21 years. Write a word name for 232.164.

**Data:** bleacherreport.com

**6.** *Stock Price.*    Walmart's stock price was recently $79.96 per share. Write a word name for 79.96.

**Data:** New York Stock Exchange

Write a word name for the number in each sentence.

**7.** One gallon of paint is equal to 3.785 L of paint.

**8.** *Water Weight.* One gallon of water weighs 8.35 lb.

Write each number as a fraction and, if possible, as a mixed numeral. Do not simplify.

**9.** 7.3

**10.** 4.9

**11.** 21.67

**12.** −57.32

**13.** −2.703

**14.** 0.079

**15.** 0.0109

**16.** 1.0008

**17.** −4.0003

**18.** −9.012

**19.** −0.0207

**20.** −0.00104

**21.** 70.00105

**22.** 60.0403

Write decimal notation for each number.

**23.** $\dfrac{3}{10}$

**24.** $\dfrac{73}{10}$

**25.** $-\dfrac{59}{100}$

**26.** $-\dfrac{67}{100}$

**27.** $\dfrac{3798}{1000}$

**28.** $\dfrac{780}{1000}$

**29.** $\dfrac{78}{10,000}$

**30.** $\dfrac{56,788}{100,000}$

**31.** $\dfrac{-18}{100,000}$

**32.** $\dfrac{-2347}{100}$

**33.** $\dfrac{486,197}{1,000,000}$

**34.** $\dfrac{8,953,074}{1,000,000}$

**35.** $7\dfrac{13}{1000}$

**36.** $4\dfrac{909}{1000}$

**37.** $-8\dfrac{431}{1000}$

**38.** $-49\dfrac{32}{1000}$

**39.** $2\dfrac{1739}{10,000}$

**40.** $9243\dfrac{1}{10}$

**41.** $8\dfrac{953,073}{1,000,000}$

**42.** $2256\dfrac{3059}{10,000}$

Which number is larger?

**43.** 0.06,  0.58

**44.** 0.008,  0.8

**45.** 0.403,  0.410

**46.** 42.06,  42.1

**47.** −5.046,  −5.043

**48.** −324.19,  −325.19

**49.** 234.07,  235.07

**50.** 0.99999,  1

**51.** 0.007,  $\dfrac{7}{100}$

**52.** $\dfrac{73}{10}$,  0.73

**53.** −0.872,  −0.873

**54.** −0.8437,  −0.84384

**55.** 0.23  **56.** 0.85  **57.** −0.372  **58.** −0.261

**59.** 2.951  **60.** 7.532  **61.** −327.2347  **62.** −8.7493

Round to the nearest hundredth.

**63.** 0.893  **64.** 0.675  **65.** −0.6666  **66.** −7.5252

**67.** 0.9952  **68.** 207.9976  **69.** −0.03488  **70.** −9.27481

Round to the nearest thousandth.

**71.** 0.5724  **72.** 0.6666  **73.** 17.0015  **74.** 123.4562

**75.** −20.20202  **76.** −0.10346  **77.** 9.98487  **78.** 67.100602

Round 809.47321 to the nearest:

**79.** Tenth.  **80.** Thousandth.  **81.** Hundredth.  **82.** One.

## Skill Maintenance

Add or subtract, as indicated.

**83.** $\begin{array}{r} 6\,8\,1 \\ +\,1\,4\,9 \end{array}$ [1.2a]

**84.** $\dfrac{681}{1000} + \dfrac{149}{1000}$ [4.2a]

**85.** $\begin{array}{r} 2\,6\,7 \\ -\ \ \,8\,5 \end{array}$ [1.3a]

**86.** $\dfrac{267}{100} - \dfrac{85}{100}$ [4.3a]

**87.** $\dfrac{37}{55} - \dfrac{49}{55}$ [4.3a]

**88.** $-\dfrac{29}{34} + \dfrac{14}{34}$ [4.2a]

**89.** $\begin{array}{r} 3\,4{,}9\,0\,3 \\ -\ \ 1{,}9\,4\,5 \end{array}$ [1.3a]

**90.** $\begin{array}{r} 4\,9\,3\,7 \\ +\,5\,7\,8\,9 \end{array}$ [1.2a]

## Synthesis

**91.** Arrange the following numbers in order from smallest to largest.

−0.989, −0.898, −1.009, −1.09, −0.098

**92.** Arrange the following numbers in order from smallest to largest.

−2.018, −2.1, −2.109, −2.0119, −2.108

*Truncating.* There are other methods of rounding decimal notation. A computer often uses a method called **truncating**. To truncate we drop off decimal places right of the rounding place, which is the same as changing all digits to the right of the rounding place to zeros. For example, rounding 6.78093456285102 to the ninth decimal place, using truncating, gives us 6.780934562. Use truncating to round each of the following to the fifth decimal place, that is, the hundred-thousandth place.

**93.** 6.78346123  **94.** 6.783461902  **95.** 99.999999999  **96.** 0.030303030303

# 5.2

## OBJECTIVES

**a** Add using decimal notation.

**b** Subtract using decimal notation.

**c** Add and subtract negative decimals.

**d** Combine like terms with decimal coefficients.

Add.

1.      0.8 4 7
    + 1 0.0 7

2.      2 . 1
        0 . 7 3
    + 3 1 . 3 6 8

Add.

3.  0.02 + 4.3 + 0.649

4.  0.12 + 3.006 + 0.4357

5.  0.4591 + 0.2374 + 8.70894

# Addition and Subtraction of Decimals

## a   ADDITION

Adding with decimal notation is similar to adding whole numbers. First, we line up the decimal points so that we can add corresponding place-value digits. Then we add digits from the right. For example, we add the thousandths, then the hundredths, and so on, carrying if necessary. If desired, we can write extra zeros to the right of the decimal point so that the number of places is the same in all of the addends.

**EXAMPLE 1**   Add: 56.314 + 17.78.

$$\begin{array}{r} 5\;6\,.\,3\;1\;4 \\ +\;1\;7\,.\,7\;8\;0 \\ \end{array}$$   Lining up the decimal points in order to add

Writing an extra zero to the right of the decimal point

$$\begin{array}{r} 5\;6\,.\,3\;1\;\boxed{4} \\ +\;1\;7\,.\,7\;8\;\boxed{0} \\ \hline \boxed{4} \end{array}$$   Adding thousandths

$$\begin{array}{r} 5\;6\,.\,3\;\boxed{1}\;4 \\ +\;1\;7\,.\,7\;\boxed{8}\;0 \\ \hline \boxed{9}\;4 \end{array}$$   Adding hundredths

$$\begin{array}{r} \overset{1}{5}\;6\,.\,\boxed{3}\;1\;4 \\ +\;1\;7\,.\,\boxed{7}\;8\;0 \\ \hline \boxed{.}\;0\;9\;4 \end{array}$$   Adding tenths

We get 10 tenths = 1 one + 0 tenths, so we carry the 1 to the ones column. Writing a decimal point in the answer

$$\begin{array}{r} \overset{1}{5}\;\overset{1}{6}\,.\,3\;1\;4 \\ +\;1\;\boxed{7}\,.\,7\;8\;0 \\ \hline \boxed{4}\,.\,0\;9\;4 \end{array}$$   Adding ones

We get 14 ones = 1 ten + 4 ones, so we carry the 1 to the tens column.

$$\begin{array}{r} \overset{1}{\boxed{5}}\;\overset{1}{6}\,.\,3\;1\;4 \\ +\;1\;7\,.\,7\;8\;0 \\ \hline \boxed{7}\;4\,.\,0\;9\;4 \end{array}$$   Adding tens

◀ **Do Exercises 1 and 2.**

**EXAMPLE 2**   Add: 3.42 + 0.237 + 14.1.

$$\begin{array}{r} 3\,.\,4\;2\;0 \\ 0\,.\,2\;3\;7 \\ +\;1\;4\,.\,1\;0\;0 \\ \hline 1\;7\,.\,7\;5\;7 \end{array}$$

Lining up the decimal points and writing extra zeros

Adding

◀ **Do Exercises 3–5.**

*Answers*

**1.** 10.917  **2.** 34.198  **3.** 4.969
**4.** 3.5617  **5.** 9.40544

Now we consider the addition $3456 + 19.347$. Keep in mind that any whole number has an "unwritten" decimal point at the right that can be followed by zeros. For example, 3456 can also be written 3456.000. When adding, we can always write in the decimal point and extra zeros if desired.

**EXAMPLE 3**   Add: $3456 + 19.347$.

$$
\begin{array}{r}
\overset{1}{\phantom{0}}\\
3456.000\\
+\quad 19.347\\
\hline
3475.347
\end{array}
$$

Writing in the decimal point and extra zeros
Lining up the decimal points
Adding

**Do Exercises 6 and 7.** ▶

## b   SUBTRACTION

Subtracting with decimal notation is similar to subtracting whole numbers. First, we line up the decimal points so that we can subtract corresponding place-value digits. Then we subtract digits from the right. For example, we subtract the thousandths, then the hundredths, the tenths, and so on, borrowing if necessary.

**EXAMPLE 4**   Subtract: $56.314 - 17.78$.

$$
\begin{array}{r}
56.314\\
-17.780\\
\end{array}
$$
Lining up the decimal points in order to subtract
Writing an extra 0

$$
\begin{array}{r}
56.314\\
-17.780\\
\hline
4
\end{array}
$$
Subtracting thousandths

$$
\begin{array}{r}
2\ 11\\
56.3\cancel{1}4\\
-17.780\\
\hline
34
\end{array}
$$
Borrowing tenths to subtract hundredths

Subtracting hundredths

$$
\begin{array}{r}
12\\
5\ \ 2\ 11\\
5\cancel{6}.3\cancel{1}4\\
-17.780\\
\hline
.534
\end{array}
$$
Borrowing ones to subtract tenths

Subtracting tenths; writing a decimal point

$$
\begin{array}{r}
15\ 12\\
4\ \cancel{5}\ \cancel{2}\ 11\\
\cancel{5}6.3\cancel{1}4\\
-17.780\\
\hline
8.534
\end{array}
$$
Borrowing tens to subtract ones

Subtracting ones

$$
\begin{array}{r}
15\ 12\\
4\ \cancel{5}\ \cancel{2}\ 11\\
\cancel{5}6.3\cancel{1}4\\
-17.780\\
\hline
38.534
\end{array}
$$
Subtracting tens

Check by adding:
$$
\begin{array}{r}
1\ 1\ 1\\
38.534\\
+17.780\\
\hline
56.314
\end{array}
$$
The answer checks because this is the top number in the subtraction.

**Do Exercises 8 and 9.** ▶

---

Add.

**6.** $789 + 123.67$

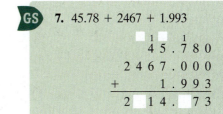

**GS  7.** $45.78 + 2467 + 1.993$

$$
\begin{array}{r}
\overset{1}{\phantom{0}}\ \ \ \ \overset{1}{\phantom{0}}\\
45.780\\
2467.000\\
+\quad\quad 1.993\\
\hline
2\ \square\ 4.\square\ 73
\end{array}
$$

*Subtract whole numbers.*   [1.3a]
Subtract.

**1.** $\begin{array}{r} 236 \\ -109 \\ \hline \end{array}$

**2.** $\begin{array}{r} 4023 \\ -1667 \\ \hline \end{array}$

**Answers: 1.** 127  **2.** 2356

MyLab Math
VIDEO

Subtract.

**GS  8.** $37.428 - 26.674$

$$
\begin{array}{r}
6\ \ 3\\
37.428\\
-26.674\\
\hline
1\ \square.7\ \square 4
\end{array}
$$

**9.** $\begin{array}{r} 0.347 \\ -0.008 \\ \hline \end{array}$

***Answers***
**6.** 912.67  **7.** 2514.773  **8.** 10.754
**9.** 0.339
***Guided Solutions:***
**7.** 1, 1; 5, 7  **8.** 13, 12; 0, 5

Subtract.

**10.** $2.9 - 0.36$

**11.** $0.43 - 0.18762$

**12.** $5.27 - 0.00008$

Subtract.

**13.** $1277 - 82.78$

**14.** $5 - 0.0089$

$$\begin{array}{r} 5 .\overset{10}{\cancel{0}}\cancel{0}\cancel{0}\cancel{0} \\ -\ 0 . 0\ 0\ 8\ 9 \\ \hline 4 .\ \ \ 9\ 1\ \ \end{array}$$

**EXAMPLE 5**  Subtract: $23.08 - 5.0053$.

$$\begin{array}{r} \overset{1\ \ 13}{2\ \cancel{3}} . 0\ \overset{7\ \ 9\ \ 10}{\cancel{8}\ \cancel{0}\ \cancel{0}} \\ -\ 5 . 0\ 0\ 5\ 3 \\ \hline 1\ 8 . 0\ 7\ 4\ 7 \end{array}$$   Writing two extra zeros to the right of the last digit

Subtracting

◀ **Do Exercises 10–12.**

When subtraction involves an integer, the "unwritten" decimal point can be written in. Extra zeros can then be written in to the right of the decimal point.

**EXAMPLE 6**  Subtract: $456 - 2.467$.

$$\begin{array}{r} 4\ 5\ \overset{5\ \ 9\ \ 9\ \ 10}{\cancel{6}.\cancel{0}\ \cancel{0}\ \cancel{0}} \\ -\ \ \ \ 2 . 4\ 6\ 7 \\ \hline 4\ 5\ 3 . 5\ 3\ 3 \end{array}$$   Writing in the decimal point and extra zeros

Subtracting

◀ **Do Exercises 13 and 14.**

## c ADDING AND SUBTRACTING WITH NEGATIVES

Negative decimals are added or subtracted like negative integers.

> To add a negative number and a positive number:
>
> **a)** Determine the sign of the number with the greater absolute value.
> **b)** Subtract the smaller absolute value from the larger one.
> **c)** The answer is the difference from part (b) with the sign from part (a).

**EXAMPLE 7**  Add: $-13.82 + 4.69$.

**a)** Since $|-13.82| > |4.69|$, the sign of the number with the greater absolute value is negative.

**b)**
$$\begin{array}{r} 1\ 3 . 8\ \overset{7\ \ 12}{\cancel{8}\ \cancel{2}} \\ -\ \ \ 4 . 6\ 9 \\ \hline 9 . 1\ 3 \end{array}$$   Finding the difference of the absolute values

**c)** The answer is negative: $-13.82 + 4.69 = -9.13$.

◀ **Do Exercises 15 and 16.**

Add.

**15.** $7.42 + (-9.38)$

**16.** $-4.201 + 7.36$

> To add two negative numbers:
>
> **a)** Add the absolute values.
> **b)** Make the answer negative.

**17.** Add: $-7.49 + (-5.8)$.

**EXAMPLE 8**  Add: $-2.306 + (-3.125)$.

**a)**
$$\begin{array}{r} 2 . 3\ \overset{1}{0}\ 6 \\ +\ 3 . 1\ 2\ 5 \\ \hline 5 . 4\ 3\ 1 \end{array} \Big\}$$   $|-2.306| = 2.306$ and $|-3.125| = 3.125$

Adding the absolute values

**b)** $-2.306 + (-3.125) = -5.431$   The answer is negative.

◀ **Do Exercise 17.**

To subtract, we add the opposite of the number being subtracted.

**EXAMPLE 9**  Subtract: $-3.1 - 4.8$.

$$-3.1 - 4.8 = -3.1 + (-4.8) \qquad \text{Adding the opposite of 4.8}$$
$$= -7.9 \qquad \text{The sum of two negative numbers is negative.}$$

**EXAMPLE 10**  Subtract: $-7.9 - (-8.5)$.

$$-7.9 - (-8.5) = -7.9 + 8.5 \qquad \text{Adding the opposite of } -8.5$$
$$= 0.6 \qquad \text{Subtracting absolute values. The answer is positive since 8.5 has the larger absolute value.}$$

**Do Exercises 18–21.** ▶

Do Exercises 18–21. ▶

### d COMBINING LIKE TERMS

Recall that like, or similar, terms have exactly the same variable factors. To combine like terms, we add or subtract coefficients to form an equivalent expression.

**EXAMPLE 11**  Combine like terms: $3.2x + 4.6x$.

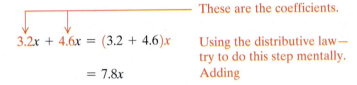
These are the coefficients.

$$3.2x + 4.6x = (3.2 + 4.6)x \qquad \text{Using the distributive law— try to do this step mentally.}$$
$$= 7.8x \qquad \text{Adding}$$

A similar procedure is used when subtracting like terms.

**EXAMPLE 12**  Combine like terms: $4.13a - 7.56a$.

$$4.13a - 7.56a = (4.13 - 7.56)a \qquad \text{Using the distributive law}$$
$$= (4.13 + (-7.56))a \qquad \text{Adding the opposite of 7.56}$$
$$= -3.43a \qquad \text{Subtracting absolute values. The coefficient is negative since } |-7.56| > |4.13|.$$

When more than one pair of like terms is present, we can rearrange the terms and then simplify.

**EXAMPLE 13**  Combine like terms: $5.7x - 3.9y - 2.4x + 4.5y$.

$$5.7x - 3.9y - 2.4x + 4.5y$$
$$= 5.7x + (-3.9y) + (-2.4x) + 4.5y \qquad \text{Rewriting as addition}$$
$$= 5.7x + (-2.4x) + (-3.9y) + 4.5y \qquad \text{Using the commutative law to rearrange}$$
$$= 3.3x + 0.6y \qquad \text{Combining like terms}$$

With practice, you will be able to perform many of the above steps mentally.

**Do Exercises 22–24.** ▶

Do Exercises 22–24. ▶

Subtract.

**18.** $9.25 - 13.41$

**19.** $-5.72 - 4.19$

**20.** $9.8 - (-2.6)$

**21.** $-5.9 - (-3.2)$

---

**CALCULATOR CORNER**

***Addition and Subtraction with Decimal Notation***  To use a calculator to add and subtract with decimal notation, we use the $\boxed{\cdot}$, $\boxed{+}$, $\boxed{-}$, and $\boxed{=}$ keys. To find $47.046 - 28.193$, for example, we press $\boxed{4}\,\boxed{7}\,\boxed{\cdot}\,\boxed{0}\,\boxed{4}\,\boxed{6}$ $\boxed{-}\,\boxed{2}\,\boxed{8}\,\boxed{\cdot}\,\boxed{1}\,\boxed{9}\,\boxed{3}\,\boxed{=}$. The display reads $\boxed{18.853}$, so $47.046 - 28.193 = 18.853$.

**EXERCISES:**  Use a calculator to perform the indicated operations.

**1.**
$$\begin{array}{r} 274.159 \\ + 43.486 \end{array}$$

**2.** $3.4 + 45 + 0.68$

**3.**
$$\begin{array}{r} 52.34 \\ - 18.51 \end{array}$$

**4.** $6.09 - 5.1$

**5.** $246 + (-3.07)$

**6.** $-12.7 - (-1.008)$

---

Combine like terms.

**22.** $5.8x - 2.1x$

**23.** $-5.9a + 7.6a$

**24.** $-4.8y + 7.5 + 2.1y - 2.1$

***Answers***
**18.** $-4.16$  **19.** $-9.91$  **20.** $12.4$
**21.** $-2.7$  **22.** $3.7x$  **23.** $1.7a$
**24.** $-2.7y + 5.4$

## ✓ Check Your Understanding

**Reading Check**   Complete each subtraction and its check by selecting a number from the list at the right.

**RC1.**
```
    2 3.7
  -  1.8 7 6
  ┌─────────┐
  └─────────┘
```
Check:
```
    2 1.8 2 4
  +  1.8 7 6
  ┌─────────┐
  └─────────┘
```

| |
|---|
| 23.7 |
| 187.623 |
| 21.824 |
| 40.9 |
| 1.876 |
| 146.723 |

**RC2.**
```
    1 8 7.6 2 3
  -    4 0.9
  ┌─────────────┐
  └─────────────┘
```
Check:
```
    1 4 6.7 2 3
  + ┌─────────────┐
    └─────────────┘
    1 8 7.6 2 3
```

**Concept Check**   Determine whether each result will be positive or negative, without performing the addition or subtraction.

**CC1.** $-4.03 + 1.58$

**CC2.** $-3.6087 + 4.1$

**CC3.** $-5.3 + (-2.6)$

**CC4.** $-1.3 - (-9.8)$

**CC5.** $11.29 - (-32.6)$

**CC6.** $-5.7 - (-0.8723)$

---

### a   Add.

**1.**
```
  426.25
+  38.12
```

**2.**
```
  641.823
+  14.915
```

**3.**
```
  659.403
+916.612
```

**4.**
```
  875.795
+324.862
```

**5.**
```
    9.104
+123.456
```

**6.**
```
    3.409
+ 81.001
```

**7.** $2.006 + 5.817$

**8.** $0.8096 + 0.7856$

**9.** $20.7 + 30.0124$

**10.** $0.263 + 0.8$

**11.** $1.06 + 9$

**12.** $12 + 18.08$

**13.** $0.34 + 3.5 + 0.127 + 768$

**14.** $2.3 + 0.729 + 23$

**15.** $17 + 3.24 + 0.256 + 0.3689$

**16.**
$$\begin{array}{r} 47.8 \\ 219.852 \\ 43.59 \\ + 666.713 \\ \hline \end{array}$$

**17.**
$$\begin{array}{r} 2.703 \\ 78.33 \\ 28.0009 \\ + 118.4341 \\ \hline \end{array}$$

**18.**
$$\begin{array}{r} 13.72 \\ 9.112 \\ 6542.7908 \\ + 23.901 \\ \hline \end{array}$$

**b**   Subtract.

**19.**
$$\begin{array}{r} 47.596 \\ - \ 6.215 \\ \hline \end{array}$$

**20.**
$$\begin{array}{r} 11.345 \\ - \ 2.105 \\ \hline \end{array}$$

**21.**
$$\begin{array}{r} 51.31 \\ - \ 2.29 \\ \hline \end{array}$$

**22.**
$$\begin{array}{r} 37.45 \\ - \ 6.32 \\ \hline \end{array}$$

**23.**
$$\begin{array}{r} 3.6 \\ - 0.036 \\ \hline \end{array}$$

**24.**
$$\begin{array}{r} 28.0 \\ - \ 0.28 \\ \hline \end{array}$$

**25.**
$$\begin{array}{r} 92.341 \\ - \ 6.42 \\ \hline \end{array}$$

**26.**
$$\begin{array}{r} 0.346 \\ - 0.0346 \\ \hline \end{array}$$

**27.**
$$\begin{array}{r} 3.0074 \\ - 1.3408 \\ \hline \end{array}$$

**28.**
$$\begin{array}{r} 32.7978 \\ - \ 0.0592 \\ \hline \end{array}$$

**29.**
$$\begin{array}{r} 6.07 \\ - 2.0078 \\ \hline \end{array}$$

**30.**
$$\begin{array}{r} 1.0 \\ - 0.9999 \\ \hline \end{array}$$

**31.** $30.24 - 0.241$

**32.** $100.12 - 0.112$

**33.** $34.07 - 30.7$

**34.** $36.2 - 16.28$

**35.** $8.45 - 7.405$

**36.** $3.801 - 2.81$

**37.** $6.003 - 2.3$

**38.** $1 - 0.0098$

**39.** $2 - 1.0908$

**40.** $100 - 0.34$

**41.** $624 - 18.79$

**42.** $7.48 - 2.6$

**43.** $57.803 - 4.6$

**44.** $25.008 - 12.4$

**45.** $263.7 - 102.08$

**46.** $19 - 1.198$

**47.** $45 - 0.999$

**48.** $10.05 - 0.392$

**c**  Add or subtract, as indicated.

**49.** $-5.02 + 1.73$

**50.** $-4.31 + 7.66$

**51.** $12.9 - 15.4$

**52.** $27.2 - 31.9$

**53.** $-2.9 + (-4.3)$

**54.** $-7.49 - 1.82$

**55.** $-4.301 + 7.68$

**56.** $-5.952 + 7.98$

**57.** $-12.9 - 3.7$

**58.** $-8.7 - 12.4$

**59.** $-2.1 - (-4.6)$

**60.** $-4.3 - (-2.5)$

**61.** $14.301 + (-17.82)$

**62.** $13.45 + (-18.701)$

**63.** $7.201 - (-2.4)$

**64.** $2.901 - (-5.7)$

**65.** $96.9 + (-21.4)$

**66.** $43.2 + (-10.9)$

**67.** $-3 - (-12.7)$

**68.** $-4.5 - (-7)$

**69.** $-4.9 - 5.392$

**70.** $89.3 - 100$

**71.** $14.7 - 15$

**72.** $-7.201 - 1.9$

**d**  Combine like terms.

**73.** $1.8x + 3.9x$

**74.** $7.9x + 1.3x$

**75.** $17.59a - 12.73a$

**76.** $23.28a - 15.79a$

**77.** $15.2t + 7.9 + 5.9t$

**78.** $29.5t - 4.8 + 7.6t$

**79.** $5.217x - 8.134x$

**80.** $6.317t - 9.429t$

**81.** $4.906y - 7.1 + 3.2y$

**82.** $9.108y + 4.2 + 3.7y$

**83.** $4.8x + 1.9y - 5.7x + 1.2y$

**84.** $3.2r - 4.1t - 5.6t + 1.9r$

**85.** $4.9 - 3.9t - 6 - 4.5t$

**86.** $5 + 9.7x - 7.2 - 12.8x$

## Skill Maintenance

Multiply.  [3.4b]

**87.** $\dfrac{3}{5} \cdot \dfrac{4}{7}$

**88.** $\dfrac{2}{9} \cdot \dfrac{7}{5}$

**89.** $\dfrac{3}{10} \cdot \dfrac{21}{100}$

Evaluate.  [2.6a]

**90.** $8 - 2x^2$, for $x = 3$

**91.** $5 - 3x^2$, for $x = -2$

**92.** $7 + 2x^2 \div 3$, for $x = 6$

## Synthesis

Combine like terms.

**93.** ▦ $-3.928 - 4.39a + 7.4b - 8.073 + 2.0001a - 9.931b - 9.8799a + 12.897b$

**94.** ▦ $79.02x + 0.0093y - 53.14z - 0.02001y - 37.987z - 97.203x - 0.00987y$

**95.** ▦ $39.123a - 42.458b - 72.457a + 31.462b - 59.491 + 37.927a$

**96.** Ryan presses the wrong key when using a calculator and adds 235.7 instead of subtracting it. The incorrect answer is 817.2. What is the correct answer?

**97.** Alicia presses the wrong key when using a calculator and subtracts 349.2 instead of adding it. The incorrect answer is $-836.9$. What is the correct answer?

**98.** ▦ Find the errors, if any, in the balances in this checkbook.

| 20___ | | RECORD ALL CHARGES OR CREDITS THAT AFFECT YOUR ACCOUNT | | | | | | |
|---|---|---|---|---|---|---|---|---|
| DATE | CHECK NUMBER | TRANSACTION DESCRIPTION | √ T | (−) PAYMENT/ DEBIT | | (+ OR −) OTHER | (+) DEPOSIT/ CREDIT | BALANCE FORWARD 2767 73 |
| 8/16 | 432 | Burch Laundry | | 23 56 | | | | 2744 16 |
| 8/19 | 433 | Rogers TV | | 20 49 | | | | 2764 65 |
| 8/20 | | Deposit | | | | | 85 00 | 2848 65 |
| 8/21 | 434 | Galaxy Records | | 48 60 | | | | 2801 05 |
| 8/22 | 435 | Electric Works | | 267 95 | | | | 2533 09 |
| | | | | | | | | |

Find *a*.

**99.**
$$
\begin{array}{r}
9\,3\,.\,a\,4\,3 \\
-\ 8\,7\,.\,9\,6\,9 \\
\hline
5\,.\,2\,7\,4
\end{array}
$$

**100.**
$$
\begin{array}{r}
4\,8\,1\,.\,a\,2\,4 \\
-\ \ 7\,2\,.\,9\,7\,8 \\
\hline
4\,0\,8\,.\,3\,4\,6
\end{array}
$$

## 5.3

### OBJECTIVES

a  Multiply using decimal notation.

b  Convert from notation like 45.7 million to standard notation, and convert between dollars and cents.

c  Evaluate algebraic expressions using decimal notation.

# Multiplication of Decimals

## a  MULTIPLICATION

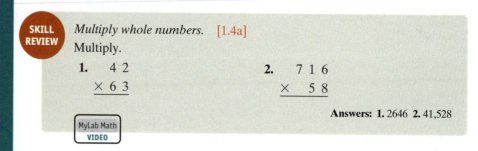

**SKILL REVIEW**

*Multiply whole numbers.*  [1.4a]

Multiply.

1.
$$\begin{array}{r} 4\ 2 \\ \times\ 6\ 3 \\ \hline \end{array}$$

2.
$$\begin{array}{r} 7\ 1\ 6 \\ \times\ \ \ 5\ 8 \\ \hline \end{array}$$

**Answers: 1.** 2646 **2.** 41,528

MyLab Math
VIDEO

To develop an understanding of decimal multiplication, consider $2.3 \times 1.12$. To find this product, we first convert each factor to fraction notation. Next, we multiply the numerators and then divide by the product of the denominators.

$$2.3 \times 1.12 = \frac{23}{10} \times \frac{112}{100} = \frac{23 \times 112}{10 \times 100} = \frac{2576}{1000} = 2.576$$

Note that the number of decimal places in the product is the sum of the numbers of decimal places in the factors.

$$\begin{array}{r} 1.1\ 2 \\ \times\ \ \ 2.3 \\ \hline 2.5\ 7\ 6 \end{array}$$  
(2 decimal places)  
(1 decimal place)  
(3 decimal places)

Now consider

$$0.011 \times 15.0002 = \frac{11}{1000} \times \frac{150{,}002}{10{,}000} = \frac{1{,}650{,}022}{10{,}000{,}000} = 0.1650022.$$

Again, note that the number of decimal places in the product is the sum of the numbers of decimal places in the factors.

$$\begin{array}{r} 1\ 5.0\ 0\ 0\ 2 \\ \times\ \ \ \ \ \ \ 0.0\ 1\ 1 \\ \hline 0.1\ 6\ 5\ 0\ 0\ 2\ 2 \end{array}$$  
(4 decimal places)  
(3 decimal places)  
(7 decimal places)

To multiply using decimals:

$0.8 \times 0.43$

a) Ignore the decimal points for the moment and multiply as though both factors were integers.

$$\begin{array}{r} {}^{2}\ \ \\ 0.4\ 3 \\ \times\ \ \ 0.8 \\ \hline 3\ 4\ 4 \end{array}$$  
Ignore the decimal points for now.

b) Place the decimal point in the result. The number of decimal places in the product is the sum of the numbers of places in the factors. (Count places from the right.)

$$\begin{array}{r} 0.4\ 3 \\ \times\ \ \ 0.8 \\ \hline 0.3\ 4\ 4 \end{array}$$  
(2 decimal places)  
(1 decimal place)  
(3 decimal places)

**EXAMPLE 1** Multiply: $8.3 \times 74.6$.

**a)** We ignore the decimal points and multiply as though both factors were integers.

$$
\begin{array}{r}
7\,4.6 \\
\times \qquad 8.3 \\
\hline
2\,2\,3\,8 \\
5\,9\,6\,8\,0 \\
\hline
6\,1\,9\,1\,8
\end{array}
$$

**b)** We place the decimal point in the result. The number of decimal places in the product is the sum of the numbers of decimal places in the factors, $1 + 1$, or 2.

$$
\begin{array}{rl}
7\,4.6 & (1 \text{ decimal place}) \\
\times \qquad 8.3 & (1 \text{ decimal place}) \\
\hline
2\,2\,3\,8 & \\
5\,9\,6\,8\,0 & \\
\hline
6\,1\,9.1\,8 & (2 \text{ decimal places})
\end{array}
$$

**Do Exercise 1.** ▶

**EXAMPLE 2** Multiply: $0.0032 \times 2148$.

$$
\begin{array}{rl}
2\,1\,4\,8 & (0 \text{ decimal places}) \\
\times\,0.0\,0\,3\,2 & (4 \text{ decimal places}) \\
\hline
4\,2\,9\,6 & \\
6\,4\,4\,4\,0 & \\
\hline
6.8\,7\,3\,6 & (4 \text{ decimal places})
\end{array}
$$

**EXAMPLE 3** Multiply: $-0.104 \times 0.86$.

We multiply the absolute values.

$$
\begin{array}{rl}
0.8\,6 & (2 \text{ decimal places}) \\
\times\quad 0.1\,0\,4 & (3 \text{ decimal places}) \\
\hline
3\,4\,4 & \\
8\,6\,0\,0 & \\
\hline
0.0\,8\,9\,4\,4 & (5 \text{ decimal places}) \qquad \text{Writing an extra zero}
\end{array}
$$

Since the product of a negative number and a positive number is negative, the answer is $-0.08944$.

**Do Exercises 2–4.** ▶

## Multiplying by 0.1, 0.01, 0.001, and So On

Now let's consider some special kinds of products. The first involves multiplying by a tenth, hundredth, thousandth, and so on. We can see a pattern in the following products.

$$0.1 \times 38 = \frac{1}{10} \times 38 = \frac{38}{10} = 3.8$$

$$0.01 \times 38 = \frac{1}{100} \times 38 = \frac{38}{100} = 0.38$$

$$0.001 \times 38 = \frac{1}{1000} \times 38 = \frac{38}{1000} = 0.038$$

Note in each case that the product is *smaller* than 38. That is, the decimal point in each product is farther to the left than the unwritten decimal point in 38. Also, each product can be obtained from 38 by moving the decimal point.

---

**1.** Multiply.

$$
\begin{array}{r}
8\,5.4 \\
\times \quad 6.2 \\
\hline
\end{array}
$$

Multiply.

**2.**
$$
\begin{array}{r}
1\,2\,3\,4 \\
\times\,0.0\,0\,4\,1 \\
\hline
\end{array}
$$

**GS** **3.** $42.65 \times 0.804$

$$
\begin{array}{r}
4\,2.6\,5 \\
\times \quad 0.8\,0\,4 \\
\hline
1\,\square\,0\,6\,0 \\
3\,4\,1\,2\,0 \quad 0 \\
\hline
3\,4.\,\square\,9\,0\,\square\,0
\end{array}
$$

**4.** $5.2014 \times (-2.41)$

To multiply any number by 0.1, 0.01, 0.001, and so on,

**a)** count the number of decimal places in the tenth, hundredth, or thousandth, and so on, and

$$0.001 \times 34.45678$$

→ 3 places

**b)** move the decimal point in the other number that many places to the left.

$$0.001 \times 34.45678 = 0.034.45678$$

Move 3 places to the left.

$$0.001 \times 34.45678 = 0.03445678$$

---

**Multiply.**

**5.** $0.1 \times 3.48$

**6.** $0.01 \times 3.48$

**7.** $(-0.001) \times 60.2$

**8.** $0.0001 \times 57$

---

**Multiply.**

**9.** $10 \times 3.48$

**10.** $100 \times 3.48$

**11.** $1000 \times (-83.9)$

**12.** $10,000 \times 57.043$

---

**EXAMPLES** Multiply.

**4.** $0.1 \times 14.605 = 1.4605$   1.4.605

**5.** $0.01 \times 14.605 = 0.14605$

**6.** $0.001 \times (-87) = -0.087$   We write an extra zero.

**7.** $0.0001 \times 23.9 = 0.00239$   We write two extra zeros.

◀ **Do Exercises 5–8.**

## Multiplying by 10, 100, 1000, and So On

Next, let's consider multiplying by 10, 100, 1000, and so on. We see a pattern in the following.

$$10 \times 97.34 = 973.4$$
$$100 \times 97.34 = 9734$$
$$1000 \times 97.34 = 97,340$$

Note in each case that the product is *larger* than 97.34. That is, the decimal point in each product is farther to the right than the decimal point in 97.34. Also, each product can be obtained from 97.34 by moving the decimal point.

---

To multiply any number by 10, 100, 1000, and so on,

**a)** count the number of zeros, and

$$1000 \times 34.45678$$

→ 3 zeros

**b)** move the decimal point in the other number that many places to the right.

$$1000 \times 34.45678 = 34.456.78$$

Move 3 places to the right.

$$1000 \times 34.45678 = 34,456.78$$

---

**EXAMPLES** Multiply.

**8.** $10 \times 14.605 = 146.05$   14.6.05

**9.** $100 \times 89.43 = 8943$

**10.** $1000 \times (-2.4167) = -2416.7$

**11.** $10,000 \times 7.52 = 75,200$   7.5200. We write two extra zeros.

◀ **Do Exercises 9–12.**

## b  NAMING LARGE NUMBERS; MONEY CONVERSION

### Naming Large Numbers

We often see notation like the following in newspapers and magazines and on television and the Internet.

- In 2017, the Internal Revenue Service processed 104.9 million income tax refunds totaling $290.43 billion.
  **Data:** Internal Revenue Service

- There are 3.04 trillion trees on the planet.
  **Data:** *Nature;* AmericanGrove.org

- From 1956 to 2016, the United States spent approximately $128.9 billion on the interstate highway system. Today, the cost to build such a system would be $234 billion.
  **Data:** Federal Highway Administration

- Americans spent an estimated $69.4 billion on pets in 2017.
  **Data:** American Pet Products Association

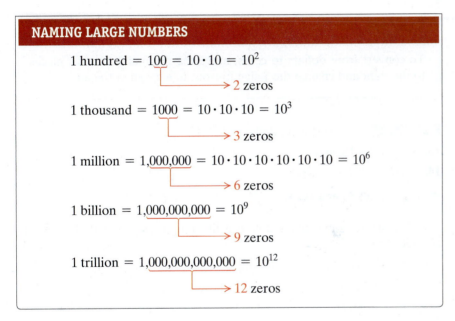

---

**NAMING LARGE NUMBERS**

1 hundred = 1$\underline{00}$ = 10 · 10 = $10^2$
       → 2 zeros

1 thousand = 1$\underline{000}$ = 10 · 10 · 10 = $10^3$
       → 3 zeros

1 million = 1,$\underline{000,000}$ = 10 · 10 · 10 · 10 · 10 · 10 = $10^6$
       → 6 zeros

1 billion = 1,$\underline{000,000,000}$ = $10^9$
       → 9 zeros

1 trillion = 1,$\underline{000,000,000,000}$ = $10^{12}$
       → 12 zeros

---

To convert a large number to standard notation, we proceed as follows.

**EXAMPLE 12** *Robocalls.* The number of robocalls received in the United States in 2017 totaled more than 2.5 billion a month. In June of that year, Atlanta's area code 404 received 50.5 million robocalls, more than any other area code. Convert 2.5 billion and 50.5 million to standard notation.
**Data:** YouMail

We convert 2.5 billion to standard notation.

2.5 billion = 2.5 × 1 billion

= 2.5 × 1,$\underline{000,000,000}$
      → 9 zeros

= 2,500,000,000    Moving the decimal point 9 places to the right

**13.** The largest building in the world is the Pentagon, which has 3.7 million square feet of floor space. Convert 3.7 million to standard notation.

**14.** *Spending on Health Care.* The United States spent $3.2 trillion on health care in 2015. Convert 3.2 trillion to standard notation.

**Data:** money.cnn.com, December 2, 2016

Convert from dollars to cents.

**15.** $15.69

$15.69 = 15.69 × $1

= 15.69 × ☐ ¢

= ☐ ¢

**16.** $0.17

Convert from cents to dollars.

**17.** 35¢

**18.** 577¢

*Answers*

**13.** 3,700,000  **14.** 3,200,000,000,000
**15.** 1569¢  **16.** 17¢  **17.** $0.35  **18.** $5.77

*Guided Solution:*

**15.** 100, 1569

---

We convert 50.5 million to standard notation.

50.5 million = 50.5 × 1 million

= 50.5 × 1,000,000

→ 6 zeros

= 50,500,000    Moving the decimal point 6 places to the right

◀ **Do Exercises 13 and 14.**

## Money Conversion

Converting from dollars to cents is like multiplying by 100. To see why, consider $19.43.

$19.43 = 19.43 × $1    We think of $19.43 as 19.43 × 1 dollar, or 19.43 × $1.

= 19.43 × 100¢    Substituting 100¢ for $1: $1 = 100¢

= 1943¢    Multiplying

---

**DOLLARS TO CENTS**

To convert from dollars to cents, move the decimal point two places to the right and change the $ sign in front to a ¢ sign at the end.

---

**EXAMPLES**  Convert from dollars to cents.

**13.** $189.64 = 18,964¢

**14.** $0.75 = 75¢

◀ **Do Exercises 15 and 16.**

Converting from cents to dollars is like multiplying by 0.01. To see why, consider 65¢.

65¢ = 65 × 1¢    We think of 65¢ as 65 × 1 cent, or 65 × 1¢.

= 65 × $0.01    Substituting $0.01 for 1¢: 1¢ = $0.01

= $0.65    Multiplying

---

**CENTS TO DOLLARS**

To convert from cents to dollars, move the decimal point two places to the left and change the ¢ sign at the end to a $ sign in front.

---

**EXAMPLES**  Convert from cents to dollars.

**15.** 395¢ = $3.95

**16.** 8503¢ = $85.03

◀ **Do Exercises 17 and 18.**

## c  EVALUATING

Algebraic expressions are often evaluated using numbers written in decimal notation.

**EXAMPLE 17**  Evaluate $Prt$ for $P = 780$, $r = 0.12$, and $t = 0.5$.

This product can be used to determine the interest paid on $780, borrowed at 12 percent simple interest, for half a year. We substitute and carry out the calculation.

$$Prt = 780 \cdot 0.12 \cdot 0.5 = 780 \cdot 0.06 = 46.8 \quad \text{This would represent } \$46.80.$$

**Do Exercise 19.** ▶

**19.** Evaluate $lwh$ for $l = 3.2$, $w = 2.6$, and $h = 0.8$. (This is the formula for the volume of a rectangular box.)

**EXAMPLE 18**  Find the perimeter of a stamp that is 3.25 cm long and 2.5 cm wide.

Recall that the perimeter, $P$, of a rectangle of length $l$ and width $w$ is given by the formula

$$P = 2l + 2w.$$

Thus, we evaluate $2l + 2w$ for $l = 3.25$ and $w = 2.5$.

$$2l + 2w = 2 \cdot 3.25 + 2 \cdot 2.5$$
$$= 6.5 + 5.0 \quad \text{Remember the rules for order of operations.}$$
$$= 11.5$$

The perimeter is 11.5 cm.

**Do Exercise 20.** ▶

**20.** Find the area of the stamp in Example 18.

**EXAMPLE 19**  *Multiple Births.*  The expression $0.278t + 30.31$ can be used to predict the rate of twin births, in twin births per 1000 births, in the United States $t$ years after 2000. Predict the rate of twin births in 2020.
**Data:** Centers for Disease Control and Prevention

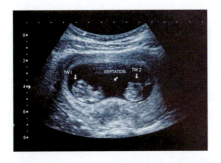

2020 is 20 years after 2000, so we evaluate $0.278t + 30.31$ for $t = 20$.

$$0.278t + 30.31 = 0.278 \cdot 20 + 30.31$$
$$= 5.56 + 30.31$$
$$= 35.87$$

In 2020, there will be approximately 35.87 twin births per 1000 births in the United States.

**Do Exercise 21.** ▶

**21.** Evaluate $6.28rh + 3.14r^2$ for $r = 1.5$ and $h = 5.1$. (This is the formula for the surface area of an open can.)

*Answers*
**19.** 6.656  **20.** 8.125 sq cm  **21.** 55.107

## ✓ Check Your Understanding

**Reading Check** Choose from the list below the factor that completes each multiplication.

**a)** 0.1  **b)** 0.01  **c)** 0.001  **d)** 0.0001  **e)** 0.00001  **f)** 0.000001

**RC1.** $6.287 \times \boxed{\phantom{x}} = 0.00006287$

**RC2.** $6.287 \times \boxed{\phantom{x}} = 0.0006287$

**RC3.** $6.287 \times \boxed{\phantom{x}} = 0.6287$

**RC4.** $6.287 \times \boxed{\phantom{x}} = 0.06287$

**RC5.** $6.287 \times \boxed{\phantom{x}} = 0.000006287$

**RC6.** $6.287 \times \boxed{\phantom{x}} = 0.006287$

**Concept Check** Match each expression with an equivalent expression from the list below.

**a)** 3800¢  **b)** 380  **c)** 0.038  **d)** \$3.80  **e)** 38,000  **f)** \$0.38

**CC1.** $0.001 \times 38$ _____

**CC2.** $1000 \times 38$ _____

**CC3.** 38¢ _____

**CC4.** \$38 _____

**CC5.** 380¢ _____

**CC6.** $10 \times 38$ _____

### a Multiply.

**1.**
$$\begin{array}{r} 6.8 \\ \times\ 7 \\ \hline \end{array}$$

**2.**
$$\begin{array}{r} 5.7 \\ \times 0.9 \\ \hline \end{array}$$

**3.**
$$\begin{array}{r} 0.84 \\ \times\ \ 8 \\ \hline \end{array}$$

**4.**
$$\begin{array}{r} 7.3 \\ \times 0.6 \\ \hline \end{array}$$

**5.**
$$\begin{array}{r} 6.3 \\ \times\ 0.0\ 4 \\ \hline \end{array}$$

**6.**
$$\begin{array}{r} 7.8 \\ \times\ 0.0\ 9 \\ \hline \end{array}$$

**7.**
$$\begin{array}{r} 8\ 7 \\ \times\ 0.0\ 0\ 6 \\ \hline \end{array}$$

**8.**
$$\begin{array}{r} 2\ 5.9 \\ \times\ 0.0\ 0\ 5 \\ \hline \end{array}$$

**9.** $10 \times 42.63$

**10.** $100 \times 2.8793$

**11.** $-1000 \times 783.686852$

**12.** $-0.34 \times 1000$

**13.** $-7.8 \times 100$

**14.** $0.00238 \times (-10)$

**15.** $0.1 \times 79.18$

**16.** $0.01 \times 789.235$

**17.** $0.001 \times 97.68$

**18.** $8976.23 \times 0.001$

**19.** $28.7 \times (-0.01)$

**20.** $0.0325 \times (-0.1)$

**21.**
$$\begin{array}{r} 2.7\ 3 \\ \times\ \ 1\ 6 \\ \hline \end{array}$$

**22.**
$$\begin{array}{r} 8.2\ 7 \\ \times\ \ \ 5.4 \\ \hline \end{array}$$

**23.**
$$\begin{array}{r} 0.9\ 8\ 4 \\ \times\ 0.0\ 3\ 1 \\ \hline \end{array}$$

**24.**
$$\begin{array}{r} 7.4\ 8\ 9 \\ \times\ \ \ \ 1.7 \\ \hline \end{array}$$

**25.** $(-37.4)(-2.4)$

**26.** $569(-1.05)$

**27.** $749(-0.43)$

**28.** $(-0.876)(-0.0204)$

**29.** 
$$
\begin{array}{r}
0.87 \\
\times\ \ 64 \\
\hline
\end{array}
$$

**30.** 
$$
\begin{array}{r}
7.25 \\
\times\ \ 60 \\
\hline
\end{array}
$$

**31.** 
$$
\begin{array}{r}
46.50 \\
\times\ \ 75 \\
\hline
\end{array}
$$

**32.** 
$$
\begin{array}{r}
8.24 \\
\times\ 703 \\
\hline
\end{array}
$$

**33.** $(-0.231)(-0.5)$

**34.** $(-12.3)(-1.08)$

**35.** $9.42 \times (-1000)$

**36.** $-7.6 \times 1000$

**37.** $-95.3 \times (-0.0001)$

**38.** $-4.23 \times (-0.001)$

 Convert from dollars to cents.

**39.** $57.06

**40.** $49.85

**41.** $0.95

**42.** $0.49

**43.** $0.01

**44.** $0.09

Convert from cents to dollars.

**45.** 72¢

**46.** 52¢

**47.** 2¢

**48.** 5¢

**49.** 6399¢

**50.** 5238¢

**51.** *U.S. National Debt.*   In 2017, the national debt of the United States reached $19.973 trillion. With a population of 325.7 million, this is about $61,300 per person. Convert 19.973 trillion and 325.7 million to standard notation.

**Data:** USdebtclock.org

**52.** *China's National Debt.*   In 2017, the national debt of China reached $2.032 trillion. China's population is 1.393 billion, so this debt is about $1460 per person. Convert 2.032 trillion and 1.393 billion to standard notation.

**Data:** USdebtclock.org

**53.** *Areas of Oceans.*   The area of the Pacific Ocean is 60.061 million sq mi. The area of the Atlantic Ocean is 29.638 million sq mi. Convert 60.061 million and 29.638 million to standard notation.

**Data:** *The World Almanac 2017*; International Hydrographic Commission

**54.** *Confined Feeding Operations.*   Most of the 41.5 million chickens sold each year in Indiana are raised in confined feeding operations (CFOs). There are approximately 1800 CFOs operating in Indiana. Convert 41.5 million to standard notation.

**Data:** Purdue College of Agriculture, *Envision*, Spring 2017

**55.** *Imports of Fruits and Vegetables.*   In 2015, 44% of U.S. imports of fruits and vegetables, with a total value of $10.4 billion, came from Mexico. Convert 10.4 billion to standard notation.

**Data:** Congressional Research Service; U.S. International Trade Commission

**56.** *Bottled Water.*   In 2016, 12.8 billion gallons of bottled water were sold in the United States, making bottled water the most popular beverage choice that year. Convert 12.8 billion to standard notation.

**Data:** Beverage Marketing Corporation

Evaluate.

**57.** $P + Prt$, for $P = 10,000$, $r = 0.04$, and $t = 2.5$
(*A formula for adding interest*)

**58.** $6.28r(h + r)$, for $r = 10$ and $h = 17.2$
(*Surface area of a cylinder*)

**59.** $vt + 0.5at^2$, for $v = 10$, $t = 1.5$, and $a = 9.8$
(*A physics formula*)

**60.** $4lh + 2h^2$, for $l = 3.5$ and $h = 1.2$
(*Surface area of a square prism*)

Find **(a)** the perimeter and **(b)** the area of a rectangular room with the given dimensions.

**61.** 12.5 ft long, 9.5 ft wide

**62.** 10.25 ft long, 8 ft wide

**63.** 8.4 m wide, 10.5 m long

**64.** 8.2 yd long, 6.4 yd wide

*Nursing.* The expression $0.0375t + 2.2$ can be used to predict the number of registered nurses, in millions, in the United States $t$ years after 2000. Estimate the number of registered nurses in the United States in the year indicated.

**Data:** Bureau of Labor Statistics, U.S. Dept. of Labor

**65.** 2015

**66.** 2020

## Skill Maintenance

Divide.

**67.** $-162 \div 6$  [2.5a]

**68.** $-216 \div (-6)$  [2.5a]

**69.** $-1035 \div (-15)$  [2.5a]

**70.** $-423 \div 3$  [2.5a]

**71.** $17\overline{)20,006}$  [1.5a]

**72.** $675 \div (-25)$  [2.5a]

## Synthesis

Consider the following names for large numbers in addition to those already discussed in this section:

$$1 \text{ quadrillion} = 1,000,000,000,000,000 = 10^{15};$$
$$1 \text{ quintillion} = 1,000,000,000,000,000,000 = 10^{18};$$
$$1 \text{ sextillion} = 1,000,000,000,000,000,000,000 = 10^{21};$$
$$1 \text{ septillion} = 1,000,000,000,000,000,000,000,000 = 10^{24}.$$

Find each of the following. Express the answer with a name that is a power of 10.

**73.** (1 trillion) $\cdot$ (1 billion)

**74.** (1 million) $\cdot$ (1 billion)

**75.** (1 trillion) $\cdot$ (1 trillion)

**76.** Is a billion millions the same as a million billions? Explain.

**77.** In Great Britain, France, and Germany, a billion means a million millions. Write standard notation for the British number 6.6 billion.

**78.** One light-year (ly) is $9.46 \times 10^{12}$ km. The star Regulus is 85 ly from the earth. How many billions of kilometers (km) from the earth is Regulus?

**Data:** *The Cambridge Factfinder*, 4th ed

**79.** 🖩 *Electric Bills*  At one time, electric bills from the Central Vermont Public Service Corporation consisted of a "customer charge" of $0.374 per day plus an "energy charge" of $0.1174 per kilowatt-hour (kWh) for the first 250 kWh used and $0.09079 per kilowatt-hour for each kilowatt-hour in excess of 250. From April 20 to May 20, the Coy-Bergers used 480 kWh of electricity. What was their bill for the period?

| **Central Vermont Public Service Corporation** | Coy-Berger R.R. 1 Braintree, VT 05060-9601 | Account No. **54879582** Meter No. **3621458** |
|---|---|---|

**ELECTRIC SERVICE**

| | kWh | |
|---|---|---|
| Current Meter Reading, 05/20 (Actual) | | 22571 |
| Previous Meter Reading, 04/20 (Actual) | − | 22091 |
| Amount of electricity Used in 30 Days | kWh | 480 |

Cost of Electricity Used for 30 Days Ending 05/20
| | | |
|---|---|---|
| Customer Charge . . . . . . . . . $0.374 per day | | $ |
| Energy Charge . . . . . . . | kWh × $0.11740 | $ |
| | kWh × $0.09079 | |

# Division of Decimals

### a | DIVISION

#### Whole-Number Divisors

**SKILL REVIEW**

*Divide whole numbers.* [1.5a]
Divide.

**1.** $5\overline{)245}$

**2.** $23\overline{)1978}$

**Answers: 1.** 49 **2.** 86

MyLab Math
VIDEO

**OBJECTIVES**

a Divide using decimal notation.

b Simplify expressions using the rules for order of operations.

To divide by a whole number,

a) place the decimal point directly above the decimal point in the dividend, and

b) divide as though dividing whole numbers.

$$
\begin{array}{r}
0.8\,4 \leftarrow \text{Quotient} \\
\text{Divisor} \rightarrow 7\,\overline{)\,5.8\,8} \leftarrow \text{Dividend} \\
5\,6 \\
\hline
2\,8 \\
2\,8 \\
\hline
0 \leftarrow \text{Remainder}
\end{array}
$$

**EXAMPLE 1**  Divide: $379.2 \div 8$.

$$
\begin{array}{r}
4\,7.4 \\
8\,\overline{)\,3\,7\,9.2} \\
3\,2 \\
\hline
5\,9 \\
5\,6 \\
\hline
3\,2 \\
3\,2 \\
\hline
0
\end{array}
$$

Place the decimal point.

Divide as though dividing whole numbers.

**EXAMPLE 2**  Divide: $82.08 \div 24$.

$$
\begin{array}{r}
3.4\,2 \\
2\,4\,\overline{)\,8\,2.0\,8} \\
7\,2 \\
\hline
1\,0\,0 \\
9\,6 \\
\hline
4\,8 \\
4\,8 \\
\hline
0
\end{array}
$$

Place the decimal point.

Divide as though dividing whole numbers.

Do Exercises 1–3. ▶

Divide.

**1.** $9\overline{)5.4}$

**2.** $15\overline{)22.5}$

**3.** $82\overline{)38.54}$

*Answers*
**1.** 0.6 **2.** 1.5 **3.** 0.47

We can think of a whole-number dividend as having a decimal point at the end, with as many zeros as we wish after the decimal point. For example, $12 = 12. = 12.0 = 12.00 = 12.000$, and so on. We can also add zeros after the last digit in the decimal portion of a number: $3.6 = 3.60 = 3.600$, and so on.

**EXAMPLE 3**   Divide: $30 \div 8$.

$$
\begin{array}{r}
3.\phantom{0} \\
8\,\overline{)\,3\,0.} \\
\underline{2\,4} \\
6
\end{array}
$$
Place the decimal point and divide to find how many ones.

$$
\begin{array}{r}
3.\phantom{0} \\
8\,\overline{)\,3\,0.0} \\
\underline{2\,4.}\!\downarrow \\
6\,0
\end{array}
$$
Write an extra zero.

$$
\begin{array}{r}
3.7 \\
8\,\overline{)\,3\,0.0} \\
\underline{2\,4}\phantom{.} \\
6\,0 \\
\underline{5\,6} \\
4
\end{array}
$$
Divide to find how many tenths.

$$
\begin{array}{r}
3.7\phantom{0} \\
8\,\overline{)\,3\,0.0\,0} \\
\underline{2\,4.}\phantom{00} \\
6\,0\phantom{0} \\
\underline{5\,6}\!\downarrow \\
4\,0
\end{array}
$$
Write another zero.

$$
\begin{array}{r}
3.7\,5 \\
8\,\overline{)\,3\,0.0\,0} \\
\underline{2\,4}\phantom{.00} \\
6\,0\phantom{0} \\
\underline{5\,6}\phantom{0} \\
4\,0 \\
\underline{4\,0} \\
0
\end{array}
$$
Divide to find how many hundredths.

Check:
$$
\begin{array}{r}
\phantom{0}^{6\ 4} \\
3.7\,5 \\
\times\phantom{000}8 \\
\hline
3\,0.0\,0
\end{array}
$$

**EXAMPLE 4**   Divide: $-4.5 \div 250$.

We first consider $4.5 \div 250$.

$$
\begin{array}{r}
0.0\,1\,8 \\
2\,5\,0\,\overline{)\,4.5\,0\,0} \\
\underline{2\,5\,0}\phantom{00} \\
2\,0\,0\,0 \\
\underline{2\,0\,0\,0} \\
0
\end{array}
$$

Check:
$$
\begin{array}{r}
0.0\,1\,8 \\
\times\phantom{00}2\,5\,0 \\
\hline
9\,0\,0 \\
3\,6\,0\,0 \\
\hline
4.5\,0\,0
\end{array}
$$

Since a negative number divided by a positive number is negative, the answer is $-0.018$.

◀ **Do Exercises 4–6.**

Divide.

**4.** $25\overline{)8}$     **5.** $-15 \div 4$

**6.** $2.15 \div 86$   **GS**

$$
\begin{array}{r}
0.\ \boxed{\phantom{0}}\,2\,\boxed{\phantom{0}} \\
8\,6\,\overline{)\,2.1\,5\,0} \\
\underline{1\,7\,2}\phantom{0} \\
4\,3\,\boxed{\phantom{0}} \\
\underline{4\,3\,0} \\
0
\end{array}
$$

*Answers*
**4.** 0.32   **5.** −3.75   **6.** 0.025
*Guided Solution:*
**6.** 0, 5, 0

## Divisors That Are Not Whole Numbers

Consider the division

$$0.24\overline{)8.208}$$

We write the division as $\dfrac{8.208}{0.24}$. Then we multiply by 1 to change to a whole-number divisor:

The division $0.24\overline{)8.208}$ is the same as $24\overline{)820.8}$.

$$\frac{8.208}{0.24} = \frac{8.208}{0.24} \times \frac{100}{100} = \frac{820.8}{24}.$$

The divisor is now a whole number.

---

To divide when the divisor is not a whole number:

**a)** move the decimal point (multiply by 10, 100, and so on) to make the divisor a whole number;

$$0.2\,4\,\overline{)8.2\,0\,8}$$

Move 2 places to the right.

**b)** move the decimal point in the dividend the same number of places (multiply the same way); and

$$0.2\,4\,\overline{)8.2\,0\,8}$$

Move 2 places to the right.

**c)** place the decimal point directly above the new decimal point in the dividend and divide as though dividing whole numbers.

$$
\begin{array}{r}
3\,4.2 \\
0.2\,4\,\overline{)8.2\,0_\wedge 8} \\
\underline{7\,2\phantom{0000}} \\
1\,0\,0\phantom{00} \\
\underline{9\,6\phantom{00}} \\
4\,8 \\
\underline{4\,8} \\
0
\end{array}
$$

(The new decimal point in the dividend is indicated by a caret.)

---

**EXAMPLE 5** Divide: $5.848 \div 8.6$.

$$8.6\,\overline{)5.8\,4\,8}$$

Multiply the divisor by 10. (Move the decimal point 1 place.) Multiply the same way in the dividend. (Move 1 place.)

$$
\begin{array}{r}
0.6\,8 \\
8.6\,\overline{)5.8_\wedge 4\,8} \\
\underline{5\,1\,6\phantom{0}} \\
6\,8\,8 \\
\underline{6\,8\,8} \\
0
\end{array}
$$

Place a decimal point above the new decimal point in the dividend and then divide.

*Note:* $\dfrac{5.848}{8.6} = \dfrac{5.848}{8.6} \cdot \dfrac{10}{10} = \dfrac{58.48}{86}$.

**Do Exercises 7–9.** ▶

---

Divide.

**GS** **7.** $0.375 \div 0.25$

$$\frac{0.375}{0.25} = \frac{0.375}{0.25} \times \frac{\boxed{\phantom{00}}}{100}$$

$$= \frac{37.5}{\boxed{\phantom{00}}}$$

$$
\begin{array}{r}
1.\boxed{\phantom{0}} \\
0.2\,5\,\overline{)0.3\,7_\wedge 5} \\
\underline{2\,5\phantom{00}} \\
1\,2\boxed{\phantom{0}} \\
\underline{1\,2\,5} \\
0
\end{array}
$$

**8.** $0.8\,3\,\overline{)4.0\,6\,7}$

**9.** $3.5\,\overline{)4\,4.8}$

---

*Answers*
**7.** 1.5  **8.** 4.9  **9.** 12.8
*Guided Solution:*
**7.** 100, 25; 5, 5

**EXAMPLE 6**  Divide: $-12 \div (-0.64)$.

Note first that a negative number divided by a negative number is positive. To find the quotient, we consider $12 \div 0.64$.

$$0.64\overline{)12.}$$

Place a decimal point at the end of the whole number.

$$0.64\overline{)12.00}$$

Multiply the divisor by 100. (Move the decimal point 2 places.) Multiply the same way in the dividend. (Move 2 places after adding extra zeros.)

$$
\begin{array}{r}
1\ 8.7\ 5 \\
0.64\overline{)1\ 2.0\ 0\ 0\ 0} \\
\underline{6\ 4} \\
5\ 6\ 0 \\
\underline{5\ 1\ 2} \\
4\ 8\ 0 \\
\underline{4\ 4\ 8} \\
3\ 2\ 0 \\
\underline{3\ 2\ 0} \\
0
\end{array}
$$

Place a decimal point above the new decimal point in the dividend and then divide.

We have $-12 \div (-0.64) = 18.75$.

◀ **Do Exercises 10 and 11.**

Divide.

**10.**  $1.6\overline{)2.5}$

**11.**  $-9 \div 0.03$

## Dividing by 10, 100, 1000, and So On

We can divide quickly by a ten, hundred, or thousand, or by a tenth, hundredth, or thousandth. Each procedure we use is based on multiplying by 1. Consider the following example:

$$\frac{23.789}{1000} = \frac{23.789}{1000} \cdot \frac{0.001}{0.001} = \frac{0.023789}{1} = 0.023789.$$

We are dividing by a number greater than 1: The result is *smaller* than 23.789.

To divide by 10, 100, 1000, and so on,

**a)** count the number of zeros in the divisor, and

$$\frac{713.49}{100}$$

2 zeros

**b)** write the quotient by moving the decimal point in the dividend that number of places to the left.

$$\frac{713.49}{100} = \frac{713.49}{100.} = \frac{7.1349}{1.00} = 7.1349$$

2 places to the left to change 100 to 1

**EXAMPLE 7**  Divide: $\dfrac{0.0104}{10}$.

$$\frac{0.0104}{10} = \frac{0.0104}{10.} = \frac{0.00104}{1.0} = 0.00104$$

1 zero     1 place to the left to change 10 to 1

*Answers*
**10.** 1.5625   **11.** −300

**EXAMPLE 8**  Divide $-213.75$ by 100.

$$\frac{-213.75}{100} = \frac{-213.75}{100.} = \frac{-2.1375}{1.00} = -2.1375$$

    2 zeros      2 places to the left

The answer is $-2.1375$.

Do Exercises 12 and 13. ▶

Divide.

**12.** $\dfrac{0.176}{100}$

**13.** $\dfrac{-98.47}{1000}$

## Dividing by 0.1, 0.01, 0.001, and So On

Now consider the following example:

$$\frac{23.789}{0.01} = \frac{23.789}{0.01} \cdot \frac{100}{100} = \frac{2378.9}{1} = 2378.9.$$

We are dividing by a number less than 1: The result is *larger* than 23.789.

---

To divide by 0.1, 0.01, 0.001, and so on,

**a)** count the number of decimal places in the divisor, and

$$\frac{713.49}{0.001}$$

    3 places ←

**b)** write the quotient by moving the decimal point in the dividend that number of places to the right.

$$\frac{713.49}{0.001} = \frac{713,490}{0.001} = \frac{713,490}{1} = 713,490$$

    3 places to the right to change 0.001 to 1

---

**EXAMPLE 9**  Divide: $\dfrac{67.8}{0.1}$.

$$\frac{67.8}{0.1} = \frac{67.8}{0.1} = \frac{678}{1} = 678$$

  1 place   1 place to the right to change 0.1 to 1

The answer is 678.

Do Exercises 14 and 15. ▶

Divide.

**14.** $\dfrac{-6.7832}{0.1}$

**15.** $\dfrac{12.78}{0.01}$

## b  ORDER OF OPERATIONS: DECIMAL NOTATION

The rules for order of operations apply when simplifying expressions involving decimal notation.

---

**RULES FOR ORDER OF OPERATIONS**

1. Do all calculations within grouping symbols first.
2. Evaluate all exponential expressions.
3. Do all multiplications and divisions in order from left to right.
4. Do all additions and subtractions in order from left to right.

---

*Answers*
**12.** 0.00176  **13.** −0.09847  **14.** −67.832
**15.** 1278

**EXAMPLE 10**   Simplify: $2.5 \times 25 \div 25,000 \times 250$.

$$2.5 \times 25 \div 25,000 \times 250 = 62.5 \div 25,000 \times 250$$
$$= 0.0025 \times 250$$
$$= 0.625.$$

Doing all multiplications and divisions in order from left to right

**EXAMPLE 11**   Simplify: $(5 - 0.06) \div 2 + 3.42 \times 0.1$.

$$(5 - 0.06) \div 2 + 3.42 \times 0.1 = 4.94 \div 2 + 3.42 \times 0.1$$

Working inside the parentheses

$$= 2.47 + 0.342$$

Multiplying and dividing in order from left to right

$$= 2.812$$

Simplify.

**16.** $625 \div 62.5 \times 25 \div 6250$
$$= \boxed{\phantom{00}} \times 25 \div 6250$$
$$= \boxed{\phantom{00}} \div 6250$$
$$= 0.04$$

**17.** $0.25 \cdot (1 + 0.08) - 0.0274$

**18.** $[(19.7 - 17.2)^2 + 3] \div (-1.25)$

**EXAMPLE 12**   Simplify: $13 - [5.4(1.3^2 + 0.21) \div 0.6]$.

$$13 - [5.4(1.3^2 + 0.21) \div 0.6]$$
$$= 13 - [5.4(1.69 + 0.21) \div 0.6]$$
$$= 13 - [5.4 \times 1.9 \div 0.6]$$

Working in the inner-most parentheses first

$$= 13 - [10.26 \div 0.6]$$

Multiplying

$$= 13 - 17.1$$

Dividing

$$= -4.1$$

◀ **Do Exercises 16–18.**

**EXAMPLE 13**   *Population Density.*   The table below shows the number of residents per square mile in the six New England states. Find the average of the numbers of residents per square mile for this group of states.

**19.** *Population Density.*   The table below shows the number of residents per square mile in five northwestern states. Find the average of the numbers of residents per square mile for this group of states.

| STATE | RESIDENTS PER SQUARE MILE |
|---|---|
| Washington | 107.9 |
| Oregon | 42.0 |
| Idaho | 20.0 |
| Montana | 7.1 |
| Wyoming | 6.0 |

DATA: 2015 U.S. Census Bureau

| STATE | NUMBER OF RESIDENTS PER SQUARE MILE |
|---|---|
| Maine | 43.1 |
| New Hampshire | 148.6 |
| Vermont | 67.9 |
| Massachusetts | 871.1 |
| Rhode Island | 1021.5 |
| Connecticut | 741.6 |

DATA: 2015 U.S. Census Bureau

The **average** of a set of numbers is the sum of the numbers divided by the number of addends. We find the sum of the population densities per square mile and divide it by the number of addends, 6:

$$\frac{43.1 + 148.6 + 67.9 + 871.1 + 1021.5 + 741.6}{6} = \frac{2893.8}{6} = 482.3.$$

Thus, the average of the numbers of residents per square mile for these six states is 482.3.

◀ **Do Exercise 19.**

**Answers**

**16.** 0.04   **17.** 0.2426   **18.** −7.4
**19.** 36.6 residents per square mile
***Guided Solution:***
**16.** 10, 250

### ✓ Check Your Understanding

**Reading Check**  Choose from the list below the divisor that completes each division.

**a)** 0.1　　　**b)** 0.01　　　**c)** 0.001　　　**d)** 0.0001　　　**e)** 0.00001　　　**f)** 0.000001

**RC1.** $\dfrac{40.345}{\square} = 4{,}034{,}500$

**RC2.** $\dfrac{40.345}{\square} = 4034.5$

**RC3.** $\dfrac{40.345}{\square} = 403{,}450$

**RC4.** $\dfrac{40.345}{\square} = 403.45$

**RC5.** $\dfrac{40.345}{\square} = 40{,}345$

**RC6.** $\dfrac{40.345}{\square} = 40{,}345{,}000$

**Concept Check**  Name the operation that should be performed first in evaluating each expression. Do not calculate.

**CC1.** $(2 - 0.04) \div 4 + 8.5$ _____

**CC2.** $0.02 + 2.06 \div 0.01$ _____

**CC3.** $5 \times 2.1 + 0.1 - 8^3$ _____

**CC4.** $18.2 - (4.1 + 6.9)$ _____

**CC5.** $16 - 9 \div 3 + 7.3$ _____

**CC6.** $4(10 - 5) \times 14.2$ _____

---

**a**　Divide.

**1.** $2\overline{)5.98}$

**2.** $5\overline{)13.5}$

**3.** $4\overline{)95.12}$

**4.** $8\overline{)25.92}$

**5.** $12\overline{)84.96}$

**6.** $23\overline{)25.07}$

**7.** $15\overline{)18}$

**8.** $30\overline{)54}$

**9.** $5.4 \div (-6)$

**10.** $3.6 \div (-4)$

**11.** $-30 \div 0.005$

**12.** $-100 \div 0.0002$

**13.** $0.06\overline{)8.4}$

**14.** $0.04\overline{)1.68}$

**15.** $2.6\overline{)104}$

**16.** $3.2\overline{)192}$

**17.** $1.8 \div (-12)$

**18.** $6 \div (-15)$

**19.** $8.5\overline{)27.2}$

**20.** $6.2\overline{)46.5}$

**21.** $-31.59 \div 8.1$

**22.** $-39.06 \div 4.2$

**23.** $-5 \div (-8)$

**24.** $-7 \div (-8)$

**25.** $0.47\overline{)0.1222}$

**26.** $0.54\overline{)0.27}$

**27.** $0.032\overline{)0.07488}$

**28.** $0.017\overline{)1.581}$

**29.** $-24.969 \div 82$

**30.** $-25.221 \div 42$

**31.** $\dfrac{213.4567}{100}$

**32.** $\dfrac{769.3265}{1000}$

**33.** $\dfrac{-23.59}{10}$

**34.** $\dfrac{-83.57}{10}$

**35.** $\dfrac{1.0237}{0.001}$

**36.** $\dfrac{3.4029}{0.001}$

**37.** $\dfrac{-92.36}{0.01}$

**38.** $\dfrac{-56.78}{0.001}$

**39.** $\dfrac{0.8172}{10}$

**40.** $\dfrac{0.5678}{1000}$

**41.** $\dfrac{0.97}{0.1}$

**42.** $\dfrac{0.97}{0.001}$

**43.** $\dfrac{52.7}{-1000}$

**44.** $\dfrac{8.9}{-100}$

**45.** $\dfrac{75.3}{-0.001}$

**46.** $\dfrac{63.47}{-0.1}$

**47.** $\dfrac{-75.3}{1000}$

**48.** $\dfrac{23,001}{100}$

**b**  Simplify.

**49.** $14 \times (82.6 + 67.9)$

**50.** $(26.2 - 14.8) \times 12$

**51.** $0.003 + 3.03 \div (-0.01)$

**52.** $42 \times (10.6 + 0.024)$

**53.** $(4.9 - 18.6) \times 13$

**54.** $4.2 \times 5.7 + 0.7 \div 3.5$

**55.** $210.3 - 4.24 \times 1.01$

**56.** $-7.32 + 0.04 \div 0.1^2$

**57.** $0.04 \times 0.1 \div 0.4 \times 50$

**58.** $30 \div 0.2 \times 0.4 \div 10$

**59.** $12 \div (-0.03) - 12 \times 0.03^2$

**60.** $(5 - 0.04)^2 \div 4 + 8.7 \times 0.4$

**61.** $(4 - 2.5)^2 \div 100 + 0.1 \times 6.5$

**62.** $4 \div 0.4 - 0.1 \times 5 + 0.1^2$

**63.** $6 \times 0.9 - 0.1 \div 4 + 0.2^3$

**64.** $5.5^2 \times [(6 - 7.8) \div 0.06 + 0.12]$

**65.** $12^2 \div (12 + 2.4) - [(2 - 2.4) \div 0.8]$

**66.** $0.01 \times \{[(4 - 0.25) \div 2.5] - (4.5 - 4.025)\}$

**67.** *Mountain Peaks in Colorado.* The elevations of four mountain peaks in Colorado are listed in the table below. Find the average elevation of these peaks.

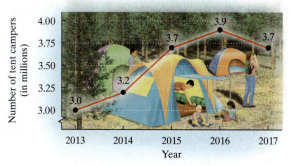

| MOUNTAIN PEAK | ELEVATION (in feet) |
|---|---|
| Mount Elbert | 14,440 |
| Mount Evans | 14,271 |
| Pikes Peak | 14,115 |
| Crested Butte | 12,168 |

**68.** *Geographical Mobility.* The table below shows the numbers of people who moved in the United States for the years 2012–2016. Find the average number of people who moved per year during this period.

| YEAR | NUMBER WHO MOVED (in millions) |
|---|---|
| 2012 | 36.5 |
| 2013 | 35.9 |
| 2014 | 35.7 |
| 2015 | 36.3 |
| 2016 | 35.1 |

DATA: U.S. Census Bureau

**69.** *Camping in National Parks.* The graph below shows the numbers of tent campers in Park Service campgrounds in the National Park system from 2013 to 2017. Find the average number of tent campers per year during this period.

**Camping in Park Service Campgrounds**

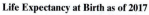

DATA: U.S. National Park Service

**70.** *Life Expectancy.* The information in the following bar graph shows the six highest life expectancies in the world. Determine the average life expectancy for these countries.

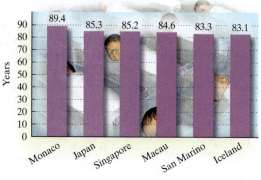

DATA: *CIA World Factbook*

The following table lists the lengths of the longest railway tunnels in the world. Use the table for Exercises 71 and 72.

| TUNNEL | Gotthard Base, Switzerland | Seikan, Japan | Yulhyeon, South Korea | Channel, France/U.K. | Songshan Lake, China |
|---|---|---|---|---|---|
| LENGTH, IN MILES | 35.483 | 33.46 | 32.5 | 31.35 | 24.117 |
| LENGTH, IN KILOMETERS | 57.104 | 53.85 | 52.3 | 50.45 | 38.813 |

DATA: worldatlas.com

**71.** Find the average length of the tunnels, in miles.

**72.** Find the average length of the tunnels, in kilometers.

## Skill Maintenance

**73.** Add: $10\frac{1}{2} + 4\frac{5}{8}$.  [4.6a]

**74.** Subtract: $10\frac{1}{2} - 4\frac{5}{8}$.  [4.6b]

Evaluate.  [1.9b]

**75.** $7^3$

**76.** $2^6$

Solve.  [1.7b]

**77.** $235 = 5 \cdot z$

**78.** $q + 31 = 72$

## Synthesis

Calculate each of the following.

**79.** ▦ $7.434 \div (-1.2) \times 9.5 + 1.47^2$

**80.** ▦ $-9.46 \times 2.1^2 \div 3.5 + 4.36$

**81.** ▦ $9.0534 - 2.041^2 \times 0.731 \div 1.043^2$

**82.** ▦ $23.042(7 - 4.037 \times 1.46 - 0.932^2)$

Solve.

**83.** $439.57 \times 0.01 \div 1000 \cdot x = 4.3957$

**84.** $5.2738 \div 0.01 \times 1000 \div t = 52.738$

**85.** $0.0329 \div 0.001 \times 10^4 \div x = 3290$

**86.** $-4.302 \times 0.1^2 \div 0.001 \cdot t = -430.2$

**87.** ▦ *Television Ratings.*  A television rating point represents 1,150,000 households. The 2018 Super Bowl was viewed in approximately 49.6 million households. How many rating points did the game receive? Round to the nearest tenth.

Data: Sportsmediawatch.com

**88.** *Size of Country.*  The world's largest country, Russia, has an area of approximately 6.6 million square miles. The smallest country, Vatican City, has an area of approximately 0.2 square mile. How many times larger is Russia?

Data: U.S. Census Bureau, International Data Base

▦ *Electric Bills.*  At one time, electric bills from the Central Vermont Public Service Corporation consisted of a "customer charge" of $0.374 per day plus an "energy charge" of $0.1174 per kilowatt-hour (kWh) for the first 250 kWh used and $0.09079 per kilowatt-hour for each kilowatt-hour in excess of 250.

**89.** From August 20 to September 20, the Kaufmans' bill was $59.10. How many kilowatt-hours of electricity did they use? (Round to the nearest kilowatt-hour.)

**90.** From July 20 to August 20, the McGuires' bill was $70. How many kilowatt-hours of electricity did they use? (Round to the nearest kilowatt-hour.)

# Mid-Chapter Review

## Concept Reinforcement

Determine whether each statement is true or false.

_____ **1.** In the number 308.00567, the digit 6 names the tens place. [5.1a]

_____ **2.** When writing a word name for decimal notation, we write the word "and" for the decimal point. [5.1a]

_____ **3.** On the number line, $-2.3$ is to the left of $-2.2$. [5.1c]

## Guided Solutions

 Fill in each blank with the number that creates a correct statement or solution.

**4.** Evaluate $P(1 + r)$ for $P = 5000$ and $r = 0.045$. [5.3c]

$$P(1 + r) = \boxed{\phantom{00}}\ (1 + \boxed{\phantom{00}}\ ) \qquad \text{Substituting}$$
$$= 5000(\boxed{\phantom{00}}\ ) \qquad \text{Adding within parentheses}$$
$$= \boxed{\phantom{00}} \qquad \text{Multiplying}$$

**5.** Simplify: $5.6 + 4.3 \times (6.5 - 0.25)^2$. [5.4b]

$$5.6 + 4.3 \times (6.5 - 0.25)^2 = 5.6 + 4.3 \times (\boxed{\phantom{00}})^2 \qquad \text{Carrying out the operation inside parentheses}$$
$$= 5.6 + 4.3 \times \boxed{\phantom{00}} \qquad \text{Evaluating the exponential expression}$$
$$= 5.6 + \boxed{\phantom{00}} \qquad \text{Multiplying}$$
$$= \boxed{\phantom{00}} \qquad \text{Adding}$$

## Mixed Review

**6.** *Indianapolis 500.* Tony Kanaan of Brazil won the 2013 Indianapolis 500 with a record average speed of 187.433 mph. The 2017 winner, Takuma Sato of Japan, won with an average speed of 155.395 mph. The difference in these average speeds is 32.038 mph. Write a word name for 32.038. [5.1a]

**Data:** Indianapolis Motor Speedway

**7.** *Top Movie in Theater Revenue.* *Avatar*, released in 2009, is the world's top-grossing movie with respect to box-office revenue. As of 2017, its box-office revenue had reached \$2.784 billion. Convert 2.784 billion to standard notation. [5.3b]

**Data:** Box-Office Mojo

Write each number as a fraction and, if possible, as a mixed numeral. [5.1b]

**8.** 4.53

**9.** 0.287

Which number is larger? [5.1c]

**10.** 0.07, 0.13

**11.** $-5.2, -5.09$

Write decimal notation. [5.1b]

**12.** $\dfrac{7}{10}$

**13.** $\dfrac{639}{100}$

**14.** $-35\dfrac{67}{100}$

**15.** $8\dfrac{2}{1000}$

Round 28.4615 to the nearest: [5.1d]

**16.** Thousandth.

**17.** Hundredth.

**18.** Tenth.

**19.** One.

Add. [5.2a], [5.2c]

20.
$$\begin{array}{r} 47.638 \\ +\ 2.457 \\ \hline \end{array}$$

21.
$$\begin{array}{r} 15.6 \\ 234.729 \\ 3.08 \\ +961.453 \\ \hline \end{array}$$

22. $-10.5 + 0.27$

23. $16 + 0.34 + 1.9$

Subtract. [5.2b], [5.2c]

24.
$$\begin{array}{r} 321.57 \\ -\ 49.38 \\ \hline \end{array}$$

25.
$$\begin{array}{r} 5.6 \\ -0.007 \\ \hline \end{array}$$

26. $34.3 - 18.75$

27. $-6.9 - 13$

Multiply. [5.3a]

28.
$$\begin{array}{r} 4.6 \\ \times 0.9 \\ \hline \end{array}$$

29.
$$\begin{array}{r} 15.3 \\ \times 6.07 \\ \hline \end{array}$$

30. $100 \times 81.236$

31. $0.1 \times (-0.483)$

Divide. [5.4a]

32. $-20.24 \div (-4)$

33. $21.76 \div 6.8$

34. $76.3 \div 0.1$

35. $914.036 \div 1000$

36. Convert $20.45 to cents. [5.3b]

37. Convert 147¢ to dollars. [5.3b]

38. Combine like terms: $3.08x - 7.1 - 4.3x$. [5.2d]

39. Evaluate $2(l + w)$ for $l = 1.3$ and $w = 0.8$. [5.3c]

Simplify. [5.4b]

40. $6.594 + 0.5318 \div 0.01$

41. $7.3 \times 4.6 - 0.8 \div 3.2$

## Understanding Through Discussion and Writing

42. A classmate rounds 236.448 to the nearest one and gets 237. Explain the possible error. [5.1d]

43. Explain the error in the following:
Subtract.
$$73.089 - 5.0061 = 2.3028 \quad [5.2b]$$

44. Explain why $10 \div 0.2 = 100 \div 2$. [5.4a]

45. Kayla made these two computational mistakes:
$$0.247 \div 0.1 = 0.0247; \quad 0.247 \div 10 = 2.47.$$
In each case, how could you convince her that a mistake has been made? [5.4a]

Copyright © 2020 Pearson Education, Inc.

# Using Fraction Notation with Decimal Notation

# 5.5

## OBJECTIVES

**a**   Use division to convert fraction notation to decimal notation.

**b**   Round numbers named by repeating decimals.

**c**   Convert certain fractions to decimal notation by using equivalent fractions.

**d**   Simplify expressions that contain both fraction notation and decimal notation.

## a   USING DIVISION TO FIND DECIMAL NOTATION

Recall that $\frac{a}{b}$ means $a \div b$. Thus, using division, we can express *any* fraction as a decimal. This means that any *rational* number (ratio of integers) can be written as a decimal.

**EXAMPLE 1**   Find decimal notation for $\frac{3}{20}$.

Because $\frac{3}{20}$ means $3 \div 20$, we can perform long division.

$$
\begin{array}{r}
0.1\,5 \\
20\,)\overline{3.0\,0} \\
\underline{2\,0}\phantom{0} \\
1\,0\,0 \\
\underline{1\,0\,0} \\
0
\end{array}
$$

> We are finished when the remainder is 0.

We have $\frac{3}{20} = 0.15$.

**EXAMPLE 2**   Find decimal notation for $\frac{-7}{8}$.

Since $\frac{-7}{8}$ means $-7 \div 8$ and a negative number divided by a positive number is negative, we know that the decimal will be negative. We divide 7 by 8 and make the result negative.

$$
\begin{array}{r}
0.8\,7\,5 \\
8\,)\overline{7.0\,0\,0} \\
\underline{6\,4}\phantom{00} \\
6\,0\phantom{0} \\
\underline{5\,6}\phantom{0} \\
4\,0 \\
\underline{4\,0} \\
0
\end{array}
$$

Thus, $\frac{-7}{8} = -0.875$.

**Do Exercises 1 and 2.** ▶

When division with decimals ends with a remainder of 0, or *terminates*, as in Examples 1 and 2, the result is called a **terminating decimal**. If the division does not terminate, the result will be a **repeating decimal**.

Find decimal notation.

**1.** $\dfrac{2}{5}$        **2.** $\dfrac{-5}{8}$

*Answers*
**1.** 0.4   **2.** −0.625

Find decimal notation.

**3.** $\dfrac{1}{6}$

$$\dfrac{1}{6} = \boxed{\phantom{0}} \div 6$$

$$
\begin{array}{r}
0\,.\,1\phantom{0}\boxed{\phantom{0}}\,6 \\
\boxed{\phantom{0}}\overline{)1\,.\,0\ 0\ 0} \\
6\phantom{00000} \\
\hline
4\ 0\phantom{000} \\
3\ 6\phantom{000} \\
\hline
\boxed{\phantom{0}}\ 0\phantom{0} \\
3\ 6 \\
\hline
\boxed{\phantom{0}}
\end{array}
$$

$$\dfrac{1}{6} = 0.1666\ldots = 0.1\overline{6}$$

**4.** $\dfrac{2}{3}$

Find decimal notation.

**5.** $\dfrac{5}{11}$    **6.** $-\dfrac{12}{11}$

**EXAMPLE 3**  Find decimal notation for $\frac{5}{6}$.

Since $\frac{5}{6}$ means $5 \div 6$, we have

$$
\begin{array}{r}
0\,.\,8\ 3\ 3 \\
6\overline{)5\,.\,0\ 0\ 0} \\
4\ 8\phantom{0000} \\
\hline
2\ 0\phantom{00} \\
1\ 8\phantom{00} \\
\hline
2\ 0\phantom{0} \\
1\ 8 \\
\hline
2
\end{array}
$$

Since 2 keeps reappearing as a remainder, the digits repeat and will continue to do so; therefore,

$$\dfrac{5}{6} = 0.83333\ldots.$$   The dots indicate an endless sequence of repeating digits in the quotient.

When there is a repeating pattern, we often use an overbar to indicate the repeating part, in this case, only the 3.

$$\dfrac{5}{6} = 0.8\overline{3}$$

◀ **Do Exercises 3 and 4.**

**EXAMPLE 4**  Find decimal notation for $-\frac{4}{11}$.

Since $-\frac{4}{11}$ is negative, we divide 4 by 11 and make the result negative.

$$
\begin{array}{r}
0\,.\,3\ 6\ 3\ 6 \\
1\,1\overline{)4\,.\,0\ 0\ 0\ 0} \\
3\ 3\phantom{00000} \\
\hline
7\ 0\phantom{000} \\
6\ 6\phantom{000} \\
\hline
4\ 0\phantom{00} \\
3\ 3\phantom{00} \\
\hline
7\ 0\phantom{0} \\
6\ 6 \\
\hline
4
\end{array}
$$

Since 7 and 4 keep repeating as remainders, the sequence of digits "36" repeats in the quotient, and

$$\dfrac{4}{11} = 0.363636\ldots, \quad \text{or} \quad 0.\overline{36}.$$

Thus, $-\frac{4}{11} = -0.\overline{36}$.

◀ **Do Exercises 5 and 6.**

When a fraction is written in simplified form, we can tell from the denominator whether its decimal notation will repeat or terminate.

For a fraction in simplified form,

• if the denominator has a prime factor other than 2 or 5, the decimal notation repeats.

• if the denominator has no prime factor other than 2 or 5, the decimal notation terminates.

**Answers**

**3.** $0.1\overline{6}$  **4.** $0.\overline{6}$  **5.** $0.\overline{45}$  **6.** $-1.\overline{09}$

*Guided Solution:*

**3.** 1, 6, 6, 4, 4

**EXAMPLE 5** Find decimal notation for $\frac{3}{7}$.

Because 7 is not a product of 2's and/or 5's, we expect a repeating decimal.

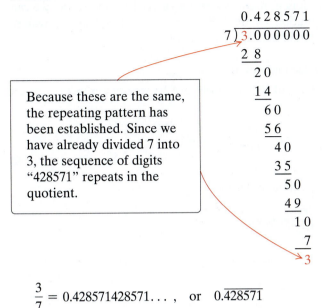

$$
\frac{3}{7} = 0.428571428571\ldots, \quad \text{or} \quad 0.\overline{428571}
$$

**Do Exercise 7.** ▶

## b  ROUNDING REPEATING DECIMALS

The repeating part of a decimal can be so long that it will not fit on a calculator display. For example, when $\frac{5}{97}$ is written as a decimal, its repeating part is 96 digits long! Most calculators round repeating decimals to 9 or 10 decimal places.

In applied problems, repeating decimals are rounded to get approximate answers. To round a repeating decimal, we can extend the decimal notation at least one place past the rounding digit and then round as before.

**EXAMPLES** Round each to the nearest tenth, hundredth, and thousandth.

| | Nearest tenth | Nearest hundredth | Nearest thousandth |
|---|---|---|---|
| **6.** $0.8\overline{3} = 0.83333\ldots$ | 0.8 | 0.83 | 0.833 |
| **7.** $3.\overline{09} = 3.090909\ldots$ | 3.1 | 3.09 | 3.091 |
| **8.** $-4.1\overline{763} = -4.1763763\ldots$ | $-4.2$ | $-4.18$ | $-4.176$ |

**Do Exercises 8–11.** ▶

**EXAMPLE 9**  *Gas Mileage.* A car travels 457 mi on 16.4 gal of gasoline. The ratio of number of miles driven to amount of gasoline used is *gas mileage*. Find the gas mileage, and convert the ratio to decimal notation rounded to the nearest tenth.

$$
\frac{\text{Miles driven}}{\text{Gasoline used}} = \frac{457}{16.4} \approx 27.86 \qquad \textcolor{red}{\text{Dividing to 2 decimal places}}
$$
$$
\approx 27.9 \qquad \textcolor{red}{\text{Rounding to 1 decimal place}}
$$

The gas mileage is about 27.9 miles per gallon, or 27.9 mpg.

**Do Exercise 12.** ▶

---

**7.** Find decimal notation for $\frac{5}{7}$.

Round each to the nearest tenth, hundredth, and thousandth.

**8.** $0.\overline{6}$       **9.** $0.6\overline{08}$

**10.** $-7.3\overline{49}$     **11.** $2.6\overline{891}$

**12.** *Gas Mileage.* A car travels 380 mi on 15.7 gal of gasoline. Find the gas mileage, and convert the ratio to decimal notation rounded to the nearest tenth.

***Answers***
**7.** $0.\overline{714285}$   **8.** 0.7; 0.67; 0.667
**9.** 0.6; 0.61; 0.608   **10.** $-7.3$; $-7.35$; $-7.349$
**11.** 2.7; 2.69; 2.689   **12.** 24.2 miles per gallon

Multiply by a form of 1 to find decimal notation for each number.

**13.** $\dfrac{4}{5}$

**14.** $-\dfrac{9}{20}$

**15.** $\dfrac{7}{200}$

**16.** $\dfrac{33}{25}$

## c   MORE WITH CONVERSIONS

Fractions like $\frac{3}{10}$ or $-\frac{71}{1000}$ can be converted to decimal notation without using long division. When a denominator is a factor of 10, 100, and so on, we can convert to decimal notation by finding (perhaps mentally) an equivalent fraction in which the denominator is a power of 10.

**EXAMPLE 10**   Find decimal notation for $-\frac{7}{500}$.

Since $500 \cdot 2 = 1000$, and 1000 is a power of 10, we use $\frac{2}{2}$ as an expression for 1 and multiply.

$$-\frac{7}{500} = -\frac{7}{500} \cdot \frac{2}{2} = -\frac{14}{1000} = -0.014$$

*Think*: $1000 \div 500 = 2$, and $7 \cdot 2 = 14$.

**EXAMPLE 11**   Find decimal notation for $\frac{9}{25}$.

$$\frac{9}{25} = \frac{9}{25} \cdot \frac{4}{4} = \frac{36}{100} = 0.36$$

Using $\frac{4}{4}$ for 1 to get a denominator of 100

As a check, we can divide.

$$
\begin{array}{r}
0.3\ 6 \\
2\,5\overline{)9.0\ 0} \\
7\ 5 \\
\hline
1\ 5\ 0 \\
1\ 5\ 0 \\
\hline
0
\end{array}
$$

Note that multiplication by 1 is much faster.

**EXAMPLE 12**   Find decimal notation for $\frac{7}{4}$.

$$\frac{7}{4} = \frac{7}{4} \cdot \frac{25}{25} = \frac{175}{100} = 1.75$$

Using $\frac{25}{25}$ for 1 to get a denominator of 100. You might also note that 7 quarters is $1.75.

◀ **Do Exercises 13–16.**

## d   CALCULATIONS WITH FRACTION AND DECIMAL NOTATION TOGETHER

Fraction notation and decimal notation can occur together in a calculation. In such cases, there are at least three ways in which we might proceed.

**EXAMPLE 13**   Calculate: $\frac{2}{3} \times 0.576$.

**Method 1:**   Perhaps the quickest method is to treat 0.576 as $\frac{0.576}{1}$. Then we multiply 0.576 by 2 and divide the result by 3.

$$\frac{2}{3} \times 0.576 = \frac{2}{3} \times \frac{0.576}{1}$$

$$= \frac{2 \times 0.576}{3 \times 1} = \frac{1.152}{3}$$

$$= 0.384$$

$$
\begin{array}{r}
0.3\ 8\ 4 \\
3\overline{)1.1\ 5\ 2} \\
9 \\
\hline
2\ 5 \\
2\ 4 \\
\hline
1\ 2 \\
1\ 2 \\
\hline
0
\end{array}
$$

*Answers*

**13.** 0.8   **14.** −0.45   **15.** 0.035   **16.** 1.32

**Method 2:** A second way to do this calculation is to convert the fraction notation to decimal notation so that both numbers are in decimal notation. Since $\frac{2}{3}$ converts to repeating decimal notation, it is first rounded to some chosen decimal place. We choose three decimal places because 0.576 has three decimal places. Then, using decimal notation, we multiply.

$$\frac{2}{3} \times 0.576 = 0.\overline{6} \times 0.576 \approx 0.667 \times 0.576 = 0.384192$$

**Method 3:** A third method is to convert the decimal notation to fraction notation so that both numbers are in fraction notation. The answer can be left in fraction notation and simplified, or we can convert back to decimal notation and, if appropriate, round.

$$\frac{2}{3} \times 0.576 = \frac{2}{3} \cdot \frac{576}{1000} = \frac{2 \cdot 576}{3 \cdot 1000}$$

$$= \frac{2 \cdot 2 \cdot 2 \cdot 2 \cdot 2 \cdot 2 \cdot 2 \cdot 3 \cdot 3}{2 \cdot 2 \cdot 2 \cdot 3 \cdot 5 \cdot 5 \cdot 5} \quad \text{Factoring}$$

$$= \frac{2 \cdot 2 \cdot 2 \cdot 3}{2 \cdot 2 \cdot 2 \cdot 3} \cdot \frac{2 \cdot 2 \cdot 2 \cdot 2 \cdot 3}{5 \cdot 5 \cdot 5} \quad \text{Removing a factor equal to 1: } \frac{2 \cdot 2 \cdot 2 \cdot 3}{2 \cdot 2 \cdot 2 \cdot 3} = 1$$

$$= \frac{2 \cdot 2 \cdot 2 \cdot 2 \cdot 3}{5 \cdot 5 \cdot 5} = \frac{48}{125}, \text{ or } 0.384$$

Note that we get an exact answer with methods 1 and 3, but method 2 gives an approximation since we rounded decimal notation for $\frac{2}{3}$.

**Do Exercises 17 and 18.** ▶

**EXAMPLE 14** *Boating.* A triangular sail on a single-sail day cruiser is 3.4 m wide and 4.2 m tall. Find the area of the sail.

1. **Familiarize.** We make a drawing and recall that the formula for the area, $A$, of a triangle with base $b$ and height $h$ is $A = \frac{1}{2}bh$.

2. **Translate.** We substitute 3.4 for $b$ and 4.2 for $h$.

$$A = \frac{1}{2}bh$$
$$= \frac{1}{2}(3.4)(4.2) \quad \text{Substituting}$$

3. **Solve.** We simplify as follows.

$$A = \frac{1}{2}(3.4)(4.2)$$
$$= \frac{3.4}{2}(4.2) \quad \text{Multiplying } \frac{1}{2} \text{ and } \frac{3.4}{1}$$
$$= 1.7(4.2) \quad \text{Dividing}$$
$$= 7.14 \quad \text{Multiplying}$$

4. **Check.** To check, we repeat the calculations using the commutative law. We also rewrite $\frac{1}{2}$ as 0.5.

$$\frac{1}{2}(3.4)(4.2) = 0.5(4.2)(3.4) = (2.1)(3.4) = 7.14$$

Our answer checks.

5. **State.** The area of the sail is 7.14 m² (square meters).

**Do Exercise 19.** ▶

---

Calculate.

 **17.** $\frac{3}{4} \times 0.62$.

**Method 1:**

$$\frac{3}{4} \times 0.62 = \frac{3}{4} \times \frac{0.62}{\Box}$$
$$= \frac{\Box}{4} = 0.465$$

**Method 2:**

$$\frac{3}{4} \times 0.62 = \Box \times 0.62$$
$$= 0.465$$

**Method 3:**

$$\frac{3}{4} \times 0.62 = \frac{3}{4} \cdot \frac{62}{\Box}$$
$$= \frac{\Box}{400}$$
$$= \frac{\Box}{200} = 0.465$$

**18.** $\frac{1}{3} \times 0.384 + \frac{5}{8} \times 0.6784$

**19.** Find the area of a triangular window that is 3.25 ft wide and 2.6 ft tall.

**Answers**

**17.** $\frac{93}{200}$, or 0.465   **18.** 0.552   **19.** 4.225 ft²

**Guided Solution:**

**17.** 1, 1.86; 0.75; 100, 186, 93

## ✓ Check Your Understanding

**Reading Check**   Determine whether the decimal notation for each fraction is terminating or repeating.

**RC1.** $\dfrac{4}{9}$ _____

**RC2.** $\dfrac{3}{32}$ _____

**RC3.** $\dfrac{39}{40}$ _____

**RC4.** $\dfrac{7}{12}$ _____

**RC5.** $\dfrac{2}{11}$ _____

**RC6.** $\dfrac{80}{125}$ _____

**Concept Check**   Choose from the list on the right the form of 1 that completes each multiplication.

**CC1.** $\dfrac{1}{5} \cdot \square = \dfrac{2}{10} = 0.2$

**CC2.** $\dfrac{27}{40} \cdot \square = \dfrac{675}{1000} = 0.675$

**CC3.** $\dfrac{11}{20} \cdot \square = \dfrac{55}{100} = 0.55$

**CC4.** $\dfrac{13}{2500} \cdot \square = \dfrac{52}{10,000} = 0.0052$

**a)** $\dfrac{25}{25}$

**b)** $\dfrac{4}{4}$

**c)** $\dfrac{2}{2}$

**d)** $\dfrac{5}{5}$

---

**a** , **c**   Find decimal notation for each number.

**1.** $\dfrac{3}{8}$

**2.** $\dfrac{3}{5}$

**3.** $\dfrac{-1}{2}$

**4.** $\dfrac{-1}{4}$

**5.** $\dfrac{3}{25}$

**6.** $\dfrac{7}{20}$

**7.** $\dfrac{9}{40}$

**8.** $\dfrac{3}{40}$

**9.** $\dfrac{13}{25}$

**10.** $\dfrac{17}{25}$

**11.** $\dfrac{-17}{20}$

**12.** $\dfrac{-13}{20}$

**13.** $-\dfrac{9}{16}$

**14.** $-\dfrac{5}{16}$

**15.** $\dfrac{7}{5}$

**16.** $\dfrac{3}{2}$

**17.** $\dfrac{28}{25}$

**18.** $\dfrac{31}{20}$

**19.** $\dfrac{11}{-8}$

**20.** $\dfrac{17}{-10}$

**21.** $-\dfrac{39}{40}$

**22.** $-\dfrac{17}{40}$

**23.** $\dfrac{121}{200}$

**24.** $\dfrac{32}{125}$

**25.** $\dfrac{8}{15}$     **26.** $\dfrac{7}{9}$     **27.** $\dfrac{1}{3}$     **28.** $\dfrac{1}{9}$

**29.** $\dfrac{-4}{3}$     **30.** $\dfrac{-8}{9}$     **31.** $\dfrac{7}{6}$     **32.** $\dfrac{7}{11}$

**33.** $-\dfrac{14}{11}$     **34.** $-\dfrac{7}{11}$     **35.** $\dfrac{-5}{12}$     **36.** $\dfrac{-11}{12}$

**37.** $\dfrac{127}{500}$     **38.** $\dfrac{83}{500}$     **39.** $\dfrac{4}{33}$     **40.** $\dfrac{5}{33}$

**41.** $\dfrac{-12}{55}$     **42.** $\dfrac{-5}{22}$     **43.** $\dfrac{4}{7}$     **44.** $\dfrac{2}{7}$

**b**   Round each to the nearest tenth, hundredth, and thousandth.

**45.** $0.\overline{18}$     **46.** $0.\overline{83}$     **47.** $0.2\overline{7}$     **48.** $3.5\overline{4}$

For Exercises 49–60, round the decimal notation for each number to the nearest tenth, hundredth, and thousandth.

**49.** $\dfrac{4}{11}$     **50.** $\dfrac{3}{11}$     **51.** $-\dfrac{5}{3}$     **52.** $-\dfrac{19}{16}$

**53.** $\dfrac{-8}{17}$     **54.** $\dfrac{-7}{13}$     **55.** $\dfrac{7}{12}$     **56.** $\dfrac{2}{15}$

**57.** $\dfrac{29}{-150}$     **58.** $\dfrac{37}{-150}$     **59.** $\dfrac{7}{-9}$     **60.** $\dfrac{5}{-13}$

*Gas Mileage.*   In each of Exercises 61–64, find the gas mileage rounded to the nearest tenth.

**61.** 285 mi; 18 gal                          **62.** 396 mi; 17 gal

**63.** 324.8 mi; 18.2 gal                      **64.** 264.8 mi; 12.7 gal

**65.** For this set of people, what is the ratio, in decimal notation rounded to the nearest thousandth, where appropriate, of:

   **a)** people wearing skirts to the total number of people?
   **b)** people wearing skirts to people wearing trousers?
   **c)** people wearing trousers to the total number of people?
   **d)** people wearing trousers to people wearing skirts?

**66.** For this set of pennies and quarters, what is the ratio, in decimal notation rounded to the nearest thousandth, where appropriate, of:

   **a)** pennies to quarters?
   **b)** quarters to pennies?
   **c)** pennies to total number of coins?
   **d)** total number of coins to pennies?

**d**   Calculate and write the result as a decimal.

**67.** $\dfrac{7}{8} \times 12.64$

**68.** $\dfrac{4}{5} \times 384.8$

**69.** $6.84 \div 2\dfrac{1}{2}$

**70.** $8\dfrac{1}{2} \div 2.125$

**71.** $\dfrac{47}{9}(-79.95)$

**72.** $\dfrac{7}{11}(-2.7873)$

**73.** $\dfrac{1}{2} - 0.5$

**74.** $3\dfrac{1}{8} - 2.75$

**75.** $\left(\dfrac{1}{6}\right)0.0765 + \left(\dfrac{3}{4}\right)0.1124$

**76.** $\left(\dfrac{2}{5}\right)6384.1 - \left(\dfrac{5}{8}\right)156.56$

**77.** $\dfrac{3}{4} \times 2.56 - \dfrac{7}{8} \times 3.94$

**78.** $\dfrac{2}{5} \times 3.91 - \dfrac{7}{10} \times 4.15$

**79.** $5.2 \times 1\dfrac{7}{8} \div 0.4$

**80.** $4\dfrac{3}{4} \times 0.5 \div 0.1$

Solve.

**81.** Find the area of a triangular shawl that is 1.8 m long and 1.2 m wide.

**82.** Find the area of a triangular sign that is 1.5 m wide and 1.5 m tall.

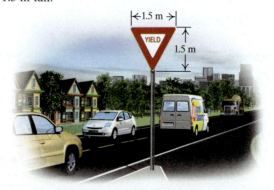

**83.** Find the area of a triangular stamp that is 3.4 cm wide and 3.4 cm tall.

**84.** Find the area of a triangular reflector that is 7.4 cm wide and 9.1 cm tall.

**85.** Find the area of the kite shown on the left.

**86.** Find the area of the kite shown on the right.

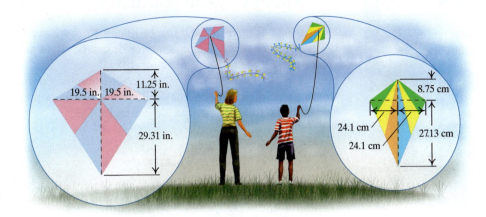

## Skill Maintenance

Solve.   [4.6c]

**87.** A recipe for bread calls for $\frac{2}{3}$ cup of water, $\frac{1}{4}$ cup of milk, and $\frac{1}{8}$ cup of oil. How many cups of liquid ingredients does the recipe call for?

**88.** A board $\frac{7}{10}$ in. thick is glued to a board $\frac{3}{5}$ in. thick. The glue is $\frac{3}{100}$ in. thick. How thick is the result?

## Synthesis

**89.** ▦ Find decimal notation for $\frac{1}{7}, \frac{2}{7}, \frac{3}{7}, \frac{4}{7}$, and $\frac{5}{7}$. Observe the pattern and predict the decimal notation for $\frac{6}{7}$. Check your answer on a calculator.

**90.** ▦ Find decimal notation for $\frac{1}{9}, \frac{1}{99}$, and $\frac{1}{999}$. Observe the pattern and predict the decimal notation for $\frac{1}{9999}$. Check your answer on a calculator.

The formula $A = \pi r^2$ is used to find the area, $A$, of a circle with radius $r$. For Exercises 91 and 92, find the area of a circle with the given radius, using $\frac{22}{7}$ for $\pi$. For Exercises 93 and 94, use 3.14 for $\pi$ or a calculator with a $\pi$ key.

**91.** $r = 2.1$ cm    **92.** $r = 1.4$ cm    **93.** ▦ $r = \dfrac{3}{4}$ ft    **94.** ▦ $r = 4\dfrac{1}{2}$ yd

# 5.6

## OBJECTIVE

a Estimate sums, differences, products, and quotients.

---

# Estimating

## a ESTIMATING SUMS, DIFFERENCES, PRODUCTS, AND QUOTIENTS

Estimating has many uses. It can be done before we even attempt a problem in order to get an idea of the answer. It can be done afterward as a check, even when we are using a calculator. In many situations, an estimate is all we need. We usually estimate by rounding the numbers so that there are one or two nonzero digits, depending on how accurate we want our estimate to be. Consider the following prices for Examples 1–3.

$269.88

$174.99

$319.95

**EXAMPLE 1**  Estimate by rounding to the nearest ten the total cost of one ladder and one electric toothbrush.

We are estimating the sum

$319.95 + $174.99 = total cost.

The estimate, found by rounding the addends to the nearest ten, is

$320 + $170 = $490.    (Estimated total cost)

◀ Do Exercise 1.

**EXAMPLE 2**  About how much more does the ladder cost than the bar stool? Estimate by rounding to the nearest ten.

We are estimating the difference

$319.95 − $269.88 = price difference.

The estimate, found by rounding each price to the nearest ten, is

$320 − $270 = $50.    (Estimated price difference)

◀ Do Exercise 2.

1. Estimate by rounding to the nearest ten the total cost of one ladder and one bar stool. Which of the following is an appropriate estimate?

**a)** $600  **b)** $570
**c)** $590  **d)** $57

2. About how much more does the ladder cost than the electric toothbrush? Estimate by rounding to the nearest ten. Which of the following is an appropriate estimate?

**a)** $100  **b)** $140
**c)** $14  **d)** $150

*Answers*
**1.** (c)  **2.** (d)

**EXAMPLE 3** Estimate the total cost of four bar stools.

We are estimating the product

$4 \times \$269.88 =$ total cost.

The estimate can be found by rounding 269.88 to the nearest ten:

$4 \times \$270 = \$1080$.

**Do Exercise 3.** ▶

**EXAMPLE 4** A student government group is planning an event for first-year students. Since the local weather is often rainy, the group decides to purchase umbrellas bearing the university logo for door prizes. The umbrellas cost $39.99 each. About how many umbrellas can be purchased for $356?

We are estimating the quotient

$\$356 \div \$39.99$.

Since we want a whole-number estimate, we need to round appropriately. Rounding $39.99 to the nearest one, we get $40. Since $356 is close to $360, which is a multiple of $40, we estimate

$\$360 \div \$40 = 9$.

The answer is about 9 umbrellas.

**Do Exercise 4.** ▶

When estimating, we usually look for numbers that are easy to work with. For example, if multiplying, we might round 0.43 to 0.5 and 8.9 to 10, because 0.5 and 10 are convenient numbers to multiply.

**EXAMPLE 5** Estimate: $4.8 \times 52$. Do not find the actual product. Which of the following is an appropriate estimate?

**a)** 25          **b)** 250          **c)** 2500          **d)** 360

We round 4.8 to the nearest one and 52 to the nearest ten:

$5 \times 50 = 250$.     (Estimated product)

Thus, an appropriate estimate is (b).

Other estimates that we might have used in Example 5 are

$5 \times 52 = 260$     or     $4.8 \times 50 = 240$.

The estimate in Example 5, $5 \times 50 = 250$, is the easiest to do because the factors have the fewest nonzero digits.

**Do Exercises 5–10.** ▶

**EXAMPLE 6** Estimate: $82.08 \div 24$. Which of the following is an appropriate estimate?

**a)** 400          **b)** 16          **c)** 40          **d)** 4

This is about $80 \div 20$, so the answer is about 4. Thus, an appropriate estimate is (d).

3. Estimate the total cost of six electric toothbrushes. Which of the following is an appropriate estimate?

**a)** $875          **b)** $1020
**c)** $1620          **d)** $10,200

4. Refer to the umbrella price in Example 4. About how many umbrellas can be purchased for $675?

Estimate each product. Do not find the actual product. Which of the choices is an appropriate estimate?

5. $2.4 \times 8$
  **a)** 16          **b)** 34
  **c)** 125          **d)** 5

6. $24 \times 0.6$
  **a)** 200          **b)** 5
  **c)** 110          **d)** 20

7. $0.86 \times 0.432$
  **a)** 0.04          **b)** 0.4
  **c)** 1.1          **d)** 4

8. $0.82 \times 0.1$
  **a)** 800          **b)** 8
  **c)** 0.08          **d)** 80

9. $0.12 \times 18.248$
  **a)** 180          **b)** 1.8
  **c)** 0.018          **d)** 18

10. $24.234 \times 5.2$
  **a)** 200          **b)** 120
  **c)** 12.5          **d)** 234

*Answers*

**3.** (b)   **4.** 17 umbrellas   **5.** (a)   **6.** (d)
**7.** (b)   **8.** (c)   **9.** (b)   **10.** (b)

Estimate each quotient. Which of the choices is an appropriate estimate?

**11.** $59.78 \div 29.1$
a) 200
b) 20
c) 2
d) 0.2

**12.** $82.08 \div 2.4$
a) 40
b) 4.0
c) 400
d) 0.4

**13.** $0.1768 \div 0.08$
a) 8
b) 10
c) 2
d) 20

**14.** Estimate: $0.0069 \div 0.15$. Which of the following is an appropriate estimate?
a) 0.5
b) 50
c) 0.05
d) 0.004

**Answers**
**11.** (c) **12.** (a) **13.** (c) **14.** (c)

**EXAMPLE 7** Estimate: $94.18 \div 3.2$. Which of the following is an appropriate estimate?

a) 30
b) 300
c) 3
d) 60

This is about $90 \div 3$, so the answer is about 30. Thus, an appropriate estimate is (a).

**EXAMPLE 8** Estimate: $0.0156 \div 1.3$. Which of the following is an appropriate estimate?

a) 0.2
b) 0.002
c) 0.02
d) 20

This is about $0.02 \div 1$, so the answer is about 0.02. Thus, an appropriate estimate is (c).

◀ **Do Exercises 11–13.**

In some cases, it is easier to estimate a quotient directly rather than by rounding the divisor and the dividend.

**EXAMPLE 9** Estimate: $0.0074 \div 0.23$. Which of the following is an appropriate estimate?

a) 0.3
b) 0.03
c) 300
d) 3

We estimate 3 for a quotient. We check by multiplying.

$$0.23 \times 3 = 0.69$$

We make the estimate smaller. We estimate 0.3 and check by multiplying.

$$0.23 \times 0.3 = 0.069$$

We make the estimate smaller. We estimate 0.03 and check by multiplying.

$$0.23 \times 0.03 = 0.0069$$

This is about 0.0074, so the quotient is about 0.03. Thus, an appropriate estimate is (b).

◀ **Do Exercise 14.**

---

**5.6** **Exercise Set**

FOR EXTRA HELP  **MyLab Math**

## ✓ Check Your Understanding

**Reading and Concept Check** Match each calculation with the most appropriate estimate from the list below.

a) 0.8
b) 0.08
c) 8
d) 80
e) 800
f) 8000

**RC1.** $0.1003 \times 0.8$ _____

**RC2.** $38.41 + 41.777$ _____

**RC3.** $0.00152 \times 4025$ _____

**RC4.** $1632 \div 1.9$ _____

**RC5.** $9.054 - 8.3111$ _____

**RC6.** $162{,}105 \times 0.0496$ _____

**a** For Exercises 1–6, use the prices shown below to estimate, by rounding to the nearest ten, each sum, difference, product, or quotient.

**$449.99**  **$164.95**  **$239.78**

1. About how much more does the stove cost than the luggage set?
   a) $300    b) $200    c) $210    d) $690

2. About how much more does the luggage set cost than the riding toy?
   a) $400    b) $80    c) $40    d) $70

3. Estimate the total cost of six riding toys.
   a) $960    b) $980    c) $900    d) $1440

4. Estimate the total cost of four stoves.
   a) $1760    b) $1800    c) $1600    d) $1080

5. About how many luggage sets can be purchased for $2400?
   a) 120    b) 100    c) 10    d) 12

6. About how many riding toys can be purchased for $3200?
   a) 16    b) 20    c) 2    d) 200

Estimate by rounding as directed.

7. $0.02 + 1.31 + 0.34$; nearest tenth

8. $0.88 + 2.07 + 1.54$; nearest one

9. $6.03 + 0.007 + 0.214$; nearest one

10. $1.11 + 8.888 + 99.94$; nearest one

11. $52.367 + 1.307 + 7.324$; nearest one

12. $12.9882 + 1.2115$; nearest tenth

13. $2.678 - 0.445$; nearest tenth

14. $12.9882 - 1.0115$; nearest one

15. $198.67432 - 24.5007$; nearest ten

Estimate. Indicate which of the choices is an appropriate estimate.

16. $234.12321 - 200.3223$
    a) 600    b) 60
    c) 300    d) 30

17. $49 \times 7.89$
    a) 400    b) 40
    c) 4    d) 0.4

18. $7.4 \times 8.9$
    a) 95    b) 63
    c) 124    d) 6

**19.** $98.4 \times 0.083$
  a) 80      b) 12
  c) 8      d) 0.8

**20.** $78 \times 5.3$
  a) 400      b) 800
  c) 40      d) 8

**21.** $3.6 \div 4$
  a) 10      b) 1
  c) 0.1      d) 0.01

**22.** $0.0713 \div 1.94$
  a) 3.5      b) 0.35
  c) 0.035      d) 35

**23.** $74.68 \div 24.7$
  a) 9      b) 3
  c) 12      d) 120

**24.** $914 \div 0.921$
  a) 10      b) 100
  c) 1000      d) 1

**25.** *Fence Posts.* A zoo plans to construct a fence around a proposed exhibit featuring animals of the Great Plains. The perimeter of the area to be fenced is 1760 ft. Estimate the number of wooden fence posts needed if the posts are placed 8.625 ft apart.

**26.** *McDonald's Stock.* Recently, McDonald's stock sold for $158.81 per share. Estimate how many shares could be purchased for $4800.

**27.** *After-School Supplies.* Elijah wants to buy 35 boxes of crayons at $1.89 per box for the after-school program that he runs. Estimate the total cost of the crayons.

**28.** *Batteries.* Charlotte buys 6 packages of AAA batteries at $8.79 per package. Estimate the total cost of the purchase.

## Skill Maintenance

Simplify. [1.9c]

**29.** $2^4 \div 4 - 2$

**30.** $3 \cdot 60 - (12 + 3)$

**31.** $200 + 40 \div 4$

Solve. [1.7b]

**32.** $p + 14 = 83$

**33.** $50 = 5 \cdot t$

**34.** $270 + y = 800$

## Synthesis

The following were done on a calculator. Estimate to determine whether the decimal point was placed correctly.

**35.** $19.7236 - 1.4738 \times 4.1097 = 1.366672414$

**36.** $28.46901 \div 4.9187 - 2.5081 = 3.279813473$

**37.** ▦ Use one of $+$, $-$, $\times$, and $\div$ in each blank to make a true sentence.
  a) $(0.37 \,\square\, 18.78) \,\square\, 2^{13} = 156{,}876.8$

  b) $2.56 \,\square\, 6.4 \,\square\, 51.2 \,\square\, 17.4 = 312.84$

# Solving Equations

We now use the addition and division principles to solve equations involving decimals.

## a  EQUATIONS WITH ONE VARIABLE TERM

 **SKILL REVIEW**

*Solve equations that require use of both the addition principle and the division principle.* [2.8e]

Solve.

**1.** $5x + 7 = -3$        **2.** $-2x - 1 = -9$

**Answers: 1.** $-2$  **2.** $4$

 MyLab Math
**VIDEO**

Recall that equations like $5x + 7 = -3$ are normally solved by first "undoing" the addition and then "undoing" the multiplication. This reverses the order of operations in which we add last and multiply first.

**EXAMPLE 1**  Solve: $0.5x + 5 = 8$.

$$0.5x + 5 = 8$$
$$0.5x + 5 - 5 = 8 - 5 \qquad \text{Subtracting 5 from both sides}$$
$$0.5x = 3 \qquad \text{Simplifying}$$
$$\frac{0.5x}{0.5} = \frac{3}{0.5} \qquad \text{Dividing both sides by 0.5}$$
$$x = 6 \qquad \text{Simplifying}$$

Check:  $\begin{array}{r|l} 0.5x + 5 = 8 \\ \hline 0.5(6) + 5 \;?\; 8 \\ 3 + 5 \\ 8 \;\big|\; 8 \quad \text{TRUE} \end{array}$

The solution is 6.

**EXAMPLE 2**  Solve: $4.2x + 3.7 = -26.12$.

$$4.2x + 3.7 = -26.12$$
$$4.2x + 3.7 - 3.7 = -26.12 - 3.7 \qquad \text{Subtracting 3.7 from both sides}$$
$$4.2x = -29.82 \qquad \text{Simplifying}$$
$$\frac{4.2x}{4.2} = \frac{-29.82}{4.2} \qquad \text{Dividing both sides by 4.2}$$
$$x = -7.1 \qquad \text{Simplifying}$$

Check:  $\begin{array}{r|l} 4.2x + 3.7 = -26.12 \\ \hline 4.2(-7.1) + 3.7 \;?\; -26.12 \\ -29.82 + 3.7 \\ -26.12 \;\big|\; -26.12 \quad \text{TRUE} \end{array}$

The solution is $-7.1$.

**Do Exercises 1–4.** ▶

Solve.

**1.** $6x + 7.4 = 11$

**2.** $0.2 - 0.1x = 1.4$

**3.** $7.4t + 1.25 = 27.89$

**4.** $-5.7 + 4.8x = -14.82$

*Answers*
**1.** 0.6  **2.** $-12$  **3.** 3.6  **4.** $-1.9$

## b  EQUATIONS WITH TWO OR MORE VARIABLE TERMS

Some equations have variable terms on both sides. To solve such an equation, we first use the addition principle to get all variable terms on one side of the equation and all constant terms on the other side.

**EXAMPLE 3**  Solve: $10x - 7 = 2x + 13$.

We begin by subtracting $2x$ from (or adding $-2x$ to) each side. This will group all variable terms on one side of the equation.

$$10x - 7 - 2x = 2x + 13 - 2x \qquad \text{Adding } -2x \text{ to both sides}$$
$$8x - 7 = 13 \qquad \text{Combining like terms}$$

We use the addition principle to isolate all constant terms on one side and then the division principle to isolate $x$.

$$8x - 7 = 13$$
$$8x - 7 + 7 = 13 + 7 \qquad \text{Adding 7 to both sides}$$
$$8x = 20 \qquad \text{Simplifying (combining like terms)}$$
$$\frac{8x}{8} = \frac{20}{8} \qquad \text{Dividing both sides by 8}$$
$$x = 2.5$$

Check:
$$\begin{array}{c|c} \multicolumn{2}{c}{10x - 7 = 2x + 13} \\ \hline 10(2.5) - 7 \ ? \ 2(2.5) + 13 \\ 25 - 7 & 5 + 13 \\ 18 & 18 \qquad \text{TRUE} \end{array}$$

The solution is 2.5.

Sometimes it may be easier to combine all variable terms on the right side and all constant terms on the left side.

**EXAMPLE 4**  Solve: $11 - 3t = 7t + 8$.

We can combine all variable terms on the right side by adding $3t$ to both sides.

$$11 - 3t = 7t + 8$$
$$11 - 3t + 3t = 7t + 8 + 3t \qquad \text{Adding } 3t \text{ to both sides}$$
$$11 = 10t + 8 \qquad \text{Combining like terms}$$
$$11 - 8 = 10t + 8 - 8 \qquad \text{Subtracting 8 from both sides}$$
$$3 = 10t$$
$$\frac{3}{10} = \frac{10t}{10} \qquad \text{Dividing both sides by 10}$$
$$0.3 = t$$

Check:
$$\begin{array}{c|c} \multicolumn{2}{c}{11 - 3t = 7t + 8} \\ \hline 11 - 3(0.3) \ ? \ 7(0.3) + 8 \\ 11 - 0.9 & 2.1 + 8 \\ 10.1 & 10.1 \qquad \text{TRUE} \end{array}$$

The solution is 0.3.

Note that in Example 4 the variable appears on the right side of the last equation. It does not matter whether the variable is isolated on the right or left side. What is important is that you have a clear direction to your work as you proceed from step to step.

**Do Exercises 5–7.** ▶

**EXAMPLE 5** Solve: $5(x + 1) = 7x + 12$.

$$5(x + 1) = 7x + 12$$

$5 \cdot x + 5 \cdot 1 = 7x + 12$    Using the distributive law to remove parentheses

$5x + 5 = 7x + 12$    Simplifying

$5x + 5 - 7x = 7x + 12 - 7x$    Subtracting $7x$ from both sides

$-2x + 5 = 12$    Simplifying

$-2x + 5 - 5 = 12 - 5$    Subtracting 5 from both sides

$-2x = 7$

$$\dfrac{-2x}{-2} = \dfrac{7}{-2}$$    Dividing both sides by $-2$

$x = -3.5$

Check:
$$\begin{array}{c|c} 5(x + 1) = 7x + 12 \\ \hline 5(-3.5 + 1) \;?\; 7(-3.5) + 12 \\ 5(-2.5) \;\big|\; -24.5 + 12 \\ -12.5 \;\big|\; -12.5 \end{array}$$    Check in the original equation.

TRUE

The solution is $-3.5$.

**Do Exercise 8.** ▶

**EXAMPLE 6** Solve: $9(x - 3) + 7 = 5x - 47$.

We use the distributive law and combine like terms before using the addition and division principles.

$$9(x - 3) + 7 = 5x - 47$$

$9x - 27 + 7 = 5x - 47$    Using the distributive law

$9x - 20 = 5x - 47$    Simplifying

$9x - 20 - 5x = 5x - 47 - 5x$    Subtracting $5x$ from both sides

$4x - 20 = -47$    Simplifying

$4x - 20 + 20 = -47 + 20$    Adding 20 to both sides

$4x = -27$    Simplifying

$$\dfrac{4x}{4} = -\dfrac{27}{4}$$    Dividing both sides by 4

$x = -6.75$

Check:
$$\begin{array}{c|c} 9(x - 3) + 7 = 5x - 47 \\ \hline 9(-6.75 - 3) + 7 \;?\; 5(-6.75) - 47 \\ 9(-9.75) + 7 \;\big|\; -33.75 - 47 \\ -87.75 + 7 \;\big|\; -80.75 \\ -80.75 \;\big|\; -80.75 \end{array}$$

TRUE

The solution is $-6.75$.

**Do Exercise 9.** ▶

Solve.

**5.** $10t - 3 = 4t + 18$

 **6.** $8 + 4x = 9x - 3$

$8 + 4x - 4x = 9x - 3 - \boxed{\phantom{0}}$

$8 = \boxed{\phantom{0}} - 3$

$8 + 3 = 5x - 3 + \boxed{\phantom{0}}$

$\boxed{\phantom{0}} = 5x$

$\dfrac{11}{5} = \dfrac{5x}{\boxed{\phantom{0}}}$

$\boxed{\phantom{0}} = x$

**7.** $2.1x - 45.3 = 17.3x + 23.1$

 **8.** Solve: $3(x + 5) = 20 - x$.

$3(x + 5) = 20 - x$

$3x + \boxed{\phantom{0}} = 20 - x$

$3x + 15 + x = 20 - x + \boxed{\phantom{0}}$

$\boxed{\phantom{0}} + 15 = 20$

$4x + 15 - \boxed{\phantom{0}} = 20 - 15$

$4x = \boxed{\phantom{0}}$

$\dfrac{4x}{4} = \dfrac{5}{\boxed{\phantom{0}}}$

$x = \boxed{\phantom{0}}$

**9.** Solve: $8(x - 2) - 15 = 4x + 2$.

*Answers*
**5.** 3.5  **6.** 2.2  **7.** $-4.5$
**8.** 1.25  **9.** 8.25
*Guided Solutions:*
**6.** $4x, 5x, 3, 11, 5, 2.2$
**8.** $15, x, 4x, 15, 5, 4, 1.25$

✓ **Check Your Understanding**

**Reading Check**  Determine whether each statement is true or false.

_____ **RC1.** Equations involving decimals never have integer solutions.

_____ **RC2.** An equation that does not involve decimals may have a decimal solution.

_____ **RC3.** In the final step of a solution of an equation, the variable may appear on the right side or on the left side of the equation.

_____ **RC4.** You should be able to give a reason for each step in your solution.

**Concept Check**  Determine whether the given number is a solution of the equation.

**CC1.** $1.2x + 10 = 4; -5$

**CC2.** $-1.5t - 1.4 = -14.9; -9$

**CC3.** $2.5a - 3.8 = 4.3a + 1.6; 3$

**CC4.** $2(1.3x + 4.2) = 1.6x - 5; -13.4$

**a**  Solve. Remember to check.

**1.** $5x = 27$

**2.** $4x = 75$

**3.** $16 \cdot y = 3.2$

**4.** $36 \cdot y = 14.76$

**5.** $-1.5t = 9.36$

**6.** $-1.2t = -11.4$

**7.** $x + 15.7 = 3.1$

**8.** $x + 13.9 = 4.2$

**9.** $1.25 = 3.8 + x$

**10.** $-1.3 = 4.05 + x$

**11.** $x - 0.37 = 1.6$

**12.** $x - 0.4 = 12$

**13.** $y - 9.8 = -1.42$

**14.** $y - 6 = -0.2$

**15.** $-9 = t - 3.1$

**16.** $-23 = t - 0.01$

**17.** $5x - 8 = 22$

**18.** $4x - 7 = 13$

**19.** $6.9x - 8.4 = 4.02$

**20.** $7.1x - 9.3 = 8.45$

**21.** $21.6 + 4.1t = 6.43$

**22.** $12.4 + 3.7t = 2.04$

**23.** $-26.25 = 7.5x + 9$

**24.** $-43.72 = 8.7x + 5$

**25.** $-4.2x + 3.04 = -4.1$

**26.** $-2.9x - 2.24 = -17.9$

**27.** $-3.05 = 7.24 - 3.5t$

**28.** $-4.62 = 5.68 - 2.5t$

**29.** $3 - 1.2y = -2.4$

**30.** $6 - 3.5y = 12.3$

**b** Solve. Remember to check.

**31.** $9x - 2 = 5x + 34$

**32.** $8x - 5 = 6x + 9$

**33.** $2x + 6 = 7x - 10$

**34.** $3x + 4 = 11x - 6$

**35.** $5y - 3 = 4 + 9y$

**36.** $6y - 5 = 8 + 10y$

**37.** $5.9x + 67 = 7.6x + 16$

**38.** $2.1x + 42 = 5.2x - 20$

**39.** $7.8a + 2 = 2.4a + 19.28$

**40.** $7.5a - 5.16 = 3.1a + 12$

**41.** $6(x + 2) = 4x + 30$

**42.** $5(x + 3) = 3x + 23$

**43.** $5(x + 3) = 15x - 6$

**44.** $2(x + 3) = 4x - 11$

**45.** $7a - 9 = 15(a - 3)$

**46.** $2a - 7 = 12(a - 3)$

**47.** $1.5(y - 6) = 1.3 - y$

**48.** $2.3(5 - y) = 0.2y + 1.7$

**49.** $2.9(x + 8.1) = 7.8x - 3.95$

**50.** $2(x + 7.3) = 6x - 0.83$

**51.** $-6.21 - 4.3t = 9.8(t + 2.1)$

**52.** $-7.37 - 3.2t = 4.9(t + 6.1)$

**53.** $4(x - 2) - 9 = 2x + 9$

**54.** $9(x - 4) + 13 = 4x - 23$

**55.** $2(4y - 1.8) + 0.4 = 8(2y - 0.4)$

**56.** $3(1.2y + 5) = 2(2.5y - 7) + 1$

**57.** $43(7 - 2x) + 34 = 50(x - 4.1) + 744$

**58.** $34(5 - 3.5x) = 12(3x - 8) + 653.5$

## Skill Maintenance

Find the area of each figure. [3.6b]

**59.**
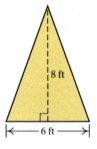
4 m, 7 m

**60.**
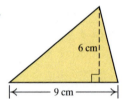
6 cm, 9 cm

**61.**
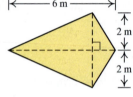
5 in., 5 in.

**62.**
8 ft, 6 ft

**63.**
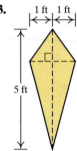
1 ft  1 ft, 5 ft

**64.**
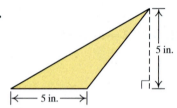
6 m, 2 m, 2 m

**65.** Subtract: $\dfrac{3}{25} - \dfrac{7}{10}$.   [4.3a]

**66.** Simplify: $\dfrac{0}{-18}$.   [3.3b]

**67.** Add: $-17 + 24 + (-9)$.   [2.2a]

**68.** Solve: $3x - 10 = 14$.   [2.8e]

## Synthesis

Solve.

**69.** ▦ $7.035(4.91x - 8.21) + 17.401 = 23.902x - 7.372815$

**70.** ▦ $8.701(3.4 - 5.1x) - 89.321 = 5.401x + 74.65787$

**71.** $5(x - 4.2) + 3[2x - 5(x + 7)] = 39 + 2(7.5 - 6x) + 3x$

**72.** $14(2.5x - 3) + 9x + 5 = 4(3.25 - x) + 2[5x - 3(x + 1)]$

**73.** ▦ $3.5(4.8x - 2.9) + 4.5 = 9.4x - 3.4(x - 1.9)$

**74.** ▦ $4.19 - 1.8(4.5x - 6.4) = 3.1(9.8 + x)$

# Applications and Problem Solving

## a TRANSLATING TO ALGEBRAIC EXPRESSIONS

To translate problems to equations, we need to be able to translate phrases to algebraic expressions. Certain key words in phrases help direct the translation.

| KEY WORDS | SAMPLE PHRASE OR SENTENCE | TRANSLATION |
|---|---|---|
| **Addition (+)** | | |
| added to | 350 lb was added to the car's weight. | $w + 350$ |
| sum of | The sum of a number and 10 | $n + 10$ |
| plus | 13 plus some number | $13 + x$ |
| more than | 176 more than last year's enrollment | $r + 176$ |
| increased by | The original estimate, increased by 50 | $y + 50$ |
| **Subtraction (−)** | | |
| subtracted from | 2 oz was subtracted from the bag's weight. | $w - 2$ |
| difference of | The difference of two prices | $p - q$ |
| minus | A construction crew of size c, minus 4 workers | $c - 4$ |
| less than | 18 less than the number of volunteers last month | $v - 18$ |
| decreased by | An essay's grade, decreased by 5 points | $g - 5$ |
| **Multiplication (·)** | | |
| multiplied by | The number of mints in a box, multiplied by 5 | $5 \cdot m$ |
| product of | The product of two numbers | $a \cdot b$ |
| times | 10 times Jon's age | $10j$ |
| twice | Twice the wholesale price | $2w$ |
| of | $\frac{1}{2}$ of the number of pages assigned | $\frac{1}{2}p$ |
| **Division (÷)** | | |
| divided by | A 16-oz bag of almonds, divided by 5 | $16 \div 5$ |
| quotient of | The quotient of 60 and 3 | $60 \div 3$ |
| divided into | 6 divided into the cost of the meal | $c \div 6$ |
| ratio of | The ratio of 456 to the number of gallons of gasoline | $456/g$ |

It is helpful to choose a descriptive variable to represent the unknown. For example, $w$ suggests weight and $g$ suggests the number of gallons of gasoline.

The following tips are helpful in translating phrases to algebraic expressions.

> **TIPS FOR TRANSLATING PHRASES TO ALGEBRAIC EXPRESSIONS**
> ....................................................................................................
> - Replace the unknown in the statement with a specific number before translating using a variable.
> - Write down what each variable represents.
> - Check the translation using another number to see if it matches the phrase.
> - Be especially careful with order when subtracting and dividing.

**EXAMPLE 1**  Translate each phrase to an algebraic expression.

**a)** A number added to 5

**b)** A number subtracted from 5

**a)** Let $n$ represent the number. Then the phrase "a number added to 5" can be translated $5 + n$, or $n + 5$.

**b)** Let $n$ represent the number. Then the phrase "a number subtracted from 5" can be translated $5 - n$.

◀ **Do Exercises 1 and 2.**

.............................. **Caution!** ..............................

Note that in Example 1, there are two correct translations for part (a) and only one correct translation for part (b). Recall that numbers can be added in either order, so $5 + n$ and $n + 5$ represent the same value. However, order is important in subtraction, so $5 - n$ and $n - 5$ do NOT represent the same value.

.....................................................................................................

**EXAMPLE 2**  Translate each phrase to an algebraic expression.

**a)** Than's age increased by six

**b)** Half of some number

**c)** Twice the cost

**d)** Seven more than twice the weight

**e)** Fifteen divided by a number

**f)** A number divided by fifteen

**g)** Six less than the product of two numbers

**h)** Nine times the sum of two numbers

| | *Phrase* | *Variable(s)* | *Algebraic Expression* |
|---|---|---|---|
| **a)** | Than's age increased by six | Let $a$ = Than's age. | $a + 6$, or $6 + a$ |
| **b)** | Half of some number | Let $n$ = the number. | $\frac{1}{2}n$, or $\frac{n}{2}$, or $n \div 2$ |
| **c)** | Twice the cost | Let $c$ = the cost. | $2c$ |
| **d)** | Seven more than twice the weight | Let $w$ = the weight. | $2w + 7$, or $7 + 2w$ |

Translate to an algebraic expression.

**1.** Twenty more than a number

**2.** Twenty less than a number

**Answers**

**1.** Let $n$ = the number; $n + 20$, or $20 + n$
**2.** Let $n$ = the number; $n - 20$

**e)** Fifteen divided by a number

Let $x$ = the number.

$15 \div x$, or $\dfrac{15}{x}$

**f)** A number divided by fifteen

Let $x$ = the number.

$x \div 15$, or $\dfrac{x}{15}$

**g)** Six less than the product of two numbers

Let $m$ and $n$ = the numbers.

$mn - 6$

**h)** Nine times the sum of two numbers

Let $a$ and $b$ = the numbers.

$9(a + b)$

**Do Exercises 3–5.** ▶

Translate to an algebraic expression.

**3.** The height of a tree, decreased by 50

**4.** Four less than ten times a number

**5.** Fourteen more than the product of the hourly rate and the number of hours worked

## b  SOLVING APPLIED PROBLEMS

**EXAMPLE 3**  *Canals.*  The Panama Canal in Panama is 50.7 mi long. The Suez Canal in Egypt is 119.9 mi long. How much longer is the Suez Canal?

Panama Canal    Suez Canal

**1. Familiarize.**  We let $l$ = the distance in miles that the length of the longer canal differs from the length of the shorter canal.

**2. Translate.**  We translate as follows, using the given information:

| Length of Panama Canal, the shorter canal | plus | Additional length | is | Length of Suez Canal, the longer canal |
|:---:|:---:|:---:|:---:|:---:|
| ↓ | ↓ | ↓ | ↓ | ↓ |
| 50.7 mi | + | $l$ | = | 119.9 mi. |

**3. Solve.**  We solve the equation by subtracting 50.7 mi from both sides:

$$50.7 + l = 119.9$$
$$50.7 + l - 50.7 = 119.9 - 50.7$$
$$l = 69.2.$$

**4. Check.**  We can check by adding.

```
   5 0 . 7
+  6 9 . 2
 1 1 9 . 9
```

The answer checks.

**5. State.**  The Suez Canal is 69.2 mi longer than the Panama Canal.

**Do Exercise 6.** ▶

**6.** *Credit-Card Transactions.* There were a total of 30.7 billion U.S. credit-card transactions in 2012. The number of transactions in 2015 was 33.8 billion. How many more credit-card transactions were there in 2015 than in 2012?

**Data:** Federal Reserve; creditcardforum.com

*Answers*

**3.** Let $h$ = the height of a tree; $h - 50$
**4.** Let $x$ = the number; $10x - 4$
**5.** Let $r$ = the hourly rate and $w$ = the number of hours worked; $rw + 14$, or $14 + rw$  **6.** 3.1 billion transactions

**EXAMPLE 4**  *Chromebook Purchase.*  A school system spent $25,293.96 for 102 Chromebooks for its seventh and eighth graders. How much did each Chromebook cost?

**1. Familiarize.**  We let $c = $ the cost of each Chromebook.

**2. Translate.**  We translate as follows:

| Number of Chromebooks purchased | times | Cost of each Chromebook | is | Total cost of purchase |
|:---:|:---:|:---:|:---:|:---:|
| 102 | · | $c$ | = | $25,293.96. |

**3. Solve.**  We solve the equation by dividing both sides by 102.

$$\frac{102 \cdot c}{102} = \frac{25{,}293.96}{102}$$

$$c = \frac{25{,}293.96}{102}$$

$$c = 247.98$$

**4. Check.**  We check by estimating:

$$25{,}293.96 \div 102 \approx 25{,}000 \div 100 = 250.$$

Since 250 is close to 247.98, the answer is probably correct.

**5. State.**  The cost of each Chromebook was $247.98.

◀ **Do Exercise 7.**

**7.** *Mileage Rates.*  For a recent year, the Internal Revenue Service allowed a tax deduction of 53.5¢ per mile driven for business purposes. What deduction, in dollars, was allowed for driving 8607 mi during the year?

## Multistep Problems

**EXAMPLE 5**  *Tracking a Bank Balance.*  Revenue of U.S. banks from overdraft fees exceeded $30 billion in 2016. (**Data:** consumersunion.org)

To avoid overdrafts and to track her spending, Maggie keeps a running account of her banking transactions. She checks her balance online, but because of pending amounts that post later, she keeps a separate record. Maggie had $2432.27 in her account. She used her debit card to pay her rent of $835 and make purchases of $14.13, $38.60, and $205.98. She then deposited her weekly pay of $748.35. What was her balance after these transactions?

**1. Familiarize.**  We first find the total of the debits. Then we find how much is left in the account after the debits are deducted. Finally, we add to this amount the deposit to find the balance in the account after all the transactions.

**2, 3. Translate and Solve.**  We let $d = $ the total amount of the debits. We are combining amounts: $835 + $14.13 + $38.60 + $205.98 = d$. To solve the equation, we add.

| | |
|---:|:---|
| 8 3 5.0 0 | First debit |
| 1 4.1 3 | Second debit |
| 3 8.6 0 | Third debit |
| + 2 0 5.9 8 | Fourth debit |
| 1 0 9 3.7 1 | Total debits |

Thus, $d = 1093.71$.

Now let $a$ = the amount in the account after the debits are deducted. We subtract: \$2432.27 − \$1093.71 = $a$.

  2432.27  Original amount
− 1093.71  Total debits
  1338.56  New amount

Thus, $a$ = 1338.56.

Finally, we let $f$ = the amount in the account after the paycheck is deposited.

| Amount after debits | plus | Amount of deposit | is | Final amount |
|---|---|---|---|---|
| ↓ | ↓ | ↓ | ↓ | ↓ |
| 1338.56 | + | 748.35 | = | $f$ |

To solve the equation, we add.

  1338.56  Balance after debits
+  748.35  Paycheck deposit
  2086.91  Final amount

**4. Check.** We repeat the computations.

**5. State.** Maggie had \$2086.91 in her account after all the transactions.

**Do Exercise 8.** ▶

**8.** *Bank Balance.* Stephen had \$915.22 in his checking account. He used his debit card to pay a charge card minimum payment of \$36 and to make purchases of \$67.50, \$178.23, and \$429.05. He then deposited his weekly pay of \$570.91. How much was in his account after these transactions?

**EXAMPLE 6** *Gas Mileage.* Ava filled her gas tank and noted that the odometer read 67,507.8. After the next fill-up, the odometer read 68,006.1. It took 16.5 gal to fill the tank. How many miles per gallon did Ava's car get?

**1. Familiarize.** We first make a drawing.

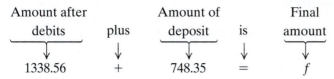

$n$ miles, 16.5 gallons

This is a two-step problem. We let $n$ = the number of miles driven and $m$ = the number of miles per gallon.

**2, 3. Translate and Solve.** First, we find the number of miles that have been driven between fill-ups. We translate and solve as follows:

| First reading | plus | Number of miles driven | is | Second reading |
|---|---|---|---|---|
| ↓ | ↓ | ↓ | ↓ | ↓ |
| 67,507.8 | + | $n$ | = | 68,006.1. |

To solve the equation, we subtract 67,507.8 from both sides:

$n$ = 68,006.1 − 67,507.8
  = 498.3.

```
  68,006.1
− 67,507.8
    498.3
```

Next, we divide the total number of miles driven by the number of gallons. This gives us $m$, that is, the gas mileage. The division that corresponds to the situation is 498.3 ÷ 16.5 = $m$.

Thus, $m$ = 30.2.

```
        3 0.2
16.5.)4 9 8.3.0
      4 9 5
        3 3 0
        3 3 0
            0
```

*Answer*
**8.** \$775.35

**4. Check.** To check, we first multiply the number of miles per gallon by the number of gallons to find the number of miles driven:

$16.5 \times 30.2 = 498.3.$

Then we add 498.3 to 67,507.8 to find the new odometer reading:

$67,507.8 + 498.3 = 68,006.1.$

The gas mileage of 30.2 checks.

**5. State.** Ava's car got 30.2 miles per gallon.

◀ **Do Exercise 9.**

Example 7 involves a formula giving the area of a circle.

> In any circle, a **diameter** is a segment that passes through the center of the circle with endpoints on the circle. A **radius** is a segment with one endpoint on the center and the other endpoint on the circle. The area, $A$, of a circle with radius of length $r$ is given by
>
> $$A = \pi \cdot r^2,$$
>
> where $\pi \approx 3.14$.
> The length $r$ of a radius of a circle is half the length $d$ of a diameter:
>
> $$r = \frac{1}{2}d.$$

**EXAMPLE 7** The Northfield Tap and Die Company stamps 6-cm-wide discs out of metal squares that are 6 cm by 6 cm. How much metal remains after the disc has been punched out?

**1. Familiarize.** We make, and label, a drawing. We let $a$ = the amount of metal remaining, in square centimeters, and list the relevant formulas.

For a square with sides of length $s$, $Area = s^2$.
For a circle with radius of length $r$, $Area = \pi \cdot r^2$, where $\pi \approx 3.14$.

The radius $r$ of a circle is half of its diameter: $r = \frac{1}{2}d$.

We also list the known information about the square and the disc:

Square: The side $s = 6$ cm.
Disc: The diameter $d = 6$ cm.

The radius $r = \frac{1}{2}(6\text{ cm}) = 3$ cm.

**2. Translate.** To find the amount left over, we subtract the area of the disc from the area of the square.

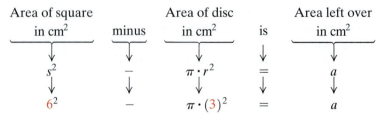

The shaded area is the amount of metal remaining.

9. *Gas Mileage.* John filled his gas tank and noted that the odometer read 38,320.8. After the next fill-up, the odometer read 38,735.5. It took 14.5 gal to fill the tank. How many miles per gallon did John get?

*Answer*
**9.** 28.6 mpg

**3. Solve.** We simplify as follows.

$$6^2 - 3.14(3)^2 \approx a$$
$$36 - 3.14 \cdot 9 \approx a$$
$$36 - 28.26 \approx a$$
$$7.74 \approx a$$

**4. Check.** We can repeat our calculation as a check. Note that 7.74 is less than the area of the disc, which in turn is less than the area of the square. This agrees with the impression given by our drawing.

**5. State.** The amount of material left over is 7.74 cm².

**Do Exercise 10.** ▶

**10.** Suppose that an 8-in.-wide disc is punched out of an 8-in. by 8-in. sheet of metal. How much material is left over?

**EXAMPLE 8** *Photo Books.* Online applications allow users to create their own photo books. On one photo book app, the price of a large hard-cover 20-page book is $24.99. Additional pages are 79 cents per page. Marta has $35 to spend on a book. What is the greatest number of pages she can put in the book?

**1. Familiarize.** Suppose that Marta put 30 pages in the book. She would have to pay an additional price per page for 30 − 20, or 10, pages. The price would be the base price plus the charge for extra pages, or

$$\$24.99 + \$0.79 \cdot 10 = \$24.99 + \$7.90 = \$32.89.$$

This familiarizes us with the way in which price is calculated. Note that we convert 79 cents to $0.79 so that only one unit, dollars, is used. Note also that Marta can put more than 30 pages in the book.

We let $p =$ the number of extra pages Marta adds to the 20-page book. Note that the total number of pages in the book is then $20 + p$.

**2. Translate.** The problem can be rephrased and translated as follows.

| Base price of book | plus | Cost per page | times | Number of extra pages | is | Cost of book |
|:---:|:---:|:---:|:---:|:---:|:---:|:---:|
| ↓ | ↓ | ↓ | ↓ | ↓ | ↓ | ↓ |
| $24.99 | + | $0.79 | · | $p$ | = | $35 |

**3. Solve.** We solve the equation.

$$24.99 + 0.79p = 35$$
$$0.79p = 10.01 \qquad \text{Subtracting 24.99 from both sides}$$
$$p = \frac{10.01}{0.79} \qquad \text{Dividing both sides by 0.79}$$
$$p \approx 12.7 \qquad \text{Rounding}$$

**4. Check.** We check in the original problem. Since Marta cannot pay for parts of a page, we must round the answer *down* to 12 in order to keep the cost under $35. This makes the total length of the book 20 + 12, or 32, pages. If Marta adds 12 pages to the book, the additional cost will be 12 times $0.79, or $9.48. If we add $9.48 to the base price of $24.99, we get $34.47, which is just under the $35 Marta has to spend.

**5. State.** With $35, Marta can make a 32-page book.

**Do Exercise 11.** ▶

**11. Bike Rentals.** Mike's Bikes rents mountain bikes. The shop charges $4.00 insurance for each rental plus $6.00 per hour. For how many hours can a person rent a bike with $25.00?

*Answers*
**10.** 13.76 in² **11.** 3.5 hr

## Problems with More Than One Unknown

**EXAMPLE 9** *Multi-sport Recreation.* Around the Bend Expeditions offers a two-day canyon trip combining biking and canoeing. Those participating will ride mountain bikes for 12 miles longer than the distance paddled. The total length of the trip is 43 miles. How long is the biking portion of the trip, and how long is the canoeing portion of the trip?

1. **Familiarize.** We first list the quantities we are asked to find:

   Length of biking portion, in miles

   Length of canoeing portion, in miles

   We will want to represent both quantities using only one variable. To do so, we use the second sentence in the problem: *Those participating will ride mountain bikes for 12 miles longer than the distance paddled.* Since the biking portion of the trip is described in terms of the canoeing portion, we let $x$ = the length of the canoeing portion, in miles. Then the length of the biking portion, also in miles, is $x + 12$.

   Length of biking portion, in miles:    $x + 12$
   Length of canoeing portion, in miles:   $x$

2. **Translate.** We use the total length of the trip to translate to an equation. Note that all lengths are in miles.

   | Length of biking portion | plus | length of canoeing portion | is | Total length of trip |
   |:---:|:---:|:---:|:---:|:---:|
   | ↓ | ↓ | ↓ | ↓ | ↓ |
   | $(x + 12)$ | $+$ | $x$ | $=$ | $43$ |

3. **Solve.** We solve the equation.

   $$\begin{aligned}
   (x + 12) + x &= 43 \\
   x + 12 + x &= 43 \qquad \text{Removing parentheses} \\
   2x + 12 &= 43 \qquad \text{Combining like terms} \\
   2x &= 31 \qquad \text{Subtracting 12 from both sides} \\
   x &= 15.5 \qquad \text{Dividing both sides by 2}
   \end{aligned}$$

   Recall that we are looking for two quantities. We will use the value of $x$ to find both of them. We return to the list of unknowns in the *Familiarize* step.

   Length of biking portion, in miles:    $x + 12 = 15.5 + 12 = 27.5$
   Length of canoeing portion, in miles:   $x = 15.5$

4. **Check.** There are two statements in the problem to verify. First, since $27.5 - 15.5 = 12$, the biking portion is 12 miles longer than the canoeing portion. Second, since $27.5 + 15.5 = 43$, the total length of the trip is 43 miles. The answer checks.

5. **State.** The biking portion of the trip is 27.5 miles, and the canoeing portion is 15.5 miles.

◀ **Do Exercise 12.**

**12.** Holly and Lorenzo worked together on a project for a psychology class. Holly spent twice as much time on the project as Lorenzo did. They spent a total of 10.5 hours on the project. How much time did each person work on the project?

*Answer*

**12.** Holly: 7 hr; Lorenzo: 3.5 hr

# Translating for Success

1. *Gas Mileage.* Art filled his SUV's gas tank and noted that the odometer read 38,271.8. At the next fill-up, the odometer read 38,677.9. It took 28.4 gal to fill the tank. How many miles per gallon did the SUV get?

2. *Dimensions of a Parking Lot.* A store's parking lot is a rectangle that measures 85.2 ft by 52.3 ft. What is the area of the parking lot?

3. *Game Snacks.* Three students pay $18.40 for snacks at a football game. What is each student's share of the cost?

4. *Electrical Wiring.* An electrician needs 1314 ft of wiring cut into $2\frac{1}{2}$-ft pieces. How many pieces will she have?

5. *College Tuition.* Wayne needs $4638 for the fall semester's tuition. On the day of registration, he has only $3092. How much does he need to borrow?

---

The goal of these matching questions is to practice step (2), Translate, of the five-step problem-solving process. Translate each word problem to an equation and select a correct translation from equations A–O.

**A.** $2\frac{1}{2} \cdot n = 1314$

**B.** $18.4 \times 3.87 = n$

**C.** $n = 85.2 \times 52.3$

**D.** $19 - (-4) = n$

**E.** $3 \times 18.40 = n$

**F.** $2\frac{1}{2} \cdot 1314 = n$

**G.** $3092 + n = 4638$

**H.** $18.4 \cdot n = 3.87$

**I.** $\dfrac{406.1}{28.4} = n$

**J.** $52.3 \cdot n = 85.2$

**K.** $n = 19 + (-4)$

**L.** $52.3 + n = 85.2$

**M.** $3092 + 4638 = n$

**N.** $3 \cdot n = 18.40$

**O.** $85.2 + 52.3 = n$

*Answers on page A-11*

---

6. *Cost of Gasoline.* What is the cost, in dollars, of 18.4 gal of gasoline at $3.87 per gallon?

7. *Temperature.* At noon, the temperature in Pierre was 19°F. At midnight, the temperature had fallen to −4°F. By how many degrees had the temperature fallen?

8. *Acres Planted.* This season Sam planted 85.2 acres of corn and 52.3 acres of soybeans. Find the total number of acres that he planted.

9. *Amount Inherited.* Tara inherited $2\frac{1}{2}$ times as much as her cousin. Her cousin received $1314. How much did Tara receive?

10. *Travel Funds.* The athletic department needs travel funds of $4638 for the tennis team and $3092 for the golf team. What is the total amount needed for travel?

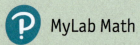
## ✓ Check Your Understanding

**Reading Check** Complete each step in the five-step problem-solving strategy with the correct word from the list on the right.

Five-Step Problem-Solving Strategy

RC1. _____ yourself with the problem situation.

RC2. _____ the problem to an equation.

RC3. _____ the equation.

RC4. _____ the solution.

RC5. _____ the answer using a complete sentence.

Solve

Familiarize

State

Translate

Check

**Concept Check** Choose from the list on the right the most appropriate step to solve each equation.

CC1. $y + 100.2 = 412.5$ _____

CC2. $100.2 + 412.5 = y$ _____

CC3. $100.2 \times y = 412.5$ _____

CC4. $y = 100.2 \times 412.5$ _____

a) Carry out the addition.

b) Divide by 100.2 on both sides.

c) Carry out the multiplication.

d) Subtract 100.2 on both sides.

**a** Translate to an algebraic expression. Choice of variables used may vary.

1. Five more than Ron's age

2. The product of four and $t$

3. 6 more than $b$

4. 7 more than Jen's weight

5. 9 less than $c$

6. 4 less than $d$

7. A number decreased by 16

8. A number increased by 20

9. 8 times Nate's speed

10. The ratio of a number and 100

11. $x$ divided by 17

12. 100 divided by the hourly rate

13. 20 added to half of a number

14. 18 subtracted from twice a number

15. 20 less than 4 times a number

16. 35 more than 10 times a number

17. The sum of the box's length and width

18. The Cessna's speed minus the wind speed

**19.** 10 more than the product of the rate and the time

**20.** 50 less than the product of the length and width

**21.** The sum of 10 times a number and the number

**22.** The difference of a number and twice the number

**23.** 5 times the difference of two numbers

**24.** One fourth of the sum of two numbers

 **b** Solve.

*Licensed Drivers.* The chart below shows the numbers of U.S. licensed drivers in 2018 by age.

| AGE | LICENSED DRIVERS (in millions) |
| --- | --- |
| 19 and under | 9.743 |
| 20–39 | 73.765 |
| 40–59 | 69.874 |
| 60–79 | 28.679 |
| 80 and over | 5.561 |

DATA: Federal Highway Administration; U.S. Department of Transporation

**25.** How many more licensed drivers were ages 40–59 than ages 60–79?

**26.** How many licensed drivers were 59 and under?

**27.** *Gasoline Cost.* What is the cost, in dollars, of 20.4 gal of gasoline at 324.9 cents per gallon? (324.9 cents = $3.249) Round the answer to the nearest cent.

**28.** *Gasoline Cost.* What is the cost, in dollars, of 15.3 gal of gasoline at 389.9 cents per gallon? (389.9 cents = $3.899) Round the answer to the nearest cent.

**29.** *Cost of Bottled Water.* The cost of a year's supply of a popular brand of bottled water, based on consumption of 64 oz per day, at a price of $3.99 for a six-pack of half-liter bottles is $918.82. This is $918.31 more than the cost of drinking the same amount of tap water for a year. What is the cost of drinking tap water for a year?

**Data:** American Water Works Association

**30.** *Body Temperature.* Normal body temperature is 98.6°F. During an illness, a patient's temperature rose 4.2°. What was the new temperature?

**31.** *Lottery Winnings.* The largest lottery jackpot in the United States totaled $1,600,000,000 and was shared equally by 3 winners. How much was each winner's share? Round to the nearest cent.

**Data:** money.cnn.com

**32.** *Lunch Costs.* A group of 4 students pays $47.84 for lunch and splits the cost equally. What is each person's share?

**33.** *Odometer Reading.* The Binford family's odometer reads 22,456.8 at the beginning of a trip. The family's driving directions tell them that they will be driving 234.7 mi. What will the odometer read at the end of the trip?

**34.** *Miles Driven.* Petra bought gasoline when the odometer read 14,296.3. At the next gasoline purchase, the odometer read 14,515.8. How many miles had been driven?

**35. Gas Mileage.** Peggy filled her van's gas tank and noted that the odometer read 26,342.8. At the next filling, the odometer read 26,736.7. It took 19.5 gal to fill the tank. How many miles per gallon did the van get?

**36. Gas Mileage.** Henry filled his Honda's gas tank and noted that the odometer read 18,943.2. At the next filling, the odometer read 19,306.2. It took 13.2 gal to fill the tank. How many miles per gallon did the car get?

**37. Record Movie Openings.** The movie *Star Wars: The Force Awakens* took in $247.97 million on its first weekend. This topped the previous record for opening-weekend revenue, set by *Star Wars: The Last Jedi*, by $27.96 million. How much did *Star Wars: The Last Jedi* take in on its opening weekend?

**Data:** boxofficemojo.com

**38. Boston Marathon.** Kenya swept the 2017 Boston Marathon. Geoffrey Kirui won the men's division in 2 hr 9.617 min. His time was 12.25 min shorter than Edna Kiplagat's winning time for the women's division. What was Edna Kiplagat's winning time?

**Data:** bostonmagazine.com

**39.** Andrew bought a DVD of the movie *Horton Hears a Who* for his nephew for $23.99 plus $1.68 sales tax. He paid for it with a $50 bill. How much change did he receive?

**40.** Claire bought a copy of the book *Make Way for Ducklings* for her daughter for $16.95 plus $0.85 sales tax. She paid for it with a $20 bill. How much change did she receive?

**41. Health Care.** Phil injects 38 units of insulin each day for a week. Each unit is 0.01 cc (cubic centimeter). How many cc's of insulin does he use in a week?

**42. Chemistry.** The water in a filled tank weighs 748.45 lb. One cubic foot of water weighs 62.5 lb. How many cubic feet of water does the tank hold?

**43. Loose Change.** In 2016, passengers at John F. Kennedy International Airport left $70,615 in loose change at airport checkpoints. This was $25,804 more than passengers left that year at Los Angeles International Airport. How much change was left at the Los Angeles airport?

**Data:** Transportation Security Administration

**44. Advertising.** In 2017, global digital ad spending was greater than TV ad spending for the first time. Advertisers spent $208.82 billion on digital advertising, which was $30.34 billion more than was spent on TV advertising. How much was spent on TV advertising?

**Data:** recode.net

**45. *Egg Costs.*** A restaurant owner bought 20 dozen eggs for $25.80. Find the cost of each egg to the nearest tenth of a cent (thousandth of a dollar).

**46. *Weight Loss.*** A person weighing 170 lb burns 8.6 calories per minute while mowing a lawn. One must burn about 3500 calories in order to lose 1 lb. How many pounds would be lost by mowing for 2 hr? Round to the nearest tenth.

**47. *Pole Vault Pit.*** Find the area and the perimeter of the landing area of the pole vault pit shown here.

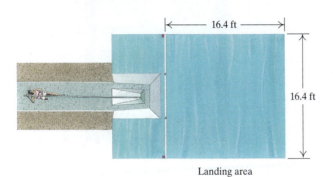

16.4 ft

16.4 ft

Landing area

**48. *Stamp.*** Find the area and the perimeter of the stamp shown here.

FOREVER

2.5 cm

3.25 cm

**49. *Study Cards.*** An instructor allows her students to bring one 7.6-cm by 12.7-cm index card to the final exam, with notes of any sort written on the card. If both sides of the card are used, how much area is available for notes?

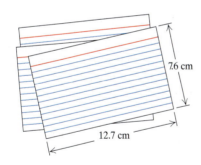

7.6 cm

12.7 cm

**50. *Stamps.*** Find the total area of the stamps shown.

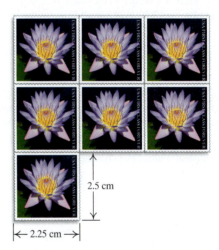

2.5 cm

2.25 cm

Find the length *d* in each figure.

**51.**

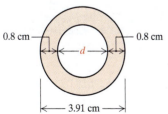

0.8 cm — — 0.8 cm

*d*

3.91 cm

**52.**

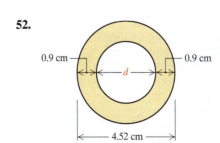

0.9 cm — — 0.9 cm

*d*

4.52 cm

**53. Carpentry.** A round, 6-ft-wide hot tub is being built into a 12-ft by 30-ft rectangular deck. How much decking is needed for the surface of the deck?

12 ft   6 ft   30 ft

**54. Travel Poster.** Sam is decorating his dorm room with a travel poster. The dimensions of the poster are as shown. How much area is not devoted to the picture?

15.7 in.

22.2 in.   27.4 in.

19.3 in.

THE GALAPAGOS ISLANDS

**55.** Lot A measures 250.1 ft by 302.7 ft. Lot B measures 389.4 ft by 566.2 ft. What is the total area of the two lots?

**56.** A 4-ft by 4-ft tablecloth is cut from a round tablecloth that is 6 ft wide. Find the area of the cloth left over.

**57. Food Consumption per Day.** The average American eats 688.6 lb of fruits and vegetables in a year. What is the average consumption in one day? (Use 1 year = 365 days.) Round to the nearest tenth of a pound.

**Data:** visualeconomics.com; niftyhomestead.com

**58. College Spending.** In 2017, the 56 million U.S. college students planned to spend $54.18 billion on back-to-school purchases. What was the average expenditure per college student?

**Data:** nrf.com

**59. Motor Vehicle Production.** In 2015, worldwide production of motor vehicles was 90.781 million. The table below lists production for the top six vehicle producing nations. Find the average vehicle production for the top three countries. Round to the nearest thousandth of a million.

| COUNTRY | MOTOR VEHICLES PRODUCED (in millions) |
|---|---|
| China | 24.503 |
| United States | 12.100 |
| Japan | 9.278 |
| Germany | 6.033 |
| South Korea | 4.556 |
| India | 4.126 |

DATA: International Organization of Motor Vehicle Manufacturers (OICA)

**60. Dow Jones Industrial Average.** The five largest one-day point gains in the Dow Jones Industrial Average are listed in the table below. Find the average of the five greatest point gains. Round the answer to the nearest hundredth of a point.

| DATE | DJIA HIGHEST POINT GAINS |
|---|---|
| 10/13/2008 | 936.42 |
| 10/28/2017 | 889.35 |
| 8/26/2017 | 619.07 |
| 11/13/2008 | 552.60 |
| 3/16/2000 | 499.19 |

DATA: Dow Jones & Co. Inc.

**61. *In-app Purchases.*** A popular game costs $2.99 to install and $1.29 per level to unlock. One month, Anna spent $22.34 to install and play the game. How many levels did she unlock?

**62. *Service Calls.*** JoJo's Service Center charges $30 for a house call plus $37.50 for each hour the job takes. For how long has a repairperson worked on a house call if the bill comes to $123.75?

**63.** Frank has been sent to the store with $40 to purchase 6 lb of cheese at $4.79 a pound and as many bottles of seltzer, at $0.64 a bottle, as possible. How many bottles of seltzer should Frank buy?

**64.** Janice has been sent to the store with $30 to purchase 5 pt of strawberries at $2.49 a pint and as many bags of chips, at $1.39 a bag, as possible. How many bags of chips should Janice buy?

**65. *Endangered Species.*** Part of the giant panda's habitat is protected by the Chinese government. There are 0.4 million more acres unprotected than there are protected acres. The total size of the giant panda's habitat is 5.4 million acres. How many acres are protected and how many are not protected?

**Data:** www.worldwildlife.org

**66. *Hours of Daylight.*** On December 22, Fairbanks, Alaska, has 16.6 more hours of darkness than it has daylight. How many hours of daylight and how many hours of darkness are there in Fairbanks on that day? [*Hint*: There are 24 hours in a day.]

**Data:** "Alaska's Winter Daylight," by Kimi Ross, at www.bellaonline.com

**67. *Vacation Spending.*** Emily spent three times as much on lodging as she did for food on a recent vacation. She spent a total of $261.20 for food and lodging. How much did she spend for each?

**68. *Homework.*** Ian spends twice as long on his science homework as he does on his history homework. One week, he spent a total of 13.5 hours on science and history homework. How much time did he spend on each subject?

**69. *Construction Pay.*** A construction worker is paid $18.50 per hour for the first 40 hr of work, and time and a half, or $27.75 per hour, for any overtime exceeding 40 hr per week. One week she works 46 hr. How much is her pay?

**70. *Summer Work.*** Zachary worked 53 hr during a week one summer. He earned $7.50 per hour for the first 40 hr and $11.25 per hour for overtime (hours exceeding 40). How much did Zachary earn during the week?

**71. *Loan Payment.*** In order to make money on loans, financial institutions are paid back more money than they loan. Suppose you borrow $120,000 to buy a house and agree to make monthly payments of $880.52 for 30 years. How much do you pay back altogether? How much more do you pay back than the amount of the loan?

**72. *Loan Payment.*** In order to make money on loans, financial institutions are paid back more money than they loan. Suppose you borrow $270,000 to buy a house and agree to make monthly payments of $1105.73 for 30 years. How much do you pay back altogether? How much more do you pay back than the amount of the loan?

## Skill Maintenance

Simplify, if possible.

**73.** $\dfrac{0}{-13}$ [3.3b]

**74.** $\dfrac{12}{0}$ [3.3b]

**75.** $\dfrac{-76}{-76}$ [3.3b]

**76.** Add: $-\dfrac{4}{5} + \dfrac{7}{10}$. [4.2b]

**77.** Subtract: $\dfrac{8}{11} - \dfrac{4}{3}$. [4.3a]

**78.** Add: $4\dfrac{1}{3} + 2\dfrac{1}{2}$. [4.6a]

## Synthesis

**79.** A poster that is 61.8 cm by 73.2 cm includes a 2-cm border. What is the area of the poster inside the border?

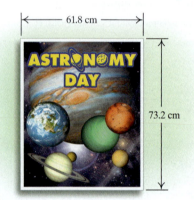

**80.** ▦ A French press coffeepot requires no filters, but costs $34.95. Kenny could buy a plastic drip cone for $4.49, but the cone requires filters which cost $0.04 per pot dripped. How many pots of coffee must Kenny make for the French press pot to be the more economical purchase?

**81.** If the daily rental for a car is $18.90 plus a certain price per mile, and Lindsey must drive 190 mi in one day and still stay within a $55.00 budget, what is the highest price per mile that Lindsey can afford?

**82.** ▦ A 25-ft by 30-ft yard contains an 8-ft-wide, round fountain. How many 1-lb bags of grass seed should be purchased to seed the lawn if 1 lb of seed covers 300 ft²?

**83.** Find the shaded area. What assumptions must you make?

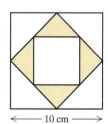

**84.** You can drive from home to work using either of two routes:

> *Route A:* Via interstate highway, 7.6 mi, with a speed limit of 65 mph.

> *Route B:* Via a country road, 5.6 mi, with a speed limit of 50 mph.

Assuming you drive at the posted speed limit, how much time can you save by taking the faster route?

## Vocabulary Reinforcement

Complete each statement with the correct word from the list at the right.

1. A _____ decimal occurs when we convert a fraction to a decimal and the denominator of the fraction has at least one factor other than 2 or 5. [5.5a]

2. A _____ decimal occurs when we convert a fraction to a decimal and the denominator of the fraction has only 2's or 5's, or both, as factors. [5.5a]

3. One _____ = 1,000,000,000. [5.3b]

4. One _____ = 1,000,000. [5.3b]

5. One _____ = 1,000,000,000,000. [5.3b]

6. The _____ consist of the integers, ..., −3, −2, −1, 0, 1, 2, 3, ..., and fractions like $\frac{-1}{2}$, $\frac{4}{5}$, and $\frac{31}{25}$. [5.1a]

trillion

million

billion

rational numbers

repeating

terminating

## Concept Reinforcement

Determine whether each statement is true or false.

_____ **1.** One thousand billion is one trillion. [5.3b]

_____ **2.** The number of decimal places in the product of two numbers is the product of the numbers of places in the factors. [5.3a]

_____ **3.** When we divide a positive number by 0.1, 0.01, 0.001, and so on, the quotient is larger than the dividend. [5.4a]

_____ **4.** For a fraction with a factor other than 2 or 5 in its denominator, decimal notation terminates. [5.5a]

_____ **5.** An estimate found by rounding to the nearest tenth is usually more accurate than one found by rounding to the nearest hundredth. [5.6a]

## Study Guide

**Objective 5.1b**  Convert between decimal notation and fraction notation.

**Example**  Write fraction notation for 5.347.

$$5.347 \qquad 5.347. \qquad \frac{5347}{1000}$$

3 decimal places   Move 3 places   3 zeros
to the right.

**Practice Exercise**

1. Write fraction notation for 50.93.

**Example**  Write decimal notation for $\frac{29}{1000}$.

$$\frac{29}{1000} \qquad 0.029. \qquad \frac{29}{1000} = 0.029$$

3 zeros   Move 3 places
to the left.

**Practice Exercise**

2. Write decimal notation for $\frac{817}{10}$.

**Example** Write decimal notation for $4\frac{63}{100}$.

$$4\frac{63}{100} = 4 + \frac{63}{100} = 4 \text{ and } \frac{63}{100} = 4.63$$

**Practice Exercise**

**3.** Write decimal notation for $42\frac{159}{1000}$.

---

**Objective 5.1d** Round decimal notation to the nearest thousandth, hundredth, tenth, one, ten, hundred, or thousand.

---

**Example** Round 19.7625 to the nearest hundredth.

Locate the digit in the hundredths place, 6. Consider the next digit to the right, 2. Since that digit, 2, is 4 or lower, the original digit does not change.

19.7625
↓
19.76

**Practice Exercise**

**4.** Round 153.346 to the nearest hundredth.

---

**Objective 5.2a** Add using decimal notation.

---

**Example** Add: 14.26 + 63.589.

$$\begin{array}{r} \overset{1}{1\,4.2\,6\,0} \\ +6\,3.5\,8\,9 \\ \hline 7\,7.8\,4\,9 \end{array}$$  **Writing an extra zero**

**Practice Exercise**

**5.** Add: 5.54 + 33.071.

---

**Objective 5.2b** Subtract using decimal notation.

---

**Example** Subtract: 67.345 − 24.28.

$$\begin{array}{r} \overset{2\ 14}{6\,7.3\,\cancel{4}\,5} \\ -2\,4.2\,8\,0 \\ \hline 4\,3.0\,6\,5 \end{array}$$  **Writing an extra zero**

**Practice Exercise**

**6.** Subtract: 221.04 − 13.192.

---

**Objective 5.3a** Multiply using decimal notation.

---

**Example** Multiply: $1.8 \times 0.04$.

$$\begin{array}{r} 1.8 \quad (1 \text{ decimal place}) \\ \times 0.04 \quad (2 \text{ decimal places}) \\ \hline 0.072 \quad (3 \text{ decimal places}) \end{array}$$

**Example** Multiply: $0.001 \times 87.1$.

$0.001 \times 87.1 \qquad 0.087.1$

3 decimal places    Move 3 places to the left.
We write an extra zero.

$0.001 \times 87.1 = 0.0871$

**Example** Multiply: $63.4 \times 100$.

$63.4 \times 100 \qquad 63.40.$

2 zeros    Move 2 places to the right.
We write an extra zero.

$63.4 \times 100 = 6340$

**Practice Exercise**

**7.** Multiply: $5.46 \times 3.5$.

**Practice Exercise**

**8.** Multiply: $17.6 \times 0.01$.

**Practice Exercise**

**9.** Multiply: $1000 \times 60.437$.

**Objective 5.4a** Divide using decimal notation.

**Example** Divide: $21.35 \div 6.1$.

$$
\begin{array}{r}
3.5 \\
6.1\,\overline{)\,2\ 1.3\!\wedge\!5} \\
1\ 8\ 3 \\
\hline
3\ 0\ 5 \\
3\ 0\ 5 \\
\hline
0
\end{array}
$$

**Example** Divide: $\dfrac{16.7}{1000}$.

$$\frac{16.7}{1000} = \frac{016.7}{1\,000.} = \frac{0.0167}{1.0} = 0.0167$$

3 zeros   Move 3 places to the left
to change 1000 to 1.

$$\frac{16.7}{1000} = 0.0167$$

**Example** Divide: $\dfrac{42.93}{0.001}$.

$$\frac{42.93}{0.001} = \frac{42.930}{0.001} = \frac{42{,}930}{1.} = 42{,}930$$

3 decimal   Move 3 places to the right
places    to change 0.001 to 1.

$$\frac{42.93}{0.001} = 42{,}930$$

**Practice Exercise**

**10.** Divide: $26.64 \div 3.6$.

**Practice Exercise**

**11.** Divide: $\dfrac{4.7}{100}$.

**Practice Exercise**

**12.** Divide: $\dfrac{156.9}{0.01}$.

## Review Exercises

Convert the number in each sentence to standard notation.   [5.3b]

**1.** Russia has the largest total area of any country in the world, at 6.59 million square miles.

**2.** Americans eat more than 3.1 billion lb of chocolate each year.
**Data:** Chocolate Manufacturers' Association

Write a word name.   [5.1a]
**3.** 3.47          **4.** 0.031

**5.** 27.0001       **6.** 0.9

Write in fraction notation and, if possible, as a mixed numeral.   [5.1b]
**7.** 0.09          **8.** $-4.561$

**9.** $-0.089$      **10.** 3.0227

Write in decimal notation.   [5.1b]
**11.** $\dfrac{34}{1000}$          **12.** $\dfrac{42{,}603}{10{,}000}$

**13.** $27\dfrac{91}{100}$          **14.** $-867\dfrac{6}{1000}$

Which number is larger?   [5.1c]
**15.** 0.034, 0.0185     **16.** $-0.91$, $-0.19$

**17.** 0.741, 0.6943     **18.** 1.038, 1.041

Round 17.4287 to the nearest:  [5.1d]

**19.** Tenth.                     **20.** Hundredth.

**21.** Thousandth.               **22.** One.

Perform the indicated operation.

**23.**    2 3 6.2 3 1          **24.**    3 7.6 4 5
             2 6 3.4                     − 8.4 9 7   [5.2b]
        +     0.1 9 8   [5.2a]

**25.** 219.3 + 2.8 + 7   [5.2a]   **26.** 745.0109 − 59.959
                                            [5.2b]

**27.** −37.8 + (−19.5)   [5.2c]   **28.** −7.52 − (−9.89)
                                            [5.2c]

**29.**      4 8             **30.** −3.7 (0.29)   [5.3a]
        × 0.2 7   [5.3a]

**31.**    2 4.6 8          **32.** 2 5 )‾8 0‾   [5.4a]
        × 1 0 0 0   [5.3a]

**33.** 11.52 ÷ (−7.2)   [5.4a]   **34.** $\dfrac{276.3}{1000}$   [5.4a]

**35.** 3.056 × 0.001   [5.3a]   **36.** $\dfrac{-4.38}{0.001}$   [5.4a]

Combine like terms.   [5.2d]

**37.** $3.7x - 5.2y - 1.5x - 3.9y$

**38.** $7.94 - 3.89a + 4.63 + 1.05a$

**39.** Evaluate: $P - Prt$ for $P = 1000$, $r = 0.05$, and
$t = 1.5$. (*A formula for depreciation*)   [5.3c]

**40.** Simplify: $9 - 3.2(-1.5) + 5.2^2$.   [5.4b]

**41.** Convert 1549 cents to dollars.   [5.3b]

**42.** Round $248.\overline{27}$ to the nearest hundredth.   [5.5b]

Multiply by a form of 1 to find decimal notation for each
number.   [5.5c]

**43.** $\dfrac{13}{5}$                **44.** $\dfrac{32}{25}$

Use division to find decimal notation for each number.   [5.5a]

**45.** $\dfrac{13}{4}$                **46.** $-\dfrac{7}{6}$

**47.** Calculate: $\dfrac{4}{15} \times 79.05$.   [5.5d]

Solve. Remember to check.

**48.** $t - 4.3 = -7.5$   **49.** $4.1x + 5.6 = -6.7$
        [5.7a]                        [5.7a]

**50.** $6x - 11 = 8x + 4$   **51.** $3(x + 2) = 5x - 7$
        [5.7b]                        [5.7b]

Solve.   [5.8b]

**52.** ***100-Meter Record.***   The fastest speed clocked
for a cheetah running a distance of 100 m is
5.95 sec. The men's world record for the 100-m
dash is held by Jamaican Usain Bolt. His time was
3.63 sec more than the cheetah's. What is the men's
100-m record held by Usain Bolt?

**Data:** "Cheetahs on the Edge," by Roff Smith, *National Geographic*,
November 2012.

**53.** Stacia, a coronary intensive care nurse, earned
$1482.36 during a recent 40-hr week. What was her
hourly wage? Round to the nearest cent.

**54.** ***Landscaping.***   A rectangular yard is 20 ft by 15 ft.
The yard is covered with grass except for a circular
flower garden with an 8-ft diameter. Find the area
of grass in the yard.

**55.** Derek had $1034.46 in his bank account. He used
his debit card to buy a Wii system for $249.99. How
much was left in his account?

**56.** ***Credit Card Processing.***   Pay Right charges $150
for software, $8.95 a month for service, and 21¢ per
online transaction. Timeless Treasures paid $178.90
for its first month of service. How many transactions
did the store process that month?

**57.** ***Recycling.***   Lisa volunteers at a local recycling
center. One Saturday, she processed 130 more
pounds of newspaper than of glass. She processed a
total of 261.4 pounds of newspaper and glass. How
many pounds of each did she process?

**58.** *Gas Mileage.* Inge wants to estimate mileage per gallon. With an odometer reading 36,057.1, she fills up. At 36,217.6 mi, the tank is refilled with 11.1 gal. Find the mileage per gallon. Round to the nearest tenth.

**59.** *Books in Libraries.* The table below lists the numbers of books, in millions, held in the five largest public libraries in the United States. Find the average number of books per library. Round to the nearest tenth.

| LIBRARY | NUMBER OF BOOKS (in millions) |
|---|---|
| Library of Congress | 34.5 |
| Boston Public Library | 19.1 |
| Harvard University | 16.8 |
| New York Public Library | 16.3 |
| University of Illinois–Urbana | 13.2 |

DATA: American Library Association

**60.** *Scanning Posters.* A high school club needs to scan posters designed by students and load them onto a flash drive. The copy center charges $12.99 for the flash drive and $1.09 per square foot for scanning. If the club needs to scan 13 posters at 3 sq ft per poster, what will the total cost be?

**61.** One pound of lean boneless ham contains 4.5 servings. It costs $5.99 per pound. What is the cost per serving? Round to the nearest cent.

**62.** *Construction.* A rectangular room measures 14.5 ft by 16.25 ft. How many feet of crown molding are needed to go around the top of the room? How many square feet of bamboo tiles are needed for the floor of the room?

**63.** *Automobile Leases.* LyDia is leasing a Ford Fusion. She paid $2359 when she signed the lease, and she pays $219 each month. So far, she has paid $5425. For how many months has she been leasing the car?

**64.** Estimate the quotient $82.304 \div 17.287$ by rounding to the nearest ten. [5.6a]

**A.** 0.4      **B.** 4
**C.** 40      **D.** 400

**65.** Translate to an algebraic expression: 15 less than twice the price. [5.8a]

**A.** $2p - 15$      **B.** $15 - 2p$
**C.** $2 + p - 15$      **D.** $2(p - 15)$

## Synthesis

**66.** ▦ In each of the following, use $+$, $-$, $\times$, or $\div$ in each blank to make a true sentence. [5.4b]

a) $2.56 - 6.4 \,\square\, 51.2 - 17.4 + 89.7 = 119.66$

b) $(11.12 \,\square\, 0.29)3^4 = 877.23$

**67.** Arrange from smallest to largest:

$$-\frac{2}{3}, \quad -\frac{15}{19}, \quad -\frac{11}{13}, \quad \frac{-5}{7}, \quad \frac{-13}{15}, \quad \frac{-17}{20}.$$

[5.1c], [5.5a]

**68.** Use the fact that $\frac{1}{3} = 0.\overline{3}$ to find repeating decimal notation for 1. Explain how you got your answer. [5.5a]

**69.** ▦ Sal's sells Sicilian pizza as a 17-in. by 20-in. pie for $15 or as an 18-in.-diameter round pie for $14. Which is a better buy and why? [5.8b]

# Understanding Through Discussion and Writing

**1.** Describe in your own words a procedure for converting from decimal notation to fraction notation. [5.1b]

**2.** A student insists that $346.708 \times 0.1 = 3467.08$. How could you convince him that a mistake had been made without checking the multiplication on a calculator? [5.3a]

**3.** When is long division *not* the fastest way to convert from fraction notation to decimal notation? [5.5a]

**4.** Consider finding decimal notation for $\frac{44}{125}$. Discuss as many ways as you can for finding such notation and give the answer. [5.5a]

# CHAPTER

## 5   Test

For
Extra
Help

For step-by-step test solutions, access the Chapter Test Prep Videos in MyLab Math.

1. **Projected World Population.** The world population is projected to reach 9.8 billion in 2050. Convert 9.8 billion to standard notation.

   **Data:** United Nations, Department of Economics and Social Affairs

2. Write a word name for 123.0047.

Write in fraction notation and, if possible, as a mixed numeral.

3. $-0.91$

4. 2.769

Write in decimal notation.

5. $\dfrac{74}{1000}$

6. $-\dfrac{37{,}047}{10{,}000}$

7. $756\dfrac{9}{100}$

8. $91\dfrac{703}{1000}$

Which number is larger?

9. 0.07, 0.162

10. 8.049, 8.0094

11. $-0.09, -0.9$

Round 5.6783 to the nearest

12. One.

13. Hundredth.

14. Thousandth.

15. Tenth.

Perform the indicated operation.

16.
$$\begin{array}{r} 402.3 \\ 2.81 \\ +\ \ \ 0.109 \\ \hline \end{array}$$

17.
$$\begin{array}{r} 0.125 \\ \times\ \ 0.24 \\ \hline \end{array}$$

18.
$$\begin{array}{r} 213.45 \\ \times\ \ 0.001 \\ \hline \end{array}$$

19.
$$\begin{array}{r} 52.091 \\ -\ \ 7.345 \\ \hline \end{array}$$

20. $342.9 + 8.1 + 5.37$

21. $-9.5 + 7.3$

22. $2 - 0.0054$

23. $1000 \times 73.962$

24. $4\overline{)19}$

25. $3.3\overline{)100.32}$

26. $\dfrac{-346.82}{1000}$

27. $\dfrac{346.82}{0.01}$

**28.** Convert $179.82 to cents.

**29.** Combine like terms:
$$4.1x + 5.2 - 3.9y + 5.7x - 9.8.$$

**30.** Evaluate: $2l + 4w + 2h$ for $l = 2.4$, $w = 1.3$, and $h = 0.8$.
(*The total girth of a postal package*)

**31.** Simplify: $20 \div 5(-2)^2 - 8.4$.

Multiply by a form of 1 to find decimal notation for each number.

**32.** $\dfrac{8}{5}$

**33.** $\dfrac{21}{4}$

Use division to find decimal notation for each number.

**34.** $-\dfrac{7}{16}$

**35.** $\dfrac{14}{9}$

**36.** Round $1.\overline{5}$ to the nearest hundredth.

Calculate.

**37.** $3 \div (-0.3) \cdot 2 - 1.5^2$

**38.** $(8 - 1.23) \div 4 + 5.6 \times 0.02$

**39.** $\dfrac{3}{8} \times 45.6 - \dfrac{1}{5} \times 36.9$

Solve. Remember to check.

**40.** $17y - 3.12 = -58.2$

**41.** $9t - 4 = 6t + 26$

**42.** $4 + 2(x - 3) = 7x - 9$

**43.** *Scanning Blueprints.* A building contractor needs to scan and load blueprints onto a flash drive. The copy center charges $10.99 for the flash drive and $1.19 per square foot for scanning. If the contractor needs to scan 5 blueprints at 6 sq ft per print, what will the total cost be?

**44.** *Gas Mileage.* Tina wants to estimate the gas mileage in her economy car. At 76,843 mi, she fills the tank with 14.3 gal of gasoline. At 77,310 mi, she fills the tank with 16.5 gal of gasoline. Find the mileage per gallon. Round to the nearest tenth.

**45.** *Checking Account Balance.* Nicholas had a balance of $820 in his checking account before making purchases of $123.89, $56.68, and $46.98 with his debit card. What was the balance after the purchases had been made?

**46.** The office manager for the Drake, Smith, and Hartner law firm buys 7 cases of copy paper at $41.99 per case. What is the total cost?

**47.** *Life Expectancy.* Life expectancies at birth for seven Asian countries are listed in the table below. Find the average life expectancy for this group of countries. Round to the nearest tenth.

| COUNTRY | LIFE EXPECTANCY (in years) |
|---|---|
| Japan | 84.74 |
| South Korea | 80.04 |
| People's Republic of China | 75.41 |
| Russia | 70.47 |
| North Korea | 70.11 |
| India | 68.13 |
| Afghanistan | 50.87 |

DATA: *The CIA World Factbook 2017*

**48.** About how many gallons of gasoline, at $2.749 per gallon, can be bought with $20?

**A.** 1 gallon      **B.** 3 gallons

**C.** 7 gallons      **D.** 12 gallons

## Synthesis

**49.** Use one of the words *sometimes*, *never*, or *always* to complete each of the following.

  **a)** The product of two numbers greater than 0 and less than 1 is _____ less than 1.

  **b)** The product of two numbers greater than 1 is _____ less than 1.

  **c)** The product of a number greater than 1 and a number less than 1 is _____ equal to 1.

  **d)** The product of a number greater than 1 and a number less than 1 is _____ equal to 0.

**50.** Silver's Health Club charges a $79 membership fee and $42.50 a month. Allise has a coupon that will allow her to join the club for $299 for six months. How much will Allise save if she uses the coupon?

**51.** *Travel Costs.* Roundtrip airfare between Burlington, VT, and Newark, NJ, often costs $359. One estimate of the true cost of driving is $1.35 per mile. Is it more economical to fly or drive the 600 mi for **(a)** an individual; **(b)** a couple; **(c)** a family of 4?

**Data:** commutesolutions.org

300 miles

Traveling by plane: $359 round trip

Traveling by car: $1.35 per mile

300 miles

VT **Burlington**   ME Portland

NY Albany

CT

Hartford

NYC

PA **Newark**

Philadelphia   NJ

Convert to fraction notation.

**1.** $2\frac{2}{9}$

**2.** 3.051

Find decimal notation.

**3.** $-\frac{7}{5}$

**4.** $\frac{6}{11}$

**5.** Determine whether 43 is prime, composite, or neither.

**6.** Determine whether 2,053,752 is divisible by 5.

Calculate.

**7.** $48 + 12 \div 4 - 10 \times 2 + 6892 \div 4$

**8.** $4.7 - \{0.1[1.2(3.95 - 1.65) + 1.5 \div 2.5]\}$

Round to the nearest hundredth.

**9.** 584.973

**10.** $218.\overline{5}$

**11.** Estimate the product $16.392 \times 9.715$ by rounding to the nearest one.

**12.** Estimate by rounding to the nearest tenth:
$2.714 + 4.562 - 3.31 - 0.0023$.

**13.** Evaluate $a \div 3 \cdot b$ for $a = 18$ and $b = 2$.

**14.** Combine like terms:
$-4p + 9 + 11p - 17$.

Add and simplify.

**15.**
$$\begin{array}{r} 2\frac{1}{4} \\ + \ 3\frac{4}{5} \\ \hline \end{array}$$

**16.**
$$\begin{array}{r} 34{,}921 \\ 93{,}092 \\ + 11{,}103 \\ \hline \end{array}$$

**17.** $\frac{1}{6} + \frac{2}{3} + \frac{8}{9}$

**18.** $143.9 + 2.053$

Subtract and simplify.

**19.** $723{,}041 - 12{,}904$

**20.** $5.903 - 19$

**21.** $5\frac{1}{7} - 4\frac{3}{7}$

**22.** $\frac{9}{10} - \frac{10}{11}$

Multiply and simplify.

**23.** $-\frac{3}{8} \cdot \frac{4}{9}$

**24.**
$$\begin{array}{r} 2\,5\,3\,2 \\ \times \ 2\,1\,0\,0 \\ \hline \end{array}$$

**25.**
$$\begin{array}{r} 2\,3.9 \\ \times \ \ \ 0.2 \\ \hline \end{array}$$

**26.**
$$\begin{array}{r} 2\,7.9\,4\,3\,1 \\ \times \ \ \ \ \ \ \ 0.0\,0\,1 \\ \hline \end{array}$$

Divide and simplify.

**27.** $16.5\overline{)3\,5.0\,1\,3}$

**28.** $26\overline{)4\,7{,}9\,1\,8}$

**29.** $13.8621 \div 0.001$

**30.** $-\dfrac{4}{9} \div \left(-\dfrac{8}{15}\right)$

Solve.

**31.** $3(x - 7) = 5x + 2$

**32.** $-75 \cdot x = 2100$

**33.** $2.4x - 7.1 = 2.05$

**34.** $1062 + y = 368{,}313$

**35.** $t + \dfrac{5}{6} = \dfrac{8}{9}$

**36.** $-\dfrac{7}{8} \cdot t = \dfrac{7}{16}$

**37.** *First Languages.*  The table below lists the numbers of first-language speakers (people who acquired the language as infants) for the top five languages spoken worldwide. How many people in the world are first-language speakers of Spanish, Arabic, or Hindi?

| LANGUAGE | FIRST-LANGUAGE SPEAKERS (in millions) |
|---|---|
| Chinese | 1213 |
| Spanish | 329 |
| English | 328 |
| Arabic | 221 |
| Hindi | 182 |

**DATA:** Living Tongues Institute for Endangered Languages; UNESCO; SIL International

**38.** Refer to the table in Exercise 37. How many more people in the world are first-language speakers of Chinese than first-language speakers of English?

**39.** After Lauren made a $450 down payment on a sofa, $\dfrac{3}{10}$ of the total cost was paid. How much did the sofa cost?

**40.** Joshua's tuition was $3600. He obtained a loan for $\dfrac{2}{3}$ of the tuition. How much was the loan?

**41.** The balance in Elliott's checking account is $314.79. After a check is written for $56.02, what is the balance in the account?

**42.** A clerk in Leah's Delicatessen sold $1\frac{1}{2}$ lb of ham, $2\frac{3}{4}$ lb of turkey, and $2\frac{1}{4}$ lb of roast beef. How many pounds of meat were sold altogether?

**43.** A triangular sail has a height of 16 ft and a base of 11 ft. Find its area.

**44.** A 4-in. by 5-in. rectangle is punched from a round piece of steel that is 9 in. wide. How much steel will be left over?

Simplify.

**45.** $\left(\dfrac{3}{4}\right)^2 - \dfrac{1}{8} \cdot \left(3 - 1\dfrac{1}{2}\right)^2$

**46.** $1.2 \times 12.2 \div 0.1 \times 3.6$

## Synthesis

**47.** Using a manufacturer's coupon, Lucy bought 2 cartons of orange juice and received a third carton free. The price of each carton was $3.59. What was the cost per carton with the coupon? Round to the nearest cent.

**48.** A carton of gelatin mix packages weighs $15\frac{3}{4}$ lb. Each package weighs $1\frac{3}{4}$ oz. How many packages are in the carton? (1 lb = 16 oz)

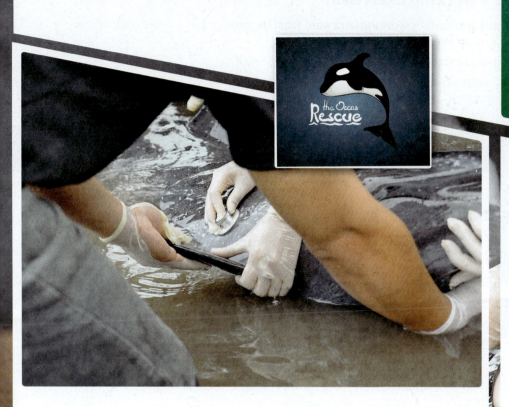

# Ratio and Proportion

Because of factors such as disease, injury, and weather, marine mammals can become stranded on land. According to the National Marine Fisheries Service, which coordinates monitoring and rescue of stranded marine mammals, over 59,000 marine mammals were stranded in the United States in the years 2001–2009. The mammals counted included cetaceans (whales, dolphins, and porpoises) and pinnipeds (seals and walruses). The accompanying graph illustrates where each type of mammal was stranded. Over 120 organizations in the United States respond to marine mammal strandings, and 32 of these are equipped to rehabilitate rescued marine mammals.

### Marine Mammal Strandings, 2001–2009

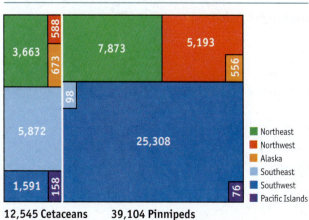

3,663 | 588 | 7,873 | 5,193
673 |
98 | 556
5,872 | 25,308
1,591 | 158 | 76

- Northeast
- Northwest
- Alaska
- Southeast
- Southwest
- Pacific Islands

**12,545 Cetaceans    39,104 Pinnipeds**

DATA: NOAA

In Example 7 of Section 6.1, we will calculate ratios of marine mammals that were stranded at Cape Hatteras National Seashore.

STUDYING FOR SUCCESS   *Working Exercises*

☐ Don't try to solve a homework problem by working backward from the answer given at the back of the text. Remember: quizzes and tests have no answer section!

☐ Check answers to odd-numbered exercises at the back of the text.

☐ Work some even-numbered exercises, whose answers are not provided, as practice. Check your answers later with a friend or your instructor.

## 6.1

### OBJECTIVES

**a** Find fraction notation for ratios.

**b** Simplify ratios.

# Introduction to Ratios

## a RATIOS

### RATIO

A **ratio** is the quotient of two quantities.

The average wind speed in Chicago is 10.4 mph. The average wind speed in Boston is 12.5 mph. The *ratio* of average wind speed in Chicago to average wind speed in Boston is written using colon notation,

Chicago wind speed → 10.4 : 12.5, ← Boston wind speed

or fraction notation,

$\dfrac{10.4}{12.5}$ ← Chicago wind speed
        ← Boston wind speed

We read both forms of notation as "the ratio of 10.4 to 12.5."

### RATIO NOTATION

The **ratio** of $a$ to $b$ is expressed by the fraction notation $\dfrac{a}{b}$, where $a$ is the numerator and $b$ is the denominator, or by the colon notation $a : b$.

**EXAMPLE 1**   Find the ratio of 7 to 8.

The ratio is $\dfrac{7}{8}$,   or   $7 : 8$.

**1.** Find the ratio of 5 to 11.

**2.** Find the ratio of 57.3 to 86.1.

**3.** Find the ratio of $6\frac{3}{4}$ to $7\frac{2}{5}$.

**EXAMPLE 2**   Find the ratio of 31.4 to 100.

The ratio is $\dfrac{31.4}{100}$,   or   $31.4 : 100$.

**EXAMPLE 3**   Find the ratio of $4\frac{2}{3}$ to $5\frac{7}{8}$. You need not simplify.

The ratio is $\dfrac{4\frac{2}{3}}{5\frac{7}{8}}$,   or   $4\frac{2}{3} : 5\frac{7}{8}$.

◀ **Do Exercises 1–3.**

**Answers**

**1.** $\dfrac{5}{11}$, or $5 : 11$   **2.** $\dfrac{57.3}{86.1}$, or $57.3 : 86.1$

**3.** $\dfrac{6\frac{3}{4}}{7\frac{2}{5}}$, or $6\frac{3}{4} : 7\frac{2}{5}$

In most of our work, we will use fraction notation for ratios.

**EXAMPLE 4**  *Media Usage.*  In 2016, American adults spent an average of 5.5 hr per week on social media and 25 hr per week on all media. Find the ratio of average time spent on social media to average time spent on all media.
**Data:** 2016 Nielsen Social Media Report

The ratio is $\dfrac{5.5}{25}$.

**EXAMPLE 5**  *Record Rainfall.*  The greatest rainfall ever recorded in the United States during a 12-month period was 739 in. in Kukui, Maui, Hawaii, from December 1981 to December 1982. What is the ratio of amount of rainfall, in inches, to time, in months? of time, in months, to amount of rainfall, in inches?
**Data:** *Time Almanac*

The ratio of amount of rainfall, in inches, to time, in months, is

$$\dfrac{739}{12}. \quad \begin{array}{l}\leftarrow \text{Rainfall} \\ \leftarrow \text{Time}\end{array}$$

The ratio of time, in months, to amount of rainfall, in inches, is

$$\dfrac{12}{739}. \quad \begin{array}{l}\leftarrow \text{Time} \\ \leftarrow \text{Rainfall}\end{array}$$

**EXAMPLE 6**  Refer to the triangle below.

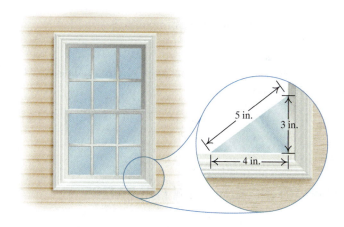

**a)** What is the ratio of the length of the longest side to the length of the shortest side?

$$\dfrac{5}{3} \quad \begin{array}{l}\leftarrow \text{Longest side} \\ \leftarrow \text{Shortest side}\end{array}$$

**b)** What is the ratio of the length of the shortest side to the length of the longest side?

$$\dfrac{3}{5} \quad \begin{array}{l}\leftarrow \text{Shortest side} \\ \leftarrow \text{Longest side}\end{array}$$

Do Exercises 4–6. ▶

**4.** *Record Snowfall.* The greatest snowfall recorded in North America during a 24-hr period was 76 in. in Silver Lake, Colorado, on April 14–15, 1921. What is the ratio of amount of snowfall, in inches, to time, in hours?

**Data:** U.S. Army Corps of Engineers

**5.** *Coffee Drinks.* A 16-oz café mocha with whole milk contains 360 calories. A 16-oz iced, blended cappucino contains 240 calories. What is the ratio of the number of calories in the café mocha to the number of calories in the cappucino? What is the ratio of the number of calories in the cappucino to the number of calories in the café mocha?

**Data:** medbroadcast.com

**GS** **6.** In the triangle below, what is the ratio of the length of the shortest side to the length of the longest side?

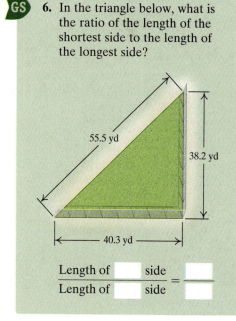

$$\dfrac{\text{Length of }\boxed{\phantom{xx}}\text{ side}}{\text{Length of }\boxed{\phantom{xx}}\text{ side}} = \dfrac{\boxed{\phantom{xx}}}{\boxed{\phantom{xx}}}$$

**Answers**

**4.** $\dfrac{76}{24}$  **5.** $\dfrac{360}{240}, \dfrac{240}{360}$  **6.** $\dfrac{38.2}{55.5}$

**Guided Solution:**
**6.** $\dfrac{\text{shortest}}{\text{longest}}, \dfrac{38.2}{55.5}$

**Marine Mammal Strandings**

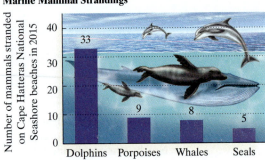

Number of mammals stranded on Cape Hatteras National Seashore beaches in 2015

Dolphins 33, Porpoises 9, Whales 8, Seals 5

DATA: nps.gov

**EXAMPLE 7** *Marine Mammal Strandings.* In 2015, a total of 57 marine mammals were stranded on the beaches of Cape Hatteras National Seashore in North Carolina. The bar graph at left shows how many dolphins, porpoises, whales, and seals were stranded.

a) What is the ratio of the number of dolphins stranded to the number of whales stranded?

b) What is the ratio of the number of cetaceans (dolphins, porpoises, and whales) stranded to the number of pinnipeds (seals) stranded?

c) What is the ratio of the number of mammals stranded that were not dolphins, porpoises, whales, or seals to the total number of mammals stranded?

a) The ratio of the number of dolphins stranded to the number of whales stranded is

$$\frac{33}{8}. \quad \leftarrow \text{Dolphins} \atop \leftarrow \text{Whales}$$

b) The total number of cetaceans (dolphins, porpoises, and whales) is

$$33 + 9 + 8 = 50.$$

The ratio of these mammals to the number of pinnipeds (seals) is

$$\frac{50}{5}. \quad \leftarrow \text{Dolphins, porpoises, and whales} \atop \leftarrow \text{Seals}$$

c) The total number of dolphins, porpoises, whales, and seals stranded is

$$33 + 9 + 8 + 5 = 55.$$

Thus, the number of other types of mammals stranded is

$$57 - 55 = 2.$$

The ratio of the number of other types of mammals to the total number of mammals stranded is

$$\frac{2}{57}. \quad \leftarrow \text{Mammals that were not listed in graph} \atop \leftarrow \text{Total number of mammals stranded}$$

◀ **Do Exercise 7.**

**7.** *NBA Playoffs.* In the final game of the 2017 NBA Playoffs, the Golden State Warriors made a total of 67 baskets. Of these, 19 were free throws, 34 were two-point field goals, and the remainder were three-point field goals.

**Data:** nba.com

a) What was the ratio of the number of two-point field goals to the number of free throws?

b) What was the ratio of the number of three-point field goals to the total number of baskets?

## b SIMPLIFYING NOTATION FOR RATIOS

**SKILL REVIEW**

*Simplify fraction notation.* [3.5b]
Simplify.

1. $\dfrac{16}{64}$

2. $\dfrac{40}{24}$

Answers: 1. $\dfrac{1}{4}$ 2. $\dfrac{5}{3}$

MyLab Math
VIDEO

*Answers*

7. (a) $\dfrac{34}{19}$; (b) $\dfrac{14}{67}$

Sometimes a ratio can be simplified. Simplifying provides a means of finding other numbers with the same ratio.

**EXAMPLE 8** Find the ratio of 6 to 8. Then simplify to find two other numbers in the same ratio.

We write the ratio in fraction notation and then simplify:

$$\frac{6}{8} = \frac{2 \cdot 3}{2 \cdot 4} = \frac{2}{2} \cdot \frac{3}{4} = 1 \cdot \frac{3}{4} = \frac{3}{4}.$$

Thus, 3 and 4 have the same ratio as 6 and 8. We can express this by saying "6 is to 8 as 3 is to 4."

**Do Exercise 8.** ▶

**EXAMPLE 9** Find the ratio of 2.4 to 10. Then simplify to find two other numbers in the same ratio.

We first write the ratio in fraction notation. Next, we multiply by 1 to clear the decimal from the numerator. Then we simplify.

$$\frac{2.4}{10} = \frac{2.4}{10} \cdot \frac{10}{10} = \frac{24}{100} = \frac{4 \cdot 6}{4 \cdot 25} = \frac{4}{4} \cdot \frac{6}{25} = \frac{6}{25}$$

Thus, 2.4 is to 10 as 6 is to 25.

**Do Exercises 9 and 10.** ▶

**EXAMPLE 10** An HDTV screen that measures approximately 46 in. diagonally has a width of 40 in. and a height of $22\frac{1}{2}$ in. Find the ratio of width to height and simplify.

The ratio is 
$$\frac{40}{22\frac{1}{2}} = \frac{40}{22.5} = \frac{40}{22.5} \cdot \frac{10}{10} = \frac{400}{225}$$
$$= \frac{25 \cdot 16}{25 \cdot 9} = \frac{25}{25} \cdot \frac{16}{9}$$
$$= \frac{16}{9}.$$

Thus, we can say that the ratio of width to height is 16 to 9, which can also be expressed as 16 : 9.

**Do Exercise 11.** ▶

8. Find the ratio of 18 to 27. Then simplify to find two other numbers in the same ratio.

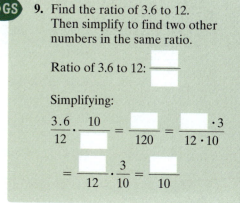

**GS** 9. Find the ratio of 3.6 to 12. Then simplify to find two other numbers in the same ratio.

Ratio of 3.6 to 12: ———

Simplifying:

$$\frac{3.6}{12} \cdot \frac{10}{\boxed{\phantom{0}}} = \frac{\boxed{\phantom{0}}}{120} = \frac{\boxed{\phantom{0}} \cdot 3}{12 \cdot 10}$$
$$= \frac{\boxed{\phantom{0}}}{12} \cdot \frac{3}{10} = \frac{\boxed{\phantom{0}}}{10}$$

10. Find the ratio of 1.2 to 1.5. Then simplify to find two other numbers in the same ratio.

11. An HDTV screen that measures 44 in. diagonally has a width of 38.4 in. and a height of 21.6 in. Find the ratio of height to width and simplify.

***Answers***

**8.** 18 is to 27 as 2 is to 3.
**9.** 3.6 is to 12 as 3 is to 10.
**10.** 1.2 is to 1.5 as 4 is to 5. **11.** $\frac{9}{16}$

***Guided Solution:***
**9.** $\frac{3.6}{12}$; 10, 36, 12, 12, 3

## ✓ Check Your Understanding

**Reading Check** Determine whether each statement is true or false.

_____ **RC1.** A ratio is a quotient.

_____ **RC2.** If there are 2 teachers and 27 students in a classroom, the ratio of students to teachers is 27 : 2.

_____ **RC3.** The ratio 6 : 7 can also be written $\frac{7}{6}$.

_____ **RC4.** The numbers 2 and 3 are in the same ratio as the numbers 4 and 9.

**Concept Check** Choose from each list all pairs of numbers that are in the same ratio as the given ratio.

**CC1.** $\frac{3}{4}$

**a)** $\frac{6}{8}$      **b)** $\frac{9}{16}$      **c)** 30 : 40      **d)** 0.3 to 0.4      **e)** 4 : 5

**CC2.** 1.6 to 10

**a)** $\frac{8}{5}$      **b)** 16 to 100      **c)** 4 : 25      **d)** $\frac{16}{1}$      **e)** $\frac{8}{50}$

**a**    Find fraction notation for each ratio. You need not simplify.

**1.** 4 to 5

**2.** 3 to 2

**3.** 178 to 572

**4.** 329 to 967

**5.** 0.4 to 12

**6.** 2.3 to 22

**7.** 3.8 to 7.4

**8.** 0.6 to 0.7

**9.** 56.78 to 98.35

**10.** 456.2 to 333.1

**11.** $8\frac{3}{4}$ to $9\frac{5}{6}$

**12.** $10\frac{1}{2}$ to $43\frac{1}{4}$

**13.** *Book Sales.* In 2015, there were 653 million print books and 204 million traditionally published e-books sold in the United States. What is the ratio of the number of print books to the number of e-books? of the number of e-books to the number of print books?

**Data:** 2015 U.S. Book Industry Year-end Review, Nielsen

**14.** *Silicon in the Earth's Crust.* Every 100 tons of the earth's crust contains about 28 tons of silicon. What is the ratio of the weight of silicon to the weight of crust? of the weight of crust to the weight of silicon?

**Data:** *The Handy Science Answer Book*

**National Parks.**   Of the 59 national parks in the United States, 35 are located in the seven states included in the following bar graph. Use the graph for Exercises 15–18. You need not simplify the ratios.

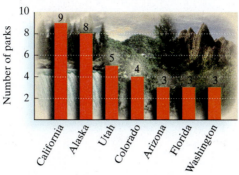

DATA: National Park Service

**15.** What is the ratio of the number of national parks in the seven states listed to the total number of national parks?

**16.** What is the ratio of the number of national parks in Alaska and California to the total number of national parks?

**17.** What is the ratio of the number of national parks in Alaska and California to the number of national parks in the other 48 states?

**18.** What is the ratio of the number of national parks in the seven states listed to the number of national parks in the other 43 states?

**19.** *Tax Freedom Day.*   Of the 365 days in 2017, the average American worked 113 days to pay his or her federal, state, and local taxes. Find the ratio of the number of days worked to pay taxes in 2017 to the number of days in the year.

**Data:** Tax Foundation

**20.** *Careers in Medicine.*   The number of jobs for nurses is expected to increase by 439,300 between 2014 and 2024. During the same decade, the number of jobs for physicians is expected to increase by 99,300. Find the ratio of the increase in jobs for physicians to the increase in jobs for nurses.

**Data:** U.S. Bureau of Labor Statistics

**21.** *Field Hockey.*   A diagram of the playing area for field hockey is shown below. What is the ratio of width to length? of length to width?

**Data:** *Sports: The Complete Visual Reference*

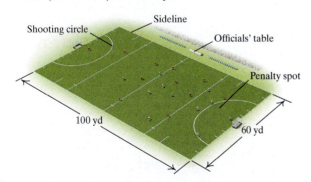

**22.** *The Leaning Tower of Pisa.*   The Leaning Tower of Pisa was reopened to the public in 2001 following a 10-yr stabilization project. The 184.5-ft tower now leans about 13 ft from its base. What is the ratio of the distance that it leans to its height? of its height to the distance that it leans?

**Data:** CNN

**b** Find the ratio of the first number to the second and simplify.

**23.** 4 to 6

**24.** 6 to 10

**25.** 18 to 24

**26.** 28 to 36

**27.** 4.8 to 10

**28.** 5.6 to 10

**29.** 2.8 to 3.6

**30.** 4.8 to 6.4

**31.** 20 to 30

**32.** 40 to 60

**33.** 56 to 100

**34.** 42 to 100

**35.** 128 to 256

**36.** 232 to 116

**37.** 0.48 to 0.64

**38.** 0.32 to 0.96

**39.** For this rectangle, find the ratios of length to width and of width to length.

478 ft

213 ft

**40.** For this right triangle, find the ratios of shortest side length to longest side length and of longest length to shortest length.

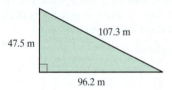

107.3 m

47.5 m

96.2 m

## Skill Maintenance

Add. Simplify if possible.

**41.** 18,468 + 390,082   [1.2a]

**42.** 24 + 3.006   [5.2a]

**43.** 4.2 + 28.07 + 365   [5.2a]

**44.** $\dfrac{1}{3} + \dfrac{3}{10}$   [4.2b]

**45.** $\dfrac{3}{8} + \dfrac{7}{12}$   [4.2b]

**46.** $4\dfrac{5}{6} + 9\dfrac{1}{2}$   [4.6a]

Subtract. Simplify if possible.

**47.** 982,001 − 39,782   [1.3a]

**48.** 12.5 − 9.9   [5.2b]

**49.** 96 − 1.625   [5.2b]

**50.** $\dfrac{9}{16} - \dfrac{1}{2}$   [4.3a]

**51.** $8\dfrac{2}{3} - 6\dfrac{4}{5}$   [4.6b]

**52.** $11 - 6\dfrac{1}{3}$   [4.6b]

## Synthesis

**53.** Find the ratio of $3\dfrac{3}{4}$ to $5\dfrac{7}{8}$ and simplify.

*Fertilizer.* Exercises 54 and 55 refer to a common lawn fertilizer known as "5, 10, 15." This mixture contains 5 parts of potassium for every 10 parts of phosphorus and 15 parts of nitrogen. (This is often denoted 5 : 10 : 15.)

**54.** Find and simplify the ratio of potassium to nitrogen and of nitrogen to phosphorus.

**55.** Simplify the ratio 5 : 10 : 15.

# Rates and Unit Prices

**a** RATES

> **SKILL REVIEW**
>
> *Divide using decimal notation.* [5.4a]
> Divide.
>
> **1.** $15\overline{)634.5}$        **2.** $2.4\overline{)6}$
>
> **Answers: 1.** 42.3 **2.** 2.5
>
> MyLab Math
> VIDEO

A 2018 Toyota Camry can travel 408 mi on 12 gal of gasoline. Let's consider the ratio of miles to gallons:

**Data:** cars.com

$$\frac{408 \text{ mi}}{12 \text{ gal}} = \frac{408}{12}\frac{\text{miles}}{\text{gallon}} = \frac{34}{1}\frac{\text{miles}}{\text{gallon}}$$

$$= 34 \text{ miles per gallon} = 34 \text{ mpg.}$$

"per" means "division," or "for each."

The ratio

$$\frac{408 \text{ mi}}{12 \text{ gal}}, \quad \text{or} \quad \frac{408}{12}\frac{\text{mi}}{\text{gal}}, \quad \text{or} \quad 34 \text{ mpg,}$$

is called a **rate**.

---

> **RATE**
>
> When a ratio is used to compare two different kinds of measure, we call it a **rate**.

---

Suppose David's car travels 574 mi on 16.8 gal of gasoline. Is the gas mileage (mpg) of his car better than that of the Toyota Camry above? To determine this, it helps to convert the ratio to decimal notation and perhaps round. Thus, we have

$$\frac{574 \text{ miles}}{16.8 \text{ gallons}} = \frac{574}{16.8} \text{ mpg} \approx 34.167 \text{ mpg.}$$

Since $34.167 > 34$, David's car gets better gas mileage than the Toyota Camry does.

**EXAMPLE 1** Abby harvested 8800 bushels of corn from 55 acres. What is the rate in bushels per acre?

$$\frac{8800 \text{ bushels}}{55 \text{ acres}} = \frac{8800}{55}\frac{\text{bushels}}{\text{acre}} = 160\frac{\text{bushels}}{\text{acre}}$$

**EXAMPLE 2**   It takes 60 oz of grass seed to seed 3000 sq ft of lawn. What is the rate in ounces per square foot?

$$\frac{60 \text{ oz}}{3000 \text{ sq ft}} = \frac{1}{50}\frac{\text{oz}}{\text{sq ft}}, \quad \text{or} \quad 0.02\frac{\text{oz}}{\text{sq ft}}$$

**EXAMPLE 3**   Martina bought 5 lb of organic russet potatoes for $4.99. What was the rate in cents per pound?

$$\frac{\$4.99}{5 \text{ lb}} = \frac{499 \text{ cents}}{5 \text{ lb}} = 99.8\cancel{c}/\text{lb}$$

**EXAMPLE 4**   *Hourly Wage.*   In 2017, Walmart sales associates earned, on average, $408.80 for a 40-hr work week. What was the rate of pay per hour?
**Data:** payscale.com

The rate of pay is the ratio of money earned to length of time worked, or

$$\frac{\$408.80}{40 \text{ hr}} = \frac{408.80}{40}\frac{\text{dollars}}{\text{hr}} = 10.22\frac{\text{dollars}}{\text{hr}}, \quad \text{or} \quad \$10.22 \text{ per hr.}$$

**EXAMPLE 5**   *Ratio of Strikeouts to Home Runs.*   In the 2016 baseball season, Nelson Cruz of the Seattle Mariners had 159 strikeouts and 43 home runs. What was his strikeout to home-run rate?
**Data:** Major League Baseball

$$\frac{159 \text{ strikeouts}}{43 \text{ home runs}} = \frac{159}{43}\frac{\text{strikeouts}}{\text{home runs}} = \frac{159}{43} \text{ strikeouts per home run}$$

$$\approx 3.698 \text{ strikeouts per home run}$$

◀ **Do Exercises 1–5.**

## b  UNIT PRICING

> **UNIT PRICE**
>
> A **unit price**, or **unit rate**, is the ratio of price to the number of units.

**EXAMPLE 6**   *Unit Price of Pears.*   Ruth bought a 15.2-oz can of pears for $1.30. What is the unit price in cents per ounce?

$$\text{Unit price} = \frac{\text{Price}}{\text{Number of units}}$$

$$= \frac{\$1.30}{15.2 \text{ oz}} = \frac{130 \text{ cents}}{15.2 \text{ oz}} = \frac{130}{15.2}\frac{\text{cents}}{\text{oz}}$$

$$\approx 8.553 \text{ cents per ounce}$$

◀ **Do Exercise 6.**

---

A ratio of distance traveled to time is called a *speed*. What is the rate, or speed, in miles per hour?

**1.** 45 mi, 9 hr

**2.** 120 mi, 10 hr

What is the rate, or speed, in feet per second?

**3.** 2200 ft, 2 sec

**4.** 52 ft, 13 sec   **GS**

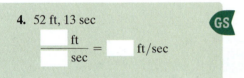

**5.** *Babe Ruth.*   In his baseball career, Babe Ruth had 1330 strikeouts and 714 home runs. What was his home-run to strikeout rate?

   **Data:** Major League Baseball

**6.** **Unit Price of Pasta Sauce.**   **GS**
Gregory bought a 26-oz jar of pasta sauce for $2.79. What is the unit price in cents per ounce?

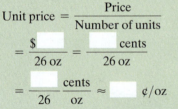

**Answers**

**1.** 5 mi/hr, or 5 mph   **2.** 12 mi/hr, or 12 mph   **3.** 1100 ft/sec   **4.** 4 ft/sec
**5.** $\frac{714}{1330}$ home run per strikeout ≈ 0.537 home run per strikeout
**6.** 10.731¢/oz

**Guided Solutions:**

**4.** $\frac{52 \text{ ft}}{13 \text{ sec}} = 4 \text{ ft/sec}$   **6.** 2.79, 279, 279, 10.731

Unit prices enable us to do comparison shopping and determine the best buy for a product on the basis of price. It is often helpful to change all prices to cents so that we can compare unit prices more easily.

**EXAMPLE 7** *Unit Price of Salad Dressing.* At the request of his customers, Angelo started bottling and selling the rosemary and lemon salad dressing that he serves in his café. The dressing is sold in containers of four sizes as listed in the table below. Compute the unit price of each size of container and determine which size is the best buy on the basis of unit price alone.

| Size | Price | Unit Price |
|------|-------|------------|
| 10 oz | $2.49 | |
| 16 oz | $3.59 | |
| 20 oz | $4.09 | |
| 32 oz | $6.79 | |

We compute the unit price of each size and fill in the chart:

$$10 \text{ oz:} \quad \frac{\$2.49}{10 \text{ oz}} = \frac{249 \text{ cents}}{10 \text{ oz}} = \frac{249}{10} \frac{\text{cents}}{\text{oz}} = 24.900¢/\text{oz};$$

$$16 \text{ oz:} \quad \frac{\$3.59}{16 \text{ oz}} = \frac{359 \text{ cents}}{16 \text{ oz}} = \frac{359}{16} \frac{\text{cents}}{\text{oz}} \approx 22.438¢/\text{oz};$$

$$20 \text{ oz:} \quad \frac{\$4.09}{20 \text{ oz}} = \frac{409 \text{ cents}}{20 \text{ oz}} = \frac{409}{20} \frac{\text{cents}}{\text{oz}} = 20.450¢/\text{oz};$$

$$32 \text{ oz:} \quad \frac{\$6.79}{32 \text{ oz}} = \frac{679 \text{ cents}}{32 \text{ oz}} = \frac{679}{32} \frac{\text{cents}}{\text{oz}} \approx 21.219¢/\text{oz}.$$

| Size | Price | Unit Price |
|------|-------|------------|
| 10 oz | $2.49 | 24.900¢/oz |
| 16 oz | $3.59 | 22.438¢/oz |
| 20 oz | $4.09 | 20.450¢/oz<br>Lowest unit price |
| 32 oz | $6.79 | 21.219¢/oz |

On the basis of unit price alone, we see that the 20-oz container is the best buy.

**Do Exercise 7.** ▶

Although we often think that "bigger is cheaper," this is not always the case, as we see in Example 7. In addition, even when a larger package has a lower unit price than a smaller package, it still might not be the best buy for you. For example, some of the food in a large package could spoil before it is used, or you might not have room to store a large package.

**7.** *Cost of Mayonnaise.* Complete the following table for Hellmann's mayonnaise sold on an online shopping site. Which size has the lowest unit price?

**Data:** Peapod.com

| Size | Price | Unit Price |
|------|-------|------------|
| 8 oz | $2.79 | |
| 15 oz | $3.49 | |
| 30 oz | $4.99 | |

**Answers**

**7.** 34.875¢/oz; 23.267¢/oz; 16.633¢/oz; the 30-oz size has the lowest unit price.

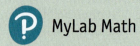
## ✓ Check Your Understanding

**Reading Check** Determine whether each statement is true or false.

_____ **RC1.** A rate is a ratio.

_____ **RC2.** It is always less expensive per unit to buy in bulk.

_____ **RC3.** The rate $0.50 per mile is a unit rate.

_____ **RC4.** The rate $5.99 per 10 oz is a unit rate.

**Concept Check** Complete each sentence with the correct quantity from the list at the right, given the following information.

Jason wrote an 8-page essay in 9 hr.

Keri graded 8 essays in 3 hr.

**CC1.** Jason wrote at a rate of ____ pages/hr.

**CC2.** Jason wrote at a rate of ____ hr/page.

**CC3.** Keri graded at a rate of ____ hr/essay.

**CC4.** Keri graded at a rate of ____ essays/hr.

a) $\frac{8}{3}$

b) $\frac{8}{9}$

c) $\frac{9}{8}$

d) $\frac{3}{8}$

**a** In Exercises 1–4, find each rate, or speed, as a ratio of distance to time. Round to the nearest hundredth where appropriate.

**1.** 120 km, 3 hr

**2.** 18 mi, 9 hr

**3.** 217 mi, 29 sec

**4.** 443 m, 48 sec

**5.** *Population Density of Monaco.* Monaco is a tiny country on the Mediterranean coast of France. It has an area of 0.75 sq mi and a population of 38,499 people. What is the rate of number of people per square mile? The rate of people per square mile is called the *population density.* Monaco has the highest population density of any country in the world.

**Data:** World Bank

**6.** *Population Density of Australia.* The continent of Australia, with the island state of Tasmania, has an area of 2,967,893 sq mi and a population of 24,127,159 people. What is the rate of number of people per square mile? The rate of people per square mile is called the *population density.* Australia has one of the lowest population densities in the world.

**Data:** World Bank

**7. *Ford F-150: City Driving.*** A 2017 Ford F-150 6-cylinder will travel 212.5 mi on 12.5 gal of gasoline in city driving. What is the rate in miles per gallon?

**Data:** fueleconomy.gov

**8. *Honda CR-V: City Driving.*** A 2017 Honda CR-V AWD 1.5L will travel 391.5 mi on 14.5 gal of gasoline in city driving. What is the rate in miles per gallon?

**Data:** fueleconomy.gov

**9. *Honda CR-V: Highway Driving.*** A 2017 Honda CR-V AWD 1.5L will travel 643.5 mi on 19.5 gal of gasoline in highway driving. What is the rate in miles per gallon?

**Data:** fueleconomy.gov

**10. *Ford F-150: Highway Driving.*** A 2017 Ford F-150 6-cylinder will travel 297 mi on 13.5 gal of gasoline in highway driving. What is the rate in miles per gallon?

**Data:** fueleconomy.gov

**11.** A car is driven 500 mi in 20 hr. What is the rate in miles per hour? in hours per mile?

**12.** A student eats 3 hamburgers in 15 min. What is the rate in hamburgers per minute? in minutes per hamburger?

**13. *Broadway Musicals.*** In the 17 years from 1987 through 2003, the musical *Les Misérables* was performed on Broadway 6680 times. What was the average rate of performances per year?

**Data:** broadwaymusicalhome.com

**14. *Employment Growth.*** In the 10 years from 2014 to 2024, the number of jobs for physical therapist assistants is expected to grow by 51,400. What is the expected average rate of growth in jobs per year?

**Data:** U.S. Bureau of Labor Statistics

**15. *Speed of Light.*** Light travels 186,000 mi in 1 sec. What is its rate, or speed, in miles per second?

**Data:** *The Handy Science Answer Book*

**16. *Speed of Sound.*** Sound travels 1100 ft in 1 sec. What is its rate, or speed, in feet per second?

**Data:** *The Handy Science Answer Book*

**17.** Impulses in nerve fibers travel 310 km in 2.5 hr. What is the rate, or speed, in kilometers per hour?

**18.** A black racer snake can travel 4.6 km in 2 hr. What is its rate, or speed, in kilometers per hour?

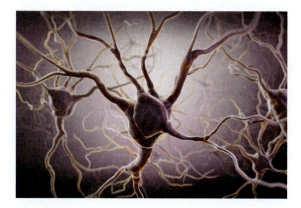

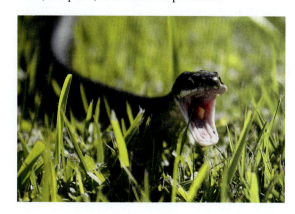

**19.** *Lawn Watering.* Watering a lawn adequately requires 623 gal of water for every 1000 ft². What is the rate in gallons per square foot?

**20.** A car is driven 200 km on 40 L of gasoline. What is the rate in kilometers per liter?

**21.** *Elephant Heart Rate.* The heart of an elephant, at rest, beats an average of 1500 beats in 60 min. What is the rate in beats per minute?

**Data:** *The Handy Science Answer Book*

**22.** *Human Heart Rate.* The heart of a human, at rest, beats an average of 4200 beats in 60 min. What is the rate in beats per minute?

**Data:** *The Handy Science Answer Book*

**b** Find each unit price in Exercises 23–30. Then, in each exercise, determine which size is the better buy based on unit price alone.

**23.** *Hidden Valley Ranch Dressing.*

| Size | Price | Unit Price |
|------|-------|------------|
| 16 oz | $4.19 | |
| 20 oz | $5.29 | |

**24.** *Miracle Whip.*

| Size | Price | Unit Price |
|------|-------|------------|
| 32 oz | $5.29 | |
| 48 oz | $8.29 | |

**25.** *Cascade Powder Detergent.*

| Size | Price | Unit Price |
|------|-------|------------|
| 45 oz | $5.90 | |
| 75 oz | $8.59 | |

**26.** *Bush's Homestyle Baked Beans.*

| Size | Price | Unit Price |
|------|-------|------------|
| 16 oz | $2.99 | |
| 28 oz | $4.51 | |

**27.** *Jif Creamy Peanut Butter.*

| Size | Price | Unit Price |
|------|-------|------------|
| 18 oz | $3.28 | |
| 28 oz | $4.89 | |

**28.** *Hills Brothers Coffee.*

| Size | Price | Unit Price |
|------|-------|------------|
| 26 oz | $9.99 | |
| 39 oz | $14.99 | |

**29.** *Campbell's Condensed Tomato Soup.*

| Size | Price | Unit Price |
|------|-------|------------|
| 10.7 oz | $1.09 | |
| 26 oz | $2.69 | |

**30.** *Nabisco Saltines.*

| Size | Price | Unit Price |
|------|-------|------------|
| 16 oz | $3.59 | |
| 32 oz | $5.19 | |

Find the unit price of each brand in Exercises 31–34. Then, in each exercise, determine which brand is the better buy based on unit price alone.

**31.** *Vanilla Ice Cream.*

| Brand | Size | Price |
|-------|------|-------|
| B | 32 oz | $5.99 |
| E | 48 oz | $6.99 |

**32.** *Orange Juice.*

| Brand | Size | Price |
|-------|------|-------|
| M | 54 oz | $4.79 |
| T | 59 oz | $5.99 |

**33.** *Tomato Ketchup.*

| Brand | Size | Price |
|-------|------|-------|
| A | 24 oz | $2.49 |
| B | 36 oz | $3.29 |
| H | 46 oz | $3.69 |

**34.** *Yellow Mustard.*

| Brand | Size | Price |
|-------|------|-------|
| F | 14 oz | $1.29 |
| G | 19 oz | $1.99 |
| P | 20 oz | $2.49 |

## Skill Maintenance

Multiply. Simplify, if possible.

**35.** $25 \times 462$  [1.4a]

**36.** $8.4 \times 80.892$  [5.3a]

**37.** $0.01 \times 274.568$  [5.3a]

**38.** $\dfrac{50}{9} \cdot \dfrac{6}{5}$  [3.6a]

**39.** $3\dfrac{4}{5} \cdot 2\dfrac{1}{4}$  [4.7a]

**40.** $4\dfrac{1}{10} \cdot 3\dfrac{1}{3}$  [4.7a]

Divide. Simplify, if possible.

**41.** $4000 \div 32$  [1.5a]

**42.** $95 \div 10$  [5.4a]

**43.** $80.892 \div 8.4$  [5.4a]

**44.** $\dfrac{50}{9} \div \dfrac{6}{5}$  [3.7b]

**45.** $200 \div 1\dfrac{1}{3}$  [4.7b]

**46.** $4\dfrac{6}{7} \div \dfrac{1}{4}$  [4.7b]

## Synthesis

**47.** Manufacturers sometimes change the sizes of their containers in such a way that the consumer thinks the price of a product has been lowered when, in reality, a higher unit price is being charged.

Some aluminum juice cans are now concave (curved in) on the bottom. Suppose the volume of the can in the figure has been reduced from 6 fl oz to 5.5 fl oz, and the price of each can has been reduced from 65¢ to 60¢. Find the unit price of each container in cents per ounce.

$\frac{5}{16}$ in.

$1\frac{13}{16}$ in.

$2\frac{1}{16}$ in.

# 6.3

## OBJECTIVES

**a** Determine whether two pairs of numbers are proportional.

**b** Solve proportions.

# Proportions

## a PROPORTIONS

**SKILL REVIEW**

*Test to determine whether two fractions are equivalent.* [3.5c]
Use = or ≠ for ☐ to write a true sentence.

**1.** $\dfrac{18}{12} \,\square\, \dfrac{9}{6}$

**2.** $\dfrac{4}{7} \,\square\, \dfrac{5}{8}$

**Answers: 1.** = **2.** ≠

When two pairs of numbers, such as 3, 2 and 6, 4, have the same ratio, we say that they are **proportional**. The equation

$$\frac{3}{2} = \frac{6}{4}$$

states that the pairs 3, 2 and 6, 4 are proportional. Such an equation is called a **proportion**. We sometimes read $\frac{3}{2} = \frac{6}{4}$ as "3 is to 2 as 6 is to 4."

Since ratios can be written using fraction notation, we can use the test for equality of fractions to determine whether two ratios are the same.

---

### A TEST FOR EQUALITY OF FRACTIONS

Two fractions are equal if their cross products are equal.

---

**EXAMPLE 1** Determine whether 1, 2 and 3, 6 are proportional.

We can use cross products.

$$1 \cdot 6 = 6 \qquad \overset{?}{\underset{}{\frac{1}{2} = \frac{3}{6}}} \qquad 2 \cdot 3 = 6$$

Since the cross products are the same, $6 = 6$, we know that $\frac{1}{2} = \frac{3}{6}$, so the numbers are proportional.

**EXAMPLE 2** Determine whether 2, 5 and 4, 7 are proportional.

We can use cross products.

$$2 \cdot 7 = 14 \qquad \overset{?}{\underset{}{\frac{2}{5} = \frac{4}{7}}} \qquad 5 \cdot 4 = 20$$

Since the cross products are not the same, $14 \neq 20$, we know that $\frac{2}{5} \neq \frac{4}{7}$, so the numbers are not proportional.

◀ **Do Exercises 1–3.**

---

Determine whether the two pairs of numbers are proportional.

**1.** 3, 4 and 6, 8

**2.** 1, 4 and 10, 39

**3.** 1, 2 and 20, 39   **GS**

We compare cross products.

$1 \cdot \fbox{\phantom{xx}} = \fbox{\phantom{xx}}$   $\dfrac{1}{2} \,?\, \dfrac{20}{39}$

$2 \cdot \fbox{\phantom{xx}} = \fbox{\phantom{xx}}$

Since $39 \neq 40$, the numbers $\underline{\phantom{xxxx}}$ proportional.
are/are not

---

**Answers**

**1.** Yes  **2.** No  **3.** No

**Guided Solution:**
**3.** 39, 39, 20, 40; are not

**EXAMPLE 3**   Determine whether 3.2, 4.8 and 0.16, 0.24 are proportional.

We can use cross products.

$$3.2 \times 0.24 = 0.768 \qquad \overset{?}{\underset{4.8}{\frac{3.2}{4.8}} = \frac{0.16}{0.24}} \qquad 4.8 \times 0.16 = 0.768$$

Since the cross products are the same, $0.768 = 0.768$, we know that $\frac{3.2}{4.8} = \frac{0.16}{0.24}$, so the numbers are proportional.

**Do Exercises 4 and 5.** ▶

Determine whether the two pairs of numbers are proportional.

**4.** 6.4, 12.8 and 5.3, 10.6

**5.** 6.8, 7.4 and 3.4, 4.2

**EXAMPLE 4**   Determine whether $4\frac{2}{3}, 5\frac{1}{2}$ and $8\frac{7}{8}, 16\frac{1}{3}$ are proportional.

We can use cross products.

$$4\frac{2}{3} \cdot 16\frac{1}{3} = \frac{14}{3} \cdot \frac{49}{3} \qquad \overset{?}{\frac{4\frac{2}{3}}{5\frac{1}{2}} = \frac{8\frac{7}{8}}{16\frac{1}{3}}} \qquad 5\frac{1}{2} \cdot 8\frac{7}{8} = \frac{11}{2} \cdot \frac{71}{8}$$

$$= \frac{686}{9} \qquad\qquad\qquad\qquad\qquad = \frac{781}{16}$$

$$= 76\frac{2}{9} \qquad\qquad\qquad\qquad\qquad = 48\frac{13}{16}$$

Since the cross products are not the same, $76\frac{2}{9} \neq 48\frac{13}{16}$, we know that the numbers are not proportional.

**Do Exercise 6.** ▶

**6.** Determine whether $4\frac{2}{3}, 5\frac{1}{2}$ and 14, $16\frac{1}{2}$ are proportional.

## b   SOLVING PROPORTIONS

One way to solve a proportion is to use cross products. Then we can divide on both sides to get the variable alone.

**EXAMPLE 5**   Solve the proportion $\frac{x}{3} = \frac{4}{6}$.

$$\frac{x}{3} = \frac{4}{6}$$

$x \cdot 6 = 3 \cdot 4$   Equating cross products (finding cross products and setting them equal)

$$\frac{x \cdot 6}{6} = \frac{3 \cdot 4}{6}$$   Dividing by 6 on both sides

$$x = \frac{3 \cdot 4}{6} = \frac{12}{6} = 2$$

We can check that 2 is the solution by replacing $x$ with 2 and finding cross products:

$$2 \cdot 6 = 12 \qquad \overset{?}{\frac{2}{3} = \frac{4}{6}} \qquad 3 \cdot 4 = 12.$$

Since the cross products are the same, it follows that $\frac{2}{3} = \frac{4}{6}$. Thus, the pairs of numbers 2, 3 and 4, 6 are proportional, and 2 is the solution of the proportion.

**Do Exercise 7.** ▶

**7.** Solve: $\frac{x}{63} = \frac{2}{9}$.

To solve $\dfrac{x}{a} = \dfrac{c}{d}$ for $x$, equate *cross products* and divide on both sides to get $x$ alone.

---

**8.** Solve: $\dfrac{x}{9} = \dfrac{5}{4}$. Write a mixed numeral for the answer.

$$\frac{x}{9} = \frac{5}{4}$$

$$x \cdot 4 = 9 \cdot \boxed{\phantom{0}}$$

$$\frac{x \cdot 4}{\boxed{\phantom{0}}} = \frac{9 \cdot 5}{\boxed{\phantom{0}}}$$

$$x = \frac{\boxed{\phantom{0}}}{4} = \frac{\boxed{\phantom{0}}}{\boxed{\phantom{0}}} \frac{\boxed{\phantom{0}}}{4}$$

**EXAMPLE 6**  Solve: $\dfrac{x}{7} = \dfrac{5}{3}$. Write a mixed numeral for the answer.

$$\frac{x}{7} = \frac{5}{3}$$

$$x \cdot 3 = 7 \cdot 5 \qquad \text{Equating cross products}$$

$$\frac{x \cdot 3}{3} = \frac{7 \cdot 5}{3} \qquad \text{Dividing by 3}$$

$$x = \frac{7 \cdot 5}{3} = \frac{35}{3}, \text{ or } 11\frac{2}{3}$$

The solution is $11\frac{2}{3}$.

◀ **Do Exercise 8.**

**EXAMPLE 7**  Solve: $\dfrac{7.7}{15.4} = \dfrac{y}{2.2}$.

$$\frac{7.7}{15.4} = \frac{y}{2.2}$$

$$7.7 \times 2.2 = 15.4 \times y \qquad \text{Equating cross products}$$

$$\frac{7.7 \times 2.2}{15.4} = \frac{15.4 \times y}{15.4} \qquad \text{Dividing by 15.4}$$

$$\frac{7.7 \times 2.2}{15.4} = y$$

$$\frac{16.94}{15.4} = y \qquad \text{Multiplying}$$

$$1.1 = y \qquad \text{Dividing: } 15.4\overline{)16.94}$$

The solution is 1.1.

**9.** Solve: $\dfrac{21}{5} = \dfrac{n}{2.5}$.

**10.** Solve: $\dfrac{6}{x} = \dfrac{25}{11}$. Write decimal notation for the answer.

**EXAMPLE 8**  Solve: $\dfrac{8}{x} = \dfrac{5}{3}$. Write decimal notation for the answer.

$$\frac{8}{x} = \frac{5}{3}$$

$$8 \cdot 3 = x \cdot 5 \qquad \text{Equating cross products}$$

$$\frac{8 \cdot 3}{5} = x \qquad \text{Dividing by 5}$$

$$\frac{24}{5} = x \qquad \text{Multiplying}$$

$$4.8 = x \qquad \text{Dividing: } 5\overline{)24}$$

The solution is 4.8.

◀ **Do Exercises 9 and 10.**

*Answers*

**8.** $11\frac{1}{4}$  **9.** 10.5  **10.** 2.64

*Guided Solution:*
**8.** 5, 4, 4, 45, 11, 1

**EXAMPLE 9**  Solve: $\dfrac{3.4}{4.93} = \dfrac{10}{n}$.

$$\frac{3.4}{4.93} = \frac{10}{n}$$

$3.4 \times n = 4.93 \times 10$     Equating cross products

$\dfrac{3.4 \times n}{3.4} = \dfrac{4.93 \times 10}{3.4}$     Dividing by 3.4

$n = \dfrac{4.93 \times 10}{3.4}$

$n = \dfrac{49.3}{3.4}$     Multiplying

$n = 14.5$     Dividing

The solution is 14.5.

**Do Exercise 11.** ▶

**11.** Solve: $\dfrac{0.4}{0.9} = \dfrac{4.8}{t}$.

**EXAMPLE 10**  Solve: $\dfrac{4\frac{2}{3}}{5\frac{1}{2}} = \dfrac{14}{x}$.

$$\frac{4\frac{2}{3}}{5\frac{1}{2}} = \frac{14}{x}$$

$4\dfrac{2}{3} \cdot x = 5\dfrac{1}{2} \cdot 14$     Equating cross products

$\dfrac{14}{3} \cdot x = \dfrac{11}{2} \cdot 14$     Converting to fraction notation

$\dfrac{\frac{14}{3} \cdot x}{\frac{14}{3}} = \dfrac{\frac{11}{2} \cdot 14}{\frac{14}{3}}$     Dividing by $\dfrac{14}{3}$

$x = \dfrac{11}{2} \cdot 14 \div \dfrac{14}{3}$

$x = \dfrac{11}{2} \cdot 14 \cdot \dfrac{3}{14}$     Multiplying by the reciprocal of the divisor

$x = \dfrac{11 \cdot 3}{2}$     Simplifying by removing a factor equal to 1: $\dfrac{14}{14} = 1$

$x = \dfrac{33}{2}$, or $16\dfrac{1}{2}$

The solution is $\dfrac{33}{2}$, or $16\frac{1}{2}$.

**Do Exercise 12.** ▶

**12.** Solve:

$$\frac{8\frac{1}{3}}{x} = \frac{10\frac{1}{2}}{3\frac{3}{4}}.$$

✓ **Check Your Understanding**

**Reading Check** Complete each statement with the appropriate word from the following list.

cross products proportion proportional ratio

**RC1.** The quotient of two quantities is their _____.

**RC2.** Two pairs of numbers that have the same ratio are _____.

**RC3.** A _____ states that two pairs of numbers have the same ratio.

**RC4.** For the equation $\dfrac{2}{x} = \dfrac{3}{y}$, the _____ are $2y$ and $3x$.

**Concept Check** Fill in the blank to make each statement correct.

**CC1.** The equation $\dfrac{1}{2} = \dfrac{5}{10}$ states that the pairs 1, 2 and _____ are proportional.

**CC2.** The equation $\dfrac{3}{10} = \dfrac{0.3}{1}$ states that the pairs _____ and 0.3, 1 are proportional.

**CC3.** The pairs 4, 6 and 10, 15 are proportional. This means that $\dfrac{4}{6} = \dfrac{\square}{\square}$.

**CC4.** The pairs 3, 5 and 0.9, 1.5 are proportional. This means that $\dfrac{3}{\square} = \dfrac{0.9}{\square}$.

**CC5.** The pairs $\dfrac{1}{3}, \dfrac{2}{9}$ and 3, 2 are proportional. This means that $\dfrac{\square}{\frac{2}{9}} = \dfrac{\square}{2}$.

**a** Determine whether the two pairs of numbers are proportional.

**1.** 5, 6 and 7, 9

**2.** 7, 5 and 6, 4

**3.** 1, 2 and 10, 20

**4.** 7, 3 and 21, 9

**5.** 2.4, 3.6 and 1.8, 2.7

**6.** 4.5, 3.8 and 6.7, 5.2

**7.** $5\dfrac{1}{3}, 8\dfrac{1}{4}$ and $2\dfrac{1}{5}, 9\dfrac{1}{2}$

**8.** $2\dfrac{1}{3}, 3\dfrac{1}{2}$ and 14, 21

**b** Solve.

**9.** $\dfrac{18}{4} = \dfrac{x}{10}$

**10.** $\dfrac{x}{45} = \dfrac{20}{25}$

**11.** $\dfrac{x}{8} = \dfrac{9}{6}$

**12.** $\dfrac{8}{10} = \dfrac{n}{5}$

**13.** $\dfrac{t}{12} = \dfrac{5}{6}$

**14.** $\dfrac{12}{4} = \dfrac{x}{3}$

**15.** $\dfrac{2}{5} = \dfrac{8}{n}$

**16.** $\dfrac{10}{6} = \dfrac{5}{x}$

**17.** $\dfrac{n}{15} = \dfrac{10}{30}$

**18.** $\dfrac{2}{24} = \dfrac{x}{36}$

**19.** $\dfrac{16}{12} = \dfrac{24}{x}$

**20.** $\dfrac{8}{12} = \dfrac{20}{x}$

**21.** $\dfrac{6}{11} = \dfrac{12}{x}$

**22.** $\dfrac{8}{9} = \dfrac{32}{n}$

**23.** $\dfrac{20}{7} = \dfrac{80}{x}$

**24.** $\dfrac{36}{x} = \dfrac{9}{5}$

**25.** $\dfrac{12}{9} = \dfrac{x}{7}$

**26.** $\dfrac{x}{20} = \dfrac{16}{15}$

**27.** $\dfrac{x}{13} = \dfrac{2}{9}$

**28.** $\dfrac{8}{11} = \dfrac{x}{5}$

**29.** $\dfrac{100}{25} = \dfrac{20}{n}$

**30.** $\dfrac{35}{125} = \dfrac{7}{m}$

**31.** $\dfrac{6}{y} = \dfrac{18}{15}$

**32.** $\dfrac{15}{y} = \dfrac{3}{4}$

**33.** $\dfrac{x}{3} = \dfrac{0}{9}$

**34.** $\dfrac{x}{6} = \dfrac{1}{6}$

**35.** $\dfrac{1}{2} = \dfrac{7}{x}$

**36.** $\dfrac{2}{5} = \dfrac{12}{x}$

**37.** $\dfrac{1.2}{4} = \dfrac{x}{9}$

**38.** $\dfrac{x}{11} = \dfrac{7.1}{2}$

**39.** $\dfrac{8}{2.4} = \dfrac{6}{y}$

**40.** $\dfrac{3}{y} = \dfrac{5}{4.5}$

**41.** $\dfrac{t}{0.16} = \dfrac{0.15}{0.40}$

**42.** $\dfrac{0.12}{0.04} = \dfrac{t}{0.32}$

**43.** $\dfrac{0.5}{n} = \dfrac{2.5}{3.5}$

**44.** $\dfrac{6.3}{0.9} = \dfrac{0.7}{n}$

**45.** $\dfrac{1.28}{3.76} = \dfrac{4.28}{y}$

**46.** $\dfrac{10.4}{12.4} = \dfrac{6.76}{t}$

**47.** $\dfrac{7}{\frac{1}{4}} = \dfrac{28}{x}$

**48.** $\dfrac{5}{\frac{1}{3}} = \dfrac{3}{x}$

**49.** $\dfrac{\frac{1}{5}}{\frac{1}{10}} = \dfrac{\frac{1}{10}}{x}$

**50.** $\dfrac{\frac{1}{4}}{\frac{1}{2}} = \dfrac{\frac{1}{2}}{x}$

**51.** $\dfrac{y}{\frac{3}{5}} = \dfrac{\frac{7}{12}}{\frac{14}{15}}$

**52.** $\dfrac{\frac{5}{8}}{\frac{5}{4}} = \dfrac{y}{\frac{3}{2}}$

**53.** $\dfrac{x}{1\frac{3}{5}} = \dfrac{2}{15}$

**54.** $\dfrac{1}{7} = \dfrac{x}{4\frac{1}{2}}$

**55.** $\dfrac{2\frac{1}{2}}{3\frac{1}{3}} = \dfrac{x}{4\frac{1}{4}}$

**56.** $\dfrac{3\frac{1}{2}}{y} = \dfrac{6\frac{1}{2}}{4\frac{2}{3}}$

**57.** $\dfrac{5\frac{1}{5}}{6\frac{1}{6}} = \dfrac{y}{3\frac{1}{2}}$

**58.** $\dfrac{10\frac{3}{8}}{12\frac{2}{3}} = \dfrac{5\frac{3}{4}}{y}$

## Skill Maintenance

**59.** *Golf Courses.* The number of golf courses in the United States has declined steadily since its peak in the early 2000s. In 2015, there were 15,372 golf courses in the United States. This was 680 fewer courses than the peak number. What was the peak number of golf courses in the United States? [1.8a]

**Data:** espn.com

**60.** *Bird Feeders.* After Raena poured a 4-lb bag of birdseed into her new bird feeder, the feeder was $\frac{2}{3}$ full. How much seed does the feeder hold when it is full? [3.8b]

**61.** Mariah bought $\frac{3}{4}$ lb of cheese at the city market and gave $\frac{1}{2}$ lb of it to Lindsay. How much cheese did Mariah have left? [4.3c]

**62.** David bought $\frac{3}{4}$ lb of fudge and gave $\frac{1}{2}$ of it to Chris. How much fudge did he give to Chris? [3.6b]

**63.** Rocky is $187\frac{1}{10}$ cm tall and his daughter is $180\frac{3}{4}$ cm tall. How much taller is Rocky? [4.6c]

**64.** A serving of fish steak (cross section) is generally $\frac{1}{2}$ lb. How many servings can be prepared from a cleaned $18\frac{3}{4}$-lb tuna? [4.7d]

**65.** *Expense Needs.* Aaron has $34.97 to spend on a book that costs $49.95, a cap that costs $14.88, and a sweatshirt that costs $29.95. How much more money does Aaron need to make these purchases? [5.8b]

**66.** *Gas Mileage.* Joanna filled her van's gas tank and noted that the odometer read 42,598.2. After the next fill-up, the odometer read 42,912.1. It took 14.6 gal to fill the tank. How many miles per gallon did the van get? [5.8b]

## Synthesis

 Solve.

**67.** $\dfrac{1728}{5643} = \dfrac{836.4}{x}$

**68.** $\dfrac{328.56}{627.48} = \dfrac{y}{127.66}$

**69.** *Strikeouts per Home Run.* Baseball Hall-of-Famer Babe Ruth had 1330 strikeouts and 714 home runs in his career. Hall-of-Famer Mike Schmidt had 1883 strikeouts and 548 home runs in his career. Find the rate of strikeouts per home run for each player. (These rates were considered among the highest in the history of the game and yet each player was voted into the Hall of Fame.)

# Mid-Chapter Review

## Concept Reinforcement

Determine whether each statement is true or false.

_____ **1.** A ratio can be written in fraction notation or in colon notation. [6.1a]

_____ **2.** A rate is a ratio. [6.2a]

_____ **3.** The largest size package of an item always has the lowest unit price. [6.2b]

_____ **4.** If $\frac{x}{t} = \frac{y}{s}$, then $xy = ts$. [6.3b]

## Guided Solutions

 **5.** What is the rate, or speed, in miles per hour? [6.2a]

120 mi, 2 hr

$$\frac{120 \,\boxed{\phantom{x}}}{\boxed{\phantom{x}}\text{ hr}} = \frac{120 \,\boxed{\phantom{x}}}{\boxed{\phantom{x}}\text{ hr}} = \boxed{\phantom{x}}\text{ mi/hr}$$

**6.** Solve: $\frac{x}{4} = \frac{3}{6}$. [6.3b]

$$\frac{x}{4} = \frac{3}{6}$$

$x \cdot \boxed{\phantom{x}} = \boxed{\phantom{x}} \cdot 3$      Equating cross products

$\dfrac{x \cdot 6}{\boxed{\phantom{x}}} = \dfrac{4 \cdot 3}{\boxed{\phantom{x}}}$      Dividing on both sides

$x = \boxed{\phantom{x}}$      Simplifying

## Mixed Review

Find fraction notation for each ratio. [6.1a]

**7.** 4 to 7

**8.** 313 to 199

**9.** 35 to 17

**10.** 59 to 101

Find the ratio of the first number to the second and simplify. [6.1b]

**11.** 8 to 12

**12.** 25 to 75

**13.** 32 to 28

**14.** 100 to 76

**15.** 112 to 56

**16.** 15 to 3

**17.** 2.4 to 8.4

**18.** 0.27 to 0.45

Find each rate, or speed, as a ratio of distance to time. Round to the nearest hundredth where appropriate. [6.2a]

**19.** 243 mi, 4 hr
**20.** 146 km, 3 hr
**21.** 65 m, 5 sec
**22.** 97 ft, 6 sec

**23.** *Record Snowfall.* The greatest recorded snowfall in a single storm occurred in the Mt. Shasta Ski Bowl in California, when 189 in. fell during a seven-day storm in 1959. What is the rate in inches per day? [6.2a]

Data: U.S. Army Corps of Engineers

**24.** *Free Throws.* During the 2016–2017 NBA basketball season, Isaiah Thomas of the Boston Celtics attempted 649 free throws and made 590 of them. What is the rate in number of free throws made to number of free throws attempted? Round to the nearest thousandth. [6.2a]

Data: National Basketball Association

**25.** Jerome bought an 18-oz jar of grape jelly for $2.09. What is the unit price in cents per ounce? [6.2b]

**26.** Martha bought 12 oz of deli honey ham for $5.99. What is the unit price in cents per ounce? [6.2b]

Determine whether the two pairs of numbers are proportional. [6.3a]

**27.** 3, 7 and 15, 35
**28.** 9, 7 and 7, 5
**29.** 2.4, 1.5 and 3.2, 2.1
**30.** $1\frac{3}{4}, 1\frac{1}{3}$ and $8\frac{3}{4}, 6\frac{2}{3}$

Solve. [6.3b]

**31.** $\dfrac{9}{15} = \dfrac{x}{20}$

**32.** $\dfrac{x}{24} = \dfrac{30}{18}$

**33.** $\dfrac{12}{y} = \dfrac{20}{15}$

**34.** $\dfrac{2}{7} = \dfrac{10}{y}$

**35.** $\dfrac{y}{1.2} = \dfrac{1.1}{0.6}$

**36.** $\dfrac{0.24}{0.02} = \dfrac{y}{0.36}$

**37.** $\dfrac{\frac{1}{4}}{x} = \dfrac{\frac{1}{8}}{\frac{1}{4}}$

**38.** $\dfrac{1\frac{1}{2}}{3\frac{1}{4}} = \dfrac{7\frac{1}{2}}{x}$

## Understanding Through Discussion and Writing

**39.** Can every ratio be written as the ratio of some number to 1? Why or why not? [6.1a]

**40.** What can be concluded about a rectangle's width if the ratio of length to perimeter is 1 to 3? Make some sketches and explain your reasoning. [6.1a, b]

**41.** Instead of equating cross products, a student solves $\dfrac{x}{7} = \dfrac{5}{3}$ by multiplying on both sides by the least common denominator, 21. Is his approach a good one? Why or why not? [6.3b]

**42.** An instructor predicts that a student's test grade will be proportional to the amount of time the student spends studying. What is meant by this? Write an example of a proportion that represents the grades of two students and their study times. [6.3b]

# Applications of Proportions

# 6.4

## OBJECTIVE

a   Solve applied problems involving proportions.

## a   APPLICATIONS AND PROBLEM SOLVING

Proportions have applications in such diverse fields as business, chemistry, health sciences, and home economics, as well as in many areas of daily life. Proportions are useful in making predictions.

**EXAMPLE 1**   *Predicting Total Distance.*   Donna drives her delivery van 800 mi in 3 days. At this rate, how far will she drive in 15 days?

1.   **Familiarize.**   We let $d$ = the distance traveled in 15 days.

2.   **Translate.**   We translate to a proportion. We make each side the ratio of distance to time, with distance in the numerator and time in the denominator.

$$\text{Distance in 15 days} \rightarrow \frac{d}{15} = \frac{800}{3} \leftarrow \text{Distance in 3 days}$$
$$\text{Time} \rightarrow \qquad\qquad\qquad \leftarrow \text{Time}$$

It may help to verbalize the proportion above as "the unknown distance $d$ is to 15 days as the known distance 800 mi is to 3 days."

3.   **Solve.**   Next, we solve the proportion:

$$d \cdot 3 = 15 \cdot 800 \qquad \text{Equating cross products}$$

$$\frac{d \cdot 3}{3} = \frac{15 \cdot 800}{3} \qquad \text{Dividing by 3 on both sides}$$

$$d = \frac{15 \cdot 800}{3}$$

$$d = 4000. \qquad \text{Multiplying and dividing}$$

4.   **Check.**   We substitute into the proportion and check cross products:

$$\frac{4000}{15} = \frac{800}{3};$$

$$4000 \cdot 3 = 12{,}000; \qquad 15 \cdot 800 = 12{,}000.$$

The cross products are the same, so the answer checks.

5.   **State.**   At the given rate, Donna will drive 4000 mi in 15 days.

**Do Exercise 1.** ▶

Problems involving proportions can be translated in more than one way. For Example 1, any one of the following is also a correct translation:

$$\frac{15}{d} = \frac{3}{800}, \qquad \frac{15}{3} = \frac{d}{800}, \qquad \frac{800}{d} = \frac{3}{15}.$$

Equating the cross products in each proportion gives us the equation $d \cdot 3 = 15 \cdot 800$, which is the equation we obtained in Example 1.

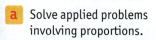

**SKILL REVIEW**   *Solve equations containing decimals.*   [5.7a]
Solve.

1.   $120 \cdot 25 = 100 \cdot n$

2.   $0.3 \times w = 1.68 \times 0.4$

**Answers: 1.** 30  **2.** 2.24

MyLab Math
VIDEO

1.   *Burning Calories.*   The readout on Mary's treadmill indicates that she burns 108 calories when she walks for 24 min. How many calories will she burn if she walks at the same rate for 30 min?

*Answer*

1.   135 calories

**EXAMPLE 2** *Recommended Dosage.* To control a fever, a doctor suggests that a child who weighs 28 kg be given 320 mg of a liquid pain reliever. If the dosage is proportional to the child's weight, how much of the medication is recommended for a child who weighs 35 kg?

**1. Familiarize.** We let $t$ = the number of milligrams of the liquid pain reliever recommended for a child who weighs 35 kg.

**2. Translate.** We translate to a proportion, keeping the amount of medication in the numerators.

$$\text{Medication suggested} \rightarrow \frac{320}{28} = \frac{t}{35} \leftarrow \text{Medication suggested}$$
$$\text{Child's weight} \rightarrow \qquad\qquad \leftarrow \text{Child's weight}$$

**3. Solve.** Next, we solve the proportion:

$$320 \cdot 35 = 28 \cdot t \qquad \text{Equating cross products}$$
$$\frac{320 \cdot 35}{28} = \frac{28 \cdot t}{28} \qquad \text{Dividing by 28 on both sides}$$
$$\frac{320 \cdot 35}{28} = t$$
$$400 = t. \qquad \text{Multiplying and dividing}$$

**4. Check.** We substitute into the proportion and check cross products:

$$\frac{320}{28} = \frac{400}{35};$$
$$320 \cdot 35 = 11{,}200; \qquad 28 \cdot 400 = 11{,}200.$$

The cross products are the same, so the answer checks.

**5. State.** The dosage for a child who weighs 35 kg is 400 mg.

◀ Do Exercise 2.

**EXAMPLE 3** *Purchasing Tickets.* Carey bought 8 tickets to an international food festival for $52. How many tickets could she purchase with $90?

**1. Familiarize.** We let $n$ = the number of tickets that can be purchased with $90.

**2. Translate.** We translate to a proportion, keeping the number of tickets in the numerators.

$$\text{Tickets} \rightarrow \frac{8}{52} = \frac{n}{90} \leftarrow \text{Tickets}$$
$$\text{Cost} \rightarrow \qquad\qquad \leftarrow \text{Cost}$$

**3. Solve.** Next, we solve the proportion:

$$52 \cdot n = 8 \cdot 90 \qquad \text{Equating cross products}$$

$$\frac{52 \cdot n}{52} = \frac{8 \cdot 90}{52} \qquad \text{Dividing by 52 on both sides}$$

$$n = \frac{8 \cdot 90}{52}$$

$$n \approx 13.8. \qquad \text{Multiplying and dividing}$$

Because it is impossible to buy a fractional part of a ticket, we must round our answer *down* to 13.

**4. Check.** As a check, we use a different approach: We find the cost per ticket and then divide $90 by that price. Since $52 \div 8 = 6.50$ and $90 \div 6.50 \approx 13.8$, we have a check.

**5. State.** Carey could purchase 13 tickets with $90.

**Do Exercise 3.** ▶

**3.** *Purchasing Shirts.* If 2 shirts can be bought for $47, how many shirts can be bought with $200?

**EXAMPLE 4** *Waist-to-Hip Ratio.* To reduce the risk of heart disease, it is recommended that a man's waist-to-hip ratio be 0.9 or lower. Mac's hip measurement is 40 in. To meet the recommendation, what should his waist measurement be?
**Data:** Mayo Clinic

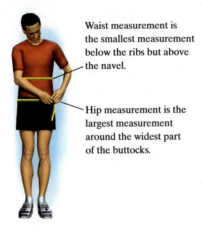

Waist measurement is the smallest measurement below the ribs but above the navel.

Hip measurement is the largest measurement around the widest part of the buttocks.

**1. Familiarize.** Note that $0.9 = \dfrac{9}{10}$. We let $w$ = Mac's waist measurement.

**2. Translate.** We translate to a proportion as follows.

$$\begin{array}{c}\text{Waist measurement} \rightarrow \\ \text{Hip measurement} \rightarrow\end{array} \frac{w}{40} = \frac{9}{10} \begin{array}{c}\nwarrow \text{Recommended} \\ \swarrow \text{waist-to-hip ratio}\end{array}$$

**3. Solve.** Next, we solve the proportion:

$$w \cdot 10 = 40 \cdot 9 \qquad \text{Equating cross products}$$

$$\frac{w \cdot 10}{10} = \frac{40 \cdot 9}{10} \qquad \text{Dividing by 10 on both sides}$$

$$w = \frac{40 \cdot 9}{10}$$

$$w = 36. \qquad \text{Multiplying and dividing}$$

*Answer*

**3.** 8 shirts

**4. Waist-to-Hip Ratio.** It is recommended that a woman's waist-to-hip ratio be 0.85 or lower. Martina's hip measurement is 40 in. To meet the recommendation, what should her waist measurement be?

**Data:** Mayo Clinic

**4. Check.** As a check, we divide 36 by 40: $36 \div 40 = 0.9$. This is the desired ratio, so the answer checks.

**5. State.** Mac's recommended waist measurement is 36 in. or less.

◀ **Do Exercise 4.**

**EXAMPLE 5** *Construction Plans.* Architects make blueprints for construction projects. These are scale drawings in which lengths are in proportion to actual sizes. The Hennesseys are adding a rectangular deck to their house. The architect's blueprint is rendered such that $\frac{3}{4}$ in. on the drawing is actually 2.25 ft on the deck. The width of the deck on the drawing is 4.3 in. How wide is the deck in reality?

28.5 ft    w    l    4.3 in.

1. **Familiarize.** We let $w$ = the width of the deck.

2. **Translate.** Then we translate to a proportion, using 0.75 for $\frac{3}{4}$ in.

$$\text{Measure on drawing} \rightarrow \frac{0.75}{2.25} = \frac{4.3}{w} \leftarrow \text{Width on drawing} \\ \text{Measure on deck} \rightarrow \phantom{} \phantom{} \leftarrow \text{Width on deck}$$

3. **Solve.** Next, we solve the proportion:

$$0.75 \times w = 2.25 \times 4.3 \quad \text{Equating cross products}$$

$$\frac{0.75 \times w}{0.75} = \frac{2.25 \times 4.3}{0.75} \quad \text{Dividing by 0.75 on both sides}$$

$$w = \frac{2.25 \times 4.3}{0.75}$$

$$w = 12.9.$$

4. **Check.** We substitute into the proportion and check cross products:

$$\frac{0.75}{2.25} = \frac{4.3}{12.9};$$

$$0.75 \times 12.9 = 9.675; \quad 2.25 \times 4.3 = 9.675.$$

The cross products are the same, so the answer checks.

5. **State.** The width of the deck is 12.9 ft.

◀ **Do Exercise 5.**

**5. Construction Plans.** In Example 5, the length of the actual deck is 28.5 ft. What is the length of the deck on the blueprint?

*Answers*

**4.** 34 in. or less   **5.** 9.5 in.

**EXAMPLE 6** *Estimating a Wildlife Population.* Scientists often use proportions to estimate the size of a wildlife population. They begin by collecting and marking, or tagging, a portion of the population. This tagged sample is released and mingles with the entire population. At a later date, the scientists collect a second sample from the population. The proportion of tagged individuals in the second sample is estimated to be the same as the proportion of tagged individuals in the entire population.

The marking can be done by using actual tags or by identifying individuals in other ways. For example, marine biologists can identify an individual whale by the patterns on its tail. Recently, scientists have begun using DNA to identify individuals in populations. For example, to identify individual bears in the grizzly bear population of the Northern Continental Divide ecosystem in Montana, geneticists use DNA from fur samples left on branches near the bears' feeding areas.

In one recent large-scale study in this ecosystem, biologists identified 545 individual grizzly bears. If later a sample of 30 bears contains 25 of the previously identified individuals, estimate the total number of bears in the ecosystem.

**Data:** Northern Divide Grizzly Bear Project

1. **Familiarize.** We let $B =$ the total number of bears in the ecosystem. We assume that the ratio of the number of identified bears to the total number of bears in the ecosystem is the same as the ratio of the number of identified bears in the later sample to the total number of bears in the later sample.

2. **Translate.** We translate to a proportion as follows.

$$\text{Identified bears} \rightarrow \frac{545}{B} = \frac{25}{30} \leftarrow \text{Identified bears in sample}$$
$$\text{Total number of bears} \rightarrow \qquad\qquad \leftarrow \text{Number of bears in sample}$$

3. **Solve.** Next, we solve the proportion:

$545 \cdot 30 = B \cdot 25$     Equating cross products

$\dfrac{545 \cdot 30}{25} = \dfrac{B \cdot 25}{25}$     Dividing by 25 on both sides

$\dfrac{545 \cdot 30}{25} = B$

$654 = B.$     Multiplying and dividing

4. **Check.** We substitute into the proportion and check cross products:

$\dfrac{545}{654} = \dfrac{25}{30};$

$545 \cdot 30 = 16{,}350; \qquad 654 \cdot 25 = 16{,}350.$

The cross products are the same, so the answer checks..

5. **State.** We estimate that there are 654 bears in the ecosystem.

**Do Exercise 6.** ▶

**GS** 6. **Estimating a Deer Population.** To determine the number of deer in a forest, a conservationist catches 153 deer, tags them, and releases them. Later, 62 deer are caught, and it is found that 18 of them are tagged. Estimate how many deer are in the forest.

1. **Familiarize.** Let $D =$ the number of deer in the forest.

2. **Translate.**

$$\frac{153}{D} = \frac{\boxed{\phantom{xx}}}{\boxed{\phantom{xx}}}$$

3. **Solve.**

$153 \cdot 62 = D \cdot \boxed{\phantom{xx}}$

$\boxed{\phantom{xx}} = D$

4. **Check.** The cross products are the same, so the answer checks.

5. **State.** There are about $\boxed{\phantom{xx}}$ deer in the forest.

*Answer*

**6.** 527 deer

*Guided Solution:*

**6.** $\dfrac{18}{62}$; 18, 527; 527

# Translating for Success

**1. Calories in Cereal.** There are 140 calories in a $1\frac{1}{2}$-cup serving of Brand A cereal. How many calories are there in 6 cups of the cereal?

**2. Calories in Cereal.** There are 140 calories in 6 cups of Brand B cereal. How many calories are there in a $1\frac{1}{2}$-cup serving of the cereal?

**3. Gallons of Gasoline.** Jared's SUV traveled 310 mi on 15.5 gal of gasoline. At this rate, how many gallons would be needed to travel 465 mi?

**4. Gallons of Gasoline.** Elizabeth's fuel-efficient car traveled 465 mi on 15.5 gal of gasoline. At this rate, how many gallons will be needed to travel 310 mi?

**5. Perimeter.** What is the perimeter of a rectangular field that measures 83.7 m by 62.4 m?

*The goal of these matching questions is to practice step (2), Translate, of the five-step problem-solving process. Translate each word problem to an equation and select a correct translation from equations A–O.*

**A.** $\dfrac{310}{15.5} = \dfrac{465}{x}$

**B.** $180 = 1\frac{1}{2} \cdot x$

**C.** $x = 71\frac{1}{8} - 76\frac{1}{2}$

**D.** $71\frac{1}{8} \cdot x = 74$

**E.** $74 \cdot 71\frac{1}{8} = x$

**F.** $x = 83.7 + 62.4$

**G.** $71\frac{1}{8} + x = 76\frac{1}{2}$

**H.** $x = 1\frac{2}{3} \cdot 180$

**I.** $\dfrac{140}{6} = \dfrac{x}{1\frac{1}{2}}$

**J.** $x = 2(83.7 + 62.4)$

**K.** $\dfrac{465}{15.5} = \dfrac{310}{x}$

**L.** $x = 83.7 \cdot 62.4$

**M.** $x = 180 \div 1\frac{2}{3}$

**N.** $\dfrac{140}{1\frac{1}{2}} = \dfrac{x}{6}$

**O.** $x = 1\frac{2}{3} \div 180$

*Answers on page A-13*

**6. Electric Bill.** Last month Todd's electric bills for his two rental properties were $83.70 and $62.40. What was the total electric bill for the two properties?

**7. Package Tape.** A postal service center uses rolls of package tape that each contain 180 ft of tape. If it takes an average of $1\frac{2}{3}$ ft of tape per package, how many packages can be taped with one roll?

**8. Online Price.** Jane spent $180 buying an area rug in a department store. Later, she saw the same rug for sale online and realized she had paid $1\frac{1}{2}$ times the online price. What was the online price?

**9. Heights of Sons.** Henry's three sons play basketball on three different college teams. Jeff's, Jason's, and Jared's heights are 74 in., $71\frac{1}{8}$ in., and $76\frac{1}{2}$ in., respectively. How much taller is Jared than Jason?

**10. Area of a Lot.** Bradley bought a lot that measured 74 yd by $71\frac{1}{8}$ yd. What was the area of the lot?

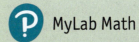
## ✓ Check Your Understanding

**Reading and Concept Check** Complete each proportion to form a correct translation of the following problem.

Christy ran 4 marathons in the first 6 months of the year. At this rate, how many marathons will she run in the first 8 months of the year? Let $m = $ the number of marathons she will run in 8 months.

**RC1.** $\dfrac{4}{6} = \dfrac{\square}{\square}$      **RC2.** $\dfrac{6}{4} = \dfrac{\square}{\square}$      **RC3.** $\dfrac{4}{m} = \dfrac{\square}{\square}$      **RC4.** $\dfrac{m}{4} = \dfrac{\square}{\square}$

**a**   Solve.

**1.** *Study Time and Test Grades.* An English instructor asserted that students' test grades are directly proportional to the amount of time spent studying. Lisa studies for 9 hr for a particular test and gets a score of 75. At this rate, for how many hours would she have had to study to get a score of 92?

**2.** *Study Time and Test Grades.* A mathematics instructor asserted that students' test grades are directly proportional to the amount of time spent studying. Brent studies for 15 hr for a final exam and gets a score of 75. At this rate, what score would he have received if he had studied for 18 hr?

**3.** *Movies.* If *The Hobbit: An Unexpected Journey* is played at the rate preferred by director Peter Jackson, a moviegoer sees 600 frames in $12\frac{1}{2}$ sec. How many frames does a moviegoer see in 160 sec?

**4.** *Sugaring.* When 20 gal of maple sap are boiled down, the result is $\frac{1}{2}$ gal of maple syrup. How much sap is needed to produce 9 gal of syrup?

**Data:** University of Maine

**5.** *Gas Mileage.* A 2017 Ford Mustang 8-cylinder with an automatic transmission will travel 372 mi on 15.5 gal of gasoline in highway driving.

   **a)** How many gallons of gasoline will it take to drive 2690 mi from Boston to Phoenix?

   **b)** How far can the car be driven on 140 gal of gasoline?

**Data:** fueleconomy.gov

**6.** *Gas Mileage.* A 2017 Honda Accord 2.4L with an automatic transmission will travel 594 mi on 16.5 gal of premium gasoline in highway driving.

   **a)** How many gallons of gasoline will it take to drive 1650 mi from Pittsburgh to Albuquerque?

   **b)** How far can the car be driven on 130 gal of gasoline?

**Data:** fueleconomy.gov

**7. _Overweight Americans._**   A recent study determined that of every 100 American adults, 71 are overweight or obese. It is estimated that the U.S. population will be about 359 million in 2030. At the given rate, how many Americans will be considered overweight or obese in 2030?

**Data:** newsweek.com; census.gov

**8. _Prevalence of Diabetes._**   A recent study determined that of every 1000 Americans age 65 or older, 252 have been diagnosed with diabetes. It is estimated that there will be about 75 million Americans in this age group in 2030. At the given rate, how many in this age group will be diagnosed with diabetes in 2030?

**Data:** American Diabetes Association

**9. _Quality Control._**   A quality-control inspector examined 100 lightbulbs and found 7 of them to be defective. At this rate, how many defective bulbs will there be in a lot of 2500?

**10. _Grading._**   A high school math teacher must grade 32 reports on famous mathematicians. She can grade 4 reports in 90 min. At this rate, how long will it take her to grade all 32 reports?

**11. _Recommended Dosage._**   To control an infection, Dr. Okeke prescribes a dosage of 200 mg of Rocephin every 8 hr for an infant weighing 15.4 lb. At this rate, what would the dosage be for an infant weighing 20.2 lb?

**12. _Metallurgy._**   In Ethan's white gold ring, the ratio of nickel to gold is 3 to 13. If the ring contains 4.16 oz of gold, how much nickel does it contain?

**13. _Painting._**   Fred uses 3 gal of paint to cover 1275 ft² of siding. How much siding can Fred paint with 7 gal of paint?

**14. _Waterproofing._**   Bonnie can waterproof 450 ft² of decking with 2 gal of sealant. How many gallons of the sealant should Bonnie buy for a 1200-ft² deck?

**15. _Publishing._**   Every 6 pages of an author's manuscript correspond to 5 published pages. How many published pages will a 540-page manuscript become?

**16. _Turkey Servings._**   An 8-lb turkey breast contains 36 servings of meat. How many pounds of turkey breast would be needed for 54 servings?

**17. _Mileage._** Jean bought a new car. In the first 8 months, it was driven 9000 mi. At this rate, how many miles will the car be driven in 1 year?

**18. _Coffee Production._** Coffee beans from 18 trees are required to produce enough coffee each year for a person who drinks 2 cups of coffee per day. Firefighters at the Sugar Creek Station brew 15 cups of coffee each day. How many coffee trees are required for this each year?

**19. _Cap'n Crunch's Peanut Butter Crunch® Cereal._** The nutritional chart on the side of a box of Quaker Cap'n Crunch's Peanut Butter Crunch® cereal states that there are 110 calories in a $\frac{3}{4}$-cup serving. How many calories are there in 6 cups of the cereal?

**20. _Rice Krispies® Cereal._** The nutritional chart on the side of a box of Kellogg's Rice Krispies® cereal states that there are 130 calories in a $1\frac{1}{4}$-cup serving. How many calories are there in 5 cups of the cereal?

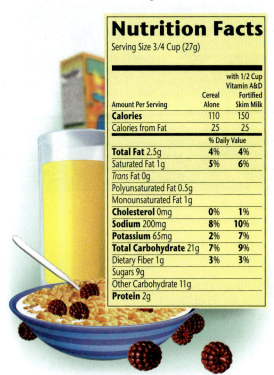

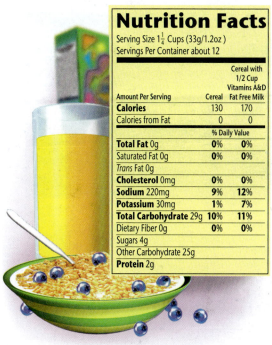

**21. _Lefties._** In a class of 40 students, on average, 6 will be left-handed. If a class includes 9 "lefties," how many students would you estimate are in the class?

**22. _Class Size._** A college advertises that its student-to-faculty ratio is 27 to 2. If 81 students register for Introductory Spanish, how many sections of the course would you expect to see offered?

**23. _Painting._** Helen can paint 950 ft$^2$ with 2 gal of paint. How many 1-gal cans does she need in order to paint a 30,000-ft$^2$ wall?

**24. _Snow to Water._** Under typical conditions, $1\frac{1}{2}$ ft of snow will melt to 2 in. of water. How many inches of water will result when $5\frac{1}{2}$ ft of snow melts?

**25. _Gasoline Mileage._** Nancy's van traveled 84 mi on 6.5 gal of gasoline. At this rate, how many gallons would be needed to travel 126 mi?

**26. _Bicycling._** Roy bicycled 234 mi in 14 days. At this rate, how far would Roy bicycle in 42 days?

**27. Grass-Seed Coverage.** It takes 60 oz of grass seed to seed 3000 ft$^2$ of lawn. At this rate, how much would be needed for 5000 ft$^2$ of lawn?

**28. Grass-Seed Coverage.** In Exercise 27, how much seed would be needed for 7000 ft$^2$ of lawn?

**29. Estimating a Whale Population.** To determine the number of humpback whales in a population, a marine biologist, using tail markings, identifies 27 individual whales. Several weeks later, 40 whales from the population are sighted at random. Of the 40 sighted, 12 are among the 27 originally identified. Estimate the number of whales in the population.

**30. Estimating a Trout Population.** To determine the number of trout in a lake, a conservationist catches 112 trout, tags them, and throws them back into the lake. Later, 82 trout are caught, and it is found that 32 of them are tagged. Estimate how many trout there are in the lake.

**31. Map Scaling.** On a road atlas map, 1 in. represents 16.6 mi. If two cities are 3.5 in. apart on the map, how far apart are they in reality?

**32. Map Scaling.** On a map, $\frac{1}{4}$ in. represents 50 mi. If two cities are $3\frac{1}{4}$ in. apart on the map, how far apart are they in reality?

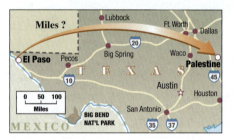

**33. Currency Exchange.** On July 31, 2017, 1 U.S. dollar was worth about 0.845 euro.

a) How much were 50 U.S. dollars worth in euros on that day?

b) How much would a car that costs 8640 euros cost in U.S. dollars?

**34. Currency Exchange.** On July 31, 2017, 1 U.S. dollar was worth about 17.8 Mexican pesos.

a) How much were 150 U.S. dollars worth in Mexican pesos on that day?

b) While traveling in Mexico at that time, Jake bought a watch that cost 3600 Mexican pesos. How much did it cost in U.S. dollars?

**35.** *Soccer Goals.*   After playing 21 games in the 2017 Major League Soccer season, Nemanja Nikolic of the Chicago Fire had scored 16 goals.

**a)** At this rate, how many games would it take him to score 19 goals?

**b)** At this rate, how many goals would Nikolic score in the whole 30-game season?

**Data:** Major League Soccer

**36.** *Home Runs.*   After playing 95 games in the 2017 Major League Baseball season, Bryce Harper of the Washington Nationals had 27 home runs.

**a)** At this rate, how many games would it take him to hit 30 home runs?

**b)** At this rate, how many home runs would Harper hit in the entire 162-game season?

**Data:** Major League Baseball

## Skill Maintenance

Solve.

**37.** $12 \cdot x = 1944$   [1.7b]

**38.** $6807 = m + 2793$   [1.7b]

**39.** $t + 4.25 = 8.7$   [5.7a]

**40.** $112.5 \cdot p = 45$   [5.7a]

**41.** $3.7 + y = 18$   [5.7a]

**42.** $0.078 = 0.3 \cdot t$   [5.7a]

**43.** $c + \dfrac{4}{5} = \dfrac{9}{10}$   [4.3b]

**44.** $\dfrac{5}{6} = \dfrac{2}{3} \cdot x$   [3.8a]

## Synthesis

**45.** 🔢 Carney College is expanding from 850 to 1050 students. To avoid any rise in the student-to-faculty ratio, the faculty of 69 professors must also be increased. How many new faculty positions should be created?

**46.** 🔢 In recognition of her outstanding work, Sheri's salary has been increased from $26,000 to $29,380. Tim is earning $23,000 and is requesting a proportional raise. How much more should he ask for?

**47.** *Baseball Statistics.*   Cy Young, one of the greatest baseball pitchers of all time, gave up an average of 2.63 earned runs every 9 innings. Young pitched 7356 innings, more than anyone else in the history of baseball. How many earned runs did he give up?

**48.** 🔢 *Real-Estate Values.*   According to Coldwell Banker Real Estate Corporation, a home selling for $189,000 in Austin, Texas, would sell for $486,300 in Key West, Florida. How much would a $350,000 home in Key West sell for in Austin? Round to the nearest $1000.

**Data:** Coldwell Banker Real Estate Corporation

**49.** 🔢 The ratio 1 : 3 : 2 is used to estimate the relative costs of a CD player, receiver, and speakers when shopping for a sound system. That is, the receiver should cost three times the amount spent on the CD player and the speakers should cost twice the amount spent on the CD player. If you had $900 to spend, how would you allocate the money, using this ratio?

# Geometric Applications

## a PROPORTIONS AND SIMILAR TRIANGLES

**SKILL REVIEW**

*Solve proportions.* [6.3b]

Solve.

**1.** $\dfrac{7}{x} = \dfrac{8}{3}$

**2.** $\dfrac{2}{1\frac{1}{2}} = \dfrac{p}{\frac{1}{4}}$

Answers: **1.** 2.625 **2.** $\dfrac{1}{3}$

MyLab Math
VIDEO

MyLab Math
ANIMATION

The following triangles appear to have the same shape, but their sizes are different. These are examples of **similar triangles**. For similar triangles, the corresponding sides of the triangles have the same ratio. That is, the following proportion is true.

$$\frac{a}{d} = \frac{b}{e} = \frac{c}{f}$$

**EXAMPLE 1** The triangles below are similar triangles. Find the missing length $x$.

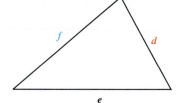

The ratio of $x$ to 9 is the same as the ratio of 24 to 8 or 21 to 7. We get the proportions

$$\frac{x}{9} = \frac{24}{8} \quad \text{and} \quad \frac{x}{9} = \frac{21}{7}.$$

We can solve either one of these proportions. We use the first one:

$$\frac{x}{9} = \frac{24}{8}$$

$x \cdot 8 = 9 \cdot 24$     Equating cross products

$\dfrac{x \cdot 8}{8} = \dfrac{9 \cdot 24}{8}$     Dividing by 8 on both sides

$x = 27.$     Simplifying

The missing length $x$ is 27. Other proportions could also be used.

◀ **Do Exercise 1.**

---

**1.** This pair of triangles is similar. Find the missing length $x$. **GS**

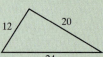

The ratio of $x$ to 20 is the same as the ratio of 9 to ☐.

$$\frac{x}{20} = \frac{9}{☐}$$

$$x \cdot ☐ = 20 \cdot 9$$

$$\frac{x \cdot 12}{☐} = \frac{20 \cdot 9}{☐}$$

$$x = \frac{☐}{12} = ☐$$

*Answer*

**1.** 15

*Guided Solution:*

**1.** 12; 12, 12, 12, 12, 180, 15

**Similar triangles** have the same shape. The lengths of their corresponding sides have the same ratio—that is, they are proportional.

**EXAMPLE 2** How high is a flagpole that casts a 56-ft shadow at the same time that a 6-ft man casts a 5-ft shadow?

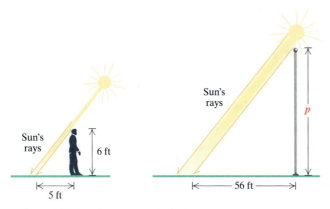

If we use the sun's rays to represent the third side of the triangle in our drawing of the situation, we see that we have similar triangles. Let $p =$ the height of the flagpole. The ratio of 6 to $p$ is the same as the ratio of 5 to 56. Thus, we have the proportion

Height of man $\rightarrow \dfrac{6}{p} = \dfrac{5}{56}$ $\leftarrow$ Length of shadow of man
Height of pole $\rightarrow$ $\leftarrow$ Length of shadow of pole

*Solve:* $6 \cdot 56 = p \cdot 5$     Equating cross products

$\dfrac{6 \cdot 56}{5} = \dfrac{p \cdot 5}{5}$     Dividing by 5 on both sides

$\dfrac{6 \cdot 56}{5} = p$     Simplifying

$67.2 = p.$

The height of the flagpole is 67.2 ft.

**Do Exercise 2.** ▶

**2.** How high is a flagpole that casts a 45-ft shadow at the same time that a 5.5-ft woman casts a 10-ft shadow?

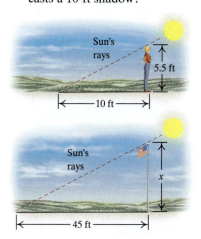

**EXAMPLE 3** *Rafters of a House.* Carpenters use similar triangles to determine the length of rafters for a house. They first choose the pitch of the roof, or the ratio of the rise to the run. Then using a triangle with that ratio, they calculate the length of the rafters needed for the house. Loren is making rafters for a roof with a 6/12 pitch on a house that is 30 ft wide. Using a rafter guide, Loren finds that the rafter length corresponding to the 6/12 pitch is 13.4.

*Answer*

**2.** 24.75 ft

Find the length $x$ of the rafters for this house to the nearest tenth of a foot.

We have the proportion

$$\underset{\substack{\text{Length of rafter} \\ \text{in 6/12 triangle}}}{\searrow} \quad \underset{\substack{\text{Length of rafter} \\ \text{on the house}}}{\nearrow} \quad \frac{13.4}{x} = \frac{12}{15}. \quad \overset{\substack{\text{Run in 6/12} \\ \text{triangle}}}{\nwarrow} \quad \overset{\substack{\text{Run in similar} \\ \text{triangle on the house}}}{\nwarrow}$$

*Solve:* $13.4 \cdot 15 = x \cdot 12$     Equating cross products

$\dfrac{13.4 \cdot 15}{12} = \dfrac{x \cdot 12}{12}$     Dividing by 12 on both sides

$\dfrac{13.4 \cdot 15}{12} = x$

$16.8 \text{ ft} \approx x.$     Rounding to the nearest tenth of a foot

The length $x$ of the rafters for the house is about 16.8 ft.

◀ **Do Exercise 3.**

**3.** *Rafters of a House.* Referring to Example 3, find the length $y$ of the rise of the rafters of the house to the nearest tenth of a foot.

---

## b   PROPORTIONS AND OTHER GEOMETRIC SHAPES

When one geometric figure is a magnification of another, the figures are similar. Thus, the corresponding lengths are proportional.

**EXAMPLE 4** The sides in the photographs below are proportional. Find the width of the larger photograph.

2.5 cm    ←— 3.5 cm —→|    |←———— 10.5 cm ————→|

We let $x =$ the width of the photograph. Then we translate to a proportion.

$$\underset{\substack{\text{Smaller width} \rightarrow}}{\text{Larger width} \rightarrow} \quad \frac{x}{2.5} = \frac{10.5}{3.5} \quad \overset{\substack{\leftarrow \text{Smaller length}}}{\leftarrow \text{Larger length}}$$

*Solve:* $x \times 3.5 = 2.5 \times 10.5$     Equating cross products

$\dfrac{x \times 3.5}{3.5} = \dfrac{2.5 \times 10.5}{3.5}$     Dividing by 3.5 on both sides

$x = \dfrac{2.5 \times 10.5}{3.5}$     Simplifying

$x = 7.5.$

Thus, the width of the larger photograph is 7.5 cm.

◀ **Do Exercise 4.**

**4.** The sides in the photographs below are proportional. Find the width of the larger photograph.

6 cm    |←— 10 cm —→|

$x$    |←———— 35 cm ————→|

**EXAMPLE 5** A scale model of an addition to an athletic facility is 12 cm wide at the base and rises to a height of 15 cm. If the actual base is to be 52 ft, what will the actual height of the addition be?

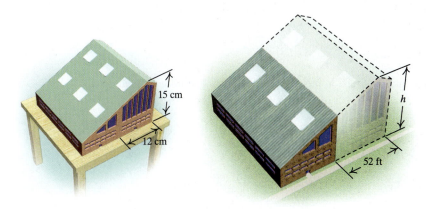

15 cm

12 cm

h

52 ft

We let $h$ = the height of the addition. Then we translate to a proportion.

Width in model → $\dfrac{12}{52} = \dfrac{15}{h}$ ← Height in model
Actual width → $\phantom{\dfrac{12}{52}}$ ← Actual height

*Solve:* $12 \cdot h = 52 \cdot 15$      Equating cross products

$\dfrac{12 \cdot h}{12} = \dfrac{52 \cdot 15}{12}$      Dividing by 12 on both sides

$h = \dfrac{52 \cdot 15}{12} = 65.$

Thus, the height of the addition will be 65 ft.

**Do Exercise 5.** ▶

**GS** 5. Refer to the figure in Example 5. If a skylight on the model is 3 cm wide, how wide will an actual skylight be?

Let $w$ = the width of an actual skylight.

$$\frac{12}{52} = \frac{\boxed{\phantom{xx}}}{w}$$

$$12 \cdot w = 52 \cdot \boxed{\phantom{xx}}$$

$$w = \boxed{\phantom{xx}}$$

The width of an actual skylight will be 13 $\boxed{\phantom{xx}}$.

**EXAMPLE 6** *Bicycle Design.* Two important dimensions to consider when buying a bicycle are *stack* and *reach*, as illustrated in the diagram at the right. The Country Racer bicycle comes in six different frame sizes, each proportional to the others. In the smallest frame size, the stack is 50 cm and the reach is 37.5 cm. In the largest size, the stack is 60 cm. What is the reach in the largest frame size of the Country Racer?

Stack

Reach

We let $r$ = the reach in the largest frame size. Then we translate to a proportion.

Stack in smallest frame → $\dfrac{50}{37.5} = \dfrac{60}{r}$ ← Stack in largest frame
Reach in smallest frame → $\phantom{\dfrac{50}{37.5}}$ ← Reach in largest frame

*Solve:*

$50 \cdot r = 37.5 \cdot 60$      Equating cross products

$\dfrac{50 \cdot r}{50} = \dfrac{37.5 \cdot 60}{50}$      Dividing by 50 on both sides

$r = \dfrac{37.5 \cdot 60}{50} = 45.$

Thus, the reach in the largest frame size is 45 cm.

**Do Exercise 6.** ▶

6. Refer to Example 6. In another frame size of the Country Racer, the reach is 42 cm. What is the stack?

*Answers*
**5.** 13 ft   **6.** 56 cm
*Guided Solution:*
**5.** 3, 3, 13; ft

✔ **Check Your Understanding**

**Reading and Concept Check**  Complete each proportion based on the following similar triangles.

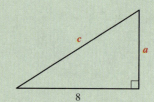

**RC1.** $\dfrac{a}{3} = \dfrac{c}{\square}$

**RC2.** $\dfrac{\square}{4} = \dfrac{c}{5}$

**RC3.** $\dfrac{3}{\square} = \dfrac{5}{c}$

**RC4.** $\dfrac{8}{4} = \dfrac{a}{\square}$

**a**  The triangles in each exercise are similar. Find the missing lengths.

**1.**

**2.**
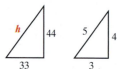

**3.**

**4.**
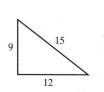

**5.**
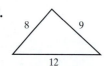

**6.**
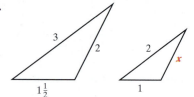

**7.**
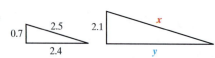

**8.**
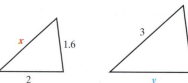

**9.** When a tree 8 m high casts a shadow 5 m long, how long a shadow is cast by a person 2 m tall?

**10.** How high is a flagpole that casts a 42-ft shadow at the same time that a $5\frac{1}{2}$-ft woman casts a 7-ft shadow?

**11.** How high is a tree that casts a 27-ft shadow at the same time that a 4-ft fence post casts a 3-ft shadow?

**12.** How high is a tree that casts a 32-ft shadow at the same time that an 8-ft light pole casts a 9-ft shadow?

**13.** Find the height $h$ of the wall.

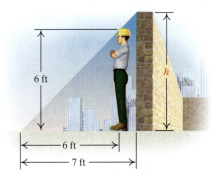

6 ft

$h$

6 ft

7 ft

**14.** Find the length $L$ of the lake. Assume that the ratio of $L$ to 120 yd is the same as the ratio of 720 yd to 30 yd.

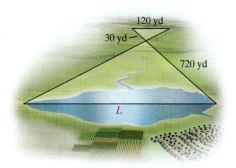

120 yd

30 yd

720 yd

$L$

**15.** Find the distance across the river. Assume that the ratio of $d$ to 25 ft is the same as the ratio of 40 ft to 10 ft.

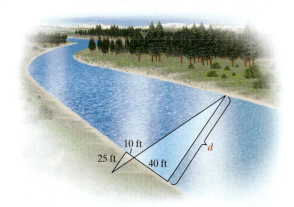

10 ft

$d$

25 ft

40 ft

**16.** To measure the height of a hill, a string is stretched from level ground to the top of the hill. A 3-ft stick is placed under the string, touching it at point $P$, a distance of 5 ft from point $G$, where the string touches the ground. The string is then detached and found to be 120 ft long. How high is the hill?

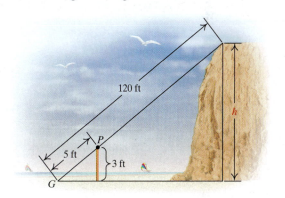

120 ft

$P$

5 ft

3 ft

$h$

$G$

**b**　In each of Exercises 17–26, the sides in each pair of figures are proportional. Find the missing lengths.

**17.**

6

$x$

9

6

**18.**

$x$

5

7

14

**19.**

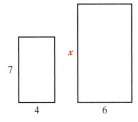

$x$

7

4

6

**20.**

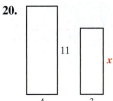

11

$x$

4

3

**21.**

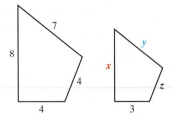

**22.**

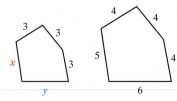

**23.**

**24.**

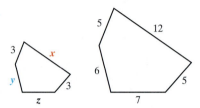

**25.**

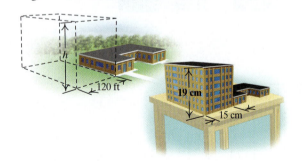

**26.**

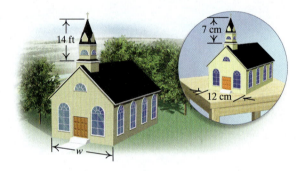

**27.** A scale model of an addition to a medical clinic is 15 cm wide at the base and rises to a height of 19 cm. If the actual base is to be 120 ft, what will the actual height of the addition be?

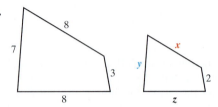

**28.** Refer to the figure in Exercise 27. If a large window on the model is 3 cm wide, how wide will the actual window be?

## Skill Maintenance

Determine whether each number is prime, composite, or neither.  [3.2b]

**29.** 83                              **30.** 28

Find the prime factorization of each number.  [3.2c]

**31.** 808                    **32.** 93

Use = or ≠ for ☐ to write a true sentence.  [3.5c]

**33.** $\dfrac{12}{8}$ ☐ $\dfrac{6}{4}$           **34.** $\dfrac{4}{7}$ ☐ $\dfrac{5}{9}$

Use < or > for ☐ to write a true sentence.  [4.2c]

**35.** $\dfrac{7}{12}$ ☐ $\dfrac{11}{15}$           **36.** $\dfrac{1}{6}$ ☐ $\dfrac{2}{11}$

Simplify.

**37.** $\left(\dfrac{1}{2}\right)^2 + \dfrac{2}{3}\cdot 4\dfrac{1}{2}$  [4.8a]

**38.** $18.3 + 2.5 \times 4.2 - (2.6 - 0.3^2)$  [5.4b]

**39.** $9 \times 15 - [2^3\cdot 6 - (2\cdot 5 + 3\cdot 10)]$  [1.9d]

**40.** $2600 \div 13 - 5^3$  [1.9c]

## Synthesis

*Hockey Goals.*   An official hockey goal is 6 ft wide. To make scoring more difficult, goalies often position themselves far in front of the goal to "cut down the angle." In Exercises 41 and 42, suppose that a slapshot is attempted from point *A* and that the goalie is 2.7 ft wide. Determine how far from the goal the goalie should be located if point *A* is the given distance from the goal. (*Hint*: First find how far the goalie should be from point *A*.)

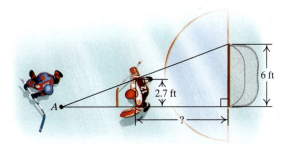

**41.**  25 ft                              **42.** 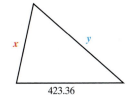 35 ft

**43.** A miniature air conditioning unit is to be built for the model referred to in Exercise 27. An actual unit is 10 ft high. How high should the model unit be?

 Solve. Round the answer to the nearest thousandth.

**44.** $\dfrac{8664.3}{10{,}344.8} = \dfrac{x}{9776.2}$

**45.** $\dfrac{12.0078}{56.0115} = \dfrac{789.23}{y}$

 The triangles in each exercise are similar triangles. Find the lengths not given.

**46.**

**47.**

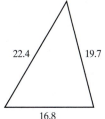

## Vocabulary Reinforcement

Complete each statement with the correct term from the list on the right. Some of the choices may not be used.

1. A ratio is the _____ of two quantities. [6.1a]

2. Similar triangles have the same _____. [6.5a]

3. To solve $\dfrac{x}{a} = \dfrac{c}{d}$ for $x$, equate the _____ and divide on both sides to get $x$ alone on one side. [6.3b]

4. A(n) _____ is a ratio used to compare two different kinds of measure. [6.2a]

5. A(n) _____ states that two pairs of numbers have the same ratio. [6.3a]

6. A unit price is the ratio of _____ to the number of units. [6.2b]

| |
|---|
| cross products |
| price |
| product |
| proportion |
| quantity |
| quotient |
| rate |
| shape |

## Concept Reinforcement

Determine whether each statement is true or false.

_____ 1. When we simplify a ratio like $\frac{8}{12}$, we find two other numbers in the same ratio. [6.1b]

_____ 2. The proportion $\dfrac{a}{b} = \dfrac{c}{d}$ can also be written as $\dfrac{c}{a} = \dfrac{d}{b}$. [6.3a]

_____ 3. Similar triangles must be the same size. [6.5a]

_____ 4. Lengths of corresponding sides of similar figures have the same ratio. [6.5b]

## Study Guide

**Objective 6.1a**  Find fraction notation for ratios.

| **Example**  Find the ratio of 7 to 18. | **Practice Exercise** |
|---|---|
| Write a fraction with a numerator of 7 and a denominator of 18: $\dfrac{7}{18}$. | 1. Find the ratio of 17 to 3. |

**Objective 6.1b**  Simplify ratios.

| **Example**  Simplify the ratio 8 to 2.5. | **Practice Exercise** |
|---|---|
| $\dfrac{8}{2.5} = \dfrac{8}{2.5} \cdot \dfrac{10}{10} = \dfrac{80}{25} = \dfrac{5 \cdot 16}{5 \cdot 5}$ $= \dfrac{5}{5} \cdot \dfrac{16}{5} = \dfrac{16}{5}$ | 2. Simplify the ratio 3.2 to 2.8. |

**Objective 6.2a**   Give the ratio of two different measures as a rate.

**Example**   A driver travels 156 mi on 6.5 gal of gas. What is the rate in miles per gallon?

$$\frac{156 \text{ mi}}{6.5 \text{ gal}} = \frac{156}{6.5} \frac{\text{mi}}{\text{gal}} = 24 \frac{\text{mi}}{\text{gal}}, \text{ or } 24 \text{ mpg}$$

**Practice Exercise**

3. A student earned $120 for working 16 hr. What was the rate of pay per hour?

---

**Objective 6.2b**   Find unit prices and use them to compare purchases.

**Example**   A 16-oz can of tomatoes costs $1.00. A 20-oz can costs $1.23. Which has the lower unit price?

$$16 \text{ oz}: \frac{\$1.00}{16 \text{ oz}} = \frac{100 \text{ cents}}{16 \text{ oz}} = 6.25\text{¢/oz}$$

$$20 \text{ oz}: \frac{\$1.23}{20 \text{ oz}} = \frac{123 \text{ cents}}{20 \text{ oz}} = 6.15\text{¢/oz}$$

Thus, the 20-oz can has the lower unit price.

**Practice Exercise**

4. A 28-oz jar of Brand A spaghetti sauce costs $2.79. A 32-oz jar of Brand B spaghetti sauce costs $3.29. Find the unit price of each brand and determine which is a better buy based on unit price alone.

---

**Objective 6.3a**   Determine whether two pairs of numbers are proportional.

**Example**   Determine whether 3, 4 and 7, 9 are proportional.

$$3 \cdot 9 = 27 \qquad \overset{?}{\frac{3}{4} = \frac{7}{9}} \qquad 4 \cdot 7 = 28$$

Since the cross products are not the same ($27 \neq 28$), $\frac{3}{4} \neq \frac{7}{9}$ and the numbers are not proportional.

**Practice Exercise**

5. Determine whether 7, 9 and 21, 27 are proportional.

---

**Objective 6.3b**   Solve proportions.

**Example**   Solve: $\frac{3}{4} = \frac{y}{7}$.

$$3 \cdot 7 = 4 \cdot y \qquad \text{Equating cross products}$$

$$\frac{3 \cdot 7}{4} = \frac{4 \cdot y}{4} \qquad \text{Dividing by 4 on both sides}$$

$$\frac{21}{4} = y$$

The solution is $\frac{21}{4}$.

**Practice Exercise**

6. Solve: $\frac{9}{x} = \frac{8}{3}$.

---

**Objective 6.4a**   Solve applied problems involving proportions.

**Example**   Martina bought 3 tickets to a campus theater production for $16.50. How much would 8 tickets cost?

We translate to a proportion.

$$\text{Tickets} \rightarrow \frac{3}{16.50} = \frac{8}{c} \leftarrow \text{Tickets}$$
$$\text{Cost} \rightarrow \phantom{\frac{3}{16.50}} \phantom{=} \phantom{\frac{8}{c}} \leftarrow \text{Cost}$$

$$3 \cdot c = 16.50 \cdot 8 \qquad \text{Equating cross products}$$

$$c = \frac{16.50 \cdot 8}{3}$$

$$c = 44$$

Eight tickets would cost $44.

**Practice Exercise**

7. On a map, $\frac{1}{2}$ in. represents 50 mi. If two cities are $1\frac{3}{4}$ in. apart on the map, how far apart are they in reality?

**Objective 6.5a**   Find lengths of sides of similar triangles using proportions.

**Example**   The triangles below are similar. Find the missing length $x$.

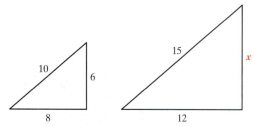

The ratio of 6 to $x$ is the same as the ratio of 8 to 12 (and also as the ratio of 10 to 15). We write and solve a proportion:

$$\frac{6}{x} = \frac{8}{12}$$

$$6 \cdot 12 = x \cdot 8$$

$$\frac{6 \cdot 12}{8} = \frac{x \cdot 8}{8}$$

$$9 = x.$$

The missing length is 9. (We could also have used other proportions, including $\frac{6}{x} = \frac{10}{15}$, to find $x$.)

**Practice Exercise**

8. The triangles below are similar. Find the missing length $y$.

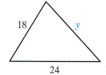

## Review Exercises

Write fraction notation for each ratio. Do not simplify.   [6.1a]

**1.** 47 to 84

**2.** 46 to 1.27

**3.** 83 to 100

**4.** 0.72 to 197

**5.** At Preston Seafood Market, 12,480 lb of tuna and 16,640 lb of salmon were sold one year.   [6.1a, b]

   **a)** Write fraction notation for the ratio of tuna sold to salmon sold.

   **b)** Write fraction notation for the ratio of salmon sold to the total number of pounds of both kinds of fish sold.

Find the ratio of the first number to the second number and simplify.   [6.1b]

**6.** 9 to 12

**7.** 3.6 to 6.4

**8.** *Gas Mileage.*   The 2017 Chevrolet Malibu 1.5L will travel 522 mi on 14.5 gal of gasoline in highway driving. What is the rate in miles per gallon?   [6.2a]

**Data:** fueleconomy.gov

**9.** *Flywheel Revolutions.*   A certain flywheel makes 472,500 revolutions in 75 min. What is the rate of spin in revolutions per minute?   [6.2a]

**10.** A lawn requires 319 gal of water for every 500 ft². What is the rate in gallons per square foot?   [6.2a]

**11.** *Calcium Supplement.*   The price for a particular calcium supplement is $18.99 for 300 tablets. Find the unit price in cents per tablet.   [6.2b]

**12.** Raquel bought a 24-oz loaf of 12-grain bread for $4.69. Find the unit price in cents per ounce.   [6.2b]

**13. Vegetable Oil.** Find the unit price. Then determine which size has the lowest unit price. [6.2b]

| Size | Price | Unit Price |
|------|-------|------------|
| 32 oz | $4.79 | |
| 48 oz | $5.99 | |
| 64 oz | $9.99 | |

Determine whether the two pairs of numbers are proportional. [6.3a]

**14.** 9, 15 and 36, 60      **15.** 24, 37 and 40, 46.25

Solve. [6.3b]

**16.** $\dfrac{8}{9} = \dfrac{x}{36}$      **17.** $\dfrac{6}{x} = \dfrac{48}{56}$

**18.** $\dfrac{120}{\frac{3}{7}} = \dfrac{7}{x}$      **19.** $\dfrac{4.5}{120} = \dfrac{0.9}{x}$

Solve. [6.4a]

**20. Quality Control.** A factory manufacturing computer circuits found 3 defective circuits in a lot of 65 circuits. At this rate, how many defective circuits can be expected in a lot of 585 circuits?

**21. Exchanging Money.** On July 31, 2017, 1 U.S. dollar was worth about 1.246 Canadian dollars.

  **a)** How much were 250 U.S. dollars worth in Canada on that day?

  **b)** While traveling in Canada that day, Jamal saw a sweatshirt that cost 50 Canadian dollars. How much would it cost in U.S. dollars?

**22.** A train travels 448 mi in 7 hr. At this rate, how far will it travel in 13 hr?

**23. Movies Released.** In the first 5 months of 2017, there were 164 movies released to theaters in the United States. At this rate, how many movies would be released in 2017?

**Data:** wildaboutmovies.com

**24. Trash Production.** In the United States, 5 people generate, on average, 22 lb of trash each day. The population of New York City is 8,537,673. How many pounds of trash are produced in New York City in one day?

**Data:** U.S. Environmental Protection Agency; New York City Department of City Planning

**25. Thanksgiving Dinner.** A traditional turkey dinner for 8 people cost about $39.90 in a recent year. How much would it cost to serve a turkey dinner for 14 people?

**Data:** American Farm Bureau Federation

**26. Lawyers per Capita.** In Massachusetts, there are about 6.4 lawyers for every 1000 people. The population of Boston is 673,200. How many lawyers would you expect there to be in Boston?

**Data:** American Bar Association; U.S. Census Bureau

Each pair of triangles in Exercises 27 and 28 is similar. Find the missing length(s). [6.5a]

**27.**

**28.**

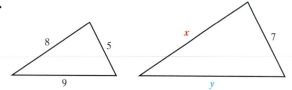

**29.** How high is a billboard that casts a 25-ft shadow at the same time that an 8-ft sapling casts a 5-ft shadow?   [6.5a]

**30.** The lengths in the figures below are proportional. Find the missing lengths.   [6.5b]

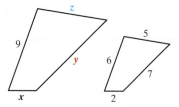

**31.** *Turkey Servings.* A 25-lb turkey serves 18 people. Find the rate in servings per pound.   [6.2a]

**A.** 0.36 serving/lb    **B.** 0.72 serving/lb
**C.** 0.98 serving/lb    **D.** 1.39 servings/lb

**32.** If 3 dozen eggs cost $5.04, how much will 5 dozen eggs cost?   [6.4a]

**A.** $6.72    **B.** $6.96
**C.** $8.40    **D.** $10.08

## Synthesis

**33.** *Paper Towels.* Find the unit price and determine which item would be the best buy based on unit price alone.   [6.2b]

| PACKAGE | PRICE | UNIT PRICE PER SHEET |
|---|---|---|
| 8 rolls, 60 (2 ply) sheets per roll | $8.67 | |
| 15 rolls, 60 (2 ply) sheets per roll | $14.92 | |
| 6 big rolls, 165 (2 ply) sheets per roll | $13.95 | |

**34.** 🖩 The following triangles are similar. Find the missing lengths.   [6.5a]

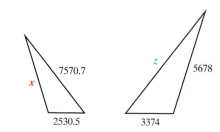

**35.** Shine-and-Glo Painters uses 2 gal of finishing paint for every 3 gal of primer. Each gallon of finishing paint covers 450 ft². If a surface of 4950 ft² needs both primer and finishing paint, how many gallons of each should be purchased?   [6.4a]

## Understanding Through Discussion and Writing

**1.** If you were a college president, which would you prefer: a low or high faculty-to-student ratio? Why?   [6.1a]

**2.** Can unit prices be used to solve proportions that involve money? Explain why or why not.   [6.2b], [6.4a]

**3.** Write a proportion problem for a classmate to solve. Design the problem so that the solution is "Leslie would need 16 gal of gasoline in order to travel 368 mi."   [6.4a]

**4.** Is it possible for two triangles to have two pairs of sides that are proportional without the triangles being similar? Why or why not?   [6.5a]

CHAPTER

**6**   **Test**

For Extra Help

For step-by-step test solutions, access the Chapter Test Prep Videos in MyLab Math.

Write fraction notation for each ratio. Do not simplify.

**1.** 85 to 97

**2.** 0.34 to 124

Find the ratio of the first number to the second number and simplify.

**3.** 18 to 20

**4.** 0.75 to 0.96

**5.** What is the rate in feet per second?

10 feet, 16 seconds

**6.** *Ham Servings.* A 12-lb shankless ham contains 16 servings. What is the rate in servings per pound?

**7.** *Gas Mileage.* Jeff's convertible will travel 464 mi on 14.5 gal of gasoline in highway driving. What is the rate in miles per gallon?

**8.** *Bagged Salad Greens.* Ron bought a 16-oz bag of salad greens for $2.49. Find the unit price in cents per ounce.

**9.** The table below lists prices for concentrated liquid laundry detergent. Find the unit price of each size in cents per ounce. Then determine which has the lowest unit price.

| Size | Price | Unit Price |
|------|-------|-----------|
| 40 oz | $6.59 | |
| 50 oz | $6.99 | |
| 100 oz | $11.49 | |
| 150 oz | $24.99 | |

Determine whether the two pairs of numbers are proportional.

**10.** 7, 8 and 63, 72

**11.** 1.3, 3.4 and 5.6, 15.2

Solve.

**12.** $\dfrac{9}{4} = \dfrac{27}{x}$

**13.** $\dfrac{150}{2.5} = \dfrac{x}{6}$

**14.** $\dfrac{x}{100} = \dfrac{27}{64}$

**15.** $\dfrac{68}{y} = \dfrac{17}{25}$

Solve.

**16.** *Distance Traveled.* An ocean liner traveled 432 km in 12 hr. At this rate, how far would it travel in 42 hr?

**17.** *Time Loss.* A watch loses 2 min in 10 hr. At this rate, how much will it lose in 24 hr?

**18. *Map Scaling.*** On a map, 3 in. represents 225 mi. If two cities are 7 in. apart on the map, how far apart are they in reality?

**19. *Tower Height.*** A birdhouse on a pole that is 3 m high casts a shadow 5 m long. At the same time, the shadow of a tower is 110 m long. How high is the tower?

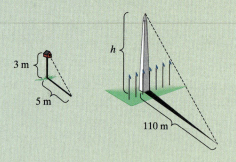

**20. *Charity Work.*** Kayla is crocheting hats for a charity. She can make 8 hats from 12 packages of yarn.

**a)** How many hats can she make from 20 packages of yarn?

**b)** How many packages of yarn does she need to make 20 hats?

**21. *Calories Burned.*** Kevin burned 200 calories while playing ultimate frisbee for $\frac{3}{4}$ hr. How many calories would he burn if he played for $1\frac{1}{5}$ hr?

The sides in each pair of figures are proportional. Find the missing lengths.

**22.**

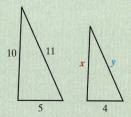

**23.**

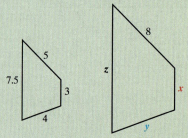

**24.** Lucita walks $4\frac{1}{2}$ mi in $1\frac{1}{2}$ hr. What is her rate in miles per hour?

**A.** $\frac{1}{3}$ mph      **B.** $1\frac{1}{2}$ mph

**C.** 3 mph      **D.** $4\frac{1}{2}$ mph

## Synthesis

Solve.

**25.** $\dfrac{x + 3}{4} = \dfrac{5x + 2}{8}$

**26.** Nancy wants to win a gift card from the campus bookstore by guessing the number of marbles in an 8-gal jar. She knows that there are 128 oz in a gallon. She goes home and fills an 8-oz jar with 46 marbles. How many marbles should she guess are in the 8-gal jar?

Calculate and simplify.

**1.**
$$\begin{array}{r} 2\,7.6\,8 \\ 3.0\,1\,9 \\ +\ 4\,8\,3.2\,9\,7 \\ \hline \end{array}$$

**2.**
$$\begin{array}{r} 2\,\frac{1}{3} \\ +\ 4\,\frac{5}{12} \\ \hline \end{array}$$

**3.** $\dfrac{6}{35} + \dfrac{5}{28}$

**4.**
$$\begin{array}{r} 4\,0.2 \\ -\ \ \ 9.7\,0\,9 \\ \hline \end{array}$$

**5.** $-32 - (-15)$

**6.** $\dfrac{3}{20} - \dfrac{4}{15}$

**7.**
$$\begin{array}{r} 3\,7.6\,4 \\ \times\ \ \ \ \ 5.9 \\ \hline \end{array}$$

**8.** $-43\,(15)$

**9.** $2\dfrac{1}{3} \cdot 1\dfrac{2}{7}$

**10.** $2.3\,\overline{)\,9\,8.9}$

**11.** $-306 \div (-6)$

**12.** $\dfrac{7}{11} \div \dfrac{14}{33}$

**13.** Write expanded notation: 30,074.

**14.** Write a word name for 120.07.

Which number is larger?

**15.** 0.7, 0.698

**16.** $-0.799, -0.8$

**17.** Find the prime factorization of 144.

**18.** Find the LCM of 18 and 30.

**19.** What part is shaded?

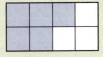

**20.** Simplify: $\dfrac{90}{144}$.

Calculate.

**21.** $\dfrac{3}{5} \times 9.53$

**22.** $7.2 \div 0.4(-1.5) + (1.2)^2$

**23.** Evaluate $\dfrac{t-7}{w}$ for $t = -3$ and $w = -2$.

**24.** Write fraction notation for the ratio 0.3 to 15. Do not simplify.

**25.** Determine whether the pairs 3, 9 and 25, 75 are proportional.

**26.** What is the rate in meters per second?

660 meters, 12 seconds

**27.** *Unit Prices.* An 8-oz can of pineapple chunks costs $0.99. A 24.5-oz jar of pineapple chunks costs $3.29. Which has the lower unit price?

Solve.

**28.** $\dfrac{14}{25} = \dfrac{x}{54}$

**29.** $-423 = 16 \cdot t$

**30.** $\dfrac{2}{3} \cdot y = \dfrac{16}{27}$

**31.** $9x - 7 = -43$

**32.** $34.56 + n = 67.9$

**33.** $2(x - 3) + 9 = 5x - 6$

**34.** Ramona's recipe for fettuccini alfredo has 520 calories in 1 cup. How many calories are there in $\frac{3}{4}$ cup?

**35.** A machine can stamp out 925 washers in 5 min. An order is placed for 1295 washers. How long will it take to stamp them out?

**36.** A 46-oz juice can contains $5\frac{3}{4}$ cups of juice. A recipe calls for $3\frac{1}{2}$ cups of juice. How many cups are left over?

**37.** It takes a carpenter $\frac{2}{3}$ hr to hang a door. How many doors can the carpenter hang in 8 hr?

**38.** *Car Travel.* A car travels 337.62 mi in 8 hr. How far does it travel in 1 hr?

**39.** *Shuttle Orbits.* A space shuttle made 16 orbits a day during an 8.25-day mission. How many orbits were made during the entire mission?

**40.** How many even prime numbers are there?
- **A.** 5
- **B.** 3
- **C.** 2
- **D.** 1
- **E.** None

**41.** The gas mileage of a car is 28.16 mpg. How many gallons per mile is this?
- **A.** $\dfrac{704}{25}$
- **B.** $\dfrac{25}{704}$
- **C.** $\dfrac{2816}{100}$
- **D.** $\dfrac{250}{704}$
- **E.** None

## Synthesis

**42.** A soccer goalie wishing to block an opponent's shot moves toward the shooter to reduce the shooter's view of the goal. If the goalie can only defend a region 10 ft wide, how far in front of the goal should the goalie be? (See the figure at right.)

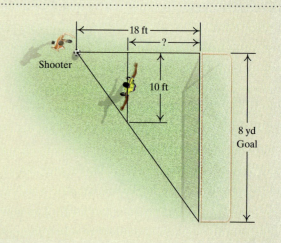

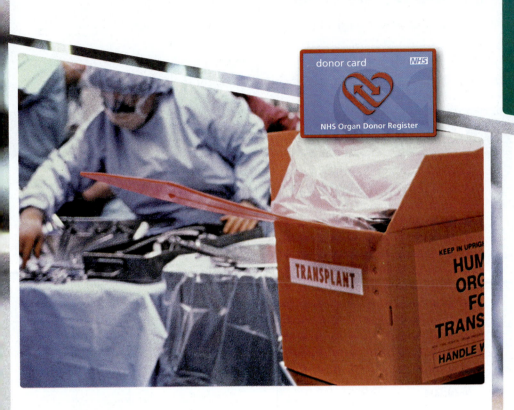

# Percent Notation

In August 2017, there were approximately 116,000 men, women, and children in the United States on the organ transplant waiting list. Every 10 minutes another person is added to the list, and 20 people die each day waiting for a transplant. A total of 33,611 transplants were performed in 2016. The graph at right shows percents for transplants of selected organs.

*DATA: organdonor.gov; U.S. Department of Health and Human Services*

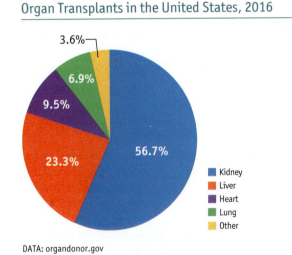

**Organ Transplants in the United States, 2016**

3.6%
6.9%
9.5%
23.3%
56.7%

- Kidney
- Liver
- Heart
- Lung
- Other

DATA: organdonor.gov

In Exercise 1 of Exercise Set 7.4, we will find the number of liver transplants and the number of heart transplants that were performed in the United States in 2016.

439

# 7.1

## OBJECTIVES

**a** Write three kinds of notation for a percent.

**b** Convert between percent notation and decimal notation.

**c** Convert between fraction notation and percent notation.

# Percent Notation

## a UNDERSTANDING PERCENT NOTATION

In 2017, 54% of the world's population lived in urban areas. What does 54% mean? It means that of every 100 people on Earth in 2017, 54 lived in urban areas. Thus, 54% is a ratio of 54 to 100, or $\frac{54}{100}$. (See Figure 1 below.) In the United States in 2017, approximately 83% of the population was urban. (See Figure 2 below.)

**Data:** worldometers.info; un.org; citywise.net

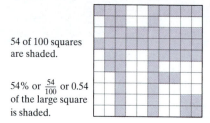

54 of 100 squares are shaded.

54% or $\frac{54}{100}$ or 0.54 of the large square is shaded.

**FIGURE 1**

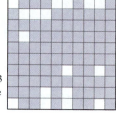

83 of 100 squares are shaded.

83% or $\frac{83}{100}$ or 0.83 of the large square is shaded.

**FIGURE 2**

We encounter percent notation frequently. Here are some examples.

- Humans start 84% of the wildfires in the United States.
  **Data:** *Proceedings of the National Academy of Sciences*

- Over an average lifetime, a person spends 29.7% of the time sitting and 0.69% of the time exercising.
  **Data:** CensusWide/Reebok poll of 18,000 people in nine countries

- In 2015, 13.3% of the people living in the United States had been born in another country.
  **Data:** U.S. Census Bureau's Current Population Surveys (Annual Social and Economic Supplements); U.S. Department of Commerce

Percent notation is often represented using a circle graph, or pie chart, to show how the parts of a quantity are related. For example, the circle graph at left illustrates the percents of consumer transactions by payment type.

**Consumer Transactions**

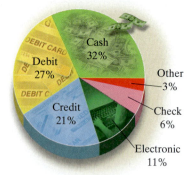

DATA: Federal Reserve Bank of San Francisco

| PERCENT NOTATION |
| --- |
| The notation **n**% means "*n* per hundred." |

This definition leads us to the following equivalent ways of defining percent notation.

---

**NOTATION FOR n%**

**Percent notation, n%,** can be expressed using:

a ratio → $n\% = \text{the ratio of } n \text{ to } 100 = \dfrac{n}{100}$,

fraction notation → $n\% = n \times \dfrac{1}{100}$, or

decimal notation → $n\% = n \times 0.01$.

---

**EXAMPLE 1** Write three kinds of notation for 67.8%.

Using a ratio: $\qquad\qquad 67.8\% = \dfrac{67.8}{100}$     A ratio of 67.8 to 100

Using fraction notation: $\quad 67.8\% = 67.8 \times \dfrac{1}{100}$     Replacing % with $\times \dfrac{1}{100}$

Using decimal notation: $\quad 67.8\% = 67.8 \times 0.01$     Replacing % with $\times 0.01$

**Do Exercises 1–4.** ▶

## b CONVERTING BETWEEN PERCENT NOTATION AND DECIMAL NOTATION

**SKILL REVIEW**

*Multiply using decimal notation.* [5.3a]
Multiply.
1. $68.3 \times 0.01$        2. $3013 \times 2.4$

           Answers: 1. 0.683   2. 7231.2

MyLab Math
VIDEO

To write decimal notation for a number like 78%, we can think of percent notation as a ratio and write

$\qquad 78\% = \dfrac{78}{100}$     Using the definition of percent as a ratio

$\qquad\qquad = 0.78.$     Dividing

Similarly,

$\qquad 4.9\% = \dfrac{4.9}{100} = 0.049.$

We can also convert 78% to decimal notation by replacing "%" with "× 0.01" and writing

$\qquad 78\% = 78 \times 0.01$     Replacing % with × 0.01

$\qquad\qquad = 0.78.$        Multiplying

Similarly,

$\qquad 4.9\% = 4.9 \times 0.01 = 0.049.$

---

Fluid milk consumption per capita dropped 16.2% from 2005 to 2016.

**Data:** U.S. Department of Agriculture

Write three kinds of notation for each percent, as in Example 1.

1. 70%          2. 23.4%

3. 100%       4. 0.6%

---

It is thought that the Roman emperor Augustus began percent notation by taxing goods sold at a rate of $\frac{1}{100}$. In time, the symbol "%" evolved by interchanging the parts of the symbol "100" to make "0/0" and then simplifying "0/0" to "%."

---

***Answers***

1. $\dfrac{70}{100}$; $70 \times \dfrac{1}{100}$; $70 \times 0.01$

2. $\dfrac{23.4}{100}$; $23.4 \times \dfrac{1}{100}$; $23.4 \times 0.01$

3. $\dfrac{100}{100}$; $100 \times \dfrac{1}{100}$; $100 \times 0.01$

4. $\dfrac{0.6}{100}$; $0.6 \times \dfrac{1}{100}$; $0.6 \times 0.01$

Dividing by 100 amounts to moving the decimal point two places to the left, which is the same as multiplying by 0.01. This leads us to a quick way to convert from percent notation to decimal notation: We drop the percent symbol and move the decimal point two places to the left.

| To convert from percent notation to decimal notation, | 36.5% |
|---|---|
| **a)** replace the percent symbol % with × 0.01, and | 36.5 × 0.01 |
| **b)** multiply by 0.01, which means move the decimal point two places to the left. | 0.36.5   Move 2 places to the left.<br>36.5% = 0.365 |

**EXAMPLE 2**   Find decimal notation for 99.44%.

**a)** Replace the percent symbol with × 0.01.      99.44 × 0.01

**b)** Move the decimal point two places to the left      0.99.44.

Thus, 99.44% = 0.9944.

**EXAMPLE 3**   The interest rate on a $2\frac{1}{2}$-year certificate of deposit is $2\frac{3}{8}$%. Find decimal notation for $2\frac{3}{8}$%.

**a)** Convert $2\frac{3}{8}$ to decimal notation and replace the percent symbol with × 0.01.      $2\frac{3}{8}$%<br>2.375 × 0.01

**b)** Move the decimal point two places to the left.      0.02.375

Thus, $2\frac{3}{8}$% = 0.02375.

◀ **Do Exercises 5–9.**

Find decimal notation.

**5.** 34%          **6.** 78.9%

**7.** $3\frac{1}{2}$%

Find decimal notation for the percent notation(s) in each sentence.

**8.** *Energy Use.* It is projected that by 2030, the United States will use 14.9% of the world's energy. India is projected to use only 6.7% of the world's energy.

**Data:** Energy Information Administration; Organization for Economic Cooperation and Development (OECD)

**9.** *Blood Alcohol Level.* A blood alcohol level of 0.08% is the standard used by the most states as the legal limit for drunk driving.

To convert 0.38 to percent notation, we can first write fraction notation, as follows:

$$0.38 = \frac{38}{100}$$   Converting to fraction notation

$$= 38\%.$$   Using the definition of percent as a ratio

Note that 100% = 100 × 0.01 = 1. Thus, to convert 0.38 to percent notation, we can multiply by 1, using 100% as a symbol for 1.

$$0.38 = 0.38 × 1$$
$$= 0.38 × 100\%$$
$$= 0.38 × 100 × 0.01$$   Replacing 100% with 100 × 0.01
$$= (0.38 × 100) × 0.01$$   Using the associative law of multiplication
$$= 38 × 0.01$$
$$= 38\%$$   Replacing × 0.01 with %

Even more quickly, since 0.38 = 0.38 × 100%, we can simply multiply 0.38 by 100 and write the % symbol.

To convert from decimal notation to percent notation, we multiply by 100%. That is, we move the decimal point two places to the right and write a percent symbol.

**Answers**

**5.** 0.34   **6.** 0.789   **7.** 0.035
**8.** 0.149; 0.067   **9.** 0.0008

| To convert from decimal notation to percent notation, multiply by 100%. That is, | $0.675 = 0.675 \times 100\%$ |
|---|---|
| **a)** move the decimal point two places to the right and | $0.67.5$    Move 2 places to the right. |
| **b)** write a % symbol. | $67.5\%$ <br> $0.675 = 67.5\%$ |

**EXAMPLE 4** Of the time off that employees take as sick leave, 0.21 is actually used for family issues. Find percent notation for 0.21.
**Data:** CCH Inc.

**a)** Move the decimal point two places to the right.     $0.21.$

**b)** Write a % symbol.     $21\%$

Thus, $0.21 = 21\%$.

**EXAMPLE 5** Find percent notation for 5.6.

**a)** Move the decimal point two places to the right, adding an extra zero.     $5.60.$

**b)** Write a % symbol.     $560\%$

Thus, $5.6 = 560\%$.

**EXAMPLE 6** Of those who play golf, 0.149 play 8–24 rounds per year. Find percent notation for 0.149.
**Data:** U.S. Golf Association

**a)** Move the decimal point two places to the right.     $0.14.9$

**b)** Write a % symbol.     $14.9\%$

Thus, $0.149 = 14.9\%$.

**EXAMPLE 7** Find percent notation for 0.00325.

**a)** Move the decimal point two places to the right.     $0.00.325$

**b)** Write a % symbol.     $0.325\%$

Thus, $0.00325 = 0.325\%$.

**Do Exercises 10–15.** ▶

Find percent notation.

**10.** 0.24      **11.** 3.47

**12.** 1      **13.** 0.005

Find percent notation for the decimal notation(s) in each sentence.

**14.** *Women in Congress.* In 2016, 0.196 of the members of the United States Congress were women.

     **Data:** Center for American Women in Politics (CAWP)

**15.** *Lacrosse.* In 2016, 824,947 Americans played lacrosse. Of this number of participants, 0.383 were high school students and 0.051 were college students.

     **Data:** uslacrosse.org

---

### c CONVERTING BETWEEN FRACTION NOTATION AND PERCENT NOTATION

**SKILL REVIEW**

*Convert from fraction notation to decimal notation.* [5.5a]
Find decimal notation.

**1.** $\dfrac{11}{16}$        **2.** $\dfrac{5}{9}$

**Answers: 1.** $0.6875$   **2.** $0.\overline{5}$

MyLab Math
**VIDEO**

To convert from fraction notation to percent notation,

$$\frac{3}{5}$$ Fraction notation

**a)** find decimal notation by division, and

$$\begin{array}{r} 0.6 \\ 5\overline{)\,3.0} \\ \underline{3\ 0} \\ 0 \end{array}$$

**b)** convert the decimal notation to percent notation.

$0.6 = 0.60 = 60\%$   Percent notation

$$\frac{3}{5} = 60\%$$

---

**EXAMPLE 8** *Programming Languages.* In a survey of over 64,000 software developers, $\frac{5}{8}$ of the developers said that they programmed in JavaScript. Find percent notation for $\frac{5}{8}$.

**Data:** insights.stackoverflow.com

**a)** Find decimal notation by division.

$$\begin{array}{r} 0.6\ 2\ 5 \\ 8\overline{)\,5.0\ 0\ 0} \\ \underline{4\ 8} \\ 2\ 0 \\ \underline{1\ 6} \\ 4\ 0 \\ \underline{4\ 0} \\ 0 \end{array} \qquad \frac{5}{8} = 0.625$$

**b)** Convert the decimal notation to percent notation.

$0.62.5$

$$\frac{5}{8} = 62.5\%, \text{ or } 62\frac{1}{2}\% \qquad 0.5 = \frac{1}{2}$$

**EXAMPLE 9** Find percent notation for $\frac{1}{6}$.

**a)** Find decimal notation by division.

$$\begin{array}{r} 0.1\ 6\ 6 \\ 6\overline{)\,1.0\ 0\ 0} \\ \underline{6} \\ 4\ 0 \\ \underline{3\ 6} \\ 4\ 0 \\ \underline{3\ 6} \\ 4 \end{array}$$

We get a repeating decimal: $0.16\overline{6}$.

**b)** Convert the decimal notation to percent notation.

$0.16.\overline{6}$

$$\frac{1}{6} = 16.\overline{6}\%, \text{ or } 16\frac{2}{3}\% \qquad 0.\overline{6} = \frac{2}{3}$$

◀ **Do Exercises 16–19.**

---

Find percent notation.

**16.** $\dfrac{1}{4}$   **17.** $\dfrac{3}{8}$

**18.** Water is the single most abundant chemical in the body. The human body is about $\frac{2}{3}$ water. Find percent notation for $\frac{2}{3}$.

**19.** Find percent notation: $\dfrac{5}{6}$.

*Answers*

**16.** 25%   **17.** 37.5%, or $37\frac{1}{2}\%$

**18.** $66.\overline{6}\%$, or $66\frac{2}{3}\%$

**19.** $83.\overline{3}\%$, or $83\frac{1}{3}\%$

In some cases, division is not the fastest way to convert a fraction to percent notation. The following are some optional ways in which this conversion might be done.

**EXAMPLE 10** Find percent notation for $\frac{69}{100}$.

We use the definition of percent as a ratio.

$$\frac{69}{100} = 69\%$$

**EXAMPLE 11** Find percent notation for $\frac{17}{20}$.

We want to multiply by 1 to get 100 in the denominator. We think of what we must multiply 20 by in order to get 100. That number is 5, so we multiply by 1 using $\frac{5}{5}$.

$$\frac{17}{20} \cdot \frac{5}{5} = \frac{85}{100} = 85\%$$

Note that this shortcut works only when the denominator of a simplified fraction is a factor of 100.

**EXAMPLE 12** Find percent notation for $\frac{18}{25}$.

$$\frac{18}{25} = \frac{18}{25} \cdot \frac{4}{4} = \frac{72}{100} = 72\%$$

**Do Exercises 20–23.** ▶

The method used in Example 10 is reversed when we convert from percent notation to fraction notation.

| | | |
|---|---|---|
| To convert from percent notation to fraction notation, | 30% | **Percent notation** |
| **a)** use the definition of percent as a ratio, and | $\dfrac{30}{100}$ | |
| **b)** simplify, if possible. | $\dfrac{3}{10}$ | **Fraction notation** |
| | $30\% = \dfrac{3}{10}$ | |

**EXAMPLE 13** Find fraction notation for 75%.

$$75\% = \frac{75}{100} \quad \text{Using the definition of percent}$$

$$= \frac{3 \cdot 25}{4 \cdot 25}$$

$$= \frac{3}{4} \cdot \frac{25}{25} \quad \text{Simplifying by removing a factor equal to 1: } \frac{25}{25} = 1$$

$$= \frac{3}{4}$$

Find percent notation.

20. $\dfrac{57}{100}$

**GS** 21. $\dfrac{19}{25} = \dfrac{19}{25} \cdot \dfrac{4}{\square}$

$$= \dfrac{76}{\square} = \square \,\%$$

22. $\dfrac{7}{10}$     23. $\dfrac{1}{4}$

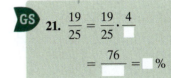

**CALCULATOR CORNER**

***Converting from Fraction Notation to Percent Notation*** A calculator can be used to convert from fraction notation to percent notation. We simply perform the division on the calculator and then use the percent key. To convert $\frac{17}{40}$ to percent notation, for example, we press

[1][7][÷][4][0][2nd][%], or
[1][7][÷][4][0][SHIFT][%].

The display reads [ 42.5 ], so $\frac{17}{40} = 42.5\%$.

**EXERCISES:** Use a calculator to find percent notation. Round to the nearest hundredth of a percent, where appropriate.

1. $\dfrac{13}{25}$     2. $\dfrac{5}{13}$

3. $\dfrac{43}{39}$     4. $\dfrac{12}{7}$

5. $\dfrac{217}{364}$     6. $\dfrac{2378}{8401}$

**EXAMPLE 14**  Find fraction notation for 62.5%.

$$62.5\% = \frac{62.5}{100}$$  Using the definition of percent

$$= \frac{62.5}{100} \times \frac{10}{10}$$  Multiplying by 1 to eliminate the decimal point in the numerator

$$= \frac{625}{1000}$$

$$= \frac{5 \cdot 125}{8 \cdot 125} = \frac{5}{8} \cdot \frac{125}{125}$$  Simplifying

$$= \frac{5}{8}$$

Find fraction notation.

**24.** 60%

**25.** 3.25%

$$= \frac{3.25}{\boxed{\phantom{00}}} = \frac{3.25}{100} \times \frac{\boxed{\phantom{00}}}{100}$$

$$= \frac{325}{\boxed{\phantom{00}}} = \frac{13 \times \boxed{\phantom{00}}}{400 \times 25}$$

$$= \frac{13}{400} \times \frac{25}{25} = \frac{\boxed{\phantom{00}}}{400}$$

**26.** $66\frac{2}{3}\%$     **27.** $12\frac{1}{2}\%$

**EXAMPLE 15**  Find fraction notation for $16\frac{2}{3}\%$.

$$16\frac{2}{3}\% = \frac{50}{3}\%$$  Converting from the mixed numeral to fraction notation

$$= \frac{50}{3} \times \frac{1}{100}$$  Using the definition of percent

$$= \frac{50 \cdot 1}{3 \cdot 50 \cdot 2} = \frac{1}{3 \cdot 2} \cdot \frac{50}{50}$$  Simplifying

$$= \frac{1}{6}$$

◀ **Do Exercises 24–27.**

The table below lists fraction, decimal, and percent equivalents that are used so often it would speed up your work if you memorized them. For example, $\frac{1}{3} = 0.\overline{3}$, so we say that the **decimal equivalent** of $\frac{1}{3}$ is $0.\overline{3}$, or that $0.\overline{3}$ has the **fraction equivalent** $\frac{1}{3}$. This table also appears on the inside back cover of the book.

**Fraction, Decimal, and Percent Equivalents**

| FRACTION NOTATION | $\frac{1}{10}$ | $\frac{1}{8}$ | $\frac{1}{6}$ | $\frac{1}{5}$ | $\frac{1}{4}$ | $\frac{3}{10}$ | $\frac{1}{3}$ | $\frac{3}{8}$ | $\frac{2}{5}$ | $\frac{1}{2}$ | $\frac{3}{5}$ | $\frac{5}{8}$ | $\frac{2}{3}$ | $\frac{7}{10}$ | $\frac{3}{4}$ | $\frac{4}{5}$ | $\frac{5}{6}$ | $\frac{7}{8}$ | $\frac{9}{10}$ | $\frac{1}{1}$ |
|---|---|---|---|---|---|---|---|---|---|---|---|---|---|---|---|---|---|---|---|---|
| DECIMAL NOTATION | 0.1 | 0.125 | $0.16\overline{6}$ | 0.2 | 0.25 | 0.3 | $0.33\overline{3}$ | 0.375 | 0.4 | 0.5 | 0.6 | 0.625 | $0.66\overline{6}$ | 0.7 | 0.75 | 0.8 | $0.83\overline{3}$ | 0.875 | 0.9 | 1 |
| PERCENT NOTATION | 10% | 12.5%, or $12\frac{1}{2}\%$ | $16.\overline{6}\%$, or $16\frac{2}{3}\%$ | 20% | 25% | 30% | $33.\overline{3}\%$, or $33\frac{1}{3}\%$ | 37.5%, or $37\frac{1}{2}\%$ | 40% | 50% | 60% | 62.5%, or $62\frac{1}{2}\%$ | $66.\overline{6}\%$, or $66\frac{2}{3}\%$ | 70% | 75% | 80% | $83.\overline{3}\%$, or $83\frac{1}{3}\%$ | 87.5%, or $87\frac{1}{2}\%$ | 90% | 100% |

**EXAMPLE 16**  Find fraction notation for $16.\overline{6}\%$.

We can use the table above or recall that $16.\overline{6}\% = 16\frac{2}{3}\% = \frac{1}{6}$. We can also recall from our work with repeating decimals in Chapter 5 that $0.\overline{6} = \frac{2}{3}$. Then we have $16.\overline{6}\% = 16\frac{2}{3}\%$ and can proceed as in Example 15.

◀ **Do Exercises 28 and 29.**

Find fraction notation.

**28.** $33.\overline{3}\%$     **29.** $83.\overline{3}\%$

*Answers*

**24.** $\frac{3}{5}$  **25.** $\frac{13}{400}$  **26.** $\frac{2}{3}$

**27.** $\frac{1}{8}$  **28.** $\frac{1}{3}$  **29.** $\frac{5}{6}$

*Guided Solution:*

**25.** 100, 100, 10,000, 25, 13

### ✓ Check Your Understanding

**Reading Check**   Fill in each blank with either "left" or "right."

**RC1.** To convert from decimal notation to percent notation, move the decimal point two places to the _____ and write a % symbol.

**RC2.** To convert from percent notation to decimal notation, replace the % symbol with × 0.01 and multiply by 0.01, which means move the decimal point two places to the _____.

**Concept Check**   Find percent notation for each shaded area.

**CC1.** 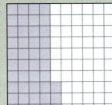    **CC2.**     **CC3.** 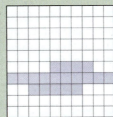    **CC4.**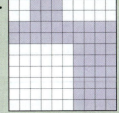

---

**a**   Write three kinds of notation, as in Example 1 on p. 441.

**1.** 90%     **2.** 58.7%     **3.** 12.5%     **4.** 130%

**b**   Find decimal notation.

**5.** 67%     **6.** 17%     **7.** 45.6%     **8.** 76.3%

**9.** 59.01%     **10.** 30.02%     **11.** 10%     **12.** 80%

**13.** 1%     **14.** 100%     **15.** 200%     **16.** 300%

**17.** 0.1%     **18.** 0.4%     **19.** 0.09%     **20.** 0.12%

**21.** 0.18%     **22.** 5.5%     **23.** 23.19%     **24.** 87.99%

**25.** $56\frac{1}{2}$%     **26.** $61\frac{3}{4}$%     **27.** $14\frac{7}{8}$%     **28.** $93\frac{1}{8}$%

For Exercises 29–32, find decimal notation for the percent notation.

| | GENERATION | AGE IN 2015 | PERCENT WITH AT LEAST ONE TATTOO |
|---|---|---|---|
| **29.** | Millennials | 18–34 | 47% |
| **30.** | Gen X | 35–50 | 36% |
| **31.** | Baby Boomers | 51–69 | 13% |
| **32.** | Mature | 70+ | 10% |

DATA: The Harris Poll (an online survey of 2225 U.S. adults, October 14–19, 2015)

Find decimal notation for the percent notation(s) in each exercise.

**33.** *Citations for Truck Drivers.* From October 2016 through August 2017, about 17% of law enforcement citations of truck drivers were for driving log violations, 6.7% were for speeding, 6.2% were for failure to wear a seat belt, and 5.5% were for driving beyond the 8-hour limit without a break.

**Data:** Federal Motor Carrier Safety Administration

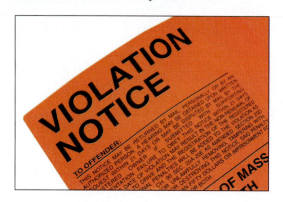

**34.** *Panama Canal Expansion.* After $5.4 billion was spent to expand the Panama Canal, larger transit vessels can now move through the locks. Prior to the opening of the new locks in June 2016, the monthly tonnage record was 30.4 million. In January of 2017, a new monthly tonnage record of 36.1 million was set. This was an 18.75% increase over the previous record.

**Data:** Panama Canal Authority

Find percent notation.

**35.** 0.47     **36.** 0.87     **37.** 0.03     **38.** 0.01     **39.** 8.7

**40.** 4     **41.** 0.334     **42.** 0.889     **43.** 12     **44.** 16.8

**45.** 0.4     **46.** 0.5     **47.** 0.006     **48.** 0.008     **49.** 0.017

**50.** 0.024     **51.** 0.2718     **52.** 0.8911     **53.** 0.0239     **54.** 0.00073

Find percent notation for the decimal notation(s) in each sentence.

**55. *Parts for the Auto Industry.*** In 2015, 0.76 of the seating and interior trim for the U.S. auto industry was produced in Canada or Mexico.

**Data:** Center for Automotive Research

**56. *Landline Phones.*** In 2012, 0.35 of Americans no longer had a landline phone. By 2016, 0.54 no longer had a landline phone.

**Data:** Influence Central

**57. *Residents Age 14 or Younger.*** In Egypt, 0.319 of the residents are age 14 or younger. In the United States, 0.190 of the residents are age 14 or younger.

**Data:** *The CIA World Factbook 2017*

**58. *Graduation Rates.*** In 2015, the high school graduation rate in the United States was 0.832. The dropout rate was 0.059.

**Data:** National Center for Educational Statistics

**59. *Growth in Digital Play.*** A recent survey showed that children prefer playing with touch screen devices to playing with construction blocks and puzzles. Of the 300 parents surveyed, 0.62 said that their children "often" or "very often" play with touch screens; only 0.49 gave the same responses for construction blocks and only 0.38 for puzzles.

**Data:** Michael Cohen Group: *The Wall Street Journal,* "Lego Hits Brick Wall As Digital Play Grows" by Saabira Chaudhuri

**60. *High Blood Pressure.*** Approximately 0.251 of American men ages 35–44 have high blood pressure. For the same age range, 0.190 of women have high blood pressure.

**Data:** Centers for Disease Control and Prevention

**C** Find percent notation.

**61.** $\dfrac{41}{100}$    **62.** $\dfrac{36}{100}$    **63.** $\dfrac{5}{100}$    **64.** $\dfrac{1}{100}$    **65.** $\dfrac{2}{10}$    **66.** $\dfrac{7}{10}$

**67.** $\dfrac{3}{10}$    **68.** $\dfrac{9}{10}$    **69.** $\dfrac{1}{2}$    **70.** $\dfrac{3}{4}$    **71.** $\dfrac{7}{8}$    **72.** $\dfrac{1}{8}$

**73.** $\dfrac{4}{5}$    **74.** $\dfrac{2}{5}$    **75.** $\dfrac{2}{3}$    **76.** $\dfrac{1}{3}$    **77.** $\dfrac{1}{6}$    **78.** $\dfrac{5}{6}$

**79.** $\dfrac{3}{16}$    **80.** $\dfrac{11}{16}$    **81.** $\dfrac{13}{16}$    **82.** $\dfrac{7}{16}$    **83.** $\dfrac{4}{25}$    **84.** $\dfrac{17}{25}$

**85.** $\dfrac{1}{20}$    **86.** $\dfrac{31}{50}$    **87.** $\dfrac{17}{50}$    **88.** $\dfrac{11}{20}$

Find percent notation for the fractions in each sentence.

**89.** *Taco Filling Preferences.* In the United States, $\frac{12}{25}$ of people surveyed choose beef as their favorite taco filling. Only $\frac{3}{20}$ choose chicken as their favorite taco filling.

**Data:** Food Network

**90.** *Popular Automobile Colors.* The three most popular colors for automobiles in Asia in 2016 were white, black, and silver. Of all automobiles in Asia, $\frac{12}{25}$ were white, $\frac{4}{25}$ black, and $\frac{1}{10}$ silver.

**Data:** AXALTA Coating Systems, "Global Automotive 2016 Color Popularity Report"

In Exercises 91–96, write percent notation for the fractions in the pie chart below.

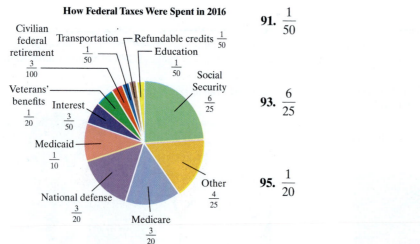

How Federal Taxes Were Spent in 2016

DATA: Committee for a Responsible Federal Budget

**91.** $\frac{1}{50}$

**92.** $\frac{3}{50}$

**93.** $\frac{6}{25}$

**94.** $\frac{4}{25}$

**95.** $\frac{1}{20}$

**96.** $\frac{3}{100}$

Find fraction notation. Simplify, if possible.

**97.** 85%

**98.** 55%

**99.** 62.5%

**100.** 12.5%

**101.** $33\frac{1}{3}\%$

**102.** $83\frac{1}{3}\%$

**103.** $16.\overline{6}\%$

**104.** $66.\overline{6}\%$

**105.** 7.25%

**106.** 4.85%

**107.** 0.8%

**108.** 0.2%

**109.** $25\frac{3}{8}\%$

**110.** $48\frac{7}{8}\%$

**111.** $78\frac{2}{9}\%$

**112.** $16\frac{5}{9}\%$

**113.** $64\frac{7}{11}\%$

**114.** $73\frac{3}{11}\%$

**115.** 150%

**116.** 110%

**117.** 0.0325%

**118.** 0.419%

**119.** $33.\overline{3}\%$

**120.** $83.\overline{3}\%$

In Exercises 121–126, find fraction notation for the percent notations in the table below.

**U.S. POPULATION BY SELECTED AGE CATEGORIES**

(Data have been rounded to the nearest percent.)

| AGE CATEGORY | PERCENT OF POPULATION |
|---|---|
| 0–18 years | 24% |
| 19–25 years | 9% |
| 26–34 years | 12% |
| 35–54 years | 26% |
| 55–64 years | 13% |
| 65+ years | 15% |

DATA: The Henry J. Kaiser Family Foundation; U.S. Census Bureau's *Current Population Survey* (Annual Social and Economic Supplements), March 2017

**121.** 24%

**122.** 26%

**123.** 9%

**124.** 12%

**125.** 15%

**126.** 13%

Find fraction notation for the percent notation in each sentence.

**127.** A $\frac{3}{4}$-cup serving of Post Selects Great Grains® cereal provides 6% of the recommended daily amount of potassium.

**Data:** Kraft Foods Global, Inc.

**128.** A 1.8-oz serving of Frosted Mini-Wheats®, Blueberry Muffin, with $\frac{1}{2}$ cup of fat-free milk provides 35% of the recommended daily amount of Vitamin $B_{12}$.

**Data:** Kellogg, Inc.

**129.** *Smoking Cigarettes.* In 2015, 7.4% of Americans with an undergraduate college degree smoked cigarettes.

**Data:** Centers for Disease Control and Prevention

**130.** *Smoking Cigarettes.* In the United States in 2015, 16.7% of men 18 years of age or older smoked cigarettes, and 13.6% of women in the same age group smoked cigarettes.

**Data:** Centers for Disease Control and Prevention

Complete each table.

**131.**

| Fraction Notation | Decimal Notation | Percent Notation |
|---|---|---|
| $\frac{1}{8}$ | | 12.5%, or 12$\frac{1}{2}$% |
| $\frac{1}{6}$ | | |
| | | 20% |
| | 0.25 | |
| | | 33.$\overline{3}$%, or 33$\frac{1}{3}$% |
| | | 37.5%, or 37$\frac{1}{2}$% |
| | | 40% |
| $\frac{1}{2}$ | | |

**132.**

| Fraction Notation | Decimal Notation | Percent Notation |
|---|---|---|
| $\frac{3}{5}$ | | |
| | 0.625 | |
| $\frac{2}{3}$ | | |
| | 0.75 | 75% |
| $\frac{4}{5}$ | | |
| $\frac{5}{6}$ | | 83.$\overline{3}$%, or 83$\frac{1}{3}$% |
| $\frac{7}{8}$ | | 87.5%, or 87$\frac{1}{2}$% |
| | | 100% |

**133.**

| Fraction Notation | Decimal Notation | Percent Notation |
|---|---|---|
| | 0.5 | |
| $\frac{1}{3}$ | | |
| | | 25% |
| | | $16.\overline{6}\%$, or $16\frac{2}{3}\%$ |
| | 0.125 | |
| $\frac{3}{4}$ | | |
| | $0.8\overline{3}$ | |
| $\frac{3}{8}$ | | |

**134.**

| Fraction Notation | Decimal Notation | Percent Notation |
|---|---|---|
| | | 40% |
| | | 62.5%, or $62\frac{1}{2}\%$ |
| | 0.875 | |
| $\frac{1}{1}$ | | |
| | 0.6 | |
| | $0.\overline{6}$ | |
| $\frac{1}{5}$ | | |

## Skill Maintenance

Solve.

**135.** $13 \cdot x = 910$ [1.7b]

**136.** $15 \cdot y = 75$ [1.7b]

**137.** $0.05 \times b = 20$ [5.7a]

**138.** $3 = 0.16 \times b$ [5.7a]

**139.** $\dfrac{24}{37} = \dfrac{15}{x}$ [6.3b]

**140.** $\dfrac{17}{18} = \dfrac{x}{27}$ [6.3b]

**141.** $\dfrac{9}{10} = \dfrac{x}{5}$ [6.3b]

**142.** $\dfrac{7}{x} = \dfrac{4}{5}$ [6.3b]

## Synthesis

Find percent notation for each shaded area.

**143.**

**144.**

Write percent notation.

**145.** $2.5\overline{74631}$

**146.** $\dfrac{54}{999}$

Write decimal notation.

**147.** $\dfrac{729}{7}\%$

**148.** $\dfrac{19}{12}\%$

# Solving Percent Problems Using Percent Equations

## a TRANSLATING TO EQUATIONS

To solve a problem involving percents, it is helpful to translate first to an equation. To distinguish the method discussed in this section from that of Section 7.3, we will call these *percent equations*.

### KEY WORDS IN PERCENT TRANSLATIONS

"**Of**" translates to "·" or "×".    "**Is**" translates to "=".

"**What**" translates to any letter.    "**%**" translates to "$\times \frac{1}{100}$" or "$\times 0.01$".

**EXAMPLES**  Translate each of the following.

**1.** 23%  of  5  is  what?

  ↓      ↓    ↓   ↓    ↓

  0.23  ·  5  =  $a$       This is a *percent equation*.

**2.** What  is  11%  of  49?

  ↓      ↓    ↓     ↓    ↓

  $a$  =  0.11  ·  49       Any letter can be used.

> **Do Exercises 1 and 2.** ▶

Translate to an equation. Do not solve.

**1.** 12% of 50 is what?

**2.** What is 40% of 60?

**EXAMPLES**  Translate each of the following.

**3.** 3  is  10%  of  what?

  ↓   ↓    ↓     ↓    ↓

  3  =  0.10  ·  $b$

············· **Caution!** ···············

Don't forget to translate percent notation to decimal notation!

**4.** 45%  of  what  is  23?

  ↓      ↓     ↓     ↓   ↓

  0.45  ×  $b$  =  23

> **Do Exercises 3 and 4.** ▶

Translate to an equation. Do not solve.

**3.** 45 is 20% of what?

**4.** 120% of what is 60?

**EXAMPLES**  Translate each of the following.

**5.** 10  is  what percent  of  20?

  ↓    ↓        ↓          ↓   ↓

  10  =      $p$       ×  20

**6.** What percent  of  50  is  7?

       ↓          ↓    ↓    ↓   ↓

       $p$     ·  50  =  7

> **Do Exercises 5 and 6.** ▶

Translate to an equation. Do not solve.

**5.** 16 is what percent of 40?

**6.** What percent of 84 is 10.5?

Each percent equation in Examples 1–6 can be written in this form:

$$\textbf{Amount = Percent number} \times \textbf{Base.}$$

In each case, one of the three pieces of information is missing.

*Answers*

**1.** $0.12 \cdot 50 = a$  **2.** $a = 0.40 \cdot 60$
**3.** $45 = 0.20 \cdot b$  **4.** $1.20 \times b = 60$
**5.** $16 = p \times 40$  **6.** $p \cdot 84 = 10.5$

For fiscal year 2017, the U.S. government budgeted $835 billion for defense (including military defense, veterans affairs, and foreign policy). Of that amount, 72.2% was budgeted for military defense. What was budgeted for military defense? (See Example 7.)

**Data:** usgovernmentspending.com

**7.** Solve:

What is 12% of $50?

**8.** Solve:

64% of 55 is what?

## b SOLVING PERCENT PROBLEMS

In solving percent problems, we use the *Translate* and *Solve* steps in the problem-solving strategy used throughout this text.

Percent problems are actually of three different types. Although the method we present does *not* require that you be able to identify which type you are solving, it is helpful to know them. Each of the three types of percent problems depends on which of the three pieces of information is missing.

**1.** Finding the *amount* (the result of taking the percent)

*Example:*     **What**   is   25%   of   60?

*Translation:*     $a$   =   0.25   $\cdot$   60

**2.** Finding the *base* (the number you are taking the percent of)

*Example:*     15   is   25%   of   **what?**

*Translation:*     15   =   0.25   $\cdot$   $b$

**3.** Finding the *percent number* (the percent itself)

*Example:*     15   is   **what percent**   of   60?

*Translation:*     15   =   $p$   $\cdot$   60

### Finding the Amount

**EXAMPLE 7**   What is 72.2% of $835,000,000,000?

*Translate:* $a = 0.722 \times 835{,}000{,}000{,}000$.

*Solve:* The letter is by itself. To solve the equation, we multiply:

$$a = 0.722 \times 835{,}000{,}000{,}000 = 602{,}870{,}000{,}000.$$

Thus, $602,870,000,000 is 72.2% of $835,000,000,000. The answer is $602,870,000,000.

◀ **Do Exercise 7.**

**EXAMPLE 8**   120% of 42 is what?

*Translate:* $1.20 \times 42 = a$.

*Solve:* The letter is by itself. To solve the equation, we carry out the calculation:

$$a = 1.2 \times 42 \qquad 120\% = 1.2$$
$$a = 50.4.$$

Thus, 120% of 42 is 50.4. The answer is 50.4.

◀ **Do Exercise 8.**

*Answers*

**7.** $6   **8.** 35.2

## Finding the Base

**EXAMPLE 9** 8% of what is 32?

*Translate:* $0.08 \cdot b = 32$.

*Solve:* This time the letter is *not* by itself. To solve the equation, we divide both sides by 0.08:

$$\frac{0.08 \cdot b}{0.08} = \frac{32}{0.08} \quad \text{Dividing}$$

$$b = \frac{32}{0.08}$$

$$b = 400.$$

Thus, 8% of 400 is 32. The answer is 400.

**EXAMPLE 10** $3 is 16% of what?

*Translate:*

| \$3 | is | 16% | of | what? |
|----|----|-----|----|-------|
| ↓ | ↓ | ↓ | ↓ | ↓ |
| 3 | = | 0.16 | · | $b$ |

*Solve:* To solve the equation, we divide both sides by 0.16:

$$\frac{3}{0.16} = \frac{0.16 \cdot b}{0.16} \quad \text{Dividing both sides by 0.16}$$

$$\frac{3}{0.16} = b \quad \quad 16\% = 0.16$$

$$18.75 = b.$$

Thus, \$3 is 16% of \$18.75. The answer is \$18.75.

**Do Exercises 9 and 10.** ▶

## Finding the Percent Number

When finding the percent number, you *must* remember to convert to percent notation after you have solved the equation.

**EXAMPLE 11** 374,000 is what percent of 561,000?

*Translate:*

| 374,000 | is | what percent | of | 561,000? |
|---------|----|--------------|----|----------|
| ↓ | ↓ | ↓ | ↓ | ↓ |
| 374,000 | = | $p$ | × | 561,000 |

*Solve:* To solve the equation, we divide both sides by 561,000 and convert the result to percent notation:

$$374,000 = p \times 561,000$$

$$\frac{374,000}{561,000} = \frac{p \times 561,000}{561,000} \quad \text{Dividing}$$

$$0.66\overline{6} = p \quad \quad \text{Converting to decimal notation}$$

$$66\tfrac{2}{3}\% = p. \quad \quad \text{Converting to percent notation}$$

Thus, 374,000 is $66\tfrac{2}{3}\%$ of 561,000. The answer is $66\tfrac{2}{3}\%$.

A survey of a group of people found that 8% of the group, or 32 people, chose cookies and cream as their favorite milkshake. How many people were surveyed? (See Example 9.)

Solve.

**GS** 9. 20% of what is 45?

| 20% | of | what | is | 45? |
|-----|----|------|----|-----|
| ↓ | ↓ | ↓ | ↓ | ↓ |
| ▢ | · | $b$ | = | ▢ |

$$\frac{0.20 \cdot b}{0.20} = \frac{45}{\boxed{\phantom{0}}}$$

$$\boxed{\phantom{0}} = \frac{45}{0.20}$$

$$b = \boxed{\phantom{0}}$$

10. \$60 is 120% of what?

Of the 561,000 single-family homes sold in 2016, 374,000 had a two-car garage. What percent of the houses had a two-car garage?
Data: U.S. Census Bureau

***Answers***
**9.** 225  **10.** \$50
***Guided Solution:***
**9.** 0.20, 45, 0.20, $b$, 225

Solve.

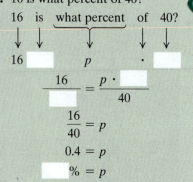

**11.** 16 is what percent of 40?

16  is  what percent  of  40?

$$\frac{16}{\boxed{\phantom{00}}} = \frac{p \cdot \boxed{\phantom{00}}}{40}$$

$$\frac{16}{40} = p$$

$$0.4 = p$$

$$\boxed{\phantom{00}} \% = p$$

**12.** What percent of $84 is $10.50?

**EXAMPLE 12** What percent of $50 is $16?

*Translate:* What percent of $50 is $16?

$$p \quad \times \quad 50 \quad = \quad 16$$

*Solve:* To solve the equation, we divide both sides by 50 and convert the result to percent notation:

$$\frac{p \times 50}{50} = \frac{16}{50} \qquad \text{Dividing by 50 on both sides}$$

$$p = \frac{16}{50}$$

$$p = 0.32$$

$$p = 32\%. \qquad \text{Converting to percent notation}$$

Thus, 32% of $50 is $16. The answer is 32%.

◀ **Do Exercises 11 and 12.**

.............................................. **Caution!** ..............................................

When a question asks "what percent?", be sure to give the answer in percent notation.

........................................................................................................................

*Answers*

**11.** 40% **12.** 12.5%, or $12\frac{1}{2}$%

*Guided Solution:*

**11.** =, 40, 40, 40, 40

---

**CALCULATOR CORNER**

****Using Percents in Computations**** Many calculators have a %⃞ key that can be used in computations. For example, to find 11% of 49, we press 1⃞ 1⃞ 2nd⃞ %⃞ ×⃞ 4⃞ 9⃞ =⃞, or 4⃞ 9⃞ ×⃞ 1⃞ 1⃞ SHIFT⃞ %⃞ =⃞. The display reads ⃞5.39, so 11% of 49 is 5.39.

In Example 9, we performed the computation 32/8%. To use the %⃞ key in this computation, we press 3⃞ 2⃞ ÷⃞ 8⃞ 2nd⃞ %⃞ =⃞, or 3⃞ 2⃞ ÷⃞ 8⃞ SHIFT⃞ %⃞ =⃞. The result is 400.

We can also use the %⃞ key to find the percent number in a problem. In Example 11, for instance, we answered the question "374,000 is what percent of 561,000?" On a calculator, we press 3⃞ 7⃞ 4⃞ 0⃞ 0⃞ 0⃞ ÷⃞ 5⃞ 6⃞ 1⃞ 0⃞ 0⃞ 0⃞ 2nd⃞ %⃞ =⃞, or 3⃞ 7⃞ 4⃞ 0⃞ 0⃞ 0⃞ ÷⃞ 5⃞ 6⃞ 1⃞ 0⃞ 0⃞ 0⃞ SHIFT⃞ %⃞ =⃞. The result is $66.\overline{6}$, so 374,000 is $66.\overline{6}$% of 561,000.

**EXERCISES:** Use a calculator to find each of the following.

**1.** What is 12.6% of $40?

**2.** 0.04% of 28 is what?

**3.** 8% of what is 36?

**4.** $45 is 4.5% of what?

**5.** 23 is what percent of 920?

**6.** What percent of $442 is $53.04?

## ✓ Check Your Understanding

**Reading Check** Match each question with the correct translation from the list on the right.

**RC1.** 18 is 40% of what? _____

**RC2.** What percent of 45 is 18? _____

**RC3.** What is 40% of 45? _____

**RC4.** 0.5% of 1200 is what? _____

**RC5.** 6 is what percent of 1200? _____

**RC6.** 6 is 0.5% of what? _____

**a)** $6 = 0.005 \cdot b$

**b)** $6 = p \cdot 1200$

**c)** $18 = 0.40 \cdot b$

**d)** $0.005 \cdot 1200 = a$

**e)** $p \cdot 45 = 18$

**f)** $a = 0.40 \cdot 45$

**Concept Check** Choose from the list on the right the most appropriate first step in solving each equation.

**CC1.** $5\% \times b = 400$

**CC2.** $40 = p \times 400$

**CC3.** $100 = 10\% \times b$

**CC4.** $p \times 850 = 85$

**a)** Divide by 100 on both sides

**b)** Divide by 850 on both sides

**c)** Divide by 400 on both sides

**d)** Divide by 85 on both sides

**e)** Divide by 0.05 on both sides

**f)** Divide by 0.1 on both sides

---

**a** Translate to an equation. Do not solve.

**1.** What is 32% of 78?

**2.** 98% of 57 is what?

**3.** 89 is what percent of 99?

**4.** What percent of 25 is 8?

**5.** 13 is 25% of what?

**6.** 21.4% of what is 20?

**b** Translate to an equation and solve.

**7.** What is 85% of 276?

**8.** What is 74% of 53?

**9.** 150% of 30 is what?

**10.** 100% of 13 is what?

**11.** What is 6% of $300?

**12.** What is 4% of $45?

**13.** 3.8% of 50 is what?

**14.** $33\frac{1}{3}\%$ of 480 is what?

(*Hint*: $33\frac{1}{3}\% = \frac{1}{3}$.)

**15.** $39 is what percent of $50?

**16.** $16 is what percent of $90?

**17.** 20 is what percent of 10?

**18.** 60 is what percent of 20?

**19.** What percent of $300 is $150?

**20.** What percent of $50 is $40?

**21.** What percent of 80 is 100?

**22.** What percent of 60 is 15?

**23.** 20 is 50% of what?

**24.** 57 is 20% of what?

**25.** 40% of what is $16?

**26.** 100% of what is $74?

**27.** 56.32 is 64% of what?

**28.** 71.04 is 96% of what?

**29.** 70% of what is 14?

**30.** 70% of what is 35?

**31.** What is $62\frac{1}{2}$% of 10?

**32.** What is $35\frac{1}{4}$% of 1200?

**33.** What is 8.3% of $10,200?

**34.** What is 9.2% of $5600?

**35.** 2.5% of what is 30.4?

**36.** 8.2% of what is 328?

## Skill Maintenance

Write fraction notation. [5.1b]

**37.** 0.9375      **38.** 0.125

Write decimal notation. [5.1b]

**39.** $\frac{3}{10}$      **40.** $\frac{17}{1000}$

Simplify. [1.9c]

**41.** $3 + (8 - 6) \cdot 2$

**42.** $2 \cdot 7 - (5 + 1)$

## Synthesis

 Solve.

**43.** $2496 is 24% of what amount?

     Estimate _____

     Calculate _____

**44.** What is 38.2% of $52,345.79?

     Estimate _____

     Calculate _____

**45.** *Recyclables.* It is estimated that 40% to 50% of all trash is recyclable. If a community produces 270 tons of trash, how much of this trash is recyclable?

**46.** *Batting.* An all-star baseball player gets a hit in 30% to 35% of his at-bats. If an all-star had 520 to 580 at-bats, how many hits would he have had?

**47.** 40% of $18\frac{3}{4}$% of $25,000 is what?

# Solving Percent Problems Using Proportions*

### OBJECTIVES

**a** Translate percent problems to proportions.

**b** Solve basic percent problems using proportions.

## a TRANSLATING TO PROPORTIONS

A percent is a ratio of some number to 100. For example, 6% is the ratio $\frac{6}{100}$. Since 9,660,000 is 6% of 161,000,000, the numbers 9,660,000 and 161,000,000 have the same ratio as 6 and 100.

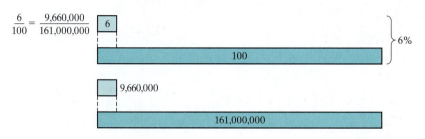

$$\frac{6}{100} = \frac{9,660,000}{161,000,000}$$

To solve a percent problem using a proportion, we translate as follows:

$$\text{Number} \longrightarrow \frac{N}{100} = \frac{a}{b} \longleftarrow \text{Amount}$$
$$100 \longrightarrow \qquad \qquad \longleftarrow \text{Base}$$

> You might find it helpful to read this as "part is to whole as part is to whole."

For example, "60% of 25 is 15" translates to

$$\frac{60}{100} = \frac{15}{25} \begin{array}{l} \longleftarrow \text{Amount} \\ \longleftarrow \text{Base} \end{array}$$

A clue for translating is that the base, $b$, corresponds to 100 and usually follows the wording "percent of." Also, $N\%$ always translates to $N/100$. Another aid in translating is to make a comparison drawing. To do this, we start with the percent side and list 0% at the top and 100% near the bottom. Then we estimate where the specified percent—in this case, 60%—is located. The corresponding quantities are then filled in. The base—in this case, 25— always corresponds to 100%, and the amount—in this case, 15—corresponds to the specified percent.

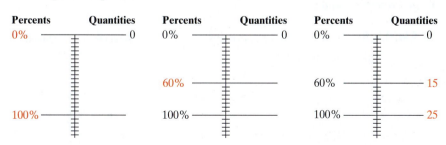

The proportion can then be read easily from the drawing: $\frac{60}{100} = \frac{15}{25}$.

In the United States, approximately 6% of the labor force is age 65 or older. In 2017, there were approximately 161,000,000 people in the labor force. This means that about 9,660,000 workers were age 65 or older.

**Data:** U.S. Department of Labor; U.S. Bureau of Labor Statistics

---

*Note: This section presents an alternative method for solving basic percent problems. You can use either equations or proportions to solve percent problems, but you might prefer one method over the other, or your instructor may direct you to use one method over the other.

**EXAMPLE 1**    Translate to a proportion.

23% of 5 is what?

$$\frac{23}{100} = \frac{a}{5}$$

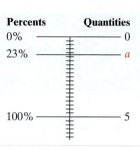

**EXAMPLE 2**    Translate to a proportion.

What is 124% of 49?

$$\frac{124}{100} = \frac{a}{49}$$

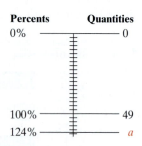

Translate to a proportion.
Do not solve.

1. 12% of 50 is what?

2. What is 40% of 60?

3. 130% of 72 is what?

◀ **Do Exercises 1–3.**

**EXAMPLE 3**    Translate to a proportion.

3 is 10% of what?

$$\frac{10}{100} = \frac{3}{b}$$

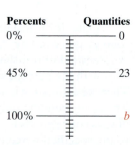

**EXAMPLE 4**    Translate to a proportion.

45% of what is 23?

$$\frac{45}{100} = \frac{23}{b}$$

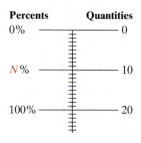

Translate to a proportion.
Do not solve.

4. 45 is 20% of what?

5. 120% of what is 60?

◀ **Do Exercises 4 and 5.**

**EXAMPLE 5**    Translate to a proportion.

10 is what percent of 20?

$$\frac{N}{100} = \frac{10}{20}$$

| Percents | Quantities |
|---|---|
| 0% | 0 |
| N% | 10 |
| 100% | 20 |

*Answers*

1. $\frac{12}{100} = \frac{a}{50}$   2. $\frac{40}{100} = \frac{a}{60}$   3. $\frac{130}{100} = \frac{a}{72}$

4. $\frac{20}{100} = \frac{45}{b}$   5. $\frac{120}{100} = \frac{60}{b}$

**EXAMPLE 6** Translate to a proportion.

What percent of 50 is 7?

$$\frac{N}{100} = \frac{7}{50}$$

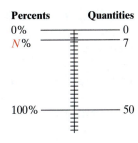

**Do Exercises 6 and 7.**

Translate to a proportion.
Do not solve.

**6.** 16 is what percent of 40?

**7.** What percent of 84 is 10.5?

## b SOLVING PERCENT PROBLEMS

*Solve proportions.* [6.3b]

Solve.

**1.** $\dfrac{3}{100} = \dfrac{27}{b}$

**2.** $\dfrac{4.3}{20} = \dfrac{N}{100}$

**Answers: 1.** 900 **2.** 21.5

MyLab Math
VIDEO

After a percent problem has been translated to a proportion, we solve as in Section 6.3.

**EXAMPLE 7** 5% of what is $20?

*Translate:* $\dfrac{5}{100} = \dfrac{20}{b}$

*Solve:* $5 \cdot b = 100 \cdot 20$     Equating cross products

$\dfrac{5 \cdot b}{5} = \dfrac{100 \cdot 20}{5}$     Dividing by 5

$b = \dfrac{2000}{5}$

$b = 400$     Simplifying

Thus, 5% of $400 is $20. The answer is $400.

**Do Exercise 8.**

**EXAMPLE 8** 120% of 42 is what?

*Translate:* $\dfrac{120}{100} = \dfrac{a}{42}$

*Solve:* $120 \cdot 42 = 100 \cdot a$     Equating cross products

$\dfrac{120 \cdot 42}{100} = \dfrac{100 \cdot a}{100}$     Dividing by 100

$\dfrac{5040}{100} = a$

$50.4 = a$     Simplifying

Thus, 120% of 42 is 50.4. The answer is 50.4.

**Do Exercises 9 and 10.**

**GS 8.** Solve: 20% of what is $45?

$$\frac{20}{\boxed{\phantom{00}}} = \frac{\boxed{\phantom{00}}}{b}$$

$20 \cdot b = 100 \cdot \boxed{\phantom{0}}$

$\dfrac{20 \cdot b}{20} = \dfrac{100 \cdot 45}{\boxed{\phantom{0}}}$

$b = \dfrac{4500}{20}$

$b = \boxed{\phantom{0}}$

Thus, 20% of $\boxed{\phantom{0}}$ is $45.

Solve.

**GS 9.** 64% of 55 is what?

$$\frac{64}{100} = \frac{a}{\boxed{\phantom{0}}}$$

$\boxed{\phantom{0}} \cdot 55 = 100 \cdot a$

$\dfrac{64 \cdot 55}{100} = \dfrac{100 \cdot a}{\boxed{\phantom{0}}}$

$\dfrac{\boxed{\phantom{0}}}{100} = a$

$\boxed{\phantom{0}} = a$

Thus, 64% of 55 is $\boxed{\phantom{0}}$.

**10.** What is 12% of 50?

***Answers***

**6.** $\dfrac{N}{100} = \dfrac{16}{40}$ **7.** $\dfrac{N}{100} = \dfrac{10.5}{84}$ **8.** $225

**9.** 35.2 **10.** 6

***Guided Solutions:***

**8.** 100, 45, 45, 20, 225, 225

**9.** 55, 64, 100, 3520, 35.2, 35.2

**EXAMPLE 9**   210 is $10\frac{1}{2}$% of what?

*Translate:* $\dfrac{210}{b} = \dfrac{10.5}{100}$     $10\frac{1}{2}\% = 10.5\%$

*Solve:* $210 \cdot 100 = b \cdot 10.5$     Equating cross products

$\dfrac{210 \cdot 100}{10.5} = \dfrac{b \cdot 10.5}{10.5}$     Dividing both sides by 10.5

$\dfrac{21{,}000}{10.5} = b$     Multiplying and simplifying

$2000 = b$     Dividing

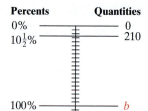

Thus, 210 is $10\frac{1}{2}$% of 2000. The answer is 2000.

◀ **Do Exercise 11.**

**11.** Solve:

 60 is 120% of what?

---

**12.** Solve:

 $12 is what percent of $40?

$\dfrac{12}{40} = \dfrac{N}{\boxed{\phantom{xx}}}$

$\boxed{\phantom{xx}} \cdot 100 = 40 \cdot N$

$\dfrac{12 \cdot 100}{\boxed{\phantom{xx}}} = \dfrac{40 \cdot N}{40}$

$\dfrac{\boxed{\phantom{xx}}}{40} = N$

$30 = N$

Thus, $12 is 30 $\boxed{\phantom{xx}}$ of $40.

**GS**

**EXAMPLE 10**   $10 is what percent of $20?

*Translate:* $\dfrac{10}{20} = \dfrac{N}{100}$

*Solve:* $10 \cdot 100 = 20 \cdot N$     Equating cross products

$\dfrac{10 \cdot 100}{20} = \dfrac{20 \cdot N}{20}$     Dividing both sides by 20

$\dfrac{1000}{20} = N$     Multiplying and simplifying

$50 = N$     Dividing

Thus, $10 is 50% of $20. The answer is 50%.

> Always "look before you leap." Many students can solve this problem mentally: $10 is half, or 50%, of $20.

◀ **Do Exercise 12.**

**EXAMPLE 11**   What percent of 50 is 16?

*Translate:* $\dfrac{N}{100} = \dfrac{16}{50}$

*Solve:* $50 \cdot N = 100 \cdot 16$     Equating cross products

$\dfrac{50 \cdot N}{50} = \dfrac{100 \cdot 16}{50}$     Dividing both sides by 50

$N = \dfrac{1600}{50}$     Multiplying and simplifying

$N = 32$     Dividing

Thus, 32% of 50 is 16. The answer is 32%.

> Note when solving percent problems using proportions that $N$ is a percent and needs only a % sign.

**13.** Solve:

 What percent of 84 is 10.5?

◀ **Do Exercise 13.**

················· **Caution!** ·················

Don't forget to add the % sign when the problem asks "What percent . . . ?"

**Answers**
**11.** 50   **12.** 30%   **13.** 12.5%
*Guided Solution:*
**12.** 100, 12, 40, 1200; %

## ✓ Check Your Understanding

**Reading Check**  Match each question with the correct translation from the list on the right.

**RC1.** 70 is 35% of what? _____

**RC2.** 70 is what percent of 200? _____

**RC3.** What is 35% of 200? _____

**RC4.** 74.8 is 110% of what? _____

**RC5.** What percent of 68 is 74.8? _____

**RC6.** 110% of 68 is what? _____

**a)** $\dfrac{110}{100} = \dfrac{a}{68}$   **b)** $\dfrac{70}{b} = \dfrac{35}{100}$

**c)** $\dfrac{a}{200} = \dfrac{35}{100}$   **d)** $\dfrac{74.8}{68} = \dfrac{N}{100}$

**e)** $\dfrac{70}{200} = \dfrac{N}{100}$   **f)** $\dfrac{74.8}{b} = \dfrac{110}{100}$

**Concept Check**  Select from the list on the right an equation equivalent to the given proportion.

**CC1.** $\dfrac{64}{b} = \dfrac{16}{100}$ _____

**CC2.** $\dfrac{6.5}{13} = \dfrac{N}{100}$ _____

**a)** $13 \cdot 100 = 6.5 \cdot N$
**b)** $64 \cdot 100 = b \cdot 16$
**c)** $64 \cdot b = 16 \cdot 100$
**d)** $6.5 \cdot 100 = 13 \cdot N$
**e)** $6.5 \cdot 13 = N \cdot 100$
**f)** $64 \cdot 16 = b \cdot 100$

---

**a**  Translate to a proportion. Do not solve.

**1.** What is 37% of 74?

**2.** 66% of 74 is what?

**3.** 4.3 is what percent of 5.9?

**4.** What percent of 6.8 is 5.3?

**5.** 14 is 25% of what?

**6.** 133% of what is 40?

**b**  Translate to a proportion and solve.

**7.** What is 76% of 90?

**8.** What is 32% of 70?

**9.** 70% of 660 is what?

**10.** 80% of 920 is what?

**11.** What is 130% of 352?

**12.** What is 225% of 83?

**13.** 4.8% of 60 is what?

**14.** 63.1% of 80 is what?

**15.** $24 is what percent of $96?

**16.** $14 is what percent of $70?

**17.** 102 is what percent of 100?

**18.** 103 is what percent of 100?

**19.** What percent of $480 is $120?

**20.** What percent of $80 is $60?

**21.** What percent of 160 is 150?

**22.** What percent of 33 is 11?

**23.** $18 is 25% of what?

**24.** $75 is 20% of what?

**25.** 60% of what is 54?

**26.** 80% of what is 96?

**27.** 65.12 is 74% of what?

**28.** 63.7 is 65% of what?

**29.** 80% of what is 16?

**30.** 80% of what is 10?

**31.** What is $62\frac{1}{2}$% of 40?

**32.** What is $43\frac{1}{4}$% of 2600?

**33.** What is 9.4% of $8300?

**34.** What is 8.7% of $76,000?

**35.** 80.8 is $40\frac{2}{5}$% of what?

**36.** 66.3 is $10\frac{1}{5}$% of what?

## Skill Maintenance

Solve. [6.3b]

**37.** $\dfrac{x}{188} = \dfrac{2}{47}$

**38.** $\dfrac{15}{x} = \dfrac{3}{800}$

**39.** $\dfrac{75}{100} = \dfrac{n}{20}$

**40.** $\dfrac{612}{t} = \dfrac{72}{244}$

Solve.

**41.** A recipe for muffins calls for $\frac{1}{2}$ qt of buttermilk, $\frac{1}{3}$ qt of skim milk, and $\frac{1}{16}$ qt of oil. How many quarts of liquid ingredients does the recipe call for? [4.2d]

**42.** The Ferristown School District purchased $\frac{3}{4}$ ton (T) of clay. If the clay is to be shared equally among the district's 6 art departments, how much will each art department receive? [3.8b]

## Synthesis

Solve.

**43.** ▦ What is 8.85% of $12,640?

Estimate _____
Calculate _____

**44.** ▦ 78.8% of what is 9809.024?

Estimate _____
Calculate _____

**45.** 30% of 80 is what percent of 120?

**46.** 40% of what is the same as 30% of 200?

**47.** ▦ What percent of 90 is the same as 26% of 135?

**48.** ▦ What percent of 80 is the same as 76% of 150?

# Mid-Chapter Review

## Concept Reinforcement

Determine whether each statement is true or false.

_____ **1.** When converting decimal notation to percent notation, move the decimal point two places to the right and write a percent symbol. [7.1b]

_____ **2.** The symbol % is equivalent to $\times$ 0.10. [7.1a]

_____ **3.** Of the numbers $\frac{1}{10}$, 1%, 0.1%, and $\frac{1}{100}$, the smallest number is 0.1%. [7.1b, c]

## Guided Solutions

 Fill in each blank with the number that creates a correct statement or solution. [7.1b, c]

**4.** $\dfrac{1}{2}\% = \dfrac{1}{2} \cdot \dfrac{1}{\boxed{\phantom{0}}} = \dfrac{1}{\boxed{\phantom{0}}}$

**5.** $\dfrac{80}{1000} = \dfrac{\boxed{\phantom{0}}}{100} = \boxed{\phantom{0}}\%$

**6.** $5.5\% = \dfrac{\boxed{\phantom{0}}}{100} = \dfrac{\boxed{\phantom{0}}}{1000} = \dfrac{11}{\boxed{\phantom{0}}}$

**7.** $0.375 = \dfrac{\boxed{\phantom{0}}}{1000} = \dfrac{\boxed{\phantom{0}}}{100} = \boxed{\phantom{0}}\%$

**8.** Solve: 15 is what percent of 80? [7.2b]

$15 = p \times \boxed{\phantom{0}}$     Translating

$\dfrac{15}{\boxed{\phantom{0}}} = \dfrac{p \times \boxed{\phantom{0}}}{\boxed{\phantom{0}}}$     Dividing on both sides

$\dfrac{15}{\boxed{\phantom{0}}} = p$     Simplifying

$\boxed{\phantom{0}} = p$     Dividing

$\boxed{\phantom{0}}\% = p$     Converting to percent notation

## Mixed Review

Find decimal notation. [7.1b]

**9.** 28%

**10.** 0.15%

**11.** $5\dfrac{3}{8}\%$

**12.** 240%

Find percent notation. [7.1b, c]

**13.** 0.71

**14.** $\dfrac{9}{100}$

**15.** 0.3891

**16.** $\dfrac{3}{16}$

**17.** 0.005

**18.** $\dfrac{37}{50}$

**19.** 6

**20.** $\dfrac{5}{6}$

Find fraction notation. Simplify.  [7.1c]

**21.** 85%  **22.** 0.048%  **23.** $22\frac{3}{4}$%  **24.** $16.\overline{6}$%

Write percent notation for the shaded area.  [7.1c]

**25.**

**26.**

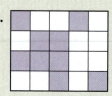

Solve.  [7.2b], [7.3b]

**27.** 25% of what is 14.5?

**28.** 220 is what percent of 1320?

**29.** What is 3.2% of 80,000?

**30.** $17.50 is 35% of what?

**31.** What percent of $800 is $160?

**32.** 130% of $350 is what?

**33.** Arrange the following numbers from smallest to largest.  [7.1b, c]

$$\frac{1}{2}\%, \ 5\%, \ 0.275, \ \frac{13}{100}, \ 1\%, \ 0.1\%, \ 0.05\%, \ \frac{3}{10}, \ \frac{7}{20}, \ 10\%$$

**34.** Solve: 8.5 is $2\frac{1}{2}$% of what?  [7.2b], [7.3b]

   **A.** 3.4         **B.** 21.25
   **C.** 0.2125      **D.** 340

**35.** Solve: $102,000 is what percent of $3.6 million?
   [7.2b], [7.3b]

   **A.** $2.8\overline{3}$ million     **B.** $2.8\overline{3}$%

   **C.** $0.028\overline{3}$%      **D.** $28.\overline{3}$%

# Understanding Through Discussion and Writing

**36.** Is it always best to convert from fraction notation to percent notation by first finding decimal notation? Why or why not?  [7.1c]

**37.** Suppose we know that 40% of 92 is 36.8. What is a quick way to find 4% of 92? 400% of 92? Explain. [7.2b], [7.3b]

**38.** In solving Example 10 in Section 7.3 a student simplifies $\frac{10}{20}$ before solving. Is this a good idea? Why or why not?  [7.3b]

**39.** What do the following have in common? Explain. [7.1b, c]

$$\frac{23}{16}, \ 1\frac{875}{2000}, \ 1.4375, \ \frac{207}{144}, \ 1\frac{7}{16}, \ 143.75\%, \ 1\frac{4375}{10,000}$$

# Applications of Percent

## 7.4
### OBJECTIVE

a  Solve applied problems involving percent.

## a  APPLIED PROBLEMS INVOLVING PERCENT

Applied problems involving percent are not always stated in a manner easily translated to an equation. In such cases, it is helpful to rephrase the problem before translating. Sometimes it also helps to make a drawing.

**EXAMPLE 1**  *Transportation to Work.*  In the United States, there were about 148,000,000 workers in 2015. Approximately 76.6% of those workers drove to work alone. How many workers drove to work alone?

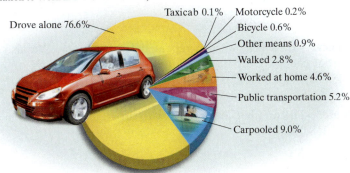

**Transportation to Work in the United States, 2015**

Drove alone 76.6%
Taxicab 0.1%
Motorcycle 0.2%
Bicycle 0.6%
Other means 0.9%
Walked 2.8%
Worked at home 4.6%
Public transportation 5.2%
Carpooled 9.0%

DATA: U.S. Department of Transportation

1. **Familiarize.**  We can simplify the pie chart shown above to help familiarize ourselves with the problem. We let $a$ = the total number of workers who drove to work alone.

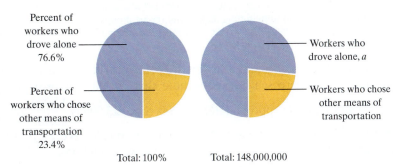

**Transportation to Work in the United States, 2015**

Percent of workers who drove alone 76.6%

Percent of workers who chose other means of transportation 23.4%

Total: 100%

Workers who drove alone, $a$

Workers who chose other means of transportation

Total: 148,000,000

**2. Translate.** There are two ways in which we can translate this problem.

*Percent equation (see Section 7.2):*

$$\underbrace{\text{What number}}_{a} \quad \underset{=}{\text{is}} \quad \underbrace{76.6\%}_{0.766} \quad \underset{\cdot}{\text{of}} \quad \underbrace{148{,}000{,}000?}_{148{,}000{,}000}$$

*Proportion (see Section 7.3):*

$$\frac{76.6}{100} = \frac{a}{148{,}000{,}000}$$

**3. Solve.** We now have two ways in which to solve the problem.

*Percent equation (see Section 7.2):*

$$a = 0.766 \cdot 148{,}000{,}000 = 113{,}368{,}000$$

*Proportion (see Section 7.3):*

$$\frac{76.6}{100} = \frac{a}{148{,}000{,}000}$$

$$76.6 \cdot 148{,}000{,}000 = 100 \cdot a \qquad \textcolor{red}{\text{Equating cross products}}$$

$$\frac{76.6 \cdot 148{,}000{,}000}{100} = \frac{100 \cdot a}{100} \qquad \textcolor{red}{\text{Dividing both sides by 100}}$$

$$\frac{11{,}336{,}800{,}000}{100} = a$$

$$113{,}368{,}000 = a \qquad \textcolor{red}{\text{Simplifying}}$$

**1. *Transportation to Work.*** There were about 148,000,000 workers in the United States in 2015. Approximately 9.0% of them carpooled to work. How many workers carpooled to work?

**Data:** U.S. Census Bureau; American Community Survey

**4. Check.** To check, we can repeat the calculations. We also can do a partial check by estimating. Since 76.6% is about 75%, or $\frac{3}{4}$, and $\frac{3}{4}$ of 148,000,000 is 111,000,000, which is close to 113,368,000, our answer is reasonable.

**5. State.** The number of workers who drove to work alone in 2015 was 113,368,000.

◀ **Do Exercise 1.**

**EXAMPLE 2** *Rescuing Sea Lions.* Marine mammal rescue groups, such as Pacific Marine Mammal Center and the Marine Mammal Care Center, gather valuable knowledge of the ocean as they rescue, rehabilitate, and release marine mammals along the coast of California. From 1975 through October 13, 2015, 20,400 marine mammals were rescued along this coastline, and of this number 12,837 were sea lions. What percent of the rescued marine mammals were sea lions?

**Data:** The Marine Mammal Center; ocregister.com, "Tangled whales and stranded sea lions off California prompt legislation to help fund marine-life rescue groups" by Erika Ritchie; *Earth Island Journal*, September 22, 2017, "Marine Mammals Are Suffering from a Life-threatening Toxin off California's Coast" by Jeremy Miller

**1. Familiarize.** The question asks for the percent of the rescued marine mammals that were sea lions. We note that 20,400 is approximately 20,000 and 12,837 is approximately 13,000. Since 13,000 is $\frac{13{,}000}{20{,}000}$, or $\frac{13}{20}$, or 65% of 20,000, our answer should be close to 65%. We let $p$ = the percent of the rescued marine mammals that were sea lions.

*Answer*

**1.** 13,320,000 workers

**2. Translate.** There are two ways in which we can translate this problem.

*Percent equation:*

$$\underbrace{12{,}837}\ \underbrace{\text{is}}\ \underbrace{\text{what percent}}\ \underbrace{\text{of}}\ \underbrace{20{,}400?}$$
$$12{,}837\ =\ p\ \cdot\ 20{,}400$$

*Proportion:*

$$\frac{N}{100} = \frac{12{,}837}{20{,}400}$$

For proportions, $N\% = p$.

**3. Solve.** We now have two ways in which to solve the problem.

*Percent equation:*

$$12{,}837 = p \cdot 20{,}400$$

$$\frac{12{,}837}{20{,}400} = \frac{p \cdot 20{,}400}{20{,}400} \qquad \text{Dividing both sides by 20,400}$$

$$\frac{12{,}837}{20{,}400} = p$$

$$0.629 \approx p \qquad \text{Finding decimal notation and rounding to the nearest thousandth}$$

$$62.9\% \approx p \qquad \text{Remember to find percent notation.}$$

Note here that the solution, $p$, includes the % symbol.

*Proportion:*

$$\frac{N}{100} = \frac{12{,}837}{20{,}400}$$

$$N \cdot 20{,}400 = 100 \cdot 12{,}837 \qquad \text{Equating cross products}$$

$$\frac{N \cdot 20{,}400}{20{,}400} = \frac{1{,}283{,}700}{20{,}400} \qquad \text{Dividing both sides by 20,400}$$

$$N = \frac{1{,}283{,}700}{20{,}400}$$

$$N \approx 62.9 \qquad \text{Dividing and rounding to the nearest tenth}$$

We use the solution of the proportion to express the answer to the problem as 62.9%. Note that in the proportion method, $N\%$ is equal to $p$ in the percent equation above.

**4. Check.** To check, we note that the answer 62.9% is close to 65%, as estimated in the *Familiarize* step.

**5. State.** About 62.9% of the rescued marine mammals were sea lions.

**Do Exercise 2.** ▶

**2.** *Presidential Assassinations in Office.* Of the 43 different U.S. presidents, 4 have been assassinated while in office. These were James A. Garfield, William McKinley, Abraham Lincoln, and John F. Kennedy. What percent have been assassinated in office?

*Answer*

**2.** About 9.3%

## ✓ Check Your Understanding

**Reading Check** Determine whether each statement is true or false.

_____ **RC1.** Applied problems involving percent can be translated using either equations or proportions.

_____ **RC2.** We can estimate to familiarize ourselves with a problem and to check to see if our answer is reasonable.

_____ **RC3.** Drawings such as pie charts are helpful when familiarizing ourselves with applied problems involving percent.

_____ **RC4.** When solving a percent problem using a proportion, $N$ is already written in percent notation.

**Concept Check** Choose from the list at right an appropriate estimate for each question.

**CC1.** 8.9 is what percent of 91?

**CC2.** 126.7 is what percent of 130?

**CC3.** 12 is what percent of 996?

**CC4.** 19.7 is what percent of 79.6?

**CC5.** 13.8 is what percent of 6.85?

**CC6.** 2.3 is what percent of 1987?

**a)** $\frac{1}{10}\%$

**b)** 1%

**c)** 10%

**d)** 25%

**e)** 100%

**f)** 200%

---

**a** Solve.

**1.** *Organ Transplants.* In 2016, there were 33,611 organ transplants in the United States. Approximately 23.3% were liver transplants and 9.5% were heart transplants. How many liver transplants and how many heart transplants were performed in 2016?

**Data:** organdonor.gov; U.S. Department of Health and Human Services

**2.** *Mississippi River.* The Mississippi River, which extends from its source, at Lake Itasca in Minnesota, to the Gulf of Mexico, is 2348 mi long. Approximately 77% of the river is navigable. How many miles of the river are navigable?

**Data:** National Oceanic and Atmospheric Administration

Mississippi River

**3.** Carrie earns $43,200 one year and then receives an 8% raise in salary. What is Carrie's new salary?

**4.** Chad earns $28,600 one year and then receives a 5% raise in salary. What is Chad's new salary?

**5.** *Test Results.*  On a test, Juan got 85%, or 119, of the items correct. How many items were on the test?

**6.** *Test Results.*  On a test, Maj Ling got 86%, or 81.7, of the items correct. (There was partial credit on some items.) How many items were on the test?

**7.** *Farmland.*  In Kansas, 47,000,000 acres are farmland. About 5% of all the farm acreage in the United States is in Kansas. What is the total number of acres of farmland in the United States?

**Data:** U.S. Department of Agriculture; National Agricultural Statistics Service

**8.** 🖩 *World Population.*  World population is increasing by 1.12% each year. In 2016, it was 7.47 billion. What will the population be in 2020?

**Data:** U.S. Census Bureau; International Data Base

**9.** *Car Depreciation.*  A car generally depreciates 25% of its original value in the first year. Lonnie's car is worth $27,300 after the first year. What was its original cost?

**10.** *Car Depreciation.*  Given normal use, an American-made car will depreciate 25% of its original value the first year and 14% of its remaining value in the second year. What is the value of a car at the end of the second year if its original cost was $36,400? $28,400? $26,800?

**11.** *Test Results.*  On a test of 80 items, Pedro got 93% correct. (There was partial credit on some items.) How many items did he get correct? incorrect?

**12.** *Test Results.*  On a test of 40 items, Christina got 91% correct. (There was partial credit on some items.) How many items did she get correct? incorrect?

**13.** *Housing Expenditure.*  The average American family spends approximately 41% of its income on housing. If a family's yearly income is $83,400, how much is its housing expenditure?

**Data:** U.S. Census Bureau; U.S. Bureau of Labor Statistics

**14.** *Transportation Expenditure.*  The average American family spends approximately 16% of its income on transportation. If a family's yearly income is $69,700, how much is its transportation expenditure?

**Data:** U.S. Census Bureau; U.S. Bureau of Labor Statistics

**15.** *Women in the Workforce in Saudi Arabia.*  In 2016, there were 13,102,100 men and women in the workforce in Saudi Arabia. Of this number, about 1,978,400 were women. What percent of the workforce were women?

**Data:** data.worldbank.org

**16.** *Women in the Workforce in the United States.*  In 2017, there were approximately 161,000,000 men and women in the workforce in the United States. Of this number, about 74,437,700 were women. What percent of the workforce were women?

**Data:** data.worldbank.org

**17.** *Tipping.*  For a party of 8 or more, some restaurants add an 18% tip to the bill. What is the total amount charged for a party of 10 if the cost of the meal, without tip, is $195?

**18.** *Tipping.*  Elaine adds a 15% tip when charging a meal to a credit card. What is the total amount charged to her card if the cost of the meal, without tip, is $18? $34? $49?

**19.** A lab technician has 540 mL of a solution of alcohol and water; 8% of the solution is alcohol. How many milliliters are alcohol? water?

**20.** A lab technician has 680 mL of a solution of water and acid; 3% of the solution is acid. How many milliliters are acid? water?

**21.** *Wasting Food.* As world population increases and the number of acres of farmland decreases, food production and packaging must be improved and less food wasted. In the United States, consumers waste 17 lb of every 90 lb of fruits and vegetables grown. What percent of the total supply of fruits and vegetables is wasted in the home?

**22.** *Credit-Card Debt.* Michael has disposable monthly income of $3400. Each month, he pays $470 toward his credit-card debt. What percent of his disposable income is allotted to paying off credit-card debt?

Monthly Disposable Income: $3400

Credit-card payments: $470

Fruit and Vegetable Supply

Consumed

Lost during picking, sorting, storing, shipping, juice production, canning, and baking

Wasted in the home

DATA: *National Geographic*, March 2016, pp. 38–39, "Waste not, Want not" by Elizabeth Royte

**23.** *U.S. Armed Forces.* There were 1,304,184 people in the United States in active military service in 2016. The numbers for the four branches are listed in the table below. What percent of the total does each branch represent? Round the answers to the nearest tenth of a percent.

**U.S. ARMED FORCES: 2016**

| TOTAL | 1,304,184* |
|---|---|
| AIR FORCE | 315,786 |
| ARMY | 474,472 |
| NAVY | 330,556 |
| MARINES | 183,370 |

*Includes National Guard, Reserve, and retired regular personnel on extended or continuous active duty. Excludes Coast Guard.

DATA: U.S. Department of Defense; U.S. Census Bureau

**24.** *Living Veterans.* There were 18,496,937 living veterans in the United States in 2016. Numbers for various age groups are listed in the table below. What percent of the total does each age group represent? Round the answers to the nearest tenth of a percent.

**LIVING VETERANS BY AGE: 2016**

| TOTAL | 18,496,937 |
|---|---|
| UNDER 35 YEARS OLD | 1,617,066 |
| 35–54 YEARS OLD | 4,354,478 |
| 55–64 YEARS OLD | 3,297,230 |
| 65–74 YEARS OLD | 4,959,939 |
| 75 YEARS OLD AND OLDER | 4,268,224 |

DATA: U.S. Census Bureau; 2016 American Community Survey

## Skill Maintenance

Convert to decimal notation. [5.1b], [5.5a]

**25.** $\dfrac{25}{11}$

**26.** $\dfrac{11}{25}$

**27.** $\dfrac{27}{8}$

**28.** $\dfrac{43}{9}$

**29.** $\dfrac{23}{25}$

## Synthesis

**30.** A couple uses a coupon to get $10 off a restaurant meal. Before subtracting $10, however, the restaurant adds a 20% tip. If the amount of the check is $40.40, how much would the meal alone (without tip) have cost without the coupon?

**31.** If $p$ is 120% of $q$, then $q$ is what percent of $p$?

# Percent Increase or Decrease

## 7.5

### OBJECTIVE

a   Solve applied problems involving percent increase or percent decrease.

## a   APPLIED PROBLEMS INVOLVING PERCENT INCREASE OR PERCENT DECREASE

**SKILL REVIEW**

*Divide using decimal notation.*   [5.4a]

Divide.

**1.** $2.25 \div 3.75$          **2.** $3.6 \div 24$

**Answers: 1.** 0.6 **2.** 0.15

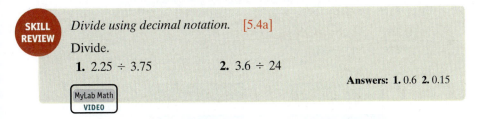

MyLab Math
**VIDEO**

Percent is often used to state increase or decrease. Let's consider an example of each, using the price of a car as the original number.

### Percent Increase

One year a car sold for $20,000. The manufacturer decides to raise the price of the following year's model by 5%. The increase is $0.05 \times \$20,000$, or $1000. The new price is $20,000 + $1000, or $21,000. Note that the new price is 105% of the *former* price.

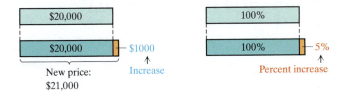

MyLab Math
**ANIMATION**

New price: $21,000

Increase ↑

Former price: $20,000

The increase, $1000, is 5% of the *former* price, $20,000. The *percent increase* is 5%.

### Percent Decrease

Abigail buys a new car for $20,000. After one year, the car depreciates in value by 25%. The decrease is $0.25 \times \$20,000$, or $5000. This lowers the value of the car to $20,000 − $5000, or $15,000. Note that the new value is 75% of the original price. If Abigail decides to sell the car after one year, $15,000 might be the most she could expect to get for it.

Original price: $20,000

Decrease ↓

New value: $15,000

The decrease, $5000, is 25% of the *original* price, $20,000. The *percent decrease* is 25%.

**Do Exercises 1 and 2.** ▶

When a quantity is decreased by a certain percent, we say that there has been a **percent decrease**.

1. *Percent Increase.*   The price of a car is $36,875. The price is increased by 4%.

    a)  How much is the increase?

    b)  What is the new price?

2. *Percent Decrease.*   The value of a car is $36,875. The car depreciates in value by 25% after one year.

    a)  How much is the decrease?

    b)  What is the depreciated value of the car?

*Answers*
**1. (a)** $1475; **(b)** $38,350
**2. (a)** $9218.75; **(b)** $27,656.25

**EXAMPLE 1** *Dow Jones Industrial Average.* The Dow Jones Industrial Average (DJIA) plunged from 11,143 to 10,365 on September 29, 2008. This was the largest one-day point drop in its history. What was the percent decrease?

**Data:** *The Wall Street Journal,* Market Data Center, October 1, 2017

1. **Familiarize.** We first determine the amount of decrease and then make a drawing.

| | |
|---|---|
| 11,143 | Opening average |
| −10,365 | Closing average |
| 778 | Decrease |

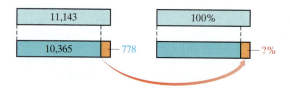

We are asking this question: The decrease is what percent of the opening average? We let $p$ = the percent decrease.

2. **Translate.** There are two ways in which we can translate this problem.

*Percent equation:*

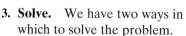

$$778 = p \cdot 11{,}143$$

*Proportion:*

$$\frac{N}{100} = \frac{778}{11{,}143}$$

For proportions, $N\% = p$.

| Percents | Quantities |
|---|---|
| 0% | 0 |
| $N\%$ | 778 |
| 100% | 11,143 |

3. **Solve.** We have two ways in which to solve the problem.

*Percent equation:*

$$778 = p \cdot 11{,}143$$

$$\frac{778}{11{,}143} = \frac{p \cdot 11{,}143}{11{,}143} \qquad \text{Dividing both sides by 11,143}$$

$$\frac{778}{11{,}143} = p$$

$$0.07 \approx p$$

$$7\% \approx p \qquad \text{Converting to percent notation}$$

*Proportion:*

$$\frac{N}{100} = \frac{778}{11{,}143}$$

$$11{,}143 \times N = 100 \times 778 \qquad \text{Equating cross products}$$

$$\frac{11{,}143 \times N}{11{,}143} = \frac{100 \times 778}{11{,}143} \qquad \text{Dividing both sides by 11,143}$$

$$N = \frac{77{,}800}{11{,}143}$$

$$N \approx 7$$

---

**CALCULATOR CORNER**

**Percent Increase or Decrease**
On many calculators, there is a fast way to increase or decrease a number by any given percentage. For example, the result of taking 5% of $20,000 and adding it to $20,000 can be found by pressing

$\boxed{2}\ \boxed{0}\ \boxed{0}\ \boxed{0}\ \boxed{0}\ \boxed{+}$

$\boxed{5}\ \boxed{\text{SHIFT}}\ \boxed{\%}\ \boxed{=}$.

The displayed result will be

$\boxed{21000}$

**EXERCISES**

1. Use a calculator with a $\boxed{\%}$ key to confirm your answers to Margin Exercises 1 and 2.

2. The selling price of Lisa's $87,000 condominium was reduced by 8%. Find the new price.

We use the solution of the proportion to express the answer to the problem as 7%. Note that $N\%$ is equal to $p$ in the percent equation above.

4. **Check.** To check, we note that, with a 7% decrease, the closing Dow average should be 93% of the opening average. Since

$$93\% \times 11{,}143 = 0.93 \times 11{,}143 \approx 10{,}363,$$

and 10,363 is close to 10,365, our answer checks. (Remember that we rounded to get 7%.)

5. **State.** The percent decrease in the DJIA was approximately 7%.

**Do Exercise 3.** ▶

When a quantity is increased by a certain percent, we say that this is a **percent increase**.

**EXAMPLE 2** *Spending on Restaurant Meals.* The average American household spent $2787 on restaurant meals and takeout in 2014. The average amount spent increased to $3008 in 2015. What was the percent increase in the amount spent on restaurant meals and takeout?

**Data:** Fool.com, "Here's What the Average American Spends on Restaurants and Takeout" by Maurie Backman, January 1, 2017

1. **Familiarize.** We first determine the increase in the amount spent and then make a drawing.

$$
\begin{array}{rl}
\$3008 & \text{Amount spent in 2015} \\
-\ 2787 & \text{Amount spent in 2014} \\
\hline
\$221 & \text{Increase}
\end{array}
$$

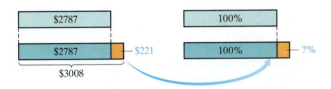

We are asking this question: The increase is what percent of the *original* amount spent? We let $p$ = the percent increase.

2. **Translate.** There are two ways in which we can translate this problem.

*Percent equation:*

$$
\begin{array}{ccccc}
221 & \text{is} & \text{what percent} & \text{of} & 2787? \\
\downarrow & \downarrow & \downarrow & \downarrow & \downarrow \\
221 & = & p & \times & 2787
\end{array}
$$

*Proportion:*

$$\frac{N}{100} = \frac{221}{2787}$$

For proportions, $N\% = p$.

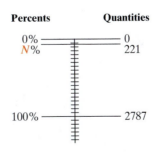

3. *Volume of Mail.* The volume of U.S. first-class single piece mail decreased from about 42.3 billion pieces in 2007 to 19.7 billion pieces in 2016. What was the percent decrease?

**Data:** U.S. Postal Service

*Answer*

**3.** About 53.4%

**3. Solve.** We have two ways in which to solve the problem.

*Percent equation:*

$$221 = p \times 2787$$

$$\frac{221}{2787} = \frac{p \times 2787}{2787} \qquad \text{Dividing by 2787 on both sides}$$

$$\frac{221}{2787} = p$$

$$0.079 \approx p$$

$$7.9\% \approx p \qquad \text{Converting to percent notation}$$

*Proportion:*

$$\frac{N}{100} = \frac{221}{2787}$$

$$2787 \times N = 100 \times 221 \qquad \text{Equating cross products}$$

$$\frac{2787 \times N}{2787} = \frac{100 \times 221}{2787} \qquad \text{Dividing by 2787 on both sides}$$

$$N = \frac{22{,}100}{2787}$$

$$N \approx 7.9$$

We use the solution of the proportion to express the answer to the problem as 7.9%. Note that $N\%$ is equal to $p$ in the percent equation above.

**4. Check.** To check, we take 7.9% of 2787:

$$7.9\% \times 2787 = 0.079 \times 2787 \approx 220.$$

Since 220 is close to 221, our answer checks. (Remember that we rounded to get 7.9%.)

**5. State.** The percent increase in the amount spent on restaurant meals and takeout was about 7.9%.

◀ **Do Exercise 4.**

The percent increase or decrease is *always* based on the original amount. To find a percent increase or decrease, we need to know (1) the amount of increase or decrease and (2) the original amount.

$$\text{Percent increase or decrease} = \frac{\text{Amount of increase or decrease}}{\text{Original amount}}$$

---

**4.** *Spending on Groceries.* The average American household spent $3971 on groceries in 2014. The average amount spent increased to $4015 in 2015. What was the percent increase in the amount spent on groceries?

**Data:** Fool.com, "Here's What the Average American Spends on Restaurants and Takeout" by Maurie Backman, January 1, 2017

---

*Answer*

**4.** About 1.1%

# Translating for Success

1. **Distance Walked.** After a knee replacement, Alex walked $\frac{1}{8}$ mi each morning and $\frac{1}{5}$ mi each afternoon. How much farther did he walk in the afternoon?

2. **Stock Prices.** A stock sold for $5 per share on Monday and only $2.125 per share on Friday. What was the percent decrease from Monday to Friday?

3. **SAT Score.** After attending a class titled "Improving Your SAT Scores," Jacob raised his total score from 884 to 1040. What was the percent increase?

4. **Change in Population.** The population of a small farming community decreased from 1040 to 884. What was the percent decrease?

5. **Lawn Mowing.** During the summer, brothers Steve and Rob earned money for college by mowing lawns. The largest lawn that they mowed was $2\frac{1}{8}$ acres. Steve can mow $\frac{1}{5}$ acre per hour, and Rob can mow only $\frac{1}{8}$ acre per hour. Working together, how many acres did they mow per hour?

The goal of these matching questions is to practice step (2), Translate, of the five-step problem-solving process. Translate each word problem to an equation and select a correct translation from equations A–O.

**A.** $x + \dfrac{1}{5} = \dfrac{1}{8}$

**B.** $250 = x \cdot 1040$

**C.** $884 = x \cdot 1040$

**D.** $\dfrac{250}{16.25} = \dfrac{1000}{x}$

**E.** $156 = x \cdot 1040$

**F.** $16.25 = 250 \cdot x$

**G.** $\dfrac{1}{5} + \dfrac{1}{8} = x$

**H.** $2\dfrac{1}{8} = x \cdot 5$

**I.** $5 = 2.875 \cdot x$

**J.** $\dfrac{1}{8} + x = \dfrac{1}{5}$

**K.** $1040 = x \cdot 884$

**L.** $\dfrac{250}{16.25} = \dfrac{x}{1000}$

**M.** $2.875 = x \cdot 5$

**N.** $x \cdot 884 = 156$

**O.** $x = 16.25 \cdot 250$

*Answers on page A-14*

6. **Land Sale.** Cole sold $2\frac{1}{8}$ acres of the 5 acres he inherited from his uncle. What percent of his land did he sell?

7. **Travel Expenses.** A magazine photographer is reimbursed 16.25¢ per mile for business travel, up to 1000 mi per week. In a recent week, he traveled 250 mi. What was the total reimbursement for travel?

8. **Trip Expenses.** The total expenses for Claire's recent business trip were $1040. She put $884 on her credit card and paid the balance in cash. What percent did she place on her credit card?

9. **Cost of Copies.** During the first summer session at a community college, the campus copy center advertised 250 copies for $16.25. At this rate, what is the cost of 1000 copies?

10. **Cost of Insurance.** Following a rise in the cost of health insurance, 250 of a company's 1040 employees canceled their insurance. What percent of the employees canceled their insurance?

## ✓ Check Your Understanding

**Reading and Concept Check** Complete the table by filling in the missing numbers and words.

| | Original Price | New Price | Change | Increase or Decrease? | Percent Increase or Decrease |
|---|---|---|---|---|---|
| **RC1.** | $50 | $40 | $ _____ | _____ | $\dfrac{\text{Change}}{\text{Original}} = \dfrac{\$\ \ }{\$\ \ } = $ _____ % |
| **RC2.** | $60 | $75 | $ _____ | _____ | $\dfrac{\text{Change}}{\text{Original}} = \dfrac{\$\ \ }{\$\ \ } = $ _____ % |
| **RC3.** | $360 | $480 | $ _____ | _____ | $\dfrac{\text{Change}}{\text{Original}} = \dfrac{\$\ \ }{\$\ \ } = $ _____ % |
| **RC4.** | $4000 | $2400 | $ _____ | _____ | $\dfrac{\text{Change}}{\text{Original}} = \dfrac{\$\ \ }{\$\ \ } = $ _____ % |

**a** Solve.

1. *Mortgage Payment Increase.* A monthly mortgage payment increases from $840 to $882. What is the percent increase?

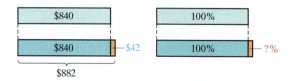

2. *Savings Increase.* The amount in a savings account increased from $200 to $216. What was the percent increase?

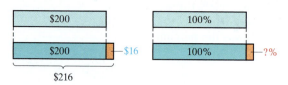

3. A person on a diet goes from a weight of 160 lb to a weight of 136 lb. What is the percent decrease?

4. During a sale, a dress decreased in price from $90 to $72. What was the percent decrease?

5. *Nurse Practitioners.* In 2016, there were 222,000 nurse practitioners in the United States. It is projected that there will be 22,000 more nurse practitioners by 2025. What is the projected percent increase?

**Data:** *AARP Bulletin,* July–August 2017

6. *Overdraft Fees.* Consumers are paying record amounts of fees for overdrawing their bank accounts. In 2016, the largest 628 banks collected $11,410,000,000 in overdraft fees, which is $250,000,000 more than they collected in 2015. What was the percent increase?

**Data:** money.cnn.com; Consumer Financial Protection Bureau

7. *Tax-Refund Fraud.* There were 376,000 reports of tax-refund fraud reported to the Internal Revenue Service in 2016. This number is 390,000 less than the number of reports of tax-refund fraud in 2014. What was the percent decrease?

**Data:** *The Wall Street Journal,* 3/4–5/2017, "Tax-ID Theft Drops" by Laura Saunders

8. *Arctic Ice.* The Arctic sea surface freezes in the winter, but the part of it that melts in the summer is increasing. In 1979, 2,780,000 square miles of the Arctic sea were still ice-covered in September. Only 1,820,000 square miles were still ice-covered in September 2016. What was the percent decrease?

**Data:** National Snow and Ice Data Center; *National Geographic,* April 2017, "Climate Change"

9. **Insulation.** A roll of unfaced fiberglass insulation has a retail price of $28.79. For two weeks, it is on sale for $20.49. What is the percent decrease?

10. **Set of Weights.** A 300-lb weight set retails for $199.95. For its grand opening, a sporting goods store reduced the price to $154.95. What is the percent decrease?

11. **Shopping Malls in the United States.** With the increase in online shopping, the percent growth in the number of shopping malls in the United States has been decreasing. The table below lists the number of malls in selected years. What was the percent increase in the number of malls from 1977 to 1987? from 2007 to 2017?

| YEAR | NUMBER OF MALLS IN THE UNITED STATES |
|------|--------------------------------------|
| 1977 | 576 |
| 1987 | 874 |
| 1997 | 1043 |
| 2007 | 1165 |
| 2017 | 1211 |

DATA: CoStar Group

12. **Milestones in the DJIA.** The table below lists the dates when the Dow Jones Industrial Average reached specified 1000-point milestones and the closing DJIA values on those dates. What was the percent increase in the DJIA from the first date it closed above 10,000 to the first date it closed above 12,000? from the first date it closed above 20,000 to the first date it closed above 22,000?

| FIRST CLOSE ABOVE | DATE | CLOSING DJIA |
|-------------------|------|--------------|
| 10,000 | 3/29/1999 | 10,000.78 |
| 12,000 | 10/19/2006 | 12,011.73 |
| 14,000 | 7/19/2007 | 14,000.41 |
| 16,000 | 11/21/2013 | 16,009.99 |
| 18,000 | 12/23/2014 | 18,024.17 |
| 20,000 | 1/25/2017 | 20,068.51 |
| 22,000 | 8/2/2017 | 22,016.24 |

DATA: us.spindices.com/indexology; macrotrends.net; marketwatch.com

13. **Credit Card Debt.** The average credit card debt per household in the United States was $11,248 in 2008, $8158 in 2016, and $8683 in 2018. What was the percent decrease from 2008 to 2016? What was the percent increase from 2016 to 2018? (Note: Credit card debt does not include credit card balances that are paid off each month.)

Data: magnifymoney.com; nerdwallet.com

14. **Centenarians.** In 1980, there were 32,194 centenarians in the United States. The number of centenarians increased to 53,364 by 2010. It is projected that there will be 604,000 centenarians by 2060. What is the percent increase from 1980 to 2010? from 2010 to 2060?

Data: *The Wall Street Journal*, 7/21/2017, "The Life Insurance Isn't So Permanent" by Leslie Scism; U.S. Census Bureau Projections

15. **Two-by-Four.** A cross-section of a standard, or nominal, "two-by-four" actually measures $1\frac{1}{2}$ in. by $3\frac{1}{2}$ in. The rough board is 2 in. by 4 in. but is planed and dried to the finished size. What percent of the wood is removed in planing and drying?

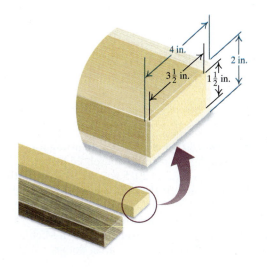

16. **Strike Zone.** In baseball, the *strike zone* is normally a 17-in. by 30-in. rectangle. Some batters give the pitcher an advantage by swinging at pitches thrown out of the strike zone. By what percent is the area of the strike zone increased if a 2-in. border is added to the outside?

Data: Major League Baseball

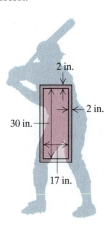

**Building Permits.** The table below shows data for the number of U.S. building permits issued for privately owned single housing units in May for the years 2005–2016. For Exercises 17–20, find the change in the number of building permits issued and the percent increase or percent decrease for the given years.

| YEAR | BUILDING PERMITS FOR SINGLE HOUSING UNITS ISSUED IN MAY | YEAR | BUILDING PERMITS FOR SINGLE HOUSING UNITS ISSUED IN MAY |
|---|---|---|---|
| 2005 | 153,673 | 2011 | 39,215 |
| 2006 | 145,063 | 2012 | 49,621 |
| 2007 | 105,295 | 2013 | 62,413 |
| 2008 | 61,147 | 2014 | 59,144 |
| 2009 | 38,841 | 2015 | 62,282 |
| 2010 | 40,099 | 2016 | 70,156 |

DATA: U.S. Census Bureau

**17.** From 2007 to 2008

**18.** From 2012 to 2013

**19.** From 2015 to 2016

**20.** From 2008 to 2009

**21.** *Increase in Population.* The population of Utah in 2010 was 2,763,885. It is projected that in 2020 the population of Utah will be 3,253,024. What will the percent increase be?

**Data:** worldpopulationreview.com; U.S. Census Bureau

**22.** *Decrease in Population.* The population of West Virginia in 2010 was 1,852,994. It is projected that the population of West Virginia will be 1,821,013 in 2020. What will the percent decrease be?

**Data:** worldpopulationreview.com; U.S. Census Bureau

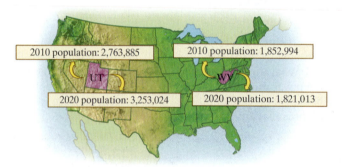

2010 population: 2,763,885

2010 population: 1,852,994

2020 population: 3,253,024

2020 population: 1,821,013

## Skill Maintenance

Find the perimeter of each polygon. [2.7b]

**23.**

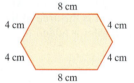

8 cm
4 cm   4 cm
4 cm   4 cm
8 cm

**24.**

10 ft
3 ft
5 ft
3 ft
12 ft

**25.**

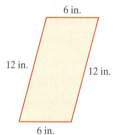

6 in.
12 in.   12 in.
6 in.

**26.**

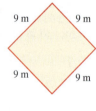

9 m   9 m
9 m   9 m

## Synthesis

**27.** 🖩 A worker receives raises of 3%, 6%, and then 9%. By what percent has the original salary increased?

**28.** *Median Household Income.* After adjusting for inflation, the change in median household income from its peak to its level in 2015 was −17.4% for both Michigan and Nevada. If the 2015 median household income was $54,203 for Michigan and $52,008 for Nevada, how much more, in dollars, did median household income decrease from its peak in Michigan than in Nevada? Round to the nearest dollar.

**Data:** advisorperspectives.com

# Sales Tax, Commission, and Discount

## a  SALES TAX

Sales tax computations represent a special type of percent increase problem. The sales tax rate in Colorado is 2.9%. This means that the tax is 2.9% of the purchase price. Suppose the purchase price of a canoe is $749.95. The sales tax is then 2.9% of $749.95, or $0.029 \times \$749.95$, or $21.74855, which is about $21.75.

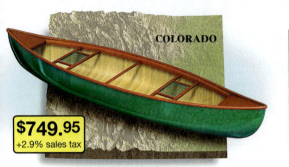

**$749.95**
+2.9% sales tax

COLORADO

**BILL:**

| | |
|---|---|
| Purchase price | = $749.95 |
| Sales tax (2.9% of $749.95) | = + 21.75 |
| Total price | $771.70 |

The total that you would pay for the canoe is the purchase price plus the sales tax:

$749.95 + $21.75, or $771.70.

---

### SALES TAX

**Sales tax** = Sales tax rate × Purchase price

**Total price** = Purchase price + Sales tax

---

**EXAMPLE 1**  *Maine Sales Tax.*   The sales tax rate in Maine is 5.5%. How much tax is charged on the purchase of 3 shrubs at $42.99 each? What is the total price?

**a)** We first find the purchase price of the 3 shrubs. It is

$3 \times \$42.99 = \$128.97$.

**b)** The sales tax on items costing $128.97 is

$$\underbrace{\text{Sales tax rate}}_{5.5\%} \quad \times \quad \underbrace{\text{Purchase price}}_{128.97,}$$

or $0.055 \times 128.97$, or $7.09335$. Thus, the tax is $7.09 (rounded to the nearest cent).

**c)** The total price is given by the purchase price plus the sales tax:

$128.97 + $7.09, or $136.06.

Maine

**$42.99**
each

plus 5.5%
sales tax

To check, note that the total price is the purchase price plus 5.5% of the purchase price. Thus, the total price is 105.5% of the purchase price. Since $1.055 \times 128.97 \approx 136.06$, we have a check. The sales tax is $7.09, and the total price is $136.06.

**1.** *Texas Sales Tax.* The sales tax rate in Texas is 6.25%. In Texas, how much tax is charged on the purchase of an ultrasound toothbrush that sells for $139.95? What is the total price?

**2.** **Wyoming Sales Tax.** In her hometown, Laramie, Wyoming, Samantha buys 4 copies of *Hillbilly Elegy* by J. D. Vance for $18.95 each. The sales tax rate in Wyoming is 4%. How much sales tax will Samantha be charged? What is the total price?

GS

Purchase price = 4 × $⬚
             = $⬚
Sales tax = ⬚% × 4 × $⬚
       = 0.04 × $⬚
       = $3.032
       ≈ $⬚
Total price = $75.80 + $⬚
       = $⬚

**3.** The sales tax on the purchase of a set of holiday dishes that costs $449 is $26.94. What is the sales tax rate?

**4.** The sales tax on the purchase of a pair of designer jeans is $4.84, and the sales tax rate is 5.5%. Find the purchase price (the price before the tax is added).

*Answers*
**1.** $8.75; $148.70  **2.** $3.03, $78.83
**3.** 6%  **4.** $88
*Guided Solution:*
**2.** 18.95, 75.80; 4, 18.95, 75.80, 3.03; 3.03, 78.83

◀ **Do Exercises 1 and 2.**

**EXAMPLE 2** The sales tax on the purchase of an eReader that costs $199 is $13.93. What is the sales tax rate?

**$199.00**
plus $13.93
sales tax

We rephrase and translate as follows:

*Rephrase:* Sales tax is what percent of purchase price?
*Translate:* $13.93 = r · 199.

To solve the equation, we divide both sides by 199:

$$\frac{13.93}{199} = \frac{r \cdot 199}{199}$$

$$\frac{13.93}{199} = r$$

$$0.07 = r$$

$$7\% = r.$$

The sales tax rate is 7%.

◀ **Do Exercise 3.**

**EXAMPLE 3** The sales tax on the purchase of a stone-top firepit is $12.74, and the sales tax rate is 8%. Find the purchase price (the price before the tax is added).

We rephrase and translate as follows:

*Rephrase:* Sales tax is 8% of what?
*Translate:* 12.74 = 0.08 · b.

To solve, we divide both sides by 0.08:

$$\frac{12.74}{0.08} = \frac{0.08 \cdot b}{0.08}$$

$$\frac{12.74}{0.08} = b$$

$$159.25 = b.$$

**Price: ?**
$12.74 tax @ 8%

The purchase price is $159.25.

◀ **Do Exercise 4.**

## b COMMISSION

When you work for a **salary**, you receive the same amount of money each week or month. When you work for a **commission**, you are paid a percent of the total sales for which you are responsible.

> **COMMISSION**
>
> **Commission** = Commission rate × Sales

**EXAMPLE 4** *Membership Sales.* A salesperson's commission rate on fitness club memberships is 3%. What is the commission on the sale of $8300 worth of fitness club memberships?

$$Commission = Commission\ rate \times Sales$$
$$C = 3\% \times 8300$$
$$C = 0.03 \times 8300$$
$$C = 249$$

The commission is $249.

**Do Exercise 5.** ▶

**5.** Isabella's commission rate is 15%. What commission does she earn on the sale of $9260 worth of exercise equipment?

**EXAMPLE 5** *Earth-Moving Equipment Sales.* Gavin earns a commission of $20,800 for selling $320,000 worth of earth-moving equipment. What is the commission rate?

$$Commission = Commission\ rate \times Sales$$
$$20{,}800 = r \times 320{,}000$$

*Answers*

**5.** $1389

To solve this equation, we divide by 320,000 on both sides:

$$\frac{20,800}{320,000} = \frac{r \times 320,000}{320,000}$$

$$0.065 = r$$

$$6.5\% = r.$$

The commission rate is 6.5%.

◀ **Do Exercise 6.**

**6.** Grayson earns a commission of $2040 for selling $17,000 worth of concert tickets. What is the commission rate?

**EXAMPLE 6** *Cruise Vacations.* Taylor's commission rate is 5.6%. She received a commission of $2457 on cruise vacation packages that she sold in November. How many dollars worth of cruise vacations did she sell?

$$\text{Commission} = \text{Commission rate} \times \text{Sales}$$
$$2457 = 5.6\% \times S, \text{ or}$$
$$2457 = 0.056 \times S$$

To solve this equation, we divide both sides by 0.056:

$$\frac{2457}{0.056} = \frac{0.056 \times S}{0.056}$$

$$\frac{2457}{0.056} = S$$

$$43,875 = S.$$

Taylor sold $43,875 worth of cruise vacation packages.

◀ **Do Exercise 7.**

**7.** Mila's commission rate is 7.5%. **GS** She receives a commission of $2970 from the sale of winter ski passes. How many dollars worth of ski passes did she sell?

$$\$2970 = \boxed{\phantom{xx}} \% \times S$$
$$\$2970 = 0.075 \times S$$
$$\frac{\$2970}{\boxed{\phantom{xx}}} = \frac{0.075 \times S}{0.075}$$
$$\$\boxed{\phantom{xx}} = S$$

**c  DISCOUNT**

Suppose that the regular price of a rug is $60, and the rug is on sale at 25% off. Since 25% of $60 is $15, the sale price is $60 − $15, or $45. We call $60 the **original**, or **marked**, **price**, 25% the **rate of discount**, $15 the **discount**, and $45 the **sale price**. Note that discount problems are a type of percent decrease problem.

---

**DISCOUNT AND SALE PRICE**

**Discount** = Rate of discount × Original price
**Sale price** = Original price − Discount

---

**EXAMPLE 7**  A leather sofa marked $2379 is on sale at $33\frac{1}{3}\%$ off. What is the discount? the sale price?

Leather sofa
$2379 original price
Save $33\frac{1}{3}\%$

**a)** *Discount* = *Rate of discount* × *Original price*

$$D \quad = \quad 33\frac{1}{3}\% \quad \times \quad 2379$$

$$D \quad = \quad \frac{1}{3} \quad \times \quad 2379$$

$$D = \frac{2379}{3} = 793$$

**b)** *Sale price* = *Original price* − *Discount*

$$S \quad = \quad 2379 \quad - \quad 793$$

$$S = 1586$$

The discount is $793, and the sale price is $1586.

**Do Exercise 8.** ▶

**EXAMPLE 8**  The price of a wooden playset is marked down from $1650 to $1353. What is the rate of discount?

We first find the discount by subtracting the sale price from the original price:

$$1650 - 1353 = 297.$$

The discount is $297.

Next, we use the equation for discount:

*Discount* = *Rate of discount* × *Original price*

$$297 \quad = \quad r \quad \times \quad 1650$$

To solve, we divide both sides by 1650:

$$\frac{297}{1650} = \frac{r \times 1650}{1650}$$

$$\frac{297}{1650} = r$$

$$0.18 = r$$

$$18\% = r.$$

The rate of discount is 18%.

To check, note that an 18% rate of discount means that the buyer pays 82% of the original price:

$$0.82 \times \$1650 = \$1353.$$

**Do Exercise 9.** ▶

8. A computer marked $660 is on sale at $16\frac{2}{3}\%$ off. What is the discount? the sale price?

9. The price of a winter coat is reduced from $75 to $60. Find the rate of discount.

*Answers*
**8.** $110; $550   **9.** 20%

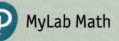
✓ **Check Your Understanding**

**Reading and Concept Check**   Complete each definition with the word *price*, *rate*, or *tax*.

**RC1.** Commission = Commission _____ × Sales

**RC2.** Discount = _____ of discount × Original price

**RC3.** Sale price = Original _____ − Discount

**RC4.** Sales tax = Sales _____ rate × Purchase price

**RC5.** Total price = Purchase price + Sales _____

**a**   Solve.

**1.** *Wyoming Sales Tax.*   The sales tax rate in Wyoming is 4%. How much sales tax would be charged on a fireplace screen with doors that costs $239?

**2.** *Kansas Sales Tax.*   The sales tax rate in Kansas is 6.5%. How much sales tax would be charged on a fireplace screen with doors that costs $239?

**3.** *New Mexico Sales Tax.*   The sales tax rate in New Mexico is 5.125%. How much sales tax is charged on a camp stove that sells for $129.95?

**4.** *Ohio Sales Tax.*   The sales tax rate in Ohio is 5.75%. How much sales tax is charged on a pet carrier that sells for $39.99?

**5.** *California Sales Tax.*   The sales tax rate in California is 7.25%. How much sales tax is charged on a purchase of 4 contour foam travel pillows at $39.95 each? What is the total price?

**6.** *Illinois Sales Tax.*   The sales tax rate in Illinois is 6.25%. How much sales tax is charged on a purchase of 3 wet-dry vacs at $60.99 each? What is the total price?

**7.** The sales tax is $30 on the purchase of a diamond ring that sells for $750. What is the sales tax rate?

**8.** The sales tax is $48 on the purchase of a dining room set that sells for $960. What is the sales tax rate?

**9.** The sales tax is $13.68 on the purchase of a patio set that sells for $456. What is the sales tax rate?

**10.** The sales tax is $35.80 on the purchase of a refrigerator-freezer that sells for $895. What is the sales tax rate?

**11.** The sales tax on the purchase of a new fishing boat is $904, and the sales tax rate is 8%. What is the purchase price (the price before tax is added)?

**12.** The sales tax on the purchase of a used car is $434, and the sales tax rate is 7%. What is the purchase price?

**13.** The sales and use tax rate in New York City is 4.875% for the city plus 4% for the state. Find the total amount paid for 6 boxes of chocolates at $19.95 each.

**14.** The sales tax rate in Nashville, Tennessee, is 2.25% for Davidson County plus 7% for the state. Find the total amount paid for 2 ladders at $39 each.

**15.** The sales tax rate in Pittsburgh, Pennsylvania, is 1% for Allegheny County plus 6% for the state. Find the total amount paid for 5 NFL tickets at $266 each.

**16.** The sales tax rate in Miami, Florida, is 1% for Dade County plus 6% for the state. Find the total amount paid for 2 tires at $49.95 each.

**17.** The sales tax rate in Champaign, Illinois, is 1.25% for the county, 1.5% for the city, and 6.25% for the state. Find the total amount paid for 3 dolls at $138 each.

**18.** The sales tax rate in Atlanta, Georgia, is 1.5% for the city, 3% for Fulton County, and 4% for the state. Find the total amount paid for 6 bicycle helmets at $39.99 each.

**b** Solve.

**19.** Benjamin's commission rate is 21%. What commission does he earn on the sale of $12,500 worth of windows?

**20.** Olivia's commission rate is 6%. What commission does she earn on the sale of $45,000 worth of lawn irrigation systems?

**21.** Alyssa earns $408 for selling $3400 worth of shoes. What is the commission rate?

**22.** Joshua earns $120 for selling $2400 worth of television sets. What is the commission rate?

**23. *Real Estate Commission.*** A real estate agent's commission rate is 7%. She receives a commission of $12,950 from the sale of a home. How much did the home sell for?

**24. *Clothing Consignment Commission.*** A clothing consignment shop's commission rate is 40%. The shop receives a commission of $552. How many dollars worth of clothing were sold?

**25.** A real estate commission rate is 8%. What is the commission from the sale of a piece of land for $68,000?

**26.** A real estate commission rate is 6%. What is the commission from the sale of a $98,000 home?

**27.** David earns $1147.50 for selling $7650 worth of car parts. What is the commission rate?

**28.** Jayla earns $280.80 for selling $2340 worth of tee shirts. What is the commission rate?

**29.** Laila's commission is increased according to how much she sells. She receives a commission of 4% for the first $1000 of sales and 7% for the amount over $1000. What is her total commission on sales of $5500?

**30.** Malik's commission is increased according to how much he sells. He receives a commission of 5% for the first $2000 of sales and 8% for the amount over $2000. What is his total commission on sales of $6200?

**c** Complete the table below by filling in the missing numbers.

|  | Marked Price | Rate of Discount | Discount | Sale Price |
|---|---|---|---|---|
| **31.** | $300 | 10% |  |  |
| **32.** | $2000 | 40% |  |  |
| **33.** | $17 | 15% |  |  |
| **34.** | $20 | 25% |  |  |
| **35.** |  | 10% | $12.50 |  |
| **36.** |  | 15% | $65.70 |  |
| **37.** | $600 |  | $240 |  |
| **38.** | $12,800 |  | $1920 |  |

**39.** Find the marked price and the rate of discount for the surfboard in this ad.

Save
**$120.**00
Surfboard Now
**$180.**00

**40.** Find the marked price and the rate of discount for the steel log rack in this ad.

Closeout
**$35**

Save
**$15**

**41.** Find the discount and the rate of discount for the pinball machine in this ad.

Best price of the year!
Now only
**$3150**
Was $3999

**42.** Find the discount and the rate of discount for the amaryllis in this ad.

Sale
3-in-1 Amaryllis
**$37.95**
Was $42.95

## Skill Maintenance

Solve.  [6.3b]

**43.** $\dfrac{x}{12} = \dfrac{24}{16}$

**44.** $\dfrac{7}{2} = \dfrac{11}{x}$

Convert to standard notation.  [5.3b]

**47.** 4.03 trillion

**48.** 5.8 million

Solve.  [5.7a]

**45.** $0.64 \cdot x = 170$

**46.** $29.44 = 25.6 \times y$

**49.** 42.7 million

**50.** 6.09 billion

## Synthesis

**51.** ▦ Sara receives a 10% commission on the first $5000 in sales and 15% on all sales beyond $5000. If Sara receives a commission of $2405, how much did she sell? Use a calculator and trial and error if you wish.

**52.** Elijah collects baseball memorabilia. He bought two autographed plaques, but then became short of funds and had to sell them quickly for $200 each. On one, he made a 20% profit, and on the other, he lost 20%. Did he make or lose money on the sale?

### OBJECTIVES

**a** Solve applied problems involving simple interest.

**b** Solve applied problems involving compound interest.

**c** Solve applied problems involving interest rates on credit cards.

# Simple Interest and Compound Interest; Credit Cards

## a SIMPLE INTEREST

 **SKILL REVIEW**

*Convert between percent notation and decimal notation.* [7.1b]
Find decimal notation.

**1.** $34\frac{5}{8}\%$        **2.** $5\frac{1}{4}\%$

Answers: **1.** 0.34625   **2.** 0.0525

MyLab Math
**VIDEO**

Suppose you put $1000 into an investment for 1 year. The $1000 is called the **principal**. If the **interest rate** is 5%, in addition to the principal, you get back 5% of the principal, which is

     5% of 1000,   or   $0.05 \cdot \$1000$,   or   $50.00.

The $50.00 is called **simple interest**. It is, in effect, the price that a financial institution pays for the use of the money over time.

> **SIMPLE INTEREST FORMULA**
>
> The **simple interest** $I$ on principal $P$, invested for $t$ years at interest rate $r$, is given by
>
> $$I = P \cdot r \cdot t.$$

---

**1.** What is the simple interest on $4300 invested at an interest rate of 4% for 1 year?

$I = P \cdot r \cdot t$

$\phantom{I}= \$4300 \times \boxed{\phantom{00}} \% \times 1$

$\phantom{I}= \$\boxed{\phantom{000}} \times 0.04 \times 1$

$\phantom{I}= \$\boxed{\phantom{000}}$

**EXAMPLE 1**  What is the simple interest on $2500 invested at an interest rate of 6% for 1 year?

We use the formula $I = P \cdot r \cdot t$:

$$I = P \cdot r \cdot t = \$2500 \cdot 6\% \cdot 1$$
$$= \$2500 \cdot 0.06 = \$150.$$

The simple interest for 1 year is $150.

◀ **Do Exercise 1.**

**EXAMPLE 2** What is the simple interest on a principal of $2500 invested at an interest rate of 6% for 3 months?

We use the formula $I = P \cdot r \cdot t$ and express 3 months as a fraction of a year:

$$I = P \cdot r \cdot t = \$2500 \cdot 6\% \cdot \frac{3}{12} = \$2500 \cdot 6\% \cdot \frac{1}{4}$$
$$= \frac{\$2500 \cdot 0.06}{4} = \$37.50.$$

The simple interest for 3 months is $37.50.

**2.** What is the simple interest on a principal of $4300 invested at an interest rate of 4% for 9 months?

◀ **Do Exercise 2.**

**Answers**

**1.** $172   **2.** $129

**Guided Solution:**

**1.** 4, 4300, 172

When time is given in days, we generally divide it by 365 to express the time as a fractional part of a year.

**EXAMPLE 3**  To pay for a shipment of lawn furniture, Patio by Design borrows $8000 at $9\frac{3}{4}\%$ for 60 days. Find **(a)** the amount of simple interest that is due and **(b)** the total amount that must be paid after 60 days.

**a)** We express 60 days as a fractional part of a year:

$$I = P \cdot r \cdot t = \$8000 \cdot 9\frac{3}{4}\% \cdot \frac{60}{365}$$

$$= \$8000 \cdot 0.0975 \cdot \frac{60}{365} \approx \$128.22.$$

The interest due for 60 days is $128.22.

**b)** The total amount to be paid after 60 days is the principal plus the interest:

$$\$8000 + \$128.22 = \$8128.22.$$

The total amount due is $8128.22.

**Do Exercise 3.** ▶

GS **3.** The Glass Nook borrows $4800 at $5\frac{1}{2}\%$ for 30 days. Find **(a)** the amount of simple interest due and **(b)** the total amount that must be paid after 30 days.

**a)**  $I = P \cdot r \cdot t$

$= \$4800 \cdot 5\frac{1}{2}\% \cdot \dfrac{\boxed{\phantom{00}}}{365}$

$= \$4800 \cdot 0.055 \cdot \dfrac{30}{365}$

$\approx \$\boxed{\phantom{00}}$

**b)**  Total amount

$= \$4800 + \boxed{\phantom{00}}$

$= \boxed{\phantom{00}}$

## b   COMPOUND INTEREST

When interest is paid *on interest*, we call it **compound interest**. This is the type of interest usually paid on investments. Suppose you have $5000 in a savings account at 6%. In 1 year, the account will contain the original $5000 plus 6% of $5000. Thus, the total in the account after 1 year will be

106% of $5000,   or   $1.06 \cdot \$5000$,   or   $5300.

Now suppose that the total of $5300 remains in the account for another year. At the end of this second year, the account will contain the $5300 plus 6% of $5300. The total in the account will thus be

106% of $5300,·   or   $1.06 \cdot \$5300$,   or   $5618.

Note that in the second year, interest is also earned on the first year's interest. When this happens, we say that interest is **compounded annually**.

**EXAMPLE 4**  Find the amount in an account if $3000 is invested at 4%, compounded annually, for 2 years.

**a)** After 1 year, the account will contain 104% of $3000:

$1.04 \cdot \$3000 = \$3120.$

**b)** At the end of the second year, the account will contain 104% of $3120:

$1.04 \cdot \$3120 = \$3244.80.$

The amount in the account after 2 years is $3244.80.

**Do Exercise 4.** ▶

**4.** Find the amount in an account if $3000 is invested at 2%, compounded annually, for 2 years.

*Answers*
**3. (a)** $21.70; **(b)** $4821.70   **4.** $3121.20
*Guided Solution:*
**3. (a)** 30, 21.70; **(b)** $21.70, $4821.70

Suppose that the interest in Example 4 were **compounded semiannually**—that is, every half year. Interest would then be calculated twice a year at a rate of 4% ÷ 2, or 2% each time. The approach used in Example 4 can then be adapted, as follows.

After the first $\frac{1}{2}$ year, the account will contain 102% of $3000:

$$1.02 \cdot \$3000 = \$3060.$$

After a second $\frac{1}{2}$ year (1 full year), the account will contain 102% of $3060:

$$1.02 \cdot \$3060 = \$3121.20.$$

After a third $\frac{1}{2}$ year ($1\frac{1}{2}$ full years), the account will contain 102% of $3121.20:

$$1.02 \cdot \$3121.20 = \$3183.624$$
$$\approx \$3183.62 \qquad \text{Rounding to the nearest cent}$$

Finally, after a fourth $\frac{1}{2}$ year (2 full years), the account will contain 102% of $3183.62:

$$1.02 \cdot \$3183.62 = \$3247.2924$$
$$\approx \$3247.29 \qquad \text{Rounding to the nearest cent}$$

Let's summarize our results and look at them another way:

End of 1st $\frac{1}{2}$ year → $1.02 \times 3000 = 3000 \times (1.02)^1$;

End of 2nd $\frac{1}{2}$ year → $1.02 \times (1.02 \times 3000) = 3000 \times (1.02)^2$;

End of 3rd $\frac{1}{2}$ year → $1.02 \times (1.02 \times 1.02 \times 3000) = 3000 \times (1.02)^3$;

End of 4th $\frac{1}{2}$ year → $1.02 \times (1.02 \times 1.02 \times 1.02 \times 3000) = 3000 \times (1.02)^4$.

Note that each multiplication was by 1.02 and that

$$\$3000 \cdot 1.02^4 \approx \$3247.30. \qquad \text{Using a calculator and rounding to the nearest cent}$$

Notice that when we round only once, the amount in the account, $3247.30, is one cent more than the amount calculated above, $3247.29.

We have illustrated the following result.

---

### COMPOUND INTEREST FORMULA

If a principal $P$ has been invested at interest rate $r$, compounded $n$ times a year, in $t$ years it will grow to an amount $A$ given by

$$A = P \cdot \left(1 + \frac{r}{n}\right)^{n \cdot t}.$$

---

Let's apply this formula to confirm our preceding discussion, where the amount invested is $P = \$3000$, the interest rate is 4%, the number of years is $t = 2$, and the number of compounding periods each year is $n = 2$. Substituting into the compound interest formula, we have

$$A = P \cdot \left(1 + \frac{r}{n}\right)^{n \cdot t} = 3000 \cdot \left(1 + \frac{4\%}{2}\right)^{2 \cdot 2}$$

$$= \$3000 \cdot \left(1 + \frac{0.04}{2}\right)^4 = \$3000(1.02)^4$$

$$= \$3000 \times 1.08243216 \approx \$3247.30.$$

*Compound Interest* A calculator is useful in computing compound interest. Not only does it perform computations quickly but it also eliminates the need to round until the computation is completed. This minimizes round-off errors that occur when rounding is done at each stage of the computation. We must keep order of operations in mind when computing compound interest.

To find the amount due on a $20,000 loan made for 25 days at 11% interest, compounded daily, we compute $20{,}000\left(1 + \dfrac{0.11}{365}\right)^{25}$.

To do this on a calculator, we press
[2][0][0][0][0][×][(][(][1][+][.]
[1][1][÷][3][6][5][)][$y^x$] (or [^])
[2][5][=]. The result is $20,151.23, rounded to the nearest cent.

Some calculators have business keys that allow such computations to be done more quickly.

#### EXERCISES

1. Find the amount due on a $16,000 loan made for 62 days at 13% interest, compounded daily.

2. An investment of $12,500 is made for 90 days at 8.5% interest, compounded daily. How much is the investment worth after 90 days?

**EXAMPLE 5** The Ibsens invest $4000 in an account paying $3\frac{5}{8}\%$, compounded quarterly. Find the amount in the account after $2\frac{1}{2}$ years.

The compounding is quarterly, so $n$ is 4. We substitute $4000 for $P$, $3\frac{5}{8}\%$, or 0.03625, for $r$, 4 for $n$, and $2\frac{1}{2}$, or $\frac{5}{2}$, for $t$ and compute $A$:

$$A = P \cdot \left(1 + \frac{r}{n}\right)^{n \cdot t} = \$4000 \cdot \left(1 + \frac{3\frac{5}{8}\%}{4}\right)^{4 \cdot 5/2}$$

$$= \$4000 \cdot \left(1 + \frac{0.03625}{4}\right)^{10}$$

$$= \$4000(1.0090625)^{10}$$

$$\approx \$4377.65.$$

The amount in the account after $2\frac{1}{2}$ years is $4377.65.

**Do Exercise 5.** ▶

**5.** A couple invests $7000 in an account paying $6\frac{3}{8}\%$, compounded semiannually. Find the amount in the account after $1\frac{1}{2}$ years.

## c  CREDIT CARDS

According to nerdwallet.com, the average credit-card debt among U.S. households with such debt was $16,883 per household in 2016.

The money you obtain through the use of a credit card is not "free" money. There is a price (interest) to be paid for the convenience of using a credit card. A balance carried on a credit card is a type of loan. Comparing interest rates is essential if one is to become financially responsible. A small change in an interest rate can make a large difference in the cost of a loan. When you make a payment on a credit-card balance, do you know how much of that payment is interest and how much is applied to reducing the principal?

**EXAMPLE 6** *Credit Cards.* After the holidays, Evan has a balance of $3216.28 on a credit card with an annual percentage rate (APR) of 19.7%. He decides not to make additional purchases with this card until he has paid off the balance.

**a)** Many credit cards require a minimum monthly payment of 2% of the balance. At this rate, what is Evan's minimum payment on a balance of $3216.28? Round the answer to the nearest dollar.

**b)** Find the amount of interest and the amount applied to reduce the principal in the minimum payment found in part (a).

**c)** If Evan had transferred his balance to a card with an APR of 12.5%, how much of his first payment would be interest and how much would be applied to reduce the principal?

**d)** Compare the amounts for 12.5% from part (c) with the amounts for 19.7% from part (b).

We solve as follows.

**a)** We multiply the balance of $3216.28 by 2%:

$$0.02 \cdot \$3216.28 = \$64.3256.$$

Evan's minimum payment, rounded to the nearest dollar, is $64.

*Answer*

5. $7690.94

**6. Credit Card.** After the holidays, Samantha has a balance of $4867.59 on a credit card with an annual percentage rate (APR) of 21.3%. She decides not to make additional purchases with this card until she has paid off the balance.

**a)** Many credit cards require a minimum monthly payment of 2% of the balance. What is Samantha's minimum payment on a balance of $4867.59? Round the answer to the nearest dollar.

**b)** Find the amount of interest and the amount applied to reduce the principal in the minimum payment found in part (a).

**c)** If Samantha had transferred her balance to a card with an APR of 13.6%, how much of her first payment would be interest and how much would be applied to reduce the principal?

**d)** Compare the amounts for 13.6% from part (c) with the amounts for 21.3% from part (b).

**b)** The amount of interest on $3216.28 at 19.7% for one month* is given by

$$I = P \cdot r \cdot t = \$3216.28 \cdot 0.197 \cdot \frac{1}{12} \approx \$52.80.$$

We subtract to find the portion of the first payment applied to reduce the principal:

$$\text{Amount applied to} \atop \text{reduce the principal} = \text{Minimum payment} - {\text{Interest for} \atop \text{the month}}$$

$$= \$64 - \$52.80$$

$$= \$11.20.$$

Thus, the principal of $3216.28 is decreased by only $11.20 with the first payment. (Evan still owes $3205.08.)

**c)** The amount of interest on $3216.28 at 12.5% for one month is

$$I = P \cdot r \cdot t = \$3216.28 \cdot 0.125 \cdot \frac{1}{12} \approx \$33.50.$$

We subtract to find the amount applied to reduce the principal in the first payment:

$$\text{Amount applied to} \atop \text{reduce the principal} = \text{Minimum payment} - {\text{Interest for} \atop \text{the month}}$$

$$= \$64 - \$33.50$$

$$= \$30.50.$$

Thus, the principal of $3216.28 would have decreased by $30.50 with the first payment. (Evan would still owe $3185.78.)

**d)** Let's organize the information for both rates in the following table.

| BALANCE BEFORE FIRST PAYMENT | FIRST MONTH'S PAYMENT | %APR | AMOUNT OF INTEREST | AMOUNT APPLIED TO PRINCIPAL | BALANCE AFTER FIRST PAYMENT |
|---|---|---|---|---|---|
| $3216.28 | $64 | 19.7% | $52.80 | $11.20 | $3205.08 |
| 3216.28 | 64 | 12.5 | 33.50 | 30.50 | 3185.78 |

Difference in balance after first payment ⟶ $19.30

At 19.7%, the interest is $52.80 and the principal is decreased by $11.20. At 12.5%, the interest is $33.50 and the principal is decreased by $30.50. Comparing the amounts of interest, we see that the interest at 19.7% is $52.80 − $33.50, or $19.30, greater than the interest at 12.5%. Comparing the reductions in principal, we see that the principal is decreased by $30.50 − $11.20, or $19.30, more with the 12.5% rate than with the 19.7% rate.

◀ **Do Exercise 6.**

*Answers*

**6. (a)** $97; **(b)** interest: $86.40; amount applied to principal: $10.60; **(c)** interest: $55.17; amount applied to principal: $41.83; **(d)** At 13.6%, the principal was reduced by $31.23 more than at the 21.3% rate. The interest at 13.6% is $31.23 less than at 21.3%.

*Actually, the interest on a credit card is computed daily with a rate called a daily percentage rate (DPR). The DPR for Example 6 would be 19.7%/365 ≈ 0.054%. When no payments or additional purchases are made during a month, the difference in total interest for the month is minimal and we will not deal with it here.

It is interesting to compare how long it takes to pay off the balance in Example 6 if Evan continues to pay $64 each month, compared to how long it takes if he pays double that amount, or $128, each month. Financial consultants frequently tell clients that if they want to take control of their debt, they should pay double the minimum payment.

| RATE | PAYMENT | NUMBER OF PAYMENTS TO PAY OFF DEBT | TOTAL PAID BACK | INTEREST COST OF PURCHASES |
|------|---------|-----------------------------------|-----------------|----------------------------|
| 19.7% | $64 | 107, or 8 yr 11 mo | $6848 | $3631.72 |
| 19.7 | 128 | 33, or 2 yr 9 mo | 4224 | 1007.72 |
| 12.5 | 64 | 72, or 6 yr | 4608 | 1391.72 |
| 12.5 | 128 | 29, or 2 yr 5 mo | 3712 | 495.72 |

As with most loans, if you pay an extra amount toward the principal with each payment, the length of the loan can be greatly reduced. Note that at the rate of 19.7%, it will take Evan almost 9 years to pay off his debt if he pays only $64 per month and does not make additional purchases. If he transfers his balance to a card with a 12.5% rate and pays $128 per month, he can eliminate his debt in approximately $2\frac{1}{2}$ years. You can see how debt can get out of control if you continue to make purchases and pay only the minimum payment each month. The debt will never be eliminated.

---

## 7.7 Exercise Set

FOR EXTRA HELP  MyLab Math

### ✓ Check Your Understanding

**Reading Check** In the simple interest formula, $I = P \cdot r \cdot t$, $t$ must be expressed in years. Convert each length of time to years. Use 1 year = 12 months = 365 days.

**RC1.** 6 months      **RC2.** 40 days      **RC3.** 285 days

**RC4.** 9 months      **RC5.** 3 months      **RC6.** 4 months

**Concept Check** Use the data listed in Tables 1 and 2 below to answer Exercises CC1 and CC2.

**TABLE 1**

| PRINCIPAL | RATE OF INTEREST | COMPOUNDED | TIME | AMOUNT IN THE ACCOUNT |
|-----------|------------------|------------|------|-----------------------|
| $10,000 | $4\frac{1}{2}\%$ | Semiannually | 3 yr | $11,428.25 |
| $10,000 | $4\frac{1}{2}\%$ | Quarterly | 3 yr | $11,436.74 |
| $10,000 | $4\frac{1}{2}\%$ | Monthly | 3 yr | $11,442.48 |

**CC1.** How much more interest is earned from a $10,000 investment for 3 years at $4\frac{1}{2}\%$ when interest is compounded monthly rather than semiannually?

**TABLE 2**

| PRINCIPAL | RATE OF INTEREST | COMPOUNDED | TIME | AMOUNT IN THE ACCOUNT |
|-----------|------------------|------------|------|-----------------------|
| $10,000 | 6% | Quarterly | 3 yr | $11,956.18 |
| $10,000 | $6\frac{1}{2}\%$ | Quarterly | 3 yr | $12,134.08 |
| $10,000 | 7% | Quarterly | 3 yr | $12,314.39 |

**CC2.** How much more interest is earned from a $10,000 investment for 3 years when interest is compounded quarterly at $6\frac{1}{2}\%$ rather than 6%?

Find the simple interest.

| | Principal | Rate of Interest | Time | Simple Interest |
|---|---|---|---|---|
| 1. | $200 | 4% | 1 year | |
| 2. | $200 | 7.7% | 1 year | |
| 3. | $4300 | 10.56% | $\frac{1}{4}$ year | |
| 4. | $80,000 | $6\frac{3}{4}$% | $\frac{1}{12}$ year | |
| 5. | $20,000 | $4\frac{5}{8}$% | 6 months | |
| 6. | $8000 | 9.42% | 2 months | |
| 7. | $50,000 | $5\frac{3}{8}$% | 3 months | |
| 8. | $100,000 | $3\frac{1}{4}$% | 9 months | |

Solve. Assume that simple interest is being calculated in each case.

9. Mia's Boutique borrows $10,000 at 9% for 60 days. Find **(a)** the amount of interest due and **(b)** the total amount that must be paid after 60 days.

10. Mason's Drywall borrows $8000 at 10% for 90 days. Find **(a)** the amount of interest due and **(b)** the total amount that must be paid after 90 days.

11. Animal Instinct Pet Supply borrows $6500 at $5\frac{1}{4}$% for 90 days. Find **(a)** the amount of interest due and **(b)** the total amount that must be paid after 90 days.

12. Andante's Cafe borrows $4500 at $12\frac{1}{2}$% for 60 days. Find **(a)** the amount of interest due and **(b)** the total amount that must be paid after 60 days.

13. Cameron's Garage borrows $5600 at 10% for 30 days. Find **(a)** the amount of interest due and **(b)** the total amount that must be paid after 30 days.

14. Shear Delights Hair Salon borrows $3600 at 4% for 30 days. Find **(a)** the amount of interest due and **(b)** the total amount that must be paid after 30 days.

**b** Interest is compounded annually. Find the amount in the account after the given length of time. Round to the nearest cent.

| | Principal | Rate of Interest | Time | Amount in the Account |
|---|---|---|---|---|
| **15.** | $400 | 5% | 2 years | |
| **16.** | $450 | 4% | 2 years | |
| **17.** | $2000 | 2.5% | 4 years | |
| **18.** | $4000 | 3.7% | 4 years | |
| **19.** | $4300 | 10.56% | 6 years | |
| **20.** | $8000 | 9.42% | 6 years | |
| **21.** | $20,000 | $6\frac{5}{8}$% | 25 years | |
| **22.** | $100,000 | $5\frac{7}{8}$% | 30 years | |

Interest is compounded semiannually. Find the amount in the account after the given length of time. Round to the nearest cent.

| | Principal | Rate of Interest | Time | Amount in the Account |
|---|---|---|---|---|
| **23.** | $4000 | 2% | 1 year | |
| **24.** | $1000 | 3% | 1 year | |
| **25.** | $20,000 | 8.8% | 4 years | |
| **26.** | $40,000 | 7.7% | 4 years | |
| **27.** | $5000 | 10.56% | 6 years | |
| **28.** | $8000 | 9.42% | 8 years | |
| **29.** | $20,000 | $7\frac{5}{8}$% | 25 years | |
| **30.** | $100,000 | $4\frac{7}{8}$% | 30 years | |

Solve.

**31.** The Aguilar family invests $4000 in an account paying 6%, compounded monthly. How much is in the account after 5 months?

**32.** Ryan and Alicia invest $2500 in an account paying 3%, compounded monthly. How much is in the account after 6 months?

**33.** Bryn and Andy invest $1200 in an account paying 10%, compounded quarterly. How much is in the account after 1 year?

**34.** The O'Hares invest $6000 in an account paying 8%, compounded quarterly. How much is in the account after 18 months?

**35.** Emilio loans his niece's business $20,000 for 50 days at 6% interest, compounded daily. How much is Emilio owed after 50 days?

**36.** Elsa loans her nephew's business $25,000 for 40 days at 5% interest, compounded daily. How much is Elsa owed after 40 days?

Solve.

**37. *Credit Cards.*** Aiden has a balance of $1278.56 on a credit card with an annual percentage rate (APR) of 19.6%. The minimum payment required in the current statement is $25.57. Find the amount of interest and the amount applied to reduce the principal in this payment and the balance after this payment.

**38. *Credit Cards.*** Ella has a balance of $1834.90 on a credit card with an annual percentage rate (APR) of 22.4%. The minimum payment required in the current statement is $36.70. Find the amount of interest and the amount applied to reduce the principal in this payment and the balance after this payment.

**39. *Credit Cards.*** Hailey has a balance of $4876.54 on a credit card with an annual percentage rate (APR) of 21.3%.

a) Many credit cards require a minimum monthly payment of 2% of the balance. What is Hailey's minimum payment on a balance of $4876.54? Round the answer to the nearest dollar.
b) Find the amount of interest and the amount applied to reduce the principal in the minimum payment found in part (a).
c) If Hailey had transferred her balance to a card with an APR of 12.6%, how much of her payment would be interest and how much would be applied to reduce the principal?
d) Compare the amounts for 12.6% from part (c) with the amounts for 21.3% from part (b).

**40. *Credit Cards.*** Luke has a balance of $5328.88 on a credit card with an annual percentage rate (APR) of 18.7%.

a) Many credit cards require a minimum monthly payment of 2% of the balance. What is Luke's minimum payment on a balance of $5328.88? Round the answer to the nearest dollar.
b) Find the amount of interest and the amount applied to reduce the principal in the minimum payment found in part (a).
c) If Luke had transferred his balance to a card with an APR of 13.2%, how much of his payment would be interest and how much would be applied to reduce the principal?
d) Compare the amounts for 13.2% from part (c) with the amounts for 18.7% from part (b).

## Skill Maintenance

**41.** Find the LCM of 32 and 50.   [4.1a]

**42.** Find the prime factorization of 228.   [3.2c]

Divide and simplify.   [3.7b]

**43.** $\dfrac{6}{125} \div \dfrac{8}{15}$

**44.** $\dfrac{16}{105} \div \dfrac{5}{14}$

Multiply and simplify.   [3.6a]

**45.** $\dfrac{4}{15} \times \dfrac{3}{20}$

**46.** $\dfrac{8}{21} \times \dfrac{49}{800}$

**47.** Simplify: $4^3 - 6^2 \div 2^2$.   [1.9c]

**48.** Solve: $x + \dfrac{2}{5} = \dfrac{9}{10}$.   [4.3b]

## Synthesis

***Effective Yield.*** The *effective yield* is the yearly rate of simple interest that corresponds to a rate for which interest is compounded two or more times a year. For example, if $P$ is invested at 12%, compounded quarterly, we multiply $P$ by $(1 + 0.12/4)^4$, or $1.03^4$. Since $1.03^4 \approx 1.126$, the 12% compounded quarterly corresponds to an effective yield of approximately 12.6%. In Exercises 49 and 50, find the effective yield for the indicated account.

**49.** ▦ The account pays 9% compounded monthly.

**50.** ▦ The account pays 10% compounded daily.

## Vocabulary Reinforcement

Complete each statement with the appropriate word or phrase from the list on the right. Some of the choices will not be used.

1. When a quantity is decreased by a certain percent, we say that this is a _____. [7.5a]

2. The _____ interest $I$ on principal $P$, invested for $t$ years at interest rate $r$, is given by $I = P \cdot r \cdot t$. [7.7a]

3. Sale price = Original price − _____. [7.6c]

4. Commission = Commission rate × _____. [7.6b]

5. Discount = _____ of discount × Original price. [7.6c]

6. When a quantity is increased by a certain percent, we say that this is a _____. [7.5a]

discount

rate

sales

commission

price

principal

percent increase

percent decrease

simple

compound

## Concept Reinforcement

Determine whether each statement is true or false.

_____ 1. A fixed principal invested for 4 years will earn more interest when interest is compounded quarterly than when interest is compounded semiannually. [7.7b]

_____ 2. Of the numbers $0.5\%$, $\frac{5}{1000}\%$, $\frac{1}{2}\%$, $\frac{1}{5}$, and $0.\overline{1}$, the largest number is $0.\overline{1}$. [7.1b, c]

_____ 3. If principal A equals principal B and principal A is invested for 2 years at 4%, compounded quarterly, while principal B is invested for 4 years at 2%, compounded semiannually, the interest earned from each investment is the same. [7.7b]

## Study Guide

**Objective 7.1b**   Convert between percent notation and decimal notation.

**Example**   Find percent notation for 1.3.

We move the decimal point two places to the right and write a percent symbol:

$\qquad 1.3 = 130\%.$

**Example**   Find decimal notation for $12\frac{3}{4}\%$.

We convert $12\frac{3}{4}$ to a decimal, move the decimal point two places to the left, and drop the percent symbol:

$\qquad 12\frac{3}{4}\% = 12.75\% = 0.1275.$

**Practice Exercise**

1. Find percent notation for 0.082.

**Practice Exercise**

2. Find decimal notation for $62\frac{5}{8}\%$.

**Objective 7.1c**   Convert between fraction notation and percent notation.

**Example**   Find percent notation for $\frac{5}{12}$.

$$\begin{array}{r} 0.4\ 1\ 6 \\ 1\ 2\ \overline{)\ 5.0\ 0\ 0} \\ \underline{4\ 8} \\ 2\ 0 \\ \underline{1\ 2} \\ 8\ 0 \\ \underline{7\ 2} \\ 8 \end{array}$$

$\frac{5}{12} = 0.41\overline{6} = 41.\overline{6}\%$, or $41\frac{2}{3}\%$

**Example**   Find fraction notation for 9.5%.

$$9.5\% = \frac{9.5}{100} = \frac{9.5}{100} \cdot \frac{10}{10} = \frac{95}{1000}$$

$$= \frac{5 \cdot 19}{5 \cdot 200}$$

$$= \frac{5}{5} \cdot \frac{19}{200}$$

$$= \frac{19}{200}$$

**Objective 7.2b**   Solve basic percent problems using percent equations.

**Example**   165 is what percent of 3300?

We have

$$165 = p \cdot 3300 \qquad \text{Translating to a percent equation}$$

$$\frac{165}{3300} = \frac{p \cdot 3300}{3300}$$

$$\frac{165}{3300} = p$$

$$0.05 = p$$

$$5\% = p.$$

Thus, 165 is 5% of 3300.

**Objective 7.3b**   Solve basic percent problems using proportions.

**Example**   18% of what is 1296?

$$\frac{18}{100} = \frac{1296}{b} \qquad \text{Translating to a proportion}$$

$$18 \cdot b = 100 \cdot 1296$$

$$\frac{18 \cdot b}{18} = \frac{129{,}600}{18}$$

$$b = 7200$$

Thus, 18% of 7200 is 1296.

**Objective 7.5a** Solve applied problems involving percent increase or percent decrease.

**Example** The average price of a hotel room in New York City in 2016 was \$252. The price of a hotel room in New York City in 2015 averaged \$260. What was the percent decrease?

**Data:** Hotels.com, March 7, 2017

We first determine the amount of decrease.

$$\$260 - \$252 = \$8$$

Then we translate to a percent equation or a proportion and solve.

*Rewording:* \$8 is what percent of \$260?

*Percent Equation:*

$$8 = p \cdot 260$$

$$\frac{8}{260} = \frac{p \cdot 260}{260}$$

$$\frac{8}{260} = p$$

$$0.031 \approx p$$

$$3.1\% \approx p$$

*Proportion:*

$$\frac{N}{100} = \frac{8}{260}$$

$$260 \cdot N = 100 \cdot 8$$

$$\frac{260 \cdot N}{260} = \frac{100 \cdot 8}{260}$$

$$N = \frac{800}{260}$$

$$N \approx 3.1$$

The percent decrease was about 3.1%.

**Practice Exercise**

7. The average price of a hotel room in Las Vegas in 2016 was \$138. The price of a hotel room in Las Vegas in 2015 averaged \$131. What was the percent increase?

**Data:** Hotels.com, March 7, 2017

---

**Objective 7.6a** Solve applied problems involving sales tax and percent.

**Example** The sales tax is \$34.23 on the purchase of a flat-screen high-definition television that costs \$489. What is the sales tax rate?

*Rephrase:* Sales tax is what percent of purchase price?

*Translate:* $34.23 = r \cdot 489$

*Solve:*

$$\frac{34.23}{489} = \frac{r \cdot 489}{489}$$

$$\frac{34.23}{489} = r$$

$$0.07 = r$$

$$7\% = r$$

The sales tax rate is 7%.

**Practice Exercise**

8. The sales tax is \$1102.20 on the purchase of a new car that costs \$18,370. What is the sales tax rate?

**Objective 7.6b** Solve applied problems involving commission and percent.

**Example** A real estate agent's commission rate is $6\frac{1}{2}\%$. She received a commission of \$17,160 on the sale of a home. For how much did the home sell?

*Rephrase:* Commission is $6\frac{1}{2}\%$ of what selling price?

*Translate:* $17{,}160 = 6\frac{1}{2}\% \times S$

*Solve:* $17{,}160 = 0.065 \times S$

$$\frac{17{,}160}{0.065} = \frac{0.065 \times S}{0.065}$$

$$264{,}000 = S$$

The home sold for \$264,000.

**Practice Exercise**

9. A real estate agent's commission rate is 7%. He received a commission of \$12,950 on the sale of a home. For how much did the home sell?

---

**Objective 7.7a** Solve applied problems involving simple interest.

**Example** To meet its payroll, a business borrows \$5200 at $4\frac{1}{4}\%$ for 90 days. Find the amount of simple interest that is due and the total amount that must be paid after 90 days.

$$I = P \cdot r \cdot t = \$5200 \cdot 4\tfrac{1}{4}\% \cdot \frac{90}{365}$$

$$= \$5200 \cdot 0.0425 \cdot \frac{90}{365}$$

$$\approx \$54.49$$

The interest due for 90 days $=$ \$54.49.

The total amount due $=$ \$5200 $+$ \$54.49 $=$ \$5254.49.

**Practice Exercise**

10. A student borrows \$2500 for tuition at $5\frac{1}{2}\%$ for 60 days. Find the amount of simple interest that is due and the total amount that must be paid after 60 days.

---

**Objective 7.7b** Solve applied problems involving compound interest.

**Example** Find the amount in an account if \$3200 is invested at 5%, compounded semiannually, for $1\frac{1}{2}$ years.

$$A = P \cdot \left(1 + \frac{r}{n}\right)^{n \cdot t}$$

$$= \$3200\left(1 + \frac{0.05}{2}\right)^{2 \cdot \frac{3}{2}}$$

$$= \$3200(1.025)^3$$

$$= \$3446.05$$

The amount in the account after $1\frac{1}{2}$ years is \$3446.05.

**Practice Exercise**

11. Find the amount in an account if \$6000 is invested at $4\frac{3}{4}\%$, compounded quarterly, for 2 years.

---

# Review Exercises

Find percent notation. [7.1b]

**1.** 1.7

**2.** 0.065

Find decimal notation for the percent notation(s) in each sentence. [7.1b]

**3.** In the 2015–2016 school year, about 20.4 million students, including 1,043,839 foreign students, were enrolled in U.S. colleges and universities. Approximately 31.5% of the foreign students were from China.

**Data:** Institute of International Education

**4.** Of all active physicians, 13.3% specialize in internal medicine and 6.7% specialize in pediatrics.

Data: AMA Physician Masterfile, December 2015

Find percent notation. [7.1c]

**5.** $\dfrac{3}{8}$          **6.** $\dfrac{1}{3}$

Find fraction notation. [7.1c]

**7.** 24%          **8.** 6.3%

Translate to a percent equation. Then solve. [7.2a, b]

**9.** 30.6 is what percent of 90?

**10.** 63 is 84% of what?

**11.** What is $38\frac{1}{2}$% of 168?

Translate to a proportion. Then solve. [7.3a, b]

**12.** 24% of what is 16.8?

**13.** 42 is what percent of 30?

**14.** What is 10.5% of 84?

Solve. [7.4a], [7.5a]

**15.** *Favorite Ice Creams.* According to a survey, 8.9% of those interviewed chose chocolate as their favorite ice cream flavor and 4.2% chose butter pecan. At this rate, of 2000 first-year college students, how many would choose chocolate as their favorite ice cream? butter pecan?

Data: International Ice Cream Association

**16.** *Physicians Specializing in Psychiatry.* Of the 860,939 active physicians in the United States in 2015, 37,736 specialized in psychiatry. What percent of the active physicians specialized in psychiatry? Round the answer to the nearest tenth of a percent.

Data: AMA Physician Masterfile, December 2015

**17.** *Water Output.* The average person loses 200 mL of water per day by sweating. This is 8% of the total output of water from the body. How much is the total output of water?

Data: Elaine N. Marieb, *Essentials of Human Anatomy and Physiology,* 6th ed. Boston: Addison Wesley Longman, Inc., 2000

**18.** *Test Scores.* After Sheila got a 75 on a math test, she was allowed to go to the math lab and take a retest. She increased her score to 84. What was the percent increase?

**19.** *Test Scores.* James got an 80 on a math test. By taking a retest in the math lab, he increased his score by 15%. What was his new score?

Solve. [7.6a, b, c]

**20.** A state charges a meals tax of $7\frac{1}{2}$%. What is the meals tax charged on a dinner party costing $320?

**21.** In a certain state, a sales tax of $453.60 is collected on the purchase of a used car for $7560. What is the sales tax rate?

**22.** Kim earns $753.50 for selling $6850 worth of televisions. What is the commission rate?

**23.** What is the rate of discount of this stepladder?

SPECIAL VALUE!

Now **$67** Was $82

8' Aluminum Stepladder

**24.** An air conditioner has a marked price of $350. It is placed on sale at 12% off. What are the discount and the sale price?

**25.** The price of a printer is marked down from $305 to $262.30. What is the rate of discount?

**26.** An insurance salesperson receives a 0.7% commission. If $150,000 worth of life insurance is sold, what is the commission?

Solve. [7.7a, b, c]

**27.** What is the simple interest on $1800 at 6% for $\frac{1}{3}$ year?

**28.** The Dress Shack borrows $24,000 at 10% simple interest for 60 days. Find **(a)** the amount of interest due and **(b)** the total amount that must be paid after 60 days.

**29.** What is the simple interest on a principal of $2200 at an interest rate of 5.5% for 1 year?

**30.** The Garcias invest $7500 in an investment account paying an annual interest rate of 4%, compounded monthly. How much is in the account after 3 months?

**31.** Find the amount in an investment account if $8000 is invested at 2.4%, compounded annually, for 2 years.

**32.** *Credit Cards.* At the end of her junior year of college, Kasha has a balance of $6428.74 on a credit card with an annual percentage rate (APR) of 18.7%. She decides not to make additional purchases with this card until she has paid off the balance.

**a)** Many credit cards require a minimum payment of 2% of the balance. At this rate, what is Kasha's minimum payment on a balance of $6428.74? Round the answer to the nearest dollar.

**b)** Find the amount of interest and the amount applied to reduce the principal in the minimum payment found in part (a).

**c)** If Kasha had transferred her balance to a card with an APR of 13.2%, how much of her payment would be interest and how much would be applied to reduce the principal?

**d)** Compare the amounts for 13.2% from part (c) with the amounts for 18.7% from part (b).

**33.** A fishing boat listed at $16,500 is on sale at 15% off. What is the sale price? [7.6c]

**A.** $14,025　　　　　　**B.** $2475

**C.** 85%　　　　　　　**D.** $14,225

**34.** Find the amount in a money market account if $10,500 is invested at 6%, compounded semiannually, for $1\frac{1}{2}$ years. [7.7b]

**A.** $11,139.45　　　　**B.** $12,505.67

**C.** $11,473.63　　　　**D.** $10,976.03

## Synthesis

**35.** Mike's Bike Shop reduces the price of a bicycle by 40% during a sale. By what percent must the store increase the sale price, after the sale, to get back to the original price? [7.6c]

# Understanding Through Discussion and Writing

**1.** Which is the better deal for a consumer and why: a discount of 40% or a discount of 20% followed by another of 22%? [7.6c]

**2.** Which is better for a wage earner, and why: a 10% raise followed by a 5% raise a year later or a 5% raise followed by a 10% raise a year later? [7.5a]

**3.** Ollie bought a microwave oven during a 10%-off sale. The sale price that Ollie paid was $162. To find the original price, Ollie calculates 10% of $162 and adds that to $162. Is this correct? Why or why not? [7.6c]

**4.** You take 40% of 50% of a number. What percent of the number could you take to obtain the same result making only one multiplication? Explain your answer. [7.4a]

**5.** A firm must choose between borrowing $5000 at 10% for 30 days and borrowing $10,000 at 8% for 60 days. Give arguments in favor of and against each option. [7.7a]

**6.** On the basis of the mathematics presented in Section 7.7, discuss what you have learned about interest rates and credit cards. [7.7c]

CHAPTER

7

Test

For
Extra
Help

For step-by-step test solutions, access the Chapter Test Prep Videos in MyLab Math.

1. **Multivitamin-Mineral Supplements.** In 2013, 31.9% of U.S. adults took multivitamin-mineral supplements. Find decimal notation for 31.9%.

   **Data:** *JAMA Internal Medicine,* Regan L. Bailey et al.

2. **Gravity.** The gravity of Mars is 0.38 as strong as Earth's. Find percent notation for 0.38.

   **Data:** www.marsinstitute.info/epo/mermarsfacts.html

3. Find percent notation for $\frac{11}{8}$.

4. Find fraction notation for 65%.

5. Translate to a percent equation. Then solve.

   What is 40% of 55?

6. Translate to a proportion. Then solve.

   What percent of 80 is 65?

Solve.

7. **Paying for Health Care.** The U.S. health care bill is $3.2 trillion annually. The pie chart below shows the percents of the bill paid by various payers. How much of the bill is paid by private health insurance? Medicaid? Round the answers to the nearest hundredth of a trillion.

   **Data:** money.cnn.com, December 2016

8. **Batting Average.** Ben Zobrist, second baseman for the Chicago Cubs, got 142 hits during the 2016 baseball season. This was about 27.1% of his at-bats. How many at-bats did he have?

   **Data:** Major League Baseball

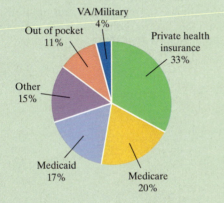

VA/Military
4%
Out of pocket
11%
Private health
insurance
33%
Other
15%
Medicaid
17%
Medicare
20%

DATA: money.cnn.com, December 2016

9. **Foreign Adoptions.** The number of foreign children adopted by Americans declined from 5647 in 2015 to 5370 in 2016. Find the percent decrease.

   **Data:** U.S. State Department; *USA TODAY,* August 13, 2016

10. There are about 7,432,000,000 people living in the world today, and approximately 4,436,000,000 live in Asia. What percent of people live in Asia?

    **Data:** Population Division/International Programs Center, U.S. Census Bureau, U.S. Dept. of Commerce

11. **Oklahoma Sales Tax** The sales tax rate in Oklahoma is 4.5%. How much tax is charged on a purchase of $560? What is the total price?

12. Noah's commission rate is 15%. What is his commission on the sale of $4200 worth of merchandise?

13. The marked price of a ceiling fan is $200, and the item is on sale at 20% off. What are the discount and the sale price?

14. What is the simple interest on a principal of $120 at the interest rate of 7.1% for 1 year?

15. A city orchestra invests $5200 at 6% simple interest. How much is in the account after $\frac{1}{2}$ year?

16. Find the amount in an account if $1000 is invested at $5\frac{3}{8}$%, compounded annually, for 2 years.

**17.** The Suarez family invests $10,000 at an annual interest rate of 4.9%, compounded monthly. How much is in the account after 3 years?

**18.** *Job Opportunities.* The table below lists job opportunities in 2014 and projected increases for 2024. Complete the table by filling in the missing numbers.

| Occupation | Total Employment in 2014 | Projected Employment in 2024 | Change | Percent of Increase |
|---|---|---|---|---|
| Interpreters and translators | 61,000 | 78,500 | 17,500 | 28.7% |
| Home health aides | 913,500 | | 348,400 | |
| Wind turbine technicians | 4400 | 9200 | | |
| Kindergarten and elementary teachers | | 1,605,200 | 87,800 | |
| Electricians | 628,800 | | 85,900 | |

DATA: *Occupational Outlook Handbook*

**19.** Find the discount and the rate of discount for the television in this ad.

19" LCD HDTV

$299⁹⁹

was $349⁹⁹

**20.** *Credit Cards.* Jayden has a balance of $2704.27 on a credit card with an annual percentage rate of 16.3%. The minimum payment required on the current statement is $54. Find the amount of interest and the amount applied to reduce the principal in this payment and the balance after this payment.

**21.** 0.75% of what number is 300?

**A.** 2.25  **B.** 40,000  **C.** 400  **D.** 225

## Synthesis

**22.** By selling a home without using a realtor, Juan and Marie can avoid paying a 7.5% commission. They receive an offer of $180,000 from a potential buyer. In order to give a comparable offer, for what price would a realtor need to sell the house? Round to the nearest hundred.

**23.** Karen's commission rate is 3%. She invests her commission from the sale of $124,000 worth of merchandise at an interest rate of 4%, compounded quarterly. How much is Karen's investment worth after 6 months?

1. **Fatal Medical Errors.** Medical error ranks third behind heart disease and cancer as a cause of death in the United States. In a recent year, the number of deaths from medical error was 362,894 less than the number of deaths from heart disease. Find the number of deaths from medical error. Use the information in the table below.

**ANNUAL DEATHS IN THE UNITED STATES**

| CAUSE OF DEATH | NUMBER OF DEATHS |
|---|---|
| Heart disease | 614,348 |
| Cancer | 591,699 |
| Medical error | ? |
| Accidents | 136,053 |
| Stroke | 133,103 |
| Alzheimer's | 93,541 |

DATA: Johns Hopkins School of Medicine; National Center for Health Statistics; *The BMJ; AARP Bulletin,* July/August 2016

2. Find percent notation: 0.269.

3. Find percent notation: $\dfrac{9}{8}$.

4. Find decimal notation: $\dfrac{13}{6}$.

5. Write fraction notation for the ratio 5 to 0.5.

6. Find the rate in kilometers per hour: 350 km, 15 hr.

Use $<$, $>$, or $=$ for ☐ to write a true sentence.

7. $\dfrac{5}{7}$ ☐ $\dfrac{6}{8}$

8. $\dfrac{6}{14}$ ☐ $\dfrac{15}{25}$

Estimate the sum or the difference by first rounding to the nearest hundred.

9. $263{,}961 + 32{,}090 + 127.89$

10. $73{,}510 - 23{,}450$

Calculate.

11. $46 - [4(6 + 4 \div 2) + 2 \times 3 - 5]$

12. Combine like terms: $5x - 9 - 7x - 5$.

Compute and simplify.

13. $\dfrac{6}{5} + 1\dfrac{5}{6}$

14. $-46.9 + 2.84$

15. 
$$\begin{array}{r} 4\ 8\ 7{,}0\ 9\ 4 \\ 6{,}9\ 3\ 6 \\ +\ \ \ 2\ 1{,}1\ 2\ 0 \\ \hline \end{array}$$

16. $35 - 34.98$

17. $3\dfrac{1}{3} - 2\dfrac{2}{3}$

18. $-\dfrac{8}{9} - \dfrac{6}{7}$

19. $\dfrac{7}{9} \cdot \dfrac{3}{14}$

20. $(-32)(-4)(-3)$

21. 
$$\begin{array}{r} 4\ 6.0\ 1\ 2 \\ \times\ \ \ \ \ 0.0\ 3 \\ \hline \end{array}$$

22. $6\dfrac{3}{5} \div 4\dfrac{2}{5}$

23. $431.2 \div 35.2$

24. $15\overline{)1850}$

Solve.

25. $36 \cdot x = 3420$

26. $y + 142.87 = 151$

27. $\dfrac{3}{7}x - 5 = 16$

28. $\dfrac{3}{4} + x = \dfrac{5}{6}$

29. $3(x - 7) + 2 = 12x - 3$

30. $\dfrac{16}{n} = \dfrac{21}{11}$

**31. Museum Attendance.** Students benefit from their school's membership in the Children's Museum of Indianapolis. Currently 112 schools have memberships, and in 2016, 102,000 students visited with their schools. This number represented about 8% of the total number of visits that year. How many visits were there to the Children's Museum of Indianapolis in 2016?

Data: Children's Museum of Indianapolis

**32. Salary of Farmer.** In 2016, the median pay for a veterinarian was $22,410 higher than the median pay for a farmer. If the pay for a veterinarian was $88,770, what was the pay for a farmer?

Data: U.S. Bureau of Labor Statistics; *Occupational Outlook Handbook*

**33.** At one point during the 2016–2017 NBA season, the Utah Jazz had won 16 out of 26 games. At this rate, how many games would they have won in the entire season of 82 games?

Data: National Basketball Association

**34. Shirts.** A total of $424.75 was paid for 5 equally priced shirts at an upscale men's store. How much did each shirt cost?

**35. Unit Price.** A 200-oz bottle of liquid laundry detergent costs $14.99. What is the unit price?

**36.** Tiana walked $\frac{7}{10}$ mi to school and then $\frac{8}{10}$ mi to the library. How far did she walk?

**37.** On a map, 1 in. represents 80 mi. How much does $\frac{3}{4}$ in. represent?

**38. Compound Interest.** The Bakers invest $8500 in an investment account paying 4%, compounded monthly. How much is in the account after 5 years?

**39. Ribbons.** How many pieces of ribbon $1\frac{4}{5}$ yd long can be cut from a length of ribbon 9 yd long?

**40. Physical Therapists.** In 2016, 210,900 people were employed as physical therapists. It is projected that by 2024, this number will increase by 71,800. Find the projected percent increase in the number of physical therapists and the number of physical therapists in 2024.

Data: U.S. Bureau of Labor Statistics; *Occupational Outlook Handbook*

**41.** Subtract and simplify: $\frac{14}{25} - \frac{3}{20}$.

A. $\frac{11}{500}$      B. $\frac{11}{5}$

C. $\frac{41}{100}$      D. $\frac{205}{500}$

**42.** The population of the state of Louisiana decreased from 4,468,976 in 2000 to 4,287,768 in 2006. What was the percent decrease?

Data: U.S. Census Bureau

A. 4.1%      B. 4.2%

C. 104%      D. 95.9%

## Synthesis

**43.** How many successive 10% discounts are necessary to lower the price of an item to below 50% of its original price?

On a trip through the mountains, a Dodge Neon traveled 240 mi on $7\frac{1}{2}$ gal of gasoline. Going across the plains, the same car averaged 36 miles per gallon.

**44.** What was the percent increase or decrease in miles per gallon when the car left the mountains for the plains?

**45.** How many miles per gallon did the Dodge average over the entire trip if it used 5 gal of gas to cross the plains?

# Data, Graphs, and Statistics

Traffic congestion in the United States in 2014 cost commuters $160 billion, wasted 3.1 billion gallons of fuel, and resulted in 6.9 billion hours of delay, or 42 hours of delay per auto commuter. The average commuter delay nearly doubled from 1986 to 2014, as shown in the accompanying line graph. Traffic congestion not only wastes time and fuel, but also causes increases in stress and pollution and delays in response of emergency vehicles.

*DATA: 2015 Urban Mobility Scorecard, The Texas A&M Transportation Institute and INRIX; usatoday.com*

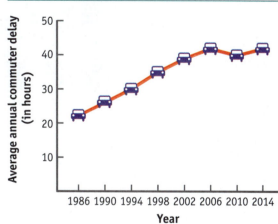

**Traffic Congestion Delays**

In Exercises 26–30 of Exercise Set 8.2, we will use a bar graph to compare average commuting times for various cities, and in Exercise 8 of Exercise Set 8.5, we will find the mean, median, and mode of hours of traffic delay.

509

# 8.1

## OBJECTIVES

**a** Extract and interpret data from tables.

**b** Extract and interpret data from graphs.

# Interpreting Data from Tables and Graphs

We use tables and graphs to display data and to communicate information about the data. For example, the following table and graphs display data on the resting heart rate for several mammals. Examine each method of presentation. Which method do you like best, and why?

**Table**

|  | MOUSE | GIRAFFE | CAT | HUMAN | HORSE | ELEPHANT |
|---|---|---|---|---|---|---|
| Average resting heart rate (in beats per minute) | 500 | 170 | 130 | 70 | 35 | 28 |

DATA: elephantnaturepark.org; vetmedicine.about.com; giraffeconservation.org; learningabouthorses.com

**Pictograph**

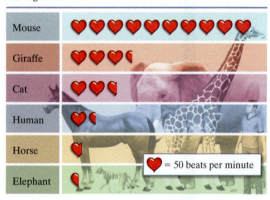

**Bar Graph**

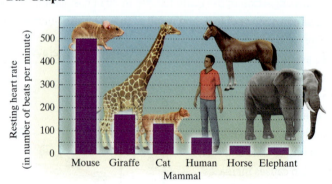

Comparing the table and the graphs reveals that the exact data values are most easily read in a table. The fastest and slowest heart rates can be determined easily from the graphs. The graph below communicates additional information about the data. The size of each mammal in this graph indicates the heart rate, not the actual size of the animal. The unexpected relative sizes of the mammals in the graph emphasize the fact that many small animals have faster heart rates than larger animals.

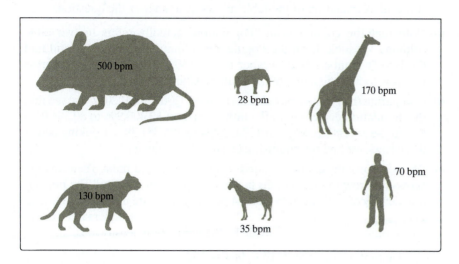

500 bpm
28 bpm
170 bpm
70 bpm
130 bpm
35 bpm

## a  READING AND INTERPRETING TABLES

A **table** is often used to present data in rows and columns.

**EXAMPLE 1**  *Population Density.*  The following table lists populations and land areas of 10 countries.

| COUNTRY | LAND AREA (in square miles) | POPULATION | | POPULATION DENSITY (per square mile) | |
| --- | --- | --- | --- | --- | --- |
| | | 2008 | 2017 | 2008 | 2017 |
| Australia | 2,941,299 | 20,434,176 | 23,232,413 | 7 | 8 |
| Brazil | 3,265,077 | 190,010,647 | 207,353,391 | 58 | 64 |
| China | 3,600,947 | 1,321,851,888 | 1,379,302,771 | 367 | 383 |
| Finland | 117,558 | 5,238,460 | 5,518,371 | 45 | 47 |
| Germany | 134,836 | 82,400,996 | 80,594,017 | 611 | 598 |
| India | 1,147,955 | 1,129,866,154 | 1,281,935,911 | 984 | 1117 |
| Japan | 144,689 | 127,433,494 | 126,451,398 | 881 | 874 |
| Kenya | 219,789 | 36,913,721 | 47,615,739 | 168 | 217 |
| Mexico | 742,490 | 108,700,891 | 124,574,795 | 146 | 168 |
| United States | 3,537,439 | 301,139,947 | 326,625,791 | 85 | 92 |

DATA: *World Almanac 2008*; U.S. Census Bureau

**a)** Which country had the largest population in 2017?

**b)** In which country or countries did the population decrease from 2008 to 2017?

**c)** What was the percent increase in population density in India from 2008 to 2017?

**d)** Find the average land area of the four largest countries in the table.

Careful examination of the table allows us to answer the questions.

**a)** Note that the column head "Population" actually refers to two table columns. We look down the Population column headed "2017" and find the largest number. That number is 1,379,302,771. Then we look across that row to find the name of the country: China.

**b)** Comparing the Population columns headed "2008" and "2017," we see that the population decreased in the fifth row (from 82,400,996 to 80,594,017) and in the seventh row (from 127,433,494 to 126,451,398). Looking across these rows, we find the countries: Germany and Japan.

**c)** We look down the column headed "Country" and find India. Then we look across that row to the columns headed "Population Density." The population density of India in 2008 was 984 people per square mile. It increased to 1117 people per square mile in 2017. To find the percent increase, we find the amount of increase and divide by the population density in 2008.

Amount of Increase: $1117 - 984 = 133$

Percent Increase: $\dfrac{133}{984} \approx 0.135 = 13.5\%$

The population density of India increased by about 13.5% from 2008 to 2017.

**d)** By looking down the column headed "Land Area," we determine that the four largest countries in the table are Australia, Brazil, China, and the United States. We find the average land area of these countries:

$$\dfrac{2{,}941{,}299 + 3{,}265{,}077 + 3{,}600{,}947 + 3{,}537{,}439}{4} = \dfrac{13{,}344{,}762}{4}$$

$$= 3{,}336{,}190.5 \text{ sq mi.}$$

◀ **Do Exercises 1–5.**

---

Use the table in Example 1 to answer Margin Exercises 1–5.

**1.** Which country has the smallest land area?

**2.** What was the population of Mexico in 2017?

**3.** What was the percent decrease **GS** in population density in Germany from 2008 to 2017?

The amount of the decrease in population density is

$$611 - 598 = \boxed{\phantom{00}}.$$

The percent decrease is

$$\dfrac{13}{\boxed{\phantom{00}}} \approx 0.021, \text{ or } \boxed{\phantom{00}}\ \%.$$

**4.** Which country had the greatest increase in population from 2008 to 2017?

**5.** Find the average population density of these countries in 2017.

---

**b** **READING AND INTERPRETING GRAPHS**

*Pictographs* (or *picture graphs*) are another way to show information. Instead of actually listing the amounts to be considered, a **pictograph** uses symbols to represent the amounts. A pictograph includes a *key* that tells what each symbol represents.

---

*Answers*

**1.** Finland  **2.** 124,574,795  **3.** About 2.1%
**4.** India, with an increase of 152,069,757
**5.** 356.8 people per square mile

*Guided Solution:*
**3.** 13; 611; 2.1

**EXAMPLE 2**  *Roller Coasters.*  The following pictograph shows the number of roller coasters listed in the Roller Coaster Data Base for six continents. Below the graph is a key that tells you that each  represents 100 roller coasters.

**Roller Coasters of the World**

| Africa | |
| Asia | |
| Australia | |
| Europe | |
| North America | |
| South America | |

 = 100 roller coasters

**a)** Which continent has the greatest number of roller coasters?

**b)** About how many roller coasters are there in Australia?

**c)** How many more roller coasters are there in Europe than in North America?

    We can determine the answers by reading the pictograph.

**a)** The continent with the most symbols is Asia, so Asia has the greatest number of roller coasters.

**b)** The pictograph shows about $\frac{1}{4}$ symbol for Australia. Since each symbol represents 100 roller coasters, there are about $\frac{1}{4} \times 100$, or 25, roller coasters in Australia.

**c)** From the graph, we see that there are about $11\frac{3}{4} \times 100$, or 1175, roller coasters in Europe and about $8\frac{3}{4} \times 100$, or 875, roller coasters in North America. Thus, there are about $1175 - 875$, or 300, more roller coasters in Europe than in North America. We could also estimate this difference by noting that Europe has 3 more symbols than North America does, and $3 \times 100 = 300$.

**Do Exercises 6–8.** ▶

    When representing data with graphs, we must be sure that the areas of regions of the graph are proportional to the numbers that the regions represent. For example, in pictographs, each symbol is the same size, and the number of symbols is proportional to the actual data values. Thus, the total area of the symbols is proportional to the data values. This *area principle* is illustrated in the following example.

Use the pictograph in Example 2 to answer Margin Exercises 6–8.

**6.** Which continent has the smallest number of roller coasters?

**7.** About how many roller coasters are there in Asia?

**GS** **8.** How many more roller coasters are there in South America than in Africa?

The graph shows about $1\frac{3}{4}$ symbols for South America.

This represents ☐ roller coasters.

The graph shows about $\frac{3}{4}$ symbol for Africa.

This represents ☐ roller coasters.

There are about ☐ more roller coasters in South America than in Africa.

**EXAMPLE 3** *Electricity Generation.* The following graph illustrates the different methods used in the United States to generate electricity. Each darker (smaller) circle represents the amount of electricity generated in 1 year, in billions of kilowatt-hours (kWh). Some methods of electricity generation are more efficient than others. The lighter circle surrounding each darker circle represents the amount of electricity that could have been generated if the method were 100% efficient.

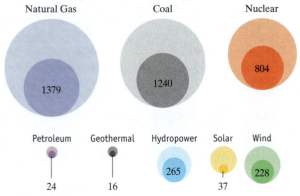

DATA: U.S. Energy Information Administration; brighthubengineering.com

a) How much electricity is generated annually with wind?

b) Which method generates the least electricity?

c) Which method of electricity generation is the most efficient?

d) Is solar generation of electricity more or less efficient than wind generation?

We use the information in the graph to answer the questions. Note that the area of the lighter circle that is not covered by the darker circle represents the amount of energy lost or wasted during the generation process.

a) From the darker green circle labeled "Wind," we see that 228 billion kWh of electricity is generated annually with wind.

b) The smallest darker circle is labeled "Geothermal," so the least electricity is generated by geothermal methods.

c) The darker circle that most nearly fills its outer circle is labeled "Hydropower," so hydropower generation is the most efficient.

d) The darker yellow circle labeled "Solar" occupies less of the area of its surrounding circle than does the darker green circle labeled "Wind." Thus, solar generation of electricity is less efficient than wind generation.

◀ **Do Exercises 9–12.**

We often use **circle graphs**, also called **pie charts**, to show the percent of a quantity in each of several categories. Circle graphs can also be used very effectively to show visually the *ratio* of one category to another. In either case, it is quite often necessary to use mathematics to find the actual amounts represented for each specific category.

Use the graph in Example 3 to answer Margin Exercises 9–12.

**9.** How much electricity is generated annually from natural gas?

**10.** Which method generates the most electricity?

**11.** Solar, geothermal, hydropower, and wind are all considered renewable energy sources. How much electricity is generated annually from renewable sources?

**12.** Is nuclear generation of electricity more or less efficient than hydropower generation?

**Answers**

**9.** 1379 billion kWh **10.** Natural gas
**11.** 546 billion kWh **12.** Nuclear generation of electricity is less efficient than hydropower generation.

**EXAMPLE 4** *Endangered Species.* According to the International Union for Conservation of Nature, seven species of whales are endangered or near-threatened. The following circle graph shows the approximate percentage of the entire population of endangered or near-threatened whales that each species represents.

**Endangered or Near-Threatened Whales**

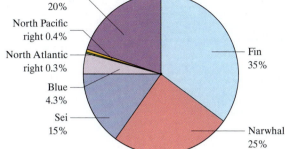

Beluga 20%
North Pacific right 0.4%
North Atlantic right 0.3%
Blue 4.3%
Sei 15%
Fin 35%
Narwhal 25%

**a)** Which species has the greatest population?

**b)** The total number of whales in these seven species is about 300,000. How many blue whales are there?

**c)** What percent of the total population of endangered or near-threatened whales are right whales?

We look at the sections of the graph to find the answers.

**a)** The largest section (or *sector*) of the graph represents 35% of the population and corresponds to fin whales.

**b)** The section representing blue whales is 4.3% of the circle. Since 4.3% of 300,000 is 12,900, there are approximately 12,900 blue whales.

**c)** There are two kinds of right whales represented on the graph: North Pacific right whales and North Atlantic right whales. We add the percents corresponding to these whales:

$$0.4\% + 0.3\% = 0.7\%.$$

Do Exercises 13–15. ▶

Use the circle graph in Example 4 to answer Margin Exercises 13–15.

**13.** Which species accounts for 20% of the entire population of endangered or near-threatened whales?

**14.** What percent of the population of endangered or near-threatened whales are fin whales or sei whales?

**15.** The total number of whales in these seven species is about 300,000. How many fin whales are there?

*Answers*
**13.** Beluga whales  **14.** 50%
**15.** 105,000 fin whales

---

**8.1** **Exercise Set**

 MyLab Math

✔ **Check Your Understanding**

**Reading Check** Determine whether each statement is true or false.

_____ **RC1.** There is only one correct way to represent a set of data.

_____ **RC2.** It is usually easy to read exact amounts from a pictograph.

_____ **RC3.** A circle graph follows the area principle.

_____ **RC4.** If the same data were displayed in a table and in a pictograph, we would have to use the pictograph to determine a maximum or a minimum.

**Concept Check**  The following statements refer to the graph on the right. Determine whether each statement is true or false.

_____ **CC1.** Anita spent 100% of her disposable income on music, clothing, electronics, and dining out.

_____ **CC2.** Anita spent more than half of her disposable income on clothing.

_____ **CC3.** Anita spent about $\frac{1}{4}$ of her disposable income on music.

_____ **CC4.** Anita spent about the same amount on electronics as she spent on dining out and music combined.

_____ **CC5.** If Anita has $100 in disposable income, she spends about $50 on electronics.

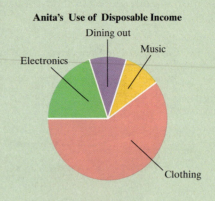

**Anita's Use of Disposable Income**

a  *Heat Index.*   In warm weather, a person can feel hot because of reduced heat loss from the skin caused by higher humidity. The **temperature–humidity index**, or **apparent temperature**, is what the temperature would have to be with no humidity in order to give the same heat effect. The following table lists the apparent temperatures for various actual temperatures and relative humidities. Use this table for Exercises 1–12.

| ACTUAL TEMPERATURE (°F) | RELATIVE HUMIDITY | | | | | | | | | |
|---|---|---|---|---|---|---|---|---|---|---|
| | 10% | 20% | 30% | 40% | 50% | 60% | 70% | 80% | 90% | 100% |
| | APPARENT TEMPERATURE (°F) | | | | | | | | | |
| 75° | 75 | 77 | 79 | 80 | 82 | 84 | 86 | 88 | 90 | 92 |
| 80° | 80 | 82 | 85 | 87 | 90 | 92 | 94 | 97 | 99 | 102 |
| 85° | 85 | 88 | 91 | 94 | 97 | 100 | 103 | 106 | 108 | 111 |
| 90° | 90 | 93 | 97 | 100 | 104 | 107 | 111 | 114 | 118 | 121 |
| 95° | 95 | 99 | 103 | 107 | 111 | 115 | 119 | 123 | 127 | 131 |
| 100° | 100 | 105 | 109 | 114 | 118 | 123 | 127 | 132 | 137 | 141 |
| 105° | 105 | 110 | 115 | 120 | 125 | 131 | 136 | 141 | 146 | 151 |

In Exercises 1–4, find the apparent temperature for the given actual temperature and humidity combinations.

**1.** 80°, 60%          **2.** 90°, 70%          **3.** 85°, 90%          **4.** 95°, 80%

**5.** Which temperature–humidity combinations give an apparent temperature of 100°?

**6.** Which temperature–humidity combinations give an apparent temperature of 111°?

**7.** At a relative humidity of 50%, what actual temperatures give an apparent temperature above 100°?

**8.** At a relative humidity of 90%, what actual temperatures give an apparent temperature above 100°?

**9.** At an actual temperature of 95°, what relative humidities give an apparent temperature above 100°?

**10.** At an actual temperature of 85°, what relative humidities give an apparent temperature above 100°?

**11.** At an actual temperature of 85°, what is the difference in humidities required to raise the apparent temperature from 94° to 108°?

**12.** At an actual temperature of 80°, what is the difference in humidities required to raise the apparent temperature from 87° to 102°?

**Planets.**   Use the following table, which lists information about the planets, for Exercises 13–18.

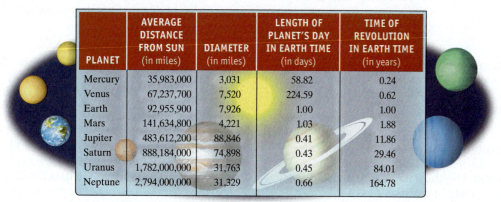

| PLANET | AVERAGE DISTANCE FROM SUN (in miles) | DIAMETER (in miles) | LENGTH OF PLANET'S DAY IN EARTH TIME (in days) | TIME OF REVOLUTION IN EARTH TIME (in years) |
|---|---|---|---|---|
| Mercury | 35,983,000 | 3,031 | 58.82 | 0.24 |
| Venus | 67,237,700 | 7,520 | 224.59 | 0.62 |
| Earth | 92,955,900 | 7,926 | 1.00 | 1.00 |
| Mars | 141,634,800 | 4,221 | 1.03 | 1.88 |
| Jupiter | 483,612,200 | 88,846 | 0.41 | 11.86 |
| Saturn | 888,184,000 | 74,898 | 0.43 | 29.46 |
| Uranus | 1,782,000,000 | 31,763 | 0.45 | 84.01 |
| Neptune | 2,794,000,000 | 31,329 | 0.66 | 164.78 |

DATA: *The Handy Science Answer Book*, Gale Research, Inc.

**13.** Find the average distance from the sun to Jupiter.

**14.** How long is a day on Venus?

**15.** Which planet has a time of revolution of 164.78 years?

**16.** Which planet has a diameter of 4221 mi?

**17.** About how many Earth diameters equal one Jupiter diameter?

**18.** How much longer is the longest time of revolution than the shortest?

**Nutrition Facts.**   Most foods are required by law to come with factual information regarding nutrition, like that in the following table of nutrition facts from a box of breakfast cereal. Use the nutrition data for Exercises 19–24 on the next page.

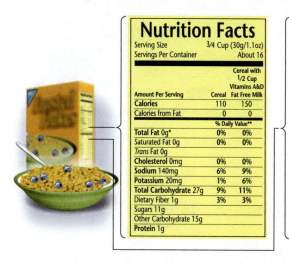

**19.** Suppose your breakfast consists of $1\frac{1}{2}$ cups of cereal with 1 cup of fat-free milk. How many calories do you consume?

**20.** Suppose your breakfast consists of $1\frac{1}{2}$ cups of cereal with 1 cup of fat-free milk. What percent of the daily value of dietary fiber do you consume?

**21.** A nutritionist recommends that you look for foods that provide 10% or more of the daily value of vitamin C. Do you get that with 1 serving of cereal and $\frac{1}{2}$ cup of fat-free milk?

**22.** Suppose you are trying to limit your daily caloric intake to 2000 calories. How many servings of cereal alone would it take to exceed 2000 calories?

**23.** Suppose your breakfast consists of $1\frac{1}{2}$ cups of cereal with 1 cup of fat-free milk. How much sodium do you consume? (*Hint*: Use the data listed in the first footnote below the table of nutrition facts.)

**24.** Suppose your breakfast consists of $1\frac{1}{2}$ cups of cereal with 1 cup of fat-free milk. How much protein do you consume? (*Hint*: Use the data listed in the first footnote below the table of nutrition facts.)

**b** *Rhino Population.* The rhinoceros is considered one of the world's most endangered animals. The total number of rhinos worldwide is approximately 33,200. The following pictograph shows the populations of the five remaining rhino species. Located in the graph is a key that tells you that each symbol represents 500 rhinos. Use the pictograph for Exercises 25–30.

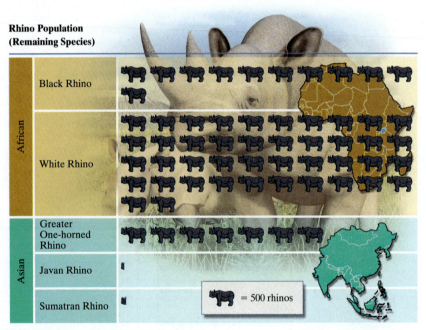

DATA: savetherhino.org

**25.** Which species has the greatest number of rhinos?

**26.** Which species has the least number of rhinos?

**27.** How many more black rhinos are there than greater one-horned rhinos?

**28.** How many more white rhinos are there than black rhinos?

**29.** How many times as large is the white rhino population than the greater one-horned rhino population?

**30.** How many more African rhinos are there than Asian rhinos?

**Personal Consumption Expenditures.** The following graph shows the amounts of personal consumption expenditures, in dollars per person per year, in the United States, for four years. The graph also shows the amounts spent on food and on financial services and insurance for those years, labeled as percents of the personal consumption expenditures. Use the graph for Exercises 31–38.

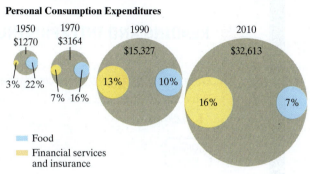

**Personal Consumption Expenditures**

DATA: *TIME*, October 10, 2011, p. 32; infoplease.com; U.S. Census Bureau

**31.** How much were personal consumption expenditures per person in 1950?

**32.** How much were personal consumption expenditures per person in 2010?

**33.** For which of the years shown was more spent on food than on financial services and insurance?

**34.** For which of the years shown was more spent on financial services and insurance than on food?

**35.** How much per person was spent on food in 1990?

**36.** How much per person was spent on financial services and insurance in 1970?

**37. a)** How much less, as a percent of personal consumption expenditures, was spent on food in 2010 than in 1950?
**b)** How much more, in dollars, was spent on food in 2010 than in 1950?

**38. a)** How much more, as a percent of personal consumption expenditures, was spent on financial services and insurance in 2010 than in 1950?
**b)** How much more, in dollars, was spent on financial services and insurance in 2010 than in 1950?

**Foreign Students.** The circle graph below shows the foreign countries sending the most students to the United States to attend colleges and universities. Use this graph for Exercises 39–44.

**39.** What percent of foreign students are from South Korea?

**40.** Together, what percent of foreign students are from China and Taiwan?

**41.** In 2016, there were approximately 1,040,000 foreign students studying at colleges and universities in the United States. According to the data in the graph, how many were from India?

**42.** In 2016, there were approximately 1,040,000 foreign students studying in the United States. How many were from Saudi Arabia?

**43.** Which country accounted for 3% of the foreign students?

**44.** Which country accounted for 16% of the foreign students?

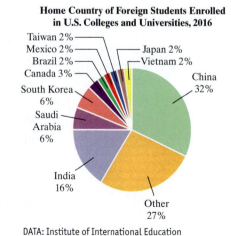

**Home Country of Foreign Students Enrolled in U.S. Colleges and Universities, 2016**

DATA: Institute of International Education

# 8.2

## OBJECTIVES

a Extract and interpret data from bar graphs.

b Draw bar graphs.

c Extract and interpret data from line graphs.

d Draw line graphs.

# Interpreting and Drawing Bar Graphs and Line Graphs

## a READING AND INTERPRETING BAR GRAPHS

**SKILL REVIEW**

*Given a pair of numbers in decimal notation, tell which is larger.* [5.1c]
Which number is larger?

**1.** 0.078, 0.1            **2.** 36.4, 9.875

                                          **Answers: 1.** 0.1 **2.** 36.4

MyLab Math
VIDEO

A **bar graph** is convenient for showing comparisons because you can tell at a glance which quantity is the largest or smallest. A *scale* is usually included with a bar graph so that estimates of values can be made with some accuracy. Bar graphs may be drawn horizontally or vertically, and sometimes a double bar graph is used to make comparisons.

**EXAMPLE 1** *Coffee and Tea Consumption.* The following horizontal bar graph is a double bar graph, showing per capita consumption, in pounds per person per year, of both coffee and tea for several countries.

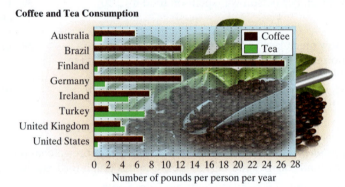

DATA: caffeineinformer.com; irishexaminer.com; joe.ie; msn.com

a) Which country has the highest per capita coffee consumption?

b) What is the per capita tea consumption in Germany?

c) In which countries do people consume more pounds of tea than coffee each year?

d) In which two countries do people consume about the same amount of coffee?

e) In which countries is per capita coffee consumption greater than 10 pounds per year?

  We use the graph to answer the questions.

a) The longest brown bar is for Finland. Thus, Finland has the highest coffee consumption per capita.

b) We look to the right along the green bar associated with Germany. Since it ends halfway between 1 and 2, we estimate Germany's per capita tea consumption to be 1.5 pounds per year.

c) The green bars are longer than the brown bars for Turkey and the United Kingdom, so people in Turkey and in the United Kingdom consume, on average, more pounds of tea than coffee each year.

**d)** The brown bars are the same length for Brazil and Germany; both countries have a per capita consumption of about 12 pounds of coffee per year. The people in Brazil and Germany, on average, consume about the same amount of coffee per year.

**e)** We move across the horizontal scale to 10. From there we move up, noting any brown bars that are longer than 10 units. We see that per capita coffee consumption is greater than 10 pounds per year in Brazil, Finland, and Germany.

**Do Exercises 1–3.** ▶

## b   DRAWING BAR GRAPHS

**EXAMPLE 2**   *Population by Age.*   Listed below are U.S. population data for selected age groups. Make a vertical bar graph of the data.

| AGE GROUP | PERCENT OF POPULATION |
|-----------|----------------------|
| Under 5 years | 7% |
| 5 to 17 years | 17% |
| 18 to 24 years | 10% |
| 25 to 44 years | 27% |
| 45 to 64 years | 26% |
| 65 years and over | 13% |

DATA: U.S. Census Bureau

First, we indicate the age groups in six equally spaced intervals on the horizontal scale and give the horizontal scale the title "Age category." (See the figure on the left below.)

Next, we scale the vertical axis. To do so, we look over the data and note that it ranges from 7% to 27%. We start the vertical scaling at 0, labeling the marks by 5's from 0 to 30. We give the vertical scale the title "Percent of population" and the graph the overall title "U.S. Population by Age."

Finally, we draw vertical bars to show the various percents, as shown in the figure on the right below.

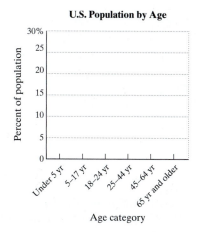

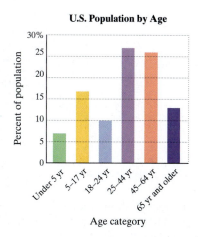

**Do Exercise 4.** ▶

Use the bar graph in Example 1 to answer Margin Exercises 1–3.

**1.** What is the per capita coffee consumption in the United Kingdom?

**2.** In which countries is per capita tea consumption less than 2 pounds per year?

**3.** How many more pounds of coffee are consumed per person in the United States than pounds of tea?

**4.** *Planetary Moons.*   Make a horizontal bar graph to show the numbers of moons orbiting the various planets.

| PLANET | MOONS |
|--------|-------|
| Earth | 1 |
| Mars | 2 |
| Jupiter | 63 |
| Saturn | 60 |
| Uranus | 27 |
| Neptune | 13 |

DATA: National Aeronautics and Space Administration

*Answers*

**1.** About 3.7 pounds per year
**2.** Australia, Brazil, Finland, Germany, and the United States   **3.** About 6 pounds per person
**4.**

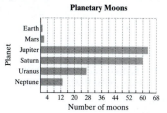

## c   READING AND INTERPRETING LINE GRAPHS

**Line graphs** are often used to show a change over time as well as to indicate patterns or trends.

**EXAMPLE 3**   *Gold.*   The following line graph shows the average price of gold, in dollars per ounce, for various years from 1970 to 2015.

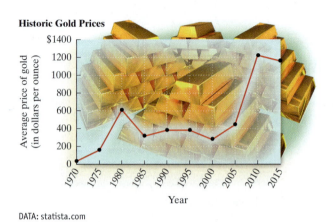

**Historic Gold Prices**

DATA: statista.com

a) For which year before 2000 was the average price of gold the highest?

b) Between which years did the average price of gold decrease?

c) For which year was the average price of gold about $1225 per ounce?

d) By how much did the average price of gold decrease from 2010 to 2015?

We look at the graph to answer the questions.

a) Before 2000, the highest point on the graph corresponds to 1980. The highest average price of gold was about $610 per ounce in 1980.

b) Reading the graph from left to right, we see that the average price of gold decreased from 1980 to 1985, from 1995 to 2000, and from 2010 to 2015.

c) We look from left to right along a line at $1225 per ounce. We see that the average price of gold was about $1225 per ounce in 2010.

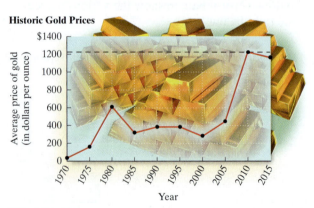

**Historic Gold Prices**

DATA: statista.com

d) The graph shows that the average price of gold was about $1225 per ounce in 2010 and about $1150 per ounce in 2015. Thus, the average price of gold decreased by $1225 − $1150 = $75 per ounce.

◀ **Do Exercises 5–7.**

---

Use the line graph in Example 3 to answer Margin Exercises 5–7.

**5.** For which year after 1980 was the average price of gold the lowest?

**6.** Between which years did the average price of gold increase by about $800 per ounce?

**7.** For which years was the average price of gold less than $400 per ounce?   **GS**

We look from left to right along a line at $ ____ per ounce. The points on the graph that are below this line correspond to the years 1970, 1975, 1985, 1990, 1995, and ____ .

*Answers*

**5.** 2000   **6.** Between 2005 and 2010
**7.** 1970, 1975, 1985, 1990, 1995, 2000

*Guided Solution:*
**7.** 400; 2000

## d  DRAWING LINE GRAPHS

**EXAMPLE 4** *Temperature in Enclosed Vehicle.* The temperature inside an enclosed vehicle increases rapidly with time. Listed in the table below are the inside temperatures of an enclosed vehicle for specified elapsed times when the outside temperature is 80°F. Make a line graph of the data.

| ELAPSED TIME | TEMPERATURE IN ENCLOSED VEHICLE WITH OUTSIDE TEMPERATURE 80°F |
|---|---|
| 10 min | 99° |
| 20 min | 109° |
| 30 min | 114° |
| 40 min | 118° |
| 50 min | 120° |
| 60 min | 123° |

DATA: General Motors; Jan Null, Golden Gate Weather Services

First, we indicate the 10-min elapsed time intervals on the horizontal scale and give the horizontal scale the title "Elapsed time (in minutes)." (See the figure on the left below.) Next, we scale the vertical axis by 10's beginning with 80 to show the number of degrees and give the vertical scale the title "Temperature (in degrees)." The jagged line at the base of the vertical scale indicates that an unused portion of the scale has been omitted. We also give the graph the overall title "Temperature in Enclosed Vehicle with Outside Temperature 80°F."

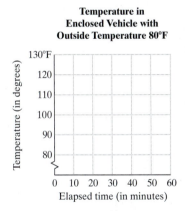

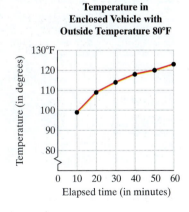

Next, we mark the temperature at the appropriate level above each elapsed time. (See the figure on the right above.) Then we draw line segments connecting the points. The rapid change in temperature can be observed easily from the graph.

**Do Exercise 8.** ▶

**8.** *Phone Sales.* Listed below are the numbers of Apple iPhones sold worldwide for the years 2007–2016. Make a line graph of the data.

| YEAR | NUMBER OF iPHONES SOLD (in millions) |
|---|---|
| 2007 | 1.39 |
| 2008 | 11.63 |
| 2009 | 20.73 |
| 2010 | 39.99 |
| 2011 | 72.29 |
| 2012 | 125.05 |
| 2013 | 150.26 |
| 2014 | 169.22 |
| 2015 | 231.22 |
| 2016 | 211.88 |

DATA: statista.com

*Answer*

**8.**

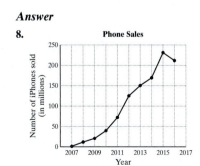

### ✓ Check Your Understanding

**Reading Check**   Determine whether each statement is true or false.

_____ **RC1.** Bar graphs may be drawn horizontally or vertically.

_____ **RC2.** A double bar graph indicates two amounts for each category.

_____ **RC3.** A line graph is always used to show trends over time.

_____ **RC4.** Some data could be illustrated using either a line graph or a bar graph.

**Concept Check**   Determine whether a line graph would be an appropriate way to display the information described.

**CC1.** The number of each kind of tree present in a nature preserve

**CC2.** The number of students enrolled in a communication class for the years between 2010 and 2018

**CC3.** The price of a stock at the end of each week

**CC4.** The percent of a state's economy represented by each economic sector

**a**   *Chocolate Desserts.*   The following horizontal bar graph shows the average caloric content of various kinds of chocolate desserts. Use the bar graph for Exercises 1–8.

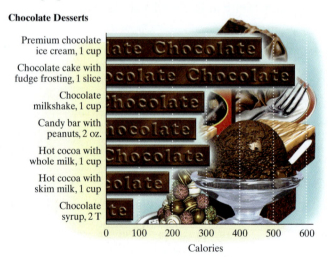

**Chocolate Desserts**

1. Estimate how many calories there are in 1 cup of hot cocoa with skim milk.

2. Estimate how many calories there are in a 2-oz candy bar with peanuts.

3. Which dessert has the highest caloric content?

4. Which dessert has the lowest caloric content?

5. Which dessert contains about 460 calories?

6. Which desserts contain about 300 calories?

7. How many more calories are there in 1 cup of hot cocoa made with whole milk than in 1 cup of hot cocoa made with skim milk?

8. If Emily drinks a 4-cup chocolate milkshake, how many calories does she consume?

**Bearded Irises.** A gardener planted six varieties of bearded iris in a new garden on campus. Students from the horticulture department were assigned to record data on the range of heights for each variety. The vertical bar graph below shows their results. The length of the light green shaded portion of each bar and the blossom illustrates the range of heights for a variety. For example, the range of heights for the miniature dwarf bearded iris is 2 in. to 9 in. Use the graph for Exercises 9–16.

9. Which variety of iris has a minimum height of 17 in.?

10. Which variety of iris has a maximum height of 28 in.?

11. What is the range of heights for the border bearded iris?

12. What is the range of heights for the standard dwarf bearded iris?

13. Which variety of iris has the smallest range in heights?

14. Which irises have a maximum height less than 16 in.?

15. What is the difference between the maximum heights of the tallest iris and the shortest iris?

16. Which irises have a range in heights less than 10 in.?

**Bearded Irises**

DATA: irises.org

**Bachelor's Degrees.** The graph at right provides data on the numbers of bachelor's degrees conferred on men and on women in selected fields. Use the bar graph for Exercises 17–20.

17. In what fields were more bachelor's degrees conferred on men than on women?

18. How many more bachelor's degrees were conferred on women in education than in communication?

19. How many more bachelor's degrees were conferred on women than on men in communication?

20. In what fields was the total number of bachelor's degrees conferred on men and on women greater than 50,000?

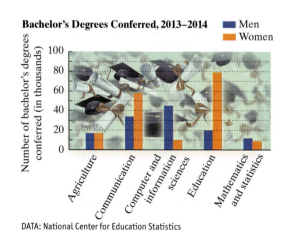

DATA: National Center for Education Statistics

**b**

**21. Cost of Living Index.** The following table lists the cost of living index for several cities. The national average of this index is 100. An index greater than 100 indicates that the cost of living is higher than average, and an index less than 100 indicates that the cost of living is lower than average. Make a horizontal bar graph to illustrate the data.

| CITY | COST OF LIVING INDEX |
|------|------|
| Chicago | 116.9 |
| Denver | 99.4 |
| New York City | 185.8 |
| Juneau | 136.5 |
| Indianapolis | 87.2 |
| San Diego | 132.3 |
| Salt Lake City | 100.6 |

DATA: U.S. Census Bureau

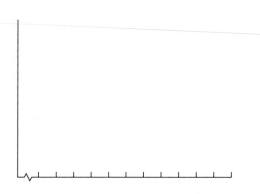

Use the data and the bar graph you created in Exercise 21 to do Exercises 22–25.

**22.** Which city has the highest cost of living index?

**23.** In which cities is the cost of living index less than 100?

**24.** In which cities is the cost of living approximately the national average?

**25.** How much higher is the cost of living index in New York City than in Chicago?

**26. Commuting Time.** The following table lists the average commuting time to work in six U.S. cities. Make a vertical bar graph to illustrate the data.

| CITY | COMMUTING TIME (in minutes) |
|------|------|
| New York City | 39.7 |
| Chicago | 33.7 |
| Los Angeles | 29.9 |
| Phoenix | 24.7 |
| Indianapolis | 22.6 |
| Oklahoma City | 20.7 |

DATA: University of Michigan Transportation Research Institute

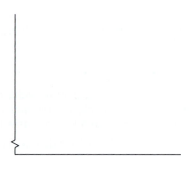

Use the data and the bar graph you created in Exercise 26 to do Exercises 27–30.

**27.** Which city has the longest commuting time?

**28.** Which city has the shortest commuting time?

**29.** The average commuting time to work in the United States is 26 minutes. What cities in this list have a longer than average commuting time?

**Data:** U.S. Census Bureau

**30.** How much time does a worker in New York City spend, on average, commuting to work in one 5-day work week?

**C** *Facebook Stock.* The line graph below shows the price per share of Facebook stock when it was first offered in May 2012 and at the beginning of each month for the remainder of that year. Use the graph for Exercises 31–34.

**Stock Performance of Facebook**

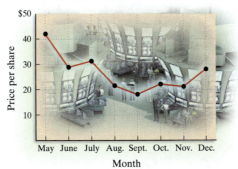

DATA: finance.yahoo.com

**31.** Estimate the opening price per share of Facebook stock in May 2012.

**32.** How much higher was the opening price of Facebook stock than its price at the beginning of September?

**33.** Between which months did the price of Facebook stock increase?

**34.** Between which months was the decrease in the price of Facebook stock the greatest?

*Monthly Loan Payment.* Suppose you borrow $110,000 at an interest rate of $5\frac{1}{2}$% to buy a condominium. The following graph shows the monthly payment required to pay off the loan, depending on the length of the loan. Use the graph for Exercises 35–42.

**$110,000 Loan Repayment**

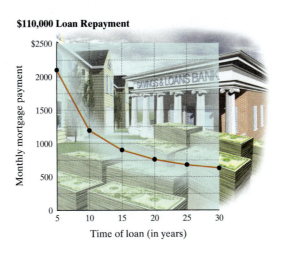

**35.** Estimate the monthly payment for a loan of 15 years.

**36.** Estimate the monthly payment for a loan of 25 years.

**37.** What time period corresponds to a monthly payment of about $760?

**38.** What time period corresponds to a monthly payment of about $625?

**39.** By how much does the monthly payment decrease when the loan period is increased from 10 years to 20 years?

**40.** By how much does the monthly payment decrease when the loan period is increased from 5 years to 20 years?

**41.** For a 10-year loan, there are 120 monthly payments. In all, how much will you pay back for a 10-year loan?

**42.** For a 20-year loan, there are 240 monthly payments. In all, how much will you pay back for a 20-year loan?

d

**43. Longevity Beyond Age 65.** The data in the table below indicate how many years beyond age 65 a male who is 65 in the given year could expect to live. Draw a line graph using the horizontal axis to scale "Year."

| YEAR | AVERAGE NUMBER OF YEARS MEN ARE ESTIMATED TO LIVE BEYOND AGE 65 |
|------|-----------------------------------------------------------------|
| 1980 | 14 |
| 1990 | 15 |
| 2000 | 15.9 |
| 2010 | 16.4 |
| 2020 | 16.9 |
| 2030 | 17.5 |

DATA: 2000 Social Security Report

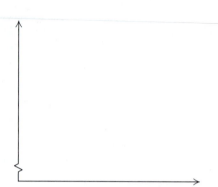

**44.** What was the percent increase in longevity (years beyond 65) between 1980 and 2000?

**45.** What is the expected percent increase in longevity between 1980 and 2030?

**46.** What is the expected percent increase in longevity between 2020 and 2030?

**47.** What is the expected percent increase in longevity between 2000 and 2030?

## Skill Maintenance

Solve.

**48.** $32 + n = 115$  [1.7b]

**49.** $x \cdot \dfrac{2}{3} = \dfrac{8}{9}$  [3.8b]

**50.** $y + \dfrac{5}{8} = \dfrac{11}{12}$  [4.3b]

**51.** $5 \cdot x = 11.3$  [5.7a]

**52.** $t + 4.752 = 11.1$  [5.7a]

**53.** $\dfrac{9}{10} = \dfrac{x}{8}$  [6.3b]

**54.** 51.2 is 64% of what?
[7.2a, b], [7.3a, b]

**55.** What is $4\dfrac{1}{2}$% of 20?
[7.2a, b], [7.3a, b]

**56.** 120 is what percent of 80?
[7.2a, b], [7.3a, b]

Calculate.

**57.** $3 \times [11 + (18 - 10) \div 2^3 - 5]$  [1.9d]

**58.** $2.56 \div (4 - 3.84) + 6.3 \times 0.2$  [5.4b]

**59.** $\dfrac{9}{10} \div \dfrac{1}{2} \cdot \dfrac{1}{3} - \left(\dfrac{1}{4} - \dfrac{1}{6}\right)$  [4.8a]

**60.** $6.25 \times 7\dfrac{1}{5}$  [5.5d]

# Ordered Pairs and Equations in Two Variables

## OBJECTIVES

**a** Plot a point, given its coordinates. Find coordinates, given a point.

**b** Determine the quadrant in which a point lies.

**c** Determine whether an ordered pair is a solution of an equation with two variables.

By using two perpendicular number lines as **axes** (pronounced ăk′sēz; singular: **axis**), we can use points to represent solutions of certain equations. First we look at graphing points using such axes.

## a POINTS AND ORDERED PAIRS

When two number lines are used as axes, a grid can be formed. Just as a location in a city might be given as the intersection of an avenue and a side street, a point on a plane might be described as the intersection of a vertical line and a horizontal line on a grid. For the point shown in the figure below, these lines pass through 3 on the horizontal axis and 4 on the vertical axis. Thus, the **first coordinate** of this point is 3 and the **second coordinate** is 4. **Ordered pair** notation, (3, 4), provides a quick way of stating this.

········· **Caution!** ·········

When writing an ordered pair, you should *always* list the coordinate from the horizontal axis first.

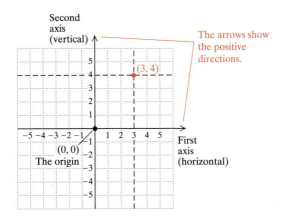

The point (0, 0), where the axes cross each other, is called the **origin**. To graph, or *plot*, the point (3, 4), we can begin at the origin and move horizontally (along the first axis) to the number 3. From there, we move up 4 units vertically and make a "dot."

**Do Exercises 1 and 2.** ▶

Plot these points on the graph below.

**1.** (2, 5)     **2.** (4, 1)

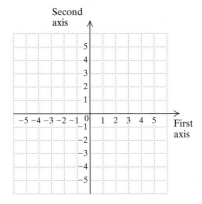

**EXAMPLE 1** Plot the points $(-5, 2)$ and $(2, -5)$.

To plot $(-5, 2)$, we locate $-5$ on the first, or horizontal, axis. From there we go up 2 units and make a dot.

To plot $(2, -5)$, we locate 2 on the first, or horizontal, axis. Then we go down 5 units and make a dot. Note that the order of the numbers within a pair is important: $(2, -5) \neq (-5, 2)$.

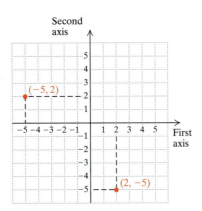

*Answers*
**1 and 2.**

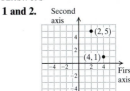

Plot these points on the graph below.

**3.** $(-2, 5)$      **4.** $(-3, -4)$

**5.** $(5, -3)$      **6.** $(-2, -1)$

**7.** $(0, -3)$      **8.** $(2\frac{1}{2}, 0)$

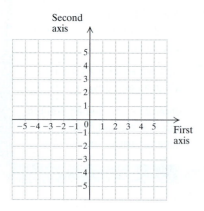

**9.** Determine the coordinates of points $A$, $B$, $C$, $D$, $E$, $F$, and $G$ on the graph below.

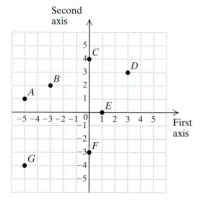

*Answers*

**3–8.**

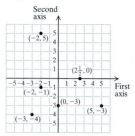

**9.** $A$: $(-5, 1)$; $B$: $(-3, 2)$; $C$: $(0, 4)$; $D$: $(3, 3)$; $E$: $(1, 0)$; $F$: $(0, -3)$; $G$: $(-5, -4)$

◀ **Do Exercises 3–8.**

To determine the coordinates of a given point, we first look directly above or below the point to find the point's horizontal coordinate. Then we look to the left or right of the point to identify the vertical coordinate.

**EXAMPLE 2**   Determine the coordinates of points $A$, $B$, $C$, $D$, $E$, $F$, and $G$.

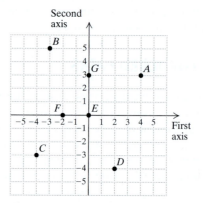

We look below point $A$ to see that its first coordinate is 4. Looking to the left of point $A$, we find that its second coordinate is 3. Thus, the coordinates of point $A$ are $(4, 3)$. The coordinates of the other points are

$B$: $(-3, 5)$;     $C$: $(-4, -3)$;     $D$: $(2, -4)$;

$E$: $(0, 0)$;        $F$: $(-2, 0)$;      $G$: $(0, 3)$.

◀ **Do Exercise 9.**

**b**   **QUADRANTS**

The axes divide the plane into four regions, or **quadrants**. For any point in region I (the *first quadrant*), both coordinates are positive. For any point in region II (the *second quadrant*), the first coordinate is negative and the second coordinate is positive. In region III (the *third quadrant*), both coordinates are negative. In region IV (the *fourth quadrant*), the first coordinate is positive and the second coordinate is negative.

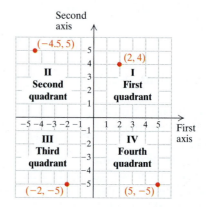

As the figure above illustrates, the point $(2, 4)$ is in the first quadrant, $(-4.5, 5)$ is in the second quadrant, $(-2, -5)$ is in the third quadrant, and $(5, -5)$ is in the fourth quadrant.

Do Exercises 10–15.

## C SOLUTIONS OF EQUATIONS

**SKILL REVIEW**

*Evaluate an algebraic expression by substitution.* [2.6a]

**1.** Evaluate $2x + 3y$ for $x = 5$ and $y = -3$.

**2.** Evaluate $x - y$ for $x = 2$ and $y = -10$.

Answers: **1.** 1 **2.** 12

MyLab Math
VIDEO

The coordinate system we have just introduced is called the **Cartesian** coordinate system, in honor of the mathematician and philosopher René Descartes (1596–1650). Legend has it that Descartes hit upon the idea of the coordinate system after watching a fly stop several times on the ceiling over his bed. We can use this coordinate system as a method of presenting solutions of equations containing two variables. Equations like $3x + 2y = 8$ have ordered pairs as solutions. In Section 8.4, we will find solutions and graph them. Here we simply practice checking to see if an ordered pair is a solution.

To determine whether an ordered pair is a solution of an equation, we normally substitute the first coordinate for the variable that comes first alphabetically and the second coordinate for the variable that is last alphabetically. The letters $x$ and $y$ are used most often.

**EXAMPLE 3** Determine whether the ordered pair $(2, 1)$ is a solution of the equation $3x + 2y = 8$.

$$\frac{3x + 2y = 8}{\begin{array}{c|c} 3 \cdot 2 + 2 \cdot 1 \ ? \ 8 \\ 6 + 2 \\ 8 & 8 \ \text{TRUE} \end{array}}$$

Substituting 2 for $x$ and 1 for $y$ (alphabetical order of variables)

Since the equation becomes true, $(2, 1)$ is a solution.

In a similar manner, we can show that $(0, 4)$ and $(4, -2)$ are also solutions of $3x + 2y = 8$. In fact, there are an infinite number of solutions of $3x + 2y = 8$.

**EXAMPLE 4** Determine whether the ordered pair $(-2, 3)$ is a solution of the equation $2t = 4s - 8$.

We substitute:

$$\frac{2t = 4s - 8}{\begin{array}{c|c} 2 \cdot 3 \ ? \ 4(-2) - 8 \\ 6 & -8 - 8 \\ 6 & -16 \quad \text{FALSE} \end{array}}$$

Using alphabetical order, substituting $-2$ for $s$ and 3 for $t$

Since the equation becomes false, $(-2, 3)$ is not a solution.

**Unless it is stated otherwise, the coordinates of an ordered pair correspond alphabetically to the variables.**

Do Exercises 16 and 17.

---

**10.** What can you say about the coordinates of a point in the third quadrant?

**11.** What can you say about the coordinates of a point in the fourth quadrant?

In which quadrant is each point located?

**GS** **12.** $(5, 3)$

To plot the point $(5, 3)$, we locate ▢ on the horizontal axis and go up ▢ units. We are now in the ▢ quadrant, or quadrant I.

**13.** $(-6, -4)$     **14.** $(10, -14)$

**15.** $\left(-13, 9\frac{1}{2}\right)$

**GS** **16.** Determine whether $(5, 1)$ is a solution of $y = 2x + 3$.

We substitute ▢ for $x$ and ▢ for $y$.

$$\frac{y = 2x + 3}{\begin{array}{c|c} 1 \ ? \ 2 \cdot 5 + 3 \\ 1 & ▢ \end{array}}$$

Since $1 = 13$ is ▢,
true/false
$(5, 1)$ ▢ a solution.
is/is not

**17.** Determine whether $(-13.6, 25.4)$ is a solution of $3x + 2y = 10$.

**Answers**
**10.** Both are negative numbers.
**11.** The first, or horizontal, coordinate is positive; the second, or vertical, coordinate is negative.
**12.** I  **13.** III  **14.** IV  **15.** II  **16.** No
**17.** Yes

**Guided Solutions:**
**12.** 5, 3; first
**16.** 5, 1; 13; false, is not

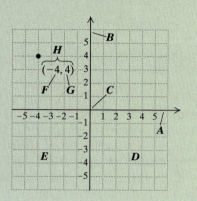
## ✓ Check Your Understanding

**Reading and Concept Check** Match the letter from the graph with the most appropriate term. Not all letters will be used.

RC1. _____ Third quadrant

RC2. _____ Origin

RC3. _____ Second coordinate

RC4. _____ Vertical axis

RC5. _____ Ordered pair

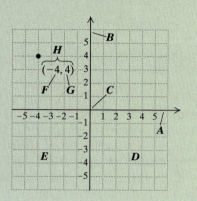

**a** Plot each group of points on the given graph below.

**1.** $(4, 4)$  $(-2, 4)$  $(5, -3)$  $(-5, -5)$  $(0, 4)$  $(0, -4)$
$(3, 0)$  $(-4, 0)$

**2.** $(2, 5)$  $(-1, 3)$  $(3, -2)$  $(-2, -4)$  $(0, 4)$  $(0, -5)$
$(5, 0)$  $(-5, 0)$

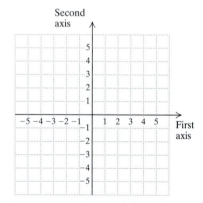

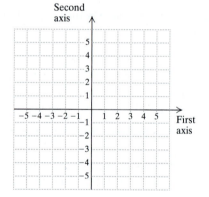

**3.** $(-2, -4)$  $(5, -4)$  $(0, 3\frac{1}{2})$  $(4, 3\frac{1}{2})$  $(-1, -3)$  $(-1, 5)$
$(4, -1)$  $(-2, 0)$

**4.** $(-3, -1)$  $(5, 1)$  $(-1, -5)$  $(0, 0)$  $(0, 1)$  $(-4, 0)$
$(2, 3\frac{1}{2})$  $(4\frac{1}{2}, -2)$

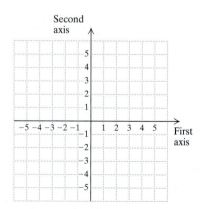

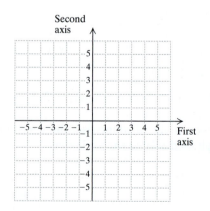

Determine the coordinates of points $A$, $B$, $C$, $D$, $E$, and $F$.

**5.**

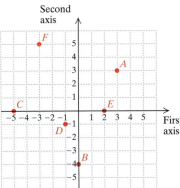

**6.**

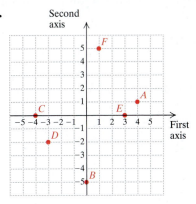

**7.**

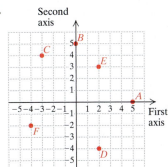

**8.**

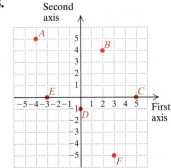

**b**  In which quadrant is each point located?

**9.** $(-5, 3)$

**10.** $(-12, 1)$

**11.** $(100, -1)$

**12.** $\left(35\frac{1}{2}, -2\frac{1}{2}\right)$

**13.** $(-6.5, -1.9)$

**14.** $(-3.4, -5.9)$

**15.** $\left(3\frac{7}{10}, 9\frac{1}{11}\right)$

**16.** $(1895, 1492)$

Complete each sentence using the words *positive* and *negative* and/or the numeral I, II, III, or IV.

**17.** In quadrant IV, first coordinates are always _____ and second coordinates are always _____.

**18.** In quadrant III, first coordinates are always _____ and second coordinates are always _____.

**19.** In quadrant _____, both coordinates are always negative.

**20.** In quadrant _____, both coordinates are always positive.

**21.** In quadrants I and _____, the first coordinate is always _____.

**22.** In quadrants II and _____, the second coordinate is always _____.

**c** Determine whether each ordered pair is a solution of the given equation.

**23.** $(4, 3)$; $y = 2x - 5$

**24.** $(1, 7)$; $y = 2x + 5$

**25.** $(2, -3)$;   $3x - y = 4$

**26.** $(-1, 4)$;   $2x + y = 6$

**27.** $(-2, -1)$;   $3x + 2y = -8$

**28.** $(0, -4)$;   $4x + 2y = -9$

**29.** $(5, -4)$;   $3x + y = 19$

**30.** $(-1, 7)$;   $x - y = -8$

**31.** $\left(2\dfrac{1}{3}, 6\right)$;   $2y - 3x = 5$

**32.** $\left(1\dfrac{1}{4}, 3\right)$;   $2y - 4x = 1$

**33.** $(2.4, 0.7)$; $y = 5x - 11.3$

**34.** $(1.8, 7.4)$; $y = 3x + 2$

## Skill Maintenance

Solve.

**35.** $3x - 4 = 17$   [2.8e]

**36.** $5(x - 2) = 3x - 4$   [5.7b]

**37.** $-\dfrac{1}{9}t = \dfrac{2}{3}t$   [3.8b]

**38.** Simplify: $\dfrac{90}{51}$.   [3.5b]

**39.** Combine like terms:   [4.6b]
$$7\dfrac{2}{11}a - 5\dfrac{1}{3}a.$$

**40.** Simplify:   [2.7a]
$$3(x - 5) + 4x - 9.$$

## Synthesis

Determine whether each ordered pair is a solution of the given equation.

**41.** ▦ $(-2.37, 1.23)$; $5.2x + 6.1y = -4.821$

**42.** ▦ $(4.16, -9.35)$; $6.5x - 7.2y = -94.35$

In Exercises 43–46, determine the quadrant(s) in which the point could be located.

**43.** The first coordinate is positive.

**44.** The second coordinate is negative.

**45.** The first and second coordinates are equal.

**46.** The first coordinate is the opposite of the second coordinate.

**47.** The points $(-1, 1)$, $(4, 1)$, and $(4, -5)$ are three vertices of a rectangle. Find the coordinates of the fourth vertex.

**48.** A parallelogram is a four-sided polygon with two pairs of parallel sides. Two examples are shown below. Three parallelograms share the vertices $(-2, -3)$, $(-1, 2)$, and $(4, -3)$. Find the fourth vertex of each parallelogram.

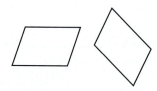

# Graphing Linear Equations

Once we can find a few ordered pairs that make an equation true, we can graph the equation.

## a FINDING SOLUTIONS

To find a solution of an equation with two variables, we first replace one variable with some number choice and then solve the resulting equation for the other variable.

**EXAMPLE 1** Find a solution of $x + y = 7$. Let $x = 5$.

If $x$ is 5, then the solution will be an ordered pair $(5, y)$. We find $y$ by substituting 5 for $x$ in $x + y = 7$:

$$5 + y = 7.$$

We solve for $y$ as follows:

$$5 + y = 7$$

$$5 + y - 5 = 7 - 5 \qquad \text{Subtracting 5 from both sides}$$

$$y = 2.$$

Since $5 + 2 = 7$, the ordered pair $(5, 2)$ is a solution of $x + y = 7$.

**Do Exercise 1.** ▶

**EXAMPLE 2** Complete these solutions of $2x + 3y = 8$: $(\square, 2)$; $(-2, \square)$.

Recall that each solution will be of the form $(x, y)$. Thus, to complete the pair $(\square, 2)$, we replace $y$ with 2 and solve for $x$:

$$2x + 3y = 8$$

$$2x + 3 \cdot 2 = 8 \qquad \text{Substituting 2 for } y$$

$$2x + 6 = 8$$

$$2x + 6 - 6 = 8 - 6 \qquad \text{Subtracting 6 from both sides}$$

$$2x = 2$$

$$\tfrac{1}{2} \cdot 2x = \tfrac{1}{2} \cdot 2 \qquad \text{Multiplying both sides by } \tfrac{1}{2}$$

$$x = 1.$$

Thus, $(1, 2)$ is a solution of $2x + 3y = 8$. Since $2(1) + 3(2) = 8$, the solution checks.

To complete the pair $(-2, \square)$, we replace $x$ with $-2$ and solve for $y$:

$$2x + 3y = 8$$

$$2(-2) + 3y = 8 \qquad \text{Substituting } -2 \text{ for } x$$

$$-4 + 3y = 8$$

$$3y = 12 \qquad \text{Adding 4 to both sides}$$

$$y = 4. \qquad \text{Dividing both sides by 3}$$

Thus, $(-2, 4)$ is also a solution of $2x + 3y = 8$. Since $2(-2) + 3(4) = 8$, the solution checks.

**Do Exercise 2.** ▶

---

**OBJECTIVES**

a Find solutions of equations in two variables.

b Graph linear equations in two variables.

c Graph equations for horizontal lines or vertical lines.

---

**SKILL REVIEW**

*Solve equations that require use of both the addition principle and the division principle.* [2.8e]

Solve.

**1.** $3x + 2 \cdot 4 = 23$

**2.** $4(-1) - y = 7$

**Answers: 1.** 5 **2.** $-11$

MyLab Math
**VIDEO**

---

**GS 1.** Find a solution of $x - y = 3$. Let $y = 5$.

Substitute 5 for $y$ and solve for $x$.

$$x - y = 3$$

$$x - \boxed{\phantom{5}} = 3$$

$$x - 5 + 5 = 3 + \boxed{\phantom{5}}$$

$$x = \boxed{\phantom{5}}$$

One solution of $x - y = 3$ is $(\boxed{\phantom{5}}, 5)$.

**2.** Complete these solutions of $5x + y = 10$: $(1, \square)$; $(\square, -5)$.

*Answers*

**1.** $(8, 5)$ **2.** $(1, 5)$; $(3, -5)$

*Guided Solution:*
**1.** 5, 5, 8; 8

---

***Graphing Equations*** Although equations can be graphed using graphing calculators or computer software, *it is still necessary to understand how equations are graphed by hand.* Graphing calculators are valuable for checking work and for graphing more challenging equations.

The part of the grid that shows in the calculator's screen is called the *window*. For now, the "Standard" window extending from $-10$ to $10$ on both the $x$- and the $y$-axis will suffice. The standard window is usually chosen within the ZOOM menu.

The keystrokes that follow may vary with the calculator used. To graph $y = x + 2$, we press $\boxed{Y=}$ and then $\boxed{X, T, \theta, n}$ $\boxed{+}$ $\boxed{2}$ $\boxed{GRAPH}$. A TRACE key can be used to move a cursor along the line. Near the bottom of the window, the coordinates of the cursor appear, as shown below.

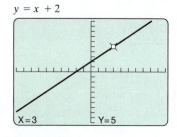
$y = x + 2$

X=3     Y=5

**EXERCISES:** Use graphing technology to graph each of the following.

1. $y = \frac{2}{3}x + 1$ (Example 8)
2. $y = x + 1$
   (Margin Exercise 10)
3. $y = -2x + 1$
   (Margin Exercise 11)
4. $y = \frac{3}{5}x$
   (Margin Exercise 12)

3. Find three solutions of
   $x + 2y = 5$. Answers may vary.

4. Find three solutions of
   $y = -2x + 1$. Answers may vary.

**Answers**
3. $(1, 2), (5, 0), (3, 1)$
4. $(0, 1), (2, -3), (-2, 5)$

**EXAMPLE 3** Find three solutions of $2x - y = 5$.
We are free to use *any* number as a replacement for either $x$ or $y$. To find one solution, we select 1 as a replacement for $x$. We then solve for $y$:

| | |
|---|---|
| $2x - y = 5$ | We are looking for an ordered pair $(1, \Box)$. |
| $2 \cdot 1 - y = 5$ | Substituting 1 for $x$ |
| $2 - y = 5$ | |
| $-y = 3$ | Subtracting 2 from both sides |
| $-1y = 3$ | Recall that $-a = -1 \cdot a$. |
| $y = -3.$ | Dividing both sides by $-1$ |

Thus, $(1, -3)$ is one solution of $2x - y = 5$.
To find a second solution, we choose to replace $y$ with 0 and solve for $x$:

| | |
|---|---|
| $2x - y = 5$ | We are looking for an ordered pair $(\Box, 0)$. |
| $2x - 0 = 5$ | Substituting 0 for $y$ |
| $2x = 5$ | Simplifying |
| $x = 2.5.$ | Dividing both sides by 2 |

Thus, $(2.5, 0)$ is a second solution of $2x - y = 5$.
To find a third solution, we can replace $x$ with 0 and solve for $y$:

| | |
|---|---|
| $2x - y = 5$ | We are looking for an ordered pair $(0, \Box)$. |
| $2 \cdot 0 - y = 5$ | Substituting 0 for $x$ |
| $0 - y = 5$ | |
| $-y = 5$ | |
| $-1y = 5$ | Try to do this step mentally. |
| $y = -5.$ | Dividing both sides by $-1$ |

The pair $(0, -5)$ is a third solution of $2x - y = 5$.
Note that three different choices for $x$ or $y$ would have given three different solutions. There are an infinite number of ordered pairs that are solutions, so it is unlikely for two students to have solutions that match entirely.

---

To find a solution of an equation with two variables:

1. Choose a replacement for one variable.
2. Solve for the other variable.
3. Write the solution as an ordered pair that reflects the alphabetical order of the variables.

---

◀ **Do Exercises 3 and 4.**

## b  GRAPHING EQUATIONS

Equations like those considered in Examples 1–3 are in the form $Ax + By = C$. All equations that can be written this way are said to be **linear** because the solutions of such an equation, when graphed, form a straight line. An equation $Ax + By = C$ is called the **standard form** of a linear equation. When the line representing the solutions is drawn, we say that we have *graphed* the equation.

Since solutions of $Ax + By = C$ are written in the form $(x, y)$, we label the horizontal axis as the $x$-axis and the vertical axis as the $y$-axis.

**EXAMPLE 4** Graph: $2x - y = 5$.

We first need to calculate several solutions of $2x - y = 5$. In Example 3, we found that $(1, -3)$, $(2.5, 0)$, and $(0, -5)$ are solutions of the equation.

Next, we plot the points. As expected, the points describe a straight line. We draw the line using a straightedge and label it with the equation, as shown on the right below.

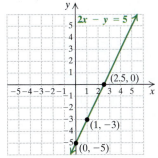

Note that two points are enough to determine a line, but we generally calculate and graph at least three ordered pairs before drawing each line. If the points do not all line up, we know that a mistake has been made.

**Do Exercise 5.** ▶

Linear equations are not always written in standard form. Equations like $y = 2x$ or $y = x + 2$ are also linear. To find solutions of equations like $y = 2x$, we usually choose values for $x$ and then calculate $y$.

**EXAMPLE 5** Graph: $y = 2x$.

First, we find some ordered pairs that are solutions. To find three ordered pairs, we can choose any three values for $x$ and then calculate the corresponding values for $y$. One good choice is 0; we also choose $-2$ and 3.

If $x$ is $0$, then $y = 2x = 2 \cdot 0 = 0$.          Thus, $(0, 0)$ is a solution.
If $x$ is $-2$, then $y = 2x = 2(-2) = -4$.          Thus, $(-2, -4)$ is a solution.
If $x$ is $3$, then $y = 2x = 2 \cdot 3 = 6$.          Thus, $(3, 6)$ is a solution.

We can compute additional pairs if we wish and list the ordered pairs that are solutions in a table.

| $x$ | $y$<br>$y = 2x$ | $(x, y)$ |
|-----|-----|-----|
| 0 | 0 | $(0, 0)$ |
| $-2$ | $-4$ | $(-2, -4)$ |
| 3 | 6 | $(3, 6)$ |
| 1 | 2 | $(1, 2)$ |

Next, we plot these points. We draw the line, or graph, with a ruler and label it $y = 2x$.

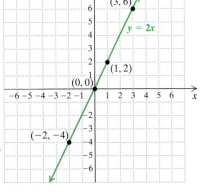

Substitute for $x$.
Compute the value of $y$.
Form the ordered pair $(x, y)$.
Plot the points.
Draw and label the graph.

**Do Exercises 6 and 7.** ▶

**5.** Graph $x + 2y = 5$. Use the results from Margin Exercise 3.

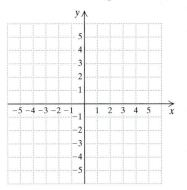

Graph.

**6.** $y = 3x$

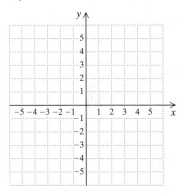

**7.** $y = \frac{1}{2}x$

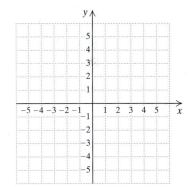

*Answers*

**5.**

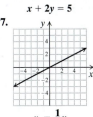

$x + 2y = 5$

**6.**

$y = 3x$

**7.**

$y = \frac{1}{2}x$

Graph.

**8.** $y = -x$ (or $y = -1 \cdot x$)

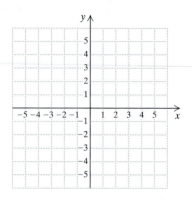

**9.** $y = -2x$

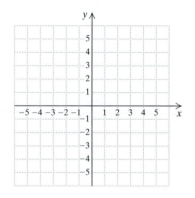

**EXAMPLE 6** Graph: $y = -3x$.

We make a table of solutions. Then we plot the points, draw the line, and label the line $y = -3x$.

If $x$ is 0, then $y = -3 \cdot 0 = 0$.

If $x$ is 1, then $y = -3 \cdot 1 = -3$.

If $x$ is $-2$, then $y = -3(-2) = 6$.

If $x$ is 2, then $y = -3 \cdot 2 = -6$.

| $x$ | $y$<br>$y = -3x$ | $(x, y)$ |
|-----|------|----------|
| 0 | 0 | $(0, 0)$ |
| 1 | $-3$ | $(1, -3)$ |
| $-2$ | 6 | $(-2, 6)$ |
| 2 | $-6$ | $(2, -6)$ |

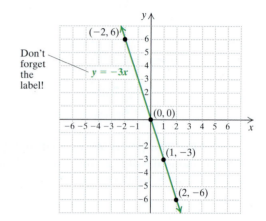

◀ **Do Exercises 8 and 9.**

**EXAMPLE 7** Graph: $y = x + 2$.

We make a table of solutions. Then we plot the points, draw the line, and label it.

If $x$ is 0, then $y = 0 + 2 = 2$.

If $x$ is 1, then $y = 1 + 2 = 3$.

If $x$ is $-1$, then $y = -1 + 2 = 1$.

If $x$ is 3, then $y = 3 + 2 = 5$.

| $x$ | $y$<br>$y = x + 2$ | $(x, y)$ |
|-----|------|----------|
| 0 | 2 | $(0, 2)$ |
| 1 | 3 | $(1, 3)$ |
| $-1$ | 1 | $(-1, 1)$ |
| 3 | 5 | $(3, 5)$ |

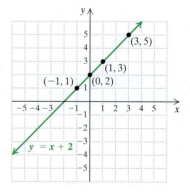

The values of $x$ in these examples were *chosen*. Different choices for $x$ would yield different points, but the same line.

**For linear equations, tables can be formed using any numbers for $x$.**

*Answers*

**8.**

**9.**

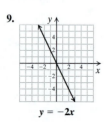

$y = -x$      $y = -2x$

**EXAMPLE 8** Graph: $y = \frac{2}{3}x + 1$.

We make a table of solutions, plot the points, and draw and label the line. For this table, we selected multiples of 3 as $x$-values to avoid fraction values for $y$.

If $x$ is 6, then $y = \frac{2}{3} \cdot 6 + 1 = 4 + 1 = 5$.
If $x$ is 3, then $y = \frac{2}{3} \cdot 3 + 1 = 2 + 1 = 3$.
If $x$ is 0, then $y = \frac{2}{3} \cdot 0 + 1 = 0 + 1 = 1$.
If $x$ is $-3$, then $y = \frac{2}{3} \cdot (-3) + 1 = -2 + 1 = -1$.

| $x$ | $y = \frac{2}{3}x + 1$ | $(x, y)$ |
|-----|------------------------|----------|
| 6 | 5 | $(6, 5)$ |
| 3 | 3 | $(3, 3)$ |
| 0 | 1 | $(0, 1)$ |
| $-3$ | $-1$ | $(-3, -1)$ |

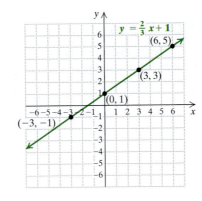

**Do Exercises 10–12.** ▶

## c | GRAPHING HORIZONTAL LINES OR VERTICAL LINES

Any equation in the form $Ax + By = C$ is linear, provided $A$ and $B$ are not both zero. If $A$ is 0 and $B$ is nonzero, there is no $x$-term and the graph is a horizontal line. If $B$ is 0 and $A$ is nonzero, there is no $y$-term and the graph is a vertical line.

**EXAMPLE 9** Graph: $y = 3$.

We regard $y = 3$ as $0 \cdot x + y = 3$. No matter what number we choose for $x$, we find that $y$ must be 3 if the equation is to be true.

Choose any number for $x$. →

| $x$ | $y = 3$ | $(x, y)$ |
|-----|---------|----------|
| $-2$ | 3 | $(-2, 3)$ |
| 0 | 3 | $(0, 3)$ |
| 4 | 3 | $(4, 3)$ |

All pairs have 3 as the $y$-coordinate.

$y$ must be 3.

When we plot $(-2, 3)$, $(0, 3)$ and $(4, 3)$ and connect the points, we obtain a horizontal line. Any ordered pair of the form $(x, 3)$ is a solution, so the line is 3 units above the $x$-axis, as shown in the graph at the top of the next page.

Graph.

**10.** $y = x + 1$

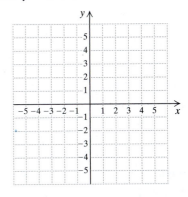

**11.** $y = -2x + 1$

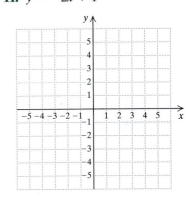

**12.** $y = \frac{3}{5}x$

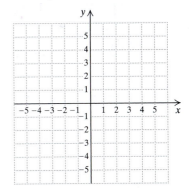

*Answers*

**10.**
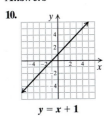
$y = x + 1$

**11.**

$y = -2x + 1$

**12.**
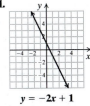
$y = \frac{3}{5}x$

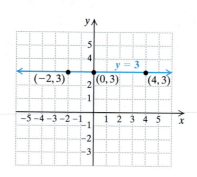

## EXAMPLE 10  Graph: $x = -4$.

We regard $x = -4$ as $x + 0 \cdot y = -4$ and make a table with $-4$ in every row in the $x$-column.

Graph.

**13.** $y = 4$

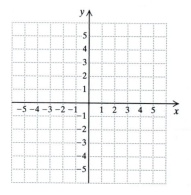

**14.** $x = 5$

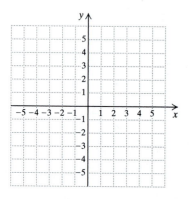

$x$ must be $-4$. →

| $x$ | | |
|---|---|---|
| $x = -4$ | $y$ | $(x, y)$ |
| $-4$ | $-5$ | $(-4, -5)$ |
| $-4$ | $1$ | $(-4, 1)$ |
| $-4$ | $3$ | $(-4, 3)$ |

All pairs have $-4$ as the $x$-coordinate.

Choose any number for $y$.

When we plot $(-4, -5)$, $(-4, 1)$, and $(-4, 3)$ and connect them, we obtain a vertical line. Any ordered pair of the form $(-4, y)$ is a solution, so the line is 4 units left of the $y$-axis.

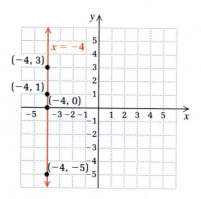

*Answers*

**13.**

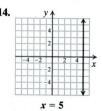

$y = 4$

**14.**

$x = 5$

◀ **Do Exercises 13 and 14.**

### HORIZONTAL LINES AND VERTICAL LINES

The graph of $y = b$ is a horizontal line.
The graph of $x = a$ is a vertical line.

## ✓ Check Your Understanding

**Reading Check** Determine whether each statement is true or false.

_____ **RC1.** A linear equation in two variables has at most one solution.

_____ **RC2.** Every solution of $y = 3x - 7$ is an ordered pair.

_____ **RC3.** The graph of $y = 3x - 7$ represents all solutions of the equation.

_____ **RC4.** If a point is on the graph of $y = 3x - 7$, the corresponding ordered pair is a solution of the equation.

_____ **RC5.** To find a solution of $y = 3x - 7$, we can choose any value for $x$ and calculate the corresponding $y$-value.

_____ **RC6.** The graph of every equation is a straight line.

**Concept Check** Determine whether the graph of each linear equation is a horizontal line, a vertical line, or a slanted line.

**CC1.** $y = x + 2$　　　**CC2.** $y = 2$　　　**CC3.** $y = 2x$　　　**CC4.** $x = 2$

**a** For each equation, use the indicated value to find an ordered pair that is a solution.

**1.** $x + y = 8$; let $x = 5$

**2.** $x + y = 5$; let $x = 4$

**3.** $2x + y = 7$; let $x = 3$

**4.** $x + 2y = 9$; let $y = 4$

**5.** $y = 3x - 1$; let $x = 5$

**6.** $y = 2x + 7$; let $x = 3$

**7.** $x + 3y = 1$; let $x = 10$

**8.** $5x + y = 7$; let $y = -8$

**9.** $2x + 5y = 15$; let $x = 0$

**10.** $5x + 2y = 18$; let $x = 0$

**11.** $3x - 2y = 8$; let $y = -1$

**12.** $2x - 5y = 12$; let $y = -2$

For each equation, complete the given ordered pairs.

**13.** $x + y = 4$; $(\square, 3)$; $(-1, \square)$

**14.** $x - y = 6$; $(\square, 2)$; $(9, \square)$

**15.** $x - y = 4$; $(\square, 3)$; $(10, \square)$

**16.** $x + y = 10$; $(\square, 8)$; $(3, \square)$

**17.** $2x + 3y = 30$; $(0, \square)$; $(\square, 0)$

**18.** $3x + 2y = 24$; $(0, \square)$; $(\square, 0)$

**19.** $3x + 5y = 14$; $(3, \square)$; $(\square, 4)$

**20.** $4x + 3y = 11$; $(5, \square)$; $(\square, 2)$

**21.** $y = 4x$; $(\square, 4)$; $(-2, \square)$

**22.** $y = 6x$; $(\square, 6)$; $(-2, \square)$

**23.** $2x + 5y = 3$; $(0, \square)$; $(\square, 0)$

**24.** $5x + 7y = 9$; $(0, \square)$; $(\square, 0)$

For each equation, find three solutions. Answers may vary.

**25.** $x + y = 9$

**26.** $x + y = 19$

**27.** $y = 4x$

**28.** $y = 5x$

**29.** $3x + y = 13$

**30.** $x + 5y = 12$

**31.** $y = 3x - 1$

**32.** $y = 2x + 5$

**33.** $y = -7x$

**34.** $y = -4x$

**35.** $4 + y = x$

**36.** $3 + y = x$

**37.** $3x + 2y = 12$

**38.** $2x + 3y = 18$

**39.** $y = \frac{1}{3}x + 2$

**40.** $y = \frac{1}{2}x + 5$

 **b** Graph each equation.

**41.** $x + y = 6$

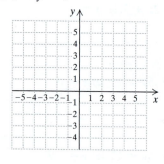

**42.** $x + y = 4$

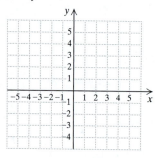

**43.** $x - 1 = y$

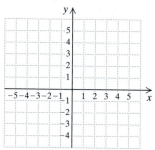

**44.** $x - 2 = y$

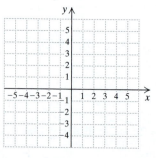

**45.** $y = x - 4$

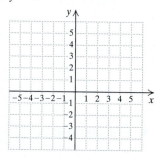

**46.** $y = x - 5$

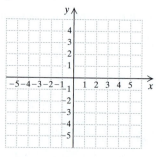

**47.** $y = \dfrac{1}{3}x$

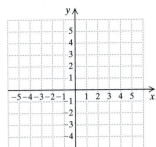

**48.** $y = -\dfrac{1}{3}x$

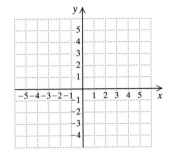

**49.** $y = x$

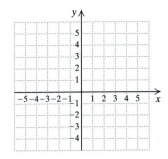

**50.** $y = x - 3$

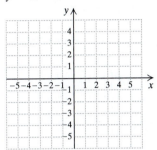

**51.** $y = 2x - 1$

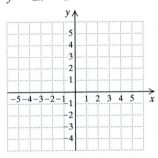

**52.** $y = 2x - 3$

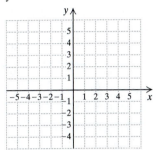

**53.** $y = 2x + 1$

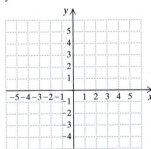

**54.** $y = 3x + 1$

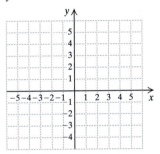

**55.** $y = \dfrac{2}{5}x$

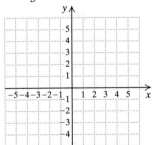

**56.** $y = \dfrac{3}{4}x$

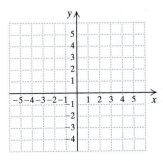

**57.** $y = -x + 4$

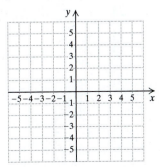

**58.** $y = -x + 5$

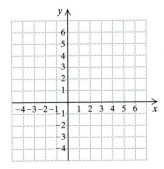

**59.** $y = \dfrac{2}{3}x + 1$

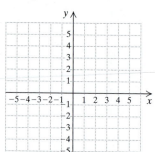

**60.** $y = \dfrac{2}{5}x - 1$

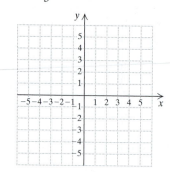

**C**  Graph.

**61.** $y = 2$

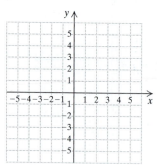

**62.** $y = 1$

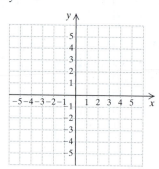

**63.** $x = 2$

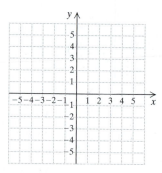

**64.** $x = 3$

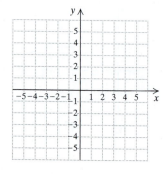

**65.** $x = -3$

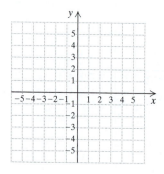

**66.** $x = -1$

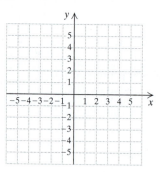

**67.** $y = -4$

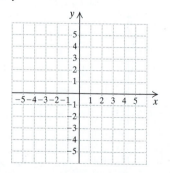

**68.** $y = -2$

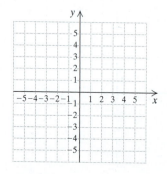

69. The tunes on Miles Davis's classic *Kind of Blue* album are approximately 9 min, $9\frac{1}{2}$ min, $5\frac{1}{2}$ min, $11\frac{1}{2}$ min, and $9\frac{1}{2}$ min long. Find the average length of a tune on that album. [4.6c], [4.7d]

70. The books on Sherry's nightstand are 243, 410, 352, and 274 pages long. What is the average length of a book on the nightstand? [5.8b]

71. A recipe for a batch of chili calls for $\frac{3}{4}$ cup of red wine vinegar. How much vinegar is needed to make $2\frac{1}{2}$ batches of chili? [4.7d]

Simplify.

72. $-\dfrac{49}{77}$ [3.5b]

73. $-8 - 5^2 \cdot 2(3 - 4)$ [2.5b]

74. $\dfrac{3}{10}\left(-\dfrac{25}{12}\right)$ [3.4b]

## Synthesis

Find three solutions of each equation. Then graph the equation.

75. $21x - 70y = -14$

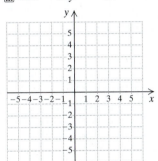

76. $25x + 80y = 100$

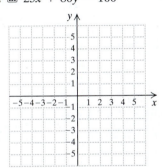

77. $50x + 75y = 180$

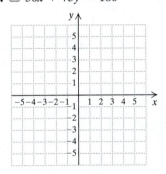

78. Use the graph in Example 4 to find three solutions of $2x - y = 5$. Do not use the ordered pairs already listed.

79. List all solutions of $x + y = 6$ that use only whole numbers.

80. Graph three solutions of $y = |x|$ in the second quadrant and another three solutions in the first quadrant.

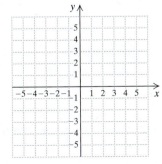

*To the Student and the Instructor*: Exercises marked with a symbol are meant to be solved using a graphing calculator.

81. Use a graphing calculator to graph each of the following.

    **(a)** $y = -0.63x + 2.8$     **(b)** $y = 2.3x - 4.1$

## Concept Reinforcement

Determine whether each statement is true or false.

_____ **1.** The key on a pictograph indicates what each symbol represents.  [8.1b]

_____ **2.** The solution of an equation in two variables is written as an ordered pair.  [8.3c]

_____ **3.** The graph of the equation $x = 5$ is a horizontal line.  [8.4c]

## Guided Solutions

 Fill in each blank with the number that creates a correct solution.

**4.** Determine whether the ordered pair $(2, -1)$ is a solution of the equation $3x + y = 5$.  [8.3c]

$$\frac{3x + y = 5}{3 \cdot \boxed{\phantom{0}} + (\boxed{\phantom{0}}) \; ? \; 5}$$

$$\boxed{\phantom{0}} + (\boxed{\phantom{0}}) \; \Big| $$

$$\Big| \; 5 \quad \text{TRUE}$$

Since $3x + y = 5$ becomes true, $(2, -1)$ is a solution.

**5.** Find a solution of $x - y = 6$. Let $x = 1$.  [8.4a]

$$x - y = 6$$
$$\boxed{\phantom{0}} - y = 6$$
$$-y = \boxed{\phantom{0}}$$
$$y = \boxed{\phantom{0}}$$

Thus, $(1, \boxed{\phantom{0}})$ is a solution of $x - y = 6$.

## Mixed Review

***Do Not Call Registry.*** Consumers who do not wish to receive marketing calls can list their telephone numbers in the National Do Not Call Registry. If a telemarketer violates the do-not-call rules, a consumer can register a complaint. The following table lists the number of complaints received for various years. Use the table for Exercises 6 and 7.

| FISCAL YEAR | NUMBER OF COMPLAINTS RECEIVED (in millions) |
|---|---|
| 2004 | 0.6 |
| 2007 | 1.3 |
| 2010 | 1.6 |
| 2013 | 3.7 |
| 2016 | 5.3 |

DATA: Federal Trade Commission

**6.** Draw a line graph, giving the horizontal scale the title "Fiscal year."  [8.2d]

**7.** What was the percent increase in complaints from 2004 to 2016?  [8.1a]

In which quadrant is each point located?   [8.3b]

**8.** $(-2, -12)$

**9.** $(-5, 6)$

**10.** $\left(\dfrac{1}{2}, -8\right)$

**11.** Determine whether the ordered pair $(0, -5)$ is a solution of $5x + y = 5$.   [8.3c]

**12.** Find a solution of $y = x + 3$. Let $x = -10$.   [8.4a]

**13.** Complete this solution of $3x - y = 7$: $(-1, \square)$.   [8.4a]

**14.** Find three solutions of $2x + y = 5$. Answers may vary.   [8.4a]

Graph each equation.   [8.4b], [8.4c]

**15.** $x + y = 3$

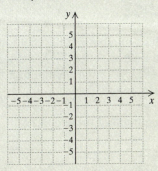

**16.** $y = x - 2$

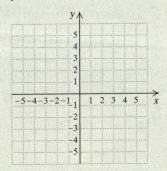

**17.** $y = 4x$

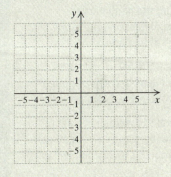

**18.** $y = 2x - 4$

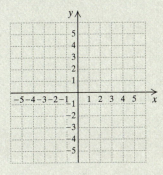

**19.** $y = -3$

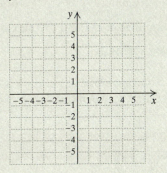

**20.** $x = 1$

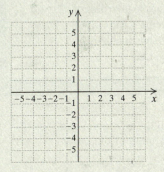

## Understanding Through Discussion and Writing

**21.** Under what conditions will the points $(a, b)$ and $(b, a)$ be in the same quadrant?   [8.3b]

**22.** In which quadrant, if any, is the point $(5, 0)$? Why?   [8.3b]

**23.** To graph a linear equation, a student plots three points and discovers that the points do not line up with each other. What should the student do next?   [8.4b]

**24.** What is the greatest number of quadrants that a line can pass through? Why?   [8.3b], [8.4b]

- [ ] Review regularly. A good way to do this is by doing the Skill Maintenance exercises found in each exercise set.
- [ ] Try creating your own glossary. Understanding terminology is essential for success in any math course.
- [ ] Memorizing is a helpful tool in the study of mathematics. Ask your instructor what you are expected to have memorized for tests.

# 8.5

## OBJECTIVES

a  Find the minimum, the maximum, and the range of a set of numbers.

b  Find the mean of a set of numbers and solve applied problems involving means.

c  Find the median of a set of numbers and solve applied problems involving medians.

d  Find the mode of a set of numbers and solve applied problems involving modes.

e  Find the quartiles of a set of numbers and write the five-number summary of a set of numbers.

**1.** Find the minimum and the maximum of these data.

25, 18, 13, 7, 6

*Answer*

**1.** Minimum: 6; Maximum: 25

# Descriptive Statistics

A **statistic** is a number describing a set of data. Statistics can describe characteristics such as the *spread* and *center* of a set of data.

## a  MINIMUM, MAXIMUM, AND RANGE

The **minimum** of a set of numbers is the smallest number in the set, and the **maximum** of a set of numbers is the largest number in the set. When a set of numbers is arranged in order from smallest to largest, the minimum is the first number listed and the maximum is the last number listed.

**EXAMPLE 1**  Find the minimum and the maximum of these data.

2, 15, 4, 7, 6, 14, 15

We first rearrange the numbers in order from smallest to largest.

2, 4, 6, 7, 14, 15, 15

  ↑                   ↑

Minimum         Maximum

The minimum is 2 and the maximum is 15.

◀ **Do Exercise 1.**

The **range** of a set of data is one measure of spread. It is defined as the difference between the maximum and the minimum of the set.

### RANGE

To find the **range** of a set of numbers, subtract the smallest number in the set from the largest number in the set.

Range = Maximum − Minimum

**EXAMPLE 2**  The following table shows the selling price of houses in two counties. Find the range in selling price for each county.

| PINE COUNTY | PRAIRIE COUNTY |
|---|---|
| $125,380 | $235,600 |
| $263,150 | $110,800 |
| $105,410 | $798,400 |
| $143,900 | $267,800 |
|  | $153,100 |

We rearrange each set of numbers in order from smallest to largest, determine the minimum and the maximum, and then find the range.

Pine County:

   105,410,   125,380,   143,900,   263,150

Minimum: 105,410
Maximum: 263,150
Range: 263,150 − 105,410 = 157,740   Range = Maximum − Minimum

Prairie County:

   110,800,   153,100,   235,600,   267,800,   798,400

Minimum: 110,800
Maximum: 798,400
Range: 798,400 − 110,800 = 687,600   Range = Maximum − Minimum

The range of selling price for Pine County is $157,740, and the range of selling price for Prairie County is $687,600. Note that the range describes the *spread* of the data.

Do Exercise 2. ▶

**2.** Find the range of these data.
95%, 74%, 100%, 72%, 86%, 81%

## b MEANS

One type of statistic is a *center point*, or *measure of central tendency*, that characterizes the data. The most common kind of center point is the **arithmetic** (pronounced ăr′ĭth-mĕt′-ĭk) **mean**, or simply the **mean**. This center point is often referred to as the *average*.

> ### MEAN
>
> To find the **mean** of a set of numbers, add the numbers and then divide by the number of items of data.

**EXAMPLE 3**   On a 4-day trip, a car was driven the following numbers of miles: 240, 302, 280, 320. What was the mean number of miles per day?

$$\frac{240 + 302 + 280 + 320}{4} = \frac{1142}{4}, \text{ or } 285.5$$

The car was driven a mean of 285.5 mi per day. Had the car been driven exactly 285.5 mi each day, the same total distance (1142 mi) would have been traveled.

*Answer*

**2.** 28%

Find the mean.

**3.** 14, 175, 36

**4.** 75, 36.8, 95.7, 12.1

**5.** In the first five games of the season, a basketball player scored 26, 21, 13, 14, and 23 points. Find the average number of points scored per game.

**6.** *Home-Run Batting Average.* Babe Ruth hit 714 home runs in 22 seasons in the major leagues. What was his average number of home runs per season? Round to the nearest tenth.
**Data:** Major League Baseball

**EXAMPLE 4** *Gas Mileage.* The 2017 Volkswagen Jetta with a 1.4 L engine is estimated to travel 520 miles on the highway on 13 gal of diesel fuel. What is the expected mean number of miles per gallon (mpg)—that is, what is the fuel mileage for highway driving?
**Data:** vw.com

We divide the total number of miles, 520, by the total number of gallons, 13:

$$\frac{520 \text{ mi}}{13 \text{ gal}} = 40 \text{ mpg}.$$

The Jetta's expected mean is 40 mi per gallon for highway driving.

◀ **Do Exercises 3–6.**

In a *weighted average*, more importance, or *weight*, is assigned to some values than to others. For example, a course syllabus may include the following description:

| COURSE COMPONENT | WEIGHT FOR GRADE |
|---|---|
| Quizzes | 20 |
| Homework | 30 |
| Tests | 50 |

If Allison has scored 70% on quizzes, 100% on homework, and 92% on tests, she cannot calculate her course grade by averaging 70, 100, and 92, because each category is weighted differently. Instead, she must multiply each percentage by its weight, add the results, and divide by the total of the weights:

$$\text{Course grade} = \frac{70 \cdot 20 + 100 \cdot 30 + 92 \cdot 50}{20 + 30 + 50}$$

$$= \frac{9000}{100} = 90.$$

Allison's course grade is 90%.

A grade point average is another example of a weighted average.

**EXAMPLE 5** *Grade Point Average.* In many schools, students are assigned grade point values for grades obtained. The **grade point average**, or **GPA**, is the average of the grade point values for each credit hour taken. At Meg's college, grade point values are assigned as follows:

A: 4.0   B: 3.0   C: 2.0   D: 1.0   F: 0.0.

Meg earned the following grades for one semester. What was her grade point average?

| COURSE | GRADE | NUMBER OF CREDIT HOURS IN COURSE |
|---|---|---|
| Colonial History | B | 3 |
| Basic Mathematics | A | 4 |
| English Literature | A | 3 |
| French | C | 4 |
| Time Management | D | 1 |

*Answers*

**3.** 75 **4.** 54.9 **5.** 19.4 points per game
**6.** 32.5 home runs per season

To find the GPA, we first multiply the grade point value for each grade by the number of credit hours in the course to determine the number of *quality points,* and then add. Here each grade is weighted by the number of credit hours in the course.

| | | |
|---|---|---|
| Colonial History | $3.0 \cdot 3 =$ | 9 |
| Basic Mathematics | $4.0 \cdot 4 =$ | 16 |
| English Literature | $4.0 \cdot 3 =$ | 12 |
| French | $2.0 \cdot 4 =$ | 8 |
| Time Management | $1.0 \cdot 1 =$ | $\underline{\phantom{0}1}$ |
| | | 46 (Total quality points) |

The total number of credit hours taken is $3 + 4 + 3 + 4 + 1$, or 15. We divide the number of quality points, 46, by the number of credit hours, 15, and round to the nearest tenth:

$$\text{GPA} = \frac{46}{15} \approx 3.1.$$

Meg's grade point average was 3.1.

**Do Exercises 7 and 8.** ▶

**EXAMPLE 6** *Grading.* To get a B in math, Geraldo must score an average of 80 on five tests. On the first four tests, his scores were 79, 88, 64, and 78. What is the lowest score that Geraldo can get on the last test and still get a B?

We can find the total of the five scores needed as follows:

$$80 + 80 + 80 + 80 + 80 = 5 \cdot 80, \quad \text{or} \quad 400.$$

The total of the scores on the first four tests is

$$79 + 88 + 64 + 78 = 309.$$

Thus, Geraldo needs to get at least

$$400 - 309, \quad \text{or} \quad 91,$$

in order to get a B. We can check this as follows:

$$\frac{79 + 88 + 64 + 78 + 91}{5} = \frac{400}{5}, \quad \text{or} \quad 80.$$

**Do Exercise 9.** ▶

## c  MEDIANS

Another type of center-point statistic is the *median.* Medians are useful when we wish to de-emphasize unusually extreme numbers. For example, suppose a small class scored as follows on an exam: 78, 81, 82, 56, 84.

Let's first list the scores in order from smallest to largest:

$$56, \quad 78, \quad 81, \quad 82, \quad 84.$$

↑
Middle score

The middle score—in this case, 81—is called the **median**. Note that because of the extremely low score of 56, the average of the scores is 76.2. In this example, the median may be a more appropriate center-point statistic.

 **7.** Soha's sociology professor included the following in the course syllabus:

| COURSE COMPONENT | WEIGHT FOR GRADE |
|---|---|
| Participation | 15 |
| Book reports | 25 |
| Research paper | 40 |

Soha received 88% on her research paper and 92% on her book reports, and she anticipates a score of 100% for participation. What is her course grade?

Course grade

$$= \frac{100 \cdot 15 + 92 \cdot \boxed{\phantom{00}} + 88 \cdot \boxed{\phantom{00}}}{15 + 25 + 40}$$

$$= \frac{7320}{\boxed{\phantom{00}}} = \boxed{\phantom{00}}$$

Soha's course grade is $\boxed{\phantom{00}}$ %.

**8.** *Grade Point Average.* Alex earned the following grades one semester.

| GRADE | NUMBER OF CREDIT HOURS IN COURSE |
|---|---|
| B | 3 |
| C | 4 |
| C | 4 |
| A | 2 |

What was Alex's grade point average? Assume that the grade point values are 4.0 for an A, 3.0 for a B, and so on. Round to the nearest tenth.

**9.** *Grading.* To get an A in math, Rosa must score an average of 90 on four tests. On the first three tests, her scores were 80, 100, and 86. What is the lowest score that Rosa can get on the last test and still get an A?

*Answers*

**7.** 91.5%  **8.** 2.5  **9.** 94

*Guided Solution:*

**7.** 25, 40, 80, 91.5; 91.5

Find the median.

**10.** 17, 13, 18, 14, 19

**11.** 20, 14, 13, 19, 16, 18, 17

**12.** 78, 81, 83, 91, 103, 102, 122, 119, 88

**EXAMPLE 7**   What is the median of this set of numbers?

99,  870,  91,  98,  106,  90,  98

We first rearrange the numbers in order from smallest to largest. Then we locate the middle number, 98.

90,  91,  98,  98,  99,  106,  870

↑
Middle number

The median is 98.

◀ **Do Exercises 10–12.**

> ### MEDIAN
>
> Once a set of data is listed in order, from smallest to largest, the **median** is the middle number if there is an odd number of data items. If there is an even number of items, the median is the number that is the average of the two middle numbers.

**EXAMPLE 8**   What is the median of this set of numbers?

69,  80,  61,  63,  62,  65

We first rearrange the numbers in order from smallest to largest. There is an even number of numbers. We look for the middle two, which are 63 and 65. The median is halfway between 63 and 65, the number 64.

61,  62,  63,  65,  69,  80

The average of the middle numbers is
$$\frac{63 + 65}{2} = \frac{128}{2}, \text{ or } 64.$$

└── The median is 64.

Find the median.

**13.** *Salaries of Part-Time Typists.*   $3300, $4000, $3900, $3600, $3800, $3400

**14.** 68, 34, 67, 69, 34, 70   **GS**

Rearrange the numbers in order from smallest to largest:

34, 34, ____, 68, ____, 70.

The middle numbers are ____ and 68.

The average of 67 and 68 is ____ .

The median is ____ .

**EXAMPLE 9**   *Salaries.*   The following are the salaries of the four highest-paid players in the National Hockey League. What is the median of the salaries?

| PLAYER | SALARY |
|--------|--------|
| Patrick Kane | $13,800,000 |
| Jonathan Toews | $13,800,000 |
| Anze Kopitar | $13,000,000 |
| Jamie Benn | $13,000,000 |

DATA: spotrac.com

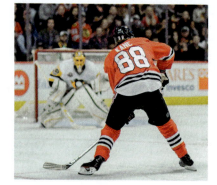

We rearrange the numbers in order from smallest to largest:

$13,000,000,   $13,000,000,   $13,800,000,   $13,800,000

The two middle numbers are $13,000,000 and $13,800,000. Their average is $13,400,000. Thus, the median salary is $13,400,000.

◀ **Do Exercises 13 and 14.**

*Answers*
**10.** 17   **11.** 17   **12.** 91
**13.** $3700   **14.** 67.5
*Guided Solution:*
**14.** 67, 69; 67, 67.5, 67.5

## d  MODES

The final type of center-point statistic we will consider is the *mode*.

---

### MODE

The **mode** of a set of data is the number or numbers that occur most often. If each number occurs the same number of times, there is *no* mode.

---

**EXAMPLE 10**  Find the mode of these data.

17,  13,  18,  17,  14,  19

To find the mode, it is helpful to first rearrange the numbers in order from smallest to largest.

13,  14,  17,  17,  18,  19

The number that occurs most often is 17. Thus, the mode is 17.

**EXAMPLE 11**  Find the mode of these data.

5,  5,  11,  11,  13,  13

The numbers in this set of data are 5, 11, and 13. Each occurs twice, so all the numbers are equally represented. There is *no mode*.

A set of data has just one average (mean) and just one median, but it can have more than one mode.

**EXAMPLE 12**  Find the modes of these data.

33,  34,  34,  34,  35,  36,  37,  37,  37,  38,  39,  40

There are two numbers that occur most often, 34 and 37. Thus, the modes are 34 and 37.

**Do Exercises 15–18.** ▶

Which center-point statistic is best for a particular situation? If someone is bowling, the *mean* from several games is a good indicator of that person's ability. If someone is applying for a job, the *median* salary at that business is often most indicative of what people are earning there, because although executives tend to make a high salary, there are few of them. For similar reasons, the selling price of homes is usually reported as a *median* price. Finally, if someone is reordering stock for a clothing store, the *mode* of the sizes sold is probably the most important statistic.

---

Find the modes of these data.

**15.** 23, 45, 45, 45, 78

**16.** 34, 34, 67, 67, 68, 70

**GS** **17.** 24, 89, 13, 28, 67, 27
Rearrange the numbers in order from smallest to largest.
13, 24, ____ , 28, 67, ____ .
Each number occurs ____ time.
There is no mode.

**18.** In a lab, Gina determined the mass, in grams, of each of five eggs:

15g, 19g, 19g, 14g, 18g.

**a)** What is the mean?
**b)** What is the median?
**c)** What is the mode?

---

## e QUARTILES AND FIVE-NUMBER SUMMARIES

We can think of the median of a set of data as a number that divides the data in half. One half, or 50%, of the numbers in the set of data are less than the median, and one half, or 50% of the numbers are greater than the median. In a similar fashion, **quartiles** divide a set of data into fourths, or **quarters**.

25% of data | 25% of data | 25% of data | 25% of data

First
quartile

Second
quartile, or median

Third
quartile

To find the quartiles of a set of data, first list the numbers in order from smallest to largest. Then find the median of the data. This is also the second quartile and "splits" the data in half. The first quartile is then the median of the set of numbers that are less than the second quartile, and the third quartile is the median of the set of numbers that are greater than the second quartile.

**EXAMPLE 13**   Find the quartiles of these data.

$$1, \quad 2, \quad 5, \quad 9, \quad 17, \quad 26, \quad 35, \quad 40, \quad 50, \quad 55$$

The numbers are already listed in order from smallest to largest. There is an even number of numbers. The middle two numbers are 17 and 26. The median, or second quartile, is the average of 17 and 26, or

$$\frac{17 + 26}{2} = \frac{43}{2} = 21.5.$$

$$1, \quad 2, \quad 5, \quad 9, \quad 17, \quad 26, \quad 35, \quad 40, \quad 50, \quad 55$$

21.5

Second quartile

The first quartile is the median of the set of numbers that are less than the second quartile:

$$1, \quad 2, \quad 5, \quad 9, \quad 17.$$

First quartile

The third quartile is the median of the set of numbers that are greater than the second quartile:

$$26, \quad 35, \quad 40, \quad 50, \quad 55.$$

Third quartile

Thus, the first quartile is 5, the second quartile (median) is 21.5, and the third quartile is 40.

◀ **Do Exercise 19**

In Example 13, the median of the set of data was not part of the data because there were an even number of numbers in the set. If the median *is* one of the numbers in the set of data, it is not considered when finding the first and third quartiles.

**19.** Find the quartiles of these data.
3, 5, 5, 7, 11, 12, 14, 15,
15, 15, 26, 30

*Answer*

**19.** First quartile: 6; second quartile: 13;
third quartile: 15

The quartiles of a set of data together with the minimum and the maximum are often listed as a **five-number summary** of the data.

---

### FIVE-NUMBER SUMMARY

The five-number summary of a set of data consists of the following statistics:

Minimum
First quartile
Median (or second quartile)
Third quartile
Maximum

---

**EXAMPLE 14**   Find the five-number summary of these data.

6,  8,  9,  1,  6,  4,  8,  10,  9,  12,  9,  13,  13

First, list the numbers in order from smallest to largest. We identify the minimum, 1, and the maximum, 13. The median is the middle number, 9.

1,  4,  6,  6,  8,  8,  9,  9,  9,  10,  12,  13,  13
↑                        ↑                        ↑
Minimum         Median,             Maximum
                      or second quartile

The first quartile is the median of the set of numbers that are less than the second quartile:

1,  4,  6,  6,  8,  8.   $\dfrac{6 + 6}{2} = 6$
            6
            ↑
      First quartile

The third quartile is the median of the set of numbers that are greater than the second quartile:

9,  9,  10,  12,  13,  13.   $\dfrac{10 + 12}{2} = 11$
        11
        ↑
  Third quartile

The five-number summary is then

Minimum: 1
First quartile: 6
Median: 9
Third quartile: 11
Maximum: 13.

**Do Exercise 20.** ▶

**20.** Find the five-number summary of these data.

3, 11, 16, 19, 7, 21, 4, 12, 8, 6, 5

*Answer*

**20.** Minimum: 3; first quartile: 5; median: 8; third quartile: 16; maximum: 21

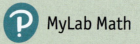
### ✓ Check Your Understanding

**Reading Check**  Complete each sentence with the appropriate word from the list on the right. Not all choices will be used.

**RC1.** A mean is a(n) _____.

**RC2.** To find the _____ of a set of numbers, add the numbers and then divide by the number of items of data.

**RC3.** To find the weighted average of a set of numbers, multiply each number by its _____, add the results, and divide by the total of the weights.

**RC4.** The _____ of a set of numbers is the number or numbers that occur most often.

> mean
>
> median
>
> mode
>
> statistic
>
> weight

**Concept Check**  For the given set of data, choose from the list on the right the appropriate number for each statistic. Choices may be used more than once or not at all.

78, 85, 74, 92, 68, 88, 100, 76, 85

**CC1.** Minimum

**CC2.** Maximum

**CC3.** Range

**CC4.** Median

**CC5.** First quartile

**CC6.** Second quartile

**a)** 32

**b)** 90

**c)** 85

**d)** 75

**e)** 68

**f)** 100

---

**a**  For each set of numbers, find the minimum, the maximum, and the range.

**1.** 3, 7, 10, 16, 25, 38

**2.** 1, 6, 15, 97

**3.** 12, 16, 38, 112, 5

**4.** 8, 7, 5, 9, 6, 11

**5.** 2, 3, 3, 2, 2, 3, 3

**6.** 8, 3, 8, 7, 3, 3, 6

**b**, **c**, **d**   For each set of numbers, find the mean, the median, and any modes that exist.

**7.** *Smithsonian Museum Visitors.*   The following table lists the number of visitors to the 8 most popular Smithsonian museums in 2016. What is the mean number of visitors for the 8 museums? the median? the mode?

| MUSEUMS | NUMBER OF VISITORS |
|---|---|
| National Air and Space Museum | 7,500,000 |
| National Museum of Natural History | 7,100,000 |
| National Museum of American History | 3,800,000 |
| National Zoo | 2,700,000 |
| National Air and Space Museum's Steven F. Udvar–Hazy Center | 1,600,000 |
| Donald W. Reynolds Center for American Art and Portraiture | 1,200,000 |
| National Museum of the American Indian | 1,100,000 |
| Smithsonian Institution Building | 1,100,000 |

DATA: newsdesk.si.edu

**8.** *Congestion.*   The following table lists the annual number of hours of traffic delay per auto commuter for 8 U.S. cities. What is the mean delay time? the median? the mode?

| CITY | NUMBER OF HOURS OF DELAY PER AUTO COMMUTER |
|---|---|
| Washington, DC | 82 |
| Los Angeles, CA | 80 |
| San Francisco, CA | 78 |
| New York City, NY | 74 |
| Boston, MA | 64 |
| Seattle, WA | 63 |
| Chicago, IL | 61 |
| Houston, TX | 61 |

DATA: *2015 Annual Urban Mobility Report,* Texas A&M Transportation Institute

**9.** 17, 19, 29, 18, 14, 29

**10.** 72, 83, 85, 88, 92

**11.** 5, 37, 20, 20, 35, 5, 25

**12.** 13, 32, 25, 27, 13

**13.** 4.3, 7.4, 1.2, 5.7, 8.3

**14.** 13.4, 13.4, 12.6, 42.9

**15.** 234, 228, 234, 229, 234, 278

**16.** $29.95, $28.79, $30.62, $28.79, $29.95

**17.** *Gas Mileage.*   The 2017 Kia Optima LX does 396 mi of highway driving on 11 gal of gasoline. What is the mean number of miles expected per gallon—that is, what is the gas mileage?

**Data:** Kia.com

**18.** *Gas Mileage.*   The 2017 Chevrolet Malibu with a 1.5 L engine does 243 mi of city driving on 9 gal of gasoline. What is the mean number of miles expected per gallon—that is, what is the gas mileage?

**Data:** fueleconomy.gov

**Grade Point Average.**   The tables in Exercises 19 and 20 show the grades of a student for one semester. In each case, find the grade point average. Assume that the grade point values are 4.0 for an A, 3.0 for a B, and so on. Round to the nearest tenth.

19.

| GRADE | NUMBER OF CREDIT HOURS IN COURSE |
|-------|----------------------------------|
| B | 4 |
| A | 5 |
| D | 3 |
| C | 4 |

20.

| GRADE | NUMBER OF CREDIT HOURS IN COURSE |
|-------|----------------------------------|
| A | 5 |
| C | 4 |
| F | 3 |
| B | 5 |

21. *Brussels Sprouts.*   The following prices per stalk of Brussels sprouts were found at five farmers' markets:

$3.99,   $4.49,   $4.99,   $3.99,   $3.49.

What was the mean price per stalk? the median price? the mode?

22. *Mangoes.*   The most popular fruit in the world is the mango, which is grown in over 2000 varieties. The following prices per pound of mangoes were found at five supermarkets:

$2.49,   $1.59,   $2.29,   $2.49,   $2.29.

What was the mean price per pound? the median price? the mode?

23. *Grading.*   To get a B in math, Rich must score an average of 80 on five tests. His scores on the first four tests were 80, 74, 81, and 75. What is the lowest score that Rich can get on the last test and still receive a B?

24. *Grading.*   To get an A in math, Cybil must score an average of 90 on five tests. Her scores on the first four tests were 90, 91, 81, and 92. What is the lowest score that Cybil can get on the last test and still receive an A?

25. *Length of Pregnancy.*   Marta was pregnant 270 days, 259 days, and 272 days for her first three pregnancies. In order for Marta's average pregnancy to equal the worldwide average of 266 days, how long must her fourth pregnancy last?

**Data:** Vardaan Hospital, Dr. Rekha Khandelwal, M.S.

26. *Male Height.*   Jason's brothers are 174 cm, 180 cm, 179 cm, and 172 cm tall. The average male is 176.5 cm tall. How tall is Jason if he and his brothers have an average height of 176.5 cm?

27. *Median Home Prices.*   The following table lists the selling prices of homes in two counties during one month.
   a) Find the median home price for each county.
   b) Which county had the lower median home price?

| JEFFERSON COUNTY | HAMILTON COUNTY |
|------------------|-----------------|
| $122,587 | $387,262 |
| 138,291 | 146,989 |
| 121,103 | 262,105 |
| 768,407 | 253,289 |
| 532,194 | 112,681 |
| 129,683 | 127,092 |
| 278,104 | 131,612 |
| 110,329 | |

28. *Median Salaries.*   The following table lists salaries for two small companies.
   a) Find the median salary for each company.
   b) Which company has the higher median salary?

| VALUE SERVICES | DEPENDABLE CARE |
|----------------|-----------------|
| $ 48,267 | $18,242 |
| 32,193 | 21,607 |
| 189,607 | 98,322 |
| 56,189 | 87,212 |
| 28,394 | 56,812 |
| 152,693 | 42,394 |
| 42,681 | 50,112 |
| | 52,987 |

**29. Movie Ticket Sales.** The following table lists the numbers of movie tickets sold annually, in billions, from 2006 to 2016.

a) Find the average number of tickets sold for the 8 years from 2006 to 2013.
b) Find the average number of tickets sold for the 8 years from 2009 to 2016.
c) On average, were more tickets sold per year from 2006 to 2013 or from 2009 to 2016?

**30. Movies Released.** The following table lists the numbers of movies issued in wide release by the six major studios annually from 2009 to 2016.

a) Find the average number of movies released for the 5 years from 2009 to 2013.
b) Find the average number of movies released for the 5 years from 2012 to 2016.
c) On average, were more movies released from 2009 to 2013 or from 2012 to 2016?

| YEAR | NUMBER OF MOVIE TICKETS SOLD (in billions) | YEAR | NUMBER OF MOVIE TICKETS SOLD (in billions) |
|---|---|---|---|
| 2006 | 1.40 | 2012 | 1.39 |
| 2007 | 1.42 | 2013 | 1.34 |
| 2008 | 1.36 | 2014 | 1.27 |
| 2009 | 1.42 | 2015 | 1.34 |
| 2010 | 1.33 | 2016 | 1.30 |
| 2011 | 1.28 | | |

DATA: the-numbers.com

| YEAR | NUMBER OF MOVIES RELEASED | YEAR | NUMBER OF MOVIES RELEASED |
|---|---|---|---|
| 2009 | 110 | 2013 | 78 |
| 2010 | 93 | 2014 | 88 |
| 2011 | 101 | 2015 | 92 |
| 2012 | 90 | 2016 | 93 |

DATA: the-numbers.com

**e** Find the quartiles for each set of numbers.

**31.** 4, 8, 10, 16, 12, 25, 30, 32, 14, 28

**32.** 13, 1, 17, 26, 18, 15, 12, 4

**33.** 12, 2, 3, 6, 7, 11, 5, 2, 10

**34.** 2, 3, 7, 3, 7, 5, 2, 9, 9, 4, 6

Find the five-number summary for each set of numbers.

**35.** 88, 73, 62, 90, 94, 98, 82, 87, 77, 79, 77

**36.** 12, 97, 32, 16, 83, 11, 10, 62, 9, 48, 53, 13, 28

**37.** 3.9, 3.8, 1.1, 1.2, 2.7, 4.0, 2.8, 3.4, 3.2, 3.7

**38.** $1.48, $2.95, $3.67, $1.22, $3.51, $4.96, $3.52, $3.99

## Skill Maintenance

Multiply.

**39.** $12.86 \times 17.5$ [5.3a]

**40.** $222 \times 0.5678$ [5.3a]

**41.** $\frac{4}{5} \cdot \frac{3}{28}$ [3.6a]

**42.** $\frac{28}{45} \cdot \frac{3}{2}$ [3.6a]

## Synthesis

**43.** The ordered set of data 18, 21, 24, $a$, 36, 37, $b$ has a median of 30 and an average of 32. Find $a$ and $b$.

**44. Hank Aaron.** Hank Aaron averaged $34\frac{7}{22}$ home runs per year over a 22-year career. After 21 years, Aaron had averaged $35\frac{10}{21}$ home runs per year. How many home runs did Aaron hit in his final year?

**45. Price Negotiations.** Amy offers $6400 for a used Ford Taurus advertised at $8000. The first offer from Jim, the car's owner, is to "split the difference" and sell the car for $(6400 + 8000) \div 2$, or $7200. Amy's second offer is to split the difference between Jim's offer and her first offer. Jim's second offer is to split the difference between Amy's second offer and his first offer. If this pattern continues and Amy accepts Jim's third (and final) offer, how much will she pay for the car?

# Frequency Distributions and Histograms

## OBJECTIVES

**a** Interpret and create frequency tables.

**b** Interpret and construct stem-and-leaf plots.

**c** Interpret and construct histograms.

**SKILL REVIEW**

*Solve percent problems.* [7.2b], [7.3b]

Translate to an equation and solve. Round to the nearest tenth of a percent.

**1.** 15 is what percent of 40?

**2.** What percent of 820 is 129?

**Answers: 1.** $37\frac{1}{2}\%$, or 37.5%  **2.** 15.7%

MyLab Math
VIDEO

The **frequency** of an item in a set of data is the number of times that item appears in the set. A **frequency distribution** describes the frequency patterns in a set of data. In this section, we will look at frequency distributions described by frequency tables, stem-and-leaf plots, and histograms.

## a FREQUENCY TABLES

A **frequency table** gives the number of times a value or values within a range appear in a set of data.

**EXAMPLE 1** *Major League Baseball.* The following list gives the winners of the Major League World Series for the years 2004–2016. Summarize the data using a frequency table.

Boston Red Sox, Chicago White Sox, St. Louis Cardinals, Boston Red Sox, Philadelphia Phillies, New York Yankees, San Francisco Giants, St. Louis Cardinals, San Francisco Giants, Boston Red Sox, San Francisco Giants, Kansas City Royals, Chicago Cubs

**Data:** espn.go.com

We list the different teams in the first column of the frequency table. Then we go through the data sequentially, writing a tally mark in the second column every time the corresponding team name appears. Finally, we write the number of tally marks in the third column.

| TEAM | TALLY MARKS | FREQUENCY |
|---|---|---|
| Boston Red Sox | /// | 3 |
| Chicago White Sox | / | 1 |
| St. Louis Cardinals | // | 2 |
| Philadelphia Phillies | / | 1 |
| New York Yankees | / | 1 |
| San Francisco Giants | /// | 3 |
| Kansas City Royals | / | 1 |
| Chicago Cubs | / | 1 |

◀ **Do Exercise 1.**

**1.** The following list gives the champion women's gymnastic teams for Olympic games from 1984 through 2016. Summarize the data using a frequency table.

Romania, Soviet Union, Soviet Union, United States, Romania, Romania, China, United States, United States

*Answer*

**1.**

| Team | Tally Marks | Frequency |
|---|---|---|
| Romania | ||| | 3 |
| Soviet Union | || | 2 |
| United States | ||| | 3 |
| China | | | 1 |

Sometimes we can better visualize a frequency distribution by recording the frequency with which values appear in classes of equal width.

**EXAMPLE 2** *Test Scores.* The following list gives test scores for a final exam in a history class. Summarize the data using a frequency table.

83, 87, 64, 49, 98, 73, 77, 75, 82, 68, 50, 93, 88

Listing each unique score and its frequency would not give us a good picture of the frequency distribution. Using a range of values for each row of the table makes more sense here. There are several good choices for intervals to use. Here we note that the minimum score is 49 and the maximum score is 98, so we decide to group scores in the 40s, 50s, 60s, and so on. Each of these intervals is called a *class*.

| TEST SCORE | TALLY MARKS | FREQUENCY |
|------------|-------------|-----------|
| 40–49 | I | 1 |
| 50–59 | I | 1 |
| 60–69 | II | 2 |
| 70–79 | III | 3 |
| 80–89 | IIII | 4 |
| 90–99 | II | 2 |

**Do Exercise 2.** ▶

2. The following list gives the number of words per discussion post for a sociology assignment. Complete the frequency table.

52, 753, 967, 134, 228, 365, 547, 862, 197, 678

| WORD COUNT | TALLY MARKS | FREQUENCY |
|------------|-------------|-----------|
| 0–250 | | |
| 251–500 | | |
| 501–750 | | |
| 751–1000 | | |

If the data consist of two variables, such as a person's eye color and hair color, a summary of the data may take the form of a **two-way frequency table**, or **contingency table.**

**EXAMPLE 3** *Survey Results.* A survey of students in a community college asked whether their high school was rural, suburban, or urban and whether it was small (fewer than 1000 students) or large (1000 students or more). The results of the survey are summarized in the following two-way frequency table.

| | RURAL | SUBURBAN | URBAN | TOTALS |
|---|-------|----------|-------|--------|
| SMALL | 251 | 150 | 452 | 853 |
| LARGE | 123 | 695 | 793 | 1611 |
| TOTALS | 374 | 845 | 1245 | 2464 |

a) How many students came from large suburban high schools?

b) How many students came from suburban high schools?

c) How many students were represented in the survey?

d) What percent of students in the survey came from large suburban high schools? Round to the nearest tenth of a percent.

e) What percent of students from suburban high schools came from large schools? Round to the nearest tenth of a percent.

*Answer*

2.

| Word Count | Tally Marks | Frequency |
|------------|-------------|-----------|
| 0–250 | IIII | 4 |
| 251–500 | I | 1 |
| 501–750 | II | 2 |
| 751–1000 | III | 3 |

**Survey Results.** A survey of community college students asked how far they traveled to class and what mode of transportation they used: public or private. The results of the survey are summarized in the following two-way frequency table. Use the table for Margin Exercises 3–7.

|  | FEWER THAN 5 MILES | 5–10 MILES | MORE THAN 10 MILES | TOTALS |
|---|---|---|---|---|
| PUBLIC | 267 | 543 | 57 | 867 |
| PRIVATE | 98 | 726 | 435 | 1259 |
| TOTALS | 365 | 1269 | 492 | 2126 |

**3.** How many students were represented in the survey?

**4.** How many students rode public transportation to class?

**5.** How many students traveled fewer than 5 miles using private transportation?

**6.** What percent of students in the survey rode public transportation to class?

**7.** What percent of students who used public transportation rode more than 10 miles?

We use the two-way frequency table to answer the questions.

**a)** To find the number of students who came from high schools that are both large and suburban, we locate the row in the table labeled "Large" and move across that row to the column labeled "Suburban." There were 695 students who came from large suburban high schools.

**b)** The number of students who came from suburban high schools is the sum of the number of students who came from small suburban schools and the number of students who came from large suburban schools. This sum can be found by locating the column labeled "Suburban" and moving down the column to the row labeled "Totals." There were 845 students who came from suburban high schools.

**c)** The number of students represented in the survey is the entry in the bottom right corner of the table. It is the sum of the entries in the table that are not in the "Totals" row or in the "Totals" column. It is also the sum of the entries to its left in the "Totals" row, and it is the sum of the entries above it in the "Totals" column. There were 2464 students represented in the survey.

**d)** We translate to an equation.

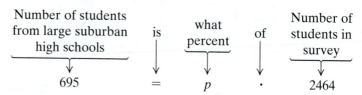

We then solve for $p$ and write the result using percent notation.

$$695 = p \cdot 2464$$

$$\frac{695}{2464} = \frac{p \cdot 2464}{2464} \qquad \text{Dividing by 2464 on both sides}$$

$$\frac{695}{2464} = p$$

$$0.282 \approx p \qquad \text{Rounding decimal notation to the nearest thousandth so that percent notation will be rounded to the nearest tenth of a percent}$$

$$28.2\% \approx p \qquad \text{Finding percent notation}$$

Approximately 28.2% of students in the survey came from large suburban high schools.

**e)** The phrase "of students from suburban high schools" tells us that we are focusing on the column labeled "Suburban." We translate to an equation.

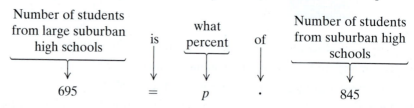

We then solve for $p$ and write the result using percent notation.

$$695 = p \cdot 845$$

$$\frac{695}{845} = \frac{p \cdot 845}{845} \qquad \text{Dividing by 845 on both sides}$$

$$\frac{695}{845} = p$$

$$0.822 \approx p \qquad \text{Rounding decimal notation to the nearest thousandth}$$

$$82.2\% \approx p \qquad \text{Finding percent notation}$$

Approximately 82.2% of students from suburban high schools came from large schools.

**Do Exercises 3–7 on the preceding page.** ▶

## b STEM-AND-LEAF PLOTS

Using ranges of values to present data in a frequency table provides a way to describe the distribution of the data, but in the process the individual data values are lost. A **stem-and-leaf plot** helps us visualize the frequency distribution by preserving the data values. In a stem-and-leaf plot, the **leaves** consist of the rightmost digit of each data value, and the **stems** consist of the remaining digit or digits. It is important that each leaf represent the same place value—for example, ones or tenths. A key included with the plot indicates the place values represented.

**EXAMPLE 4** *ACT Scores.* The ACT is a standardized test used to assess college readiness. The following stem-and-leaf plot shows the average ACT composite score for each of 51 states and districts.

| 17 | 7 |
| 18 | 4 5 7 |
| 19 | 1 1 5 9 9 9 |
| 20 | 0 0 0 1 2 2 2 3 3 3 4 5 6 6 7 8 |
| 21 | 1 1 4 7 9 9 |
| 22 | 0 1 2 3 6 7 |
| 23 | 0 1 1 1 3 3 4 6 6 9 |
| 24 | 5 5 8 |

Key: $17 \mid 7 = 17.7$

**Data:** *The World Almanac and Book of Facts 2017*

**a)** Find the minimum, the maximum, and the range of the average ACT scores.

**b)** Find the median of the average ACT scores.

We use the stem-and-leaf plot to answer the questions.

**a)** The data are arranged in order in a stem-and-leaf plot, so the minimum is the first number listed and the maximum is the last number listed. The minimum average ACT score is 17.7 and the maximum is 24.8. The range is then $24.8 - 17.7 = 7.1$.

**b)** There are 51 scores listed. Since 51 is an odd number, the median is the middle score, or the 26th score listed. Beginning at 17.7, we count the leaves, in order, until we reach the 26th leaf, which is the 8 in the row corresponding to the stem 20. Thus, the median score is 20.8.

**Do Exercises 8 and 9.** ▶

*Public School Revenue.* The following stem-and-leaf plot shows the percent of revenue for public elementary and high schools that came from the U.S. federal government in 2012–2013. The plot includes data for each of 51 states and districts. Use the plot for Margin Exercises 8 and 9.

| 4 | 4 4 |
| 5 | 5 7 7 |
| 6 | 0 3 7 |
| 7 | 1 3 6 8 9 9 |
| 8 | 0 6 6 6 6 7 7 |
| 9 | 0 2 2 4 6 7 8 |
| 10 | 0 1 7 7 |
| 11 | 2 7 8 8 9 |
| 12 | 1 1 3 4 6 6 9 |
| 13 | 0 6 9 |
| 14 | |
| 15 | 0 2 2 |
| 16 | 1 |

Key: $4 \mid 4 = 4.4\%$

**Data:** *The World Almanac and Book of Facts 2017*

**8.** Find the minimum, the maximum, and the range of the percent of revenue from the federal government.

**9.** Find the median of the percent of revenue from the federal government.

*Answers*

**8.** Minimum: 4.4%; maximum: 16.1%; range: 11.7%  **9.** 9.6%

To construct a stem-and-leaf plot, we use the following procedure.

---

**CONSTRUCTING A STEM-AND-LEAF PLOT**

1. Determine the place value of the leaves.
2. Arrange the data in order from smallest to largest.
3. Determine the minimum stem and the maximum stem.
4. Draw a vertical line and list the stems to the left of the line. Begin with the minimum stem and list all consecutive whole numbers until the maximum stem is reached.
5. To the right of the vertical line, list all leaves corresponding to each stem, using numerical order. Write the leaves in columns so that all first leaves align, all second leaves align, and so on.
6. Add a key to the plot to indicate what value each stem and leaf represents.

---

**EXAMPLE 5** *Test Scores.* Construct a stem-and-leaf plot for the following list of test scores.

78, 62, 47, 83, 84, 83, 98, 97, 74, 76, 68, 79, 100

We follow the procedure described above.

1. Since the rightmost digit of each score is in the ones place, the place value of the leaves is ones.

2. We arrange the data in order from smallest to largest.

47, 62, 68, 74, 76, 78, 79, 83, 83, 84, 97, 98, 100

3. The minimum value is 47. The leaf is 7, so the minimum stem is 4. The maximum value is 100. The leaf is 0, so the maximum stem is 10.

4. We draw a vertical line and list all consecutive whole numbers, beginning with 4 and ending with 10, as shown in Figure 1.

```
 4 |              4 | 7            4 | 7
 5 |              5 |              5 |
 6 |              6 | 2 8          6 | 2 8
 7 |              7 | 4 6 8 9      7 | 4 6 8 9
 8 |              8 | 3 3 4        8 | 3 3 4
 9 |              9 | 7 8          9 | 7 8
10 |             10 | 0           10 | 0
                                  Key: 4 | 7 = 47
```

**FIGURE 1**          **FIGURE 2**          **FIGURE 3**

5. To the right of the vertical line, we list all leaves corresponding to each stem, aligning the leaves in columns, as shown in Figure 2.

6. Finally, we add a key, as shown in Figure 3. We write the first stem, a vertical line, and the first leaf, and we indicate that this represents the value 47.

◀ Do Exercise 10.

---

**10. Class Size.** Construct a stem-and-leaf plot for the following list of class sizes.

15, 36, 22, 107, 9, 17, 19, 45, 18, 27, 24

*Answer*

**10.**
```
 0 | 9
 1 | 5 7 8 9
 2 | 2 4 7
 3 | 6
 4 | 5
 5 |
 6 |
 7 |
 8 |
 9 |
10 | 7
Key: 0|9 = 9
```

## c HISTOGRAMS

MyLab Math
ANIMATION

A **histogram** is another type of graph that we use to visualize frequency distributions.

**EXAMPLE 6** *Fuel Economy.* Listed below are the fuel economy ratings, in miles per gallon, for combined city and highway driving for all midsize car models from a recent year sold in the United States.

23, 20, 21, 21, 28, 24, 21, 22, 20, 20, 21, 19, 28, 26, 24, 23, 24, 20, 17, 26,
19, 17, 16, 14, 13, 29, 29, 21, 21, 22, 24, 22, 22, 20, 23, 23, 22, 16, 14, 21,
21, 19, 22, 21, 29, 31, 30, 33, 30, 29, 27, 33, 31, 30, 28, 26, 24, 30, 28, 22,
23, 24, 32, 29, 27, 22, 17, 19, 21, 18, 17, 23, 24, 31, 32, 27, 13, 13, 47, 29,
28, 26, 26, 25, 28, 43, 43, 29, 28, 25, 25, 22, 30, 32, 32, 32, 31, 32, 30, 30,
20, 19, 18, 29, 21, 22, 20, 20, 21, 19, 18, 23, 26, 28, 28, 29, 29, 30, 26, 26,
23, 21, 31, 24, 40, 19, 18, 19, 18, 20, 26, 25, 21, 22, 45, 21, 25, 24, 31, 32,
21, 22, 23, 20, 19, 23, 22, 23, 22, 26, 25, 31, 25, 22, 30, 34, 33, 34, 24, 27,
20, 50, 50, 24, 28, 40, 41, 40, 25, 25, 34, 23, 35, 26, 25, 21, 23

It is difficult to make sense of the 177 numbers in this data set, so the data are displayed below in a histogram.

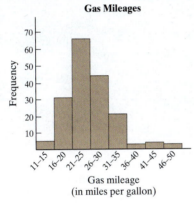

**Gas Mileages**

Gas mileage
(in miles per gallon)

DATA: fueleconomy.gov

a) In which class of gas mileages did the greatest number of midsize models fall?

b) About how many midsize models had gas mileages that were less than 16 mpg?

c) About how many more midsize models had gas mileages between 26 mpg and 30 mpg than between 31 mpg and 35 mpg?

We use the histogram to answer the questions.

a) The tallest rectangle in the histogram is above the class 21–25, so the range 21 mpg to 25 mpg included the greatest number of midsize models.

b) The rectangle corresponding to 11–15 is 5 units high, so 5 midsize models had gas mileages that were less than 16 mpg.

c) From the histogram, we estimate that about 44 midsize models had gas mileages in the 26–30 range and about 21 models had gas mileages in the 31–35 range. Thus, about 44 − 21, or 23, more midsize models had gas mileages between 26 mpg and 30 mpg than between 31 mpg and 35 mpg.

**Do Exercises 11–13.** ▶

The following histogram illustrates test grades for a class of 100 students. Use the histogram for Margin Exercises 11–13.

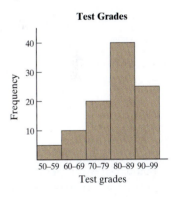

**Test Grades**

Test grades

11. Which range of grades included the greatest number of students?

12. About how many students received a test grade between 90 and 99?

13. About how many more students received a grade between 90 and 99 than a grade between 50 and 59?

*Answers*

**11.** 80–89 **12.** About 25 students
**13.** About 20 students

To construct a histogram, we first create a frequency table for classes of the data. The histogram consists of rectangles whose width is the class width and whose heights correspond to the frequencies. The rectangles forming the histogram should touch.

**EXAMPLE 7** *Major League Baseball.* The following list gives the numbers of games played in by the players listed on the roster of the 2016 Chicago Cubs team. Construct a histogram representing the frequency distribution of the number of games played.

> 1, 2, 2, 3, 3, 5, 5, 7, 8, 8, 11, 14, 16, 16, 17, 17, 26, 28, 29, 29, 31, 32, 33, 34, 35, 47, 48, 51, 54, 55, 67, 68, 74, 76, 81, 86, 86, 107, 125, 142, 142, 147, 151, 155, 155

**Data:** baseball-reference.com

The data are arranged in order from smallest to largest, so we can see that the minimum is 1 and the maximum is 155. We decide to use 8 classes, each of width 20, for the frequency table, as shown on the left below.

| GAMES PLAYED | FREQUENCY |
|---|---|
| 1–20 | 16 |
| 21–40 | 9 |
| 41–60 | 5 |
| 61–80 | 4 |
| 81–100 | 3 |
| 101–120 | 1 |
| 121–140 | 1 |
| 141–160 | 6 |

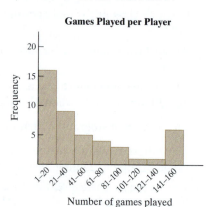

For the histogram, we label the horizontal axis "Number of games played" and write the range of values represented by each class. We label the vertical axis "Frequency." We draw rectangles of equal width, whose sides touch and whose heights correspond to the frequencies in the table, as shown on the right above.

◀ **Do Exercise 14.**

**14.** The following list gives the number of words per discussion post for a sociology assignment. Use the frequency table from Margin Exercise 2 to construct a histogram.

> 52, 753, 967, 134, 228, 365, 547, 862, 197, 678

*Answer*

**14.**

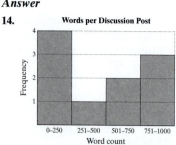

### ✓ Check Your Understanding

**Reading Check** Determine whether each statement is true or false.

_____ **RC1.** Stem-and-leaf plots and histograms are both visual representations of frequency distributions.

_____ **RC2.** A stem-and-leaf plot preserves the data values.

_____ **RC3.** When constructing a histogram, you should leave a small space between the rectangles.

_____ **RC4.** All class intervals for a data set should have the same width.

_____ **RC5.** We can read the minimum and maximum values of a data set directly from a stem-and-leaf plot.

**Concept Check** Results of a survey on phones and computers owned by students were summarized in a two-way frequency table. Use the highlighted row or column of the table to answer the question. Round to the nearest tenth of a percent.

**CC1.** What percent of PC owners have an iPhone?

|  | PC | MAC | LINUX | TOTALS |
|---|---|---|---|---|
| **iPHONE** | 23 | 112 | 2 | 137 |
| **ANDROID** | 42 | 20 | 1 | 63 |
| **TOTALS** | 65 | 132 | 3 | 200 |

**CC2.** What percent of iPhone owners have a PC?

|  | PC | MAC | LINUX | TOTALS |
|---|---|---|---|---|
| **iPHONE** | 23 | 112 | 2 | 137 |
| **ANDROID** | 42 | 20 | 1 | 63 |
| **TOTALS** | 65 | 132 | 3 | 200 |

**a** Summarize each set of data in Exercises 1–4 using a frequency table.

1. *Tennis.* The four major annual tennis tournaments are called the Grand Slam tournaments. The following list gives the Grand Slam men's singles champions for 2014–2017.

> Stan Wawrinka, Rafael Nadal, Novak Djokovic, Marin Cilic, Novak Djokovic, Stan Wawrinka, Novak Djokovic, Novak Djokovic, Novak Djokovic, Novak Djokovic, Andy Murray, Stan Wawrinka, Roger Federer, Rafael Nadal, Roger Federer, Rafael Nadal

**Data:** espn.com

**2.** *Baby Names.*   The following list gives the most popular names for baby girls in the United States for 2001–2016.

Emily, Emily, Emily, Emily, Emily, Emily, Emily, Emma, Isabella, Isabella, Sophia, Sophia, Sophia, Emma, Emma, Emma

**Data:** Social Security Administration

**3.** 4, 4, 7, 6, 0, 4, 2, 4, 4, 7, 5, 7, 7, 4, 0, 6, 6, 6, 7, 0, 4, 5, 5

**4.** 31, 28, 31, 30, 31, 30, 31, 31, 30, 31, 30, 31

**5.** *College Tuition.*   The following list gives the annual tuition for the 25 most expensive colleges in the United States. Complete the frequency table.

$52,666, $55,161, $52,491, $52,550, $52,945, $50,982, $49,062, $52,283, $51,438, $51,024, $50,358, $52,385, $52,760, $50,855, $50,430, $52,476, $51,614, $50,547, $52,002, $51,464, $52,430, $50,394, $51,548, $49,073, $50,910

**Data:** businessinsider.com

| CLASS | TALLY MARKS | FREQUENCY |
| --- | --- | --- |
| $49,000–$49,999 | | |
| $50,000–$50,999 | | |
| $51,000–$51,999 | | |
| $52,000–$52,999 | | |
| $53,000–$53,999 | | |
| $54,000–$54,999 | | |
| $55,000–$55,999 | | |

**6.** *College Enrollment.*   The following list gives the number of students enrolled in each college that is part of the Southeastern Conference. Complete the frequency table.

33,724, 27,845, 37,098, 1,641, 29,727, 23,212, 36,130, 20,873, 23,212, 12,567, 50,645, 35,424, 26,754, 31,524, 27,287

**Data:** collegeraptor.com

| CLASS | TALLY MARKS | FREQUENCY |
| --- | --- | --- |
| 0–9,999 | | |
| 10,000–19,999 | | |
| 20,000–29,999 | | |
| 30,000–39,999 | | |
| 40,000–49,999 | | |
| 50,000–59,999 | | |

Results of a survey concerning beverage choices are summarized in the following two-way frequency table. Use the table for Exercises 7–14. Where appropriate, round to the nearest tenth of a percent.

|  | REGULAR SOFT DRINKS | DIET SOFT DRINKS | NO SOFT DRINKS | TOTALS |
|---|---|---|---|---|
| COFFEE | 265 | 165 | 423 | 853 |
| NO COFFEE | 348 | 98 | 112 | 558 |
| TOTALS | 613 | 263 | 535 | 1411 |

7. How many people are represented in the survey?

8. How many people surveyed drink coffee?

9. How many people surveyed drink diet soft drinks?

10. How many people surveyed drink no coffee and no soft drinks?

11. What percent of all people surveyed drink coffee and diet soft drinks?

12. What percent of all people surveyed drink no coffee but drink regular soft drinks?

13. What percent of people surveyed who drink coffee do not drink soft drinks?

14. What percent of people surveyed who drink diet soft drinks also drink coffee?

The following two-way frequency table shows how many students in each of three sections of a study skills class passed the class, failed the class, or were given an incomplete for the class. Use the table to answer Exercises 15–20.

|  | SECTION A | SECTION B | SECTION C | TOTALS |
|---|---|---|---|---|
| PASSED | 24 | 22 | 19 | 65 |
| FAILED | 3 | 5 | 2 | 10 |
| INCOMPLETE | 0 | 1 | 4 | 5 |
| TOTALS | 27 | 28 | 25 | 80 |

15. What percent of all students represented were those who failed Section B?

16. What percent of all students represented were those who passed Section A?

17. What percent of students in Section C failed?

18. What percent of students who failed were in Section B?

19. What percent of all students represented passed the class?

20. What percent of students in Section C received an incomplete?

Interpret each of the following stem-and-leaf plots.

**21.** The following stem-and-leaf plot gives the percent of 2014 college graduates who graduated with student debt for each of 49 states.

```
4 | 6 6 7 7 8
5 | 0 4 4 4 5 5 5 6 7 8 8 9 9 9
6 | 0 0 0 1 1 1 2 2 2 2 2 3 4 5 5 5 5 7 7 7 8 8 8 8 9 9
7 | 0 0 0 2 6
```
Key: 4|6 = 4.6%

**Data:** *The World Almanac and Book of Facts 2017*

**a)** Find the minimum, the maximum, and the range of the percent of graduates with student debt.

**b)** Find the median of the percent of graduates with student debt.

**22.** The following stem-and-leaf plot gives the percent of public school students in eighth grade who scored at or above a basic level of science knowledge in 2011, for each of 51 states and districts.

```
2 | 2
3 |
4 | 7
5 | 2 4 5 6 7 8 8
6 | 0 1 1 2 2 2 3 3 3 3 4 4 4 6 7 8 8 9 9
7 | 0 1 1 2 2 2 2 3 3 3 4 5 5 6 6 7 7 8 9 9
8 | 0 0 2
```
Key: 2|2 = 22%

**Data:** *The World Almanac and Book of Facts 2017*

**a)** Find the minimum, the maximum, and the range of the percent of eighth-grade students who scored at or above a basic level of science knowledge.

**b)** Find the median of the percent of eighth-grade students who scored at or above a basic level of science knowledge.

*Highest Mountains.* The following stem-and-leaf plots give the heights of the 20 highest mountains in North America and in South America. Use the plots to answer Exercises 23–26.

**North America's Highest Mountains**

```
14 | 8 8
15 | 0 3 4 8 9
16 | 2 4 4 5
17 | 0 1 2 4 7
18 | 0 5
19 | 5
20 | 3
```
14|8 = 14,800 ft

DATA: summitpost.org

**South America's Highest Mountains**

```
21 | 1 1 1 2 4 5 6 7 7 7 8 9
22 | 0 1 1 1 2 3 6 8
```
21|1 = 21,100 ft

DATA: andes.org.uk

**23.** Which of the two continents contains the highest mountain?

**24.** For which of the two continents is there a larger range of heights in its 20 highest mountains?

**25.** Find the mean of the heights of the 20 highest mountains in North America.

**26.** Find the mean of the heights of the 20 highest mountains in South America.

**27. Highest Bridges.** The following list gives the distance, in meters, from the road surface to the water or ground below for the 20 highest bridges in the world. Create a stem-and-leaf plot to represent the data. *Hint:* Let each leaf represent the digit in the tens place.

570, 500, 360, 340, 370, 320, 490, 390, 320, 430,
330, 360, 400, 330, 370, 460, 390, 320, 380, 340

**Data:** highestbridges.com

Key:

**28. Unemployment Rates.** In August 2017, the United States unemployment rate was 4.4%. The following list gives the unemployment rates for the 26 states with rates statistically different from the nation's rate. Create a stem-and-leaf plot to represent the data.

7.2, 3.5, 5.1, 2.4, 6.4, 2.6, 2.9, 3.5, 3.3, 3.9, 5.4, 3.8,
3.8, 5.3, 3.9, 2.8, 2.7, 6.3, 2.3, 5.4, 3.3, 3.3, 3.5, 3.0,
3.8, 3.4

**Data:** U.S. Department of Labor, Bureau of Labor Statistics

Key:

**29. Average Temperature.** The following list gives the average annual temperature, in degrees Fahrenheit, for each of 50 states. Create a stem-and-leaf plot to represent the data.

63, 27, 60, 60, 59, 45, 49, 55, 71, 64, 70, 44, 52, 52,
48, 54, 57, 66, 41, 54, 48, 44, 41, 63, 55, 43, 49, 50,
44, 53, 53, 45, 59, 40, 51, 60, 48, 49, 50, 62, 45, 58,
65, 49, 43, 55, 48, 52, 43, 42

**Data:** currentresults.com

Key:

**30. Average Rainfall.** The following list gives the average annual rainfall, in inches, for each of 50 states. Create a stem-and-leaf plot to represent the data.

58, 23, 14, 51, 22, 16, 50, 46, 55, 51, 64, 19, 39, 42,
34, 29, 49, 60, 42, 45, 48, 33, 27, 59, 42, 15, 24, 10,
43, 47, 15, 42, 50, 18, 39, 40, 27, 43, 48, 50, 20, 54,
29, 12, 43, 44, 38, 45, 33, 13

**Data:** currentresults.com

Key:

**C** **Basketball.** The following histogram illustrates the number of points scored per game by the Los Angeles Lakers during a recent basketball season. Use the graph for Exercises 31–34.

**Los Angeles Lakers Regular Season Points per Game**

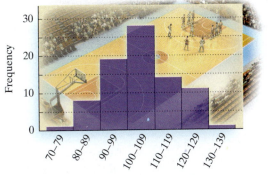

Number of points scored per game

DATA: National Basketball Association

**31.** In how many games did the Lakers score 90–99 points?

**32.** In what point range did the highest number of Laker scores lie?

**33.** In what point range(s) did the lowest number of Laker scores lie?

**34.** In how many more games did the Lakers score 100–109 points than 90–99 points?

**Licensed Drivers.** The following histogram illustrates the percent of residents who are licensed drivers for each of 50 states. Use the graph for Exercises 35–38.

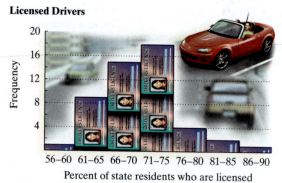

**Licensed Drivers**

Frequency (y-axis): 4, 8, 12, 16, 20

Percent of state residents who are licensed (x-axis): 56–60  61–65  66–70  71–75  76–80  81–85  86–90

DATA: U.S. Department of Transportation, Federal Highway Administration

**35.** In how many states are 61% to 75% of residents licensed drivers?

**36.** In how many states are 76% to 90% of all residents licensed drivers?

**37.** What can you conclude about the median of the data?

**38.** What can you conclude about the range of the data?

**39.** Construct a histogram representing the data in Exercise 3.

**40.** Construct a histogram representing the data in Exercise 4.

**41.** *College Tuition.* Use the frequency table in Exercise 5 to construct a histogram representing the annual tuition data for the 25 most expensive colleges in the United States.

**42.** *College Enrollment.* Use the frequency table in Exercise 6 to construct a histogram representing the enrollment data for colleges in the Southeastern Conference.

## Skill Maintenance

Simplify.

**43.** $\left(\dfrac{3}{5}\right)^2 - \dfrac{1}{10} \cdot 2\dfrac{1}{2}$  [4.8a]

**44.** $1000 \div 100 \div 2 \cdot 5$  [1.9c]

**45.** $6.1 - 4.32 + 0.8$  [5.4b]

**46.** $\dfrac{\frac{1}{2}}{\frac{1}{8}}$  [4.8b]

**47.** $2\dfrac{1}{3} + 1\dfrac{1}{4} - \left(\dfrac{1}{2}\right)^2$  [4.8a]

**48.** $5 \cdot (1 + 3)^2 - 10 - 2(5 - 4)$  [1.9c]

## Synthesis

Two-sided stem-and-leaf plots can be used to compare two sets of data. In a two-sided plot, the stems are listed in the middle. Leaves from one set of data extend to the right, as before, and data values are read from left to right. Leaves from the other set of data extend to the left, and data values are read from right to left.

The following two-sided stem-and-leaf plot shows the percents of 15-year-olds in 39 countries who believe they are too fat. Use the plot to answer Exercises 49–52.

```
                                          Boys      Girls
                                    88 │ 0 │
                            988655410 │ 1 │ 5 6 8
     9988776655444333322221111110 │ 2 │ 0 7 8 9
                                      │ 3 │ 3 4 4 5 5 6 7 8 8 9 9
                                      │ 4 │ 1 1 3 3 5 5 5 6 7 8 9
                                      │ 5 │ 0 0 1 1 1 2 2 2 2 4
                                       Key: 1 │ 5 = 15%
```

**49.** What is the minimum percent of girls who believe they are too fat?

**50.** What is the maximum percent of boys who believe they are too fat?

**51.** Find the five-number summary of the data for boys.

**52.** Find the five-number summary of the data for girls.

**53.** Create a two-sided stem-and-leaf plot for the heights of the highest mountains in North America and South America, given before Exercises 23–26.

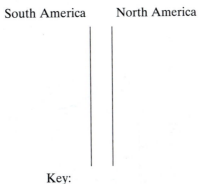

South America    North America

Key:

**Predictions and Probability**

### OBJECTIVES

a  Make predictions from a set of data using interpolation or extrapolation.

b  Use a tree diagram to count the number of possible outcomes.

c  Calculate the probability of an event occurring.

### a  MAKING PREDICTIONS

Sometimes we use data to make predictions or estimates of missing data points. One process for doing so is called **interpolation**. Interpolation enables us to estimate missing "in-between values" on the basis of known information.

**EXAMPLE 1**  *Monthly Mortgage Payments.*  When money is borrowed and then repaid in monthly installments, the payment amount increases as the total number of payments decreases. The table below lists the size of a monthly payment when $110,000 is borrowed (at 9% interest) for various lengths of time. Use interpolation to estimate the monthly payment on a 35-yr loan.

| YEAR | MONTHLY PAYMENT |
|------|-----------------|
| 5 | $2283.42 |
| 10 | 1393.43 |
| 15 | 1115.69 |
| 20 | 989.70 |
| 25 | 923.12 |
| 30 | 885.08 |
| 35 | ? |
| 40 | 848.50 |

To use interpolation, we first plot the points and look for a trend. It seems reasonable to draw a line between the points corresponding to 30 and 40. We can "zoom in" to better visualize the situation. To estimate the second coordinate that is paired with 35, we trace a vertical line up from 35 to the graph and then left to the vertical axis. Thus, we estimate the value to be $867. We can also estimate this value by averaging $885.08 and $848.50.

$$\frac{\$885.08 + \$848.50}{2} = \$866.79 \approx \$867$$

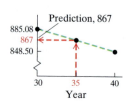

When we estimate in this manner to find an in-between value, we are *interpolating*. Real-world information about the data might tell us that an estimate found in this way is unreliable. For example, data from the stock market might be too erratic for interpolation.

**Do Exercise 1.** ▶

We often analyze data with the intention of going "beyond" the data. One process for doing so is called **extrapolation**.

**EXAMPLE 2** *Daily Mobile Device Use.* The data in the following table and graphs show the average number of minutes per day that Americans spent using mobile devices for several years. Use extrapolation to estimate the average number of minutes per day that Americans will spend using mobile devices in 2019.

| YEAR | AVERAGE DAILY MOBILE DEVICE USE (in minutes) |
|------|------|
| 2015 | 177 |
| 2016 | 186 |
| 2017 | 195 |
| 2018 | 203 |
| 2019 | ? |

DATA: statista.com

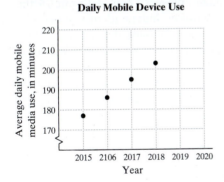

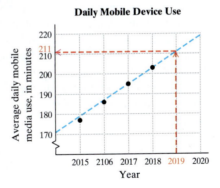

First, we analyze the data and note that they increase from 2015 through 2018. Then we draw a "representative" line through the data and beyond. We choose to draw the line through the two most recent points. To estimate a value for 2019, we draw a vertical line up from 2019 until it hits the representative line. We go to the left and read off a value—about 211. Thus, we estimate the average daily mobile device use of Americans to be about 211 minutes in 2019. When we estimate in this way to find a "go-beyond value," we are *extrapolating*. Estimates found with this method vary depending on the "representative" line chosen.

**Do Exercise 2.** ▶

In calculus and statistics, other methods of interpolation and extrapolation are developed. The two basic concepts remain unchanged, but more complicated methods of determining what line "best fits" the given data are used. These methods are often computer-based.

**1.** *Study Time and Test Scores.* A professor gathered the following data comparing study time and test scores. Use interpolation to estimate the test score received when study time is 21 hr.

| STUDY TIME (in hours) | TEST SCORE (in percent) |
|------|------|
| 19 | 83 |
| 20 | 85 |
| 21 | ? |
| 22 | 91 |
| 23 | 93 |

**2.** *Daily TV Use.* The following table shows the average number of minutes per day that Americans spent watching TV for several years. Use extrapolation to estimate the average number of minutes per day that Americans will spend watching TV in 2019.

| YEAR | AVERAGE DAILY TV USE (in minutes) |
|------|------|
| 2015 | 251 |
| 2016 | 245 |
| 2017 | 240 |
| 2018 | 235 |
| 2019 | ? |

DATA: statista.com

*Answers*

**1.** 88%  **2.** About 230 min

## b   TREE DIAGRAMS

To calculate a probability, we first need to be able to count how many ways a given event can occur. In many applications, a *tree diagram* can help us to count the number of possible outcomes of an *experiment*.

Rolling a die is an example of what is called an **experiment** in probability. A probability experiment is a procedure that has more than one clearly defined outcome. We do not know before conducting an experiment what the outcome will be. The outcomes of rolling a standard six-sided die are ⚀, ⚁, ⚂, ⚃, ⚄, and ⚅. Flipping a coin is also an experiment. The outcomes of flipping a coin are heads and tails.

We can use tree diagrams to illustrate the outcomes of the experiments "rolling a die" and "flipping a coin" as shown in the following figure.

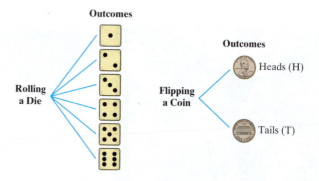

**EXAMPLE 3**   The popular children's game "Rock, Paper, Scissors" can be thought of as an experiment. In this game, players begin with one of their hands forming a closed fist. Each player chooses, independently of the other, whether to leave the hand closed (rock), open the hand flat (paper), or open two fingers (scissors). Draw a tree diagram to illustrate the outcomes of the experiment.

A tree diagram begins on the left at one point. A branch is drawn to the right for each outcome. The experiment has three outcomes, as shown in the following figure.

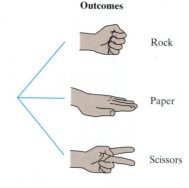

Outcomes

Rock

Paper

Scissors

**3.** A bag contains one red marble and one blue marble. Draw a tree diagram to illustrate the outcomes of selecting one marble from the bag.

◀ **Do Exercise 3.**

*Answer*

**3.**
   ⟨ Red
      Blue

When two procedures are performed together, all possible outcomes for the second procedure are drawn at the end of each branch of the first procedure. We follow each branch to its end to find the outcome represented by that part of the diagram.

**EXAMPLE 4**  Draw a tree diagram to illustrate two players playing "Rock, Paper, Scissors." List all possible outcomes.

We begin with the diagram drawn in Example 3. For each outcome for the first player, there are three possible outcomes for the second player.

| First Player | Second Player | Outcomes |
|---|---|---|
| Rock | Rock | ⟹ Rock, Rock |
| | Paper | ⟹ Rock, Paper |
| | Scissors | ⟹ Rock, Scissors |
| Paper | Rock | ⟹ Paper, Rock |
| | Paper | ⟹ Paper, Paper |
| | Scissors | ⟹ Paper, Scissors |
| Scissors | Rock | ⟹ Scissors, Rock |
| | Paper | ⟹ Scissors, Paper |
| | Scissors | ⟹ Scissors, Scissors |

There are a total of nine possible outcomes for two players playing "Rock, Paper, Scissors."

**Do Exercise 4.** ▶

**EXAMPLE 5**  A bag contains four marbles: one red, one blue, one purple, and one green. Draw a tree diagram to illustrate an experiment consisting of selecting a marble from the bag and then flipping a coin. List all possible outcomes.

There are four outcomes when selecting a marble, and two when flipping a coin. We draw a tree diagram and list outcomes at the end of each branch.

| Choosing a Marble | Flipping a Coin | Outcomes |
|---|---|---|
| Red | H | ⟹ Red, H |
| | T | ⟹ Red, T |
| Blue | H | ⟹ Blue, H |
| | T | ⟹ Blue, T |
| Purple | H | ⟹ Purple, H |
| | T | ⟹ Purple, T |
| Green | H | ⟹ Green, H |
| | T | ⟹ Green, T |

There are a total of eight possible outcomes for the experiment of first selecting a marble and then flipping a coin.

**Do Exercise 5.** ▶

4. Draw a tree diagram to illustrate an experiment consisting of flipping a coin twice. How many outcomes are possible?

5. Using the bag of marbles described in Example 5, draw a tree diagram to illustrate an experiment consisting of flipping a coin and then selecting a marble from the bag. How many outcomes are possible?

*Answers*

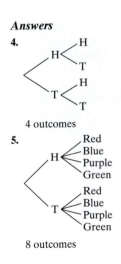

4.

4 outcomes

5.

8 outcomes

## C   PROBABILITY

Using probability, we can attach a numerical value to the likelihood that a specific event will occur.

Suppose we flip a fair coin. Because the coin is just as likely to land heads as it is to land tails, we say that the *probability* of it landing heads is $\frac{1}{2}$. Similarly, if we roll a fair die (plural: dice), we are as likely to roll a ▨ as we are to roll a ▨, ▨, ▨, ▨, or ▨. Because of this, we say that the probability of rolling a ▨ is $\frac{1}{6}$.

**EXAMPLE 6**   A die is about to be rolled. Find the probability that a number greater than 4 will be rolled.

Since ▨, ▨, ▨, ▨, ▨, and ▨ are all equally likely to be rolled, and since two of these possibilities involve numbers greater than 4, we have

$$\begin{array}{l}\text{The probability of rolling} \\ \text{a number greater than 4}\end{array} = \dfrac{2}{6} \begin{array}{l}\leftarrow \text{Number of ways to roll a 5 or 6} \\ \leftarrow \text{Number of (equally likely) possible} \\ \quad\ \text{outcomes}\end{array}$$

$$= \dfrac{1}{3}.$$

A coin landing heads and rolling a ▨ are examples of "events." To find the probability, or likelihood, of an event occurring, we use the following principle.

---

### THE PRIMARY PRINCIPLE OF PROBABILITY

If an event $E$ can occur $m$ ways out of $n$ equally likely possible outcomes, then

$$\text{The probability of } E \text{ occurring } = \dfrac{m}{n}.$$

---

The probability of any event is greater than or equal to 0 and less than or equal to 1.

**6.** Find the probability of the arrow in Example 7 landing on red or blue.   **GS**

Probability of landing on red or blue

$$= \dfrac{\begin{array}{l}\text{Number of ways to} \\ \text{land on red or blue}\end{array}}{\begin{array}{l}\text{Number of ways to} \\ \text{land on a space}\end{array}}$$

$$= \dfrac{\boxed{\phantom{0}}}{12}$$

$$= \dfrac{\boxed{\phantom{0}}}{3}$$

**EXAMPLE 7**   A spinner such as the one shown below is used by flicking the arrow so that it rotates quickly. Find the probability that the arrow will come to rest on a green space. (Assume that the arrow will not rest on a line.)

There are 12 sections of equal size, so there are 12 equally likely possible outcomes. Since 4 of the possibilities are green, we have

$$\begin{array}{l}\text{Probability of} \\ \text{landing on green}\end{array} = \dfrac{\text{Number of ways to land on green}}{\text{Number of ways to land on a space}}$$

$$= \dfrac{4}{12} = \dfrac{1}{3}.$$

◀ **Do Exercise 6.**

*Answer*

6. $\dfrac{2}{3}$

*Guided Solution*
6. 8, 2

**EXAMPLE 8** A cloth bag contains 20 equal-size marbles: 5 are red, 7 are blue, and 8 are yellow. A marble is randomly selected. Find the probability that **(a)** a red marble is selected; **(b)** a blue marble is selected; **(c)** a yellow marble is selected; **(d)** a black marble is selected.

**a)** Since all 20 marbles are equally likely to be selected, we have

$$\frac{\text{The probability of}}{\text{selecting a red marble}} = \frac{\text{Number of ways to select a red marble}}{\text{Number of ways to select any marble}}$$

$$= \frac{5}{20} = \frac{1}{4}, \text{ or } 0.25$$

**b)** $$\frac{\text{The probability of}}{\text{selecting a blue marble}} = \frac{\text{Number of ways to select a blue marble}}{\text{Number of ways to select any marble}}$$

$$= \frac{7}{20}, \text{ or } 0.35$$

**c)** $$\frac{\text{The probability of}}{\text{selecting a yellow marble}} = \frac{\text{Number of ways to select a yellow marble}}{\text{Number of ways to select any marble}}$$

$$= \frac{8}{20} = \frac{2}{5}, \text{ or } 0.4$$

**d)** $$\frac{\text{The probability}}{\text{of selecting a black marble}} = 0 \qquad \textcolor{red}{\text{There are 0 ways to select a black marble; } \frac{0}{20} = 0.}$$

**Do Exercise 7.** ▶

**7.** A school play is attended by 500 people: 250 children, 100 seniors, and 150 (nonsenior) adults. After everyone has been seated, one audience member is selected at random. Find the probability of each of the following.

**a)** A child is selected.

**b)** A senior is selected.

**c)** A (nonsenior) adult is selected.

A standard deck of 52 playing cards is made up as shown below.

A deck of 52 cards

King  Queen  Jack  10  9  8  7  6  5  4  3  2  Ace

**EXAMPLE 9** A card is randomly selected from a well-shuffled (mixed) deck of cards. Find the probability that **(a)** the card is a jack; **(b)** the card is a club.

**a)** $$\frac{\text{The probability of}}{\text{selecting a jack}} = \frac{\text{Number of ways to select a jack}}{\text{Number of ways to select any card}}$$

$$= \frac{4}{52} = \frac{1}{13}$$

**b)** $$\frac{\text{The probability of}}{\text{selecting a club}} = \frac{\text{Number of ways to select a club}}{\text{Number of ways to select any card}}$$

$$= \frac{13}{52} = \frac{1}{4}$$

**Do Exercise 8.** ▶

**8.** A card is randomly selected from a well-shuffled deck of cards. Find the probability of each of the following.

**a)** The card is a diamond.

**b)** The card is a king or queen.

*Answers*

**7. (a)** $\frac{1}{2}$, or 0.5; **(b)** $\frac{1}{5}$, or 0.2; **(c)** $\frac{3}{10}$, or 0.3

**8. (a)** $\frac{1}{4}$, or 0.25; **(b)** $\frac{2}{13}$

# Translating
# for Success

1. *Vacation Miles.* The Saenz family drove their new van 13,640.8 mi in the first year. Of this total, 2018.2 mi were driven while on vacation. How many nonvacation miles did they drive?

2. *Rail Miles.* Of the recent $15\frac{1}{2}$ million passenger miles on a rail passenger line, 80% were transportation-to-work miles. How many rail miles, in millions, were transportation to work?

3. *Sales Tax Rate.* The sales tax on the purchase of 10 bath towels that cost $129.50 is $8.42. What is the sales tax rate?

4. *Water Level.* During heavy rains in early spring, the water level in a pond rose 0.5 in. every 35 min. How much did the water rise in 90 min?

5. *Marathon Training.* At one point in his daily training routine for a marathon, Rocco had run $15\frac{1}{2}$ mi. This was 80% of the distance he intended to run that day. How far did Rocco plan to run?

The goal of these matching questions is to practice step (2), Translate, of the five-step problem-solving process. Translate each word problem to an equation and select a correct translation from equations A–O.

**A.** $8.42 \cdot x = 129.50$

**B.** $x = 80\% \cdot 15\frac{1}{2}$

**C.** $x = \dfrac{84 - 68}{84}$

**D.** $2018.2 + x = 13{,}640.8$

**E.** $\dfrac{5}{100} = \dfrac{x}{3875}$

**F.** $2018.2 = x \cdot 13{,}640.8$

**G.** $4\frac{1}{6} \cdot 73 = x$

**H.** $\dfrac{x}{5} = \dfrac{100}{3875}$

**I.** $15\frac{1}{2} = 80\% \cdot x$

**J.** $8.42 = x \cdot 129.50$

**K.** $\dfrac{0.5}{35} = \dfrac{x}{90}$

**L.** $x \cdot 4\frac{1}{6} = 73$

**M.** $x = \dfrac{84 - 68}{68}$

**N.** $x = 8.42\% \cdot 129.50$

**O.** $0.5 \times 35 = 90 \cdot x$

Answers on page A-18

6. *Vacation Miles.* The Ning family drove 2018.2 mi on their summer vacation. If they put a total of 13,640.8 mi on their new van during that year, what percent were vacation miles?

7. *Sales Tax.* The sales tax rate is 8.42%. Salena purchased 10 pillows at $12.95 each. How much tax was charged on this purchase?

8. *Charity Donations.* Rachel donated $5 to her favorite charity for each $100 she earned. One month, she earned $3875. How much did she donate that month?

9. *Tuxedos.* Emil Tailoring Company purchased 73 yd of fabric for a new line of tuxedos. How many tuxedos can be produced if it takes $4\frac{1}{6}$ yd of fabric for each tuxedo?

10. *Percent Increase.* In a calculus-based physics course, Mime got 68% on the first exam and 84% on the second. What was the percent increase in her score?

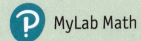
✓ **Check Your Understanding**

**Reading Check**   Choose from the list on the right the number that matches each description.

**RC1.** The probability of selecting a red marble from a bag containing 2 blue marbles

**RC2.** The probability of selecting a red marble from a bag containing 2 red marbles

**RC3.** The probability of selecting a red marble from a bag containing 2 blue marbles and 2 red marbles

**RC4.** Not a probability

**a)** 0
**b)** $\frac{1}{2}$
**c)** 1
**d)** 2

**Concept Check**   Follow the steps in each exercise to find the probability that the event described will occur when a card is randomly selected from a well-shuffled deck of cards. Refer, if necessary, to the figure on p. 579 that illustrates a standard deck of 52 playing cards.

**CC1.** Find the probability of selecting a red 4.

   **a)** How many ways can we select a red 4? In other words, how many red cards are numbered 4?

   **b)** How many ways can we select any card? In other words, how many cards are there in all?

   **c)** What is the probability of selecting a red 4?

**CC2.** Find the probability of selecting a 4 of hearts.

   **a)** How many ways can we select a 4 of hearts?

   **b)** How many ways can we select any card?

   **c)** What is the probability of selecting a 4 of hearts?

---

**a**   Use interpolation or extrapolation to find the missing data values.

**1.** *Study Time and Grades.*   A math instructor asked her students to keep track of how much time each spent studying the chapter on decimal notation. They collected the information together with test scores from that chapter's test. The data are given in the following table. Estimate the missing value.

| STUDY TIME (in hours) | TEST GRADE |
|---|---|
| 13 | 80 |
| 15 | 85 |
| 17 | 80 |
| 19 | ? |
| 21 | 86 |
| 23 | 91 |

**2.** *Maximum Heart Rate.*   A person's maximum heart rate depends on his or her sex, age, and resting heart rate. The following table relates resting heart rate and maximum heart rate for a 30-yr-old woman. Estimate the missing value.

| RESTING HEART RATE (in beats per minute) | MAXIMUM HEART RATE (in beats per minute) |
|---|---|
| 58 | 173 |
| 65 | 178 |
| 70 | ? |
| 78 | 185 |
| 85 | 188 |

DATA: American Heart Association

**3. Global Recorded Digital Music Revenue.**

| YEAR | DIGITAL MUSIC REVENUE (in billions of dollars) |
|------|-----|
| 2012 | 5.4 |
| 2013 | 5.7 |
| 2014 | 6.0 |
| 2015 | 6.6 |
| 2016 | 7.8 |
| 2017 | ? |

DATA: ifpi.org

**4. Global Recorded Physical Music Revenue.**

| YEAR | PHYSICAL MUSIC REVENUE (in billions of dollars) |
|------|-----|
| 2012 | 7.6 |
| 2013 | 6.8 |
| 2014 | 6.1 |
| 2015 | 5.8 |
| 2016 | 5.4 |
| 2017 | ? |

DATA: ifpi.org

**5. International Students.**

| YEAR | NEW INTERNATIONAL STUDENT ENROLLMENT IN U.S. INSTITUTIONS |
|------|-----|
| 2011–2012 | 228,000 |
| 2012–2013 | 251,000 |
| 2013–2014 | 270,000 |
| 2014–2015 | ? |
| 2015–2016 | 301,000 |

DATA: Institute of International Education

**6. Baseball Ticket Price.**

| YEAR | AVERAGE MAJOR LEAGUE BASEBALL TICKET PRICE |
|------|-----|
| 2014 | $27.93 |
| 2015 | 29.94 |
| 2016 | 31.00 |
| 2017 | ? |
| 2018 | 32.44 |

DATA: statista.com

**7. Mobile Ad Spending.**

| YEAR | MOBILE AD SPENDING (in billions of dollars) |
|------|-----|
| 2016 | 40.5 |
| 2017 | 49.8 |
| 2018 | 57.7 |
| 2019 | 65.8 |
| 2020 | ? |

DATA: smartinsights.com

**8. Vegetable Production.**

| YEAR | AREA PLANTED IN SQUASH IN THE UNITED STATES (in acres) |
|------|-----|
| 2015 | 41,250 |
| 2016 | 37,100 |
| 2017 | 35,500 |
| 2018 | ? |

DATA: U.S. Department of Agriculture

**b**  Use the following figures for Exercises 9–14 on the following page.

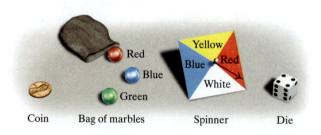

Coin          Bag of marbles          Spinner          Die

**9.** Draw a tree diagram and list the outcomes of selecting one marble from the bag.

**10.** Draw a tree diagram and list the outcomes of spinning the spinner.

**11.** An experiment consists of flipping the coin and then selecting one marble from the bag. Draw a tree diagram and list the outcomes of the experiment.

**12.** An experiment consists of spinning the spinner and then flipping the coin. Draw a tree diagram and list the outcomes of the experiment.

**13.** An experiment consists of rolling the die and then spinning the spinner. How many outcomes are there of the experiment?

**14.** An experiment consists of spinning the spinner and then selecting one marble from the bag. How many outcomes are there of the experiment?

**C**    Find each of the following probabilities.

***Rolling a Die.***   In Exercises 15–20, assume that one die is rolled.

**15.** Find the probability that a ⚃ is rolled.

**16.** Find the probability that a ⚁ is rolled.

**17.** Find the probability that an odd number is rolled.

**18.** Find the probability that a number greater than 2 is rolled.

**19.** Find the probability that a 7 is rolled.

**20.** Find the probability that a ⚀, ⚁, ⚂, ⚃, ⚄, or ⚅ is rolled.

***Playing Cards.***   In Exercises 21–26, assume that one card is randomly selected from a well-shuffled deck (see p. 579).

**21.** Find the probability that the card is the jack of spades.

**22.** Find the probability that the card is a picture card (jack, queen, or king).

**23.** Find the probability that an 8 or a 6 is selected.

**24.** Find the probability that a black 5 is selected.

**25.** Find the probability that a red picture card (jack, queen, or king) is selected.

**26.** Find the probability that a 10 is selected.

*Candy Colors.* A box of candy was found to contain the following numbers of gumdrops.

| | |
|---|---|
| Strawberry | 7 |
| Lemon | 8 |
| Orange | 9 |
| Cherry | 4 |
| Lime | 5 |
| Grape | 6 |

In Exercises 27–30, assume that one of the gumdrops is randomly chosen from the box.

**27.** Find the probability that a cherry gumdrop is selected.

**28.** Find the probability that an orange gumdrop is selected.

**29.** Find the probability that the gumdrop is *not* lime.

**30.** Find the probability that the gumdrop is *not* lemon.

## Skill Maintenance

**31.** *Album Sales.* There were 200.5 million music albums sold in 2016. Of these, 104.8 million were CDs. What percent of albums sold were CDs? Round to the nearest tenth of a percent. [7.4a]

**Data:** *Nielsen Music Year-End Report U.S. 2016*

**32.** *Digital Downloads.* There were 964.3 million songs downloaded in 2015, and 723.7 million songs downloaded in 2016. What was the percent decrease from 2015 to 2016? Round to the nearest tenth of a percent. [7.5a]

**Data:** *Nielsen Music Year-End Report U.S. 2016*

## Synthesis

**33.** A coin is flipped twice. What is the probability that two heads occur?

**34.** A coin is flipped twice. What is the probability that one head and one tail occur?

**35.** A die is rolled twice. What is the probability that a ▨ is rolled twice?

**36.** A day is chosen randomly during a leap year. What is the probability that the day is in July?

**37.** In the game "Rock, Paper, Scissors," described in Example 3 of this section, if players make identical selections, there is no winner. If players make different selections, a winner is determined by the following rules.

- Rock wins against Scissors. (Rock smashes Scissors.)
- Paper wins against Rock. (Paper covers Rock.)
- Scissors wins against Paper. (Scissors cuts Paper.)

Use the tree diagram in Example 4 to find each probability, assuming that all selections are random.

**a)** The probability that player 1 wins

**b)** The probability that player 2 wins

**c)** The probability that neither player wins

## Vocabulary Reinforcement

Choose the term from the list on the right that best completes each sentence. Not every term will be used.

1. A(n) _____ presents data in rows and columns. [8.1a]

2. The _____ is the point where the axes cross. [8.3a]

3. A(n) _____ uses symbols to represent amounts. [8.1b]

4. The _____ of a set of data is the number or numbers that occur most often. [8.5d]

5. The _____ of a set of data is the sum of the numbers in the set divided by the number of items of data. [8.5b]

mean

median

mode

table

pictograph

histogram

origin

quadrant

## Concept Reinforcement

Determine whether each statement is true or false.

_____ 1. To find the mean of a set of numbers, add the numbers and then multiply by the number of items of data. [8.5b]

_____ 2. If each number in a set of data occurs the same number of times, there is no mode. [8.5d]

_____ 3. If there is an odd number of items in a set of data, the middle number is the median. [8.5c]

## Study Guide

**Objective 8.1a** Extract and interpret data from tables.

**Example** Which oatmeal listed below has the greatest number of calories?

| PRODUCT | COST | CALORIES | FAT (g) | FIBER (g) | SUGARS (g) |
|---|---|---|---|---|---|
| Quaker Quick-1 Minute | 0.19 | 150 | 3.0 | 4 | 1 |
| Kashi Heart to Heart Golden Brown Maple | 0.44 | 160 | 2.0 | 5 | 12 |
| Quaker Organic Maple & Brown Sugar | 0.54 | 150 | 2.0 | 3 | 12 |
| Nature's Path Organic Maple Nut | 0.47 | 200 | 4.0 | 4 | 12 |

PER PACKET (instant) OR SERVING (longer-cooking)

DATA: *Consumer Reports*, November 2008

We look down the column headed "Calories" and find the largest number, 200. The name of the oatmeal in that row is Nature's Path Organic Maple Nut.

**Practice Exercises**

1. Which oatmeal has the greatest cost per serving? What is that cost?

2. How many grams of sugar are in the Kashi oatmeal?

**Objective 8.2a** Extract and interpret data from bar graphs.

**Example** The horizontal bar graph below shows the building costs of selected stadiums. Estimate how much more Lucas Oil Stadium cost than Invesco Field cost.

**Cost of Sports Stadiums**

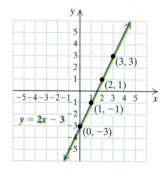

Cost (in millions of dollars)

DATA: National Football League

We estimate each cost.
*Lucas Oil Stadium:* $720 million; *Invesco Field:*
$360 million
*Difference:* $720 million − $360 million =
$360 million

**Practice Exercises**

**3.** Which stadium cost less than $100 million?

**4.** Estimate how much more Invesco Field cost than Bank of America Stadium did.

---

**Objective 8.3c** Determine whether an ordered pair is a solution of an equation with two variables.

**Example** Determine whether the ordered pair $(-2, 1)$ is a solution of the equation $x + 5y = 3$.

$$x + 5y = 3$$
$$\overline{-2 + 5 \cdot 1 \,?\, 3}$$
$$-2 + 5 \,\bigg|$$
$$3 \,\bigg|\, 3 \quad \text{TRUE}$$

Since the equation becomes true for $x = -2$ and $y = 1$, $(-2, 1)$ is a solution.

**Practice Exercise**

**5.** Determine whether the ordered pair $(2, -4)$ is a solution of the equation $3x + y = 10$.

---

**Objective 8.4b** Graph linear equations in two variables.

**Example** Graph: $y = 2x - 3$.

We make a table of solutions. Then we plot the points, draw the line, and label it.

| $x$ | $y$ | $(x, y)$ |
|-----|-----|----------|
| 0 | −3 | $(0, -3)$ |
| 1 | −1 | $(1, -1)$ |
| 2 | 1 | $(2, 1)$ |
| 3 | 3 | $(3, 3)$ |

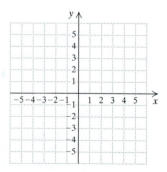

**Practice Exercise**

**6.** Graph: $y = x - 4$.

**Objectives 8.5b, c, d**   Find the mean, the median, and the mode of a set of numbers.

**Example**   Find the mean, the median, and the mode:

$$2.6, \quad 3.5, \quad 61.8, \quad 10.4, \quad 3.5, \quad 21.6, \quad 10.4, \quad 3.5.$$

*Mean:* We add the numbers and divide by the number of data items:

$$\frac{2.6 + 3.5 + 61.8 + 10.4 + 3.5 + 21.6 + 10.4 + 3.5}{8}$$

$$= \frac{117.3}{8} = 14.6625.$$

*Median:* We first rearrange the numbers from smallest to largest:

$$2.6, \quad 3.5, \quad 3.5, \quad 3.5, \quad 10.4, \quad 10.4, \quad 21.6, \quad 61.8.$$

The median is halfway between the middle two numbers, which are 3.5 and 10.4. The average of these middle numbers is 6.95.

*Mode:* Since 3.5 occurs the most often, it is the mode.

**Practice Exercise**

**7.** Find the mean, the median, and the mode of this set of numbers:

$$8, 13, 1, 4, 8, 7, 15.$$

---

**Objective 8.5e**   Find the quartiles of a set of numbers and write the five-number summary of a set of numbers.

**Example**   Find the five-number summary:

$$3, \quad 7, \quad 2, \quad 9, \quad 9, \quad 2, \quad 11, \quad 2, \quad 5, \quad 8.$$

We list the numbers in order from left to right and find the median, or second quartile. The first quartile is the median of the numbers that are less than the second quartile. The third quartile is the median of the numbers that are greater than the second quartile.

$$2, \quad 2, \quad 2, \quad 3, \quad 5, \quad 7, \quad 8, \quad 9, \quad 9, \quad 11$$

First quartile    Median, 6    Third quartile

The five-number summary includes these three statistics as well as the minimum and the maximum.

Minimum: 2

First quartile: 2

Median: 6

Third quartile: 9

Maximum: 11

**Practice Exercise**

**8.** Find the five-number summary for the set of numbers.

$$10, \quad 85, \quad 2, \quad 4, \quad 32, \quad 33, \quad 8, \quad 65, \quad 58$$

---

**Objective 8.7c**   Calculate the probability of an event occurring.

**Example**   A company randomly selects a month of the year for an annual party. What is the probability that a month whose name begins with J is chosen?

There are 12 months in a year, so there are 12 equally likely possible outcomes. There are 3 months whose names begin with J: January, June, and July. We have

$$\text{The probability that the name of the month begins with J} = \frac{\text{Number of months whose names begin with J}}{\text{Number of months in the year}}$$

$$= \frac{3}{12} = \frac{1}{4}.$$

**Practice Exercise**

**9.** A month of the year is randomly selected. What is the probability that the name of the month begins with A?

## Review Exercises

*Internet Usage and Smartphone Ownership.*  The table below lists the percents of the populations of several countries who use the Internet and who own a smartphone. Use this table for Exercises 1–3.  [8.1a]

| COUNTRY | PERCENT USING INTERNET | PERCENT OWNING SMARTPHONE |
|---------|------------------------|---------------------------|
| South Korea | 94% | 88% |
| United States | 89 | 72 |
| Poland | 69 | 41 |
| China | 65 | 58 |
| Ukraine | 60 | 27 |
| Mexico | 54 | 35 |
| India | 22 | 17 |

DATA: pewglobal.org

**1.** What percent of China's population owns a smartphone?

**2.** In what countries does less than 50% of the population own a smartphone?

**3.** In which country do approximately twice as many people use the Internet as own a smartphone?

*College Costs.*  The circle graph below shows the various cost categories for a full-time resident student at an Oklahoma regional university and the percent of the total college cost represented by each category. Use this graph for Exercises 4–6.  [8.1b]

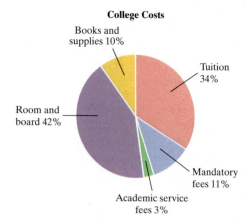

**College Costs**

Books and supplies 10%
Tuition 34%
Room and board 42%
Mandatory fees 11%
Academic service fees 3%

DATA: okcollegestart.org

**4.** What percent of college costs is tuition?

**5.** Which category accounts for the greatest part of the total college costs?

**6.** In a recent year, the total college cost for a full-time resident student at an Oklahoma regional university was $11,500. How much did a student pay for room and board?

Find the mean.  [8.5b]

**7.** 26, 34, 53, 41

**8.** 0.2, 1.7, 1.9, 2.4, 3.1

**9.** $2, $14, $17, $17, $21, $29

Find the median.  [8.5c]

**10.** 7, 11, 14, 17, 18

**11.** 4.6, 5.2, 5.4, 9.8

**12.** $2, $17, $21, $29, $14, $17

Find the mode.  [8.5d]

**13.** 26, 34, 43, 26, 51

**14.** 17, 7, 11, 11, 14, 17, 18

**15.** 20, 10, 20, 50, 20, 60

**16.** *Gas Mileage.*  A 2017 Mazda 3 does 336 mi of city driving on 12 gal of gasoline. What is the gas mileage?  [8.5b]

**17.** *Grade Point Average.*  Find the grade point average for one semester given the following grades. Assume the grade point values are 4.0 for A, 3.0 for B, and so on. Round to the nearest tenth.  [8.5b]

| COURSE | GRADE | NUMBER OF CREDIT HOURS IN COURSE |
|--------|-------|----------------------------------|
| Math | A | 5 |
| English | B | 3 |
| Computer Science | C | 4 |
| Spanish | B | 3 |
| College Skills | B | 1 |

**Tornadoes.** The bar graph below shows the total number of tornadoes that occurred in the United States from 2014 through 2016, by month. Use the graph for Exercises 18–21. [8.2a]

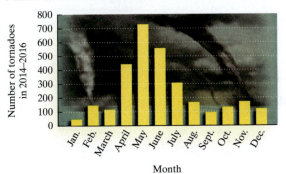

Number of Tornadoes in 2014–2016

DATA: spc.noaa.gov

**18.** Which month had the greatest number of tornadoes?

**19.** How many tornadoes occurred in February?

**20.** How many more tornadoes occurred in May than in June?

**21.** Do more tornadoes occur in the winter or in the spring?

**First-Class Postage.** The table below lists the cost of first-class postage in various years. Use the table for Exercises 22 and 23.

| YEAR | FIRST-CLASS POSTAGE |
|---|---|
| 2001 | 34¢ |
| 2002 | 37 |
| 2006 | 39 |
| 2007 | 41 |
| 2008 | 42 |
| 2009 | 44 |
| 2012 | 45 |
| 2013 | 46 |
| 2014 | 49 |
| 2016 | 47 |
| 2017 | 49 |

DATA: U.S. Postal Service

**22.** Make a vertical bar graph of the data. [8.2b]

**23.** Make a line graph of the data. [8.2d]

**Homelessness.** The line graph below shows the average number of homeless children in New York City's shelter system each night for various years. Use the graph for Exercises 24–27. [8.2c]

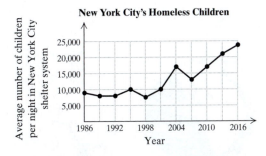

New York City's Homeless Children

DATA: NYC Department of Homeless Services and Human Resources Administration; NYC Stat, shelter census reports; Coalition for the Homeless

**24.** During which year after 1990 were there the fewest children in the shelter system?

**25.** How many children were in the shelter system each night in 2001?

**26.** In which years were there about 17,000 children each night in the shelter system?

**27.** By how much did the number of children in the shelter system each night increase from 2007 to 2016?

**Governors' Salaries.** The histogram below shows the numbers of state governors in the United States who receive annual salaries in the given ranges. Use the graph for Exercises 28 and 29. [8.6c]

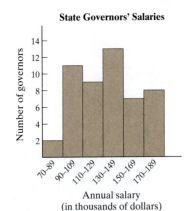

State Governors' Salaries

DATA: knowledgecenter.csg.org

**28.** Which salary range has the smallest number of governors?

**29.** How many governors make less than $130,000?

**30.** An experiment consists of spinning the spinner shown below and then selecting a card at random from the hand shown. Draw a tree diagram to illustrate the experiment, and list all outcomes. [8.7b]

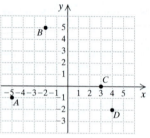

Determine the coordinates for each point. [8.3a]

**31.** *A*

**32.** *B*

**33.** *C*

**34.** *D*

Plot each point on the graph below. [8.3a]

**35.** (2, 5)     **36.** (0, −3)     **37.** (−4, −2)

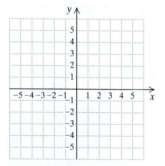

In which quadrant is each point located? [8.3b]

**38.** (3, −8)     **39.** (−20, −14)     **40.** $\left(4\frac{9}{10}, 1\frac{3}{10}\right)$

**41.** Complete these solutions of $2x + 4y = 10$: (1, □); (□, −2). [8.4a]

Graph on a plane. [8.4b, c]

**42.** $y = 2x - 5$

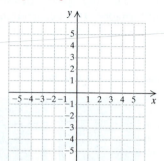

**43.** $y = -\frac{3}{4}x$

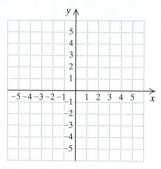

**44.** $x + y = 4$

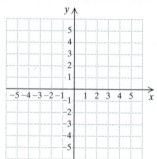

**45.** $x = -5$

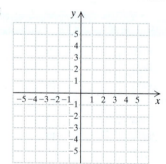

**46.** $y = 6$

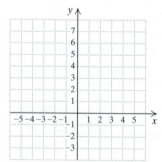

The following two-way frequency table displays the results of a survey of instructors of three subjects concerning their use of online homework. Use the table for Exercises 47–49. [8.6a]

|  | MATHEMATICS | BUSINESS | HISTORY | TOTALS |
|---|---|---|---|---|
| ASSIGN ONLINE HOMEWORK | 98 | 87 | 12 | 197 |
| NO ONLINE HOMEWORK | 42 | 23 | 18 | 83 |
| TOTALS | 140 | 110 | 30 | 280 |

**47.** How many business teachers are represented in the table?

**48.** How many history teachers assign online homework?

**49.** What percent of mathematics teachers do not assign online homework?

*Vegetable Consumption.* In 2015, the per capita consumption of fresh vegetables was greater than 2 lb per person for 16 vegetables. The following list gives the per capita consumption for those vegetables. Use the data for Exercises 50–52.

20.5, 18.9, 13.5, 11.1, 11, 8.8, 8.6, 7.6, 7.5, 7.4, 6.3, 5.4, 4.6, 3.1, 3, 2.4

**50.** Find the five-number summary for the data. [8.5e]

**51.** Construct a stem-and-leaf plot for the data. [8.6b]

**52.** Construct a histogram for the data. Use the following classes:

2.0–5.9
6.0–9.9
10.0–13.9
14.0–17.9
18.0–21.9  [8.6c]

A deck of 52 playing cards is thoroughly shuffled and a card is randomly selected. [8.7c]

**53.** Find the probability that the 5 of clubs is selected.

**54.** Find the probability that a red card is selected.

**55.** Use extrapolation and the graph for Exercises 24–27 to estimate the number of children in New York City's shelter system each night in 2018. [8.7a]

**56.** What is the mean of this set of data?

$$\frac{1}{2}, \frac{1}{3}, \frac{1}{4}, \frac{1}{5} \quad [8.5b]$$

**A.** $\frac{77}{240}$  **B.** $\frac{1}{3}$  **C.** $\frac{7}{24}$  **D.** $\frac{1}{4}$

**57.** Find the mode(s) of this set of data.

6, 9, 6, 8, 8, 5, 10, 5, 9, 10  [8.5d]

**A.** 8  **B.** 5, 6, 8, 9, 10
**C.** 9  **D.** No mode exists.

## Synthesis

**58.** The ordered set of data 298, 301, 305, *a*, 323, *b*, 390 has a median of 316 and a mean of 326. Find *a* and *b*. [8.5b, c]

**59.** A typing pool consists of four senior typists who earn $12.35 per hour and nine other typists who earn $11.15 per hour. Find the mean hourly wage. [8.5b]

# Understanding Through Discussion and Writing

**1.** Find a real-world situation that fits this equation:

$$T = \frac{20,500 + 22,800 + 23,400 + 26,000}{4}. \quad [8.5b]$$

**2.** Can bar graphs always, sometimes, or never be converted to line graphs? Why? [8.2b, d]

**3.** Is it possible for the mean of a set of numbers to be greater than all but one of the numbers in the set? Why or why not? [8.5b]

**4.** Is it possible for the median of a set of four numbers to be one of the numbers in the set? Why or why not? [8.5c]

**5.** One way the U.S. Census Bureau reports income is by giving a 3-yr mean of median incomes. Why would this statistic be chosen? [8.5b, c]

**6.** Would a company considering expansion be more interested in interpolation or extrapolation? Why? [8.7a]

CHAPTER

**8**   Test

For
Extra
Help

For step-by-step test solutions, access the Chapter Test Prep Videos
in MyLab Math.

This table lists the number of calories burned during various walking activities. Use it for Exercises 1 and 2.

| WALKING ACTIVITY | CALORIES BURNED IN 30 MIN | | |
|---|---|---|---|
| | 110 LB | 132 LB | 154 LB |
| Walking | | | |
| Fitness (5 mph) | 183 | 213 | 246 |
| Mildly energetic (3.5 mph) | 111 | 132 | 159 |
| Strolling (2 mph) | 69 | 84 | 99 |
| Hiking | | | |
| 3 mph with 20-lb load | 210 | 249 | 285 |
| 3 mph with 10-lb load | 195 | 228 | 264 |
| 3 mph with no load | 183 | 213 | 246 |

**1.** Which activity provides the greatest benefit in burned calories for a person who weighs 132 lb?

**2.** What activities would allow a person weighing 154 lb to burn at least 250 calories every 30 min?

*Waste Generated.* The number of pounds of waste generated per person per year varies greatly among countries around the world. In the pictograph at right, each symbol represents approximately 100 lb of waste. Use the pictograph for Exercises 3–6.

**3.** In which country does each person generate 1300 lb of waste per year?

**4.** In which countries does each person generate more than 1500 lb of waste per year?

**5.** How many pounds of waste per person per year are generated in Canada?

**6.** How many more pounds of waste per person per year are generated in the United States than in Mexico?

**Amount of Waste Generated (per person per year)**

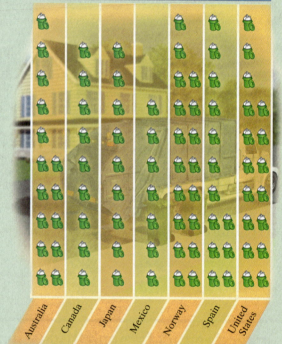

DATA: OECD, Key Environmental Indicators

 = 100 pounds

*Hurricanes.* The following line graph shows the numbers of Atlantic hurricanes for the years 2000–2016. Use the graph for Exercises 7–12.

**Atlantic Hurricanes**

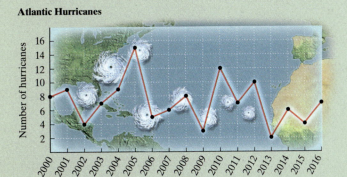

DATA: National Hurricane Center

**7.** What year had the greatest number of Atlantic hurricanes?

**8.** In what year were there 3 Atlantic hurricanes?

**9.** How many hurricanes were there in 2012?

**10.** How many more hurricanes were there in 2005 than in 2006?

**11.** Find the mean number of hurricanes per year for the years 2008–2012.

**12.** Use extrapolation to estimate the number of Atlantic hurricanes in 2017.

*Book Circulation.* The table below lists the average number of books checked out per day of the week for a branch library. Use this table for Exercises 13 and 14.

| DAY | NUMBER OF BOOKS CHECKED OUT |
|---|---|
| Sunday | 210 |
| Monday | 160 |
| Tuesday | 240 |
| Wednesday | 270 |
| Thursday | 310 |
| Friday | 275 |
| Saturday | 420 |

**13.** Make a vertical bar graph of the data.

**14.** Make a line graph of the data.

In which quadrant is each point located?

**15.** $\left(-\frac{1}{2}, 7\right)$

**16.** $(-5, -6)$

Determine the coordinates of each point.

**17.** $A$

**18.** $B$

**19.** $C$

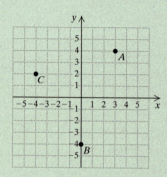

**20.** Complete the following solution of the equation $y - 3x = -10$: ( ⬚ , 2).

Graph.

**21.** $y = 2x - 2$

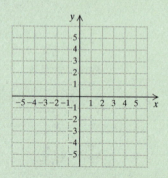

**22.** $y = -\frac{3}{2}x$

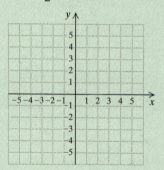

**23.** $x = -2$

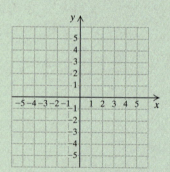

Find the mean.

**24.** 45, 49, 52, 54

**25.** 1, 2, 3, 4, 5

**26.** 3, 17, 17, 18, 18, 20

Find the median and any modes that exist.

**27.** 45, 47, 54, 54

**28.** 1, 2, 3, 4, 5

**29.** 20, 17, 17, 18, 3, 18

**30. Grades.** To get a C in chemistry, Ted must score an average of 70 on four tests. His scores on the first three tests were 68, 71, and 65. What is the lowest score he can get on the last test and still get a C?

**31. Grade Point Average.** Find the grade point average for one semester given the following grades. Assume the grade point values are 4.0 for A, 3.0 for B, and so on. Round to the nearest tenth.

| COURSE | GRADE | NUMBER OF CREDIT HOURS IN COURSE |
| --- | --- | --- |
| Introductory Algebra | B | 3 |
| English | A | 3 |
| Business | C | 4 |
| Spanish | B | 3 |
| Typing | B | 2 |

**Interest Rates.** The following is a list of interest rates, in percent, offered by various banks for a $1000 6-month deposit. Use the data for Exercises 32–35.

> 1.3, 1.3, 0.9, 0.8, 0.7, 0.9, 0.9, 1.5, 1.4, 1.0, 1.2, 1.2, 0.8, 0.8, 0.7, 1.4, 1.2

**32.** Find the five-number summary.

**33.** Construct a stem-and-leaf plot.

Key:

**34.** Complete the frequency table.

| CLASS | TALLY MARKS | FREQUENCY |
| --- | --- | --- |
| 0.7–0.9 | | |
| 1.0–1.2 | | |
| 1.3–1.5 | | |

**35.** Use the frequency table in Exercise 34 to construct a histogram.

Results of a survey of the number of hours students work each week are summarized in the following two-way frequency table. Use the table for Exercises 36 and 37.

| | FEWER THAN 10 HOURS | 10–20 HOURS | MORE THAN 20 HOURS | TOTALS |
| --- | --- | --- | --- | --- |
| FULL-TIME STUDENTS | 36 | 21 | 18 | 75 |
| PART-TIME STUDENTS | 2 | 12 | 31 | 45 |
| TOTALS | 38 | 33 | 49 | 120 |

**36.** How many full-time students are working more than 20 hours a week?

**37.** What percent of full-time students are working more than 20 hours a week?

**38.** A bag contains a white cube, a black cube, and a red cube. Draw a tree diagram to illustrate an experiment consisting of selecting a cube from the bag and then flipping a coin. List the outcomes of the experiment.

**39.** A bag of mini chocolate bars contains 11 milk chocolate bars, 8 dark chocolate bars, and 10 chocolate bars with peanuts. All bars are the same shape and size. If a bar is drawn at random, find the probability that it is dark chocolate.

**40.** A standard six-sided die is about to be rolled. Find the probability that a number greater than 4 will be rolled.

**A.** $\dfrac{1}{2}$      **B.** $\dfrac{1}{3}$      **C.** $\dfrac{1}{4}$      **D.** $\dfrac{1}{6}$

## Synthesis

**41.** The ordered set of data 69, 71, 73, $a$, 78, 98, $b$ has a median of 74 and a mean of 82. Find $a$ and $b$.

Graph.

**42.** $\dfrac{1}{4}x + 3\dfrac{1}{2}y = 1$

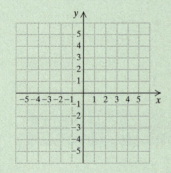

**43.** $\dfrac{5}{6}x - 2\dfrac{1}{3}y = 1$

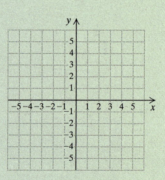

**44.** Find the area of a rectangle whose vertices are $(-3, 1)$, $(5, 1)$, $(5, 8)$, and $(-3, 8)$.

**1.** *Insects.*  There are 1.4 billion insects for every human being. Write standard notation for 1.4 billion.

**Data:** National Geographic

**2.** *Gas Mileage.*  A 2017 Honda CR-V does 364 mi of city driving on 13 gal of gasoline. What is the gas mileage?

**3.** In 402,513, what does the digit 5 mean?

**4.** Evaluate: $3 + 5^3$.

**5.** Find all the factors of 60.

**6.** Round 52.045 to the nearest tenth.

**7.** Convert to fraction notation: $3\frac{3}{10}$.

**8.** Convert from cents to dollars: 210¢.

**9.** Find $-x$ when $x = -9$

**10.** Find percent notation for $\frac{7}{20}$.

**11.** Determine whether 11, 30 and 4, 12 are proportional.

**12.** Evaluate $2x - y$ for $x = 3$ and $y = 8$.

Perform the indicated operation and, if possible, simplify.

**13.** $\frac{-2}{15} + \frac{3}{10}$

**14.** $-18 + (-21)$

**15.** $\frac{14}{15} - \frac{3}{5}$

**16.** $350 - 24.57$

**17.** $3\frac{3}{7} \cdot 4\frac{3}{8}$

**18.** $(17.4)(-2.43)$

**19.** $\frac{13}{15} \div \frac{26}{27}$

**20.** $-1334.183 \div (-21.4)$

Solve.

**21.** $\frac{5}{8} = \frac{6}{x}$

**22.** $\frac{2}{5} \cdot y = \frac{3}{10}$

**23.** $\frac{3}{8}x + 2 = 11$

**24.** $3(x - 5) = 7x + 2$

Solve.

**25.** *Energy Consumption.*  In a recent year, American utility companies generated 1240 billion kilowatt-hours (kWh) of electricity using coal, 804 billion using nuclear power, 1379 billion using natural gas, 608 billion using renewable sources, and 49 billion using other methods. How many kilowatt-hours of electricity were produced that year?

**Data:** U.S. Energy Information Administration

**26.** A piece of fabric $1\frac{3}{4}$ yd long is cut into 7 equal strips. What is the length of each strip?

**27.** A recipe calls for $\frac{3}{4}$ cup of sugar. How much sugar should be used for $\frac{1}{2}$ of the recipe?

**28.** *Billionaires.*  The number of billionaires in the world increased from 1810 in 2016 to 2043 in 2017. What was the percent increase?

**Data:** Forbes

**29. Peanut Products.** In any given year, the average American eats 2.7 lb of peanut butter, 1.5 lb of salted peanuts, 1.2 lb of peanut candy, 0.7 lb of in-shell peanuts, and 0.1 lb of peanuts in other forms. How many pounds of peanuts and products containing peanuts does the average American eat in a year?

**30.** A landscaper bought 22 evergreen trees for $210. What was the cost of each tree? Round to the nearest cent.

**31.** A factory manufacturing valves for engines was discovered to have made 4 defective valves in a lot of 18 valves. At this rate, how many defective valves can be expected in a lot of 5049 valves?

**32.** A salesperson earns $182 selling $2600 worth of electronic equipment. What is the commission rate?

**33.** Find the perimeter and the area of the rectangle.

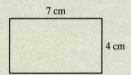

7 cm

4 cm

**34.** In which quadrant is the point $(-4, 9)$ located?

**35.** Graph: $y = \dfrac{1}{2}x - 4$.

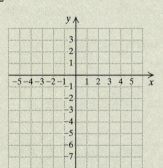

**36.** Combine like terms: $6x + 4y - 8x - 3y$.

**37.** Multiply: $5(2a - 3b + 1)$.

**Video Game Industry.** The following table lists the monthly retail revenue of the U.S. video game industry from August 2016 through July 2017. Use the table for Exercises 38–40.

| NUMBER OF MONTHS AFTER JULY 2016 | U.S. VIDEO GAME REVENUE (in billions) |
|---|---|
| 1 | $0.57 |
| 2 | 0.84 |
| 3 | 0.86 |
| 4 | 1.97 |
| 5 | 2.78 |
| 6 | 0.61 |
| 7 | 0.72 |
| 8 | 1.36 |
| 9 | 0.64 |
| 10 | 0.54 |
| 11 | 0.77 |
| 12 | 0.59 |

DATA: statista.com

**38.** Find the mean and the median of these revenues.

**39.** Make a vertical bar graph of the data.

**40.** Make a line graph of the data.

## Synthesis

**41.** Simplify:

$$\left(\frac{3}{4}\right)^2 - \frac{1}{8} \cdot \left(3 - 1\frac{1}{2}\right)^2.$$

**42.** Add and write the answer as a mixed numeral:

$$-5\frac{42}{100} + \frac{355}{100} + \frac{89}{10} + \frac{17}{1000}.$$

**43.** A square with sides parallel to the axes has the point $(2, 3)$ at its center. Find the coordinates of the square's vertices if each side is 8 units long.

# Geometry and Measurement

The table below lists the 5 longest railway tunnels in the world. The Gotthard Base Tunnel through the Swiss Alps is the longest and deepest railway tunnel. Construction of the Gotthard Base Tunnel took almost two decades and cost over $10 billion. The tracks could not be installed until 28 million tons of rock had been excavated. The tunnel can carry 315 passenger and freight trains a day.

### Longest Railway Tunnels

| Tunnel (year completed) | Length |
|---|---|
| Gotthard Base, Switzerland (2016) | Two tubes: 57.104 km, 57.017 km |
| Seikan, Japan (1988) | 53.85 km |
| Yulhyeon, South Korea (2016) | 52.3 km |
| Channel, France/United Kingdom (1994) | 50.45 km |
| Songshan Lake, China (2016) | 38.813 km |

DATA: *USA TODAY*, May, 29, 2016, "World's Longest Rail Tunnel to Open," by Helena Bachmann; *The World Almanac* 2017; Alp Transit Gotthard Ltd.

We will convert the length of the Gotthard Base Tunnel from kilometers to miles in Example 16 in Section 9.1.

# Systems of Linear Measurement

## 9.1

### OBJECTIVES

**a** Convert from one American unit of length to another.

**b** Convert from one metric unit of length to another.

**c** Convert between American and metric units of length.

Length, or distance, is one kind of measure. To find lengths, we start with some **unit segment** and assign to it a measure of 1. Suppose $\overline{AB}$ below is a unit segment. Let's measure segment $\overline{CD}$, using $\overline{AB}$ as our unit segment.

Unit segment:   $A \vdash\!\!\!\!\!\rule[0.4ex]{2em}{0.4pt}\!\!\!\!\!\dashv B$
$\underbrace{\qquad}_{1}$

$C \vdash\!\!\!\rule[0.4ex]{10em}{0.4pt}\!\!\!\dashv D$
$\underbrace{\quad}_{1}\underbrace{\quad}_{1}\underbrace{\quad}_{1}\underbrace{\quad}_{1}$

Since we can place 4 unit segments end to end along $\overline{CD}$, the measure of $\overline{CD}$ is 4.

Sometimes we need to use parts of units. For example, the measure of the segment $\overline{MN}$ below is $1\frac{1}{2}$. We place one unit segment and one half-unit segment end to end.

$M \vdash\!\!\!\rule[0.4ex]{6em}{0.4pt}\!\!\!\dashv N$
$\underbrace{\quad}_{1}\underbrace{\ }_{\frac{1}{2}}$

Use the unit below to measure the length of each segment or object.

$A \vdash\!\!\!\!\!\rule[0.4ex]{2em}{0.4pt}\!\!\!\!\!\dashv B$
$\underbrace{\quad}_{1}$

1. $\vdash\!\!\!\rule[0.4ex]{7em}{0.4pt}\!\!\!\dashv$

2. $\vdash\!\!\!\rule[0.4ex]{10em}{0.4pt}\!\!\!\dashv$

3.

4.

◀ **Do Exercises 1–4.**

### a   AMERICAN MEASURES

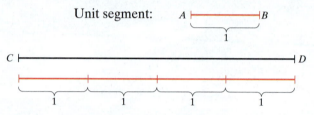

**SKILL REVIEW**

*Convert between mixed numerals and fraction notation.*   [4.5a]

1. Convert $6\frac{3}{8}$ to fraction notation.   2. Convert $\frac{96}{5}$ to a mixed numeral.

Answers: **1.** $\frac{51}{8}$ **2.** $19\frac{1}{5}$

MyLab Math
VIDEO

American units of length are related as follows.

| AMERICAN UNITS OF LENGTH | |
|---|---|
| 12 inches (in.) = 1 foot (ft) | 3 feet = 1 yard (yd) |
| 36 inches = 1 yard | 5280 feet = 1 mile (mi) |

*Answers*

**1.** 2   **2.** 3   **3.** $1\frac{1}{2}$   **4.** $2\frac{1}{2}$

(Actual size, in inches)

We can visualize comparisons of the units as follows:

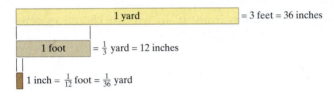

The symbols ′ and ″ are also used for feet and inches, as in 13 in. = 13″ and 27 ft = 27′. American units have also been called "English," or "British–American," because at one time they were used in both North America and Great Britain. Today, both Canada and England have officially converted to the metric system.

To change from certain American units to others, we make substitutions. Such a substitution is usually helpful when we are converting from a *larger* unit to a *smaller* one.

**EXAMPLE 1** Complete: $7\frac{1}{3}$ yd = _____ in.

$$7\frac{1}{3} \text{ yd} = 7\frac{1}{3} \times 1 \text{ yd} \qquad \text{We think of } 7\frac{1}{3} \text{ yd as } 7\frac{1}{3} \times \text{yd, or } 7\frac{1}{3} \times 1 \text{ yd.}$$

$$= 7\frac{1}{3} \times 36 \text{ in.} \qquad \text{Substituting 36 in. for 1 yd}$$

$$= \frac{22}{3} \times 36 \text{ in.}$$

$$= 264 \text{ in.}$$

**Do Exercises 5–7.** ▶

Sometimes it helps to use multiplying by 1 in making conversions. For example, 12 in. = 1 ft, so

$$\frac{12 \text{ in.}}{1 \text{ ft}} = 1 \quad \text{and} \quad \frac{1 \text{ ft}}{12 \text{ in.}} = 1.$$

If we divide 12 in. by 1 ft or 1 ft by 12 in., we get 1 because the lengths are the same. Let's first use multiplying by 1 to convert from *smaller* units to *larger* units.

**EXAMPLE 2** Complete: 48 in. = _____ ft.

To convert from "in." to "ft," we multiply by 1 using a symbol for 1 with "in." on the bottom and "ft" on the top to eliminate inches and to convert to feet:

$$48 \text{ in.} = \frac{48 \text{ in.}}{1} \times \frac{1 \text{ ft}}{12 \text{ in.}} \qquad \text{Multiplying by 1 using } \frac{1 \text{ ft}}{12 \text{ in.}} \text{ to eliminate in.}$$

$$= \frac{48 \text{ in.}}{12 \text{ in.}} \times 1 \text{ ft}$$

$$= \frac{48}{12} \times \frac{\text{in.}}{\text{in.}} \times 1 \text{ ft}$$

$$= 4 \times 1 \text{ ft} \qquad \text{The } \frac{\text{in.}}{\text{in.}} \text{ acts like 1, so we can omit it.}$$

$$= 4 \text{ ft.}$$

Complete.

**5.** 8 yd = _____ in.

**GS 6.** $2\frac{5}{6}$ yd = _____ ft

$$2\frac{5}{6} \text{ yd} = 2\frac{5}{6} \times 1 \text{ yd}$$

$$= \frac{\boxed{\phantom{00}}}{6} \times \boxed{\phantom{00}} \text{ ft}$$

$$= \frac{17}{\boxed{\phantom{00}}} \text{ ft}$$

$$= 8\frac{1}{2} \text{ ft}$$

**7.** 3.8 mi = _____ in.

**Answers**

**5.** 288  **6.** $8\frac{1}{2}$  **7.** 240,768

*Guided Solution:*

**6.** 17, 3, 2

Complete.

**8.** 72 in. = _____ ft

$$72 \text{ in.} = \frac{72 \text{ in.}}{1} \cdot \frac{\boxed{\phantom{x}}}{12 \text{ in.}}$$

$$= \frac{\boxed{\phantom{x}}}{12} \cdot 1 \text{ ft}$$

$$= \boxed{\phantom{x}} \text{ ft}$$

GS

**9.** 24 ft = _____ yd

Complete.

**10.** 35 ft = _____ yd

**11.** 26,400 ft = _____ mi

**12.** 2640 ft = _____ mi

Complete.

**13.** 6 mi = _____ ft

**14.** *Pedestrian Paths.* There are 23 mi of pedestrian paths in Central Park in New York City. Convert 23 miles to yards.

**Answers**

**8.** 6  **9.** 8  **10.** $11\frac{2}{3}$, or $11.\overline{6}$  **11.** 5

**12.** $\frac{1}{2}$, or 0.5  **13.** 31,680  **14.** 40,480 yd

**Guided Solution:**

**8.** 1 ft, 72, 6

---

We can also look at this conversion as "canceling" units:

$$48 \text{ in.} = \frac{48 \text{ in.}}{1} \cdot \frac{1 \text{ ft}}{12 \text{ in.}} = \frac{48}{12} \cdot 1 \text{ ft} = 4 \text{ ft}.$$

◀ **Do Exercises 8 and 9.**

In Examples 3–5, we will use only the "canceling" method.

**EXAMPLE 3**  Complete: 70 ft = _____ yd.

Since we are converting from "ft" to "yd," we choose a symbol for 1 with "yd" on the top and "ft" on the bottom:

$$70 \text{ ft} = 70 \text{ ft} \cdot \frac{1 \text{ yd}}{3 \text{ ft}} \qquad \text{If it helps, write 70 ft as } \frac{70 \text{ ft}}{1}.$$

$$= \frac{70}{3} \cdot 1 \text{ yd} = 23.\overline{3} \text{ yd, or } 23\frac{1}{3} \text{ yd.}$$

**EXAMPLE 4**  Complete: 23,760 ft = _____ mi.

We choose a symbol for 1 with "mi" on the top and "ft" on the bottom:

$$23{,}760 \text{ ft} = 23{,}760 \text{ ft} \cdot \frac{1 \text{ mi}}{5280 \text{ ft}} \qquad 5280 \text{ ft} = 1 \text{ mi, so } \frac{1 \text{ mi}}{5280 \text{ ft}} = 1$$

$$= \frac{23{,}760}{5280} \cdot 1 \text{ mi}$$

$$= 4.5 \text{ mi.}$$

◀ **Do Exercises 10–12.**

**EXAMPLE 5**  *Illuminated Bridge.*  The steel towers of the George Washington Bridge between Fort Lee, New Jersey, and New York City are lit on special holidays. There are 760 light fixtures within the interiors of the towers. The lighting requires 7 mi of steel conduit and 31 mi of wiring. This bridge was first illuminated on July 4, 2000. Convert 7 mi and 31 mi to yards.

**Data:** untappedcities.com, "Daily What?! The George Washington Bridge Gets Lit Up Like the Empire State Building (Sometimes)," by Michelle Young, 2/17/2015; "George Washington Bridge Interesting Facts," by the Port Authority of New York and New Jersey

We have

$$7 \text{ mi} = 7 \text{ mi} \cdot \frac{5280 \text{ ft}}{1 \text{ mi}} \cdot \frac{1 \text{ yd}}{3 \text{ ft}}$$

$$= \frac{7 \cdot 5280}{1 \cdot 3} \cdot 1 \text{ yd}$$

$$= 12{,}320 \text{ yd;}$$

$$31 \text{ mi} = 31 \text{ mi} \cdot \frac{5280 \text{ ft}}{1 \text{ mi}} \cdot \frac{1 \text{ yd}}{3 \text{ ft}}$$

$$= \frac{31 \cdot 5280}{1 \cdot 3} \cdot 1 \text{ yd}$$

$$= 54{,}560 \text{ yd.}$$

The illuminated lights require 12,320 yd of steel conduit and 54,560 yd of wiring.

◀ **Do Exercises 13 and 14.**

## b THE METRIC SYSTEM

Because the **metric system** is based on powers of 10, it allows for easy conversion between units. The metric system does not use inches, feet, pounds, and so on, but the units for time and electricity are the same as those in the American system.

The basic unit of length is the **meter.** It is just over a yard. In fact, 1 meter ≈ 1.1 yd.

(Comparative sizes are shown.)

| 1 Meter |
|---|

| 1 Yard |
|---|

The other units of length are multiples of the length of a meter:

10 times a meter,   100 times a meter,   1000 times a meter,   and so on,

or fractions of a meter:

$\frac{1}{10}$ of a meter,   $\frac{1}{100}$ of a meter,   $\frac{1}{1000}$ of a meter,   and so on.

You should memorize the names and abbreviations for metric units of length. Think of *kilo*- for 1000, *hecto*- for 100, *deka*- for 10, *deci*- for $\frac{1}{10}$, *centi*- for $\frac{1}{100}$, and *milli*- for $\frac{1}{1000}$. (The units dekameter and decimeter are not used often.) We will also use these prefixes when considering units of area, capacity, and mass.

### Thinking Metric

To familiarize yourself with metric units, consider the following.

| | |
|---|---|
| 1 kilometer (1000 meters) | is a bit more than $\frac{1}{2}$ mile (≈0.6 mi). |
| 1 meter | is just over a yard (≈1.1 yd). |
| 1 centimeter (0.01 meter) | is a little more than the width of a jumbo paperclip (≈0.4 in.). |
| 1 millimeter (0.001 meter) | is about the diameter of a paperclip wire. |

---

**METRIC UNITS OF LENGTH**

1 *kilo*meter (km) = 1000 meters (m)
1 *hecto*meter (hm) = 100 meters (m)
1 *deka*meter (dam) = 10 meters (m)
1 meter (m)
1 *deci*meter (dm) = $\frac{1}{10}$ meter (m)
1 *centi*meter (cm) = $\frac{1}{100}$ meter (m)
1 *milli*meter (mm) = $\frac{1}{1000}$ meter (m)

---

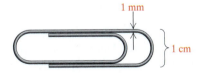

1 inch is 2.54 centimeters.

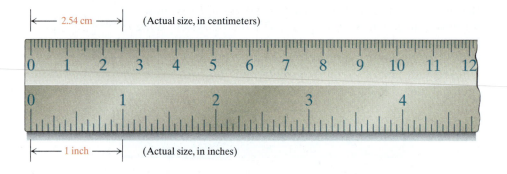

2.54 cm (Actual size, in centimeters)

1 inch (Actual size, in inches)

Using a centimeter ruler, measure each object.

**15.**

The millimeter (mm) is used to measure small distances, especially in industry.

1 mm

3 mm

**16.**

Centimeters (cm) are used for body dimensions and clothing sizes.

163 cm
(64.2 in.)
(5 ft 4 in.)

RELAXED FIT
97 cm/81 cm
(38 in./32 in.)

**17.**

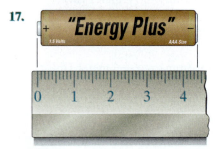

"Energy Plus"
1.5 Volts    AAA Size

◀ **Do Exercises 15–17.**

The meter (m) is used for expressing dimensions of large objects—say, the height of Hoover Dam, 221.4 m, or the distance around a standard athletic track, 400 m—and for expressing somewhat smaller dimensions like the length and width of an organic vegetable garden.

400 m around

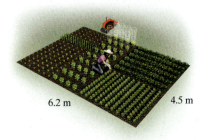

6.2 m     4.5 m

*Answers*

**15.** 2 cm, or 20 mm

**16.** 2.3 cm, or 23 mm

**17.** 4.4 cm, or 44 mm

Kilometers (km) are used for longer distances, mostly in cases where miles are used. 1 mile is about 1.6 km.

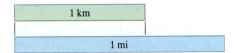

**Do Exercises 18–23.** ▶

Metric conversions from larger units to smaller ones are often most easily made using substitution.

**EXAMPLE 6**   Complete: 4 km = _____ m.

$$4 \text{ km} = 4 \cdot 1 \text{ km}$$

$$= 4 \cdot 1000 \text{ m} \qquad \text{Substituting 1000 m for 1 km}$$

$$= 4000 \text{ m} \qquad \text{Multiplying by 1000}$$

**Do Exercises 24 and 25.** ▶

Since

$$\frac{1}{10} \text{ m} = 1 \text{ dm}, \qquad \frac{1}{100} \text{ m} = 1 \text{ cm}, \quad \text{and} \quad \frac{1}{1000} \text{ m} = 1 \text{ mm},$$

we can use the following conversions.

---

**METRIC CONVERSIONS**

1 m = 10 dm,     1 m = 100 cm,   and   1 m = 1000 mm

---

Remembering these equations will help you to write forms of 1 when canceling to make conversions.

**EXAMPLE 7**   Complete: 93.4 m = _____ cm.

To convert from "m" to "cm," we multiply by 1 using a symbol for 1 with "m" on the bottom and "cm" on the top. This process introduces centimeters and at the same time eliminates meters.

$$93.4 \text{ m} = 93.4 \text{ m} \cdot \frac{100 \text{ cm}}{1 \text{ m}} \qquad \text{Multiplying by 1 using } \frac{100 \text{ cm}}{1 \text{ m}}$$

$$= 93.4 \text{ m} \cdot \frac{100 \text{ cm}}{1 \text{ m}} = 93.4 \cdot 100 \cdot 1 \text{ cm} = 9340 \text{ cm}$$

Complete with mm, cm, m, or km.

**18.** A stick of gum is 7 _____ long.

**19.** New Orleans is 2941 _____ from San Diego.

**20.** A penny is 1 _____ thick.

**21.** The halfback ran 7 _____.

**22.** The book is 3 _____ thick.

**23.** The desk is 2 _____ long.

Complete.

**GS 24.** 23 km = _____ m
$$23 \text{ km} = 23 \cdot 1 \text{ km}$$
$$= 23 \cdot \boxed{\phantom{000}} \text{ m}$$
$$= \boxed{\phantom{000}} \text{ m}$$

**25.** 4 hm = _____ m

**Complete.**

**26.** 1.78 m = _____ cm

**27.** 9.04 m = _____ mm

**Complete.**

**28.** 7814 m = _____ km

**29.** 7814 m = _____ dam $\quad$ **GS**

$$7814 \text{ m} = 7814 \text{ m} \cdot \frac{1 \text{ dam}}{\boxed{\phantom{xx}}}$$

$$= \frac{7814}{\boxed{\phantom{xx}}} \cdot \frac{\text{m}}{\text{m}} \cdot 1 \text{ dam}$$

$$= \boxed{\phantom{xx}} \text{ dam}$$

**30.** 9.67 mm = _____ cm

**31.** 89 km = _____ cm

**EXAMPLE 8** Complete: 0.248 m = _____ mm.

We are converting from "m" to "mm," so we choose a symbol for 1 with "mm" on the top and "m" on the bottom:

$$0.248 \text{ m} = 0.248 \text{ m} \cdot \frac{1000 \text{ mm}}{1 \text{ m}} = 0.248 \cdot 1000 \cdot 1 \text{ mm} = 248 \text{ mm}.$$

◀ **Do Exercises 26 and 27.**

**EXAMPLE 9** Complete: 2347 m = _____ km.

We multiply by 1 using $\frac{1 \text{ km}}{1000 \text{ m}}$:

$$2347 \text{ m} = 2347 \text{ m} \cdot \frac{1 \text{ km}}{1000 \text{ m}} = \frac{2347}{1000} \cdot 1 \text{ km} = 2.347 \text{ km}.$$

Since 1 m = 1000 mm and 1 m = 100 cm, we know that 1000 mm = 100 cm. More simply, 10 mm = 1 cm.

**EXAMPLE 10** Complete: 8.42 mm = _____ cm.

We can multiply by 1 using either $\frac{1 \text{ cm}}{10 \text{ mm}}$ or $\frac{100 \text{ cm}}{1000 \text{ mm}}$. Both expressions for 1 will eliminate mm and leave cm:

$$8.42 \text{ mm} = 8.42 \text{ mm} \cdot \frac{1 \text{ cm}}{10 \text{ mm}} = \frac{8.42}{10} \cdot 1 \text{ cm} = 0.842 \text{ cm}.$$

◀ **Do Exercises 28–31.**

## Mental Conversion

Changing from one unit to another in the metric system amounts to the movement of a decimal point. That is because the metric system is based on 10. Let's find a faster way to convert. Look at the following table.

| 1000 m | 100 m | 10 m | 1 m | 0.1 m | 0.01 m | 0.001 m |
|--------|-------|------|-----|-------|--------|---------|
| 1 km | 1 hm | 1 dam | 1 m | 1 dm | 1 cm | 1 mm |

Each place in the table has a value $\frac{1}{10}$ that of the place to the left or 10 times that of the place to the right. Thus, moving one place in the table corresponds to moving one decimal place.

**EXAMPLE 11** Complete: 35.7 mm = _____ cm.

*Think*: Centimeters is the next larger unit after millimeters. Thus, we move the decimal point one place to the left.

35.7 $\quad$ 3.5.7 $\quad$ 35.7 mm = 3.57 cm $\qquad$ Converting to a larger unit shifts the decimal point to the left.

---

*Answers*

**26.** 178  **27.** 9040  **28.** 7.814
**29.** 781.4  **30.** 0.967  **31.** 8,900,000

*Guided Solution:*

**29.** 10 m, 10, 781.4

**EXAMPLE 12** Complete: 1.886 km = _____ cm.

*Think*: To go from km to cm in the table is a move of five places to the right. Thus, we move the decimal point five places to the right.

| 1000 m | 100 m | 10 m | 1 m | 0.1 m | 0.01 m | 0.001 m |
|--------|-------|------|-----|-------|--------|---------|
| 1 km   | 1 hm  | 1 dam | 1 m | 1 dm | 1 cm   | 1 mm    |

5 places to the right

1.886   1.88600.   1.886 km = 188,600 cm

**EXAMPLE 13** Complete: 3 m = _____ cm.

*Think*: To go from m to cm in the table is a move of two places to the right. Thus, we move the decimal point two places to the right.

| 1000 m | 100 m | 10 m | 1 m | 0.1 m | 0.01 m | 0.001 m |
|--------|-------|------|-----|-------|--------|---------|
| 1 km   | 1 hm  | 1 dam | 1 m | 1 dm | 1 cm   | 1 mm    |

2 places to the right

3     3.00.     3 m = 300 cm

You should try to make metric conversions mentally as much as possible. The fact that conversions can be done so easily is an important advantage of the metric system.

> The most commonly used metric units of length are km, m, cm, and mm. We have purposely used these more often than the others in the exercises and examples.

**Do Exercises 32–35.** ▶

Complete. Try to do this mentally using the table.

**32.** 6780 m = _____ km

**33.** 9.74 cm = _____ mm

**34.** 1 mm = _____ cm

**35.** 845.1 mm = _____ dm

**c  CONVERTING BETWEEN AMERICAN UNITS AND METRIC UNITS**

We can make conversions between American units and metric units by substituting based on the following table. The conversions shown are all approximations except for the first one; 1 in. is defined to be 2.54 cm.

| AMERICAN | METRIC |
|----------|--------|
| 1 in.    | 2.54 cm |
| 1 ft     | 0.305 m |
| 1 yd     | 0.914 m |
| 1 mi     | 1.609 km |
| 0.621 mi | 1 km |
| 1.094 yd | 1 m |
| 3.281 ft | 1 m |
| 39.370 in. | 1 m |

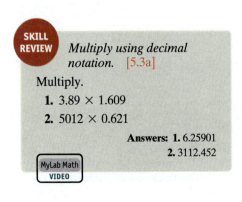

**SKILL REVIEW**   *Multiply using decimal notation.* [5.3a]

Multiply.

**1.** 3.89 × 1.609

**2.** 5012 × 0.621

**Answers: 1.** 6.25901
           **2.** 3112.452

MyLab Math
VIDEO

*Answers*

**32.** 6.78   **33.** 97.4   **34.** 0.1   **35.** 8.451

Complete.

**36.** 100 yd = _____ m
(The length of a football field excluding the end zones)

**37.** 2.5 mi = _____ km
(The length of the tri-oval track at Daytona International Speedway)

**38.** 2383 km = _____ mi
(The distance from St. Louis to Phoenix)

**39.** The Pacific Coast Highway, which is part of California State Route 1, is 655.8 mi long. Find this length in kilometers, rounded to the nearest tenth.

**40.** The height of the Stratosphere Tower in Las Vegas, Nevada, is 1149 ft. Find the height in meters.

**41.** Complete:
3.175 mm = _____ in.
(The thickness of a quarter)

*Answers*
**36.** 91.4   **37.** 4.0225   **38.** 1479.843
**39.** 1055.2 km   **40.** 350.445 m
**41.** 0.125

**EXAMPLE 14**   Complete: 11 in. = _____ cm.
(The wingspan of the world's largest butterfly, the Queen Alexandra)
**Source:** *Top 10 of Everything 2013*

$$11 \text{ in.} = 11 \cdot 1 \text{ in.}$$
$$= 11 \cdot 2.54 \text{ cm} \qquad \text{Substituting 2.54 cm for 1 in.}$$
$$= 27.94 \text{ cm}$$

This answer would probably be rounded to the nearest one: 28 cm.

**EXAMPLE 15**   Complete: 26.2 mi = _____ km.
(The length of the Olympic marathon)

Since we are given that $1 \text{ mi} \approx 1.609 \text{ km}$, we can convert using substitution.

$$26.2 \text{ mi} = 26.2 \cdot 1 \text{ mi}$$
$$\approx 26.2 \cdot 1.609 \text{ km} \qquad \text{Converting from mi to km}$$
$$\approx 42.16 \text{ km}$$

We can also convert by multiplying by 1.

$$26.2 \text{ mi} \approx \frac{26.2 \text{ mi}}{1} \cdot \frac{1 \text{ km}}{0.621 \text{ mi}}$$
$$\approx 42.19 \text{ km}$$

Note that, since the conversion factors given in the table are approximations, the answers found by the two methods differ slightly.

**EXAMPLE 16**   *Gotthard Base Tunnel.*   The Gotthard Base Tunnel, completed in 2016, is the longest railway tunnel in the world. The length of one tube of this tunnel through the Swiss Alps is 57.104 km. Convert 57.104 km to miles.
**Data:** Alp Transit Gotthard Ltd., *The World Almanac 2017*

We let $T$ = the length of the tunnel. To convert kilometers to miles, we substitute 0.621 mi for 1 km.

$$T = 57.104 \text{ km}$$
$$= 57.104 \cdot 1 \text{ km}$$
$$\approx 57.104 \cdot 0.621 \text{ mi} \qquad \text{Substituting 0.621 mi for 1 km}$$
$$\approx 35.46 \text{ mi}$$

◀ **Do Exercises 36–40.**

**EXAMPLE 17**   Complete: 0.10414 mm = _____ in.
(The thickness of a $1 bill)

In this case, we must make two substitutions or multiply by two forms of 1 since the table on the preceding page does not provide a direct way to convert from millimeters to inches. Here we choose to multiply by forms of 1.

$$0.10414 \text{ mm} = 0.10414 \cdot 1 \text{ mm} \cdot \frac{1 \text{ cm}}{10 \text{ mm}} = 0.010414 \text{ cm}$$

$$= 0.010414 \cdot 1 \text{ cm} \cdot \frac{1 \text{ in.}}{2.54 \text{ cm}} = 0.0041 \text{ in.}$$

◀ **Do Exercise 41.**

## ✔ Check Your Understanding

**Reading Check**   Complete each sentence using > or < for ☐.

**RC1.** 1 in. ☐ 1 m

**RC2.** 1 m ☐ 1 ft

**RC3.** 1 cm ☐ 1 in.

**RC4.** 1 mi ☐ 1 km

**RC5.** 1 yd ☐ 1 m

**RC6.** 1 km ☐ 1 ft

**Concept Check**   Fill in each blank with "left" or "right."

**CC1.** To convert 36.71 mm to centimeters (cm), move the decimal point 1 place to the _____.

**CC2.** To convert 802.7 km to millimeters (mm), move the decimal point 6 places to the _____.

**a**   Complete.

**1.** 1 yd = _____ ft

**2.** 1 ft = _____ in.

**3.** 1 in. = _____ ft

**4.** 1 mi = _____ ft

**5.** 1 mi = _____ yd

**6.** 1 ft = _____ yd

**7.** 3 yd = _____ ft

**8.** 4 yd = _____ in.

**9.** 84 in. = _____ ft

**10.** 18 in. = _____ ft

**11.** 48 ft = _____ yd

**12.** 29 ft = _____ yd

**13.** 5 mi = _____ yd

**14.** 5 mi = _____ ft

**15.** 63 in. = _____ ft

**16.** 19 ft = _____ yd

**17.** 11,616 ft = _____ mi

**18.** 5.2 yd = _____ ft

**19.** 7.1 mi = _____ ft

**20.** 31,680 ft = _____ mi

**21.** $7\frac{1}{2}$ ft = _____ yd

**22.** 360 in. = _____ yd

**23.** 36 in. = _____ ft

**24.** 7.2 ft = _____ in.

**25.** 1760 yd = _____ mi

**26.** 3520 yd = _____ mi

**27.** 45 in. = _____ yd

**28.** $6\frac{1}{3}$ yd = _____ in.

**29.** 25 mi = _____ ft

**30.** 240 in. = _____ ft

**31.** 2 mi = _____ in.

**32.** 63,360 in. = _____ mi

Complete. Do as much as possible mentally.

**33.** **a)** 1 km = _____ m
    **b)** 1 m = _____ km

**34.** **a)** 1 hm = _____ m
    **b)** 1 m = _____ hm

**35.** **a)** 1 dam = _____ m
    **b)** 1 m = _____ dam

**36.** **a)** 1 dm = _____ m
    **b)** 1 m = _____ dm

**37.** **a)** 1 cm = _____ m
    **b)** 1 m = _____ cm

**38.** **a)** 1 mm = _____ m
    **b)** 1 m = _____ mm

**39.** 8.3 km = _____ m

**40.** 27 km = _____ m

**41.** 98 cm = _____ m

**42.** 0.789 cm = _____ m

**43.** 8921 m = _____ km

**44.** 8664 m = _____ km

**45.** 32.17 m = _____ km

**46.** 4.733 m = _____ km

**47.** 289 m = _____ cm

**48.** 869 m = _____ cm

**49.** 477 cm = _____ m

**50.** 6.27 mm = _____ m

**51.** 6.88 m = _____ cm

**52.** 6.88 m = _____ dm

**53.** 1 mm = _____ cm

**54.** 1 cm = _____ km

**55.** 1 km = _____ cm

**56.** 2 km = _____ cm

**57.** 14.2 cm = _____ mm

**58.** 25.3 cm = _____ mm

**59.** 8.2 mm = _____ cm

**60.** 9.7 mm = _____ cm

**61.** 4500 mm = _____ cm

**62.** 8,000,000 m = _____ km

**63.** 0.024 mm = _____ m

**64.** 60,000 mm = _____ dam

**65.** 6.88 m = _____ dam

**66.** 7.44 m = _____ hm

**67.** 2.3 dam = _____ dm

**68.** 9 km = _____ hm

Complete the following table.

| | Object | Millimeters (mm) | Centimeters (cm) | Meters (m) |
|---|---|---|---|---|
| 69. | Width of a football field | | 4844 | |
| 70. | Length of a football field | | | 109.09 |
| 71. | Width of a calculator | | | 0.085 |
| 72. | Width of a credit card | 56 | | |
| 73. | Thickness of an index card | 0.27 | | |
| 74. | Thickness of a piece of cardboard | | 0.23 | |
| 75. | Height of One World Trade Center, New York, New York | | | 541.3 |
| 76. | Height of The Gateway Arch, St. Louis, Missouri | 192,000 | | |

**C** Complete. Answers may vary slightly, depending on the conversion factor used.

**77.** 10 km = _____ mi
(A common running distance)

**78.** 5 mi = _____ km
(A common running distance)

**79.** 14 in. = _____ cm
(A common paper length)

**80.** 400 m = _____ yd
(A common race distance)

**81.** 65 mph = _____ km/h
(A common speed limit in the United States)

**82.** 100 km/h = _____ mph
(A common speed limit in Canada)

**83.** 94 ft = _____ m
(The length of an NCAA basketball court)

**84.** 1.91 m = _____ in.
(The height of Jeremy Lin of the NBA)

**85.** 180 cm = _____ in.
(A common snowboard length)

**86.** 0.25 in. = _____ mm
(The thickness of an eraser on a pencil)

**87.** 7.6 mm = _____ in.
(The thickness of an iPhone 5s)

**88.** 70 in. = _____ cm
(A common height for a man)

Complete the following table. Answers may vary, depending on the conversion factor used.

| | Object | Yards (yd) | Centimeters (cm) | Inches (in.) | Meters (m) | Millimeters (mm) |
|---|---|---|---|---|---|---|
| 89. | Width of a piece of typing paper | | | $8\frac{1}{2}$ | | |
| 90. | Length of a piece of typing paper | | 27.94 | | | |
| 91. | Length of the Channel Tunnel connecting France and England | | | | 50,450 | |
| 92. | Height of Jin Mao Tower, Shanghai | 460 | | | | |

## Skill Maintenance

Solve.

**93.** $-7x - 9x = 24$   [5.7b]

**94.** $-2a + 9 = 5a + 23$   [5.7b]

**95.** If 3 calculators cost $43.50, how much would 7 calculators cost?   [6.4a]

**96.** A principal of $500 is invested at a rate of 8.9% for 1 year. Find the simple interest.   [7.7a]

Convert to percent notation.

**97.** 0.47   [7.1b]

**98.** $\dfrac{7}{20}$   [7.1c]

**99.** A living room is 12 ft by 16 ft. Find the perimeter and area of the room.   [1.2b], [1.4b]

**100.** A bedroom measures 10 ft by 12 ft. Find the perimeter and area of the room.   [1.2b], [1.4b]

## Synthesis

In Exercises 101–104, each sentence is incorrect. Insert or move a decimal point to make the sentence correct.

**101.** When my right arm is extended, the distance from my left shoulder to the end of my right hand is 10 m.

**102.** The height of the Shanghai World Financial Center is 49.2 m.

**103.** A stack of ten quarters is 140 cm high.

**104.** The width of an adult's hand is 112 cm.

**105.** *Noah's Ark.*   It is thought that the biblical measure called a *cubit* was equal to about 18 in.: 1 cubit $\approx$ 18 in. The dimensions of Noah's ark are given as follows: "The length of the ark shall be three hundred cubits, the breadth of it fifty cubits, and the height of it thirty cubits" (*Holy Bible, King James Version,* Gen. 6:15). What were the dimensions of Noah's ark in inches? in feet?

**106.** *Goliath's Height.*   The biblical measure called a *span* was considered to be half of a cubit (1 cubit $\approx$ 18 in.; see Exercise 105). The giant Goliath's height "was six cubits and a span" (*Holy Bible, King James Version,* 1 Sam. 17:4). What was the height of Goliath in inches? in feet?

Complete. Answers may vary, depending on the conversion factor used.

**107.** ▦ 2 mi = _____ cm

**108.** ▦ 10 km = _____ in.

**109.** ▦ The current world record for the women's 100-m dash is 10.49 sec, set by Florence Griffith-Joyner in Indianapolis, Indiana, on July 16, 1988. How fast is this in miles per hour? Round to the nearest tenth of a mile per hour.

**Data:** International Association of Athletics Federations

**110.** ▦ The current world record for the men's 100-m dash is 9.58 sec, set by Usain Bolt of Jamaica on August 16, 2009, in Berlin, Germany. How fast is this in miles per hour? Round to the nearest hundredth of a mile per hour.

**Data:** International Association of Athletics Federations

**111.** Develop a substitution to convert from inches to millimeters.

**112.** Develop a substitution to convert from millimeters to inches. How does it relate to the answer for Exercise 111?

# Converting Units of Area

**9.2**

## a  AMERICAN UNITS

**OBJECTIVES**

**a** Convert from one American unit of area to another.

**b** Convert from one metric unit of area to another.

**SKILL REVIEW**

*Evaluate exponential notation.*   [1.9b]

Evaluate.

**1.** $9^2$

**2.** $10^3$

Answers: **1.** 81 **2.** 1000

MyLab Math
VIDEO

It is often necessary to convert units of area. First we will convert from one American unit of area to another.

**EXAMPLE 1**  Complete: $1 \text{ yd}^2 = $ _____ $\text{ft}^2$.

We recall that 1 yd = 3 ft and make a sketch. Note that $1 \text{ yd}^2 = 9 \text{ ft}^2$. The same result can be found as follows:

$1 \text{ yd}^2 = 1 \cdot (3 \text{ ft})^2$   Substituting 3 ft for 1 yd

$= 3 \text{ ft} \cdot 3 \text{ ft}$

$= 9 \text{ ft}^2$.   Note that $\text{ft} \cdot \text{ft} = \text{ft}^2$.

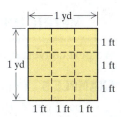

**EXAMPLE 2**  Complete: $2 \text{ ft}^2 = $ _____ $\text{in}^2$.

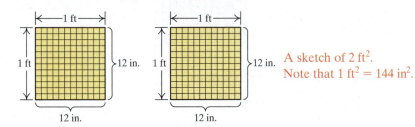

A sketch of $2 \text{ ft}^2$.
Note that $1 \text{ ft}^2 = 144 \text{ in}^2$.

$2 \text{ ft}^2 = 2 \cdot (12 \text{ in.})^2$   Substituting 12 in. for 1 ft

$= 2 \cdot 12 \text{ in.} \cdot 12 \text{ in.} = 288 \text{ in}^2$   Note that $\text{in.} \cdot \text{in.} = \text{in}^2$.

**Do Exercises 1–3.** ▶

Complete.

**1.** $1 \text{ ft}^2 = $ _____ $\text{in}^2$

**2.** $10 \text{ ft}^2 = $ _____ $\text{in}^2$

**3.** $7 \text{ yd}^2 = $ _____ $\text{ft}^2$

American units of area are related as follows.

### AMERICAN UNITS OF AREA

1 square yard $(\text{yd}^2) = 9$ square feet $(\text{ft}^2)$

1 square foot $(\text{ft}^2) = 144$ square inches $(\text{in}^2)$

1 square mile $(\text{mi}^2) = 640$ acres

1 acre $= 43{,}560 \text{ ft}^2$

················ **Caution!** ················

Be aware of when you are converting units of length and when you are converting units of area.

$1 \text{ yd} = 3 \text{ ft} \Rightarrow 1 \text{ yd}^2 = 9 \text{ ft}^2$

$1 \text{ ft} = 12 \text{ in.} \Rightarrow 1 \text{ ft}^2 = 144 \text{ in}^2$

*Answers*
**1.** 144 **2.** 1440 **3.** 63

**EXAMPLE 3**  Complete: 36 ft² = _____ yd².

To convert from "ft²" to "yd²" we write 1 with yd² on top and ft² on the bottom.

$$36 \text{ ft}^2 = 36 \text{ ft}^2 \cdot \frac{1 \text{ yd}^2}{9 \text{ ft}^2} \qquad \text{Multiplying by 1 using } \frac{1 \text{ yd}^2}{9 \text{ ft}^2}$$

$$= \frac{36}{9} \cdot \text{yd}^2 = 4 \text{ yd}^2$$

When converting from larger units to smaller units, we can substitute directly from the table.

**EXAMPLE 4**  Complete: 7 mi² = _____ acres.

$$7 \text{ mi}^2 = 7 \cdot 640 \text{ acres} \qquad \text{Substituting 640 acres for 1 mi}^2$$
$$= 4480 \text{ acres}$$

◀ Do Exercises 4 and 5.

**b**  **METRIC UNITS**

We next convert from one metric unit of area to another.

**EXAMPLE 5**  Complete: 1 cm² = _____ mm².

Since 1 cm = 10 mm, a square centimeter will have sides of length 10 mm.

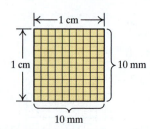

The area of the square can be written

$$(1 \text{ cm})(1 \text{ cm}) = 1 \text{ cm}^2$$

or

$$(10 \text{ mm})(10 \text{ mm}) = 100 \text{ mm}^2.$$

Thus, 1 cm² = 100 mm².

**EXAMPLE 6**  Complete: 1 km² = _____ m².

$$1 \text{ km}^2 = (1 \text{ km})(1 \text{ km})$$
$$= (1000 \text{ m})(1000 \text{ m}) \qquad \text{Substituting 1000 m for 1 km}$$
$$= 1{,}000{,}000 \text{ m}^2 \qquad \text{Note that m} \cdot \text{m} = \text{m}^2.$$

**EXAMPLE 7**  Complete: 1 cm² = _____ m².

$$1 \text{ cm}^2 = (1 \text{ cm})(1 \text{ cm})$$
$$= \left(\frac{1}{100} \text{ m}\right)\left(\frac{1}{100} \text{ m}\right) \qquad \text{Substituting } \frac{1}{100} \text{ m for 1 cm}$$
$$= \frac{1}{10{,}000} \text{ m}^2 = 0.0001 \text{ m}^2$$

◀ Do Exercises 6 and 7.

---

Complete.

**4.** 360 in² = _____ ft²   **GS**

$$360 \text{ in}^2 = 360 \text{ in}^2 \cdot \frac{1 \text{ ft}^2}{\boxed{\phantom{00}}}$$

$$= \frac{360}{\boxed{\phantom{00}}} \cdot \frac{\text{in}^2}{\text{in}^2} \cdot 1 \text{ ft}^2$$

$$= \boxed{\phantom{00}} \text{ ft}^2$$

**5.** 5 mi² = _____ acres

Complete.

**6.** 1 m² = _____ mm²

**7.** 1 mm² = _____ cm²

*Answers*
**4.** 2.5  **5.** 3200  **6.** 1,000,000  **7.** 0.01
*Guided Solution:*
**4.** 144 in², 144, 2.5

## Mental Conversion

Let's compare conversions of units of length to the conversions of units of area in Examples 6 and 7.

**Converting units of length**

**1 km → m**
1 km = 1000 m
Move the decimal point
3 places to the right.

**1 cm → m**
1 cm = 0.01 m
Move the decimal point
2 places to the left.

**Converting units of area**

**1 km² → m²**
1 km² = 1,000,000 m²
Move the decimal point
6 places to the right.

**1 cm² → m²**
1 cm² = 0.0001 m²
Move the decimal point
4 places to the left.

In general, a metric area conversion requires moving the decimal point twice as many places as the corresponding length conversion. We can use a table as before and multiply the number of places we move by 2 to determine the number of places to move the decimal point.

**EXAMPLE 8** Complete: $3.48 \text{ cm}^2 = $ _____ $\text{mm}^2$.

*Think:* To go from cm to mm in the table is a move of 1 place to the right.

| 1000 m | 100 m | 10 m | 1 m | 0.1 m | 0.01 m | 0.001 m |
|--------|-------|------|-----|-------|--------|---------|
| 1 km | 1 hm | 1 dam | 1 m | 1 dm | 1 cm | 1 mm |

1 move to the right

So we move the decimal point 2 · 1, or 2, places to the right.

$$3.48 \qquad 3.48. \qquad 3.48 \text{ cm}^2 = 348 \text{ mm}^2$$

2 places to the right

**EXAMPLE 9** Complete: $586.78 \text{ mm}^2 = $ _____ $\text{m}^2$.

*Think*: To go from mm to m in the table is a move of 3 places to the left.

| 1000 m | 100 m | 10 m | 1 m | 0.1 m | 0.01 m | 0.001 m |
|--------|-------|------|-----|-------|--------|---------|
| 1 km | 1 hm | 1 dam | 1 m | 1 dm | 1 cm | 1 mm |

3 moves to the left

So we move the decimal point 2 · 3, or 6, places to the left.

$$586.78 \qquad 0.000586.78 \qquad 586.78 \text{ mm}^2 = 0.00058678 \text{ m}^2$$

6 places to the left

**Do Exercises 8–10.** ▶

Complete.

**8.** $2.88 \text{ m}^2 = $ _____ $\text{cm}^2$

**9.** $4.3 \text{ mm}^2 = $ _____ $\text{cm}^2$

**10.** $678,000 \text{ m}^2 = $ _____ $\text{km}^2$

*Answers*
**8.** 28,800   **9.** 0.043   **10.** 0.678

## ✓ Check Your Understanding

**Reading Check**   Determine whether each equation is true or false.

**RC1.** _____ $9 \text{ ft}^2 = 1 \text{ yd}^2$

**RC2.** _____ $1 \text{ mi}^2 = 640 \text{ ft}^2$

**RC3.** _____ $1 \text{ acre} = 43{,}560 \text{ ft}^2$

**RC4.** _____ $36 \text{ in}^2 = 1 \text{ yd}^2$

**RC5.** _____ $1 \text{ km}^2 = 1{,}000{,}000 \text{ m}^2$

**RC6.** _____ $1 \text{ ft}^2 = 144 \text{ in}^2$

**Concept Check**   Choose from the list on the right the symbol for 1 that can be used to complete each unit conversion.

**CC1.** $216 \text{ in}^2 = 216 \text{ in}^2 \cdot \boxed{\phantom{xx}} = 1.5 \text{ ft}^2$

**CC2.** $21{,}780 \text{ ft}^2 = 21{,}780 \text{ ft}^2 \cdot \boxed{\phantom{xx}} = 0.5 \text{ acre}$

**a)** $\dfrac{1 \text{ mi}^2}{464 \text{ acres}}$     **b)** $\dfrac{1 \text{ acre}}{43{,}560 \text{ ft}^2}$

**c)** $\dfrac{43{,}560 \text{ ft}^2}{1 \text{ acre}}$     **d)** $\dfrac{144 \text{ in.}}{1 \text{ ft}^2}$

**e)** $\dfrac{1 \text{ yd}^2}{9 \text{ ft}^2}$     **f)** $\dfrac{1 \text{ ft}^2}{144 \text{ in}^2}$

**a**   Complete.

**1.** $1 \text{ ft}^2 = $ _____ $\text{in}^2$

**2.** $1 \text{ yd}^2 = $ _____ $\text{ft}^2$

**3.** $1 \text{ mi}^2 = $ _____ acres

**4.** $1 \text{ acre} = $ _____ $\text{ft}^2$

**5.** $1 \text{ in}^2 = $ _____ $\text{ft}^2$

**6.** $1 \text{ ft}^2 = $ _____ $\text{yd}^2$

**7.** $5 \text{ yd}^2 = $ _____ $\text{ft}^2$

**8.** $4 \text{ ft}^2 = $ _____ $\text{in}^2$

**9.** $7 \text{ ft}^2 = $ _____ $\text{in}^2$

**10.** $2 \text{ acres} = $ _____ $\text{ft}^2$

**11.** $432 \text{ in}^2 = $ _____ $\text{ft}^2$

**12.** $54 \text{ ft}^2 = $ _____ $\text{yd}^2$

**13.** $22 \text{ yd}^2 = $ _____ $\text{ft}^2$

**14.** $40 \text{ ft}^2 = $ _____ $\text{in}^2$

**15.** $15 \text{ ft}^2 = $ _____ $\text{in}^2$

**16.** $144 \text{ ft}^2 = $ _____ $\text{yd}^2$

**17.** $20 \text{ mi}^2 = $ _____ acres

**18.** $576 \text{ in}^2 = $ _____ $\text{ft}^2$

**19.** $1 \text{ mi}^2 = $ _____ $\text{ft}^2$

**20.** $1 \text{ mi}^2 = $ _____ $\text{yd}^2$

**21.** $720 \text{ in}^2 = $ _____ $\text{ft}^2$

**22.** $27 \text{ ft}^2 = $ _____ $\text{yd}^2$

**23.** $3 \text{ ft}^2 = $ _____ $\text{yd}^2$

**24.** $72 \text{ in}^2 = $ _____ $\text{ft}^2$

**25.** 1 acre = _____ mi$^2$

**26.** 4 acres = _____ ft$^2$

**27.** 40.3 mi$^2$ = _____ acres

**28.** 1080 in$^2$ = _____ ft$^2$

**29.** 216 in$^2$ = _____ ft$^2$

**30.** 69 ft$^2$ = _____ yd$^2$

**b**   Complete.

**31.** 19 km$^2$ = _____ m$^2$

**32.** 39 km$^2$ = _____ m$^2$

**33.** 6.31 m$^2$ = _____ cm$^2$

**34.** 2.7 m$^2$ = _____ mm$^2$

**35.** 6.5432 mm$^2$ = _____ cm$^2$

**36.** 8.38 cm$^2$ = _____ mm$^2$

**37.** 349 cm$^2$ = _____ m$^2$

**38.** 125 mm$^2$ = _____ m$^2$

**39.** 250,000 mm$^2$ = _____ cm$^2$

**40.** 5900 mm$^2$ = _____ cm$^2$

**41.** 472,800 m$^2$ = _____ km$^2$

**42.** 1.37 cm$^2$ = _____ mm$^2$

## Skill Maintenance

In Exercises 43 and 44, find the simple interest. [7.7a]

**43.** On $2000 at an interest rate of 8% for 1.5 years

**44.** On $2000 at an interest rate of 5.3% for 2 years

In Exercises 45 and 46, find **(a)** the amount of simple interest due and **(b)** the total amount that must be paid back. [7.7a]

**45.** A firm borrows $6400 at 8.4% for 150 days.

**46.** A firm borrows $4200 at 11% for 30 days.

## Synthesis

Complete.

**47.** 🖩 1 m$^2$ = _____ ft$^2$

**48.** 🖩 1 in$^2$ = _____ cm$^2$

**49.** 🖩 2 yd$^2$ = _____ m$^2$

**50.** 🖩 1 acre = _____ m$^2$

**51.** *Aalsmeer Flower Auction.* The fourth-largest building in the world in terms of floor space houses the Aalsmeer Flower Auction in Aalsmeer, Netherlands. It covers approximately 990,000 m$^2$. Each day, over 20 million flowers are sold there. Convert 990,000 m$^2$ to square feet.

**Data:** www.amsterdamlogue.com; www.youTube.com/Aalsmeer, Netherlands, Flower Auction

**52.** *The Palazzo.* The largest building in the United States in terms of floor space is the Palazzo, a hotel and casino on the Las Vegas Strip in Paradise, Nevada. It contains approximately 6,948,980 ft$^2$. Convert 6,948,980 ft$^2$ to square meters. Round the answer to the nearest thousand.

**Data:** real-estate.knoji.com

**53.** Janie's Rubik's Cube has 54 cm² of area. Each side of Norm's cube is twice as wide as Janie's. Find the area of Norm's cube.

**54.**  In order to remodel an office, a carpenter needs to purchase carpeting, at $8.45 a square yard, and molding for the base of the walls, at $0.87 a foot. If the room is 9 ft by 12 ft, with a 3-ft-wide doorway, what will the materials cost?

Find the area of the shaded region of each figure. Give the answer in square feet. (Figures are not drawn to scale.)

**55.**
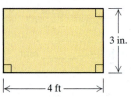
3 in.
4 ft

**56.**
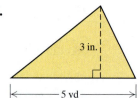
3 in.
5 yd

**57.**
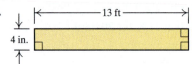
13 ft
4 in.

**58.**
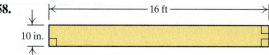
16 ft
10 in.

Find the area of the shaded region of each figure.

**59.**
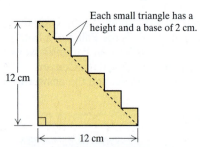
Each small triangle has a height and a base of 2 cm.
12 cm
12 cm

**60.**
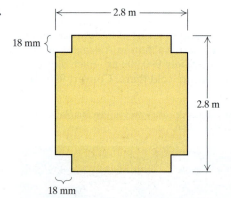
2.8 m
18 mm
2.8 m
18 mm

**61.**  A 30-ft by 60-ft ballroom is to be remodeled by placing an 18-ft by 42-ft dance floor in the middle and carpeting the rest of the room. The new dance floor is to be laid in tiles that are 8-in. by 8-in. squares. How many such tiles are needed? What percent of the area is the dance floor?

# Parallelograms, Trapezoids, and Circles

We have already discussed how to find the perimeter of polygons and the area of squares, rectangles, and triangles. In this section, we find the area of *parallelograms*, *trapezoids*, and *circles*. We also calculate the perimeter, or *circumference*, of a circle.

### OBJECTIVES

**a** Find the area of a parallelogram or trapezoid.

**b** Find the circumference, area, radius, or diameter of a circle, given the length of a radius or diameter.

## **a** PARALLELOGRAMS AND TRAPEZOIDS

A **parallelogram** is a four-sided figure with two pairs of parallel sides, as shown below.

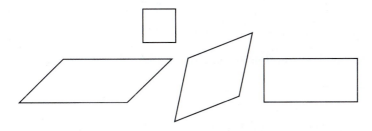

To understand how to find the area of a parallelogram, consider the one below.

If we cut off a piece and move it to the other end, we get a rectangle.

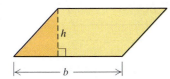 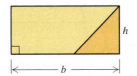

We can find the area by multiplying the length of a **base** $b$ by $h$, the **height**.

---

**SKILL REVIEW**

*Calculate using fraction and decimal notation together.* [5.5d]

Calculate.

**1.** $\dfrac{1}{2} \times 16.243$

**2.** $0.5 \times \dfrac{3}{8}$

**Answers: 1.** $8.1215$ **2.** $0.1875$, or $\dfrac{3}{16}$

MyLab Math
**VIDEO**

---

### AREA OF A PARALLELOGRAM

The **area of a parallelogram** is the product of the length of a base $b$ and the height $h$:

$$A = b \cdot h.$$

**EXAMPLE 1** Find the area of this parallelogram.

$A = b \cdot h$
$\quad = 7 \text{ km} \cdot 5 \text{ km}$
$\quad = 35 \text{ km}^2$

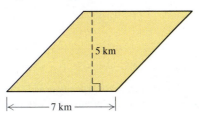

Find the area.

**1.**

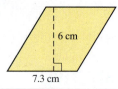

6 cm

7.3 cm

**2.**

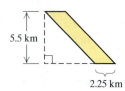

5.5 km

2.25 km

**EXAMPLE 2** Find the area of this parallelogram.

$$A = b \cdot h$$
$$= (1.2 \text{ m}) \cdot (6 \text{ m})$$
$$= 7.2 \text{ m}^2$$

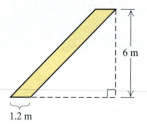

6 m

1.2 m

◀ **Do Exercises 1 and 2.**

A **trapezoid** is a polygon with four sides, two of which, the **bases**, are parallel to each other.*

To understand how to find the area of a trapezoid, consider the one below.

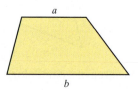

$a$

$b$

Think of cutting out another just like it and placing the second one like this:

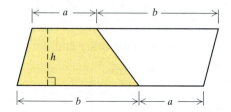

The resulting figure is a parallelogram with an area of

$$h \cdot (a + b). \qquad \text{The base of the parallelogram has length } a + b.$$

The trapezoid we started with has half the area of the parallelogram, or

$$\frac{1}{2} \cdot h \cdot (a + b).$$

---

### AREA OF A TRAPEZOID

The **area of a trapezoid** is half the product of the height and the sum of the lengths of the parallel sides, or the product of the height and the average length of the bases:

$$A = \frac{1}{2} \cdot h \cdot (a + b) = h \cdot \frac{a + b}{2}.$$

$a$

$h$

$b$

---

**Answers**

**1.** 43.8 cm² **2.** 12.375 km²

**620** CHAPTER 9 Geometry and Measurement

**EXAMPLE 3** Find the area of this trapezoid.

$$A = \frac{1}{2} \cdot h \cdot (a + b)$$

$$= \frac{1}{2} \cdot 7 \text{ cm} \cdot (12 + 18) \text{ cm}$$

$$= \frac{7 \cdot 30}{2} \cdot \text{ cm}^2 = \frac{7 \cdot 15 \cdot 2}{1 \cdot 2} \text{ cm}^2$$

$$= 105 \text{ cm}^2 \quad \text{Removing a factor equal to } 1 \ (\tfrac{2}{2} = 1) \text{ and multiplying}$$

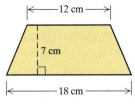

**Do Exercises 3 and 4.** ▶

## b CIRCLES

### Radius and Diameter

At right is a circle with center $O$. Segment $\overline{AC}$ is a *diameter*. A **diameter** is a segment that passes through the center of the circle and has endpoints on the circle. Segment $\overline{OB}$ is a *radius*. A **radius** is a segment with one endpoint on the center and the other endpoint on the circle. The words *radius* and *diameter* are also used to represent the lengths of a circle's radius and diameter, respectively.

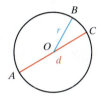

---

**DIAMETER AND RADIUS**

Suppose that $d$ is the diameter of a circle and $r$ is the radius. Then

$$d = 2 \cdot r \quad \text{and} \quad r = \frac{d}{2}.$$

---

**EXAMPLE 4** Find the length of a radius of this circle.

$$r = \frac{d}{2}$$

$$= \frac{12 \text{ m}}{2}$$

$$= 6 \text{ m}$$

The radius is 6 m.

**EXAMPLE 5** Find the length of a diameter of this circle.

$$d = 2 \cdot r$$

$$= 2 \cdot \frac{1}{4} \text{ ft}$$

$$= \frac{1}{2} \text{ ft}$$

The diameter is $\frac{1}{2}$ ft.

**Do Exercises 5 and 6.** ▶

Find the area.

**3.**

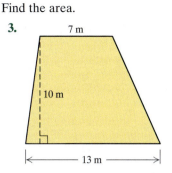

7 m
10 m
13 m

**4.**

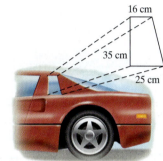

16 cm
35 cm
25 cm

**5.** Find the length of a radius.

20 m

**6.** Find the length of a diameter.

$2\frac{1}{2}$ ft

*Answers*

**3.** 100 m²   **4.** 717.5 cm²   **5.** 10 m   **6.** 5 ft

## Circumference

The perimeter of a circle is called its **circumference**. The circumference of a circle depends on its diameter. Suppose that we take a dinner plate and measure the circumference $C$ and diameter $d$ with a tape measure. For the specific plate shown, we have

$$\frac{C}{d} = \frac{33.7 \text{ in.}}{10.75 \text{ in.}} \approx 3.1.$$

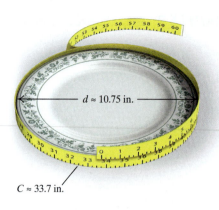

$d \approx 10.75$ in.

$C \approx 33.7$ in.

If we did this with plates and circles of several sizes, the result would always be a number close to 3.1. For any circle, if we divide the circumference $C$ by the diameter $d$, we get the same number. We call this number $\pi$ (pi). The number $\pi$ is a nonterminating, nonrepeating decimal. It is impossible to precisely express $\pi$ as a decimal or a fraction.

---

### CIRCUMFERENCE AND DIAMETER

The circumference $C$ of a circle of diameter $d$ is given by

$$C = \pi \cdot d.$$

The number $\pi$ is about 3.14, or about $\frac{22}{7}$.

---

**EXAMPLE 6** Find the circumference of this circle. Use 3.14 for $\pi$.

$C = \pi \cdot d$
$\approx 3.14 \cdot 6 \text{ cm}$
$= 18.84 \text{ cm}$

6 cm

The circumference is about 18.84 cm.

◀ **Do Exercise 7.**

Since $d = 2 \cdot r$, where $r$ is the length of a radius, it follows that

$$C = \pi \cdot d = \pi \cdot (2 \cdot r).$$

---

### CIRCUMFERENCE AND RADIUS

The circumference $C$ of a circle of radius $r$ is given by

$$C = 2 \cdot \pi \cdot r.$$

---

**EXAMPLE 7** Find the circumference of this circle. Use $\frac{22}{7}$ for $\pi$.

$C = 2 \cdot \pi \cdot r$
$\approx 2 \cdot \frac{22}{7} \cdot 70 \text{ in.}$
$= 2 \cdot 22 \cdot \frac{70}{7} \text{ in.}$
$= 44 \cdot 10 \text{ in.} = 440 \text{ in.}$

70 in.

The circumference is about 440 in.

---

**7.** Find the circumference of the circle. Use 3.14 for $\pi$. **GS**

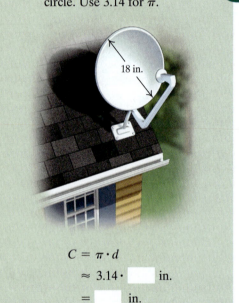

18 in.

$C = \pi \cdot d$
$\approx 3.14 \cdot \boxed{\phantom{00}} \text{ in.}$
$= \boxed{\phantom{00}} \text{ in.}$

*Answer*
**7.** 56.52 in.
*Guided Solution:*
**7.** 18, 56.52

**EXAMPLE 8**  Find the perimeter of this figure. Use 3.14 for $\pi$.

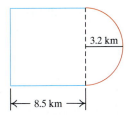

We let $P$ = the perimeter. Note that the figure has three straight edges (line segments) and one curved edge. The curved edge is half the circumference of a circle of radius 3.2 km. The top and bottom straight edges have lengths of 8.5 km. The left edge is the same length as twice the radius of the semicircle on the right, or 6.4 km.

$$\underbrace{\text{Perimeter}}\quad \text{is}\quad \underbrace{\overset{\text{length of}}{\text{straight edges}}}\quad \text{plus}\quad \underbrace{\overset{\text{length of}}{\text{curved edge}}}$$

$$P \;=\; 2(8.5\text{ km}) + 6.4\text{ km} \;+\; \frac{1}{2}\cdot 2\cdot \pi\cdot 3.2\text{ km}$$

$$\approx 17\text{ km} + 6.4\text{ km} + 1\cdot 3.14\cdot 3.2\text{ km}$$

$$= 23.4\text{ km} + 10.048\text{ km}$$

$$= 33.448\text{ km}$$

The perimeter is about 33.448 km.

**Do Exercises 8–10.** ▶

## Area

To understand how to find the area of a circle with radius $r$, think of cutting half a circular region into small slices and arranging them as shown below. Note that half of the circumference is half of $2\pi r$, or simply $\pi r$.

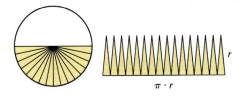

Then imagine slicing the other half of the circular region and arranging the pieces in with the others as shown below.

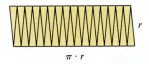

The thinner the slices, the closer this comes to being a rectangle. Its length is $\pi r$ and its width is $r$, so its area is

$$(\pi \cdot r)\cdot r.$$

This is the area of a circle.

---

**8.** Find the circumference of this circle. Use 3.14 for $\pi$.

**9.** Find the circumference of this bicycle wheel. Use $\frac{22}{7}$ for $\pi$.

**10.** Find the perimeter of this figure. Use 3.14 for $\pi$.

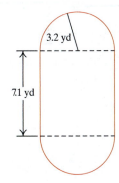

*Answers*
**8.** 15.7 m   **9.** 220 cm   **10.** 34.296 yd

**11.** Find the area of this circle. Use $\frac{22}{7}$ for $\pi$. **GS**

5 km

$A = \pi \cdot r \cdot r$

$\approx \frac{22}{7} \cdot 5 \text{ km} \cdot \boxed{\phantom{0}} \text{ km}$

$= \frac{22}{7} \cdot 25 \text{ km}^2$

$= \frac{550}{7} \text{ km}^2$

$= \boxed{\phantom{00}}\frac{4}{7} \text{ km}^2$

**12.** Find the area of this circle. Use 3.14 for $\pi$.

20.8 cm

**13.** Find the area of the shaded region. Use 3.14 for $\pi$. (*Hint*: The figure consists of a square and two semicircles, or one complete circle.)

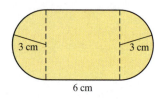

3 cm        3 cm

6 cm

*Answers*

**11.** $78\frac{4}{7} \text{ km}^2$  **12.** 339.6 cm$^2$
**13.** 64.26 cm$^2$
*Guided Solution:*
**11.** 5, 78

**EXAMPLE 9**  Find the area of this circle. Use $\frac{22}{7}$ for $\pi$.

$A = \pi \cdot r^2 = \pi \cdot r \cdot r$

$\approx \frac{22}{7} \cdot 14 \text{ cm} \cdot 14 \text{ cm}$

$= \frac{22}{7} \cdot \frac{7 \cdot 2}{1} \text{ cm} \cdot 14 \text{ cm}$

$= 616 \text{ cm}^2$   *Note: $r^2 \neq 2r$.*

14 cm

The area is about 616 cm$^2$.

**EXAMPLE 10**  Find the area of this circle. Use 3.14 for $\pi$. Round to the nearest hundredth.

The diameter is 4.2 m; the radius is 4.2 m ÷ 2, or 2.1 m.

$A = \pi \cdot r \cdot r$

$\approx 3.14 \cdot 2.1 \text{ m} \cdot 2.1 \text{ m}$

$= 3.14 \cdot 4.41 \text{ m}^2$

$= 13.8474 \text{ m}^2 \approx 13.85 \text{ m}^2$

4.2 m

The area is about 13.85 m$^2$.

◀ Do Exercises 11 and 12.

We can often add or subtract areas to determine the area of a region.

**EXAMPLE 11**  Find the area of the shaded region.

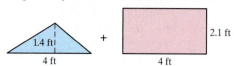

1.4 ft

2.1 ft

4 ft

The region consists of a triangle and a rectangle, as illustrated below. Note that the base of the triangle is the same as the length of the rectangle. We calculate the areas separately and add to find the area of the shaded region.

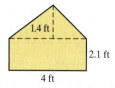

1.4 ft

4 ft    +    4 ft    2.1 ft

$\text{Area} = \frac{1}{2} \cdot 4 \text{ ft} \cdot 1.4 \text{ ft} + 4 \text{ ft} \cdot 2.1 \text{ ft}$

$= 2.8 \text{ ft}^2 + 8.4 \text{ ft}^2$

$= 11.2 \text{ ft}^2$

◀ Do Exercise 13.

**EXAMPLE 12** *Area of Cake Pans.* Tyler can make either a 9-in. round cake or a 9-in. square cake for a party. If he makes the square cake, how much more area will he have on the top for decorations?

The area of the square is

$$A = s \cdot s$$
$$= 9 \text{ in.} \cdot 9 \text{ in.} = 81 \text{ in}^2.$$

The diameter of the circle is 9 in., so the radius is 9 in./2, or 4.5 in. The area of the circle is

$$A = \pi \cdot r \cdot r$$
$$\approx 3.14 \cdot 4.5 \text{ in.} \cdot 4.5 \text{ in.} = 63.585 \text{ in}^2.$$

The area of the top of the square cake is larger by about

$$81 \text{ in}^2 - 63.585 \text{ in}^2, \quad \text{or} \quad 17.415 \text{ in}^2.$$

**Do Exercise 14.** ▶

---

## AREA AND CIRCUMFERENCE FORMULAS

Area of a parallelogram:   $A = b \cdot h$

Area of a trapezoid:   $A = \dfrac{1}{2} \cdot h \cdot (a + b)$

or $A = h \cdot \dfrac{a + b}{2}$

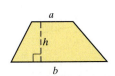

Circumference of a circle:   $C = \pi \cdot d$
or $C = 2 \cdot \pi \cdot r$

Area of a circle:   $A = \pi \cdot r^2$

*Note*: $d = 2r$ and $\pi \approx 3.14 \approx \frac{22}{7}$.

---

·········· **Caution!** ··········

Remember that circumference is always measured in linear units like ft, m, cm, yd, and so on, but area is measured in square units like ft², m², cm², yd², and so on.

·······················································

**14.** Which is larger and by how much: a 10-ft-square flower bed or a 12-ft-diameter circular flower bed?

*Answer*
**14.** A 12-ft-diameter flower bed is 13.04 ft² larger.

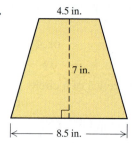
### ✓ Check Your Understanding

**Reading Check**  Complete each statement with the correct word from the list on the right. A word may be used more than once or not at all.

**RC1.** The _____ of a circle is half the length of its diameter.

**RC2.** The _____ of a circle is found by multiplying its diameter by $\pi$.

**RC3.** The _____ of a circle is found by multiplying its radius by $2\pi$.

**RC4.** The _____ of a circle is found by multiplying the square of its radius by $\pi$.

> area
> circumference
> diameter
> radius

**Concept Check**  Select from choices (a)–(d) the closest approximation of the circumference of the circle that has the given diameter or radius.

**CC1.** $d = 12$ in.

**a)** 3.768 in.  **b)** 18.84 in.

**c)** 37.68 in.  **d)** 188.4 in.

**CC2.** $r = 4.5$ m

**a)** 28.26 m  **b)** 14.13 m

**c)** 1.413 m  **d)** 282.6 m

---

**a**  Find the area of each parallelogram or trapezoid.

**1.**
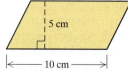
5 cm
10 cm

**2.**

4 cm
4 cm

**3.**
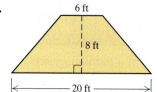
6 ft
8 ft
20 ft

**4.**
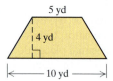
5 yd
4 yd
10 yd

**5.**
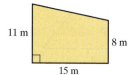
11 m
8 m
15 m

**6.**
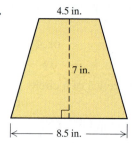
4.5 in.
7 in.
8.5 in.

**7.**
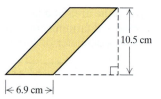
10.5 cm
6.9 cm

**8.**
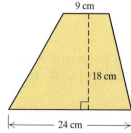
9 cm
18 cm
24 cm

**9.**
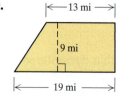
13 mi
9 mi
19 mi

**10.**

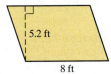

5.2 ft

8 ft

**11.**

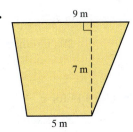

9 m

7 m

5 m

**12.**

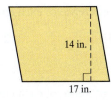

14 in.

17 in.

**13.**

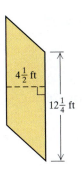

$4\frac{1}{2}$ ft

$12\frac{1}{4}$ ft

**14.**

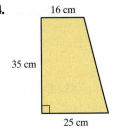

16 cm

35 cm

25 cm

**15.**

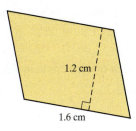

1.2 cm

1.6 cm

**16.**

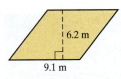

6.2 m

9.1 m

**17.**

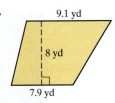

9.1 yd

8 yd

7.9 yd

**18.**

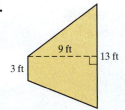

9 ft

3 ft

13 ft

**b**  Find the length of a diameter of each circle.

**19.**

7 cm

**20.**

8 m

**21.**

$\frac{7}{8}$ in.

**22.**

$8\frac{2}{3}$ mi

Find the length of a radius of each circle.

**23.**

20 ft

**24.**

10 in.

**25.**

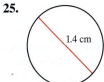

1.4 cm

**26.**

60.9 km

Find the circumference of each circle in Exercises 19–22. Use $\frac{22}{7}$ for $\pi$.

**27.** Exercise 19          **28.** Exercise 20          **29.** Exercise 21          **30.** Exercise 22

Find the circumference of each circle in Exercises 23–26. Use 3.14 for $\pi$.

**31.** Exercise 23          **32.** Exercise 24          **33.** Exercise 25          **34.** Exercise 26

Find the area of each circle in Exercises 19–22. Use $\frac{22}{7}$ for $\pi$.

**35.** Exercise 19          **36.** Exercise 20          **37.** Exercise 21          **38.** Exercise 22

Find the area of each circle in Exercises 23–26. Use 3.14 for $\pi$.

**39.** Exercise 23          **40.** Exercise 24          **41.** Exercise 25          **42.** Exercise 26

Solve. Use 3.14 for $\pi$ unless otherwise specified.

**43.** *Pond Edging.*  Quiet Designs plans to incorporate a circular pond with a diameter of 30 ft in a landscape design. The pond will be edged using stone pavers. How many feet of pavers will be needed?

**44.** *Gypsy-Moth Tape.*  To protect an elm tree in her backyard, Laura decides to attach gypsy moth caterpillar tape around the trunk. The tree has a 1.1-ft diameter. What length of tape is needed?

**45.** *Areas of Pizza Pans.*  How much larger is a pizza made in a 16-in.-square pizza pan than a pizza made in a 16-in.-diameter circular pan?

16 in.

16 in.

16 in.

**46.** *Dimensions of a Penny.*  A penny has a 1-cm radius. What is its diameter? its circumference? its area?

1 cm

**47.** *Earth.*  The circumference of the earth at the equator is 24,901 mi. What is the diameter of the earth at the equator? the radius?

**48.** *Dimensions of a Quarter.*  The circumference of a quarter is 7.85 cm. What is the diameter? the radius? the area?

**49. *Oil Spill.*** The radius of an oil slick is 150 mi. How much area is covered by the slick?

**50. *Radio.*** A college radio station is allowed by the FCC to broadcast over an area with a radius of 6 mi. How much area is this?

**51. *Masonry.*** The Harris-Regency Hotel plans to install a 1-yd-wide walk around a circular swimming pool. The diameter of the pool is 20 yd. What is the area of the walk?

**52. *Trampoline.*** The standard backyard trampoline has a diameter of 14 ft. What is its area?

**Data:** International Trampoline Industry Association, Inc.

**53. *Circumference of a Baseball Bat.*** In Major League Baseball, the diameter of the barrel of a bat cannot be more than $2\frac{3}{4}$ in., and the diameter of the bat handle cannot be less than $\frac{16}{19}$ in. Find the maximum circumference of the barrel of a bat and the minimum circumference of the bat handle. Use $\frac{22}{7}$ for $\pi$.

**Data:** Major League Baseball

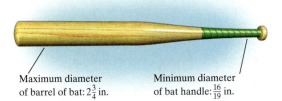

Maximum diameter of barrel of bat: $2\frac{3}{4}$ in.

Minimum diameter of bat handle: $\frac{16}{19}$ in.

**54. *Roller-Rink Floor.*** A roller-rink floor is shown below. Each end is a semicircle. What is the total area? If hardwood flooring costs $62.50 per square meter, how much will the flooring cost?

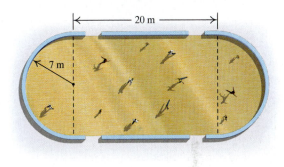

20 m

7 m

Find the perimeter of each figure. Use 3.14 for $\pi$.

**55.**

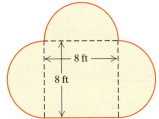

8 ft

8 ft

**56.**

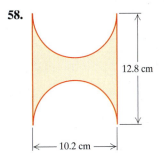

← 8 in. → ← 8 in. → ← 8 in. → ← 8 in. →

**57.**

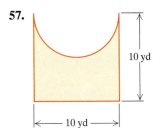

10 yd

10 yd

**58.**

12.8 cm

10.2 cm

Find the area of the shaded region in each figure. Use 3.14 for $\pi$.

**59.**

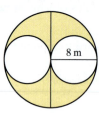

8 m

**60.**

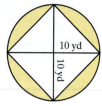

10 yd

10 yd

**61.**

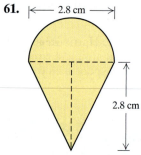

2.8 cm

2.8 cm

**62.**

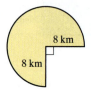

8 km

8 km

**63.**

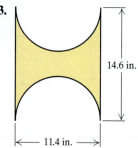

14.6 in.

11.4 in.

**64.**

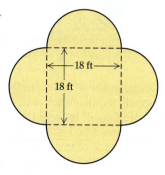

18 ft

18 ft

**65.**

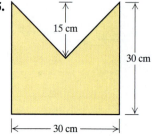

15 cm

30 cm

30 cm

**66.**

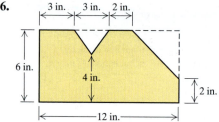

3 in.   3 in.   2 in.

6 in.

4 in.

2 in.

12 in.

**67.**

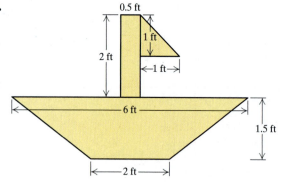

0.5 ft

1 ft

2 ft

1 ft

6 ft

1.5 ft

2 ft

**68.**

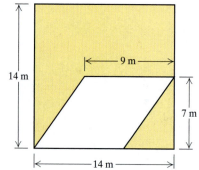

9 m

14 m

7 m

14 m

## Skill Maintenance

Convert to fraction notation.   [7.1c]

**69.** 9.25%

**70.** $87\frac{1}{2}\%$

Convert to percent notation.   [7.1c]

**71.** $\frac{5}{4}$

**72.** $\frac{2}{3}$

**73.** The weight of a human brain is 2.5% of total body weight. A person weighs 200 lb. What does his brain weigh?   [7.4a]

**74.** Jack's commission is increased according to how much he sells. He receives a commission of 6% for the first $3000 and 10% on the amount over $3000. What is the total commission on sales of $8500?   [7.6b]

## Synthesis

**75.** ▦ Calculate the surface area of an unopened steel can that has a height of 3.5 in. and a diameter of 2.5 in. (*Hint*: Make a sketch and "unroll" the sides of the can.) Use 3.14 for $\pi$.

**76.** ▦ The distance from Kansas City to Indianapolis is 500 mi. A car was driven this distance using tires with a radius of 14 in. How many revolutions of each tire occurred on the trip? Use $\frac{22}{7}$ for $\pi$.

**77.** *Sports Marketing.* Tennis balls are often packed vertically, three in a can, one on top of another. Without using a calculator, determine the larger measurement: the can's circumference or the can's height.

**78.** ▦ The sides of a cake box are trapezoidal, as shown in the figure. Determine the surface area of the box.

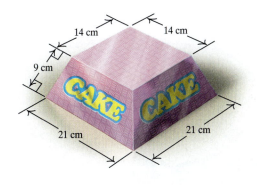

**79.** The radius of one circle is twice another circle's radius. If the area of the smaller circle is $A$, what is the area of the larger circle?

**80.** The radius of one circle is twice another circle's radius. If the circumference of the smaller circle is $C$, what is the circumference of the larger circle?

**81.** A cake recipe calls for a 9-inch round baking pan. If only square pans are available, what size pan should be used for the cake? Round to the nearest inch.

**82.** Peggy fenced a square garden with sides of 12 ft. If she decides to use the fence around a circular garden instead, what size garden can she fence? Which garden has the greater area?

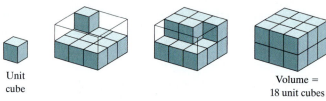

## 9.4

### OBJECTIVES

**a** Find the volume of a rectangular solid, a cylinder, and a sphere.

**b** Convert from one unit of capacity to another.

**c** Solve applied problems involving volume and capacity.

**MyLab Math**

**ANIMATION**

# Volume and Capacity

## a VOLUME

The **volume** of an object is the number of unit cubes needed to fill it.

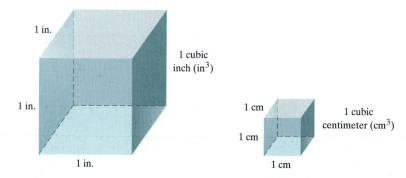

Two unit cubes commonly used to measure volume are shown below.

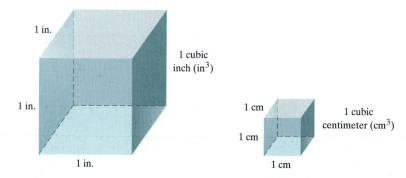

### Rectangular Solids

**EXAMPLE 1** Find the volume.

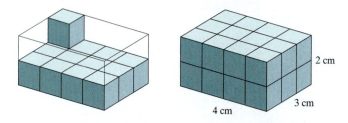

The figure is made up of 2 layers of 12 cubes each, so its volume is 24 cubic centimeters (cm³). Note that $24 = 4 \cdot 3 \cdot 2$.

◀ **Do Exercise 1.**

**1.** Find the volume.

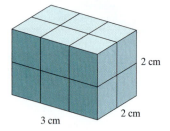

---

### VOLUME OF A RECTANGULAR SOLID

The **volume of a rectangular solid** is found by multiplying length by width by height:

$$V = \underbrace{l \cdot w}_{\substack{\uparrow \\ \text{Area} \\ \text{of base}}} \cdot \underset{\substack{\uparrow \\ \text{Height}}}{h}.$$

---

*Answer*

**1.** 12 cm³

**EXAMPLE 2** *Volume of a Safety Deposit Box.* Tricia rents a safety deposit box at her bank. The dimensions of the box are 18 in. × 10.5 in. × 5 in. Find the volume of this rectangular solid.

$$V = l \cdot w \cdot h$$
$$= 18 \text{ in.} \cdot 10.5 \text{ in.} \cdot 5 \text{ in.}$$
$$= 945 \text{ in}^3$$

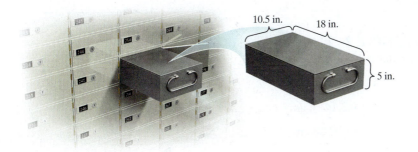

**Do Exercises 2 and 3.** ▶

Volumes are described in units such as cubic centimeters (cm³) and cubic inches (in³). *Dimensional analysis* is an excellent way of determining the correct units for an answer.

- If measurements of length are added, use a one-dimensional unit of length: 3 ft + 2 ft + 7 ft = 12 ft.
- If two measurements of length are multiplied, use a two-dimensional unit of area: 8 ft · 7 ft = 56 ft².
- If three measurements of length are multiplied, use a three-dimensional unit of volume: 3 m · 2 m · 4 m = 24 m³.

## Cylinders

A rectangular solid is shown below. Note that we can think of the volume as the product of the area of the base times the height:

$$V = l \cdot w \cdot h$$
$$= (l \cdot w) \cdot h$$
$$= (\text{Area of the base}) \cdot h$$
$$= B \cdot h,$$

Area of base $= B = l \cdot w$

where B represents the area of the base.

Like rectangular solids, **circular cylinders** have bases of equal area that lie in parallel planes. The bases of circular cylinders are circular regions.

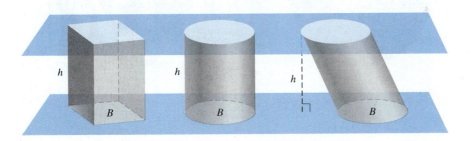

**2. Carry-on Luggage.** A carry-on bag measures 23 in. by 10 in. by 13 in. Find the volume of this solid.

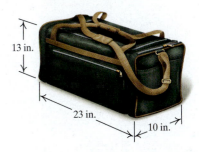

**3. Cord of Wood.** A cord of wood measures 4 ft by 4 ft by 8 ft. What is the volume of a cord of wood?

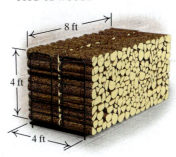

**4.** Find the volume of the cylinder. Use 3.14 for $\pi$.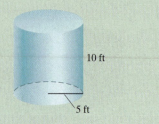

10 ft

5 ft

$V = \pi \cdot r^2 \cdot h$

$\approx 3.14 \cdot 5 \text{ ft} \cdot 5 \text{ ft} \cdot \boxed{\phantom{00}} \text{ ft}$

$= 3.14 \cdot 250 \text{ ft}^3$

$= \boxed{\phantom{00}} \text{ ft}^3$

**5.** Find the volume of the cylinder. Use $\frac{22}{7}$ for $\pi$.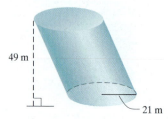

49 m

21 m

**6.** Find the volume of the sphere. Use $\frac{22}{7}$ for $\pi$.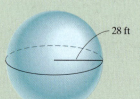

28 ft

$V = \frac{4}{3} \cdot \pi \cdot r^3$

$\approx \frac{4}{3} \cdot \frac{22}{7} \cdot (\boxed{\phantom{00}} \text{ ft})^3$

$= \frac{4}{3} \cdot \frac{22}{7} \cdot \boxed{\phantom{00}} \text{ ft}^3$

$= \frac{275{,}968}{3} \text{ ft}^3$

$= \boxed{\phantom{00}} \frac{1}{3} \text{ ft}^3$

**7.** The radius of a standard-sized golf ball is 2.1 cm. Find its volume. Use 3.14 for $\pi$.

**Answers**

**4.** 785 ft$^3$  **5.** 67,914 m$^3$  **6.** 91,989$\frac{1}{3}$ ft$^3$
**7.** 38.77272 cm$^3$
**Guided Solutions:**
**4.** 10, 785  **6.** 28, 21,952, 91,989

The volume of a circular cylinder is the product of the area of the base and the height. The height is always measured perpendicular to the base.

---

**VOLUME OF A CIRCULAR CYLINDER**

The **volume of a circular cylinder** is the product of the area of the base $B$ and the height $h$:

$$V = B \cdot h, \quad \text{or} \quad V = \pi \cdot r^2 \cdot h.$$

---

**EXAMPLE 3** Find the volume of this circular cylinder. Use 3.14 for $\pi$.

$V = B \cdot h = \pi \cdot r^2 \cdot h$

$\approx 3.14 \cdot 4 \text{ cm} \cdot 4 \text{ cm} \cdot 12 \text{ cm}$

$= 602.88 \text{ cm}^3$

12 cm

4 cm

◀ **Do Exercises 4 and 5.**

## Spheres

A **sphere** is the three-dimensional counterpart of a circle. It is the set of all points in space that are a given distance (the radius) from a given point (the center). The volume of a sphere depends on its radius.

---

**VOLUME OF A SPHERE**

The **volume of a sphere** of radius $r$ is given by

$$V = \frac{4}{3} \cdot \pi \cdot r^3.$$

---

**EXAMPLE 4** *Bowling Ball.* The radius of a standard-sized bowling ball is 4.2915 in. Find the volume of a standard-sized bowling ball (disregarding the finger holes). Round to the nearest hundredth of a cubic inch. Use 3.14 for $\pi$.

$r = 4.2915$ in.

$V = \frac{4}{3} \cdot \pi \cdot r^3 \approx \frac{4}{3} \cdot 3.14 \cdot (4.2915 \text{ in.})^3$

$\approx 330.90 \text{ in}^3$      Using a calculator

◀ **Do Exercises 6 and 7.**

## b CAPACITY

### American Units

To answer a question like "How much soda is in the bottle?" we need measures of **capacity**. American units of capacity are fluid ounces, cups, pints, quarts, and gallons. These units are related as follows.

| AMERICAN UNITS OF CAPACITY | |
| --- | --- |
| 1 gallon (gal) = 4 quarts (qt) | 1 pt = 2 cups |
| | = 16 fluid ounces (fl oz) |
| 1 qt = 2 pints (pt) | 1 cup = 8 fluid oz |

Fluid ounces, abbreviated fl oz, are often referred to as ounces, or oz.

**EXAMPLE 5** Complete: 24 qt = _____ gal.

Since we are converting from a *smaller* unit to a *larger* unit, we multiply by 1 using 1 gal in the numerator and 4 qt in the denominator:

$$24 \text{ qt} = 24 \text{ qt} \cdot \frac{1 \text{ gal}}{4 \text{ qt}} = \frac{24}{4} \cdot 1 \text{ gal} = 6 \text{ gal.}$$

**EXAMPLE 6** Complete: 9 gal = _____ oz.

The box above does not list how many ounces are in 1 gal. We convert gallons to quarts, quarts to pints, and pints to ounces, using the relationships given in the box.

Since we are converting from a *larger* unit to a *smaller* unit, we use substitution:

$9 \text{ gal} = 9 \cdot 1 \text{ gal} = 9 \cdot 4 \text{ qt} = 36 \text{ qt}$     Substituting 4 qt for 1 gal

$\phantom{9 \text{ gal}} = 36 \cdot 1 \text{ qt} = 36 \cdot 2 \text{ pt} = 72 \text{ pt}$     Substituting 2 pt for 1 qt

$\phantom{9 \text{ gal}} = 72 \cdot 1 \text{ pt} = 72 \cdot 16 \text{ oz} = 1152 \text{ oz.}$     Substituting 16 oz for 1 pt

Thus, 9 gal = 1152 oz.

**Do Exercises 8 and 9.** ▶

### Metric Units

One unit of capacity in the metric system is a **liter**. A liter is just a bit more than a quart. It is defined as follows.

1 liter ≈ 1.06 quarts

1 liter        1 quart

**SKILL REVIEW**

*Convert from one metric unit of length to another.* [9.1b]

Complete.

1. 42.7 cm = _____ mm

2. 42.7 mm = _____ cm

     **Answers: 1.** 427  **2.** 4.27

MyLab Math
VIDEO

Complete.

**GS**   **8.** 80 qt = _____ gal

$$80 \text{ qt} = 80 \text{ qt} \cdot \frac{1 \text{ gal}}{\boxed{\phantom{xx}}}$$

$$= \frac{80}{\boxed{\phantom{xx}}} \cdot 1 \text{ gal}$$

$$= \boxed{\phantom{xx}} \text{ gal}$$

**9.** 5 gal = _____ pt

**Answers**

**8.** 20   **9.** 40

*Guided Solution:*

**8.** 4 qt, 4, 20

## METRIC UNITS OF CAPACITY

> 1 liter (L) = 1000 cubic centimeters (1000 cm$^3$)
>
> The script letter $\ell$ is also used for "liter."

The metric prefixes are also used with liters. The most common is **milli-**. The milliliter (mL) is, then, $\frac{1}{1000}$ liter.

> 1 L = 1000 mL = 1000 cm$^3$;
> 0.001 L = 1 mL = 1 cm$^3$.

A common unit for drug dosage is the milliliter (mL) or the cubic centimeter (cm$^3$). The notation "cc" is also used for cubic centimeter, especially in medicine. The milliliter and the cubic centimeter represent the same measure of capacity. A milliliter is about $\frac{1}{5}$ of a teaspoon.

3 cm$^3$

5 mL

> 1 mL = 1 cm$^3$ = 1 cc

Volumes for which quarts and gallons are used are expressed in liters. Large volumes are expressed using measures of cubic meters (m$^3$).

◀ **Do Exercises 10–13.**

Complete with mL or L.

**10.** The patient received an injection of 2 _____ of penicillin.

**11.** There are 250 _____ in a coffee cup.

**12.** The gas tank holds 80 _____.

**13.** Bring home 8 _____ of milk.

**EXAMPLE 7**   Complete: 4.5 L = _____ mL.

$$4.5\text{ L} = 4.5 \cdot 1\text{ L} = 4.5 \cdot 1000\text{ mL} \qquad \text{Substituting 1000 mL for 1 L}$$
$$= 4500\text{ mL}$$

We can also convert units of capacity by multiplying by a form of 1.

**EXAMPLE 8**   Complete: 280 mL = _____ L.

$$280\text{ mL} = 280\text{ mL} \cdot \frac{1\text{ L}}{1000\text{ mL}}$$
$$= \frac{280}{1000}\text{ L}$$
$$= 0.28\text{ L}$$

Complete.

**14.** 0.97 L = _____ mL    **GS**

$$0.97\text{ L} = 0.97 \cdot 1\text{ L}$$
$$= 0.97 \cdot \boxed{\phantom{00}}\text{ mL}$$
$$= \boxed{\phantom{00}}\text{ mL}$$

**15.** 8990 mL = _____ L

*Answers*

**10.** mL   **11.** mL   **12.** L   **13.** L   **14.** 970
**15.** 8.99

*Guided Solution:*
**14.** 1000, 970

◀ **Do Exercises 14 and 15.**

## c SOLVING APPLIED PROBLEMS

**EXAMPLE 9** At a self-service gas station, 89-octane gasoline sells for 102.6¢ a liter. Estimate the price of 1 gal in dollars.

Since 1 liter is about 1 quart and there are 4 quarts in a gallon, the price of a gallon is about 4 times the price of a liter.

$$4 \cdot 102.6¢ = 410.4¢ = \$4.104$$

Thus, 89-octane gasoline sells for about $4.10 a gallon.

**Do Exercise 16.** ▶

**EXAMPLE 10** *Propane Gas Tank.* A propane gas tank is shaped like a circular cylinder with half of a sphere at each end. Find the volume of the tank if the cylindrical section is 5 ft long with a 4-ft diameter. Use 3.14 for $\pi$.

**1. Familiarize.** We first make a drawing.

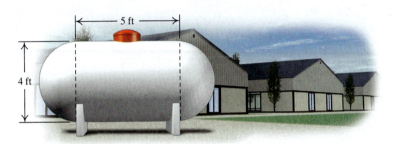

**2. Translate.** This is a two-step problem. We first find the volume of the cylindrical portion. Then we find the volume of the two ends and add. Note that the radius is 2 ft and that together the two ends make a sphere. We let $V =$ the total volume.

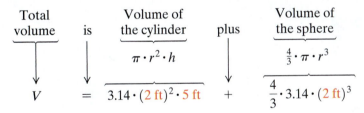

**3. Solve.** The volume of the cylinder is approximately

$$3.14 \cdot (2 \text{ ft})^2 \cdot 5 \text{ ft} = 3.14 \cdot 2 \text{ ft} \cdot 2 \text{ ft} \cdot 5 \text{ ft}$$
$$= 62.8 \text{ ft}^3.$$

The volume of the two ends is approximately

$$\frac{4}{3} \cdot 3.14 \cdot (2 \text{ ft})^3 = \frac{4}{3} \cdot 3.14 \cdot 2 \text{ ft} \cdot 2 \text{ ft} \cdot 2 \text{ ft}$$
$$\approx 33.5 \text{ ft}^3.$$

The total volume is approximately

$$62.8 \text{ ft}^3 + 33.5 \text{ ft}^3 = 96.3 \text{ ft}^3.$$

**4. Check.** We can repeat the calculations. The answer checks.

**5. State.** The volume of the tank is about 96.3 ft³.

**Do Exercise 17.** ▶

---

**16.** At a gas station, the price of 87-octane gasoline is 96.7 cents a liter. Estimate the price of 1 gal in dollars.

---

### CALCULATOR CORNER

***Volumes Using Pi*** Many calculators have a $\boxed{\pi}$ key that can be used to give a more precise value of $\pi$ than 3.14.

When we use a $\boxed{\pi}$ key to find the volume of the circular cylinder in Example 3, the result is approximately 603.19. Note that this is slightly different from the result found using 3.14 for $\pi$.

**EXERCISES:**

**1.** Use a calculator with a $\boxed{\pi}$ key to find the volume of a cylinder with radius 1.2 cm and height 0.25 cm.

**2.** Use a calculator with a $\boxed{\pi}$ key to find the volume of a sphere with radius $\frac{3}{8}$ in.

---

**17.** *Medicine Capsule.* A cold capsule is 8 mm long and 4 mm in diameter. Find the volume of the capsule. Use 3.14 for $\pi$. (*Hint*: First find the length of the cylindrical section.)

---

*Answers*
**16.** $3.87   **17.** 83.7$\overline{3}$ mm³

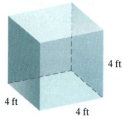
## ✓ Check Your Understanding

**Reading Check** Match each formula with the correct phrase from the list on the right.

**RC1.** _____ $V = l \cdot w \cdot h$

**RC2.** _____ $V = \pi \cdot r^2 \cdot h$

**RC3.** _____ $V = \dfrac{4}{3} \cdot \pi \cdot r^3$

a) the volume of a cylinder

b) the volume of a rectangular solid

c) the volume of a sphere

**Concept Check** Select from choices (a)–(d) the closest approximation of the volume of the rectangular solid that has the given dimensions.

**CC1.** $l = 10.1\,\text{ft}$, $w = 5.6\,\text{ft}$, $h = 100\,\text{ft}$

  a) $56\,\text{ft}^3$        b) $600\,\text{ft}^3$

  c) $6000\,\text{ft}^3$     d) $56{,}000\,\text{ft}^3$

**CC2.** $l = 0.21\,\text{m}$, $w = 100\,\text{cm}$, $h = 10\,\text{m}$

  a) $210{,}000\,\text{cm}^3$    b) $2{,}000{,}000\,\text{cm}$

  c) $2\,\text{m}^3$          d) $20\,\text{m}^2$

**a**   Find each volume. Use 3.14 for $\pi$ in Exercises 9–12. Use $\dfrac{22}{7}$ for $\pi$ in Exercises 13 and 14.

**1.**

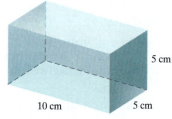

10 cm   5 cm   5 cm

**2.**

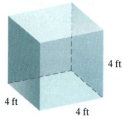

4 ft   4 ft   4 ft

**3.**

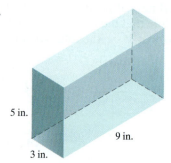

5 in.   3 in.   9 in.

**4.**

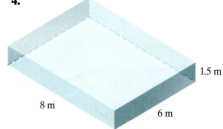

8 m   6 m   1.5 m

**5.**

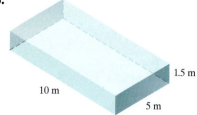

10 m   5 m   1.5 m

**6.**

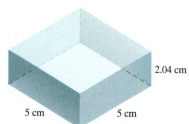

5 cm   5 cm   2.04 cm

**7.**

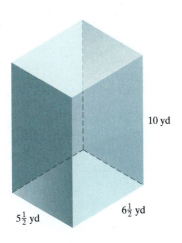

10 yd

$5\frac{1}{2}$ yd    $6\frac{1}{2}$ yd

**8.**

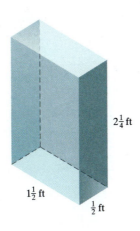

$2\frac{1}{4}$ ft

$1\frac{1}{2}$ ft    $\frac{1}{2}$ ft

**9.**

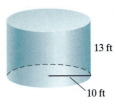

13 ft

10 ft

**10.**

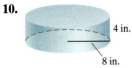

4 in.

8 in.

**11.**

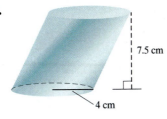

7.5 cm

4 cm

**12.**

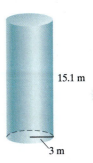

15.1 m

3 m

**13.**

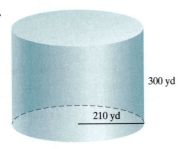

300 yd

210 yd

**14.**

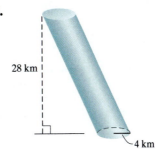

28 km

4 km

Find each volume. Use 3.14 for $\pi$ in Exercises 15–18. Use $\frac{22}{7}$ for $\pi$ in Exercises 19 and 20.

**15.**

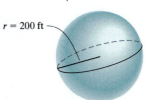

$r = 100$ in.

**16.**

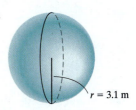

$r = 200$ ft

**17.**

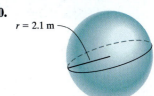

$r = 3.1$ m

**18.**

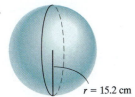

$r = 15.2$ cm

**19.**

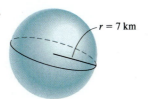

$r = 7$ km

**20.**

$r = 2.1$ m

**b** Complete.

**21.** 1 L = _____ mL = _____ cm³

**22.** _____ L = 1 mL = _____ cm³

**23.** 59 L = _____ mL

**24.** 714 L = _____ mL

**25.** 49 mL = _____ L

**26.** 43 mL = _____ L

**27.** 27.3 L = _____ cm³

**28.** 49.2 L = _____ cm³

**29.** 5 gal = _____ pt

**30.** 48 oz = _____ pt

**31.** 10 qt = _____ oz

**32.** 2 gal = _____ cups

**33.** 24 oz = _____ cups

**34.** 20 cups = _____ pt

**35.** 10 gal = _____ qt

**36.** 5 gal = _____ cups

**37.** 3 gal = _____ cups

**38.** 72 oz = _____ cups

**39.** 15 pt = _____ gal

**40.** 9 qt = _____ gal

**c** Solve.

**41.** *Oak Log.* An oak log has a diameter of 12 cm and a length (height) of 42 cm. Find the volume. Use 3.14 for $\pi$.

**42.** *Ladder Rung.* A rung of a ladder is 2 in. in diameter and 16 in. long. Find the volume. Use 3.14 for $\pi$.

**43.** *Times Square Ball.* The current Times Square Ball, located in New York City's Times Square, is an icosahedral geodesic sphere with a diameter of 12 ft. Find the approximate volume of the sphere. Use 3.14 for $\pi$, and round to the nearest cubic foot.

**Data:** timessquarenyc.org

**44.** *Culinary Arts.* Raena often makes individual soufflés in cylindrical baking dishes called *ramekins*. The diameter of each ramekin is 3.5 in., and the height is 1.75 in. Find the approximate volume of a ramekin. Use 3.14 for $\pi$.

**45.** *Volume of a Trash Can.* The diameter of the base of a cylindrical trash can is 0.7 yd. The height is 1.1 yd. Find the volume. Use 3.14 for $\pi$ and round the answer to the nearest hundredth.

**46.** *Tennis Ball.* The diameter of a tennis ball is 6.5 cm. Find the volume. Use 3.14 for $\pi$ and round the answer to the nearest hundredth.

**47.** ▦ *Volume of Earth.* The radius of the earth is about 3980 mi. Find the volume of the earth. Use 3.14 for $\pi$. Round to the nearest ten thousand cubic miles.

**48.** ▦ *Astronomy.* The radius of the largest moon of Uranus is about 789 km. Find the volume of this satellite. Use $\frac{22}{7}$ for $\pi$. Round to the nearest ten thousand cubic kilometers.

**49.** *Oceanography.* The Deep-ESP (Environmental Sample Processor) is designed to study deep-sea environments. Its titanium pressure housing allows the ESP sphere to descend over 2 miles below the ocean surface. If the diameter of the sphere is 40 in., what is its volume? Use 3.14 for $\pi$ and round to the nearest cubic inch.

**Data:** "Extreme Life," by Henry Bortman, 06/15/09, on www.astrobio.net

**50.** *Gas Pipeline.* The 638-mi Rockies Express-East gas pipeline from Colorado to Ohio is constructed with 80-ft sections of steel pipe with a radius of 21 in., or $1\frac{3}{4}$ ft. Find the volume of one section. Use $\frac{22}{7}$ for $\pi$.

**Data:** Rockies Express Pipeline

**51.** ▦ A sphere with diameter 1 m is circumscribed by a cube. How much greater is the volume of the cube than the volume of the sphere? Use 3.14 for $\pi$.

**52.** ▦ *Golf-Ball Packaging.* The box shown is just big enough to hold 3 golf balls. If the radius of a golf ball is 2.1 cm, how much air surrounds the three balls? Use 3.14 for $\pi$.

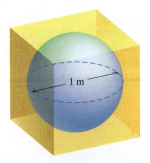

1 m

**53.** *Architecture.* The Westhafen Tower in Frankfort, Germany, is a cylindrical building with a height of 110 m and a radius of 21 m. Find the volume of the building. Use 3.14 for $\pi$. Round to the nearest cubic meter.

**54.** *Weather Forecasting.* Every day, the National Weather Service launches spherical weather balloons from 100 locations in the United States. Each balloon can rise up to 115,000 ft and travel up to 200 miles. Find the volume of a 6-ft-diameter weather balloon. Use 3.14 for $\pi$.

**Data:** Kaysam Worldwide, Inc.

**55.** *Metallurgy.* If all the gold in the world could be gathered together, it would form a cube 18 yd on a side. Find the volume of the world's gold.

**56.** ▦ The width of a dollar bill is 2.3125 in., the length is 6.0625 in., and the thickness is 0.0041 in. Find the volume occupied by 1 million one-dollar bills.

**57.** *Toys.* Toy stores often sell capsules that dissolve in water allowing a toy inside the capsule to expand. One such capsule is 40 mm long with a diameter of 8 mm.

 **a)** What is the volume of the capsule? Use 3.14 for $\pi$.

 **b)** The manufacturer claims that the toy in the capsule will expand 600%. What is the volume of the toy after expansion?

**58.** *Silo.* A silo is a circular cylinder with half a sphere on the top of the cylinder. If the diameter of the cylinder is 6 m and the height of the cylinder is 13 m, find the volume of the silo. Use 3.14 for $\pi$.

## Skill Maintenance

**59.** Find the simple interest on $600 at 8% for $\frac{1}{2}$ yr.   [7.7a]

**60.** Find the simple interest on $5000 at 7% for $\frac{1}{2}$ yr.   [7.7a]

**61.** If 9 pens cost $8.01, how much would 12 pens cost?   [6.4a]

**62.** Solve: $9(x - 1) = 3x + 5$.   [5.7b]

**63.** Solve: $-5y + 3 = -12y - 4$.   [5.7b]

**64.** A barge travels 320 km in 15 days. At this rate, how far will it travel in 21 days?   [6.4a]

**65.** Evaluate $\frac{9}{5}C + 32$ for $C = 15$.   [4.7c]

**66.** Evaluate $\frac{5}{9}(F - 32)$ for $F = 50$.   [4.7c]

## Synthesis

**67.** 🖩 *Truck Rental.*   The storage compartment of a moving truck is 9.83 ft by 5.67 ft by 5.83 ft, with an "attic" measuring 1.5 ft by 5.67 ft by 2.5 ft. Find the total volume of the compartment.

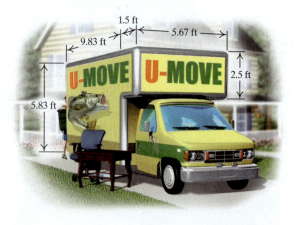

**68.** *Tennis-Ball Packaging.*   Tennis balls are generally packaged in circular cylinders that hold 3 balls each. The diameter of a tennis ball is 6.5 cm. If the tennis balls were "liquid" and could be poured into the cylinder, how many tennis balls would the cylinder hold?

**69.** The volume of a ball is $36\pi$ cm³. Find the dimensions of a rectangular box that is just large enough to hold the ball.

**70.** 🖩 The volume of a basketball is $2304\pi$ cm³. Find the volume of a cube-shaped box that is just large enough to hold the ball.

**71.** 🖩 A 2-cm-wide stream of water passes through a 30-m-long garden hose. At the instant that the water is turned off, how many liters of water are in the hose? Use 3.141593 for $\pi$.

**72.** *Remarkable Feat.*   In 1982, Larry Walters captured the world's imagination by riding a lawn chair attached to 42 helium-filled weather balloons to an altitude of 16,000 ft, before using a BB gun to pop a few balloons and safely descend. Walters used balloons measuring approximately 7 ft in diameter. Find the total volume of the balloons used. Use $\frac{22}{7}$ for $\pi$.

**Data:** MarkBarry.com

**73.** *Conservation.*   Many people leave the water running while brushing their teeth. Suppose that one person wastes 32 oz of water in such a manner each day. How much water, in gallons, would that person waste in a week? in 30 days? in a year? If each of 330 million Americans wastes water this way, estimate how much water is wasted in a year.

**74.** The volumes occupied by all the world's gold and by a million dollar bills are described in Exercises 55 and 56. How many million dollar bills would fit into the volume of all the world's gold?

## Concept Reinforcement

Determine whether each statement is true or false.

_____ **1.** Distances that are measured in miles in the American system would probably be measured in meters in the metric system. [9.1c]

_____ **2.** One meter is slightly more than one yard. [9.1c]

_____ **3.** One kilometer is longer than one mile. [9.1c]

_____ **4.** The area of a parallelogram with base 8 cm and height 5 cm is the same as the area of a rectangle with length 8 cm and width 5 cm. [9.3a]

_____ **5.** The exact value of the ratio $C/d$ is $\pi$. [9.3b]

## Guided Solutions

**GS** Fill in each blank with the number or unit that creates a correct solution.

**6.** Complete: $16\frac{2}{3}$ yd = _____ ft. [9.1a]

$$16\frac{2}{3}\text{ yd} = 16\frac{2}{3}\cdot 1\ \square\ = \frac{50}{3}\cdot 3\ \square\ = 50\ \square$$

**7.** Complete: 10,200 mm = _____ ft. [9.1c]

$$10,200\text{ mm} = 10,200\text{ mm}\cdot\frac{1\ \square}{1000\ \square} \approx 10.2\ \square\cdot\frac{3.281\ \square}{1\ \square} = 33.4662\ \square$$

**8.** Find the circumference and the area. Use 3.14 for $\pi$. [9.3b]

10.2 in.

$C = \pi\cdot d$        $A = \pi\cdot r\cdot r$

$C \approx\ \square\ \cdot\ \square\ $ in.      $A \approx\ \square\ \cdot\ \square\ $ in. $\cdot\ \square\ $ in.

$C =\ \square\ $ in.          $A =\ \square\ $ in$^{\square}$

## Mixed Review

Complete. [9.1a, b, c], [9.2a, b]

**9.** $5\frac{1}{2}$ mi = _____ yd

**10.** 840 in. = _____ ft

**11.** 24.05 cm = _____ dm

**12.** 0.15 m = _____ km

**13.** 3753 ft = _____ yd

**14.** 26,400 ft = _____ mi

**15.** 1800 m = _____ cm

**16.** 0.007 km = _____ cm

**17.** 2.5 yd = _____ m

**18.** 36 m = _____ ft

**19.** 6 in. = _____ cm

**20.** 100 mi = _____ km

**21.** $5 \text{ ft}^2$ = _____ $\text{in}^2$

**22.** $10 \text{ yd}^2$ = _____ $\text{ft}^2$

**23.** $60 \text{ ft}^2$ = _____ $\text{yd}^2$

**24.** $48 \text{ in}^2$ = _____ $\text{ft}^2$

**25.** 2 acres = _____ $\text{ft}^2$

**26.** $1.5 \text{ mi}^2$ = _____ acres

**27.** Arrange from smallest to largest:

100 in., 430 ft, $\frac{1}{100}$ mi, 3.5 ft, 6000 ft, 1000 in., 2 yd. [9.1a]

**28.** Arrange from largest to smallest:

3240 cm, 300 m, 250 dm, 150 hm, 33,000 mm, 310 dam, 13 km. [9.1b]

Find the area. [9.3a]

Find the circumference and the area. Use 3.14 for $\pi$. [9.3b]

**29.**

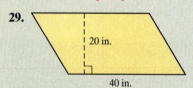

20 in.
40 in.

**30.**

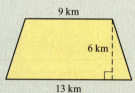

9 km
6 km
13 km

**31.**

7 in.

**32.**

8.6 cm

Find the volume. Use 3.14 for $\pi$ in Exercises 34–36. [9.4a]

**33.**

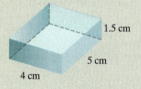

1.5 cm
5 cm
4 cm

**34.**

15 ft
6 ft

**35.**

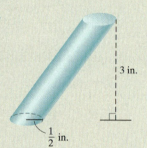

3 in.
$\frac{1}{2}$ in.

**36.**

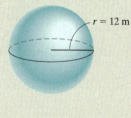

r = 12 m

Complete. [9.4b]

**37.** 7.4 L = _____ mL

**38.** $62 \text{ cm}^3$ = _____ mL

**39.** 0.5 mL = _____ L

**40.** 3 gal = _____ qt

**41.** 18 cups = _____ pt

**42.** 32 oz = _____ cups

## Understanding Through Discussion and Writing

**43.** A student makes the following error:

23 in. = 23 · (12 ft) = 276 ft.

Explain the error. [9.1a]

**44.** Explain why a 16-in.-diameter pizza that costs $16.25 is a better buy than a 10-in.-diameter pizza that costs $7.85. [9.3b]

**45.** Explain how the area of a triangle can be found by considering the area of a parallelogram. [9.3a]

**46.** The radius of one circle is twice another circle's radius. Is the area of the first circle twice the area of the other circle? Why or why not? [9.3b]

# Angles and Triangles

## 9.5

### a MEASURING ANGLES

We see a real-world application of *angles* of various types in the different back postures of the bicycle riders illustrated below.

**Style of Biking Determines Cycling Posture**

**Road**
About 180° flat

Riders prefer a more aerodynamic flat-back position.

**Mountain**
About 45°

Riders prefer a semi-upright position to help lift the front wheel over obstacles.

**Comfort**
About 90°

Riders prefer an upright position that lessens stress on the lower back and neck.

DATA: USA TODAY research

### OBJECTIVES

**a** Name an angle in six different ways and measure an angle with a protractor.

**b** Classify an angle as right, straight, acute, or obtuse.

**c** Identify complementary, supplementary, and vertical angles and find the measure of a complement or a supplement of a given angle.

**d** Classify a triangle as equilateral, isosceles, or scalene, and as right, obtuse, or acute.

**e** Given two of the angle measures of a triangle, find the third.

An **angle** is a set of points consisting of two **rays**, or half-lines, with a common endpoint. The endpoint is called the **vertex** of the angle. The rays are called the **sides** of the angle.

Ray $BA$, or $\vec{BA}$

Ray $BC$, or $\vec{BC}$

Vertex

The angle above can be named

angle $ABC$,     angle $CBA$,     angle $B$,     $\angle ABC$,     $\angle CBA$,     or    $\angle B$.

Note that the vertex is written in the middle of the name. If there is only one angle with a given vertex in a drawing, the angle may be named using simply its vertex.

**Do Exercises 1 and 2.** ▶

Name the angle in six different ways.

**1.**

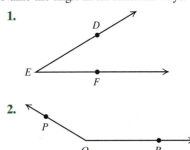

**2.**

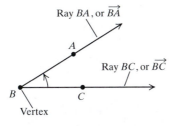

***Answers***

**1.** Angle $DEF$, angle $FED$, angle $E$, $\angle DEF$, $\angle FED$, or $\angle E$   **2.** Angle $PQR$, angle $RQP$, angle $Q$, $\angle PQR$, $\angle RQP$, or $\angle Q$

**3.** Give another name for ∠1.

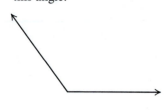

Angles may also be numbered. In the following figure, another name for ∠ABC is ∠1, and another name for ∠CBD is ∠2.

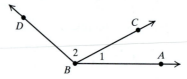

◀ **Do Exercise 3.**

To measure angles, we start with some predetermined angle and assign to it a measure of 1. We call it a *unit angle*. Suppose ∠U, shown below, is a unit angle. Let's measure ∠DEF. If we made 3 copies of ∠U, they would "fill up" ∠DEF. Thus, the measure of ∠DEF would be 3 units.

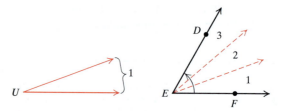

The unit most commonly used for angle measure is the degree. Below is such a unit. Its measure is 1 degree, or 1°. There are 360 degrees in a circle.

A 1° angle:

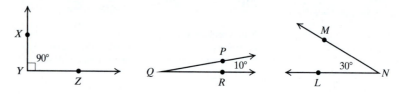

Here are some other angles with their degree measures.

To indicate the *measure* of ∠XYZ, we write $m \angle XYZ = 90°$. The symbol ⌐ is sometimes drawn on a figure to indicate a 90° angle.

A device called a **protractor** is used to measure angles. Protractors often have two scales (inside and outside). In the center of the protractor is a vertex indicator such as ▲ or a small hole. To measure an angle like ∠Q below, we place the protractor's ▲ at the vertex and line up one of the angle's sides at 0°. Then we check where the angle's other side crosses the scale. In the figure below, the side $\overrightarrow{QR}$ lines up with 0° on the *inside* scale, so we check where the angle's other side crosses the *inside* scale. We see that $m \angle Q = 145°$.

**4.** Use a protractor to measure this angle.

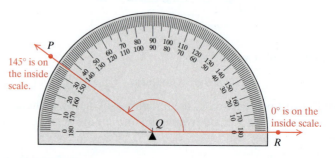

145° is on the inside scale.

0° is on the inside scale.

◀ **Do Exercise 4.**

Let's find the measure of $\angle ABC$. This time we line up one of the angle's sides, $\overrightarrow{BC}$, with 0° on the *outside* scale. Then we check where the angle's other side, $\overrightarrow{BA}$, crosses the *outside* scale. We see that $m \angle ABC = 42°$.

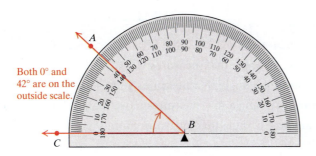

Both 0° and 42° are on the outside scale.

Do Exercise 5. ▶

A protractor can be used to draw a circle graph.

**EXAMPLE 1** *Transportation.* According to a recent poll, 45% of adults believe that flying is the safest mode of transportation, 39% believe that cars are the safest, and 16% believe that trains are the safest. Draw a circle graph to represent these figures.

**Data:** Marist Institute for Public Opinion

Every circle graph contains a total of 360°. Thus,

45% of the circle is a 0.45(360°), or 162°, angle;

39% of the circle is a 0.39(360°), or 140.4°, angle;

16% of the circle is a 0.16(360°), or 57.6°, angle.

We begin by drawing a 162° angle. Beginning at the center of the circle, we draw a horizontal segment to the circle. That segment is one side of the angle. We use a protractor to mark off a 162° angle. From that mark, we draw a segment to the center of the circle to complete the angle. This section of the circle graph we label with both the percent (45%) and the type of transportation (Airplanes).

From the second segment drawn, we repeat the above procedure to draw a 140.4° angle. Since protractors are marked in units of 1°, we must approximate this angle. This section we label with 39% and Cars.

The remainder of the circle represents Trains and should be a 57.6° angle; we measure to confirm this, and label the section with 16% and Trains.

Finally, we give a title to the graph: Safest Mode of Transportation.

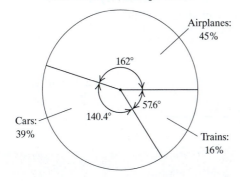

**Safest Mode of Transportation**

Airplanes: 45%

162°

57.6°

140.4°

Cars: 39%

Trains: 16%

Do Exercise 6. ▶

**5.** Use a protractor to measure this angle.

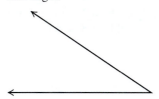

**6. Lengths of Engagement of Married Couples.** The data below list the percentages of married couples who were engaged for a certain time period before marriage. Use this information to draw a circle graph.

**Data:** Bruskin Goldring Research

| Less than 1 yr: | 24% |
|---|---|
| 1–2 yr: | 21% |
| More than 2 yr: | 35% |
| Never engaged: | 20% |

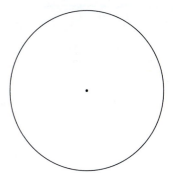

## b CLASSIFYING ANGLES

The following are ways in which we classify angles.

Classify each angle as right, straight, acute, or obtuse. Use a protractor if necessary.

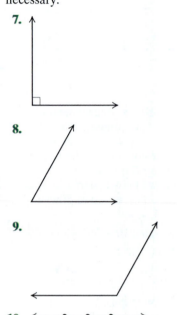

**7.**

**8.**

**9.**

**10.**

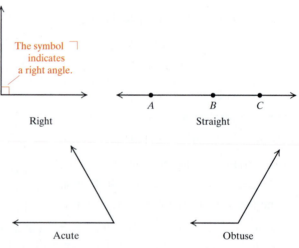

The symbol ⌐ indicates a right angle.

Right

Straight

Acute

Obtuse

◀ **Do Exercises 7–10.**

## c COMPLEMENTARY, SUPPLEMENTARY, AND VERTICAL ANGLES

Certain pairs of angles share special properties.

### Complementary Angles

When the sum of the measures of two angles is 90°, the angles are said to be **complementary**. For example, in the figure below, ∠1 and ∠2 are complementary. If two angles are complementary, each is an acute angle. Note that when complementary angles are adjacent to each other (that is, they have a side in common), they form a right angle.

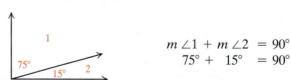

$$m \angle 1 + m \angle 2 = 90°$$
$$75° + 15° = 90°$$

*Answers*

**7.** Right  **8.** Acute  **9.** Obtuse
**10.** Straight

**EXAMPLE 2** Identify each pair of complementary angles.

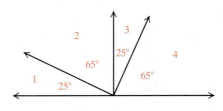

| $\angle 1$ and $\angle 2$ $25° + 65° = 90°$ | $\angle 2$ and $\angle 3$ |
|---|---|
| $\angle 1$ and $\angle 4$ | $\angle 3$ and $\angle 4$ |

**EXAMPLE 3** Find the measure of a complement of a 39° angle.

$90° - 39° = 51°$

The measure of a complement of a 39° angle is 51°.

**Do Exercises 11–14.** ▶

## Supplementary Angles

Next, consider $\angle 1$ and $\angle 2$ as shown below. Because the sum of their measures is 180°, $\angle 1$ and $\angle 2$ are said to be **supplementary**. Note that when supplementary angles are adjacent, they form a straight angle.

$$m\angle 1 + m\angle 2 = 180°$$
$$30° + 150° = 180°$$

---

**SUPPLEMENTARY ANGLES**

Two angles are **supplementary** if the sum of their measures is 180°. Each angle is called a **supplement** of the other.

---

**EXAMPLE 4** Identify each pair of supplementary angles.

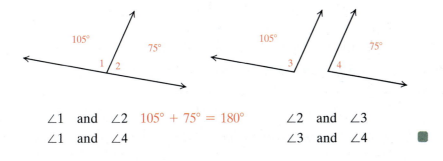

| $\angle 1$ and $\angle 2$ $105° + 75° = 180°$ | $\angle 2$ and $\angle 3$ |
|---|---|
| $\angle 1$ and $\angle 4$ | $\angle 3$ and $\angle 4$ |

**11.** Identify each pair of complementary angles.

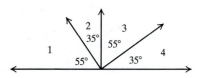

Find the measure of a complement of each angle.

**12.**

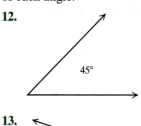

**13.**

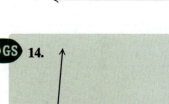

**GS** **14.**

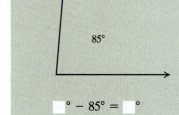

$\square° - 85° = \square°$

*Answers*
**11.** $\angle 1$ and $\angle 2$; $\angle 1$ and $\angle 4$; $\angle 2$ and $\angle 3$; $\angle 3$ and $\angle 4$   **12.** 45°   **13.** 72°   **14.** 5°
*Guided Solution:*
**14.** 90, 5

**15.** Identify each pair of supplementary angles.

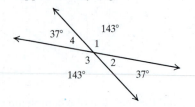

Find the measure of a supplement of an angle with each given measure.

**16.** 38°          **17.** 157°

**18.** 90°

**19.** 71°          **GS**

$$\boxed{\phantom{000}}° - 71° = \boxed{\phantom{000}}°$$

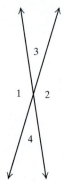

**20.** Identify each pair of vertical angles.

**EXAMPLE 5**  Find the measure of a supplement of a 112° angle.

$$180° - 112° = 68°$$

The measure of a supplement of a 112° angle is 68°.

◀ **Do Exercises 15–19.**

## Vertical Angles

When two lines intersect, four angles are formed. The pairs of angles that do not share any side in common are said to be **vertical** (or *opposite*) angles. Thus, in the drawing below, $\angle 1$ and $\angle 3$ are vertical angles, as are $\angle 4$ and $\angle 2$. Note that $m\angle 1 = m\angle 3$ and $m\angle 4 = m\angle 2$.

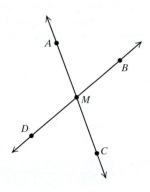

**EXAMPLE 6**  Identify each pair of vertical angles.

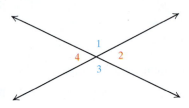

$\angle AMB$ and $\angle CMD$ are vertical angles.
$\angle BMC$ and $\angle DMA$ are vertical angles.

◀ **Do Exercise 20.**

Note in the figure in Example 6 that $\angle DMA$ and $\angle AMB$ are supplementary; that is,

$$m\angle DMA + m\angle AMB = 180°.$$

Also, $\angle CMD$ and $\angle DMA$ are supplementary, so

$$m\angle CMD + m\angle DMA = 180°.$$

Therefore,

$$m\angle DMA + m\angle AMB = m\angle CMD + m\angle DMA$$
$$m\angle AMB = m\angle CMD. \quad \text{Subtracting } m\angle DMA \text{ from both sides}$$

This shows that vertical angles have the same measure.

**Answers**

**15.** $\angle 1$ and $\angle 2$; $\angle 1$ and $\angle 4$; $\angle 2$ and $\angle 3$; $\angle 3$ and $\angle 4$  **16.** 142°  **17.** 23°  **18.** 90°
**19.** 109°  **20.** $\angle 1$ and $\angle 2$; $\angle 3$ and $\angle 4$

**Guided Solution:**
**19.** 180, 109

## VERTICAL ANGLES

Two angles are **vertical** if they are formed by two intersecting lines and have no side in common. Vertical angles have the same measure.

**Do Exercise 21.** ▶

If two angles have the same measure, we say that they are **congruent**, denoted by the symbol ≅. In the figure below, the measures of angles *WVZ* and *XVY* are equal, and the angles are congruent:

$$m \angle WVZ = m \angle XVY$$
$$\angle WVZ \cong \angle XVY.$$

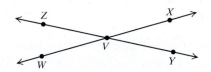

Note that we do not write that angles are equal: The *measures are equal* and the *angles are congruent*.

**Do Exercise 22.** ▶

### d TRIANGLES

A **triangle** is a polygon made up of three segments, or sides. Consider these triangles. The triangle with vertices *A*, *B*, and *C* can be named △*ABC*.

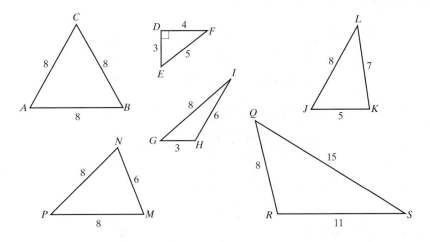

We can classify triangles according to sides and according to angles.

## TYPES OF TRIANGLES

**Equilateral triangle:** All sides are the same length.

**Isosceles triangle:** Two or more sides are the same length.

**Scalene triangle:** All sides are of different lengths.

**Right triangle:** One angle is a right angle.

**Obtuse triangle:** One angle is an obtuse angle.

**Acute triangle:** All three angles are acute.

**Do Exercises 23–26.** ▶

**21.** Complete:

$$m \angle ANC = \underline{\hspace{1cm}};$$
$$m \angle ANB = \underline{\hspace{1cm}}.$$

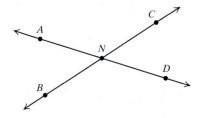

**22.** Complete:

$$\angle PMQ \cong \underline{\hspace{1cm}};$$
$$\angle SMP \cong \underline{\hspace{1cm}}.$$

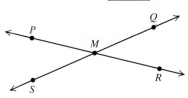

**23.** Which triangles on this page are

   **a)** equilateral?

   **b)** isosceles?

   **c)** scalene?

**24.** Are all equilateral triangles isosceles?

**25.** Are all isosceles triangles equilateral?

**26.** Which triangles on this page are

   **a)** right triangles?

   **b)** obtuse triangles?

   **c)** acute triangles?

*Answers*

**21.** *m* ∠*BND* or *m* ∠*DNB*; *m* ∠*CND* or *m* ∠*DNC*
**22.** ∠*SMR* or ∠*RMS*; ∠*QMR* or ∠*RMQ*
**23.** **(a)** △*ABC*; **(b)** △*ABC*, △*MPN*; **(c)** △*DEF*, △*GHI*, △*JKL*, △*QRS*
**24.** Yes **25.** No **26.** **(a)** △*DEF*; **(b)** △*GHI*, △*QRS*; **(c)** △*ABC*, △*MPN*, △*JKL*

SECTION 9.5  Angles and Triangles  **651**

**SUM OF THE ANGLE MEASURES OF A TRIANGLE**

The sum of the angle measures of every triangle is 180°. To see this, note that we can think of cutting apart a triangle as shown on the left below. If we reassemble the pieces, we see that a straight angle is formed.

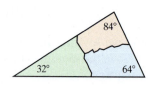
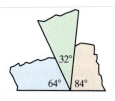

$$64° + 32° + 84° = 180°$$

**27.** Find $m \angle P + m \angle Q + m \angle R$.

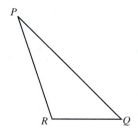

---

**SUM OF THE ANGLE MEASURES OF A TRIANGLE**

In any $\triangle ABC$, the sum of the measures of the angles is 180°:

$$m \angle A + m \angle B + m \angle C = 180°.$$

---

◀ **Do Exercise 27.**

If we know the measures of two angles of a triangle, we can calculate the measure of the third angle.

**EXAMPLE 7** Find the missing angle measure.

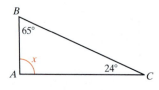

$$m \angle A + m \angle B + m \angle C = 180°$$
$$x + 65° + 24° = 180°$$
$$x + 89° = 180°$$
$$x = 180° - 89° \qquad \text{Subtracting 89° from both sides}$$
$$x = 91°$$

Thus, $m \angle A = 91°$.

◀ **Do Exercise 28.**

**28.** Find the missing angle measure.

*Answers*

**27.** 180°  **28.** 64°

**Exercise Set**

**P** **MyLab Math**

**RC1.** _____ An angle whose measure is 90°

**RC2.** _____ An angle whose measure is 180°

**RC3.** _____ An angle whose measure is greater than 0° and less than 90°

**RC4.** _____ An angle whose measure is greater than 90° and less than 180°

**RC5.** _____ A pair of angles whose measures add to 90°

**RC6.** _____ A pair of angles whose measures add to 180°

**a)** acute angle

**b)** complementary angles

**c)** straight angle

**d)** supplementary angles

**e)** obtuse angle

**f)** right angle

**a**   Name each angle in six different ways.

**1.**

**2.**

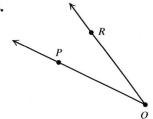

Give another name for ∠1 in each figure.

**3.**

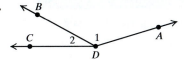

**4.**

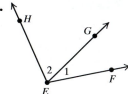

Use a protractor to measure each angle.

**5.**

**6.**

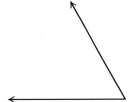

**7.**

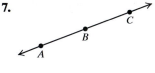

**8.**

**9.**

**10.**

Use the given information and a protractor to draw a circle graph.

**11.** *Credit Score.* The FICO Credit Score is based on five types of information, as shown in the table below.

| PERSONAL INFORMATION | PERCENT |
|---|---|
| Payment history | 35% |
| Debt level | 30% |
| Length of credit history | 15% |
| Credit inquiries | 10% |
| Mix of credit | 10% |

DATA: credit.about.com

**12.** *Snacking Habits.* Some adults snack heavily and some not at all. The table below lists the frequency of snacking of American adults.

| FREQUENCY OF SNACKING | PERCENT |
|---|---|
| Never | 10% |
| Occasionally | 45% |
| Moderately | 35% |
| Heavily | 10% |

DATA: Market Facts for Hershey Foods

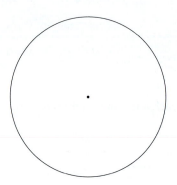

**13.** *Wind Energy.* The table below shows how wind-turbine capacity is distributed among countries.

| COUNTRY | PERCENT |
|---|---|
| China | 26% |
| United States | 20% |
| Germany | 12% |
| Spain | 9% |
| India | 7% |
| Rest of the world | 26% |

DATA: Earth Policy Institute

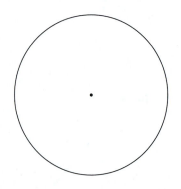

**14.** *Causes of Spinal Cord Injuries.* The table below lists the causes of spinal cord injury.

| CAUSES | PERCENT |
|---|---|
| Motor vehicle accidents | 44% |
| Acts of violence | 24% |
| Falls | 22% |
| Sports | 8% |
| Other | 2% |

DATA: National Spinal Cord Injury Association

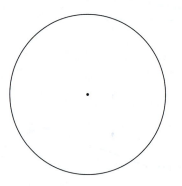

**15.–20.** Classify each of the angles in Exercises 5–10 as right, straight, acute, or obtuse.

**21.–24.** Classify each of the angles in Margin Exercises 1, 2, 4, and 5 as right, straight, acute, or obtuse.

**c**  Identify two pairs of vertical angles for each figure.

**25.**

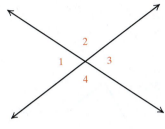

**26.**

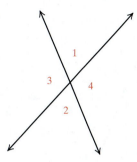

**27.**

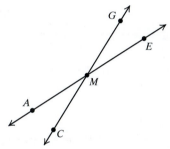

**28.**

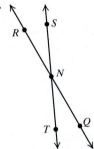

Complete.

**29.** Refer to Exercise 25.

   $m \angle 2 =$ _____

   $m \angle 3 =$ _____

**30.** Refer to Exercise 26.

   $m \angle 4 =$ _____

   $m \angle 2 =$ _____

**31.** Refer to Exercise 27.

   $\angle AMC \cong$ _____

   $\angle AMG \cong$ _____

**32.** Refer to Exercise 28.

   $\angle RNS \cong$ _____

   $\angle TNR \cong$ _____

Find the measure of a complement of an angle with the given measure.

**33.** 11°    **34.** 83°    **35.** 67°    **36.** 5°

**37.** 58°    **38.** 32°    **39.** 29°    **40.** 54°

Find the measure of a supplement of an angle with the given measure.

**41.** 3°    **42.** 54°    **43.** 139°    **44.** 13°

**45.** 75°    **46.** 128°    **47.** 104°    **48.** 49°

**d** Classify each triangle as equilateral, isosceles, or scalene. Then classify it as right, obtuse, or acute.

**49.**

**50.**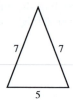

**51.**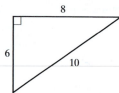

**52.**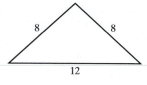

**53.**

**54.**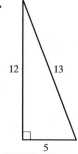

**55.**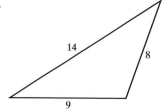

**56.**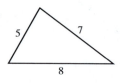

**e** Find each missing angle measure.

**57.**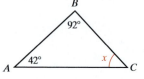

**58.**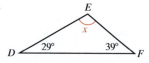

**59.**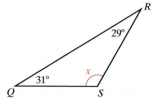

**60.**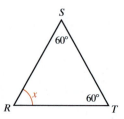

**61.**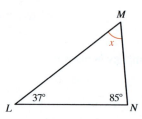

**62.**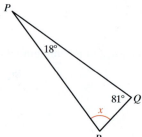

## Skill Maintenance

Find the simple interest. [7.7a]

| | Principal | Rate of Interest | Time | Simple Interest |
|---|---|---|---|---|
| **63.** | $2000 | 8% | 1 year | |
| **64.** | $750 | 6% | $\frac{1}{2}$ year | |
| **65.** | $4000 | 7.4% | $\frac{1}{2}$ year | |
| **66.** | $200,000 | 6.7% | $\frac{1}{12}$ year | |

Interest is compounded semiannually. Find the amount in the account after the given length of time. Round to the nearest cent. [7.7b]

| | Principal | Rate of Interest | Time | Amount in the Account |
|---|---|---|---|---|
| **67.** | $25,000 | 6% | 5 years | |
| **68.** | $150,000 | $6\frac{7}{8}$% | 15 years | |
| **69.** | $150,000 | 7.4% | 20 years | |
| **70.** | $160,000 | 7.4% | 20 years | |

## Synthesis

**71.** ▦ In the figure, $m \angle 1 = 79.8°$ and $m \angle 3 = 33.07°$. Find $m \angle 2$, $m \angle 4$, $m \angle 5$, and $m \angle 6$.

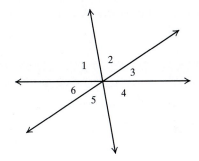

**72.** ▦ In the figure, $m \angle 2 = 42.17°$ and $m \angle 6 = 81.9°$. Find $m \angle 1$, $m \angle 3$, $m \angle 4$, and $m \angle 5$.

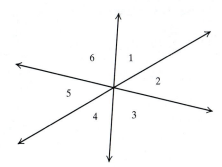

**73.** Find $m \angle ACB$, $m \angle CAB$, $m \angle EBC$, $m \angle EBA$, $m \angle AEB$, and $m \angle ADB$ in the rectangle shown below.

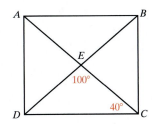

**74.** The angles in the figure are supplementary. Find the measure of each angle.

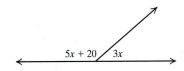

## 9.6

### OBJECTIVES

**a** Simplify square roots of squares, such as $\sqrt{25}$.

**b** Approximate square roots.

**c** Given the lengths of any two sides of a right triangle, find the length of the third side.

**d** Solve applied problems involving right triangles.

# Square Roots and the Pythagorean Theorem

## a SQUARE ROOTS

**SQUARE ROOT**

If a number is a product of a factor times itself, then that factor is a **square root** of the number. (If $c^2 = a$, then $c$ is a square root of $a$.)

The number 36 has two square roots, 6 and $-6$: we have $6 \cdot 6 = 36$ and $(-6) \cdot (-6) = 36$.

**EXAMPLE 1** Find the square roots of 25.

The square roots of 25 are 5 and $-5$, because $5^2 = 25$ and $(-5)^2 = 25$. ■

·················· **Caution!** ··················

To find the *square* of a number, multiply the number by itself. To find a *square root* of a number, find a number that, when squared, gives the original number. For example, $16^2 = 16 \cdot 16 = 256$, and $\sqrt{16} = 4$ because $4 \cdot 4 = 16$.

◀ **Do Exercises 1–12.**

Every positive number has two square roots. However, the symbol $\sqrt{\phantom{n}}$ (called a **radical sign**) represents only the positive square root of the number underneath. Thus, $\sqrt{9}$ means 3, not $-3$.

**RADICAL SIGN**

If $n$ is a positive number, $\sqrt{n}$ means the positive square root of $n$. "The" square root of $n$ means $\sqrt{n}$.

**EXAMPLES** Simplify.

**2.** $\sqrt{36} = 6$    The square root of 36 is 6 because $6^2 = 36$ and 6 is positive.

**3.** $\sqrt{144} = 12$    Note that $12^2 = 144$.

◀ **Do Exercises 13–22.**

---

Find each square.

**1.** $9^2$

**2.** $10^2$

**3.** $11^2$

**4.** $12^2$

**5.** $13^2$

**6.** $14^2$

**7.** $15^2$

**8.** $16^2$

Find all square roots. Use the results of Exercises 1–8 above, if necessary.

**9.** 100

**10.** 81

**11.** 49

**12.** 196

Simplify. Use the results of Exercises 1–8 above, if necessary.

**13.** $\sqrt{49}$

**14.** $\sqrt{16}$

**15.** $\sqrt{121}$

**16.** $\sqrt{100}$

**17.** $\sqrt{81}$

**18.** $\sqrt{64}$

**19.** $\sqrt{225}$

**20.** $\sqrt{169}$

**21.** $\sqrt{1}$

**22.** $\sqrt{0}$

*Answers*

**1.** 81  **2.** 100  **3.** 121  **4.** 144
**5.** 169  **6.** 196  **7.** 225  **8.** 256
**9.** $-10, 10$  **10.** $-9, 9$  **11.** $-7, 7$
**12.** $-14, 14$  **13.** 7  **14.** 4  **15.** 11
**16.** 10  **17.** 9  **18.** 8  **19.** 15
**20.** 13  **21.** 1  **22.** 0

## b APPROXIMATING SQUARE ROOTS

Many square roots can't be written as whole numbers or fractions. For example, $\sqrt{2}$, $\sqrt{3}$, $\sqrt{39}$, and $\sqrt{70}$ cannot be precisely represented in decimal notation. To see this, consider the following decimal approximations for $\sqrt{2}$. Each gives a closer approximation, but none is exactly $\sqrt{2}$:

$\sqrt{2} \approx 1.4$     because    $(1.4)^2 = 1.96$;

$\sqrt{2} \approx 1.41$     because    $(1.41)^2 = 1.9881$;

$\sqrt{2} \approx 1.414$     because    $(1.414)^2 = 1.999396$;

$\sqrt{2} \approx 1.4142$     because    $(1.4142)^2 = 1.99996164$.

Decimal approximations like these are commonly found by using a calculator.

**EXAMPLE 4** Use a calculator to approximate $\sqrt{3}$, $\sqrt{27}$, and $\sqrt{180}$ to three decimal places.

We use a calculator to find each square root. Since the calculator displays more than three decimal places, we round to three places.

$$\sqrt{3} \approx 1.732, \quad \sqrt{27} \approx 5.196, \quad \sqrt{180} \approx 13.416$$

As a check, note that $1 \cdot 1 = 1$ and $2 \cdot 2 = 4$, so we expect $\sqrt{3}$ to be between 1 and 2. Similarly, we expect $\sqrt{27}$ to be between 5 and 6 and $\sqrt{180}$ to be between 13 and 14.

**Do Exercises 23–26.** ▶

## c THE PYTHAGOREAN THEOREM

**SKILL REVIEW**

*Evaluate exponential notation.*    [1.9b]

Evaluate.

**1.** $5^2$                                **2.** $8^2$

Answers: **1.** 25   **2.** 64

MyLab Math
**VIDEO**

A **right triangle** is a triangle with a 90° angle, as shown here. In a right triangle, the longest side is called the **hypotenuse**. It is the side opposite the right angle. The other two sides are called **legs**. We generally use the letters $a$ and $b$ for the lengths of the legs and $c$ for the length of the hypotenuse. They are related as follows.

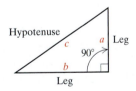

---

**THE PYTHAGOREAN THEOREM**

In any right triangle, if $a$ and $b$ are the lengths of the legs and $c$ is the length of the hypotenuse, then

$$a^2 + b^2 = c^2, \quad \text{or}$$

$$(\text{Leg})^2 + (\text{Other leg})^2 = (\text{Hypotenuse})^2.$$

The equation $a^2 + b^2 = c^2$ is called the **Pythagorean equation**.*

---

\* The *converse* of the Pythagorean theorem is also true. That is, if $a^2 + b^2 = c^2$, then the triangle is a right triangle.

Use a calculator to approximate to three decimal places.

**23.** $\sqrt{5}$              **24.** $\sqrt{78}$

**25.** $\sqrt{168}$         **26.** $\sqrt{321}$

**Answers**

**23.** 2.236   **24.** 8.832   **25.** 12.961

**26.** 17.916

The Pythagorean theorem is named for the Greek mathematician Pythagoras (569?–500? B.C.E.). We can think of this relationship as adding areas.

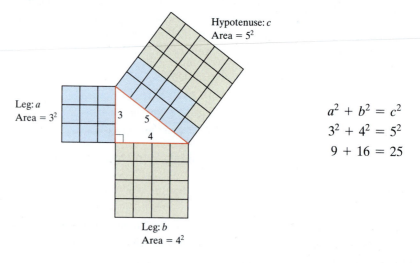

Hypotenuse: $c$
Area $= 5^2$

Leg: $a$
Area $= 3^2$

Leg: $b$
Area $= 4^2$

$$a^2 + b^2 = c^2$$
$$3^2 + 4^2 = 5^2$$
$$9 + 16 = 25$$

If we know the lengths of any two sides of a right triangle, we can use the Pythagorean equation to determine the length of the third side.

**EXAMPLE 5** Find the length of the hypotenuse of this right triangle.

We substitute in the Pythagorean equation:

$$a^2 + b^2 = c^2$$
$$6^2 + 8^2 = c^2 \qquad \text{Substituting}$$
$$36 + 64 = c^2$$
$$100 = c^2.$$

The solution of this equation is the square root of 100, which is 10:

$$c = \sqrt{100} = 10.$$

◀ **Do Exercise 27.**

·········· **Caution!** ··········

Before applying the Pythagorean theorem, make sure you have a right triangle. Then determine which side of the triangle is the hypotenuse, or $c$. The hypotenuse is always the longest side of a right triangle and is always opposite the right angle.

**EXAMPLE 6** Find the length $b$ for the right triangle shown. Give an exact answer and an approximation to three decimal places.

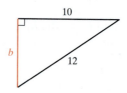

Recall that the leg opposite the right angle is the hypotenuse. Thus, for this triangle, $c = 12$. We substitute in the Pythagorean equation:

$$a^2 + b^2 = c^2$$
$$10^2 + b^2 = 12^2 \qquad \text{Substituting}$$
$$100 + b^2 = 144.$$

**27.** Find the length of the hypotenuse **GS** of this right triangle.

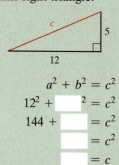

$$a^2 + b^2 = c^2$$
$$12^2 + \boxed{\phantom{x}}^2 = c^2$$
$$144 + \boxed{\phantom{x}} = c^2$$
$$\boxed{\phantom{x}} = c^2$$
$$\boxed{\phantom{x}} = c$$

**Answer**

**27.** $c = 13$

**Guided Solution:**

**27.** 5, 25, 169, 13

Next, we solve for $b^2$ and then $b$:

$$100 + b^2 - 100 = 144 - 100$$  Subtracting 100 from both sides
$$b^2 = 144 - 100$$
$$b^2 = 44$$  Solving for $b^2$

*Exact answer*:  $b = \sqrt{44}$  Solving for $b$
*Approximation*: $b \approx 6.633$.  Using a calculator

**Do Exercises 28–30.** ▶

## d | APPLICATIONS

**EXAMPLE 7** *Height of Ladder.* A 12-ft ladder leans against a building. The bottom of the ladder is 7 ft from the building. How high is the top of the ladder? Give an exact answer and an approximation to the nearest tenth of a foot.

1. **Familiarize.** We first make a drawing. In it we see a right triangle. We let $h =$ the unknown height. We also note that the ladder forms the hypotenuse of the triangle, so $c = 12$.

2. **Translate.** We substitute 7 for $a$, $h$ for $b$, and 12 for $c$ in the Pythagorean equation:

$$a^2 + b^2 = c^2$$  Pythagorean equation
$$7^2 + h^2 = 12^2.$$

3. **Solve.** We solve for $h^2$ and then $h$:

$$49 + h^2 = 144$$  $7^2 = 49$ and $12^2 = 144$
$$49 + h^2 - 49 = 144 - 49$$  Subtracting 49 from both sides
$$h^2 = 144 - 49$$
$$h^2 = 95$$

*Exact answer*:  $h = \sqrt{95}$  Solving for $h$
*Approximation*:  $h \approx 9.7$ ft.

4. **Check.**  $7^2 + (\sqrt{95})^2 = 49 + 95 = 144 = 12^2$.
5. **State.** The top of the ladder is $\sqrt{95}$ ft, or about 9.7 ft, from the ground.

**Do Exercise 31.** ▶

For each right triangle, find the length of the leg not given. Give an exact answer and an approximation to three decimal places.

28.

29.

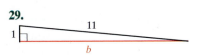

30.

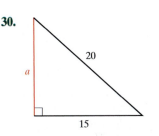

31. How long is a guy wire reaching from the top of an 18-ft pole to a point on the ground 10 ft from the pole? Give an exact answer and an approximation to the nearest tenth of a foot.

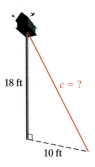

**Answers**
28. $a = \sqrt{75}$; $a \approx 8.660$  29. $b = \sqrt{120}$; $b \approx 10.954$  30. $a = \sqrt{175}$; $a \approx 13.229$
31. $\sqrt{424}$ ft $\approx 20.6$ ft

# Translating for Success

1. **Servings of Pork.** An 8-lb pork roast contains 37 servings of meat. How many pounds of pork would be needed for 55 servings?

2. **Height of a Ladder.** A 14.5-ft ladder leans against a house. The bottom of the ladder is 9.4 ft from the house. How high is the top of the ladder?

3. **Cruise Cost.** A group of 6 college students pays $4608 for a spring break cruise. What is each person's share?

4. **Sales Tax Rate.** The sales tax is $14.95 on the purchase of a new ladder that costs $299. What is the sales tax rate?

5. **Volume of a Sphere.** Find the volume of a sphere whose radius is 7.2 cm.

*The goal of these matching questions is to practice step (2), Translate, of the five-step problem-solving process. Translate each word problem to an equation and select a correct translation from equations A–O.*

A. $x = 17.2 \text{ km} \cdot \dfrac{1000 \text{ m}}{1 \text{ km}}$

B. $6 \cdot x = \$4608$

C. $x = \dfrac{4}{3} \cdot \pi \cdot 6^2 \cdot (7.2)$

D. $x = \pi \cdot \left(5\dfrac{1}{2} \div 2\right)^2 \cdot 7$

E. $x = 6\% \times 5 \times \$14.95$

F. $x = \pi \cdot 5\dfrac{1}{2} \cdot 7$

G. $(9.4)^2 + x^2 = (14.5)^2$

H. $\$14.95 = x \cdot \$299$

I. $x = 2(14.5 + 9.4)$

J. $(9.4 + 14.5)^2 = x$

K. $\dfrac{8}{37} = \dfrac{x}{55}$

L. $x = 17.2 \text{ km} \cdot \dfrac{1 \text{ km}}{1000 \text{ m}}$

M. $x = 6 \cdot \$4608$

N. $8 \cdot 37 = 55 \cdot x$

O. $x = \dfrac{4}{3} \cdot \pi \cdot (7.2)^3$

*Answers on page A-22*

6. **Inheritance.** Six children each inherit $4608 from their mother's estate. What is the total inheritance?

7. **Sales Tax.** Erica buys 5 pairs of earrings at $14.95 each. The sales tax rate is 6%. How much sales tax will be charged?

8. **Bridge Length.** The Vasco Da Gama Bridge in Portugal is 17.2 km long. Convert this distance to meters.

9. **Volume of a Storage Tank.** The diameter of a cylindrical grain-storage tank is $5\frac{1}{2}$ yd. Its height is 7 yd. Find its volume.

10. **Perimeter of a Photo.** A rectangular photo is 14.5 cm by 9.4 cm. What is the perimeter of the photo?

## ✓ Check Your Understanding

**Reading Check**  Determine whether each statement is true or false.

_____ **RC1.** 10 is a square root of 100.

_____ **RC2.** $\sqrt{3} = 9$.

_____ **RC3.** In a right triangle, the side opposite the right angle is the hypotenuse.

_____ **RC4.** In a right triangle, the sum of the lengths of the legs is the length of the hypotenuse.

**Concept Check**  In Exercises CC1–CC4, the lengths of the three sides of a right triangle are given. Identify the lengths of the legs and the length of the hypotenuse.

**CC1.** 30, 34, 16

**CC2.** 6, $\sqrt{37}$, 1

**CC3.** $2\sqrt{22}, \sqrt{11}, \sqrt{77}$

**CC4.** 10, 16, $2\sqrt{39}$

**a**  Find both square roots for each number listed.

**1.** 16

**2.** 9

**3.** 121

**4.** 49

**5.** 169

**6.** 144

**7.** 2500

**8.** 3600

Simplify.

**9.** $\sqrt{64}$

**10.** $\sqrt{4}$

**11.** $\sqrt{81}$

**12.** $\sqrt{49}$

**13.** $\sqrt{225}$

**14.** $\sqrt{121}$

**15.** $\sqrt{625}$

**16.** $\sqrt{900}$

**17.** $\sqrt{400}$

**18.** $\sqrt{169}$

**19.** $\sqrt{10,000}$

**20.** $\sqrt{1,000,000}$

**b**  Approximate each number to the nearest thousandth. Use a calculator.

**21.** $\sqrt{48}$

**22.** $\sqrt{17}$

**23.** $\sqrt{8}$

**24.** $\sqrt{7}$

**25.** $\sqrt{3}$

**26.** $\sqrt{6}$

**27.** $\sqrt{12}$

**28.** $\sqrt{18}$

**29.** $\sqrt{19}$

**30.** $\sqrt{75}$

**31.** $\sqrt{110}$

**32.** $\sqrt{10}$

**c**  Find the length of the third side of each right triangle. Give an exact answer and, when appropriate, an approximation to the nearest thousandth.

**33.**

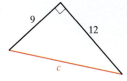

9, 12, $c$

**34.**

8, 15, $c$

**35.**

7, 7, $c$

**36.**

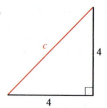

**37.**

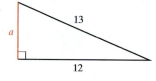

**38.**

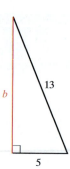

**39.**

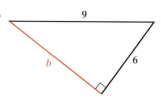

**40.**

For each right triangle, find the length of the side not given. Assume that $c$ represents the length of the hypotenuse. Give an exact answer and, when appropriate, an approximation to the nearest thousandth.

**41.** $a = 10, b = 24$

**42.** $a = 5, b = 12$

**43.** $a = 9, c = 15$

**44.** $a = 18, c = 30$

**45.** $a = 4, b = 5$

**46.** $a = 5, b = 6$

**47.** $a = 1, c = 32$

**48.** $b = 1, c = 20$

**d**  In Exercises 49–56, give an exact answer and an approximation to the nearest tenth.

**49.** A 30-ft string of lights reaches from the top of a pole to a point on the ground 16 ft from the base of the pole. How tall is the pole?

**50.** A 25-ft wire reaches from the top of a telephone pole to a point on the ground 18 ft from the base of the pole. How tall is the pole?

**51.** *Softball Diamond.*  A slow-pitch softball diamond is actually a square 65 ft on a side. How far is it from home plate to second base?

**52.** *Baseball Diamond.*  A baseball diamond is actually a square 90 ft on a side. How far is it from home plate to second base?

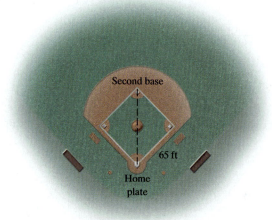

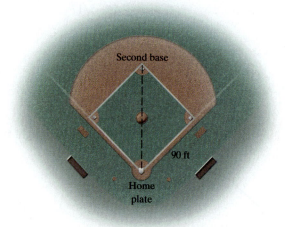

**53. *Great Pyramid.*** The Pyramid of Cheops is 146 m high. The distance at ground level from the center of the pyramid to the middle of one of the faces is 115 m, as shown below. What is the slant height of a side of the pyramid?

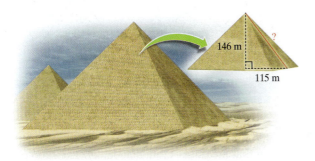

**54. *Construction.*** In order to support a masonry wall, Matthew erects braces at a height of 12 ft on the wall. The braces are anchored to the ground 15 ft from the base of the wall. How long are the braces?

**55.** An airplane is flying at an altitude of 4100 ft. The slanted distance directly to the airport is 15,100 ft. How far is the airplane horizontally from the airport?

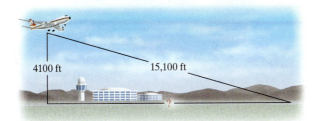

**56.** A surveyor had poles located at points *P*, *Q*, and *R* around a lake. The distances that the surveyor was able to measure are marked on the drawing. What is the distance from *P* to *R* across the lake?

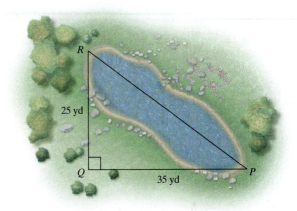

## Skill Maintenance

Solve.

**57.** Food expenses account for 26% of the average family's budget. A family makes $1800 one month. How much do they spend for food? [7.4a]

**58.** Blakely County has a population that is increasing by 4% each year. This year the population is 180,000. What will it be next year? [7.5a]

**59.** Dexter College has a student body of 1850 students. Of these, 17.5% are seniors. How many students are seniors? [7.4a]

**60.** The price of a cell phone was reduced from $70 to $61.60. Find the percent decrease in price. [7.5a]

Simplify. [1.9b]

**61.** $2^3$

**62.** $5^3$

**63.** $10^3$

**64.** $10^4$

Simplify. [1.9c, d]

**65.** $90 \div 15 \cdot 2 - (1 + 2)^2$

**66.** $10^3 - \{2 \times [5 \times 3 - (4 + 2)]\}$

**67.** ▦ Find the area of the trapezoid shown. Round to the nearest hundredth.

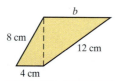

**68.** Which of the triangles below has the larger area? If the areas are the same, state so.

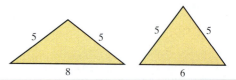

**69.** ▦ Caiden's new TV has a screen that measures $31\frac{3}{4}$ in. by $56\frac{1}{2}$ in. Determine the diagonal measurement of the screen. Round your answer to the nearest tenth of an inch.

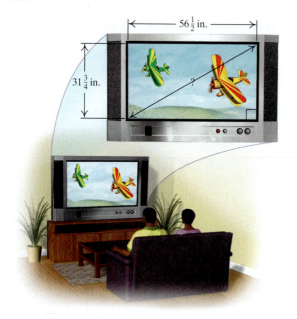

**70.** A Philips 42-in. plasma television has a rectangular screen that measures 42 in. diagonally. The ratio of width to height is 16 to 9. Find the width and the height of the screen.

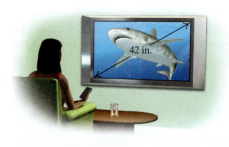

**71.** A conventional 19-in. television set has a rectangular screen that measures 19 in. diagonally. The ratio of width to height in a conventional television set is 4 to 3. Find the width and the height of the screen.

**72.** ▦ A cube is circumscribed by a sphere with a 1-m diameter. How much more volume is in the sphere?

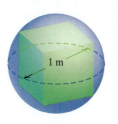

# Weight, Mass, and Temperature

## a WEIGHT: THE AMERICAN SYSTEM

**OBJECTIVES**

**a** Convert from one American unit of weight to another.

**b** Convert from one metric unit of mass to another.

**c** Convert temperatures from Celsius to Fahrenheit and from Fahrenheit to Celsius.

**SKILL REVIEW**

*Multiply an integer and a fraction.* [3.4a]
Multiply.

**1.** $\frac{3}{4} \cdot 16$

**2.** $2480 \times \frac{1}{2000}$

Answers: **1.** 12 **2.** $\frac{31}{25}$, or 1.24

MyLab Math
VIDEO

The American units of weight are as follows.

| AMERICAN UNITS OF WEIGHT |
|---|
| 1 lb = 16 ounces (oz) |
| 1 ton (T) = 2000 pounds (lb) |

The term "ounce" used here for weight is different from the "ounce" we used for capacity in Section 9.4.

**EXAMPLE 1** A well-known hamburger is called a "quarter-pounder." Find its name in ounces: a " _____ ouncer."

$$\frac{1}{4} \text{ lb} = \frac{1}{4} \cdot 1 \text{ lb}$$

$$= \frac{1}{4} \cdot 16 \text{ oz} \qquad \text{Substituting 16 oz for 1 lb}$$

$$= 4 \text{ oz}$$

A "quarter-pounder" could also be called a "four-ouncer."

**EXAMPLE 2** Complete: 15,360 lb = _____ T.

$$15{,}360 \text{ lb} = 15{,}360 \text{ lb} \cdot \frac{1 \text{ T}}{2000 \text{ lb}} \qquad \text{Multiplying by 1}$$

$$= \frac{15{,}360}{2000} \text{ T}$$

$$= 7.68 \text{ T} \qquad \text{Dividing by 2000}$$

**Do Exercises 1–3.** ▶

Complete.

**1.** 5 lb = _____ oz

**2.** 8640 lb = _____ T

**3.** 1 T = _____ oz

## b MASS: THE METRIC SYSTEM

There is a difference between **mass** and **weight**, but the terms are often used interchangeably. People sometimes use the word "weight" when, technically, they are referring to "mass." Weight is related to the force of the earth's gravity. The farther you are from the center of the earth, the less you weigh. Your mass, on the other hand, stays the same no matter where you are.

*Answers*
**1.** 80 **2.** 4.32 **3.** 32,000

## METRIC UNITS OF MASS

1 metric ton (t) = 1000 kilograms (kg)

1 *kilo*gram (kg) = 1000 grams (g)

1 *hecto*gram (hg) = 100 grams (g)

1 *deka*gram (dag) = 10 grams (g)

1 gram (g)

1 *deci*gram (dg) = $\frac{1}{10}$ gram (g)

1 *centi*gram (cg) = $\frac{1}{100}$ gram (g)

1 *milli*gram (mg) = $\frac{1}{1000}$ gram (g)

The basic unit of mass is the **gram** (g), which is the mass of 1 cubic centimeter (1 cm³) of water. Since a cubic centimeter is small, a gram is a small unit of mass.

$$1\text{ g} = 1\text{ gram} = \text{the mass of 1 cm}^3 \text{ of water}$$

1 g = 1 cm³ of water

The metric units of mass are listed to the left. The prefixes are the same as those for length.

### Thinking Metric

One gram is about the mass of 1 raisin or 1 package of artificial sweetener. Since 1 kg is about 2.2 lb, 1000 kg is about 2200 lb, or 1 metric ton (t), which is about 10% more than 1 American ton (T), which is 2000 lb.

1 gram

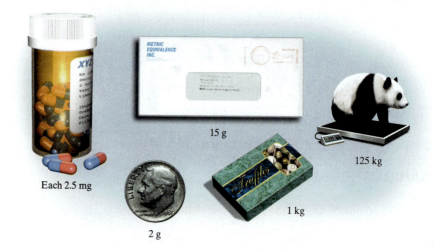

1 kilogram of grapes          1 pound of grapes

Small masses, such as dosages of medicine and vitamins, may be measured in milligrams (mg). The gram (g) is used for objects ordinarily measured in ounces, such as the mass of a letter, a piece of candy, or a coin.

Each 2.5 mg          2 g          15 g          1 kg          125 kg

Complete with mg, g, kg, or t.

**4.** A laptop computer has a mass of 6 _____ .

**5.** Eric has a body mass of 85.4 _____ .

**6.** This is a 3- _____ vitamin.

**7.** A pen has a mass of 12 _____.

**8.** A sport utility vehicle has a mass of 3 _____.

The kilogram (kg) is used for larger food packages and for body masses. The metric ton (t) is used for very large masses, such as the mass of an automobile, a truckload of gravel, or an airplane.

◀ **Do Exercises 4–8.**

*Answers*

**4.** kg   **5.** kg   **6.** mg   **7.** g   **8.** t

## Changing Units Mentally

As before, changing from one metric unit to another requires only the movement of a decimal point. We use this table.

| 1000 g | 100 g | 10 g | 1 g | 0.1 g | 0.01 g | 0.001 g |
|--------|-------|------|-----|-------|--------|---------|
| 1 kg | 1 hg | 1 dag | 1 g | 1 dg | 1 cg | 1 mg |

**EXAMPLE 3**  Complete: 8 kg = _____ g.

*Think:* A kilogram is 1000 times the mass of a gram. To go from kg to g in the table is a move of three places to the right. Thus, we move the decimal point three places to the right.

| 1000 g | 100 g | 10 g | 1 g | 0.1 g | 0.01 g | 0.001 g |
|--------|-------|------|-----|-------|--------|---------|
| 1 kg | 1 hg | 1 dag | 1 g | 1 dg | 1 cg | 1 mg |

3 places to the right

8.0    8.000.    8 kg = 8000 g

**EXAMPLE 4**  Complete: 4235 g = _____ kg.

*Think:* There are 1000 grams in 1 kilogram. To go from g to kg in the table is a move of three places to the left. Thus, we move the decimal point three places to the left.

| 1000 g | 100 g | 10 g | 1 g | 0.1 g | 0.01 g | 0.001 g |
|--------|-------|------|-----|-------|--------|---------|
| 1 kg | 1 hg | 1 dag | 1 g | 1 dg | 1 cg | 1 mg |

3 places to the left

4235.0    4.235.0    4235 g = 4.235 kg

**EXAMPLE 5**  Complete: 6.98 cg = _____ mg.

*Think:* One centigram has the mass of 10 milligrams. To go from cg to mg is a move of one place to the right. Thus, we move the decimal point one place to the right.

| 1000 g | 100 g | 10 g | 1 g | 0.1 g | 0.01 g | 0.001 g |
|--------|-------|------|-----|-------|--------|---------|
| 1 kg | 1 hg | 1 dag | 1 g | 1 dg | 1 cg | 1 mg |

1 place to the right

6.98    6.9.8    6.98 cg = 69.8 mg

The most commonly used metric units of mass are kg, g, and mg. We have intentionally used those more often than the others in the exercises.

**Do Exercises 9–12.** ▶

Complete.

**9.** 6.2 kg = _____ g

**10.** 304.8 cg = _____ g

**11.** 7.7 cg = _____ mg

**12.** 2344 mg = _____ cg

*Answers*

**9.** 6200  **10.** 3.048  **11.** 77  **12.** 234.4

## C TEMPERATURE

### Estimated Conversions

Below are two temperature scales: **Fahrenheit** for American measure and **Celsius**, used internationally and in science.

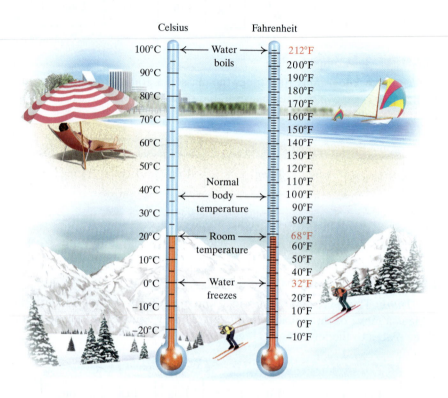

By laying a straightedge horizontally between the scales, we can make an approximate conversion from one measure of temperature to another and get an idea of how the temperature scales compare.

**EXAMPLES** Convert to Celsius using the scales shown above. Approximate to the nearest ten degrees.

**6.** 212°F (Boiling point of water)     100°C     This is exact.
**7.** 32°F (Freezing point of water)     0°C     This is exact.
**8.** 105°F     40°C     This is approximate.

◀ **Do Exercises 13–15.**

**EXAMPLES** Make an approximate conversion to Fahrenheit using the scales shown above.

**9.** 44°C (Hot bath)     110°F     This is approximate.
**10.** 20°C (Room temperature)     68°F     This is exact.
**11.** 83°C     180°F     This is approximate.

◀ **Do Exercises 16–18.**

Convert to Celsius. Use a straightedge and the scales shown on this page. Approximate to the nearest ten degrees.

**13.** 180°F (Brewing coffee)

**14.** 25°F (Cold day)

**15.** −10°F (Miserably cold day)

Convert to Fahrenheit. Use a straightedge and the scales shown on this page. Approximate to the nearest ten degrees.

**16.** 25°C (Nice day at the park)

**17.** 40°C (Temperature of a patient with a high fever)

**18.** 10°C (Cold bath)

*Answers*

**13.** 80°C **14.** 0°C **15.** −20°C
**16.** 80°F **17.** 100°F **18.** 50°F

## Exact Conversions

A formula allows us to make exact conversions from Celsius to Fahrenheit.

Convert to Fahrenheit.

**19.** 80°C

**20.** 35°C

> ### CELSIUS TO FAHRENHEIT
>
> $$F = \frac{9}{5} \cdot C + 32, \quad \text{or} \quad F = 1.8 \cdot C + 32$$
>
> $$\left(\text{Multiply the Celsius temperature by } \frac{9}{5}, \text{ or } 1.8, \text{ and add 32.}\right)$$

**EXAMPLES**   Convert to Fahrenheit.

**12.** 0°C (Freezing point of water)

$$F = \frac{9}{5} \cdot C + 32 = \frac{9}{5} \cdot 0 + 32 = 0 + 32 = 32$$

Thus, 0°C = 32°F.

**13.** 37°C (Normal body temperature)

$$F = 1.8 \cdot C + 32 = 1.8 \cdot 37 + 32 = 66.6 + 32 = 98.6$$

Thus, 37°C = 98.6°F.

Check the answers to Examples 12 and 13 using the scales on p. 670.

**Do Exercises 19 and 20.** ▶

A second formula allows us to make exact conversions from Fahrenheit to Celsius.

> ### FAHRENHEIT TO CELSIUS
>
> $$C = \frac{5}{9} \cdot (F - 32), \quad \text{or} \quad C = \frac{F - 32}{1.8}$$
>
> $$\left(\text{Subtract 32 from the Fahrenheit temperature and multiply by } \frac{5}{9} \text{ or}\right.$$
> $$\left.\text{divide by 1.8.}\right)$$

**EXAMPLES**   Convert to Celsius.

**14.** 212°F (Boiling point of water)

$$C = \frac{5}{9} \cdot (F - 32) = \frac{5}{9} \cdot (212 - 32) = \frac{5}{9} \cdot 180 = 100$$

Thus, 212°F = 100°C.

**15.** 77°F

$$C = \frac{F - 32}{1.8} = \frac{77 - 32}{1.8} = \frac{45}{1.8} = 25$$

Thus, 77°F = 25°C.

Check the answers to Examples 14 and 15 using the scales on p. 670.

**Do Exercises 21 and 22.** ▶

### CALCULATOR CORNER

***Temperature Conversions***
Temperature conversions can be done quickly using a calculator. To convert 37°C to Fahrenheit, for example, we press ⬚1⬚ ⬚·⬚ ⬚8⬚ ⬚×⬚ ⬚3⬚ ⬚7⬚ ⬚+⬚ ⬚3⬚ ⬚2⬚ ⬚=⬚. The calculator displays ⬚98.6⬚ , so 37°C = 98.6°F. We can convert 212°F to Celsius by pressing ⬚(⬚ ⬚2⬚ ⬚1⬚ ⬚2⬚ ⬚−⬚ ⬚3⬚ ⬚2⬚ ⬚)⬚ ⬚÷⬚ ⬚1⬚ ⬚·⬚ ⬚8⬚ ⬚=⬚. The display reads ⬚100⬚ , so 212°F = 100°C. Note that we must use parentheses when converting from Fahrenheit to Celsius in order to get the correct result.

**EXERCISES:**   Use a calculator to convert each temperature to Fahrenheit.

**1.** 5°C

**2.** 50°C

Use a calculator to convert each temperature to Celsius.

**3.** 68°F

**4.** 113°F

Convert to Celsius.

**GS 21.** 95°F

$$C = \frac{5}{9}(F - 32)$$

$$= \frac{5}{9}(\boxed{\phantom{000}} - 32)$$

$$= \frac{5}{9} \cdot \boxed{\phantom{000}} = \boxed{\phantom{000}}$$

Thus, 95°F = $\boxed{\phantom{00}}$ °C.

**22.** 113°F

***Answers***
**19.** 176°F  **20.** 95°F  **21.** 35°C  **22.** 45°C
***Guided Solution:***
**21.** 95, 63, 35; 35

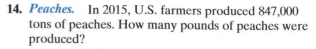
## ✓ Check Your Understanding

**Reading Check**   Determine whether each statement is true or false.

**RC1.** _____ 400 g > 40 dg    **RC2.** _____ 5 hg < 400 g    **RC3.** _____ 0.5 kg = 500 g

**RC4.** _____ 48 oz = 4 lb    **RC5.** _____ 7500 lb < 3.5 T    **RC6.** _____ 800 cg > 6 g

**Concept Check**   Select the temperature in each pair that is higher.

**CC1.** −10°C, −10°F                  **CC2.** 80°C, 200°F

**CC3.** 100°F, 30°C                  **CC4.** 0°F, 0°C

---

**a**   Complete.

**1.** 1 T = _____ lb        **2.** 1 lb = _____ oz        **3.** 6000 lb = _____ T

**4.** 8 T = _____ lb        **5.** 4 lb = _____ oz        **6.** 10 lb = _____ oz

**7.** 6.32 T = _____ lb        **8.** 8.07 T = _____ lb        **9.** 3200 oz = _____ T

**10.** 6400 oz = _____ T        **11.** 80 oz = _____ lb        **12.** 960 oz = _____ lb

**13.** *Pecans.*   In 2015, U.S. farmers produced 254,290,000 pounds of pecans. How many tons of pecans were produced?

**Data:** National Agricultural Statistics Service, U.S. Department of Agriculture

**14.** *Peaches.*   In 2015, U.S. farmers produced 847,000 tons of peaches. How many pounds of peaches were produced?

**Data:** National Agricultural Statistics Service, U.S. Department of Agriculture

**b**   Complete.

**15.** 1 kg = _____ g        **16.** 1 hg = _____ g        **17.** 1 g = _____ kg

**18.** 1 dg = _____ g        **19.** 1 cg = _____ g        **20.** 1 mg = _____ g

**21.** 1 g = _____ mg

**22.** 1 g = _____ cg

**23.** 1 g = _____ dg

**24.** 25 kg = _____ g

**25.** 234 kg = _____ g

**26.** 9403 g = _____ kg

**27.** 5200 g = _____ kg

**28.** 1.506 kg = _____ g

**29.** 897 mg = _____ kg

**30.** 45 cg = _____ g

**31.** 7.32 kg = _____ g

**32.** 0.0025 cg = _____ mg

**33.** 8492 g = _____ kg

**34.** 9466 g = _____ kg

**35.** 585 mg = _____ cg

**36.** 96.1 mg = _____ cg

**37.** 8 kg = _____ cg

**38.** 0.06 kg = _____ mg

**39.** 1 t = _____ kg

**40.** 2 t = _____ kg

**41.** 3.4 cg = _____ dag

**42.** 115 mg = _____ g

**43.** 60.3 kg = _____ t

**44.** 15.68 kg = _____ t

**C** Convert to Celsius. Round the answer to the nearest ten degrees. Use the scales on p. 670.

**45.** 178°F

**46.** 195°F

**47.** 140°F

**48.** 107°F

**49.** 68°F

**50.** 45°F

**51.** 10°F

**52.** 120°F

Convert to Fahrenheit. Round the answer to the nearest ten degrees. Use the scales on p. 670.

**53.** 80°C

**54.** 93°C

**55.** 58°C

**56.** 33°C

**57.** −10°C

**58.** −5°C

**59.** 5°C

**60.** 15°C

Convert to Fahrenheit. Use the formula $F = \frac{9}{5} \cdot C + 32$, or $F = 1.8C + 32$.

**61.** 30°C    **62.** 85°C    **63.** 40°C    **64.** 90°C

**65.** 2°C    **66.** 8°C    **67.** −1°C    **68.** −15°C

**69.** 3000°C (The melting point of iron)    **70.** 1000°C (The melting point of gold)

Convert to Celsius. Use the formula $C = \frac{5}{9} \cdot (F - 32)$ or $C = \dfrac{F - 32}{1.8}$.

**71.** 77°F    **72.** 59°F    **73.** 131°F    **74.** 140°F

**75.** 178°F    **76.** 110°F    **77.** 5°F    **78.** −4°F

**79.** 98.6°F (Normal body temperature)    **80.** 104°F (High-fevered body temperature)

**81.** *Highest Temperatures.* The highest temperature ever recorded in the world is 56.7°C in Furnace Creek (Death Valley), California, on July 10, 1913. The highest temperature ever recorded in Africa is 131.0°F in Kebili, Tunisia, on July 7, 1931.

**Data:** infoplease.com

**a)** Convert each temperature to the other scale.
**b)** How much higher in degrees Fahrenheit was the world record than the African record?

**82.** *Boiling Point and Altitude.* The boiling point of water actually changes with altitude. The boiling point is 212°F at sea level, but lowers about 1°F for every 500 ft of increase in altitude above sea level.

**Data:** *The Handy Geography Answer Book; The New York Times Almanac*

**a)** What is the boiling point at an elevation of 1500 ft above sea level?
**b)** The elevation of Tucson is 2564 ft above sea level and that of Phoenix is 1117 ft. What is the boiling point in each city?
**c)** How much lower is the boiling point in Denver, whose elevation is 5280 ft, than in Tucson?
**d)** What is the boiling point at the top of Mt. McKinley (Denali) in Alaska, the highest point in the United States, at 20,320 ft?

## Skill Maintenance

Graph.  [8.4b]

**83.** $y = -3x$

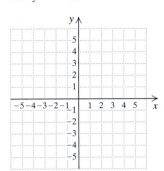

**84.** $y = \frac{1}{3}x$

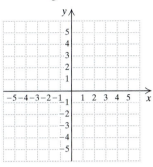

**85.** $y = x + 1$

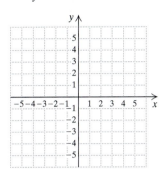

**86.** $y = 3x - 4$

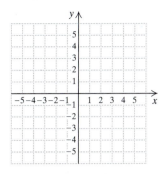

Solve.  [1.8a]

**87.** *Morel Mushrooms.*   During the spring, Kate's Country Market sold 43 pounds of fresh morel mushrooms. The mushrooms sold for $22 a pound. Find the total amount Kate took in from the sale of the mushrooms.

**88.** Sandy can type 62 words per minute. How long will it take her to type 12,462 words?

## Synthesis

**89.** A box of gelatin-mix packages weighs $15\frac{3}{4}$ lb. Each package weighs $1\frac{3}{4}$ oz. How many packages are in the box?

**90.** At $1.59 a dozen, the cost of eggs is $1.06 per pound. How much does an egg weigh?

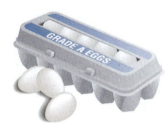

**91.** Use the formula $F = \frac{9}{5} \cdot C + 32$ to find the temperature that is the same for both the Fahrenheit and Celsius scales.

**92.** Which represents a bigger change in temperature: a drop of 5°F or a drop of 5°C Why?

**93. *Chemistry.*** Another temperature scale often used is the **Kelvin** scale. Conversions from Celsius to Kelvin can be carried out using the formula

$$K = C + 273.$$

A chemistry textbook describes an experiment in which a reaction takes place at a temperature of 400 Kelvin. A student wishes to perform the experiment, but has only a Fahrenheit thermometer. At what Fahrenheit temperature will the reaction take place?

**94.** ▦ A large egg is about $5\frac{1}{2}$ cm tall with a diameter of 4 cm. Estimate the mass of such an egg by averaging the volumes of two spheres. (*Hint*: 1 cc of water has a mass of 1 g.)

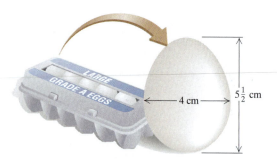

Complete. Use 453.6 g = 1 lb. Round to four decimal places.

**95.** ▦ 1 lb = _____ kg

**96.** ▦ 1 g = _____ lb

**97.** ▦ *Track and Field.* In shot put, a woman's shot weighs 8.8 lb and has a 4.5-in. diameter. Find its mass per cubic centimeter, given that 1 lb = 453.6 g.

**Data:** National Collegiate Athletic Association

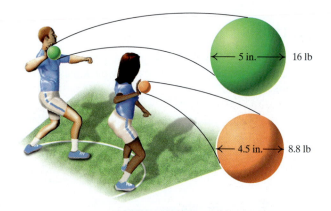

**98.** ▦ *Track and Field.* In shot put, a man's shot weighs 16 lb and has a 5-in. diameter. Find its mass per cubic centimeter, given that 1 lb = 453.6 g.

**99. *Large Diamonds.*** A **carat** (also spelled **karat**) is a unit of weight for precious stones; 1 carat = 200 mg. The Golden Jubilee Diamond weighs 545.67 carats and is the largest faceted diamond in the world. The Hope Diamond, located at the Smithsonian National Museum of Natural History, weighs 45.52 carats.

**Data:** Cape Town Diamond Museum; Smithsonian

**a)** How many grams does the Golden Jubilee Diamond weigh?

**b)** How many grams does the Hope Diamond weigh?

**c)** ▦ Given that 1 lb = 453.6 g, how many ounces does each diamond weigh?

# Medical Applications

## a MEASUREMENTS AND MEDICINE

**OBJECTIVE**

a Make conversions and solve applied problems concerning medical dosages.

Measurements play a critical role in health care. Doctors, nurses, aides, technicians, and others all need to use the proper units and perform the proper calculations to assure the best possible care of patients.

Because of the ease with which conversions can be made and its extensive use in science—among other reasons—the metric system is the primary system of measurement in medicine.

**EXAMPLE 1** *Medical Dosage.* A physician ordered 3.5 L of 5% dextrose in water (abbreviated D5W) to be administered over a 24-hr period. How many milliliters were ordered?

We convert 3.5 L to milliliters:

$$3.5 \text{ L} = 3.5 \cdot 1 \text{ L}$$

$$= 3.5 \cdot 1000 \text{ mL} \qquad \text{Substituting}$$

$$= 3500 \text{ mL}.$$

The physician ordered 3500 mL of D5W.

**Do Exercise 1.**

**1.** *Medical Dosage.* A physician ordered 2400 mL of 0.9% saline solution to be administered intravenously over a 24-hr period. How many liters were ordered?

Liquids at a pharmacy are often labeled in liters or milliliters. Thus, if a physician's prescription is given in ounces, it must be converted. For conversion, a pharmacist knows that 1 oz ≈ 29.57 mL.*

**EXAMPLE 2** *Prescription Size.* A prescription calls for 3 oz of theophylline, a drug commonly used for children with asthma. For how many milliliters is the prescription?

We convert as follows:

$$3 \text{ oz} = 3 \cdot 1 \text{ oz}$$

$$\approx 3 \cdot 29.57 \text{ mL} \qquad \text{Substituting}$$

$$= 88.71 \text{ mL}.$$

The prescription calls for 88.71 mL of theophylline.

**Do Exercise 2.**

**2.** *Prescription Size.* A prescription calls for 2 oz of theophylline.
**a)** For how many milliliters is the prescription?
**b)** For how many liters is the prescription?

---

*In practice, most pharmacists use 30 mL as an approximation of 1 oz.

*Answers*
**1.** 2.4 L   **2. (a)** About 59.14 mL;
**(b)** about 0.059 L

**EXAMPLE 3** *Pill Splitting.* Chlorthalidone is a commonly prescribed drug used to treat hypertension. Tania's physician directs her to reduce her dosage from 25 mg to 12.5 mg. Tania's original prescription contained 30 tablets, each 25 mg.

**a)** How many milligrams of chlorthalidone, in total, were in the original prescription?

**b)** How many 12.5-mg doses can Tania obtain from the original prescription?

**a)** The original prescription contained 30 tablets, each containing 25 mg of chlorthalidone. To find the total amount in the prescription, we multiply:

$$30 \text{ tablets} \cdot 25 \text{ mg/tablet} = 750 \text{ mg.}$$

The original prescription contained 750 mg of chlorthalidone.

**b)** Since 12.5 mg is half of 25 mg ($25 \div 2 = 12.5$), Tania's original 30 doses can each be split in half, yielding $30 \cdot 2$, or 60, doses at 12.5 mg per dose.

◀ **Do Exercise 3.**

Another metric unit that is used in medicine is the microgram (mcg). It is defined as follows.

---

**MICROGRAM**

$$1 \text{ microgram} = 1 \text{ mcg} = \frac{1}{1{,}000{,}000} \text{ g} = 0.000001 \text{ g}$$

$$1{,}000{,}000 \text{ mcg} = 1 \text{ g}$$

---

One microgram is one-millionth of a gram, so one million micrograms is one gram. A microgram is also one-thousandth of a milligram, so one thousand micrograms is one milligram.

**EXAMPLE 4** Complete: 1 mg = _____ mcg.
We convert to grams and then to micrograms:

$$1 \text{ mg} = 0.001 \text{ g}$$
$$= 0.001 \cdot 1 \text{ g}$$
$$= 0.001 \cdot 1{,}000{,}000 \text{ mcg} \qquad \text{Substituting 1,000,000 mcg for 1 g}$$
$$= 1000 \text{ mcg.}$$

◀ **Do Exercise 4.**

**EXAMPLE 5** *Medical Dosage.* Sublingual nitroglycerin comes in 0.4-mg tablets. How many micrograms are in each tablet?
**Data:** Steven R. Smith, M.D.

We are to complete 0.4 mg = _____ mcg. Thus,

$$0.4 \text{ mg} = 0.4 \cdot 1 \text{ mg}$$
$$= 0.4 \cdot 1000 \text{ mcg} \qquad \text{Substituting 1000 mcg for 1 mg (from Example 4)}$$
$$= 400 \text{ mcg.}$$

We can also do this problem in a manner similar to that used in Example 4.

◀ **Do Exercise 5.**

**3.** *Pill Splitting.* If Tania's physician originally prescribed 14 tablets that were each 25 mg, how many milligrams were in the original prescription? How many 12.5-mg doses could be obtained from the original prescription?

**4.** Complete:

1 mcg = _____ mg.

1 mcg = ☐ g
   = 0.000001 · 1 g
   = 0.000001 · ☐ mg
   = ☐ mg

**5.** *Medical Dosage.* A physician prescribes 500 mcg of alprazolam, an antianxiety medication. How many milligrams is this dosage?
**Data:** Steven R. Smith, M.D.

*Answers*
**3.** 350 mg; 28 doses   **4.** 0.001 mg
**5.** 0.5 mg
*Guided Solution:*
**4.** 0.000001, 1000, 0.001

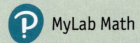
### ✓ Check Your Understanding

**Reading and Concept Check**  Determine whether each statement is true or false.

_____ **RC1.** For medicines in liquid form, 1 oz is about 30 mL.

_____ **RC2.** One microgram is one-thousandth of a milligram.

_____ **RC3.** One microgram is one million grams.

_____ **RC4.** There are 1000 micrograms in a milligram and 1000 milligrams in a gram.

**a**  *Medical Dosage.*  Solve each of the following. (None of these medications should be taken without consulting your own physician.)

**1.** An emergency-room physician orders 2.0 L of Ringer's lactate to be administered over 2 hr for a patient in shock. How many milliliters is this?

**2.** Ingrid received 84 mL per hour of normal saline solution. How many liters did Ingrid receive in a 24-hr period?

**3.** To battle hypertension and prostate enlargement, Rick is directed to take 4 mg of doxazosin each day for 30 days. How many grams is this?

**4.** To battle high cholesterol, Kit is directed to take 40 mg of atorvastatin for 60 days. How many grams is this?

**5.** Cephalexin is an antibiotic that frequently is prescribed in 500-mg tablets. Dr. Bouvier prescribes 2 g of cephalexin per day for a patient with a skin abscess. How many 500-mg tablets would have to be taken in order to achieve this daily dosage?

**6.** Quinidine gluconate is a liquid mixture, part medicine and part water, which is administered intravenously. There are 80 mg of quinidine gluconate in each cubic centimeter (cc) of the liquid mixture. Dr. Nassat orders 500 mg of quinidine gluconate to be administered daily to a patient with malaria. How much of the solution would have to be administered in order to achieve the recommended daily dosage?

**7.** Albuterol is a medication used for the treatment of asthma. It comes in an inhaler that contains 17 mg of albuterol mixed with a liquid. One actuation (inhalation) from the mouthpiece delivers a 90-mcg dose of albuterol.

**a)** Dr. Martinez orders 2 inhalations 4 times per day. How many micrograms of albuterol does the patient inhale per day?

**b)** How many actuations/inhalations are contained in one inhaler?

**c)** Delia is going away for 4 months and wants to take enough albuterol to last for that time. Her physician has prescribed 2 inhalations 4 times per day. Estimate how many inhalers Delia will need to take with her for the 4-month period.

**8.** Amoxicillin is a common antibiotic prescribed for children. It is a liquid suspension composed of part amoxicillin and part water. In one formulation of amoxicillin suspension, there are 250 mg of amoxicillin in 5 cc of the liquid suspension. Dr. Scarlotti prescribes 400 mg per day for a 2-year-old child with an ear infection. How much of the amoxicillin liquid suspension would the child's parent need to administer in order to achieve the recommended daily dosage of amoxicillin?

**9.** Dr. Norris tells a patient to purchase 0.5 L of hydrogen peroxide. Commercially, hydrogen peroxide is found on the shelves in bottles that hold 4 oz, 8 oz, and 16 oz. Which bottle comes closest to filling the prescription?

**10.** Dr. Lopez wants a patient to receive 3 L of a normal glucose solution in a 24-hr period. How many milliliters per hour should the patient receive?

**11.** Mirtazapine is a commonly prescribed antianxiety drug. Joanne has a 14-tablet supply of 30-mg tablets and her physician now directs her to reduce her dosage to 15 mg.

  **a)** How many grams were originally prescribed?

  **b)** How many 15-mg doses can Joanne obtain from the original prescription?

**12.** Nefazodone is a commonly prescribed antidepressant. Chad has a 90-tablet supply of 200-mg tablets when his physician directs him to cut his dosage to 100 mg.

  **a)** How many grams are in Chad's current supply?

  **b)** How many 100-mg doses can Chad obtain from his supply?

**13.** Amoxicillin is an antibiotic obtainable in a liquid suspension form, part medication and part water, and is frequently used to treat infections in infants. One formulation of the drug contains 125 mg of amoxicillin per 5 mL of liquid. A pediatrician orders 150 mg per day for a 4-month-old child with an ear infection. How much of the amoxicillin suspension would the parent need to administer to the infant in order to achieve the recommended daily dose?

**14.** Diphenhydramine HCL is an antihistamine available in liquid form, part medication and part water. One formulation contains 25 mg of medication in 5 mL of liquid. An allergist orders 40-mg doses for a high school student. How many milliliters should be in each dose?

Complete.

**15.** 1 mg = _____ mcg

**16.** 1 mcg = _____ mg

**17.** 325 mcg = _____ mg

**18.** 0.45 mg = _____ mcg

**19.** Dr. Djihn prescribes 0.25 mg of alprazolam, an antianxiety medication. How many micrograms are in this dose?

**20.** Dr. Kramer prescribes 0.4 mg of alprazolam, an antianxiety medication. How many micrograms are in this dose?

**21.** Digoxin is a medication used to treat heart problems. A cardiologist orders 0.125 mg of digoxin to be taken once daily. How many micrograms of digoxin are there in the daily dosage?

**22.** Digoxin is a medication used to treat heart problems. An internist orders 0.25 mg of digoxin to be taken once a day. How many micrograms of digoxin are there in the daily dosage?

**23.** Triazolam is a medication used for the short-term treatment of insomnia. A physician advises her patient to take one of the 0.125-mg tablets each night for 7 nights. How many milligrams of triazolam will the patient have ingested over that 7-day period? How many micrograms?

**24.** Clonidine is a medication used to treat high blood pressure. The usual starting dose of clonidine is one 0.1-mg tablet twice a day. If a patient is started on this dose by his physician, how many total milligrams of clonidine will the patient have taken before he returns to see his physician 14 days later? How many micrograms?

## Skill Maintenance

Subtract.  [1.3a]

**25.**
$$\begin{array}{r} 5789 \\ -2431 \\ \hline \end{array}$$

**26.**
$$\begin{array}{r} 8429 \\ -1015 \\ \hline \end{array}$$

**27.**
$$\begin{array}{r} 4097 \\ -3243 \\ \hline \end{array}$$

**28.**
$$\begin{array}{r} 8390 \\ -2056 \\ \hline \end{array}$$

Simplify.  [2.7a]

**29.** $7x + 9 - 2x - 1$

**30.** $8x + 12 - 2x - 7$

**31.** $8t - 5 - t - 4$

**32.** $9r - 6 - r - 4$

## Synthesis

**33.** ▦ A patient is directed to take 200 mg of nefazodone three times a day for one week, then 200 mg twice a day for a week, and then 100 mg three times a day for a week.

**a)** How many grams of medication are used altogether?
**b)** What is the average dosage size?

**34.** ▦ A patient is directed to take 200 mg of bupropion twice a day for a week, then 200 mg in the evening and 100 mg in the morning for a week, and then 100 mg twice a day for a week.

**a)** How many grams of medication are used altogether?
**b)** What is the average dosage size?

**35.** Naproxen sodium is a painkiller that lasts approximately 12 hours. A typical dose is one 220-mg tablet. Ibuprofen is a similar painkiller that lasts about 6 hours and has a typical dosage of two 200-mg tablets. Which is the better buy: a bottle containing 44 g of naproxen sodium costing $11 or a bottle containing 72 g of ibuprofen costing $11.24? Why?

## Conversions and Formulas

| | |
|---|---|
| *American Units of Length:* | 12 in. = 1 ft; 3 ft = 1 yd; 36 in. = 1 yd; 5280 ft = 1 mi |
| *Metric Units of Length:* | 1 km = 1000 m; 1 hm = 100 m; 1 dam = 10 m; 1 dm = 0.1 m; 1 cm = 0.01 m; 1 mm = 0.001 m |
| *American–Metric Conversion:* | 1 m ≈ 39.370 in.; 1 m ≈ 3.281 ft; 1 ft ≈ 0.305 m; 1 in. ≈ 2.54 cm; 1 km ≈ 0.621 mi; 1 mi ≈ 1.609 km; 1 m ≈ 1.094 yd; 1 yd ≈ 0.914 m |
| *American Units of Area:* | $1 \text{ yd}^2 = 9 \text{ ft}^2$; $1 \text{ ft}^2 = 144 \text{ in}^2$; $1 \text{ mi}^2 = 640$ acres; 1 acre = 43,560 $\text{ft}^2$ |
| *American Units of Capacity:* | 1 gal = 4 qt; 1 qt = 2 pt; 1 pt = 16 oz; 1 pt = 2 cups; 1 cup = 8 oz |
| *Metric Units of Capacity:* | 1 L = 1000 mL = 1000 $\text{cm}^3$ = 1000 cc |
| *American–Metric Conversion:* | 1 oz ≈ 29.57 mL |
| *American Units of Weight:* | 1 T = 2000 lb; 1 lb = 16 oz |
| *Metric Units of Mass:* | 1 t = 1000 kg; 1 kg = 1000 g; 1 hg = 100 g; 1 dag = 10 g; 1 dg = 0.1 g; 1 cg = 0.01 g; 1 mg = 0.001 g; 1 mcg = 0.000001 g |
| *Temperature Conversion:* | $F = \dfrac{9}{5} \cdot C + 32$, or $F = 1.8 \cdot C + 32$; $C = \dfrac{5}{9} \cdot (F - 32)$, or $C = \dfrac{F - 32}{1.8}$ |

| | | | |
|---|---|---|---|
| *Area of a Parallelogram:* | $A = b \cdot h$ | *Area of a Circle:* | $A = \pi \cdot r \cdot r$, or $A = \pi \cdot r^2$ |
| *Area of a Trapezoid:* | $A = \dfrac{1}{2} \cdot h \cdot (a + b)$ | *Volume of a Rectangular Solid:* | $V = l \cdot w \cdot h$ |
| | | *Volume of a Circular Cylinder:* | $V = \pi \cdot r^2 \cdot h$ |
| *Radius and Diameter of a Circle:* | $d = 2 \cdot r$, or $r = \dfrac{d}{2}$ | *Volume of a Sphere:* | $V = \dfrac{4}{3} \cdot \pi \cdot r^3$ |
| *Circumference of a Circle:* | $C = \pi \cdot d$, or $C = 2 \cdot \pi \cdot r$ | *Sum of Angle Measures of a Triangle:* | $m \angle A + m \angle B + m \angle C = 180°$ |
| | | *Pythagorean Equation:* | $a^2 + b^2 = c^2$ |

## Vocabulary Reinforcement

Complete each statement with the correct word from the list at the right. Some of the choices may not be used and some may be used more than once.

1. A parallelogram is a four-sided figure with two pairs of _____ sides.   [9.3a]

2. A(n) _____ is the set of all points in space that are a given distance from a given point.   [9.4a]

3. The _____ of a circle is half the length of its diameter.   [9.3b]

4. Two angles are _____ if the sum of their measures is 180°.   [9.5c]

5. A(n) _____ triangle has all sides of different lengths.   [9.5d]

6. The _____ of a right triangle is the side opposite the right angle.   [9.6c]

circumference

radius

isosceles

scalene

parallel

sphere

cylinder

hypotenuse

leg

complementary

supplementary

# Concept Reinforcement

Determine whether each statement is true or false.

_____ 1. To convert mm$^2$ to cm$^2$, move the decimal point 2 places to the left. [9.2b]

_____ 2. Since 1 yd = 3 ft, we multiply by 3 to convert square yards to square feet. [9.2a]

_____ 3. The number $\pi$ is greater than 3.14 and $\frac{22}{7}$. [9.3b]

_____ 4. The acute angles of a right triangle are complementary. [9.5b, c, e]

_____ 5. The length of the hypotenuse of a right triangle is greater than the length of either of its legs. [9.6c]

_____ 6. You would probably use your furnace when the temperature was 40°C. [9.7c]

# Study Guide

**Objective 9.1a** Convert from one American unit of length to another.

**Example** Complete: 126 in. = _____ yd.

$$126 \text{ in.} = \frac{126 \text{ in.}}{1} \cdot \frac{1 \text{ yd}}{36 \text{ in.}} = 3.5 \text{ yd}$$

**Practice Exercise**

1. Complete: 7 ft = _____ yd.

**Objective 9.1b** Convert from one metric unit of length to another.

**Example** Complete: 38 km = _____ cm.

To go from km to cm, we move the decimal point 5 places to the right.

38   38.00000.   38 km = 3,800,000 cm

**Practice Exercise**

2. Complete: 4.6 cm = _____ km.

**Objective 9.1c** Convert between American and metric units of length.

**Example** Complete: 42 ft = _____ m.
(Note: 1 ft ≈ 0.305 m.)

42 ft = 42 · 1 ft ≈ 42 · 0.305 m = 12.81 m

**Practice Exercise**

3. Complete: 10 m = _____ yd.
(Note: 1 m ≈ 1.094 yd.)

**Objective 9.2a** Convert from one American unit of area to another.

**Example** Complete: 14,400 in$^2$ = _____ ft$^2$.

$$14,400 \text{ in}^2 = \frac{14,400 \text{ in}^2}{1} \cdot \frac{1 \text{ ft}^2}{144 \text{ in}^2} = 100 \text{ ft}^2$$

**Practice Exercise**

4. Complete: 81 ft$^2$ = _____ yd$^2$.

**Objective 9.2b** Convert from one metric unit of area to another.

**Example** Complete: 9.6 m$^2$ = _____ cm$^2$.

To go from m$^2$ to cm$^2$, we move the decimal point 2 × 2, or 4, places to the right.

9.6   9.6000.   9.6 m$^2$ = 96,000 cm$^2$

**Practice Exercise**

5. Complete: 52.4 cm$^2$ = _____ mm$^2$.

**Objective 9.3a**  Find the area of a parallelogram or trapezoid.

**Examples**  Find the area of this parallelogram.

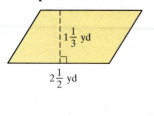

1⅓ yd

2½ yd

$$A = b \cdot h$$

$$= 2\frac{1}{2} \text{ yd} \cdot 1\frac{1}{3} \text{ yd}$$

$$= \frac{5}{2} \cdot \frac{4}{3} \cdot \text{yd} \cdot \text{yd}$$

$$= \frac{20}{6} \text{ yd}^2 = \frac{10}{3} \text{ yd}^2,$$

$$\text{or } 3\frac{1}{3} \text{ yd}^2$$

**Practice Exercises**

**6.** Find the area of this parallelogram.

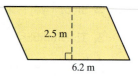

2.5 m

6.2 m

Find the area of this trapezoid.

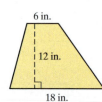

6 in.

12 in.

18 in.

$$A = \frac{1}{2} \cdot h \cdot (a + b)$$

$$= \frac{1}{2} \cdot 12 \text{ in.} \cdot (6 \text{ in.} + 18 \text{ in.})$$

$$= \frac{12 \cdot 24}{2} \text{ in}^2 = 144 \text{ in}^2$$

**7.** Find the area of this trapezoid.

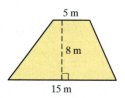

5 m

8 m

15 m

---

**Objective 9.3b**  Find the circumference, area, radius, or diameter of a circle, given the length of a radius or diameter.

**Examples**  Find the circumference of this circle. Use 3.14 for $\pi$.

4 ft

$$C = \pi \cdot d, \quad \text{or} \quad 2 \cdot \pi \cdot r$$

$$\approx 2 \cdot 3.14 \cdot 4 \text{ ft}$$

$$= 25.12 \text{ ft}$$

**Practice Exercises**

**8.** Find the circumference of this circle. Use 3.14 for $\pi$.

6 in.

Find the area of this circle. Use $\frac{22}{7}$ for $\pi$.

21 mm

$$A = \pi \cdot r \cdot r, \quad \text{or} \quad \pi \cdot r^2$$

$$\approx \frac{22}{7} \cdot 21 \text{ mm} \cdot 21 \text{ mm}$$

$$= \frac{22 \cdot 21 \cdot 21}{7} \text{ mm}^2 = 1386 \text{ mm}^2$$

**9.** Find the area of this circle. Use $\frac{22}{7}$ for $\pi$.

14 cm

---

**Objective 9.4a**  Find the volume of a rectangular solid, a cylinder, and a sphere.

**Examples**  Find the volume of this rectangular solid.

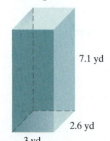

7.1 yd

2.6 yd

3 yd

$$V = l \cdot w \cdot h$$

$$= 3 \text{ yd} \cdot 2.6 \text{ yd} \cdot 7.1 \text{ yd}$$

$$= 3 \cdot 2.6 \cdot 7.1 \text{ yd}^3$$

$$= 55.38 \text{ yd}^3$$

**Practice Exercises**

**10.** Find the volume of this rectangular solid.

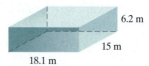

6.2 m

15 m

18.1 m

Find the volume of this circular cylinder. Use 3.14 for $\pi$.

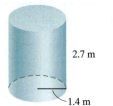

$$V = B \cdot h, \quad \text{or} \quad \pi \cdot r^2 \cdot h$$
$$\approx 3.14 \cdot 1.4 \text{ m} \cdot 1.4 \text{ m} \cdot 2.7 \text{ m}$$
$$= 16.61688 \text{ m}^3$$

2.7 m

1.4 m

**11.** Find the volume of this circular cylinder. Use $\frac{22}{7}$ for $\pi$.

$5\frac{2}{5}$ ft

$1\frac{1}{3}$ ft

Find the volume of this sphere. Use $\frac{22}{7}$ for $\pi$.

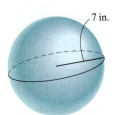

7 in.

$$V = \frac{4}{3} \cdot \pi \cdot r^3$$
$$\approx \frac{4}{3} \cdot \frac{22}{7} \cdot 7 \text{ in.} \cdot 7 \text{ in.} \cdot 7 \text{ in.}$$
$$= 1437\frac{1}{3} \text{ in}^3$$

**12.** Find the volume of this sphere. Use 3.14 for $\pi$.

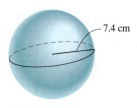

7.4 cm

---

**Objective 9.4b** Convert from one unit of capacity to another.

**Example** Complete: 6 gal = _____ pt.

$$6 \text{ gal} = 6 \cdot 1 \text{ gal}$$
$$= 6 \cdot 4 \text{ qt}$$
$$= 24 \cdot 1 \text{ qt}$$
$$= 24 \cdot 2 \text{ pt} = 48 \text{ pt}$$

**Practice Exercise**

**13.** Complete: 16 qt = _____ cups.

---

**Objective 9.5c** Find the measure of a complement or a supplement of a given angle.

**Example** Find the measure of a complement and a supplement of an angle that measures 65°.

The measure of the complement of an angle of 65° is 90° − 65°, or 25°.

The measure of the supplement of an angle of 65° is 180° − 65°, or 115°.

**Practice Exercise**

**14.** Find the measure of a complement and a supplement of an angle that measures 38°.

---

**Objective 9.5e** Given two of the angle measures of a triangle, find the third.

**Example** Find the missing angle measure.

$$m\angle A + m\angle B + m\angle C = 180°$$
$$x + 130° + 28° = 180°$$
$$x + 158° = 180°$$
$$x = 180° - 158°$$
$$x = 22°$$
The measure of $\angle A$ is 22°.

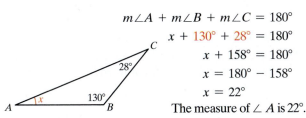

**Practice Exercise**

**15.** Find the missing angle measure.

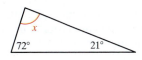

$x$

72°       21°

**Objective 9.6c** Given the lengths of any two sides of a right triangle, find the length of the third side.

**Example** Find the length of the third side of this triangle. Give an exact answer and an approximation to three decimal places.

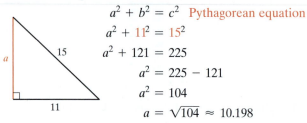

$$a^2 + b^2 = c^2 \quad \text{Pythagorean equation}$$
$$a^2 + 11^2 = 15^2$$
$$a^2 + 121 = 225$$
$$a^2 = 225 - 121$$
$$a^2 = 104$$
$$a = \sqrt{104} \approx 10.198$$

**Practice Exercise**

16. Find the length of the third side of this right triangle. Give an exact answer and an approximation to three decimal places.

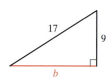

---

**Objective 9.7a** Convert from one American unit of weight to another.

**Example** Complete: 4020 oz = _____ lb.

$$4020 \text{ oz} = 4020 \text{ o\!z} \times \frac{1 \text{ lb}}{16 \text{ o\!z}} = 251.25 \text{ lb}$$

**Practice Exercise**

17. Complete: 10,280 lb = _____ T.

---

**Objective 9.7b** Convert from one metric unit of mass to another.

**Example** Complete: 5.62 cg = _____ g.

To go from cg to g, we move the decimal point 2 places to the left.

5.62    0.05.62    5.62 cg = 0.0562 g

**Practice Exercise**

18. Complete: 9.78 mg = _____ g.

---

**Objective 9.7c** Convert temperatures from Celsius to Fahrenheit and from Fahrenheit to Celsius.

**Examples** Convert 18°C to Fahrenheit and 95°F to Celsius.

$$F = \frac{9}{5}C + 32 = 1.8 \cdot 18 + 32$$
$$= 32.4 + 32 = 64.4$$

Thus, 18°C = 64.4°F.

$$C = \frac{5}{9} \cdot (F - 32) = \frac{5}{9} \cdot (95 - 32)$$
$$= \frac{5}{9} \cdot 63 = 35$$

Thus, 95°F = 35°C.

**Practice Exercises**

19. Convert 68°C to Fahrenheit.

20. Convert 104°F to Celsius.

---

**Objective 9.8a** Make conversions and solve applied problems concerning medical dosages.

**Example** A physician ordered 320 mL of 5% dextrose in water (D5W) to be administered intravenously over 4 hr. How many liters of D5W is this?

We convert 320 mL to liters:

$$320 \text{ mL} = 320 \cdot 1 \text{ mL}$$
$$= 320 \cdot 0.001 \text{ L}$$
$$= 0.32 \text{ L}.$$

The physician ordered 0.32 L of D5W.

**Practice Exercise**

21. A physician orders 0.5 L of normal saline solution. How many milliliters are ordered?

## Review Exercises

Complete.

**1.** 10 ft = _____ yd
[9.1a]

**2.** $\frac{5}{6}$ yd = _____ in.
[9.1a]

**3.** 1.7 mm = _____ cm
[9.1b]

**4.** 2 yd = _____ in.
[9.1a]

**5.** 4 km = _____ cm
[9.1b]

**6.** 14 in. = _____ ft
[9.1a]

**7.** 200 m = _____ yd
[9.1c]

**8.** 20 mi = _____ km
[9.1c]

**9.** 5 lb = _____ oz
[9.7a]

**10.** 3 g = _____ kg
[9.7b]

**11.** 50 qt = _____ gal
[9.4b]

**12.** 28 gal = _____ pt
[9.4b]

**13.** 60 mL = _____ L
[9.4b]

**14.** 0.4 L = _____ mL
[9.4b]

**15.** 0.7 T = _____ lb
[9.7a]

**16.** 0.2 g = _____ mg
[9.7b]

**17.** 4.7 kg = _____ g
[9.7b]

**18.** 4 cg = _____ g
[9.7b]

**19.** 4 $yd^2$ = _____ $ft^2$
[9.2a]

**20.** 0.7 $km^2$ = _____ $m^2$
[9.2b]

**21.** 1008 $in^2$ = _____ $ft^2$
[9.2a]

**22.** 570 $cm^2$ = _____ $m^2$
[9.2b]

**23.** Find the circumference of a circle of radius 5 m. Use 3.14 for $\pi$.   [9.3b]

**24.** Find the length of a radius of the circle.   [9.3b]

$\frac{28}{11}$ in.

**25.** Find the length of a diameter of the circle.   [9.3b]

12 m

**26.** *Track and Field.*   Track meets are held on a track similar to the one shown below. Find the shortest distance around the track. Use 3.14 for $\pi$.   [9.3b]

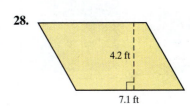
85.56 yd
85.56 yd

Find the area of each figure in Exercises 27–32. [9.3a, b]

**27.**
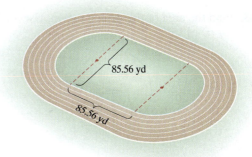
7 in.
3 in.
5 in.

**28.**
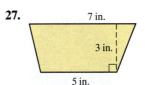
4.2 ft
7.1 ft

**29.**

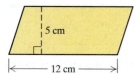
5 cm
12 cm

**30.**

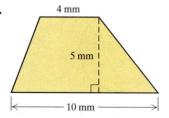
4 mm
5 mm
10 mm

**31.** Use $\frac{22}{7}$ for $\pi$.

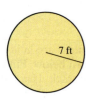
7 ft

**32.** Use 3.14 for $\pi$.

20 cm

**33.** Find the area of the shaded region. Use 3.14 for $\pi$.   [9.3b]

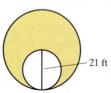
21 ft

**34.** A "Norman" window is designed with dimensions as shown. Find its area. Use 3.14 for $\pi$.   [9.3b]

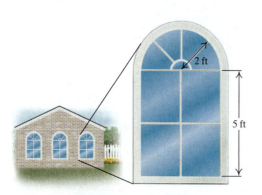
2 ft
5 ft

Use a protractor to measure each angle.   [9.5a]

**35.**

**36.**

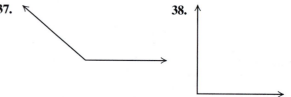
P   Q   R

**37.**

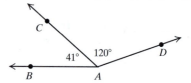

**38.**

**39–42.** Classify each of the angles in Exercises 35–38 as right, straight, acute, or obtuse.   [9.5b]

**43.** Find the measure of a complement of $\angle BAC$.   [9.5c]

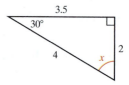
C
41°   120°   D
B   A

**44.** Find the measure of a supplement of a 44° angle.   [9.5c]

Use the following triangle for Exercises 45–47.

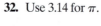
3.5
30°
4   x   2

**45.** Find the missing angle measure.   [9.5e]

**46.** Classify the triangle as equilateral, isosceles, or scalene.   [9.5d]

**47.** Classify the triangle as right, obtuse, or acute.   [9.5d]

Find the volume of each figure. Use 3.14 for $\pi$.  [9.4a]

**48.**

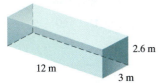

2.6 m
12 m
3 m

**49.**

14 ft
3 ft   4.6 ft

**50.**

90 cm
10 cm

**51.**

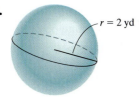

$r = 2$ yd

**52.**

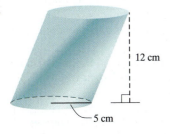

12 cm
5 cm

**53.** Simplify: $\sqrt{64}$.  [9.6a]

**54.** ▦ Use a calculator to approximate $\sqrt{83}$ to three decimal places.  [9.6b]

For each right triangle, find the length of the side not given. Find an exact answer and an approximation to three decimal places. Assume that $c$ represents the length of the hypotenuse.  [9.6c]

**55.** $a = 15, b = 25$        **56.** $a = 4, c = 10$

**57.**

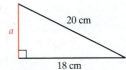

$c$
5 ft
8 ft

**58.**

20 cm
$a$
18 cm

Solve.  [9.6d]

**59.** A wire 24 ft long reaches from the top of a pole to a point on the ground 16 ft from the base of the pole. How tall is the pole? Round to the nearest tenth of a foot.

**60.** *Construction.*  Chloe is designing rafters for a house. The rise of each rafter will be 6 ft and the run 12 ft. What is the rafter length? Round to the nearest hundredth of a foot.

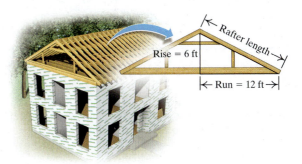

Rafter length
Rise = 6 ft
Run = 12 ft

**61.** Convert 45°C to Fahrenheit.  [9.7c]

**62.** Convert 68°F to Celsius.  [9.7c]

*Medical Dosage.* Solve. [9.8a]

63. Amoxicillin is an antibiotic obtainable in a liquid suspension form, part medication and part water, and is frequently used to treat infections in infants. One formulation of the drug contains 125 mg of amoxicillin per 5 mL of liquid. A pediatrician orders 200 mg per day for a child with an ear infection. How much of the amoxicillin suspension would the parent need to administer to the child in order to achieve the recommended daily dose?

64. An emergency-room physician orders 3 L of Ringer's lactate to be administered over 4 hr for a patient suffering from shock and severe low blood pressure. How many milliliters is this?

65. A physician prescribes 0.25 mg of alprazolam, an antianxiety medication. How many micrograms are in this dose?

66. Find the measure of a supplement of a $20\frac{3}{4}^\circ$ angle. [9.5c]

   **A.** $339\frac{1}{4}^\circ$          **B.** $159\frac{1}{4}^\circ$

   **C.** $69\frac{1}{4}^\circ$           **D.** $70\frac{1}{4}^\circ$

67. Complete: 0.16 gal = _____ cups. [9.4b]
   **A.** 1.28             **B.** 2.56
   **C.** 0.64             **D.** 160

## Synthesis

68. *Running Record.* The world record for running the 200-m dash is 19.19 sec, set by Usain Bolt of Jamaica in the 2009 World Championships in Berlin. What would his record time be if he had run at the same rate in a 200-yd dash? [9.1c]

   **Data:** iaaf.com

69. It is known that 1 gal of water weighs 8.3453 lb. Which weighs more: an ounce of pennies or an ounce (as capacity) of water? Explain. [9.4b], [9.7a]

70. A community center has a rectangular swimming pool that is 50 ft wide, 100 ft long, and 10 ft deep. The pool is filled with water to a line that is 1 ft from the top. Water costs $2.25 per 1000 ft³. How much does it cost to fill the pool? [9.4c]

71. ▦ One lap around a standard running track is 440 yd. A marathon is 26 mi, 385 yd long. How many laps around a track does a marathon require? [9.1a]

72. ▦ Find the area of the largest round pizza that can be baked in a 35-cm by 50-cm pan. Use 3.14 for $\pi$. [9.3b]

# Understanding Through Discussion and Writing

1. Give at least two reasons why someone might prefer the use of grams to the use of ounces. [9.7a, b]

2. Is it possible for a triangle to contain two 90° angles? Why or why not? [9.5e]

3. Which is larger and why: one square meter or one square yard? [9.1c], [9.2a, b]

4. Explain a procedure that could be used to determine the measure of an angle's supplement from the measure of the angle's complement. [9.5c]

5. Explain how the Pythagorean theorem can be used to prove that a triangle is a *right* triangle. [9.6c]

6. Which occupies more volume: two spheres, each with radius $r$, or one sphere with radius $2r$? Explain why. [9.4c]

**CHAPTER**

**9**  **Test**

For Extra Help

For step-by-step test solutions, access the Chapter Test Prep Videos in MyLab Math.

Complete.

**1.** 8 ft = _____ in.

**2.** 280 cm = _____ m

**3.** 2 yd² = _____ ft²

**4.** 5 km = _____ m

**5.** 9.1 mm = _____ cm

**6.** 4520 m² = _____ km²

**7.** 2983 mL = _____ L

**8.** 3.8 kg = _____ g

**9.** 10 gal = _____ oz

**10.** 0.69 L = _____ mL

**11.** 9 lb = _____ oz

**12.** 4.11 T = _____ lb

**13.** Find the length of a radius of this circle.

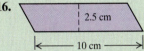

16 cm

**14.** Find the area of a circle of radius 4 m. Use 3.14 for π.

**15.** Find the circumference of a circle of radius 14 ft. Use $\frac{22}{7}$ for π.

Find each area. Use 3.14 for π.

**16.**

2.5 cm

10 cm

**17.**

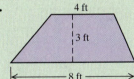

4 ft

3 ft

8 ft

**18.**

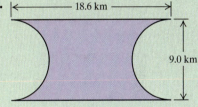

18.6 km

9.0 km

**19.** Find the measure of a complement and a supplement of ∠*CAD*.

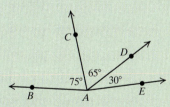

*C*

*D*

65°

75°  30°

*B*  *A*  *E*

Use a protractor to measure each angle.

**20.**

**21.**

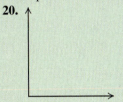

**22.** Classify the angle in Exercise 20 as right, straight, acute, or obtuse.

**23.** Classify the angle in Exercise 21 as right, straight, acute, or obtuse.

Use the following triangle for Exercises 24–26.

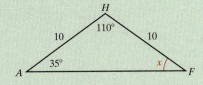

*H*

110°

10  10

35°  *x*

*A*  *F*

**24.** Find the missing angle measure.

**25.** Classify the triangle as equilateral, isosceles, or scalene.

**26.** Classify the triangle as right, obtuse, or acute.

**27.** A twelve-box rectangular carton of 12-oz juice boxes measures $10\frac{1}{2}$ in. by 8 in. by 5 in. What is the volume of the carton?

In Exercises 28–30, find the volume for each figure. Use 3.14 for $\pi$.

**28.**

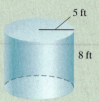

5 ft

8 ft

**29.**

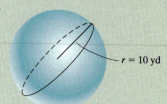

$r = 10$ yd

**30.**

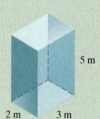

5 m

2 m    3 m

**31.** How long must a wire be in order to reach from the top of a 13-m antenna to a point on the ground 9 m from the base of the antenna? Round to the nearest tenth of a meter.

For each right triangle, find the length of the side not given. Find an exact answer and an approximation to three decimal places.

**32.**

$c$

1

1

**33.**

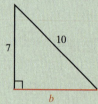

7

10

$b$

**34.** Find $\sqrt{121}$.

**35.** Convert 32°F to Celsius.

**36.** An antihistamine solution contains 25 mg of medication in 5 mL of liquid. Pat is directed to take a 30-mg dose of the antihistamine. How many mL's of the solution should Pat take?

**A.** 6 mL        **B.** 55 mL
**C.** 125 mL      **D.** 150 mL

## Synthesis

**37.** The measure of $\angle SMC$ is three times that of its complement. Find the measure of the supplement of $\angle SMC$.

**38.** 📷 A *board foot* is the amount of wood in a piece 12 in. by 12 in. by 1 in. A carpenter places the following order for a certain kind of lumber:

25 pieces: 2 in. by 4 in. by 8 ft;
32 pieces: 2 in. by 6 in. by 10 ft;
24 pieces: 2 in. by 8 in. by 12 ft.

The price of this type of lumber is $225 per thousand board feet. What is the total cost of the carpenter's order?

Find the volume of the solid. (Note that the solids are not drawn in perfect proportion.) Give the answer in cubic feet. Use 3.14 for $\pi$.

**39.**

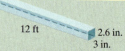

12 ft

2.6 in.
3 in.

**40.**

18 ft

$\frac{3}{4}$ in.

Solve.

1. **The Arctic Sea.** The Arctic sea surface freezes each winter, but the area that is still ice-covered in September has been decreasing. In 1979, 2.78 million square miles of the Arctic sea surface remained ice-covered in September. In 2016, only 1.82 million square miles were ice-covered in September. Find standard notation for 2.78 million and for 1.82 million.

**Data:** *National Geographic,* April 2017, "Ice Is Melting Fast"; National Snow and Ice Data Center

2. **Firefighting.** During a fire, the firefighters get a 1-ft layer of water on the 25-ft by 60-ft first floor of a 5-floor building. Water weighs $62\frac{1}{2}$ lb per cubic foot. What is the total weight of the water on the floor?

Calculate.

3. $1\frac{1}{2} + 2\frac{2}{3}$

4. $120.5 - 32.98$

5. $-27{,}148 \div 22$

6. $8^3 + 45 \cdot 24 - 9^2 \div 3$

7. $\left(\frac{1}{4}\right)^2 \div \left(\frac{1}{2}\right)^3 \times 2^4 + (10.3)(4)$

8. $14 \div [33 \div 11 + 8 \times 2 - (15 - 3)]$

Find fraction notation.

9. 1.209          10. 17%

Use $<$, $>$, or $=$ for $\square$ to write a true sentence.

11. $\frac{5}{6} \square \frac{7}{8}$          12. $\frac{15}{18} \square \frac{10}{12}$

Complete.

13. 2.5 yd = _____ in.     14. 6 oz = _____ lb

15. 15°C = _____ °F     16. 0.087 L = _____ mL

17. 3 yd$^2$ = _____ ft$^2$     18. 17 cm = _____ m

Solve.

19. $x + \frac{3}{4} = \frac{7}{8}$          20. $\frac{3}{x} = \frac{7}{10}$

21. $1 - 7x = 4 - (x + 9)$     22. $-15x = 240$

23. Find the perimeter and the area of the parallelogram.

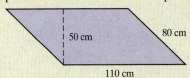

24. Find the diameter, the circumference, and the area of this circle. Use $\frac{22}{7}$ for $\pi$.

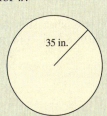

**25.** Find the volume of this sphere. Use $\frac{22}{7}$ for $\pi$.

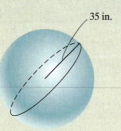
35 in.

**26.** Find the mean: 49, 53, 60, 62, 69.

Solve.

**27.** What is the simple interest on $8000 at 4.2% for $\frac{1}{4}$ year?

**28.** What is the amount in an account after 25 years if $8000 is invested at 4.2%, compounded annually?

**29.** How long must a rope be in order to reach from the top of an 8-m tree to a point on the ground 15 m from the bottom of the tree?

**30.** The sales tax on an office supply purchase of $5.50 is $0.33. What is the sales tax rate?

**31.** A bolt of fabric in a fabric store has $10\frac{3}{4}$ yd on it. A customer purchases $8\frac{5}{8}$ yd. How many yards remain on the bolt?

**32.** What is the cost, in dollars, of 15.6 gal of gasoline at 329.9¢ per gallon? Round to the nearest cent.

**33.** A box of powdered milk that makes 20 qt costs $4.99. A box that makes 8 qt costs $1.99. Which size has the lower unit price?

**34.** It is $\frac{7}{10}$ km from Maria's dormitory to the library. Maria started to walk from the dorm to the library, changed her mind after going $\frac{1}{4}$ of the distance, and returned to the dorm. How far did she walk?

**35.** Find the missing angle measure.

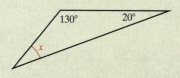
130°  20°
x

**36.** Combine like terms: $12a - 7 - 3a - 9$.

**37.** Graph: $y = -\dfrac{1}{3}x + 2$.

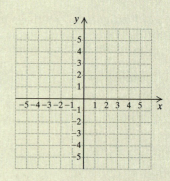

## Synthesis

Find the volume in cubic feet. Use 3.14 for $\pi$.

**38.**

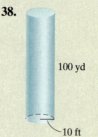
100 yd

10 ft

**39.**

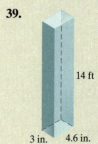
14 ft

3 in.   4.6 in.

# Exponents and Polynomials

The common octopus, *Octopus vulgaris*, is most often found in temperate and tropical waters at a depth of 100 to 150 meters. This species can reach a length of 1 meter, legs included. The octopus has three hearts and no bones. Octopuses can learn and are probably the most intelligent of all invertebrates. One-third of an octopus's neurons are located in its brain, and the remaining two-thirds of its neurons are in its arms, which are lined with suckers. The table at left shows a comparison of the number of neurons in a human, an octopus, a mouse, and a pond snail.

### Comparing Number of Neurons

| | Number of neurons |
|---|---|
| **Pond snail** | 10,000 |
| **Mouse** | 80,000,000 |
| **Octopus** | 500,000,000 |
| **Human** | 86,000,000,000 |

DATA: *National Geographic*, November 2016, p. 70, "The Power of Eight," by Olivia Judson

DATA: *animaldiversity.org; arkive.org; National Geographic, November 2016, p. 70, "The Power of Eight," Olivia Judson; Roger Hanlon, Marine Biological Laboratory; Guy Levy and Benny Hochner, Hebrew University of Jerusalem; Cliff Ragsdale, University of Chicago*

We will calculate the number of neurons in the arms of an *Octopus vulgaris* in Exercise 36 of Exercise Set 10.3.

☐ Browse through each chapter, reviewing highlighted or boxed information and noting important formulas.
☐ Attend any exam tutoring sessions offered by your college or university.
☐ Retake the chapter tests that you took in class, or take the chapter tests in the text.
☐ Work through the Cumulative Review for Chapters 1–10 as a sample final exam.

# 10.1 Integers as Exponents

## OBJECTIVES

**a** Evaluate algebraic expressions containing whole-number exponents.

**b** Express exponential expressions involving negative exponents as equivalent expressions containing positive exponents.

We have already used the numbers 2, 3, . . ., as exponents. In this section, we consider 1, 0, and negative integers as exponents.

### a ONE AND ZERO AS EXPONENTS

Look for a pattern in the following:

$$8 \cdot 8 \cdot 8 \cdot 8 = 8^4 \qquad \text{We divide by 8 each time.}$$
$$8 \cdot 8 \cdot 8 = 8^3$$
$$8 \cdot 8 = 8^2$$
$$8 = 8^?$$
$$1 = 8^?.$$

The exponents decrease by 1 each time. Continuing the pattern, we have

$$8 = 8^1 \quad \text{and} \quad 1 = 8^0.$$

We make the following definitions.

---

**SKILL REVIEW**

*Evaluate exponential notation.* [1.9b]

Evaluate.

1. $10^2$   2. $(-10^2)$

Answers: 1. 100  2. 100

MyLab Math
VIDEO

---

.......... **Caution!** ..........

Don't confuse the powers 1 and 0. Be careful: $8^0 = 1$, but $8^1 = 8$.

---

| THE EXPONENTS 1 AND 0 |
|---|

$b^1 = b$, for any number $b$.
$b^0 = 1$, for any nonzero number $b$.

---

**EXAMPLE 1**  Evaluate $23^1$, $-23^1$, and $(-23)^1$.

$$23^1 = 23;$$
$$-23^1 = -23; \qquad \text{We read } -23^1 \text{ as "the opposite of } 23^1.\text{"}$$
$$(-23)^1 = -23 \qquad \text{We read } (-23)^1 \text{ as "negative 23 to the first."}$$

**EXAMPLE 2**  Evaluate $3^0$, $(-3)^0$, and $-3^0$.

$$3^0 = 1; \qquad\qquad\qquad\qquad \text{3 is the base; 0 is the exponent.}$$
$$(-3)^0 = 1; \qquad\qquad\qquad\quad -3 \text{ is the base; 0 is the exponent.}$$
$$-3^0 = -1 \cdot 3^0 = -1 \cdot 1 = -1 \qquad \text{Note that } -3^0 \neq (-3)^0.$$

◀ Do Exercises 1–6.

**EXAMPLE 3**  Evaluate $m^0 + 5$ for $m = 9$.

$$m^0 + 5 = 9^0 + 5 = 1 + 5 = 6$$

---

Evaluate.

1. $16^1$   2. $-16^1$

3. $(-35)^1$   4. $7^0$

5. $(-9)^0$   6. $-8^0$

**Answers**

1. 16  2. −16  3. −35  4. 1  5. 1
6. −1

**EXAMPLE 4** Evaluate $(3x + 2)^0$ for $x = -5$.

We substitute $-5$ for $x$ and follow the rules for order of operations:

$$(3x + 2)^0 = (3(-5) + 2)^0 \quad \text{Substituting}$$
$$= (-15 + 2)^0 \quad \text{Multiplying}$$
$$= (-13)^0 = 1.$$

**Do Exercises 7 and 8.** ▶

**7.** Evaluate $(2x - 9)^0$ for $x = 3$.

**8.** Evaluate $t^0 - 4$ for $t = 7$.

## b NEGATIVE INTEGERS AS EXPONENTS

The pattern used to help define the exponents 1 and 0 can also be used to define negative-integer exponents:

$$
\begin{array}{l|l}
8 \cdot 8 \cdot 8 = 8^3 & \text{We divide by 8 each time.} \\
8 \cdot 8 = 8^2 & \\
8 = 8^1 & \\
1 = 8^0 & \\
\dfrac{1}{8} = 8^? & \\
\dfrac{1}{8 \cdot 8} = 8^?. &
\end{array}
$$

The exponents decrease by 1 each time. To continue the pattern, we would say that

$$\frac{1}{8} = 8^{-1} \quad \text{and} \quad \frac{1}{8 \cdot 8} = 8^{-2}.$$

Thus, if we are to preserve the above pattern, we must have

$$\frac{1}{8^1} = 8^{-1} \quad \text{and} \quad \frac{1}{8^2} = 8^{-2}.$$

This leads to our definition of negative exponents.

### NEGATIVE EXPONENTS

For any nonzero numbers $a$ and $b$, and any integer $n$,

$$a^{-n} = \frac{1}{a^n} \quad \text{and} \quad \left(\frac{a}{b}\right)^{-n} = \left(\frac{b}{a}\right)^n.$$

(A base raised to a negative exponent is equal to the reciprocal of the base raised to a positive exponent.)

**EXAMPLES** Write an equivalent expression using a positive exponent. Then simplify.

**5.** $4^{-2} = \dfrac{1}{4^2} = \dfrac{1}{16}$    Note that $4^{-2}$ represents a *positive* number.

**6.** $(-3)^{-2} = \dfrac{1}{(-3)^2} = \dfrac{1}{(-3)(-3)} = \dfrac{1}{9}$

**7.** $m^{-3} = \dfrac{1}{m^3}$

**8.** $ab^{-1} = a \cdot b^{-1} = a\left(\dfrac{1}{b^1}\right) = a\left(\dfrac{1}{b}\right) = \dfrac{a}{b}$    Think of $a$ as $\dfrac{a}{1}$ if you wish.

*Answers*

**7.** 1   **8.** −3

Write an equivalent expression with positive exponents. Then simplify.

**9.** $4^{-3}$     **10.** $5^{-2}$

**11.** $2^{-4}$     **12.** $(-2)^{-3}$

**13.** $\left(\dfrac{5}{3}\right)^{-2}$

**14.** $xy^{-2} = x \cdot y^{-2}$

$= x \cdot \dfrac{1}{\boxed{\phantom{xx}}}$

$= \dfrac{x}{\boxed{\phantom{xx}}}$

Write an equivalent expression with negative exponents.

**15.** $\dfrac{1}{9^2}$     **16.** $\dfrac{7}{x^4}$

Write an equivalent expression with positive exponents.

**17.** $\dfrac{m^3}{n^{-5}}$     **18.** $\dfrac{ab}{c^{-1}}$

Write an equivalent expression with positive exponents.

**19.** $\dfrac{a^4}{b^{-6}}$     **20.** $\dfrac{x^7 y}{z^{-4}}$

**21.** $\dfrac{a^4 b^{-7}}{c^{-3}} = \dfrac{a^4 c^{\boxed{\phantom{x}}}}{b^{\boxed{\phantom{x}}}}$  GS

*Answers*

**9.** $\dfrac{1}{4^3}; \dfrac{1}{64}$  **10.** $\dfrac{1}{5^2}; \dfrac{1}{25}$  **11.** $\dfrac{1}{2^4}; \dfrac{1}{16}$

**12.** $\dfrac{1}{(-2)^3}; -\dfrac{1}{8}$  **13.** $\left(\dfrac{3}{5}\right)^2; \dfrac{9}{25}$

**14.** $x \cdot \dfrac{1}{y^2}; \dfrac{x}{y^2}$  **15.** $9^{-2}$  **16.** $7x^{-4}$  **17.** $m^3 n^5$

**18.** $abc$  **19.** $a^4 b^6$  **20.** $x^7 y z^4$  **21.** $\dfrac{a^4 c^3}{b^7}$

*Guided Solutions:*

**14.** $y^2, y^2$  **21.** $3, 7$

**9.** $\left(\dfrac{5}{6}\right)^{-2} = \left(\dfrac{6}{5}\right)^2$     $\left(\dfrac{a}{b}\right)^{-n} = \left(\dfrac{b}{a}\right)^n$

$= \dfrac{6}{5} \cdot \dfrac{6}{5} = \dfrac{36}{25}$

.................................................... **Caution!** ....................................................

Note in Example 5 that

$4^{-2} \neq 4(-2)$   and   $4^{-2} \neq \dfrac{1}{-4^2}.$

In general, $a^{-n} \neq a(-n)$. The negative exponent also does not indicate that a negative number is in the result.

....................................................

◀ **Do Exercises 9–14.**

**EXAMPLES**   Write an equivalent expression using a negative exponent.

**10.** $\dfrac{1}{7^2} = 7^{-2}$     Reading $a^{-n} = \dfrac{1}{a^n}$ from right to left: $\dfrac{1}{a^n} = a^{-n}$

**11.** $\dfrac{5}{x^8} = 5 \cdot \dfrac{1}{x^8} = 5x^{-8}$

◀ **Do Exercises 15 and 16.**

Consider an expression like

$$\dfrac{a^2}{b^{-3}},$$

in which the denominator is a negative power. We can simplify as follows:

$\dfrac{a^2}{b^{-3}} = \dfrac{a^2}{\dfrac{1}{b^3}}$     Rewriting $b^{-3}$ as $\dfrac{1}{b^3}$

$= a^2 \cdot \dfrac{b^3}{1}$     To divide by a fraction, we multiply by its reciprocal.

$= a^2 b^3.$

◀ **Do Exercises 17 and 18.**

Our work above indicates that to divide by a base raised to a negative power, we can instead multiply by the opposite power of the same base. This will shorten our work.

**EXAMPLES**   Write an equivalent expression using positive exponents.

**12.** $\dfrac{x^3}{y^{-2}} = x^3 y^2$     Instead of dividing by $y^{-2}$, multiply by $y^2$.

**13.** $\dfrac{a^2 b^5}{c^{-6}} = a^2 b^5 c^6$     Instead of dividing by $c^{-6}$, multiply by $c^6$.

**14.** $\dfrac{x^{-2} y}{z^{-3}} = x^{-2} y z^3 = \dfrac{y z^3}{x^2}$

◀ **Do Exercises 19–21.**

## ✔ Check Your Understanding

**Reading Check**  Determine whether each statement is true or false.

_____ **RC1.** The expression $a^0$ is equal to 1 for any number $a$.

_____ **RC2.** When a positive number is raised to a negative exponent, the result is negative.

_____ **RC3.** The expressions $-6^0$ and $(-6)^0$ are equivalent.

_____ **RC4.** If $a > 1$, then $a^{-1} < 1$.

**Concept Check**  Choose the number from the following list that makes each statement true. Not every number will be used.

$$-2 \qquad -1 \qquad 0 \qquad 1 \qquad 2$$

**CC1.** $7^{\square} = 1$  **CC2.** $7^{\square} = 7$  **CC3.** $7^{-2} = \dfrac{1}{7^{\square}}$  **CC4.** $\dfrac{1}{7} = 7^{\square}$

---

**a**  Evaluate.

**1.** $4^0$

**2.** $17^0$

**3.** $3.14^1$

**4.** $2.67^1$

**5.** $(-19.57)^1$

**6.** $(-34.6)^0$

**7.** $(-98.6)^0$

**8.** $(-98.6)^1$

**9.** $x^0, x \neq 0$

**10.** $a^0, a \neq 0$

**11.** $x^1$, for $x = 97$

**12.** $a^1$, for $a = -10$

**13.** $(3x - 17)^0$, for $x = 10$

**14.** $(7x - 45)^0$, for $x = 8$

**15.** $(5x - 3)^1$, for $x = 4$

**16.** $(35 - 4x)^1$, for $x = 8$

**17.** $(4m - 19)^0$, for $m = 3$

**18.** $(9 - 2x)^0$, for $x = 5$

**19.** $3x^0 + 4$, for $x = -2$

**20.** $7x^0 + 6$, for $x = -3$

**21.** $(3x)^0 + 4$, for $x = -2$

**22.** $(7x)^0 + 6$, for $x = -3$

**23.** $(5 - 3x^0)^1$, for $x = 19$

**24.** $(5x^1 - 29)^0$, for $x = 4$

---

**b**  Write an equivalent expression with a positive exponent. Then simplify, if possible.

**25.** $3^{-2}$

**26.** $2^{-3}$

**27.** $10^{-4}$

**28.** $5^{-6}$

**29.** $t^{-4}$

**30.** $x^{-2}$

**31.** $7^{-1}$

**32.** $10^{-1}$

**33.** $(-5)^{-2}$

**34.** $(-4)^{-3}$

**35.** $(-x)^{-2}$

**36.** $(-n)^{-2}$

Write an equivalent expression using positive exponents. Then, if possible, simplify.

**37.** $3x^{-7}$

**38.** $-6y^{-2}$

**39.** $mn^{-5}$

**40.** $cd^{-1}$

**41.** $xy^{-1}$

**42.** $pw^{-3}$

**43.** $-7a^{-9}$

**44.** $9p^{-4}$

**45.** $\dfrac{1}{x^{-3}}$

**46.** $\dfrac{1}{x^{-1}}$

**47.** $\dfrac{x}{y^{-4}}$

**48.** $\dfrac{r}{t^{-7}}$

**49.** $\dfrac{r^5}{t^{-3}}$

**50.** $\dfrac{x^7}{y^{-5}}$

**51.** $\dfrac{xy}{z^{-1}}$

**52.** $\dfrac{ab}{c^{-6}}$

**53.** $\dfrac{a^8b}{c^{-2}}$

**54.** $\dfrac{x^7y}{t^{-6}}$

**55.** $\dfrac{x^2y^3}{z^{-8}}$

**56.** $\dfrac{x^8y^{10}}{z^{-1}}$

**57.** $\dfrac{x^{-4}y^3}{t^{-2}}$

**58.** $\dfrac{ab^{-6}}{c^{-2}}$

**59.** $\dfrac{cd^{-4}}{a^{-1}}$

**60.** $\dfrac{x^3y^{-8}}{z^{-4}}$

**61.** $\left(\dfrac{2}{5}\right)^{-2}$

**62.** $\left(\dfrac{3}{7}\right)^{-2}$

**63.** $\left(\dfrac{5}{a}\right)^{-3}$

**64.** $\left(\dfrac{x}{3}\right)^{-4}$

Write an equivalent expression using a negative exponent.

**65.** $\dfrac{1}{7^3}$

**66.** $\dfrac{1}{5^2}$

**67.** $\dfrac{9}{x^3}$

**68.** $\dfrac{4}{y^2}$

## Skill Maintenance

**69.** When used for a singles match, a regulation tennis court is 27 ft by 78 ft. Find its perimeter.  [2.7b]

**70.** The Floral Doctor's delivery van traveled 147 mi on 10.5 gal of gas. How many miles per gallon did the van get?  [6.2a]

**71.** Ramon's new truck gets 21 mpg. This is 20% more than the mileage his old truck got. What mileage did the old truck get?  [7.5a]

**72.** A 5% sales tax is added to the price of a two-speed washing machine. If the machine is priced at $399, find the total amount paid.  [7.6a]

**73.** Of the 8 fish Mac caught, 3 were trout. What percentage were not trout?  [7.4a]

**74.** The diameter of a compact disc is 12 cm. What is its circumference? (Use 3.14 for $\pi$.)  [9.3b]

**75.** Multiply: $-57 \cdot 48$.  [2.4a]

**76.** Multiply: $(-72)(-46)$.  [2.4a]

## Synthesis

Simplify.

**77.** $3 + 4^{-1} \cdot 10^2$

**78.** $(2 + 3)^{-2} \cdot 10^2 \div 2^{-1}$

**79.** $4^2 \div 2^{-1} \cdot 5^1 - (12 \cdot 10)^0$

**80.** $(42 - 6 \cdot 7)^8 \cdot (13 - 5 \cdot 4)^{-3}$

**81.** ▦ Evaluate $\dfrac{3^x}{3^{x-1}}$ for $x = -4$ and then for $x = -40$.

**82.** ▦ Evaluate $\dfrac{5^x}{5^{x+1}}$ for $x = -3$ and then for $x = -30$.

# Working with Exponents

There are several rules that we can use to simplify exponential expressions. We first consider multiplying powers with like bases. Throughout this section, we assume that a variable in a denominator is not zero.

a  Use the product rule to multiply exponential expressions with like bases.

b  Use the quotient rule to divide exponential expressions with like bases.

c  Use the power rule to raise powers to powers.

d  Raise a product to a power and a quotient to a power.

## a  MULTIPLYING POWERS WITH LIKE BASES

When we multiply expressions like $a^3$ and $a^2$, we can use the associative law to combine the factors:

$$a^3 \cdot a^2 = \underbrace{(a \cdot a \cdot a)}_{3 \text{ factors}} \underbrace{(a \cdot a)}_{2 \text{ factors}} = \underbrace{a \cdot a \cdot a \cdot a \cdot a}_{5 \text{ factors}} = a^5.$$

Note that the exponent in $a^5$ is the sum of the exponents in $a^3$ and $a^2$. Similarly,

$$b^4 \cdot b^3 = (b \cdot b \cdot b \cdot b)(b \cdot b \cdot b) = b^7, \quad \text{where} \quad 4 + 3 = 7.$$

Adding the exponents gives the correct result.

---

### THE PRODUCT RULE

For any nonzero number $a$ and any integers $m$ and $n$,

$$a^m \cdot a^n = a^{m+n}.$$

(To multiply powers with the same base, keep the base and add the exponents.)

---

.............. **Caution!** ..............

Use the product rule correctly.

$$2^4 \cdot 2^{10} = 2^{14}$$
$$2^4 \cdot 2^{10} \neq 4^{14} \qquad \text{Incorrect!}$$
$$2^4 \cdot 2^{10} \neq 2^{40} \qquad \text{Incorrect!}$$

**EXAMPLES**  Multiply and simplify. Here "simplify" means to express the product as one base with a positive exponent whenever possible.

**1.** $3^6 \cdot 3^2 = 3^{6+2}$      Adding exponents: $a^m \cdot a^n = a^{m+n}$

$\quad = 3^8$

**2.** $x^4 \cdot x^5 = x^{4+5} = x^9$

**3.** $5 \cdot 5^6 \cdot 5^2 = 5^1 \cdot 5^6 \cdot 5^2$      Writing 5 as $5^1$

$\quad = 5^{1+6+2}$

$\quad = 5^9$

**4.** $x^3 \cdot x^{-8} = x^{3+(-8)}$      The product rule works with negative exponents.

$\quad = x^{-5}$

$\quad = \dfrac{1}{x^5}$      Writing with a positive exponent

**5.** $(a^3 b^2)(a^3 b^5) = (a^3 a^3)(b^2 b^5)$      Using the associative law and the commutative law

$\quad = a^6 b^7$      Adding exponents

Do Exercises 1–5. ▶

Simplify.

**1.** $x^4 \cdot x^9$          **2.** $y^{10} \cdot y$

**3.** $2^4 \cdot 2^2 \cdot 2^3$      **4.** $7^{-1} \cdot 7^4$

**5.** $(xy^3)(x^4 y^5)$

*Answers*

**1.** $x^{13}$  **2.** $y^{11}$  **3.** $2^9$  **4.** $7^3$  **5.** $x^5 y^8$

# b  DIVIDING POWERS WITH LIKE BASES

Recall that any expression that is divided or multiplied by 1 is unchanged. This, together with the fact that anything (besides 0) divided by itself is 1, leads to a rule for division:

$$\frac{a^5}{a^2} = \frac{a \cdot a \cdot a \cdot a \cdot a}{a \cdot a} = \frac{a \cdot a \cdot a}{1} \cdot \frac{a \cdot a}{a \cdot a} = \frac{a \cdot a \cdot a}{1} \cdot 1 = a^3.$$

Note that the exponent in $a^3$ is the difference of the exponents in $\frac{a^5}{a^2}$; that is, $5 - 2 = 3$. Similarly,

$$\frac{x^4}{x^3} = \frac{x \cdot x \cdot x \cdot x}{x \cdot x \cdot x} = \frac{x}{1} \cdot \frac{x \cdot x \cdot x}{x \cdot x \cdot x} = \frac{x}{1} \cdot 1 = x^1, \text{ or } x.$$

Subtracting the exponents gives the correct result: $4 - 3 = 1$.

> **Caution!**
>
> Use the quotient rule correctly.
>
> $$\frac{7^{10}}{7^2} = 7^8$$
>
> $$\frac{7^{10}}{7^2} \neq 1^8 \qquad \text{Incorrect!}$$
>
> $$\frac{7^{10}}{7^2} \neq 7^5 \qquad \text{Incorrect!}$$

---

**THE QUOTIENT RULE**

For any nonzero number $a$ and any integers $m$ and $n$,

$$\frac{a^m}{a^n} = a^{m-n}.$$

(To divide powers with the same base, keep the base and subtract the exponent of the denominator from the exponent of the numerator.)

---

**EXAMPLES**  Divide and simplify. Here "simplify" means to write the expression as one base with a positive exponent whenever possible.

**6.** $\dfrac{4^7}{4^2} = 4^{7-2}$        Subtracting exponents

$\quad\quad = 4^5$

**7.** $\dfrac{x^9}{x^{-3}} = x^{9-(-3)} = x^{12}$     Use parentheses when subtracting negative exponents.

**8.** $\dfrac{p}{p^5} = \dfrac{p^1}{p^5}$        Writing $p$ as $p^1$

$\quad\quad = p^{1-5}$

$\quad\quad = p^{-4}$

$\quad\quad = \dfrac{1}{p^4}$       Writing with a positive exponent

**9.** $\dfrac{3^{10}}{3^{10}} = 3^{10-10} = 3^0 = 1$    $a^0 = 1$

**10.** $\dfrac{c^8 d^{12}}{c^3 d^{10}} = \dfrac{c^8}{c^3} \cdot \dfrac{d^{12}}{d^{10}}$

$\quad\quad = c^{8-3} d^{12-10}$

$\quad\quad = c^5 d^2$

◀ **Do Exercises 6–10.**

Simplify.

**6.** $\dfrac{x^5}{x^3}$          **7.** $\dfrac{a^8}{a}$

**8.** $\dfrac{x^{11}}{x^{11}}$         **9.** $\dfrac{x^2 y^5}{x y^4}$

**10.** $\dfrac{4^{-3}}{4^{10}} = 4^{\square - \square}$      **GS**

$\quad\quad = 4^{\square}$

$\quad\quad = \dfrac{1}{4^{\square}}$

**Answers**

**6.** $x^2$ **7.** $a^7$ **8.** $1$ **9.** $xy$ **10.** $\dfrac{1}{4^{13}}$

*Guided Solution:*

**10.** $-3, 10, -13, 13$

## c | RAISING A POWER TO A POWER

Consider an expression like $(7^2)^4$:

$$(7^2)^4 = (7^2)(7^2)(7^2)(7^2) \qquad \text{There are four factors of } 7^2.$$
$$= (7\cdot7)(7\cdot7)(7\cdot7)(7\cdot7)$$
$$= 7\cdot7\cdot7\cdot7\cdot7\cdot7\cdot7\cdot7 \qquad \text{Using an associative law}$$
$$= 7^8.$$

Note that the exponent in $7^8$ is the product of the exponents in $(7^2)^4$. Similarly,

$$(y^5)^3 = y^5 \cdot y^5 \cdot y^5 \qquad \text{There are three factors of } y^5.$$
$$= (y\cdot y\cdot y\cdot y\cdot y)(y\cdot y\cdot y\cdot y\cdot y)(y\cdot y\cdot y\cdot y\cdot y)$$
$$= y^{15};$$

or $\qquad (y^5)^3 = y^{5\cdot3} = y^{15}.$

---

### THE POWER RULE

For any nonzero number $a$ and any integers $m$ and $n$,

$$(a^m)^n = a^{mn}.$$

(To raise a power to a power, multiply the exponents and leave the base unchanged.)

---

**EXAMPLES** Simplify. Express the answer using positive exponents.

**11.** $(2^4)^3 = 2^{4\cdot3}$    Multiplying exponents
$$= 2^{12}$$

**12.** $(y^{-3})^8 = y^{-3\cdot8} = y^{-24} = \dfrac{1}{y^{24}}$    **13.** $(x^{-2})^{-10} = x^{(-2)(-10)} = x^{20}$

**Do Exercises 11–13.** ▶

Simplify.

**11.** $(3^4)^5$

**12.** $(5^{-2})^8$

**13.** $(t^{-8})^{-5}$

## d | RAISING A PRODUCT OR A QUOTIENT TO A POWER

When an expression inside parentheses is raised to a power, the inside expression is the base. Let's compare $2a^3$ and $(2a)^3$.

$2a^3 = 2\cdot a\cdot a\cdot a$    The base is $a$.      $(2a)^3 = (2a)(2a)(2a)$    The base is $2a$.
$$= (2\cdot2\cdot2)(a\cdot a\cdot a)$$
$$= 2^3 a^3$$
$$= 8a^3$$

We see that $2a^3$ and $(2a)^3$ are *not* equivalent. Note too that $(2a)^3$ can be simplified by cubing each factor in $2a$. This leads to the following rule for raising a product to a power.

---

### RAISING A PRODUCT TO A POWER

For any nonzero numbers $a$ and $b$ and any integer $n$,

$$(ab)^n = a^n b^n.$$

(To raise a product to a power, raise each factor to that power.)

---

*Answers*

**11.** $3^{20}$    **12.** $\dfrac{1}{5^{16}}$    **13.** $t^{40}$

Simplify.

**14.** $(x^3y^8)^5$

**15.** $(-2ab^4)^2$

GS

$$= (-2)\ \square\ (a)\ \square\ (b^4)\ \square$$
$$= \square\ a\ \square\ b\ \square$$

**EXAMPLES** Simplify.

**14.** $(3a^4)^2 = (3^1a^4)^2$  Writing 3 as $3^1$

$\qquad = (3^1)^2 \cdot (a^4)^2$  Raising each factor to the second power

$\qquad = 3^2 \cdot a^8 = 9a^8$  Multiplying exponents and simplifying

**15.** $(-5x^5y^8)^3 = (-5)^3(x^5)^3(y^8)^3$  Raising each factor to the third power

$\qquad = -125x^{15}y^{24}$

◀ **Do Exercises 14 and 15.**

There is a similar rule for raising a quotient to a power.

---

**RAISING A QUOTIENT TO A POWER**

For any nonzero numbers $a$ and $b$ and any integer $n$,

$$\left(\frac{a}{b}\right)^n = \frac{a^n}{b^n}.$$

(To raise a quotient to a power, raise the numerator to the power and divide by the denominator raised to the power.)

---

**EXAMPLES**

**16.** $\left(\dfrac{x}{7}\right)^2 = \dfrac{x^2}{7^2} = \dfrac{x^2}{49}$  Squaring the numerator and the denominator

**17.** $\left(\dfrac{5}{a^4}\right)^3 = \dfrac{5^3}{(a^4)^3}$  Raising a quotient to a power

$\qquad = \dfrac{5^3}{a^{12}} = \dfrac{125}{a^{12}}$  Using the power rule (multiplying exponents) and simplifying

◀ **Do Exercises 16 and 17.**

In the following summary of definitions and rules, we assume that no denominators are 0 and $0^0$ is not considered.

Simplify.

**16.** $\left(\dfrac{a}{b}\right)^8$

**17.** $\left(\dfrac{x^4}{y^5}\right)^2$

---

**DEFINITIONS AND PROPERTIES OF EXPONENTS**

For any integers $m$ and $n$,

| | |
|---|---|
| 1 as an exponent: | $a^1 = a$ |
| 0 as an exponent: | $a^0 = 1$ |
| Negative exponents: | $a^{-n} = \dfrac{1}{a^n}$ |
| The product rule: | $a^m \cdot a^n = a^{m+n}$ |
| The quotient rule: | $\dfrac{a^m}{a^n} = a^{m-n}$ |
| The power rule: | $(a^m)^n = a^{mn}$ |
| Raising a product to a power: | $(ab)^n = a^nb^n$ |
| Raising a quotient to a power: | $\left(\dfrac{a}{b}\right)^n = \dfrac{a^n}{b^n}$ |

---

**Answers**

**14.** $x^{15}y^{40}$  **15.** $4a^2b^8$  **16.** $\dfrac{a^8}{b^8}$  **17.** $\dfrac{x^8}{y^{10}}$

**Guided Solution:**

**15.** 2, 2, 2, 4, 2, 8

## ✓ Check Your Understanding

**Reading Check** Choose from the list on the right the word that best completes each sentence. Words may be used more than once or not at all.

**RC1.** When simplifying $6^4 6^7$, we _____ 4 and 7.

**RC2.** When simplifying $(3^8)^2$, we _____ 8 and 2.

**RC3.** When simplifying $7^{-3} 7^{-5}$, we _____ −3 and −5.

**RC4.** When simplifying $\dfrac{4^9}{4^{-6}}$, we _____ 9 and −6.

**RC5.** When simplifying $(-2)^4 (-2)^{-10}$, we _____ 4 and −10.

**RC6.** When simplifying $\dfrac{5^{-4}}{5}$, we _____ −4 and 1.

**RC7.** When simplifying $(y^{-3})^{-9}$, we _____ −3 and −9.

| | |
|---|---|
| add | |
| subtract | |
| multiply | |
| divide | |

**Concept Check** Match each expression with an equivalent expression from the column on the right.

**CC1.** $4^5 \cdot 4^3$ _____

**CC2.** $(4^5)^3$ _____

**CC3.** $\dfrac{4^5}{4^3}$ _____

**CC4.** $\left(\dfrac{4}{5}\right)^3$ _____

**CC5.** $(4 \cdot 5)^3$ _____

**a)** $4^{5-3}$

**b)** $4^{5 \cdot 3}$

**c)** $4^{5+3}$

**d)** $4^3 \cdot 5^3$

**e)** $\dfrac{4^3}{5^3}$

---

**a** Simplify. Assume that no denominator is 0.

**1.** $3^4 \cdot 3^2$

**2.** $4^5 \cdot 4^7$

**3.** $x^2 \cdot x^{12}$

**4.** $y^3 \cdot y^9$

**5.** $a^6 \cdot a$

**6.** $x \cdot x^4$

**7.** $8 \cdot 8^5$

**8.** $2^{10} \cdot 2$

**9.** $5^{17} \cdot 5^{21}$

**10.** $4^{32} \cdot 4^{16}$

**11.** $x^2 \cdot x \cdot x^5$

**12.** $y \cdot y^4 \cdot y^8$

**13.** $a^{-5} \cdot a^8$

**14.** $c^{12} \cdot c^{-9}$

**15.** $p^6 \cdot p^{-5}$

**16.** $t^{-4} \cdot t^5$

**17.** $x^{-5} \cdot x^{-7}$

**18.** $y^{-3} \cdot y^{-4}$

**19.** $5^4 \cdot 5^{-10}$

**20.** $6^{-12} \cdot 6^2$

**21.** $(a^3 b^5)(ab^9)$

**22.** $(x^2 y^{10})(x^5 y)$

**23.** $a^{-4} \cdot a^9 \cdot a^{-6}$

**24.** $x^{-11} \cdot x^4 \cdot x^6$

**b** Simplify. Assume that no denominator is 0.

**25.** $\dfrac{8^9}{8^2}$

**26.** $\dfrac{3^{11}}{3^3}$

**27.** $\dfrac{x^4}{x^3}$

**28.** $\dfrac{x^{12}}{x^{11}}$

**29.** $\dfrac{a^{10}}{a}$

**30.** $\dfrac{c^{12}}{c}$

**31.** $\dfrac{x^2}{x^5}$

**32.** $\dfrac{y^4}{y^{12}}$

**33.** $\dfrac{7^8}{7^8}$

**34.** $\dfrac{10^6}{10^6}$

**35.** $\dfrac{t^2}{t^{-1}}$

**36.** $\dfrac{n^{-5}}{n^8}$

**37.** $\dfrac{c^{-2}}{c^{-6}}$

**38.** $\dfrac{t^{-2}}{t}$

**39.** $\dfrac{x^4 y^{12}}{xy^7}$

**40.** $\dfrac{u^9 v^{15}}{u^8 v}$

**c**   Simplify. Assume that no denominator is 0.

**41.** $(x^2)^5$

**42.** $(y^5)^7$

**43.** $(2^{-3})^6$

**44.** $(3^{-2})^4$

**45.** $(7^{-1})^{-6}$

**46.** $(9^{-4})^{-3}$

**47.** $(y^{12})^{-1}$

**48.** $(x^{-18})^{-1}$

**d**   Simplify. Assume that no denominator is 0.

**49.** $(x^2 y^5)^2$

**50.** $(a^3 b^4)^2$

**51.** $(2x^2 y)^3$

**52.** $(3xy^3)^2$

**53.** $(-2a^2 b^5)^3$

**54.** $(-5a^4 b^2)^3$

**55.** $(-9x^2 y^2)^2$

**56.** $(-10x^5 y)^2$

**57.** $\left(\dfrac{a}{b}\right)^7$

**58.** $\left(\dfrac{x}{y}\right)^{10}$

**59.** $\left(\dfrac{a^2}{2}\right)^3$

**60.** $\left(\dfrac{x^4}{5}\right)^2$

**61.** $\left(\dfrac{x^4}{y^5}\right)^3$

**62.** $\left(\dfrac{a^6}{b^{10}}\right)^4$

**63.** $\left(\dfrac{2^4}{3^5}\right)^2$

**64.** $\left(\dfrac{5^9}{3^5}\right)^4$

## Skill Maintenance

**65.** A sidewalk of uniform width is built around three sides of a store, as shown in the figure. What is the area of the sidewalk?   [1.8a]

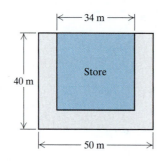

**66.** A real estate agent's commission rate is 6%. The agent receives a commission of $7380 on the sale of a home. For how much did the home sell?   [7.6b]

## Synthesis

Simplify. Assume that no denominator is 0.

**67.** $(3y)^{10}(3y)^6$

**68.** $(x^2 y^3 z^8)^0$

**69.** $\left(\dfrac{x}{y}\right)^{-1}$

**70.** $a^x \cdot a^{2x}$

**71.** $\dfrac{x^{3y}}{x^{2y}}$

**72.** $(a^{-3} b^2)^{-4}$

**73.** $\dfrac{a^{33}}{a^{3(11)}}$

**74.** $\left(\dfrac{1}{a}\right)^{-n}$

# Scientific Notation

**a  WRITING SCIENTIFIC NOTATION**

We can write numbers using different types of notation, such as fraction notation, decimal notation, and percent notation. Now we study another, **scientific notation,** which makes use of exponential notation. Scientific notation is especially useful when calculations involve very large or very small numbers. The following are examples of scientific notation:

- The number of eggs a coconut octopus may lay: $1 \times 10^5 = 100,000$

- The length of an *E. coli* bacterium: $2 \times 10^{-6}$ m $= 0.000002$ m

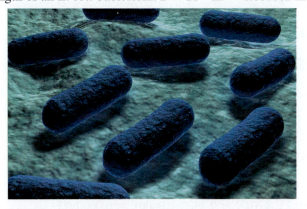

---

**SCIENTIFIC NOTATION**

**Scientific notation** for a number is an expression of the type

$$M \times 10^n,$$

where $n$ is an integer, $M$ is greater than or equal to 1 and less than 10 ($1 \le M < 10$), and $M$ is expressed in decimal notation. The notation $10^n$ is also considered to be scientific notation when $M = 1$.

---

You should try to make conversions to scientific notation mentally as much as possible. To do so, note the following.

---

In scientific notation, a positive power of 10 indicates a large number (greater than or equal to 10) and a negative power of 10 indicates a small number (between 0 and 1).

---

Converting from scientific notation to decimal notation involves multiplying by a power of 10. To convert $M \times 10^n$ to decimal notation, we move the decimal point.

- When $n$ is positive, we move the decimal point right $n$ places.
- When $n$ is negative, we move the decimal point left $|n|$ places.

**EXAMPLES** Convert each number to decimal notation.

**1.** $7.893 \times 10^5 = 789{,}300$     The decimal point moves right 5 places.

*Think*: Positive exponent indicates large number.

*Check*: $7.893 \times 10^5 = 7.893 \times 100{,}000 = 789{,}300$

**2.** $4.7 \times 10^{-8} = 0.000000047$     The decimal point moves left 8 places.

*Think*: Negative exponent indicates small number.

*Check*: $4.7 \times 10^{-8} = 4.7 \times 0.00000001 = 0.000000047$

◀ **Do Exercises 1 and 2.**

We reverse the process when converting from decimal notation to scientific notation. To convert a number to scientific notation, we write it in the form $M \times 10^n$, where $M$ has exactly one nonzero digit to the immediate left of the decimal point. Compare the following.

| *Scientific Notation* | *NOT Scientific Notation* | |
|---|---|---|
| $1.98 \times 10^{12}$ | $19.8 \times 10^{11}$ | 19.8 is greater than 10. |
| $2.007 \times 10^{-4}$ | $0.2007 \times 10^{-3}$ | 0.2007 is less than 1. |

◀ **Do Exercises 3–6.**

**EXAMPLES** Convert each number to scientific notation.

**3.** $78{,}000 = 78{,}000. = 7.8000 \times 10^4 = 7.8 \times 10^4$

                           4 places

*Check*: $7.8 \times 10^4 = 7.8 \times 10{,}000 = 78{,}000$

**4.** $0.0000923 = 000009.23 \times 10^{-5} = 9.23 \times 10^{-5}$

                    5 places

*Check*: $9.23 \times 10^{-5} = 9.23 \times 0.00001 = 0.0000923$

◀ **Do Exercises 7 and 8.**

## b   MULTIPLYING AND DIVIDING USING SCIENTIFIC NOTATION

### Multiplying

Consider the product $400 \cdot 2000 = 800{,}000$. In scientific notation, this is

$$(4 \times 10^2) \cdot (2 \times 10^3) = (4 \cdot 2)(10^2 \cdot 10^3) = 8 \times 10^5.$$

By applying the commutative and associative laws, we found this product by multiplying $4 \cdot 2$, to get 8, and $10^2 \cdot 10^3$, to get $10^5$ (adding exponents).

---

Convert each number to decimal notation.

**1.** $6.893 \times 10^{11}$

**2.** $5.67 \times 10^{-5}$

Determine whether each number is written in scientific notation.

**3.** $8.06 \times 10^3$

**4.** $0.49 \times 10^{12}$

**5.** $10.608 \times 10^{-3}$

**6.** $1.008 \times 10^{-6}$

Convert each number to scientific notation.

**7.** $0.000517$

**8.** $523{,}000{,}000$

---

**Answers**

**1.** 689,300,000,000   **2.** 0.0000567   **3.** Yes
**4.** No   **5.** No   **6.** Yes   **7.** $5.17 \times 10^{-4}$
**8.** $5.23 \times 10^8$

**EXAMPLE 5** Multiply: $(1.8 \times 10^6) \cdot (2.3 \times 10^{-4})$.

We apply the commutative and associative laws to get

$$(1.8 \times 10^6) \cdot (2.3 \times 10^{-4}) = (1.8 \cdot 2.3) \times (10^6 \cdot 10^{-4})$$
$$= 4.14 \times 10^{6+(-4)}$$
$$= 4.14 \times 10^2.$$

**EXAMPLE 6** Multiply: $(3.1 \times 10^5) \cdot (4.5 \times 10^{-3})$.

$$(3.1 \times 10^5) \cdot (4.5 \times 10^{-3}) = (3.1 \times 4.5)(10^5 \cdot 10^{-3})$$
$$= 13.95 \times 10^2 \qquad \text{Not scientific notation;}$$
$$\text{13.95 is greater than 10.}$$
$$= (1.395 \times 10^1) \times 10^2 \qquad \text{Substituting}$$
$$\text{1.395} \times 10^1 \text{ for 13.95}$$
$$= 1.395 \times (10^1 \times 10^2) \qquad \text{Using the associative}$$
$$\text{law}$$
$$= 1.395 \times 10^3 \qquad \text{Adding exponents;}$$
$$\text{the answer is now in}$$
$$\text{scientific notation.}$$

**Do Exercises 9 and 10.** ▶

## Dividing

Consider the quotient $800{,}000 \div 400 = 2000$. In scientific notation, this is

$$(8 \times 10^5) \div (4 \times 10^2) = \frac{8 \times 10^5}{4 \times 10^2} = \frac{8}{4} \times 10^{5-2} = 2 \times 10^3.$$

**EXAMPLE 7** Divide: $(3.41 \times 10^5) \div (1.1 \times 10^{-3})$.

$$(3.41 \times 10^5) \div (1.1 \times 10^{-3}) = \frac{3.41 \times 10^5}{1.1 \times 10^{-3}}$$
$$= \frac{3.41}{1.1} \times \frac{10^5}{10^{-3}}$$
$$= 3.1 \times 10^{5-(-3)}$$
$$= 3.1 \times 10^8$$

**EXAMPLE 8** Divide: $(3.2 \times 10^{-7}) \div (8.0 \times 10^6)$.

$$(3.2 \times 10^{-7}) \div (8.0 \times 10^6) = \frac{3.2 \times 10^{-7}}{8.0 \times 10^6}$$
$$= \frac{3.2}{8.0} \times \frac{10^{-7}}{10^6}$$
$$= 0.4 \times 10^{-13} \qquad \text{Not scientific}$$
$$\text{notation; 0.4 is less}$$
$$\text{than 1.}$$
$$= (4.0 \times 10^{-1}) \times 10^{-13} \qquad \text{Substituting}$$
$$\text{4.0} \times 10^{-1} \text{ for 0.4}$$
$$= 4.0 \times (10^{-1} \times 10^{-13}) \qquad \text{Using the}$$
$$\text{associative law}$$
$$= 4.0 \times 10^{-14} \qquad \text{Adding exponents}$$

**Do Exercises 11 and 12.** ▶

Multiply and write scientific notation for the result.

**9.** $(1.12 \times 10^{-8})(5 \times 10^{-7})$

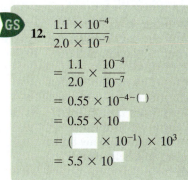

**GS 10.** $(9.1 \times 10^{-17})(8.2 \times 10^3)$
$$= (9.1 \times 8.2)(10^{-17} \cdot 10^3)$$
$$= \boxed{\phantom{xx}} \times 10^{-14}$$
$$= (\boxed{\phantom{xx}} \times 10^1) \times 10^{-14}$$
$$= 7.462 \times 10^{\boxed{\phantom{x}}}$$

Divide and write scientific notation for each result.

**11.** $\dfrac{4.2 \times 10^5}{2.1 \times 10^2}$

**GS 12.** $\dfrac{1.1 \times 10^{-4}}{2.0 \times 10^{-7}}$
$$= \frac{1.1}{2.0} \times \frac{10^{-4}}{10^{-7}}$$
$$= 0.55 \times 10^{-4-(\phantom{x})}$$
$$= 0.55 \times 10^{\boxed{\phantom{x}}}$$
$$= (\boxed{\phantom{xx}} \times 10^{-1}) \times 10^3$$
$$= 5.5 \times 10^{\boxed{\phantom{x}}}$$

*Answers*
**9.** $5.6 \times 10^{-15}$  **10.** $7.462 \times 10^{-13}$
**11.** $2.0 \times 10^3$  **12.** $5.5 \times 10^2$
*Guided Solutions:*
**10.** $74.62, 7.462, -13$  **12.** $-7, 3, 5.5, 2$

## C  APPLICATIONS WITH SCIENTIFIC NOTATION

**EXAMPLE 9**  *Distance from the Sun to Earth.*  Light from the sun traveling at a rate of 300,000 km/s (kilometers per second) reaches Earth in 499 sec. Find the distance, expressed in scientific notation, from the sun to Earth.

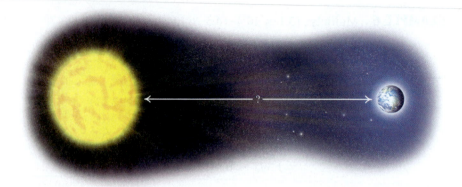

The time it takes light to reach Earth from the sun is $4.99 \times 10^2$ sec. The speed is $3.0 \times 10^5$ km/s. Recall that distance can be expressed in terms of speed and time as

Distance = Speed × Time.

We substitute $3.0 \times 10^5$ for the speed and $4.99 \times 10^2$ for the time:

$$\text{Distance} = (3.0 \times 10^5)(4.99 \times 10^2) \qquad \text{Substituting}$$
$$= 14.97 \times 10^7 \qquad \text{Note that 14.97 is greater than 10.}$$
$$= (1.497 \times 10^1) \times 10^7$$
$$= 1.497 \times (10^1 \times 10^7) \qquad \text{Converting to scientific notation}$$
$$= 1.497 \times 10^8 \text{ km.}$$

Thus, the distance from the sun to Earth is $1.497 \times 10^8$ km.

**13.** *Niagara Falls Water Flow.* On the Canadian side, the amount of water that spills over Niagara Falls in 1 min during the summer is about

$$1.3088 \times 10^8 \text{ L.}$$

How much water spills over the falls in one day? Express the answer in scientific notation.

**14.** *DNA.* The width of a DNA (deoxyribonucleic acid) double helix is about $2 \times 10^{-9}$ m. If its length, fully stretched, is $5 \times 10^{-2}$ m, how many times longer is the helix than it is wide?

**EXAMPLE 10**  *Social Networking.*  The social networking site LinkedIn allows registered users to upload information about their professional careers. Users can also verify, or endorse, the skills of other users. In March 2018, the 550 million LinkedIn users had 10 billion endorsements. On average, how many endorsements did each user have?
**Data:** expandedramblings.com

In order to find the average number of endorsements per LinkedIn user, we divide the total number of endorsements by the number of users. We first write each number using scientific notation:

$$550 \text{ million} = 550,000,000 = 5.5 \times 10^8,$$
$$10 \text{ billion} = 10,000,000,000 = 1.0 \times 10^{10}.$$

We then divide $1.0 \times 10^{10}$ by $5.5 \times 10^8$:

$$\frac{1.0 \times 10^{10}}{5.5 \times 10^8} = \frac{1.0}{5.5} \times \frac{10^{10}}{10^8}$$
$$\approx 0.18 \times 10^2 = (1.8 \times 10^{-1}) \times 10^2 = 1.8 \times 10.$$

On average, each user has $1.8 \times 10$, or 18, endorsements.

◀ Do Exercises 13 and 14.

**Answers**
**13.** $1.884672 \times 10^{11}$ L
**14.** The length of the helix is $2.5 \times 10^7$ times its width.

# Translating for Success

**1. Test Items.** On a test of 90 items, Sally got 80% correct. How many items did she get correct?

**2. Suspension Bridge.** The San Francisco/Oakland Bay suspension bridge is 0.4375 mi long. Convert this distance to yards.

**3. Population Growth.** Brookdale's growth rate per year is 0.9%. If the population was 1,500,000 in 2015, what is the population in 2016?

**4. Roller Coaster Drop.** The Manhattan Express Roller Coaster at the New York–New York Hotel and Casino, Las Vegas, Nevada, has a 144-ft drop. The California Screamin' Roller Coaster at Disney California Adventure, Anaheim, California, has a 32.635-m drop. How much larger in meters is the drop of the Manhattan Express than the drop of the California Screamin'?

**5. Driving Distance.** Nate drives the company car 675 mi in 15 days. At this rate, how far will he drive in 20 days?

---

*The goal of these matching questions is to practice step (2), Translate, of the five-step problem-solving process. Translate each word problem to an equation and select a correct translation from equations A–O.*

**A.** $32.635 \text{ m} + x = 144 \text{ ft} \cdot \dfrac{0.305 \text{ m}}{1 \text{ ft}}$

**B.** $x = 0.4375 \text{ mi} \times \dfrac{5280 \text{ ft}}{1 \text{ mi}} \times \dfrac{12 \text{ in.}}{1 \text{ ft}}$

**C.** $80\% \cdot x = 90$

**D.** $x = 0.89 \text{ km} \cdot \dfrac{1000 \text{ m}}{1 \text{ km}} \cdot \dfrac{1 \text{ m}}{3.281 \text{ ft}}$

**E.** $x = 80\% \cdot 90$

**F.** $x = 420 \text{ m} + 75 \text{ ft} \cdot \dfrac{0.305 \text{ m}}{1 \text{ ft}}$

**G.** $\dfrac{x}{20} = \dfrac{675}{15}$

**H.** $x = 0.4375 \text{ mi} \times \dfrac{5280 \text{ ft}}{1 \text{ mi}} \times \dfrac{1 \text{ yd}}{3 \text{ ft}}$

**I.** $x = 0.89 \text{ km} \cdot \dfrac{0.621 \text{ mi}}{1 \text{ km}} \cdot \dfrac{5280 \text{ ft}}{1 \text{ mi}}$

**J.** $\dfrac{x}{15} = \dfrac{675}{20}$

**K.** $x = 420 \text{ m} \cdot \dfrac{3.281 \text{ ft}}{1 \text{ m}} + 75 \text{ ft}$

**L.** $144 \text{ ft} + x = 32.635 \text{ m} \cdot \dfrac{1 \text{ ft}}{0.305 \text{ m}}$

**M.** $x = 1{,}500{,}000 - 0.9\% \, (1{,}500{,}000)$

**N.** $20 \cdot x = 675$

**O.** $x = 1{,}500{,}000 + 0.9\%(1{,}500{,}000)$

*Answers on page A-24*

---

**6. Test Items.** Jason answered correctly 90 items on a recent test. These items represented 80% of the total number of questions. How many items were on the test?

**7. Population Decline.** Flintville's growth rate per year is −0.9%. If the population was 1,500,000 in 2015, what is the population in 2016?

**8. Bridge Length.** The Tatara Bridge in Onomichi-Imabari, Japan, is 0.89 km long. Convert this distance to feet.

**9. Height of Tower.** The Willis Tower in Chicago is 75 ft taller than the Jin Mao Building in Shanghai. The height of the Jin Mao Building is 420 m. What is the height of the Willis Tower in feet?

**10. Gasoline Usage.** Nate's company car gets 20 miles to the gallon in city driving. How many gallons will it use in 675 mi of city driving?

## ✓ Check Your Understanding

**Reading Check**  Determine whether each statement is true or false.

_____ **RC1.** To convert a number less than 1 to scientific notation, move the decimal point to the right.

_____ **RC2.** A positive exponent in scientific notation indicates a number greater than or equal to 10.

_____ **RC3.** The number $12.856 \times 10^{16}$ is written in scientific notation.

_____ **RC4.** The notations $\frac{1}{2}$, $0.5$, $2^{-1}$, and $5 \times 10^{-1}$ all represent the same number.

**Concept Check**  State whether scientific notation for each of the following numbers includes a positive power of 10 or a negative power of 10.

**CC1.** The distance from Earth to the moon, in feet

**CC2.** The width of a classroom, in miles

**CC3.** The mass of the moon, in grams

**CC4.** The mass of a hydrogen atom, in grams

---

**a**  Convert each number to scientific notation.

**1.** 28,000,000,000

**2.** 4,900,000,000,000

**3.** 907,000,000,000,000,000

**4.** 168,000,000,000,000

**5.** 0.00000304

**6.** 0.000000000865

**7.** 0.000000018

**8.** 0.00000000002

**9.** 100,000,000,000

**10.** 0.0000001

**11.** *Population of the United States.*  It is estimated that the population of the United States will be 419,854,000 in 2050. Convert 419,854,000 to scientific notation.

**Data:** U.S. Census Bureau

**12.** *Microprocessors.*  The minimum feature size of a microprocessor is the transistor gate length. In 2011, the transistor gate length for a new microprocessor was about 0.000000028 m. Convert 0.000000028 to scientific notation.

**13.** *Wavelength of Light.*  The wavelength of red light is 0.00000068 m. Convert 0.00000068 to scientific notation.

**14.** *Gym Memberships.*  In a recent year, Americans spent $19,000,000,000 on gym memberships. Convert 19,000,000,000 to scientific notation.

Convert each number to decimal notation.

**15.** $8.74 \times 10^7$     **16.** $1.85 \times 10^8$     **17.** $5.704 \times 10^{-8}$     **18.** $8.043 \times 10^{-4}$

**19.** $10^7$     **20.** $10^6$     **21.** $10^{-5}$     **22.** $10^{-8}$

 Multiply or divide and write scientific notation for each result.

**23.** $(3 \times 10^4)(2 \times 10^5)$     **24.** $(3.9 \times 10^8)(1.2 \times 10^{-3})$     **25.** $(5.2 \times 10^5)(6.5 \times 10^{-2})$

**26.** $(7.1 \times 10^{-7})(8.6 \times 10^{-5})$     **27.** $(9.9 \times 10^{-6})(8.23 \times 10^{-8})$     **28.** $(1.123 \times 10^4) \times 10^{-9}$

**29.** $\dfrac{8.5 \times 10^8}{3.4 \times 10^{-5}}$     **30.** $\dfrac{5.6 \times 10^{-2}}{2.5 \times 10^5}$     **31.** $(3.0 \times 10^6) \div (6.0 \times 10^9)$

**32.** $(1.5 \times 10^{-3}) \div (1.6 \times 10^{-6})$     **33.** $\dfrac{7.5 \times 10^{-9}}{2.5 \times 10^{12}}$     **34.** $\dfrac{4.0 \times 10^{-3}}{8.0 \times 10^{20}}$

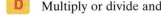 Solve.

**35.** *Coral Reefs.*   There are 10 million bacteria per square centimeter of coral in a coral reef. The coral reefs near the Hawaiian Islands cover 14,000 km². How many bacteria are there in Hawaii's coral reefs? Express the answer in scientific notation.

**Data:** livescience.com; U.S. Geological Survey

**36.** *Nervous System of an Octopus.*   The common octopus, *Octopus vulgaris*, has 500 million neurons in its nervous system, and approximately 67% of its neurons are in its arms. How many neurons are there in the arms of an octopus? Write the answer in scientific notation.

**Data:** *National Geographic*, November 2016, p. 70, "The Power of Eight," Olivia Judson; Roger Hanlon, Marine Biological Laboratory; Guy Levy and Benny Hochner, Hebrew University of Jerusalem; Cliff Ragsdale, University of Chicago

*Space Travel.*   Use the following information for Exercises 37 and 38.

| APPROXIMATE DISTANCE FROM EARTH TO: | |
| --- | --- |
| Moon | 240,000 miles |
| Mars | 35,000,000 miles |
| Pluto | 2,670,000,000 miles |

**37.** *Time to Reach Mars.*   Suppose it takes about 3 days for a space vehicle to travel from Earth to the moon. About how long would it take the same space vehicle traveling at the same speed to reach Mars? Express the answer in scientific notation.

**38.** *Time to Reach Pluto.*   Suppose it takes about 3 days for a space vehicle to travel from Earth to the moon. About how long would it take the same space vehicle traveling at the same speed to reach the dwarf planet Pluto? Express the answer in scientific notation.

**39. Spam.** In 2017, approximately $1.5 \times 10^{11}$ spam e-mails were sent worldwide per day. If there were $5 \times 10^9$ active e-mail accounts in 2017, how many spam e-mails, on average, were sent per day to each account?

Data: marketingprofs.com; statista.com

**40. Data Transmission.** Researchers in the United Kingdom have used an optical communication system to transmit data at a rate of 1.125 terabits per second. The first 19 Marvel Cinematic movies, released beginning in 2008, have a total run-time of 41 hours and consume about 720 gigabits of data. Using the optical communication system described, how long would it take to transfer all 19 movies? (*Note:* 1 gigabit $= 10^9$ bits and 1 terabit $= 10^{12}$ bits.)

Data: engineering.com; businessinsider.com; Netflix

**41. Gold Leaf.** Gold can be milled into a very thin film called *gold leaf*. This film is so thin that it took only 43 oz of gold to cover the dome of Georgia's state capitol building. The gold leaf used was $5 \times 10^{-6}$ m thick. In contrast, a U.S. penny is $1.55 \times 10^{-3}$ m thick. How many sheets of gold leaf are in a stack that is the height of a penny?

Data: georgiaencyclopedia.org

**42. Relative Size.** An influenza virus is about $1.2 \times 10^{-7}$ m in diameter. A staphylococcus bacterium is about $1.5 \times 10^{-6}$ m in diameter. How many influenza viruses would it take, laid side by side, to equal the diameter of the bacterium?

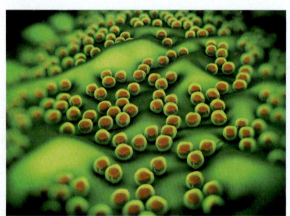

## Skill Maintenance

Solve.

**43.** $2x - 4 - 5x + 8 = x - 3$  [5.7b]

**44.** $4(x - 3) + 5 = 6(x + 2) - 8$  [5.7b]

**45.** $2 + \frac{3}{4}t = \frac{1}{2}$  [4.4b]

**46.** $\frac{5}{16} = \frac{3}{x}$  [6.3b]

## Synthesis

**47.** 🖩 Carry out the indicated operations. Express the result in scientific notation.

$$\frac{(5.2 \times 10^6)(6.1 \times 10^{-11})}{1.28 \times 10^{-3}}$$

**48.** Find the reciprocal and express it in scientific notation.

$$6.25 \times 10^{-3}$$

**49.** Find the LCM for $6.4 \times 10^8$ and $1.28 \times 10^4$.

**50. Stars.** It is estimated that there are 10 billion trillion stars in the known universe. Express the number of stars in scientific notation. *Note:* 1 billion $= 10^9$; 1 trillion $= 10^{12}$.

# Mid-Chapter Review

## Concept Reinforcement

Determine whether each statement is true or false.

_____ **1.** Raising a number to a negative power may not result in a negative number. [10.1b]

_____ **2.** Any number raised to the zero power is zero. [10.1a]

_____ **3.** In scientific notation, if the exponent is negative, the number is very large. [10.3a]

## Guided Solutions

 Fill in each blank with the number that creates a correct statement or solution.

**4.** $y^2 \cdot y^{10} = y^{\boxed{\phantom{x}} +10} = y^{\boxed{\phantom{x}}}$  [10.2a]

**5.** $\dfrac{y^2}{y^{10}} = y^{\boxed{\phantom{x}} -10} = y^{\boxed{\phantom{x}}} = \dfrac{1}{y^{\boxed{\phantom{x}}}}$  [10.2b]

**6.** Convert to scientific notation.

$4{,}032{,}000{,}000{,}000 = \boxed{\phantom{x}} \times 10^{\boxed{\phantom{x}}}$  [10.3a]

**7.** Convert to scientific notation.

$0.000038 = \boxed{\phantom{x}} \times 10^{\boxed{\phantom{x}}}$  [10.3a]

## Mixed Review

Evaluate. [10.1a]

**8.** $(-12)^0$

**9.** $(-12)^1$

**10.** $-12^0$

**11.** $(12 - 6t)^1$ for $t = 5$

Write an equivalent expression with a positive exponent. [10.1b]

**12.** $x^{-7}$

**13.** $3^{-1}$

**14.** $9x^{-6}$

**15.** $\dfrac{1}{x^{-4}}$

Simplify. Assume that no denominator is 0. [10.2a, b, c, d]

**16.** $3^4 \cdot 3^{11}$

**17.** $6 \cdot 6^9$

**18.** $x^{-6} \cdot x^{17}$

**19.** $y^{-2} \cdot y \cdot y^{-3}$

**20.** $(x^2 y^8)(x^4 y^3)$

**21.** $\dfrac{8^{11}}{8^2}$

**22.** $\dfrac{x^8}{x^9}$

**23.** $\dfrac{d^5}{d^5}$

**24.** $\dfrac{n^4}{n^{-1}}$

**25.** $\dfrac{a^{12}b^{11}}{a^4b^7}$

**26.** $(x^4)^8$

**27.** $(y^{-2})^{-6}$

**28.** $(x^{-3})^1$

**29.** $(5^4)^{-12}$

**30.** $(x^3y^4)^5$

**31.** $(3x^2y^4)^2$

**32.** $(-2ab^5)^3$

**33.** $\left(\dfrac{c}{d}\right)^{10}$

**34.** $\left(\dfrac{2}{x^3}\right)^2$

**35.** $\left(\dfrac{a^4}{b^8}\right)^7$

Convert each number to scientific notation. [10.3a]

**36.** 0.000907

**37.** 431,080,000,000

Convert each number to decimal notation. [10.3a]

**38.** $2.08 \times 10^8$

**39.** $4.1 \times 10^{-6}$

**40.** The wavelength of a radio wave is given by the velocity divided by the frequency. The velocity of radio waves is approximately 300,000,000 m/sec, and the frequency of Rick's ham radio repeater is 1,200,000,000 cycles per sec. What is the wavelength of Rick's repeater frequency? Express the answer in scientific notation. [10.3c]

**41.** A good length for a radio antenna is one fourth of the wavelength of the frequency it is designed to receive. What should the length of such a *quarterwave* antenna be for the frequency described in Exercise 40? [10.3c]

**42.** Multiply. Write scientific notation for the result: $(4.5 \times 10^{12})(5.2 \times 10^{20})$. [10.3b]

  **A.** $23.4 \times 10^{240}$          **B.** $23.4 \times 10^{32}$
  **C.** $2.34 \times 10^{33}$           **D.** $2.34 \times 10^{31}$

**43.** Divide. Write scientific notation for the result:
$\dfrac{2.4 \times 10^{-6}}{4.8 \times 10^{-20}}$. [10.3b]

  **A.** $5.0 \times 10^{13}$            **B.** $0.5 \times 10^{14}$
  **C.** $0.5 \times 10^{-26}$           **D.** $5.0 \times 10^{-15}$

## Understanding Through Discussion and Writing

**44.** Under what circumstances is $a^0 > a^1$? [10.1a]

**45.** Consider the expresssion $x^{-3}$. When evaluated, will the expression ever be negative? Explain. [10.1b]

**46.** Which number is larger and why: $5^{-8}$ or $6^{-8}$? Do not use a calculator. [10.1b]

**47.** Emma can give a measurement using the unit km or the unit mm. Which measurement would require the larger exponent if scientific notation were used? Why? [10.3c]

# Addition and Subtraction of Polynomials

## 10.4

### OBJECTIVES

**a** Add polynomials.

**b** Find the opposite of a polynomial.

**c** Subtract polynomials.

**d** Evaluate a polynomial.

A *term* is a number, a variable, a product of numbers and/or variables, or a quotient of numbers and/or variables. Thus, expressions like

$$5x^2, \quad -34, \quad \frac{3}{4}ab^2, \quad xy^3z^5, \quad \text{and} \quad \frac{7n}{m}$$

are terms. A term is called a **monomial** if there is no division by a variable expression. All of the terms above, except for $\frac{7n}{m}$, are monomials. Monomials are used to form **polynomials** like the following:

$$a^2b + c^3, \quad 5y + 3, \quad 3x^2 + 2x - 5, \quad -7a^3 + \tfrac{1}{2}a, \quad 37p^4, \quad x, \quad 0.$$

---

> ### POLYNOMIAL
>
> A **polynomial** is a monomial or a combination of sums and/or differences of monomials.

---

Polynomials can be classified by number of terms. A monomial is a polynomial with one term. A polynomial with two terms is called a **binomial**, and a polynomial with three terms is called a **trinomial.**

In a monomial, the number multiplied by the variable or variables is called the **coefficient**. The coefficient of $-6a^2b$ is $-6$. The coefficient of $\frac{1}{3}x^4$ is $\frac{1}{3}$. Since $x^2 = 1 \cdot x^2$, the coefficient of $x^2$ is 1.

**Do Exercises 1 and 2.** ▶

**1.** What is the coefficient of $19x^3$?

**GS** **2.** What is the coefficient of $-a^2$? Since $-a^2 = \boxed{\phantom{0}} \cdot a^2$, the coefficient of $-a^2$ is $\boxed{\phantom{0}}$.

## a  ADDING POLYNOMIALS

Addition is commutative and associative; that is, we can reorder and regroup addends without changing the sum. We use these properties when adding polynomials in order to pair "like" terms. Recall that when two terms have the same variable(s) raised to the same power(s), they are like terms and can be combined.

**EXAMPLE 1**  Add: $(5x^3 + 4x^2 + 3x) + (2x^3 + 5x^2 - x)$.

$(5x^3 + 4x^2 + 3x) + (2x^3 + 5x^2 - x)$     $5x^3$ and $2x^3$ are like terms; $4x^2$ and $5x^2$ are like terms; $3x$ and $-x$ are like terms.

$= (5x^3 + 2x^3) + (4x^2 + 5x^2) + (3x - x)$     Using the commutative and associative laws to pair like terms

$= 7x^3 + 9x^2 + 2x$     Combining like terms. Remember that $x = 1x$.

**SKILL REVIEW**  *Combine like terms.* [2.7a]

Combine like terms.

**1.** $8x^2 - 3x + 4x - x^2$

**2.** $2x^3 - 6 - 10 - 3x^3 + 9x$

       **Answers: 1.** $7x^2 + x$    **2.** $-x^3 + 9x - 16$

**MyLab Math**
**VIDEO**

*Answers*

**1.** 19    **2.** $-1$

*Guided Solution:*

**2.** $-1, -1$

**EXAMPLE 2**  Add: $(3a^2 + 7a^2b) + (5a^2 - 6ab^2)$.

$$(3a^2 + 7a^2b) + (5a^2 - 6ab^2) = (3a^2 + 5a^2) + 7a^2b - 6ab^2$$
$$= 8a^2 + 7a^2b - 6ab^2 \qquad \text{Combining like terms}$$

Add.

**3.** $(7a^2 + 2a + 8) + (2a^2 + a - 9)$

**4.** $(5x^2y + 3x^2 + 4) + (2x^2y + 4x)$ **GS**

$$= (5x^2y + \boxed{\phantom{xx}}) + 3x^2 + 4 + 4x$$
$$= \boxed{\phantom{xx}} + 3x^2 + 4x + 4$$

**5.** $(2a^3 + 17) + (2a^2 - 9a)$

**EXAMPLE 3**  Add: $(7x^2 + 5) + (5x^3 + 4x)$.

$$(7x^2 + 5) + (5x^3 + 4x) = 7x^2 + 5 + 5x^3 + 4x \qquad \text{There are no like terms here.}$$
$$= 5x^3 + 7x^2 + 4x + 5 \qquad \text{Rearranging the order}$$

Note in Example 3 that we wrote the answer so that the powers of $x$ decrease as we read from left to right. This **descending order** is the traditional way of expressing an answer.

◀ **Do Exercises 3–5.**

**b**  **OPPOSITES OF POLYNOMIALS**

To subtract a number, we can add its opposite. We can similarly subtract a polynomial by adding its opposite. To check if two polynomials are opposites, recall that 5 and $-5$ are opposites because $5 + (-5) = 0$.

> **THE OPPOSITE OF A POLYNOMIAL**
>
> Two polynomials are **opposites**, or **additive inverses**, of each other if their sum is zero.

The opposite of $-9x^2 + x - 7$ is $9x^2 - x + 7$ because

$$(-9x^2 + x - 7) + (9x^2 - x + 7) = -9x^2 + 9x^2 + x + (-x) + (-7) + 7$$
$$= 0.$$

This can be said using algebraic symbolism:

$$\text{The opposite of} \quad (-9x^2 + x - 7) \quad \text{is} \quad 9x^2 - x + 7.$$
$$- \qquad (-9x^2 + x - 7) \quad = \quad 9x^2 - x + 7$$

Note that in the opposite polynomials, the terms in each pair of like terms are opposites.

$$-9x^2 \text{ and } 9x^2 \text{ are opposites.}$$
$$x \text{ and } -x \text{ are opposites.}$$
$$-7 \text{ and } 7 \text{ are opposites.}$$

> **TO FIND THE OPPOSITE OF A POLYNOMIAL**
>
> We can find an equivalent polynomial for the opposite, or additive inverse, of a polynomial by replacing each term with its opposite—that is, by *changing the sign of every term.*

*Answers*
**3.** $9a^2 + 3a - 1$
**4.** $7x^2y + 3x^2 + 4x + 4$
**5.** $2a^3 + 2a^2 - 9a + 17$
*Guided Solution:*
**4.** $2x^2y, 7x^2y$

**EXAMPLE 4**   Find two equivalent expressions for the opposite of

$4x^5 - 7x^3 - 8x + \frac{5}{6}$.

**a)** $-\left(4x^5 - 7x^3 - 8x + \frac{5}{6}\right)$   This is one expression for the opposite of
$4x^5 - 7x^3 - 8x + \frac{5}{6}$.

**b)** $-4x^5 + 7x^3 + 8x - \frac{5}{6}$   Changing the sign of every term

Thus, $-\left(4x^5 - 7x^3 - 8x + \frac{5}{6}\right)$ is equivalent to $-4x^5 + 7x^3 + 8x - \frac{5}{6}$, and
each is the opposite of the original polynomial $4x^5 - 7x^3 - 8x + \frac{5}{6}$.

**Do Exercises 6–9.** ▶

**EXAMPLE 5**   Simplify: $-\left(-7x^4 - \frac{5}{9}x^3 + 8x^2 - x + 67\right)$.

$-\left(-7x^4 - \frac{5}{9}x^3 + 8x^2 - x + 67\right) = 7x^4 + \frac{5}{9}x^3 - 8x^2 + x - 67$

**Do Exercises 10–12.** ▶

## c   SUBTRACTING POLYNOMIALS

We can now subtract a polynomial by adding the opposite of that polynomial. That is, for any polynomials $P$ and $Q$, $P - Q = P + (-Q)$.

**EXAMPLE 6**   Subtract:

$(9x^5 + x^3 - 2x^2 + 4) - (2x^5 + x^4 - 4x^3 - 3x^2)$.

We have

$(9x^5 + x^3 - 2x^2 + 4) - (2x^5 + x^4 - 4x^3 - 3x^2)$

$= (9x^5 + x^3 - 2x^2 + 4) + [-(2x^5 + x^4 - 4x^3 - 3x^2)]$   Adding the opposite

$= (9x^5 + x^3 - 2x^2 + 4) + [-2x^5 - x^4 + 4x^3 + 3x^2]$   Changing the sign of every term

$= 9x^5 + x^3 - 2x^2 + 4 - 2x^5 - x^4 + 4x^3 + 3x^2$

$= 7x^5 - x^4 + 5x^3 + x^2 + 4.$   Combining like terms

**Do Exercises 13 and 14.** ▶

To shorten our work, we often begin by changing the sign of each term in the polynomial being subtracted.

**EXAMPLE 7**   Subtract: $(5a^4 - 7a^3 + 5a^2b) - (-3a^4 + 4a^2b + 6)$.

We have

$(5a^4 - 7a^3 + 5a^2b) - (-3a^4 + 4a^2b + 6)$

$= 5a^4 - 7a^3 + 5a^2b + 3a^4 - 4a^2b - 6$   Removing all parentheses; changing the sign of every term in the second polynomial

$= 8a^4 - 7a^3 + a^2b - 6.$   Combining like terms

**Do Exercise 15.** ▶

---

Find two equivalent expressions for the opposite of each polynomial.

**6.** $12x^4 - 3x^2 + 4x$

**7.** $-4x^4 + 3x^2 - 4x$

**8.** $-13x^6 + 2x^4 - 3x^2 + x - \frac{5}{13}$

**9.** $-8a^3b + 5ab^2 - 2ab$

Simplify.

**10.** $-(4x^3 - 6x + 3)$

**11.** $-(5x^3y + 3x^2y^2 - 7xy^3)$

**12.** $-\left(14x^{10} - \frac{1}{2}x^5 + 5x^3 - x^2 + 3x\right)$

Subtract.

**13.** $(7x^3 + 2x + 4) - (5x^3 - 4)$

**14.** $(-3x^2 + 5x - 4) - (-4x^2 + 11x - 2)$

**15.** Subtract:

$(7x^3 + 3x^2 - xy) - (5x^3 + 3xy + 2)$.

***Answers***

**6.** $-(12x^4 - 3x^2 + 4x); -12x^4 + 3x^2 - 4x$
**7.** $-(-4x^4 + 3x^2 - 4x); 4x^4 - 3x^2 + 4x$
**8.** $-\left(-13x^6 + 2x^4 - 3x^2 + x - \frac{5}{13}\right);$

$13x^6 - 2x^4 + 3x^2 - x + \frac{5}{13}$
**9.** $-(-8a^3b + 5ab^2 - 2ab);$
$8a^3b - 5ab^2 + 2ab$
**10.** $-4x^3 + 6x - 3$
**11.** $-5x^3y - 3x^2y^2 + 7xy^3$
**12.** $-14x^{10} + \frac{1}{2}x^5 - 5x^3 + x^2 - 3x$
**13.** $2x^3 + 2x + 8$
**14.** $x^2 - 6x - 2$
**15.** $2x^3 + 3x^2 - 4xy - 2$

## d EVALUATING POLYNOMIALS AND APPLICATIONS

It is important to keep in mind that when we add or subtract polynomials, we are *not* solving an equation. Rather, we are finding an equivalent expression that is usually more concise. One reason we do this is to make it easier to evaluate.

**EXAMPLE 8** Evaluate both $(5x^3 + 4x^2 + 3x) + (2x^3 + 5x^2 - x)$ and $7x^3 + 9x^2 + 2x$ for $x = 2$ (see Example 1).

**a)** When $x$ is replaced by 2 in $(5x^3 + 4x^2 + 3x) + (2x^3 + 5x^2 - x)$, we have

$$5 \cdot 2^3 + 4 \cdot 2^2 + 3 \cdot 2 + 2 \cdot 2^3 + 5 \cdot 2^2 - 2,$$

or $\quad 5 \cdot 8 + 4 \cdot 4 + 3 \cdot 2 + 2 \cdot 8 + 5 \cdot 4 - 2,$

or $\quad 40 + 16 + 6 + 16 + 20 - 2,$ which is 96.

**b)** Similarly, when $x$ is replaced by 2 in $7x^3 + 9x^2 + 2x$, we have

$$7 \cdot 2^3 + 9 \cdot 2^2 + 2 \cdot 2,$$

or $\quad 7 \cdot 8 + 9 \cdot 4 + 4,$

or $\quad 56 + 36 + 4.$ As expected, this is also 96.

Note how much easier it is to evaluate the simplified sum in part (b) rather than the original expression.

◀ **Do Exercise 16.**

Polynomials are frequently evaluated in real-world situations.

**EXAMPLE 9** *Athletics.* In a sports league of $n$ teams in which all teams play each other twice, the total number of games played is given by the polynomial

$$n^2 - n.$$

A women's softball league has 10 teams. If each team plays every other team twice, what is the total number of games played?

We evaluate the polynomial for $n = 10$:

$$n^2 - n = 10^2 - 10 = 100 - 10 = 90.$$

The league plays 90 games.

◀ **Do Exercises 17–19.**

---

**16.** Evaluate each expression for $a = 2$. (See Margin Exercise 3.)

**a)** $(7a^2 + 2a + 8) + (2a^2 + a - 9)$

**b)** $9a^2 + 3a - 1$

**17.** In the situation of Example 9, how many games are played in a league with 12 teams?

Wendy is pedaling down a hill. Her distance from the top of the hill, in meters, can be approximated by

$$\frac{1}{2}t^2 + 3t,$$

where $t$ is the number of seconds she has been pedaling and $t < 30$.

**18.** How far has Wendy traveled in 4 sec?

**19.** How far has Wendy traveled in 10 sec?

*Answers*

**16. (a)** 41; **(b)** 41   **17.** 132 games
**18.** 20 m   **19.** 80 m

## ✓ Check Your Understanding

**Reading Check**  Determine whether each statement is true or false.

_____ **RC1.** To find the opposite of a polynomial, we need change only the sign of the first term.

_____ **RC2.** We can subtract a polynomial by adding its opposite.

_____ **RC3.** The sum of two monomials is always a monomial.

_____ **RC4.** The coefficient of $3x^7$ is 7.

**Concept Check**  Simplify.

**CC1.** $x^2 - (x^2)$

**CC2.** $x^2 - (-x^2)$

**CC3.** $x^2 - (x^2 - x)$

**CC4.** $x^2 + x - (x^2 + x)$

**CC5.** $x^2 + x - (x^2 - x)$

**CC6.** $x^2 - x - (-x^2 - x)$

**a**  Add.

**1.** $(3x + 7) + (-7x + 3)$

**2.** $(6x + 1) + (-7x + 2)$

**3.** $(-9x + 7) + (x^2 + x - 2)$

**4.** $(x^2 - 5x + 4) + (5x - 9)$

**5.** $(x^2 - 7) + (x^2 + 7)$

**6.** $(x^3 + x^2) + (2x^3 - 5x^2)$

**7.** $(6t^4 + 4t^3 - 1) + (5t^2 - t + 1)$

**8.** $(5t^2 - 3t + 12) + (2t^2 + 8t - 30)$

**9.** $(2 + 4x + 6x^2 + 7x^3) + (5 - 4x + 6x^2 - 7x^3)$

**10.** $(3x^4 - 6x - 5x^2 + 5) + (6x^2 - 4x^3 - 1 + 7x)$

**11.** $(9x^8 - 7x^4 + 2x^2 + 5) + (8x^7 + 4x^4 - 2x)$

**12.** $(4x^5 - 6x^3 - 9x + 1) + (6x^3 + 9x^2 + 9x)$

**13.** $\left(\frac{1}{2}t^3 - \frac{3}{4}t^2 + \frac{1}{3}\right) + \left(\frac{3}{2}t^3 - \frac{1}{4}t^2 + \frac{1}{6}t\right)$

**14.** $\left(\frac{2}{3}t^5 - t^4 + \frac{1}{2}t^2 + 4\right) + \left(\frac{7}{3}t^5 + \frac{1}{2}t^4 - \frac{1}{2}t^2 + \frac{1}{3}t\right)$

**15.** $(9x^3 - 3x + x^2 - 10) + (3x - x^2 - 9x^3 + 10)$

**16.** $(5 - x^4 + 3x^2 + 2x) + (x^4 - 3x^2 - 2x - 5)$

**17.** $(8a^3b^2 + 5a^2b^2 + 6ab^2) + (5a^3b^2 - a^2b^2 - 4a^2b)$

**18.** $(6x^3y^3 - 4x^2y^2 + 3xy^2) + (x^3y^3 + 7x^3y^2 - 2xy^2)$

**19.** $(17.5abc^3 + 4.3a^2bc) + (-4.9a^2bc - 5.2abc)$

**20.** $(23.9x^3yz - 19.7x^2y^2z) + (-14.6x^3yz - 8x^2yz)$

---

**b** Find two equivalent expressions for the opposite of each polynomial.

**21.** $-5x$

**22.** $x^2 - 3x$

**23.** $-x^2 + 13x - 7$

**24.** $-7x^3 - x^2 - x$

**25.** $12x^4 - 3x^3 + 3$

**26.** $4x^3 - 6x^2 - 8x + 1$

Simplify.

**27.** $-(3x - 5)$

**28.** $-(-2x + 4)$

**29.** $-(4x^2 - 3x + 2)$

**30.** $-(-6a^3 + 2a^2 - 9a + 1)$

**31.** $-(-4x^4 + 6x^2 + \frac{3}{4}x - 8)$

**32.** $-(-5x^4 + 4x^3 - x^2 + 0.9)$

---

**c** Subtract.

**33.** $(3x + 2) - (-4x + 3)$

**34.** $(6x + 1) - (-7x + 2)$

**35.** $(9t^2 + 7t + 5) - (5t^2 + 7t - 1)$

**36.** $(8t^2 - 5t + 7) - (3t^2 - 5t - 8)$

**37.** $(-8x + 2) - (x^2 + x - 3)$

**38.** $(x^2 - 5x + 4) - (8x - 9)$

**39.** $(7a^2 + 5a - 9) - (8a^2 + 7)$

**40.** $(8a^2 - 6a + 5) - (2a^2 - 19a)$

**41.** $(8x^4 + 3x^3 - 1) - (4x^2 - 3x + 5)$

**42.** $(-4x^2 + 2x) - (3x^3 - 5x^2 + 3)$

**722**    **CHAPTER 10**    Exponents and Polynomials

**43.** $(1.2x^3 + 4.5x^2 - 3.8x) - (-3.4x^3 - 4.7x^2 + 23)$

**44.** $(0.5x^4 - 0.6x^2 + 0.7) - (2.3x^4 + 1.8x - 3.9)$

**45.** $\left(\frac{5}{8}x^3 - \frac{1}{4}x - \frac{1}{3}\right) - \left(-\frac{1}{8}x^3 + \frac{1}{4}x - \frac{1}{3}\right)$

**46.** $\left(\frac{1}{5}x^3 + 2x^2 - 0.1\right) - \left(-\frac{2}{5}x^3 + 2x^2 + 0.01\right)$

**47.** $(9x^3y^3 + 8x^2y^2 + 7xy) - (3x^3y^3 - 2x^2y + 6xy)$

**48.** $(3x^4y + 2x^3y - 7x^2y) - (5x^4y + 2x^2y^2 - 8x^2y)$

**d** Evaluate each polynomial for $x = 4$.

**49.** $-7x + 5$

**50.** $-3x + 1$

**51.** $2x^2 - 5x + 7$

**52.** $3x^2 + x + 7$

**53.** $x^3 - 5x^2 + x$

**54.** $7 - x + 3x^2$

Evaluate each polynomial for $x = -1$.

**55.** $2x + 9$

**56.** $6 - 2x$

**57.** $x^2 - 2x + 1$

**58.** $5x - 6 + x^2$

**59.** $-3x^3 + 7x^2 - 3x - 2$

**60.** $-2x^3 - 5x^2 + 4x + 3$

**61.** *Skydiving.* During the first 13 sec of a jump, the distance $S$, in feet, that a skydiver falls in $t$ seconds can be approximated by the polynomial equation

$$S = 11.12t^2.$$

In 2009, 108 U.S. skydivers fell headfirst in formation from a height of 18,000 ft. How far had they fallen 10 sec after having jumped from the plane?

**Data:** telegraph.co.uk

**62.** *SCAD Thrill Ride.* The distance $s$, in feet, traveled by a body falling freely from rest in $t$ seconds is approximated by the polynomial equation

$$s = 16t^2.$$

The SCAD thrill ride is a 2.5-sec free fall into a net. How far does the diver fall?

**Data:** scadfreefall.co.uk

*Daily Accidents.* The average number of accidents per day involving drivers who are $a$ years old is approximated by the polynomial

$$0.4a^2 - 40a + 1039.$$

**63.** Evaluate the polynomial for $a = 18$ to find the daily number of accidents involving 18-year-old drivers.

**64.** Evaluate the polynomial for $a = 20$ to find the daily number of accidents involving 20-year-old drivers.

**Total Revenue.** Cutting Edge Electronics is marketing a new kind of phone. *Total revenue* is the total amount of money taken in. The firm determines that when it sells $x$ phones, it takes in

$$280x - 0.4x^2 \text{ dollars.}$$

**65.** What is the total revenue from the sale of 75 phones?

**66.** What is the total revenue from the sale of 100 phones?

**Total Cost.** Cutting Edge Electronics determines that the total cost of producing $x$ phones is given by

$$5000 + 0.6x^2 \text{ dollars.}$$

**67.** What is the total cost of producing 500 phones?

**68.** What is the total cost of producing 650 phones?

**Minutes of Daylight.** The number of minutes of daylight in Chicago, on a date $n$ days after December 21, can be approximated by

$$-0.01096n^2 + 4n + 548.$$

**69.** ▦ Determine the number of minutes of daylight in Chicago 92 days after December 21.

**70.** ▦ Determine the number of minutes of daylight in Chicago 123 days after December 21.

**Medicine.** Ibuprofen is a medication used to relieve pain. The polynomial

$$0.5t^4 + 3.45t^3 - 96.65t^2 + 347.7t$$

can be used to estimate the number of milligrams of ibuprofen in the bloodstream $t$ hours after 400 mg of the medication has been swallowed, where $0 \leq t \leq 6$.

**Data:** Dr. P. Carey, Burlington, VT

**71.** ▦ Determine the number of milligrams of ibuprofen in the bloodstream 1 hr after the medication has been swallowed.

**72.** ▦ Determine the number of milligrams of ibuprofen in the bloodstream 5 hr after the medication has been swallowed.

# Skill Maintenance

**73.** A 10-lb fish serves 7 people. What is the ratio of servings to pounds?   [6.1a]

**74.** A bicycle salesperson's commission rate is 22%. A commission of $783.20 is received. How many dollars' worth of bicycles were sold?   [7.6b]

**75.** The sales tax rate in Pennsylvania is 6%. How much sales tax would be paid in Pennsylvania on the purchase of a laptop computer for $1350?   [7.6a]

**76.** Find the area of a rectangle that is 6.5 m by 4 m.   [5.8b]

**77.** Find the area of a circle with radius 20 cm. Use 3.14 for $\pi$.   [9.3b]

**78.** Melanie earned $4740 for working 12 weeks. What was the rate of pay per week?   [6.2a]

Write the prime factorization for each number.   [3.2c]

**79.** 168

**80.** 192

**81.** 735

**82.** 117

# Synthesis

*Minutes of Daylight.* The number of minutes of daylight in Los Angeles, on a date $n$ days after December 21, can be approximated by

$$-0.0085n^2 + 3.1014n + 593.$$

**83.** ▦ Determine the number of minutes of daylight in Los Angeles on Ground Hog Day (February 2).

**84.** ▦ How much more daylight is available in Chicago than in Los Angeles on July 4? (See Exercises 69 and 70.)

**85.** ▦ *Medicine.* The polynomial

$$0.5t^4 + 3.45t^3 - 96.65t^2 + 347.7t$$

used in Exercises 71 and 72 describes the number of milligrams of ibuprofen in the bloodstream $t$ hours after 400 mg of the medication has been swallowed. To visualize this, make a line graph, with the number of hours on the horizontal axis and the amount of ibuprofen in the bloodstream on the vertical axis. Plot points for $t = 0, 1, 2, 3, 4, 5,$ and 6.

**86.** ▦ *Daily Accidents.* The polynomial

$$0.4a^2 - 40a + 1039$$

from Exercises 63 and 64 can be used to estimate the average number of accidents per day involving drivers who are $a$ years old. To visualize this, make a vertical bar graph, with the age of drivers on the horizontal axis and the average number of accidents per day on the vertical axis. Show bars for $a = 20, 30, 40, 50, 60, 70,$ and 80.

**87.** *Total Profit.* Total profit is defined as total revenue minus total cost. Find a polynomial giving the total profit for Cutting Edge Electronics, described in Exercises 65–68, when $x$ phones are produced and sold.

**88.** *Total Profit.* Use the polynomial found in Exercise 87 to find **(a)** the total profit when 200 phones are produced and sold and **(b)** the total profit when 500 phones are produced and sold.

Perform the indicated operations and simplify.

**89.** $(7y^2 - 5y + 6) - (3y^2 + 8y - 12) + (8y^2 - 10y + 3)$

**90.** $(3x^2 - 4x + 6) - (-2x^2 + 4) + (-5x - 3)$

**91.** $(-y^4 - 7y^3 + y^2) + (-2y^4 + 5y - 2) - (-6y^3 + y^2)$

**92.** $(-4 + x^2 + 2x^3) - (-6 - x + 3x^3) - (-x^2 - 5x^3)$

**93.** Complete: $9x^4 + \underline{\quad} + 5x^2 - 7x^3 + \underline{\quad} - 9 + \underline{\quad} = 12x^4 - 5x^3 + 5x^2 - 16.$

**94.** Complete: $8t^4 + \underline{\quad} - 2t^3 + \underline{\quad} - 2t^2 + t - \underline{\quad} - 3 + \underline{\quad} = 8t^4 + 7t^3 - 3t + 4.$

# Introduction to Multiplying and Factoring Polynomials

## a  MULTIPLYING MONOMIALS

Recall that the area of a square with sides of length $x$ is $x^2$.

$x$

Area $= x^2$

$x$

If a rectangle is 3 times as long as it is wide, we can represent its width by $x$ and its length by $3x$.

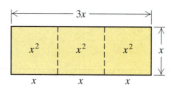

$3x$

$x^2$   $x^2$   $x^2$   $x$

$x$   $x$   $x$

Area $= 3x^2$

The area, $3x^2$, is the product of $3x$ and $x$. This product can be found using an associative law:

$$(3x)x = 3(xx) = 3x^2.$$

To find other products of monomials, we may need to use a commutative law as well.

**EXAMPLE 1**  Multiply: $(4x)(5x)$.

$$
\begin{aligned}
(4x)(5x) &= 4 \cdot x \cdot 5 \cdot x && \text{Using an associative law}\\
&= 4 \cdot 5 \cdot x \cdot x && \text{Using a commutative law}\\
&= (4 \cdot 5)(x \cdot x) && \text{Using an associative law}\\
&= 20x^2
\end{aligned}
$$

The multiplication in Example 1 can be regarded as finding the area of a rectangle of width $4x$ and length $5x$. Note that the area consists of 20 squares, each of which has area $x^2$.

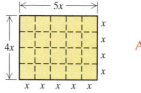

$5x$

$x$
$x$
$4x$   $x$
$x$

$x$ $x$ $x$ $x$ $x$

Area $= 20x^2$

◀ Do Exercises 1 and 2.

Usually the steps in Example 1 are combined: We multiply coefficients and we multiply variables.

Multiply.

**1.** $(6a)(3a)$

**2.** $(-7x)(2x)$

$$
\begin{aligned}
&= -7 \cdot x \cdot 2 \cdot x\\
&= -7 \cdot \boxed{\phantom{2}} \cdot x \cdot x\\
&= (-7 \cdot 2)(x \cdot x)\\
&= \boxed{\phantom{xx}}
\end{aligned}
$$

***Answers***

**1.** $18a^2$  **2.** $-14x^2$

***Guided Solution:***

**2.** $2, -14x^2$

**EXAMPLES**  Multiply.

**2.** $(5x)(6x) = (5 \cdot 6)(x \cdot x)$    Multiplying the coefficients and the variables
$$= 30x^2$$    Simplifying

**3.** $(3x)(-y) = (3x)(-1y)$    Rewriting $-y$ as $-1y$
$$= (3)(-1)(x \cdot y)$$
$$= -3xy$$

**Do Exercises 3–5.** ▶

**Multiply.**

**3.** $(4a)(12a)$

**4.** $(-m)(5m)$

**5.** $(-6a)(-7b)$

We can use the product rule when multiplying monomials.

**EXAMPLES**  Multiply and simplify.

**4.** $x^2 \cdot x^5 = x^{2+5}$    Adding exponents
$$= x^7$$

**5.** $(3a^4)(5a^2) = (3 \cdot 5)(a^4 \cdot a^2)$   Multiplying coefficients;
$$= 15a^6$$   adding exponents

**6.** $(-4x^2y^3)(3x^6y^7) = (-4 \cdot 3)(x^2 \cdot x^6)(y^3 \cdot y^7)$
$$= -12x^8y^{10}$$

**Do Exercises 6–9.** ▶

**Multiply.**

**6.** $a^5 \cdot a^4$

**7.** $(2x^8)(4x^5)$

**8.** $(-7m^4)(-5m^7)$

**9.** $(3a^5b^4)(5a^2b^8)$

## b   MULTIPLYING A MONOMIAL AND ANY POLYNOMIAL

The product of $x$ and $x + 2$ can be visualized as the area of a rectangle with width $x$ and length $x + 2$, as illustrated in the figure at right.

The distributive law is used to find products of polynomials algebraically.

**EXAMPLE 7**  Multiply: $2x$ and $5x + 3$.

$$2x(5x + 3) = 2x \cdot 5x + 2x \cdot 3$$    Using the distributive law
$$= 10x^2 + 6x$$    Multiplying each pair of monomials

Area $= x(x + 2)$
$$= x \cdot x + x \cdot 2$$
$$= x^2 + 2x$$

**EXAMPLE 8**  Multiply: $5x(2x^2 - 3x + 4)$.

$$5x(2x^2 - 3x + 4) = 5x \cdot 2x^2 - 5x \cdot 3x + 5x \cdot 4$$    Note that $x \cdot x^2 = x^1 \cdot x^2 = x^3$.
$$= 10x^3 - 15x^2 + 20x$$

**EXAMPLE 9**  Multiply: $-3r^2s(2r^3s^2 - 5rs)$.

$$-3r^2s(2r^3s^2 - 5rs) = -3r^2s \cdot 2r^3s^2 - (-3r^2s)5rs$$
$$= -6r^5s^3 + 15r^3s^2$$

**Do Exercises 10–12.** ▶

**Multiply.**

**10.** $4x$ and $3x + 5$

**11.** $3a(2a^2 - 5a + 7)$

**12.** $4a^3b^2(2a^2 + 5b^4)$

*Answers*

**3.** $48a^2$   **4.** $-5m^2$   **5.** $42ab$   **6.** $a^9$
**7.** $8x^{13}$   **8.** $35m^{11}$   **9.** $15a^7b^{12}$
**10.** $12x^2 + 20x$   **11.** $6a^3 - 15a^2 + 21a$
**12.** $8a^5b^2 + 20a^3b^6$

## **c** FACTORING

**Factoring** is the reverse of multiplying. We use the distributive law, beginning with a sum or a difference of terms that contain a common factor:

$$ab + ac = a(b + c) \quad \text{and} \quad rs - rt = r(s - t).$$

---

> ### FACTORING
>
> To **factor** an expression is to find an equivalent expression that is a product.

To *factor* an expression like $10y + 15$, we find an equivalent expression that is a product. To do this, we look to see if the terms have a factor in common. If there *is* a common factor, we can "factor it out" using the distributive law. Note the following:

The prime factorization of $10y$ is $2 \cdot 5 \cdot y$.

The prime factorization of 15 is $3 \cdot 5$.

We factor out the common factor, 5:

$$10y + 15 = 5 \cdot 2y + 5 \cdot 3 \qquad \text{Try to do this step mentally.}$$
$$= 5(2y + 3). \qquad \text{Using the distributive law}$$

We generally factor out the *largest* common factor. This is the product of all prime factors common to all terms.

**EXAMPLE 10**    Factor $12a - 30$.

The prime factorization of $12a$ is $2 \cdot 2 \cdot 3 \cdot a$.

The prime factorization of 30 is $2 \cdot 3 \cdot 5$.

Both factorizations include a factor of 2 and a factor of 3. Thus, 2 is a common factor, 3 is a common factor, and $2 \cdot 3$ is a common factor. The largest common factor is $2 \cdot 3$, or 6:

$$12a - 30 = 6(2a - 5). \qquad \text{Try to go directly to this step.} \qquad ■$$

**EXAMPLE 11**    Factor $9x + 27y - 9$.

The prime factorization of $9x$ is $3 \cdot 3 \cdot x$.

The prime factorization of $27y$ is $3 \cdot 3 \cdot 3 \cdot y$.

The prime factorization of 9 is $3 \cdot 3$.

$$9x + 27y - 9 = 9 \cdot x + 9 \cdot 3y - 9 \cdot 1 \qquad \text{The largest common factor is } 3 \cdot 3, \text{ or } 9.$$
$$= 9(x + 3y - 1) \qquad\qquad\qquad ■$$

In Example 11, the 1 in the factorization is necessary. To see this, reverse the factorization process by multiplying. This provides a check for the answer.

$$9(x + 3y - 1) = 9 \cdot x + 9 \cdot 3y - 9 \cdot 1 = 9x + 27y - 9$$

---

Factorizations can always be checked by multiplying.

---

Note in Example 11 that although $3(3x + 9y - 3)$ is also equivalent to $9x + 27y - 9$, it is not factored "completely." However, we can complete the process by factoring out another factor of 3:

$$9x + 27y - 9 = 3(3x + 9y - 3) = 3 \cdot 3(x + 3y - 1) = 9(x + 3y - 1).$$

Remember to factor out the *largest common factor*.

**EXAMPLES** Factor. Try to write just the answer.

**12.** $-3x + 6y - 9z = -3(x - 2y + 3z)$

We generally factor out a negative factor when the first coefficient is negative. We might also factor as $-3x + 6y - 9z = 3(-x + 2y - 3z)$.

**13.** $18z - 12x - 24 = 6(3z - 2x - 4)$

The largest common factor is $2 \cdot 3$.

$18z = 2 \cdot 3 \cdot 3 \cdot z;$
$12x = 2 \cdot 2 \cdot 3 \cdot x;$
$24 = 2 \cdot 2 \cdot 2 \cdot 3$

Check:  $6(3z - 2x - 4) = 6 \cdot 3z - 6 \cdot 2x - 6 \cdot 4 = 18z - 12x - 24$ ∎

*Remember*: An expression is factored when it is written as a product.

**Do Exercises 13–16.** ▶

**EXAMPLE 14** Factor each of the following.

**a)** $10x^6 + 15x^2$

**b)** $8xy^3 - 6xy^2 + 4xy$

**a)** The prime factorization of $10x^6$ is $2 \cdot 5 \cdot x \cdot x \cdot x \cdot x \cdot x \cdot x$.

The prime factorization of $15x^2$ is $3 \cdot 5 \cdot x \cdot x$.

$$10x^6 + 15x^2 = 5x^2 \cdot 2x^4 + 5x^2 \cdot 3 \quad \text{The largest common factor is } 5x^2.$$
$$= 5x^2(2x^4 + 3)$$

**b)** The prime factorization of $8xy^3$ is $2 \cdot 2 \cdot 2 \cdot x \cdot y \cdot y \cdot y$.

The prime factorization of $6xy^2$ is $2 \cdot 3 \cdot x \cdot y \cdot y$.

The prime factorization of $4xy$ is $2 \cdot 2 \cdot x \cdot y$.

$$8xy^3 - 6xy^2 + 4xy = 2xy \cdot 4y^2 - 2xy \cdot 3y + 2xy \cdot 2 \quad \text{The largest common factor is } 2xy.$$
$$= 2xy(4y^2 - 3y + 2)$$

The checks are left for the student. ∎

The largest common factor can be determined by considering the coefficients and the variables separately. The largest common factor of the coefficients is found using prime factorizations. The largest common variable factors can be found by examining the exponents.

When a variable appears in every term of a polynomial, the *largest* common factor of that variable is the *smallest* of the powers of that variable in the polynomial.

**Do Exercises 17–19.** ▶

Factor.

**13.** $6z - 12$

**14.** $3x - 6y + 12$

**15.** $16a - 36b + 42$

**16.** $-12x + 32y - 16z$

Factor.

**17.** $5a^3 + 10a$

**18.** $14x^3 - 7x^2 + 21x$

**GS** **19.** $9a^2b - 6ab^2$

$9a^2b = 3 \cdot 3 \cdot a \cdot a \cdot \boxed{\phantom{x}}$

$6ab^2 = 2 \cdot 3 \cdot a \cdot b \cdot \boxed{\phantom{x}}$

The largest common factor is

$3 \cdot a \cdot \boxed{\phantom{x}}.$

$9a^2b - 6ab^2$

$= 3ab \cdot \boxed{\phantom{x}} - 3ab \cdot \boxed{\phantom{x}}$

$= 3ab(\boxed{\phantom{x}} - \boxed{\phantom{x}})$

*Answers*

**13.** $6(z - 2)$  **14.** $3(x - 2y + 4)$
**15.** $2(8a - 18b + 21)$
**16.** $-4(3x - 8y + 4z)$  **17.** $5a(a^2 + 2)$
**18.** $7x(2x^2 - x + 3)$  **19.** $3ab(3a - 2b)$

*Guided Solution:*

**19.** $b; b; b; 3a, 2b, 3a, 2b$

## ✓ Check Your Understanding

**Reading Check**  Determine whether each statement is true or false.

_____ **RC1.** When multiplying the monomials $4x^2$ and $5x^3$, we multiply the coefficients.

_____ **RC2.** When multiplying the monomials $4x^2$ and $5x^3$, we multiply the exponents.

_____ **RC3.** After we have multiplied $a(b + c)$, there will be two terms in the product.

_____ **RC4.** The product of two monomials may be a binomial.

**Concept Check**  Match each expression with an equivalent expression from the column on the right. Choices may be used more than once or not at all.

**CC1.** $8x \cdot 2x$  _____

**CC2.** $(-16x)(-x)$  _____

**CC3.** $2x(8x - 1)$  _____

**CC4.** $8x(2x - 1)$  _____

**a)** $16x^2$

**b)** $-16x^2$

**c)** $16x^2 - 1$

**d)** $16x^2 - 2x$

**e)** $16x^2 - 8x$

---

**a**  Multiply.

**1.** $(4a)(7a)$

**2.** $(7x)(6x)$

**3.** $(-4x)(15x)$

**4.** $(-9a)(10a)$

**5.** $(7x^5)(4x^3)$

**6.** $(10a^2)(3a^2)$

**7.** $(-0.1x^6)(0.7x^3)$

**8.** $(0.3x^3)(-0.4x^6)$

**9.** $(5x^2y^3)(7x^4y^9)$

**10.** $(9a^5b^4)(2a^4b^7)$

**11.** $(4a^3b^4c^2)(3a^5b^4)$

**12.** $(7x^3y^5z^2)(8x^3z^4)$

**13.** $(3x^2)(-4x^3)(2x^6)$

**14.** $(-2y^5)(10y^4)(-3y^3)$

**b**  Multiply.

**15.** $3x(-x + 7)$

**16.** $2x(4x - 6)$

**17.** $-3x(x - 2)$

**18.** $-9x(-x - 1)$

**19.** $x^2(x^3 + 1)$

**20.** $-2x^3(x^2 - 1)$

**21.** $5x(2x^2 - 6x + 1)$  **22.** $-4x(2x^3 - 6x^2 - 5x + 1)$  **23.** $4xy(3x^2 + 2y)$

**24.** $7xy(3x^2 - 6y^2)$  **25.** $3a^2b(4a^5b^2 - 3a^2b^2)$  **26.** $4a^2b^2(2a^3b - 5ab^2)$

**C**  Factor. Check by multiplying.

**27.** $2x + 8$  **28.** $3x + 12$  **29.** $7a - 35$  **30.** $9a - 18$

**31.** $28x + 21y$  **32.** $8x - 10y$  **33.** $9a - 27b + 81$  **34.** $5x + 10 + 15y$

**35.** $18 - 6m$  **36.** $28 - 4y$  **37.** $-16 - 8x + 40y$  **38.** $-35 + 14x - 21y$

**39.** $9x^5 + 9x$  **40.** $5x^6 + 5x$  **41.** $a^3 - 8a^2$  **42.** $a^5 - 9a^2$

**43.** $8x^3 - 6x^2 + 2x$  **44.** $9x^4 - 12x^3 + 3x$  **45.** $12a^4b^3 + 18a^5b^2$  **46.** $15a^5b^2 + 20a^2b^3$

## Skill Maintenance

Graph.  [8.4b, c]

**47.** $y = x - 5$  **48.** $y = -2x + 8$  **49.** $x = 2$  **50.** $y = \dfrac{2}{3}x$

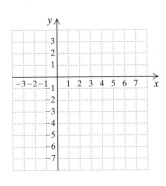  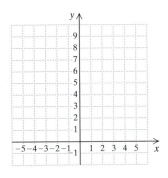  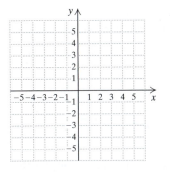  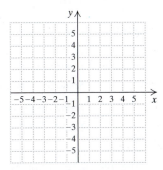

## Synthesis

Factor.

**51.** ▦ $391x^{391} + 299x^{299}$  **52.** ▦ $703a^{437} + 437a^{703}$

**53.** $84a^7b^9c^{11} - 42a^8b^6c^{10} + 49a^9b^7c^8$

**54.** Draw a figure similar to those preceding Examples 1 and 7 to show that $2x \cdot 3x = 6x^2$.

## Vocabulary Reinforcement

Complete each statement with the correct word or number from the column at the right. Some of the choices may not be used.

1. Any nonzero number raised to the power 0 is equal to _____.
   [10.1a]

2. The _____ rule states that when multiplying with exponential notation, if the bases are the same, we keep the base and add the exponents.   [10.2a]

3. The number $7.9 \times 10^{-3}$ is written in _____ notation.   [10.3a]

4. A base raised to a negative exponent is equal to the _____ of the base raised to a positive exponent.   [10.1b]

5. The _____ rule states that when raising a power to a power, we multiply the exponents and leave the base unchanged.   [10.2c]

6. The _____ of $3x^2$ is 3.   [10.4a]

| |
|---|
| 0 |
| 1 |
| coefficient |
| exponent |
| fraction |
| scientific |
| power |
| product |
| quotient |
| reciprocal |

## Concept Reinforcement

_____  1. If $x^7$ is divided by $x^7$, the result is the same as $x^0$.   [10.1a], [10.2b]

_____  2. We subtract a polynomial by adding its opposite.   [10.4c]

_____  3. If a polynomial is written as a product, it is factored.   [10.5c]

## Study Guide

**Objective 10.1a**   Evaluate algebraic expressions containing whole-number exponents.

**Examples**   Evaluate $-6^1$.
Since $6^1 = 6$, $-6^1 = -6$.

Evaluate $(-5)^0$.
Since $b^0 = 1$ for all $b \neq 0$, $(-5)^0 = 1$.

**Practice Exercises**

Evaluate.

1. $(-3)^1$

2. $12^0$

**Objective 10.1b**   Express exponential expressions involving negative exponents as equivalent expressions containing positive exponents.

**Example**   Write an expression equivalent to $x^{-10}$ using a positive exponent.

$$x^{-10} = \frac{1}{x^{10}}$$

**Practice Exercise**

3. Write an expression equivalent to $3^{-12}$ using a positive exponent.

**Objective 10.2a**   Use the product rule to multiply exponential expressions with like bases.

**Example**   Simplify $x^2 \cdot x^8$.

$x^2 \cdot x^8 = x^{2+8} = x^{10}$

**Practice Exercise**

4. Simplify $y \cdot y^{11}$.

**Objective 10.2b**   Use the quotient rule to divide exponential expressions with like bases.

**Example**   Simplify $\dfrac{7^{12}}{7^{10}}$.

$$\dfrac{7^{12}}{7^{10}} = 7^{12-10} = 7^2 = 49$$

**Practice Exercise**

5.  Simplify $\dfrac{x^9}{x^2}$.

---

**Objective 10.2c**   Use the power rule to raise powers to powers.

**Example**   Simplify $(x^3)^{-5}$.

$$(x^3)^{-5} = x^{3(-5)} = x^{-15} = \dfrac{1}{x^{15}}$$

**Practice Exercise**

6.  Simplify $(3^{-2})^{-6}$.

---

**Objective 10.2d**   Raise a product to a power and a quotient to a power.

**Example**   Simplify $(a^2b)^9$.
$$(a^2b)^9 = (a^2)^9(b)^9 = a^{18}b^9$$

**Practice Exercise**

7.  Simplify $(3x^7)^2$.

---

**Objective 10.3a**   Convert between scientific notation and decimal notation.

**Examples**   Convert 96,000,000,000 to scientific notation.
$$96{,}000{,}000{,}000 = 9.6 \times 10^{10}$$
Convert $6.02 \times 10^{-5}$ to decimal notation.
$$6.02 \times 10^{-5} = 0.0000602$$

**Practice Exercises**

8.  Convert 0.000803 to scientific notation.

9.  Convert $3.48 \times 10^3$ to decimal notation.

---

**Objective 10.4a**   Add polynomials.

**Example**   Add: $(3x^3 - 7x^2 - 9) + (2x^3 + x^2 - 1)$.
$$(3x^3 - 7x^2 - 9) + (2x^3 + x^2 - 1)$$
$$= 3x^3 + 2x^3 - 7x^2 + x^2 - 9 - 1$$
$$= 5x^3 - 6x^2 - 10$$

**Practice Exercise**

10.  Add: $(2a^2 - 5a + 3) + (a^3 + a - 5)$.

---

**Objective 10.4c**   Subtract polynomials.

**Example**   Subtract:
$$(9y^3 - 8y^2 + 11y) - (4y^3 - 2y^2 - 12y).$$
$$(9y^3 - 8y^2 + 11y) - (4y^3 - 2y^2 - 12y)$$
$$= 9y^3 - 8y^2 + 11y - 4y^3 + 2y^2 + 12y$$
$$= 5y^3 - 6y^2 + 23y$$

**Practice Exercise**

11.  Subtract: $(x^2 - x - 1) - (7x^2 - 6x + 2)$.

---

**Objective 10.5a**   Multiply monomials.

**Example**   Multiply: $(-3m^2)(9m^5)$.
$$(-3m^2)(9m^5) = (-3 \cdot 9)(m^2 \cdot m^5)$$
$$= -27m^7$$

**Practice Exercise**

12.  Multiply: $(-2x^3)(-7x)$.

**Objective 10.5b** Multiply a monomial and any polynomial.

**Example** Multiply: $2x^3(3x^4 - 5)$.

$$2x^3(3x^4 - 5) = 2x^3 \cdot 3x^4 - 2x^3 \cdot 5$$
$$= 6x^7 - 10x^3$$

**Practice Exercise**

**13.** Multiply: $5x^2(x^3 + 2x^2 - 3x + 9)$.

**Objective 10.5c** Use the distributive law to factor.

**Example** Factor $6x^3 - 20x^2 + 2x$.

$$6x^3 - 20x^2 + 2x = 2x(3x^2 - 10x + 1)$$

**Practice Exercise**

**14.** Factor $20a^4 - 30a$.

## Review Exercises

Evaluate. [10.1a]

**1.** $(-53)^0$

**2.** $46^1$

**3.** $(5x + 7)^1$ for $x = -2$

**4.** $(3x - 2)^0$ for $x = 5$

Write an equivalent expression using positive exponents. Then simplify, if possible. [10.1b]

**5.** $12^{-2}$

**6.** $8a^{-7}$

**7.** $\dfrac{x^{-3}}{y^5 z^{-6}}$

**8.** $\left(\dfrac{4}{5}\right)^{-2}$

**9.** Write an expression equivalent to $\dfrac{1}{x^7}$ using a negative exponent. [10.1b]

Simplify. [10.2a, b, c, d]

**10.** $x^4 \cdot x^{11}$

**11.** $\dfrac{x^{16}}{x^4}$

**12.** $(3^4)^{11}$

**13.** $(x^2 y^4)^3$

**14.** $\dfrac{x^3}{x^{11}}$

**15.** $\dfrac{u^2 v^8}{uv^7}$

**16.** $x^{-5} \cdot x^{-6}$

**17.** $(3x^4)^2$

**18.** $a^{-2} \cdot a \cdot a^7$

**19.** $\left(\dfrac{y^2}{2}\right)^5$

**20.** Write scientific notation for 42,700,000. [10.3a]

**21.** Write scientific notation for 0.0001924. [10.3a]

Simplify. Write the answer in scientific notation. [10.3b]

**22.** $(5.1 \times 10^6)(2.3 \times 10^4)$

**23.** $\dfrac{1.6 \times 10^2}{6.4 \times 10^{18}}$

**24. *Pizza Consumption.*** Each person in the United States eats an average of 46 slices of pizza per year. The U.S. population is projected to be about 360 million in 2030. At this rate, how many slices of pizza would be consumed in 2030? Express the answer in scientific notation. [10.3c]

**Data:** Packaged Facts; U.S. Census Bureau

Perform the indicated operation. [10.4a, c]
**25.** $(-4x + 9) + (7x - 15)$

**26.** $(7x^4 - 5x^3 + 3x - 5) + (x^3 - 4x + 2)$

**27.** $(9a^5 + 8a^3 + 4a + 7) - (a^5 - 4a^3 + a^2 - 2)$

**28.** $(8a^3b^3 + 9a^2b^3) - (3a^3b^3 - 2a^2b^3 + 7)$

**29.** Find two equivalent expressions for the opposite of $12x^3 - 4x^2 + 9x - 3$. [10.4b]

**30.** Evaluate $5t^3 + t$ for $t = -2$. [10.4d]

**31.** The altitude, in feet, of a falling golf ball $t$ seconds after it reaches the peak of its flight can be estimated by $-16t^2 + 200$. Find the ball's altitude 3 sec after it has reached its peak. [10.4d]

Multiply.
**32.** $(5x^3)(6x^4)$ [10.5a]     **33.** $(-2xy)(9x^2y^4)$ [10.5a]

**34.** $3x(6x^3 - 4x - 1)$     **35.** $2a^4b(7a^3b^3 + 5a^2b^3)$
[10.5b]                        [10.5b]

Factor. [10.5c]
**36.** $45x^3 - 10x$     **37.** $7a - 35b - 49ac$

**38.** Evaluate $6 - 2x - x^2$ for $x = -1$. [10.4d]

    **A.** 3             **B.** 5
    **C.** 7             **D.** 9

**39.** Factor out the largest common factor: $6x^3y - 9x^2y^5$. [10.5c]

    **A.** $3x^2y(2x - 3y^4)$     **B.** $3(2x^2y - 3x^2y^5)$
    **C.** $3x^2(2xy - 3y^5)$     **D.** $y(6x^3 - 9x^2y^4)$

## Synthesis

**40.** Simplify: $-3x^5 \cdot 3x^3 - x^6(2x)^2 + (3x^4)^2 + (2x^4)^2 - 40x^2(x^3)^2$. [10.4a, c]

Factor. [10.5c]
**41.** $39a^3b^7c^6 - 130a^2b^5c^8 + 52a^4b^6c^5$

**42.** $w^5x^6y^4z^5 - w^7x^3y^7z^3 + w^6x^2y^5z^6 - w^6x^7y^3z^4$

**43.** $10a^4b^{-5} + 12a^7b^{-3}$

## Understanding Through Discussion and Writing

**1.** Can $x^{-2}$ represent a negative number? Why or why not? [10.1b]

**2.** Is it true that if $a > b$, then $a^{-1} < b^{-1}$? Explain your answer. [10.1b]

**3.** Adi claims that $(3x^{-5})(-4x^{-2}) = -x^{10}$. What mistake(s) is she probably making? [10.1b], [10.5a]

**4.** Using the quotient rule, explain why $7^0$ is 1. [10.1a], [10.2b]

**5.** Is every term a monomial? Why or why not? [10.4a]

**6.** If all of a polynomial's coefficients are prime, is it still possible to factor the polynomial? Why or why not? [10.5c]

CHAPTER

**10**  **Test**

For
Extra
Help

For step-by-step test solutions, access the Chapter Test Prep Videos
in MyLab Math.

Evaluate.

**1.** $193^1$

**2.** $-9^0$

**3.** $(3x - 7)^0$ for $x = 2$

Write an equivalent expression with positive exponents. Then simplify, if possible.

**4.** $5^{-3}$

**5.** $\dfrac{5a^{-3}}{b^{-2}}$

**6.** $\left(\dfrac{3}{5}\right)^{-3}$

Simplify. Express the answer using positive exponents.

**7.** $a^{12} \cdot a^{13}$

**8.** $\dfrac{x}{x^2}$

**9.** $(5x^3y^4)^2$

**10.** $\left(\dfrac{3}{x^9}\right)^3$

**11.** $x^{-1} \cdot x^{-7} \cdot x$

**12.** $\dfrac{a^{10}b^{12}}{a^4b^{10}}$

**13.** Write scientific notation for 0.00047.

**14.** Write scientific notation for 8,250,000.

**15.** Find the product and write the answer using scientific notation:
$(3.2 \times 10^{-8})(5.7 \times 10^{-9})$.

**16.** Add: $(12a^3 - 9a^2 + 8) + (6a^3 + 4a^2 - a)$.

**17.** Find two equivalent expressions for the opposite of $-9a^4 + 7b^2 - ab + 3$.

**18.** Subtract: $(12x^4 + 7x^2 - 6) - (9x^4 + 8x^2 + 5)$.

**19.** The height, in meters, of a ball $t$ sec after it has been thrown is approximated by $-4.9t^2 + 15t + 2$. How high is the ball 2 sec after it has been thrown?

Multiply.

**20.** $(-5x^4y^3)(2x^2y^5)$

**21.** $2a(5a^2 - 4a + 3)$

Factor.

**22.** $35x^6 - 25x^3 + 15x^2$

**23.** $6ab - 9bc + 12ac$

**24.** Americans throw away 2.5 million plastic beverage bottles every hour. How many bottles do they throw away in a week? Give your answer in scientific notation.

**Data:** grabstats.com

**A.** $6.0 \times 10^7$ bottles     **B.** $1.75 \times 10^7$ bottles

**C.** $2.52 \times 10^9$ bottles     **D.** $4.2 \times 10^8$ bottles

## Synthesis

**25.** The polynomial

$$0.041h - 0.018A - 2.69$$

can be used to estimate the lung capacity, in liters, of a female of height $h$, in centimeters, and age $A$, in years. Find the lung capacity of a 30-yr-old woman who is 150 cm tall.

**26.** Simplify: $x^{8y} \cdot x^{2y} \cdot x^{-y}$.

This exam reviews the entire textbook. Many answers to exercises (and to real-world problems) can be written in several kinds of notation. For this exam, here is the guideline we follow: Use the notation given in the problem. That is, if the problem is given using mixed numerals, give the answer as a mixed numeral. If the problem is given in decimal notation, give the answer in decimal notation.

**1. *Area.*** There are 5.1 million km² of forest in the Amazon jungle. Find standard notation for 5.1 million.

**Data:** outbackbrazil.com

Add and, if possible, simplify.

**2.**
$$\begin{array}{r} 4\,9\,0\,3 \\ 5\,2\,7\,8 \\ 6\,3\,9\,1 \\ +\,4\,5\,1\,3 \\ \hline \end{array}$$

**3.**
$$\begin{array}{r} 5\frac{4}{9} \\ +\,3\frac{1}{3} \\ \hline \end{array}$$

**4.** $-29 + 53$

**5.** $-543 + (-219)$

**6.** $-34.56 + 2.783 + 0.433 + (-13.02)$

**7.** $(4x^5 + 7x^4 - 3x^2 + 9) + (6x^5 - 8x^4 + 2x^3 - 7)$

Subtract and, if possible, simplify.

**8.**
$$\begin{array}{r} 6\,7\,4 \\ -\,4\,3\,1 \\ \hline \end{array}$$

**9.** $-4x - 13x$

**10.** $\dfrac{2}{5} - \dfrac{7}{8}$

**11.**
$$\begin{array}{r} 4\frac{1}{3} \\ -\,1\frac{5}{8} \\ \hline \end{array}$$

**12.**
$$\begin{array}{r} 2\,0.0 \\ -\ \ \ 0.0\,0\,2\,7 \\ \hline \end{array}$$

**13.** $(7x^3 + 2x^2 - x) - (5x^3 - 3x^2 - 8x)$

Multiply and, if possible, simplify.

**14.**
$$\begin{array}{r} 2\,9\,7 \\ \times\ \ 1\,6 \\ \hline \end{array}$$

**15.** $349 \cdot (-213)$

**16.** $2\dfrac{3}{4} \cdot 1\dfrac{2}{3}$

**17.** $-\dfrac{9}{7} \cdot \dfrac{14}{15}$

**18.** $12 \cdot \dfrac{5}{6}$

**19.**
$$\begin{array}{r} 3\,4.0\,9 \\ \times\ \ \ \ 7.6 \\ \hline \end{array}$$

**20.** $3(8x - 5)$

**21.** $(9a^3b^2)(3a^5b)$

**22.** $7x^2(3x^3 - 2x + 8)$

Divide and simplify. State the answer using a remainder when appropriate.

**23.** $6\,\overline{)\,3\,4\,3\,8}$

**24.** $34\,\overline{)\,1\,9\,1\,4}$

Divide and, if possible, simplify.

**25.** $\dfrac{4}{5} \div \left(-\dfrac{8}{15}\right)$     **26.** $-2\dfrac{1}{3} \div (-30)$

**27.** $2.7 \,\overline{)\, 1\ 0\ 5 . 3}$

Simplify.

**28.** $10 \div 2 \times 20 - 5^2$     **29.** $\dfrac{|3^2 - 5^2|}{2 - 2 \cdot 5}$

**30.** Write exponential notation: $14 \cdot 14 \cdot 14$.

**31.** Round 68,489 to the nearest thousand.

**32.** Round $21.\overline{83}$ to the nearest hundredth.

**33.** Determine whether 1368 is divisible by 3.

**34.** Find all the factors of 15.

**35.** Find the LCM of 15 and 35.

**36.** Simplify $\dfrac{24}{33}$.

**37.** Convert to a mixed numeral: $-\dfrac{18}{5}$.

**38.** Use $<$ or $>$ for $\square$ to write a true sentence:
$-17 \,\square\, -29$.

**39.** Use $<$ or $>$ for $\square$ to write a true sentence:
$\dfrac{4}{7} \,\square\, \dfrac{3}{5}$.

**40.** Which number is greater: $-1.001$ or $-0.9976$?

**41.** Evaluate $\dfrac{a^2 - b}{3}$ for $a = -9$ and $b = -6$.

Factor.

**42.** $40 - 5t$

**43.** $18a^3 - 15a^2 + 6a$

**44.** What part is shaded?

Write decimal notation for each number.

**45.** $\dfrac{429}{10,000}$     **46.** $-\dfrac{13}{25}$

**47.** $\dfrac{8}{9}$     **48.** $7\%$

Write each number in fraction notation.

**49.** $6.71$     **50.** $-7\dfrac{1}{4}$

**51.** $40\%$

Write each number in percent notation.

**52.** $\dfrac{17}{20}$     **53.** $1.5$

**54.** Estimate the sum $9.389 + 4.2105$ to the nearest tenth.

Solve.

**55.** $234 + y = 789$

**56.** $3.9a = 249.6$

**57.** $\dfrac{2}{3} \cdot t = \dfrac{5}{6}$

**58.** $\dfrac{8}{17} = \dfrac{36}{x}$

**59.** $7x - 9 = 26$

**60.** $-2(x - 5) = 3x + 12$

*Egg Consumption.* The line graph below shows egg consumption per person in the United States for recent years. Use the graph for Exercises 61–64.

**Egg Consumption**

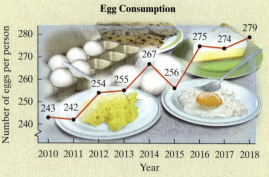

DATA: statista.com

**61.** Find the lowest egg consumption and the year(s) in which it occurred.

**62.** Find the highest egg consumption and the year(s) in which it occurred.

**63.** Find the mean, the median, and the mode(s) of the egg consumptions.

**64.** What was the percent increase in egg consumption from 2011 to 2018?

**65.** *Dead Sea.* The lowest point in the world is the Dead Sea on the border of Israel and Jordan. It is 1312 ft below sea level. Convert 1312 ft to yards; to meters.

**Data:** *The Handy Geography Answer Book*

**66.** In Sam's writing lab, 3 of the 20 students are left-handed. If a student is randomly selected, what is the probability that he or she is left-handed?

**67.** Find the missing angle measure.

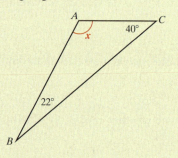

**68.** *Closest Star.* Excluding the sun, the closest star to Earth is Proxima Centauri, which is 4.3 light-years away (one light-year $= 5.88 \times 10^{12}$ mi). How far, in miles, is Proxima Centauri from Earth? Express the answer in scientific notation.

**69.** On Monday morning, a bolt of fabric contained $8\frac{1}{4}$ yd. Madison sold $3\frac{5}{8}$ yd from the bolt. How much fabric remains?

**70.** From Indira's income of $32,000, amounts of $6400 and $1600 are paid for federal and state taxes. How much remains after these taxes have been paid?

**71.** A toddler walks $\frac{3}{5}$ km per hour. At this rate, how far would the child walk in $\frac{1}{2}$ hr?

**72.** Eight gallons of paint covers 2000 ft$^2$. How much paint is needed to cover 3250 ft$^2$?

**73.** What is the simple interest on $4000 principal at 8% for $\frac{3}{4}$ yr?

**74.** The population of Bridgeton is 29,000 this year and is increasing at a rate of 4% per year. What will the population be next year?

**75.** *Tree Height.* While visiting the Grove of the Patriarchs in Mt. Rainier National Park, Abi photographed one of the tallest trees in the grove. Standing 100 ft from the base of the tree, she estimated that the distance between her and the top of the tree was about 300 ft. About how tall is the tree? Round to the nearest foot.

**76.** *Medical Dosage.* A doctor suggests that a child who weighs 24 kg be given 42 mg of phenytoin sodium. If the dosage is proportional to the child's weight, how much phenytoin sodium is recommended for a child who weighs 32 kg?

**77.** *Earth vs. Jupiter.* The mass of Earth is about $6 \times 10^{21}$ metric tons. The mass of Jupiter is about $1.908 \times 10^{24}$ metric tons. About how many times the mass of Earth is the mass of Jupiter? Express the answer in scientific notation.

Earth

Jupiter

**78.** At the start of a trip, the odometer on the Oquendos' minivan read 27,428.6 mi, and at the end of the trip, the reading was 27,914.5 mi. How long was the trip?

**79.** Eight identical dresses cost a total of $679.68. What is the cost of each dress?

**80.** Eighteen ounces of a fruit "smoothie" cost $3.06. Find the unit price in cents per ounce.

**81.** Luis paid $35 a day plus 15¢ a mile for a van rental. If his one-day van rental cost $68, how many miles did he drive?

**82.** Baldacci Real Estate received $5880 commission on the sale of an $84,000 home. What was the rate of commission?

**83.** *Falling Distance.* The distance, in feet, traveled by a body falling freely from rest in $t$ seconds is approximated by the polynomial $16t^2$. A stone is dropped from a cliff and takes 8 sec to hit the ground. How high is the cliff?

$16t^2$

Evaluate.

**84.** $18^2$

**85.** $37^0$

**86.** $42^1$

**87.** $\sqrt{121}$

Write an equivalent expression with a positive exponent. Then simplify, if possible.

**88.** $4^{-3}$

**89.** $\left(\dfrac{5}{4}\right)^{-2}$

Express each of the following in scientific notation.

**90.** 4,357,000

**91.** $(6.2 \times 10^7)(4.3 \times 10^{-23})$

Complete.

**92.** $\dfrac{1}{3}$ yd = _____ in.

**93.** 5.8 km = _____ m

**94.** 10 lb = _____ oz

**95.** 8190 mL = _____ L

**96.** 3917 mm = _____ cm

**97.** 60,000 g = _____ kg

**98.** 2.3 g = _____ mg

**99.** 28 qt = _____ gal

**100.** 10 yd$^2$ = _____ ft$^2$

**101.** 200 cm$^2$ = _____ m$^2$

The data in the following table show the percent of people who eat salad a certain number of times per week.

| NUMBER OF SALADS PER WEEK | PERCENT |
|---|---|
| None | 3% |
| 2 or fewer | 37% |
| 3–6 | 47% |
| At least one a day | 13% |

DATA: Market Facts for the Association of Dressings and Sauces

**102.** Make a bar graph of the data

**103.** Make a circle graph of the data.

**104.** Plot the following points:

$(-5, 2), (4, 0), (3, -4), (0, 2).$

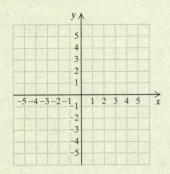

**105.** Graph: $y = -\dfrac{1}{3}x.$

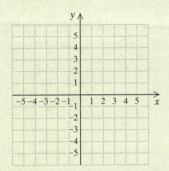

**106.** Graph: $y = 3.$

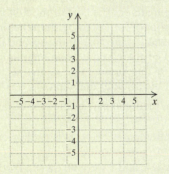

**107.** These triangles are similar. Find the missing lengths.

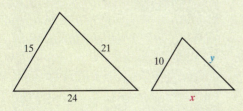

**108.** A rectangular mirror measures 20 in. by 24 in. Find its perimeter.

Find the area of each figure. Use 3.14 for $\pi$.

**109.**

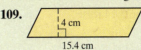

4 cm

15.4 cm

**110.**

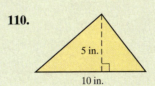

5 in.

10 in.

**111.**

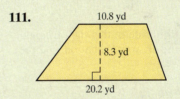

10.8 yd

8.3 yd

20.2 yd

**112.**

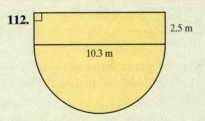

2.5 m

10.3 m

**113.** Find the diameter, the circumference, and the area of this circle. Use 3.14 for $\pi$.

10.4 in.

**114.** Find the volume.

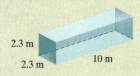

2.3 m
2.3 m
10 m

Find the volume of each shape. Use 3.14 for $\pi$.

**115.**

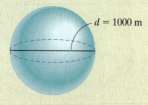

16 ft

4 ft

**116.**

$d = 1000$ m

**117.** Find the length of the third side of this right triangle. Give an exact answer and an approximation to three decimal places.

11 ft

$a$

6 ft

*Test Scores.* The following list gives the test scores for a midterm exam in a course on world literature. Use the data for Exercises 118–121.

72, 68, 93, 98, 87, 89, 74, 76, 60, 79,
82, 87, 85, 85, 96, 88, 71, 63, 92, 84

**118.** Find the five-number summary for the test scores.

**119.** Complete the following frequency table for the test scores.

| Test Scores | Frequency |
|---|---|
| 60—69 | |
| 70—79 | |
| 80—89 | |
| 90—99 | |

**120.** Construct a stem-and-leaf plot for the test scores.

**121.** Construct a histogram using the frequency table in Exercise 119.

**122.** Find the measure of a complement of $\angle CBA$, if $m \angle CBA = 40°$.

   **A.** 140°           **B.** 40°
   **C.** 50°            **D.** 90°

**123.** Convert 100°C to Fahrenheit.

   **A.** $37\frac{7}{9}$°F        **B.** 237.6°F
   **C.** 180°F          **D.** 212°F

**124.** A month of the year is randomly selected. What is the probability that the name of the month contains an *r* in its spelling?

   **A.** $\frac{1}{2}$           **B.** $\frac{1}{3}$
   **C.** $\frac{2}{3}$           **D.** 8

## Synthesis

**125.** A housing development is constructed on a dead-end road that runs along a river and ends in a cul-de-sac, as shown in the figure. The property owners agree to share the cost of maintaining the road in the following manner. The cost of the first fifth of the road in front of lot 1 is to be shared equally among all five lot owners. The cost of the second fifth in front of lot 2 is to be shared equally among the owners of lots 2–5, and so on. Assume that all five sections of the road cost the same to maintain.

**a)** What fractional part of the cost is paid by each owner?

**b)** What percent of the cost is paid by each owner?

**c)** If lots 3, 4, and 5 were all owned by the same person, what percent of the cost of maintenance would this person pay?

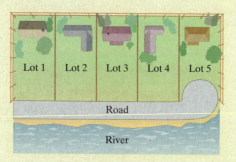

Lot 1   Lot 2   Lot 3   Lot 4   Lot 5

Road

River

# D

**Developmental Units**

A   ADDITION

S   SUBTRACTION

M   MULTIPLICATION

D   DIVISION

## OBJECTIVES

**a** Add any two of the numbers 0, 1, 2, 3, 4, 5, 6, 7, 8, 9.

**b** Find certain sums of three numbers such as $1 + 7 + 9$.

**c** Add two whole numbers when carrying is not necessary.

**d** Add two whole numbers when carrying is necessary.

# Addition

## a   BASIC ADDITION

The sum $3 + 4$ can be found by counting out a set of 3 objects and a separate set of 4 objects, putting them together, and counting all the objects.

| A set of 3 | + | A set of 4 | = | A set of 7 |
|---|---|---|---|---|

The numbers to be added are called **addends**. The result is the **sum**.

$$3 + 4 = 7$$

Addend   Addend   Sum

**EXAMPLES**   Add. Think of putting sets of objects together.

**1.** $5 + 6 = 11$

$$\begin{array}{r} 5 \\ + 6 \\ \hline 11 \end{array}$$

**2.** $8 + 5 = 13$

$$\begin{array}{r} 8 \\ + 5 \\ \hline 13 \end{array}$$

We can also do these problems by counting up from one of the numbers. For example, in Example 2, we start at 8 and count up 5 times: 9, 10, 11, 12, 13.

◀ **Do Exercises 1–6.**

What happens when we add 0? Think of a set of 5 objects. If we add 0 objects to it, we still have 5 objects. Similarly, if we have a set with 0 objects in it and add 5 objects to it, we have a set with 5 objects. Thus,

$$5 + 0 = 5 \quad \text{and} \quad 0 + 5 = 5.$$

---

### ADDITION OF 0

Adding 0 to a number does not change the number:

$$a + 0 = 0 + a = a.$$

We say that 0 is the **additive identity**.

---

Add; think of joining sets of objects.

**1.** $4 + 5$   **2.** $5 + 2$

**3.**
$$\begin{array}{r} 9 \\ + 5 \end{array}$$
**4.**
$$\begin{array}{r} 8 \\ + 8 \end{array}$$

**5.**
$$\begin{array}{r} 9 \\ + 7 \end{array}$$
**6.**
$$\begin{array}{r} 7 \\ + 9 \end{array}$$

Add.

**7.** $8 + 0$   **8.** $0 + 8$

**9.**
$$\begin{array}{r} 7 \\ + 0 \end{array}$$
**10.**
$$\begin{array}{r} 46 \\ + 0 \end{array}$$

**11.** $0 + 13$   **12.** $58 + 0$

**EXAMPLES**   Add.

**3.** $0 + 9 = 9$

$$\begin{array}{r} 0 \\ + 9 \\ \hline 9 \end{array}$$

**4.** $0 + 0 = 0$

$$\begin{array}{r} 0 \\ + 0 \\ \hline 0 \end{array}$$

**5.** $97 + 0 = 97$

$$\begin{array}{r} 97 \\ + 0 \\ \hline 97 \end{array}$$

◀ **Do Exercises 7–12.**

**Answers**

**1.** 9   **2.** 7   **3.** 14   **4.** 16
**5.** 16   **6.** 16   **7.** 8   **8.** 8
**9.** 7   **10.** 46   **11.** 13   **12.** 58

Your objective for this part of the section is to be able to add any two of the numbers 0, 1, 2, 3, 4, 5, 6, 7, 8, 9. Adding 0 is easy. The rest of the sums are listed in this table. Memorize the table by saying it to yourself over and over or by using flash cards.

| + | 1 | 2 | 3 | 4 | 5 | 6 | 7 | 8 | 9 |
|---|---|---|---|---|---|---|---|---|---|
| 1 | 2 | 3 | 4 | 5 | 6 | 7 | 8 | 9 | 10 |
| 2 | 3 | 4 | 5 | 6 | 7 | 8 | 9 | 10 | 11 |
| 3 | 4 | 5 | 6 | 7 | 8 | 9 | 10 | 11 | 12 |
| 4 | 5 | 6 | 7 | 8 | 9 | 10 | 11 | 12 | 13 |
| 5 | 6 | 7 | 8 | 9 | 10 | 11 | 12 | 13 | 14 |
| 6 | 7 | 8 | 9 | 10 | 11 | 12 | 13 | 14 | 15 |
| 7 | 8 | 9 | 10 | 11 | 12 | 13 | 14 | 15 | 16 |
| 8 | 9 | 10 | 11 | 12 | 13 | 14 | 15 | 16 | 17 |
| 9 | 10 | 11 | 12 | 13 | 14 | 15 | 16 | 17 | 18 |

6 + 7 = 13
Find 6 at the left, and 7 at the top.

7 + 6 = 13
Find 7 at the left, and 6 at the top.

It is very important that you *memorize* the basic addition facts! If you do not, you will always have trouble with addition.

Note the following.

$$3 + 4 = 7 \qquad 7 + 6 = 13 \qquad 7 + 2 = 9$$
$$4 + 3 = 7 \qquad 6 + 7 = 13 \qquad 2 + 7 = 9$$

We can add whole numbers in any order. This is the *commutative law of addition*. Because of this law, you need to learn only about half the table above, as shown by the shading.

**Do Exercises 13 and 14.** ▶

## b  CERTAIN SUMS OF THREE NUMBERS

To add $3 + 5 + 4$, we can add 3 and 5, then 4. We can also add 5 and 4, then add 3. Either way, we get 12.

**EXAMPLE 6**  Add from the top mentally.

$$\begin{array}{r} 1 \\ 7 \\ + 9 \\ \hline \end{array}$$

We first add 1 and 7, getting 8. Then we add 8 and 9, getting 17.

1
7 → 8
+ 9   9 → 17
17 ←

**EXAMPLE 7**  Add from the top mentally.

$$\begin{array}{r} 2 \\ 4 \\ + 8 \\ \hline 14 \end{array}$$

4 → 6
8 → 14
14 ←

**Do Exercises 15–18.** ▶

Complete the table.

**13.**

| + | 1 | 2 | 3 | 4 | 5 |
|---|---|---|---|---|---|
| 1 |   |   | 4 |   |   |
| 2 |   |   |   |   |   |
| 3 |   |   |   | 7 |   |
| 4 |   |   |   |   |   |
| 5 |   |   |   |   |   |

**14.**

| + | 6 | 5 | 7 | 4 | 9 |
|---|---|---|---|---|---|
| 7 |   |   | 14 |   |   |
| 9 |   |   |   |   |   |
| 5 |   |   | 9 |   |   |
| 8 |   |   |   |   |   |
| 4 |   |   |   |   |   |

Add from the top mentally.

**15.**
$$\begin{array}{r} 1 \\ 6 \\ + 9 \\ \hline \end{array}$$

**16.**
$$\begin{array}{r} 2 \\ 3 \\ + 4 \\ \hline \end{array}$$

**17.**
$$\begin{array}{r} 6 \\ 1 \\ + 4 \\ \hline \end{array}$$

**18.**
$$\begin{array}{r} 5 \\ 2 \\ + 8 \\ \hline \end{array}$$

*Answers*

**13.**

| + | 1 | 2 | 3 | 4 | 5 |
|---|---|---|---|---|---|
| 1 | 2 | 3 | 4 | 5 | 6 |
| 2 | 3 | 4 | 5 | 6 | 7 |
| 3 | 4 | 5 | 6 | 7 | 8 |
| 4 | 5 | 6 | 7 | 8 | 9 |
| 5 | 6 | 7 | 8 | 9 | 10 |

**14.**

| + | 6 | 5 | 7 | 4 | 9 |
|---|---|---|---|---|---|
| 7 | 13 | 12 | 14 | 11 | 16 |
| 9 | 15 | 14 | 16 | 13 | 18 |
| 5 | 11 | 10 | 12 | 9 | 14 |
| 8 | 14 | 13 | 15 | 12 | 17 |
| 4 | 10 | 9 | 11 | 8 | 13 |

**15.** 16   **16.** 9   **17.** 11   **18.** 15

## c ADDITION (NO CARRYING)

We now move to a gradual, conceptual development of addition of whole numbers. To add larger numbers, we can add the ones first, then the tens, then the hundreds, and so on.

**EXAMPLE 8**  Add: 5722 + 3234.

```
  5 7 2 2      Add ones.
+ 3 2 3 4
          6
```

```
  5 7 2 2      Add tens.
+ 3 2 3 4
        5 6
```

```
  5 7 2 2      Add hundreds.
+ 3 2 3 4
      9 5 6
```

```
  5 7 2 2      Add thousands.
+ 3 2 3 4
    8 9 5 6
```

This is for explanation.

```
  5 7 2 2
+ 3 2 3 4
  8 9 5 6
```

You should write only this.

◀ **Do Exercises 19–22.**

Add.

**19.**   2 4
        + 3 5

**20.**   3 4 6
        + 2 0 3

**21.**   8 3 2 7
        + 1 6 5 2

**22.**   3 4 6 1
        + 2 0 3 5

## d ADDITION (WITH CARRYING)

### Carrying Tens, Hundreds, and Thousands

**EXAMPLE 9**  Add: 18 + 27.

```
  1 8      Add ones.   Think:      8
+ 2 7                           +  7
      ?                           1 5
```

15 ones = 10 ones + 5 ones
         = 1 ten + 5 ones

```
   1
  1 8
+ 2 7
      5
```

Write 5 in the ones column. Then write 1 above the tens to indicate that we are *carrying* 1 ten from the ones column.

```
   1
  1 8
+ 2 7      Add tens.
  4 5
```

◀ **Do Exercises 23 and 24.**

We can use money to help explain Example 9.

18¢ → 1 dime and 8 pennies
+27¢ → 2 dimes and 7 pennies
        3 dimes and 15 pennies
              or
45¢ ← 4 dimes and 5 pennies

Add.

**23.**   1 9
        + 3 7

**24.**   4 6
        + 3 9

**EXAMPLE 10**  Add: 256 + 391.

$$
\begin{array}{r}
2\;5\;6 \\
+\;3\;9\;1 \\
\hline
7
\end{array}
$$
Add ones.

$$
\begin{array}{r}
{}^{1}2\;5\;6 \\
+\;3\;9\;1 \\
\hline
4\;7
\end{array}
$$
Add tens. We get 14 tens.
Now 14 tens = 10 tens + 4 tens = 1 hundred + 4 tens.
Write 4 in the tens column and 1 above the hundreds.

$$
\begin{array}{r}
{}^{1}2\;5\;6 \\
+\;3\;9\;1 \\
\hline
6\;4\;7
\end{array}
$$
Add hundreds.

**Do Exercises 25 and 26.** ▶

The carrying in Example 10 is like exchanging 14 dimes for 1 dollar bill and 4 dimes.

Add.

**25.**  $\begin{array}{r} 3\;4\;1 \\ +\;4\;8\;8 \\ \hline \end{array}$  **26.**  $\begin{array}{r} 7\;3\;0 \\ +\;2\;9\;6 \\ \hline \end{array}$

**EXAMPLE 11**  Add: 4803 + 3792.

$$
\begin{array}{r}
4\;8\;0\;3 \\
+\;3\;7\;9\;2 \\
\hline
5
\end{array}
$$
Add ones.

$$
\begin{array}{r}
4\;8\;0\;3 \\
+\;3\;7\;9\;2 \\
\hline
9\;5
\end{array}
$$
Add tens.

$$
\begin{array}{r}
{}^{1}4\;8\;0\;3 \\
+\;3\;7\;9\;2 \\
\hline
5\;9\;5
\end{array}
$$
Add hundreds. We get 15 hundreds. Now 15 hundreds = 10 hundreds + 5 hundreds = 1 thousand + 5 hundreds. Write 5 in the hundreds column and 1 above the thousands.

$$
\begin{array}{r}
{}^{1}4\;8\;0\;3 \\
+\;3\;7\;9\;2 \\
\hline
8\;5\;9\;5
\end{array}
$$
Add thousands.

**Do Exercise 27.** ▶

**TO THE STUDENT**

If you had trouble with Section 1.2 and have studied Developmental Unit A, you should go back and work through Section 1.2 after completing Exercise Set A.

**27.** Add.

$$
\begin{array}{r}
7\;8\;5\;0 \\
+\;4\;8\;4\;8 \\
\hline
\end{array}
$$

## Carrying More Than Once

**EXAMPLE 12**  Add: 767 + 993.

$$
\begin{array}{r}
7\;{}^{1}6\;7 \\
+\;\;9\;9\;3 \\
\hline
0
\end{array}
$$
Add ones. We get 10 ones. Now 10 ones = 1 ten + 0 ones. Write 0 in the ones column and 1 above the tens.

$$
\begin{array}{r}
{}^{1}7\;{}^{1}6\;7 \\
+\;\;9\;9\;3 \\
\hline
6\;0
\end{array}
$$
Add tens. We get 16 tens. Now 16 tens = 1 hundred + 6 tens. Write 6 in the tens column and 1 above the hundreds.

$$
\begin{array}{r}
{}^{1}7\;{}^{1}6\;7 \\
+\;\;9\;9\;3 \\
\hline
1\;7\;6\;0
\end{array}
$$
Add hundreds. We get 17 hundreds.

**Do Exercises 28 and 29.** ▶

Add.

**28.**  $\begin{array}{r} 7\;9\;8\;9 \\ +\;5\;6\;7\;2 \\ \hline \end{array}$  **29.**  $\begin{array}{r} 5\;6{,}7\;8\;9 \\ +\;1\;4{,}5\;3\;9 \\ \hline \end{array}$

*Answers*
**25.** 829  **26.** 1026  **27.** 12,698
**28.** 13,661  **29.** 71,328

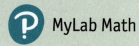
**a**  Add. Try to do these mentally. If you have trouble, think of putting sets of objects together.

| | | | | | |
|---|---|---|---|---|---|
| **1.** 8 <br> + 9 | **2.** 8 <br> + 7 | **3.** 6 <br> + 7 | **4.** 9 <br> + 5 | **5.** 5 <br> + 7 | **6.** 5 <br> + 6 |
| **7.** 9 <br> + 8 | **8.** 9 <br> + 7 | **9.** 8 <br> + 4 | **10.** 9 <br> + 1 | **11.** 8 <br> + 2 | **12.** 3 <br> + 8 |
| **13.** 0 <br> + 7 | **14.** 4 <br> + 3 | **15.** 2 <br> + 9 | **16.** 0 <br> + 0 | **17.** 3 <br> + 0 | **18.** 9 <br> + 9 |
| **19.** 8 <br> + 6 | **20.** 3 <br> + 7 | **21.** 2 <br> + 2 | **22.** 7 <br> + 7 | **23.** 6 <br> + 5 | **24.** 7 <br> + 8 |
| **25.** 8 <br> + 8 | **26.** 8 <br> + 1 | **27.** 5 <br> + 8 | **28.** 5 <br> + 9 | **29.** 4 <br> + 7 | **30.** 6 <br> + 1 |

**31.** 6 + 7  **32.** 7 + 7  **33.** 3 + 9  **34.** 6 + 0  **35.** 6 + 4

**36.** 9 + 3  **37.** 5 + 5  **38.** 5 + 3  **39.** 1 + 1  **40.** 4 + 5

**41.** 9 + 4  **42.** 0 + 8  **43.** 4 + 6  **44.** 2 + 7  **45.** 3 + 7

**46.** 3 + 3  **47.** 5 + 8  **48.** 3 + 6  **49.** 4 + 4  **50.** 4 + 7

**b**  Add from the top mentally.

| | | | | |
|---|---|---|---|---|
| **51.** 1 <br> 8 <br> + 3 | **52.** 1 <br> 7 <br> + 5 | **53.** 3 <br> 2 <br> + 5 | **54.** 4 <br> 3 <br> + 5 | **55.** 1 <br> 7 <br> + 9 |
| **56.** 5 <br> 2 <br> + 6 | **57.** 4 <br> 5 <br> + 1 | **58.** 1 <br> 9 <br> + 6 | **59.** 1 <br> 8 <br> + 7 | **60.** 1 <br> 6 <br> + 8 |

**c** Add.

61.
```
  23
+ 16
```

62.
```
  54
+ 35
```

63.
```
  67
+ 20
```

64.
```
  496
+ 503
```

65.
```
  700
+ 200
```

66.
```
  801
+  67
```

67.
```
  666
+ 333
```

68.
```
  523
+ 325
```

69.
```
  747
+ 130
```

70.
```
  8250
+ 9430
```

71.
```
  6552
+ 4321
```

72.
```
  3406
+ 1293
```

73.
```
  7340
+ 3527
```

74.
```
  4825
+ 5070
```

75.
```
  2073
+ 1925
```

76.
```
  9111
+ 9111
```

77.
```
  7889
+ 9000
```

78.
```
  52,433
+ 12,056
```

79.
```
  43,723
+ 56,276
```

80.
```
  51,670
+ 26,107
```

**d** Add.

81.
```
  38
+  8
```

82.
```
  17
+  9
```

83.
```
  17
+ 38
```

84.
```
  95
+  6
```

85.
```
  862
+ 781
```

86.
```
  613
+ 799
```

87.
```
  355
+ 491
```

88.
```
  280
+ 348
```

89.
```
  814
+ 390
```

90.
```
  274
+ 333
```

91.
```
  9990
+   10
```

92.
```
  999
+  11
```

93.
```
  999
+ 111
```

94.
```
  839
+ 388
```

95.
```
  909
+ 202
```

96.
```
  808
+ 909
```

97.
```
  8718
+ 1420
```

98.
```
  3854
+ 2700
```

99.
```
  4828
+ 1283
```

100.
```
  6995
+ 1432
```

101.
```
  9889
+    1
```

102.
```
  6889
+ 4723
```

103.
```
  9128
+ 1997
```

104.
```
  8898
+ 6645
```

105.
```
  9989
+ 6785
```

106.
```
  46,889
+ 21,786
```

107.
```
  23,448
+ 10,989
```

108.
```
  67,658
+ 98,786
```

109.
```
  77,548
+ 23,767
```

110.
```
  44,684
+  4,765
```

# Subtraction

### S

## OBJECTIVES

**a** Find basic differences such as $5 - 3$, $13 - 8$, and so on.

**b** Subtract one whole number from another when borrowing is not necessary.

**c** Subtract one whole number from another when borrowing is necessary.

## a  BASIC SUBTRACTION

Subtraction can be explained by taking away part of a set.

**EXAMPLE 1**  Subtract: $7 - 3$.

We can do this by counting out 7 objects and then taking away 3 of them. Then we count the number that remain: $7 - 3 = 4$.

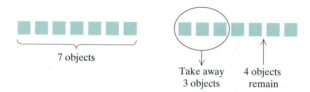

7 objects

Take away 3 objects          4 objects remain

We could also do this mentally by starting at 7 and counting down 3 times: 6, 5, 4.

**EXAMPLE 2**  Subtract: $11 - 6$. Think of "take away."

$$11 - 6 = 5 \qquad \textit{Take away:} \text{ "11 take away 6 is 5."}$$

$$\begin{array}{r} 11 \\ -\ 6 \\ \hline 5 \end{array}$$

◀ **Do Exercises 1–4.**

Subtract.

**1.** $10 - 6$    **2.** $11 - 4$

**3.** $\begin{array}{r} 16 \\ -\ 8 \\ \hline \end{array}$    **4.** $\begin{array}{r} 10 \\ -\ 7 \\ \hline \end{array}$

The addition table in Developmental Unit A will enable you to subtract also. First, recall how addition and subtraction are related.

*An addition*:

4          +          3          =          7

*Two related subtractions*:

**a)**

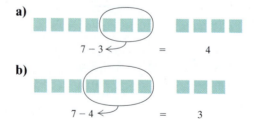

$7 - 3$          =          4

**b)**

$7 - 4$          =          3

Since we know that

$$4 + 3 = 7, \qquad \text{A basic addition fact}$$

we also know the two subtraction facts:

$$7 - 3 = 4 \quad \text{and} \quad 7 - 4 = 3.$$

*Answers*
**1.** 4  **2.** 7  **3.** 8  **4.** 3

**EXAMPLE 3** From $8 + 9 = 17$, write two subtraction facts.

**a)** The addend 8 is subtracted from the sum 17.

$8 + 9 = 17.$   The related sentence is   $17 - 8 = 9.$

**b)** The addend 9 is subtracted from the sum 17.

$8 + 9 = 17.$   The related sentence is   $17 - 9 = 8.$

**Do Exercises 5 and 6.** ▶

We can use the idea that subtraction is defined in terms of addition to think of subtraction as "how much more."

**EXAMPLE 4** Find: $13 - 6$.

To find $13 - 6$, we ask, "6 plus what number is 13?"

$6 + \square = 13$

Using the addition table at the right, we find 13 inside the table and 6 along the left side. Then we read the answer, 7, from the top. Thus, we have

$13 - 6 = 7.$

Strive to do this kind of thinking as fast as you can mentally, without having to use the table.

**Do Exercises 7–10.** ▶

## b  SUBTRACTION (NO BORROWING)

We now move to a gradual, conceptual development of subtraction of whole numbers. To subtract larger numbers, we can subtract the ones first, then the tens, then the hundreds, and so on.

**EXAMPLE 5** Subtract: $5787 - 3214$.

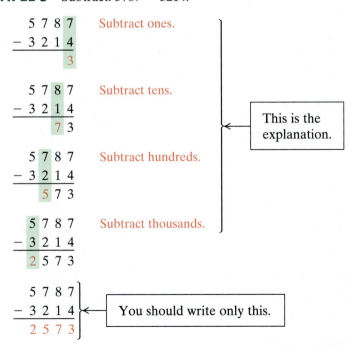

For each addition fact, write two subtraction facts.

**5.** $8 + 4 = 12$

**6.** $6 + 7 = 13$

| + | 1 | 2 | 3 | 4 | 5 | 6 | 7 | 8 | 9 |
|---|---|---|---|---|---|---|---|---|---|
| 1 | 2 | 3 | 4 | 5 | 6 | 7 | 8 | 9 | 10 |
| 2 | 3 | 4 | 5 | 6 | 7 | 8 | 9 | 10 | 11 |
| 3 | 4 | 5 | 6 | 7 | 8 | 9 | 10 | 11 | 12 |
| 4 | 5 | 6 | 7 | 8 | 9 | 10 | 11 | 12 | 13 |
| 5 | 6 | 7 | 8 | 9 | 10 | 11 | 12 | 13 | 14 |
| 6 | 7 | 8 | 9 | 10 | 11 | 12 | 13 | 14 | 15 |
| 7 | 8 | 9 | 10 | 11 | 12 | 13 | 14 | 15 | 16 |
| 8 | 9 | 10 | 11 | 12 | 13 | 14 | 15 | 16 | 17 |
| 9 | 10 | 11 | 12 | 13 | 14 | 15 | 16 | 17 | 18 |

Subtract. Try to do these mentally.

**7.** $14 - 6$    **8.** $12 - 5$

**9.**   $\begin{array}{r} 13 \\ -\ 4 \\ \hline \end{array}$    **10.**   $\begin{array}{r} 11 \\ -\ 7 \\ \hline \end{array}$

Subtract.

**11.**   $\begin{array}{r} 78 \\ -\ 64 \\ \hline \end{array}$    **12.**   $\begin{array}{r} 29 \\ -\ 9 \\ \hline \end{array}$

**13.**   $\begin{array}{r} 542 \\ -\ 301 \\ \hline \end{array}$    **14.**   $\begin{array}{r} 6896 \\ -\ 4871 \\ \hline \end{array}$

***Answers***

**5.** $12 - 8 = 4;\ 12 - 4 = 8$
**6.** $13 - 6 = 7;\ 13 - 7 = 6$   **7.** 8
**8.** 7   **9.** 9   **10.** 4   **11.** 14   **12.** 20
**13.** 241   **14.** 2025

**Do Exercises 11–14.** ▶

## C SUBTRACTION (WITH BORROWING)

We now consider subtraction when borrowing, or regrouping, is necessary.

### Borrowing from the Tens Place

**EXAMPLE 6**  Subtract: 37 − 18.

$$\begin{array}{r} 3\,7 \\ -\,1\,8 \\ \hline ? \end{array}$$

Try to subtract ones: 7 − 8 is not a whole number.

$$\begin{array}{r} {}^{2}\phantom{3}{}^{17}\phantom{7} \\ \cancel{3}\ \cancel{7} \\ -\,1\ 8 \\ \hline \end{array}$$

Borrow a ten. That is, 1 ten = 10 ones, and 10 ones + 7 ones = 17 ones. Write 2 above the tens column and 17 above the ones. We regard 37 as 20 + 17.

$$\begin{array}{r} {}^{2}\phantom{3}{}^{17}\phantom{7} \\ \cancel{3}\ \cancel{7} \\ -\,1\ 8 \\ \hline 9 \end{array}$$

Subtract ones.

> You should write only this.
>
> $$\begin{array}{r} {}^{2}\ {}^{17} \\ \cancel{3}\ \cancel{7} \\ -\,1\ 8 \\ \hline 1\ 9 \end{array}$$

$$\begin{array}{r} {}^{2}\phantom{3}{}^{17}\phantom{7} \\ \cancel{3}\ \cancel{7} \\ -\,1\ 8 \\ \hline 1\ 9 \end{array}$$

Subtract tens.

◀ Do Exercises 15 and 16.

> The borrowing in Example 6 is like exchanging 3 dimes and 7 pennies for 2 dimes and 17 pennies.

Subtract.

**15.**  $\begin{array}{r} 4\,6 \\ -\,2\,9 \\ \hline \end{array}$

**16.**  $\begin{array}{r} 7\,4 \\ -\,3\,8 \\ \hline \end{array}$

### Borrowing Hundreds

**EXAMPLE 7**  Subtract: 538 − 275.

$$\begin{array}{r} 5\,3\,8 \\ -\,2\,7\,5 \\ \hline 3 \end{array}$$

Subtract ones.

$$\begin{array}{r} 5\,3\,8 \\ -\,2\,7\,5 \\ \hline ?\,3 \end{array}$$

Try to subtract tens: 3 tens − 7 tens is not a whole number.

$$\begin{array}{r} {}^{4}\phantom{5}{}^{13}\phantom{3} \\ \cancel{5}\ \cancel{3}\ 8 \\ -\,2\ 7\ 5 \\ \hline 3 \end{array}$$

Borrow a hundred. That is, 1 hundred = 10 tens, and 10 tens + 3 tens = 13 tens. Write 4 above the hundreds column and 13 above the tens.

$$\begin{array}{r} {}^{4}\phantom{5}{}^{13}\phantom{3} \\ \cancel{5}\ \cancel{3}\ 8 \\ -\,2\ 7\ 5 \\ \hline 6\,3 \end{array}$$

Subtract tens.

> You should write only this.
>
> $$\begin{array}{r} {}^{4}\ {}^{13} \\ \cancel{5}\ \cancel{3}\ 8 \\ -\,2\ 7\ 5 \\ \hline 2\ 6\ 3 \end{array}$$

$$\begin{array}{r} {}^{4}\phantom{5}{}^{13}\phantom{3} \\ \cancel{5}\ \cancel{3}\ 8 \\ -\,2\ 7\ 5 \\ \hline 2\,6\,3 \end{array}$$

Subtract hundreds.

◀ Do Exercises 17 and 18.

> The borrowing in Example 7 is like exchanging 5 dollars and 3 dimes for 4 dollars and 13 dimes.

Subtract.

**17.**  $\begin{array}{r} 6\,4\,6 \\ -\,1\,9\,2 \\ \hline \end{array}$

**18.**  $\begin{array}{r} 7\,3\,3 \\ -\,4\,8\,3 \\ \hline \end{array}$

*Answers*

**15.** 17  **16.** 36  **17.** 454  **18.** 250

## Borrowing More Than Once

Sometimes we must borrow more than once.

**EXAMPLE 8**   Subtract: 672 − 394.

$$
\begin{array}{r}
6\ \overset{6}{\cancel{7}}\ \overset{12}{\cancel{2}} \\
-\ 3\ 9\ 4 \\
\hline
8
\end{array}
$$
Borrowing a ten to subtract ones

$$
\begin{array}{r}
5\ \overset{16}{\cancel{6}}\ \overset{6}{\cancel{7}}\ \overset{12}{\cancel{2}} \\
-\ 3\ 9\ 4 \\
\hline
2\ 7\ 8
\end{array}
$$
Borrowing a hundred to subtract tens

**Do Exercises 19 and 20.** ▶

Subtract.

**19.**  $\begin{array}{r} 5\ 6\ 3 \\ -\ 1\ 8\ 7 \\ \hline \end{array}$   **20.**  $\begin{array}{r} 7\ 3\ 3 \\ -\ 4\ 8\ 8 \\ \hline \end{array}$

**EXAMPLE 9**   Subtract: 6357 − 1769.

$$
\begin{array}{r}
6\ 3\ \overset{4}{\cancel{5}}\ \overset{17}{\cancel{7}} \\
-\ 1\ 7\ 6\ 9 \\
\hline
8
\end{array}
$$
7 − 9 is not a whole number. We borrow a ten.

$$
\begin{array}{r}
6\ \overset{2}{\cancel{3}}\ \overset{14}{\cancel{4}}\ \overset{17}{\cancel{5}\ 7} \\
-\ 1\ 7\ 6\ 9 \\
\hline
8\ 8
\end{array}
$$
4 tens minus 6 tens is not a whole number. We borrow a hundred.

$$
\begin{array}{r}
5\ \overset{12}{\cancel{6}}\ \overset{2}{\cancel{3}}\ \overset{14}{\cancel{4}}\ \overset{17}{\cancel{5}\ 7} \\
-\ 1\ 7\ 6\ 9 \\
\hline
4\ 5\ 8\ 8
\end{array}
$$
2 hundreds minus 7 hundreds is not a whole number. We borrow a thousand.

We can always check by adding the answer to the number being subtracted.

**EXAMPLE 10**   Subtract: 8341 − 2673. Check by adding.

We check by adding 5668 and 2673.

$$
\begin{array}{r}
7\ \overset{12}{\cancel{2}}\ \overset{13}{\cancel{3}}\ \overset{11}{\cancel{4}\ 1} \\
8\ \cancel{3}\ \cancel{4}\ \cancel{1} \\
-\ 2\ 6\ 7\ 3 \\
\hline
5\ 6\ 6\ 8
\end{array}
\qquad
\begin{array}{l}
Check: \\
\begin{array}{r}
\overset{1}{5}\ \overset{1}{6}\ \overset{1}{6}\ 8 \\
+\ 2\ 6\ 7\ 3 \\
\hline
8\ 3\ 4\ 1
\end{array}
\end{array}
$$

**Do Exercises 21 and 22.** ▶

Subtract. Check by adding.

**21.**  $\begin{array}{r} 4\ 2\ 3\ 6 \\ -\ 1\ 6\ 7\ 9 \\ \hline \end{array}$   **22.**  $\begin{array}{r} 7\ 5\ 4\ 1 \\ -\ 3\ 8\ 6\ 7 \\ \hline \end{array}$

## Zeros in Subtraction

Note the following:

50 is 5 tens;

70 is 7 tens.

Then

100 is 10 tens;

200 is 20 tens.

Complete.

**23.**  80 = _____ tens

**24.**  60 = _____ tens

**25.**  300 = _____ tens

**26.**  900 = _____ tens

**Do Exercises 23–26.** ▶

**Answers**

**19.** 376  **20.** 245  **21.** 2557  **22.** 3674
**23.** 8  **24.** 6  **25.** 30  **26.** 90

Complete.

**27.** $5000 = $ _____ tens

**28.** $9000 = $ _____ tens

**29.** $5380 = $ _____ tens

**30.** $6770 = $ _____ tens

Subtract.

**31.**
```
  6 0
- 1 8
```

**32.**
```
  4 8 0
- 2 5 6
```

Subtract.

**33.**
```
  6 0 2
- 4 6 4
```

**34.**
```
  4 0 8
- 3 6 4
```

Subtract.

**35.**
```
  4 0 0 6
- 1 2 3 8
```

**36.**
```
  9 0 0 1
- 7 8 0 4
```

Subtract.

**37.**
```
  3 0 0 0
- 1 7 5 4
```

**38.**
```
  8 0 1 7
- 3 2 8 9
```

Also,

230 is 2 hundreds + 3 tens

or 20 tens + 3 tens

or 23 tens.

Similarly,

1000 is 100 tens;

2000 is 200 tens;

4670 is 467 tens.

◀ **Do Exercises 27–30.**

**EXAMPLE 11**   Subtract: $50 - 37$.

```
  4 10
  5 0      We have 5 tens. We keep 4 of them in the tens column and
- 3 7      put 1 ten, or 10 ones, with the ones.
  1 3
```

◀ **Do Exercises 31 and 32.**

**EXAMPLE 12**   Subtract: $803 - 547$.

```
  7 9 13
  8 0 3     We have 8 hundreds, or 80 tens. We keep 79 tens and
- 5 4 7     put 1 ten, or 10 ones, with the ones.
  2 5 6
```

◀ **Do Exercises 33 and 34.**

**EXAMPLE 13**   Subtract: $9003 - 2789$.

```
  8 9 9 13
  9 0 0 3    We have 9 thousands, or 900 tens. We keep 899 tens
- 2 7 8 9    and put 1 ten, or 10 ones, with the ones.
  6 2 1 4
```

◀ **Do Exercises 35 and 36.**

**EXAMPLES**   Subtract.

**14.**
```
  4 9 9 10
  5 0 0 0
- 2 8 6 1
  2 1 3 9
```

**15.**
```
        10
  4 9 0 13
  5 0 1 3     We have 5 thousands, or
- 1 8 5 7     49 hundreds and 10 tens.
  3 1 5 6
```

◀ **Do Exercises 37 and 38.**

---

**TO THE STUDENT**

If you had trouble with Section 1.3 and have studied Developmental Unit S, you should go back and work through Section 1.3 after completing Exercise Set S.

---

*Answers*

**27.** 500  **28.** 900  **29.** 538  **30.** 677
**31.** 42  **32.** 224  **33.** 138  **34.** 44
**35.** 2768  **36.** 1197  **37.** 1246  **38.** 4728

**a** Subtract. Try to do these mentally.

1. 
$$\begin{array}{r} 7 \\ -\ 0 \\ \hline \end{array}$$

2. 
$$\begin{array}{r} 8 \\ -\ 8 \\ \hline \end{array}$$

3. 
$$\begin{array}{r} 7 \\ -\ 7 \\ \hline \end{array}$$

4. 
$$\begin{array}{r} 8 \\ -\ 3 \\ \hline \end{array}$$

5. 
$$\begin{array}{r} 5 \\ -\ 2 \\ \hline \end{array}$$

6. 
$$\begin{array}{r} 1\ 6 \\ -\ 8 \\ \hline \end{array}$$

7. 
$$\begin{array}{r} 1\ 7 \\ -\ 9 \\ \hline \end{array}$$

8. 
$$\begin{array}{r} 1\ 2 \\ -\ 6 \\ \hline \end{array}$$

9. 
$$\begin{array}{r} 1\ 1 \\ -\ 4 \\ \hline \end{array}$$

10. 
$$\begin{array}{r} 1\ 2 \\ -\ 9 \\ \hline \end{array}$$

11. 
$$\begin{array}{r} 1\ 4 \\ -\ 7 \\ \hline \end{array}$$

12. 
$$\begin{array}{r} 1\ 8 \\ -\ 9 \\ \hline \end{array}$$

13. 
$$\begin{array}{r} 1\ 3 \\ -\ 7 \\ \hline \end{array}$$

14. 
$$\begin{array}{r} 1\ 5 \\ -\ 9 \\ \hline \end{array}$$

15. 
$$\begin{array}{r} 9 \\ -\ 7 \\ \hline \end{array}$$

16. $7 - 3$

17. $4 - 1$

18. $2 - 0$

19. $3 - 3$

20. $6 - 3$

21. $7 - 6$

22. $9 - 8$

23. $10 - 3$

24. $6 - 6$

25. $11 - 7$

26. $12 - 8$

27. $5 - 0$

28. $4 - 0$

29. $13 - 9$

30. $14 - 9$

31. $11 - 2$

32. $12 - 3$

33. $16 - 9$

34. $18 - 9$

35. $11 - 5$

36. $10 - 4$

37. $10 - 8$

38. $14 - 8$

39. $15 - 8$

40. $10 - 2$

**b** Subtract.

41. 
$$\begin{array}{r} 6\ 4 \\ -\ 3\ 1 \\ \hline \end{array}$$

42. 
$$\begin{array}{r} 5\ 5 \\ -\ 3\ 4 \\ \hline \end{array}$$

43. 
$$\begin{array}{r} 5\ 4\ 8 \\ -\ 3\ 0\ 1 \\ \hline \end{array}$$

44. 
$$\begin{array}{r} 5\ 9\ 6 \\ -\ 4\ 0\ 3 \\ \hline \end{array}$$

45. 
$$\begin{array}{r} 7\ 0\ 0 \\ -\ 2\ 0\ 0 \\ \hline \end{array}$$

**46.**  765
       − 111

**47.**  525
       − 323

**48.**  747
       − 130

**49.**  988
       − 700

**50.**  9450
       − 8230

**51.**  6552
       − 4321

**52.**  7547
       − 3421

**53.**  5875
       − 2111

**54.**  38,695
       − 37,004

**55.**  67,899
       − 66,673

**56.**  99,999
       −      1

**57.**  56,780
       − 56,770

**58.**  42,111
       − 32,010

**59.**  77,654
       − 66,611

**60.**  23,456
       − 12,345

**C** Subtract.

**61.**  93
       − 28

**62.**  42
       − 13

**63.**  86
       − 78

**64.**  98
       − 89

**65.**  625
       − 317

**66.**  735
       − 609

**67.**  853
       − 236

**68.**  961
       − 747

**69.**  787
       − 698

**70.**  6769
       − 2367

**71.**  6431
       − 2876

**72.**  7654
       − 1765

**73.**  5246
       − 2859

**74.**  6328
       − 2679

**75.**  7641
       − 3809

**76.**  8743
       −  599

**77.**  12,647
       −  4,897

**78.**  16,222
       −  5,777

**79.**  46,781
       − 12,988

**80.**  470
       − 189

**81.**  690
       − 235

**82.**  703
       − 132

**83.**  6406
       −  258

**84.**  2309
       −  109

**85.**  3406
       − 1293

**86.**  6807
       − 3059

**87.**  8000
       − 2794

**88.**  8002
       − 6543

**89.**  38,000
       − 37,695

**90.**  16,043
       − 11,588

# Multiplication

## a  BASIC MULTIPLICATION

To multiply, we begin with two numbers, called **factors**, and get a third number, called a **product**. Multiplication of whole numbers can be explained by counting. The product $3 \times 5$ can be found by counting out 3 sets of 5 objects each, joining them (in a rectangular array if desired), and counting all the objects.

$$3 \times 5 = 15$$

Factor  Factor  Product

We can also think of multiplication as repeated addition.

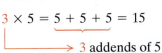

$$3 \times 5 = 5 + 5 + 5 = 15$$

3 addends of 5

**EXAMPLES**  Multiply. Think either of putting sets of objects together in a rectangular array or of repeated addition.

**1.** $5 \times 6 = 30$

$$\begin{array}{r} 6 \\ \times\ 5 \\ \hline 30 \end{array}$$

**2.** $8 \times 4 = 32$

$$\begin{array}{r} 4 \\ \times\ 8 \\ \hline 32 \end{array}$$

**Do Exercises 1–4.** ▶

## Multiplying by 0

How do we multiply by 0? Consider $4 \cdot 0$. Using repeated addition, we see that

$$4 \cdot 0 = 0 + 0 + 0 + 0 = 0.$$

4 addends of 0

We can also think of $4 \cdot 0$ as 4 sets with 0 objects in each set, so the total is 0.

Consider $0 \cdot 4$. Using repeated addition, this is 0 addends of 4, which is 0. Using sets, this is 0 sets with 4 objects in each set, which is 0. Thus, we have the following.

| MULTIPLICATION BY 0 |
| :--- |
| Multiplying by 0 gives 0. |

**EXAMPLES**  Multiply.

**3.** $13 \times 0 = 0$

$$\begin{array}{r} 0 \\ \times\ 13 \\ \hline 0 \end{array}$$

**4.** $0 \cdot 11 = 0$

$$\begin{array}{r} 11 \\ \times\ 0 \\ \hline 0 \end{array}$$

**5.** $0 \cdot 0 = 0$

$$\begin{array}{r} 0 \\ \times\ 0 \\ \hline 0 \end{array}$$

**Do Exercises 5 and 6.** ▶

Multiply. Think of joining sets in a rectangular array or of repeated addition.

**1.** $7 \cdot 8$ (The dot, $\cdot$, means the same as $\times$.)

**2.** $\begin{array}{r} 9 \\ \times\ 4 \\ \hline \end{array}$

**3.** $4 \cdot 7$

**4.** $\begin{array}{r} 7 \\ \times\ 6 \\ \hline \end{array}$

Multiply.

**5.** $8 \cdot 0$

**6.** $\begin{array}{r} 17 \\ \times\ 0 \\ \hline \end{array}$

***Answers***

**1.** 56  **2.** 36  **3.** 28  **4.** 42
**5.** 0  **6.** 0

## Multiplying by 1

How do we multiply by 1? Consider $5 \cdot 1$. Using repeated addition, we see that

$$5 \cdot 1 = \underbrace{1 + 1 + 1 + 1 + 1}_{} = 5.$$

$\longrightarrow$ 5 addends of 1

We can also think of $5 \cdot 1$ as 5 sets with 1 object in each set, for a total of 5 objects.

Consider $1 \cdot 5$. Using repeated addition, this is 1 addend of 5, which is 5. Using sets, this is 1 set of 5 objects, which is again 5 objects. Thus, we have the following.

---

### MULTIPLICATION BY 1

Multiplying a number by 1 does not change the number:

$$a \cdot 1 = 1 \cdot a = a.$$

We say that 1 is the **multiplicative identity**.

---

**EXAMPLES** Multiply.

**6.** $13 \cdot 1 = 13$

$$\begin{array}{r} 1 \\ \times\ 13 \\ \hline 13 \end{array}$$

**7.** $1 \cdot 7 = 7$

$$\begin{array}{r} 7 \\ \times\ 1 \\ \hline 7 \end{array}$$

**8.** $1 \cdot 1 = 1$

$$\begin{array}{r} 1 \\ \times\ 1 \\ \hline 1 \end{array}$$

◀ **Do Exercises 7 and 8.**

You should be able to multiply any two of the numbers 0, 1, 2, 3, 4, 5, 6, 7, 8, 9. Multiplying by 0 and 1 is easy. The rest of the products are listed in the following table.

| × | 2 | 3 | 4 | 5 | 6 | 7 | 8 | 9 |
|---|---|---|---|---|---|---|---|---|
| 2 | 4 | 6 | 8 | 10 | 12 | 14 | 16 | 18 |
| 3 | 6 | 9 | 12 | 15 | 18 | 21 | 24 | 27 |
| 4 | 8 | 12 | 16 | 20 | 24 | 28 | 32 | 36 |
| 5 | 10 | 15 | 20 | 25 | 30 | 35 | 40 | 45 |
| 6 | 12 | 18 | 24 | 30 | 36 | 42 | 48 | 54 |
| 7 | 14 | 21 | 28 | 35 | 42 | 49 | 56 | 63 |
| 8 | 16 | 24 | 32 | 40 | 48 | 56 | 64 | 72 |
| 9 | 18 | 27 | 36 | 45 | 54 | 63 | 72 | 81 |

$5 \times 7 = 35$
Find 5 at the left, and 7 at the top.

$8 \cdot 4 = 32$
Find 8 at the left, and 4 at the top.

It is *very* important that you have the basic multiplication facts *memorized*. If you do not, you will always have trouble with multiplication.

The *commutative law of multiplication* says that we can multiply numbers in any order. Thus, you need to learn only about half the table, as shown by the shading.

◀ **Do Exercises 9 and 10.**

---

Multiply.

**7.** $8 \cdot 1$

**8.**
$$\begin{array}{r} 2\ 3 \\ \times\ \ \ 1 \\ \hline \end{array}$$

Complete the table.

**9.**

| × | 2 | 3 | 4 | 5 |
|---|---|---|---|---|
| 2 |  |  |  |  |
| 3 |  | 12 |  |  |
| 4 |  |  |  |  |
| 5 |  | 15 |  |  |
| 6 |  |  |  |  |

**10.**

| × | 6 | 7 | 8 | 9 |
|---|---|---|---|---|
| 5 |  |  |  |  |
| 6 |  |  | 48 |  |
| 7 |  |  |  |  |
| 8 |  | 56 |  |  |
| 9 |  |  |  |  |

*Answers*

**7.** 8  **8.** 23

**9.**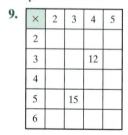

| × | 2 | 3 | 4 | 5 |
|---|---|---|---|---|
| 2 | 4 | 6 | 8 | 10 |
| 3 | 6 | 9 | 12 | 15 |
| 4 | 8 | 12 | 16 | 20 |
| 5 | 10 | 15 | 20 | 25 |
| 6 | 12 | 18 | 24 | 30 |

**10.**

| × | 6 | 7 | 8 | 9 |
|---|---|---|---|---|
| 5 | 30 | 35 | 40 | 45 |
| 6 | 36 | 42 | 48 | 54 |
| 7 | 42 | 49 | 56 | 63 |
| 8 | 48 | 56 | 64 | 72 |
| 9 | 54 | 63 | 72 | 81 |

## b  MULTIPLYING BY 10, 100, AND 1000

We now move to a gradual, conceptual development of multiplication of whole numbers. We begin by considering multiplication by 10, 100, and 1000.

### Multiplying by 10

We know that

$$50 = 5 \text{ tens} \qquad 340 = 34 \text{ tens} \qquad \text{and} \qquad 2340 = 234 \text{ tens}$$
$$= 5 \cdot 10, \qquad \qquad = 34 \cdot 10, \qquad \qquad = 234 \cdot 10.$$

Turning this around, we see that to multiply any whole number by 10, all we need to do is write a 0 on the end of the number.

> **MULTIPLICATION BY 10**
>
> To multiply a whole number by 10, write 0 on the end of the number.

**EXAMPLES**  Multiply.

**9.** $10 \cdot 6 = 60$

**10.** $10 \cdot 47 = 470$

**11.** $10 \cdot 583 = 5830$

**Do Exercises 11–15.** ▶

Let's find $4 \cdot 90$. This is $4 \cdot (9 \text{ tens})$, or 36 tens. The procedure is the same as multiplying 4 and 9 and writing a 0 on the end. Thus, $4 \cdot 90 = 360$.

**EXAMPLES**  Multiply.

**12.** $5 \cdot 70 = 350$

$5 \cdot 7$, then write a 0

**13.** $8 \cdot 80 = 640$

**14.** $5 \cdot 60 = 300$

**Do Exercises 16 and 17.** ▶

### Multiplying by 100

Note the following:

$$300 = 3 \text{ hundreds} \qquad 4700 = 47 \text{ hundreds} \qquad \text{and} \qquad 56{,}800 = 568 \text{ hundreds}$$
$$= 3 \cdot 100, \qquad \qquad = 47 \cdot 100, \qquad \qquad = 568 \cdot 100.$$

Turning this around, we see that to multiply any whole number by 100, all we need to do is write two 0's on the end of the number.

> **MULTIPLICATION BY 100**
>
> To multiply a whole number by 100, write two 0's on the end of the number.

Multiply.

**11.** $10 \cdot 7$   **12.** $10 \cdot 45$

**13.** $10 \cdot 273$   **14.** $10 \cdot 10$

**15.** $10 \cdot 100$

Multiply.

**16.** $\begin{array}{r} 70 \\ \times\ 8 \\ \hline \end{array}$   **17.** $\begin{array}{r} 60 \\ \times\ 6 \\ \hline \end{array}$

*Answers*

**11.** 70  **12.** 450  **13.** 2730  **14.** 100
**15.** 1000  **16.** 560  **17.** 360

Multiply.

**18.** $100 \cdot 7$     **19.** $100 \cdot 23$

**20.** $100 \cdot 723$     **21.** $100 \cdot 100$

**22.** $100 \cdot 1000$

Multiply.

**23.** $\begin{array}{r} 7\,0\,0 \\ \times\ \ \ \ 8 \\ \hline \end{array}$     **24.** $\begin{array}{r} 4\,0\,0 \\ \times\ \ \ \ 4 \\ \hline \end{array}$

Multiply.

**25.** $1000 \cdot 9$     **26.** $1000 \cdot 852$

**27.** $1000 \cdot 10$     **28.** $3 \cdot 4000$

**29.** $9 \cdot 8000$

**EXAMPLES**  Multiply.

**15.** $100 \cdot 6 = 600$

**16.** $100 \cdot 39 = 3900$

**17.** $100 \cdot 448 = 44{,}800$

◀ **Do Exercises 18–22.**

Let's find $4 \cdot 900$. This is $4 \cdot (9 \text{ hundreds})$, or 36 hundreds. The procedure is the same as multiplying 4 and 9 and writing two 0's on the end. Thus, $4 \cdot 900 = 3600$.

**EXAMPLES**  Multiply.

**18.** $6 \cdot 800 = 4800$     $6 \cdot 8$, then write 00

**19.** $9 \cdot 700 = 6300$

**20.** $5 \cdot 500 = 2500$

◀ **Do Exercises 23 and 24.**

## Multiplying by 1000

Note the following:

$$6000 = 6 \text{ thousands} \quad \text{and} \quad 19{,}000 = 19 \text{ thousands}$$
$$= 6 \cdot 1000 \quad\quad\quad\quad\quad\quad = 19 \cdot 1000.$$

Turning this around, we see that to multiply any whole number by 1000, all we need to do is write three 0's on the end of the number.

> **MULTIPLYING BY 1000**
>
> To multiply a whole number by 1000, write three 0's on the end of the number.

**EXAMPLES**  Multiply.

**21.** $1000 \cdot 8 = 8000$

**22.** $2000 \cdot 13 = 26{,}000$     $2 \cdot 13$, then write 000

**23.** $1000 \cdot 567 = 567{,}000$

◀ **Do Exercises 25–29.**

## Multiplying Multiples by Multiples

Let's multiply 50 and 30. This is $50 \cdot (3 \text{ tens})$, or 150 tens, or 1500. The procedure is the same as multiplying 5 and 3 and writing two 0's on the end.

## TO MULTIPLY MULTIPLES OF TENS, HUNDREDS, THOUSANDS, AND SO ON:

**a)** Multiply the one-digit numbers.

**b)** Count the number of zeros.

**c)** Write that many 0's on the end of the product.

**EXAMPLES**  Multiply.

**24.**
```
      80      1 zero at end
 ×    60      1 zero at end
    4800
      ↑———— 6 · 8, then write 00
```

**25.**
```
      800     2 zeros at end
 ×     60     1 zero at end
    48,000
       ↑———— 6 · 8, then write 000
```

**26.**
```
      800     2 zeros at end
 ×    600     2 zeros at end
    480,000
      ↑———— 6 · 8, then write 0,000
```

**27.**
```
      800     2 zeros at end
 ×     50     1 zero at end
    40,000
      ↑———— 5 · 8, then write 000
```

**Do Exercises 30–33.** ▶

Multiply.

**30.**
```
  9 0 0 0
 ×      6
```

**31.**
```
    8 0
 × 7 0
```

**32.**
```
    8 0 0
 ×    7 0
```

**33.**
```
    6 0 0
 ×    3 0
```

## c  MULTIPLYING LARGER NUMBERS

The product $3 \times 24$ can be represented as

$$3 \times (2 \text{ tens} + 4) = (2 \text{ tens} + 4) + (2 \text{ tens} + 4) + (2 \text{ tens} + 4)$$
$$= 6 \text{ tens} + 12$$
$$= 6 \text{ tens} + 1 \text{ ten} + 2$$
$$= 7 \text{ tens} + 2$$
$$= 72.$$

```
We multiply the 4 ones by 3, getting      12
We multiply the 2 tens by 3, getting    + 60
                   Then we add:            72
```

**EXAMPLE 28**  Multiply: $3 \times 24$.

```
    2 4     We use the approach described above.
 ×    3
    1 2 ←—Multiply the 4 ones by 3.
    6 0 ←—Multiply the 2 tens by 3.
    7 2 ←—Add.
```

**Do Exercises 34–36.** ▶

Multiply.

**34.**
```
    1 4
 ×    2
```

**35.**
```
    5 8
 ×    2
```

**36.**
```
    3 7
 ×    4
```

**EXAMPLE 29**  Multiply: $5 \times 734$.

```
      7 3 4
 ×        5
        2 0 ←— Multiply the 4 ones by 5.
      1 5 0 ←— Multiply the 3 tens by 5.
    3 5 0 0 ←— Multiply the 7 hundreds by 5.
    3 6 7 0 ←— Add.
```

**Do Exercises 37 and 38.** ▶

Multiply.

**37.**
```
    8 2 3
 ×      6
```

**38.**
```
    1 3 4 8
 ×        5
```

***Answers***
**30.** 54,000  **31.** 5600  **32.** 56,000
**33.** 18,000  **34.** 28  **35.** 116
**36.** 148  **37.** 4938  **38.** 6740

Let's look at Example 29 again. Instead of writing each product on a separate line, we can use a shorter form.

**EXAMPLE 30**   Multiply: $5 \times 734$.

$$\begin{array}{r} \overset{2}{7\ 3\ 4} \\ \times \quad\quad 5 \\ \hline 0 \end{array}$$

Multiply the ones by 5: $5 \cdot (4 \text{ ones}) = 20 \text{ ones} = 2 \text{ tens} + 0 \text{ ones}$. Write 0 in the ones column and 2 above the tens.

$$\begin{array}{r} \overset{1}{7}\ \overset{2}{3}\ 4 \\ \times \quad\quad 5 \\ \hline 7\ 0 \end{array}$$

Multiply 3 tens by 5 and add 2 tens: $5 \cdot (3 \text{ tens}) = 15 \text{ tens}$; $15 \text{ tens} + 2 \text{ tens} = 17 \text{ tens} = 1 \text{ hundred} + 7 \text{ tens}$. Write 7 in the tens column and 1 above the hundreds.

$$\begin{array}{r} \overset{1}{7}\ \overset{2}{3}\ 4 \\ \times \quad\quad 5 \\ \hline 3\ 6\ 7\ 0 \end{array}$$

Multiply the 7 hundreds by 5 and add 1 hundred: $5 \cdot (7 \text{ hundreds}) = 35 \text{ hundreds}$; $35 \text{ hundreds} + 1 \text{ hundred} = 36 \text{ hundreds}$.

$$\begin{array}{r} \overset{1}{7}\ \overset{2}{3}\ 4 \\ \times \quad\quad 5 \\ \hline 3\ 6\ 7\ 0 \end{array}$$

You should write only this.

◀ **Do Exercises 39–42.**

Multiply using the short form.

**39.**
$$\begin{array}{r} 5\ 8 \\ \times \quad 2 \\ \hline \end{array}$$

**40.**
$$\begin{array}{r} 3\ 7 \\ \times \quad 4 \\ \hline \end{array}$$

**41.**
$$\begin{array}{r} 8\ 2\ 3 \\ \times \quad\ \ 6 \\ \hline \end{array}$$

**42.**
$$\begin{array}{r} 1\ 3\ 4\ 8 \\ \times \quad\quad 5 \\ \hline \end{array}$$

### d   MULTIPLYING BY MULTIPLES OF 10, 100, AND 1000

To multiply 327 by 50, we multiply by 10 (write a 0) and then multiply 327 by 5.

$$\begin{array}{r} 3\ 2\ 7 \\ \times \quad 5\ 0 \\ \hline 1\ 6\,,3\ 5\ 0 \end{array}$$

Write a 0.
Multiply $5 \cdot 327$.

**EXAMPLE 31**   Multiply: $400 \times 289$.

$$\begin{array}{r} 2\ 8\ 9 \\ \times\ 4\ 0\ 0 \\ \hline 0\ 0 \end{array}$$

Write two 0's.

$$\begin{array}{r} 2\ 8\ 9 \\ \times \quad 4\ 0\ 0 \\ \hline 1\ 1\ 5\,,6\ 0\ 0 \end{array}$$

Multiply 4 and 289:

$$\begin{array}{r} \overset{3}{2}\ \overset{3}{8}\ 9 \\ \times \quad\quad 4 \\ \hline 1\ 1\ 5\ 6 \end{array}$$

$$\begin{array}{r} \overset{3}{2}\ \overset{3}{8}\ 9 \\ \times \quad 4\ 0\ 0 \\ \hline 1\ 1\ 5\,,6\ 0\ 0 \end{array}$$

Try to write only this.

◀ **Do Exercises 43–45.**

Multiply.

**43.**
$$\begin{array}{r} 7\ 4\ 6 \\ \times \quad\ \ 8 \\ \hline \end{array}$$

**44.**
$$\begin{array}{r} 7\ 4\ 6 \\ \times \quad\ 8\ 0 \\ \hline \end{array}$$

**45.**
$$\begin{array}{r} 7\ 4\ 6 \\ \times\ 8\ 0\ 0 \\ \hline \end{array}$$

**TO THE STUDENT**

If you had trouble with Section 1.4 and have studied Developmental Unit M, you should go back and work through Section 1.4 after completing Exercise Set M.

*Answers*

**39.** 116   **40.** 148   **41.** 4938   **42.** 6740
**43.** 5968   **44.** 59,680   **45.** 596,800

**a** Multiply. Try to do these mentally.

1.  $\begin{array}{r} 3 \\ \times\ 4 \\ \hline \end{array}$

2.  $\begin{array}{r} 6 \\ \times\ 0 \\ \hline \end{array}$

3.  $\begin{array}{r} 7 \\ \times\ 1 \\ \hline \end{array}$

4.  $\begin{array}{r} 0 \\ \times\ 2 \\ \hline \end{array}$

5.  $\begin{array}{r} 10 \\ \times\ 1 \\ \hline \end{array}$

6.  $\begin{array}{r} 6 \\ \times\ 5 \\ \hline \end{array}$

7.  $\begin{array}{r} 5 \\ \times\ 2 \\ \hline \end{array}$

8.  $\begin{array}{r} 9 \\ \times\ 7 \\ \hline \end{array}$

9.  $\begin{array}{r} 9 \\ \times\ 6 \\ \hline \end{array}$

10.  $\begin{array}{r} 2 \\ \times\ 6 \\ \hline \end{array}$

11.  $\begin{array}{r} 7 \\ \times\ 0 \\ \hline \end{array}$

12.  $\begin{array}{r} 8 \\ \times\ 9 \\ \hline \end{array}$

13.  $\begin{array}{r} 1 \\ \times\ 8 \\ \hline \end{array}$

14.  $\begin{array}{r} 8 \\ \times\ 0 \\ \hline \end{array}$

15.  $\begin{array}{r} 4 \\ \times\ 7 \\ \hline \end{array}$

16.  $\begin{array}{r} 3 \\ \times\ 8 \\ \hline \end{array}$

17.  $\begin{array}{r} 5 \\ \times\ 9 \\ \hline \end{array}$

18.  $\begin{array}{r} 2 \\ \times\ 9 \\ \hline \end{array}$

19.  $\begin{array}{r} 0 \\ \times\ 7 \\ \hline \end{array}$

20.  $\begin{array}{r} 5 \\ \times\ 7 \\ \hline \end{array}$

21.  $\begin{array}{r} 9 \\ \times\ 5 \\ \hline \end{array}$

22.  $\begin{array}{r} 5 \\ \times\ 8 \\ \hline \end{array}$

23.  $\begin{array}{r} 0 \\ \times\ 0 \\ \hline \end{array}$

24.  $\begin{array}{r} 2 \\ \times\ 8 \\ \hline \end{array}$

25. $5 \cdot 5$

26. $9 \cdot 9$

27. $1 \cdot 1$

28. $0 \cdot 0$

29. $2 \cdot 2$

30. $6 \cdot 6$

31. $1 \cdot 8$

32. $0 \cdot 1$

33. $3 \cdot 9$

34. $2 \cdot 9$

35. $6 \cdot 0$

36. $10 \cdot 1$

37. $6 \cdot 8$

38. $9 \cdot 6$

39. $12 \cdot 0$

40. $9 \cdot 8$

41. $3 \cdot 5$

42. $1 \cdot 8$

43. $1 \cdot 9$

44. $2 \cdot 1$

45. $8 \cdot 4$

46. $3 \cdot 2$

47. $5 \cdot 3$

48. $1 \cdot 6$

49. $4 \cdot 2$

50. $4 \cdot 5$

51. $5 \cdot 4$

52. $4 \cdot 4$

53. $5 \cdot 2$

54. $8 \cdot 0$

**b** Multiply.

**55.**
$$\begin{array}{r} 1\,0 \\ \times\ \ 8 \\ \hline \end{array}$$

**56.**
$$\begin{array}{r} 7 \\ \times\,1\,0 \\ \hline \end{array}$$

**57.**
$$\begin{array}{r} 2\,0 \\ \times\ \ 8 \\ \hline \end{array}$$

**58.**
$$\begin{array}{r} 3\,0 \\ \times\ \ 7 \\ \hline \end{array}$$

**59.**
$$\begin{array}{r} 4\,5 \\ \times\,1\,0 \\ \hline \end{array}$$

**60.**
$$\begin{array}{r} 7\,8 \\ \times\,1\,0 \\ \hline \end{array}$$

**61.**
$$\begin{array}{r} 8\,0 \\ \times\ \ 7 \\ \hline \end{array}$$

**62.**
$$\begin{array}{r} 9\,0 \\ \times\ \ 4 \\ \hline \end{array}$$

**63.**
$$\begin{array}{r} 1\,0\,0 \\ \times\ \ \ 8 \\ \hline \end{array}$$

**64.**
$$\begin{array}{r} 1\,0\,0 \\ \times\ \ \ 3 \\ \hline \end{array}$$

**65.**
$$\begin{array}{r} 1\,0\,0 \\ \times\ \ \ 9 \\ \hline \end{array}$$

**66.**
$$\begin{array}{r} 1\,0\,0 \\ \times\ \ 1\,0 \\ \hline \end{array}$$

**67.**
$$\begin{array}{r} 3\,4\,5\,7 \\ \times\ \ 1\,0\,0 \\ \hline \end{array}$$

**68.**
$$\begin{array}{r} 4\,0\,0 \\ \times\ \ \ 3 \\ \hline \end{array}$$

**69.**
$$\begin{array}{r} 7\,0\,0 \\ \times\ \ \ 7 \\ \hline \end{array}$$

**70.**
$$\begin{array}{r} 5\,0\,0 \\ \times\ \ \ 8 \\ \hline \end{array}$$

**71.**
$$\begin{array}{r} 1\,0\,0 \\ \times\,1\,0\,0 \\ \hline \end{array}$$

**72.**
$$\begin{array}{r} 1\,0\,0\,0 \\ \times\ \ \ \ 7 \\ \hline \end{array}$$

**73.**
$$\begin{array}{r} 1\,0\,0\,0 \\ \times\ \ \ \ 9 \\ \hline \end{array}$$

**74.**
$$\begin{array}{r} 1\,0\,0\,0 \\ \times\ \ \ \ 2 \\ \hline \end{array}$$

**75.**
$$\begin{array}{r} 4\,5\,7 \\ \times\,1\,0\,0\,0 \\ \hline \end{array}$$

**76.**
$$\begin{array}{r} 6\,7\,6\,9 \\ \times\,1\,0\,0\,0 \\ \hline \end{array}$$

**77.**
$$\begin{array}{r} 2\,0\,0\,0 \\ \times\ \ \ \ 9 \\ \hline \end{array}$$

**78.**
$$\begin{array}{r} 5\,0\,0\,0 \\ \times\ \ \ \ 4 \\ \hline \end{array}$$

**79.**
$$\begin{array}{r} 6\,0\,0\,0 \\ \times\ \ \ \ 8 \\ \hline \end{array}$$

**80.**
$$\begin{array}{r} 8\,0\,0\,0 \\ \times\ \ \ \ 2 \\ \hline \end{array}$$

**81.**
$$\begin{array}{r} 3\,0\,0\,0 \\ \times\ \ \ \ 2 \\ \hline \end{array}$$

**82.**
$$\begin{array}{r} 1\,0\,0\,0 \\ \times\,1\,0\,0\,0 \\ \hline \end{array}$$

**83.**
$$\begin{array}{r} 4\,0 \\ \times\,3\,0 \\ \hline \end{array}$$

**84.**
$$\begin{array}{r} 2\,0 \\ \times\,1\,0 \\ \hline \end{array}$$

**85.**
$$\begin{array}{r} 8\,0 \\ \times\,5\,0 \\ \hline \end{array}$$

**86.**
$$\begin{array}{r} 5\,0 \\ \times\,5\,0 \\ \hline \end{array}$$

**87.**
$$\begin{array}{r} 4\,0\,0 \\ \times\ \ 3\,0 \\ \hline \end{array}$$

**88.**
$$\begin{array}{r} 2\,0\,0 \\ \times\ \ 3\,0 \\ \hline \end{array}$$

**89.**
$$\begin{array}{r} 7\,0\,0 \\ \times\ \ 9\,0 \\ \hline \end{array}$$

**90.**
$$\begin{array}{r} 4\,0\,0 \\ \times\,3\,0\,0 \\ \hline \end{array}$$

**91.**
$$\begin{array}{r} 4\,0\,0\,0 \\ \times\ \ 2\,0\,0 \\ \hline \end{array}$$

**92.**
$$\begin{array}{r} 6\,0\,0\,0 \\ \times\ \ \ 2\,0 \\ \hline \end{array}$$

**93.**
$$\begin{array}{r} 4\,0\,0\,0 \\ \times\,4\,0\,0\,0 \\ \hline \end{array}$$

**94.**
$$\begin{array}{r} 8\,0\,0\,0 \\ \times\ \ \ 1\,0 \\ \hline \end{array}$$

**c** Multiply.

**95.**
$$\begin{array}{r} 4\,9 \\ \times\ \ 3 \\ \hline \end{array}$$

**96.**
$$\begin{array}{r} 7\,4 \\ \times\ \ 6 \\ \hline \end{array}$$

**97.**
$$\begin{array}{r} 5\,9\,3 \\ \times\ \ \ 5 \\ \hline \end{array}$$

**98.**
$$\begin{array}{r} 6\,0\,9 \\ \times\ \ \ 8 \\ \hline \end{array}$$

**99.**
$$\begin{array}{r} 8\,9\,9 \\ \times\ \ \ 7 \\ \hline \end{array}$$

**100.**
$$\begin{array}{r} 8\,6\,5 \\ \times\ \ \ 4 \\ \hline \end{array}$$

**101.**
$$\begin{array}{r} 8\,1\,1\,8 \\ \times\ \ \ \ 2 \\ \hline \end{array}$$

**102.**
$$\begin{array}{r} 6\,7\,5\,4 \\ \times\ \ \ \ 2 \\ \hline \end{array}$$

**103.**
$$\begin{array}{r} 4\,3{,}7\,7\,7 \\ \times\ \ \ \ \ \ 2 \\ \hline \end{array}$$

**104.**
$$\begin{array}{r} 3\,2{,}5\,6\,4 \\ \times\ \ \ \ \ \ 6 \\ \hline \end{array}$$

**d** Multiply.

**105.**
$$\begin{array}{r} 5\,8 \\ \times\,6\,0 \\ \hline \end{array}$$

**106.**
$$\begin{array}{r} 9\,3 \\ \times\,3\,0 \\ \hline \end{array}$$

**107.**
$$\begin{array}{r} 4\,2 \\ \times\,8\,0 \\ \hline \end{array}$$

**108.**
$$\begin{array}{r} 7\,8 \\ \times\,9\,0 \\ \hline \end{array}$$

**109.**
$$\begin{array}{r} 3\,4\,6 \\ \times\ \ 6\,0 \\ \hline \end{array}$$

**110.**
$$\begin{array}{r} 2\,6\,7 \\ \times\ \ 4\,0 \\ \hline \end{array}$$

**111.**
$$\begin{array}{r} 8\,9\,7 \\ \times\,4\,0\,0 \\ \hline \end{array}$$

**112.**
$$\begin{array}{r} 3\,6\,6 \\ \times\,3\,0\,0 \\ \hline \end{array}$$

**113.**
$$\begin{array}{r} 8\,3\,4 \\ \times\,7\,0\,0 \\ \hline \end{array}$$

**114.**
$$\begin{array}{r} 3\,3\,3 \\ \times\,9\,0\,0 \\ \hline \end{array}$$

**115.**
$$\begin{array}{r} 5\,6\,7\,3 \\ \times\,2\,0\,0\,0 \\ \hline \end{array}$$

**116.**
$$\begin{array}{r} 4\,6\,7\,8 \\ \times\,5\,0\,0\,0 \\ \hline \end{array}$$

**117.**
$$\begin{array}{r} 6\,7\,8\,8 \\ \times\,9\,0\,0\,0 \\ \hline \end{array}$$

**118.**
$$\begin{array}{r} 9\,1\,2\,9 \\ \times\,8\,0\,0\,0 \\ \hline \end{array}$$

# Division

## a  BASIC DIVISION

Division can be explained by arranging a set of objects in a rectangular array. This can be done in two ways. Both methods below illustrate the division $18 \div 6$.

**Method 1**  We take 18 objects and determine how many rows, each with 6 objects, we can form.

} 3 rows of 6 objects

Since there are 3 rows of 6 objects, we have $18 \div 6 = 3$.

**Method 2**  We take 18 objects and arrange them in 6 rows, then determine how many objects are in each row.

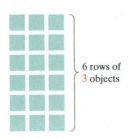
} 6 rows of 3 objects

Since there are 3 objects in each of the 6 rows, we have $18 \div 6 = 3$.

We can also use other notation for division:

$$18 \div 6 = 18/6 = \frac{18}{6} = 3 \quad \text{and} \quad 6\overline{)18}.$$

with $3$ above in the long division.

**EXAMPLES**  Divide.

1. $9\overline{)36}$ with $4$ above    *Think:* 36 objects: How many rows, each with 9 objects?
   *or:*  36 objects: How many objects in each of 9 rows?

2. $42 \div 7 = 6$

3. $\dfrac{24}{3} = 8$

**Do Exercises 1–4.** ▶

The multiplication table in Developmental Unit M will enable you to divide also. First, recall how multiplication and division are related.

*A multiplication:* $5 \cdot 4 = 20$

*Two related divisions:*

a)
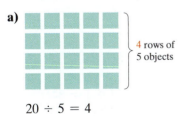
} 4 rows of 5 objects

$20 \div 5 = 4$

b)
} 5 rows of 4 objects

$20 \div 4 = 5$

**OBJECTIVES**

**a**  Find basic quotients such as $20 \div 5$, $56 \div 7$, and so on.

**b**  Divide by estimating multiples of thousands, hundreds, tens, and ones.

Divide.

1. $24 \div 6$    2. $64 \div 8$

3. $\dfrac{63}{7}$    4. $\dfrac{27}{9}$

*Answers*
1. 4  2. 8  3. 9  4. 3

Since we know that

$$5 \cdot 4 = 20, \qquad \text{A basic multiplication fact}$$

we also know two division facts:

$$20 \div 5 = 4 \quad \text{and} \quad 20 \div 4 = 5.$$

In the division $20 \div 4 = 5$, 20 is the *dividend*, 4 is the *divisor*, and 5 is the *quotient*.

**EXAMPLE 4**   From $7 \cdot 8 = 56$, write two division facts.

**a)** We have

$$7 \cdot 8 = 56 \qquad \text{Multiplication sentence}$$
$$7 = 56 \div 8. \qquad \text{Related division sentence}$$

**b)** We also have

$$7 \cdot 8 = 56 \qquad \text{Multiplication sentence}$$
$$8 = 56 \div 7. \qquad \text{Related division sentence}$$

◀ **Do Exercises 5 and 6.**

For each multiplication fact, write two division facts.

**5.** $6 \cdot 2 = 12$

**6.** $7 \times 6 = 42$

We can use the idea that division is defined in terms of multiplication to do basic divisions.

**EXAMPLE 5**   Find: $35 \div 5$.

To find $35 \div 5$, we ask, "5 times what number is 35?"

$$5 \cdot \square = 35$$

Using the multiplication table at the left, we find 35 inside the table and 5 along the left side. Then we read the answer, 7, from the top. Thus, we have

$$35 \div 5 = 7.$$

Strive to do this kind of thinking as fast as you can mentally, without having to use the table.

◀ **Do Exercises 7–10.**

| × | 2 | 3 | 4 | 5 | 6 | 7 | 8 | 9 |
|---|---|---|---|---|---|---|---|---|
| 2 | 4 | 6 | 8 | 10 | 12 | 14 | 16 | 18 |
| 3 | 6 | 9 | 12 | 15 | 18 | 21 | 24 | 27 |
| 4 | 8 | 12 | 16 | 20 | 24 | 28 | 32 | 36 |
| 5 | 10 | 15 | 20 | 25 | 30 | 35 | 40 | 45 |
| 6 | 12 | 18 | 24 | 30 | 36 | 42 | 48 | 54 |
| 7 | 14 | 21 | 28 | 35 | 42 | 49 | 56 | 63 |
| 8 | 16 | 24 | 32 | 40 | 48 | 56 | 64 | 72 |
| 9 | 18 | 27 | 36 | 45 | 54 | 63 | 72 | 81 |

Divide.

**7.** $28 \div 4$       **8.** $81 \div 9$

**9.** $\dfrac{16}{2}$       **10.** $\dfrac{54}{6}$

## Division by 1

Note that $3 \div 1 = 3$   because   $3 \cdot 1 = 3$.

**DIVISION BY 1**

Any number divided by 1 is that same number:

$$a \div 1 = \frac{a}{1} = a.$$

**EXAMPLES**   Divide.

**6.** $\dfrac{8}{1} = 8$          **7.** $6 \div 1 = 6$          **8.** $34 \div 1 = 34$

◀ **Do Exercises 11–13.**

Divide.

**11.** $6 \div 1$       **12.** $\dfrac{13}{1}$

**13.** $1 \div 1$

*Answers*

**5.** $12 \div 2 = 6$; $12 \div 6 = 2$
**6.** $42 \div 6 = 7$; $42 \div 7 = 6$   **7.** 7   **8.** 9
**9.** 8   **10.** 9   **11.** 6   **12.** 13   **13.** 1

## Division by 0

We cannot divide by 0. To see why, suppose the number 4 could be divided by 0, giving some number, $\square$, as the result. Because division is defined in terms of multiplication,

if   $4 \div 0 = \square$,   then   $4 = \square \cdot 0 = 0$.    False! $4 \neq 0$

Thus, the only number that could possibly be divided by 0 would be 0 itself. But since 0 times any number is 0, this division could give us any number we wanted. We avoid these difficulties by agreeing not to divide *any* number by 0.

---

**DIVISION BY 0**

Division by 0 is not defined. (We agree not to divide by 0.)

---

## Dividing 0 by Other Numbers

Note that $0 \div 3 = 0$   because   $0 \cdot 3 = 0$.

---

**DIVISION INTO 0**

Zero divided by any number other than 0 is 0:

$$\frac{0}{a} = 0, \quad a \neq 0.$$

---

**EXAMPLES**   Divide.

**9.** $0 \div 8 = 0$

**10.** $\dfrac{0}{9} = 0$

**11.** $\dfrac{137}{0}$ is undefined.

**12.** $12 \div 0$ is undefined.

**Do Exercises 14–21.** ▶

## Dividing a Number by Itself

Note that $3 \div 3 = 1$   because   $1 \cdot 3 = 3$.

---

**DIVISION OF A NUMBER BY ITSELF**

Any number other than 0 divided by itself is 1:

$$\frac{a}{a} = 1, \quad a \neq 0.$$

---

**EXAMPLES**   Divide.

**13.** $8 \div 8 = 1$   $1 \cdot 8 = 8$

**14.** $\dfrac{32}{32} = 1$   $1 \cdot 32 = 32$

**Do Exercises 22–27.** ▶

Divide, if possible. If not possible, write "undefined."

**14.** $\dfrac{8}{4}$

**15.** $\dfrac{5}{0}$

**16.** $\dfrac{0}{5}$

**17.** $\dfrac{0}{0}$

**18.** $12 \div 0$

**19.** $100 \div 10$

**20.** $\dfrac{5}{3-3}$

**21.** $\dfrac{8-8}{4}$

Divide.

**22.** $23 \div 23$

**23.** $\dfrac{67}{67}$

**24.** $\dfrac{41}{41}$

**25.** $17 \div 17$

**26.** $17 \div 1$

**27.** $\dfrac{54}{54}$

**Answers**

**14.** 2  **15.** Undefined  **16.** 0
**17.** Undefined  **18.** Undefined  **19.** 10
**20.** Undefined  **21.** 0  **22.** 1  **23.** 1
**24.** 1  **25.** 1  **26.** 17  **27.** 1

## b DIVIDING BY ESTIMATING MULTIPLES

We divide using a process of estimating multiples of 10, 100, 1000, and so on.

**EXAMPLE 15** Divide: $743 \div 3$.

**a)** Are there any hundreds in the quotient? To find how many hundreds, we find products of 3 and multiples of 100.

$$3 \cdot 100 = 300$$
$$3 \cdot 200 = 600$$
$$3 \cdot 300 = 900 \quad \leftarrow 743$$

```
        2 0 0
3 ) 7 4 3
    6 0 0  ← Multiply 3 × 200.
    1 4 3  ← Subtract.
```

**b)** Now look at 143 and go to the tens place. Are there any tens in the quotient?

$$3 \cdot 10 = 30$$
$$3 \cdot 20 = 60$$
$$3 \cdot 30 = 90$$
$$3 \cdot 40 = 120$$
$$3 \cdot 50 = 150 \quad \leftarrow 143$$

```
          4 0
        2 0 0
3 ) 7 4 3
    6 0 0
    1 4 3
    1 2 0  ← Multiply 3 × 40.
      2 3  ← Subtract.
```

**c)** Now look at 23 and go to the ones place. Are there any ones in the quotient?

$$3 \cdot 1 = 3$$
$$3 \cdot 2 = 6$$
$$3 \cdot 3 = 9$$
$$3 \cdot 4 = 12$$
$$3 \cdot 5 = 15$$
$$3 \cdot 6 = 18$$
$$3 \cdot 7 = 21 \quad \leftarrow 23$$
$$3 \cdot 8 = 24$$

```
          2 4 7  ← Adding:
              7     200 + 40 + 7 = 247
            4 0
          2 0 0
3 ) 7 4 3
    6 0 0
    1 4 3
    1 2 0
      2 3
      2 1  ← Multiply 3 × 7.
        2  ← Subtract. This is the
              remainder, since
              2 is less than 3.
```

We generally write only the *short form* when we divide.

```
    2 4 7      We write a 2 above the hundreds
3 ) 7 4 3      digit in the dividend to record 200.
    6 0 0      We write a 4 to record 40.
    1 4 3      We write a 7 to record 7.
    1 2 0
      2 3
      2 1
        2
```

The answer is 247 R 2.

◀ **Do Exercises 28–31.**

**TO THE STUDENT**

If you had trouble with Section 1.5 and have studied Developmental Unit D, you should go back and work through Section 1.5 after completing Exercise Set D.

Divide.

**28.** $4\overline{)385}$     **29.** $7\overline{)8846}$

Divide using the short form.

**30.** $2\overline{)648}$     **31.** $9\overline{)3758}$

*Answers*

**28.** 96 R 1   **29.** 1263 R 5   **30.** 324
**31.** 417 R 5

Ⓟ MyLab Math

**a** Divide, if possible.

**1.** $24 \div 8$    **2.** $72 \div 9$    **3.** $28 \div 7$    **4.** $22 \div 22$    **5.** $32 \div 1$

**6.** $45 \div 5$    **7.** $14 \div 2$    **8.** $40 \div 8$    **9.** $37 \div 1$    **10.** $10 \div 2$

**11.** $36 \div 4$    **12.** $12 \div 3$    **13.** $54 \div 9$    **14.** $18 \div 2$    **15.** $20 \div 4$

**16.** $16 \div 2$    **17.** $72 \div 8$    **18.** $42 \div 7$    **19.** $12 \div 4$    **20.** $8 \div 4$

**21.** $54 \div 6$    **22.** $18 \div 9$    **23.** $9 \div 3$    **24.** $28 \div 4$    **25.** $56 \div 7$

**26.** $24 \div 6$    **27.** $14 \div 2$    **28.** $14 \div 7$    **29.** $21 \div 7$    **30.** $36 \div 6$

**31.** $8 \div 8$    **32.** $32 \div 8$    **33.** $30 \div 5$    **34.** $18 \div 6$    **35.** $49 \div 7$

**36.** $81 \div 9$    **37.** $0 \div 7$    **38.** $9 \div 0$    **39.** $16 \div 0$    **40.** $42 \div 6$

**41.** $\dfrac{48}{6}$    **42.** $\dfrac{35}{5}$    **43.** $\dfrac{9}{9}$    **44.** $\dfrac{45}{9}$    **45.** $\dfrac{0}{5}$    **46.** $\dfrac{0}{8}$

**47.** $\dfrac{6}{2}$    **48.** $\dfrac{3}{3}$    **49.** $\dfrac{8}{2}$    **50.** $\dfrac{7}{1}$    **51.** $\dfrac{5}{5}$    **52.** $\dfrac{6}{1}$

**53.** $\dfrac{2}{2}$    **54.** $\dfrac{25}{5}$    **55.** $\dfrac{4}{2}$    **56.** $\dfrac{24}{3}$    **57.** $\dfrac{0}{9}$    **58.** $\dfrac{0}{4}$

**59.** $\dfrac{40}{5}$    **60.** $\dfrac{3}{1}$    **61.** $\dfrac{16}{4}$    **62.** $\dfrac{9}{0}$    **63.** $\dfrac{32}{8}$    **64.** $\dfrac{15}{15}$

**b** Divide.

**65.** $4 \overline{)277}$     **66.** $2 \overline{)399}$     **67.** $8 \overline{)737}$     **68.** $6 \overline{)831}$

**69.** $5 \overline{)105}$     **70.** $6 \overline{)708}$     **71.** $9 \overline{)820}$     **72.** $3 \overline{)965}$

**73.** $5 \overline{)8619}$     **74.** $3 \overline{)8775}$     **75.** $9 \overline{)7777}$     **76.** $8 \overline{)4179}$

**77.** $5 \overline{)4823}$     **78.** $8 \overline{)5437}$     **79.** $2 \overline{)5794}$     **80.** $7 \overline{)9298}$

**81.** $20 \overline{)875}$     **82.** $30 \overline{)987}$     **83.** $50 \overline{)6893}$     **84.** $40 \overline{)9946}$

**85.** $11 \overline{)415}$     **86.** $11 \overline{)836}$     **87.** $12 \overline{)6312}$     **88.** $12 \overline{)1970}$

# Answers

## CHAPTER 1

### Exercise Set 1.1, p. 6
**RC1.** digit  **RC2.** period  **RC3.** expanded
**RC4.** standard  **CC1.** Five million  **CC2.** Forty-two million  **CC3.** Three billion  **CC4.** Eighteen billion
**CC5.** Seven trillion  **CC6.** Forty trillion
**1.** 5 thousands  **3.** 5 hundreds  **5.** 1  **7.** 4
**9.** 4 thousands + 6 hundreds + 9 tens + 2 ones
**11.** 4 thousands + 0 hundreds + 9 tens + 0 ones, or 4 thousands +
9 tens  **13.** 9 ten thousands + 3 thousands + 9 hundreds +
8 tens + 6 ones  **15.** 4 hundred thousands + 0 ten thousands +
1 thousand + 6 hundreds + 9 tens + 0 ones, or 4 hundred
thousands + 1 thousand + 6 hundreds + 9 tens  **17.** 1 billion +
3 hundred millions + 7 ten millions + 3 millions + 5 hundred
thousands + 4 ten thousands + 1 thousand + 2 hundreds +
7 tens + 8 ones  **19.** 2 hundred millions + 5 ten millions +
8 millions + 3 hundred thousands + 1 ten thousand +
6 thousands + 0 hundreds + 5 tens + 1 one, or 2 hundred millions + 5 ten millions + 8 millions + 3 hundred thousands +
1 ten thousand + 6 thousands + 5 tens + 1 one  **21.** Eighty-
five  **23.** Eighty-eight thousand  **25.** One hundred
twenty-three thousand, seven hundred sixty-five  **27.** Seven
billion, seven hundred fifty-four million, two hundred eleven
thousand, five hundred seventy-seven  **29.** Three hundred
ninety-four thousand, two hundred forty-nine  **31.** Thirty
million, seven hundred fourteen thousand, two hundred
eighty-six  **33.** 632,896  **35.** 50,324  **37.** 2,233,812
**39.** 8,000,000,000  **41.** 40,000,000  **43.** 30,000,103
**45.** 64,186,000  **47.** 138

### Calculator Corner, p. 10
**1.** 121  **2.** 1602  **3.** 1932  **4.** 864

### Exercise Set 1.2, p. 12
**RC1.** addends  **RC2.** sum  **RC3.** 0  **RC4.** perimeter
**CC1.** 50  **CC2.** 110  **CC3.** 800  **CC4.** 1700
**CC5.** 6000  **CC6.** 10,000  **1.** 387  **3.** 164  **5.** 5198
**7.** 100  **9.** 8503  **11.** 5266  **13.** 4466  **15.** 6608
**17.** 34,432  **19.** 101,310  **21.** 230  **23.** 18,424
**25.** 31,685  **27.** 132 yd  **29.** 1661 ft  **31.** 570 ft
**33.** 8 ten thousands  **34.** Nine billion, three hundred forty-six
million, three hundred ninety-nine thousand, four hundred sixty-
eight  **35.** 1 + 99 = 100, 2 + 98 = 100, . . . , 49 + 51 = 100.
Then 49 100's = 4900 and 4900 + 50 + 100 = 5050.

### Calculator Corner, p. 15
**1.** 28  **2.** 47  **3.** 67  **4.** 119  **5.** 2128  **6.** 2593

### Exercise Set 1.3, p. 17
**RC1.** minuend  **RC2.** subtraction symbol
**RC3.** subtrahend  **RC4.** difference  **CC1.** 2  **CC2.** 93
**CC3.** 7  **CC4.** 600  **CC5.** 995  **CC6.** 1  **1.** 44

**3.** 533  **5.** 39  **7.** 14  **9.** 369  **11.** 26  **13.** 234
**15.** 417  **17.** 5382  **19.** 2778  **21.** 3069  **23.** 1089
**25.** 7748  **27.** 4144  **29.** 56  **31.** 454  **33.** 3749
**35.** 2191  **37.** 43,028  **39.** 95,974  **41.** 4418
**43.** 1305  **45.** 9989  **47.** 48,017  **49.** 1345  **50.** 924
**51.** 22,692  **52.** 10,920  **53.** Six million, three hundred
seventy-five thousand, six hundred two  **54.** 9 thousands +
1 hundred + 0 tens + 3 ones, or 9 thousands + 1 hundred +
3 ones  **55.** 3; 4

### Calculator Corner, p. 21
**1.** 448  **2.** 21,970  **3.** 6380  **4.** 39,564  **5.** 180,480
**6.** 2,363,754

### Exercise Set 1.4, p. 23
**RC1.** factors  **RC2.** product  **RC3.** 0  **RC4.** 1
**CC1.** 800  **CC2.** 4800  **CC3.** 6000  **CC4.** 80,000
**CC5.** 63,000  **CC6.** 400,000  **1.** 520  **3.** 564  **5.** 1527
**7.** 64,603  **9.** 4770  **11.** 3995  **13.** 870  **15.** 1920
**17.** 46,296  **19.** 14,652  **21.** 258,312  **23.** 798,408
**25.** 20,723,872  **27.** 362,128  **29.** 302,220  **31.** 49,101,136
**33.** 25,236,000  **35.** 20,064,048  **37.** 529,984 sq mi
**39.** 8100 sq ft  **41.** 12,685  **42.** 10,834  **43.** 8889
**44.** 254,119  **45.** 4 hundred thousands  **46.** 0
**47.** 1 ten thousand + 2 thousands + 8 hundreds + 4 tens +
7 ones  **48.** Seven million, four hundred thirty-two
thousand  **49.** 247,464 sq ft

### Calculator Corner, p. 30
**1.** 28  **2.** 123  **3.** 323  **4.** 36

### Exercise Set 1.5, p. 32
**RC1.** quotient  **RC2.** dividend  **RC3.** remainder
**RC4.** divisor  **CC1.** 2 R 1  **CC2.** 3 R 3  **CC3.** 12 R 4
**CC4.** 48 R 1  **1.** 12  **3.** 1  **5.** 22  **7.** 0  **9.** Not defined
**11.** 6  **13.** 55 R 2  **15.** 108  **17.** 307  **19.** 753 R 3
**21.** 74 R 1  **23.** 92 R 2  **25.** 1703  **27.** 987 R 5
**29.** 12,700  **31.** 127  **33.** 52 R 52  **35.** 29 R 5
**37.** 40 R 12  **39.** 90 R 22  **41.** 29  **43.** 105 R 3
**45.** 1609 R 2  **47.** 1007 R 1  **49.** 23  **51.** 107 R 1
**53.** 370  **55.** 609 R 15  **57.** 304  **59.** 3508 R 219
**61.** 8070  **63.** 1241  **64.** 66,444  **65.** 19,800  **66.** 9380
**67.** 40 ft  **68.** 99 sq ft  **69.** 54, 122; 33, 2772; 4, 8
**71.** 30 buses

### Mid-Chapter Review: Chapter 1, p. 35
**1.** False  **2.** True  **3.** True  **4.** False  **5.** True
**6.** False  **7.** Ninety-five million, four hundred six thousand,
two hundred thirty-seven

**8.**
$$\begin{array}{r} \overset{5\;9\;14}{6\,0\,4} \\ -\ 4\,9\,7 \\ \hline 1\,0\,7 \end{array}$$
  **9.** 6 hundreds  **10.** 6 ten thousands

**11.** 6 thousands   **12.** 6 ones   **13.** 2   **14.** 6   **15.** 5   **16.** 1
**17.** 5 thousands + 6 hundreds + 0 tens + 2 ones, or 5 thousands + 6 hundreds + 2 ones   **18.** 6 ten thousands + 9 thousands + 3 hundreds + 4 tens + 5 ones   **19.** One hundred thirty-six
**20.** Sixty-four thousand, three hundred twenty-five
**21.** 308,716   **22.** 4,567,216   **23.** 798   **24.** 1030   **25.** 7922
**26.** 7534   **27.** 465   **28.** 339   **29.** 1854   **30.** 4328
**31.** 216   **32.** 15,876   **33.** 132,275   **34.** 5,679,870   **35.** 253
**36.** 112 R 5   **37.** 23 R 19   **38.** 144 R 31   **39.** 25 m
**40.** 8 sq in.   **41.** When numbers are being added, it does not matter how they are grouped.   **42.** Subtraction is not commutative. For example, $5 - 2 = 3$, but $2 - 5 \neq 3$.   **43.** Answers will vary. Suppose one coat costs $150. Then the multiplication $4 \cdot \$150$ gives the cost of four coats. Or, suppose one ream of copy paper costs $4. Then the multiplication $\$4 \cdot 150$ gives the cost of 150 reams.   **44.** If we use the definition of division, $0 \div 0 = a$ such that $a \cdot 0 = 0$. We see that $a$ could be *any* number since $a \cdot 0 = 0$ for any number $a$. Thus we cannot say that $0 \div 0 = 0$. This is why we agree not to allow division by 0.

## Exercise Set 1.6, p. 43

**RC1.** True   **RC2.** False   **RC3.** False   **RC4.** True
**CC1.** 100   **CC2.** 1000   **CC3.** 700   **1.** 50   **3.** 460
**5.** 730   **7.** 900   **9.** 100   **11.** 1000   **13.** 9100
**15.** 32,800   **17.** 6000   **19.** 8000   **21.** 45,000   **23.** 373,000
**25.** $80 + 90 = 170$   **27.** $8070 - 2350 = 5720$
**29.** 220; incorrect   **31.** 890; incorrect
**33.** $7300 + 9200 = 16,500$   **35.** $6900 - 1700 = 5200$
**37.** 1600; correct   **39.** 1500; correct
**41.** $10,000 + 5000 + 9000 + 7000 = 31,000$
**43.** $92,000 - 23,000 = 69,000$   **45.** $50 \cdot 70 = 3500$
**47.** $30 \cdot 30 = 900$   **49.** $900 \cdot 300 = 270,000$
**51.** $400 \cdot 200 = 80,000$   **53.** $350 \div 70 = 5$
**55.** $8450 \div 50 = 169$   **57.** $1200 \div 200 = 6$
**59.** $8400 \div 300 = 28$   **61.** $1900   **63.** $1600; no
**65.** Answers will vary depending on the options chosen.
**67. (a)** $309,600; **(b)** $360,000   **69.** 90 people
**71.** <   **73.** >   **75.** <   **77.** >   **79.** >   **81.** >
**83.** $1,014,023 < 1,894,934$, or $1,894,934 > 1,014,023$
**85.** $10,425 > 10,038$, or $10,038 < 10,425$   **87.** 86,754
**88.** 13,589   **89.** 48,824   **90.** 4415   **91.** 1702   **92.** 17,748
**93.** 54 R 4   **94.** 208   **95.** Left to the student   **97.** Left to the student

## Exercise Set 1.7, p. 52

**RC1.** (c)   **RC2.** (a)   **RC3.** (d)   **RC4.** (b)   **CC1.** No
**CC2.** Yes   **CC3.** Yes   **CC4.** No   **1.** 14   **3.** 0   **5.** 90,900
**7.** 450   **9.** 352   **11.** 25   **13.** 29   **15.** 0   **17.** 79
**19.** 45   **21.** 8   **23.** 14   **25.** 32   **27.** 143   **29.** 17,603
**31.** 37   **33.** 1035   **35.** 66   **37.** 324   **39.** 335   **41.** 18,252
**43.** 104   **45.** 45   **47.** 4056   **49.** 2847   **51.** 15   **53.** 205
**55.** 457   **57.** 142 R 5   **58.** 142   **59.** 334   **60.** 334 R 11
**61.** <   **62.** >   **63.** >   **64.** <   **65.** 6,376,000
**66.** 6,375,600   **67.** 347

## Translating for Success, p. 62

**1.** E   **2.** M   **3.** D   **4.** G   **5.** A   **6.** O   **7.** F   **8.** K
**9.** J   **10.** H

## Exercise Set 1.8, p. 63

**RC1.** Familiarize   **RC2.** Translate   **RC3.** Solve
**RC4.** Check   **CC1.** (d)   **CC2.** (b)   **CC3.** (c)
**CC4.** (a)   **1.** 107 lb   **3.** 549 lb   **5.** 18 rows   **7.** 43 events
**9.** 2054 mi   **11.** 2,073,600 pixels   **13.** 95 milligrams
**15.** 168 hr   **17.** $999 per month   **19.** $467   **21.** $11,232
**23.** 151,500   **25.** $78   **27.** $390 per month   **29.** $24,456
**31.** 35 weeks; 2 episodes   **33.** 21 columns   **35.** 236 gal

**37. (a)** 4200 sq ft; **(b)** 268 ft   **39.** 56 cartons   **41.** 645 mi; 5 in.
**43.** $247   **45.** 525 min, or 8 hr 45 min   **47.** 99,300 jobs
**49.** 118 seats   **51.** 32 $10 bills   **53.** $400   **55.** 106 bones
**57.** 8273   **58.** 7759   **59.** 806,985   **60.** 147 R 4   **61.** 34 m
**62.** 9706 sq ft   **63.** $200 \times 600 = 120,000$   **64.** 66
**65.** 792,000 mi; 1,386,000 mi

## Calculator Corner, p. 71

**1.** 243   **2.** 15,625   **3.** 20,736   **4.** 2048

## Calculator Corner, p. 73

**1.** 49   **2.** 85   **3.** 36   **4.** 0   **5.** 73   **6.** 49

## Exercise Set 1.9, p. 75

**RC1.** exponent   **RC2.** squared   **RC3.** multiplication
**RC4.** 3   **CC1.** Multiplication   **CC2.** Division
**CC3.** Subtraction   **CC4.** Multiplication
**CC5.** Exponentiation   **1.** $3^4$   **3.** $5^2$   **5.** $7^5$   **7.** $10^3$
**9.** 49   **11.** 729   **13.** 20,736   **15.** 243   **17.** 22
**19.** 20   **21.** 100   **23.** 1   **25.** 49   **27.** 5   **29.** 434
**31.** 41   **33.** 88   **35.** 4   **37.** 303   **39.** 20   **41.** 70
**43.** 295   **45.** 32   **47.** 906   **49.** 62   **51.** 102   **53.** 32
**55.** $94   **57.** 401   **59.** 110   **61.** 7   **63.** 544   **65.** 708
**67.** 27   **69.** 452   **70.** 835   **71.** 13   **72.** 37
**73.** 4898   **74.** 100   **75.** 104,286 sq mi   **76.** 98 gal
**77.** 24; $1 + 5 \cdot (4 + 3) = 36$   **79.** 7; $12 \div (4 + 2) \cdot 3 - 2 = 4$

## Summary and Review: Chapter 1, p. 78

### Vocabulary Reinforcement

**1.** perimeter   **2.** minuend   **3.** digits; periods   **4.** dividend
**5.** factors; product   **6.** additive   **7.** associative   **8.** divisor; remainder; dividend

### Concept Reinforcement

**1.** True   **2.** True   **3.** False   **4.** False   **5.** True   **6.** False

### Study Guide

**1.** 2 thousands   **2.** 65,302   **3.** 3237   **4.** 225,036
**5.** 315 R 14   **6.** 36,000   **7.** <   **8.** 36   **9.** 216

### Review Exercises

**1.** 8 thousands   **2.** 3   **3.** 2 thousands + 7 hundreds + 9 tens + 3 ones   **4.** 5 ten thousands + 6 thousands + 0 hundreds + 7 tens + 8 ones, or 5 ten thousands + 6 thousands + 7 tens + 8 ones   **5.** 4 millions + 0 hundred thousands + 0 ten thousands + 7 thousands + 1 hundred + 0 tens + 1 one, or 4 millions + 7 thousands + 1 hundred + 1 one
**6.** Sixty-seven thousand, eight hundred nineteen   **7.** Two million, seven hundred eighty-one thousand, four hundred twenty-seven   **8.** 476,588   **9.** 1,500,000,000   **10.** 14,272
**11.** 66,024   **12.** 21,788   **13.** 98,921   **14.** 5148   **15.** 1689
**16.** 2274   **17.** 17,757   **18.** 5,100,000   **19.** 6,276,800
**20.** 506,748   **21.** 27,589   **22.** 5,331,810   **23.** 12 R 3
**24.** 5   **25.** 913 R 3   **26.** 384 R 1   **27.** 4 R 46   **28.** 54
**29.** 452   **30.** 5008   **31.** 4389   **32.** 345,800   **33.** 345,760
**34.** 346,000   **35.** 300,000   **36.** >   **37.** <
**38.** $41,300 + 19,700 = 61,000$   **39.** $38,700 - 24,500 = 14,200$
**40.** $400 \cdot 700 = 280,000$   **41.** 8   **42.** 45   **43.** 58
**44.** 0   **45.** $4^3$   **46.** 10,000   **47.** 36   **48.** 65   **49.** 233
**50.** 260   **51.** 165   **52.** $502   **53.** $484   **54.** 1982
**55.** 19 cartons   **56.** $13,585   **57.** 14 beehives
**58.** 98 sq ft; 42 ft   **59.** 137 beakers filled; 13 mL left over
**60.** $27,598   **61.** B   **62.** A   **63.** D   **64.** 8   **65.** $a = 8$, $b = 4$   **66.** 6 days

## Understanding Through Discussion and Writing

**1.** No; if subtraction were associative, then $a - (b - c) = (a - b) - c$ for any $a$, $b$, and $c$. But, for example,

$$12 - (8 - 4) = 12 - 4 = 8,$$

whereas

$$(12 - 8) - 4 = 4 - 4 = 0.$$

Since $8 \neq 0$, this example shows that subtraction is not associative. **2.** By rounding prices and estimating their sum, a shopper can estimate the total grocery bill while shopping. This is particularly useful if the shopper wants to spend no more than a certain amount. **3.** Answers will vary. Anthony is driving from Kansas City to Minneapolis, a distance of 512 mi. He stops for gas after driving 183 mi. How much farther must he drive? **4.** The parentheses are not necessary in the expression $9 - (4 \cdot 2)$. Using the rules for order of operations, the multiplication would be performed before the subtraction even if the parentheses were not present. The parentheses are necessary in the expression $(3 \cdot 4)^2$; $(3 \cdot 4)^2 = 12^2 = 144$, but $3 \cdot 4^2 = 3 \cdot 16 = 48$.

## Test: Chapter 1, p. 83

**1.** [1.1a] 5 **2.** [1.1b] 8 thousands + 8 hundreds + 4 tens + 3 ones **3.** [1.1c] Thirty-eight million, four hundred three thousand, two hundred seventy-seven
**4.** [1.2a] 9989 **5.** [1.2a] 63,791 **6.** [1.2a] 3165
**7.** [1.2a] 10,515 **8.** [1.3a] 3630 **9.** [1.3a] 1039
**10.** [1.3a] 6848 **11.** [1.3a] 5175 **12.** [1.4a] 41,112
**13.** [1.4a] 5,325,600 **14.** [1.4a] 2405 **15.** [1.4a] 534,264
**16.** [1.5a] 3 R 3 **17.** [1.5a] 70 **18.** [1.5a] 97
**19.** [1.5a] 805 R 8 **20.** [1.6a] 35,000 **21.** [1.6a] 34,530
**22.** [1.6a] 34,500 **23.** [1.6b] 23,600 + 54,700 = 78,300
**24.** [1.6b] 54,800 − 23,600 = 31,200
**25.** [1.6b] 800 · 500 = 400,000 **26.** [1.6c] > **27.** [1.6c] <
**28.** [1.7b] 46 **29.** [1.7b] 13 **30.** [1.7b] 14
**31.** [1.7b] 381 **32.** [1.8a] 83 calories **33.** [1.8a] 20 staplers
**34.** [1.8a] 1,256,615 sq mi **35. (a)** [1.2b], [1.4b] 300 in., 5000 sq in.; 264 in., 3872 sq in.; 228 in., 2888 sq in.;
**(b)** [1.8a] 2112 sq in. **36.** [1.8a] 1852 12-packs; 7 cakes left over
**37.** [1.8a] $95 **38.** [1.9a] $12^4$ **39.** [1.9b] 343
**40.** [1.9b] 100,000 **41.** [1.9c] 31 **42.** [1.9c] 98 **43.** [1.9c] 2
**44.** [1.9c] 18 **45.** [1.9d] 216 **46.** [1.9c] A **47.** [1.4b], [1.8a] 336 sq in. **48.** [1.9c] 9 **49.** [1.8a] 80 payments

## CHAPTER 2

## Exercise Set 2.1, p. 90

**RC1.** True **RC2.** True **RC3.** False **CC1.** True
**CC2.** True **CC3.** False **CC4.** False **CC5.** True
**CC6.** False **CC7.** True **CC8.** True **1.** 24; −2
**3.** 7,200,000,000,000; −460 **5.** 2073; −282 **7.** < **9.** >
**11.** > **13.** < **15.** < **17.** < **19.** > **21.** 57
**23.** 0 **25.** 24 **27.** 53 **29.** 8 **31.** 7 **33.** −7 **35.** 0
**37.** 21 **39.** −53 **41.** 1 **43.** 7 **45.** −9 **47.** −17
**49.** 23 **51.** −1 **53.** 85 **55.** −345 **57.** 0
**59.** −8 **61.** 825 **62.** 125 **63.** 7106 **64.** 4
**65.** 42 **66.** 69 **67.** > **69.** = **71.** −1, 0, 1
**73.** −100, −5, 0, $|3|$, 4, $|-6|$, $7^2$, $10^2$, $2^7$, $2^{10}$

## Calculator Corner, p. 94

**1.** 13 **2.** −8

## Exercise Set 2.2, p. 95

**RC1.** add, negative **RC2.** subtract, negative
**RC3.** opposites **RC4.** identity **CC1.** right **CC2.** left
**CC3.** left **CC4.** right **1.** −5 **3.** −4 **5.** 6 **7.** 0

**9.** −4 **11.** −12 **13.** −11 **15.** −15 **17.** 42 **19.** 5
**21.** −4 **23.** 0 **25.** 0 **27.** −9 **29.** 7 **31.** 0 **33.** 45
**35.** −3 **37.** 0 **39.** −10 **41.** −24 **43.** −5 **45.** −21
**47.** −30 **49.** 6 **51.** −21 **53.** 25 **55.** −17 **57.** 6
**59.** −65 **61.** −160 **63.** −62 **65.** −23 **67.** 6681 **68.** 73
**69.** 3 ten thousands + 9 thousands + 4 hundreds + 1 ten + 7 ones
**70.** 2352 **71.** 32 **72.** 3500 **73.** −40 **75.** −6483
**77.** All negative **79.** negative **81.** negative

## Exercise Set 2.3, p. 100

**RC1.** opposite **RC2.** opposite **RC3.** difference **CC1.** (c)
**CC2.** (b) **CC3.** (d) **CC4.** (a) **1.** −4 **3.** −7
**5.** −6 **7.** 0 **9.** −4 **11.** −7 **13.** −6 **15.** 0 **17.** 0
**19.** 14 **21.** 11 **23.** −14 **25.** 5 **27.** −7 **29.** −1
**31.** 18 **33.** −10 **35.** −3 **37.** −21 **39.** 5 **41.** −8
**43.** 12 **45.** −19 **47.** −68 **49.** −81 **51.** 116 **53.** 0
**55.** 55 **57.** 19 **59.** −62 **61.** −139 **63.** 6 **65.** 107
**67.** 219 **69.** −11,156 ft **71.** 17 lb **73.** 3780 m **75.** −3°
**77.** 18 points **79.** Profit of $4300 **81.** Balance of −$83, or $83 in debt **83.** 64 **84.** 4896 **85.** 1 **86.** 4147
**87.** 8 cans **88.** 288 oz **89.** 35 **90.** 3 **91.** 32 **92.** 165
**93.** −309,882 **95.** False; $3 - 0 \neq 0 - 3$ **97.** True
**99.** True **101.** 17 **103.** Up 15 points

## Calculator Corner, p. 107

**1.** 148,035,889 **2.** −1,419,857 **3.** −1,124,864 **4.** 1,048,576
**5.** −531,441 **6.** −117,649 **7.** −7776 **8.** −19,683

## Exercise Set 2.4, p. 108

**RC1.** negative **RC2.** positive **RC3.** positive
**RC4.** negative **CC1.** 1 **CC2.** −1 **CC3.** 1 **CC4.** −1
**1.** −16 **3.** −60 **5.** −48 **7.** −30 **9.** 15 **11.** 18
**13.** 42 **15.** 20 **17.** −120 **19.** 0 **21.** 72 **23.** −340
**25.** 400 **27.** 0 **29.** 24 **31.** 420 **33.** −70 **35.** 30
**37.** 0 **39.** −294 **41.** 36 **43.** −125 **45.** 10,000
**47.** −16 **49.** −243 **51.** 1 **53.** −121 **55.** −64
**57.** 532,500 **58.** 60,000,000 **59.** 80 **60.** 2550 **61.** 5
**62.** 48 **63.** 40 sq ft **64.** 4 trips **65.** 243 **67.** 0 **69.** 7
**71.** −2209 **73.** 130,321 **75.** −2197 **77.** 116,875
**79.** −$23 **81. (a)** Both $m$ and $n$ must be odd. **(b)** At least one of $m$ and $n$ must be even.

## Calculator Corner, p. 112

**1.** −4 **2.** −2 **3.** 787

## Exercise Set 2.5, p. 113

**RC1.** True **RC2.** True **RC3.** False **CC1.** 0
**CC2.** Undefined **CC3.** Undefined **CC4.** 0 **1.** −6
**3.** −13 **5.** −2 **7.** 4 **9.** −9 **11.** 2 **13.** −12 **15.** −8
**17.** Undefined **19.** −8 **21.** −23 **23.** 0 **25.** −19
**27.** −41 **29.** −7 **31.** −7 **33.** −334 **35.** 23 **37.** 8
**39.** 12 **41.** −1 **43.** 0 **45.** −10 **47.** −86 **49.** −9
**51.** 18 **53.** −10 **55.** −67 **57.** 10 **59.** −25 **61.** −7988
**63.** −3000 **65.** 60 **67.** 1 **69.** −37 **71.** −22 **73.** 2
**75.** 7 **77.** Undefined **79.** 3 **81.** 2 **83.** 0 **85.** 28 sq in.
**86.** 248 rooms **87.** 12 gal **88.** 27 gal **89.** 150 Cal
**90.** 672 g **91.** 4 pieces; 2 pieces **92.** 4 lozenges;
4 lozenges **93.** 0 **95.** 0 **97.** −2 **99.** 992

**101.** $\boxed{(}\ \boxed{1}\ \boxed{5}\ \boxed{x^2}\ \boxed{-}\ \boxed{5}\ \boxed{x^y}\ \boxed{3}\ \boxed{)}\ \boxed{\div}\ \boxed{(}\ \boxed{3}\ \boxed{x^2}\ \boxed{+}$
$\boxed{4}\ \boxed{x^2}\ \boxed{)}\ \boxed{=}$ **103.** 5 **105.** Positive **107.** Negative
**109.** Positive

## Mid-Chapter Review: Chapter 2, p. 116

**1.** False **2.** False **3.** True
**4.** $-x = -(-4) = 4$; $-(-x) = -(-(-4)) = -(4) = -4$

**5.** $5 - 13 = 5 + (-13) = -8$   **6.** $-6 - (-7) = -6 + 7 = 1$
**7.** $450; -79$   **8.** $-9$   **9.** $<$   **10.** $<$   **11.** $<$   **12.** $>$
**13.** 38   **14.** 18   **15.** 0   **16.** 12   **17.** 56   **18.** $-3$
**19.** 0   **20.** 49   **21.** 19   **22.** 23   **23.** $-2$   **24.** $-16$
**25.** 0   **26.** $-17$   **27.** 1   **28.** 2   **29.** $-26$   **30.** $-4$
**31.** $-13$   **32.** 16   **33.** 6   **34.** $-12$   **35.** $-36$   **36.** $-54$
**37.** 26   **38.** 82   **39.** 81   **40.** $-81$   **41.** 25   **42.** $-5$
**43.** 75   **44.** 14   **45.** 42   **46.** $-38$   **47.** $-13$   **48.** 2
**49.** $33°C$   **50.** \$44   **51.** Answers may vary. The student may be confusing distance from 0 on the number line with position on the number line. Although $-45$ is farther from 0 than $-21$, it is less than $-21$ because it is to the left of $-21$ on the number line.
**52.** Answers may vary. Subtraction of integers is not associative, as can be illustrated by an example: Compare $3 - (9 - 10) = 4$ with $(3 - 9) - 10 = -16$.   **53.** Answers may vary. If we think of the addition on the number line, we start at a negative number and move to the left. This always brings us to a point on the negative portion of the number line.   **54.** Yes; consider $m - (-n)$, where both $m$ and $n$ are positive. Then $m - (-n) = m + n$. Now $m + n$, the sum of two positive numbers, is positive.

## Calculator Corner, p. 120

**1.** 243   **2.** 1024   **3.** $-32$   **4.** $-3125$

## Exercise Set 2.6, p. 122

**RC1.** Division   **RC2.** Multiplication   **RC3.** Multiplication
**RC4.** Division   **CC1.** $3 - 2(5)$   **CC2.** $3(4 + (-7))$
**CC3.** $\dfrac{5(-4)}{2}$   **CC4.** $-(-3)^2$   **1.** \$20   **3.** $-2$
**5.** 1   **7.** 18 yr   **9.** $-7$   **11.** 14 ft   **13.** 14 ft   **15.** $-21$
**17.** $-21$   **19.** 400 ft   **21.** 22   **23.** 0   **25.** 36 handshakes
**27.** $-3$   **29.** 11   **31.** $-1100$   **33.** $\dfrac{-5}{t}; \dfrac{5}{-t}$   **35.** $\dfrac{n}{-b}; -\dfrac{n}{b}$
**37.** $\dfrac{-9}{p}; -\dfrac{9}{p}$   **39.** $\dfrac{14}{-w}; -\dfrac{14}{w}$   **41.** $-5; -5; -5$   **43.** $-27; -27; -27$
**45.** $36; -12$   **47.** $45; 45$   **49.** $216; -216$   **51.** $1; 1$
**53.** $32; -32$   **55.** $5a + 5b$   **57.** $4x + 4$   **59.** $2b + 10$
**61.** $7 - 7t$   **63.** $30x - 12$   **65.** $8x + 56 + 48y$
**67.** $-7y + 14$   **69.** $3x + 6$   **71.** $-4x + 12y + 8z$
**73.** $8a - 24b + 8c$   **75.** $4x - 12y - 28z$
**77.** $20a - 25b + 5c - 10d$   **79.** $-3m - 2n$
**81.** $-2a + 3b - 4$   **83.** $-x + y + z$   **85.** Twenty-three million, forty-three thousand, nine hundred twenty-one
**86.** 901   **87.** $5280 - 2480 = 2800$   **88.** 994   **89.** 5 in.
**90.** \$63   **91.** $698°F$   **93.** 4438   **95.** 279   **97.** 2
**99.** 2,560,000   **101.** $-32 \boxed{\times} (88 \boxed{-} 29) = -1888$
**103.** True   **105.** True

## Translating for Success, p. 129

**1.** J   **2.** G   **3.** A   **4.** K   **5.** H   **6.** N   **7.** F
**8.** M   **9.** D   **10.** C

## Exercise Set 2.7, p. 130

**RC1.** closed   **RC2.** perimeter   **RC3.** perimeter
**RC4.** square   **CC1.** $5a, a$   **CC2.** $x, 5x$ and $4, 7$
**CC3.** $x, -6x$ and $-3y, 2y$   **CC4.** $-2ab, 4ab$   **CC5.** $t^2, -t^2$
**CC6.** $6xy^2, -8xy^2$ and $9x^2y, -x^2y$   **1.** $2a, 5b, -7c$
**3.** $mn, -6n, 8$   **5.** $3x^2y, -4y^2, -2z^3$   **7.** $14x$   **9.** $-5a$
**11.** $3x + 6y$   **13.** $-13a + 62$   **15.** $-4 + 4t + 6y$
**17.** $6a + 4b - 2$   **19.** $-1 + a - 12b$   **21.** $7x^2 + 3y$
**23.** $7x^4 + y^3$   **25.** $6a^2 - 3a$   **27.** $3x^3 - 8x^2 + 4$
**29.** $9x^3y - xy^3 + 3xy$   **31.** $-4a^6 - 3b^4 + 2a^6b^4$   **33.** 10 ft
**35.** 42 km   **37.** 8 m   **39.** 210 ft   **41.** 138 ft   **43.** 36 ft
**45.** 56 in.   **47.** 260 cm   **49.** 64 ft   **51.** 17 servings
**52.** 210   **53.** 16   **54.** 13   **55.** 15   **56.** 25   **57.** 375

**58.** 16   **59.** $7x + 1$   **61.** $-29 - 3a$   **63.** $-10 - x - 27y$
**65.** \$45   **67.** 912 mm

## Exercise Set 2.8, p. 139

**RC1.** variable   **RC2.** solution   **RC3.** solution
**RC4.** variable   **CC1.** (c)   **CC2.** (a)   **CC3.** (d)   **CC4.** (b)
**1.** Equivalent equations   **3.** Equivalent expressions
**5.** Equivalent expressions   **7.** Equivalent equations
**9.** Equivalent expressions   **11.** Equivalent equations   **13.** $-3$
**15.** $-8$   **17.** 18   **19.** $-15$   **21.** 32   **23.** $-17$   **25.** 17
**27.** 0   **29.** 10   **31.** $-14$   **33.** 5   **35.** 0   **37.** $-11$
**39.** $-56$   **41.** 12   **43.** 390   **45.** 4   **47.** $-9$   **49.** $-15$
**51.** $-26$   **53.** $-5$   **55.** $-4$   **57.** $-172$   **59.** $-4$   **61.** 7
**63.** 3   **65.** $-3$   **67.** $-7$   **69.** $-8$   **71.** 8   **73.** 24
**75.** 6   **77.** 5   **79.** $-8$   **81.** 29   **82.** 7   **83.** 8   **84.** 27
**85.** 16   **86.** 26   **87.** 1   **88.** 24   **89.** 8   **91.** 0   **93.** $-5$
**95.** 29   **97.** $-20$   **99.** 1027   **101.** $-343$   **103.** 17

## Summary and Review: Chapter 2, p. 142

### Vocabulary Reinforcement

**1.** integers   **2.** absolute   **3.** opposite   **4.** not defined
**5.** substituting   **6.** constant   **7.** distributive law
**8.** addition principle   **9.** equivalent

### Concept Reinforcement

**1.** True   **2.** True   **3.** False   **4.** False

### Study Guide

**1.** (a) 17; (b) 300   **2.** 21   **3.** 14   **4.** $-90$   **5.** $-11$   **6.** $-12$
**7.** $30x - 40y - 5z$   **8.** $17a - 7b$   **9.** $-6$

### Review Exercises

**1.** $-45; 72$   **2.** $>$   **3.** $<$   **4.** $>$   **5.** 39   **6.** 23   **7.** 0
**8.** 72   **9.** 59   **10.** $-9$   **11.** $-11$   **12.** 6   **13.** $-24$
**14.** $-12$   **15.** 23   **16.** $-1$   **17.** 0   **18.** $-4$   **19.** 12
**20.** 92   **21.** $-84$   **22.** $-40$   **23.** $-3$   **24.** $-5$   **25.** 0
**26.** $-5$   **27.** 2   **28.** $-180$   **29.** $-20$   **30.** $-62$   **31.** 7
**32.** $-6, -6, -6$   **33.** $20x + 36$   **34.** $6a - 12b + 15$
**35.** $-20x - 10y$   **36.** $17a$   **37.** $6x$   **38.** $-3m + 6$
**39.** 36 in.   **40.** 100 cm   **41.** $-8$   **42.** $-9$
**43.** $-13$   **44.** 11   **45.** 15   **46.** $-7$   **47.** 8-yd gain
**48.** $-\$130$   **49.** A   **50.** B   **51.** 403 and 397
**52.** (a) $-7 + (-6) + (-5) + (-4) + (-3) + (-2) + (-1) + 0 + 1 + 2 + 3 + 4 + 5 + 6 + 7 + 8 = 8$; (b) 0
**53.** 662,582   **54.** $-88,174$   **55.** $-240$   **56.** $x < -2$
**57.** $x < 0$

### Understanding Through Discussion and Writing

**1.** We know that the product of an even number of negative numbers is positive, and the product of an odd number of negative numbers is negative. Since $(-7)^8$ is equivalent to the product of eight negative numbers, it will be a positive number. Similarly, since $(-7)^{11}$ is equivalent to the product of eleven negative numbers, it will be a negative number.   **2.** The expression $-x$ does not always represent a negative number. When $x = 0$, $-x = 0$, and when $x$ is negative, $-x$ is positive.   **3.** Jake is expecting the multiplication to be performed before the division.   **4.** The expression $-x^2$ represents a negative number, except when $x = 0$. For all other values of $x$, $x^2$ is positive, and the opposite of $x^2$ is negative.

## Test: Chapter 2, p. 147

**1.** [2.1a] $-542; 307$   **2.** [2.1b] $>$   **3.** [2.1c] 739   **4.** [2.1d] $-19$
**5.** [2.2a] $-11$   **6.** [2.2a] $-21$   **7.** [2.2a] 9   **8.** [2.3a] $-12$
**9.** [2.3a] $-15$   **10.** [2.3a] $-24$   **11.** [2.3a] 19   **12.** [2.3a] 38
**13.** [2.4b] $-64$   **14.** [2.4a] $-270$   **15.** [2.4a] 0
**16.** [2.5a] 8   **17.** [2.5a] $-8$   **18.** [2.5b] $-1$   **19.** [2.5b] 25

**20.** [2.3b] 14°F higher   **21.** [2.6a] −3
**22.** [2.6b] $14x + 21y − 7$   **23.** [2.7a] $4x − 17$
**24.** [2.7b] 20 ft   **25.** [2.8c] 5   **26.** [2.8b] −12   **27.** [2.8c] −95
**28.** [2.8e] 4   **29.** [2.6b] C   **30.** [2.7b] 66 ft
**31.** [2.5b] $35x − 7$   **32.** [2.5b] $−24x − 57$   **33.** [2.5b] 103,097
**34.** [2.5b] 1086

## Cumulative Review: Chapters 1–2, p. 149

**1.** [1.1c] 505,302   **2.** [1.1c] Five billion, three hundred eighty million, one thousand, four hundred thirty-seven
**3.** [1.2a] 18,827   **4.** [1.2a] 8857   **5.** [2.2a] 56   **6.** [2.2a] −102
**7.** [1.3a] 7846   **8.** [1.3a] 2428   **9.** [2.3a] 14   **10.** [2.3a] −43
**11.** [1.4a] 16,767   **12.** [1.4a] 8,266,500   **13.** [2.4a] −344
**14.** [2.4a] 72   **15.** [1.5a] 104   **16.** [1.5a] 62   **17.** [2.5a] 0
**18.** [2.5a] −5   **19.** [1.6a] 428,000   **20.** [1.6a] 5300
**21.** [1.6b] $749,600 + 301,400 = 1,051,000$
**22.** [1.6b] $700 × 500 = 350,000$   **23.** [2.1b] <   **24.** [2.1c] 279
**25.** [1.9c], [2.5b] 36   **26.** [1.9d], [2.5b] 2   **27.** [1.9c], [2.5b] −86
**28.** [1.9b] 125   **29.** [2.6a] 3   **30.** [2.6a] 28
**31.** [2.6b] $−2x − 10$   **32.** [2.6b] $18x − 12y + 24$
**33.** [2.2a] −2   **34.** [2.4b] −8   **35.** [2.3a] −15   **36.** [2.5a] 32
**37.** [2.3a] 13   **38.** [2.4b] −30   **39.** [2.3a] 16   **40.** [2.5b] −57
**41.** [1.7b], [2.8b] 27   **42.** [2.8c] −3   **43.** [2.8e] 15
**44.** [2.8e] −8   **45.** [1.8a] 104 yr   **46.** [1.8a] 497,051 acres
**47.** [2.3b] 285°C   **48.** [1.8a] Westside Appliance
**49.** [2.7a] $10x − 14$   **50.** [1.8a] Cases: 6; six-packs: 3; loose cans: 4   **51.** [2.5b], [2.7a] $4a$   **52.** [2.5b] −1071
**53.** [2.1c], [2.8e] ±3

# CHAPTER 3

## Calculator Corner, p. 153

**1.** Yes   **2.** No   **3.** No   **4.** Yes

## Exercise Set 3.1, p. 156

**RC1.** (c)   **RC2.** (a)   **RC3.** (d)   **RC4.** (f)
**RC5.** (b)   **RC6.** (e)   **1.** 7, 14, 21, 28, 35, 42, 49, 56, 63, 70
**3.** 20, 40, 60, 80, 100, 120, 140, 160, 180, 200   **5.** 3, 6, 9, 12, 15, 18, 21, 24, 27, 30   **7.** 12, 24, 36, 48, 60, 72, 84, 96, 108, 120
**9.** 10, 20, 30, 40, 50, 60, 70, 80, 90, 100   **11.** 25, 50, 75, 100, 125, 150, 175, 200, 225, 250   **13.** No   **15.** Yes   **17.** Yes
**19.** Yes; the sum of the digits is 12, which is divisible by 3.
**21.** No; the ones digit is not 0 or 5.   **23.** Yes; the ones digit is 0.
**25.** Yes; the sum of the digits is 18, which is divisible by 9.
**27.** No; the ones digit is not even.   **29.** No; the ones digit is not even.   **31.** 6825 is divisible by 3 and 5.   **33.** 119,117 is divisible by none of these numbers.   **35.** 127,575 is divisible by 3, 5, and 9.   **37.** 9360 is divisible by 2, 3, 5, 6, 9, and 10.
**39.** 555; 300; 36; 45,270; 711; 13,251; 8064   **41.** 300; 45,270
**43.** 300; 36; 45,270; 8064   **45.** 56; 324; 784; 200; 42; 812; 402
**47.** 55,555; 200; 75; 2345; 35; 1005   **49.** 324   **51.** 53   **52.** 5
**53.** −8   **54.** −24   **55.** 45 gal   **56.** 4320 min   **57.** 125
**58.** 16   **59.** $9^3$   **60.** $3^6$   **61.** 99,969   **63.** 30   **65.** 210
**67.** 840   **69. (a)** $999a + 99b + 9c = 9(111a + 11b + c) = 3(333a + 33b + 3c)$; therefore, $999a + 99b + 9c$ is divisible by both 9 and 3; **(b)** if $a + b + c + d$ is divisible by 9, then $a + b + c + d = 9n$ for some number $n$. Then $abcd = 999a + 99b + 9c + 9n = 9(111a + 11b + c + n)$, so $abcd$ is divisible by 9. A similar argument holds for 3.
**71.** 332,986,412 is divisible by 4 and 11.

## Exercise Set 3.2, p. 163

**RC1.** prime   **RC2.** factor   **RC3.** divisible   **RC4.** multiple
**RC5.** composite   **RC6.** factorization   **CC1.** True
**CC2.** False   **CC3.** False   **CC4.** True   **CC5.** False
**1.** No   **3.** Yes   **5.** 1, 2, 3, 6, 9, 18   **7.** 1, 2, 3, 6, 9, 18, 27, 54
**9.** 1, 2, 4   **11.** 1   **13.** 1, 2, 7, 14, 49, 98   **15.** 1, 3, 5, 15, 17, 51, 85, 255   **17.** Prime   **19.** Composite   **21.** Composite

**23.** Prime   **25.** Neither   **27.** Composite   **29.** Prime
**31.** Prime   **33.** $3·3·3$   **35.** $2·7$   **37.** $2·2·2·5$
**39.** $5·5$   **41.** $2·31$   **43.** $2·2·5·5$   **45.** $11·13$   **47.** $11·11$
**49.** $3·7·13$   **51.** $5·5·7$   **53.** $11·19$   **55.** $2·2·2·2·3·5·5$
**57.** $3·3·7·11$   **59.** $2·2·7·103$   **61.** $2·3·11·17$   **63.** −26
**64.** 256   **65.** 8   **66.** −23   **67.** 1   **68.** −98   **69.** 0   **70.** 0
**71.** $13·19·19·29$   **73.** $23·31·61·73·149$   **75.** Answers may vary. One arrangement is a three-dimensional rectangular array consisting of 2 tiers of 12 objects each, where each tier consists of a rectangular array of 4 rows with 3 objects each.
**77.**

| Product | 56 | 63 | 36 | 72 | 140 | 96 |
| --- | --- | --- | --- | --- | --- | --- |
| Factor | 7 | 7 | 2 | 2 | 10 | 8 |
| Factor | 8 | 9 | 18 | 36 | 14 | 12 |
| Sum | 15 | 16 | 20 | 38 | 24 | 20 |

| Product | 48 | 168 | 110 | 90 | 432 | 63 |
| --- | --- | --- | --- | --- | --- | --- |
| Factor | 6 | 21 | 10 | 9 | 24 | 3 |
| Factor | 8 | 8 | 11 | 10 | 18 | 21 |
| Sum | 14 | 29 | 21 | 19 | 42 | 24 |

## Exercise Set 3.3, p. 170

**RC1.** (e)   **RC2.** (d)   **RC3.** (c)   **RC4.** (b)
**RC5.** (f)   **RC6.** (a)
**CC1.** Answers may vary; shade any 11 units.

**CC2.** Answers may vary; shade any 4 units.

**1.** Numerator: 3; denominator: 4   **3.** Numerator: 7; denominator: −9   **5.** Numerator: $2x$; denominator: $3y$   **7.** $\frac{6}{12}$
**9.** $\frac{1}{8}$   **11.** $\frac{3}{4}$   **13.** $\frac{4}{8}$   **15.** $\frac{12}{12}$   **17.** $\frac{9}{8}$   **19.** $\frac{4}{3}$   **21.** $\frac{11}{9}$
**23.** $\frac{5}{8}$   **25.** $\frac{4}{7}$   **27.** $\frac{12}{16}$   **29.** $\frac{38}{16}$   **31. (a)** $\frac{2}{8}$; **(b)** $\frac{6}{8}$
**33. (a)** $\frac{3}{8}$; **(b)** $\frac{5}{8}$   **35. (a)** $\frac{5}{7}$; **(b)** $\frac{5}{2}$; **(c)** $\frac{2}{7}$; **(d)** $\frac{2}{5}$
**37. (a)** $\frac{4}{15}$; **(b)** $\frac{4}{11}$; **(c)** $\frac{11}{15}$   **39. (a)** $\frac{57}{10,000}$; **(b)** $\frac{39}{10,000}$;
**(c)** $\frac{33}{10,000}$; **(d)** $\frac{20}{10,000}$; **(e)** $\frac{16}{10,000}$; **(f)** $\frac{12}{10,000}$   **41.** $\frac{4}{7}$   **43.** 0
**45.** 7   **47.** 1   **49.** 1   **51.** 0   **53.** $19x$   **55.** 1   **57.** −87
**59.** 0   **61.** Undefined   **63.** $7n$   **65.** Undefined   **67.** −210
**68.** −322   **69.** 0   **70.** 0   **71.** 5289 libraries   **72.** 71 gal
**73.** $\frac{1}{6}$   **75.** $\frac{2}{16}$, or $\frac{1}{8}$   **77.** ⬡⬡⬢⬢⬢   **79.**
**81.** $\frac{52}{365}$   **83.** $\frac{3}{4}; \frac{1}{4}$

## Exercise Set 3.4, p. 179

**RC1.** True   **RC2.** True   **RC3.** True   **RC4.** True
**CC1.** 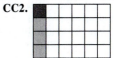   **CC2.**

**1.** $\dfrac{3}{8}$  **3.** $\dfrac{-5}{6}$, or $-\dfrac{5}{6}$  **5.** $\dfrac{14}{3}$  **7.** $-\dfrac{7}{9}$, or $\dfrac{-7}{9}$  **9.** $\dfrac{5x}{6}$

**11.** $\dfrac{-6}{5}$, or $-\dfrac{6}{5}$  **13.** $\dfrac{2a}{7}$  **15.** $\dfrac{-17m}{6}$, or $-\dfrac{17m}{6}$  **17.** $\dfrac{6}{5}$

**19.** $\dfrac{2x}{7}$  **21.** $\dfrac{4}{15}$  **23.** $\dfrac{-1}{40}$, or $\dfrac{1}{40}$  **25.** $\dfrac{2}{15}$  **27.** $\dfrac{2x}{9y}$  **29.** $\dfrac{9}{16}$

**31.** $\dfrac{14}{39}$  **33.** $\dfrac{-3}{50}$, or $-\dfrac{3}{50}$  **35.** $\dfrac{7a}{64}$  **37.** $\dfrac{100}{y}$  **39.** $\dfrac{-147}{20}$, or

$-\dfrac{147}{20}$  **41.** $\dfrac{40}{3}$ yd  **43.** $\dfrac{1}{12,324}$  **45.** $\dfrac{1}{24}$  **47.** $\dfrac{9}{20}$  **49.** $50$

**50.** $6399$  **51.** $\dfrac{71,269}{180,433}$  **53.** $\dfrac{-56}{1125}$, or $-\dfrac{56}{1125}$  **55.** $\dfrac{1}{80}$ gal

## Calculator Corner, p. 184

**1.** $\dfrac{14}{15}$  **2.** $\dfrac{7}{8}$  **3.** $\dfrac{138}{167}$  **4.** $\dfrac{7}{25}$

## Exercise Set 3.5, p. 186

**RC1.** Equivalent  **RC2.** simplify  **RC3.** common
**RC4.** cross  **CC1.** Yes  **CC2.** Yes  **CC3.** No  **CC4.** No
**CC5.** Yes  **CC6.** Yes  **1.** $\dfrac{5}{10}$  **3.** $\dfrac{-36}{-48}$  **5.** $\dfrac{35}{50}$  **7.** $\dfrac{11t}{5t}$

**9.** $\dfrac{20}{4}$  **11.** $-\dfrac{51}{54}$  **13.** $\dfrac{15}{-40}$  **15.** $\dfrac{-42}{132}$  **17.** $\dfrac{x}{8x}$  **19.** $\dfrac{-10a}{7a}$

**21.** $\dfrac{4ab}{9ab}$  **23.** $\dfrac{12b}{27b}$  **25.** $\dfrac{1}{2}$  **27.** $-\dfrac{2}{3}$  **29.** $\dfrac{2}{5}$  **31.** $3$  **33.** $\dfrac{3}{4}$

**35.** $-\dfrac{12}{7}$  **37.** $\dfrac{3}{4}$  **39.** $\dfrac{-1}{3}$  **41.** $-5$  **43.** $\dfrac{7}{8}$  **45.** $\dfrac{-2}{3}$

**47.** $\dfrac{2}{5}$  **49.** $\dfrac{9}{8}$  **51.** $\dfrac{12}{13}$  **53.** $\dfrac{17}{19}$  **55.** $\dfrac{3}{8}$  **57.** $\dfrac{3y}{2}$  **59.** $\dfrac{-9}{10b}$

**61.** $=$  **63.** $\neq$  **65.** $=$  **67.** $\neq$  **69.** $\neq$  **71.** $=$  **73.** $\neq$

**75.** $=$  **77.** $3520$  **78.** $89$  **79.** $6498$  **80.** $85$  **81.** $\dfrac{17}{29}$

**83.** $-\dfrac{29x}{15y}$  **85. (a)** $\dfrac{4}{10}=\dfrac{2}{5}$; **(b)** $\dfrac{6}{10}=\dfrac{3}{5}$  **87.** No; $\dfrac{204}{590}\neq\dfrac{213}{664}$
because $204\cdot 664\neq 590\cdot 213$.

## Mid-Chapter Review: Chapter 3, p. 188

**1.** True  **2.** False  **3.** False  **4.** True
**5.** $\dfrac{25}{25}=1$  **6.** $\dfrac{0}{9}=0$  **7.** $\dfrac{8}{1}=8$  **8.** $\dfrac{6}{13}=\dfrac{18}{39}$

**9.** $\dfrac{70}{225}=\dfrac{2\cdot 5\cdot 7}{3\cdot 3\cdot 5\cdot 5}=\dfrac{5}{5}\cdot\dfrac{2\cdot 7}{3\cdot 3\cdot 5}=1\cdot\dfrac{14}{45}=\dfrac{14}{45}$
**10.** 84; 17,576; 224; 132; 594; 504; 1632
**11.** 84; 300; 132; 500; 180  **12.** 17,576; 224; 500
**13.** 84; 300; 132; 120; 1632  **14.** 300; 180; 120
**15.** Prime  **16.** Prime  **17.** Composite  **18.** Neither
**19.** 1, 2, 4, 5, 8, 10, 16, 20, 32, 40, 80, 160; $2\cdot 2\cdot 2\cdot 2\cdot 2\cdot 5$
**20.** 1, 2, 3, 6, 37, 74, 111, 222; $2\cdot 3\cdot 37$  **21.** 1, 2, 7, 14, 49, 98;
$2\cdot 7\cdot 7$  **22.** 1, 3, 5, 7, 9, 15, 21, 35, 45, 63, 105, 315; $3\cdot 3\cdot 5\cdot 7$
**23.** $\dfrac{8}{24}$, or $\dfrac{1}{3}$  **24.** $\dfrac{5}{4}$  **25.** $\dfrac{7}{9}$  **26.** $\dfrac{8}{45}$  **27.** $\dfrac{-40}{11}$, or $-\dfrac{40}{11}$
**28.** $\dfrac{2}{5}$  **29.** $\dfrac{11}{3}$  **30.** $1$  **31.** $0$  **32.** $\dfrac{9}{31}$  **33.** $\dfrac{-9}{5}$  **34.** $\dfrac{5}{42}$
**35.** $\dfrac{21}{29}$  **36.** Undefined  **37.** $=$  **38.** $\neq$  **39.** $\dfrac{25}{200}$, or $\dfrac{1}{8}$
**40.** $\dfrac{21}{10,000}$ mi$^2$  **41.** Find the product of two prime numbers.
**42.** If we use the divisibility tests, it is quickly clear that none of the even-numbered years is a prime number. In addition, the divisibility tests for 5 and 3 show that 2001, 2005, 2007, 2013, 2015, and 2019 are not prime numbers. Then the years 2003, 2009, 2011, and 2017 can be divided by prime numbers to determine whether they are prime. When we do this, we find that 2003, 2011, and 2017 are prime numbers. If the divisibility tests

are not used, each of the numbers from 2000 to 2020 can be divided by prime numbers to determine if it is prime.  **43.** It is possible to cancel only when identical *factors* appear in the numerator and the denominator of a fraction. Situations in which it is not possible to cancel include the occurrence of identical *addends* or *digits* in the numerator and the denominator.
**44.** No; since the only factors of a prime number are the number itself and 1, two different prime numbers cannot contain a common factor (other than 1).

## Exercise Set 3.6, p. 194

**RC1.** products  **RC2.** Factor  **RC3.** 1  **RC4.** Carry out
**CC1.** $\dfrac{1}{8}$  **CC2.** $\dfrac{4}{21}$  **1.** $\dfrac{1}{3}$  **3.** $-\dfrac{1}{8}$  **5.** $\dfrac{4}{7}$  **7.** $\dfrac{1}{15}$  **9.** $-\dfrac{27}{10}$

**11.** $\dfrac{4x}{9}$  **13.** $1$  **15.** $1$  **17.** $3$  **19.** $7$  **21.** $12$  **23.** $3a$

**25.** $1$  **27.** $1$  **29.** $\dfrac{1}{5}$  **31.** $\dfrac{9}{25}$  **33.** $60n$  **35.** $-\dfrac{10}{3}$  **37.** $1$

**39.** $3$  **41.** $\dfrac{119}{750}$  **43.** $\dfrac{20}{187}$  **45.** $-\dfrac{42}{275}$  **47.** $\dfrac{16}{5x}$  **49.** $\dfrac{11}{40}$

**51.** $\dfrac{5a}{28b}$  **53.** $\dfrac{5}{8}$ in.  **55.** About \$28,700,000,000

**57.** 480 addresses  **59.** $\dfrac{1}{2}$  **61.** \$115,500  **63.** 160 mi

**65.** Food: \$8400; housing: \$10,500; clothing: \$4200; savings:
\$3000; taxes: \$8400; other expenses: \$7500  **67.** 60 in$^2$
**69.** $\dfrac{35}{4}$ mm$^2$  **71.** $\dfrac{63}{8}$ m$^2$  **73.** 92 mi$^2$  **75.** $\dfrac{7}{4}$ in$^2$  **77.** 35
**78.** 85  **79.** 125  **80.** 120  **81.** 4989  **82.** 8546  **83.** 8699
**84.** 6407  **85.** $\dfrac{129}{485}$  **87.** $\dfrac{1}{12}$  **89.** 13,380 mm$^2$

## Calculator Corner, p. 200

**1.** $\dfrac{1}{6}$  **2.** $\dfrac{20}{9}$  **3.** $-\dfrac{9}{7}$  **4.** $\dfrac{3}{2}$

## Exercise Set 3.7, p. 202

**RC1.** True  **RC2.** False  **RC3.** True  **RC4.** True
**CC1.** (g)  **CC2.** (f)  **CC3.** (c)  **CC4.** (b)  **1.** $\dfrac{3}{7}$  **3.** $\dfrac{1}{9}$

**5.** $7$  **7.** $-\dfrac{9}{8}$  **9.** $\dfrac{c}{a}$  **11.** $\dfrac{m}{-3n}$  **13.** $\dfrac{-15}{8}$  **15.** $\dfrac{1}{7m}$  **17.** $4a$

**19.** $-3z$  **21.** $\dfrac{4}{7}$  **23.** $\dfrac{4}{15}$  **25.** $4$  **27.** $-2$  **29.** $\dfrac{25}{7}$

**31.** $\dfrac{1}{8}$  **33.** $\dfrac{3}{28}$  **35.** $-8$  **37.** $\dfrac{x}{2}$  **39.** $\dfrac{1}{9x}$  **41.** $35a$  **43.** $1$

**45.** $-\dfrac{2}{3}$  **47.** $\dfrac{99}{224}$  **49.** $\dfrac{112a}{3}$  **51.** $\dfrac{14}{15}$  **53.** $\dfrac{7}{32}$  **55.** $-\dfrac{25}{12}$

**57.** $-9$  **58.** $-81$  **59.** $75$  **60.** $81$  **61.** $2$  **62.** $2$

**63.** $\dfrac{100}{9}$  **65.** $36$  **67.** $\dfrac{121}{900}$  **69.** $\dfrac{9}{19}$  **71.** $\dfrac{220}{51}$

## Translating for Success, p. 207

**1.** C  **2.** H  **3.** A  **4.** N  **5.** O  **6.** F  **7.** I  **8.** L
**9.** D  **10.** M

## Exercise Set 3.8, p. 208

**RC1.** True  **RC2.** True  **RC3.** False  **RC4.** True
**CC1.** (b)  **CC2.** (c)  **CC3.** (d)  **CC4.** (a)  **1.** 15  **3.** 9
**5.** $-45$  **7.** $\dfrac{2}{17}$  **9.** $\dfrac{12}{5}$  **11.** $-\dfrac{16}{21}$  **13.** $-\dfrac{2}{25}$  **15.** $\dfrac{1}{6}$

**17.** $-80$  **19.** $-\dfrac{1}{6}$  **21.** $-\dfrac{7}{13}$  **23.** $\dfrac{27}{31}$  **25.** $\dfrac{6}{7}$  **27.** $\dfrac{12}{5}$
**29.** $-\dfrac{7}{15}$  **31.** $\dfrac{9}{5}$  **33.** $6$  **35.** $\dfrac{10}{7}$  **37.** 960 extension cords
**39.** 32 pairs  **41.** 24 bowls  **43.** 16 L  **45.** 20 packages
**47.** $\dfrac{1}{8}$ T  **49.** 9 bees  **51.** 2400 words  **53.** 32 laps
**55.** 288 km; 108 km  **57.** $\dfrac{1}{16}$ in.  **59.** 26  **60.** $-42$  **61.** $-67$
**62.** $-65$  **63.** 20  **64.** 6  **65.** $17x$  **66.** $4a$  **67.** $7a + 3$
**68.** $4x - 7$  **69.** $\dfrac{2}{9}$  **71.** $\dfrac{7}{8}$ lb  **73.** 103 slices  **75.** 288 bees

## Summary and Review: Chapter 3, p. 212

### Vocabulary Reinforcement
**1.** multiplicative  **2.** factors  **3.** prime  **4.** denominator
**5.** equivalent  **6.** reciprocals  **7.** factorization  **8.** multiple

### Concept Reinforcement
**1.** True  **2.** False  **3.** True  **4.** True

### Study Guide
**1.** 1, 2, 4, 8, 13, 26, 52, 104  **2.** $2 \cdot 2 \cdot 2 \cdot 13$  **3.** 0, 1, 18  **4.** $\dfrac{5}{14}$
**5.** $\neq$  **6.** $\dfrac{70}{9}$  **7.** $\dfrac{7}{10}$  **8.** $\dfrac{7}{3}$ cups

### Review Exercises
**1.** 8, 16, 24, 32, 40, 48, 56, 64, 72, 80  **2.** No  **3.** No  **4.** No
**5.** Yes  **6.** Yes  **7.** 1, 2, 3, 4, 5, 6, 10, 12, 15, 20, 30, 60
**8.** 1, 2, 4, 8, 11, 16, 22, 44, 88, 176  **9.** Prime  **10.** Neither
**11.** Composite  **12.** $2 \cdot 5 \cdot 7$  **13.** $2 \cdot 2 \cdot 2 \cdot 3 \cdot 3$  **14.** $3 \cdot 3 \cdot 5$
**15.** $2 \cdot 3 \cdot 5 \cdot 5$  **16.** $2 \cdot 2 \cdot 2 \cdot 3 \cdot 3 \cdot 3 \cdot 3$  **17.** $2 \cdot 2 \cdot 2 \cdot 2 \cdot 3 \cdot 5 \cdot 5$
**18.** Numerator: 9; denominator: 7  **19.** $\dfrac{3}{5}$  **20.** $\dfrac{7}{6}$
**21.** (a) $\dfrac{3}{5}$; (b) $\dfrac{5}{3}$; (c) $\dfrac{3}{8}$  **22.** 0  **23.** 1  **24.** 48  **25.** 1
**26.** $-\dfrac{2}{3}$  **27.** $\dfrac{1}{4}$  **28.** $-1$  **29.** $\dfrac{3}{4}$  **30.** $\dfrac{2}{5}$  **31.** Undefined
**32.** $\dfrac{2}{7}$  **33.** $\dfrac{32}{225}$  **34.** $\dfrac{15}{21}$  **35.** $\dfrac{-30}{55}$
**36.** $\dfrac{16}{100} = \dfrac{4}{25}; \dfrac{17}{100} = \dfrac{17}{100}; \dfrac{48}{100} = \dfrac{12}{25}; \dfrac{4}{100} = \dfrac{1}{25}; \dfrac{15}{100} = \dfrac{3}{20}$
**37.** $\neq$  **38.** $=$  **39.** $\neq$  **40.** $=$  **41.** $\dfrac{13}{2}$  **42.** $-\dfrac{1}{7}$  **43.** 8
**44.** $\dfrac{5y}{3x}$  **45.** $\dfrac{14}{45}$  **46.** $\dfrac{3y}{7x}$  **47.** $\dfrac{2}{3}$  **48.** $-14$  **49.** $\dfrac{10}{7}$
**50.** 1  **51.** $24x$  **52.** $\dfrac{1}{4}$  **53.** $\dfrac{80}{3}$  **54.** $\dfrac{1}{15}$  **55.** $6a$  **56.** $-1$
**57.** $\dfrac{3}{4}$  **58.** $\dfrac{4}{9}$  **59.** 240  **60.** $\dfrac{-3}{10}$  **61.** 28  **62.** $\dfrac{2}{3}$
**63.** 42 m$^2$  **64.** $\dfrac{35}{2}$ ft$^2$  **65.** 9 days  **66.** About $60,794
**67.** 1000 km  **68.** $\dfrac{1}{3}$ cup  **69.** $\dfrac{1}{6}$ mi  **70.** 60 bags  **71.** D
**72.** B  **73.** $\dfrac{17}{6}$  **74.** 2, 8  **75.** $a = 11{,}176; b = 9887$
**76.** 13, 11, 101, 37

### Understanding Through Discussion and Writing
**1.** The student is probably multiplying the divisor by the reciprocal of the dividend rather than multiplying the dividend by the reciprocal of the divisor.
**2.** $9432 = 9 \cdot 1000 + 4 \cdot 100 + 3 \cdot 10 + 2 \cdot 1 =$
$9(999 + 1) + 4(99 + 1) + 3(9 + 1) + 2 \cdot 1 =$

$9 \cdot 999 + 9 \cdot 1 + 4 \cdot 99 + 4 \cdot 1 + 3 \cdot 9 + 3 \cdot 1 + 2 \cdot 1$. Since 999, 99, and 9 are each a multiple of 9, $9 \cdot 999$, $4 \cdot 99$, and $3 \cdot 9$ are multiples of 9. This leaves $9 \cdot 1 + 4 \cdot 1 + 3 \cdot 1 + 2 \cdot 1$, or $9 + 4 + 3 + 2$. If the sum of the digits, $9 + 4 + 3 + 2$, is divisible by 9, then 9432 is divisible by 9.  **3.** Taking $\frac{1}{2}$ of a number is equivalent to multiplying the number by $\frac{1}{2}$. Dividing by $\frac{1}{2}$ is equivalent to multiplying by the reciprocal of $\frac{1}{2}$, or 2. Thus taking $\frac{1}{2}$ of a number is not the same as dividing by $\frac{1}{2}$.  **4.** We first consider an object and take $\frac{4}{7}$ of it. We divide the object into 7 parts and take 4 them, as shown by the shading below in the left figure.

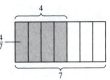
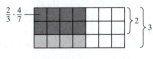

Next, we take $\frac{2}{3}$ of the shaded area in the left figure. We divide it into 3 parts and take two of them, as shown in the right figure. The entire object has been divided into 21 parts, 8 of which have been shaded twice. Thus, $\frac{2}{3} \cdot \frac{4}{7} = \frac{8}{21}$.  **5.** Since $\frac{1}{7}$ is a smaller number than $\frac{2}{3}$, there are more $\frac{1}{7}$'s in 5 than $\frac{2}{3}$'s. Thus, $5 \div \frac{1}{7}$ is a greater number than $5 \div \frac{2}{3}$.  **6.** No; in order to simplify a fraction, we must be able to remove a factor of the type $\frac{n}{n}, n \neq 0$, where $n$ is a factor that the numerator and the denominator have in common.

## Test: Chapter 3, p. 217

**1.** [3.1b] Yes  **2.** [3.1b] No  **3.** [3.2a] 1, 2, 3, 5, 6, 9, 10, 15, 18, 30, 45, 90  **4.** [3.2b] Composite  **5.** [3.2c] $2 \cdot 2 \cdot 3 \cdot 3$
**6.** [3.2c] $2 \cdot 2 \cdot 3 \cdot 5$  **7.** [3.3a] Numerator: 4; denominator: 9
**8.** [3.3a] $\dfrac{3}{4}$  **9.** [3.3a] $\dfrac{3}{7}$  **10.** [3.3a] (a) $\dfrac{180}{47}$; (b) $\dfrac{47}{93}$
**11.** [3.3b] 32  **12.** [3.3b] 1  **13.** [3.3b] 0  **14.** [3.5b] $\dfrac{-1}{3}$
**15.** [3.5b] 6  **16.** [3.5b] $\dfrac{1}{5}$  **17.** [3.3b] Undefined  **18.** [3.5b] $\dfrac{2}{3}$
**19.** [3.5c] $=$  **20.** [3.5c] $\neq$  **21.** [3.5a] $\dfrac{15}{40}$  **22.** [3.7a] $\dfrac{42}{a}$
**23.** [3.7a] $-\dfrac{1}{9}$  **24.** [3.6a] $\dfrac{5}{2}$  **25.** [3.7b] $\dfrac{8}{33}$  **26.** [3.4a] $\dfrac{3x}{8}$
**27.** [3.7b] $-\dfrac{3}{14}$  **28.** [3.7b] 18  **29.** [3.6a] $\dfrac{2}{9}$  **30.** [3.8b] $\dfrac{3}{20}$ lb
**31.** [3.6b] 125 lb  **32.** [3.8a] 64  **33.** [3.8a] $-\dfrac{7}{4}$
**34.** [3.6b] $\dfrac{91}{2}$ m$^2$  **35.** [3.3a] C  **36.** [3.6b] $\dfrac{7}{48}$ acre
**37.** [3.6a], [3.7b] $-\dfrac{7}{960}$

## Cumulative Review: Chapters 1–3, p. 219

**1.** [1.1c] Two million, fifty-six thousand, seven hundred eighty-three  **2.** [1.2a] 10,982  **3.** [2.2a] $-43$  **4.** [2.2a] $-33$
**5.** [1.3a] 2129  **6.** [2.3a] $-23$  **7.** [2.3a] $-8$  **8.** [1.4a] 16,905
**9.** [2.4a] $-312$  **10.** [3.6a] $-30x$  **11.** [3.6a] $\dfrac{7}{15}$
**12.** [1.5a] 235 R 3  **13.** [2.5a] $-17$  **14.** [3.7b] $-28$
**15.** [3.7b] $\dfrac{2}{3}$  **16.** [1.6a] 4510  **17.** [1.6b] $900 \times 500 = 450{,}000$
**18.** [2.1c] 479  **19.** [2.5b] 8  **20.** [3.2b] Composite
**21.** [2.6a] $-21$  **22.** [1.7b], [2.8b] 25  **23.** [1.7b], [2.8c] 7
**24.** [3.8a] $-45$  **25.** [2.8c] $-12$  **26.** [2.8e] $-5$  **27.** [2.8c] 10
**28.** [2.7a] $5x - 5$  **29.** [2.7a] $3x + 7y$  **30.** [3.3b] 1
**31.** [3.3b] 0  **32.** [3.3b] $63x$  **33.** [3.5b] $-\dfrac{5}{27}$  **34.** [3.7a] $\dfrac{5}{2}$

**35.** [3.7a] $\frac{1}{57}$   **36.** [3.5a] $\frac{21}{70}$   **37.** [1.8a] 8 oz   **38.** [1.8a] 8 mpg

**39.** [3.6b] 4375 students   **40.** [3.8b] 5 qt   **41.** [3.6b] $\frac{3}{5}$ mi

**42.** [2.6a], [3.6a], [3.7b] $-\frac{54}{169}$   **43.** [2.1c], [2.6a], [3.6a] $-\frac{9}{100}$

**44.** [3.6b] $468

## CHAPTER 4

### Exercise Set 4.1, p. 227
**RC1.** True   **RC2.** True   **RC3.** True   **RC4.** False
**CC1.** LCM = $2 \cdot 3 \cdot 5 = 30$
**CC2.** LCM = $2 \cdot 2 \cdot 2 \cdot 3 \cdot 5 = 120$
**CC3.** LCM = $2 \cdot 3 \cdot 5 \cdot 11 = 330$
**CC4.** LCM = $2 \cdot 2 \cdot 2 \cdot 3 \cdot 3 \cdot 5 \cdot 5 = 1800$
**CC5.** LCM = $2 \cdot 2 \cdot 3 \cdot 5 \cdot 5 = 300$
**CC6.** LCM = $2 \cdot 2 \cdot 2 \cdot 3 \cdot 5 \cdot 5 \cdot 5 \cdot 7 = 21{,}000$
**1.** 4   **3.** 50   **5.** 40   **7.** 54   **9.** 150   **11.** 120   **13.** 72
**15.** 420   **17.** 144   **19.** 180   **21.** 42   **23.** 30   **25.** 72
**27.** 60   **29.** 36   **31.** 900   **33.** 48   **35.** 210   **37.** 300
**39.** 60   **41.** $abc$   **43.** $9x^2$   **45.** $4x^3y$   **47.** $24r^3s^2t^4$
**49.** $a^3b^2c^2$   **51.** Every 60 yr   **53.** Every 420 yr   **55.** 14
**56.** $-27$   **57.** 7935   **58.** $\frac{2}{3}$   **59.** $-\frac{8}{7}$   **60.** $-167$   **61.** 2592
**63.** 54,033   **65.** 24 ft   **67.** 24 strands   **69. (a)** Not the LCM because $a^2b^5$ is not a factor of $a^3b^3$; **(b)** Not the LCM because $a^3b^2$ is not a factor of $a^2b^5$; **(c)** The LCM because both $a^3b^2$ and $a^2b^5$ are factors of $a^3b^5$ and it is the smallest such expression
**71.** 2520   **73.** 27 and 2; 27 and 6; 27 and 18

### Exercise Set 4.2, p. 235
**RC1.** True   **RC2.** False   **RC3.** False   **RC4.** True
**CC1.** $\frac{7}{20} + \frac{15}{20}$   **CC2.** $\frac{9}{24} + \frac{20}{24}$   **CC3.** $\frac{42}{60} + \frac{44}{60} + \frac{15}{60}$
**1.** $\frac{5}{9}$   **3.** 1   **5.** $\frac{2}{5}$   **7.** $\frac{13}{a}$   **9.** $-\frac{1}{2}$   **11.** $\frac{7}{9}x$   **13.** $\frac{1}{2}t$
**15.** $-\frac{9}{x}$   **17.** $\frac{7}{24}$   **19.** $-\frac{1}{10}$   **21.** $\frac{23}{24}$   **23.** $\frac{83}{20}$   **25.** $\frac{5}{24}$
**27.** $\frac{37}{100}x$   **29.** $\frac{19}{20}$   **31.** $\frac{7}{8}$   **33.** $-\frac{99}{100}$   **35.** $-\frac{1}{30}x$   **37.** $-\frac{33}{7}t$
**39.** $-\frac{17}{24}$   **41.** $\frac{3}{4}$   **43.** $\frac{437}{1000}$   **45.** $\frac{5}{4}$   **47.** $\frac{239}{78}$   **49.** $\frac{59}{90}$   **51.** $-\frac{5}{4}$
**53.** $>$   **55.** $<$   **57.** $>$   **59.** $<$   **61.** $>$   **63.** $>$
**65.** $\frac{4}{15}, \frac{3}{10}, \frac{5}{12}$   **67.** $\frac{37}{12}$ mi   **69.** $\frac{13}{16}$ lb   **71.** $\frac{27''}{32}$   **73.** $\frac{13}{12}$ lb
**75.** $\frac{7}{8}$ in.   **77.** $\frac{33}{20}$ mi   **79.** $\frac{4}{5}$ qt; $\frac{8}{5}$ qt; $\frac{2}{5}$ qt   **81.** $-13$   **82.** 4
**83.** $-8$   **84.** $-31$   **85.** $\frac{10}{3}$   **86.** 42; 42   **87.** 35   **88.** 6407
**89.** 11   **90.** 3   **91.** $\frac{9}{7}$   **92.** $-\frac{2}{5}$   **93.** $\frac{13}{30}t + \frac{31}{35}$   **95.** $7t^2 + \frac{9}{a}t$
**97.** $>$   **99.** $\frac{4}{15}$; $320   **101.** $a = 2, b = 8$

### Translating for Success, p. 243
**1.** J   **2.** E   **3.** D   **4.** B   **5.** I   **6.** N   **7.** A   **8.** C
**9.** L   **10.** F

### Exercise Set 4.3, p. 244
**RC1.** numerators; denominator   **RC2.** denominators
**RC3.** denominators   **RC4.** denominators
**CC1.** 16   **CC2.** 40   **CC3.** 132   **CC4.** 360   **1.** $\frac{2}{3}$   **3.** $-\frac{1}{4}$
**5.** $\frac{2}{a}$   **7.** $-\frac{1}{2}$   **9.** $-\frac{4}{5a}$   **11.** $\frac{2}{t}$   **13.** $\frac{13}{16}$   **15.** $-\frac{1}{3}$   **17.** $\frac{7}{10}$
**19.** $-\frac{17}{60}$   **21.** $\frac{47}{100}$   **23.** $\frac{26}{75}$   **25.** $-\frac{71}{100}$   **27.** $\frac{13}{24}$   **29.** $-\frac{29}{50}$
**31.** $-\frac{1}{24}$   **33.** $-\frac{41}{72}$   **35.** $\frac{1}{360}$   **37.** $\frac{2}{9}x$   **39.** $-\frac{7}{20}a$   **41.** $\frac{7}{9}$
**43.** $\frac{4}{11}$   **45.** $\frac{1}{15}$   **47.** $\frac{9}{8}$   **49.** $\frac{2}{15}$   **51.** $-\frac{7}{24}$   **53.** $\frac{2}{15}$   **55.** $-\frac{5}{4}$
**57.** $\frac{5}{12}$ hr   **59.** $\frac{1}{32}$ in.   **61.** $\frac{11}{20}$ lb   **63.** $\frac{7}{20}$ hr   **65.** $\frac{3}{16}$ in.
**67.** $\frac{7}{24}$ cup   **69.** $\frac{1}{4}$   **71.** 1   **72.** Not defined   **73.** Not defined
**74.** 4   **75.** $\frac{4}{21}$   **76.** $\frac{3}{2}$   **77.** 21   **78.** $\frac{1}{32}$   **79.** $\frac{1}{16}$   **81.** $-\frac{11}{10}$

**83.** $-\frac{64}{35}$   **85.** $-\frac{37}{1000}$   **87.** $\frac{21}{40}$ km   **89.** $\frac{43}{50}$   **91.** $\frac{14}{3553}$
**93.** *Day 1:* Cut off $\frac{1}{7}$ of bar and pay the contractor.
  *Day 2:* Cut off $\frac{2}{7}$ of the bar's original length and trade it for the $\frac{1}{7}$.
  *Day 3:* Give the $\frac{1}{7}$ back to the contractor.
  *Day 4:* Trade the $\frac{4}{7}$ remaining for the contractor's $\frac{3}{7}$.
  *Day 5:* Give the contractor the $\frac{1}{7}$ again.
  *Day 6:* Trade the $\frac{2}{7}$ for the $\frac{1}{7}$.
  *Day 7:* Give the contractor the $\frac{1}{7}$ again.
  This assumes that the contractor does not spend parts of the gold bar immediately.

### Exercise Set 4.4, p. 252
**RC1.** False   **RC2.** False   **RC3.** False   **RC4.** True
**CC1.** Yes   **CC2.** Yes   **CC3.** No   **1.** 3   **3.** $\frac{3}{5}$   **5.** $-1$
**7.** $\frac{27}{2}$   **9.** $\frac{1}{2}$   **11.** $\frac{3}{2}$   **13.** $-\frac{4}{3}$   **15.** $\frac{1}{2}$   **17.** $\frac{8}{7}$   **19.** $-6$
**21.** $\frac{21}{5}$   **23.** 6   **25.** $-\frac{17}{4}$   **27.** $-\frac{3}{4}$   **29.** $\frac{3}{2}$   **31.** $\frac{9}{2}$   **33.** $-\frac{10}{3}$
**35.** $-\frac{1}{5}$   **37.** $-\frac{1}{6}$   **39.** $\frac{35}{12}$   **41.** $-13$   **42.** $-8$   **43.** 18
**44.** 27   **45.** The balance has decreased $150.
**46.** $1180 profit   **47.** $\frac{5}{7m}$   **48.** $20n$   **49.** 4   **51.** $-\frac{290}{697}$
**53.** $\frac{436}{35}$   **55.** 2 cm

### Calculator Corner, p. 256
**1.** $5\frac{4}{7}$   **2.** $8\frac{2}{5}$   **3.** $1476\frac{1}{6}$   **4.** $676\frac{4}{9}$   **5.** $134\frac{1}{15}$   **6.** $12\frac{169}{454}$

### Exercise Set 4.5, p. 258
**RC1.** True   **RC2.** True   **RC3.** False   **RC4.** True
**CC1.** $3\frac{1}{12}$   **CC2.** $12\frac{3}{8}$   **CC3.** $26\frac{21}{23}$   **1.** $\frac{23}{8}$   **3.** $\frac{25}{4}$
**5.** $-\frac{161}{8}$   **7.** $\frac{51}{10}$   **9.** $\frac{103}{5}$   **11.** $-\frac{100}{3}$   **13.** $\frac{13}{8}$   **15.** $-\frac{51}{4}$
**17.** $\frac{57}{10}$   **19.** $-\frac{507}{100}$   **21.** $5\frac{1}{3}$   **23.** $7\frac{1}{2}$   **25.** $5\frac{7}{10}$   **27.** $7\frac{2}{9}$
**29.** $-5\frac{1}{2}$   **31.** $11\frac{1}{2}$   **33.** $-1\frac{1}{3}$   **35.** $61\frac{2}{5}$   **37.** $-8\frac{13}{50}$
**39.** $108\frac{5}{8}$   **41.** $906\frac{3}{7}$   **43.** $40\frac{4}{7}$   **45.** $-20\frac{2}{15}$   **47.** $-22\frac{3}{7}$
**49.** $275\frac{2}{3}$ backpacks   **51.** $4231\frac{1}{6}$ hr   **53.** $18x + 6y - 24$
**54.** $-3a + 6b - 12c$   **55.** $237\frac{19}{541}$   **57.** $8\frac{2}{3}$   **59.** $3\frac{2}{3}$   **61.** $52\frac{1}{7}$

### Mid-Chapter Review: Chapter 4, p. 260
**1.** True   **2.** True   **3.** False   **4.** False
**5.** $\dfrac{11}{42} - \dfrac{3}{35} = \dfrac{11}{2 \cdot 3 \cdot 7} - \dfrac{3}{5 \cdot 7}$
$= \dfrac{11}{2 \cdot 3 \cdot 7} \cdot \left(\dfrac{5}{5}\right) - \dfrac{3}{5 \cdot 7} \cdot \left(\dfrac{2 \cdot 3}{2 \cdot 3}\right)$
$= \dfrac{11 \cdot 5}{2 \cdot 3 \cdot 7 \cdot 5} - \dfrac{3 \cdot 2 \cdot 3}{5 \cdot 7 \cdot 2 \cdot 3}$
$= \dfrac{55}{2 \cdot 3 \cdot 5 \cdot 7} - \dfrac{18}{2 \cdot 3 \cdot 5 \cdot 7}$
$= \dfrac{55 - 18}{2 \cdot 3 \cdot 5 \cdot 7} = \dfrac{37}{210}$

**6.** $x + \frac{1}{8} = \frac{2}{3}$
$x + \frac{1}{8} - \frac{1}{8} = \frac{2}{3} - \frac{1}{8}$
$x + 0 = \frac{2}{3} \cdot \frac{8}{8} - \frac{1}{8} \cdot \frac{3}{3}$
$x = \frac{16}{24} - \frac{3}{24}$
$x = \frac{13}{24}$
The solution is $\frac{13}{24}$.

**7.**
45 and 50 — 120
50 and 80 — 720
30 and 24 — 400
18, 24, and 80 — 450
30, 45, and 50

**8.** $\frac{16}{45}$   **9.** $\frac{25}{12}$   **10.** $\frac{1}{18}$   **11.** $-\frac{19}{90}$   **12.** $\frac{7}{240}$   **13.** $-\frac{7}{24}x$
**14.** $-\frac{17}{60}$   **15.** $\frac{6}{91}$   **16.** $\frac{22}{15}$ mi   **17.** $\frac{101}{20}$ hr   **18.** $\frac{1}{5}, \frac{2}{7}, \frac{3}{10}, \frac{4}{9}$
**19.** $\frac{13}{80}$   **20.** $-\frac{8}{9}$   **21.** $17\frac{8}{15}$   **22.** C   **23.** C

**24.** No; if one number is a multiple of the other, for example, the LCM is the larger of the numbers. **25.** We multiply by 1, using the notation $n/n$, to express each fraction in terms of the least common denominator. **26.** Write $\frac{8}{5}$ as $\frac{16}{10}$ and $\frac{8}{2}$ as $\frac{40}{10}$ and since taking 40 tenths away from 16 tenths would give a result less than 0, the answer cannot possibly be $\frac{8}{3}$. You could also find the sum $\frac{8}{3} + \frac{8}{2}$ and show that it is not $\frac{8}{5}$. **27.** No; $2\frac{1}{3} = \frac{7}{3}$ but $2 \cdot \frac{1}{3} = \frac{2}{3}$.

## Exercise Set 4.6, p. 268

**RC1.** (d) **RC2.** (a) **RC3.** (b) **RC4.** (c)
**CC1.** (a), (b), (c) **CC2.** (b), (c), (d)
**1.** $11\frac{2}{5}$ **3.** $9\frac{1}{2}$ **5.** $5\frac{1}{3}$ **7.** $13\frac{7}{12}$ **9.** $12\frac{1}{10}$ **11.** $17\frac{5}{24}$
**13.** $21\frac{1}{2}$ **15.** $27\frac{7}{8}$ **17.** $7\frac{1}{5}$ **19.** $6\frac{1}{10}$ **21.** $1\frac{3}{5}$ **23.** $13\frac{1}{4}$
**25.** $15\frac{3}{8}$ **27.** $7\frac{5}{12}$ **29.** $11\frac{5}{18}$ **31.** $8\frac{3}{4}t$ **33.** $2x$ **35.** $8\frac{3}{10}y$
**37.** $11\frac{8}{9}t$ **39.** $10\frac{5}{24}x$ **41.** $2\frac{1}{3}a$ **43.** $23\frac{3}{20}$ ft **45.** $\frac{15}{16}$ in.
**47.** $1\frac{3}{10}$ in. **49.** $134\frac{1}{4}$ in. **51.** $28\frac{3}{4}$ yd **53.** $4\frac{5}{6}$ ft **55.** $5\frac{7}{8}$ in.
**57.** $20\frac{1}{8}$ in. **59.** $3\frac{4}{5}$ hr **61.** $66\frac{5}{8}$ ft **63.** $7\frac{3}{8}$ ft **65.** $-\frac{2}{5}$
**67.** $-3\frac{1}{4}$ **69.** $-3\frac{13}{15}$ **71.** $-7\frac{3}{8}$ **73.** $2\frac{7}{12}$ **75.** $-1\frac{8}{9}$
**77.** 3 ten thousands + 8 thousands + 1 hundred + 2 tens + 5 ones **78.** Two million, five thousand, six hundred eighty-nine **79.** $9^4$ **80.** 81 **81.** Yes **82.** No **83.** No **84.** Yes
**85.** No **86.** Yes **87.** Yes **88.** Yes **89.** $8568\frac{786}{1189}$
**91.** $10\frac{7}{12}$ **93.** $-28\frac{3}{8}$ **95.** $55\frac{3}{4}$ in.

## Calculator Corner, p. 277

**1.** $\frac{7}{12}$ **2.** $\frac{11}{10}$ **3.** $\frac{35}{16}$ **4.** $-\frac{3}{10}$ **5.** $10\frac{2}{15}$ **6.** $1\frac{1}{28}$
**7.** $10\frac{11}{15}$ **8.** $2\frac{91}{115}$

## Translating for Success, p. 278

**1.** O **2.** K **3.** F **4.** D **5.** H **6.** G **7.** L
**8.** E **9.** M **10.** J

## Exercise Set 4.7, p. 279

**RC1.** True **RC2.** True **RC3.** True **RC4.** True
**CC1.** $\frac{7}{3} \cdot \frac{5}{23}$ **CC2.** $\frac{73}{10} \cdot \frac{10}{17}$ **CC3.** $\frac{74}{9} \cdot \frac{16}{1}$ **CC4.** $\frac{47}{7} \cdot \frac{1}{30}$
**1.** $22\frac{2}{3}$ **3.** $1\frac{2}{3}$ **5.** $-56\frac{2}{3}$ **7.** $16\frac{1}{3}$ **9.** $-10\frac{3}{25}$ **11.** $35\frac{91}{100}$
**13.** $6\frac{1}{4}$ **15.** $1\frac{1}{5}$ **17.** $1\frac{2}{3}$ **19.** $-1\frac{1}{8}$ **21.** $1\frac{8}{43}$ **23.** $\frac{9}{40}$
**25.** $23\frac{2}{5}$ **27.** $15\frac{5}{7}$ **29.** $-27\frac{2}{9}$ **31.** $-1\frac{1}{3}$ **33.** $12\frac{1}{4}$ **35.** $8\frac{3}{20}$
**37.** About 35 mi **39.** 144 cups **41.** About $46,000,000,000
**43.** 75 mph **45.** $12\frac{4}{5}$ tiles **47.** $5\frac{1}{2}$ cups of flour, $2\frac{2}{3}$ cups
of sugar **49.** 15 mpg **51.** $62\frac{1}{2}$ sq ft **53.** 400 cu ft
**55.** $16\frac{1}{2}$ servings **57.** $441\frac{1}{4}$ sq ft **59.** $76\frac{1}{4}$ sq ft
**61.** $27\frac{5}{16}$ sq cm **63.** 68°F **65.** $13\frac{1}{4}$ in. $\times 13\frac{1}{4}$ in.: perimeter = 53 in., area = $175\frac{9}{16}$ sq in.; $13\frac{1}{4}$ in. $\times 3\frac{1}{4}$ in.: perimeter = 33 in., area = $43\frac{1}{16}$ sq in. **67.** 22 m **68.** 4500 sq yd **69.** $16\frac{25}{64}$
**71.** $\frac{4}{9}$ **73.** $r = \frac{240}{13}$, or $18\frac{6}{13}$ **75.** 88 gal

## Exercise Set 4.8, p. 288

**RC1.** (c) **RC2.** (d) **RC3.** (a) **RC4.** (b)
**CC1.** $\frac{2}{3} \div \frac{1}{6}$ **CC2.** $\frac{10}{1} \div \frac{3}{8}$ **CC3.** $\frac{1}{4} \div \frac{4}{1}$ **CC4.** $\frac{16}{5} \div \frac{9}{2}$
**1.** $\frac{7}{24}$ **3.** $\frac{3}{2}$, or $1\frac{1}{2}$ **5.** $\frac{7}{8}$ **7.** $\frac{59}{30}$, or $1\frac{29}{30}$ **9.** $\frac{7}{16}$ **11.** $\frac{3}{20}$
**13.** $-6$ **15.** $\frac{1}{36}$ **17.** $-\frac{3}{100}$ **19.** $-1$ **21.** $\frac{19}{4}$, or $4\frac{3}{4}$ **23.** $\frac{3}{8}$
**25.** $\frac{2}{9}$ **27.** $\frac{3}{11}$ **29.** $-\frac{14}{3}$, or $-4\frac{2}{3}$ **31.** $-\frac{5}{4}$, or $-1\frac{1}{4}$ **33.** $\frac{1}{100}$
**35.** $-\frac{1}{6}$ **37.** $\frac{2x}{35}$ **39.** $-\frac{3n}{28}$ **41.** $\frac{6}{7x}$ **43.** $-\frac{5}{18}$ **45.** $-\frac{7}{12}$
**47.** $\frac{1}{3}$ **49.** $\frac{37}{48}$ **51.** $\frac{25}{72}$ **53.** $\frac{103}{16}$, or $6\frac{7}{16}$ **55.** $16\frac{7}{96}$ mi
**57.** $9\frac{19}{40}$ lb **59.** 20 **60.** 2 **61.** 84 **62.** 100 **63.** 1, 2, 3, 6,
7, 14, 21, 42 **64.** No **65.** Prime: 5, 7, 23, 43; composite: 9, 14;

neither: 1 **66.** $2 \cdot 3 \cdot 5 \cdot 5$ **67.** $\frac{53}{14}$, or $3\frac{11}{14}$ **69.** $\frac{8x}{147}$ **71.** 0
**73.** $\frac{1}{2}$ **75.** $\frac{1}{2}$

## Summary and Review: Chapter 4, p. 291

### Vocabulary Reinforcement
**1.** least common multiple **2.** mixed numeral **3.** fraction
**4.** complex fraction **5.** denominators **6.** least common
multiple **7.** greatest **8.** numerators

### Concept Reinforcement
**1.** True **2.** True **3.** True **4.** False

### Study Guide
**1.** 156 **2.** $\frac{28}{45}$ **3.** $<$ **4.** $\frac{4}{35}$ **5.** $-\frac{1}{12}$ **6.** $\frac{26}{3}$ **7.** $7\frac{5}{6}$
**8.** $7\frac{27}{28}$ **9.** $14\frac{14}{25}$ **10.** About 4,500,000 **11.** $\frac{9}{2}$, or $4\frac{1}{2}$

### Review Exercises
**1.** 36 **2.** 90 **3.** 30 **4.** 1404 **5.** $\frac{7}{9}$ **6.** $\frac{9}{x}$ **7.** $-\frac{7}{15}$
**8.** $\frac{7}{16}$ **9.** $\frac{2}{9}$ **10.** $-\frac{1}{8}$ **11.** $\frac{4}{27}$ **12.** $\frac{1}{18}$ **13.** $>$ **14.** $<$
**15.** $\frac{19}{40}$ **16.** 11 **17.** $-\frac{5}{6}$ **18.** $\frac{12}{25}$ **19.** $\frac{2}{5}$ **20.** $\frac{15}{2}$ **21.** $\frac{67}{8}$
**22.** $\frac{13}{3}$ **23.** $-\frac{12}{7}$ **24.** $2\frac{1}{3}$ **25.** $-6\frac{3}{4}$ **26.** $12\frac{3}{5}$ **27.** $3\frac{1}{2}$
**28.** $-877\frac{1}{3}$ **29.** $82\frac{1}{3}$ **30.** $10\frac{2}{5}$ **31.** $11\frac{11}{15}$ **32.** $-9$ **33.** $1\frac{3}{4}$
**34.** $7\frac{7}{9}$ **35.** $4\frac{11}{15}$ **36.** $-5\frac{1}{8}$ **37.** $-14\frac{1}{4}$ **38.** $\frac{7}{9}x$ **39.** $3\frac{7}{40}a$
**40.** 16 **41.** $-3\frac{1}{2}$ **42.** $2\frac{21}{50}$ **43.** 6 **44.** $-24$ **45.** $-1\frac{7}{17}$
**46.** $\frac{1}{8}$ **47.** $\frac{9}{10}$ **48.** $13\frac{5}{7}$ **49.** $2\frac{8}{11}$ **50.** $4\frac{1}{4}$ yd **51.** $3\frac{1}{8}$ pizzas
**52.** 24 lb **53.** $63\frac{2}{3}$ pies; $19\frac{1}{3}$ pies **54.** $3\frac{3}{4}$ mi **55.** $177\frac{3}{4}$ sq in.
**56.** $50\frac{1}{4}$ sq in. **57.** $850 **58.** $4\frac{5}{6}$ ft **59.** 1 **60.** $\frac{39}{40}$ **61.** 3
**62.** $\frac{77}{240}$ **63.** $-\frac{8}{9}$ **64.** $-\frac{2x}{3}$ **65.** A **66.** D
**67.** $\frac{600}{13}$, or $46\frac{2}{13}$ **68.** $\frac{6}{3} + \frac{5}{4} = 3\frac{1}{4}$ **69.** (a) 6; (b) 10;
**(c)** $-28$; **(d)** $-1$

### Understanding Through Discussion and Writing
**1.** No; if the sum of the fraction parts of the mixed numerals is $n/n$, then the sum of the mixed numerals is an integer. For example, $1\frac{1}{5} + 6\frac{4}{5} = 7\frac{5}{5} = 8$. **2.** A wheel makes $33\frac{1}{3}$ revolutions per minute. It rotates for $4\frac{1}{2}$ min. How many revolutions does it make? Answers may vary. **3.** The student is multiplying the whole numbers to get the whole-number portion of the answer and multiplying fractions to get the fraction part of the answer. The student should have converted each mixed numeral to fraction notation, multiplied, simplified, and then converted back to a mixed numeral. The correct answer is $4\frac{6}{7}$. **4.** It might be necessary to find the least common denominator before adding or subtracting. The least common denominator is the least common multiple of the denominators. **5.** Suppose that a room has dimensions $15\frac{3}{4}$ ft by $28\frac{5}{8}$ ft. The equation $2 \cdot 15\frac{3}{4} + 2 \cdot 28\frac{5}{8} = 88\frac{3}{4}$ gives the perimeter of the room, in feet. Answers may vary. **6.** Note that $5 \cdot 3\frac{2}{7} = 5(3 + \frac{2}{7}) = 5 \cdot 3 + 5 \cdot \frac{2}{7}$. The products $5 \cdot 3$ and $5 \cdot \frac{2}{7}$ should be added rather than multiplied together. The student could also have converted $3\frac{2}{7}$ to fraction notation, multiplied, simplified, and converted back to a mixed numeral. The correct answer is $16\frac{3}{7}$.

## Test: Chapter 4, p. 297

**1.** [4.1a] 48 **2.** [4.2a] 3 **3.** [4.2b] $-\frac{5}{24}$ **4.** [4.3a] $\frac{2}{t}$
**5.** [4.3a] $\frac{1}{12}$ **6.** [4.3a] $-\frac{1}{12}$ **7.** [4.3b] $\frac{1}{4}$ **8.** [4.4a] $-\frac{12}{5}$
**9.** [4.4a] $\frac{3}{20}$ **10.** [4.2c] $>$ **11.** [4.5a] $\frac{7}{2}$ **12.** [4.5a] $-\frac{75}{8}$
**13.** [4.5a] $-8\frac{2}{9}$ **14.** [4.5b] $162\frac{7}{11}$ **15.** [4.6a] $14\frac{1}{5}$

**16.** [4.6a] $12\frac{5}{12}$  **17.** [4.6b] $4\frac{7}{24}$  **18.** [4.6d] $8\frac{4}{7}$
**19.** [4.6d] $-5\frac{7}{10}$  **20.** [4.3a] $-\frac{1}{8}x$  **21.** [4.6b] $1\frac{54}{55}a$
**22.** [4.7a] 39  **23.** [4.7a] $-18$  **24.** [4.7b] 6  **25.** [4.7b] 2
**26.** [4.7c] $19\frac{3}{5}$  **27.** [4.7c] $28\frac{1}{20}$  **28.** [4.7d] 55 mph
**29.** [4.7d] 80 books  **30.** [4.6c] **(a)** 3 in.; **(b)** $4\frac{1}{2}$ in.
**31.** [4.3c] $\frac{1}{16}$ in.  **32.** [4.8b] $6\frac{11}{36}$ ft  **33.** [4.8a] $3\frac{1}{2}$
**34.** [4.8a] $-\frac{9}{4}$, or $-2\frac{1}{4}$  **35.** [4.8b] $-\frac{2}{3}$  **36.** [4.1a] D
**37.** [4.1a] **(a)** 24, 48, 72; **(b)** 24  **38.** [4.3c] Rebecca walks
$\frac{17}{56}$ mi farther.

## Cumulative Review: Chapters 1–4, p. 299

**1.** [4.6c] $23\frac{1}{4}$ gal  **2.** [1.8a] 31 people  **3.** [3.6b] $\frac{2}{5}$ tsp; 4 tsp
**4.** [4.7d] 16 pieces  **5.** [1.8a] \$108  **6.** [4.2d] $\frac{33}{20}$ mi
**7.** [3.3a] $\frac{5}{16}$  **8.** [3.3a] $\frac{4}{3}$  **9.** [1.2a] 8982  **10.** [1.3a] 4518
**11.** [1.4a] 5004  **12.** [2.4a] $-145$  **13.** [4.2b] $\frac{5}{12}$  **14.** [4.6a] $8\frac{1}{4}$
**15.** [4.3a] $-\frac{5}{t}$  **16.** [4.6b] $1\frac{1}{6}$  **17.** [3.6a] $\frac{3}{2}$  **18.** [3.6a] $-15$
**19.** [2.3a] 16  **20.** [4.7b] $7\frac{1}{3}$  **21.** [1.5a] 715
**22.** [1.5a] 56 R 11  **23.** [4.5b] $56\frac{11}{45}$  **24.** [1.1a] 5
**25.** **(a)** [4.7d] $142\frac{1}{4}$ sq ft; **(b)** [4.6c] 54 ft  **26.** [1.6a] 38,500
**27.** [4.1a] 72  **28.** [4.8a] $\frac{1377}{100}$, or $13\frac{77}{100}$  **29.** [4.2c] $>$
**30.** [3.5c] $=$  **31.** [4.2c] $>$  **32.** [2.6a] 4  **33.** [3.5b] $\frac{4}{5}$
**34.** [3.3b] 0  **35.** [3.5b] $-32$  **36.** [4.5a] $\frac{37}{8}$
**37.** [4.5a] $-5\frac{2}{3}$  **38.** [1.7b] 93  **39.** [4.3b] $\frac{5}{9}$
**40.** [3.8a] $-\frac{12}{7}$  **41.** [4.4a] $\frac{2}{21}$  **42.** [3.1a, b], [3.2a, b, c]
Factors of 68: 1, 2, 4, 17, 34, 68
Factorization of 68: $2 \cdot 2 \cdot 17$, or $2 \cdot 34$
Prime factorization of 68: $2 \cdot 2 \cdot 17$
Numbers divisible by 6: 12, 54, 72, 300
Numbers divisible by 8: 8, 16, 24, 32, 40, 48, 64, 864
Numbers divisible by 5: 70, 95, 215
Prime numbers: 2, 3, 17, 19, 23, 31, 47, 101
**43.** [3.2b] 2003  **44.** [4.4a], [4.4b] $\frac{3}{7}$

# CHAPTER 5

## Exercise Set 5.1, p. 309

**RC1.** 5  **RC2.** 2  **RC3.** 3  **RC4.** 0  **RC5.** 8  **RC6.** 4
**RC7.** 1  **RC8.** 6  **CC1.** 0.099, 0.89999, 0.909, 0.9889, 0.99,
0.9999, 1, 1.00009  **CC2.** 2.000001, 2.0119, 2.018, 2.0302, 2.1,
2.108, 2.109
**1.** Seventeen and seven thousand one hundred seventy-eight
ten-thousandths  **3.** One hundred three and six tenths
**5.** Two hundred thirty-two and one hundred sixty-four thousandths
**7.** Three and seven hundred eighty-five thousandths
**9.** $\frac{73}{10}$; $7\frac{3}{10}$  **11.** $\frac{2167}{100}$; $21\frac{67}{100}$  **13.** $-\frac{2703}{1000}$, $-2\frac{703}{1000}$  **15.** $\frac{109}{10,000}$
**17.** $-\frac{40,003}{10,000}$; $-4\frac{3}{10,000}$  **19.** $-\frac{207}{10,000}$  **21.** $\frac{7,000,105}{100,000}$; $70\frac{105}{100,000}$
**23.** 0.3  **25.** $-0.59$  **27.** 3.798  **29.** 0.0078  **31.** $-0.00018$
**33.** 0.486197  **35.** 7.013  **37.** $-8.431$  **39.** 2.1739
**41.** 8.953073  **43.** 0.58  **45.** 0.410  **47.** $-5.043$  **49.** 235.07
**51.** $\frac{7}{100}$  **53.** $-0.872$  **55.** 0.2  **57.** $-0.4$  **59.** 3.0
**61.** $-327.2$  **63.** 0.89  **65.** $-0.67$  **67.** 1.00  **69.** $-0.03$
**71.** 0.572  **73.** 17.002  **75.** $-20.202$  **77.** 9.985  **79.** 809.5
**81.** 809.47  **83.** 830  **84.** $\frac{830}{100}$, or $\frac{83}{10}$  **85.** 182  **86.** $\frac{182}{100}$, or $\frac{91}{50}$
**87.** $-\frac{12}{55}$  **88.** $-\frac{15}{34}$  **89.** 32,958  **90.** 10,726
**91.** $-1.09$, $-1.009$, $-0.989$, $-0.898$, $-0.098$  **93.** 6.78346
**95.** 99.99999

## Calculator Corner, p. 315

**1.** 317.645  **2.** 49.08  **3.** 33.83  **4.** 0.99  **5.** 242.93
**6.** $-11.692$

## Exercise Set 5.2, p. 316

**RC1.** 21.824; 23.7  **RC2.** 146.723; 40.9  **CC1.** Negative
**CC2.** Positive  **CC3.** Negative  **CC4.** Positive
**CC5.** Positive  **CC6.** Negative  **1.** 464.37  **3.** 1576.015
**5.** 132.56, or 132.560  **7.** 7.823  **9.** 50.7124  **11.** 10.06
**13.** 771.967  **15.** 20.8649  **17.** 227.468, or 227.4680
**19.** 41.381  **21.** 49.02  **23.** 3.564  **25.** 85.921
**27.** 1.6666  **29.** 4.0622  **31.** 29.999  **33.** 3.37  **35.** 1.045
**37.** 3.703  **39.** 0.9092  **41.** 605.21  **43.** 53.203
**45.** 161.62  **47.** 44.001  **49.** $-3.29$  **51.** $-2.5$  **53.** $-7.2$
**55.** 3.379  **57.** $-16.6$  **59.** 2.5  **61.** $-3.519$  **63.** 9.601
**65.** 75.5  **67.** 9.7  **69.** $-10.292$  **71.** $-0.3$  **73.** 5.7$x$
**75.** 4.86$a$  **77.** 21.1$t$ + 7.9  **79.** $-2.917x$  **81.** 8.106$y$ − 7.1
**83.** $-0.9x$ + 3.1$y$  **85.** $-1.1$ − 8.4$t$  **87.** $\frac{12}{35}$  **88.** $\frac{14}{45}$  **89.** $\frac{63}{1000}$
**90.** $-10$  **91.** $-7$  **92.** 31  **93.** $-12.001$ − 12.2698$a$ + 10.366$b$
**95.** 4.593$a$ − 10.996$b$ − 59.491  **97.** $-138.5$  **99.** 2

## Calculator Corner, p. 322

**1.** 142.803  **2.** $-0.5076$  **3.** 7916.4  **4.** 20.4153

## Exercise Set 5.3, p. 326

**RC1.** (e)  **RC2.** (d)  **RC3.** (a)  **RC4.** (b)  **RC5.** (f)
**RC6.** (c)  **CC1.** (c)  **CC2.** (e)  **CC3.** (f)  **CC4.** (a)
**CC5.** (d)  **CC6.** (b)  **1.** 47.6  **3.** 6.72  **5.** 0.252  **7.** 0.522
**9.** 426.3  **11.** $-783,686.852$  **13.** $-780$  **15.** 7.918
**17.** 0.09768  **19.** $-0.287$  **21.** 43.68  **23.** 0.030504
**25.** 89.76  **27.** $-322.07$  **29.** 55.68  **31.** 3487.5  **33.** 0.1155
**35.** $-9420$  **37.** 0.00953  **39.** 5706¢  **41.** 95¢  **43.** 1¢
**45.** \$0.72  **47.** \$0.02  **49.** \$63.99  **51.** 19,973,000,000,000;
325,700,000  **53.** 60,061,000; 29,638,000  **55.** 10,400,000,000
**57.** 11,000  **59.** 26.025  **61.** **(a)** 44 ft; **(b)** 118.75 sq ft
**63.** **(a)** 37.8 m; **(b)** 88.2 m²  **65.** 2.7625 million nurses, or
2,762,500 nurses  **67.** $-27$  **68.** 36  **69.** 69  **70.** $-141$
**71.** 1176 R 14  **72.** $-27$  **73.** $10^{21} = 1$ sextillion
**75.** $10^{24} = 1$ septillion  **77.** 6,600,000,000,000  **79.** \$61.45

## Calculator Corner, p. 330

**1.** 14.3  **2.** 2.56  **3.** $-0.064$  **4.** 75.8

## Calculator Corner, p. 331

**1.** 28 R 2  **2.** 116 R 3  **3.** 74 R 10  **4.** 415 R 3

## Exercise Set 5.4, p. 335

**RC1.** (e)  **RC2.** (b)  **RC3.** (d)  **RC4.** (a)  **RC5.** (c)
**RC6.** (f)  **CC1.** Subtract: $2 - 0.04$  **CC2.** Divide: $2.06 \div 0.01$
**CC3.** Evaluate: $8^3$  **CC4.** Add: $4.1 + 6.9$  **CC5.** Divide: $9 \div 3$
**CC6.** Subtract: $10 - 5$  **1.** 2.99  **3.** 23.78  **5.** 7.08  **7.** 1.2
**9.** $-0.9$  **11.** $-6000$  **13.** 140  **15.** 40  **17.** $-0.15$
**19.** 3.2  **21.** $-3.9$  **23.** 0.625  **25.** 0.26  **27.** 2.34
**29.** $-0.3045$  **31.** 2.134567  **33.** $-2.359$  **35.** 1023.7
**37.** $-9236$  **39.** 0.08172  **41.** 9.7  **43.** $-0.0527$
**45.** $-75,300$  **47.** $-0.0753$  **49.** 2107  **51.** $-302.997$
**53.** $-178.1$  **55.** 206.0176  **57.** 0.5  **59.** $-400.0108$
**61.** 0.6725  **63.** 5.383  **65.** 10.5  **67.** 13,748.5 ft
**69.** 3.5 million tent campers  **71.** 31.382 mi
**73.** $15\frac{1}{8}$  **74.** $5\frac{7}{8}$  **75.** 343  **76.** 64  **77.** 47  **78.** 41
**79.** $-56.6916$  **81.** 6.254194585  **83.** 1000  **85.** 100
**87.** 43.1 points  **89.** 450 kWh

## Mid-Chapter Review: Chapter 5, p. 339

**1.** False  **2.** True  **3.** True  **4.** $P(1 + r) = 5000(1 + 0.045)$
$= 5000(1.045)$
$= 5225$

**5.** $5.6 + 4.3 \times (6.5 - 0.25)^2 = 5.6 + 4.3 \times (6.25)^2$
$= 5.6 + 4.3 \times 39.0625$
$= 5.6 + 167.96875$
$= 173.56875$
**6.** Thirty-two and thirty-eight thousandths   **7.** 2,784,000,000
**8.** $\frac{453}{100}$, $4\frac{53}{100}$   **9.** $\frac{287}{1000}$   **10.** 0.13   **11.** $-5.09$   **12.** 0.7
**13.** 6.39   **14.** $-35.67$   **15.** 8.002   **16.** 28.462   **17.** 28.46
**18.** 28.5   **19.** 28   **20.** 50.095   **21.** 1214.862   **22.** $-10.23$
**23.** 18.24   **24.** 272.19   **25.** 5.593   **26.** 15.55   **27.** $-19.9$
**28.** 4.14   **29.** 92.871   **30.** 8123.6   **31.** $-0.0483$   **32.** 5.06
**33.** 3.2   **34.** 763   **35.** 0.914036   **36.** 2045¢   **37.** \$1.47
**38.** $-1.22x - 7.1$   **39.** 4.2   **40.** 59.774   **41.** 33.33
**42.** The student probably rounded over successively from the thousandths place as follows: $236.448 \approx 236.45 \approx 236.5 \approx 237$. The student should have considered only the tenths place and rounded down. **43.** The decimal points were not lined up before the subtraction was carried out.   **44.** $10 \div 0.2 = \frac{10}{0.2} = \frac{10}{0.2} \cdot \frac{10}{10} = \frac{100}{2} = 100 \div 2$.   **45.** $0.247 \div 0.1 = \frac{247}{1000} \div \frac{1}{10} = \frac{247}{1000} \cdot \frac{10}{1} = \frac{247 \cdot 10}{10 \cdot 100} = \frac{247}{100} = 2.47 \neq 0.0247$; $0.247 \div 10 = \frac{247}{1000} \div 10 = \frac{247}{1000} \cdot \frac{1}{10} = \frac{247}{10,000} = 0.0247 \neq 2.47$

## Calculator Corner, p. 343
**1.** $-0.1\overline{6}$   **2.** $0.\overline{63}$   **3.** $6.\overline{3}$   **4.** $-57.\overline{1}$

## Calculator Corner, p. 344
**1.** 123.150432   **2.** 52.59026102

## Exercise Set 5.5, p. 346
**RC1.** Repeating   **RC2.** Terminating   **RC3.** Terminating
**RC4.** Repeating   **RC5.** Repeating   **RC6.** Terminating
**CC1.** (c)   **CC2.** (a)   **CC3.** (d)   **CC4.** (b)   **1.** 0.375
**3.** $-0.5$   **5.** 0.12   **7.** 0.225   **9.** 0.52   **11.** $-0.85$
**13.** $-0.5625$   **15.** 1.4   **17.** 1.12   **19.** $-1.375$   **21.** $-0.975$
**23.** 0.605   **25.** $0.5\overline{3}$   **27.** $0.\overline{3}$   **29.** $-1.\overline{3}$   **31.** $1.1\overline{6}$
**33.** $-1.\overline{27}$   **35.** $-0.41\overline{6}$   **37.** 0.254   **39.** $0.\overline{12}$   **41.** $-0.2\overline{18}$
**43.** $0.\overline{571428}$   **45.** 0.2; 0.18; 0.182   **47.** 0.3; 0.28; 0.278
**49.** 0.4; 0.36; 0.364   **51.** $-1.7$; $-1.67$; $-1.667$
**53.** $-0.5$; $-0.47$; $-0.471$   **55.** 0.6; 0.58; 0.583
**57.** $-0.2$; $-0.19$; $-0.193$   **59.** $-0.8$; $-0.78$; $-0.778$
**61.** 15.8 mpg   **63.** 17.8 mpg   **65.** **(a)** 0.571; **(b)** 1.333; **(c)** 0.429;
**(d)** 0.75   **67.** 11.06   **69.** 2.736   **71.** $-417.51\overline{6}$   **73.** 0
**75.** 0.09705   **77.** $-1.5275$   **79.** 24.375   **81.** 1.08 m$^2$
**83.** 5.78 cm$^2$   **85.** 790.92 in$^2$   **87.** $1\frac{1}{24}$ cups   **88.** $1\frac{33}{100}$ in.
**89.** $0.\overline{142857}$, $0.\overline{285714}$, $0.\overline{428571}$, $0.\overline{571428}$, $0.\overline{714285}$; $0.\overline{857142}$
**91.** 13.86 cm$^2$   **93.** 1.76625 ft$^2$ or 1.767145868 ft$^2$

## Exercise Set 5.6, p. 352
**RC1.** (b)   **RC2.** (d)   **RC3.** (c)   **RC4.** (e)   **RC5.** (a)
**RC6.** (f)
**1.** (c)   **3.** (a)   **5.** (c)   **7.** 1.6   **9.** 6   **11.** 60   **13.** 2.3
**15.** 180   **17.** (a)   **19.** (c)   **21.** (b)   **23.** (b)
**25.** $1800 \div 9 = 200$ posts; answers may vary
**27.** $\$2 \cdot 35 = \$70$; answers may vary   **29.** 2   **30.** 165
**31.** 210   **32.** 69   **33.** 10   **34.** 530   **35.** No   **37. (a)** $+, \times$;
**(b)** $+, \times, -$

## Calculator Corner, p. 356
Left to the student

## Exercise Set 5.7, p. 358
**RC1.** False   **RC2.** True   **RC3.** True   **RC4.** True
**CC1.** Yes   **CC2.** No   **CC3.** No   **CC4.** Yes

**1.** 5.4   **3.** 0.2   **5.** $-6.24$   **7.** $-12.6$   **9.** $-2.55$   **11.** 1.97
**13.** 8.38   **15.** $-5.9$   **17.** 6   **19.** 1.8   **21.** $-3.7$   **23.** $-4.7$
**25.** 1.7   **27.** 2.94   **29.** 4.5   **31.** 9   **33.** 3.2   **35.** $-1.75$
**37.** 30   **39.** 3.2   **41.** 9   **43.** 2.1   **45.** 4.5   **47.** 4.12
**49.** 5.6   **51.** $-1.9$   **53.** 13   **55.** 0   **57.** $-1.5$   **59.** 14 m$^2$
**60.** 27 cm$^2$   **61.** $\frac{25}{2}$ in$^2$   **62.** 24 ft$^2$   **63.** 5 ft$^2$   **64.** 12 m$^2$
**65.** $-\frac{29}{50}$   **66.** 0   **67.** $-2$   **68.** 8   **69.** 3.1   **71.** 36
**73.** 1.1212963

## Translating for Success, p. 369
**1.** I   **2.** C   **3.** N   **4.** A   **5.** G   **6.** B   **7.** D   **8.** O
**9.** F   **10.** M

## Exercise Set 5.8, p. 370
**RC1.** Familiarize   **RC2.** Translate   **RC3.** Solve
**RC4.** Check   **RC5.** State   **CC1.** (d)   **CC2.** (a)   **CC3.** (b)
**CC4.** (c)   **1.** Let $a = $ Ron's age: $a + 5$, or $5 + a$   **3.** $b + 6$, or $6 + b$   **5.** $c - 9$   **7.** Let $n = $ the number; $n - 16$
**9.** Let $s = $ Nate's speed; $8s$   **11.** $\frac{x}{17}$
**13.** Let $x = $ the number; $\frac{1}{2}x + 20$, or $20 + \frac{1}{2}x$
**15.** Let $x = $ the number; $4x - 20$   **17.** Let $l = $ the length and $w = $ the width; $l + w$, or $w + l$   **19.** Let $r = $ the rate and $t = $ the time; $rt + 10$, or $10 + rt$   **21.** Let $n = $ the number; $10n + n$, or $n + 10n$   **23.** Let $x$ and $y = $ the numbers; $5(x - y)$
**25.** 41.195 million drivers   **27.** \$66.28   **29.** \$0.51
**31.** \$533,333,333.33   **33.** 22,691.5 mi   **35.** 20.2 mpg
**37.** \$220.01 million   **39.** \$24.33   **41.** 2.66 cc   **43.** \$44,811
**45.** 10.8¢   **47.** Area: 268.96 sq ft; perimeter: 65.6 ft
**49.** 193.04 cm$^2$   **51.** 2.31 cm   **53.** 331.74 ft$^2$
**55.** 296,183.55 sq ft   **57.** 1.9 lb   **59.** 15.294 million
**61.** 15 levels   **63.** 17 bottles   **65.** 2.5 million protected acres, 2.9 million unprotected acres   **67.** \$65.30 for food, \$195.90 for lodging   **69.** \$906.50   **71.** \$316,987.20; \$196,987.20
**73.** 0   **74.** Undefined   **75.** 1   **76.** $-\frac{1}{10}$   **77.** $-\frac{20}{33}$
**78.** $6\frac{5}{6}$   **79.** 3999.76 cm$^2$   **81.** \$0.19   **83.** 25 cm$^2$. We assume that the figures are nested squares formed by connecting the midpoints of consecutive sides of the next larger square.

## Summary and Review: Chapter 5, p. 377
### Vocabulary Reinforcement
**1.** repeating   **2.** terminating   **3.** billion   **4.** million
**5.** trillion   **6.** rational numbers

### Concept Reinforcement
**1.** True   **2.** False   **3.** True   **4.** False   **5.** False

### Study Guide
**1.** $\frac{5093}{100}$   **2.** 81.7   **3.** 42.159   **4.** 153.35   **5.** 38.611
**6.** 207.848   **7.** 19.11   **8.** 0.176   **9.** 60,437   **10.** 7.4
**11.** 0.047   **12.** 15,690

### Review Exercises
**1.** 6,590,000   **2.** 3,100,000,000   **3.** Three and forty-seven hundredths   **4.** Thirty-one thousandths   **5.** Twenty-seven and one ten-thousandth   **6.** Nine tenths   **7.** $\frac{9}{100}$
**8.** $-\frac{4561}{1000}$, $-4\frac{561}{1000}$   **9.** $-\frac{89}{1000}$   **10.** $\frac{30,227}{10,000}$, $3\frac{227}{10,000}$   **11.** 0.034
**12.** 4.2603   **13.** 27.91   **14.** $-867.006$   **15.** 0.034   **16.** $-0.19$
**17.** 0.741   **18.** 1.041   **19.** 17.4   **20.** 17.43   **21.** 17.429
**22.** 17   **23.** 499.829   **24.** 29.148   **25.** 229.1   **26.** 685.0519
**27.** $-57.3$   **28.** 2.37   **29.** 12.96   **30.** $-1.073$   **31.** 24,680
**32.** 3.2   **33.** $-1.6$   **34.** 0.2763   **35.** 0.003056   **36.** $-4380$
**37.** $2.2x - 9.1y$   **38.** $-2.84a + 12.57$   **39.** 925   **40.** 40.84

**41.** $15.49  **42.** 248.27  **43.** 2.6  **44.** 1.28  **45.** 3.25
**46.** $-1.1\overline{6}$  **47.** 21.08  **48.** $-3.2$  **49.** $-3$  **50.** $-7.5$
**51.** 6.5  **52.** 9.58 sec  **53.** $37.06  **54.** 249.76 ft$^2$
**55.** $784.47  **56.** 95 transactions  **57.** 195.7 lb of newspaper, 65.7 lb of glass  **58.** 14.5 mpg  **59.** 20.0 million books
**60.** $55.50  **61.** $1.33  **62.** 61.5 ft; 235.625 sq ft
**63.** 14 months  **64.** B  **65.** A  **66.** (a) +; (b) −
**67.** $\frac{-13}{15}, \frac{-17}{20}, -\frac{11}{13}, -\frac{15}{19}, \frac{-5}{7}, -\frac{2}{3}$
**68.** $1 = 3 \cdot \frac{1}{3} = 3(0.33333333\ldots) = 0.99999999\ldots,$ or $0.\overline{9}$

**69.** The rectangular pizza, at $\frac{4.4¢}{\text{in}^2}$, is a better buy than the round pizza, which costs $\frac{5.5¢}{\text{in}^2}$.

## Understanding Through Discussion and Writing

**1.** Count the number of decimal places. Move the decimal point that many places to the right and write the result over a denominator of 1 followed by that many zeros.
**2.** $346.708 \times 0.1 = \frac{346,708}{1000} \times \frac{1}{10} = \frac{346,708}{10,000} = 34.6708 \neq 3467.08$
**3.** When the denominator of a fraction is a multiple of 10, long division is not the fastest way to convert the fraction to decimal notation. Many times this is also the case when the denominator has only 2's or 5's or both as factors.  **4.** Multiply by 1 to get a denominator that is a power of 10:

$$\frac{44}{125} = \frac{44}{125} \cdot \frac{8}{8} = \frac{352}{1000} = 0.352.$$

We can also divide to find that $\frac{44}{125} = 0.352$.

## Test: Chapter 5, p. 382

**1.** [5.3b] 9,800,000,000  **2.** [5.1a] One hundred twenty-three and forty-seven ten-thousandths  **3.** [5.1b] $-\frac{91}{100}$
**4.** [5.1b] $\frac{2769}{1000}, 2\frac{769}{1000}$  **5.** [5.1b] 0.074  **6.** [5.1b] $-3.7047$
**7.** [5.1b] 756.09  **8.** [5.1b] 91.703  **9.** [5.1c] 0.162
**10.** [5.1c] 8.049  **11.** [5.1c] $-0.09$  **12.** [5.1d] 6
**13.** [5.1d] 5.68  **14.** [5.1d] 5.678  **15.** [5.1d] 5.7
**16.** [5.2a] 405.219  **17.** [5.3a] 0.03  **18.** [5.3a] 0.21345
**19.** [5.2b] 44.746  **20.** [5.2a] 356.37  **21.** [5.2c] $-2.2$
**22.** [5.2b] 1.9946  **23.** [5.3a] 73,962  **24.** [5.4a] 4.75
**25.** [5.4a] 30.4  **26.** [5.4a] $-0.34682$  **27.** [5.4a] 34,682
**28.** [5.3b] 17,982¢  **29.** [5.2d] $9.8x - 3.9y - 4.6$
**30.** [5.3c] 11.6  **31.** [5.4b] 7.6  **32.** [5.5c] 1.6  **33.** [5.5c] 5.25
**34.** [5.5a] $-0.4375$  **35.** [5.5a] $1.\overline{5}$  **36.** [5.5b] 1.56
**37.** [5.4b] $-22.25$  **38.** [5.4b] 1.8045  **39.** [5.5d] 9.72
**40.** [5.7a] $-3.24$  **41.** [5.7b] 10  **42.** [5.7b] 1.4
**43.** [5.8b] $46.69  **44.** [5.8b] 28.3 mpg  **45.** [5.8b] $592.45
**46.** [5.8b] $293.93  **47.** [5.4b], [5.8b] 71.4 years  **48.** [5.6a] C
**49.** [5.3a] (a) always; (b) never; (c) sometimes; (d) sometimes
**50.** [5.8b] $35  **51.** [5.8b] (a) Fly; (b) fly; (c) drive

## Cumulative Review: Chapters 1–5, p. 385

**1.** [4.5a] $\frac{20}{9}$  **2.** [5.1b] $\frac{3051}{1000}$  **3.** [5.5a] $-1.4$  **4.** [5.5a] $0.5\overline{4}$
**5.** [3.2b] Prime  **6.** [3.1b] No  **7.** [1.9c] 1754
**8.** [5.4b] 4.364  **9.** [5.1d] 584.97  **10.** [5.5b] 218.56
**11.** [5.6a] 160  **12.** [5.6a] 4  **13.** [2.6a] 12  **14.** [2.7a] $7p - 8$
**15.** [4.6a] $6\frac{1}{20}$  **16.** [1.2a] 139,116  **17.** [4.2b] $\frac{31}{18}$
**18.** [5.2a] 145.953  **19.** [1.3a] 710,137  **20.** [5.2b] $-13.097$
**21.** [4.6b] $\frac{5}{7}$  **22.** [4.3a] $-\frac{1}{110}$  **23.** [3.6a] $-\frac{1}{6}$
**24.** [1.4a] 5,317,200  **25.** [5.3a] 4.78  **26.** [5.3a] 0.0279431
**27.** [5.4a] 2.122  **28.** [1.5a] 1843  **29.** [5.4a] 13,862.1
**30.** [3.7b] $\frac{5}{6}$  **31.** [5.7b] $-11.5$  **32.** [2.8c] $-28$
**33.** [5.7a] 3.8125  **34.** [1.7b] 367,251  **35.** [4.3b] $\frac{1}{18}$
**36.** [3.8a] $-\frac{1}{2}$  **37.** [1.8a] 732 million  **38.** [1.8a] 885 million
**39.** [3.8b] $1500  **40.** [3.6b] $2400  **41.** [5.8b] $258.77
**42.** [4.6c] $6\frac{1}{2}$ lb  **43.** [3.6b] 88 ft$^2$  **44.** [5.8b] 43.585 in$^2$
**45.** [4.8a] $\frac{9}{32}$  **46.** [5.4b] 527.04  **47.** [5.8b] $2.39
**48.** [4.7d] 144 packages

# CHAPTER 6

## Exercise Set 6.1, p. 392
**RC1.** True  **RC2.** True  **RC3.** False  **RC4.** False
**CC1.** (a), (c), (d)  **CC2.** (b), (c), (e)

**1.** $\frac{4}{5}$  **3.** $\frac{178}{572}$  **5.** $\frac{0.4}{12}$  **7.** $\frac{3.8}{7.4}$  **9.** $\frac{56.78}{98.35}$  **11.** $\frac{8\frac{3}{4}}{9\frac{5}{6}}$  **13.** $\frac{653}{204}, \frac{204}{653}$
**15.** $\frac{35}{59}$  **17.** $\frac{17}{42}$  **19.** $\frac{113}{365}$  **21.** $\frac{60}{100}; \frac{100}{60}$  **23.** $\frac{2}{3}$  **25.** $\frac{3}{4}$
**27.** $\frac{12}{25}$  **29.** $\frac{7}{9}$  **31.** $\frac{2}{3}$  **33.** $\frac{14}{25}$  **35.** $\frac{1}{2}$  **37.** $\frac{3}{4}$  **39.** $\frac{478}{213}; \frac{213}{478}$
**41.** 408,550  **42.** 27.006  **43.** 397.27  **44.** $\frac{19}{30}$  **45.** $\frac{23}{24}$
**46.** $14\frac{1}{3}$  **47.** 942,219  **48.** 2.6  **49.** 94.375  **50.** $\frac{1}{16}$
**51.** $1\frac{13}{15}$  **52.** $4\frac{2}{3}$  **53.** $\frac{30}{47}$  **55.** $1:2:3$

## Exercise Set 6.2, p. 398
**RC1.** True  **RC2.** False  **RC3.** True  **RC4.** False
**CC1.** (b)  **CC2.** (c)  **CC3.** (d)  **CC4.** (a)
**1.** 40 km/h  **3.** 7.48 mi/sec  **5.** 51,332 people/sq mi
**7.** 17 mpg  **9.** 33 mpg  **11.** 25 mph; 0.04 hr/mi
**13.** About 393 performances/year  **15.** 186,000 mi/sec
**17.** 124 km/h  **19.** 0.623 gal/ft$^2$  **21.** 25 beats/min
**23.** 26.188¢/oz; 26.450¢/oz; 16 oz  **25.** 13.111¢/oz; 11.453¢/oz;
75 oz  **27.** 18.222¢/oz; 17.464¢/oz; 28 oz  **29.** 10.187¢/oz;
10.346¢/oz; 10.7 oz  **31.** B: 18.719¢/oz; E: 14.563¢/oz; Brand E
**33.** A: 10.375¢/oz; B: 9.139¢/oz; H: 8.022¢/oz; Brand H
**35.** 11,550  **36.** 679.4928  **37.** 2.74568  **38.** $\frac{20}{3}$  **39.** $8\frac{11}{20}$
**40.** $13\frac{2}{3}$  **41.** 125  **42.** 9.5  **43.** 9.63  **44.** $\frac{125}{27}$  **45.** 150
**46.** $19\frac{3}{7}$  **47.** 6-oz: 10.833¢/oz; 5.5 oz: 10.909¢/oz

## Exercise Set 6.3, p. 406
**RC1.** ratio  **RC2.** proportional  **RC3.** proportion
**RC4.** cross products  **CC1.** 5, 10  **CC2.** 3, 10  **CC3.** $\frac{10}{15}$

**CC4.** $\frac{3}{5} = \frac{0.9}{1.5}$  **CC5.** $\frac{\frac{1}{3}}{\frac{2}{9}} = \frac{3}{2}$

**1.** No  **3.** Yes  **5.** Yes  **7.** No  **9.** 45  **11.** 12
**13.** 10  **15.** 20  **17.** 5  **19.** 18  **21.** 22  **23.** 28
**25.** $\frac{28}{3}$, or $9\frac{1}{3}$  **27.** $\frac{26}{9}$, or $2\frac{8}{9}$  **29.** 5  **31.** 5  **33.** 0
**35.** 14  **37.** 2.7  **39.** 1.8  **41.** 0.06  **43.** 0.7  **45.** 12.5725
**47.** 1  **49.** $\frac{1}{20}$  **51.** $\frac{3}{8}$  **53.** $\frac{16}{75}$  **55.** $\frac{51}{16}$, or $3\frac{3}{16}$  **57.** $\frac{546}{185}$, or $2\frac{176}{185}$
**59.** 16,052 golf courses  **60.** 6 lb  **61.** $\frac{1}{4}$ lb  **62.** $\frac{3}{8}$ lb
**63.** $6\frac{7}{20}$ cm  **64.** $37\frac{1}{2}$ servings  **65.** $59.81  **66.** 21.5 mpg
**67.** Approximately 2731.4  **69.** Ruth: 1.863 strikeouts per home run; Schmidt: 3.436 strikeouts per home run

## Mid-Chapter Review: Chapter 6, p. 409
**1.** True  **2.** True  **3.** False  **4.** False
**5.** $\frac{120\,\text{mi}}{2\,\text{hr}} = \frac{120}{2}\frac{\text{mi}}{\text{hr}} = 60$ mi/hr
**6.**  $\frac{x}{4} = \frac{3}{6}$
$x \cdot 6 = 4 \cdot 3$
$\frac{x \cdot 6}{6} = \frac{4 \cdot 3}{6}$
$x = 2$

**7.** $\frac{4}{7}$  **8.** $\frac{313}{199}$  **9.** $\frac{35}{17}$  **10.** $\frac{59}{101}$  **11.** $\frac{2}{3}$  **12.** $\frac{1}{3}$  **13.** $\frac{8}{7}$  **14.** $\frac{25}{19}$
**15.** $\frac{2}{1}$  **16.** $\frac{5}{1}$  **17.** $\frac{2}{7}$  **18.** $\frac{3}{5}$  **19.** 60.75 mi/hr, or 60.75 mph
**20.** 48.67 km/h  **21.** 13 m/sec  **22.** 16.17 ft/sec
**23.** 27 in./day  **24.** About 0.909 free throw made/attempt
**25.** 11.611¢/oz  **26.** 49.917¢/oz  **27.** Yes  **28.** No
**29.** No  **30.** Yes  **31.** 12  **32.** 40  **33.** 9  **34.** 35
**35.** 2.2  **36.** 4.32  **37.** $\frac{1}{2}$  **38.** $\frac{65}{4}$, or $16\frac{1}{4}$  **39.** Yes; every
ratio $\frac{a}{b}$ can be written as $\frac{\frac{a}{b}}{1}$.  **40.** By making some sketches, we
see that the rectangle's length must be twice the width.

**41.** The student's approach will work. However, when we use the approach of equating cross products, we eliminate the need to find the least common denominator.　**42.** The instructor thinks that the longer a student studies, the higher his or her grade will be. An example is the situation in which one student gets a test grade of 96 after studying for 8 hr while another student gets a score of 78 after studying for $6\frac{1}{2}$ hr. This is represented by the proportion $\frac{96}{8} = \frac{78}{6\frac{1}{2}}$.

## Translating for Success, p. 416

**1.** N　**2.** I　**3.** A　**4.** K　**5.** J　**6.** F　**7.** M　**8.** B
**9.** G　**10.** E

## Exercise Set 6.4, p. 417

**RC1.** $\frac{m}{8}$　**RC2.** $\frac{8}{m}$　**RC3.** $\frac{6}{8}$　**RC4.** $\frac{8}{6}$
**1.** 11.04 hr　**3.** 7680 frames　**5. (a)** About 112 gal;
**(b)** 3360 mi　**7.** 254.89 million, or 254,890,000　**9.** 175 bulbs
**11.** About 262 mg every 8 hr　**13.** 2975 ft²　**15.** 450 pages
**17.** 13,500 mi　**19.** 880 calories　**21.** 60 students
**23.** 64 cans　**25.** 9.75 gal　**27.** 100 oz　**29.** 90 whales
**31.** 58.1 mi　**33. (a)** 42.25 euros; **(b)** \$10,224.85
**35. (a)** 25 games; **(b)** about 23 goals　**37.** 162　**38.** 4014
**39.** 4.45　**40.** 0.4　**41.** 14.3　**42.** 0.26　**43.** $\frac{1}{10}$　**44.** $\frac{5}{4}$
**45.** 17 positions　**47.** 2150 earned runs　**49.** CD player: \$150;
receiver: \$450; speakers: \$300

## Exercise Set 6.5, p. 426

**RC1.** 5　**RC2.** 8　**RC3.** $a$　**RC4.** 3
**1.** 25　**3.** $\frac{4}{3}$, or $1\frac{1}{3}$　**5.** $x = \frac{27}{4}$, or $6\frac{3}{4}$; $y = 9$
**7.** $x = 7.5$; $y = 7.2$　**9.** 1.25 m　**11.** 36 ft　**13.** 7 ft
**15.** 100 ft　**17.** 4　**19.** $10\frac{1}{2}$　**21.** $x = 6$; $y = 5.25$; $z = 3$
**23.** $x = 5\frac{1}{3}$, or $5.\overline{3}$; $y = 4\frac{2}{3}$, or $4.\overline{6}$; $z = 5\frac{1}{3}$, or $5.\overline{3}$　**25.** 20 ft
**27.** 152 ft　**29.** Prime　**30.** Composite　**31.** $2 \cdot 2 \cdot 2 \cdot 101$, or
$2^3 \cdot 101$　**32.** $3 \cdot 31$　**33.** =　**34.** ≠　**35.** <　**36.** <
**37.** $\frac{13}{4}$, or $3\frac{1}{4}$　**38.** 26.29　**39.** 127　**40.** 75　**41.** 13.75 ft
**43.** 1.25 cm　**45.** 3681.437　**47.** $x \approx 0.35$; $y = 0.4$

## Summary and Review: Chapter 6, p. 430

### Vocabulary Reinforcement
**1.** quotient　**2.** shape　**3.** cross products　**4.** rate
**5.** proportion　**6.** price

### Concept Reinforcement
**1.** True　**2.** True　**3.** False　**4.** True

### Study Guide
**1.** $\frac{17}{3}$　**2.** $\frac{8}{7}$　**3.** \$7.50/hr　**4.** A: 9.964¢/oz; B: 10.281¢/oz;
Brand A　**5.** Yes　**6.** $\frac{27}{8}$　**7.** 175 mi　**8.** 21

### Review Exercises
**1.** $\frac{47}{84}$　**2.** $\frac{46}{1.27}$　**3.** $\frac{83}{100}$　**4.** $\frac{0.72}{197}$　**5. (a)** $\frac{12,480}{16,640}$, or $\frac{3}{4}$; **(b)** $\frac{16,640}{29,120}$,
or $\frac{4}{7}$　**6.** $\frac{3}{4}$　**7.** $\frac{9}{16}$　**8.** 36 mpg　**9.** 6300 revolutions/min
**10.** 0.638 gal/ft²　**11.** 6.33¢/tablet　**12.** 19.542¢/oz
**13.** 14.969¢/oz; 12.479¢/oz; 15.609¢/oz; 48 oz　**14.** Yes
**15.** No　**16.** 32　**17.** 7　**18.** $\frac{1}{40}$　**19.** 24　**20.** 27 defective
circuits　**21. (a)** 311.50 Canadian dollars; **(b)** 40.13 U.S.
dollars　**22.** 832 mi　**23.** About 394 movies　**24.** About
37,565,761 lb　**25.** About \$69.83　**26.** About 4308 lawyers

**27.** $x = \frac{14}{3}$, or $4\frac{2}{3}$　**28.** $x = \frac{56}{5}$, or $11\frac{1}{5}$; $y = \frac{63}{5}$, or $12\frac{3}{5}$
**29.** 40 ft　**30.** $x = 3$; $y = \frac{21}{2}$, or $10\frac{1}{2}$; $z = \frac{15}{2}$, or $7\frac{1}{2}$　**31.** B
**32.** C　**33.** 1.806¢/sheet; 1.658¢/sheet; 1.409¢/sheet; 6 big rolls
**34.** $x = 4258.5$; $z \approx 10,094.3$　**35.** Finishing paint: 11 gal;
primer: 16.5 gal

## Understanding Through Discussion and Writing
**1.** In terms of cost, a low faculty-to-student ratio is less expensive than a high faculty-to-student ratio. In terms of quality of education and student satisfaction, a high faculty-to-student ratio is more desirable. A college president must balance the cost and quality issues.　**2.** Yes; unit prices can be used to solve proportions involving money. In Example 3 of Section 6.4, for instance, we could have divided \$90 by the unit price, or the price per ticket, to find the number of tickets that could be purchased for \$90.　**3.** Leslie used 4 gal of gasoline to drive 92 mi. At the same rate, how many gallons would be needed to travel 368 mi?　**4.** Yes; consider the following pair of triangles.

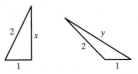

Two pairs of sides are proportional, but we can see that $x$ is shorter than $y$, so the ratio of $x$ to $y$ is clearly not the same as the ratio of 1 to 1 (or 2 to 2).

## Test: Chapter 6, p. 435
**1.** [6.1a] $\frac{85}{97}$　**2.** [6.1a] $\frac{0.34}{124}$　**3.** [6.1b] $\frac{9}{10}$　**4.** [6.1b] $\frac{25}{32}$
**5.** [6.2a] 0.625 ft/sec　**6.** [6.2a] $1\frac{1}{3}$ servings/lb
**7.** [6.2a] 32 mpg　**8.** [6.2b] About 15.563¢/oz
**9.** [6.2b] 16.475¢/oz; 13.980¢/oz; 11.490¢/oz; 16.660¢/oz; 100 oz
**10.** [6.3a] Yes　**11.** [6.3a] No　**12.** [6.3b] 12
**13.** [6.3b] 360　**14.** [6.3b] 42.1875　**15.** [6.3b] 100
**16.** [6.4a] 1512 km　**17.** [6.4a] 4.8 min　**18.** [6.4a] 525 mi
**19.** [6.5a] 66 m　**20.** [6.4a] **(a)** 13 hats; **(b)** 30 packages
**21.** [6.4a] 320 calories　**22.** [6.5a] $x = 8$; $y = 8.8$
**23.** [6.5b] $x = \frac{24}{5}$, or 4.8; $y = \frac{32}{5}$, or 6.4; $z = 12$
**24.** [6.2a] C　**25.** [6.3b] $\frac{4}{3}$, or $1.\overline{3}$
**26.** [6.4a] 5888 marbles

## Cumulative Review: Chapters 1–6, p. 437
**1.** [5.2a] 513.996　**2.** [4.6a] $6\frac{3}{4}$　**3.** [4.2b] $\frac{7}{20}$
**4.** [5.2b] 30.491　**5.** [2.3a] −17　**6.** [4.3a] $-\frac{7}{60}$
**7.** [5.3a] 222.076　**8.** [2.4a] −645　**9.** [4.7a] 3
**10.** [5.4a] 43　**11.** [2.5a] 51　**12.** [3.7b] $\frac{3}{2}$
**13.** [1.1b] 3 ten thousands + 0 thousands + 0 hundreds +
7 tens + 4 ones, or 3 ten thousands + 7 tens + 4 ones
**14.** [5.1a] One hundred twenty and seven hundredths
**15.** [5.1c] 0.7　**16.** [5.1c] −0.799
**17.** [3.2c] $2 \cdot 2 \cdot 2 \cdot 2 \cdot 3 \cdot 3$, or $2^4 \cdot 3^2$　**18.** [4.1a] 90
**19.** [3.3a] $\frac{5}{8}$　**20.** [3.5b] $\frac{5}{8}$　**21.** [5.5d] 5.718
**22.** [5.4b] −25.56　**23.** [2.6a] 5　**24.** [6.1a] $\frac{0.3}{15}$
**25.** [6.3a] Yes　**26.** [6.2a] 55 m/sec　**27.** [6.2b] 8-oz can
**28.** [6.3b] 30.24　**29.** [5.7a] −26.4375　**30.** [3.8a] $\frac{8}{9}$
**31.** [2.8e] −4　**32.** [5.7a] 33.34　**33.** [5.7b] 3
**34.** [3.6b] 390 calories　**35.** [6.4a] 7 min　**36.** [4.6c] $2\frac{1}{4}$ cups
**37.** [3.8b] 12 doors　**38.** [5.8b], [6.2a] 42.2025 mi
**39.** [5.8b] 132 orbits　**40.** [3.2b] D　**41.** [6.2a] B
**42.** [6.5a] $10\frac{1}{2}$ ft

# CHAPTER 7

## Calculator Corner, p. 442
**1.** 0.14 **2.** 0.00069 **3.** 0.438 **4.** 1.25

## Calculator Corner, p. 445
**1.** 52% **2.** 38.46% **3.** 110.26% **4.** 171.43% **5.** 59.62%
**6.** 28.31%

## Exercise Set 7.1, p. 447
**RC1.** right **RC2.** left **CC1.** 43% **CC2.** 86%
**CC3.** 19% **CC4.** 50%
**1.** $\frac{90}{100}$; $90 \times \frac{1}{100}$; $90 \times 0.01$ **3.** $\frac{12.5}{100}$; $12.5 \times \frac{1}{100}$; $12.5 \times 0.01$
**5.** 0.67 **7.** 0.456 **9.** 0.5901 **11.** 0.1 **13.** 0.01 **15.** 2
**17.** 0.001 **19.** 0.0009 **21.** 0.0018 **23.** 0.2319 **25.** 0.565
**27.** 0.14875 **29.** 0.47 **31.** 0.13 **33.** 0.17; 0.067; 0.062; 0.055
**35.** 47% **37.** 3% **39.** 870% **41.** 33.4% **43.** 1200%
**45.** 40% **47.** 0.6% **49.** 1.7% **51.** 27.18% **53.** 2.39%
**55.** 76% **57.** 31.9%; 19.0% **59.** 62%; 49%; 38%
**61.** 41% **63.** 5% **65.** 20% **67.** 30% **69.** 50%
**71.** 87.5%, or $87\frac{1}{2}$% **73.** 80% **75.** $66.\overline{6}$%, or $66\frac{2}{3}$%
**77.** $16.\overline{6}$%, or $16\frac{2}{3}$% **79.** 18.75%, or $18\frac{3}{4}$%
**81.** 81.25%, or $81\frac{1}{4}$% **83.** 16% **85.** 5% **87.** 34%
**89.** Beef: 48%; chicken: 15% **91.** 2% **93.** 24% **95.** 5%
**97.** $\frac{17}{20}$ **99.** $\frac{5}{8}$ **101.** $\frac{1}{3}$ **103.** $\frac{1}{6}$ **105.** $\frac{29}{400}$ **107.** $\frac{1}{125}$
**109.** $\frac{203}{800}$ **111.** $\frac{176}{225}$ **113.** $\frac{711}{1100}$ **115.** $\frac{3}{2}$ **117.** $\frac{13}{40,000}$ **119.** $\frac{1}{3}$
**121.** $\frac{6}{25}$ **123.** $\frac{9}{100}$ **125.** $\frac{3}{20}$ **127.** $\frac{3}{50}$ **129.** $\frac{37}{500}$
**131.**

| Fraction Notation | Decimal Notation | Percent Notation |
|---|---|---|
| $\frac{1}{8}$ | 0.125 | 12.5%, or $12\frac{1}{2}$% |
| $\frac{1}{6}$ | $0.1\overline{6}$ | $16.\overline{6}$%, or $16\frac{2}{3}$% |
| $\frac{1}{5}$ | 0.2 | 20% |
| $\frac{1}{4}$ | 0.25 | 25% |
| $\frac{1}{3}$ | $0.\overline{3}$ | $33.\overline{3}$%, or $33\frac{1}{3}$% |
| $\frac{3}{8}$ | 0.375 | 37.5%, or $37\frac{1}{2}$% |
| $\frac{2}{5}$ | 0.4 | 40% |
| $\frac{1}{2}$ | 0.5 | 50% |

**133.**

| Fraction Notation | Decimal Notation | Percent Notation |
|---|---|---|
| $\frac{1}{2}$ | 0.5 | 50% |
| $\frac{1}{3}$ | $0.\overline{3}$ | $33.\overline{3}$%, or $33\frac{1}{3}$% |
| $\frac{1}{4}$ | 0.25 | 25% |
| $\frac{1}{6}$ | $0.1\overline{6}$ | $16.\overline{6}$%, or $16\frac{2}{3}$% |
| $\frac{1}{8}$ | 0.125 | 12.5%, or $12\frac{1}{2}$% |
| $\frac{3}{4}$ | 0.75 | 75% |
| $\frac{5}{6}$ | $0.8\overline{3}$ | $83.\overline{3}$%, or $83\frac{1}{3}$% |
| $\frac{3}{8}$ | 0.375 | 37.5%, or $37\frac{1}{2}$% |

**135.** 70 **136.** 5 **137.** 400 **138.** 18.75 **139.** $\frac{185}{8}$, or 23.125
**140.** $\frac{51}{2}$, or 25.5 **141.** $\frac{9}{2}$, or 4.5 **142.** $\frac{35}{4}$, or 8.75 **143.** 20%
**145.** $257.\overline{46317}$% **147.** $1.04\overline{142857}$

## Calculator Corner, p. 456
**1.** $5.04 **2.** 0.0112 **3.** 450 **4.** $1000 **5.** 2.5% **6.** 12%

## Exercise Set 7.2, p. 457
**RC1.** (c) **RC2.** (e) **RC3.** (f) **RC4.** (d) **RC5.** (b)
**RC6.** (a) **CC1.** (e) **CC2.** (c) **CC3.** (f) **CC4.** (b)
**1.** $a = 0.32 \cdot 78$ **3.** $89 = p \cdot 99$ **5.** $13 = 0.25 \cdot b$ **7.** 234.6
**9.** 45 **11.** $18 **13.** 1.9 **15.** 78% **17.** 200% **19.** 50%
**21.** 125% **23.** 40 **25.** $40 **27.** 88 **29.** 20 **31.** 6.25
**33.** $846.60 **35.** 1216 **37.** $\frac{9375}{10,000}$, or $\frac{15}{16}$ **38.** $\frac{125}{1000}$, or $\frac{1}{8}$
**39.** 0.3 **40.** 0.017 **41.** 7 **42.** 8 **43.** $10,000 (can vary);
$10,400 **45.** 108 to 135 tons **47.** $1875

## Exercise Set 7.3, p. 463
**RC1.** (b) **RC2.** (e) **RC3.** (c) **RC4.** (f) **RC5.** (d)
**RC6.** (a) **CC1.** (b) **CC2.** (d)
**1.** $\frac{37}{100} = \frac{a}{74}$ **3.** $\frac{N}{100} = \frac{4.3}{5.9}$ **5.** $\frac{25}{100} = \frac{14}{b}$ **7.** 68.4 **9.** 462
**11.** 457.6 **13.** 2.88 **15.** 25% **17.** 102% **19.** 25%
**21.** 93.75%, or $93\frac{3}{4}$% **23.** $72 **25.** 90 **27.** 88 **29.** 20
**31.** 25 **33.** $780.20 **35.** 200 **37.** 8 **38.** 4000 **39.** 15
**40.** 2074 **41.** $\frac{43}{48}$ qt **42.** $\frac{1}{8}$ T **43.** $1170 (can vary); $1118.64
**45.** 20% **47.** 39%

## Mid-Chapter Review: Chapter 7, p. 465
**1.** True **2.** False **3.** True **4.** $\frac{1}{2}$% $= \frac{1}{2} \cdot \frac{1}{100} = \frac{1}{200}$
**5.** $\frac{80}{1000} = \frac{8}{100} = 8$% **6.** 5.5% $= \frac{5.5}{100} = \frac{55}{1000} = \frac{11}{200}$
**7.** $0.375 = \frac{375}{1000} = \frac{37.5}{100} = 37.5$%
**8.** $\quad 15 = p \times 80$
$\quad\quad \frac{15}{80} = \frac{p \times 80}{80}$
$\quad\quad \frac{15}{80} = p$
$\quad 0.1875 = p$
$\quad 18.75\% = p$
**9.** 0.28 **10.** 0.0015
**11.** 0.05375 **12.** 2.4 **13.** 71% **14.** 9% **15.** 38.91%
**16.** 18.75%, or $18\frac{3}{4}$% **17.** 0.5% **18.** 74% **19.** 600%
**20.** $83.\overline{3}$%, or $83\frac{1}{3}$% **21.** $\frac{17}{20}$ **22.** $\frac{3}{6250}$ **23.** $\frac{91}{400}$ **24.** $\frac{1}{6}$
**25.** 62.5%, or $62\frac{1}{2}$% **26.** 45% **27.** 58 **28.** $16.\overline{6}$%, or $16\frac{2}{3}$%
**29.** 2560 **30.** $50 **31.** 20% **32.** $455
**33.** 0.05%, 0.1%, $\frac{1}{2}$%, 1%, 5%, 10%, $\frac{13}{100}$, 0.275, $\frac{3}{10}$, $\frac{7}{20}$ **34.** D **35.** B
**36.** Some will say that the conversion will be done most accurately by first finding decimal notation. Others will say that it is more efficient to become familiar with some or all of the fraction and percent equivalents that appear inside the back cover and to make the conversion by going directly from fraction notation to percent notation. **37.** Since 40% $\div$ 10 = 4%, we can divide 36.8 by 10, obtaining 3.68. Since 400% $=$ 40% $\times$ 10, we can multiply 36.8 by 10, obtaining 368. **38.** Answers may vary. Some will say this is a good idea since it makes the computations in the solution easier. Others will say it is a poor idea since it adds an extra step to the solution. **39.** They all represent the same number.

## Exercise Set 7.4, p. 470
**RC1.** True **RC2.** True **RC3.** True **RC4.** False
**CC1.** (c) **CC2.** (e) **CC3.** (b) **CC4.** (d) **CC5.** (f)
**CC6.** (a)
**1.** 7831 liver transplants; 3193 heart transplants **3.** $46,656
**5.** 140 items **7.** About 940,000,000 acres **9.** $36,400
**11.** 74.4 items correct; 5.6 items incorrect **13.** $34,194
**15.** About 15.1% **17.** $230.10 **19.** Alcohol: 43.2 mL;
water: 496.8 mL **21.** About 18.9% **23.** Air Force: 24.2%;
Army: 36.4%; Navy: 25.3%; Marines: 14.1% **25.** $2.\overline{27}$ **26.** 0.44
**27.** 3.375 **28.** $4.\overline{7}$ **29.** 0.92 **31.** $83\frac{1}{3}$%, or $83.\overline{3}$%

## Calculator Corner, p. 474
**1.** Left to the student **2.** $80,040

## Translating for Success, p. 477

**1.** J  **2.** M  **3.** N  **4.** E  **5.** G  **6.** H  **7.** O  **8.** C
**9.** D  **10.** B

## Exercise Set 7.5, p. 478

**RC1.** $10, decrease, $\frac{$10}{$50}$, 20%  **RC2.** $15, increase, $\frac{$15}{$60}$, 25%
**RC3.** $120, increase, $\frac{$120}{$360}$, $33\frac{1}{3}$%  **RC4.** $1600, decrease, $\frac{$1600}{$4000}$, 40%
**1.** 5%  **3.** 15%  **5.** About 9.9%  **7.** About 50.9%
**9.** About 28.8%  **11.** 1977 to 1987: 51.7%; 2007 to 2017: 3.9%
**13.** 2008 to 2016: 27.5% decrease; 2016 to 2018: 6.4% increase
**15.** 34.375%, or $34\frac{3}{8}$%  **17.** Change: 44,148 permits; percent
decrease: about 41.9%  **19.** Change: 7874 permits; percent
increase: about 12.6%  **21.** About 17.7%  **23.** 32 cm
**24.** 33 ft  **25.** 36 in.  **26.** 36 m  **27.** About 19%

## Exercise Set 7.6, p. 486

**RC1.** rate  **RC2.** rate  **RC3.** price  **RC4.** tax  **RC5.** tax
**1.** $9.56  **3.** $6.66  **5.** $11.59; $171.39  **7.** 4%  **9.** 3%
**11.** $11,300  **13.** $130.32  **15.** $1423.10  **17.** $451.26
**19.** $2625  **21.** 12%  **23.** $185,000  **25.** $5440  **27.** 15%
**29.** $355  **31.** $30; $270  **33.** $2.55; $14.45  **35.** $125;
$112.50  **37.** 40%; $360  **39.** $300; 40%  **41.** $849; 21.2%
**43.** 18  **44.** $\frac{22}{7}$  **45.** 265.625  **46.** 1.15  **47.** 4,030,000,000,000
**48.** 5,800,000  **49.** 42,700,000  **50.** 6,090,000,000
**51.** $17,700

## Calculator Corner, p. 492

**1.** $16,357.18  **2.** $12,764.72

## Exercise Set 7.7, p. 495

**RC1.** $\frac{6}{12}$ year  **RC2.** $\frac{40}{365}$ year  **RC3.** $\frac{285}{365}$ year  **RC4.** $\frac{9}{12}$ year
**RC5.** $\frac{3}{12}$ year  **RC6.** $\frac{4}{12}$ year  **CC1.** $14.23  **CC2.** $177.90
**1.** $8  **3.** $113.52  **5.** $462.50  **7.** $671.88  **9. (a)** $147.95;
**(b)** $10,147.95  **11. (a)** $84.14; **(b)** $6584.14  **13. (a)** $46.03;
**(b)** $5646.03  **15.** $441.00  **17.** $2207.63  **19.** $7853.38
**21.** $99,427.40  **23.** $4080.40  **25.** $28,225.00  **27.** $9270.87
**29.** $129,871.09  **31.** $4101.01  **33.** $1324.58  **35.** $20,165.05
**37.** Interest: $20.88; amount applied to principal: $4.69; balance
after payment: $1273.87  **39. (a)** $98; **(b)** interest: $86.56;
amount applied to principal: $11.44; **(c)** interest: $51.20; amount
applied to principal: $46.80; **(d)** At 12.6%, the principal is
reduced by $35.36 more than at the 21.3% rate. The interest
at 12.6% is $35.36 less than at 21.3%.  **41.** 800
**42.** $2 \cdot 2 \cdot 3 \cdot 19$  **43.** $\frac{9}{100}$  **44.** $\frac{32}{75}$  **45.** $\frac{1}{25}$  **46.** $\frac{7}{300}$
**47.** 55  **48.** $\frac{1}{2}$  **49.** 9.38%

## Summary and Review: Chapter 7, p. 499

### Vocabulary Reinforcement
**1.** percent decrease  **2.** simple  **3.** discount  **4.** sales
**5.** rate  **6.** percent increase

### Concept Reinforcement
**1.** True  **2.** False  **3.** True

### Study Guide
**1.** 8.2%  **2.** 0.62625  **3.** $63.\overline{63}$%, or $63\frac{7}{11}$%  **4.** $\frac{17}{250}$
**5.** $4.1\overline{6}$%, or $4\frac{1}{6}$%  **6.** 10,000  **7.** About 5.3%  **8.** 6%
**9.** $185,000  **10.** Simple interest: $22.60; total amount due:
$2522.60  **11.** $6594.26

### Review Exercises
**1.** 170%  **2.** 6.5%  **3.** 0.315  **4.** 0.133; 0.067
**5.** 37.5%, or $37\frac{1}{2}$%  **6.** $33.\overline{3}$%, or $33\frac{1}{3}$%  **7.** $\frac{6}{25}$
**8.** $\frac{63}{1000}$  **9.** $30.6 = p \cdot 90$; 34%  **10.** $63 = 0.84 \cdot b$; 75
**11.** $a = 0.385 \cdot 168$; 64.68  **12.** $\frac{24}{100} = \frac{16.8}{b}$; 70
**13.** $\frac{42}{30} = \frac{N}{100}$; 140%  **14.** $\frac{10.5}{100} = \frac{a}{84}$; 8.82  **15.** 178 students;

84 students  **16.** 4.4%  **17.** 2500 mL  **18.** 12%  **19.** 92
**20.** $24  **21.** 6%  **22.** 11%  **23.** About 18.3%  **24.** $42; $308
**25.** 14%  **26.** $1050  **27.** $36  **28. (a)** $394.52;
**(b)** $24,394.52  **29.** $121  **30.** $7575.25  **31.** $8388.61
**32. (a)** $129; **(b)** interest: $100.18; amount applied to principal:
$28.82; **(c)** interest: $70.72; amount applied to principal:
$58.28; **(d)** At 13.2%, the principal is decreased by $29.46 more
than at the 18.7% rate. The interest at 13.2% is $29.46 less than
at 18.7%.  **33.** A  **34.** C  **35.** $66.\overline{6}$%, or $66\frac{2}{3}$%

### Understanding Through Discussion and Writing

**1.** A 40% discount is better. When successive discounts are
taken, each is based on the previous discounted price rather
than on the original price. A 20% discount followed by a
22% discount is the same as a 37.6% discount off the original
price.  **2.** Let $S =$ the original salary. After both raises
have been given, the two situations yield the same salary:
$1.05 \cdot 1.1S = 1.1 \cdot 1.05S$. However, the first situation is better for
the wage earner, because $1.1S$ is earned the first year when a
10% raise is given while in the second situation $1.05S$ is earned
that year.  **3.** No; the 10% discount was based on the original
price rather than on the sale price.  **4.** For a number $n$, 40%
of 50% of $n$ is $0.4(0.5n)$, or $0.2n$, or 20% of $n$. Thus, taking 40%
of 50% of a number is the same as taking 20% of the number.
**5.** The interest due on the 30-day loan will be $41.10 while that
due on the 60-day loan will be $131.51. This could be an argu-
ment in favor of the 30-day loan. On the other hand, the 60-day
loan puts twice as much cash at the firm's disposal for twice as
long as the 30-day loan does. This could be an argument in favor
of the 60-day loan.  **6.** Answers will vary.

## Test: Chapter 7, p. 505

**1.** [7.1b] 0.319  **2.** [7.1b] 38%  **3.** [7.1c] 137.5%  **4.** [7.1c] $\frac{13}{20}$
**5.** [7.2a, b] $a = 0.40 \cdot 55$; 22  **6.** [7.3a, b] $\frac{N}{100} = \frac{65}{80}$; 81.25%
**7.** [7.4a] Private health insurance: $1.06 trillion; Medicaid:
$0.54 trillion  **8.** [7.4a] About 524 at-bats  **9.** [7.5a] 4.9%
**10.** [7.4a] 59.7%  **11.** [7.6a] $25.20; $585.20  **12.** [7.6b] $630
**13.** [7.6c] $40; $160  **14.** [7.7a] $8.52  **15.** [7.7a] $5356
**16.** [7.7b] $1110.39  **17.** [7.7b] $11,580.07  **18.** [7.5a] Home
health aides: 1,261,900, 38.1%; wind turbine technicians: 4800,
109.1%; kindergarten and elementary teachers: 1,517,400, 5.8%;
electricians: 714,700, 13.7%  **19.** [7.6c] $50; about 14.3%
**20.** [7.7c] Interest: $36.73; amount applied to principal: $17.27;
balance after payment: $2687  **21.** [7.2a, b], [7.3a, b] B
**22.** [7.6b] $194,600  **23.** [7.6b], [7.7b] $3794.77

## Cumulative Review: Chapters 1–7, p. 507

**1.** [1.8a] 251,454  **2.** [7.1b] 26.9%  **3.** [7.1c] 112.5%,
or $112\frac{1}{2}$%  **4.** [5.5a] $2.1\overline{6}$  **5.** [6.1a] $\frac{5}{0.5}$, or $\frac{10}{1}$  **6.** [6.2a] $\frac{70 \text{ km}}{3 \text{ hr}}$,
or $23.\overline{3}$ km/hr, or $23\frac{1}{3}$ km/hr  **7.** [4.2c] <  **8.** [4.2c] <
**9.** [1.6a], [5.6a] 296,200  **10.** [1.6a] 50,000  **11.** [1.9d] 13
**12.** [2.7a] $-2x - 14$  **13.** [4.6a] $3\frac{1}{30}$  **14.** [5.2a] $-44.06$
**15.** [1.2a] 515,150  **16.** [5.2b] 0.02  **17.** [4.6b] $\frac{2}{3}$
**18.** [4.3a] $-\frac{110}{63}$  **19.** [3.6a] $\frac{1}{6}$  **20.** [2.4b] $-384$
**21.** [5.3a] 1.38036  **22.** [4.7b] $1\frac{1}{2}$  **23.** [5.4a] 12.25
**24.** [1.5a] 123 R 5  **25.** [1.7b] 95  **26.** [5.7a] 8.13  **27.** [4.4a] 49
**28.** [4.3b] $\frac{1}{12}$  **29.** [5.7b] $-\frac{16}{9}$, or $-1\frac{7}{9}$  **30.** [6.3b] $\frac{176}{21}$
**31.** [7.4a] About 1,275,000 visitors  **32.** [1.8a] $66,360
**33.** [6.4a] About 50 games  **34.** [5.8b] $84.95
**35.** [6.2b] 7.495 cents/oz  **36.** [4.2d] $\frac{3}{2}$ mi, or $1\frac{1}{2}$ mi
**37.** [6.4a] 60 mi  **38.** [7.7b] $10,378.47  **39.** [4.7d] 5 pieces
**40.** [7.5a] Percent increase: 34.0%; 282,700 physical therapists
**41.** [4.3a] C  **42.** [7.5a] A  **43.** [7.6c] 7 discounts
**44.** [7.5a] 12.5% increase  **45.** [5.8a] 33.6 mpg

# CHAPTER 8

## Exercise Set 8.1, p. 515

**RC1.** False **RC2.** False **RC3.** True **RC4.** False
**CC1.** True **CC2.** True **CC3.** False **CC4.** True
**CC5.** False
**1.** 92° **3.** 108° **5.** 85°, 60%; 90°, 40%; 100°, 10% **7.** 90°
and higher **9.** 30% and higher **11.** 90% − 40% = 50%
**13.** 483,612,200 mi **15.** Neptune **17.** 11 Earth diameters
**19.** 300 calories **21.** Yes **23.** 410 mg **25.** White rhino
**27.** About 2000 rhinos **29.** There are about six times as many
white rhinos as greater one-horned rhinos. **31.** $1270 per
person **33.** 1950 and 1970 **35.** $1532.70 per person
**37. (a)** 15% less; **(b)** $2003.51 more per person **39.** 6%
**41.** 166,400 students **43.** Canada

## Exercise Set 8.2, p. 524

**RC1.** True **RC2.** True **RC3.** False **RC4.** True
**CC1.** No **CC2.** Yes **CC3.** Yes **CC4.** No
**1.** 190 calories **3.** 1 slice of chocolate cake with fudge
frosting **5.** 1 cup of premium chocolate ice cream
**7.** About 125 calories **9.** Miniature tall bearded
**11.** 16 in. to 26 in. **13.** Tall bearded **15.** 25 in.
**17.** Computer and information sciences, mathematics and
statistics **19.** About 25,000 more degrees

**21.**

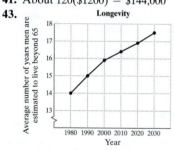

**23.** Denver and Indianapolis **25.** 68.9 **27.** New York City
**29.** New York City, Chicago, and Los Angeles **31.** $42
**33.** June to July, September to October, and November to
December **35.** About $900 **37.** 20 years **39.** About $450
**41.** About 120($1200) = $144,000

**43.**

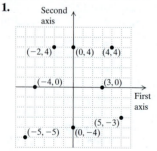

**45.** 25% **47.** 10.1% **48.** 83 **49.** $\frac{4}{3}$ **50.** $\frac{7}{24}$ **51.** 2.26
**52.** 6.348 **53.** 7.2 **54.** 80 **55.** 0.9 **56.** 150% **57.** 21
**58.** 17.26 **59.** $\frac{31}{60}$ **60.** 45

## Exercise Set 8.3, p. 532

**RC1.** *E* **RC2.** *C* **RC3.** *G* **RC4.** *B* **RC5.** *H*

**1.**

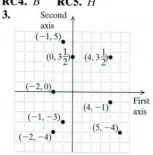

**3.**

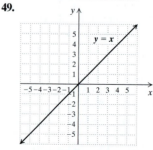

**5.** *A*: (3, 3); *B*: (0, −4); *C*: (−5, 0); *D*: (−1, −1); *E*: (2, 0); *F*: (−3, 5)
**7.** *A*: (5, 0); *B*: (0, 5); *C*: (−3, 4); *D*: (2, −4); *E*: (2, 3); *F*: (−4, −2)
**9.** II **11.** IV **13.** III **15.** I **17.** positive; negative
**19.** III **21.** IV; positive **23.** Yes **25.** No **27.** Yes
**29.** No **31.** Yes **33.** Yes **35.** 7 **36.** 3 **37.** 0 **38.** $\frac{30}{17}$
**39.** $1\frac{28}{33}a$ **40.** $7x − 24$ **41.** Yes **43.** I, IV **45.** I, III
**47.** (−1, −5)

## Calculator Corner, p. 536

**1.** $y = \frac{2}{3}x + 1$ **2.** $y = x + 1$

**3.** $y = −2x + 1$ **4.** $y = \frac{3}{5}x$

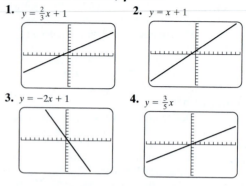

## Exercise Set 8.4, p. 541

**RC1.** False **RC2.** True **RC3.** True **RC4.** True
**RC5.** True **RC6.** False
**CC1.** Slanted line **CC2.** Horizontal line **CC3.** Slanted line
**CC4.** Vertical line
**1.** (5, 3) **3.** (3, 1) **5.** (5, 14) **7.** (10, −3) **9.** (0, 3)
**11.** (2, −1) **13.** (1, 3); (−1, 5) **15.** (7, 3); (10, 6)
**17.** (0, 10); (15, 0) **19.** (3, 1); (−2, 4) **21.** (1, 4); (−2, −8)
**23.** $(0, \frac{3}{5})$; $(\frac{3}{2}, 0)$ **25.** (0, 9), (4, 5), (10, −1)
**27.** (0, 0), (1, 4), (2, 8) **29.** (0, 13), (1, 10), (2, 7)
**31.** (0, −1), (2, 5), (−1, −4) **33.** (0, 0), (1, −7), (−2, 14)
**35.** (0, −4), (4, 0), (1, −3) **37.** (0, 6), (4, 0), (1, $\frac{9}{2}$)
**39.** (0, 2), (3, 3), (−3, 1)

**41.**

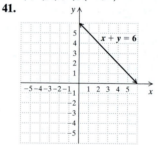

**43.**

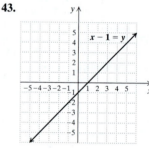

**45.**

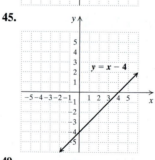

**47.**

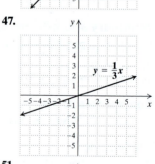

**49.**

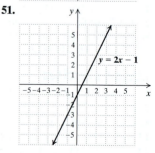

**51.**

**53.**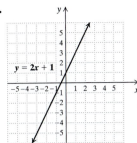

**55.**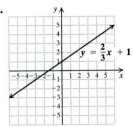

**57.**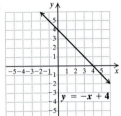

**59.**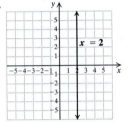

**61.**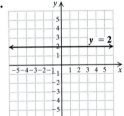

**63.**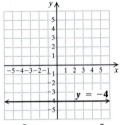

**65.**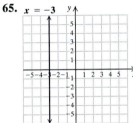

**67.**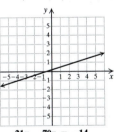

**69.** 9 min    **70.** 319.75 pages    **71.** $1\frac{7}{8}$ cups    **72.** $-\frac{7}{11}$

**73.** 42    **74.** $-\frac{5}{8}$    **75.** $(0, 0.2)$, $(-4, -1)$, $(1, 0.5)$;
answers may vary

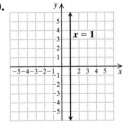
$21x - 70y = -14$

**77.** $(-3, 4.4)$, $(-3.9, 5)$, $(3, 0.4)$; answers may vary

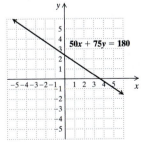
$50x + 75y = 180$

**79.** $(0, 6)$, $(1, 5)$, $(2, 4)$, $(3, 3)$, $(4, 2)$, $(5, 1)$, $(6, 0)$

**81. (a)**    $y_1 = -0.63x + 2.8$    **(b)**    $y_1 = 2.3x - 4.1$

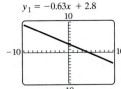    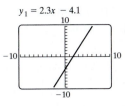

## Mid-Chapter Review: Chapter 8, p. 546

**1.** True    **2.** True    **3.** False
**4.** $3 \cdot 2 + (-1)$; $6 + (-1)$; 5
**5.** $1 - y = 6$; $-y = 5$; $y = -5$. Thus, $(1, -5)$ is a solution of
$x - y = 6$.

**6.**

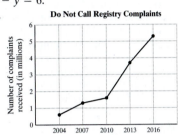

**7.** About 783%    **8.** III    **9.** II    **10.** IV    **11.** No
**12.** $(-10, -7)$    **13.** $(-1, -10)$    **14.** $(1, 3)$, $(0, 5)$, $(-1, 7)$

**15.**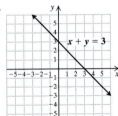

**16.**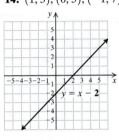

**17.**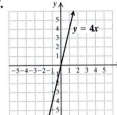

**18.**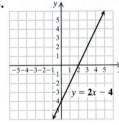

**19.**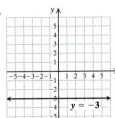

**20.**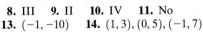

**21.** The points $(a, b)$ and $(b, a)$ will be in the same quadrant when both $a$ and $b$ are negative or when both are positive.    **22.** The point $(5, 0)$ is not in any quadrant. It lies on the $x$-axis, and the axes are not considered part of any quadrant.    **23.** The student should check each point carefully in the original equation. At least one of the points will not be a solution of the equation. After identifying any points that are not on the graph of the line, the student should calculate and plot new points to replace them. If these points do line up with each other, the student can then draw the line.    **24.** A line can pass through at most three quadrants. For example, the line $y = x - 2$ passes through quadrants I, III, and IV. A straight line cannot pass through four quadrants.

## Exercise Set 8.5, p. 556

**RC1.** statistic  **RC2.** mean  **RC3.** weight  **RC4.** mode
**CC1.** (e)  **CC2.** (f)  **CC3.** (a)  **CC4.** (c)  **CC5.** (d)
**CC6.** (c)

**1.** Minimum: 3; maximum: 38; range: 35   **3.** Minimum: 5;
maximum: 112; range: 107   **5.** Minimum: 2; maximum: 3; range: 1
**7.** Mean: 3,262,500 visitors; median: 2,150,000 visitors;
mode: 1,100,000 visitors   **9.** Mean: 21; median: 18.5; mode: 29
**11.** Mean: 21; median: 20; modes: 5, 20   **13.** Mean: 5.38;
median: 5.7; no mode exists   **15.** Mean: 239.5; median: 234;
mode: 234   **17.** 36 mpg   **19.** 2.7   **21.** Mean: $4.19;
median: $3.99; mode: $3.99   **23.** 90   **25.** 263 days
**27.** (a) Jefferson County: $133,987; Hamilton County: $146,989;
(b) Jefferson County   **29.** (a) 1.3675 billion tickets;
(b) 1.33375 billion tickets; (c) 2006 to 2013   **31.** First quartile:
10; second quartile: 15; third quartile: 28   **33.** First quartile: 2.5;
second quartile: 6; third quartile: 10.5   **35.** Minimum: 62;
first quartile: 77; median: 82; third quartile: 90; maximum: 98
**37.** Minimum: 1.1; first quartile: 2.7; median: 3.3; third
quartile: 3.8; maximum: 4.0   **39.** 225.05   **40.** 126.0516
**41.** $\frac{3}{35}$   **42.** $\frac{14}{15}$   **43.** $a = 30$; $b = 58$   **45.** $6950

## Exercise Set 8.6, p. 567

**RC1.** True  **RC2.** True  **RC3.** False  **RC4.** True
**RC5.** True  **CC1.** 35.4%  **CC2.** 16.8%

**1.**

| Winner | Frequency |
|---|---|
| Stan Wawrinka | 3 |
| Rafael Nadal | 3 |
| Novak Djokovic | 6 |
| Marin Cilic | 1 |
| Andy Murray | 1 |
| Roger Federer | 2 |

**3.**

| Number | Frequency |
|---|---|
| 0 | 3 |
| 2 | 1 |
| 4 | 7 |
| 5 | 3 |
| 6 | 4 |
| 7 | 5 |

**5.**

| Class | Tally Marks | Frequency |
|---|---|---|
| $49,000–$49,999 | \|\| | 2 |
| $50,000–$50,999 | Ⅷ \|\| | 7 |
| $51,000–$51,999 | Ⅷ | 5 |
| $52,000–$52,999 | Ⅷ Ⅷ | 10 |
| $53,000–$53,999 |  | 0 |
| $54,000–$54,999 |  | 0 |
| $55,000–$55,999 | \| | 1 |

**7.** 1411   **9.** 263   **11.** 11.7%   **13.** 49.6%
**15.** 6.25%   **17.** 8%   **19.** 81.25%
**21.** (a) Minimum: 4.6%; maximum: 7.6%; range: 3.0%;
(b) 6.1%   **23.** South America   **25.** 16,760 ft

**27.** 
```
3 | 2 2 2 3 3 4 4 6 6 7 7 8 9 9
4 | 0 3 6 9
5 | 0 7
```
        Key: 3|2 = 320 m

**29.** 
```
2 | 7
3 |
4 | 0 1 1 2 3 3 3 4 4 4 5 5 5 5 8 8 8 8 9 9 9 9
5 | 0 0 1 2 2 2 3 3 4 4 5 5 5 7 8 9 9
6 | 0 0 0 2 3 3 4 5 6
7 | 0 1
```
        Key: 2|7 = 27°F

**31.** About 19 games   **33.** 70–79 and 130–139   **35.** 42 states
**37.** The median is greater than 66% and less than 75%.

**39.**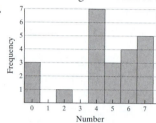

**41.**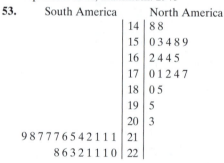

**43.** $\frac{11}{100}$   **44.** 25   **45.** 2.58   **46.** 4   **47.** $3\frac{1}{3}$   **48.** 68
**49.** 15%   **51.** Minimum: 8%; first quartile: 18%; median: 22%;
third quartile: 25%; maximum: 29%

**53.**
```
     South America          North America
                    14 | 8 8
                    15 | 0 3 4 8 9
                    16 | 2 4 4 5
                    17 | 0 1 2 4 7
                    18 | 0 5
                    19 | 5
                    20 | 3
   9 8 7 7 7 6 5 4 2 1 1 1 | 21
       8 6 3 2 1 1 1 0 | 22
```
        Key: 14|8 = 14,800 ft

## Translating for Success, p. 580

**1.** D   **2.** B   **3.** J   **4.** K   **5.** I   **6.** F   **7.** N   **8.** E
**9.** L   **10.** M

## Exercise Set 8.7, p. 581

**RC1.** (a)  **RC2.** (c)  **RC3.** (b)  **RC4.** (d)
**CC1.** (a) 2; (b) 52; (c) $\frac{2}{52} = \frac{1}{26}$  **CC2.** (a) 1; (b) 52; (c) $\frac{1}{52}$
**1.** 83   **3.** About $9 billion   **5.** About 286,000 students
**7.** About $74 billion

**9.**

Outcomes: Red, Blue, Green

**11.**

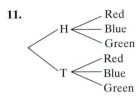

Outcomes: H, Red; H, Blue; H, Green;
T, Red; T, Blue; T, Green

**13.** 24 outcomes   **15.** $\frac{1}{6}$, or $0.1\overline{6}$   **17.** $\frac{1}{2}$, or 0.5   **19.** 0
**21.** $\frac{1}{52}$   **23.** $\frac{2}{13}$   **25.** $\frac{3}{26}$   **27.** $\frac{4}{39}$   **29.** $\frac{34}{39}$   **31.** 52.3%
**32.** 25.0%   **33.** $\frac{1}{4}$, or 0.25   **35.** $\frac{1}{36}$   **37.** (a) $\frac{1}{3}$; (b) $\frac{1}{3}$; (c) $\frac{1}{3}$

## Summary and Review: Chapter 8, p. 585

### Vocabulary Reinforcement

**1.** table   **2.** origin   **3.** pictograph   **4.** mode   **5.** mean

### Concept Reinforcement

**1.** False   **2.** True   **3.** True

### Study Guide

**1.** Quaker Organic Maple & Brown Sugar; $0.54 per serving
**2.** 12 g   **3.** Arrowhead Stadium   **4.** About $110 million more
**5.** No   **6.**

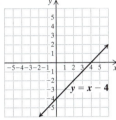

**7.** Mean: 8; median: 8; mode: 8   **8.** Minimum: 2; first quartile: 6;
median: 32; third quartile: 61.5; maximum: 85   **9.** $\frac{1}{6}$

### Review Exercises

**1.** 58%   **2.** Poland, Ukraine, Mexico, and India   **3.** Ukraine
**4.** 34%   **5.** Room and board   **6.** $4830   **7.** 38.5   **8.** 1.86
**9.** $16.67   **10.** 14   **11.** 5.3   **12.** $17   **13.** 26   **14.** 11 and 17
**15.** 20   **16.** 28 mpg   **17.** 3.1   **18.** May   **19.** About 150 tornadoes   **20.** About 175 more tornadoes   **21.** In the spring

**22.**

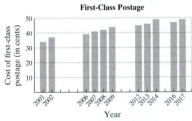

**23.**

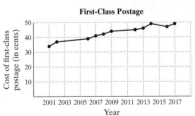

**24.** 1998   **25.** About 10,000 children   **26.** 2004 and 2010
**27.** By about 11,000 children   **28.** $70,000–$89,000

**29.** About 22 governors
**30.**

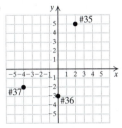

Outcomes: Blue, 2; Blue, 5; Blue, 6; Blue, 10;
White, 2; White, 5; White, 6; White, 10

**31.** $(-5, -1)$   **32.** $(-2, 5)$   **33.** $(3, 0)$   **34.** $(4, -2)$
**35.–37.**

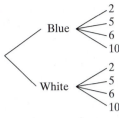

**38.** IV   **39.** III   **40.** I   **41.** $(1, 2)$; $(9, -2)$

**42.**

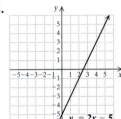

**43.**

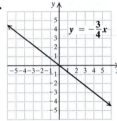

**44.**

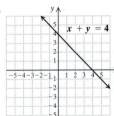

**45.**

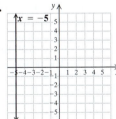

**46.**

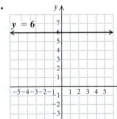

**47.** 110 teachers

**48.** 12 teachers   **49.** 30%   **50.** Minimum: 2.4 lb; first quartile:
5 lb; median: 7.55 lb; third quartile: 11.05 lb; maximum: 20.5 lb

**51.** 

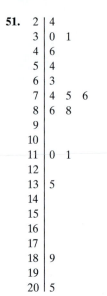

| | |
|---|---|
| 2 | 4 |
| 3 | 0  1 |
| 4 | 6 |
| 5 | 4 |
| 6 | 3 |
| 7 | 4  5  6 |
| 8 | 6  8 |
| 9 | |
| 10 | |
| 11 | 0  1 |
| 12 | |
| 13 | 5 |
| 14 | |
| 15 | |
| 16 | |
| 17 | |
| 18 | 9 |
| 19 | |
| 20 | 5 |

Key: $2 \mid 4 = 2.4$

**52.**

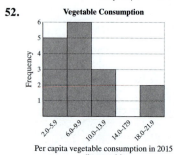

**53.** $\frac{1}{52}$  **54.** $\frac{1}{2}$  **55.** About 27,000 children  **56.** A  **57.** D
**58.** $a = 316, b = 349$  **59.** \$11.52/hr

## Understanding Through Discussion and Writing

**1.** The equation could represent a person's average income during a 4-yr period. Answers may vary.  **2.** Bar graphs that show change over time can be successfully converted to line graphs. Other bar graphs cannot be successfully converted to line graphs.  **3.** It is possible for the mean of a set of numbers to be larger than all but one number in the set. To see this, note that the mean of the set {6, 8} is 7, which is larger than all of the numbers in the set but one.  **4.** The median of a set of four numbers *can* be in the set. For example, the median of the set {11, 15, 15, 17} is 15, which is in the set.  **5.** The median income is often used instead of the mean income because it is not artificially raised by the small number of people with very high incomes. The mean of the medians over a 3-yr period would lessen the impact of a sudden drop or rise in income for one year.  **6.** A company considering expansion would probably be more interested in extrapolation than in interpolation because it will be future sales or other activity that will influence profit after expansion.

## Test: Chapter 8, p. 592

**1.** [8.1a] Hiking at 3 mph with 20-lb load  **2.** [8.1a] Hiking at 3 mph with 10-lb load, hiking at 3 mph with 20-lb load
**3.** [8.1b] Spain  **4.** [8.1b] Norway and the United States
**5.** [8.1b] 900 lb  **6.** [8.1b] 1000 lb  **7.** [8.2c] 2005
**8.** [8.2c] 2009  **9.** [8.2c] 10 hurricanes  **10.** [8.2c] 10 more hurricanes  **11.** [8.5b] About 8 hurricanes/year
**12.** [8.7a] About 10 hurricanes

**13.** [8.2b]

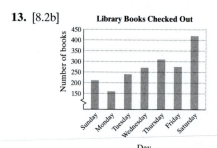

**14.** [8.2d]

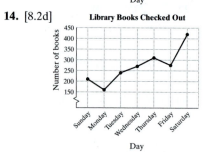

**15.** [8.3b] II  **16.** [8.3b] III  **17.** [8.3a] (3, 4)
**18.** [8.3a] (0, −4)  **19.** [8.3a] (−4, 2)  **20.** [8.4a] (4, 2)
**21.** [8.4b]  **22.** [8.4b]

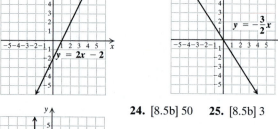

**23.** [8.4c]  **24.** [8.5b] 50  **25.** [8.5b] 3

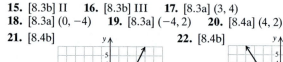

**26.** [8.5b] 15.5  **27.** [8.5c,d] Median: 50.5; mode: 54
**28.** [8.5c,d] Median: 3; no mode exists  **29.** [8.5c,d] Median: 17.5; modes: 17, 18  **30.** [8.5b] 76  **31.** [8.5b] 2.9
**32.** [8.5e] Minimum: 0.7%; first quartile: 0.8%; median: 1.0%; third quartile: 1.3%; maximum: 1.5%

**33.** [8.6b]

| | |
|---|---|
| 0 | 7  7  8  8  8  9  9  9 |
| 1 | 0  2  2  2  3  3  4  4  5 |

Key: $0 \mid 7 = 0.7\%$

**34.** [8.6a]

| Class | Tally marks | Frequency |
|---|---|---|
| 0.7–0.9 | ⅢⅢ  Ⅲ | 8 |
| 1.0–1.2 | ⅢⅠ | 4 |
| 1.3–1.5 | ⅢⅢ | 5 |

**35.** [8.6c]

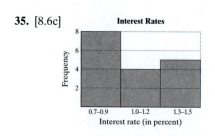

**36.** [8.6a] 18 students    **37.** [8.6a] 24%
**38.** [8.7b]

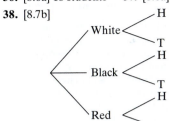

Outcomes: White, H; White, T; Black, H; Black, T; Red, H; Red, T

**39.** [8.7c] $\frac{8}{29}$    **40.** [8.7c] B    **41.** [8.5b, c] $a = 74, b = 111$

**42.** [8.4b]

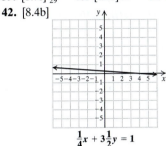

$$\frac{1}{4}x + 3\frac{1}{2}y = 1$$

**43.** [8.4b]

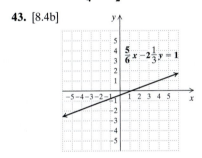

$$\frac{5}{6}x - 2\frac{1}{3}y = 1$$

**44.** [8.3a] 56 sq units

## Cumulative Review: Chapters 1–8, p. 597

**1.** [5.3b] 1,400,000,000    **2.** [8.5b] 28 mpg    **3.** [1.1a] 5 hundreds
**4.** [1.9c] 128    **5.** [3.2a] 1, 2, 3, 4, 5, 6, 10, 12, 15, 20, 30, 60
**6.** [5.1d] 52.0    **7.** [4.5a] $\frac{33}{10}$    **8.** [5.3b] \$2.10    **9.** [2.1d] 9
**10.** [7.1c] 35%    **11.** [6.3a] No    **12.** [2.6a] $-2$    **13.** [4.2b] $\frac{1}{6}$
**14.** [2.2a] $-39$    **15.** [4.3a] $\frac{1}{3}$    **16.** [5.2b] 325.43    **17.** [4.7a] 15
**18.** [5.3a] $-42.282$    **19.** [3.7b] $\frac{9}{10}$    **20.** [5.4a] 62.345
**21.** [6.3b] $9\frac{3}{5}$    **22.** [3.8a] $\frac{3}{4}$    **23.** [4.4a] 24    **24.** [5.7b] $-\frac{17}{4}$
**25.** [1.8a] 4080 billion kWh    **26.** [4.7d] $\frac{1}{4}$ yd    **27.** [3.6b] $\frac{3}{8}$ cup
**28.** [7.5a] About 12.9%    **29.** [5.8b] 6.2 lb    **30.** [5.8b] \$9.55
**31.** [6.4a] 1122 defective valves    **32.** [7.6b] 7%    **33.** [1.4b], [2.7b] 22 cm; 28 cm$^2$    **34.** [8.3b] II
**35.** [8.4b]

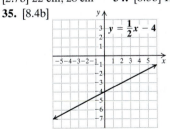

**36.** [2.7a] $-2x + y$

---

**37.** [2.6b] $10a - 15b + 5$    **38.** [8.5b, c] About \$1.02 billion; \$0.745 billion
**39.** [8.2b]

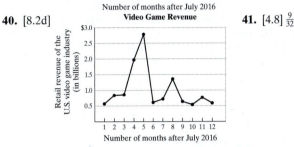

**40.** [8.2d]

Video Game Revenue

**41.** [4.8] $\frac{9}{32}$

**42.** [4.6a] $7\frac{47}{1000}$    **43.** [8.3a] $(-2, -1), (-2, 7), (6, 7), (6, -1)$

## CHAPTER 9

### Exercise Set 9.1, p. 609

**RC1.** <    **RC2.** >    **RC3.** <    **RC4.** >
**RC5.** <    **RC6.** >    **CC1.** left    **CC2.** right
**1.** 3    **3.** $\frac{1}{12}$    **5.** 1760    **7.** 9    **9.** 7    **11.** 16
**13.** 8800    **15.** $5\frac{1}{4}$, or 5.25    **17.** $2\frac{1}{5}$, or 2.2    **19.** 37,488
**21.** $2\frac{1}{2}$, or 2.5    **23.** 3    **25.** 1    **27.** $1\frac{1}{4}$, or 1.25    **29.** 132,000
**31.** 126,720    **33.** (a) 1000; (b) 0.001    **35.** (a) 10; (b) 0.1
**37.** (a) 0.01; (b) 100    **39.** 8300    **41.** 0.98    **43.** 8.921
**45.** 0.03217    **47.** 28,900    **49.** 4.77    **51.** 688    **53.** 0.1
**55.** 100,000    **57.** 142    **59.** 0.82    **61.** 450    **63.** 0.000024
**65.** 0.688    **67.** 230    **69.** 48,440; 48.44    **71.** 85; 8.5
**73.** 0.027; 0.00027    **75.** 541,300; 54,130    **77.** 6.21
**79.** 35.56    **81.** 104.585    **83.** 28.67    **85.** 70.866    **87.** 0.299

| | yd | cm | in. | m | mm |
|---|---|---|---|---|---|
| **89.** | 0.2361 | 21.59 | $8\frac{1}{2}$ | 0.2159 | 215.9 |
| **91.** | 55,192.3 | 5,045,000 | 1,986,216.5 | 50,450 | 50,450,000 |

**93.** $-\frac{3}{2}$    **94.** $-2$    **95.** \$101.50    **96.** \$44.50    **97.** 47%
**98.** 35%    **99.** 56 ft; 192 ft$^2$    **100.** 44 ft; 120 ft$^2$    **101.** 1.0 m
**103.** 1.4 cm    **105.** Length: 5400 in., or 450 ft; breadth: 900 in., or 75 ft; height: 540 in., or 45 ft    **107.** 321,800    **109.** 21.3 mph
**111.** 1 in. = 25.4 mm

### Exercise Set 9.2, p. 616

**RC1.** True    **RC2.** False    **RC3.** True    **RC4.** False
**RC5.** True    **RC6.** True    **CC1.** (f)    **CC2.** (b)
**1.** 144    **3.** 640    **5.** $\frac{1}{144}$    **7.** 45    **9.** 1008    **11.** 3    **13.** 198
**15.** 2160    **17.** 12,800    **19.** 27,878,400    **21.** 5    **23.** $\frac{1}{3}$
**25.** $\frac{1}{640}$    **27.** 25,792    **29.** $1\frac{1}{2}$    **31.** 19,000,000    **33.** 63,100
**35.** 0.065432    **37.** 0.0349    **39.** 2500    **41.** 0.4728    **43.** \$240
**44.** \$212    **45.** (a) \$220.93; (b) \$6620.93    **46.** (a) \$37.97; (b) \$4237.97    **47.** 10.76    **49.** 1.67    **51.** 10,657,311 ft$^2$
**53.** 216 cm$^2$    **55.** 1 ft$^2$    **57.** $4\frac{1}{3}$ ft$^2$    **59.** 84 cm$^2$
**61.** 1701 tiles; 42%

## Calculator Corner, p. 625

**1.** Answers will vary.   **2.** 1417.99 in.; 160,005.91 in$^2$
**3.** 1729.27 in$^2$   **4.** 125,663.71 ft$^2$

## Exercise Set 9.3, p. 626

**RC1.** radius   **RC2.** circumference   **RC3.** circumference
**RC4.** area   **CC1.** (c)   **CC2.** (a)
**1.** 50 cm$^2$   **3.** 104 ft$^2$   **5.** 142.5 m$^2$   **7.** 72.45 cm$^2$   **9.** 144 mi$^2$
**11.** 49 m$^2$   **13.** $55\frac{1}{8}$ ft$^2$   **15.** 1.92 cm$^2$   **17.** 68 yd$^2$   **19.** 14 cm
**21.** $1\frac{3}{4}$ in.   **23.** 10 ft   **25.** 0.7 cm   **27.** 44 cm   **29.** $5\frac{1}{2}$ in.
**31.** 62.8 ft   **33.** 4.396 cm   **35.** 154 cm$^2$   **37.** $2\frac{13}{32}$ in$^2$
**39.** 314 ft$^2$   **41.** 1.5386 cm$^2$   **43.** 94.2 ft   **45.** About 55.04 in$^2$
larger   **47.** About 7930.25 mi; about 3965.13 mi   **49.** 70,650 mi$^2$
**51.** 65.94 yd$^2$   **53.** Maximum circumference of barrel: $8\frac{9}{14}$ in.;
minimum circumference of handle: $2\frac{86}{133}$ in.   **55.** 45.68 ft
**57.** 45.7 yd   **59.** 100.48 m$^2$   **61.** 6.9972 cm$^2$   **63.** 64.4214 in$^2$
**65.** 675 cm$^2$   **67.** 7.5 ft$^2$   **69.** $\frac{37}{400}$   **70.** $\frac{7}{8}$   **71.** 125%
**72.** $66.\overline{6}$%, or $66\frac{2}{3}$%   **73.** 5 lb   **74.** $730   **75.** 37.2875 in$^2$
**77.** Circumference   **79.** 4$A$   **81.** An 8-inch square pan

## Calculator Corner, p. 637

**1.** 1.13 cm$^3$   **2.** 0.22 in$^3$

## Exercise Set 9.4, p. 638

**RC1.** (b)   **RC2.** (a)   **RC3.** (c)   **CC1.** (c)   **CC2.** (c)
**1.** 250 cm$^3$   **3.** 135 in$^3$   **5.** 75 m$^3$   **7.** $357\frac{1}{2}$ yd$^3$   **9.** 4082 ft$^3$
**11.** 376.8 cm$^3$   **13.** 41,580,000 yd$^3$   **15.** 4,186,666.$\overline{6}$ in$^3$
**17.** Approximately 124.725 m$^3$   **19.** $1437\frac{1}{3}$ km$^3$   **21.** 1000; 1000
**23.** 59,000   **25.** 0.049   **27.** 27,300   **29.** 40   **31.** 320
**33.** 3   **35.** 40   **37.** 48   **39.** $1\frac{7}{8}$   **41.** 4747.68 cm$^3$
**43.** 904 ft$^3$   **45.** 0.42 yd$^3$   **47.** 263,947,530,000 mi$^3$
**49.** 33,493 in$^3$   **51.** 0.477 m$^3$   **53.** 152,321 m$^3$   **55.** 5832 yd$^3$
**57.** (a) About 1875.63 mm$^3$; (b) about 11,253.78 mm$^3$   **59.** $24
**60.** $175   **61.** $10.68   **62.** $\frac{7}{3}$, or $2.\overline{3}$   **63.** $-1$   **64.** 448 km
**65.** 59   **66.** 10   **67.** About 346.2 ft$^3$   **69.** 6 cm by 6 cm by
6 cm   **71.** 9.424779 L   **73.** $1\frac{3}{4}$ gal; 7.5 gal; $91\frac{1}{4}$ gal; about
30,000,000,000 gal

## Mid-Chapter Review: Chapter 9, p. 643

**1.** False   **2.** True   **3.** False   **4.** True   **5.** True
**6.** $16\frac{2}{3}$ yd $= 16\frac{2}{3} \cdot 1$ yd $= \frac{50}{3} \cdot 3$ ft $= 50$ ft
**7.** 10,200 mm $= 10,200$ mm $\cdot \frac{1 \text{ m}}{1000 \text{ mm}} \approx 10.2$ m $\cdot \frac{3.281 \text{ ft}}{1 \text{ m}} = 33.4662$ ft
**8.** $C \approx 3.14 \cdot 10.2$ in.
$C \approx 32.028$ in.;
$A \approx 3.14 \cdot 5.1$ in. $\cdot 5.1$ in.
$A \approx 81.6714$ in$^2$
**9.** 9680   **10.** 70   **11.** 2.405   **12.** 0.00015   **13.** 1251
**14.** 5   **15.** 180,000   **16.** 700   **17.** 2.285   **18.** 118.116
**19.** 15.24   **20.** 160.9   **21.** 720   **22.** 90   **23.** $6\frac{2}{3}$
**24.** $\frac{1}{3}$   **25.** 87,120   **26.** 960   **27.** 3.5 ft, 2 yd, 100 in., $\frac{1}{100}$ mi,
1000 in., 430 ft, 6000 ft   **28.** 150 hm, 13 km, 310 dam, 300 m,
33,000 mm, 3240 cm, 250 dm   **29.** 800 in$^2$   **30.** 66 km$^2$
**31.** $C = 43.96$ in.; $A = 153.86$ in$^2$   **32.** $C = 27.004$ cm;
$A = 58.0586$ cm$^2$   **33.** 30 cm$^3$   **34.** 1695.6 ft$^3$
**35.** 2.355 in$^3$   **36.** 7234.56 m$^3$   **37.** 7400   **38.** 62
**39.** 0.0005   **40.** 12   **41.** 9   **42.** 4   **43.** The student should
have multiplied by $\frac{1}{12}$ (or divided by 12) to convert inches to feet.
The correct procedure is as follows:

$$23 \text{ in.} = 23 \text{ in.} \cdot \frac{1 \text{ ft}}{12 \text{ in.}} = \frac{23 \text{ in.}}{12 \text{ in.}} \cdot 1 \text{ ft}$$
$$= \frac{23}{12} \cdot \frac{\text{in.}}{\text{in.}} \cdot 1 \text{ ft} = \frac{23}{12} \cdot 1 \text{ ft} = \frac{23}{12} \text{ ft.}$$

**44.** The area of a 16-in-diameter pizza is approximately
3.14 $\cdot$ 8 in. $\cdot$ 8 in., or 200.96 in$^2$. At $16.25, its unit price is
$\frac{\$16.25}{200.96 \text{ in}^2}$, or about $0.08/in$^2$. The area of a 10-in.-diameter pizza
is approximately 3.14 $\cdot$ 5 in. $\cdot$ 5 in., or 78.5 in$^2$. At $7.85, its unit

price is $\frac{\$7.85}{78.5 \text{ in}^2}$, or $0.10/in$^2$. Since the 16-in.-diameter pizza has
the lower unit price, it is a better buy.   **45.** A parallelogram
with base $b$ and height $h$ can be divided into two triangles, each
with base $b$ and height $h$. Since the area of a parallelogram is
given by $A = bh$, the area of one of the triangles is given by
$A = \frac{1}{2}bh$.   **46.** No; let $r$ = radius of the smaller circle. Then
its area is $\pi \cdot r \cdot r$, or $\pi r^2$. The radius of the larger circle is $2r$,
and its area is $\pi \cdot 2r \cdot 2r$, or $4\pi r^2$, or $4 \cdot \pi r^2$. Thus, the area of the
larger circle is 4 times the area of the smaller circle.

## Exercise Set 9.5, p. 653

**RC1.** (f)   **RC2.** (c)   **RC3.** (a)   **RC4.** (e)   **RC5.** (b)
**RC6.** (d)
**1.** Angle $GHI$, angle $IHG$, angle $H$, $\angle GHI$, $\angle IHG$, or $\angle H$
**3.** $\angle ADB$ or $\angle BDA$   **5.** 10°   **7.** 180°   **9.** 90°
**11.**

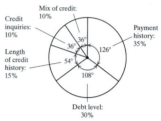

**13.**

Wind Energy

**15.** Acute   **17.** Straight   **19.** Right   **21.** Acute   **23.** Obtuse
**25.** $\angle 1$ and $\angle 3$; $\angle 2$ and $\angle 4$   **27.** $\angle GME$ (or $\angle EMG$) and
$\angle AMC$ (or $\angle CMA$); $\angle AMG$ (or $\angle GMA$) and $\angle EMC$ (or $\angle CME$)
**29.** $m\angle 4$; $m\angle 1$   **31.** $\angle GME$ or $\angle EMG$; $\angle CME$ or $\angle EMC$
**33.** 79°   **35.** 23°   **37.** 32°   **39.** 61°   **41.** 177°   **43.** 41°
**45.** 105°   **47.** 76°   **49.** Scalene; obtuse   **51.** Scalene; right
**53.** Equilateral; acute   **55.** Scalene; obtuse   **57.** 46°
**59.** 120°   **61.** 58°   **63.** $160   **64.** $22.50   **65.** $148
**66.** $1116.67   **67.** $33,597.91   **68.** $413,458.31
**69.** $641,566.26   **70.** $684,337.34   **71.** $m\angle 2 = 67.13°$;
$m\angle 4 = 79.8°$; $m\angle 5 = 67.13°$; $m\angle 6 = 33.07°$
**73.** $m\angle ACB = 50°$; $m\angle CAB = 40°$; $m\angle EBC = 50°$;
$m\angle EBA = 40°$; $m\angle AEB = 100°$; $m\angle ADB = 50°$

## Calculator Corner, p. 659

**1.** 6.6   **2.** 19.8   **3.** 1.9   **4.** 24.5   **5.** 121.2   **6.** 85.4

## Translating for Success, p. 662

**1.** K   **2.** G   **3.** B   **4.** H   **5.** O   **6.** M   **7.** E
**8.** A   **9.** D   **10.** I

## Exercise Set 9.6, p. 663

**RC1.** True   **RC2.** False   **RC3.** True   **RC4.** False
**CC1.** Legs: 16 and 30; hypotenuse: 34   **CC2.** Legs: 1 and 6;
hypotenuse: $\sqrt{37}$   **CC3.** Legs: $\sqrt{11}$ and $\sqrt{77}$; hypotenuse:
$2\sqrt{22}$   **CC4.** Legs: 10 and $2\sqrt{39}$; hypotenuse: 16

**1.** −4, 4 **3.** −11, 11 **5.** −13, 13 **7.** −50, 50 **9.** 8
**11.** 9 **13.** 15 **15.** 25 **17.** 20 **19.** 100 **21.** 6.928
**23.** 2.828 **25.** 1.732 **27.** 3.464 **29.** 4.359 **31.** 10.488
**33.** $c = 15$ **35.** $c = \sqrt{98}$; $c \approx 9.899$ **37.** $a = 5$
**39.** $b = \sqrt{45}$; $b \approx 6.708$ **41.** $c = 26$ **43.** $b = 12$
**45.** $c = \sqrt{41}$; $c \approx 6.403$ **47.** $b = \sqrt{1023}$; $b \approx 31.984$
**49.** $\sqrt{644}$ ft $\approx 25.4$ ft **51.** $\sqrt{8450}$ ft $\approx 91.9$ ft
**53.** $\sqrt{34{,}541}$ m $\approx 185.9$ m **55.** $\sqrt{211{,}200{,}000}$ ft $\approx 14{,}532.7$ ft
**57.** $468 **58.** 187,200 **59.** About 324 students **60.** 12%
**61.** 8 **62.** 125 **63.** 1000 **64.** 10,000 **65.** 3 **66.** 982
**67.** 47.80 cm² **69.** 64.8 in. **71.** Width: 15.2 in.; height: 11.4 in.

## Calculator Corner, p. 671

**1.** 41°F **2.** 122°F **3.** 20°C **4.** 45°C

## Exercise Set 9.7, p. 672

**RC1.** True **RC2.** False **RC3.** True **RC4.** False
**RC5.** False **RC6.** True **CC1.** −10°C **CC2.** 200°F
**CC3.** 100°F **CC4.** 0°C
**1.** 2000 **3.** 3 **5.** 64 **7.** 12,640 **9.** 0.1 **11.** 5
**13.** 127,145 tons **15.** 1000 **17.** $\frac{1}{1000}$, or 0.001 **19.** $\frac{1}{100}$, or 0.01
**21.** 1000 **23.** 10 **25.** 234,000 **27.** 5.2 **29.** 0.000897
**31.** 7320 **33.** 8.492 **35.** 58.5 **37.** 800,000 **39.** 1000
**41.** 0.0034 **43.** 0.0603 **45.** 80°C **47.** 60°C **49.** 20°C
**51.** −10°C **53.** 180°F **55.** 140°F **57.** 10°F **59.** 40°F
**61.** 86°F **63.** 104°F **65.** 35.6°F **67.** 30.2°F **69.** 5432°F
**71.** 25°C **73.** 55°C **75.** 81.$\overline{1}$°C **77.** −15°C **79.** 37°C
**81. (a)** 56.7°C = 134°F; 131°F = 55°C; **(b)** 3°F

**83.** 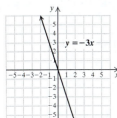 **84.**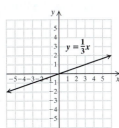

**85.** 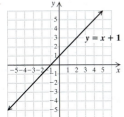 **86.**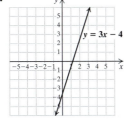

**87.** $946 **88.** 201 min, or 3 hr 21 min **89.** 144 packages
**91.** −40° **93.** 260.6°F **95.** 0.4536 **97.** About 5.1 g/cm³
**99. (a)** 109.134 g; **(b)** 9.104 g; **(c)** Golden Jubilee: 3.85 oz;
Hope: 0.321 oz

## Exercise Set 9.8, p. 679

**RC1.** True **RC2.** True **RC3.** False **RC4.** True
**1.** 2000 mL **3.** 0.12 g **5.** 4 tablets **7. (a)** 720 mcg;
**(b)** about 189 actuations; **(c)** 6 inhalers **9.** 16 oz
**11. (a)** 0.42 g; **(b)** 28 doses **13.** 6 mL **15.** 1000 **17.** 0.325
**19.** 250 mcg **21.** 125 mcg **23.** 0.875 mg; 875 mcg **25.** 3358
**26.** 7414 **27.** 854 **28.** 6334 **29.** $5x + 8$ **30.** $6x + 5$
**31.** $7t − 9$ **32.** $8r − 10$ **33. (a)** 9.1 g; **(b)** 162.5 mg
**35.** Naproxen sodium is the better buy. It costs 11 cents/day and
the ibuprofen costs 25 cents/day.

## Summary and Review: Chapter 9, p. 682

### Vocabulary Reinforcement

**1.** parallel **2.** sphere **3.** radius **4.** supplementary
**5.** scalene **6.** hypotenuse

### Concept Reinforcement

**1.** True **2.** False **3.** False **4.** True **5.** True **6.** False

### Study Guide

**1.** $\frac{7}{3}$, or $2\frac{1}{3}$, or $2.\overline{3}$ **2.** 0.000046 **3.** 10.94 **4.** 9 **5.** 5240
**6.** 15.5 m² **7.** 80 m² **8.** 37.68 in. **9.** 616 cm² **10.** 1683.3 m³
**11.** $30\frac{6}{35}$ ft³ **12.** 1696.537813 cm³ **13.** 64 **14.** Complement:
52°; supplement: 142° **15.** 87° **16.** $\sqrt{208} \approx 14.422$
**17.** 5.14 **18.** 0.00978 **19.** 154.4°F **20.** 40°C **21.** 500 mL

### Review Exercises

**1.** 3.$\overline{3}$ **2.** 30 **3.** 0.17 **4.** 72 **5.** 400,000 **6.** $1\frac{1}{6}$
**7.** 218.8 **8.** 32.18 **9.** 80 **10.** 0.003 **11.** 12.5 **12.** 224
**13.** 0.06 **14.** 400 **15.** 1400 **16.** 200 **17.** 4700
**18.** 0.04 **19.** 36 **20.** 700,000 **21.** 7 **22.** 0.057
**23.** 31.4 m **24.** $\frac{14}{11}$ in. **25.** 24 m **26.** 439.7784 yd
**27.** 18 in² **28.** 29.82 ft² **29.** 60 cm² **30.** 35 mm²
**31.** 154 ft² **32.** 314 cm² **33.** 1038.555 ft² **34.** 26.28 ft²
**35.** 54° **36.** 180° **37.** 140° **38.** 90° **39.** Acute
**40.** Straight **41.** Obtuse **42.** Right **43.** 49° **44.** 136°
**45.** 60° **46.** Scalene **47.** Right **48.** 93.6 m³
**49.** 193.2 ft³ **50.** 28,260 cm³ **51.** $33\frac{37}{75}$ yd³ **52.** 942 cm³
**53.** 8 **54.** 9.110 **55.** $c = \sqrt{850}$; $c \approx 29.155$
**56.** $b = \sqrt{84}$; $b \approx 9.165$ **57.** $c = \sqrt{89}$; $c \approx 9.434$ ft
**58.** $a = \sqrt{76}$ cm; $a \approx 8.718$ cm **59.** About 17.9 ft
**60.** About 13.42 ft **61.** 113°F **62.** 20°C **63.** 8 mL
**64.** 3000 mL **65.** 250 mcg **66.** B **67.** B **68.** 17.54 sec
**69.** 1 gal = 128 oz, so 1 oz of water (as capacity) weighs $\frac{8.3453}{128}$ lb,
or about 0.0652 lb. An ounce of pennies weighs $\frac{1}{16}$ lb, or
0.0625 lb. Thus, an ounce of water (as capacity) weighs more
than an ounce of pennies. **70.** $101.25 **71.** 104.875 laps
**72.** 961.625 cm²

### Understanding Through Discussion and Writing

**1.** Grams are more easily converted to other units of mass
than ounces. Since 1 gram is much smaller than 1 ounce,
masses that might be expressed using fractional or decimal
parts of ounces can often be expressed by whole numbers
when grams are used. **2.** No, the sum of the three angles
of a triangle is 180°. If you have two 90° angles, totaling 180°,
the third angle would be 0°. A triangle can't have an angle of
0°. **3.** Since 1 m is slightly longer than 1 yd, it follows that
1 m² is larger than 1 yd². **4.** Add 90° to the measure of the
angle's complement. **5.** Show that the sum of the squares of
the lengths of the legs is the same as the square of the length
of the hypotenuse. **6.** Volume of two spheres, each with
radius $r$: $2(\frac{4}{3}\pi r^3) = \frac{8}{3}\pi r^3$; volume of one sphere with radius
$2r$: $\frac{4}{3}\pi(2r)^3 = \frac{32}{3}\pi r^3 = 4 \cdot \frac{8}{3}\pi r^3$. The volume of the sphere with
radius $2r$ is four times the volume of the two spheres, each with
radius $r$.

## Test: Chapter 9, p. 691

**1.** [9.1a] 96 **2.** [9.1b] 2.8 **3.** [9.2a] 18 **4.** [9.1b] 5000
**5.** [9.1b] 0.91 **6.** [9.2b] 0.00452 **7.** [9.4b] 2.983
**8.** [9.7b] 3800 **9.** [9.4b] 1280 **10.** [9.4b] 690 **11.** [9.7a] 144
**12.** [9.7a] 8220 **13.** [9.3b] 8 cm **14.** [9.3b] 50.24 m²

**15.** [9.3b] 88 ft **16.** [9.3a] 25 cm² **17.** [9.3a] 18 ft²
**18.** [9.3b] 103.815 km² **19.** [9.5c] Complement: 25°;
supplement: 115° **20.** [9.5a] 90° **21.** [9.5a] 35°
**22.** [9.5b] Right **23.** [9.5b] Acute **24.** [9.5e] 35°
**25.** [9.5d] Isosceles **26.** [9.5d] Obtuse **27.** [9.4c] 420 in³
**28.** [9.4a] 628 ft³ **29.** [9.4a] 4186.$\overline{6}$ yd³ **30.** [9.4a] 30 m³
**31.** [9.6d] About 15.8 m **32.** [9.6c] $c = \sqrt{2}$; $c \approx 1.414$
**33.** [9.6c] $b = \sqrt{51}$; $b \approx 7.141$ **34.** [9.6a] 11 **35.** [9.7c] 0°C
**36.** [9.8a] A **37.** [9.5c] 112.5° **38.** [9.4c] $188.40
**39.** [9.4a] 0.65 ft³ **40.** [9.4a] 0.055 ft³

## Cumulative Review: Chapters 1–9, p. 693

**1.** [5.3b] 2,780,000; 1,820,000 **2.** [9.4c] 93,750 lb **3.** [4.6a] $4\frac{1}{6}$
**4.** [5.2b] 87.52 **5.** [2.5a] −1234 **6.** [1.9c] 1565 **7.** [5.5d] 49.2
**8.** [1.9d] 2 **9.** [5.1b] $\frac{1209}{1000}$ **10.** [7.1c] $\frac{17}{100}$ **11.** [4.2c] <
**12.** [3.5c] = **13.** [9.1a] 90 **14.** [9.7a] $\frac{3}{8}$ **15.** [9.7c] 59
**16.** [9.4b] 87 **17.** [9.2a] 27 **18.** [9.1b] 0.17 **19.** [4.3b] $\frac{1}{8}$
**20.** [6.3b] $4\frac{2}{7}$ **21.** [5.7b] 1 **22.** [2.8b], [3.8a] −16 **23.** [2.7b],
[9.3a] 380 cm; 5500 cm² **24.** [9.3b] 70 in.; 220 in.; 3850 in²
**25.** [9.4a] 179,666$\frac{2}{3}$ in³ **26.** [8.5b] 58.6 **27.** [7.7a] $84
**28.** [7.7b] $22,376.03 **29.** [9.6d] 17 m **30.** [7.6a] 6%
**31.** [4.6c] $2\frac{1}{8}$ yd **32.** [5.8b] $51.46 **33.** [6.2b] The 8-qt box
**34.** [3.6b] $\frac{7}{20}$ km **35.** [9.5e] 30° **36.** [2.7a] $9a - 16$
**37.** [8.4b]

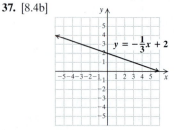

$y = -\frac{1}{3}x + 2$

**38.** [9.1a], [9.4a] 94,200 ft³ **39.** [9.1a], [9.4a] 1.342 ft³

## CHAPTER 10

### Calculator Corner, p. 697

**1.** 4.21399177; 4.21399177 **2.** 4.768371582; 4.768371582
**3.** −0.2097152; −0.2097152 **4.** −0.0484002582; −0.0484002582
**5.** 2.0736; 2.0736 **6.** 0.4932701843; 0.4932701843

### Exercise Set 10.1, p. 699

**RC1.** False **RC2.** False **RC3.** False **RC4.** True
**CC1.** 0 **CC2.** 1 **CC3.** 2 **CC4.** −1
**1.** 1 **3.** 3.14 **5.** −19.57 **7.** 1 **9.** 1 **11.** 97 **13.** 1
**15.** 17 **17.** 1 **19.** 7 **21.** 5 **23.** 2 **25.** $\frac{1}{3^2}$; $\frac{1}{9}$
**27.** $\frac{1}{10^4}$; $\frac{1}{10,000}$ **29.** $\frac{1}{t^4}$ **31.** $\frac{1}{7^1}$; $\frac{1}{7}$ **33.** $\frac{1}{(-5)^2}$; $\frac{1}{25}$
**35.** $\frac{1}{(-x)^2}$; $\frac{1}{x^2}$ **37.** $\frac{3}{x^7}$ **39.** $\frac{m}{n^5}$ **41.** $\frac{x}{y}$ **43.** $-\frac{7}{a^9}$
**45.** $x^3$ **47.** $xy^4$ **49.** $r^5t^3$ **51.** $xyz$ **53.** $a^8bc^2$
**55.** $x^2y^3z^8$ **57.** $\frac{t^2y^3}{x^4}$ **59.** $\frac{ac}{d^4}$ **61.** $\frac{25}{4}$ **63.** $\frac{a^3}{125}$
**65.** $7^{-3}$ **67.** $9x^{-3}$ **69.** 210 ft **70.** 14 mpg **71.** 17.5 mpg
**72.** $418.95 **73.** 62.5% **74.** 37.68 cm **75.** −2736
**76.** 3312 **77.** 28 **79.** 159 **81.** 3; 3

### Exercise Set 10.2, p. 705

**RC1.** add **RC2.** multiply **RC3.** add **RC4.** subtract
**RC5.** add **RC6.** subtract **RC7.** multiply

---

**CC1.** (c) **CC2.** (b) **CC3.** (a) **CC4.** (e) **CC5.** (d)
**1.** $3^6$ **3.** $x^{14}$ **5.** $a^7$ **7.** $8^6$ **9.** $5^{38}$ **11.** $x^8$ **13.** $a^3$
**15.** $p$ **17.** $\frac{1}{x^{12}}$ **19.** $\frac{1}{5^6}$ **21.** $a^4b^{14}$ **23.** $\frac{1}{a}$ **25.** $8^7$
**27.** $x$ **29.** $a^9$ **31.** $\frac{1}{x^3}$ **33.** 1 **35.** $t^3$ **37.** $c^4$ **39.** $x^3y^5$
**41.** $x^{10}$ **43.** $\frac{1}{2^{18}}$ **45.** $7^6$ **47.** $\frac{1}{y^{12}}$ **49.** $x^4y^{10}$ **51.** $8x^6y^3$
**53.** $-8a^6b^{15}$ **55.** $81x^4y^4$ **57.** $\frac{a^7}{b^7}$ **59.** $\frac{a^6}{8}$ **61.** $\frac{x^{12}}{y^{15}}$
**63.** $\frac{2^8}{3^{10}}$ **65.** 912 m² **66.** $123,000 **67.** $(3y)^{16}$ **69.** $\frac{y}{x}$
**71.** $x^y$ **73.** 1

### Calculator Corner, p. 710

**1.** $1.3545 \times 10^{-4}$ **2.** $3.6 \times 10^{12}$ **3.** $8 \times 10^{-26}$ **4.** $3 \times 10^{13}$

### Translating for Success, p. 711

**1.** E **2.** H **3.** O **4.** A **5.** G **6.** C **7.** M **8.** I
**9.** K **10.** N

### Exercise Set 10.3, p. 712

**RC1.** True **RC2.** True **RC3.** False **RC4.** True
**CC1.** Positive power of 10 **CC2.** Negative power of 10
**CC3.** Positive power of 10 **CC4.** Negative power of 10
**1.** $2.8 \times 10^{10}$ **3.** $9.07 \times 10^{17}$ **5.** $3.04 \times 10^{-6}$ **7.** $1.8 \times 10^{-8}$
**9.** $10^{11}$ **11.** $4.19854 \times 10^8$ **13.** $6.8 \times 10^{-7}$ **15.** 87,400,000
**17.** 0.00000005704 **19.** 10,000,000 **21.** 0.00001 **23.** $6 \times 10^9$
**25.** $3.38 \times 10^4$ **27.** $8.1477 \times 10^{-13}$ **29.** $2.5 \times 10^{13}$
**31.** $5.0 \times 10^{-4}$ **33.** $3.0 \times 10^{-21}$ **35.** $1.4 \times 10^{21}$ bacteria
**37.** $4.375 \times 10^2$ days **39.** $3 \times 10^1$ e-mails per day
**41.** $3.1 \times 10^2$ sheets **43.** $\frac{7}{4}$, or 1.75 **44.** $-\frac{11}{2}$, or −5.5
**45.** −2 **46.** 9.6 **47.** $2.478125 \times 10^{-1}$ **49.** $6.4 \times 10^8$

### Mid-Chapter Review: Chapter 10, p. 715

**1.** True **2.** False **3.** False **4.** $y^2 \cdot y^{10} = y^{2+10} = y^{12}$
**5.** $\frac{y^2}{y^{10}} = y^{2-10} = y^{-8} = \frac{1}{y^8}$ **6.** $4,032,000,000,000 = 4.032 \times 10^{12}$
**7.** $0.000038 = 3.8 \times 10^{-5}$ **8.** 1 **9.** −12 **10.** −1 **11.** −18
**12.** $\frac{1}{x^7}$ **13.** $\frac{1}{3}$ **14.** $\frac{9}{x^6}$ **15.** $x^4$ **16.** $3^{15}$ **17.** $6^{10}$ **18.** $x^{11}$
**19.** $\frac{1}{y^4}$ **20.** $x^6y^{11}$ **21.** $8^9$ **22.** $\frac{1}{x}$ **23.** 1 **24.** $n^5$
**25.** $a^8b^4$ **26.** $x^{32}$ **27.** $y^{12}$ **28.** $\frac{1}{x^3}$ **29.** $\frac{1}{5^{48}}$ **30.** $x^{15}y^{20}$
**31.** $9x^4y^8$ **32.** $-8a^3b^{15}$ **33.** $\frac{c^{10}}{d^{10}}$ **34.** $\frac{4}{x^6}$ **35.** $\frac{a^{28}}{b^{56}}$
**36.** $9.07 \times 10^{-4}$ **37.** $4.3108 \times 10^{11}$ **38.** 208,000,000
**39.** 0.0000041 **40.** $2.5 \times 10^{-1}$ m **41.** $6.25 \times 10^{-2}$ m, or 6.25 cm
**42.** C **43.** A **44.** When $a$ is not zero, $a^0 = 1$. Thus, $a^0 > a^1$
is equivalent to $1 > a$. So $a^0 > a^1$ for all numbers less than 1.

**45.** The expression $x^{-3}$ is equivalent to $\frac{1}{x^3}$. This expression will
be negative whenever $x$ is negative. **46.** Since $5^{-8} = \frac{1}{5^8}$ and
$6^{-8} = \frac{1}{6^8}$, we know that $5^{-8} > 6^{-8}$ because $5^8 < 6^8$.

**47.** An answer given in mm would require a larger exponent in
scientific notation because mm is a smaller unit than km.

## Exercise Set 10.4, p. 721

**RC1.** False  **RC2.** True  **RC3.** False  **RC4.** False
**CC1.** 0  **CC2.** $2x^2$  **CC3.** $x$  **CC4.** 0  **CC5.** $2x$  **CC6.** $2x^2$
**1.** $-4x + 10$  **3.** $x^2 - 8x + 5$  **5.** $2x^2$  **7.** $6t^4 + 4t^3 + 5t^2 - t$
**9.** $7 + 12x^2$  **11.** $9x^8 + 8x^7 - 3x^4 + 2x^2 - 2x + 5$
**13.** $2t^3 - t^2 + \frac{1}{6}t + \frac{1}{3}$  **15.** 0  **17.** $13a^3b^2 + 4a^2b^2 - 4a^2b + 6ab^2$
**19.** $-0.6a^2bc + 17.5abc^3 - 5.2abc$  **21.** $-(-5x); 5x$
**23.** $-(-x^2 + 13x - 7); x^2 - 13x + 7$
**25.** $-(12x^4 - 3x^3 + 3); -12x^4 + 3x^3 - 3$  **27.** $-3x + 5$
**29.** $-4x^2 + 3x - 2$  **31.** $4x^4 - 6x^2 - \frac{3}{4}x + 8$  **33.** $7x - 1$
**35.** $4t^2 + 6$  **37.** $-x^2 - 9x + 5$  **39.** $-a^2 + 5a - 16$
**41.** $8x^4 + 3x^3 - 4x^2 + 3x - 6$  **43.** $4.6x^3 + 9.2x^2 - 3.8x - 23$
**45.** $\frac{3}{4}x^3 - \frac{1}{2}x$  **47.** $6x^3y^3 + 8x^2y^2 + 2x^2y + xy$  **49.** $-23$
**51.** 19  **53.** $-12$  **55.** 7  **57.** 4  **59.** 11  **61.** 1112 ft
**63.** About 449 accidents  **65.** $18,750  **67.** $155,000
**69.** About 823 min  **71.** 255 mg  **73.** $\frac{7}{10}$ serving per pound
**74.** $3560  **75.** $81  **76.** 26 m²  **77.** 1256 cm²  **78.** $395 per
week  **79.** $2 \cdot 2 \cdot 2 \cdot 3 \cdot 7$  **80.** $2 \cdot 2 \cdot 2 \cdot 2 \cdot 2 \cdot 3$
**81.** $3 \cdot 5 \cdot 7 \cdot 7$  **82.** $3 \cdot 3 \cdot 13$  **83.** About 711 min
**85.**

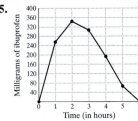

**87.** $-x^2 + 280x - 5000$  **89.** $12y^2 - 23y + 21$
**91.** $-3y^4 - y^3 + 5y - 2$  **93.** $3x^4, 2x^3, (-7)$; order of answers
may vary.

## Exercise Set 10.5, p. 730

**RC1.** True  **RC2.** False  **RC3.** True  **RC4.** False
**CC1.** (a)  **CC2.** (a)  **CC3.** (d)  **CC4.** (e)
**1.** $28a^2$  **3.** $-60x^2$  **5.** $28x^8$  **7.** $-0.07x^9$  **9.** $35x^6y^{12}$
**11.** $12a^8b^8c^2$  **13.** $-24x^{11}$  **15.** $-3x^2 + 21x$  **17.** $-3x^2 + 6x$
**19.** $x^5 + x^2$  **21.** $10x^3 - 30x^2 + 5x$  **23.** $12x^3y + 8xy^2$
**25.** $12a^7b^3 - 9a^4b^3$  **27.** $2(x + 4)$  **29.** $7(a - 5)$
**31.** $7(4x + 3y)$  **33.** $9(a - 3b + 9)$  **35.** $6(3 - m)$
**37.** $-8(2 + x - 5y)$  **39.** $9x(x^4 + 1)$  **41.** $a^2(a - 8)$
**43.** $2x(4x^2 - 3x + 1)$  **45.** $6a^4b^2(2b + 3a)$
**47.**

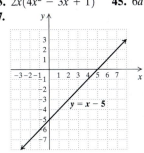

**48.**

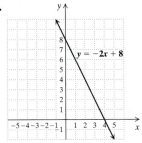

**49.**

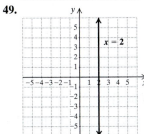

**50.**

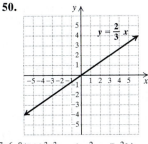

**51.** $23x^{299}(17x^{92} + 13)$  **53.** $7a^7b^6c^8(12b^3c^3 - 6ac^2 + 7a^2b)$

## Summary and Review: Chapter 10, p. 732

### Vocabulary Reinforcement

**1.** 1  **2.** product  **3.** scientific  **4.** reciprocal
**5.** power  **6.** coefficient

### Concept Reinforcement

**1.** True  **2.** True  **3.** True

### Study Guide

**1.** $-3$  **2.** 1  **3.** $\frac{1}{3^{12}}$  **4.** $y^{12}$  **5.** $x^7$  **6.** $3^{12}$  **7.** $9x^{14}$
**8.** $8.03 \times 10^{-4}$  **9.** 3480  **10.** $a^3 + 2a^2 - 4a - 2$
**11.** $-6x^2 + 5x - 3$  **12.** $14x^4$  **13.** $5x^5 + 10x^4 - 15x^3 + 45x^2$
**14.** $10a(2a^3 - 3)$

### Review Exercises

**1.** 1  **2.** 46  **3.** $-3$  **4.** 1  **5.** $\frac{1}{12^2}; \frac{1}{144}$  **6.** $\frac{8}{a^7}$  **7.** $\frac{z^6}{x^3y^5}$
**8.** $\left(\frac{5}{4}\right)^2; \frac{25}{16}$  **9.** $x^{-7}$  **10.** $x^{15}$  **11.** $x^{12}$  **12.** $3^{44}$  **13.** $x^6y^{12}$
**14.** $\frac{1}{x^8}$  **15.** $uv$  **16.** $\frac{1}{x^{11}}$  **17.** $9x^8$  **18.** $a^6$  **19.** $\frac{y^{10}}{32}$
**20.** $4.27 \times 10^7$  **21.** $1.924 \times 10^{-4}$  **22.** $1.173 \times 10^{11}$
**23.** $2.5 \times 10^{-17}$  **24.** $1.656 \times 10^{10}$ slices  **25.** $3x - 6$
**26.** $7x^4 - 4x^3 - x - 3$  **27.** $8a^5 + 12a^3 - a^2 + 4a + 9$
**28.** $5a^3b^3 + 11a^2b^3 - 7$  **29.** $-(12x^3 - 4x^2 + 9x - 3);$
$-12x^3 + 4x^2 - 9x + 3$  **30.** $-42$  **31.** 56 ft  **32.** $30x^7$
**33.** $-18x^3y^5$  **34.** $18x^4 - 12x^2 - 3x$  **35.** $14a^7b^4 + 10a^6b^4$
**36.** $5x(9x^2 - 2)$  **37.** $7(a - 5b - 7ac)$  **38.** C  **39.** A
**40.** $-40x^8$  **41.** $13a^2b^5c^5(3ab^2c - 10c^3 + 4a^2b)$
**42.** $w^5x^2y^3z^3(x^4yz^2 - w^2xy^4 + wy^2z^3 - wx^5z)$
**43.** $2a^4b^{-5}(5 + 6a^3b^2)$

### Understanding Through Discussion and Writing

**1.** Because $x^{-2} = \frac{1}{x^2}$ and because $x^2$ is never negative, it follows
that $x^{-2}$ is never negative.  **2.** It is not true that if $a > b$,
then $a^{-1} < b^{-1}$. For example, $3 > -4$, but $\frac{1}{3} \not< -\frac{1}{4}$.  **3.** Adi is
probably adding coefficients and multiplying exponents instead
of the other way around.  **4.** Consider the expression $\frac{7^2}{7^2}$.
Since this is equivalent to $\frac{49}{49}, \frac{7^2}{7^2} = 1$. And, using the quotient
rule, we have $\frac{7^2}{7^2} = 7^{2-2} = 7^0$. Therefore, $7^0 = 1$.  **5.** Not every
term is a monomial. For example, $\frac{3}{x^2}$ is a term but not a monomial.
**6.** A polynomial whose coefficients are all prime may
still be factored. For example, $3x + 3y = 3(x + y)$, and
$5x^3 + 7x^2 + 13x = x(5x^2 + 7x + 13)$.

## Test: Chapter 10, p. 736

**1.** [10.1a] 193  **2.** [10.1a] $-1$  **3.** [10.1a] 1  **4.** [10.1b] $\frac{1}{5^3}; \frac{1}{125}$
**5.** [10.1b] $\frac{5b^2}{a^3}$  **6.** [10.1b] $\left(\frac{5}{3}\right)^3; \frac{125}{27}$  **7.** [10.2a] $a^{25}$
**8.** [10.2b] $\frac{1}{x}$  **9.** [10.2d] $25x^6y^8$  **10.** [10.2d] $\frac{27}{x^{27}}$  **11.** [10.2a] $\frac{1}{x^7}$
**12.** [10.2b] $a^6b^2$  **13.** [10.3a] $4.7 \times 10^{-4}$  **14.** [10.3a] $8.25 \times 10^6$
**15.** [10.3b] $1.824 \times 10^{-16}$  **16.** [10.4a] $18a^3 - 5a^2 - a + 8$
**17.** [10.4b] $-(-9a^4 + 7b^2 - ab + 3); 9a^4 - 7b^2 + ab - 3$
**18.** [10.4c] $3x^4 - x^2 - 11$  **19.** [10.4d] 12.4 m
**20.** [10.5a] $-10x^6y^8$  **21.** [10.5b] $10a^3 - 8a^2 + 6a$
**22.** [10.5c] $5x^2(7x^4 - 5x + 3)$  **23.** [10.5c] $3(2ab - 3bc + 4ac)$
**24.** [10.3c] D  **25.** [10.4d] 2.92 L  **26.** [10.2a] $x^{9y}$

# Cumulative Review/Final Examination: Chapters 1–10, p. 738

**1.** [5.3b] 5,100,000    **2.** [1.2a] 21,085    **3.** [4.6a] $8\frac{7}{9}$

**4.** [2.2a] 24    **5.** [2.2a] −762    **6.** [5.2c] −44.364

**7.** [10.4a] $10x^5 - x^4 + 2x^3 - 3x^2 + 2$    **8.** [1.3a] 243

**9.** [2.7a] −17x    **10.** [4.3a] $-\frac{19}{40}$    **11.** [4.6b] $2\frac{17}{24}$

**12.** [5.2b] 19.9973    **13.** [10.4c] $2x^3 + 5x^2 + 7x$    **14.** [1.4a] 4752

**15.** [2.4a] −74,337    **16.** [4.7a] $4\frac{7}{12}$    **17.** [3.6a] $-\frac{6}{5}$    **18.** [3.6a] 10

**19.** [5.3a] 259.084    **20.** [2.6b] 24x − 15    **21.** [10.5a] $27a^8b^3$

**22.** [10.5b] $21x^5 - 14x^3 + 56x^2$    **23.** [1.5a] 573    **24.** [1.5a] 56 R 10

**25.** [3.7b] $-\frac{3}{2}$    **26.** [4.7b] $\frac{7}{90}$    **27.** [5.4a] 39    **28.** [1.9c] 75

**29.** [2.1c], [2.5b] −2    **30.** [1.9a] $14^3$    **31.** [1.6a] 68,000

**32.** [5.5b] 21.84    **33.** [3.1b] Yes    **34.** [3.2a] 1, 3, 5, 15

**35.** [4.1a] 105    **36.** [3.5b] $\frac{8}{11}$    **37.** [4.5a] $-3\frac{3}{5}$    **38.** [2.1b] >

**39.** [4.2c] <    **40.** [5.1c] −0.9976    **41.** [2.6a] 29

**42.** [10.5c] 5(8 − t)    **43.** [10.5c] $3a(6a^2 - 5a + 2)$    **44.** [3.3a] $\frac{3}{5}$

**45.** [5.1b] 0.0429    **46.** [5.5a] −0.52    **47.** [5.5a] $0.\overline{8}$

**48.** [7.1b] 0.07    **49.** [5.1b] $\frac{671}{100}$    **50.** [4.5a] $-\frac{29}{4}$    **51.** [7.1c] $\frac{2}{5}$

**52.** [7.1c] 85%    **53.** [7.1b] 150%    **54.** [5.6a] 13.6

**55.** [1.7b] 555    **56.** [5.7a] 64    **57.** [3.8a] $\frac{5}{4}$    **58.** [6.3b] 76.5

**59.** [2.8e] 5    **60.** [5.7b] $-\frac{2}{5}$    **61.** [8.2c] 242 eggs/person; 2011

**62.** [8.2c] 279 eggs/person; 2018    **63.** [8.5b, c, d] Mean: $260.\overline{5}$ eggs/person; median: 256 eggs/person; no mode

**64.** [7.5a] About 15.3%    **65.** [9.1a, c] $437\frac{1}{3}$ yd; about 400 m

**66.** [8.7c] $\frac{3}{20}$, or 0.15    **67.** [9.5e] 118°

**68.** [10.3c] $2.5284 \times 10^{13}$ mi    **69.** [4.6c] $4\frac{5}{8}$ yd

**70.** [1.8a] $24,000    **71.** [3.4c] $\frac{3}{10}$ km    **72.** [6.4a] 13 gal

**73.** [7.7a] $240    **74.** [7.5a] 30,160    **75.** [9.6d] About 283 ft
**76.** [9.8a] 56 mg    **77.** [10.3c] The mass of Jupiter is $3.18 \times 10^2$ times the mass of Earth.    **78.** [5.8b] 485.9 mi
**79.** [5.8b] $84.96    **80.** [6.2b] 17¢/oz    **81.** [5.8b] 220 mi
**82.** [7.6b] 7%    **83.** [10.4d] 1024 ft    **84.** [1.9b] 324

**85.** [10.1a] 1    **86.** [10.1a] 42    **87.** [9.6a] 11    **88.** [10.1b] $\frac{1}{4^3}; \frac{1}{64}$

**89.** [10.1b] $\left(\frac{4}{5}\right)^2; \frac{16}{25}$    **90.** [10.3a] $4.357 \times 10^6$

**91.** [10.3b] $2.666 \times 10^{-15}$    **92.** [9.1a] 12    **93.** [9.1b] 5800
**94.** [9.7a] 160    **95.** [9.4b] 8.19    **96.** [9.1b] 391.7    **97.** [9.7b] 60
**98.** [9.7b] 2300    **99.** [9.4b] 7    **100.** [9.2a] 90    **101.** [9.2b] 0.02
**102.** [8.2b]

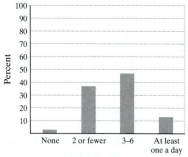

**103.** [9.5a]

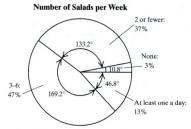

**104.** [8.3a]

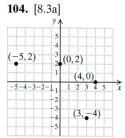

**105.** [8.4b]

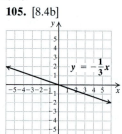

**106.** [8.4c]

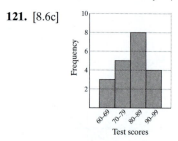

**107.** [6.5a] x = 16, y = 14    **108.** [2.7b] 88 in.
**109.** [9.3a] 61.6 cm²    **110.** [3.6b] 25 in²    **111.** [9.3a] 128.65 yd²
**112.** [9.3b] 67.390325 m²    **113.** [9.3b] Diameter: 20.8 in.; circumference: 65.312 in.; area: 339.6224 in²    **114.** [9.4a] 52.9 m³
**115.** [9.4a] 803.84 ft³    **116.** [9.4a] $523,333,333.\overline{3}$ m³
**117.** [9.6c] $\sqrt{85}$ ft; 9.220 ft    **118.** [8.5e] Minimum; 60; first quartile: 73; median: 84.5; third quartile: 88.5; maximum: 98

**119.** [8.6a]

| Test Scores | Frequency |
|---|---|
| 60–69 | 3 |
| 70–79 | 5 |
| 80–89 | 8 |
| 90–99 | 4 |

**120.** [8.6b]

```
6 | 0  3  8
7 | 1  2  4  6  9
8 | 2  4  5  5  7  7  8  9
9 | 2  3  6  8
```
Key: 6|0 = 60

**121.** [8.6c]

**122.** [9.5c] C    **123.** [9.7c] D    **124.** [8.7c] C

**125. (a)** [3.6a], [4.2d] Lot 1 pays $\frac{1}{25}$, lot 2 pays $\frac{9}{100}$, lot 3 pays $\frac{47}{300}$, lot 4 pays $\frac{77}{300}$, lot 5 pays $\frac{137}{300}$;  **(b)** [7.4a] 4%, 9%, $15\frac{2}{3}$%, $25\frac{2}{3}$%, $45\frac{2}{3}$%;  **(c)** [7.1c], [7.4a] 87%

# DEVELOPMENTAL UNITS

## Exercise Set A, p. 750

**1.** 17 **2.** 15 **3.** 13 **4.** 14 **5.** 12 **6.** 11 **7.** 17 **8.** 16
**9.** 12 **10.** 10 **11.** 10 **12.** 11 **13.** 7 **14.** 7 **15.** 11
**16.** 0 **17.** 3 **18.** 18 **19.** 14 **20.** 10 **21.** 4 **22.** 14
**23.** 11 **24.** 15 **25.** 16 **26.** 9 **27.** 13 **28.** 14 **29.** 11
**30.** 7 **31.** 13 **32.** 14 **33.** 12 **34.** 6 **35.** 10 **36.** 12
**37.** 10 **38.** 8 **39.** 2 **40.** 9 **41.** 13 **42.** 8 **43.** 10
**44.** 9 **45.** 10 **46.** 6 **47.** 13 **48.** 9 **49.** 8 **50.** 11
**51.** 12 **52.** 13 **53.** 10 **54.** 12 **55.** 17 **56.** 13 **57.** 10
**58.** 16 **59.** 16 **60.** 15 **61.** 39 **62.** 89 **63.** 87
**64.** 999 **65.** 900 **66.** 868 **67.** 999 **68.** 848 **69.** 877
**70.** 17,680 **71.** 10,873 **72.** 4699 **73.** 10,867 **74.** 9895
**75.** 3998 **76.** 18,222 **77.** 16,889 **78.** 64,489 **79.** 99,999
**80.** 77,777 **81.** 46 **82.** 26 **83.** 55 **84.** 101 **85.** 1643
**86.** 1412 **87.** 846 **88.** 628 **89.** 1204 **90.** 607
**91.** 10,000 **92.** 1010 **93.** 1110 **94.** 1227 **95.** 1111
**96.** 1717 **97.** 10,138 **98.** 6554 **99.** 6111 **100.** 8427
**101.** 9890 **102.** 11,612 **103.** 11,125 **104.** 15,543
**105.** 16,774 **106.** 68,675 **107.** 34,437 **108.** 166,444
**109.** 101,315 **110.** 49,449

## Exercise Set S, p. 757

**1.** 7 **2.** 0 **3.** 0 **4.** 5 **5.** 3 **6.** 8 **7.** 8 **8.** 6
**9.** 7 **10.** 3 **11.** 7 **12.** 9 **13.** 6 **14.** 6 **15.** 2
**16.** 4 **17.** 3 **18.** 2 **19.** 0 **20.** 3 **21.** 1 **22.** 1
**23.** 7 **24.** 0 **25.** 4 **26.** 4 **27.** 5 **28.** 4
**29.** 4 **30.** 5 **31.** 9 **32.** 9 **33.** 7 **34.** 9 **35.** 6
**36.** 6 **37.** 2 **38.** 6 **39.** 7 **40.** 8 **41.** 33 **42.** 21
**43.** 247 **44.** 193 **45.** 500 **46.** 654 **47.** 202
**48.** 617 **49.** 288 **50.** 1220 **51.** 2231 **52.** 4126
**53.** 3764 **54.** 1691 **55.** 1226 **56.** 99,998 **57.** 10
**58.** 10,101 **59.** 11,043 **60.** 11,111 **61.** 65 **62.** 29
**63.** 8 **64.** 9 **65.** 308 **66.** 126 **67.** 617 **68.** 214
**69.** 89 **70.** 4402 **71.** 3555 **72.** 5889 **73.** 2387
**74.** 3649 **75.** 3832 **76.** 8144 **77.** 7750 **78.** 10,445
**79.** 33,793 **80.** 281 **81.** 455 **82.** 571 **83.** 6148
**84.** 2200 **85.** 2113 **86.** 3748 **87.** 5206 **88.** 1459
**89.** 305 **90.** 4455

## Exercise Set M, p. 765

**1.** 12 **2.** 0 **3.** 7 **4.** 0 **5.** 10 **6.** 30 **7.** 10 **8.** 63
**9.** 54 **10.** 12 **11.** 0 **12.** 72 **13.** 8 **14.** 0 **15.** 28
**16.** 24 **17.** 45 **18.** 18 **19.** 0 **20.** 35 **21.** 45 **22.** 40
**23.** 0 **24.** 16 **25.** 25 **26.** 81 **27.** 1 **28.** 0 **29.** 4
**30.** 36 **31.** 8 **32.** 0 **33.** 27 **34.** 18 **35.** 0 **36.** 10
**37.** 48 **38.** 54 **39.** 0 **40.** 72 **41.** 15 **42.** 8 **43.** 9
**44.** 2 **45.** 32 **46.** 6 **47.** 15 **48.** 6 **49.** 8 **50.** 20
**51.** 20 **52.** 16 **53.** 10 **54.** 0 **55.** 80 **56.** 70 **57.** 160
**58.** 210 **59.** 450 **60.** 780 **61.** 560 **62.** 360 **63.** 800
**64.** 300 **65.** 900 **66.** 1000 **67.** 345,700 **68.** 1200
**69.** 4900 **70.** 4000 **71.** 10,000 **72.** 7000 **73.** 9000
**74.** 2000 **75.** 457,000 **76.** 6,769,000 **77.** 18,000
**78.** 20,000 **79.** 48,000 **80.** 16,000 **81.** 6000
**82.** 1,000,000 **83.** 1200 **84.** 200 **85.** 4000 **86.** 2500
**87.** 12,000 **88.** 6000 **89.** 63,000 **90.** 120,000
**91.** 800,000 **92.** 120,000 **93.** 16,000,000 **94.** 80,000
**95.** 147 **96.** 444 **97.** 2965 **98.** 4872 **99.** 6293
**100.** 3460 **101.** 16,236 **102.** 13,508 **103.** 87,554
**104.** 195,384 **105.** 3480 **106.** 2790 **107.** 3360
**108.** 7020 **109.** 20,760 **110.** 10,680 **111.** 358,800
**112.** 109,800 **113.** 583,800 **114.** 299,700 **115.** 11,346,000
**116.** 23,390,000 **117.** 61,092,000 **118.** 73,032,000

## Exercise Set D, p. 771

**1.** 3 **2.** 8 **3.** 4 **4.** 1 **5.** 32 **6.** 9 **7.** 7 **8.** 5
**9.** 37 **10.** 5 **11.** 9 **12.** 4 **13.** 6 **14.** 9 **15.** 5
**16.** 8 **17.** 9 **18.** 6 **19.** 3 **20.** 2 **21.** 9 **22.** 2
**23.** 3 **24.** 7 **25.** 8 **26.** 4 **27.** 7 **28.** 2 **29.** 3
**30.** 6 **31.** 1 **32.** 4 **33.** 6 **34.** 3 **35.** 7 **36.** 9
**37.** 0 **38.** Undefined **39.** Undefined **40.** 7 **41.** 8
**42.** 7 **43.** 1 **44.** 5 **45.** 0 **46.** 0 **47.** 3 **48.** 1
**49.** 4 **50.** 7 **51.** 1 **52.** 6 **53.** 1 **54.** 5 **55.** 2
**56.** 8 **57.** 0 **58.** 0 **59.** 8 **60.** 3 **61.** 4
**62.** Undefined **63.** 4 **64.** 1 **65.** 69 R 1 **66.** 199 R 1
**67.** 92 R 1 **68.** 138 R 3 **69.** 21 **70.** 118 **71.** 91 R 1
**72.** 321 R 2 **73.** 1723 R 4 **74.** 2925 **75.** 864 R 1
**76.** 522 R 3 **77.** 964 R 3 **78.** 679 R 5 **79.** 2897
**80.** 1328 R 2 **81.** 43 R 15 **82.** 32 R 27 **83.** 137 R 43
**84.** 248 R 26 **85.** 37 R 8 **86.** 76 **87.** 526 **88.** 164 R 2

# Guided Solutions

## CHAPTER 1

### Section 1.1

**8.** $2718 = 2$ thousands $+ 7$ hundreds $+ 1$ ten $+ 8$ ones

**17.** One million, eight hundred seventy-nine thousand, two hundred four

### Section 1.2

**2.**
$$
\begin{array}{r}
\overset{1\ \ 1\ \ 1}{7\ 9\ 6\ 8} \\
+\ 5\ 4\ 9\ 7 \\
\hline
1\ 3{,}4\ 6\ 5
\end{array}
$$

**5.** Perimeter $= 4$ in. $+ 5$ in. $+ 9$ in. $+ 6$ in. $+ 5$ in. $= 29$ in.

### Section 1.3

**1.**
$$
\begin{array}{r}
7\ 8\ 9\ 3 \\
-\ 4\ 0\ 9\ 2 \\
\hline
3\ 8\ 0\ 1
\end{array}
$$
Check:
$$
\begin{array}{r}
3\ 8\ 0\ 1 \\
+\ 4\ 0\ 9\ 2 \\
\hline
7\ 8\ 9\ 3
\end{array}
$$

**5.**
$$
\begin{array}{r}
\overset{4\ \ 9\ 13}{5\ 0\ 3} \\
-\ 2\ 9\ 8 \\
\hline
2\ 0\ 5
\end{array}
$$

### Section 1.4

**4.**
$$
\begin{array}{r}
\overset{1\ 2\ 4}{1\ 3\ 4\ 8} \\
\times\ \ \ \ \ \ 5 \\
\hline
6\ 7\ 4\ 0
\end{array}
$$

**20.** $A = l \cdot w$
$= 12$ ft $\cdot 8$ ft
$= 96$ sq ft

### Section 1.5

**8.** $0 \div 2$ means $0$ divided by $2$.
Since zero divided by any nonzero number is $0$, $0 \div 2 = 0$.

**9.** $7 \div 0$ means $7$ divided by $0$.
Since division by $0$ is not defined, $7 \div 0$ is not defined.

### Section 1.6

**26.** Nearest ten:
$$
\begin{array}{r}
8\ 4\ 0 \\
\times\ 2\ 5\ 0 \\
\hline
4\ 2\ \ 0\ 0\ 0 \\
1\ 6\ 8\ \ 0\ 0\ 0 \\
\hline
2\ 1\ 0{,}0\ 0\ 0
\end{array}
$$
Nearest hundred:
$$
\begin{array}{r}
8\ 0\ 0 \\
\times\ 2\ 0\ 0 \\
\hline
1\ 6\ 0{,}0\ 0\ 0
\end{array}
$$

**31.** Since 8 is to the left of 12 on the number line, $8 < 12$.

### Section 1.7

**13.**
$$x + 9 = 17$$
$$x + 9 - 9 = 17 - 9$$
$$x = 8$$
Check:
$$\frac{x + 9\ = 17}{8 + 9\ \overset{?}{=}\ 17}$$
$$17\ \big|$$
Since $17 = 17$ is true, the answer checks.
The solution is 8.

**19.**
$$\frac{144}{9} = \frac{9 \cdot n}{9}$$
$$16 = n$$
Check:
$$\frac{144 = 9 \cdot n}{144\ \overset{?}{=}\ 9 \cdot 16}$$
$$144$$
Since $144 = 144$ is true, the answer checks.
The solution is 16.

### Section 1.8

**4.** **1. Familiarize.** Let $p =$ the number of pages William still has to read.

**2. Translate.**

| Pages already read | plus | Number of pages to read | is | Total number of pages |
|---|---|---|---|---|
| ↓ | ↓ | ↓ | ↓ | ↓ |
| 86 | + | $p$ | = | 234 |

**3. Solve.**
$$86 + p = 234$$
$$86 + p - 86 = 234 - 86$$
$$p = 148$$

**4. Check.** If William reads 148 more pages, he will have read a total of $86 + 148$ pages, or 234 pages.

**5. State.** William has 148 more pages to read.

**9.** **1. Familiarize.** Let $x =$ the number of hundreds in 3500. Let $t =$ the time it takes to lose one pound.

**2. Translate.**
$$100 \cdot x = 3500$$
$$x \cdot 2 = t$$

**3. Solve.** From Example 7, we know that $x = 35$.
$$x \cdot 2 = t$$
$$35 \cdot 2 = t$$
$$70 = t$$

**4. Check.** Since $70 \div 2 = 35$, there are 35 groups of 2 min in 70 min. Thus you will burn $35 \times 100 = 3500$ calories.

**5. State.** You must swim for 70 min, or 1 hr 10 min, in order to lose one pound.

### Section 1.9

**5.** $10^4 = 10 \cdot 10 \cdot 10 \cdot 10 = 10{,}000$

**15.** $9 \times 4 - (20 + 4) \div 8 - (6 - 2)$
$= 9 \times 4 - 24 \div 8 - 4$
$= 36 - 24 \div 8 - 4$
$= 36 - 3 - 4$
$= 33 - 4$
$= 29$

**25.** $[18 - (2 + 7) \div 3] - (31 - 10 \times 2)$
$= [18 - 9 \div 3] - (31 - 10 \times 2)$
$= [18 - 3] - (31 - 20)$
$= 15 - 11$
$= 4$

# CHAPTER 2

## Section 2.1

**10.** The distance of 18 from 0 is 18, so $|18| = 18$.

**22.** $-(-x) = -(-(-2))$
   $= -(2) = -2$

## Section 2.2

**19.** $-12$ and 12 have the same absolute value. The answer is 0.

**27.** Add the positive numbers:

   $25 + 10 = 35$.

   Add the negative numbers:

   $-15 + (-5) + (-9) + (-14) = -43$.

   Finally, add the results:

   $35 + (-43) = -8$.

## Section 2.3

**11.** $2 - 8 = 2 + (-8) = -6$

**17.** $-6 - (-2) - (-4) - 12 + 3 = -6 + 2 + 4 + (-12) + 3$
   $= -6 + (-12) + 2 + 4 + 3$
   $= -18 + 9$
   $= -9$

## Section 2.4

**9.** Multiply absolute values:

   $3 \cdot 6 = 18$.

   The signs are different, so the answer is negative.

   $3(-6) = -18$

**17.** $(-4)(-5)(-2)(-3)(-1) = 20 \cdot 6 \cdot (-1)$
   $= 120 \cdot (-1)$
   $= -120$

## Section 2.5

**11.** $(-2) \cdot |3 - 2^2| + 5 = (-2) \cdot |3 - 4| + 5$
   $= (-2) \cdot |-1| + 5$
   $= (-2) \cdot 1 + 5$
   $= -2 + 5$
   $= 3$

## Section 2.6

**4.** $\dfrac{-6}{x} = -\dfrac{6}{x} = \dfrac{6}{-x}$

**17.** $6(x + y + z) = 6 \cdot x + 6 \cdot y + 6 \cdot z$
   $= 6x + 6y + 6z$

## Section 2.7

**7.** The like terms are

   $4m$ and $m$,

   $-2n^2$ and $n^2$,

   $5$ and $-9$.

   $4m + (-2n^2) + 5 + n^2 + m + (-9)$
   $= 4m + m + (-2n^2) + n^2 + 5 + (-9)$
   $= 5m + (-n^2) + (-4)$
   $= 5m - n^2 - 4$

**13.** $P = 4s$
   $= 4 \cdot 9$ in.
   $= 36$ in.

## Section 2.8

**12.** $-t = -3$
   $-1 \cdot t = -3$
   $\dfrac{-1 \cdot t}{-1} = \dfrac{-3}{-1}$
   $t = 3$

**18.** $2x - 9 = 43$
   $2x - 9 + 9 = 43 + 9$
   $2x = 52$
   $\dfrac{2x}{2} = \dfrac{52}{2}$
   $x = 26$

# CHAPTER 3

## Section 3.1

**4.**
$$\begin{array}{r} 8 \\ 2\overline{)16} \\ \underline{16} \\ 0 \end{array}$$

Since the remainder is 0, 16 is divisible by 2.

**22.** Add the digits:

   $1 + 7 + 2 + 1 + 6 = 17$.

   Since 17 is not divisible by 3, the number 17,216 is not divisible by 3.

## Section 3.2

**4.** 1 is a factor of 45.     $1 \cdot 45$
   2 is not a factor of 45.
   3 is a factor of 45.     $3 \cdot 15$
   4 is not a factor of 45.
   5 is a factor of 45.     $5 \cdot 9$
   6 is not a factor of 45.
   7 is not a factor of 45.
   8 is not a factor of 45.
   Factors of 45: 1, 3, 5, 9, 15, 45.

**9.**
$$\begin{array}{r} 7 \\ 7\overline{)49} \\ 2\overline{)98} \end{array}$$
   $98 = 2 \cdot 7 \cdot 7$

## Section 3.3

**12.** Each gallon is divided into 4 equal parts.

   The unit is $\dfrac{1}{4}$.

   There are 7 equal units shaded.

   The part that is shaded is $\dfrac{7}{4}$.

**24.** $\dfrac{4 - 4}{567} = \dfrac{0}{567} = 0$

## Section 3.4

**9.** $\dfrac{3}{8} \cdot \dfrac{5}{7} = \dfrac{3 \cdot 5}{8 \cdot 7}$
   $= \dfrac{15}{56}$

## Section 3.5

**5.** $\dfrac{4}{3} = \dfrac{4}{3} \cdot \dfrac{5}{5}$
   $= \dfrac{4 \cdot 5}{3 \cdot 5}$
   $= \dfrac{20}{15}$

**25.** $2 \cdot 20 = 40 \qquad 3 \cdot 14 = 42$

$$\dfrac{2}{3} \; \square \; \dfrac{14}{20}$$

Since $40 \neq 42,\ \dfrac{2}{3} \neq \dfrac{14}{20}$.

## Section 3.6

**1.** 
$$\dfrac{2}{3} \cdot \dfrac{7}{8} = \dfrac{2 \cdot 7}{3 \cdot 8}$$
$$= \dfrac{2 \cdot 7}{3 \cdot 2 \cdot 2 \cdot 2}$$
$$= \dfrac{2}{2} \cdot \dfrac{7}{3 \cdot 2 \cdot 2}$$
$$= 1 \cdot \dfrac{7}{3 \cdot 2 \cdot 2}$$
$$= \dfrac{7}{12}$$

**7.** 
$$A = \dfrac{1}{2} \cdot b \cdot h$$
$$= \dfrac{1}{2} \cdot 11 \text{ cm} \cdot \dfrac{12}{5} \text{ cm}$$
$$= \dfrac{1 \cdot 11 \cdot 12}{2 \cdot 5} \text{ cm}^2$$
$$= \dfrac{1 \cdot 11 \cdot 2 \cdot 2 \cdot 3}{2 \cdot 5} \text{ cm}^2$$
$$= \dfrac{66}{5} \text{ cm}^2$$

## Section 3.7

**6.** 
$$\dfrac{6}{7} \div \dfrac{3}{4} = \dfrac{6}{7} \cdot \dfrac{4}{3}$$
$$= \dfrac{6 \cdot 4}{7 \cdot 3}$$
$$= \dfrac{2 \cdot 3 \cdot 2 \cdot 2}{7 \cdot 3}$$
$$= \dfrac{3}{3} \cdot \dfrac{2 \cdot 2 \cdot 2}{7}$$
$$= \dfrac{2 \cdot 2 \cdot 2}{7}$$
$$= \dfrac{8}{7}$$

## Section 3.8

**1.** 
$$\dfrac{3}{2} \cdot \dfrac{2}{3}x = \dfrac{3}{2} \cdot 8$$
$$1x = \dfrac{3 \cdot 8}{2}$$
$$x = \dfrac{3 \cdot 2 \cdot 4}{2}$$
$$x = 12$$

**4.** 
$$-\dfrac{7}{6}\left(-\dfrac{6}{7}a\right) = -\dfrac{7}{6} \cdot \dfrac{9}{14}$$
$$a = -\dfrac{7 \cdot 3 \cdot 3}{2 \cdot 3 \cdot 2 \cdot 7}$$
$$a = -\dfrac{3}{4}$$

# CHAPTER 4

## Section 4.1

**12. 1.** $18 = 2 \cdot 3 \cdot 3$
$40 = 2 \cdot 2 \cdot 2 \cdot 5$

**2.** Select the factorization of 40:
$2 \cdot 2 \cdot 2 \cdot 5$.
This is not a multiple of 18. We need two factors of 3.

**3.** $\text{LCM} = 2 \cdot 2 \cdot 2 \cdot 5 \cdot 3 \cdot 3 = 360$

**18.** $5a^2 = 5 \cdot a^2$
$a^3 b = \quad a^3 \cdot b$
$\text{LCM} = 5 \cdot a^3 \cdot b,\ \text{or}\ 5a^3 b$

## Section 4.2

**9.** The LCD is 24.
$$\dfrac{3}{8} + \dfrac{5}{6} = \dfrac{3}{8} \cdot 1 + \dfrac{5}{6} \cdot 1$$
$$= \dfrac{3}{8} \cdot \dfrac{3}{3} + \dfrac{5}{6} \cdot \dfrac{4}{4}$$
$$= \dfrac{9}{24} + \dfrac{20}{24}$$
$$= \dfrac{29}{24}$$

**20. 1. Familiarize.** Let $T =$ the total amount of berries in the salad.

**2. Translate.** To find the total amount, we add.
$$\dfrac{7}{8} + \dfrac{3}{4} + \dfrac{5}{16} = T$$

**3. Solve.** The LCD is 16.
$$\dfrac{7}{8} \cdot \dfrac{2}{2} + \dfrac{3}{4} \cdot \dfrac{4}{4} + \dfrac{5}{16} = T$$
$$\dfrac{14}{16} + \dfrac{12}{16} + \dfrac{5}{16} = T$$
$$\dfrac{31}{16} = T$$

**4. Check.** The answer is reasonable because it is larger than any of the individual amounts.

**5. State.** There are $\dfrac{31}{16}$ qt of berries in the salad.

## Section 4.3

**6.** The LCD is 18.
$$\dfrac{5}{6} - \dfrac{1}{9} = \dfrac{5}{6} \cdot \dfrac{3}{3} - \dfrac{1}{9} \cdot \dfrac{2}{2}$$
$$= \dfrac{15}{18} - \dfrac{2}{18}$$
$$= \dfrac{13}{18}$$

**13.** 
$$\dfrac{3}{5} + t = -\dfrac{7}{8}$$
$$\dfrac{3}{5} + t - \dfrac{3}{5} = -\dfrac{7}{8} - \dfrac{3}{5}$$
$$t + 0 = -\dfrac{7}{8} \cdot \dfrac{5}{5} - \dfrac{3}{5} \cdot \dfrac{8}{8}$$
$$t = -\dfrac{35}{40} - \dfrac{24}{40}$$
$$t = \dfrac{-35 - 24}{40}$$
$$t = \dfrac{-35 + (-24)}{40}$$
$$t = \dfrac{-59}{40} = -\dfrac{59}{40}$$

## Section 4.4

**2.**
$$\frac{1}{2}x - \frac{1}{5} = \frac{7}{10}$$
$$\frac{1}{2}x - \frac{1}{5} + \frac{1}{5} = \frac{7}{10} + \frac{1}{5}$$
$$\frac{1}{2}x = \frac{7}{10} + \frac{2}{10}$$
$$\frac{1}{2}x = \frac{9}{10}$$
$$2 \cdot \frac{1}{2}x = 2 \cdot \frac{9}{10}$$
$$1x = \frac{2 \cdot 3 \cdot 3}{2 \cdot 5}$$
$$x = \frac{9}{5}$$

**6.**
$$20 = 6 - \frac{2}{3}x$$
$$3(20) = 3\left(6 - \frac{2}{3}x\right)$$
$$60 = 3 \cdot 6 - \frac{3 \cdot 2}{3}x$$
$$60 = 18 - 2x$$
$$60 - 18 = 18 - 2x - 18$$
$$42 = -2x$$
$$\frac{42}{-2} = \frac{-2x}{-2}$$
$$-21 = x$$

## Section 4.5

**6.**
$$4 \cdot 6 = 24$$
$$24 + 5 = 29$$
$$4\frac{5}{6} = \frac{29}{6}$$

**16.**
$$\begin{array}{r} 3 \\ 5\overline{)17} \\ \underline{15} \\ 2 \end{array}$$
$$\frac{17}{5} = 3\frac{2}{5}, \text{ so } \frac{-17}{5} = -3\frac{2}{5}$$

## Section 4.6

**5.**
$$8\frac{2}{3} = 8\frac{4}{6}$$
$$-5\frac{1}{2} = -5\frac{3}{6}$$
$$\rule{2cm}{0.4pt}$$
$$3\frac{1}{6}$$

**6.**
$$5 = 4\frac{3}{3}$$
$$-1\frac{1}{3} = -1\frac{1}{3}$$
$$\rule{2cm}{0.4pt}$$
$$3\frac{2}{3}$$

## Section 4.7

**3.**
$$-2 \cdot 6\frac{2}{5} = -\frac{2}{1} \cdot \frac{32}{5}$$
$$= -\frac{64}{5}$$
$$= -12\frac{4}{5}$$

**6.**
$$2\frac{1}{4} \div 1\frac{1}{5} = \frac{9}{4} \div \frac{6}{5}$$
$$= \frac{9}{4} \cdot \frac{5}{6}$$
$$= \frac{3 \cdot 3 \cdot 5}{2 \cdot 2 \cdot 3}$$
$$= \frac{3}{3} \cdot \frac{3 \cdot 5}{2 \cdot 2 \cdot 2}$$
$$= \frac{15}{8}$$
$$= 1\frac{7}{8}$$

## Section 4.8

**2.**
$$\frac{1}{3} \cdot \frac{3}{4} \div \frac{5}{8} - \frac{1}{10} = \frac{3}{12} \div \frac{5}{8} - \frac{1}{10}$$
$$= \frac{3}{12} \cdot \frac{8}{5} - \frac{1}{10}$$
$$= \frac{3 \cdot 2 \cdot 2 \cdot 2}{3 \cdot 2 \cdot 2 \cdot 5} - \frac{1}{10}$$
$$= \frac{2}{5} - \frac{1}{10}$$
$$= \frac{4}{10} - \frac{1}{10} = \frac{3}{10}$$

**7.**
$$\frac{10}{\frac{5}{8}} = 10 \div \frac{5}{8}$$
$$= 10 \cdot \frac{8}{5}$$
$$= \frac{10 \cdot 8}{5}$$
$$= \frac{2 \cdot 5 \cdot 8}{5 \cdot 1}$$
$$= 16$$

# CHAPTER 5

## Section 5.1

**7.** 0.896.  3 places
$$0.896 = \frac{896}{1000}$$

**10.** $\frac{743}{100}$   7.43.
2 zeros  2 places
$$\frac{743}{100} = 7.43$$

## Section 5.2

**7.**
$$\begin{array}{r} \scriptstyle 1\ 1\ 1\ 1 \\ 4\ 5.7\ 8\ 0 \\ 2\ 4\ 6\ 7.0\ 0\ 0 \\ +\ \ \ \ \ \ \ 1.9\ 9\ 3 \\ \hline 2\ 5\ 1\ 4.7\ 7\ 3 \end{array}$$

**8.**
$$\begin{array}{r} \scriptstyle 13 \\ \scriptstyle 6\ 3\ 12 \\ 3\ 7.4\ 2\ 8 \\ -\ 2\ 6.6\ 7\ 4 \\ \hline 1\ 0.7\ 5\ 4 \end{array}$$

**14.**
$$\begin{array}{r} \scriptstyle 4\ 9\ 9\ 9\ 10 \\ 5.0\ 0\ 0\ 0 \\ -\ 0.0\ 0\ 8\ 9 \\ \hline 4.9\ 9\ 1\ 1 \end{array}$$

## Section 5.3

**3.**
$$\begin{array}{r} 4\ 2.6\ 5 \\ \times\ \ 0.8\ 0\ 4 \\ \hline 1\ 7\ 0\ 6\ 0 \\ 3\ 4\ 1\ 2\ 0\ 0\ 0 \\ \hline 3\ 4.2\ 9\ 0\ 6\ 0 \end{array}$$

**15.**
$$\$15.69 = 15.69 \times \$1$$
$$= 15.69 \times 100¢$$
$$= 1569¢$$

## Section 5.4

**6.**
$$\begin{array}{r} 0.0\ 2\ 5 \\ 8\ 6\overline{)2.1\ 5\ 0} \\ \underline{1\ 7\ 2} \\ 4\ 3\ 0 \\ \underline{4\ 3\ 0} \\ 0 \end{array}$$

**7.** $\dfrac{0.375}{0.25} = \dfrac{0.375}{0.25} \times \dfrac{100}{100}$

$= \dfrac{37.5}{25}$

$$
\begin{array}{r}
1.5 \\
0.2\,5\,)\overline{0.3\,7\,5} \\
\underline{2\,5} \\
1\,2\,5 \\
\underline{1\,2\,5} \\
0
\end{array}
$$

**16.** $625 \div 62.5 \times 25 \div 6250$
$= 10 \times 25 \div 6250$
$= 250 \div 6250$
$= 0.04$

## Section 5.5

**3.** $\dfrac{1}{6} = 1 \div 6$

$$
\begin{array}{r}
0.1\,6\,6 \\
6\,)\overline{1.0\,0\,0} \\
\underline{6} \\
4\,0 \\
\underline{3\,6} \\
4\,0 \\
\underline{3\,6} \\
4
\end{array}
$$

$\dfrac{1}{6} = 0.1666\ldots = 0.1\overline{6}$

**17.** Method 1:
$\dfrac{3}{4} \times 0.62 = \dfrac{3}{4} \times \dfrac{0.62}{1}$
$= \dfrac{1.86}{4} = 0.465$

Method 2:
$\dfrac{3}{4} \times 0.62 = 0.75 \times 0.62$
$= 0.465$

Method 3:
$\dfrac{3}{4} \times 0.62 = \dfrac{3}{4} \cdot \dfrac{62}{100}$
$= \dfrac{186}{400}$
$= \dfrac{93}{200} = 0.465$

## Section 5.7

**6.**
$$
\begin{aligned}
8 + 4x &= 9x - 3 \\
8 + 4x - 4x &= 9x - 3 - 4x \\
8 &= 5x - 3 \\
8 + 3 &= 5x - 3 + 3 \\
11 &= 5x \\
\dfrac{11}{5} &= \dfrac{5x}{5} \\
2.2 &= x
\end{aligned}
$$

**8.**
$$
\begin{aligned}
3(x + 5) &= 20 - x \\
3x + 15 &= 20 - x \\
3x + 15 + x &= 20 - x + x \\
4x + 15 &= 20 \\
4x + 15 - 15 &= 20 - 15 \\
4x &= 5 \\
\dfrac{4x}{4} &= \dfrac{5}{4} \\
x &= 1.25
\end{aligned}
$$

# CHAPTER 6

## Section 6.1

**6.** $\dfrac{\text{Length of shortest side}}{\text{Length of longest side}} = \dfrac{38.2}{55.5}$

**9.** Ratio of 3.6 to 12: $\dfrac{3.6}{12}$

Simplifying:

$\dfrac{3.6}{12} \cdot \dfrac{10}{10} = \dfrac{36}{120} = \dfrac{12 \cdot 3}{12 \cdot 10} = \dfrac{12}{12} \cdot \dfrac{3}{10} = \dfrac{3}{10}$

## Section 6.2

**4.** $\dfrac{52 \text{ ft}}{13 \text{ sec}} = 4 \text{ ft/sec}$

**6.** Unit price $= \dfrac{\text{Price}}{\text{Number of units}}$

$= \dfrac{\$2.79}{26 \text{ oz}} = \dfrac{279 \text{ cents}}{26 \text{ oz}}$

$= \dfrac{279}{26} \dfrac{\text{cents}}{\text{oz}} \approx 10.731 \text{¢/oz}$

## Section 6.3

**3.** We compare cross products.

$1 \cdot 39 = 39 \qquad \dfrac{1}{2} \; ? \; \dfrac{20}{39}$

$2 \cdot 20 = 40$

Since $39 \ne 40$, the numbers are not proportional.

**8.** $\dfrac{x}{9} = \dfrac{5}{4}$

$x \cdot 4 = 9 \cdot 5$

$\dfrac{x \cdot 4}{4} = \dfrac{9 \cdot 5}{4}$

$x = \dfrac{45}{4} = 11\dfrac{1}{4}$

## Section 6.4

**2. 1. Familiarize.** Let $p =$ the amount of paint needed, in gallons.

**2. Translate.** $\dfrac{4}{1600} = \dfrac{p}{6000}$

**3. Solve.**

$4 \cdot 6000 = 1600 \cdot p$

$15 = p$

**4. Check.** The cross products are the same.

**5. State.** For 6000 ft$^2$, they would need 15 gal of paint.

**6. 1. Familiarize.** Let $D =$ the number of deer in the forest.

**2. Translate.** $\dfrac{153}{D} = \dfrac{18}{62}$

**3. Solve.**

$153 \cdot 62 = D \cdot 18$

$527 = D$

**4. Check.** The cross products are the same.

**5. State.** There are about 527 deer in the forest.

## Section 6.5

**1.** The ratio of $x$ to 20 is the same as the ratio of 9 to 12.

$\dfrac{x}{20} = \dfrac{9}{12}$

$x \cdot 12 = 20 \cdot 9$

$\dfrac{x \cdot 12}{12} = \dfrac{20 \cdot 9}{12}$

$x = \dfrac{180}{12} = 15$

**5.** Let $w =$ the width of an actual skylight.

$\dfrac{12}{52} = \dfrac{3}{w}$

$12 \cdot w = 52 \cdot 3$

$w = 13$

The width of an actual skylight will be 13 ft.

# CHAPTER 7

## Section 7.1

**21.** $\dfrac{19}{25} = \dfrac{19}{25} \cdot \dfrac{4}{4}$

$\quad = \dfrac{76}{100} = 76\%$

**25.** $3.25\% = \dfrac{3.25}{100} = \dfrac{3.25}{100} \times \dfrac{100}{100}$

$\quad = \dfrac{325}{10,000} = \dfrac{13 \times 25}{400 \times 25}$

$\quad = \dfrac{13}{400} \times \dfrac{25}{25} = \dfrac{13}{400}$

## Section 7.2

**9.** 20% of what is 45?

$\quad\ \downarrow\qquad\ \ \downarrow\quad\ \downarrow\ \ \downarrow\quad\ \downarrow$

$\quad 0.20\quad \cdot \quad b\quad = \quad 45$

$\dfrac{0.20 \cdot b}{0.20} = \dfrac{45}{0.20}$

$b = \dfrac{45}{0.2}$

$b = 225$

**11.** 16 is what percent of 40?

$\quad \downarrow\quad \downarrow\qquad\ \ \downarrow\qquad \downarrow\quad \downarrow$

$\quad 16\ \ =\qquad\ p\qquad \cdot\ \ 40$

$\dfrac{16}{40} = \dfrac{p \cdot 40}{40}$

$\dfrac{16}{40} = p$

$0.4 = p$

$40\% = p$

## Section 7.3

**8.** $\dfrac{20}{100} = \dfrac{45}{b}$

$20 \cdot b = 100 \cdot 45$

$\dfrac{20 \cdot b}{20} = \dfrac{100 \cdot 45}{20}$

$b = \dfrac{4500}{20}$

$b = 225$

Thus, 20% of $225 is $45.

**9.** $\dfrac{64}{100} = \dfrac{a}{55}$

$64 \cdot 55 = 100 \cdot a$

$\dfrac{64 \cdot 55}{100} = \dfrac{100 \cdot a}{100}$

$\dfrac{3520}{100} = a$

$35.2 = a$

Thus, 64% of 55 is 35.2.

**12.** $\dfrac{12}{40} = \dfrac{N}{100}$

$12 \cdot 100 = 40 \cdot N$

$\dfrac{12 \cdot 100}{40} = \dfrac{40 \cdot N}{40}$

$\dfrac{1200}{40} = N$

$30 = N$

Thus, $12 is 30% of $40.

## Section 7.6

**2.** Purchase price $= 4 \times \$18.95$

$\qquad\qquad\quad = \$75.80$

Sales tax $= 4\% \times \$75.80$

$\qquad\quad = 0.04 \times \$75.80$

$\qquad\quad = \$3.032$

$\qquad\quad \approx \$3.03$

Total price $= \$75.80 + \$3.03$

$\qquad\quad\ = \$78.83$

**7.** $\$2970 = 7.5\% \times S$

$\$2970 = 0.075 \cdot S$

$\dfrac{\$2970}{0.075} = \dfrac{0.075 \cdot S}{0.075}$

$\$39,600 = S$

## Section 7.7

**1.** $I = P \cdot r \cdot t$

$\quad = \$4300 \cdot 4\% \cdot 1$

$\quad = \$4300 \cdot 0.04 \cdot 1$

$\quad = \$172$

**3.** **a)** $I = P \cdot r \cdot t$

$\quad = \$4800 \cdot 5\frac{1}{2}\% \cdot \dfrac{30}{365}$

$\quad = \$4800 \cdot 0.055 \cdot \dfrac{30}{365}$

$\quad \approx \$21.70$

**b)** Total amount

$\quad = \$4800 + \$21.70$

$\quad = \$4821.70$

# CHAPTER 8

## Section 8.1

**3.** The amount of the decrease in population density is $611 - 598 = 13$.

The percent decrease is $\dfrac{13}{611} \approx 0.021$, or $2.1\%$.

**8.** The graph shows about $1\frac{3}{4}$ symbols for South America.

This represents 175 roller coasters.

The graph shows about $\frac{3}{4}$ symbol for Africa.

This represents 75 roller coasters.
There are about 100 more roller coasters in South America than in Africa.

## Section 8.2

**7.** We look from left to right along a line at $400 per ounce. The points on the graph that are below this line correspond to the years 1970, 1975, 1985, 1990, 1995, and 2000.

## Section 8.3

**12.** To plot the point (5, 3), we locate 5 on the horizontal axis and go up 3 units. We are now in the first quadrant, or quadrant I.

**16.** We substitute 5 for $x$ and 1 for $y$.

$\dfrac{y = 2x + 3}{1\ ?\ 2 \cdot 5 + 3}$

$1\ |\ 13$

Since $1 = 13$ is false, (5, 1) is not a solution.

## Section 8.4

**1.** Substitute 5 for $y$ and solve for $x$.

$x - y = 3$

$x - 5 = 3$

$x - 5 + 5 = 3 + 5$

$x = 8$

One solution of $x - y = 3$ is $(8, 5)$.

## Section 8.5

**7.** Course grade $= \dfrac{100 \cdot 15 + 92 \cdot 25 + 88 \cdot 40}{15 + 25 + 40}$

$\qquad\qquad\quad = \dfrac{7320}{80} = 91.5$

Soha's course grade is 91.5%.

**14.** Rearrange the numbers in order from smallest to largest:

$\qquad 34,\ 34,\ 67,\ 68,\ 69,\ 70.$

The middle numbers are 67 and 68.
The average of 67 and 68 is 67.5.
The median is 67.5.

**17.** Rearrange the numbers in order from smallest to largest.

13, 24, 27, 28, 67, 89.

Each number occurs one time.

There is no mode.

## Section 8.7

**6.** Probability of landing on red or blue

$$= \frac{\text{Number of ways to land on red or blue}}{\text{Number of ways to land on a space}}$$

$$= \frac{8}{12}$$

$$= \frac{2}{3}$$

# CHAPTER 9

## Section 9.1

**6.** $2\frac{5}{6}\,\text{yd} = 2\frac{5}{6} \cdot 1\,\text{yd}$

$$= \frac{17}{6} \cdot 3\,\text{ft}$$

$$= \frac{17}{2}\,\text{ft}$$

$$= 8\frac{1}{2}\,\text{ft}$$

**8.** $72\,\text{in.} = \frac{72\,\text{in.}}{1} \cdot \frac{1\,\text{ft}}{12\,\text{in.}}$

$$= \frac{72}{12} \cdot 1\,\text{ft}$$

$$= 6\,\text{ft}$$

**24.** $23\,\text{km} = 23 \cdot 1\,\text{km}$

$$= 23 \cdot 1000\,\text{m}$$

$$= 23{,}000\,\text{m}$$

**29.** $7814\,\text{m} = 7814\,\text{m} \cdot \dfrac{1\,\text{dam}}{10\,\text{m}}$

$$= \frac{7814}{10} \cdot \frac{\text{m}}{\text{m}} \cdot 1\,\text{dam}$$

$$= 781.4\,\text{dam}$$

## Section 9.2

**4.** $360\,\text{in}^2 = 360\,\text{in}^2 \cdot \dfrac{1\,\text{ft}^2}{144\,\text{in}^2}$

$$= \frac{360}{144} \cdot \frac{\text{in}^2}{\text{in}^2} \cdot 1\,\text{ft}^2$$

$$= 2.5\,\text{ft}^2$$

## Section 9.3

**7.** $C = \pi \cdot d$

$$\approx 3.14 \cdot 18\,\text{in.}$$

$$= 56.52\,\text{in.}$$

**11.** $A = \pi \cdot r \cdot r$

$$\approx \frac{22}{7} \cdot 5\,\text{km} \cdot 5\,\text{km}$$

$$= \frac{22}{7} \cdot 25\,\text{km}^2$$

$$= \frac{550}{7}\,\text{km}^2$$

$$= 78\frac{4}{7}\,\text{km}^2$$

## Section 9.4

**4.** $V = \pi \cdot r^2 \cdot h$

$$\approx 3.14 \cdot 5\,\text{ft} \cdot 5\,\text{ft} \cdot 10\,\text{ft}$$

$$= 3.14 \cdot 250\,\text{ft}^3$$

$$= 785\,\text{ft}^3$$

**6.** $V = \frac{4}{3} \cdot \pi \cdot r^3$

$$\approx \frac{4}{3} \cdot \frac{22}{7} \cdot (28\,\text{ft})^3$$

$$= \frac{4}{3} \cdot \frac{22}{7} \cdot 21{,}952\,\text{ft}^3$$

$$= \frac{275{,}968}{3}\,\text{ft}^3$$

$$= 91{,}989\frac{1}{3}\,\text{ft}^3$$

**8.** $80\,\text{qt} = 80\,\text{qt} \cdot \dfrac{1\,\text{gal}}{4\,\text{qt}}$

$$= \frac{80}{4} \cdot 1\,\text{gal}$$

$$= 20\,\text{gal}$$

**14.** $0.97\,\text{L} = 0.97 \cdot 1\,\text{L}$

$$= 0.97 \cdot 1000\,\text{mL}$$

$$= 970\,\text{mL}$$

## Section 9.5

**14.** $90° - 85° = 5°$

**19.** $180° - 71° = 109°$

## Section 9.6

**27.** $a^2 + b^2 = c^2$

$$12^2 + 5^2 = c^2$$

$$144 + 25 = c^2$$

$$169 = c^2$$

$$13 = c$$

## Section 9.7

**21.** $C = \frac{5}{9}(F - 32)$

$$= \frac{5}{9}(95 - 32)$$

$$= \frac{5}{9} \cdot 63 = 35$$

Thus, $95°\text{F} = 35°\text{C}$.

## Section 9.8

**4.** $1\,\text{mcg} = 0.000001\,\text{g}$

$$= 0.000001 \cdot 1\,\text{g}$$

$$= 0.000001 \cdot 1000\,\text{mg}$$

$$= 0.001\,\text{mg}$$

# CHAPTER 10

## Section 10.1

**14.** $xy^{-2} = x \cdot y^{-2}$

$$= x \cdot \frac{1}{y^2}$$

$$= \frac{x}{y^2}$$

**21.** $\dfrac{a^4 b^{-7}}{c^{-3}} = \dfrac{a^4 c^3}{b^7}$

## Section 10.2

**10.** $\dfrac{4^{-3}}{4^{10}} = 4^{-3-10}$

$$= 4^{-13}$$

$$= \frac{1}{4^{13}}$$

**15.** $(-2ab^4)^2 = (-2)^2 (a)^2 (b^4)^2$

$$= 4a^2 b^8$$

## Section 10.3

**10.** $(9.1 \times 10^{-17})(8.2 \times 10^3)$

$$= (9.1 \times 8.2)(10^{-17} \cdot 10^3)$$

$$= 74.62 \times 10^{-14}$$

$$= (7.462 \times 10^1) \times 10^{-14}$$

$$= 7.462 \times 10^{-13}$$

**12.** $\dfrac{1.1 \times 10^{-4}}{2.0 \times 10^{-7}} = \dfrac{1.1}{2.0} \times \dfrac{10^{-4}}{10^{-7}}$

$$= 0.55 \times 10^{-4-(-7)}$$

$$= 0.55 \times 10^3$$

$$= (5.5 \times 10^{-1}) \times 10^3$$

$$= 5.5 \times 10^2$$

## Section 10.4

**2.** Since $-a^2 = -1 \cdot a^2$, the coefficient of $-a^2$ is $-1$.

**4.** $(5x^2y + 3x^2 + 4) + (2x^2y + 4x)$

$= (5x^2y + 2x^2y) + 3x^2 + 4 + 4x$

$= 7x^2y + 3x^2 + 4x + 4$

## Section 10.5

**2.** $(-7x)(2x) = -7 \cdot x \cdot 2 \cdot x$

$\qquad = -7 \cdot 2 \cdot x \cdot x$

$\qquad = (-7 \cdot 2)(x \cdot x)$

$\qquad = -14x^2$

**19.** $9a^2b = 3 \cdot 3 \cdot a \cdot a \cdot b$

$\quad 6ab^2 = 2 \cdot 3 \cdot a \cdot b \cdot b$

The largest common factor is $3 \cdot a \cdot b$.

$9a^2b - 6ab^2 = 3ab \cdot 3a - 3ab \cdot 2b$

$\qquad\qquad\quad = 3ab(3a - 2b)$

# Glossary

## A

**Absolute value**  The distance that a number is from 0 on the number line

**Acute angle**  An angle whose measure is greater than $0°$ and less than $90°$

**Acute triangle**  A triangle in which all three angles are acute

**Addends**  In addition, the numbers being added

**Additive identity**  The number 0

**Additive inverse**  A number's opposite; two numbers are additive inverses of each other if their sum is zero.

**Additive inverse of a polynomial**  Two polynomials are additive inverses, or opposites, of each other if their sum is zero.

**Algebraic expression**  A number or variable, or a collection of numbers and variables, on which operations are performed

**Angle**  A set of points consisting of two rays (half-lines) with a common endpoint (vertex)

**Area**  The number of square units that fill a plane region

**Arithmetic mean**  A center point of a set of numbers found by adding the numbers and dividing by the number of items of data; also called *mean* or *average*

**Associative law of addition**  The statement that when three numbers are added, regrouping the addends gives the same sum

**Associative law of multiplication**  The statement that when three numbers are multiplied, regrouping the factors gives the same product

**Average**  A center point of a set of numbers found by adding the numbers and dividing by the number of items of data; also called the *mean*

**Axes**  Two perpendicular number lines used to identify points in a plane

## B

**Bar graph**  A graphic display of data using bars proportional in length to the numbers represented

**Base**  In exponential notation, the number being raised to a power

**Binomial**  A polynomial composed of two terms

## C

**Celsius**  A temperature scale in which water freezes at $0°$ and boils at $100°$

**Circumference**  The distance around a circle

**Coefficient**  The numeric multiplier of a variable

**Commission**  A percent of total sales paid to a salesperson

**Commutative law of addition**  The statement that when two numbers are added, changing the order in which the numbers are added does not affect the sum

**Commutative law of multiplication**  The statement that when two numbers are multiplied, changing the order in which the numbers are multiplied does not affect the product

**Complementary angles**  Two angles for which the sum of their measures is $90°$

**Complex fraction**  A fraction in which the numerator and/or denominator contains one or more fractions

**Composite number**  A natural number, other than 1, that is not prime

**Compound interest**  Interest computed on the sum of an original principal and the interest previously accrued by that principal

**Congruent angles**  Two angles that have the same measure

**Constant**  A number or letter that stands for just one number

**Contingency table**  A table that displays frequencies for more than one variable; also called a *two-way frequency table* if there are two variables

**Cross products**  Given an equation with a single fraction on each side, the products formed by multiplying the left numerator and the right denominator, and the left denominator and the right numerator

## D

**Decimal notation**  A representation of a number that uses base-10 place values and may include a decimal point

**Denominator**  The number below the fraction bar in a fraction

**Descending order**  When a polynomial is written with the powers of the variable decreasing as read from left to right, it is said to be in descending order.

**Diameter**  A segment that passes through the center of a circle and has its endpoints on the circle, or the length of such a segment

**Difference**  The result of subtracting one number from another

**Digit**  A number 0, 1, 2, 3, 4, 5, 6, 7, 8, or 9 that names a place-value location

**Discount**  The amount subtracted from the original price of an item to find the sale price

**Distributive law**  The statement that multiplying a factor by the sum of two numbers gives the same result as multiplying the factor by each of the two numbers and then adding

**Dividend**  In division, the number being divided

**Divisible**  The number $b$ is said to be divisible by another number $a$ if $b$ is a multiple of $a$.

**Divisor**  In division, the number dividing another number

## E

**Equation** A number sentence that says that the expressions on either side of the equals sign, =, represent the same number

**Equilateral triangle** A triangle in which all sides are the same length

**Equivalent equations** Equations with the same solutions

**Equivalent expressions** Expressions that have the same value for all allowable replacements

**Equivalent fractions** Fractions that represent the same number

**Evaluate** To substitute a value for each occurrence of a variable in an expression and calculate the result

**Experiment** In probability, a procedure that can be repeated and that has a defined set of outcomes

**Exponent** In expressions of the form $a^n$, the number $n$ is an exponent.

**Exponential notation** A representation of a number using a base raised to a power

**Extrapolation** The process of estimating a value that goes beyond the given data

## F

**Factor** *Verb*: to write an equivalent expression that is a product. *Noun*: a multiplier

**Factoring** Writing an expression as a product

**Factorization** A number expressed as a product of two or more numbers

**Fahrenheit** A temperature scale in which water freezes at 32° and boils at 212°

**Five-number summary** The set of statistics consisting of the minimum, the first quartile, the median, the third quartile, and the maximum of a set of data

**Fraction notation** A number written using a numerator and a denominator

**Frequency** The number of times that an item appears in a set of data

**Frequency distribution** A description of the frequency patterns in a set of data

**Frequency table** A table describing the number of times a value or values within a range appear in a set of data

## G

**Grade point average (GPA)** The average of the grade point values for each credit hour taken

## H

**Histogram** A special kind of graph that shows how often certain numbers appear in a set of data

**Hypotenuse** In a right triangle, the side opposite the right angle

## I

**Inequality** A mathematical sentence using $<$, $>$, $\leq$, $\geq$, or $\neq$

**Integers** The whole numbers and their opposites

**Interest** A percentage of an amount that has been invested or borrowed

**Interest rate** The percent at which interest is calculated on a principal

**Interpolation** The process of estimating a value between given values

**Isosceles triangle** A triangle in which two or more sides are the same length

## L

**Least common denominator (LCD)** The least common multiple of the denominators of two or more fractions

**Least common multiple (LCM)** The smallest number that is a multiple of two or more numbers

**Leaves** The rightmost digits of the data values displayed in a stem-and-leaf plot

**Legs** In a right triangle, the two sides that form the right angle

**Like terms** Terms that have exactly the same variable factors

**Line graph** A graph in which quantities are represented as points connected by straight-line segments

**Linear equation** Any equation that can be written in the form $Ax + By = C$, where $x$ and $y$ are variables

## M

**Marked price** The original price of an item

**Maximum** The largest number in a set of numbers

**Mean** A center point of a set of numbers found by adding the numbers and dividing the sum of the numbers by the number of items in the set; also called the *average*

**Median** In a set of data listed in order from smallest to largest, the middle number if there is an odd number of data items, or the average of the two middle numbers if there is an even number of data items

**Minimum** The smallest number in a set of numbers

**Minuend** The number from which another number is being subtracted

**Mixed numeral** A number represented by a whole number and a fraction less than 1

**Mode** The number or numbers that occur most often in a set of data

**Monomial** A constant, a variable, or a product of a constant and one or more variables

**Multiple of a number** The product of the number and an integer

**Multiplicative identity** The number 1

## N

**Natural numbers** The counting numbers: 1, 2, 3, 4, 5, ...

**Negative integers** Integers to the left of zero on the number line

**Numerator** The number above the fraction bar in a fraction

## O

**Obtuse angle** An angle whose measure is greater than 90° and less than 180°

**Obtuse triangle** A triangle in which one angle is an obtuse angle

**Opposite** The opposite, or additive inverse, of a number $x$ is written $-x$. Opposites are the same distance from 0 on the number line but on different sides of 0.

**Opposite of a polynomial** Two polynomials are opposites, or additive inverses, of each other if their sum is zero.

**Ordered pair** A pair of numbers of the form $(a, b)$ for which the order in which the numbers are listed is important

**Origin** The point $(0, 0)$ on a graph where the two axes intersect

**Original price** The price of an item before a discount is deducted

## P

**Palindrome prime** A prime number that remains a prime number when its digits are reversed

**Parallelogram** A four-sided polygon with two pairs of parallel sides

**Percent notation** A representation of a number as parts per 100; $n\%$

**Perimeter** The distance around an object or the sum of the lengths of its sides

**Periods** Groups of three digits, separated by commas

**Pi ($\pi$)** The number that results when the circumference of a circle is divided by its diameter; $\pi \approx 3.14$, or $\frac{22}{7}$

**Pictograph** A graphic means of displaying information using symbols to represent the amounts

**Polygon** A closed geometric figure with three or more line segments as sides

**Polynomial** A monomial or a sum of monomials

**Positive integers** Integers to the right of zero on the number line

**Prime factorization** A factorization of a composite number as a product of prime numbers

**Prime number** A natural number that has exactly two different factors: itself and 1

**Principal** An amount of money that is invested or borrowed

**Product** The result when one number is multiplied by another

**Proportion** An equation stating that two ratios are equal

**Protractor** A device used to measure and draw angles

**Purchase price** The price of an item before sales tax is added

**Pythagorean equation** The equation $a^2 + b^2 = c^2$, where $a$ and $b$ are lengths of the legs of a right triangle and $c$ is the length of the hypotenuse

## Q

**Quadrants** The four regions into which the axes divide a plane

**Quartile** One of three numbers that divide a set of data into four groups, each of which contains 25% of the data

**Quotient** The result when one number is divided by another

## R

**Radical sign** The symbol $\sqrt{\phantom{x}}$

**Radius** A segment with one endpoint on the center of a circle and the other endpoint on the circle, or the length of such a segment

**Range** The difference between the maximum and the minimum of a set of numbers

**Rate** A ratio used to compare two different kinds of measure

**Ratio** The quotient of two quantities; the ratio of $a$ to $b$ is $\frac{a}{b}$, also written $a : b$

**Rational number** Any number that can be written as the ratio of two integers $\frac{a}{b}$, where $b \neq 0$

**Ray** A part of a line consisting of one endpoint and all the points of the line on one side of the endpoint

**Reciprocal** A multiplicative inverse; two numbers are reciprocals if their product is 1

**Rectangle** A four-sided polygon with four 90° angles

**Repeating decimal** A decimal in which a block of digits repeats indefinitely

**Right angle** An angle whose measure is 90°

**Right triangle** A triangle that includes a right angle

**Rounding** Approximating the value of a number; used when estimating

## S

**Sale price** The price of an item after a discount has been deducted

**Sales tax** A tax added to the purchase price of an item

**Scalene triangle** A triangle in which all sides are of different lengths

**Scientific notation** A representation of a number written in the form $M \times 10^n$, where $n$ is an integer, $1 \leq M < 10$, and $M$ is expressed in decimal notation

**Similar triangles** Triangles in which corresponding sides are proportional and corresponding angles are congruent

**Simple interest** A percentage of an amount $P$ invested or borrowed for $t$ years, computed by calculating principal × interest rate × time

**Simplify** To rewrite an expression in an equivalent, abbreviated form

**Solution of an equation** A replacement for the variable that makes the equation true

**Solve** To find all solutions of an equation, inequality, or problem

**Sphere** The set of all points in space that are a given distance (the radius) from a given point (the center)

**Square** A four-sided polygon with four right angles and all sides of equal length

**Square root** The number $c$ is a square root of $a$ if $c^2 = a$.

**Standard form of a linear equation** An equation written in the form $Ax + By = C$

**Statistic** A number that describes a set of data

**Stem-and-leaf plot** A diagram describing the frequency distribution of a set of data in which each value in the set is listed by its leaf, consisting of its rightmost digit, associated with its stem, consisting of the remaining digits

**Stems** The digits of data values displayed in a stem-and-leaf plot, other than the rightmost digits

**Straight angle** An angle whose measure is 180°

**Substitute** To replace a variable with a number

**Subtrahend** In subtraction, the number being subtracted

**Sum** The result in addition

**Supplementary angles** Two angles for which the sum of their measures is 180°

## T

**Table** A representation of data in rows and columns

**Term** A number, a variable, or a product or a quotient of numbers and/or variables

**Terminating decimal** A decimal that can be written using a finite number of decimal places

**Total price** The sum of the purchase price of an item and the sales tax on the item

**Trapezoid** A four-sided polygon with exactly two parallel sides

**Triangle** A three-sided polygon

**Trinomial** A polynomial that is composed of three terms

**Two-way frequency table** A table that displays frequencies for two variables; also called a *contingency table*

## U

**Unit price**  The ratio of price to the number of units

**Unit rate**  The ratio of quantity to the number of units

## V

**Value**  The numerical result after a number has been substituted into an expression and the calculations have been carried out

**Variable**  A letter that represents an unknown number

**Vertex**  The common endpoint of the two rays that form an angle

**Vertical angles**  Two angles that are formed by two intersecting lines and that have no side in common

**Volume**  The number of unit cubes needed to fill an object

## W

**Whole numbers**  The natural numbers and 0: 0, 1, 2, 3, 4, 5, ...

# Index

Quarters, 554
Quartiles, 554
Quintillion, 328
Quotient, 26
   estimating, 40, 351
   as a mixed numeral, 257
   raising to a power, 704
   zeros in, 31
Quotient rule, 702, 704

**R**

Radical sign ($\sqrt{\phantom{x}}$), 658
Radius, 366, 621, 682
Raising a power to a power, 703, 704
Raising a product to a power, 703, 704
Raising a quotient to a power, 704
Range of data, 548
Rate
   commission, 483
   of discount, 484
   interest, 490
   ratio as, 395
   sales tax, 481
   unit, 396
Ratio, 167, 388
   percent as, 441
   and proportion, 402
   as a rate, 395
   simplifying, 391
Rational numbers, 302
Rays, 645
Reciprocal, 199
   and division, 200
   of zero, 199
Rectangle, 127
   area of, 22
   perimeter of, 128
Rectangular array
   in division, 26, 767
   in multiplication, 19, 759
Rectangular solid, volume of, 632, 682
Regrouping in addition, 9
Related sentences
   addition, subtraction, 752
   multiplication, division, 767
Relatively prime, 164
Remainder, 28
   finding on a calculator, 331
Removing a factor equal to one, 182
   and canceling, 184
Repeated addition, 19
Repeated subtraction, 26
Repeating decimals, 341, 342
   rounding, 343
Right angle, 127, 648
Right triangle, 651, 659
Rounding
   of decimal notation, 308, 311
     repeating, 343
   and estimating, 39
   by truncating, 311
   of whole numbers, 37

**S**

Salary, 483
Sale price, 484
Sales tax, 481
Scale on a graph, 520, 523
Scalene triangle, 651
Scientific notation, 707
   on a calculator, 710
   converting to/from decimal notation, 708
   dividing using, 709
   multiplying using, 708
Screw, pitch, 195
Second coordinate, 529
Second quartile, 554
Segment, unit, 600
Semiannually compounded interest, 492
Sentences, *see* Related sentences
Septillion, 328
Seven, divisibility by, 158
Sextillion, 328
Sides of an angle, 645
Signs of numbers, 89
   changing, 89
Similar geometric figures, 422, 424
Similar terms, 126. *See also* Like terms.
Similar triangles, 422, 423
Simple interest, 490
Simplest fraction notation, 182
Simplifying
   expressions, 71
   fraction notation, 182, 184
     complex, 286
     for ratios, 391
Six, divisibility by, 155
Solution of an equation, 48, 133, 531, 535.
   *See also* Solving equations.
Solve step in problem solving, 54
Solving equations, 48, 133
   using addition principle, 134, 241, 355
   checking solutions, 50, 134, 357
   and clearing fractions, 250
   by dividing on both sides, 51
   using division principle, 135, 355
   using multiplication principle, 204
   using the principles together, 137, 248, 356, 357
   by subtracting on both sides, 49, 241
   by trial, 48
Solving problems, 54, 61. *See also Index of Applications.*
Solving proportions, 403
Sphere, 634
   volume, 634, 682
Spread of data, 548
Square, 128
   area, 70
   perimeter, 128
Square of a number, 658
Square roots, 658
   on a calculator, 659

Standard form of a linear equation, 536
Standard notation for numbers, 4
Standard window, 536
State the answer, 54
Statistic, 548
Stem-and-leaf plot, 563
   two-sided, 573
Stems, 563
Straight angle, 648
Studying for Success, 2, 37, 86, 118, 152, 190, 222, 262, 302, 341, 388, 411, 440, 467, 510, 548, 600, 645, 696, 717
Substituting for a variable, 118
Subtraction
   basic, 752
   and borrowing, 15, 754
   checking, 15
   with decimal notation, 313, 314, 315
   definition, 14, 97
   difference, 14, 97
     estimating, 39, 350
   estimating differences, 39, 350
   of exponents, 702, 704
   using fraction notation, 239
   and "how much more," 753
   of integers, 97
   minuend, 14
   using mixed numerals, 263
   and opposites, 98, 719
   of polynomials, 719
   related addition sentence, 752
   repeated, 26
   subtrahend, 14
   as "take away," 752
   of whole numbers, 14, 15
   zeros in, 755
Subtrahend, 14
Sum, 9, 746
   of angle measures of a triangle, 652, 682
   estimating, 39, 350
   of three numbers, 747
Summary, five-number, 555
Supplement of an angle, 649
Supplementary angles, 649
Symbol
   approximately equal to ($\approx$), 40
   greater than ($>$), 42
   less than ($<$), 42
   for multiplication, 20
   not equal to ($\neq$), 42
   percent (%), 440
   radical ($\sqrt{\phantom{x}}$), 658
Systems of measures, *see* American system of measures; Metric system of measures

**T**

Table
   contingency, 561
   of data, 510, 511
   frequency, 560, 561

**I-6**   Index